METRIC UNITS FOR STRESS

$$1 = 10^0 = P_a = N/m^2$$

$$1000 = 10^3 = kP_a = kN/m^2$$

$$1,000,000 = 10^6 = MP_a = N/mm^2$$

$$10^9 = GP_a = kN/mm^2$$

BUILDING STRUCTURES

BUILDING STRUCTURES

JAMES AMBROSE

Professor of Architecture
University of Southern California

A Wiley-Interscience Publication
JOHN WILEY & SONS

New York / Chichester / Brisbane / Toronto / Singapore

Library of Congress Cataloging in Publication Data:
Ambrose, James E.
    Building structures / James Ambrose. — 2nd ed.
       p.   cm.
    Includes bibliographical references and index.
    ISBN 0-471-54060-9
    1. Structural design.   2. Structural analysis (Engineering)
  I. Title.
  TA658.A49   1993   92-16362
  624.1'7—dc20   CIP

Printed in the United States of America

Preface

This book provides a comprehensive source for study of the general topic of building structures. Its primary intent is to serve as a resource for courses on the topic in architecture schools and as a review on the topics for candidates for the examination for architectural registration.

Because of its orientation to these basic uses, the computational work is kept at a relatively simple level and the fundamental work in engineering analysis and design is not dealt with as extensively as it should be in a reference for professional structural engineers. The general style of the work here is one that is intended for persons whose primary concerns are the overall design of buildings and their construction. Involvement in the topic of structures is thus presented to develop some familiarity with the issues and with the usual processes of design development work; but the context of the broader concerns of building design are never far from mind.

The scope of the topic is large, and indeed the nine parts of this book cover much of the topic. Presentations range from consideration of fundamental concepts, through elementary applied mechanics and practical concerns for basic materials and systems, to the pragmatic consideration of some realistic design examples involving buildings of specific form and use.

It is to be expected that not all readers will be concerned with the full range of the presentations nor necessarily want to take them up in the sequence presented. The divisions of the book permit some random usage and are to a degree freestanding. Thus any of Parts 3 through 8 might be taken up first, although the fundamental materials in Parts 1 and 2 should be engaged somehow before proceeding too far.

Part 9 is by far the most important one, with the fully engaged building structure being posed as the problem for design. In a way, the rest of the book is essentially supportive of this concluding work and might be accessed only as needed. However, for a comprehensive study of the topic, a more disciplined approach is more practical.

A study manual has been prepared for use with this book. It provides supplementary study materials and recommendations for various uses for the book. It is prepared for assistance to readers who need to use the book in an essentially self-study situation with little outside help. However, the lists of significant terms, the questions, and the sample exams may also be of considerable support to students using the book as a course text.

This book is not essentially oriented for development of computer applications. However, it is to be expected that most readers will have some experience with computers, which are increasingly utilized in professional design work. A publication with demonstrations of computer applications related to this work is in preparation and will hopefully be available to interested readers shortly after this book is published.

The general attempt has been made to use current design and construction methods and the most recent codes and industry standards. However, the need for a somewhat abbreviated, simpler development of the subject makes it impractical to use truly complex investigative methods or the pragmatic shortcuts employed by experienced designers. Learning is an essential aim here, and the topics are best taken up in the simplest context and by basic hand-computational means for easier comprehension. It is to be expected, of course, that anyone expecting to advance to professional structural design work will not stop with the simple lessons presented here.

My experience in recent years with the preparation of new editions of the works of Harry Parker has strongly influenced the development of the materials for this book. The presentation here is more broadly developed, but owes much directly and in spirit to the highly distilled simplicity and conciseness of the work in Parker's several books on "simplified" structures.

I am grateful to many sources for data and reproductions of graphs and illustrations. These sources are acknowledged for their permission where the presentations occur in the book. I am especially grateful to the International Conference of Building Officials for their permission to borrow extensively from the *Uniform Building Code.*

I am also grateful for the feedback from various users of the first edition, including my own students. This has resulted in considerable reshaping of the presentation in this edition. Most notable is the reduction of size, which has been achieved in spite of an actual expansion of the topic coverage; most notably the additional cases presented in Part 9.

Finally, I am also most grateful for the assistance provided by my wife, Peggy, who continues to find ways to relieve me of tasks in the development of my writing projects. If her contributions grow much more, I may have to acknowledge her as coauthor on any future editions.

JAMES AMBROSE

Westlake Village, California
December 1992

v

Contents

Introduction

This book deals broadly with the topic of structures related to buildings. Emphasis is placed on the concerns of the working professional designers who must cope with the practical problems of figuring out how to make plans for sensible buildings. Designers' concerns range from a basic understanding of structural behaviors to the determination of the construction details for a specific type of building.

The materials in this book are arranged to present a logical sequence of study. However, it is to be expected that few readers will start at page 1 and proceed diligently to the end, as if reading a novel. The separate book parts are therefore developed as reasonably freestanding, with appropriate referencing to other parts for those readers who need some reinforcement. Additionally, at any time, the reader can use the Table of Contents, the Index, or the Glossary to seek help in understanding unfamiliar terms or ideas.

The parts of the book and their intended uses are as follows.

Part 1. A general discussion of structures and structural concerns in building design intended for basic orientation and familiarization with the subject.

Part 2. Basic development of the topics of applied mechanics (mostly statics), strength of materials, and elementary structural analysis. This provides fundamental computational relationships for use in Parts 3 through 9.

Parts 3 through 6. These deal individually with the four main structural materials and with the products and systems that are produced from them. Practical usage is emphasized and examples are illustrated in the building case studies in Part 9.

Part 7. An abbreviated treatment of soil mechanics and the design of simple foundation elements for buildings.

Part 8. General development of the problems of dealing with the effects of wind and earthquakes on building structures.

Part 9. Demonstration examples of the development of structural systems for two buildings, proceeding from preliminary planning to construction detailing. This part illustrates applications of much of the work in Parts 3 through 8.

COMPUTATIONS

In professional design firms most structural computations are done with the aid of computers. The purpose of this book is primarily to illustrate concepts, relationships, and basic processes of design. Displayed computations are limited mostly to elementary ones and are presented in step-by-step form for study purposes. Simple "hand" calculations are used to illustrate computational processes.

Because of the many assumptions and approximations that are necessary in work for structural design, final numbers used for design solutions are seldom significant beyond the third digit in terms of accuracy. The actual work for most displayed computations in this book was done on an eight-digit pocket calculator, and the results are thus sometimes displayed to eight digits to correspond to work by readers using the same form of calculator. However, answers are also frequently rounded off to three digits, which is generally all that is significant to the final work.

SYMBOLS

The following symbols are used in this book.

Symbol	Reading
$>$	Is greater than
$<$	Is less than
$\geq$	Equal to or greater than
$\leq$	Equal to or less than
$\simeq$	Is approximately equal to
6′	Six feet
6″	Six inches
6#	Six pounds
Σ	The sum of
ΔL, Δt, etc.	Change in L, t, etc.

STANDARD NOTATION

There is some general usage of standard terms and symbols, but much of the notation in structural work derives from individual references. The following list includes many commonly used letters and symbols, but specific usage is also illustrated in other parts of this book.

a	1. Moment arm; 2. acceleration; 3. increment of an area.
A	Gross (total) area of a surface.
d	Depth (height) of a spanning member: beam, truss, etc.
D	1. Diameter; 2. deflection.

e	1. Eccentricity (mislocation) from some zero reference point; 2. elongation (stretch due to tension).
E	Modulus of elasticity, direct stress.
f	Computed unit stress.
F	1. Force; 2. allowable unit stress.
g	Acceleration due to gravity.
G	Modulus of elasticity, shear stress.
h	Height
H	Horizontal component (effect) of a force.
I	Moment of inertia (bending).
J	Polar moment of inertia (torsion).
M	Moment (rotational, bending).
n	Ratio of the moduli of elasticity of two materials.
N	Number of.
P	1. Percent; 2. unit pressure.
r	Radius of gyration.
R	Radius (of a circle, etc.).
s	1. Spacing of a set of objects; 2. distance travelled (displacement); 3. strain (unit deformation).
t	1. Thickness; 2. time.
T	1. Temperature; 2. torsional moment.
V	1. Gross (total) shear force; 2. vertical component (effect) of a force.
w	1. Width; 2. unit of a uniformly distributed load.
W	1. Gross (total) value of a uniformly distributed load; 2. gross (total) weight of an object.

Greek letters used:

μ (mu)	Coefficient of friction.
ϕ (phi)	Angle.
θ (theta)	Angle.

UNITS OF MEASUREMENT

At the time of this writing, despite increasing pressures, the English system of units (now more appropriately called the U.S. system) is still widely used in the United States. This is especially so in the building industry; to the extent that many major references for data are developed primarily in U.S. units. The work in this book is presented primarily in U.S. units, although some data and some answers to computational problems are also given in SI units enclosed in brackets following the equivalent quantities in U.S. units.

For some work, such as much of that in Part 2, the units themselves are of little significance, and we have thus chosen to simplify the presentations by using only U.S. units. For use by readers, Table 1 lists the standard units of measurement in the U.S. system with the abbreviations used in this work. Table 2 yields the corresponding units in the SI system. Conversion factors used in shifting from one system to the other are given in Table 3.

TABLE 1. Units of Measurement: U.S. System

Name of Unit	Abbreviation	Use
Length		
Foot	ft	Large dimensions, building plans, beam spans
Inch	in.	Small dimensions, size of member cross sections
Area		
Square feet	ft^2	Large areas
Square inches	in.^{2}	Small areas, properties of cross sections
Volume		
Cubic feet	ft^3	Large volumes, quantities of materials
Cubic inches	in.^{3}	Small volumes
Force, mass		
Pound	lb	Specific weight, force, load
Kip		1000 lb
Pounds per foot	lb/ft	Linear load (as on a beam)
Kips per foot	kips/ft	Linear load (as on a beam)
Pounds per square foot	lb/ft^2, psf	Distributed load on a surface
Kips per square foot	kips/ft^2, ksf	Distributed load on a surface
Pounds per cubic foot	lb/ft^3, pcf	Relative density, weight
Moment		
Foot-pounds	ft-lb	Rotational or bending moment
Inch-pounds	in.-lb	Rotational or bending moment
Kip-feet	kip-ft	Rotational or bending moment
Kip-inches	kip-in.	Rotational or bending moment
Stress		
Pounds per square foot	lb/ft^2, psf	Soil pressure
Pounds per square inch	lb/in.^{2}, psi	Stresses in structures
Kips per square foot	kips/ft^2, ksf	Soil pressure
Kips per square inch	kips/in.^{2}, ksf	Stresses in structures
Temperature		
Degree Fahrenheit	°F	Temperature

TABLE 2. Units of Measurement: SI System

Name of Unit	Abbreviation	Use
Length		
Meter	m	Large dimensions, building plans, beam spans
Millimeter	mm	Small dimensions, size of member cross sections
Area		
Square meters	m^2	Large areas
Square millimeters	mm^2	Small areas, properties of cross sections
Volume		
Cubic meters	m^3	Large volumes
Cubic millimeters	mm^3	Small volumes
Mass		
Kilogram	kg	Mass of materials (equivalent to weight in U.S. system)
Kilograms per cubic meter	kg/m^3	Density
Force (load on structures)		
Newton	N	Force or load
Kilonewton	kN	1000 newtons
Stress		
Pascal	Pa	Stress or pressure (1 pascal $= 1\ N/m^2$)
Kilopascal	kPa	1000 pascals
Megapascal	MPa	1,000,000 pascals
Gigapascal	GPa	1,000,000,000 pascals
Temperature		
Degree Celsius	°C	Temperature

TABLE 3. Factors for Conversion of Units

To Convert from U.S. Units to SI Units Multiply by:	U.S. Unit	SI Unit	To Convert from SI Units to U.S. Units Multiply by:
25.4	in.	mm	0.03937
0.3048	ft	m	3.281
645.2	$in.^2$	mm^2	1.550×10^{-3}
16.39×10^3	$in.^3$	mm^3	61.02×10^{-6}
416.2×10^3	$in.^4$	mm^4	2.403×10^{-6}
0.09290	ft^2	m^2	10.76
0.02832	ft^3	m^3	35.31
0.4536	lb (mass)	kg	2.205
4.448	lb (force)	N	0.2248
4.448	kip (force)	kN	0.2248
1.356	ft-lb (moment)	N-m	0.7376
1.356	kip-ft (moment)	kN-m	0.7376
1.488	lb/ft (mass)	kg/m	0.6720
14.59	lb/ft (load)	N/m	0.06853
14.59	kips/ft (load)	kN/m	0.06853
6.895	psi (stress)	kPa	0.1450
6.895	ksi (stress)	MPa	0.1450
0.04788	psf (load or pressure)	kPa	20.93
47.88	ksf (load or pressure)	kPa	0.02093
$0.566 \times$ (°F $-$ 32)	°F	°C	$(1.8 \times$ °C$) + 32$

PART 1

BASIC CONCEPTS

The purpose of Part 1 is to provide a basis for the more intense parts that follow—a sort of pep talk before the big game. The goal is to build some motivation for a continuing effort by giving a broad view of the subject and its general relations to building design. Readers whose commitments or involvements already provide sufficient motivation for learning may not need the pep talk, but will benefit from the overview.

CHAPTER ONE

Basic Concerns

All physical objects have structures. Consequently, the design of structures is part of the general problem of design for all physical objects. It is not possible to understand fully why buildings are built the way they are without having some knowledge and understanding of the problems of their structures. Building designers cannot function in an intelligent manner in making decisions about the form and fabric of a building without some comprehension of basic concepts of structures.

1.1. SAFETY

Life safety is a major concern in structures. Two prime considerations are for resistance to fires and for a low statistical likelihood of collapse under loads. Major elements of fire resistance are

Combustibility of the Structure. If structural materials are combustible, they will contribute fuel to the fire as well as hasten the collapse of the structure.

Loss of Strength at High Temperature. This consists of a race against time, from the moment of inception of the fire to the failure of the structure—a long interval increasing the chance for the occupants to escape.

Containment of the Fire. Fires usually start at a single location, and preventing their spread is highly desirable. Walls, floors, and roofs should resist burn-through by the fire.

Major portions of building code regulations have to do with aspects of fire safety. Materials, systems, and details of construction are rated for fire resistance on the basis of experience and tests. These regulations constitute major restraints on building design with regard to selection of materials and use of details for the building construction.

Building fire safety involves much more than structural behavior. Clear exit paths, proper exits, detection and alarm systems, firefighting devices (sprinklers, standpipes, hose cabinets, etc.), and lack of toxic or highly combustible materials are also important. All of these factors will contribute to the race against time, as illustrated in Fig. 1.1.

The structure must also sustain loads, safety in this case consisting of some margin of structural capacity beyond that strictly required for the actual task. This margin is expressed by the safety factor, SF, defined as follows:

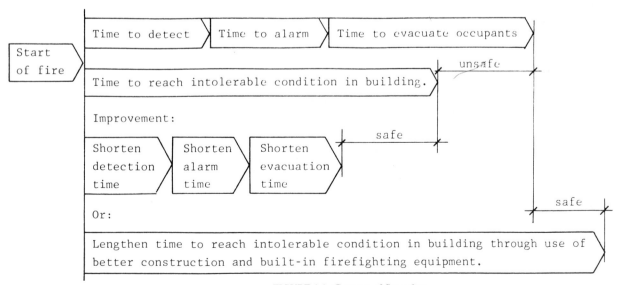

FIGURE 1.1. Concept of fire safety.

$$SF = \frac{\text{actual capacity of the structure}}{\text{required capacity of the structure}}$$

If a structure is required to carry 40,000 lb and is actually able to carry 70,000 lb before falling, the safety factor is expressed as SF = 70,000/40,000 = 1.75. Desire for safety must be tempered by practical concerns. The user of a structure may find comfort in a safety factor as high as 10, but the cost or gross size of the structure may be undesirable. Building structures are generally designed for an average safety factor of about 2. There is no particular reason for this other than experience. That is, there have been relatively few structural failures attributed to minor design or construction mistakes with the consistent use of the average safety factor of 2.

For many reasons, structural design is a highly imprecise undertaking. One should not assume, therefore, that the true safety factor in a given situation can be established with great accuracy. What the designer strives for is simply a general level of assurance of adequate performance without pushing the limits of the structure too close to the edge of failure.

There are two basic techniques for assuring the margin of safety. The method once used widely is called the *working stress* or *service load method*. With this method stress conditions under actual usage (service loads) are determined and limits for the stresses are set at some percentage of the predetermined ultimate capacities of the materials. The margin of safety is inferred from the specific percentage used for the working stresses.

A problem encountered with the working stress method is that many materials do not behave in the same manner near their ultimate failure limits as they do at working load levels. Thus prediction of failure from a stress evaluation cannot be made on the basis of a simple linear proportionality, and using a working stress of one-half of the tested ultimate stress does not truly constitute a safety factor of 2.

The other principal method for assuring safety is called the *strength design* or *factored load method*. The basis of this method is simple. The total load capacity of the structure at failure is determined, and the allowable (service) load is set at the desired level simply by dividing the failure load by the safety factor selected. The only stress conditions considered in this method are those occurring at ultimate failure. These conditions can be reasonably well-established by testing prototypes to ultimate failure.

Although life safety is certainly important, the structural designer must also deal with many other factors in developing a satisfactory solution for any building structure.

1.2. GENERAL CONCERNS

Feasibility

Structures are real and thus must use materials and products that are available and can be handled by existing craftspeople and production organizations. Building designers must have a reasonable grasp of the current inventory of available materials and products and of the usual processes for building construction. Keeping abreast of this body of knowledge is a challenge in the face of the growth of knowledge, the ever-changing state of the technology, and the market competition among suppliers and builders.

Feasibility is not just a matter of technological potentialities, but relates to the overall practicality of a structure. Just because something *can* be built is no reason that it *should* be built. Consideration must be given to the complexity of the design, dollar cost, construction time, acceptability by code-enforcing agencies, and so on.

Economy

Buildings cost a lot, and investors are seldom carefree, least of all with the cost of the structure. Except for the condition of a highly exposed structure that constitutes a major design feature, structures are usually appreciated as little as the buried piping, wiring, and other hidden service elements. Expensive structures do not often add value in the way that expensive hardware or carpet may. What is often desired is simple adequacy, and the hard-working, low-cost structure is much appreciated.

However vital, the structure represents a minor part of the total construction cost in most cases. The result is that in comparing alternative structures, the cost of the structure itself may be less important than the influence of the structure on other building costs. A particular structure may have high performance efficiency and low cost in its own right but may produce forms and details that make other aspects of the building construction difficult and expensive, with an end result that is not real economy.

Optimization

Building designers often are motivated by desires for originality and individual expression. However, they are also usually pressured to produce a practical design in terms of function and feasibility. In many instances this requires making decisions that constitute balances between conflicting or opposing considerations. The optimal or best solution is often elusive.

Obvious conflicts are those between desires for safety, quality of finishes, grandeur of spaces, and general sumptuousness on the one hand, and practical feasibility and economy on the other. All of these attributes may be important, but often changes that tend to improve one factor work to degrade others. Some rank ordering of the various attributes is generally necessary, with dollar cost usually ending up high on the list. Thus the best solution may have to be qualified in terms of the specific priorities used in the design.

Integration

Good structural design requires integration of the structure into the whole physical system of the building. It is necessary to realize the potential influences of structural design decisions on the general architectural design and on the development of the systems for power, lighting, thermal control, ventilation, water supply, waste handling, vertical transportation, firefighting, and so on. The most popular structural systems have become so in many cases largely because of their ability to accommodate the other subsystems of the building and to facilitate popular architectural forms and details.

CHAPTER TWO

Architectural Considerations

Buildings serve many purposes and take on a wide variety of forms and details to meet the requirements and fulfill the aspirations of their users. Each building is also unique in its specific combination of location, orientation, and surroundings.

2.1. USAGE REQUIREMENTS

Primary architectural functions that relate to the structure are

Need for shelter and enclosure.
Need for interior spatial definition, subdivision, and separation.
Need for unobstructed interior space.

In addition to its basic force-resistive purpose, the structure must serve to generate the building forms that relate to these basic usage functions.

Shelter and Enclosure

Exterior building surfaces usually form a barrier between the building interior and the exterior environment. This is generally required for security and privacy and often in order to protect against hostile external conditions (thermal, acoustic, air quality, precipitation, etc.). Figure 2.1 shows many potential requirements of the building skin. The skin is viewed as a selective filter that must block some things while permitting the passage of others.

In some instances, elements that serve a structural purpose also fulfill some of the filter functions of the building

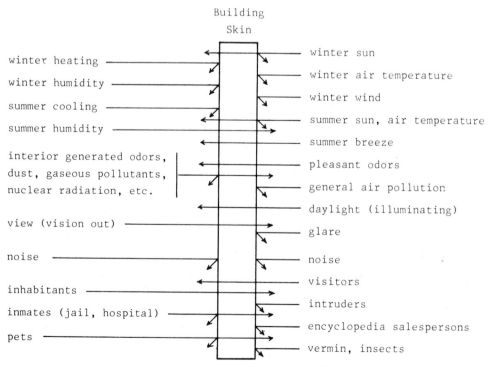

FIGURE 2.1. Functions of the exterior wall as a selective filter.

skin, and properties other than strictly structural ones must be considered in the choice of materials and details of the structure. Basic structural requirements cannot be ignored but frequently can be relatively minor as final decision criteria.

When need exists for complete enclosure, the structure must either provide it directly or facilitate the addition of other elements to provide it. Solid walls and shell domes are examples of structures that provide naturally closed surfaces (see Fig. 2.2). It may be necessary to enhance the basic structure with insulation, waterproofing, and so on, to develop all of the required skin functions, but the enclosure function is inherent in the structural system.

Frame systems, however, generate open structures that must be provided with separate skin elements to develop the enclosure function (see Fig. 2.3). In some cases the skin may interact structurally with the frame; in other cases it may add little to the basic structural behavior. An example of the latter is a heavy steel frame for a high-rise building with a thin curtain wall of light metal and glass elements.

Interior Space Division

Most buildings have interior space division with separate rooms and often separate levels. Structural elements used to develop the interior must relate to the usage requirements of individual spaces and to the needs for separation. In multiple-level buildings structural elements that form the floor for one level must simultaneously form the ceiling for spaces below. These two functions generate separate form restrictions, surface treatments, attachments, or incorporation of elements such as light fixtures, air ducts, power outlets, and plumbing fixtures. In addition, the floor–ceiling structure must provide a barrier to the transmission of sound and fire. As in the case of the building skin, the choice of construction must be made with all necessary functions in mind.

FIGURE 2.2. The self-skinning structure: a concrete shell surface. The St. Louis Planetarium. Architects: Helmuth, Obata, and Kassabaum, St. Louis.

FIGURE 2.3. Building surface developed as an applied skin: gypsum drywall, plywood, and stucco on a wood frame.

Generating Unobstructed Space

Housing of activities creates the need for producing unobstructed interior spaces. These spaces may be very small (bathrooms) or very large (sports arenas). Generating open space involves the basic structural task of spanning, illustrated in Fig. 2.4. The magnitude of the spanning problem is determined by the load and the span. As the span increases, the required structural effort increases rapidly, and options for the structural system narrow.

A particularly difficult problem is that of developing a large unobstructed space in the lower portion of a multiple-level building. As shown in Fig. 2.5, this generates a major load on the transitional spanning structure. This situation is unusual, however, and most large spanning structures consist only of roofs, for which the loads are relatively light.

2.2. ARCHITECTURAL ELEMENTS

Most buildings consist of combinations of three basic elements: walls, roofs, and floors. These elements are arranged to create both space division and unobstructed space.

Walls

Walls are usually vertical and potentially lend themselves to the task of supporting roofs and floors. Even when they do not serve as supports, they often incorporate the columns that do serve this function. Thus the design development of spanning roof and floor systems must begin with consideration of the wall systems over which they span.

Walls may be classified on the basis of their architectural and structural functions, which affects many of the

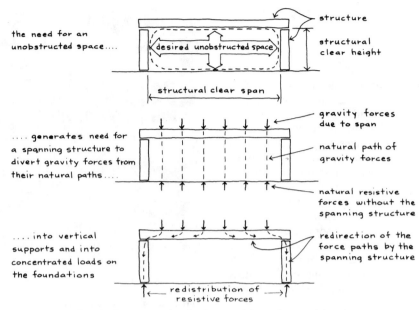

the need for an
unobstructed space....

desired unobstructed space

structure

structural
clear height

structural clear span

....generates need for
a spanning structure to
divert gravity forces from
their natural paths....

gravity forces
due to span

natural path of
gravity forces

natural resistive
forces without the
spanning structure

....into vertical
supports and into
concentrated loads on
the foundations

redirection of the
force paths by the
spanning structure

redistribution of
resistive forces

FIGURE 2.4. The structural task of generating unobstructed interior space.

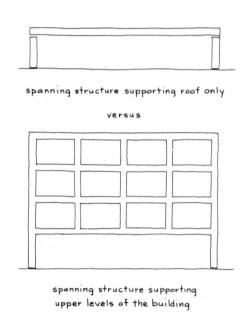

spanning structure supporting roof only

versus

spanning structure supporting
upper levels of the building

FIGURE 2.5. Load conditions for the spanning structure.

design decisions regarding choice of materials and details. Some basic categories are:

Structural Walls. These serve essential functions in the general building structural system. Bearing walls support roofs, floors, or other walls. Shear walls brace the building against horizontal forces, utilizing the stiffness of the wall in its own plane, as shown in Fig. 2.6.

Nonstructural Walls. Actually there is no such thing as a nonstructural wall, since the least that any wall must do is hold itself up. However, the term *nonstructural* is used to describe walls that do not contribute to the general struc-

tural system of the building; that is, they do not support or brace other parts of the building. When they occur on the exterior they are called *curtain walls*; on the interior they are called *partitions*.

Exterior Walls. As part of the building skin, exterior walls usually have a number of required functions, including the barrier and selective filter functions described previously. Wind produces both inward and outward (suction) pressures that the exterior walls must transmit to the building's lateral bracing system. Exterior walls are usually permanent, as opposed to some interior walls that can be relocated if they are nonstructural.

Interior Walls. Although some barrier functions are usually required of any wall, interior walls need not provide separation between the interior and exterior environments and do not sustain direct wind pressures. They may be relatively permanent, as when they enclose stairs, elevators, ducts, or toilets, but they are often essentially partitions and can be built as such.

Many walls must incorporate doors or windows or provide hiding places for items such as ducts, wiring, or piping. Walls of hollow construction provide convenient hiding places, whereas those of solid construction can present problems in this regard. Walls that are not flat and vertical can create problems with hanging objects such as pictures or drapes. Walls that are not straight in plan and walls that intersect at other than right angles can create problems in the installation of doors and windows and the arrangement of furniture (see Fig. 2.7).

Roofs

Roofs have two primary functions: they must act as skin elements and must facilitate the runoff of water from pre-

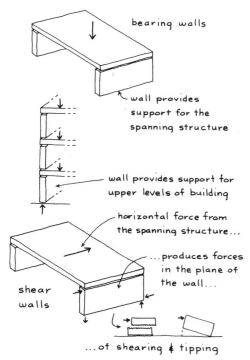

FIGURE 2.6. Structural functions of walls.

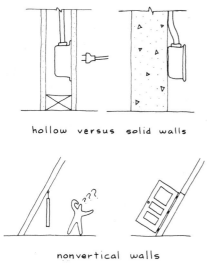

FIGURE 2.7. Problems of wall form.

cipitation. Barrier functions must be met and the roof geometry and surface must relate to the drainage problem. Whereas floors must generally be flat, roofs generally must not be, as some slope is required for drainage. The so-called flat roof must actually have some slope: typically, a minimum of $\frac{1}{4}$ in./ft, or approximately 1%. In addition, the complete drainage operation must be developed so that runoff water is collected and removed from the roof.

Floors are meant to be walked on; roofs generally are not. Thus in addition to being nonhorizontal, roofs may be constructed of materials or systems that are not rigid, the

ultimate possibility being a fabric or membrane surface held in position by tension.

Because of the freedom of geometry and lack of need for rigidity or solidity, the structural options for roofs are more numerous than those for floors. The largest enclosed, unobstructed spaces usually are those spanned only by roofs. Thus most of the dramatic and exotically formed spanning structures for buildings are those used for roofs.

Floors

Floor structures must often serve both as a floor for upper spaces and as a ceiling for lower spaces. The floor function usually requires a flat, horizontal form, limiting the choice to flat-spanning systems. Barrier requirements derive from the needs of the spaces above and below.

Most floor structures are relatively short in span, since flat-spanning systems are relatively inefficient and load magnitudes for floors are generally higher than those for roofs. Achieving large open spaces under floors is considerably more difficult than achieving such spaces under roofs.

2.3. FORM-SCALE RELATIONSHIPS

There is a great variety of types of architectural space, and therefore there are many different categories of structural problems. The breakdown in Fig. 2.8 illustrates variables of form in terms of the interior space division and of scale in terms of the span or the clear height. Other variables include the number of levels or adjacent spaces. Although this does not include all possible variations, it includes many typical cases. The following discussion deals with some of the structural problems inherent in various situations represented in Fig. 2.8.

Single Space

This type ordinarily represents the greatest degree of freedom in the choice of the structural system. The building

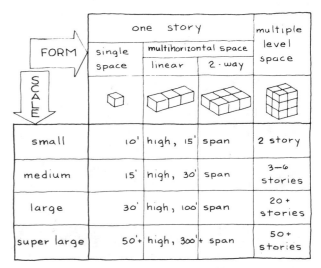

FIGURE 2.8. Form-scale relationships.

basically requires only walls and a roof, although a floor structure other than paving may be used if the building is elevated above the ground. Some possible uses for such buildings and the potential structural systems follow.

Small Scale (10 ft high, 15 ft span). This includes small sheds, cabins, and single-car garages. The range of possible structural systems is considerable, including tents, air-inflated bubbles, ice block igloos, and mud huts, as well as more ordinary construction (see Fig. 2.9).

Medium Scale (15 ft high, 30 ft span). This includes small stores, classrooms, and agricultural buildings. The 15-ft wall height is just beyond the limit for 2×4 wood studs and the 30-ft span is beyond the usual limit for solid wood joists or rafters on a horizontal span. The use of a truss, a gabled frame, or some other more efficient spanning system becomes feasible at this scale, although some flat decks or beam systems are also possible.

Large Scale (30 ft high, 100+ ft span). This includes gyms, theaters, and large showrooms. The 30-ft wall height represents a significant structural problem, usually requiring braced construction of some kind. This span is generally beyond the feasible limit for a flat-spanning beam system, and the use of a truss, arch, or some other system is usually required. Because of the size of the spanning elements, they are often quite widely spaced, requiring a secondary spanning system to fill in between the major spanning elements. Loads from the major spanning elements place heavy concentrated forces on walls, often requiring columns or piers. If the columns or piers are incorporated in the wall plane, they may serve the dual function of bracing the tall walls.

Super Large Scale (50+ ft high, 300+ ft span). This includes large convention centers and sports arenas. Walls become major structural elements, requiring considerable bracing. The spanning structure requires considerable height in the form of distance from top to bottom of truss elements, rise-to-span ratio of an arch, sag-to-span ratio of a cable, and so

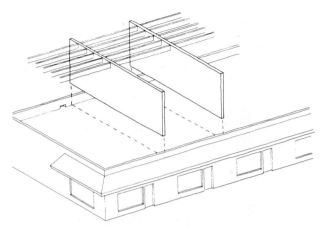

FIGURE 2.10. Multiple horizontal spaces can be produced with a large number of optional structural modules, one of the simplest being the repetition of ordinary bearing walls and roof joists.

on. Use of superefficient spanning structural systems becomes a necessity at this scale.

Multiple Horizontal Space: Linear

This category includes motels, small shopping centers, and school classroom wings. Multiplication may be done with walls that serve the dual functions of supporting the roof and dividing the interior spaces, or it may be done only in terms of multiples of the roof structural system with no interior structure as such. The roof system has somewhat less geometric freedom than that of the single-space building, and a modular system of some kind is usually indicated (see Fig. 2.10).

Although space utilization and construction simplicity generally will be obtained with the linear multiplication of rectangular plan units, there are some other possibilities, as shown in Fig. 2.11. If units are spaced by separate connecting links, more freedom can be obtained for the roof structure of the individual units.

Structural options remain essentially the same as those for the single-space building. If adjacent spaces are significantly different in height or span, it may be desirable to change the system of construction, using systems appropriate to the scale of the individual spaces.

Multiple Horizontal Space: Two-Way

This category includes factories, stores, warehouses, and large single-story offices. As with linear multiplication, the unit repetition may be done with or without interior walls, utilizing interior columns as supporting elements (see Fig. 2.12).

Constraints on plan and roof surface geometry are greater here than in linear multiplication. The relative efficiency of rectangular plan units becomes generally higher, although other possibilities exist. Modular organization and coordination become increasingly logical in the development of structural systems.

FIGURE 2.9. Small-scale, single-space buildings serve many functions, including gourmet dining.

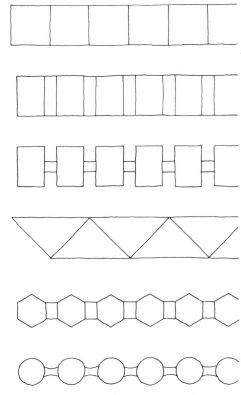

FIGURE 2.11. Linear plan multiples.

FIGURE 2.12. Two-way multiplication of horizontal spaces, achieved here at a medium scale with a common system: steel posts and beams, light steel trusses, and a light formed sheet steel deck.

Continuity in spanning structures, although also possible with linear multiplication, becomes more useful when multiplication is two-way.

Although still possible using linear multiplication, roof structures that are other than flat and horizontal become increasingly less feasible for two-way multiplication. Roof drainage becomes a major problem when the distance from the center to the edge of the building is great. The pitch required for water runoff to an edge is often not feasible, in which case more costly and complex interior roof drains are required.

Structural alternatives are generally the same as they are for the single-space building. The two-way multiplication of very large scale spaces is uncommon, and these structures tend to fall into the medium-span category. As with linear multiplication, if adjacent spaces are significantly different in size, a change in the structural system may be justified.

Multilevel Space

The jump from single to multiple levels has significant structural implications.

Need for Framed Floor Structure. This is a spanning, separating element not inherently required for the single-story building.

Need for Stacking of Support Elements. Lower elements must support upper elements as well as the spanning elements immediately above them. This works best if support elements are aligned vertically and imposes a need to coordinate building plans at the various levels.

Increased Concern for Lateral Loads. As the building becomes tall, wind and earthquake loads impose greater overturning effects as well as greater horizontal force in general, and the design of lateral bracing becomes a major problem.

Vertical Penetration of the Structure. Elevators, stairs, ducts, chimneys, piping, and wiring must be carried upward through the horizontal structure at each level, and spanning systems must accommodate the penetrations.

Increased Foundation Loads. As building height increases without an increase in plan size, the total vertical gravity load for each unit of plan area increases, eventually creating a need for a very heavy foundation.

The existence of many levels also creates a problem involving the depth of the flat-spanning systems at each level. As shown in Fig. 2.13, the depth of the structure (*A* in the illustration) is the distance from the top to the bottom of the spanning system: deck + beams + fireproofing in the example shown.

In many multistoried buildings a ceiling is hung below the floor structure, and the space between the ceiling and the underside of the floor contains various items, such as ducts, wiring, sprinkler piping, and recessed light fixtures. Architecturally, the critical depth of this construction is the total dimension from the top of the floor finish to the underside of the ceiling (*B* in the illustration).

The floor-to-floor height, from finish floor level to finish floor level, is the total construction depth plus the clear floor to ceiling height at each story. Repeated as required, the sum of these dimensions equals the total building height and volume, although only the clear space is of real value. There is thus an efficiency relationship in the ratio of clear height to total story height that constrains the depth allowed for the floor construction, and structural

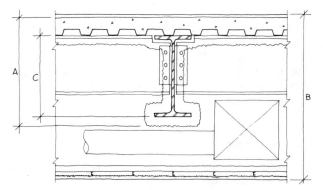

FIGURE 2.13. Dimensional relationships in floor–ceiling systems for multistory buildings. *A*: total depth of the structure. *B*: total depth of the floor–ceiling construction. *C*: net usable depth for the major spanning elements.

efficiency is often compromised in favor of other economic concerns, such as efficiency of the ducting system and total building skin surface. A critical limit for structural design is the net dimension permitted for the largest spanning elements (*C* in Fig. 2.13), which must be established cooperatively by the designers of the various building subsystems.

Sometimes it is possible to avoid placing the largest of the contained elements (usually air ducts) under the largest spanning elements. Some techniques for accomplishing this are shown in Fig. 2.14.

An important architectural aspect of the multiple-level building is the plan of the vertical supporting elements,

since these represent fixed objects around which interior spaces must be arranged. Because of the stacking required, vertical structural elements are often a constant plan condition at each level, despite possible changes in architectural requirements at the various levels. An apartment building with parking in the lower levels presents the problem of developing plans containing fixed locations of vertical structural elements that accommodate both the multiple parking spaces and the rooms of the apartments.

Vertical structural elements are usually walls or columns, situated in one of three possible ways, as shown in Fig. 2.15.

As isolated and freestanding columns or wall units in the interior of the building.

As columns or walls at the location of permanent features such as stairs, elevators, toilets, or duct shafts.

As columns or walls at the building periphery.

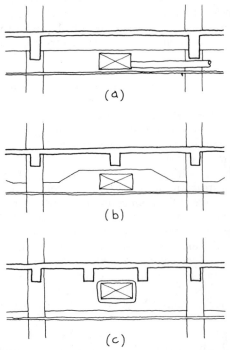

FIGURE 2.14. Accommodating air ducts in the floor–ceiling system: (*a*) by running major ducts parallel to major beams; (*b*) by varying beam depth; (*c*) by penetrating exceptionally deep beams.

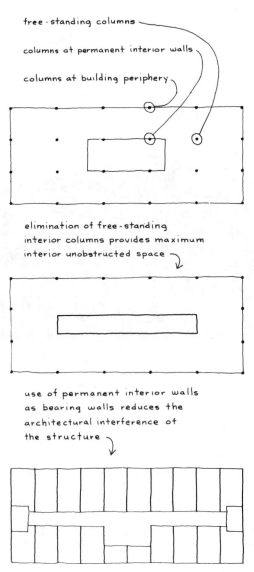

FIGURE 2.15. Development of vertical supports in multistory buildings.

Freestanding interior columns tend to be annoying for planning, because they restrict the placement of doors and hallways and are usually not desirable objects within rooms. They are clumsy to incorporate into thin walls, as shown in Fig. 2.16, producing lumps that interfere with furniture placing or door swings. Planning decisions often involve the choice of one of a number of undesirable situations.

This annoyance has motivated some designers to plan multiple-level buildings with very few, if any, freestanding interior columns. The middle plan in Fig. 2.15 shows such a solution in which vertical supports occur only at the periphery and at fixed central elements. For buildings with fixed plan modules, such as hotels, dormitories, and jails, a bearing-wall system using frequent permanent interior walls may be acceptable (see the lower plan in Fig. 2.15).

When columns are placed at the building periphery, their relationship to the building skin wall has a great bearing on the exterior appearance as well as interior planning. Figure 2.17 shows five possible locations for columns relative to the exterior wall plane, each of which has various merits and problems.

Although freestanding interior columns (Fig. 2.17a) are usually the least desirable, they may be tolerated if they are small (as in a low-rise building) and are of an unobtrusive shape (round, octagonal, etc.). In some cases they may be treated as significant features of the design. In framed structures of wood or steel the cantilevered edge implied by the interior columns usually presents a clumsy framing problem. In poured-in-place concrete structures, however, the cantilever is simply achieved and may even be advantageous to the structure in that it reduces stress on interior spans and aids the transfer of load to the columns.

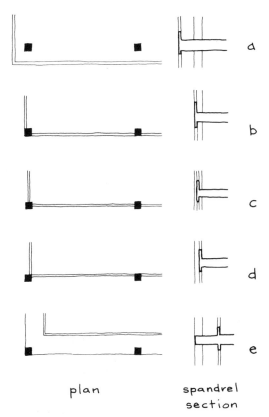

plan   spandrel
                  section

FIGURE 2.17. Relations of the structure to the building skin.

Placing columns totally outside the wall (Fig. 2.17e) eliminates both the interior planning lump and the required cantilevered edge. A continuous exterior ledge is produced and may be used as a sun shield, for window washing, as a balcony, or as an exterior corridor. However, unless some such use justifies it, the ledge may be a nuisance, creating water runoff and dirt accumulation problems. The totally exterior column also creates a potential problem with thermal expansion, as discussed in Chapter 3.

If the wall and column are joined, three possibilities exist for the usually thick column and usually thin wall, as shown in Fig. 2.17b–d. For a smooth exterior surface, the column may be flush with the outside of the wall, although this creates the same problems for planning that were discussed for interior freestanding columns. If the wall is aligned with the inside edge of the column, the interior surface will be smooth (for ease of interior planning), but the outside will be dominated by the vertical ridges of the columns. The least useful scheme would seem to be to place the column midway in the wall plane (Fig. 2.17c). Another solution, of course, is to thicken the wall sufficiently to accommodate the column—a neat architectural trick, but generally resulting in considerable wasted plan space.

In tall buildings, column sizes usually vary from top to bottom, although it is possible to achieve considerable range of strength within a fixed dimension, as shown in Fig. 2.18. Although some designers prefer the more honest expression of function represented by varying the column

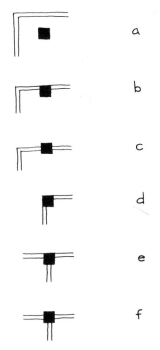

FIGURE 2.16. Interior column–wall relationships.

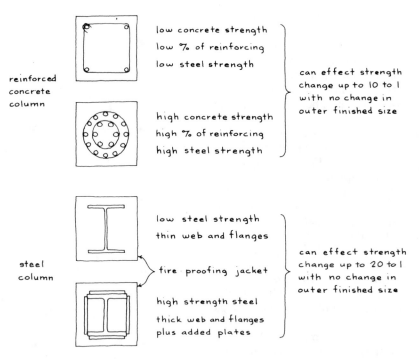

FIGURE 2.18. Variation in column strength without change in architectural finished size.

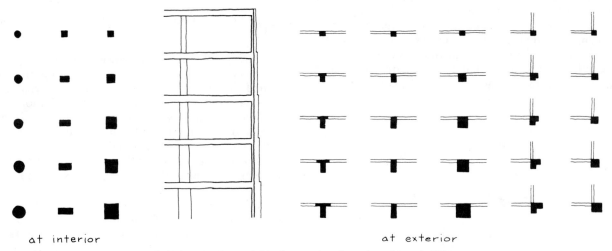

at interior   at exterior

FIGURE 2.19. Patterns of size increase for columns in multistory buildings.

size, planning is often simplified by the use of a constant column size.

Planning problems usually make it desirable to reduce column size as much as possible. If size changes for interior columns are required, the usual procedure is to have the column grow concentrically, as shown in Fig. 2.19. Exceptions are columns at the edges of stairwells or elevator shafts, where it is usually desired to keep the inside surface of the shaft aligned vertically.

For exterior columns, size changes relate to the column-to-skin relationship. If the wall is aligned with the inside edge of the column, there are several ways to let it grow in size without changing this alignment.

In very tall buildings lateral bracing often constitutes a major consideration in development of the structure and in development of architectural planning. In regions of high earthquake probability or frequent violent windstorms, the lateral bracing problem may dominate planning even for small and low-rise buildings.

2.4. THE BUILDING–GROUND RELATIONSHIP

As shown in Fig. 2.20, there are five basic variations of this relationship.

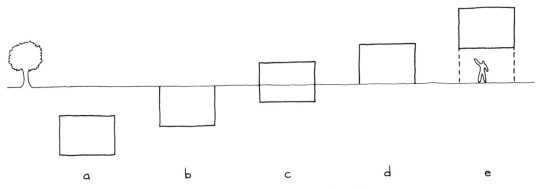

FIGURE 2.20. Building–ground relationships.

Subterranean Building

Figure 2.20*a* illustrates a situation that includes bomb shelters, subway stations, and underground parking. The insulating effect of the ground can be useful for interior thermal control in extreme climates. Exterior building surfaces must deal with soil pressure, water penetration, and deterioration conditions in general caused by constant contact with soil, which strongly limits the choice of materials and construction details for the roof, floor, and walls.

If the distance below the ground surface is great, the soil load limits the feasibility of spanning large unobstructed spaces. Exterior walls will be generally unpenetrated, although the passage of people, air, and various building services must be dealt with. There is no concern for wind, and design for earthquakes is a different matter than for the aboveground building.

Ground-Level Roof

Figure 2.20*b* illustrates a situation similar to the totally submerged building, except that the single surface exposed to air offers some possibilities for direct light and ventilation. Roof loading is less critical than in the case of a submerged building, although it is likely to have some traffic requiring the use of paving. Weight of paving plus traffic will constitute a load considerably higher than the usual one for a roof, and the feasibility of spanning large unobstructed spaces is questionable.

Limitations on construction in contact with the ground are the same as for the submerged building. The roof is likely to be penetrated with openings for air, plumbing vents, light, and passage of occupants. Some possibilities of opening such a building to dissipate the buried feeling are shown in Fig. 2.21.

Partially Submerged Building

In this case the building often consists of two structural elements: the superstructure (aboveground) and the substructure (below ground). The structure below ground has all of the problems of the submerged building and in addition must support the superstructure. If the superstructure is very tall, gravity loads will be high and the substructure will have the major task of serving to distribute the vertical load to the supporting earth materials. The horizontal forces due to wind and earthquakes must also be transferred through the substructure to the ground.

Substructures are usually built of concrete or masonry. If the superstructure is also built of concrete or masonry, there may be some continuity of the systems for the two elements. If the superstructure is built of wood or steel, the building will usually consist of two structures, one on top of the other.

The superstructure must deal with the various barrier and filter functions illustrated in Fig. 2.1. Penetrations for doors and windows must be facilitated. The superstructure is highly visible, whereas the substructure is largely not; thus the form of the superstructure is usually of much greater concern in architectural design.

Grade-Level Floor

Years ago basements were common; they were often required for the housing of a gravity heating system and for

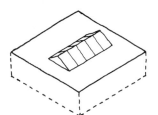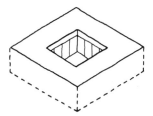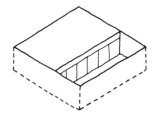

FIGURE 2.21. Opening up a building with a grade-level roof.

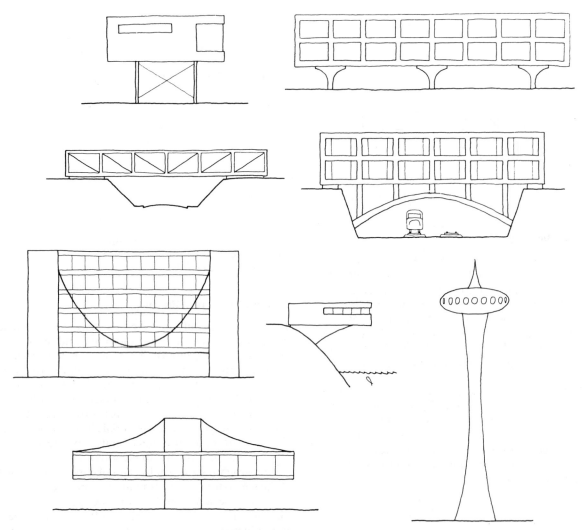

FIGURE 2.22. Buildings aboveground.

storage of the wood or coal used as an energy source. They were also useful for food storage before refrigeration, for general storage of junk, and for cowering in during windstorms. The advent of forced-circulation heating systems, refrigeration, and the high cost of construction has limited the use of basements unless they are needed for parking or housing of extensive equipment.

If there is no basement, the building is reduced to a superstructure and a foundation. If the building is short and vertical loads are low, the foundations can be minimal. If there is no frost problem and surface-level soils are adequate, the foundations may extend a very short distance below the ground surface. The floor at ground level may consist of a simple paving of the ground surface, called a *slab on grade*.

A problem that can develop with a light building having no basement is that the absence of a heavy substructure critically reduces resistance to the horizontal movement or the toppling (called *overturn*) effects caused by wind and earthquakes. Thus although only minimal foundations may be required for gravity loads, they would have to be increased in mass or ballasted with soil fill to provide the necessary anchorage for the building.

Aboveground Building

As shown in Fig. 2.22, buildings are sometimes built on legs, are cantilevered, or are suspended so that they are literally in midair. The support structures must be built on or into the ground, but the building proper has no direct contact with the ground.

The bottom floor of such a building must be designed for the barrier and filter functions usually associated only with roofs and exterior walls. In addition, its underside is often visible, becoming an unusual design problem.

Support structures may be quite modest if the buildings are not large or imposing. However, if the open span beneath the building is great or the height aboveground is considerable, the support structure may become a dominant element. Because most buildings are approached at ground level, the support structure and exposed underside of the building are important architectural design issues.

CHAPTER THREE

Structural Functions

In Chapter 2 the role of the structure is discussed in terms of the various uses and architectural design considerations for the building. Now let us consider the problems of the structure created in performing its various load-resisting functions. Basic issues to be dealt with are

The load sources and their effects.

What the structure accomplishes in terms of its performance as a supporting, spanning, or bracing element.

What happens to the structure internally as it performs its various tasks.

What is involved in determining the necessary structural elements and systems for specific required tasks.

3.1. LOAD SOURCES

The term *load* refers to any effect that results in a need for some resistive effort on the part of the structure. There are many sources for loads and many ways in which they can be classified. The principal kinds and sources of loads on building structures are the following:

Gravity

Source. Weight of the structure and other parts of the building; weight of occupants and contents; weight of snow, ice, or water on roof.

Calculation. By determination of the volume, density, and type of dispersion of items.

Application. Vertically downward, constant in magnitude.

Wind

Source. Moving air, in fluid-flow action.

Calculation. From anticipated maximum wind velocities established by local weather history.

Application. As pressure (perpendicular to surfaces) or frictional drag (parallel to surfaces); basically as an overall horizontal force effect on the building, although any surface may be affected in relation to its own individual geometry or orientation.

Earthquakes (Seismic Shock)

Source. Shaking of the ground as a result of large subterranean faults, volcanic eruptions, or underground explosions.

Calculation. By prediction of the probability of occurrence on the basis of the history of the region and records of previous seismic activity; principal force effect is the horizontally impelled inertial action of the building's own mass.

Application. Back-and-forth, up-and-down movement of the supporting ground; response of the building structure on the basis of its own dynamic properties.

Hydraulic Pressure

Source. Principally from groundwater when the free level is above the bottom of the basement.

Calculation. As fluid pressure proportional to the depth of the fluid.

Application. As horizontal pressure on walls, upward pressure on floors.

Soil Pressure (Active)

Source. Action of soil as a semifluid on objects buried in the ground.

Calculation. Usually by considering the soil as an equivalent fluid with a fluid density some fraction of the true soil density.

Application. Horizontal pressure on walls.

Thermal Change

Source. Temperature variations in the building materials from fluctuations in outdoor temperature and inside–outside temperature differences.

Calculation. From weather histories, internal design temperatures, and coefficients of expansion of the materials.

Application. Forces exerted on structure if free expansion is restrained; distortions and stresses within structure if connected parts differ in temperature.

Shrinkage. Volume reduction in concrete, in mortar joints in masonry, and in large unseasoned timber elements; may produce forces similar to those caused by thermal effects.

Vibration. In addition to seismic effects, vibrations may be caused by heavy machinery, vehicles, or high-intensity sounds.

Internal Actions. Forces may be induced by the settling of supports, slippage or loosening of connections, warping of elements, and so on.

Handling. Forces are exerted on structural elements during production, erection, transportation, storage, remodeling, and so on. These effects are not necessarily evident in the form of the finished building, but must be considered in its production.

3.2. LIVE AND DEAD LOADS

In building design a distinction is made between *live* and *dead loads.* A *dead load* is essentially a permanent load, such as the weight of permanent parts of the building construction. A *live load* is technically anything that is not permanently applied as a force on the structure. However, the specific term *live load* is generally used in building codes to refer to the assumed design loads in the form of dispersed gravity load on roof and floor surfaces as a result of the location and the particular usage of the building.

3.3. STATIC VERSUS DYNAMIC FORCES

A slightly different distinction exists between static and dynamic force effects. This distinction essentially has to do with the time-dependent character of the force. Thus the weight of the structure produces a static effect unless the structure is suddenly moved or stopped from moving, at which time a dynamic effect occurs through the inertia or momentum of the mass of the structure. The more sudden the start or stop, the greater the dynamic effect (see Fig. 3.1*a*).

Other dynamic forces are produced by ocean waves, earthquakes, blasts, sonic booms, vibration of heavy machines, and the bouncing effect of people walking. The effects of dynamic forces are very different from those of static forces. A light steel-framed building, for example, may be very strong in resisting static forces, but a dynamic loading may cause large distortions or vibrations, resulting in cracking of plaster, loosening of connections, and so on. A heavy masonry structure, although possibly not as strong as the steel frame for static load resistance, has considerable stiffness and dead weight and may thus absorb the energy of dynamic effects without any perceptible movement.

In the examples just cited, the effect of the force on the function of the structure was described. This may be dis-

tinct from the effect on the structure itself. The steel frame is flexible and responds with motions that may be objectionable. However, from a structural point of view it is probably more resistive to dynamic force than the masonry structure. Structural steel is ductile and the flexible frame dissipates some of the energy of the dynamic load through its motion, similar to a boxer rolling with a punch. Masonry, in contrast, is brittle and stiff and absorbs the energy almost entirely in the form of shock to the material. In evaluating dynamic force effects and the response of structures to them, both of these considerations must be made: the behavior of the structure itself and the effects on its usefulness in the building.

3.4. LOAD DISPERSION

Forces may be distinguished by the manner of their dispersion. Gas under pressure in a container exerts a pressure that is uniform in all directions at all points. The dead load of roofing, the weight of snow on a flat roof, and the weight of water on the flat bottom of a tank are all loads that are uniformly distributed on a surface. The weight of a beam or a cable is a load that is uniformly distributed along a line. The foot of a column or the end of a beam represents loads that are concentrated at a relatively small location (see Fig. 3.1*c*).

Randomly dispersed live loads may result in unbalanced conditions or in reversals of internal forces in the structure (see Fig. 3.1*d*). The shifting of all the passengers to one side of a ship can cause the ship to capsize. A concentration of load in one span of a beam that is continuous through several spans may result in upward deflection in adjacent spans or lifting of the beam from some supports. Because live loads are generally variable in occurrence, location, and sometimes even in direction, several combinations of them must often be investigated in order to determine the worst effects on a structure.

3.5. LOAD COMBINATIONS

A difficult judgment for the designer is that of the likelihood of the simultaneous occurrence of various force effects. Combinations must be considered carefully to determine those that cause critical situations and that have some reasonable possibility of actual simultaneous occurrence. For example, it is generally considered unreasonable to design for the simultaneous occurrence of the highest anticipated wind velocity and the strongest earthquake. It is also not possible for the wind to blow from two separate directions at the same time, although wind from all directions must be individually considered.

3.6. REACTIONS

Successful functioning of the structure in resisting various loads involves two considerations. The structure must have

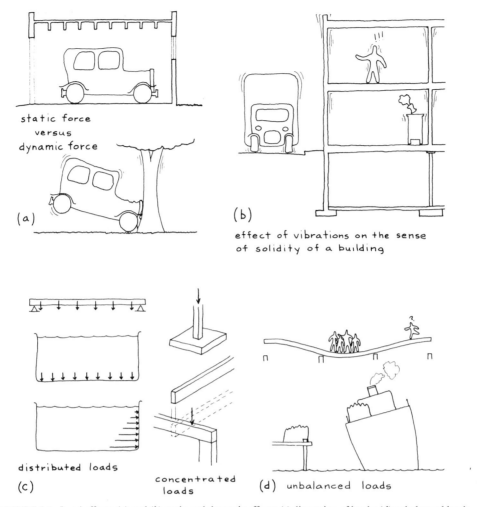

FIGURE 3.1. Load effects: (*a*) and (*b*) static and dynamic effects; (*c*) dispersion of loads; (*d*) unbalanced loads.

sufficient internal strength and stiffness to redirect the loads to its supports without developing undue stress on the materials or an undesirable amount of deformation in the form of sag, stretching, twisting, and so on. In addition, the supports for the structure must keep the structure from moving or collapsing. These support forces are called *reactions*.

Figure 3.2 shows a column supporting a load that generates a linear compressive effect. The reaction generated by the support must be equal in magnitude and opposite in sense (up versus down) to the combined load, which is the sum of the applied load plus the weight of the column. The balancing of the active loads and the reactions produces the necessary static condition for the structure. This condition is referred to as the *state of static equilibrium*.

Figure 3.2 also shows the reaction forces required for various spanning structures. For the beam the reactions consist of two vertical forces whose sum must be equal to the sum of the applied loads plus the beam weight. If the applied load is not symmetrical, these two reaction forces will not be equal, although their sum must still be equal to the load on the beam.

For the gable frame the reactions must provide horizontal as well as vertical resistance, even though the load on the structure is entirely vertical. The horizontal forces are required to keep the frame from moving outward at the supports. The net reaction forces are thus combinations of the vertical and horizontal force components required for the complete equilibrium of the structure.

The arch and the cable also require both horizontal and vertical reaction components. When the cable sag or the arch rise is low in comparison to the span, the horizontal component is very large. Thus the magnitude of the force of compression in the arch or tension in the cable may be considerable, even though the vertical load on the span is relatively small.

There is another type of reaction effort that can be visualized by considering the situation of the cantilever beam, as shown in Fig. 3.2. Since there is no reaction force at the free end of the cantilever, the supported end must develop resistance to rotation as well as to vertical movement. This rotational effect is called *moment* and has a unit that is different from that of simple direct force. Force is measured in units of pounds, tons, and so on. The moment effect is a product of force times distance, producing a unit

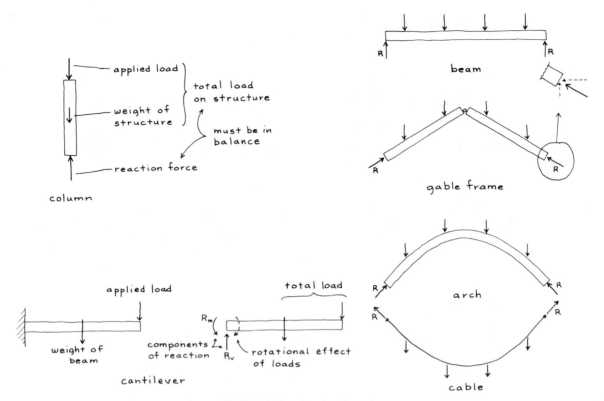

FIGURE 3.2. Development of reactions.

of pound-feet or some other combination of force and length units. The total reaction for the cantilever therefore consists of two components that cannot be combined: the vertical force (R_v) and the resisting moment (R_m).

For the rigid frame shown in Fig. 3.3 there are three possible components of the reactions. If vertical force alone is resisted at the supports, the bottoms of the frame columns will move outward and rotate, as shown in Fig. 3.3*a*. If horizontal resistance is developed, as is shown for the gable frame, the arch, and the cable in Fig. 3.2, the column bottoms can be pushed back to their original positions but will still rotate, as shown in Fig. 3.3*b*. Finally, if a moment resistance is developed at the supports, the column bottoms can be held completely in their original positions, as shown in Fig. 3.3*c*. For this total resistance to movement at the supports, the reactions must develop all three components as shown.

The applied loads and support reactions for a structure constitute the external force system, which operates on the structure. This system of forces is in some ways independent of the structure's ability to respond. That is, the external forces must be in equilibrium if the structure is to be functional, regardless of the materials, strength, stiffness, and so on, of the structure itself. However, the form of details of the structure may affect the nature of the required reactions. The span and applied loads may be the same for a beam, a gable, an arch, a cable, and a rigid frame, but the required reactions will be affected by the specific structure.

3.7. INTERNAL FORCES

In response to the external effects of loads and reactions, certain internal forces are generated within a structure as the material of the structure strives to resist the deformations induced by the external effects. These internal force effects are generated by *stresses* in the material of the structure. The stresses are actually incremental forces within the material, and they result in incremental deformations, called *strains*.

When subjected to external forces a structure sags, twists, stretches, shortens, and so on; or, to be more technical, it stresses and strains, thus assuming some new shape as the incremental strains accumulate into overall dimensional changes. Whereas stresses are not visually apparent, their accompanying strains often are.

As shown in Fig. 3.4, a person standing on a wooden plank that spans two supports will cause the plank to sag downward and assume a curved profile. The sag may be visualized as the manifestation of a strain phenomenon accompanying a stress phenomenon. In this example the principal cause of the structure's deformation is bending resistance, called *internal bending moment*. The stresses associated with this internal force action are horizontally directed compression in the upper portion of the plank and horizontally directed tension in the lower portion. Anyone could have predicted that the plank would assume a sagged profile when the person stepped on it; but we can

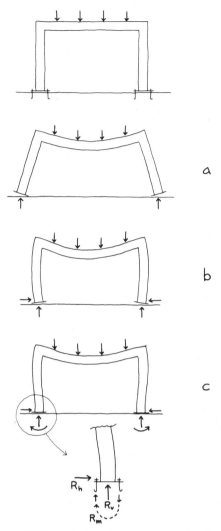

FIGURE 3.3. Reactions for a rigid frame.

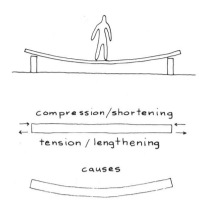

FIGURE 3.4. Development of bending.

Any structure must have certain characteristics in order to function. For purposes of structural action it must have adequate strength for an acceptable margin of safety and must have a reasonable resistance to dimensional deformation. It must also be inherently stable, both internally and externally. These three characteristics—stability, strength, and stiffness—are the principal functional requirements of structures.

3.8. STRESS AND STRAIN

Internal force actions are resisted by stresses in the material of the structure. There are three basic types of stress: tension, compression, and shear. Tension and compression are similar in nature, although opposite in sign or direction. Both tension and compression produce a linear type of strain and can be visualized as pressure effects perpendicular to the surface of a stressed cross section. Because of these similarities, both tension and compression stress are referred to as direct stresses; one is considered positive and the other negative.

Shear stress occurring in the plane of a cross section is similar to a frictional sliding effect. Strain due to shear is different from that due to tension or compression, consisting of an angular change rather than a linear shortening or lengthening.

3.9. DYNAMIC EFFECTS

Vibrations, moving loads, and sudden changes in the state of motion, such as the jolt of braking or rapid acceleration of vehicles, cause forces that result in stresses and strains in structures. The study of dynamic forces and their effects is complex, although a few of the basic concepts can be illustrated simply.

For structural investigation and design the significant distinction between static and dynamic effects has to do with the response of the structure to the loading. If the principal response of the structure can be effectively evaluated in static terms (force, stress, linear deformation, etc.), the effect on the structure is essentially static, even though the load may be time dependent in nature. If, however, the structure's response can be evaluated effectively only in

also predict the deformation as an accumulation of the strains, resulting in the shortening of the upper portion and the lengthening of the lower portion of the plank.

For the relatively thin wooden plank, the bending action and strain effects would be quite apparent. If the plank is replaced by a thick wood beam, the sag may not be visible. However, the sag and the internal bending, however slight, do exist. For structural investigation we exaggerate deformed profiles, considering structures to be considerably more flexible than they really are.

Because stress and strain are inseparable, it is possible to infer one from the other. This allows us to visualize the nature of internal force effects by imagining the exaggerated form of the deformed structure under load. Thus although stresses cannot be seen, strains can, and the nature of the accompanying stresses can be inferred. This relationship can be used as a simple visualization device or it can be used in actual laboratory testing where quantified stresses are determined by careful measurement of strains.

terms of energy capacity, work accomplished, cyclic movements, and so on, the effect of the loading is of a true dynamic character.

A critical factor in the evaluation of the structure's response to dynamic loads is the *fundamental period* of the structure's vibration. This is the time required for one full cycle of motion, in the form of a bounce or a continuing vibration. The relation of this time to the time of buildup of the load is a major factor in determining the relative degree of the dynamic effect on the structure. A structure's fundamental period may vary from a fraction of a second to several seconds, depending on the size, shape, mass, material, stiffness, support constraints, and possible presence of various damping effects.

Design for dynamic effects begins with an evaluation of possible sources and their potential effects. The response of the structure is then considered, using the variables of its mass, natural vibration, energy-absorbing capabilities, natural damping, and so on. Once the dynamic behavior is understood, the designer can consider how to manipulate the variables to improve the structure's behavior or to reduce the load effects.

3.10. DESIGN FOR STRUCTURAL BEHAVIOR

In professional design practice the investigation of structural behavior is an important part of the design process. To incorporate this investigation into the design work the designer needs to develop a number of capabilities, including the following:

1. The ability to visualize and evaluate the sources that produce loads on the structure.
2. The ability to quantify the loads and the effects they produce on the structure.
3. The ability to analyze the structure's response to the loads in terms of forces, stresses, and strains.
4. The ability to determine the structure's limits of load-carrying capacity.
5. The ability to manipulate the variables of material, form, and construction details for the structure in order to optimize its responses to loads.

Although analysis of stresses and strains is necessary in the design process, there is a sort of chicken-and-egg relationship between analysis and design. To analyze some of the structure's responses, one needs to know some of its properties, but to determine the necessary properties, one needs some results from the analysis. In simple cases it is sometimes possible to derive expressions for desired properties by the simple inversion of analytical formulas. For example, in a simple compression member, if the load produces a uniform stress on the cross section, the formula for this simple stress is

$$\text{stress} = \frac{\text{total load on the member}}{\text{area of the cross section}}$$

Therefore, if the load is known and the stress limit for the material is established, this formula can be easily con-

verted to one for finding the required area of the cross section, as follows:

$$\text{required area} = \frac{\text{total load on the member}}{\text{stress limit for the material}}$$

Most structural situations are more complex, however, and involve variables and relationships that are not so simply converted for design use. In the case of the compression member, for example, if the member is a slender column, its load capacity will be limited to some degree by the tendency to buckle. The relative stiffness of the member in resisting buckling can be determined only after the geometry of the cross section is known. Therefore, the design of such an element is a hit-or-miss situation, consisting of guessing at an appropriate cross-sectional shape and size, analyzing for its load capacity, comparing the answer to the original loading, and refining as necessary until a reasonable fit is established.

Professional designers use their own experience together with various design aids, such as tabulations of capacities of commonly used elements, to shortcut the design process. Detailed computations for structural analysis are performed in many instances only after various shortcuts and approximations have been used to establish a preliminary design of the structure.

3.11. INVESTIGATION OF STRUCTURAL BEHAVIOR

Whether for design purposes, for research, or for the study of structural behavior as a learning experience, analysis of stresses and strains is important. Analysis may be performed as a testing procedure on the actual structure with a loading applied to simulate actual usage conditions. If carefully done, this is a highly reliable procedure. However, except for some of the commonly used, simple elements of construction, it is generally not possible to perform destructive load tests on building structures built to full scale. The behavior of building structures must usually be anticipated speculatively on the basis of demonstrated performance of similar structures or on a modeling of the actions involved. The modeling can be done in the form of physical tests on scaled-down structures, but is most often done mathematically, using the current state of knowledge in the form of formulas for analysis. When the structure, the loading conditions, and the necessary formulizations are relatively simple, computations may be done "by hand." More commonly, however, computations of even routine nature are done by professional designers using computer-assisted techniques. It is imperative that the designer keep the upper hand in this situation by knowing what the computer is doing.

Sources for Further Study

M. Salvadori and R. Heller, *Structure in Architecture: The Building of Buildings*, 3rd ed., Prentice-Hall, 1986.

H. Cowan and F. Wilson, *Structural Systems*, Von Nostrand, 1981.

CHAPTER FOUR

Structural Materials

All materials—solid, liquid, or gaseous—have some structural nature. The air we breathe has a structural nature: it resists compression. Every time you ride in an automobile you are sitting on an air-supported structure. Water supports the largest human-made vehicles: huge ocean liners and battleships. Oil resists compression so strongly that it is used as the resisting element in hydraulic presses and jacks capable of developing tremendous force.

In the design of building structures, use is made of the available structural materials and the products formed from them. The discussion in this chapter deals with common structural materials and their typical uses in contemporary construction.

4.1. GENERAL CONSIDERATIONS

Broad classifications of materials can be made, such as distinctions among animal, vegetable, and mineral; between organic and inorganic materials; and among the physical states of solid, liquid, and gaseous. Various chemical and physical properties distinguish individual materials from others. In studying or designing structures, particular properties of materials are of concern. These critical properties may be split between essential structural properties and general properties.

Essential structural properties include the following:

Strength. May vary for different types of force, in different directions, at different ages, or at different amounts of temperature or moisture content.
Deformation Resistance. Degree of rigidity, elasticity, ductility; variation with time, temperature, and so on.
Hardness. Resistance to surface indentation, scratching, abrasion, or wear.
Fatigue Resistance. Time loss of strength; progressive fracture; shape change with time.
Uniformity of Physical Structure. Grain and knots in wood, cracks in concrete, shear planes in stone, effects of crystallization in metals.

General properties of interest in using and evaluating structural materials include the following:

Form. Natural, reshaped, or reconstituted.
Weight. As contributing to gravity loading of the structure.

Fire Resistance. Combustibility, conductivity, melting point, and general behavior at high temperatures.
Coefficient of Thermal Expansion. Relating to dimensional changes with change in temperature; critical when different materials are mated, as in lamination, reinforced concrete, or jacketed columns.
Durability. Resistance to weathering, rot, insects, and wear.
Workability. In producing, shaping, assembling, altering.
Appearance. Natural or reworked.
Availability and Cost.

In any given situation choices of materials must often be made on the basis of several properties—both structural and general. For any specific structural task there is seldom a single material that is superior in all respects, and the importance of various properties must often be ranked.

4.2. WOOD

Technical innovations have overcome some of the long-standing limitations of wood. Size and form limitations have been extended through glue lamination (see Fig. 4.1). Special fastening techniques have made large structures possible through better jointing. Combustibility, rot, and insect infestation can be retarded by utilizing chemical impregnations. Treatment with steam or ammonia gas can render wood highly flexible, allowing it to assume plastic forms.

Dimensional movements from changes in temperature or moisture remain a problem in wood. Fire resistance can be developed only to a degree. Although easily worked, wood elements are soft and readily damaged; thus damage due to handling and use is a problem.

Although hundreds of species exist, building structural use is limited primarily to a few softwoods: Douglas fir, southern pine, northern white pine, spruce, redwood, cedar, and hemlock. Local availability and cost are major factors in the selection of a particular species. Economy is generally achieved by using the lowest grade (quality) of material suitable for the work. Grade is influenced by the lack of knots and splits and by the particular grain characteristics of an individual piece.

FIGURE 4.1. Glued-laminated wood ribs radiate from a central core in 20°-increment spacing to form the roof of this small library. Roof deck is glued-laminated wood hung in tension from the ribs. Thomas O. Freeman Library, Lake Forest College, Lake Forest, Illinois. Architects: Perkins and Will, Chicago. Structural engineers: The Engineers Collaborative, Chicago.

FIGURE 4.2. Typical steel framing elements: rolled columns and beams, light prefabricated truss joists, formed sheet steel deck.

One technological process has produced a unique wood element: the plywood sheet. Although other sheet-form materials such as fiberboard and gypsum plasterboard have emerged to compete with plywood, it remains in wide use, both as a surfacing material and as part of a variety of structural wood-plywood fabricated products.

In the development of structural components and systems it is not always wise or possible to use a single material. Materials are often mixed, each performing its appropriate tasks, such as in reinforced concrete construction. Light, medium-span, roof trusses often have wood top and bottom chords and steel zigzag interior elements.

Because of its availability, low cost, and simple working possibilities, wood is used extensively for secondary and temporary construction—that is, for scaffolding, bracing, and forming of poured concrete. However, it is also widely used for permanent construction and is generally the structural material of choice unless its limitations preclude its use.

4.3. STEEL

Steel is used in a variety of types and forms in nearly every building. From huge columns to the smallest nails, steel is the most versatile of traditional structural materials. It is also the strongest, the most resistant to aging, and generally the most reliable in quality. Steel is a completely industrialized material and is subject to tight control of its content and the details of its forming and fabrication. It has the additional desirable qualities of being noncombustible, nonrotting, and dimensionally stable with time and moisture change.

Although the bulk material is expensive, steel can be used in small quantities because of its great strength and its forming processes, making it competitive with materials of lower bulk cost. Economy can also be achieved through mass production of standardized items. (See Fig. 4.2.)

Two principal disadvantages of steel for building structures are inherent in the material. These are its rapid heat gain and resultant loss of strength when exposed to fire, and its rusting when exposed to moisture and air or corrosive conditions. Several techniques can be used to overcome its fire sensitivity, including the use of special coatings that expand in volume when heated, forming a noncombustible surface insulation. Coatings of one kind or another are also possible means of protection against corrosion, although some special steels resist ordinary air rusting sufficiently to be left exposed without any treatment. So-called stainless steel is a special alloy both too costly and not in general having the desired properties for structural use.

In times past, encasing of steel frames in concrete was a common means of achieving fire protection. This is still done in some cases where concrete slabs or walls are used in conjunction with steel frames. It is now more common, however, to surround the steel structure with construction of masonry, plaster, plasterboard, or some other fire-resistive materials. When it is not possible to use the encasing construction for protection, a special material can be sprayed onto the steel surfaces, where it adheres and is built up to a thickness appropriate for the desired degree of insulation. Although somewhat messy, this material offers the advantage of easily covering complicated forms such as trusses and the undersides of formed sheet steel decks. If wall and ceiling construction covers the structure, the lack of visual delight is not a problem.

The vocabulary of steels in use for building structures has recently been expanded, and there is now a wide range from which to choose the correct steel for a particular situation of stress magnitude, corrosion, form of elements, or fastening technique. It is not uncommon for the various elements of the framework of a steel building to consist of a dozen or more varieties of material, with a wide spectrum of property variation. For example, the steel used in the wire of woven cable is five times as strong as that used for ordinary steel beams!

4.4. CONCRETE

In building construction, the word *concrete* is used to describe a variety of materials having one thing in common: the use of a binding agent to form a solid mass from a loose, inert aggregate. The three basic ingredients of ordinary concrete are water, a binding agent (such as cement), and a large volume of loose aggregate (such as sand and gravel). Tremendous variation of the end product is possible through the use of different binders and aggregate and with the use of special chemicals and air-void-producing foaming agents.

Ordinary concrete has several attributes, chief among which are its low bulk cost and its resistance to moisture, rot, insects, fire, and wear. Because it is formless in its mixed condition, it can be made to assume a large variety of forms. Large-scale, monolithic structures are naturally formed with this material (see Fig. 4.3).

One of concrete's chief shortcomings is its lack of tensile stress resistance. The use of inert or prestressed reinforcing is imperative for any structural function involving considerable bending or torsion. Precisely because the material is formless, its forming and finishing is often one of the major expenses in its use. Factory precasting in permanent forms is one current technique used to overcome this problem (see Fig. 4.4).

4.5. ALUMINUM

In alloyed form, aluminum is used for a large variety of structural, decorative, and functional elements in building construction. Principal advantages are its light weight (one-third that of steel) and its high resistance to corrosion. Among the disadvantages are its softness, its low stiffness, its high rate of thermal expansion, its lower resistance to fire, and its relatively high cost (see Fig. 4.5).

FIGURE 4.4. The exterior wall structure of this high-rise building consists of hollow, precast concrete units filled with sitecast concrete during the construction of the floor structure. This technique preserves many of the best qualities of both precast and sitecast construction. Beneficial Standard Life Insurance Company Building, Los Angeles. Architects: Skidmore, Owings and Merrill, San Francisco. Precast units by Rockwin Schokbeton, Santa Fe Springs, California.

Large-scale structural use in buildings is limited primarily because of cost or increased dimensional movements caused by the low stiffness of the material. This low stiffness also reduces its resistance to buckling. Small-scale structural use—wall and roof skin panels, door and window frames, and hardware—is considerable, however. Here its corrosion resistance, easy working character, and the possibilities for its forming in production are used to good advantage.

4.6. MASONRY

The term *masonry* is used to describe a variety of formations consisting of separate elements bonded together by some binding filler. The elements may be cut or rough stone, fired clay tile or bricks, or cast units of concrete. The binder is traditionally cement–lime mortar, although considerable effort is being made in experimentation with various new adhesive compounds. The resulting assemblage is similar to a concrete structure and possesses many of the same properties. A major difference is that the construction process does not usually require the same amount of temporary forming and bracing as it does for a structure of poured concrete. However, it requires considerable hand labor, which imposes some time limitations and makes the end product highly subject to the individual skill of the craftsperson.

Reinforcing techniques have been developed in recent years to extend the structural possibilities of masonry. Figure 4.6 shows a typical form of construction currently in wide use in the western United States because of its significantly improved resistance to earthquakes.

FIGURE 4.3. Reinforced concrete frame detailed to express the natural form of a structure of monolithic material. Surface texture is produced by careful detailing of the wood forms. Temple Street parking garage, New Haven, Connecticut. Architect: Paul Rudolph. Structural engineer: Henry A. Pfisterer.

FIGURE 4.5. Aluminum geodesic dome. This 144-ft-diameter dome was the first of its type produced by Kaiser Aluminum and Chemical Company. The dome consists of 575 preformed, diamond-shaped panels of thin aluminum sheets. The panels were completely erected, using a central rigging mast, in 20 working hours. Cast aluminum gussets serve the dual purposes of connecting the six panels that meet at a joint and facilitate attachment of the rigging cables. Architects: Welton Becket and Edwin Bauer (for the first dome—a 200-seat auditorium for the Hawaiian Village Hotel in Honolulu). Geodesic system patented by R. Buckminster Fuller. Photos: Werner Stoy, Honolulu, suppled by Kaiser Aluminum and Chemical Sales, Inc., Oakland, California.

FIGURE 4.6. Reinforced masonry construction. Steel reinforcing rods are placed both vertically and horizontally in the hollow voids and grout is placed around the rods, filling the voids. The result is in effect similar to the creation of a two-way rigid frame of reinforced concrete inside a wall.

Shrinkage of the mortar and thermal-expansion cracking are two major problems with masonry structures. Both necessitate extreme care in detailing, material quality control, and field inspection during construction.

4.7. PLASTICS

Plastic elements represent the widest variety of usage in building construction. The tremendous variation of material properties and formation processes provides a virtually unlimited field for the designer's imagination. Some of the principal problems with plastics are lack of fire resistance, low stiffness, high rate of thermal expansion, and some cases of chemical or physical instability with time.

Use of these materials advances steadily as they replace more traditional materials and also create entirely new functional possibilities. A few of the more important uses in building construction are

Glass Substitute. In clear or translucent form, as bubble-form skylights, windowpanes, and corrugated sheet panels (see Figs. 4.7, 5.18, and 5.24).

Coatings. Sprayed, painted, or rolled in liquid form, or laid on in films or sheets to provide protection for walls, roofs, foundation walls, and countertops.

Adhesives. The famous epoxy family of bonding and matrix binders for connecting and patching.

Formed Elements. Moldings, fixtures, panels, and hardware.

Foamed. In preformed or foamed-in-place applications, as insulation and filler for various purposes.

FIGURE 4.7. Domed plastic skylights. 180 bubble-shaped skylights in metal frames for the roof for this swimming pool at the Ambassador Motor Lodge in Minneapolis, Minnesota. The individual bubbles of Plexiglas are 90 in. square in plan. Designers: Synergetics, Inc., Raleigh, North Carolina. Photo: Rohm and Haas Company, Philadelphia.

The development in recent years of pneumatic and tension-sustained surface structures has spurred the development of various plastic membrane and fabric products for building use. Small structures may use thin plastic membranes, but for larger structures the surface material is usually a coated fabric with enhanced resistance to tension and tearing (see Fig. 5.20).

The plastic-surfaced structure can also be created by using plastic elements on a framework. A special variation of this is the steel cable structure that uses the cables to define a network and then uses translucent plastic elements to form the surface (see Fig. 5.18).

The use of plastics for building structures is still inhibited by the traditional conservatism of building regulatory agencies; by the largely unwarranted associations of cheapness and impermanence in the minds of architects, engineers, and public; by the oil shortage; and by increasing concern for the performance of plastics during fires. However, evolution of taste, custom, and design standards and the expansion of the technology will inevitably encourage its increased use.

4.8. MISCELLANEOUS MATERIALS

Glass. Ordinary glass possesses considerable strength, but has the undesirable characteristic of being brittle and subject to shattering under shock. Special treatment can increase its strength and shock resistance, but it is relatively expensive for use in large quantities. Large-scale structural use is not conceivable for this material. Considerable use is made, however, for surfacing panels as well as transparent window panes.

Fiberglass. A special use of glass is made by producing it in fibrous form, in which it is capable of realizing close to its ideal strength. This strength may exceed that of high-strength steel, and although its form restricts its usage, various structural utilizations can be achieved. One familiar use is that in which the fibers are suspended in a resin, producing fiberglass-reinforced plastic.

Paper. Paper—that is, sheet material of basically rag or wood fiber content—is used considerably in building construction. It has been replaced in many uses, however, by sheet plastic. Various coatings, laminations, impregnations, and reinforcings can be used to make the material water resistant, rot resistant, or tougher. Structural use, however, is limited to relatively minor functions, one being a forming material for poured concrete. Paper-faced plaster panels are used in "drywall."

The most ordinary of buildings will typically use a great number of different materials in its construction. A building structure is usually composed of some elements of all the basic materials. Wood- and steel-framed structures have concrete foundations; concrete and masonry structures use steel reinforcing; wood structures use a great many steel fastening devices; and so on. Structural design typically consists of matching choices from an almost endless number of possible combinations.

Sources for Further Study

H. Rosen, *Construction Materials for Architecture*, Wiley, 1985.

D. Ellison, W. Huntington, and R. Mickadeit, *Building Construction: Materials and Types of Construction*, 6th ed., Wiley, 1987.

CHAPTER FIVE

Structural Systems

The materials, products, and systems available for the construction of building structures constitute a vast inventory through which the designer must sift carefully for the appropriate selection in each case. The material in this chapter presents some of the general issues relating to this inventory.

5.1. ATTRIBUTES OF STRUCTURAL SYSTEMS

A specific structural system derives its unique character from any of a number of considerations—and probably from many of them simultaneously. Considered separately, these are the following.

1. Specific structural functions, some of which are support in compression (a pier, footing, or column); support in tension (a straight vertical hanger or a guy wire); spanning—horizontally (a beam in a floor), vertically (a sheet of window glass), or in some other position (a sloping rafter); cantilevering—vertically (a flagpole or tower) or horizontally (a balcony or canopy). A single element or system may be required to perform more than one of these functions in various situations of use.
2. The geometric form or orientation. Note the difference between the nature of the flat beam and the arch, both of which function as horizontally spanning structures. The difference is primarily one of overall structural form. Compare the arch with the draped cable, both in use as horizontally spanning elements. They are obviously different in function. The difference, however, is not one of form but of orientation to the load.
3. The material(s) of the elements.
4. The manner of joining the elements if the system consists of an articulated set of parts.
5. The manner of supporting the structure.
6. The specific loading conditions, or the forces the structure must sustain.
7. The separately imposed considerations of usage in terms of form and scale limits.
8. The limitations of form and scale of the elements and the nature of their joining imposed by the properties of the materials, the production processes, and

the need for special functions such as demountability and movement.

Structural systems occur in virtually endless variety. The designer, in attempting to find the ideal structure for a specific purpose, is faced with an exhaustive process of comparative "shopping." Most designers agree that except for a few special situations, there is no such thing as the ideal structure for any particular job. At best, the shopping can narrow the field to a few acceptable solutions.

A checklist of sorts can be used to rate the available systems for a given purpose. The following are some items that may be included in such a list.

Economy. The economy of the structure itself as well as its influence on the overall economy of the project. Special consideration may be given factors such as delay because of slow construction, adaptability to modifications, and first cost versus maintenance cost over the life of the structure.

Special Structural Requirements. Unique aspects of the structure's action, details required for development of its strength and stability, adaptability to special loading, need for symmetry or modular arrangement. Thus arches require horizontal resisting forces at their bases to resist their thrusts, tension elements must be hung *from* something, structures of thin metal parts must be stiffened for stability against buckling, and domes must have some degree of symmetry and a concentric continuity.

Problems of Design. Difficulty of performing an analysis of the structural action, ease of detailing the structure, ease of integrating the physical structure with the detail requirements of its use.

Problems of Construction. Availability of materials and of skilled labor and equipment, adaptability to unitized fabrication and assembly and to prefabrication, speed of erection, special requirements for temporary bracing or support, required skill and precision of field work.

Material and Scale Limitation. The feasibility range of size for specific systems. These vary with the development of new materials and new techniques and with experimentation of new uses for existing systems. Certain practical limits do exist and in many cases are virtually insurmountable.

FIGURE 5.1. Santa Ines Mission Church, Santa Ynez, California. Historical craft-developed building techniques are illustrated in this structure, built in the early part of the nineteenth century. Huge buttresses of adobe and brick brace the walls of the church—a reaction to the destruction of an earlier church by an earthquake. Hand-hewn wood beams and planks form the roof of the arcade, supported on one side by a series of brick arches. Although these forms and techniques are the end products of a long development of building tradition and still in effect valid, they have been made largely obsolete by contemporary industrialized technology.

5.2. CATEGORIZATION OF STRUCTURAL SYSTEMS

Structural systems can be categorized in a variety of ways. One broad differentiation is that made between solid structures, framed structures, and surface structures.

Solid structures (Fig. 5.2a) are those in which strength and stability are achieved through mass, even though the structure is not completely solid. Large piers and abutments, dams, seawalls, retaining walls, caves, and ancient burial pyramids are examples. These structures are highly resistive to forces such as those created by blasts, violent winds, wave action, and vibrations. Although their exact analysis may be highly indeterminate, the distribution of load stresses may be diffuse enough to allow simple approximations with a reasonable assurance of safety.

In *framed structures* (Fig. 5.2b) the essential structure consists of a network of assembled elements. The building, bridge, or ship is completed by filling in the voids as required between these spaced elements. Although the infilling elements may have a structural character themselves and serve to stiffen and brace the frame, they are not primary elements of the basic structure. Animal skeletons, steel beam and column systems, and trussed towers are examples. These structures are generally most adaptable to variations in form, dissymmetry of layout or detail, and the carrying of special loads. They can be cumbersome, however, if the complexity of their assembly details becomes excessive. The attachment of the infilling elements must also be facilitated.

Surface structures (Fig. 5.2c) can be very efficient because of their simultaneous twofold function as structure and enclosure and because they may be inherently very stable and strong, especially in the case of three-dimensional forms. They are, however, somewhat limited in resisting concentrated loads and in facilitating sudden discontinuities such as openings. Furthermore, the mathematics of their analysis can be extremely complicated.

Other categorizations can be made for particular types of construction or configuration of the structure. Thus we describe certain family groups such as structural walls, post and beam, arch, suspension, pneumatic, trussed, folded plate, or thin shell systems. Each of these has certain characteristics and is subject to specific material–scale limitations. Each lends itself most aptly to certain uses. A knowledge of the specific attributes of the various systems is essential to the designer, but can be gained only by exhaustive study and some experience. The inventive designer can, of course, consider new variations of the basic systems and possibly even invent systems with no existing categorical identification.

A complete presentation of all structural materials and systems and a discussion of their relative merits, potentialities, and limitations would undoubtedly fill a volume several times the size of this book. Nevertheless, a short survey of traditional systems with some commentary follows. The categorization used—for example, post and beam—has no particular significance; it is merely a convenient one for discussion.

5.3. STRUCTURAL WALLS

It seems to be a direct structural development to use the enclosing and dividing walls of a building for support and

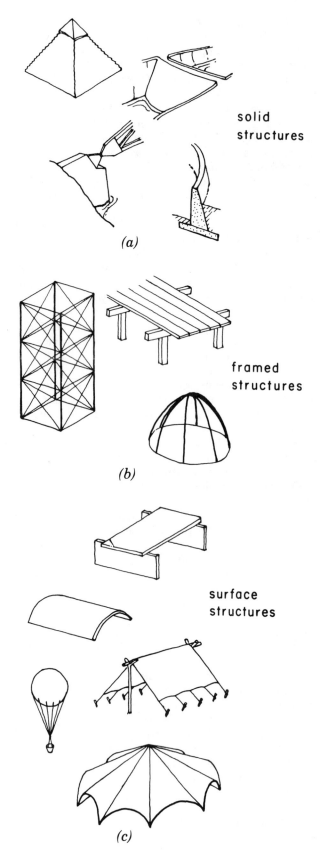

solid structures

(a)

framed structures

(b)

surface structures

(c)

FIGURE 5.2. Categorization of structures by basic constructive nature.

bracing. When this system is utilized there are usually two distinct elements in the total building structure:

The Walls. Used to provide lateral stability as well as to support the spanning elements.

Spanning Elements. Function as floors or roofs.

The spanning elements are usually structurally distinct from the walls and can be considered separately. They may consist of a variety of assemblies, from simple wooden boards and joists to complex precast concrete or steel truss units. The flat spanning system is discussed as a separate category.

Bearing walls are essentially compression elements. They may be monolithic, or they may be frameworks assembled of many pieces. They may be uninterrupted, or they may be pierced in a variety of ways (see Fig. 5.3). Holes for windows and doors may be punched in the solid wall, and as long as their heads are framed and they are arranged so as not to destroy the structural potential of the wall, the structure remains intact.

Even when not used for vertical load transfer, walls are often used to provide lateral stability. This can be achieved by the wall acting independently or in combined interaction with the frame of the building. An example of the latter is a plywood sheet attached to a series of wooden studs. Even if it does not share in vertical load development, it will function in preventing lateral collapse of the studs. This lateral bracing potential of the rigid vertical plane is often utilized in stabilizing buildings against the forces of wind or seismic shock.

Consider the simple structure shown in Fig. 5.4, which consists of a single space bounded by four walls and a flat roof. The two heavy vertical end walls in the upper illustration are capable of resisting horizontal force in a direction parallel to the plane of the walls. However, horizontal forces in a direction perpendicular to the plane of the walls would not be resisted as easily. If the other two walls of the structure are also rigid, they may, of course, function to resist horizontal forces parallel to their planes, thus providing complete support for the building. Another device, however, is that of simply turning the two end walls slightly around the corners, as shown in the lower illustration. This makes them independently stable against horizontal forces from all directions.

The device just illustrated is one technique for stabilizing the flat wall against horizontal force perpendicular to its plane. It may also be necessary to stabilize the wall against buckling under vertical loads if it is very tall and thin. In addition to folding or curving the wall in plan, some other means for achieving this transverse resistance are given below (see Fig. 5.5).

Spreading the Base. This can be done by actually fattening the wall toward its base, as with a gravity dam, or by attaching the wall rigidly to a broad footing.

Stiffening the Wall with Ribs. The wall can be thickened locally, forming monolithic ribs, or pilasters, which also

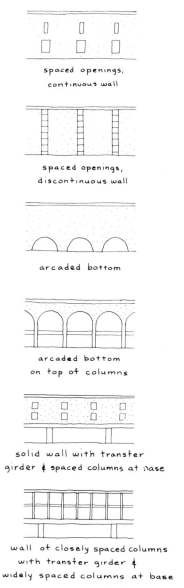

spaced openings,
continuous wall

spaced openings,
discontinuous wall

arcaded bottom

arcaded bottom
on top of columns

solid wall with transfer
girder & spaced columns at base

wall of closely spaced columns
with transfer girder &
widely spaced columns at base

FIGURE 5.3. Opening up the structural wall.

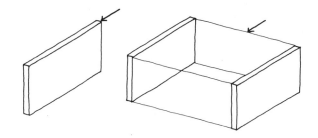

stable in one direction only

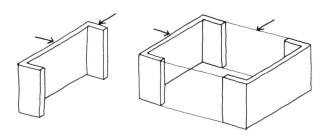

stable in both directions

FIGURE 5.4. Stability of walls.

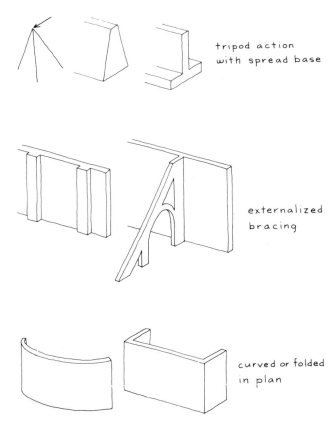

tripod action
with spread base

externalized
bracing

curved or folded
in plan

FIGURE 5.5. Means of stabilizing walls.

function to receive concentrated vertical loads on the wall.

Bracing the Wall.   External bracing devices such as buttresses, struts, and guys function strictly for lateral force resistance.

If the wall functions as a horizontal spanning element, as it does in the arcaded base structure, its basic structural behavior depends on the ratio of the span to the height of the wall. If this ratio is less than 3 or so, the wall does not act strictly as a beam but develops a corbeling or arching action, depending on the construction and material of the wall. A horizontal lintel placed over an opening in a masonry wall does not actually support the entire wall directly over it, but only some triangular portion immediately above it. If this triangular portion is eliminated, the form of the corbeled opening is anticipated, whereby the wall spans the opening effortlessly. If an arch is used for the top of the opening, it may actually function as little more than a liner for the opening.

5.4. POST-AND-BEAM SYSTEMS

Primitive cultures' use of tree trunks as building elements was the origin of this basic system. Later expansion of the vocabulary of materials into stone, masonry, concrete, and metals carried over the experience and tradition of form and detail established with wood. This same tradition, plus the real potentialities inherent in the system, keep this building technique a major part of our structural repertoire (see Fig. 5.6).

The two basic elements of the system are the post and the beam (lintel).

Post.   Essentially a linear compression element subject to crushing or buckling, depending on its relative slenderness.

Beam (Lintel).   Essentially a linear element subject to transverse load; must develop internal resistance to shear and bending and resist excessive deflection (sag).

Critical aspects of the system are the relation of length to radius of gyration (or simple thickness) of the post and the relation of depth to span of the beam. Efficiency of the geometric cross-sectional shape of the beam in bending resistance is also critical (see the discussion of bending in Chapter 3).

The stability of the system under lateral loading is critical in two different ways (see Fig. 5.7). Consider first the

FIGURE 5.6. Massive, self-stabilizing masonry columns and simply detailed wood framing produce a structure with a clear lineage of historical development. Close inspection, however, reveals quite evidently that the elements are of contemporary industrial origin. Buena Park Civic Center, Buena Park, California. Architects: Smith and Williams, South Pasadena, California.

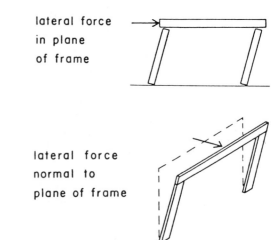

bracing required for:

lateral force
in plane
of frame

lateral force
normal to
plane of frame

FIGURE 5.7. Bracing of framed structures for lateral loads.

resistance to horizontal load in the same plane as the frame. This resistance can be provided in a number of ways, for instance, by fixing the base of the posts, using self-stabilizing posts, connecting the posts and beam rigidly (as in the legs of a table), using trussing or X-bracing, or by using a sufficiently rigid infilling panel.

Stability against horizontal loads perpendicular to the plane of the frame is a slightly different situation. Many of the same techniques of bracing can be used for this also. Another possibility is having an interaction between the frame and the infilling elements used to span from frame to frame, because the three-dimensional building usually implies a set, or series, of frames.

Some variations on the basic system are the following (see Fig. 5.8).

Use of Extended Beam Ends.   Produces overhangs, or cantilevers. This serves to reduce the degree of bending and sag at the center of the span, thus increasing the relative efficiency of the spanning element.

Rigid Attachment of Beam and Posts.   A device for producing stability in the plane of the frame, as already discussed. It achieves some reduction of bending and sag at midspan of the beam, but does so at the expense of the post—in contrast to the extended beam ends. It also produces an outward kick at the base of the post.

Rigid Attachment with Extended Beam Ends.   Combines the stabilizing advantage of the rigid attachment with the sag and bending reduction of the extended ends. If carefully designed, bending in the posts and kick at the base can be eliminated.

Widened Top of Post.   Serves to reduce the span of the beam. As the beam deflects and curves, however, its load becomes concentrated at the edge of the top of the post, thus causing bending in the post. Several variations are possible using, for example, V-shaped or Y-shaped posts.

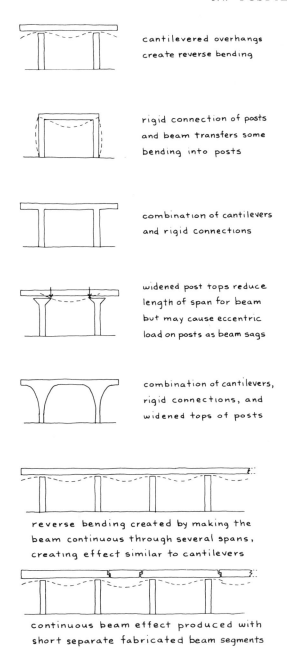

FIGURE 5.8. Variations of the post and beam.

Widened Post Top with Rigid Attachment and Extended Beam Ends. Can combine advantages of all three techniques.

Continuous (Monolithic, Multispan) Beam. Produces the same effects as the extended beam ends for the single span. Additional gain is in the tying together of the system. A variation in which internal joints are placed off the column preserves the advantages of the bending and sag reductions, but allows shorter beam segments. The latter is an advantage in wood, steel, and precast concrete structures. Poured-in-place concrete can, of course, achieve virtually any desired length of monolithic structure.

As with the wall-bearing structure, the post and beam requires the use of a secondary structural system for infilling to produce the solid surfaces of walls, floors, and roofs (see Fig. 5.9). A great variety is possible in these systems, as discussed in the section on walls and flat-spanning systems. One possible variation is to combine the post and wall monolithically, producing a series of pilasters. Similarly, the beam and flat deck may be combined monolithically, producing a continuous ribbed deck or a series of T-shaped beams.

The post and beam suggests the development of rectilinear arrangements and simple, straight forms. The beam may, however, be curved in plan, tilted from the horizontal (as a roof rafter), or have other than a flat top or bottom. Posts can be T-shaped, Y-shaped, V-shaped, or multi-

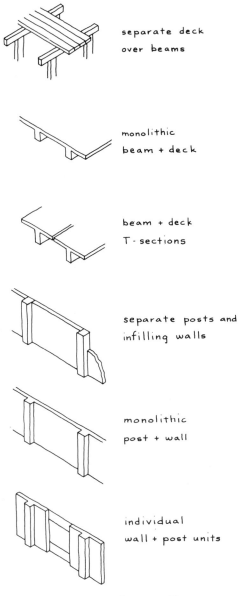

FIGURE 5.9. Infilling for post-and-beam system.

tiered. The system, in fact, lends itself to a greater degree of variation than practically any other system, which is one reason for its continuing widespread use (see Fig. 5.10).

Materials can be separately considered for the post and beam. Masonry, for example, is a possibility for the post, but highly unlikely for the beam. In the main—for structures of building scale—the materials are limited to wood, stone and masonry, concrete, and steel.

5.5. RIGID FRAMES

When the members of a linear framework are rigidly attached—that is, when the joints are capable of transferring bending between the members—the system assumes a particular character. If all joints are rigid, it is impossible to load any one member transversely without causing bending in all members. This, plus the inherent stability of the system, are its unique aspects in comparison to the simple post and beam systems. The rigid-frame action may be restricted to a single plane, or it may be extended in all directions in the three-dimensional framework (see Fig. 5.11).

The joints take on a high degree of importance in this system. In fact, in the usual case, the highest magnitude of stresses and internal force are concentrated at the joints. If the frame is assembled from separate elements, the jointing must be studied carefully for structural function and feasibility.

A popular form of rigid frame is the gabled frame, in which two elements are joined at the top of the gable peak, usually by a single hinge joint. This system is a logical one in laminated wood and is, in fact, one of the few possibilities for the rigid frame in wood. It is often used in steel or precast concrete as well.

Occasionally, the rigid-frame action is objectionable—for instance, when the beam transfers large bending to a small column or causes large curvature or kick at the base of a column. It is sometimes necessary to avoid the rigid frame action deliberately or to control it by using special joint detailing that controls the magnitude of bending or the actual joint turning, which can be transferred between the members.

5.6. FLAT-SPANNING SYSTEMS

Compared to the arch, the dome, or the draped catenary, the flat-spanning structure is very hardworking. In fact, it is exceeded only by the cantilever in this respect. Consequently, scale limits can only be overcome by various techniques that improve its efficiency. One of these is to develop the system as a two-way rather than a one-way spanning system. For a simple flat plate the load-bearing strength can thus be increased by almost 50% and the deflection reduced by an even higher degree (see Fig. 5.12).

Maximum benefit is derived from two-way spanning if the spans are equal. The more different they become, the '

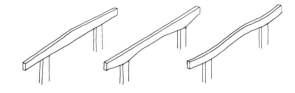

beam form variations

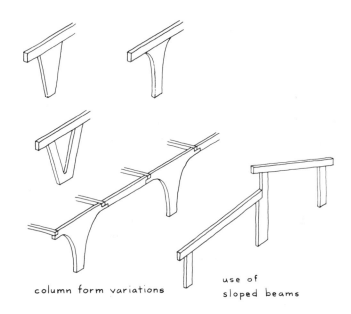

column form variations   use of sloped beams

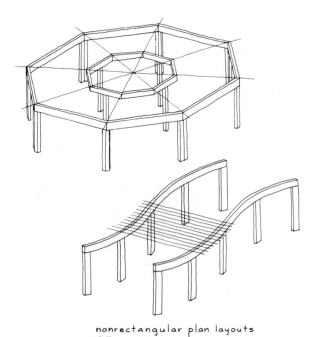

nonrectangular plan layouts

FIGURE 5.10. Form variations with the post-and-beam system.

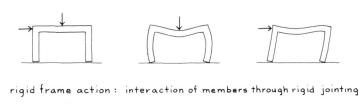

rigid frame action : interaction of members through rigid jointing

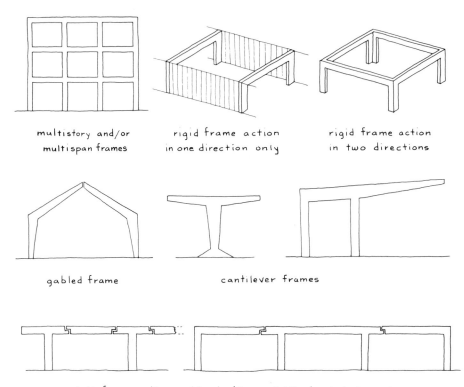

multistory and/or   rigid frame action   rigid frame action
multispan frames   in one direction only   in two directions

gabled frame   cantilever frames

rigid frame units combined with nonrigidly jointed elements

FIGURE 5.11. Rigid frames.

less the work in the long direction. At a 2:1 ratio, less than 10% of the resistance will be offered by the long span.

The other chief device for increasing efficiency is to improve the bending characteristic of the spanning elements. A simple example is the difference in effectiveness illustrated by a flat sheet of paper and one that has been pleated or corrugated. The real concept involved is that of increasing the depth of the element; the flat-sheet form has the least depth possible, whereas the sandwich panel with two sheets separated by considerable space is at the other extreme.

A critical relationship in the flat span, as in the beam, is the ratio of the span to the depth. Load capacity falls off rapidly as this ratio is pushed to its limits. Resistance to deflection is often more critical than resistance to the stresses in bending or shear.

Efficiency can also be increased by extending the element beyond its supports, by using monolithic elements continuous over several spans, or by developing bending transfer between the element and its supports (all the tricks already illustrated for the beam).

5.7.   TRUSS SYSTEMS

A framework of linear elements connected by joints can be stabilized independently by guys, struts, or rigid infilling panels. If it is internally stabilizing, or self-stabilizing, it becomes so through one of two means. The first is the use of rigid joints, as previously discussed. This results both in shear and bending in the members of the frame and usually in considerable movement or deflection of the frame under lateral loads.

The second means of stabilizing a linear framework is by arranging the members in patterns of planar triangles or spatial tetrahedrons (see Fig. 5.13). This is called *trussing*, and when the structural element produced is a flat-spanning or cantilevering planar unit, it is called simply a

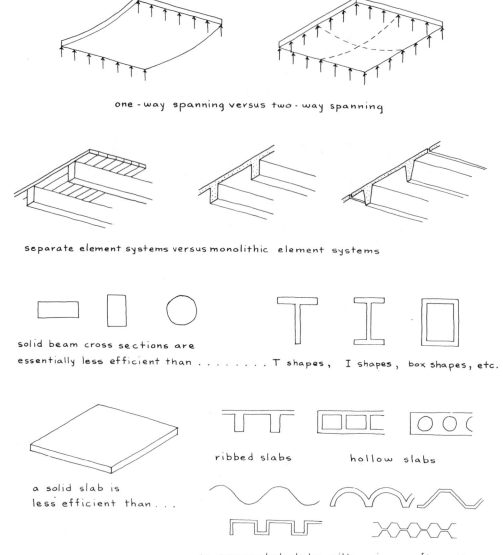

one - way spanning versus two - way spanning

separate element systems versus monolithic element systems

solid beam cross sections are
essentially less efficient than . . . . T shapes, I shapes, box shapes, etc.

a solid slab is
less efficient than . . .

ribbed slabs   hollow slabs

or corrugated slabs with various configurations

FIGURE 5.12. Elements of flat-spanning systems.

truss. The triangulated frame can also be used to produce other structural forms, such as rigid frames, arches, three-dimensional towers, and two-way flat-span systems. If the overall element has some other classification, it is referred to as a *trussed arch* or a *trussed tower.*

The triangular subdivision of the planar system, or the tetrahedral subdivision of the spatial system, produces geometric units that are nondeformable—that is, the arrangement cannot be changed without changing the length of a member or disconnecting a joint. This is one of the basic concepts at work in the truss. The other is the technique of causing widely spaced small masses of material to interact, giving the ultimate maximum efficiency in bending resistance (see the discussion of bending efficiency in Chapter 3).

The multiplicity of joints in the trussed system makes their detailing a major item in truss design. The logic of form of the linear members derives as much, if not more, from the jointing as from their function as tension-resistive or compression-resistive elements. The elimination of bending and shear in the members is, by the way, another basic concept of the truss and is actually or essentially achieved in many trusses. The less flexible the joints and the stouter the members, the less the "pure" truss action.

An almost infinite variety of truss configurations is possible. The particular configuration, the loads sustained, the scale, the material, and the jointing methods are all design considerations.

The two-way truss—often called a *space frame,* although the term is confusing—has been developed largely in steel

structural efficiency of the
truss derives from the
spatial separation of
opposed masses of material

minimum mass   maximum separation

rectilinear frames
must be stabilized by...

rigid frame action   or   trussing

common truss forms
are named for
their developers
or their geometry

Fink or W truss   Belgian truss

Pratt truss   Howe truss

a tower
composed of
four vertical
planar trusses

K truss   Warren truss

arches and rigid frames formed with
trussing instead of solid elements

or... just about any shape
can be "trussed"

FIGURE 5.13. Basic aspects of trussed structures.

(Fig. 5.14). A small-scale system developed by the Unistrut Corporation consists of small-dimension scaffolding elements assembled by simple bolting in patterns that generate from a basic modular unit in the form of a square-based pyramid.

5.8. ARCH, VAULT, AND DOME SYSTEMS

The basic concept in the arch is the development of a spanning structure through the use of only internal compression (see Fig. 5.15). The profile of the "pure" arch may actually be geometrically derived from the loading and support conditions. For a single-span arch with no fixity at the base in the form of moment resistance, with supports at the same level, and with a uniformly distributed load on the entire span, the resulting form is that of a second-degree curve, or a parabola. Various other curves—circular or elliptical—can be used, but the basic form is that of the familiar curve, convex downward, if the load is primarily gravity.

Basic considerations are the necessary horizontal forces at the base from thrust and the ratio of span to rise. As this ratio increases, the thrust increases, producing higher compression in the arch and larger horizontal forces at the support.

In the great stone arches of old the principal load was gravity—the weight of the arch itself. Although other forces existed, they were usually incidental in magnitude compared to the gravity force. In contemporary construction, the lightness of the structure has changed this situation, virtually eliminating the possibility of the pure arch.

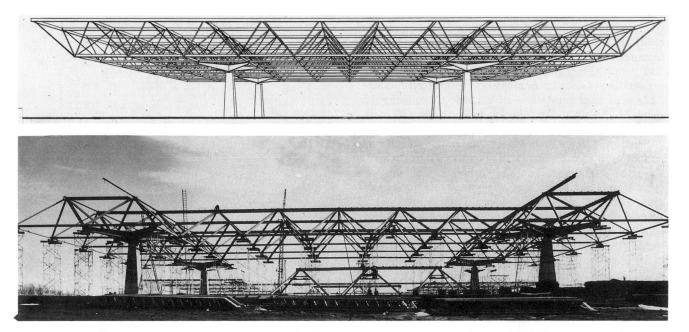

FIGURE 5.14. Two-way spanning truss system supported on only four columns. Columns have four-fingered tops to reduce the concentrated effects in the truss. Overall size of the roof is 216 × 297 ft. Pekin High School Gym, Pekin, Illinois. Architects: Foley, Hackler, Thompson, and Lee, Peoria, Illinois. Structural engineer: The Engineers Collaborative, Chicago.

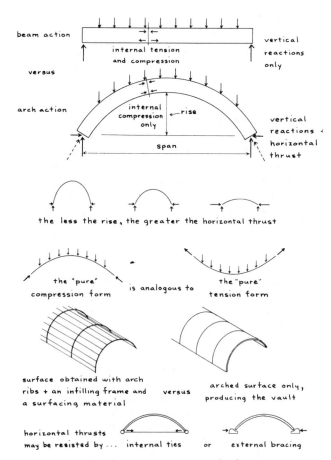

FIGURE 5.15. Basic aspects of arches.

Horizontal forces of wind or seismic movement, or even uplift from aerodynamic effects, require the consideration of more than simple gravity force in deriving the arch form and detail. Consequently, most arches today are continuous ribs of steel, laminated wood, reinforced concrete, or trussed configurations—all capable of considerable bending and shear—in addition to the basic arch compression.

The thrust of the arch—that is, the horizontal component of it—is resolved in one of two ways. The most direct way is to balance the force at one support against that at the other simply by using a tension tie across the base of the arch. This very possibly, however, destroys the internal space defined by the arch and is therefore not always acceptable. The second way is to resolve each kick separately outside the arch. This means creating a heavy abutment, or, if the arch rests on the top of a wall or a column, creating a strut or a buttress for the wall or column.

A major consideration in the structural behavior of an arch is the nature of its basic configuration. The three most common cases are those shown in Fig. 5.16, consisting of the fixed arch (Fig. 5.16a), the two-hinged arch (Fig. 5.16b), and the three-hinged arch (Fig. 5.16c).

The fixed arch occurs most commonly with reinforced concrete bridge and tunnel construction. Maintaining the fixed condition (no rotation) at the base is generally not feasible for very long span arches, so this form is more often used for short to medium spans. It may occur in the action of a series of arches built continuously with their supporting piers, as shown in Fig. 5.16a. The fixed arch is highly indeterminate in its action and is subject to internal stress and abutment forces as a result of thermal expansion and contraction.

The two-hinged arch is the most common form for long spans. The pinned base is more feasibly developed for a large arch and is not subjected to forces as a result of thermal change to the degree that the fixed support is. This arch is also indeterminate, although not to the same degree as the fixed arch.

The three-hinged arch is a popular form for medium-span building roof structures. The principal reasons for this popularity are the following.

1. The pinned bases are more easily developed than fixed ones, making shallow bearing-type foundations reasonable for the medium-span structure.
2. Thermal expansion and contraction of the arch segments will cause vertical movement at the peak pin, but have no appreciable effect on the bases. This further simplifies the foundation design.
3. Construction can often be achieved by prefabricating the two arch sections (most often of glued-laminated timber or welded steel) and connecting them at the peak in the field. The pin joint is a much easier connection to achieve under these circumstances.

For a uniformly distributed loading applied over the horizontal span of the arch, the net internal force follows a parabolic profile over the span. If the arch is formed with the profile of the centroid of its cross section following a

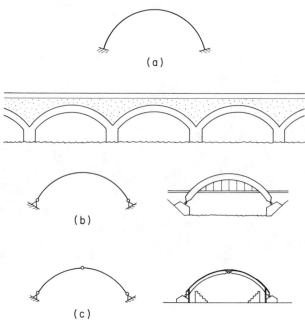

FIGURE 5.16. Types of arches.

parabola that passes through the three pins, it will be found that the net internal force in the arch is pure axial compression. This is sometimes described as a "pure" form for the arch. However, partially distributed live loads and effects of wind and earthquakes will result in other internal force effects. If the arch is exceptionally heavy (dead load predominant), the parabolic form may be of some significance. Otherwise, the general internal force condition for design will be some combination of compression plus shear and bending. Old stone arches (unreinforced masonry, essentially) work primarily in compression, with their own dead weight plus that of the construction supported being the predominant loading. Relatively long-span arches for buildings constructed of steel, laminated timber, or reinforced concrete are usually quite light with respect to the live loads and effects of wind or earthquakes. In the latter case, the arch sections are typically designed for combinations of compression plus some bending and shear.

In any event, the structural behavior of the three-hinged arch and its response to thermal change or movement of the supports is more easily predicted and controlled.

If adjacent arches are assembled side by side in a row perpendicular to their planes, the vault is produced—that is, a surface, rather than a planar rib, is obtained. If vaults intersect, complex three-dimensional forms are created in the lines of intersection. The forms resulting from intersecting vaults and the ribs placed at the intersection lines and edges were the main structural essence of the Gothic church construction.

If the single arch is rotated in place about its crown, or apex, the form generated is a dome. This structural form relates to a circle in plan, in contrast to the vault, which relates to a rectangle, or cross form.

Both vault and dome forms can be created as ribbed (that is, a set of skeleton arches with an infilling shell) or as direct shell forms. In our time few arches, vaults, or domes are made of cut stone. Reinforced concrete is probably the most obvious material for the shell forms, although ribbed systems are equally feasible in laminated wood or steel. Vaults of plywood are currently used extensively, and "bubble" dome forms in plastic are the most widely used skylight elements.

5.9. TENSION STRUCTURES

The tension suspension structure was highly developed by some primitive societies through the use of vines or strands woven from grass or shredded bamboo. These structures achieved impressive spans; foot bridges spanning 100 ft have been recorded. The development of steel, however, heralded the great span capability of this system. At first in chain and link, and later in the cable woven of drawn wire, the suspension structure quickly took over as the long-span champion.

Structurally, the single draped cable is merely the inverse of the arch in both geometry and internal force (see Fig. 5.17). The compression-arch parabola is merely

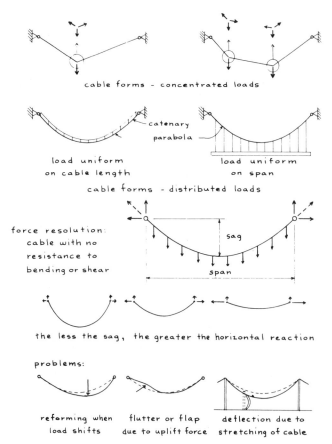

FIGURE 5.17. Basic aspects of suspended cable systems.

flipped over to produce the tension cable. Span-to-sag ratio and horizontal inward thrust at the supports have their parallels in the arch behavior.

Additional problems with the suspension element are its lack of stiffness, which causes reforming under load changes and possible fluttering or flapping, and the more difficult connection at its supports. The latter is due to basic differences between transfer of compression and tension.

Steel is obviously the principal material for this system, and the cable is the logical form. Actually, the largest spans use clusters of cables—up to 3 ft in diameter for the Golden Gate Bridge with its 4000-ft span. Although a virtually solid steel element 3 ft in diameter hardly seems flexible, one must consider the span-to-thickness ratio—approximately 1330:1. This is like a 1-in.-diameter rod over 100 ft long. One cannot anchor this size element by tying a clove hitch around a stake!

Structures can also be hung simply by tension elements. The deck of the suspension bridge, for example, is not placed directly on the cables but is hung with another system of cables. Cantilevers or spanning systems may thus be supported by hanging as well as by columns, piers, or walls.

There are many possibilities for the utilization of tension elements in structures in addition to the simple draped or vertically hung cable. Cables can be arranged in a circular, radiating pattern with an inner tension hub and an outer compression rim similar to those in a bicycle wheel (see Fig. 5.18). Cables can also be arranged in crisscrossing networks, as draped systems, or as restraining elements for air-inflated membrane surfaces.

Tension surface structures can be produced by direct tensioning, for instance, in the familiar fabric tent. A more recent innovation is that of the tension membrane surface maintained by air inflation.

5.10. SURFACE STRUCTURES

The neatness of our categorization of structural systems eventually breaks down, since variations within one system tend to produce different systems, and overlapping between categories exists. Thus the rigidly connected post and beam become the rigid frame. Surface structures are essentially those consisting of relatively thin, extensive surfaces functioning primarily by resolving only internal forces within their surfaces (see Fig. 5.19).

We have already discussed several surface structures. The wall in resisting compression, in stabilizing the building by resisting in-plane shear, and in spanning like a beam acts as a surface structure. The vault and the dome are really surface structures. These can also develop non-surface actions, however. The wall, in bending under loads perpendicular to its surface, develops out-of-plane action.

The purest surface structures are tension surfaces, since they are often made of materials incapable of any signifi-

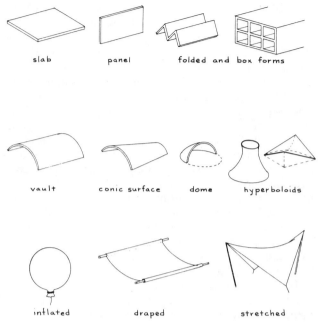

FIGURE 5.19. Basic forms of surface structures.

cant out-of-plane resistance. Thus the canvas tent, the rubber balloon, and the plastic bag are all limited in capability to tension resistance within the planes of their surfaces. The forms they assume, then, must all be completely "pure." In fact, the pure compression surface is sometimes derived by simulating it in reverse with a tension surface (see Fig. 5.20).

Compression surfaces must be more rigid than tension ones because of the possibility of buckling. This increased stiffness makes them difficult to use in a way that avoids developing out-of-plane bending and shear.

Compression resistive surface structures of curved form are also called *shells*. The egg, the light bulb, the plastic bubble, and the auto fender are all examples of shells. At the building scale the most extensively exploited material for this system has been reinforced concrete (Figs. 2.2 and 5.21). The largest structure of this type is that of an exposition hall in France with clear span of 700 ft. The structure is a concrete shell of double, or sandwich, form.

Both simple and complex geometries are possible with shells. Edges, corners, openings, and point supports are potential locations of high stress and out-of-plane bending; consequently, reinforcing is often necessary, usually consisting of monolithically cast ribs in concrete. The existence of these stiffening ribs often alters the pure surface structure character of the shells and results in complex behaviors, for example, with aspects of arch or rigid-frame action.

A special variation of the shell is the surface produced by multiple folds or pleats. The overall form of the structure may assume curvature, but the individual elements are all flat surfaces. These structures are referred to as *folded plates*. They lend themselves to execution in plywood

FIGURE 5.18. New York State's "Tent of Tomorrow" pavilion at the 1964 World's Fair. The 100-ft-high concrete columns carry an elliptical steel compression ring, 350 × 250 ft in plan. Suspended from the ring, a double layer of steel cables converge toward a steel tension ring at the center. The roof surface consists of translucent sandwich panels formed by two sheets of reinforced plastic, which are separated by an aluminum grid. The panels, approximately 3000 in number, are trapezoidal in plan and vary from 3 to 17 ft long and 3 to 4 ft wide. Architect: Phillip Johnson Associates, New York. Structural engineer: Lev Zetlin and Associates, New York. Plastic panels: Filon Corporation. Photo: Filon Corporation.

Inflated Structures

Inflation, or air pressure, can be used as a structural device in a variety of ways (see Fig. 5.22). Simple internal inflation of a totally enclosing membrane surface, for instance, in the simple rubber balloon is the most direct. This requires about the least structural material imaginable for spanning. The structure is unavoidably highly flexible, however, and dependent on the constant differential between inside and outside pressure. It is also necessarily "lumpy" in form because the surface is stretched. It has nevertheless been utilized for buildings of considerable size.

A second use of inflation is the stiffening of a structural element. This can be a sandwich or hollow ribbed structure of tension membrane material given a rigid frame character by inflation of the voids within the structural element, for example, the inflated inner tube or air mattress. The need for sealing the space enclosed by the structure is thus eliminated (see Fig. 5.20).

A third possibility is that of using a combination of inflation and simple tension draping or stretching on a frame. Thus the pillow can be suspended—its lower surface draped in tension and its upper surface maintained by inflation. An advantage in this system is the elimination of the water pocket normally formed by the draping of a surface.

FIGURE 5.20. Pneumatically stiffened structure for the Three-Rivers Art Festival, Pittsburgh, Pennsylvania, 1976. Double-surfaced elements, analogous to air mattresses, form this pavilion structure. Designers and Builders: Chrysalis East. Photo: Joseph Valerio.

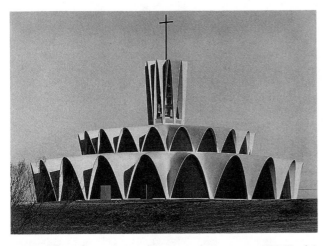

FIGURE 5.21. Multiple-element concrete shell structure. Church of the Priory of Saint Mary and Saint Louis, School for Boys, Creve Coeur, Missouri. Three tiers of thin shells produce a light, airy quality that is in sharp contrast to the traditional heavy, crude aspect of poured concrete. Architects: Helmuth, Obata, and Kassabaum, St. Louis. Structural engineers: John P. Nix, St. Louis, and Paul Weidlinger, New York. Photo: James K. Mellow, St. Louis, from Helmuth, Obata, and Kassabaum.

or sheet metal as well as concrete. There is some advantage in the ease of forming the flat surfaces and the straight-line intersections.

5.11.   SPECIAL SYSTEMS

Most of the common systems have now been itemized. Innumerable special systems are possible, each creating a new category by its unique aspects. Some of these are described briefly as follows.

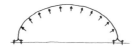 single surface - tension maintained by pressure difference between interior of building and outside

double surface - tension and stiffening produced by inflation of the structure

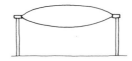 double surface - bottom draped in tension from the supports, top held up by internal inflation

cable restrained - internal pressure pushes membrane against the network of restraining cables

FIGURE 5.22. Basic forms of air-supported structures.

Lamella Frameworks

This is a system for forming arch or dome surfaces utilizing a network of perpendicular ribs that appear to be diagonal in plan. It has been used at both modest and great spans and has been executed in wood, steel, and precast concrete. One great advantage is in the repetition of similar-sized elements and joint details. Another advantage is in the use of straight-line elements to produce the curved vault surface (see Fig. 5.23).

Geodesic Domes

A few lines can scarcely do justice to this system. Developed from ideas innovated by R. Buckminster Fuller, this technique for forming hemispherical surfaces is based on spherical triangulation (see Fig. 5.24). It is also useful at both small and large scales and subject to endless variation of detail, member configuration, and materials. In addition to ordinary wood, steel, and concrete, it has been executed in plywood, plastic, cardboard, bamboo, and aluminum (see Fig. 4.5).

The chief attributes of the system are its multiplication of basic units and joints and the extreme efficiency of its internal force resolution. Its developers claim that its efficiency increases with size, making it difficult to see any basis for establishing a limiting scale.

Mast Structures

These are structures similar to trees, having single legs for vertical support and supporting one or a series of "branches." They obviously require very stable bases, well anchored against the overturning effect of horizontal forces. Their chief advantage lies in the minimum of space occupied by the base (see Fig. 5.25).

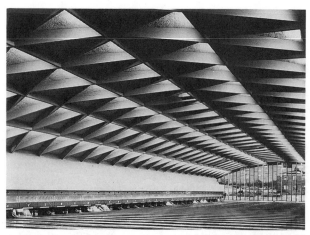

FIGURE 5.23. Wood lamella structure. Simple wood elements in a diagonal lamella pattern form this 109-ft-span roof for a bowling alley in Detroit, Michigan. Architects: Hawthorne and Schmiedeke, Detroit. Photo: American Institute of Timber Construction.

FIGURE 5.24. Geodesic dome structure, Climatron, Missouri Botanical Gardens, St. Louis. This 175-ft-diameter dome has plastic glazing of Plexiglas suspended from its frame of tubular aluminum elements in a geodesic pattern. Architects: Murphy and Mackey, St. Louis. Photo: Rohm and Haas Company, Philadelphia.

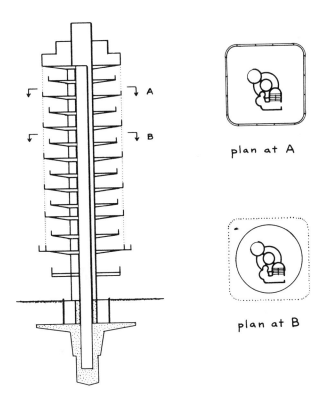

plan at A

plan at B

FIGURE 5.25. Tower structure—mast form. Laboratory Tower, Johnson Wax Company, Racine, Wisconsin. A concrete central core supports alternating round and square concrete floors in this treelike structure. The tap root foundation firmly plants the structure in the ground. Architect: Frank Lloyd Wright.

Multiple Monopode Units

Multiple mushroom-, lily-pad, or morning-glory form elements can be used to produce the one-story building of many horizontal increments. Principally developed with

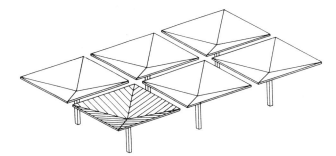

FIGURE 5.26. Monopode structural units.

reinforced concrete shell forms, this system offers savings in the repetitive use of a single form (see Fig. 5.26). An exciting example in steel at large scale is the exhibition building in Turin, Italy, by Pier Luigi Nervi.

This brief sampling does not pretend to present the complete repertoire of contemporary structural systems for buildings. The continual development of new materials, systems, and construction processes keeps this a dynamic area of endeavor. New systems are added; established ones become outmoded. Modern techniques of analysis and design make the rational, reliable design of complex systems feasible.

One seemingly inevitable trend is that toward the industrialization of the building process. This tends to emphasize those materials, systems, and processes that lend themselves to industrial production. Prefabrication, modular coordination, component systems, and machine-produced details are increasingly evident in our building structures. Although the handcrafted structure will always have a certain charm, the increasing use of industrialized products and systems will undoubtedly continue.

PART 2

INVESTIGATION OF STRUCTURES

Part 2 consists of a survey of basic concepts and procedures from the general field of applied mechanics as they have evolved in the process of investigating of the behavior of structures. The purpose of studying this material is twofold. First is the general need for a comprehensive understanding of what structures must do and how they do it. Second is the need for some factual, quantified basis for the exercise of judgment in the processes of structural design. If it is accepted that the understanding of a problem is the necessary first step in its solution, this essentially analytical study should be seen as the cornerstone of any successful design process.

Although considerable use is made of mathematics in the work in this part, it is mostly only a matter of procedural efficiency. The concepts, not the mathematical manipulations, are the critical concerns. To reinforce that assertion, most of the concepts presented in Part 2 were first developed essentially without mathematics in the discussions in Part 1. If the reader has not already done so, a quick reading of Part 1 is strongly recommended before proceeding with the material in Part 2.

CHAPTER SIX

Introduction to Structural Investigation

Investigation of structures is itself a major field of study. The material in this chapter consists of discussions of the nature, purposes, and various techniques of the investigation of structures. As in all of the work in this book, the primary focus is on material relevant to the tasks of structural design.

6.1. PURPOSE OF INVESTIGATION

Most structures exist because of some usage need. Their evaluation must therefore begin with consideration of the effectiveness with which they facilitate or satisfy the usage requirements. Three factors of this effectiveness may be considered: the structure's functionality, feasibility, and safety.

Functionality deals with the general physical relationships of the structure's form, detail, durability, fire resistance, and so on, as these relate to its use. Feasibility includes considerations of cost, the availability of materials, and the practicality of production. Safety in terms of structural actions is generally obtained in the form of some margin between the structure's capacity for resistance and the demands placed on it.

Analysis of structural behaviors serves chiefly to establish the nature of the structure's deformations (which may limit its functions) and to relate its performance to the loads that it must sustain (in order to compare capacity and demand). There are two critical phases of the structure's behavior: its working condition or typical service condition, and its ultimate response or limit at failure. We have use for both of these: the first for visualization of the deformations during use, and the second for establishing realistic margins of safety.

6.2. MEANS OF INVESTIGATION

Analysis for investigation may progress with the following considerations.

1. Determination of the structure's physical being with regard to material, form, scale, detail, orientation, location, support conditions, and internal character.
2. Determination of the demands placed on the structure: that is, the loads and their manner of application and any usage limits on deformations.
3. Determination of the structure's responses in terms of deformations and development of internal stresses.
4. Determination of the limits of the structure's capabilities.
5. Evaluation of the structure's effectiveness.

Analysis may be performed in several ways. One can visualize graphically the nature of the structure's deformation under load—through mental images or with sketches. Using available theories and techniques, one can manipulate mathematical models of the structure. Finally, one can actually load and measure responses of the structure itself or of a scaled model of the actual structure.

When reasonably precise quantitative evaluations are required, the most useful tools are direct measurements of physical responses or careful mathematical modeling with theories that have been demonstrated to be reliable. Ordinarily, mathematical modeling of some kind must precede actual construction—even that of a test model. Direct measurement is usually limited to experimental studies or to efforts to verify questionable theories or techniques.

Design of building structures is generally a highly conservative (not adventurous) field. Use of physical testing in the routine design process is rare, and most design work is based largely on experience with some measure of support from computations using commonly accepted procedures and data that have been certified as being reliable.

6.3. ASPECTS OF INVESTIGATION

The subject of structural investigation is traditionally divided into three areas of study: mechanics (statics and dynamics), strength of materials, and analysis of structural elements and systems.

Mechanics is the branch of physics that deals with the motion of physical objects and the forces that cause their motion. It is usually divided into the topics of statics and dynamics. Statics deals with the condition of bodies at rest; that is, motion is implied or impending due to the presence of forces, but at present no motion occurs. Dynamics treats the general case of forces and motions, a significant variable being time.

Although dynamics is the more general field of mechanics, we have a particular concern for statics, as this is the generally desired condition of building structures. Consideration of dynamics is mostly limited to the development of a few basic ideas that relate to the movements of the structure in the form of deformations under load. However, it is also necessary to use dynamics in a true investigation of dynamic effects, such as those caused by blasts, induced vibrations, or seismic movements.

Strength of materials deals with the resistance of materials and structural elements to deformations caused by forces. This involves relationships between the external operating forces and the reactive stresses that develop in the material to generate the necessary internal resisting forces.

As a general concern, structural analysis deals with the overall consideration of the behavior of a structure in all of the terms relating to its required tasks. What must be investigated depends on the tasks as well as the type of structure. In general, if the investigation is to be considered complete, all reasonably conceivable problems must be considered.

6.4. REALISM IN INVESTIGATION

Investigation is a fact-finding mission and its reliability is vulnerable to both the accuracy of the input data and the skills and judgment of the investigator. A difficulty arises in the academic process because of the usual procedure of dealing with "setup" problems. There is some justification for this procedure, because the study of individual ideas is made simpler when the ideas can be isolated. There is a tendency, therefore, to deal with highly controlled situations with simplified circumstances in order that the results may be restricted to certain cases. Thus problems are first designed (by the teacher) to ensure certain solutions, and the student's efforts are limited to manipulations of predetermined procedures in order to extract predictable answers.

Although the learning procedure just described has some validity, the danger lies in the student's not being aware of its contrived or artificial nature. Very often in real problem-solving situations the most difficult aspect of the work is clear identification of the problem itself. Once the problem is precisely defined, the actual investigation can often truly proceed almost automatically, assuming that one possesses the skills or the facilities for doing the work.

In this work we have largely followed the usual procedure of working with deliberately simplified conditions in order to isolate and observe the basic concepts. On some occasions the true complexity of the problem is discussed before this isolation is made. However, only in the design examples in Part 9 is the true breadth of the problem of designing building structures presented in a realistic manner. In most of the rest of the book the work is presented in limited engagements, and the reader must keep the broader picture in mind to maintain some sense of orientation.

6.5. TECHNIQUES AND AIDS FOR INVESTIGATION

The professional designer or investigator uses all the means available for accomplishment of the work. In this age mathematical modeling is aided greatly by the use of computers. However, routine problems are still often treated by use of simple hand computations or reference to data in handbook tables or graphs.

Our purpose here is essentially educational, so the emphasis is on visualization and understanding, not necessarily on efficiency of computational means. In this book, major use is made of graphical visualization, and we would like to encourage this habit on the part of the reader. The use of sketches as learning and problem-solving aids cannot be overemphasized. Two types of graphical devices are most useful: the free-body diagram and the profile of the load-deformed structure.

A free-body diagram consists of a picture of any isolated physical element together with representations of all of the forces that act externally on the element. The isolated element may be a whole structure or any fractional part of it. Consider the structure shown in Fig. 6.1. Figure 6.1a shows the entire structure, consisting of attached horizontal and vertical elements (beams and columns) that produce a planar rigid-frame bent. This may be one of a set of such

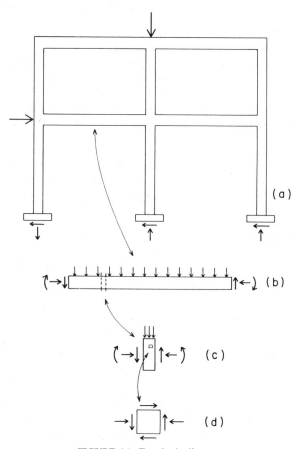

FIGURE 6.1. Free-body diagrams.

frames comprising a building structural system. The free-body diagram in Fig. 6.1*a* represents the entire structure, with forces external to it represented by arrows. The forces include the entire weight of the structure, the horizontal force of the wind, and the net forces acting on the structure's foundations.

Shown in Fig. 6.1*b* is a free-body diagram of a single beam from the framed bent. Operating on the beam are its own weight plus the forces of interaction between the ends of the beam and the columns to which it is attached. These interactions are not visible in the free-body diagram of the full frame, so one purpose for the diagram of the single beam is simply the visualization of the nature of these interactions. We can now see that the columns transmit to the ends of the beams horizontal and vertical forces as well as rotational bending actions.

In Fig. 6.1*c* we see an isolated portion of the beam length, which is produced by slicing vertical planes a short distance apart and removing the portion between them. Operating on this free body are its own weight and the actions of the beam segments on the opposite sides of the slicing planes, since it is these actions that hold the removed portion in place in the whole beam. This device (called the *cut section*) is used to visualize the internal force actions in the beam, since they are not visible in the whole beam (Fig. 6.1*b*).

Finally, in Fig. 6.1*d*, we isolate a tiny segment, or particle, of the material of the beam and visualize the external effects consisting of the interactions between this particle and those adjacent to it. This is our basic device for the visualization of stress; in this case, due to its location in the beam, the particle is subject to a combination of shear and linear compression stresses.

Figure 6.2*a* shows the exaggerated deformation of the bent under wind loading. The type of deflection of the structure and the character of bending in each member can be visualized from this figure. As shown in Fig. 6.2*b*, the character of deformation of segments and particles can also be visualized. These diagrams are very helpful in establishing the qualitative nature of the relationships between force actions and shape changes or between stresses and strains.

Another useful graphic device is the scaled plot of some mathematical relationship or of the data from some observed physical phenomenon. Considerable use is made of this technique in presenting ideas in this book. The graph in Fig. 6.3 represents the form of damped vibration of an elastic spring. It consists of the plot of the displacement ($+$ or $-s$) against elapsed time t, and represents the graph of a mathematical formula of the general form of

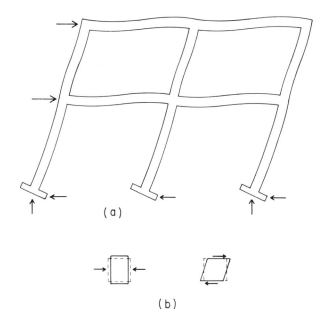

(a)

(b)

FIGURE 6.2. Visualization of structural deformation.

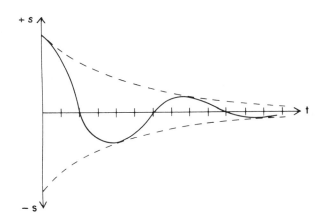

FIGURE 6.3. Graphic plot of a cyclic motion.

$$s = \frac{1}{e^t} P \sin(Qt + R)$$

Although the equation is technically sufficient for the description of the phenomenon, the graph helps in the visualization of many aspects of the relationship, such as the rate of decay of the displacement, the interval (period) of the vibration, the specific position at some specific elapsed time, and so on.

CHAPTER SEVEN

Static Forces

Chapter 7 presents basic concepts and procedures that are used in the analysis of the effects of static forces. Topics selected and procedures illustrated are limited to those that relate directly to the problems of designing ordinary building structures.

7.1. PROPERTIES OF FORCES

Static forces are those that can be dealt with adequately without considering the time-dependent aspects of their action. This limits considerations to those dealing with the following properties:

Magnitude, or the amount, of the force, which is measured in weight units such as pounds or tons.

Direction of the force, which refers to the orientation of its path or line of action. Direction is usually described by the angle that the line of action makes with some reference, such as the horizontal.

Sense of the force, which refers to the manner in which it acts along its line of action (e.g., up or down). Sense is usually expressed algebraically in terms of the sign of the force, either plus or minus.

Forces can be represented graphically in terms of these three properties by the use of an arrow, as shown in Fig. 7.1. Drawn to some scale, the length of the arrow represents the magnitude of the force. The angle of inclination of the arrow represents the direction of the force. The location of the arrow-head determines the sense of the force. This form of representation can be more than merely symbolic, since actual mathematical manipulations may be per-

formed using the vector representation that the force arrows constitute. In the work in this book arrows are used in a symbolic way for visual reference when performing algebraic computations, and in a truly representative way when performing graphical analyses.

In addition to the basic properties of magnitude, direction, and sense, some other concerns that may be significant for certain investigations are

The *position of the line of action* of the force with respect to the lines of action of other forces or to some object on which the force operates, as shown in Fig. 7.2.

The *point of application* of the force along its line of action may be of concern in analyzing for the specific effect of the force on an object; as shown in Fig. 7.3.

When forces are not resisted, they tend to produce motion. An inherent aspect of static forces is that they exist in a state of *static equilibrium*, that is, with no motion occurring. In order for static equilibrium to exist, it is necessary to have a balanced system of forces. An important considera-

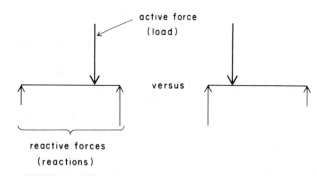

FIGURE 7.2. Effect of the location of the line of action of a force.

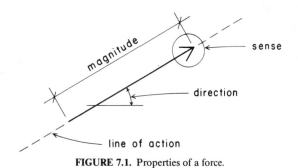

FIGURE 7.1. Properties of a force.

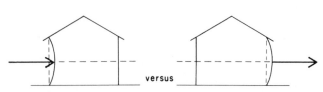

FIGURE 7.3. Effect of the position of a force along its line of action.

tion in the analysis of static forces is the nature of the geometric arrangement of the forces in a given set of forces that constitute a single system. The usual technique for classifying force systems involves consideration of whether the forces in the system are

Coplanar. All acting in a single plane, such as the plane of a vertical wall.
Parallel. All having the same direction.
Concurrent. All having their lines of action intersect at a common point.

Using these three considerations, the possible variations are given in Table 7.1 and illustrated in Fig. 7.4. Note that variation 5 in the table is really not possible, because a set of coacting forces that is parallel and concurrent cannot be noncoplanar; in fact, they all fall on a single line of action and are called *collinear.*

It is necessary to qualify a set of forces in the manner just illustrated before proceeding with any analysis, whether it is to be performed algebraically or graphically.

7.2. COMPOSITION AND RESOLUTION OF FORCES

In structural analysis it is sometimes necessary to perform either addition or subtraction of force vectors. The process of addition is called *composition*, or combining of forces. The process of subtraction is called *resolution*, or the resolving of forces into *components*. A component is any force that represents part, but not all, of the effect of the original force.

TABLE 7.1. Classification of Force Systems

System Variation	Qualifications		
	Coplanar	Parallel	Concurrent
1	Yes	Yes	Yes
2	Yes	Yes	No
3	Yes	No	Yes
4	Yes	No	No
5	No[a]	Yes	Yes
6	No	Yes	No
7	No	No	Yes
8	No	No	No

[a] Not possible if forces are parallel and concurrent.

Resolution

In Fig. 7.5 a single force is shown, acting upward toward the right. One type of component of such a force is the net horizontal effect, which is shown as F_h in Fig. 7.5a. The vector for this force can be found by determining the side of the right-angled triangle, as shown in the illustration. The magnitude of this vector may be calculated as $F(\cos \theta)$ or may be measured directly from the graphic construction if the vector for F is placed at the proper angle and has a length proportional to its actual magnitude.

If a force is resolved into two or more components, the set of components may be used to replace the original force. A single force may be resolved completely into its horizontal and vertical components, as shown in Fig. 7.5b. This is a useful type of resolution for algebraic analysis, as will be demonstrated in the examples that follow. However, for any force, there are an infinite number of potential components into which it can be resolved.

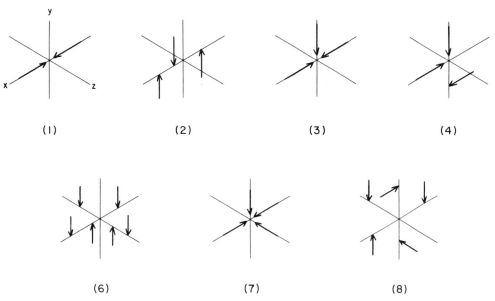

FIGURE 7.4. Classification of force systems—orthogonal reference axes.

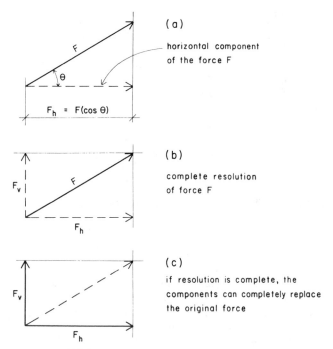

$$F_h = F(\cos \theta)$$

(a)

horizontal component
of the force F

(b)

complete resolution
of force F

(c)

if resolution is complete, the
components can completely replace
the original force

FIGURE 7.5. Resolution of a force into components.

Composition

Whether performed algebraically or graphically- the combining of forces is essentially the reverse of the process just shown for resolution of a single force into components. Consider the two forces shown in Fig. 7.6a. The combined effect of these two forces may be determined graphically by use of the parallelogram shown in Fig. 7.6b or the force triangle shown in Fig. 7.6c. The product of the addition of force vectors is called the *resultant* of the forces. Note that we could have used either of the two triangles formed in Fig. 7.6b to get the resultant. This is simply a matter of the sequence of addition, which does not affect the answer; thus if $A + B = C$, then $B + A = C$ also.

When the addition of force vectors is performed algebraically, the usual procedure is first to resolve the forces into their horizontal and vertical components (or into any mutually perpendicular set of components). The resultant is then expressed as the sum of the two sets of components. If the actual magnitude and direction of the resultant are required, they may be determined as follows (see Fig. 7.6d).

$$R = \sqrt{(\Sigma F_v)^2 + (\Sigma F_h)^2}$$

$$\tan \theta = \frac{\Sigma F_v}{\Sigma F_h}$$

When performing algebraic summations, it is necessary to use the sign (or sense) of the forces. Note in Fig. 7.6 that the horizontal components of the two forces are both in the same direction and thus have the same sense (or algebraic sign) and that the summation is one of addition of the two components. However, the vertical components of the two forces are opposite in sense and the summation consists of finding the difference between the magnitudes of the two components. Keeping track of the algebraic signs is a major concern in algebraic analyses of forces. It is necessary to establish a sign convention and to use it carefully and consistently throughout the calculations. Whenever possible, the graphic manifestation of the force analysis should be sketched and used for a reference while performing algebraic analyses, because this will usually help in keeping track of the proper sense of the forces.

When more than two forces must be added, the graphic process consists of the construction of a force polygon. This process may be visualized as the successive addition of pairs of forces in a series of force triangles, as shown in Fig. 7.7. The first pair of forces is added to produce their resultant; this resultant is then added to the third force; and so on. The process continues until the last of the forces is added to the last of the intermediate resultants to produce the final resultant for the system. This process is shown in Fig. 7.7 merely for illustration, since we do not actually need to find the intermediate resultants, but may simply add the forces in a continuous sequence, producing the single force polygon. It may be observed, however, that this is actually a composite of the individual force triangles.

For an algebraic analysis, we simply add up the two sets of components, as demonstrated previously, regardless of their number. The true magnitude and direction of the resultant can then be found using the equations previously given for R and $\tan \theta$. The equivalent process in the graphic analysis is the closing of the force polygon, the resultant being the vector represented by the line that completes the figure, extending from the tail of the first force to the head of the last force.

As mentioned previously, the sequence of the addition of the forces is actually arbitrary. Thus there is not a single

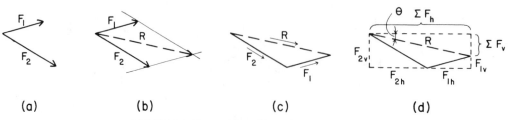

(a)   (b)   (c)   (d)

FIGURE 7.6. Composition of forces by vector addition.

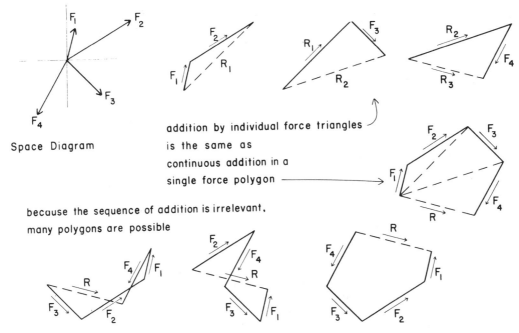

FIGURE 7.7. Composition of forces by graphic construction.

polygon that may be constructed, but rather a whole series of polygons, all producing the same resultant.

7.3. MOTION AND STATIC EQUILIBRIUM

The natural state of a static force system is one of equilibrium. This means that the resultant of any complete interactive set of static forces must be zero. For various purposes it is sometimes desirable to find the resultant combined effect of a limited number of forces, which may indeed be a net force. If a condition of equilibrium is then desired, it may be visualized in terms of producing the *equilibrant*, which is the force that will totally cancel the resultant. The equilibrant, therefore, is the force that is equal in magnitude and direction, but opposite in sense, to the resultant.

In structural design the typical force analysis problem begins with the assumption that the net effect is one of static equilibrium. Therefore, if some forces in a system are known (e.g., the loads on a structure), and some are unknown (e.g., the forces generated in members of the structure in resisting the loads), the determination of the unknown forces consists of finding out what is required to keep the whole system in equilibrium. The relationships and procedures that can be utilized for such analyses depend on the geometry or arrangement of the forces, as discussed in Sec. 7.1.

For a simple concentric, coplanar force system, the conditions necessary for static equilibrium can be stated as follows:

$\Sigma F_v = 0$ (sum of the vertical components equals zero)
$\Sigma F_h = 0$ (sum of the horizontal components equals zero)

In other words, if both components of the resultant are zero, the resultant is zero and the system is in equilibrium.

In a graphic solution for the concurrent, coplanar system, the resultant will be zero if the force polygon closes on itself, that is, if the head of the last force vector coincides with the tail of the first force vector.

Although a basic concept of statics is that no motion occurs, it is often helpful in investigation of force actions to consider the potential motion that is impending due to the effect of a force. This is especially true when considering the necessary requirements for maintaining equilibrium of a structure.

7.4. ANALYSIS OF COPLANAR, CONCURRENT FORCES

The forces that operate on individual joints in planar trusses constitute sets of coplanar, concurrent forces. The following discussion deals with the analysis of such systems, both algebraically and graphically, and introduces some of the procedures that will be used in the examples of truss analysis in this book.

In the preceding examples, forces have been identified as F_1, F_2, F_3, and so on. However, a different system of notation will be used in the work that follows. This method consists of placing a letter in each space that occurs between the forces or their lines of action, each force then being identified by the two letters that appear in the adjacent spaces. A set of five forces is shown in Fig. 7.8*a*. The common intersection point is identified as *BCGFE* and the forces are *BC*, *CG*, *GF*, *FE*, and *EB*. Note particularly that the forces have been identified by reading around the joint in a continuous clockwise manner. This is a convention that will be used throughout this book, because it has some relevance to the methods of graphic analysis that will be explained later.

In Fig. 7.8*b* a portion of a truss is shown. Reading around the joint *BCGFE* in a clockwise manner, the upper chord member between the two top-joints is read as member *CG*. Reading around the joint *CDIHG*, the same member is read as *GC*. Either designation may be used when referring to the member itself. However, if the effect of the force in the member on a joint is being identified, it is important to use the proper sequence for the two-letter designation.

In Fig. 7.9*a* a weight is shown hanging from two wires that are attached at separate points to the ceiling. The two sloping wires and the vertical wire that supports the weight directly meet at joint *CAB*. The "problem" in this case is to find the tension forces in the three wires. We refer to these forces that exist within the members of a structure as *internal forces*. In this example it is obvious that the force in the vertical wire will be the same as the magnitude of the weight: 50 lb. Thus the solution is reduced to the determination of the tension forces in the two sloping wires. This problem is presented in Fig. 7.9*b*, where the force in the vertical wire is identified in terms of both direction and magnitude, while the other two forces are identified only in terms of their directions, which must be parallel to the wires. The senses of the forces in this example are obvious, although this will not always be true in such problems.

A graphic solution of this problem can be performed by using the available information to construct a force polygon consisting of the vectors for the three forces: *BC*, *CA*, and *AB*. The process for this construction is as follows.

1. The vector for AB is totally known and can be represented as shown by the vertical arrow with its head down and its length measured in some scale to be 50.
2. The vector for force *BC* is known as to direction and must pass through the point *b* on the force polygon, as shown in Fig. 7.9*d*.
3. Similarly, we may establish that the vector for force *CA* will lie on the line shown in Fig. 7.9*e*, passing through the point *a* on the polygon.
4. Because these are the only vectors in the polygon, the point *c* is located at the intersection of these two lines, and the completed polygon is as shown in Fig. 7.9*f*, with the sense established by the continuous flow of the arrows. This "flow" is determined by reading the vectors in continuous clockwise sequence on the space diagram, starting with the vector of known sense. We thus read the direction of the arrows as flowing from *a* to *b* to *c* to *a*.

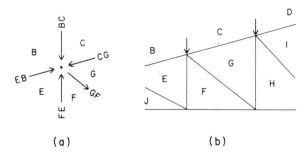

FIGURE 7.8. Force notation.

With the force polygon completed, we can determine the magnitudes for forces *BC* and *CA* by measuring their lengths on the polygon, using the same scale that was used to lay out force *AB*.

For an algebraic solution of the problem illustrated in Fig. 7.9, we first resolve the forces into their horizontal and vertical components, as shown in Fig. 7.9*g*. This increases the number of unknowns from two to four. However, we have two extra relationships that may be used in addition to the conditions for equilibrium, because the direction of forces *BC* and *CA* are known. As shown in Fig. 7.9*a*, force *BC* is at an angle with a slope of 1 vertical to 2 horizontal. Using the rule that the hypotenuse of a right triangle is related to the sides such that the square of the hypotenuse is equal to the sum of the squares of the sides, we can determine that the length of the hypotenuse of the slope triangle is

$$l = \sqrt{(1)^2 + (2)^2} = \sqrt{5} = 2.236$$

We can now use the relationships of this triangle to express the relationships of the force *BC* to its components. Thus, referring to Fig. 7.10,

$$\frac{BC_v}{BC} = \frac{1}{2.236}, \quad BC_v = \frac{1}{2.236}(BC) = 0.447(BC)$$

$$\frac{BC_h}{BC} = \frac{2}{2.236}, \quad BC_h = \frac{2}{2.236}(BC) = 0.894(BC)$$

These relationships are shown in Fig. 7.9*g* by indicating the dimensions of the slope triangle with the hypotenuse having a value of 1. Similar calculations will produce the values shown for the force *CA*. We can now express the conditions required for equilibrium. (Directions up and right are considered positive.)

$$\Sigma F_v = 0 = -50 + BC_v + CA_v$$
$$0 = -50 + 0.447(BC) + 0.707(CA) \quad (1)$$

and

$$\Sigma F_h = 0 = -BC_h + CA_h$$
$$0 = -0.894(BC) + 0.707(CA) \quad (2)$$

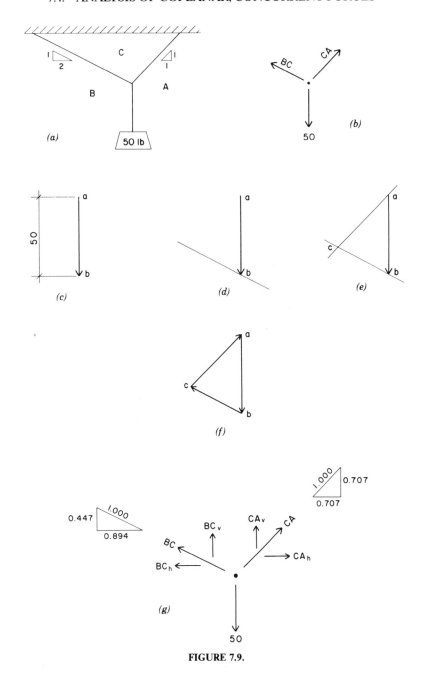

FIGURE 7.9.

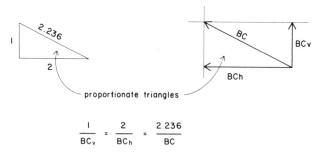

$$\frac{1}{BC_v} = \frac{2}{BC_h} = \frac{2.236}{BC}$$

FIGURE 7.10. Determination of force components.

We can eliminate *CA* from these two equations by subtracting equation (2) from equation (1) as follows:

Equation (1): $0 = -50 + 0.447(BC) + 0.707(CA)$

Equation (2): $0 = \qquad + 0.894(BC) - 0.707(CA)$

Combining: $0 = -50 + 1.341(BC)$

Then

$$BC = \frac{50}{1.341} = 37.29 \text{ lb}$$

Using equation (2) yields

$$0 = -0.894(37.29) + 0.707(CA)$$

$$CA = \frac{0.894}{0.707}(37.29) = 47.15 \text{ lb}$$

When answers obtained from algebraic solutions are compared to those obtained from graphic solutions, the level of correlation may be low, unless great care is exercised in the graphic work and a very large scale is used for the constructions. If the scale used for the graphic solution in this example is actually as small as that shown on the printed page in Fig. 7.9, it is unreasonable to expect accuracy beyond the second digit.

When the so-called method of joints is used, finding the internal forces in the members of a planar truss consists of solving a series of concurrent force systems. Figure 7.11 shows a truss with the truss form, the loads, and the reactions displayed in a *space diagram*. Below the space diagram is a figure consisting of the free-body diagrams of the individual joints of the truss. These are arranged in the same manner as they are in the truss in order to show their interrelationships. However, each joint constitutes a complete concurrent planar force system that must have its independent equilibrium. "Solving" the problem consists of determining the equilibrium conditions for all of the joints. The procedures used for this solution will be illustrated in Sec. 7.5.

7.5. GRAPHICAL ANALYSIS OF PLANAR TRUSSES

Figure 7.12 shows a single-span planar truss that is subjected to vertical gravity loads. We will use this example to illustrate the procedures for determining the internal forces in the truss, that is, the tension and compression forces in

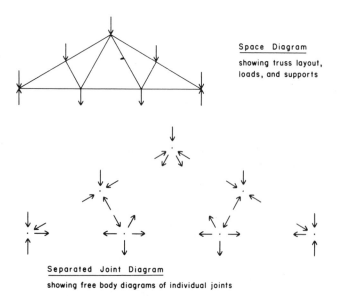

Space Diagram

showing truss layout, loads, and supports

Separated Joint Diagram

showing free body diagrams of individual joints

FIGURE 7.11. Diagrams used to represent trusses and their actions.

the individual members of the truss. The space diagram in the figure shows the truss form, the support conditions, and the loads. The letters on the space diagram identify individual forces at the truss joints, as discussed in Sec. 7.4. The sequence of placement of the letters is arbitrary, the only necessary consideration being to place a letter in each space between the loads and the individual truss members so that each force at a joint can be identified by a two-letter symbol.

The separated joint diagram in the figure provides a useful means for visualization of the complete force system at each joint as well as the interrelation of the joints through the truss members. The individual forces at each joint are designated by two-letter symbols that are obtained by simply reading around the joint in the space diagram in a clockwise direction. Note that the two-letter symbols are reversed at the opposite ends of each of the truss members. Thus the top chord member at the left end of the truss is designated as *BI* when shown in the joint at the left support (joint 1) and is designated as *IB* when shown in the first interior upper chord joint (joint 2). The purpose of this procedure will be demonstrated in the following explanation of the graphical analysis.

The third diagram in Fig. 7.12 is a composite force polygon for the external and internal forces in the truss. It is called a *Maxwell diagram* after its originator, James Maxwell, an English engineer. The construction of this diagram constitutes a complete solution for the magnitudes and senses of the internal forces in the truss. The procedure for this construction is as follows.

1. *Construct the Force Polygon for the External Forces.* Before this can be done, the values for the reactions must be found. There are graphic techniques for finding the reactions, but it is usually much simpler and faster to find them with an algebraic solution. In this example, although the truss is not symmetrical, the loading is, and it may simply be observed that the reactions are each equal to one-half of the total load on the truss, or 5000/2 = 2500 lb. Because the external forces in this case are all in a single direction, the force polygon for the external forces is actually a straight line. Using the two-letter symbols for the forces and starting with letter A at the left end, we read the force sequence by moving in a clockwise direction around the outside of the truss. The loads are thus read as *AB, BC, CD, DE, EF,* and *FG,* and the two reactions are read as *GH* and *HA.* Beginning at *A* on the Maxwell diagram, the force vector sequence for the external forces is read from *A* to *B, B* to *C, C* to *D,* and so on, ending back at *A,* which shows that the force polygon closes and the external forces are in the necessary state of static equilibrium. Note that we have pulled the vectors for the reactions off to the side in the diagram to indicate them more clearly. Note also that we have used lowercase letters for the vector ends in the Maxwell diagram, whereas uppercase letters are used on the space diagram. The alphabetic correlation is thus retained (*A* to *a*), while any possible confusion between the two diagrams is prevented. The letters on the space diagram designate spaces, while the letters on the Maxwell diagram designate points of intersection of lines.

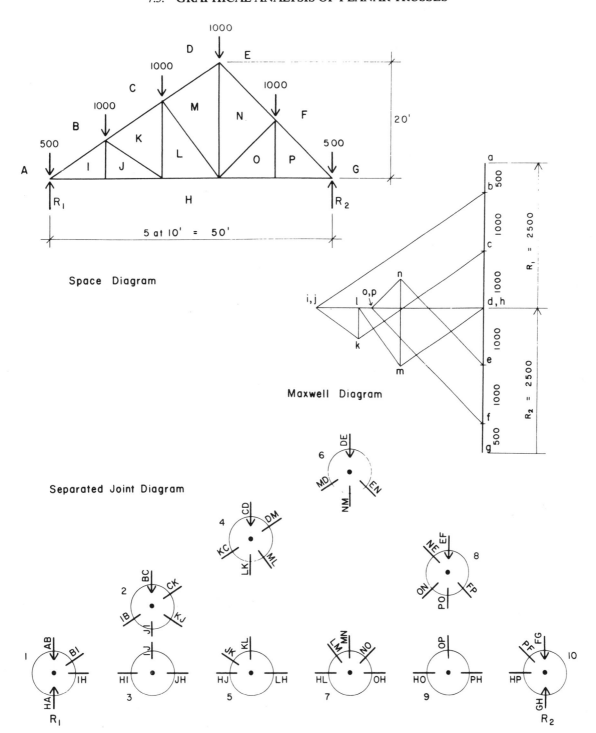

FIGURE 7.12. Graphical analysis of a truss.

2. *Construct the Force Polygons for the Individual Joints.* The graphic procedure for this consists of locating the points on the Maxwell diagram that correspond to the remaining letters, *I* through *P*, on the space diagram. When all the lettered points on the diagram are located, the complete force polygon for each joint may be read on the diagram. To locate these points, we use two relationships.

The first is that the truss members can resist only forces that are parallel to the members' positioned directions. Thus we know the directions of all the internal forces.

The second relationship is a simple one from plane geometry: A point may be located as the intersection of two lines. Consider the forces at joint 1, as shown in the separated joint diagram in Fig. 7.12. Note that there are four forces and that two of them are known (the load and the reaction) and two are unknown (the internal forces in the truss members). The force polygon for this joint, as shown on the Maxwell diagram, is read as *ABIHA*. *AB* represents the load; *BI* the force in the upper chord member; *IH* the force in the lower chord member; and *HA* the reaction. Thus the location of point *I* on the Maxwell diagram is determined by noting that *I* must be in a horizontal direction from *H* (corresponding to the horizontal position of the lower chord) and in a direction from *B* that is parallel to the position of the upper chord.

The remaining points on the Maxwell diagram are found by the same process, using two known points on the diagram to project lines of known direction whose intersection will determine the location of another point. Once all the points are located, the diagram is complete and can be used to find the magnitude and sense of each internal force. The process for construction of the Maxwell diagram typically consists of moving from joint to joint along the truss. Once one of the letters for an internal space is determined on the Maxwell diagram, it may be used as a known point for finding the letter for an adjacent space on the space diagram.

The only limitation of the process is that it is not possible to find more than one unknown point on the Maxwell diagram for any single joint. Consider joint 7 on the separated joint diagram in Fig. 7.12. If we attempt to solve this joint first, knowing only the locations of letters *A* through *H* on the Maxwell diagram, we must locate four unknown points: *L*, *M*, *N*, and *O*. This is three more unknowns than we can determine in a single step, so we must first solve for three of the unknowns by using other joints.

Solving for a single unknown point on the Maxwell diagram corresponds to finding two unknown forces at a joint, because each letter on the space diagram is used twice in the force identifications for the internal forces. Thus for joint 1 in the previous example, the letter *I* is part of the identity for forces *BI* and *IH*, as shown on the separated joint diagram. The graphic determination of single points on the Maxwell diagram, therefore, is analogous to finding two unknown quantities in an algebraic solution. As discussed previously, two unknowns are the maximum that can be solved for in the equilibrium of a coplanar, concurrent force system, which is the condition of the individual joints in the truss.

When the Maxwell diagram is completed, the internal forces can be read from the diagram as follows.

1. The magnitude is determined by measuring the length of the line in the diagram, using the scale that was used to plot the vectors for the external forces.
2. The sense of individual forces is determined by reading the forces in clockwise sequence around a single joint in the space diagram and tracing the same letter sequences on the Maxwell diagram.

Figure 7.13 shows the force system at joint 1 and the force polygon for these forces as taken from the Maxwell diagram. The forces known initially are shown as solid lines on the force polygon, and the unknown forces are shown as dashed lines. Starting with letter *A* on the force

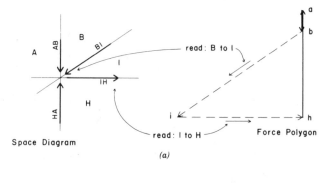

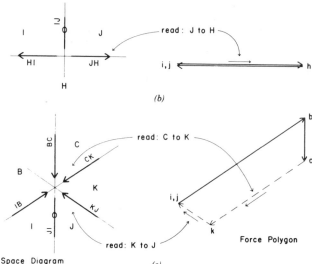

FIGURE 7.13. Analysis of truss joints.

system, we read the forces in a clockwise sequence as *AB*, *BI*, *IH*, and *HA*. On the Maxwell diagram we note that moving from *a* to *b* is moving in the order of the sense of the force, that is from tail to head of the force vector that represents the external load on the joint.

If we continue in this sequence on the Maxwell diagram, this force sense flow will be a continuous one. Thus reading from *b* to *i* on the Maxwell diagram is reading from tail to head of the force vector, which tells us that force *BI* has its head at the left end. Transferring this sense indication from the Maxwell diagram to the joint diagram indicates that force *BI* is in compression; that is, it is pushing, rather than pulling, on the joint. Reading from *i* to *h* on the Maxwell diagram shows that the arrowhead for this vector is on the right, which translates to a tension effect on the joint diagram.

Having solved for the forces at joint 1 as described, we may use the fact that we now know the forces in truss members *BI* and *IH* when we proceed to consider the adjacent joints, 2 and 3. However, we must be careful to note that the sense reverses at the opposite ends of the members in the joint diagrams. Referring to the separated joint diagram in Fig. 7.12, if the upper chord member shown as force *BI* in joint 1 is in compression, its arrowhead is at the lower left end in the diagram from joint 1, as shown in Fig. 7.13*a*. However, when the same force is shown as *IB* at joint 2, its pushing effect on the joint will be indicated by having the arrowhead at the upper right end in the diagram for joint 2. Similarly, the tension effect of the lower chord is shown in joint 1 by placing the arrowhead on the right end of the force *IH*, but the same tension force will be indicated in joint 3 by placing the arrowhead on the left end of the vector for force *HI*.

If we choose the solution sequence of solving joint 1 and then joint 2, we can transfer the known force in the upper chord to joint 2. Thus the solution for the five forces at joint 2 is reduced to finding three unknowns, since the load *BC* and the chord force *IB* are now known. However, we still cannot solve joint 2, since there are two unknown points on the Maxwell diagram (*k* and *j*) corresponding to the three unknown forces. An option, therefore, is to proceed from joint 1 to joint 3, at which there are presently only two unknown forces.

On the Maxwell diagram we can find the single unknown point *j* by projecting vector *IJ* vertically from *i* and projecting vector *JH* horizontally from point *h*. Since point *i* is also located horizontally from point *h*, we thus find that the vector *IJ* has zero magnitude, since both *i* and *j* must be on a horizontal line from *h* in the Maxwell diagram. This indicates that there is actually no stress in this truss member for this loading condition and that points *i* and *j* are coincident on the Maxwell diagram. The joint force diagram and the force polygon for joint 3 are as shown in Fig. 7.13*b*. In the joint force diagram we place a zero, rather than an arrowhead, on the vector line for *IJ* to indicate the zero stress condition. In the force polygon in Fig. 7.13*b*, we have slightly separated the two force vectors for clarity, although they are actually coincident on the same line.

Having solved for the forces at joint 3, we can next proceed to joint 2, since there now remain only two unknown forces at this joint. The forces at the joint and the force polygon for joint 2 are shown in Fig. 7.13*c*. As explained for joint 1, we read the force polygon in a sequence determined by reading in a clockwise direction around the joint: *BCK-JIB*. Following the continuous direction of the force arrows on the force polygon in this sequence, we can establish the sense for the two forces *CK* and *KJ*.

It is possible to proceed from one end and to work continuously across the truss from joint to joint to construct the Maxwell diagram in this example. The sequence in terms of locating points on the Maxwell diagram would be *i-j-k-l-m-n-o-p*, which would be accomplished by solving the joints in the following sequence: 1, 3, 2, 5, 4, 6, 7, 9, 8. However, it is advisable to minimize the error in graphic construction by working from both ends of the truss. Thus a better procedure would be to find points *i-j-k-l-m*, working from the left end of the truss, and then to find points *p-o-n-m*, working from the right end. This would result in finding two locations for the point *m*, whose separation constitutes the error in drafting accuracy.

7.6. ALGEBRAIC ANALYSIS OF PLANAR TRUSSES

Graphic solution for the internal forces in a truss using the Maxwell diagram corresponds essentially to an algebraic

solution by the *method of joints*. This method consists of solving the concentric force systems at the individual joints using simple force equilibrium equations. We will illustrate the method and the corresponding graphic solution using the previous example.

As with the graphic solution, we first determine the external forces, consisting of the loads and the reactions. We then proceed to consider the equilibrium of the individual joints, following a sequence as in the graphic solution. The limitation of this sequence, corresponding to the limit of finding only one unknown point in the Maxwell diagram, is that we cannot find more than two unknown forces at any single joint. Referring to Fig. 7.14, the solution for joint 1 is as follows.

The force system for the joint is drawn with the sense and magnitude of the known forces shown, but with the unknown internal forces represented by lines without ar-

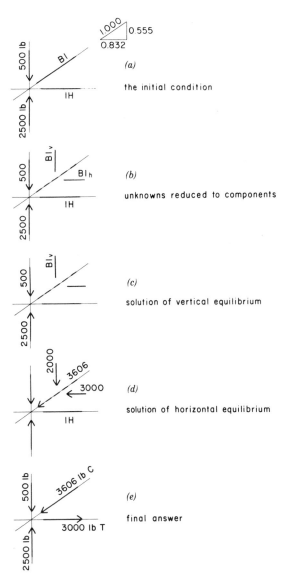

FIGURE 7.14. Algebraic solution of joint 1.

rowheads, since their senses and magnitudes initially are unknown. For forces that are not vertical or horizontal, we replace the forces with their horizontal and vertical components. We then consider the two conditions necessary for the equilibrium of the system: The sum of the vertical forces is zero and the sum of the horizontal forces is zero.

If the algebraic solution is performed carefully, the sense of the forces will be determined automatically. However, we recommend that whenever possible the sense be predetermined by simple observation of the joint conditions, as will be illustrated in the solutions.

The problem to be solved at joint 1 is as shown in Fig. 7.14a. In Fig. 7.14b the system is shown with all forces expressed as vertical and horizontal components. Note that although this now increases the number of unknowns to three (IH, BI_v, and BI_h), there is a numeric relationship between the two components of BI. When this condition is added to the two algebraic conditions for equilibrium, the number of usable relationships totals three, so that we have the necessary conditions to solve for the three unknowns.

The condition for vertical equilibrium is shown in Fig. 7.14c. Because the horizontal forces do not affect the vertical equilibrium, the balance is between the load, the reaction, and the vertical component of the force in the upper chord. Simple observation of the forces and the known magnitudes makes it obvious that force BI_v must act downward, indicating that BI is a compression force. Thus the sense of BI is established by simple visual inspection of the joint, and the algebraic equation for vertical equilibrium (with upward force considered positive) is

$$\Sigma F_v = 0 = +2500 - 500 - BI_v$$

From this equation we determine BI_v to have a magnitude of 2000 lb. Using the known relationships between BI, BI_v, and BI_h, we can determine the values of these three quantities if any one of them is known. Thus

$$\frac{BI}{1.000} = \frac{BI_v}{0.555} = \frac{BI_h}{0.832}$$

$$BI_h = \frac{0.832}{0.555}(2000) = 3000 \text{ lb}$$

$$BI = \frac{1.000}{0.555}(2000) = 3606 \text{ lb}$$

The results of the analysis to this point are shown in Figure 7.14d, from which we can observe the conditions for equilibrium of the horizontal forces. Stated algebraically (with force sense toward the right considered positive) the condition is

$$\Sigma F_h = 0 = IH - 3000$$

from which we establish that the force in IH is 3000 lb.

The final solution for the joint is then as shown in Fig. 7.14e. On this diagram the internal forces are identified as to sense by using C to indicate compression and T to indicate tension.

As with the graphic solution, we can proceed to consider the forces at joint 3. The initial condition at this joint is as shown in Fig. 7.15a, with the single known force in member HI and the two unknown forces in IJ and JH. Since the forces at this joint are all vertical and horizontal, there is no need to use components. Consideration of vertical equilibrium makes it obvious that it is not possible to have a force in member IJ. Stated algebraically, the condition for vertical equilibrium is

$$\Sigma F_v = 0 = IJ \quad \text{(since } IJ \text{ is the only vertical force)}$$

It is equally obvious that the force in JH must be equal and opposite to that in HI, since they are the only two horizontal forces. That is, stated algebraically,

$$\Sigma F_h = 0 = JH - 3000$$

The final answer for the forces at joint 3 is as shown in Fig. 7.15b. Note the convention for indicating a truss member with no internal force.

If we now proceed to consider joint 2, the initial condition is as shown in Fig. 7.16a. Of the five forces at the joint only two remain unknown. Following the procedure for joint 1, we first resolve the forces into their vertical and horizontal components, as shown in Fig. 7.16b.

Because we do not know the sense of forces CK and KJ, we may use the procedure of considering them to be positive until proven otherwise. That is, if we enter them into the algebraic equations with an assumed sense, and the solution produces a negative answer, then our assumption is wrong. However, we must be careful to be consistent with the sense of the force vectors, as the following solution will illustrate.

Let us arbitrarily assume that force CK is in compression and force KJ is in tension. If this is so, the forces and their components will be as shown in Fig. 7.16c. If we then consider the conditions for vertical equilibrium, the forces involved will be those shown in Fig. 7.16d, and the equation for vertical equilibrium will be

$$\Sigma F_v = 0 = -1000 + 2000 - CK_v - KJ_v$$

or

$$0 = +1000 - 0.555CK - 0.555KJ \qquad (1)$$

If we consider the conditions for horizontal equilibrium, the forces will be as shown in Fig. 7.16e, and the equation will be

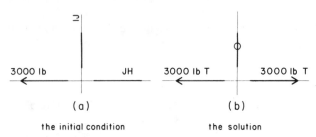

FIGURE 7.15. Algebraic solution of joint 3.

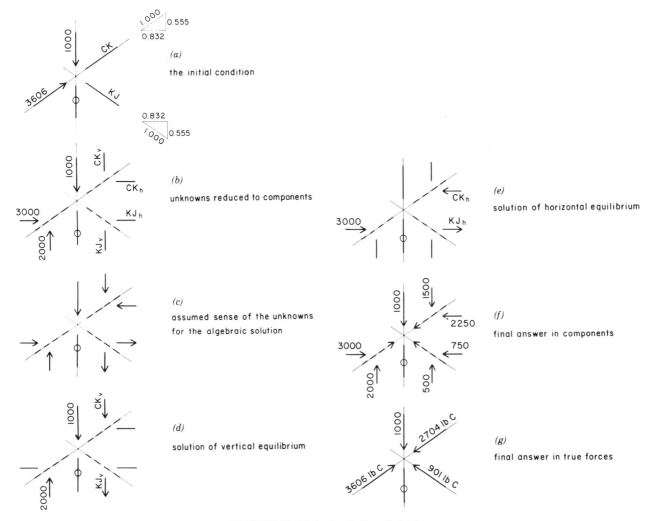

FIGURE 7.16. Algebraic solution of joint 2.

$$\Sigma F_h = 0 = +3000 - CK_h + KJ_h$$

or

$$0 = +3000 - 0.832CK + 0.832KJ \qquad (2)$$

Note the consistency of the algebraic signs and the sense of the force vectors, with positive forces considered as upward and toward the right. We may solve these two equations simultaneously for the two unknown forces as follows.

1. Multiply equation (1) by 0.832/0.555. Thus

$$0 = \frac{0.832}{0.555}(+1000) + \frac{0.832}{0.555}(-0.555CK)$$
$$+ \frac{0.832}{0.555}(-0.555KJ)$$

$$0 = +1500 - 0.832CK - 0.832KJ$$

2. Add this equation to equation (2) and solve for CK. Thus

$$0 = +1500 - 0.832CK - 0.832KJ$$
$$0 = +3000 - 0.832CK + 0.832KJ$$

Adding gives

$$0 = +4500 - 1.664CK$$

Therefore,

$$CK = \frac{4500}{1.664} = 2704 \text{ lb}$$

Note that the assumed sense of compression in CK is correct, since the algebraic solution produces a positive answer.

3. Substituting the value for CK in equation (1)

$$0 = +1000 - 0.555(2704) - 0.555(KJ)$$
$$= +1000 - 1500 - 0.555(KJ)$$

Then

$$KJ = -\frac{500}{0.555} = -901 \text{ lb}$$

Because the algebraic solution produces a negative quantity for *KJ* the assumed sense for *KJ* is wrong, and the member is actually in compression.

The final answers for the forces at joint 2 are thus as shown in Fig. 7.6*g*. To verify that equilibrium exists, however, the forces are shown in the form of their vertical and horizontal components in Fig. 7.16*f*.

When all of the internal forces have been determined for the truss, the results may be recorded or displayed in a number of ways. The most direct way is to display them on a scaled diagram of the truss, as shown in the upper part of Fig. 7.17. The force magnitudes are recorded next to each member with the sense shown as *T* for tension or *C* for compression. Zero-stress members are indicated by the conventional symbol consisting of a zero placed directly on the member.

When solving by the algebraic method of joints, the results may be recorded on a separated joint diagram as shown in the lower portion of Fig. 7.17. If the values for the vertical and horizontal components of force in sloping members are shown, it is a simple matter to verify the equilibrium of the individual joints.

7.7. VISUALIZATION OF TRUSS BEHAVIOR

It is often possible to determine the sense of the internal forces in a truss with little or no quantified calculations. Where this is so, it is useful to do so as a first step in the truss analysis. If a graphic analysis is performed, the sense determined by the preliminary inspection serves as a cross-check on the sense determined from the Maxwell diagram. If an algebraic analysis is performed, the preliminary inspection will aid greatly in keeping track of minus signs in the equilibrium equations. In addition to these practical uses, the preliminary analysis for the sense of internal forces is a good exercise in the visualization of truss behavior and of equilibrium conditions in general. The following examples will illustrate procedures that can be used for such an analysis.

Consider the truss shown in Fig. 7.18*a*. For a consideration of the sense of internal forces, we may proceed in a manner similar to that for the graphic analysis or the algebraic analysis by the method of joints. We thus consider the joints as follows.

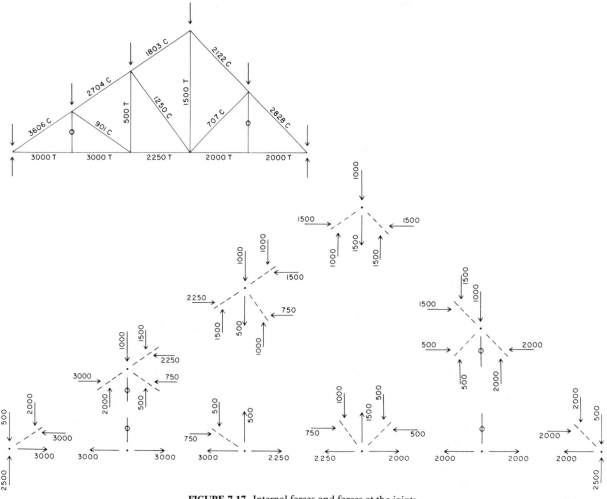

FIGURE 7.17. Internal forces and forces at the joints.

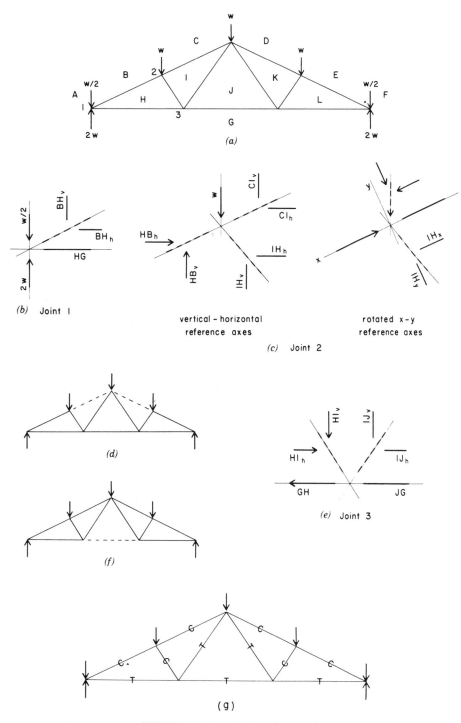

FIGURE 7.18. Visualization of truss actions.

Joint 1 (see Fig. 7.18*b*). For vertical equilibrium the vertical component in *BH* must act downward. Thus member *BH* is in compression and its horizontal component acts toward the left. For horizontal equilibrium the only other horizontal force, that in member *HG*, must act toward the right, indicating that *HG* is in tension.

Joint 2 (see Fig. 7.18*c*). Member *HB* is the same as *BH* as shown at joint 1. The known compression force in the member, therefore, is transferred to joint 2, as shown. If we use a rotated set of reference axes (*x* and *y*), it may be observed that the force in member *IH* is alone in opposing the load effect. Therefore, *IH* must be in compression. (*IH*$_y$ must oppose the *y* component of the load.)

The sense of the force in member *CI* is difficult to establish without some quantified analysis. One approach is to consider the action of the whole truss and to attempt to visualize what the result would be if these members were removed. As shown in Fig. 7.18*d*, it is fairly easy to visualize that the force in these members must be one of compression to hold the structure in place.

Joint 3 (see Fig. 7.18*e*). We first transfer the known conditions for members *GH* and *HI* from the previous analysis of joints 1 and 2. Considering vertical equilibrium, we observe that the vertical component in *IJ* must oppose that in *HI*. Therefore, member *IJ* is in tension. For member *JG* we may use the technique illustrated for member *CI* at joint 2. As shown in Fig. 7.18*f*, it may be observed that this member must act in tension for the structure to function.

Because of the symmetry of the truss in this example, the sense of all internal forces is established with the consideration of only these three joints. The results of the analysis are displayed on the truss figure shown in Fig. 7.18*g*, using *T* and *C* to indicate internal forces of tension and compression, respectively.

For a second example of this type of analysis we consider the truss shown in Fig. 7.19*a*. As before, we begin at one support and move across the truss from joint to joint.

Joint 1 (see Fig. 7.19*b*). As in the preceding example, consideration of vertical equilibrium establishes the compression in member *BI*, after which consideration of horizontal equilibrium establishes the tension in member *IH*.

Joint 2 (see Fig. 7.19*c*). Since *IJ* is the only potential vertical force, it must be zero. Since *JH* alone opposes the tension in *HI*, it must be in tension also.

Joint 3 (see Fig. 7.19*d*). With member IJ essentially nonexistent, due to its zero-stress condition, this joint is similar to joint 2 in the previous example. Thus the use of rotated reference axes may be a means of establishing the condition of compression in member *KJ*. Also, as before, we may observe the necessity for compression in member *CK* by considering the result of removing it from the truss, as illustrated in Fig. 7.19*e*.

Joint 4 (see Fig. 7.19*f*). Because of the known compression force in member *JK*, it may be observed that *KL* must be in tension. Because the two known horizontal forces are opposite in sense, it is not possible to visualize the required sense of member *LH*. As before, the device of removing the member, shown in Fig. 7.19*g*, may be used to establish the required tension in this member.

Joint 5 (see Fig. 7.19*h*). Using the rotated x-y reference axes, it is possible to establish the sense of the force in member *ML*. Because the other two *y*-direction forces have the same sense, the *y* component of *ML* must oppose them, establishing the requirement for a compression force in the member. Neither set of reference axes can be used to establish the sense of the force in member *DM*, however. As before, we can use the device

of removing the member to establish its required compression action, as shown in Fig. 7.19*i*.

Working from the other end of the truss, it is now possible to establish the sense of the force in the remaining members. The final answers for the senses in all members are as shown on the truss figure in Fig. 7.19*j*.

7.8. MOMENTS

In the analysis of concurrent forces, it is sufficient to consider only the basic vector properties of the forces: magnitude, direction, and sense. However, when forces are not concurrent, it is necessary to include the consideration of another type of force action called the *moment*, or rotational effect.

Consider the two interacting vertical forces shown in Fig. 7.20*a*. Since the forces are concurrent, the condition of equilibrium is fully established by satisfying the single algebraic equation: $\Sigma F_v = 0$. However, if the same two forces are not concurrent, as shown in Fig. 7.20*b*, the single force summation is not sufficient to establish equilibrium. In this case the force summation establishes the same fact as before: There is no net tendency for vertical motion. However, because of their separation, the forces tend to cause a counterclockwise motion in the form of a rotational effect, called the moment. The moment has three basic properties:

1. It exists in a particular plane—in this case the plane defined by the two force vectors.
2. It has a magnitude, which is expressed as the product of the force magnitude times the distance between the two vectors. In the example shown in Fig. 7.20*b*, the magnitude of the moment is $(10)(a)$. The unit for this quantity becomes a compound of the force unit and the distance unit: lb-in., kip-ft, and so on.
3. It has a sense of rotational direction. In the example the sense is counterclockwise.

Because of potential moment effects the consideration of equilibrium for noncurrent forces must include another summation: $\Sigma M = 0$. Rotational equilibrium can be established in various ways. One way is shown in Fig. 7.20*c*. In this example a second set of forces whose rotational effect counteracts that of the first set has been added. The complete equilibrium of this general coplanar force system (nonconcurrent and nonparallel) now requires the satisfaction of three summation equations:

$$\Sigma F_v = 0 = +10 - 10$$
$$\Sigma F_h = 0 = +4 - 4$$
$$\Sigma M = 0 = +10(a) - 4(2.5a)$$
$$= +10(a) - (10a)$$

Because all of these summations total zero, the system is indeed in equilibrium.

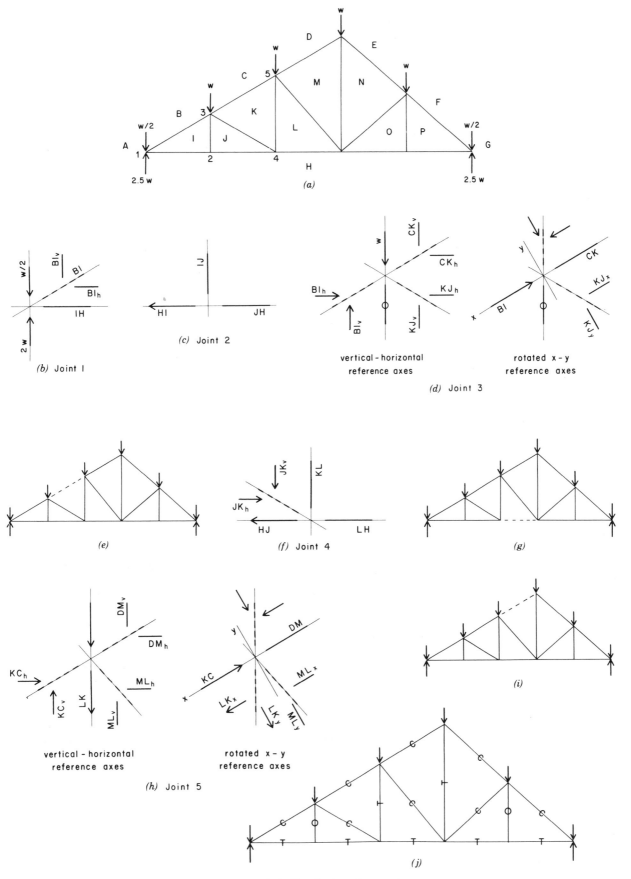

FIGURE 7.19. Visualization of truss actions.

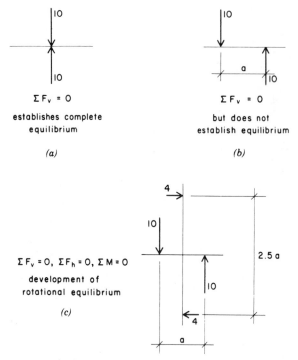

$\Sigma F_v = 0$

establishes complete
equilibrium

(a)

$\Sigma F_v = 0$

but does not
establish equilibrium

(b)

$\Sigma F_v = 0, \Sigma F_h = 0, \Sigma M = 0$

development of
rotational equilibrium

(c)

FIGURE 7.20. Consideration of rotational equilibrium.

7.9. ANALYSIS OF COPLANAR, NONCONCURRENT FORCES

Solution of equilibrium problems with nonconcurrent forces involves the application of the available algebraic summation equations. The following example illustrates the procedure for the solution of a simple parallel force system.

Figure 7.21 shows a 20-kip force applied to a beam at a point between the beam's supports. The supports must generate the two vertical forces, R_1 and R_2, in order to oppose this load. (In this case we will ignore the weight of the beam itself, which will also add load to the supports, and we will consider only the effect of the distribution of the load added to the beam.) Since there are no horizontal forces, the complete equilibrium of this force system can be established by the satisfaction of two summation equations:

$$\Sigma F_y = 0$$
$$\Sigma M = 0$$

Considering the force summation first,

$$\Sigma F_v = 0 = -20 + (R_1 + R_2)$$
(sense up considered positive)

Thus

$$R_1 + R_2 = 20$$

This yields one equation involving the two unknown quantities. If we proceed to write a moment summation involving the same two quantities, we then have two equations that can be solved simultaneously to find the two unknowns. We can simplify the algebraic task somewhat if we use the technique of making the moment summation in a way that eliminates one of the unknowns. This is done simply by using a moment reference point that lies on the line of action of one of the unknown forces. If we choose a point on the action line of R_2, as shown in Fig. 7.21*b*, the summation will be as follows:

$$\Sigma M = 0 = -20(7) + R_1(10) + R_2(0)$$
(clockwise moment plus)

Then

$$R_1(10) = 140$$
$$R_1 = 14 \text{ kips}$$

Using the relationship established from the previous force summation, we have

$$R_1 + R_2 = 20$$
$$14 + R_2 = 20$$
$$R_2 = 6 \text{ kips}$$

The solution is then as shown in Fig. 7.21*c*.

In the structure shown in Fig. 7.22*a*, the forces consist of a vertical load, a horizontal load, and some unknown reactions at the supports. Since the forces are not all parallel, we may use all equilibrium conditions in the determination of the unknown reactions. Although it is not strictly necessary, we will use the technique of finding the reactions separately for the two loads and then adding the two results to find the true reactions for the combined load.

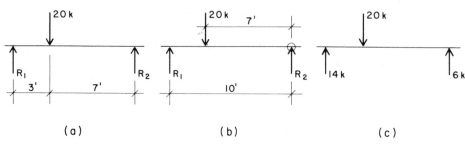

(a)   *(b)*   *(c)*

FIGURE 7.21. Analysis of a simple parallel force system.

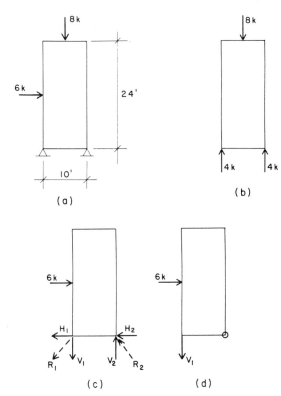

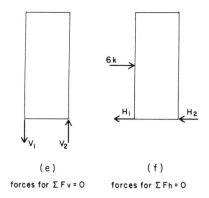

FIGURE 7.22. Analysis of a general planar force system.

The vertical load and its reactions are shown in Fig. 7.22*b*. In this case, with the symmetrically placed load, each reaction is simply one-half the total load.

For the horizontal load, the reactions will have the components shown in Fig. 7.22*c*, with the vertical reaction components developing resistance to the moment effect of the load, and the horizontal reaction components combining to resist the actual horizontal force effect. The solution for the reaction forces may be accomplished by finding these four components; then, if desired, the actual reaction forces and their directions may be found from the components, as explained in Sec. 7.2.

Let us first consider a moment summation, choosing the location of R_2 as the point of rotation. Since the action lines of V_2, H_1, and H_2 all pass through this point, their moments will be zero and the summation is reduced to dealing with the forces shown in Fig. 7.22*d*. Thus

$$\Sigma M = 0 = +6(12) - V_1(10) \quad \text{(clockwise moment considered}$$
$$V_1 = \frac{6(12)}{10} = 7.2 \text{ kips} \quad \text{positive)}$$

We next consider the summation of vertical forces, which involves only V_1 and V_2, as shown in Fig. 7.22*e*. Thus

$$\Sigma F_v = 0 = -V_1 + V_2 \quad \text{(sense up considered}$$
$$0 = -7.2 + V_2 \quad \text{positive)}$$
$$V_2 = +7.2 \text{ kips}$$

For the summation of horizontal forces the forces involved are those shown in Fig. 7.22*f*. Thus

$$\Sigma F_h = 0 = +6 - H_1 - H_2 \quad \text{(force toward right considered}$$
$$H_1 + H_2 = 6 \text{ kips} \quad \text{positive)}$$

This presents an essentially indeterminate situation that cannot be solved unless some additional relationships can be established. Some possible relationships are the following.

1. R_1 offers resistance to horizontal force, but R_2 does not. This may be the result of the relative mass or stiffness of the supporting structure or the type of connection between the supports and the structure above. If a sliding, rocking, or rolling connection is used, some minor frictional resistance may be developed, but the support is essentially without significant capability for the development of horizontal resistance. In this case $H_1 = 6$ kips and $H_2 = 0$.
2. The reverse of the preceding; R_2 offers resistance, but R_1 does not. $H_1 = 0$ and $H_2 = 6$ kips.
3. Details the construction indicate an essentially symmetrical condition for the two supports. In this case it may be reasonable to assume that the two reactions are equal. Thus $H_1 = H_2 = 3$ kips.

For this example we will assume the symmetrical condition for the supports with the horizontal force being shared equally by the two supports. Adding the results of the separate analyses we obtain the results for the combined reactions as shown in Fig. 7.22*g*. The reactions are shown both in terms of their components and in their resultant form as single forces. The magnitudes of the single force resultants are obtained as follows:

$$R_1 = \sqrt{(3)^2 + (3.2)^2} = \sqrt{19.24} = 4.386 \text{ kips}$$

$$R = \sqrt{(11.2^2 + (3)^2} = \sqrt{134.44} = 11.595 \text{ kips}$$

The directions of these forces are obtained as follows:

$$\theta_1 = \arctan \frac{3.2}{3} = \arctan 1.0667 = 46.85°$$

$$\theta_1 = \arctan \frac{11.2}{3} = \arctan 3.7333 = \text{'}5.0°$$

Note that the angles for the reactions as shown in Fig. 7.22*g* are measured as counterclockwise rotations from a right-side horizontal reference. Thus, as illustrated, the angles are actually

$$\theta_1 = 180 + 46.85 = 226.85°$$

and

$$\theta_2 = 180 - 75.0 = 105.0°$$

If this standard reference system is used, it is possible to indicate both the direction and sense of a force vector with the single value of the rotational angle. The technique is illustrated in Fig. 7.23. In Fig. 7.23*a* four forces are shown, all of which are rotated 45° from the horizontal. If we simply make the statement, "the force is at an angle of 45° from the horizontal," the situation may be any one of the four shown. If we use the reference system just described, however, we would describe the four situations as shown in Fig. 7.23*b*, and they would be identified unequivocally.

7.10. ANALYSIS OF NONCOPLANAR FORCES

Forces and structures exist in reality in a three-dimensional world. The work in preceding chapters has been limited mostly to systems of forces operating in two-dimensional planes. This is commonly done in practice, primarily for the same reasons that we have done it here: It makes both visualization and computations easier. As long as the full three-dimensional character of the forces and the structure are eventually dealt with, this approach is usually quite adequate. For visualization, as well as some computations, however, it is sometimes necessary to work with forces in noncoplanar systems. This section presents some exercises that will help in the development of an awareness of the problems of working with such force systems.

Graphical representation, visualization, and any mathematical computation all become more complex with noncoplanar systems. The following discussions rely heavily on the examples to illustrate basic concepts. The orthogonal axis system $x - y - z$ is used for ease of both visualization and computation.

Units of measurement for both forces and dimensions are of small significance in this work. Because of this, and because of the complexity of both the graphical representations and the mathematical computations, the conversions for SI units have been omitted, except for the exercise problems and answers.

Noncoplanar, Concurrent, Nonparallel Systems

Figure 7.24 shows a single force acting in such a manner that it has component actions in three dimensions. That is, it has x, y, and z components. If this force represents the resultant of a system of forces, it may be identified as follows:

$$R = \sqrt{(\Sigma F_x)^2 + (\Sigma F_y)^2 + (\Sigma F_z)^2}$$

$$\cos \theta_x = \frac{\Sigma F_x}{R}, \quad \cos \theta_y = \frac{\Sigma F_y}{R}, \quad \cos \theta_z = \frac{\Sigma F_z}{R}$$

Equilibrium for this type of system can be established by fulfilling the following conditions:

$$\Sigma F_x = 0, \quad \Sigma F_y = 0, \quad \Sigma F_z = 0$$

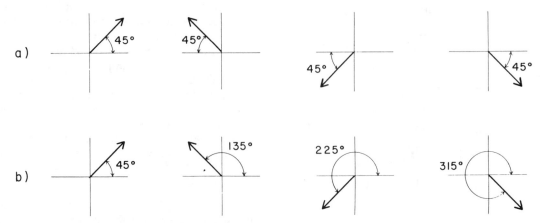

FIGURE 7.23. Reference notation for angular direction: (*a*) horizontal reference axis; (*b*) polar reference axis.

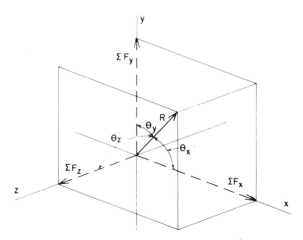

FIGURE 7.24. Components of a noncoplanar force.

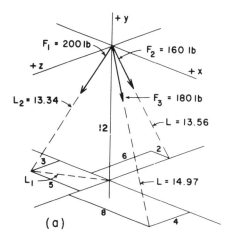

(a)

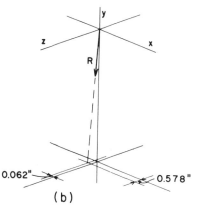

(b)

FIGURE 7.25.

Example 1. Find the resultant of the three forces shown in Fig. 7.25a.

Solution. The technique used is that of solving for the geometry of the force lines first, and then using proportionate relationships to find the force vector components. Although trigonometry could also be used, the direct use of simple geometry makes the visualization of the actions easier.

Referring to the line lengths shown in Fig. 7.25a:

$$L_1 = \sqrt{(5)^2 + (3)} = \sqrt{34} = 5.83$$
$$L_2 = \sqrt{(12)^2 + (5.83)^2} = \sqrt{178} = 13.34$$

(*Note:* To reinforce the point that the unit of measurement for dimensions is not relevant, we have omitted it.)

The other line lengths can be established in a similar manner. Their values are shown on the figure. The determination of the force components and their summation is presented in Table 7.2. Note that the signs of the components (+ and −) are established with reference to the + directions indicated for the three axes, as shown in Fig. 7.25a. Using the summations, we determine the value of the resultant as

$$R = \sqrt{(2.4)^2 + (466.1)^2 + (22.4)^2} = \sqrt{217,757} = 466.6$$

The direction of R may be established by expressing the three cosine equations, as described earlier, or by establishing its point of intersection with the x-z plane, as shown in Fig. 7.25b. Using the latter method, we have

$$\frac{\Sigma F_x}{\Sigma F_y} = \frac{x \text{ distance from } z\text{-axis}}{12} = \frac{2.4}{466.1}$$

Therefore,

$$x \text{ distance from } z\text{-axis} = \frac{2.4}{466.1}\,(12) = 0.062$$

and similarly,

$$z \text{ distance from } x\text{-axis} = \frac{22.4}{466.1}\,(12) = 0.578$$

TABLE 7.2. Summation of Force Components (Fig. 7.25)

Force	x component	y component	z component
F_1	$200 \frac{5}{13.34} = 75$ ↖	$200 \frac{12}{13.34} = 180$ ↓	$200 \frac{3}{13.34} = 45$ ↙
F_2	$160 \frac{2}{13.56} = 23.6$ ↖	$160 \frac{12}{13.56} = 141.7$ ↓	$160 \frac{6}{13.56} = 70.8$ ↗
F_3	$180 \frac{8}{14.97} = 96.2$ ↘	$180 \frac{12}{14.97} = 144.4$ ↓	$180 \frac{4}{14.97} = 48.2$ ↙
	$\Sigma F_x = 2.4 \text{ lb}$ ↖	$\Sigma F_y = 466.1 \text{ lb}$ ↓	$\Sigma F_z = 22.4 \text{ lb}$ ↙

Example 2. For the structure shown in Fig. 7.26a, find the tension in the guy wires and the compression in the mast for the loading indicated.

Solution. As in Example 1, the geometry of the wires is established first. Thus

$$L = \sqrt{(9)^2 + (12)^2 + (20)^2} = \sqrt{625} = 25$$

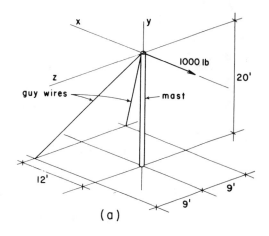

(a)

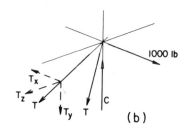

(b)

FIGURE 7.26.

Consider the concentric forces that act at the top of the mast, as shown in Fig. 7.26b. For equilibrium in the x direction

$$\Sigma F_x = 0 = 1000 - 2(T_x), \quad T_x = 500 \text{ lb}$$

Then, from the geometry of the wire,

$$\frac{T}{T_x} = \frac{25}{12}, \quad T = \frac{25}{12} (T_x) = \frac{25}{12} (500) = 1041.67 \text{ lb}$$

For the compression in the mast, consider the equilibrium of forces in the y direction. Thus

$$\Sigma F_y = 0 = C - 2(T_y), \quad C = 2(T_y)$$

where

$$\frac{T_y}{T_x} = \frac{20}{12}, \quad T_y = \frac{20}{12} (T_x) = \frac{20}{12} (500)$$

Thus

$$C = 2 \times \frac{20}{12} (500) = 1666.67 \text{ lb}$$

Example 3. Find the tension in each of the three wires in Fig. 7.27 due to the force indicated.
Solution. As before, we find the lengths of the wires as follows:

$$L_1 = \sqrt{(5)^2 + (4)^2 + (20)^2} = \sqrt{441} = 21$$
$$L_2 = \sqrt{(2)^2 + (8)^2 + (20)^2} = \sqrt{468} = 21.63$$
$$L_3 = \sqrt{(12)^2 + (20)^2} = \sqrt{544} = 23.32$$

The three equilibrium equations for the concentric forces are thus

$$\Sigma F_x = 0 = \frac{4}{21} T_1 - \frac{8}{21.63} T_2 + \frac{0}{23.32} T_3$$

$$\Sigma F_z = 0 = \frac{5}{21} T_1 + \frac{2}{21.63} T_2 - \frac{12}{23.32} T_3$$

$$\Sigma F_y = 0 = \frac{20}{21} T_1 + \frac{20}{21.63} T_2 + \frac{20}{23.32} T_3 - 1000$$

Solution of these three equations with three unknowns yields the following:

$$T_1 = 525 \text{ lb}, \quad T_2 = 271 \text{ lb}, \quad T_3 = 290 \text{ lb}$$

Noncoplanar, Noncurrent, Parallel Systems

Consider the force system shown in Fig. 7.28a. Assuming the direction of the forces to be parallel to the y-axis, the resultant can be stated as

$$R = \Sigma F_y$$

and its location in the x-z plane can be established (see Fig. 7.28a) by two moment equations, taken with respect to the x-axis and the z-axis; thus

$$L_x = \frac{\Sigma M_x}{R} \quad \text{and} \quad L_z = \frac{\Sigma M_x}{R}$$

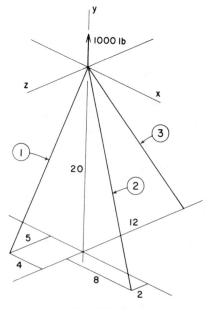

FIGURE 7.27.

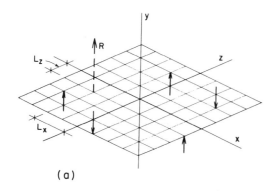

(a)

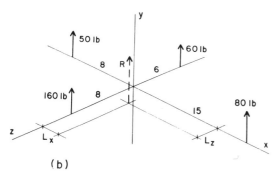

(b)

FIGURE 7.28.

Equilibrium for the system can be established by fulfilling the following conditions:

$$\Sigma F_y = 0, \quad \Sigma M_x = 0, \quad \Sigma M_z = 0$$

As with the coplanar parallel systems, the resultant may be a couple. That is, the summation of forces may be zero, but there may be a net rotational effect about the x-axis and/or the z-axis. When this is the case, the resultant couple may be visualized in terms of two component couples, one in the x-y plane (for ΣM_z) and one in the z-y plane (for ΣM_x) (see Example 5).

Example 4. Find the resultant of the system shown in Fig. 7.28b.
Solution. We first find the resultant force; thus

$$R = \Sigma F_y = 50 + 60 + 160 + 80 = 350 \text{ lb}$$

Then, for its location,

$$\Sigma M_x = +(160 \times 8) - (60 \times 6) = 920 \text{ ft-lb}$$

$$\Sigma M_z = +(50 \times 8) - (80 \times 15) = -800 \text{ ft-lb}$$

and the distances from the axes are

$$L_x = \frac{800}{350} = 2.29 \text{ ft}, \quad L_z = \frac{920}{350} = 2.63 \text{ ft}$$

Noncoplanar, Nonconcurrent, Nonparallel Systems

This is the general spatial force system with no simplifying conditions regarding its geometry. The resultant for such a system may be any of four possibilities, as follows.

1. Zero. If the system is in equilibrium.
2. A Force. If F is not zero.
3. A Couple. If M is not zero.
4. A Force plus a Couple. Which is the general case.

If the resultant is a force, its magnitude is determined as

$$R = \sqrt{(\Sigma F_x)^2 + (\Sigma F_y)^2 + (\Sigma F_z)^2}$$

and its direction by

$$\cos \theta_x = \frac{\Sigma F_x}{R}, \quad \cos \theta_y = \frac{\Sigma F_y}{R}, \quad \cos \theta_z = \frac{\Sigma F_z}{R}$$

If the resultant is a couple, it may be determined in terms of its component moments about the three axes in a procedure similar to that shown for the parallel systems. For most purposes these component moments will be sufficient, although their geometric combination into a single couple is possible.

Equilibrium for the general spatial force system can be established by fulfilling the following conditions:

$$\Sigma F_x = 0, \quad \Sigma F_y = 0, \quad \Sigma F_z = 0$$

$$\Sigma M_y = 0, \quad \Sigma M_x = 0, \quad \Sigma M_z = 0$$

The potential complexity of this system makes for a large number of possible situations. Although in theory a problem could involve the necessity for solving six simultaneous equations, simplifying conditions often reduce the complexity of the mathematical work. The larger problem is often the simple visualization and graphic representation of the force system.

CHAPTER EIGHT

Stresses and Strains

Limitations on developed stresses and strains are primary devices for the control of structural behavior. This is most true with the use of the working stress method of design, in which assigned limits are placed on stresses under service load conditions. It is less directly but just as effectively true with the use of the ultimate strength method, in which the stress and strain conditions at failure are used as a limit. Chapter 8 presents the various stress and strain considerations that are encountered in structural investigation and some of the basic techniques for their computation in ordinary situations of design.

8.1. DEVELOPMENT OF INTERNAL FORCES

Although stresses and strains result from the actions of external forces, we visualize them directly as the products of internal force actions. Thus the individual actions of tension, compression, shear, bending, and torsion are each visualized as the manifestations of a characteristic set of internal stresses and strains in the material of the structure. As discussed in Sec. 6.4, the free-body diagram and the cut section are essential tools in these visualizations.

At any particular location within a structure under load there is usually not a single internal force action, but rather some combination of actions. Consider the actions of the simple, axially loaded column and the single tension rod as shown in Figs. 8.1a and b. In these the internal force actions can be visualized in the form of simple, single effects of compression or tension.

We do indeed make use of such simple elements on occasion, but more often need to consider more complex actions. For example, for the vertical support for the sign shown in Fig. 8.1c, the internal actions developed by a combined loading of gravity plus wind include compression, shear, torsion, and two-way bending. These occur all at once, so a true analysis must somehow combine all the effects into some net stressed and strained condition.

Although true service conditions must be considered for a realistic design, it is the usual practice to visualize and analyze for individual internal actions separately. This is not necessary for the computer, but most human beings find it more feasible to handle individual actions one at a time before attempting to visualize their net effect. For initial study in particular, we will follow this procedure of

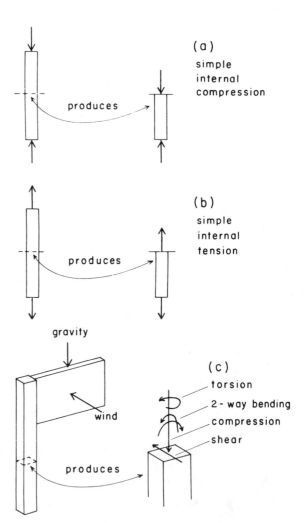

FIGURE 8.1. Development of internal forces.

divide and conquer in the following sections, before considering combined and net effects.

8.2. STRESS AND STRAIN

Stress is not truly one-dimensional or two-dimensional; it is always three-dimensional, as are the shape changes in the three-dimensional materials that experience the stress.

This idea was discussed and illustrated in Part 1, and it is important to keep the true nature of stress and strain in mind. For purposes of computation in structural investigation, and even more so in design work, however, we routinely make determinations of some simplified stress conditions. These are not done as the true representatives of the complete stress and strain actions, simply as indicators of actions.

The following sections consider the stresses and strains that can be visualized directly from the actions of individual internal force effects. Later sections in this chapter and several of the sections in Chapter 12 present issues relating to more complex considerations of combined stresses, multi-direction stresses, and so on.

It is in many ways easier to visualize strains and their accumulation in overall deformations of structures than to visualize internal stresses and force actions. For this reason we often use the graphic representation of the deformed material or structure to infer the accompanying stress or internal force actions. This is a basic technique used by the professional designer as well as the beginning student.

8.3. DIRECT STRESS

Direct stress—tension or compression—results from the action of a direct force. It is visualized as operating at right angles to a surface, produced by the force action of tension or compression that acts at a right angle to the surface. The examples in Fig. 8.2 illustrate this basic development and the simple form of the stress formulas used for computation. In Fig. 8.2a a block is being pressed down against some unyielding surface, with a pressure on the bottom of the block developed by the crushing force C. If the crushing force acts with a symmetrical orientation on the block (called an axial force) the compressive pressure effect may be visualized as compression stress, with its unit value expressed as

$$f_c = \frac{C}{A}$$

where

f_c = unit stress, expressed as force per unit area: pounds per square inch (psi), newtons per square meter (N/m²), etc.

C = total force in pounds, newtons, etc.

A = area of contact between the block and the surface that supports it

If the block is glued to the ceiling and operated on by a tension force, the determination of the tension stress on the glue joint is of the same form (see Fig. 8.2b). In a structural member, such as the column or tension rod shown in Fig. 8.1a, the development of internal direct force and the resulting direct stress is visualized in the same manner, the surface being that of a cut section at a right angle to the direct force.

The simple formula for direct stress can be used in three ways for different situations, as follows.

1. Given the crushing force C and the area of contact (or the area of the cross section of the column) A, find the magnitude of the unit direct stress. This computation takes the form presented in the illustration:

$$f = \frac{C}{A}$$

2. Given the area A and some limit for the stress f, find the limit for the compressive force. For this computation we use the form

$$C = fA$$

3. Given the limit for the stress f and the need to develop a specific amount of total compression C, find the area of contact or cross section required. In this case we use the form

$$A = \frac{C}{f}$$

The following example problems demonstrate the use of the formula for direct stress.

Example 1. A 6 × 6 wood post (actually 5.5 in. on a side [140 mm]) sustains a compression load of 20 kips [89 kN]. Find the value of the unit compressive stress in the post.
Solution: Properties of wood posts are available from Table A.2 (Appendix A); however, we may find the area of the post cross section simply as

$$A = (5.5)^2 = 30.25 \text{ in.}^2 [0.0196 \text{ m}^2]$$

The stress is then found as

$$f = \frac{C}{A} = \frac{20,000}{30.25} = 661 \text{ psi } [4.54 \text{ MPa}]$$

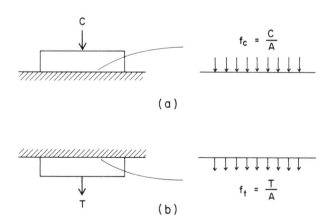

FIGURE 8.2. Development of direct stresses.

Example 2. If the post in Example 1 has a maximum limit for compression stress of 1000 psi [6.9 MPa], what is the maximum total compressive force that it can carry?
Solution. With the known values for f and A, the force is expressed as

$$C = fA = 1000(30.25)$$
$$= 30,250 \text{ lb or } 30.25 \text{ kips } [135 \text{ kN}]$$

Example 3. If the post in Example 2 must carry a load of 50 kips [222 kN], and the stress limit is 1000 psi [6.9 MPa], what area is required for the post cross section?
Solution. Using the given values for C and f, we obtain

$$A = \frac{C}{f} = \frac{50,000}{1000} = 50 \text{ in.}^2 \ [0.03226 \text{ m}^2]$$

Strain due to direct stress is visualized as a linear shortening (due to compression) or lengthening (due to tension) in the direction of the force action and the stress. Strain is ordinarily expressed as unit strain in the form of a unitless quantity, indicated as a percentage or a decimal fraction. If the stress is constant throughout the length of the member (as it approximately is in the column or tie rod), the strain will also be constant and will accumulate in the total length change of the member. Using e for the total length change (elongation) and L for the original length of the member, we express the unit strain as ε (lower case Greek epsilon) as follows:

$$\varepsilon = \frac{e}{L}$$

Relations between strains and deformations of various kinds are discussed more fully in Sec. 8.7.

8.4. SHEAR STRESS

There are three fundamentally different situations in which force actions result in the direct development of shear stress. These are as follows:

1. Stress produced by a direct shearing action (cutting, slicing effect), called *direct shear stress*.
2. Stress produced in the normal functioning of beams, called *beam shear stress*.
3. Stress produced by torsion, called *torsional shear stress*.

In this section we treat only the first case, that of direct shear stress. Shear in beams is discussed in Chapter 9 and stresses caused by torsion are discussed in Sec. 8.6.

Figure 8.3 shows two examples involving the development of direct shear stress. In Fig. 8.3a the lateral slip of the tongue-and-groove joint is resisted by the development of shear stresses at the base of the tongue. The shear stress is assumed to be of a uniform value on the developed section area, with a unit magnitude of

$$f_v = \frac{V}{A}$$

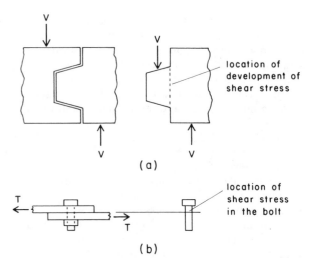

FIGURE 8.3. Development of shear stress.

in which V is the operating shear force and A is the total area of the section at the base of the tongue. It may be observed that the form of this expression is the same as that for simple direct stress of tension or compression.

In the example in Fig. 8.3b the tension forces exerted on the two overlapping steel plates tend to produce a slippage, which is resisted by the bolt that connects the plates, resulting in a slicing effect on the bolt. The expression for the shear stress on the slice-resisting cross section of the bolt is the same in form as that for the stress at the base of the tongue in Fig. 8.3a. In this case the operating force is the tension force T and the area is that of the bolt cross section.

When used in the manner shown in Fig. 8.3b, the bolt is said to be in a condition of *single shear*, since the bolt need be sliced only once for the joint to fail. In the use of the bolt in Fig. 8.4a it may be observed that the bolt must be sliced at two separate places simultaneously if the plates slip; in this case the bolt is said to be in *double shear*. Extending this concept further, it may be observed that the hinge pin shown in Fig. 8.4b must be sliced a total of eight places if the hinge is to be separated in the direction of the forces

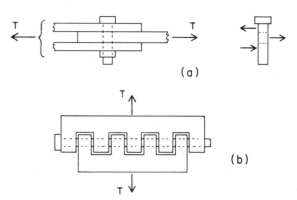

FIGURE 8.4. Development of multiple shear sections in a pinned connection.

shown. Actually, pins and bolts have other aspects of structural behavior in addition to shear. However, when considering only the shear effect, the area used in the stress formula is that of the cross section times the number of sliced sections.

The following example problems illustrate the use of the relationships for simple direct shear stress.

Example 1. A wood tongue-and-groove joint is formed as shown in Fig. 8.3a, with the width of the tongue at its base equal to 0.25 in. If the maximum stress is limited to a value of 50 psi, what is the total shear load capacity of the joint, expressed in units of pounds per running foot of joint length?
Solution. For a length of 1 ft (12 in.) of the joint the total area at the base of the tongue is

$$A = 12(0.25) = 3.0 \text{ in.}^2$$

and the limit for the load is

$$V = f_v A = 50(3.0) = 150 \text{ lb/ft [2.19 kN/m]}$$

(*Note:* With U.S. units the 1-ft length is commonly used in such computations. Obviously, other units would be used in computations with SI units.)

Example 2. A steel bolt is used as shown in Fig. 8.4a. The tension force on the joint is 20 kips [89 kN] and the limit for shear stress in the bolt is 14 ksi [100 MPa]. Find the minimum diameter required for the round bolt.
Solution. For the total resisting area we find

$$A = \frac{T}{f_v} = \frac{20}{14} = 1.429 \text{ in.}^2 \text{ [890 mm}^2\text{]}$$

Observing that the bolt is in double shear, we express the total resisting area of the two sliced cross sections of the bolt as

$$A = (2)\frac{\pi D^2}{4} = 0.5\pi D^2 = 1.571 D^2$$

Then

$$1.571 D^2 = 1.429$$

$$D = \sqrt{\frac{1.429}{1.571}} = \sqrt{0.906} = 0.954 \text{ in. [23.8 mm]}$$

Example 3. A hinge of the form shown in Fig. 8.4b is connected with a 0.25-in. [6-mm]-diameter pin. Find the value of the unit shear stress in the pin if the separating force is 2000 lb [8.9 kN].
Solution. The total resisting area of the eight sliced sections is

$$A = 8\pi R^2 = 8\pi(0.125)^2 = 0.393 \text{ in.}^2 \text{ [226 mm}^2\text{]}$$

and the unit stress is

$$f_v = \frac{2000}{0.393} = 5089 \text{ psi [39.4 MPa]}$$

8.5. BENDING STRESS

When a linear structural member is subjected to a bending moment that lies in a plane parallel to the longitudinal axis of the member, the effect is called *bending*. Such a situation is shown in Fig. 8.5a and we observe the following regarding its actions.

1. The bending tends to cause a curling up or curving of the member.
2. The curving indicates that the material on one side of the member is being lengthened due to tension, while the material on the opposite side is being shortened due to compression.
3. Due to the reversal of stress from side to side, there will be some point of transition at which the stress is zero on the member's cross section.
4. Internal bending resistance at a cross section is developed by an internal force couple, which is produced by the resultants of the tension and compression stresses.

If the member being bent has a symmetrical cross section and the moment exists in the plane containing the axis of symmetry of the section (see Fig. 8.5b), we may make the following observations.

1. The neutral stress points referred to in item 3 above will lie on an axis through the centroid of the section and perpendicular to the axis containing the bending moment. This axis at which no stress occurs is called the *neutral axis*.

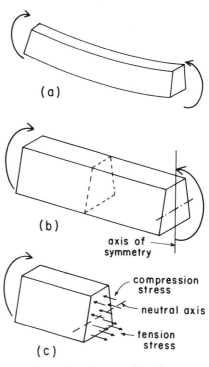

FIGURE 8.5. Development of bending stress.

2. If the material of the member is homogeneous, isotropic, and elastic (basically meaning of a single elastic material), planar cross sections of the member at right angles to its longitudinal axis will retain their planar form during bending. As the beam curves, these plane sections will rotate about their neutral axes.

3. The radius of curvature of the member at any point along its length (assuming an initial straight condition) will be inversely proportional to the bending moment at that point.

The observations just made can be verified by tests, and because of them, we can make the following derivations in terms of the stresses and strains in the member. We first observe that due to the plane section observation, the strain at any point is proportional to its distance from the neutral axis. Accepting this, if the material is elastic—meaning that stress is proportional to strain—and the stress is below the material's failure limit, the stress at any point is proportional to the distance of the point from the neutral axis. Thus (see Fig. 8.6a)

$$\frac{\varepsilon_y}{y} = K_1 \quad \text{and} \quad \frac{f_y}{y} = K_2$$

The total internal resisting moment at a section may be determined by a summation of the increments of moment resistance consisting of the effects of stresses on unit areas of the cross section. If we use a unit area that consists of a slice dy thick and y distance from the neutral axis (see Fig. 8.6b), we can assume the stress to be a constant on the unit area. Thus the increment of internal force developed on the unit area is the product of the stress times the unit area, or

$$dF = f_y \, dA$$

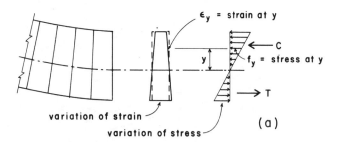

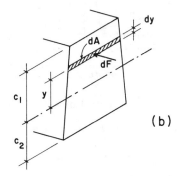

FIGURE 8.6. Bending stress on a cross section.

and the unit moment developed by this unit force is

$$dM = dF \, y = (f_y \, dA)y$$

Using the calculus, the total resisting moment of the cross section can be expressed as

$$M = \int_{c_2}^{c_1} dM = \int_{c_2}^{c_1} (f_y \, dA)y = \int_{c_2}^{c_1} f_y y \, dA$$

Since f_y/y is a constant, we can make the following transformation:

$$M = \int_{c_2}^{c_1} \frac{f_y}{y} y^2 \, dA = \frac{f_y}{y} \int_{c_2}^{c_1} y^2 \, dA$$

Astute students of geometry and calculus will recognize that the expression following the integral sign in this formula is the basic form for indication of the second moment of the area (or moment of inertia; see Appendix A) of the section about its neutral centroidal axis. We designate this basic geometric property as I and thus express the formula more simply as

$$M = \frac{f_y}{y} I$$

or, transforming it into an expression for stress, we have

$$f_y = \frac{My}{I}$$

which is the general formula for bending stress.

Referring to Fig. 8.6, we note that the maximum stress in compression will be obtained when $y = c_1$ and the maximum in tension when $y = c_2$. Further, if the section is one with symmetry about the neutral axis, c_1 will be equal to c_2. With these considerations, a simpler expression for maximum bending stress that is used commonly takes the form

$$f = \frac{Mc}{I}$$

An additional simplification can be made by use of a single term for the combination of I and c. This term is called the *section modulus*, defined as

$$S = \frac{I}{c}$$

Substituting this in the stress formula reduces it to

$$f = \frac{M}{S}$$

For beam design purposes, the formula is transposed to

$$S = \frac{M}{f}$$

When designing a beam, once the required resisting moment is determined, and the material to be used is established (yielding a value for the limit of stress), a single required property for the beam—*S*—can be found for the selection of the beam.

The following examples illustrate simple problems using the formulas for bending stress.

Example 1. A beam has a T-shaped section as shown in Fig. 8.7. A bending moment occurs in the axis of symmetry of the section, producing compression in the upper portion and tension in the lower portion of the section. If the bending moment is 10 kip-ft (14 kN-m), and the moment of inertia of the section about the neutral axis is 30.71 in.[4] [12.02 × 10^6 mm^4], find the values for maximum tension and compression stress due to bending.

Solution. For compression at the top of the section we use the distance from the neutral axis of 2.167 in. and the stress formula in the form

$$f_y = \frac{My}{I} = \frac{[10(12)]2.167}{30.71} = 8.47 \text{ ksi [63.1 MPa]}$$

Note that the moment in kip-ft must be changed to inch units, as the other values use inches.

For tension stress at the bottom we use the distance of 3.833 in.; thus

$$f_y = \frac{My}{I} = \frac{[10(12)]3.833}{30.71} = 14.98 \text{ ksi [111.6 MPa]}$$

Example 2. For the beam section in Fig. 8.7, what is the value for *S* (section modulus) that indicates the moment resistance of the section based on maximum bending stress?

Solution. If the section is limited by the maximum stress, we must use the greatest distance from the neutral axis to any edge, or 3.833 in. for the section given. Thus

$$S = \frac{I}{c} = \frac{30.71}{3.833} = 8.01 \text{ in.}^3 \text{ [125.4 × } 10^3 \text{ mm}^3]$$

8.6. TORSIONAL STRESS

When a linear element is subjected to a twisting moment that exists in a plane that is perpendicular to the longitudinal axis of the element, the effect is called *torsion*.

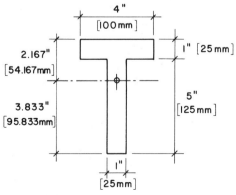

FIGURE 8.7.

Torsion is developed in a number of ways in building structures, most notably in the beams and columns of structural frameworks. However, the need to investigate stresses or deformations and to design specifically for torsional resistance is relatively uncommon in design work. In most cases structural members that experience torsion are also subject to various other actions and are often not designed primarily for the torsional effects. Nevertheless, it is important to have a general understanding of the nature of torsion and of the situations that generate it. It is to that end—the general understanding of torsion—that the following material is presented.

Torsion of Cylindrical Shafts

The stresses and strains produced by torsion are three-dimensional in nature and are usually very complex. For purpose of discussion of the phenomena, we first consider the simplest geometric case; that of a shaft with a solid or hollow cylindrical cross section.

Consider a shaft such as that shown in Fig. 8.8a. Under the action of the force couple *P—P* a twisting moment equal to *Pa* is developed. This moment is called the torque and is designated *T*. Considering the action of the shaft, we make the following observations.

1. The magnitude of *T* is constant at all cross sections along the shaft.
2. Diagonal lines in the cross sections remain straight during the twisting of the shaft. Thus the amount of movement at any point on a cross section is directly proportional to its distance from the center of the shaft. [This assumes that the material of the shaft is elastic and is analogous to the relationships observed for bending in Sec. 8.5 (Fig. 8.6).]
3. The rotation of a particular section (θ) is directly proportional to its distance from the fixed end. Thus, referring to Fig. 8.8a, we have

$$\frac{\theta_x}{\theta_L} = \frac{x}{L}$$

or

$$\frac{\theta_L}{L} = \frac{\theta_x}{x} = \text{a constant}$$

Observations 2 and 3 can be verified by tests. As a consequence of them we can say

4. The shear strain (measured as an angle and designated γ, Greek lowercase gamma, as shown in Fig. 8.8b) is the same at all points in the shaft that are the same distance from the center of the shaft and is proportional to the distance from the center.

Using these observations, we can obtain an expression for torsional shear stress. The basic technique is similar to that used in Sec. 8.5 for the derivation of the expression for bending stress. A general formula for stress would include the variable of the distance from the center of the shaft. However, because maximum stress is usually of critical

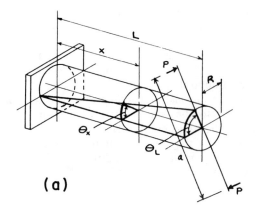

(a)

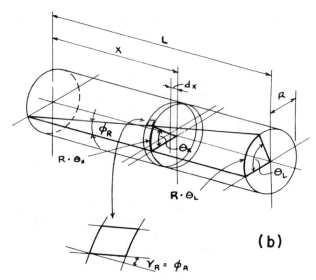

(b)

FIGURE 8.8. Development of torsion.

J = polar moment of inertia of the area of the cross section, as discussed in Sec. A.7 in Appendix A

Torsion Effects

The torsional shear condition results in a variety of effects that cause a number of different responses, depending on the material and the configuration of the member being twisted. The shear stress phenomenon is essentially three-dimensional ad includes the usual complex responses, including shear on mutually perpendicular planes and the diagonal tension and compression effects. Some common effects and responses are the following.

1. *Transverse Shear.* This tends to cause a failure by the separation of adjacent cross-sectional surfaces, as shown in Fig. 8.9a. This does not often occur in members of continuous form, but is associated with joint failures in connected members.
2. *Longitudinal Shear.* This results in longitudinal splitting and is a characteristic failure in wooden shafts (see Fig. 8.9b).
3. *Diagonal Tension.* This results in a spiral type of separation and is a typical response with brittle materials such as cast iron. This is especially true for materials weak in tension, such as stone or concrete (see Fig. 8.9c).
4. *Diagonal Compression.* This results in a spiral-form crushing.or collapse. This is typical for soft materials such as rubber or soft plastic. It is also the form of failure most common for thin-walled cylinders, occurring as a spiral-form buckling (see Fig. 8.9d).

The general case of torsion, including considerations of torsion in noncylindrical elements, is discussed more fully in Sec. 12.6.

8.7. STRAIN AND DEFORMATION

Simple relationships between stress and strain permit the use of a highly effective technique in structural investigation. This consists of visualizing the form of the structure as it is deformed by the loads acting on it. From the nature of the deformations it is usually possible to infer the particular strain conditions in the material whose gross accumulation produces the overall deformation. Finally, from the type of strain it is possible to determine the character of the stress and the type of internal force action that relates to the stress development. Because it is more difficult to "see" stress and internal force, the direct interpretation of the geometric character of the structure's deformation becomes a practical analytic device.

The visualization of the deformed structure is usually not critical to the understanding of actions of simple tension or compression members, but it is often quite useful with members subjected to bending or torsion. It was used, for example, in the derivation of the formulas for stress due

concern and will be obtained using the maximum distance—the radius of the shaft—we state the stress formula for the maximum condition as

$$f_v = \frac{TR}{J}$$

where T = twisting moment

R = shaft radius

J = polar moment of inertia of the area of the shaft cross section

For the elastic shaft, with stress proportional to strain, it is possible to establish a relationship between the twisting moment and the rotation of the shaft. The general form of this relationship is

$$\theta_x = \frac{Tx}{GJ}$$

where G = shear modulus of the material, as discussed in Sec. 12.4

this form of
failure

results from this
stress condition

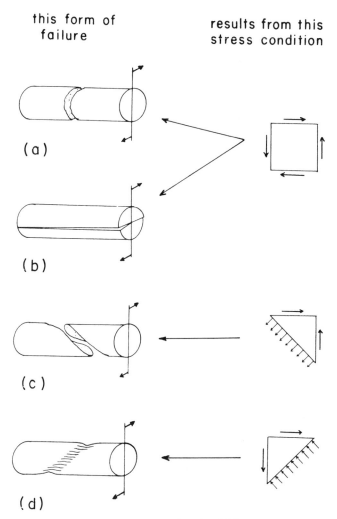

(a)

(b)

(c)

(d)

FIGURE 8.9. Forms of torsional failure.

to bending and torsion in Secs. 8.5 and 8.6. In those sections it was used both for assistance in simple visualization of the actions and as a means for establishing actual geometric relationships critical to the derivations.

Deformation of the structure itself is sometimes a design issue. In building structures, deformations due to direct stress actions of tension and compression are usually so small in dimension as not to be critical, with the possible exception of very long tension members of high-strength steel. The most common deformation "problem" is that of the deflection of members subject to bending. This may be critical for the loss of straightness, the actual dimension of displacement, the rotations of the ends of members, or the degree of curvature produced. In structural design work, computations for deformations are usually limited to these determinations.

8.8. STRESS–STRAIN RELATIONSHIPS

We have thus far discussed stress and strain as related phenomena, but have not dealt with their specific relationships in detail. We will now consider some of the aspects of these relationships as they occur with the commonly used structural materials.

The curves on the graph in Fig. 8.10 indicate three basic types of stress–strain relationships. The curves are plotted with unit strain as the horizontal variable and unit stress as the vertical variable. The curves manifest relationships involving three types of material response, as follows.

1. *Elastic Behavior.* This indicates a constant proportionality of stress to strain.
2. *Inelastic Behavior.* This is the general case when stress and strain do not remain in constant relationship over the range of stress increase. The relationship may be predictable, but is not one of constant proportionality.
3. *Plastic Behavior.* This is the case when increase of strain occurs at relatively constant stress. Plastic strain may be a natural development at some range of stress or may occur due to heat, moisture, time, or chemical change.

Referring to Fig. 8.10, curve 1 represents a material that is elastic throughout its entire range of stress. In this case stress and strain can be related by expressing the slope or the angle of the line. This is ordinarily done by defining the slope in terms of the tangent of the angle, which is simply the ratio of stress to strain. This value is called the *elastic modulus,* or more commonly the *modulus of elasticity,* and is designated as E in standard notation. Thus, as shown in Fig. 8.10,

$$E = \tan \theta_1 = \frac{\text{stress}}{\text{strain}} = \frac{f}{\varepsilon}$$

Typical values for E for common structural materials are given in Table 8.1.

Curve 2 in Fig. 8.10 represents a material for which the relation of stress to strain varies continuously over the range of stress. Although the curve shown here is a hypothetical one, many materials, including wood and concrete, behave essentially in this manner. No single value of E can be defined for such a curve. Two values that are used for E are the modulus at some specific point indicated by the tangent to the curve at that point and the average modulus for some range of stress. On the curve in Fig. 8.10, θ_P indicates the tangent modulus at point P, and θ_2 indicates the average for E over the range of stress from zero to the value at P.

Curve 3 in Fig. 8.10 represents the general form of response for ordinary structural-grade steel in tension. The curve is complex and must be discussed in detail. Referring to the letters on the curve, we note the following.

O–Q. The material is elastic in this range with an E of approximately 29,000 ksi [200 GPa]. Note that E has the same unit as stress since strain is dimensionless.

Q–R. At Q the curve begins to deviate from the straight line. Point Q establishes the *proportional limit* for the material, and as stress is increased, strain becomes increasingly inelastic. Finally, at some point (R) the ma-

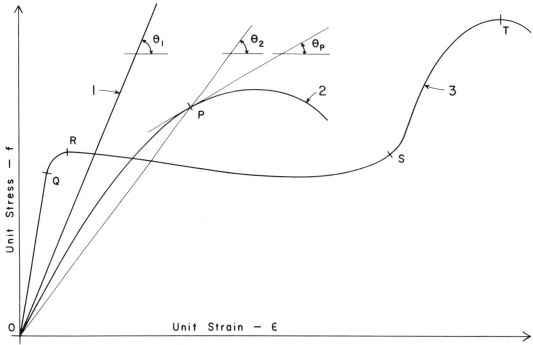

FIGURE 8.10. Forms of stress–strain behavior.

TABLE 8.1. **Values of Modulus of Elasticity (E) for Common Structural Materials**[a]

| Material | E | | Type of Stress |
	ksi	GPa	
Steel	29,000	200	Tension, compression
Aluminum (structural alloys)	10,000	70	Tension, compression
Concrete (stone aggregate)	3–5000	20–35	Compression
Wood (Douglas fir)	1.5–2000	10–14	Tension, compression parallel to grain
Plastic			
Hard (acrylic)	1000	7	Tension, compression
Soft (vinyl)	350	2.4	Tension, compression

[a] For usable (service load) range of stress of type indicated. Ignores effects of time, temperature, moisture, etc.

terial begins to deform plastically, with considerable increase in strain at an approximately constant stress resistance level near that of Q and R. The maximum stress achieved before the plastic deformation (R) is called the *yield stress*.

R–S. If the material has strain as great as that shown in the range from R to S—as much as 15–20 times the strain at Q—it is said to be highly *ductile*. As the material nears the end of this range it begins once again to develop additional stress resistance. The point at which the curve rises to the level of the stress at R establishes the end of the plastic deformation range relating to evaluation of ductility.

S–T. This range establishes the ultimate stress resistance of the material, which may be twice or more the value at R. Eventually, the material fails, usually in a brittle manner. This portion of the graph is not drawn to scale, as the strain at failure may be as much as 10–15 times that at point S.

The curves in Fig. 8.10 are significantly different and indicate different requirements for evaluation of structural responses of different materials. This is indeed the case, as will be demonstrated in the discussions of design for the materials in Parts 3 through 6.

8.9. DESIGN CONTROL OF STRESS AND STRAIN

For design purposes the control of stress is a basic means for limiting the use of structures to situations that are within their capacities by some safe margin. With the working stress method of design this is accomplished by the establishment of *allowable stresses* that are some fraction of the demonstrated capacity limit for the materials to be used. This does not consist of determining a single stress value, as there are both different types of stress (tension, shear, etc.) and different material responses (such as that related to grain direction in wood). In addition, the loading conditions may affect considerations of stress responses relating to the character of the loads (static versus dynamic, etc.), to their time duration (wind versus gravity dead load, etc.) or to the reliability of their determination.

Industry design standards and building codes are the usual sources for the stress limits used in design work. These sources and the use of criteria and data from them are discussed in detail in Parts 3 through 6.

Except in special situations, control of strain is not generally a concern. Control of stress, of course, in effect, results in control of strain. Accumulated strains in the form of gross deformations—such as the deflection of a beam—represent one type of concern for the magnitude of strain. Cracking of concrete is related to a specific amount of strain under tension, and its control must recognize that issue. The sharing of load by materials that interact—such as concrete and steel in reinforced concrete—involves consideration of shared strain conditions.

A primary concern in design is structural safety; thus the direct relation between structural strength and stress magnitudes is a vital matter. However, for a really successful design, there are many concerns in addition to the major one of structural collapse. A bouncy floor structure is undesirable, even though collapse may not be at risk. Excessive cracks in concrete are visually objectionable, even though they may be "natural" to the structural behavior and not a safety issue. This is to say that strict adherence to the legal limits for stress magnitudes may assure safety, but there are often many additional concerns in design, and it is not uncommon for stress limits to be controlled by other factors.

CHAPTER NINE

Beams

The generic name for a structural member that is used for spanning, sustains lateral loading, and develops internal force actions of bending and shear is *beam*. Such a member may have another name—girder, joist, purlin, header, lintel, deck—but its general structural functioning is described as beam action. Chapter 9 deals with the range of considerations involved in the investigation of beams.

9.1. TYPES OF BEAMS

The most commonly used beam is the *simple beam*. As shown in Fig. 9.1, this consists of a single-span beam with supports at each end, offering only vertical force resistance. Because the supports do not offer restraint to the rotation of the beam ends, the beam takes the simple curved form of deformation under load. Supports that do not restrain rotation are called *free*, *pinned*, or *simple* supports. Thus the beam in Fig. 9.1a is actually a *simply supported beam*, although the simpler reference of simple beam is commonly used.

A *cantilever* beam consists of a single-span beam with only one support, as shown in Fig. 9.1b. For stability of the beam, this must be a rotation-resisting support, called a *fixed* support or a *moment-resistive* support. Cantilevers exist less often as shown in Fig. 9.1b, than as extensions of beams over their supports, as shown at the right end of the beam in Fig. 9.1c. The beam with extended ends is called an *overhanging beam*. There may be an extension at only one end, as shown in the figure, or at both ends, the latter being called a *double overhanging beam*.

While the simple beam and the single cantilever have deformed shapes with simple single curvature, overhanging beams have inflected shapes of multiple curvature. This multiple-curved form is also present with *continuous beams*, as shown in Fig. 9.1d.

Figure 9.1e shows a single-span beam with both ends fully fixed against rotation. This is called a *restrained beam* or a *fixed-end beam* and its deformed shape takes the profile of the doubly inflected curve shown.

Visualization of the deformed shape of the beam is a useful tool in investigation. It helps to establish the charac-

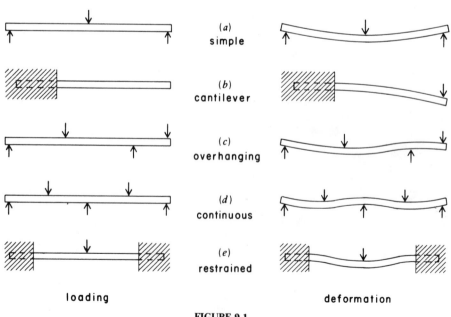

(a) simple

(b) cantilever

(c) overhanging

(d) continuous

(e) restrained

loading   deformation

FIGURE 9.1.

ter of support reactions as well as the nature of distribution of internal force effects in the beam.

It is possible for beam support conditions to approach the true situation of full fixity or complete freedom of moment restraint. Many supports, however, tend to offer partial restraint, being somewhere between the extreme conditions illustrated in Fig. 9.1. Details of the connections at the beam supports as well as the nature of the supporting structures will qualify these conditions. For initial investigation, however, we usually assume either simple or fully fixed supports, reserving judgment as to any need for adjustment until more is known about the final form and details of the structure.

9.2.   LOAD AND SUPPORT CONDITIONS

Members that serve as beams exist in a variety of situations and sustain many types of loads. The most common types of load are the following (see Fig. 9.2):

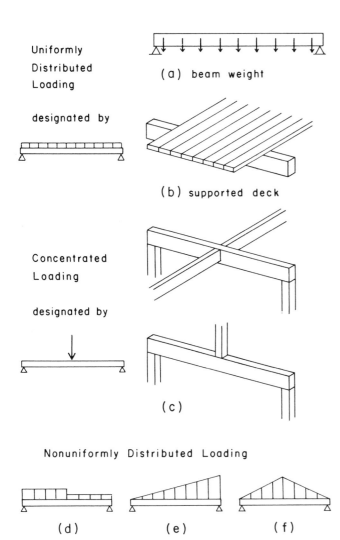

Uniformly Distributed Loading

designated by

(a) beam weight

(b) supported deck

Concentrated Loading

designated by

(c)

Nonuniformly Distributed Loading

(d)   (e)   (f)

FIGURE 9.2. Beam loading conditions.

1. *Uniformly Distributed Load.* The dead weight of the beam itself is constituted as a load that is distributed evenly along the beam length, as shown in Fig. 9.2a. This is the most common loading and is called a uniformly distributed load, or simply a uniform load. If the beam supports a roof or floor deck, as shown in Fig. 9.2b, the weight of the deck and any loads carried by the deck will usually also be uniformly distributed on the beam.

2. *Concentrated Load.* The second most common load is one in which the force is delivered to the beam at a single location, effectively as though it were concentrated at a point. In framing systems, beams that support the ends of other beams (Fig. 9.2c) sustain concentrated loads consisting of the end reactions of the supported beams.

3. *Nonuniformly Distributed Load.* Complexities of the building layout or construction sometimes result in distributed loads on beams that are not uniform in magnitude along the beam length. The loading shown in Fig. 9.2d indicates a change in the magnitude of the distributed loading over part of the beam length, possibly due to the need to support a wall over the beam for part of its length. In Fig. 9.2e the distributed load varies continuously in magnitude from zero at one end to a maximum value at the other end. This may occur in a number of ways, a common situation being the horizontal soil pressure on a retaining wall or basement wall. Another load that varies in magnitude is shown in Fig. 9.2f; this is the form of load commonly assumed for a beam that serves as a lintel over an opening in a masonry wall.

As with other structural elements, beams usually sustain combinations of loads rather than single loadings. For a given design situation it may thus be necessary to investigate the beam behavior for several loadings. This situation is discussed in Parts 3 through 6 for beams of various materials and the issue in general is discussed in the building design examples in Part 9.

Beam actions often occur in members that also serve other structural functions. Thus the design of the member must consider the interactions and net effects of the multiple force actions. In this chapter we consider only beam actions, which may be the sole function or only part of the activity for a given member.

Some aspects of beam supports are discussed in Sec. 9.1, including the identity of the support as free or fixed with regard to rotation restraint. This identity has to do with the manner of attachment of the beam to the support and with what exactly is providing support. Consider the situation illustrated in Fig. 9.2c, in which beams are shown being supported by another beam. In this case the manner of attachment and the torsional stiffness of the supporting beam are key issues in establishing the degree of restraint for the ends of the supported beams. In the same illustration, the supporting beam is shown to be supported on the tops of two columns at its ends. In this case any restraint of the beam ends depends on the rigidity of the beam-to-column connection and the bending stiffness of the columns.

In many cases the design of the supports is an extension of the design of the beam. Actually, few structural members can be designed completely as entities; each is a part of a system, and the entire system must be considered at some time in the complete design of the structure. Thus in a real design situation it is necessary to realize that while the beam behavior depends on support conditions, the requirements for the supports depend on the beam actions. In the design process there is often no single logical point at which to begin investigation that does not require some assumptions about other parts of the system. As the investigation and design proceeds, some early assumptions may be found to be in error and readjustment is required. Working designers spend a considerable amount of their time in the general activity of readjustment.

The loads and the force actions generated by the supports constitute an external force system acting on the beam. The character of the beam and the nature of this external force system will determine whether the beam is stable or unstable and whether it is statically determinate or indeterminate. The illustrations in Fig. 9.3 show the range of possibilities in these regards.

In Fig. 9.3a the single-span beam is supported at one end with a free support capable of vertical force resistance only. The beam is obviously unstable. It can be made stable by the addition of an additional reaction component: in the form of a second vertical support, or a moment resistance at the presently supported end, as shown in Fig. 9.3b.

If we add two additional support components to the beam in Fig. 9.3a, we perform an overkill of the unstable condition, producing a situation that is stable but one that is statically redundant. This is not necessarily a bad design decision, but it does have one consequence: We cannot solve for the reactions at the supports by the sole use of equations for static equilibrium. The situation is therefore described as being statically indeterminate.

A manner in which determinacy may be restored to the two-span beam in Fig. 9.3c is shown in Fig. 9.3d. This consists of producing the beam in two segments connected by a pin-type connection within the first span. This permits investigation of the beam in sequence as two statically determinate segments.

We will deal primarily with the behavior of simple statically determinate beams in this chapter, although some aspects of investigation of indeterminate beams are presented in Sec. 9.12. Structures with internal pins, including beams and frames, are treated in Chapter 14.

9.3. BEAM ACTIONS

Let us consider the behavior of beams with regard to the various actions that may need consideration in structural investigation for design.

Flexure, or Bending.   Bending is a primary beam function, involving the need for some resistance to internal moment at most cross sections of the beam. It requires the development of stresses that vary in magnitude and switch from compression to tension across the beam section, as discussed in Sec. 8.5. For most loadings, the magnitude—and possibly the sign—of the internal bending moment will vary along the beam length. Of critical concern is the greatest magnitude of internal moment, producing the requirement for maximum bending resistance by the beam.

Shear.   Internal shear as a direct result of the vertical force effect of the loads is the other primary beam function. Shear stress development is not as simple as in the case of direct shear and is discussed in Sec. 9.6. As discussed in Sec. 8.4, the vertical shear produces an equal reactive horizontal shear and diagonal tension and compression, any of which may be critical in a given situation.

Rotation and Deflection.   Beam deformation is manifested as angular change and deflection. The angular change (rotation) may be visualized as the movement of vertical plane sections or of tangents to the curved beam profile. Deflection is the distance of dislocation of points in the beam from their original, unloaded positions.

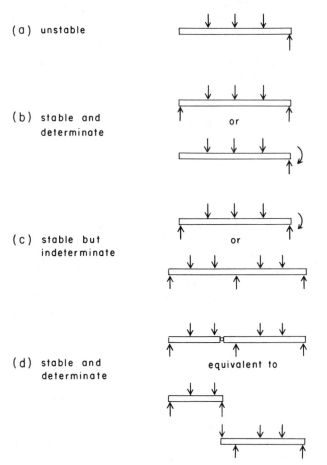

(a) unstable

(b) stable and determinate

or

(c) stable but indeterminate

or

(d) stable and determinate

equivalent to

FIGURE 9.3. Stability and determinacy of beams.

Lateral Buckling. If a beam lacks lateral stiffness and is not braced, it may buckle sideways. This is due essentially to the columnlike action of the compression side of the beam. As with other buckling phenomena, the chief determinants are the stiffness of the beam material and the slenderness of the member. The usual solution is to brace the beam; if this is not possible, the moment capacity must be reduced for design.

Rotational Buckling. If the beam lacks torsional resistance and is not adequately braced, it may be rolled over by the loads or at its ends by the support forces. As with lateral buckling, the best solution is to brace it against this type of motion. Tall and thin wood joists, such as 2 × 12s, and steel I-shaped beams with narrow flanges are examples of members especially vulnerable to this action.

Torsion. Beams may experience torsion in a variety of ways, the most common situation being a misalignment of the load, causing a combination of twisting and bending. This is not the same as torsional buckling, although the response of the beam is similar—a twisting or rolling over.

Bearing. If a beam is supported in simple bearing on top of a wall or column, the support force must be developed as a compressive bearing effect. This effect occurs in the beam and also in the support.

Other actions may occur in particular situations, but these are the common problems of most beams. Investigation is undertaken for design purposes to provide the necessary data for design decisions.

9.4. REACTIONS

For statically determinate beams the first step in the investigation of beam behavior is the determination of the effects of the supports on the beam—called the *reactions*. For the simplest case the reactions are a set of vertical forces that respond to the vertical loads on the beam, constituting with the loads a system of coplanar, parallel forces. This system yields to solution by consideration of static equilibrium if there are not more than two unknowns. Using the beam shown in Fig. 9.4, we demonstrate the usual procedure for finding the reactions in the following example.

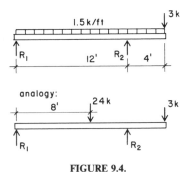

FIGURE 9.4.

Example 1. Find the reaction forces at the supports, R_1 and R_2, for the beam in Fig. 9.4.

Solution. The general mathematical technique is to write two equations involving the two unknowns, and then to solve them simultaneously. This procedure is simplest if one equation can be written that involves only one of the unknowns. We may thus consider a summation of moments of the forces about a point on the line of action of one of the reactions, as follows:

ΣM about $R_1 =$
$$(24 \times 8) + (R_2 \times 12) + (3 \times 16) = 0$$

from which

$$R_2 = \frac{192 + 48}{12} = 20 \text{ kips}$$

For a second step we may now write any equilibrium equation that includes the action of R_1 and it will be an equation with one unknown. The simplest choice for this is a summation of vertical forces; thus

$$F = 0 = R_1\uparrow + 24\downarrow + 20\uparrow + 3\downarrow$$

from which

$$R_1 = 7 \text{ kips}$$

For a check we can write another equation for R_1 to see if the system works, such as a summation of moments about R_2. Thus

M about $R_2 =$
$$(7 \times 12) + (24 \times 4) + (3 \times 4) = 0$$
$$84 + 96 + 12 = 0$$

which verifies the answer for R_1.

This is the usual form of solution for a beam with two supports loaded only with loads perpendicular to the beam.

9.5. SHEAR

Internal shear in a beam is the effort required of the beam to maintain the equilibrium of the external forces. Because the shear is itself a direct force, it is thus possible to use a simple summation of forces to establish the necessary equilibrium. Consider the beam shown in Fig. 9.5, which is the same beam that was used for the solution for reactions in Sec. 9.4 (Example 1). We therefore present it here with the given values for the reaction forces and proceed to consider the problem of determining values for the internal shear in the beam.

If we want to find the value of internal shear at some point in the beam—say at 3 ft from the left end—we cut a section through the beam at that point and remove the portion of the beam on one side—say the right side—and consider the remaining portion as a free body. This free body will be acted on by the loads and reactions that directly effect it plus the actions of the removed portion; the latter

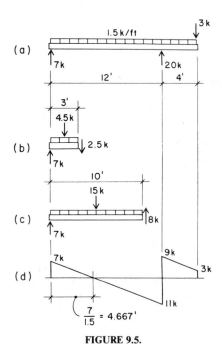

FIGURE 9.5.

representing the internal forces at the section. To find the internal shear at the section, we now consider the equilibrium of vertical forces on the free body; thus

$$\Sigma F_v = 0 = 7\uparrow + 4.5\downarrow + V$$

from which

$$V = 2.5 \text{ kips} \downarrow$$

as shown in Fig. 9.5b.

Let us consider the internal shear at a second section, say 10 ft from the left end. Using a similar summation yields

$$F_v = 0 = 7\uparrow + 15\downarrow + V$$

from which

$$V = 8 \text{ kips} \uparrow$$

as shown in Fig. 9.5c.

If we continue this process, we may find the values for internal shear at as many points as we desire, until we have a general description of the variation of shear along the beam length. Our usual technique for displaying this general form is through use of a simple graph of the shear, called the *shear diagram*. For this beam the shear diagram takes the form shown in Fig. 9.5d. Note that the shear has a sign as well as magnitude. The convention used to establish the sign for the shear graph shown in Fig. 9.5d is to proceed as demonstrated with the free bodies in Fig. 9.5b and c and to consider force up in sense as positive.

In effect, the shear diagram is simply a plot of the external vertical forces along the beam length. To produce the graph shown in Fig. 9.5d, we simply start at the left end of the beam and plot the forces as we encounter them, proceeding to the other end of the beam. We start at zero, since

the beam end is discontinuous, and there is literally nothing beyond its ends. We should also end at zero at the opposite end, which is a means of checking the correctness of our work.

Note that the shear graph passes through the zero axis, indicating a switch in the sign of the internal shear at some point between the two supports. The location of this point may be found by noting that the rate of decline of the graph is the "rate" or unit of the uniformly distributed load on the beam in this portion of its length: 1.5 kips/ft. The distance from the left end to the point of zero shear is thus the distance required for the drop from 7 kips at the rate of 1.5 kips/ft; thus

$$x = \frac{7}{1.5} = 4.667 \text{ ft}$$

The significance of the sign of the internal shear will be demonstrated in discussion of internal moment in Sec. 9.6 and consideration of internal stress conditions in Sec. 9.7.

9.6. BENDING MOMENT

Internal moment in a beam is the effort required by the beam to maintain rotational equilibrium under the action of the external forces. A technique for finding internal moments is to use the same procedures that were demonstrated in Sec. 9.5 for the finding of internal shear. To show the procedure we will use the same beam that was used in Secs. 9.4 and 9.6—shown now in Fig. 9.6 with the reactions and shear diagram complete.

At Fig. 9.6a we show the free body for the portion of the beam cut 3 ft from the left end. In Sec. 9.5 we found the internal shear at this section to be 2.5 kips with sense as shown in Fig. 9.6a. We now consider the equilibrium of moments for this free body. If we sum moments at the cut section, we write

$$\Sigma M = (7 \times 3)\curvearrowright + (4.5 \times 1.5)\curvearrowleft + M_3 = 0$$
$$M_3 = 14.25 \text{ kip-ft} \curvearrowleft$$

To verify this answer, we may write a second equation for moment about the left end; thus

$$M \text{ about } R_1 = (4.5 \times 1.5)\curvearrowright + (2.5 \times 3)\curvearrowright + M_3 = 0$$

or

$$6.75\curvearrowright + 7.5\curvearrowright + 14.25\curvearrowleft = 0$$

The value of 14.25 kip-ft for M_3 is the magnitude of the internal moment at the cut section, representing the effect of the portion of beam that was removed. Note that the moment has a sign. With respect to the cut section, this sign indicates the existence of compression in the upper part of the beam and tension in the lower part.

Let us now consider a cut section at the location of zero shear: 4.667 ft from the left end of the beam. With a similar moment summation, we will find the moment at this sec-

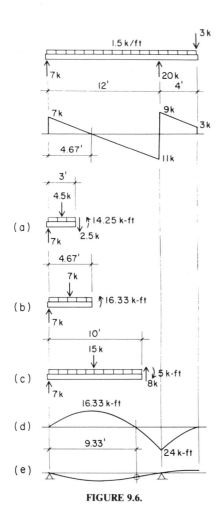

FIGURE 9.6.

tion to be 16.333 kip-ft with a sign the same as that at the previous section (see Fig. 9.6*b*).

Finally, we consider a cut section at 10 ft from the left end of the beam. The consideration of equilibrium in this case will produce an internal moment of 5 kip-ft with a sign of opposite rotation from that found at the previous two sections. This switch of moments indicates that the beam at this location has compression in the bottom portion and tension in the top (see Fig. 9.6*c*).

If we continue to investigate sections along the beam length we may eventually establish the complete pattern of moment variation along the beam length, which takes the form of the graph in Fig. 9.6*d*. This graph is called the *moment diagram*. Below the moment diagram is shown a sketch of the general form of the deflected beam (Fig. 9.6*e*).

Mathematically it can be demonstrated that the load, shear, moment, rotation, and deflection for the beam are interrelated in sequence. Thus the shear is the first integration of the load, the moment is the second integration, the rotation is the third integration, and the deflection is the fourth integration. This process can also be reversed using derivatives. Based on these relationships, plus consideration of the investigation just performed and illustrated in

Fig. 9.6, we can make the following statements with regard to investigation of beams.

1. The internal shear at any point along a beam is the sum of the loads and reactions on one side of that point. Either side may be used.
2. The internal moment at any point along a beam is the sum of moments of the loads and reactions on one side of that point. Either side may be used.
3. The change in moment between any two points on the beam is equal to the area under the shear diagram between the two points.
4. Points of maximum value on the moment diagram correspond to points of zero value on the shear diagram.
5. The sign of the internal moment is related to the type of curvature of the deflected shape.
6. Points of zero moment along the beam indicate points of change of curvature—called *inflection* or *contraflexure*—of the deflected shape.

The reader should verify these statements with the information displayed in Fig. 9.6. The shear and moment diagrams as displayed are produced by following a particular sign convention, as follows.

For the Shear Diagram. Begin from the left end, considering force up to be positive.
For the Moment Diagram. Consider internal moment that causes compression in the top of the beam positive and moment that causes tension in the top of the beam negative.

Regarding statement 3, if the sign conventions just described are used, it may be observed that areas of positive shear on the shear diagram relate to positive changes of moment, while areas of negative shear relate to negative changes of moment. This is the basis for statement 4.

Visualization of the deflected shape is a very useful device in the investigation of members subjected to bending. In most cases this visualization can be made on the basis of consideration of the loads, supports, and beam form before any other investigation is done. If so, it provides immediate clues to the character of the moment in the beam.

Relations between moment, rotation, and deflection are considered more fully in Sec. 9.8. General investigation of beams is considered further in Secs. 9.10 and 9.11.

9.7. STRESSES

Stresses in beams vary across individual cross sections. The variation of bending stress, as discussed in Sec. 8.5, occurs as shown in Fig. 9.7*a* on a simple rectangular cross section. On the same section, shear stress, as discussed in Sec. 12.4, varies as shown in Fig. 9.7*b*. The values and signs of internal shear and moment that produces these stresses also varies along the beam length. The stress condition in a

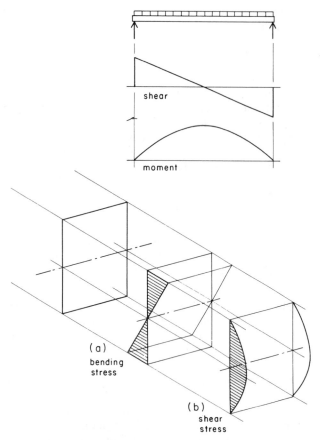

FIGURE 9.7. Distribution of stresses in a beam.

beam is thus not a single situation, but a complex series of situations.

Of critical concern are the maximum stresses. The most obvious of these are the maximum tension and compression due to bending and the maximum shear stress. For the beam shown in Fig. 9.7, consisting of a simple beam with a single, uniformly distributed load, we observe that these maximum stresses will occur as follows.

> Maximum shear will occur at the neutral axis at the beam ends.
> Maximum tension will occur at the bottom of the beam at midspan.
> Maximum compression will occur at the top of the beam at midspan.

In this example it may be noted that the maximum shear stress occurs where bending stress is zero, and the maximum bending stress occurs where the shear stress is zero. This is frequently the case with simple span beams. However, with overhanging or continuous beams negative moments at supports are frequently the maximum moments in the beam. Thus in Fig. 9.6, for example, the maximum value for shear and the maximum value for moment both occur at the right support.

Although we observe the stress conditions for shear and bending as separate phenomena, they do not truly occur as

such. In some cases, therefore, we must consider their interaction. For a beam, this type of stress combination is discussed in Sec. 12.5. There may, of course, be other stresses as well; due to torsion, axial force of tension or compression, or to two-way bending effects. In such situations the investigation can become quite complex.

Beams are used in a variety of ways and are made from different materials in many forms. The particular problems of investigation and design of different beams in different situations of use are discussed in other parts of the book. It is not our purpose here to explore all of the variations, but to concentrate on the simple fundamentals of beam behavior.

9.8. ROTATION AND DEFLECTION

There are various situations in structural design in which it is necessary to determine the actual deformation of a beam. Most often this has to do with deflection, and usually with the single value of the maximum deflection. For beams in ordinary situations deflections are usually determined through the use of derived formulas that incorporate the variables and represent the situation for a particular load and support condition.

Rotations are not often of interest in themselves, although problems in connection design or of the effects of a beam on its supports may involve their consideration. Computations for rotations are more often done as part of the procedure for some other investigation, such as the computation of deflection of a complex beam or of the sideways deflection of a rigid frame.

Rotations and deflections can be determined by integration or indirectly by the moment area method. These methods can be used for any beam, but are generally required only for unusual cases.

9.9. ANALYSIS FOR GENERAL BEHAVIOR

Let us consider a general process for the investigation of a simple beam with regard to determination of reactions, shear, moment, rotation, and deflection. For the simple beam shown in Fig. 9.8a, we observe the following.

1. Reaction forces may be found by the process of static equilibrium equations, as demonstrated in Sec. 9.4. However, for the simple beam with a symmetrical load, it can be observed that the reactions are simply each equal to one-half of the total load.

2. The shear is determined by constructing the shear diagram, recalling that it is simply a plot of the reaction and load forces. Using the convention of starting at the left end and calling force up positive, we first plot the reaction force. For the next portion of the beam—up to the first of the concentrated loads—the diagram declines at the rate of the unit load: 0.6 kip/ft. Just to the left of the first concentrated load the shear declines to a value of 8.4 kips. Then the occurrence of the 6-kip load drops it to 2.4 kips. This is all of the computation required, as we can observe that the

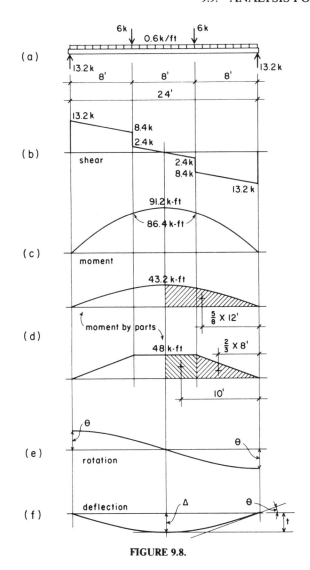

FIGURE 9.8.

concentrated load is also required. This may be found with another set of moment equations. Both the maximum moment and the moment at the concentrated load can also be found by using the shear areas, as follows:

$$M_8 = \frac{13.2 + 8.4}{2}(8) = 86.4 \text{ kip-ft}$$

$$M_{12} = 86.4 + \frac{2.4(4)}{2} = 86.4 + 4.8 = 91.2 \text{ kip-ft}$$

4. Figure 9.8e is a graph of the variation of rotation for the beam. There is no particular practical use for this figure, other than to see its relation to the other diagrams. Rotations are generally more meaningful when observed directly on the deflected shape diagram in Fig. 9.8f. From either figure we can observe that the rotation remains zero at the center of the span and attains a maximum value at the supports.

5. The maximum deflection for the symmetrically loaded simple beam will occur at the center of the span; indicated as Δ on Fig. 9.8f.

For computation of actual values of rotation or deflection, it is most practical to consider separate actions of the beam with regard to the two types of load. For this example we may find the maximum deflection by use of the formulas given in Fig. B.1 in Appendix B for cases 2 and 3. Thus

$$\Delta = \frac{5}{384} \frac{WL^3}{EI} + \frac{23}{648} \frac{PL^3}{EI}$$

$$= \frac{5}{384} \frac{[0.6(24)](24)^3}{EI} + \frac{23}{648} \frac{6(24)^3}{EI}$$

$$= \frac{2592 + 2944}{EI} = \frac{5536}{EI} \text{ kip-ft}^3$$

If the actual dimension for this deflection is required, the values for E and I can be inserted in this equation, taking care to note the necessary corrections for units. For example, if E is used in lb/in.^{2} and I in in.^{4}, the numeric value obtained would be multiplied by 1000 x (12)^3 to correct from kips to pounds and from feet to inches.

Both rotation and deflection can also be found by use of the moment area method. For this computation we do something analogous to the two-part deflection formula by drawing separate moment diagrams for the two loadings. These are shown in Fig. 9.8d. For the rotation at the end we note that the angle θ in Fig. 9.8f can be expressed as the change in rotation from the center of the span to the end. By the moment area method, this change is defined as the area under the moment diagram between the center and the end. Using the two partial moment diagrams in Fig. 9.8d, we determine this total area as

$$\frac{\theta}{EI} = \frac{2}{3}[43.2(12)] + 48(4) + \frac{1}{2}[48(8)]$$

$$\theta = \frac{345.6 + 192 + 192}{EI} = \frac{729.6}{EI} \text{ kip-ft}^2$$

diagram will be symmetrical—although opposite in sign—on the other half of the beam.

3. For the simple beam the moment diagram will be all positive in value, with zero moment at each end of the beam. Due to the symmetry of both the beam and the loads, the diagram will be symmetrical. From the form of the shear diagram we observe that the maximum moment value will occur at the center of the span, coinciding with the point where the shear diagram passes through zero. The value of the moment at the center may be found from an equation of the moments of reaction and load forces to the left of the section; thus

$$M_{12} = 13.2(12) + 0.6(12)(6) + 6(4)$$

$$= 158.4 + 43.2 + 24$$

$$= 91.2 \text{ kip-ft}$$

For actual design purposes, this may be the only information required regarding moments. For the construction of the moment diagram, however, the moment value at the

For the deflection at the center we note that the tangent to the elastic curve (deflected shape) at the center remains horizontal. We can therefore find the value for the tangential deviation at the end with respect to the center (t in Fig. 9.8f) and note that this is equal to the actual deflection at the center. By the moment area method this deviation is equal to the moment of the moment area between the two points (the center and the end) about the point of deviation (the end). Using the partial diagrams in Fig. 9.8d and the values for the three segments of area found in the computation for the rotation, we find

$$\frac{\Delta}{EI} = t = 345.6 \left(\frac{5}{8}\right)(12) + 192(10) + 192\left(\frac{2}{3}\right)(8)$$

$$\Delta = \frac{2592 + 1920 + 1024}{EI} = \frac{5536}{EI} \text{ kip-ft}^3$$

which is the same value that was obtained with the deflection formulas.

For most practical purposes the behavior of the beam is now completely described.

9.10. VISUALIZATION OF BEAM BEHAVIOR

Using the various observations and derivations from the preceding sections, it is possible to approach the problem of investigating a beam as an exercise in visual logic. That is, it is possible to produce the general forms for the shear diagram, moment diagram, and load-deflected shape for a beam by simple inspection of the load and support conditions, with recourse to minimal computation if any. The following examples—illustrated in Fig. 9.9—demonstrate this technique.

Figure 9.9a shows a simple beam with a single uniformly distributed load. The symmetrical conditions make the visualization of the deflected shape easy—a symmetrical curve with maximum deflection and zero rotation at the center of the span and equal rotations at the ends. The moment diagram will also be symmetrical—a single parabola with its maximum value at its apex in the center of the span. The shear diagram will consist of a single straight line, relating to the single unit value of the uniformly distributed load. The shear diagram is essentially symmetrical, with one-half flipped over the reference line. The reactions are equal and both act upward. All of these observations can be made without any numeric computation.

For the beam in Fig. 9.9b, an easily made computation will aid somewhat in visualization of the shear diagram. This consists of noting that the symmetry will result in the value of each reaction being equal to 1.5 times the value of a single load P. Thus the true proportions of the shear areas can be visualized. This also helps in visualizing the moment diagram, remembering that the change in moments relate to shear areas. Although not precisely the same, the general form of the deflected shape is essentially the same as that for the beam in Fig. 9.9a.

For the beam in Fig. 9.9c the most useful first step is the visualization of the form of the deflected shape. It should

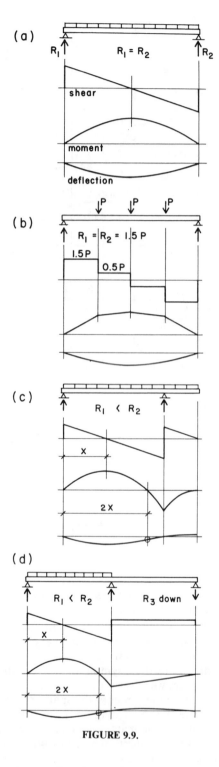

FIGURE 9.9.

be easy to observe that the cantilevered end will produce a negative moment at the right support, resulting in the inflected shape shown. It is not possible to determine the exact position of the cantilevered end without computations, but the general character of the beam profile can be observed. From the deflected shape we note that there is an inflection point somewhere between the center of the span and the right support. From the loading, the supports and beam form, and the deflected shape, we have many clues to visualize the moment diagram. We note the following.

1. The diagram will be part positive and part negative, relating to the two segments of the deflected shape.
2. The moment is zero at the location of the inflection point in the deflected shape.
3. The segments of the diagram will be parabolic curves, relating to the uniformly distributed load.
4. The moment will be zero at the beam ends, have a maximum negative value at the right support, and a maximum positive value somewhere between the left support and the inflection point.

If we now proceed to the shear diagram, we may note the following.

1. The lines of the diagram will have a constant slope, relating to the constant value of the unit load.
2. The reaction at the left support is smaller than that at the right support, which should be reflected in the proportions of the shear diagram.
3. The point of zero shear will occur nearer to the left support, corresponding to the location of the maximum positive moment. Remembering that moment change is related to shear areas, we should observe that the positive portion of the moment diagram will constitute a symmetrical parabola, and thus the apex (maximum positive moment) will be halfway between the left support and the inflection point.

The beam in Fig. 9.9d is continuous over three supports and is thus statically indeterminate. This is a problem in performing computations, but does not necessarily prevent us from visualizing its actions. The beam is shown with a uniformly distributed load in only one span. This is not a complete loading, of course, as there would be at least the beam dead load in the second span in the case of a horizontal beam. However, the beam may not be horizontal, or we may be considering only the action of the live loads. In any event, it is a loading that can be investigated.

If the support at R_3 is not capable of downward resistance (hold-down effect), the beam end will go up and the form of the deflected shape in the second span will remain a straight line—merely rotating with one end up. The remainder of the beam will then function in the same manner as the beam in Fig. 9.9a. We assume this not to be the case, but can approach the visualization of the effect of R_3 and the form of the true deflected shape by considering the adding of R_3 to the condition that would exist without it. We thus determine that R_3 acts down and the deflected shape is inflected. This can also assist in the visualization of the form of the shear diagram.

For the moment diagram we make most of the same observations that were made for the beam in Fig. 9.9c. Because there is no load in the second span, the shear value on the shear diagram remains constant throughout the span. With change of moment related to shear areas, this

helps to establish the straight-line form of the moment diagram in the second span.

Some additional sets of diagrams for beams are shown in Fig. 9.10. The reader should attempt to verify these by covering all but the beam load diagram and attempting to produce the shear, moment, and deflected shape diagrams.

Visualizations of this type are very useful in learning the various relationships of the different aspects of beam behavior. They are also truly the most important step in performing real investigations.

9.11. BUCKLING

Buckling of beams—in one form or another—is mostly a problem with beams that are relatively weak on their transverse axes, that is, the axis of the beam cross section at right angles to the axis of bending. This is not a frequent condition in concrete beams, but is a common one with beams of wood or steel or with trusses that perform beam functions. The cross sections shown in Fig. 9.11 illustrate members that are relatively susceptible to buckling in beam action.

Other than redesigning the beam section, there are two general approaches to solving the buckling "problem." The first—and often preferred—approach is to brace the member so as to effectively prevent the buckling response. To visualize where and how the bracing should be done, we must consider the various possibilities for buckling; the three major ones being those shown in Fig. 9.12.

Figure 9.12a shows the response described as *lateral* (literally, *sideways*) *buckling*. This is caused by the compression column action of the portion of the beam that develops compressive bending stress. For a simple span beam, the most critical location for this is at the midspan, if the beam is unbraced. If the beam is braced, as shown in Fig. 9.12b, the critical location is midway between bracing points. The buckling occurs in response to the sign of the moment: at the top with positive moment and the bottom with negative moment.

Bracing the beam against lateral buckling means essentially simply preventing its sideways movement. From structural theory we know that only a small force is required to restrain the beam—usually less than about 3% of the total compressive force that induces the buckling. For beams, joists, rafters, or trusses that directly support roof or floor decks, the deck will often supply sufficient bracing; subject to evaluation of the stiffness of the deck and the adequacy of the attachment to the supporting member being braced. For beams that support other beams, the supported beams will often provide bracing if they are reasonably closely spaced and adequately connected.

The other type of buckling of beams, called *torsional buckling* commonly occurs in one of two ways. The first way is at the supports, as shown in Fig. 9.12b, where the concentrated effect of the reaction force may cause a rollover effect. Bracing for this effect means simply preventing the lateral rotation at the supports. Decking, structural walls, or various parts of the general framing may be made

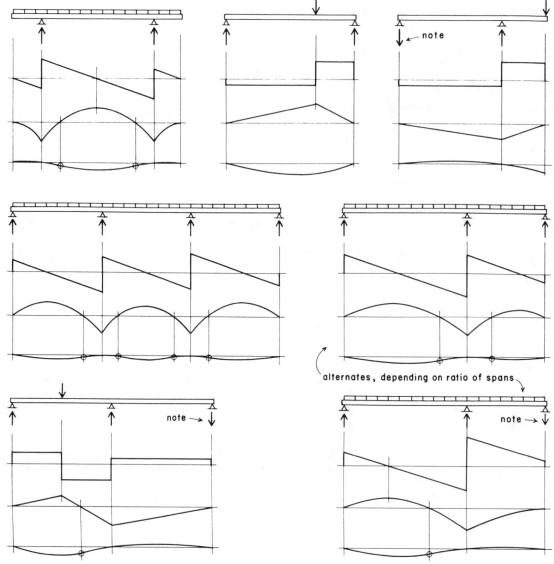

FIGURE 9.10. Visualization of beam actions.

to perform this bracing, often resulting in no need for additional elements.

The second way in which torsional buckling can occur, as shown in Fig. 9.12c, is by the rolling over of the beam at some point due to the tension force. The analogy is made in the illustration with a tied member in which the tension effect causes a sideways flip-over of the structure. Tension in the solid beam operates much like the tension tie and can cause the same phenomenon. This is critical at the same locations as the lateral buckling due to compression, and the two actions may compound the failure of the beam. Thus the torsion-weak beam must be braced for both lateral and torsional buckling in most cases, both at the same locations.

Where bracing can be utilized it may be possible to utilize the full bending resistance of the member on its strong axis (the x-axis in the sections in Fig. 9.11). However, if the

member is unbraced—or bracing must be spaced a reasonably great distance for economy—and the member is not totally incapacitated by the buckling effect, it is usually necessary to consider some reduction of its full bending capacity. Computations for such reductions are often quite complex, although use of empirical adjustments provided in design codes makes the work somewhat more feasible for ordinary situations. Examples of the use of such criteria and procedures are given in Parts 3 and 4 for wood and steel beams, respectively.

9.12. STATICALLY INDETERMINATE BEAMS

Analysis of the behavior of indeterminate structures requires use of some conditions in addition to those provided by consideration of static equilibrium. They are thus

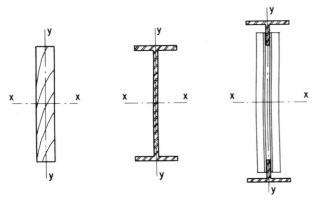

FIGURE 9.11. Beam sections lacking lateral strength.

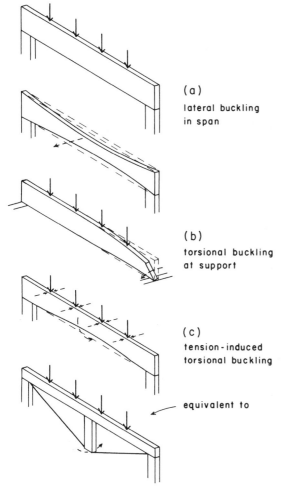

(a)
lateral buckling
in span

(b)
torsional buckling
at support

(c)
tension-induced
torsional buckling

equivalent to

FIGURE 9.12. Buckling failure of beams.

imaged as having a negative character: being *not* capable of something—analysis by statics alone. Actually, there is nothing wrong or bad about a structure that is statically indeterminate. In fact, there are several potential advantages, including reduction of the maximum bending moments, reduction of deformations, and a general redundancy of stability and internal resisting forces that enhances safety. In some cases the structure is deliberately made redundant to gain one or more of these advantages. In other cases the structure is inherently indeterminate due to the nature of the construction—such as with poured-in-place concrete structures with multiple spans.

In some cases additional behavior conditions can be established on the basis of observations of behavior that is controlled, such as fixed supports where the rotation remains zero, any support where the deflection remains zero, internal pins where the moment must be zero, and so on. It is also possible to use the fact that the beam assumes a smooth curve as its deflected shape. In most cases, however, these conditions only allow for shortcuts or for the implementation of other, more fundamental techniques, such as the classic slope-deflection method.

We will not attempt to develop the general problem of investigation of statically indeterminate structures in this book. This is an exhaustive subject, well developed in many standard reference texts. We must, however, consider some aspects of behavior of indeterminate structures so that the reader can gain some appreciation for the issues involved. In current practice really complex indeterminate structures are routinely investigated using computer-aided techniques, the software for which is readily available (for a price) to the design professions.

9.13. USE OF TABULATED FORMULAS AND DIAGRAMS

Load and support conditions for beams tend to fall into categories that are repeatedly encountered in design work. It is possible, therefore, to derive, once and for all, formulas for responses (reactions, maximum shears, maximum moments, and maximum deflections) for the most common situations. Tabulations of these, usually presented with the shear and moment diagrams displayed, can be found in many references [for example, the *AISC Manual* (Ref. 5), which presents some 42 cases]. A sampling of the most common cases is presented in this book in Appendix B.

CHAPTER TEN

Tension Elements

Tension members are used in a number of ways in building structures. Structural behavior may be simple, as in the case of a single hanger or tie rod, or extremely complex, as in the case of cable networks or restraining cables for tents and pneumatic structures. Chapter 10 contains discussions of the nature and problems of various elements and examples of investigation of several simple elements.

10.1. AXIALLY LOADED ELEMENTS

The simplest case of tension stress occurs when a linear element is subjected to tension and the tension force is aligned on an axis that coincides with the centroid of the member's cross section. A member that is loaded in this manner is said to sustain axial load. The stress is assumed to be evenly distributed on the cross section and is expressed as

$$f = \frac{T}{A}$$

In the usual case with an axially loaded member, the stress is also essentially evenly distributed over the length of the member. We may thus express the strain condition as

$$\varepsilon = \frac{f}{E}$$

and the total elongation of a member L distance in length as

$$e = \varepsilon L = \frac{f}{E} L = \frac{TL}{AE}$$

The tension force may thus be expressed as

$$T = fA \quad \text{or} \quad \frac{AEe}{L}$$

When a member is short, as in the case of a short hanger or a truss member, the usable tension capacity is usually limited by stress. However, for very long members, elongation may be critical and may limit the capacity to a value below that of the safe stress limit.

Example 1. An arch spans 100-ft [30-m] meter and is tied at its spring points by a 1-in. [25-mm]-diameter round steel rod. Find the limit for the axial tension force in the rod if stress is limited to 22 ksi [150 MPa] and total elongation is limited to 1.0 in. [25 mm].

Solution. The cross-sectional area of the rod is

$$A = \pi R^2 = 3.14(0.5)^2 = 0.785 \text{ in.}^2 \, [491 \text{ mm}^2]$$

The maximum force based on stress is

$$T = fA = 22(0.785) = 17.27 \text{ kips } [73.7 \text{ kN}]$$

and the maximum force based on elongation is

$$T = \frac{AEe}{L} = \frac{0.785(29,000)(1.0)}{100(12)} = 18.97 \text{ kips } [81.8 \text{ kN}]$$

(E values obtained from Table 8.1).

In this case the stress limit is critical, limiting T to the lower value. However, if the span were only a few feet longer, the elongation would be critical.

10.2. NET SECTION AND EFFECTIVE AREA

The development of tension in a structural member involves connecting it to something. Achieving tension-resistive connections often involves situations that reduce the load-carrying effectiveness of the tension member. Two examples of this are the bolted connection and the threaded connection.

Bolts are commonly used with members of wood or steel. Insertion of the bolts requires drilling or punching of holes in the members. If a cross section is cut in the member at the location of a bolt hole (or a row of holes), a reduced area is obtained, called the *net section*. If a stress computation is made for this area, the unit stress obtained will be higher than that at unreduced sections in the member. The total behavior of a bolted connection is more complex and involves many considerations besides the simple tension on the net section. Nevertheless, it is often the case that considerations for the connection may be the limiting factors in establishing the tension capacity for bolted tension members.

A simple tension member, extensively used, is a round steel rod with spiral threads cut in its ends. The ends are simply inserted in holes in connected members and a nut is screwed onto the threaded ends of the rod. The cutting of the threads reduces the cross section of the rod, producing a net section just as with the bolt holes. A special kind of rod is one that has forged enlargements at its ends—called *upset ends*—which have sufficient diameters so that the cutting of the threads does not produce a net section smaller than that in the main part of the rod. For building structures, upset rods are seldom used, so the tension rod is typically designed for the reduced section at the threaded ends. This consideration also applies to the use of bolts when the load is axial to the bolt rather than a shear effect on the bolt.

In steel structures it sometimes happens that the practical considerations of achieving connections makes it difficult to develop fully the tension potential of the connected tension member. Figure 10.1 shows a typical connection involving a steel angle, in which one leg of the angle is connected by welding to a supporting element. At the connection the tension force is developed only in the leg of the angle that is directly grabbed by the welds. At some distance along the member, some tension will surely be developed in the other leg; however, for a conservative design it is not unreasonable to ignore the unconnected leg and consider the connected leg to function in the manner of a simple bar. The full angle will still be effective for stiffness or other considerations, but the tension is limited by the member form and the layout of the connection.

There are many other considerations in the design of connections, which are more fully discussed in Parts 3 and 4. It is important to keep in mind, however, that the problems of achieving connections and of effectively transferring end tension forces to supporting elements is a major part of the design of tension members.

10.3. FLEXIBLE ELEMENTS

Tension elements are somewhat unique in that there is little basis for limitations on slenderness or aspect ratio. By comparison, columns with height-to-width ratios in excess of 30 or so tend to approach a condition of excessive slenderness, and beams with span-to-depth ratios over 20 or so usually have critical deflection problems. For members in tension lateral stiffness may literally be zero, as in the case of rope or chain, or it may be virtually zero and negligible, as in the case of very long wires, cables, and tie rods.

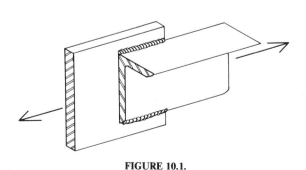

FIGURE 10.1.

The excessively slender structural element will unavoidably follow a pattern of form development that is a direct response to the only force resistance it can muster—pure tension. If it has virtually no bending, torsion, or shear capability, it cannot be made to act to develop those internal force actions. This severely limits the manner in which such elements can be used and means that the designer must precisely determine the logical pattern that loads will develop with such a structure.

The hanger rod, tie rod, or truss member that functions as a simple two-force member is no problem; it will assume a straight line form to resolve the tension force, even if it has to straighten out to do so. The only problem in such cases is to assure that connection details do not result in something other than the pure axial transfer of tension force to the member. For such members length is not related to stress problems and is limited only (if at all) by considerations for elongation, sag, vibration in a low-tension state, or available lengths from suppliers.

Tension elements that have significant stiffness may also function in pure axially stressed situations. However, they also have some capacity for other actions and may develop combined stress actions. An example of a simple case of this is discussed in Sec. 10.5.

When the superflexible tension element is used for spanning, it cannot assume the rigid, minutely deflected form of a beam. It must, instead, assume a profile that permits it to act essentially in pure tension. This profile must be "honest," that is, not necessarily one that the designer may concoct, but one that the loads and the loaded structure can actually achieve. The behavior of the spanning tension structure is considered in the next section.

10.4. SPANNING CABLES

Flexible elements may be used for spanning if they are properly supported and are permitted to assume the profile natural to resolution by pure internal tension. Although many types of elements may be used in this manner, we will discuss the case of the steel cable (actually, bridge strand) as representative of the structural type.

Consider the single-span cable shown in Fig. 10.2a, spanning horizontally and supporting only its own dead weight. The natural draped shape assumed by the cable is a catenary curve whose profile is described by the equation

$$y = \frac{a}{2} (e^{x/a} + e^{-x/a})$$

Except for cables that actually do carry only their own weight (such as electrical transmission lines), or which carry loads proportionally small with respect to their weights, this form is not particularly useful.

We deal in this section with problems in which the weight of the cable can be ignored without significant error. When this is assumed, the cable profile becomes a pure response to the static resolution of the loads. The cable will thus assume a simple parabolic form (Fig. 10.2b)

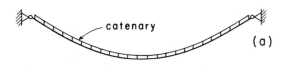

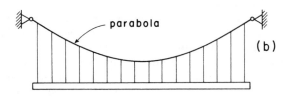

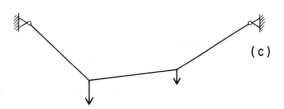

FIGURE 10.2. Cable response to loads.

when loaded with a uniformly distributed load, and a form consisting of straight segments (Fig. 10.2c) when loaded with individual concentrated loads.

Consider the cable shown in Fig. 10.3a, supporting a single concentrated load W, and having four exterior reaction components: H_1, V_1, H_2, and V_2. Without consideration of the internal nature of the structure, the analysis is indeterminate with regard to use of static equilibrium conditions alone. To analyze the structure we must use the fact that the cable is incapable of developing bending resistance, and therefore there can be no internal bending moment at any point along the cable. We may also note that individual segments of the cable operate as two-force members; thus the direction of T_1 must be the same as the slope of the left segment of the cable. Similarly, the direction of T_2 must be the same as the slope of the right segment.

Referring to the free-body diagram of the whole cable in Fig. 10.3b, if we take moments about support 2,

$$\Sigma M_2 = (W \times b) + (V_1 \times L) = 0$$

from which

$$V_1 = \frac{b}{L} W$$

Similarly, using moments about support 1, we have

$$V_2 = \frac{a}{L} W$$

Now, considering the free-body diagram of the left portion of the cable, as shown in Fig. 10.3c, we take moments about the point of the load:

$$\Sigma M = 0 = (V_1 \times a) + (H_1 \times y)$$

from which

$$H_1 = \frac{a}{y} V_1$$

Referring to the free body for the entire cable (Fig. 10.3b), we note that the two horizontal reaction components are the only horizontal forces. Therefore, $H_1 = H_2$. We have now established relationships to determine the four reaction components. The actual tension forces, T_1 and T_2, can be found from the vector combinations, thus

$$T_1 = \sqrt{(H_1)^2 + (V_1)^2} \quad \text{and} \quad T_2 = \sqrt{(H_2)^2 + (V_2)^2}$$

It may also be observed that the single load and the two cable tension forces form a simple concentric force system at the point of the load (Fig. 10.3d). This system can be analyzed graphically by construction of the force triangle shown in Fig. 10.3e. If desired, the values for the reaction components can be projected from the vectors for T_1 and T_2, as shown.

Example 1. Find the horizontal and vertical components of the reactions and the tension forces in the cable for the system shown in Fig. 10.4a.
Solution. Using the relations just derived for the structure in Fig. 10.3 yields

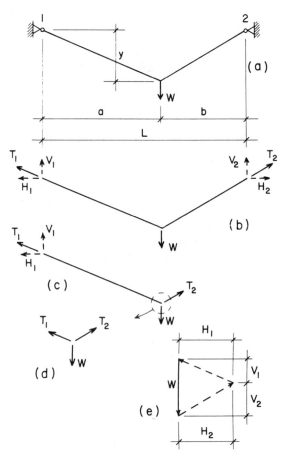

FIGURE 10.3.

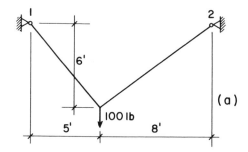

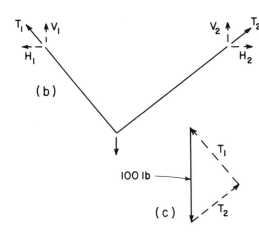

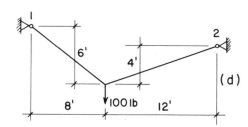

(b)

(c)

(d)

(e)

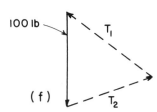

(f)

FIGURE 10.4.

$$V_1 = \frac{b}{L} W = \frac{8}{13} (100) = 61.54 \text{ lb}$$

$$V_2 = \frac{a}{L} W = \frac{5}{13} (100) = 38.46 \text{ lb}$$

$$H_1 = H_2 = \frac{a}{y} V_1 = \frac{5}{6} (61.54) = 51.28 \text{ lb}$$

$$T_1 = \sqrt{(V_1)^2 + (H_1)^2} = \sqrt{(61.54)^2 + (51.28)^2}$$
$$= 80.1 \text{ lb}$$

$$T_2 = \sqrt{(V_2)^2 + (H_2)^2} = \sqrt{(38.46)^2 + (51.28)^2}$$
$$= 64.1 \text{ lb}$$

When the two supports are not at the same elevation, the preceding problem becomes somewhat more complex. The solution is still determinate, however, and may be accomplished as follows.

Example 2. Find the horizontal and vertical components of the reactions and the tension forces in the cables for the system shown in Fig. 10.4*d*.
Solution. Referring to the free-body diagram in Fig. 10.4*e*, we note that

$$\Sigma F_v = 0 = V_1 + V_2 = 100$$

$$\Sigma F_h = 0 = H_1 + H_2 \text{ (thus } H_1 = H_2)$$

$$\Sigma M_2 = 0 = (V_1 \times 20) + (H_1 \times 2) +$$
$$(100 \times 12)$$

From the moment equation

$$10(V_1) - H_1 = 600$$

From the geometry of T_1, we observe

$$H_1 = \frac{8}{6} (V_1)$$

and substituting gives

$$10(V_1) - \frac{8}{6} (V_1) = 600$$

$$\frac{52}{6} (V_1) = 600$$

$$V_1 = \frac{6}{52} (600) = 69.23 \text{ lb}$$

Then

$$H_1 = \frac{8}{6} V_1 = \frac{8}{6} (69.23) = 92.31 \text{ lb} = H_2$$

$$V_2 = 100 - V_1 = 100 - 69.23 = 30.77 \text{ lb}$$

$$T_1 = \sqrt{(69.23)^2 + (92.31)^2} = 115.4 \text{ lb}$$

$$T_2 = \sqrt{(30.77)^2 + (92.31)^2} = 97.3 \text{ lb}$$

When a cable is loaded by more than one concentrated load, the rest position of the loaded cable must be found.

However, if the location of any single load point is known, the problem is statically determinate. Some of the relationships that are useful in this analysis are the following.

1. Internal moment is zero at all points on the cable.
2. The horizontal component of cable tension is the same at all points along the cable. (Find H and the geometry of the cable, and the cable tension can be found at any point.)
3. The angle of the cable segment at any point is the same as the direction of the internal tension vector at that point. (Find either one, and the other is determined.)

A cable loaded with a uniformly distributed load along a horizontal span (not along the cable itself), as shown in Fig. 10.5a, assumes a simple parabolic (second-degree) curve profile. Referring to the free-body diagram in Fig. 10.5b, it may be observed that the horizontal component of the internal tension is the same for all points along the cable, due to equilibrium of the horizontal forces. The vertical component of the internal force varies as the slope of the cable changes, becoming a maximum value at the support and zero at the center of the span. Thus the maximum internal tension will occur at the supports, and the minimum internal tension will occur at the center of the span.

Referring to Fig. 10.5a, we note that

$$V_1 = V_2 = \frac{wL}{2}$$

Referring to Fig. 10.5c, which is a free-body diagram of the left half of the cable, a summation of moments about the left support yields

$$\Sigma M = (H_c \times y) \ + \ \left(\frac{wL}{2} \times \frac{L}{4} \right) \ = 0$$

Thus

$$H_c = \frac{wL^2}{8y}$$

which is the general expression for horizontal force in the cable at all points.

The approximate length of the parabolic curve may be obtained by using the formula

$$S [L \left\{ 1 + \frac{8}{3} \left(\frac{y}{L} \right)^2 - \frac{32}{5} \left(\frac{y}{L} \right)^4 \right\}$$

where

S = curve length

y = maximum sag

L = horizontal span length

This formula is obtained by taking the first three significant terms of the binomial expansion of the expression obtained by integrating $ds = (dx^2 + dy^2)^{1/2}$ over the length of the cable.

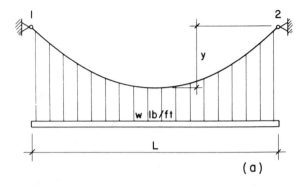

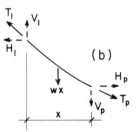

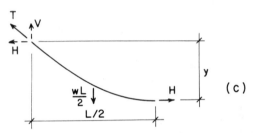

FIGURE 10.5. Behavior of the uniformly loaded cable.

10.5. COMBINED ACTION: TENSION PLUS BENDING

Various situations occur in which both axial force of tension and a bending moment exist at the same cross section in a structural member. Consider the hanger shown in Fig. 10.6a, in which a 2-in. square steel bar is welded to a plate, which is bolted to the bottom of a wood beam. A short piece of steel with a hole in it is welded to the face of the bar, and a load is hung by some means from the hole. In this situation the steel bar is subjected to combined actions of tension and bending, both of which are produced by the hung load (Fig. 10.6b). The bending occurs because the load is not applied axially to the bar; the bending moment thus produced has a magnitude of 2(5000) = 10,000 in.-lb [50(22) = 1100 kN-mm, or 1.10 kN-m].

For this simple case, the stresses due to the two phenomena are found separately and added as follows. For the direct tension effect (Fig. 10.6c)

$$f_a = \frac{N}{A} = \frac{5000}{4} = 1250 \text{ psi [8.8 MPa]}$$

For the bending we first find the section modulus of the cross section; thus

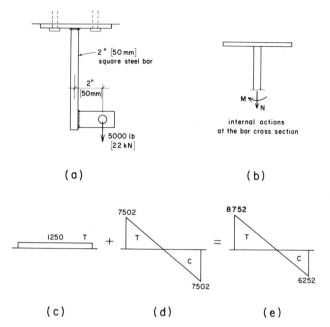

internal actions
at the bar cross section

(a)   (b)

(c)   (d)   (e)

FIGURE 10.6. Development of the combined stress.

$$S = \frac{bd^2}{6} = \frac{2(2)^2}{6} = 1.333 \text{ in.}^3 \ [20.82 \times 10^3 \text{ mm}^3]$$

Then, for the bending stress (Fig. 10.6*d*),

$$f_a = \frac{M}{S} = \frac{10,000}{1.333} = 7502 \text{ psi } [52.8 \text{ MPa}]$$

and the stress combinations are (Fig. 10.6*e*)

maximum $f = f_a + f_b = 1250 + 7502 =$
                                                    8752 psi [61.6 MPa]

minimum $f = f_a - f_b = 1250 + 7502 =$
                                                    6252 psi [44.0 MPa]

Note that although the tension stress due to the direct force is uniformly distributed on the cross section, the bending stress is not. The condition is therefore as shown in Fig. 10.6*e* and the maximum and minimum stresses just computed are the edge stresses as shown in the figure. For the algebraic work the tension stress was considered positive, which means that compressive stress is then negative. It is necessary to be aware of this algebraic sign condition when combining stresses.

CHAPTER ELEVEN

Compression Elements

Compression is developed in a number of ways in structures, including the compressive component that accompanies the development of internal bending moment and the diagonal compression due to shear. In Chapter 11 we consider elements whose primary structural purpose is the resistance of direct compression, although they may also be required to perform other structural tasks.

11.1. TYPES OF COMPRESSION ELEMENTS

A number of types of primary compression elements are used in building structures. Major ones are the following.

Columns. These are usually linear vertical elements, used when supported loads are concentrated, or when a need for open space precludes the use of walls as supports. Relative slenderness may vary over a considerable range, from stout to slender, depending on the magnitude of the loads or the material and construction of the column. In various situations columns may also be called *posts*, *piers*, or *struts*.

Piers. This term generally refers to relatively stout columns. The term is also used, however, to describe massive bridge supports, abutments, deep foundation elements cast in excavated shafts, and vertical masonry elements that are transitional in form between walls and columns. All of these elements usually resist major compression forces, but may also be required to develop other resistances, such as bending, lateral shear, or uplift.

Truss Compression Members. Truss members are usually either primarily tension or compression elements. In some situations, however, they may need to resist both types of force for different loading combinations. They may also be subjected to bending in addition to their primary truss functions, as in the case of a roof or floor deck attached directly to a top chord, or a ceiling attached directly to a bottom chord. Because compression resistance relates to slenderness, truss compression members are often somewhat heavier than members that take only tension.

Bearing Walls. When walls are used for supports, taking vertical compression, they are called bearing walls. If they are interior walls, this may be their singular structural

function and they perform essentially like columns. Exterior walls, however, are usually also designed for major lateral bending due to wind or earth pressure or for action as shear walls, resisting shear effects in the wall plane.

The Ground. Almost all buildings are supported on the ground, developing reactive compression effects—both vertical and horizontal—in the ground materials.

Short Compression Elements

The general case for axial compression capacity as related to slenderness is indicated in Fig. 11.1. The limiting conditions are those of the very stout (not slender, usually meaning "short") element that fails essentially by crushing of the material and the very slender element that has its failure precipitated by buckling. Piers and abutments are typically of the short type, but columns or truss members may also fall in this range on occasion.

When subjected primarily to axial compression force, the capacity of the short compression member is directly proportional to the mass of the material and its strength in resisting crushing stresses (Fig. 11.1, zone 1). The crushing limit may be established by column-type action, involving a generally uniformly distributed compression on the entire member cross section. However, as sometimes occurs with a pier or abutment, the transfer of the compressive load to the member may involve highly concentrated bearing stresses on some fraction of the whole member cross section. In the latter case, the load limit may be established by the compressive bearing stresses, not by the column action of the member. This issue is dealt with in discussions of the design of masonry and concrete piers in Parts 5 and 6.

Slender Compression Elements

Very slender elements that sustain compression tend to buckle (Fig. 11.1, zone 3). Buckling is a sudden lateral deflection at right angles to the direction of the compression. If the member is held in position, the buckling may serve to relieve the member of the compressive effort and the member may spring or snap back into alignment when the compressive force is removed. If the force is not removed, the member will usually quickly fail—essentially by excessive bending action. It is essentially the inability of

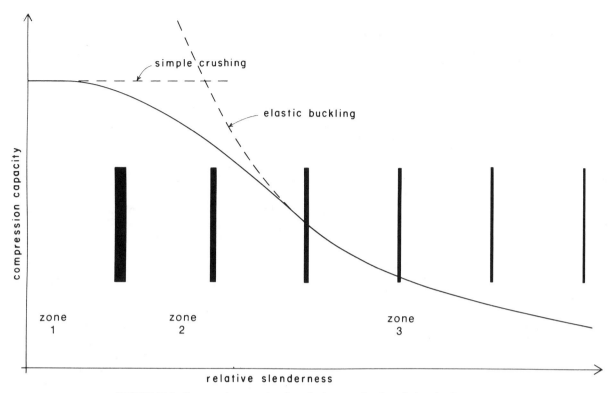

FIGURE 11.1. Compression capacity of wood columns, related to relative slenderness.

the member to resist significant bending that is the reason for its susceptibility to buckling.

The classic means for describing elastic buckling is the Euler curve, having the form

$$P = \frac{\pi^2 EI}{L^2}$$

This second-degree curve is one of the boundary conditions shown in Fig. 11.1. From the form of the equation for the Euler curve, it may be noted that the pure elastic buckling response is

1. Proportional to the stiffness of the material of the member—indicated by E, the elastic modulus of elasticity of the material for direct stress.
2. Proportional to the bending stiffness of the member as indicated by moment of inertia I of its cross-section area.
3. Inversely proportional to the member length, or actually to the second power of the length. The length in this case is an indication of potential slenderness.

The two basic limiting response mechanisms—crushing and buckling—are entirely different in nature, relating to different properties of the material and of the form of the member. The crushing limit is indicated by the straight line on the graph in Fig. 11.1; it is a constant value over the range of length it affects. The buckling load varies over the range of the member length, going to infinity for short

length (which it does not actually affect) and to zero for extremely great length.

Buckling may be affected by constraints, such as lateral bracing that effectively prevents the sideways deflection, or end connections that restrain the rotation associated with the assumption of a single curvature mode of deflection. Figure 11.2a shows the case for the member that is the general basis for the response as indicated by the Euler formula. This form of response may be altered by lateral constraints, as shown in Fig. 11.2b, that result in a multimode response. In the example shown in Fig. 11.2b the deflected form is such that the length over which buckling occurs is reduced to one-half of the column height.

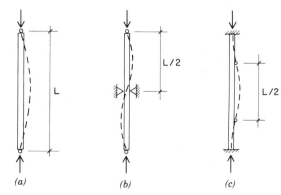

FIGURE 11.2. Effect of restraining conditions on column buckling.

By comparison to the member in Fig. 11.2a, the member in Fig. 11.2c has its ends restrained against rotation, having what is described as fixed ends. In this case the deflected (buckled) shape is a doubly inflected curve. It can be shown that the inflection points on this curve will occur at the quarter points of the height; thus the length over which the simple free buckling occurs is one half the full height of the member. Conditions such as all three cases in Fig. 11.2 occur in building structures, and modifications are used to qualify the member response on the basis of evaluation of whatever constraints are judged to occur. One means of qualifying the response is by using an *effective buckling length.* In the cases shown in Fig. 11.2b and c this modified length would be one-half of the actual length.

Member straightness in the unloaded condition can be a critical factor for the very slender member. If the member is not reasonably straight, it may be in what is virtually a prebuckled condition and buckling may occur progressively with a minor load. A deflected position can also occur due to bending induced by other loadings (Secs. 11.3 and 11.4).

A final consideration for the very slender member is the relative dynamic character of the load. Buckling is essentially a dynamic type of response, occurring as a very sudden collapse. This is not a very desirable mode of failure for building structures. Not that any failure is desirable; yet there are those that have a nature that is better than others. Brittle fracture and buckling are less preferred than slow ductile yielding or multistaged responses. If load buildup is slow, a buckling tendency may actually be detected in time for corrective measures to be taken (provide bracing or remove load). If the load buildup is quick, however, buckling will most likely occur suddenly and without warning. Truly dynamic loads are often not of a long-time occurrence, such as a surge of wind pressure (called a *gust*) or a single major movement during an earthquake. In the latter case, if the stability of the entire structure is not at risk, the instantaneous buckling of some members may not result in collapse, and the structure may snap back into a safe condition.

11.2. RELATIVE SLENDERNESS

The two limiting responses—pure crushing and pure elastic buckling—are actually only true at the extreme ends of the range of member length or slenderness. Between these limits there is a transitional range in which the actual response is some combination of the two forms of response. Except for piers and abutments, most primary compression members in building structures fall in this range. The three ranges, numbered 1, 2, and 3, are indicated on Fig. 11.1. The form of the response on the graph shows the behavior in the intermediate range (zone 2) to be simply a geometric transition from the straight horizontal line representing crushing to the second-degree curve of the Euler formula. The points of actual transition are arbitrary in the illustration, but have been repeatedly demonstrated in laboratory tests.

For the design of steel and wood columns, codes provide criteria for the evaluation of compression capacity over the full range of slenderness. The application of these criteria is explained in Chapters 18 and 23. For masonry and concrete columns some adjustment of capacity is also provided for; however, columns in these materials are actually mostly quite stout.

On Fig. 11.1 the proportions of columns corresponding to the range of slenderness are indicated. These are actually derived from present code criteria for solid wood columns of ordinary structural grade-lumber. Some minor variation will occur with other materials and with other cross-sectional forms, but the effect on the column profiles will be small. Consideration of these profiles will confirm our earlier allegation that most building structural columns tend to fall in the intermediate range.

11.3. INTERACTION: COMBINED COMPRESSION AND BENDING

There are a number of situations in which structural members are subjected to the combined effects that result in development of axial compression and internal bending. Stresses developed by these two actions are both of the direct stress type and can be directly combined for the consideration of net stress conditions. The stress investigation is considered in Sec. 11.4. However, the actions of a column and a bending member are essentially different in character and it is therefore customary to consider this combined activity by what is described as interaction.

The classic form of interaction is represented by the graph in Fig. 11.3. Referring to the notation on the graph:

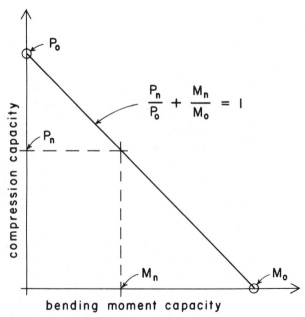

$$\frac{P_n}{P_o} + \frac{M_n}{M_o} = 1$$

FIGURE 11.3. Column interaction: compression plus bending.

The maximum axial load capacity of the member (with no bending) is P_o.

The maximum bending moment capacity of the member (without compression) is M_o.

At some compression load below P_o (indicated as P_n) the member is assumed to have some tolerance for a bending moment (indicated as M_n) in combination with the axial load.

Combinations of P_n and M_n are assumed to fall on a line connecting P_o and M_o. The equation of this line has the form

$$\frac{P_n}{P_o} + \frac{M_n}{M_o} = 1$$

A graph similar to that in Fig. 11.3 can be constructed using stresses rather than loads and moments. This is the procedure used with wood and steel members; the graph taking a simple form expressed as

$$\frac{f_a}{F_a} + \frac{f_b}{F_b} \leq 1$$

where

f_a = actual stress due to the axial load

F_a = allowable column-action stress

f_b = actual stress due to bending

F_b = allowable beam-action stress in flexure

For various reasons, real structural members do not adhere to the classic straight-line form of response as shown for interaction in Fig. 11.3. Figure 11.4 shows a form of response characteristic of reinforced concrete columns. Although there is some approximation of the pure interaction response in the midrange of combined effects, major deviation occurs at both ends of the range. At the low moment end, where almost pure compression occurs, the column is capable of developing only some percentage of

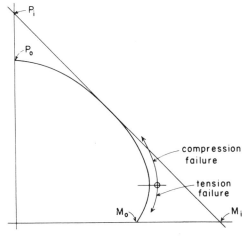

FIGURE 11.4. Interaction response: concrete columns.

the full, theoretical capacity. There are many contributing reasons for this, including the general nature of the composite (concrete and steel) material and typical inaccuracies in construction. At the high end of the moment range, where little compression occurs, failure is controlled by various procedures and criteria that tend to result in a failure that is predicated by tension yielding of the reinforcing, rather than by compression failure of the concrete. As compression is added at the low end of the axial load range, it actually tends to increase the moment capacity, up to the point where failure eventually becomes one essentially of compression.

Steel and wood members also have various deviations from the simple interaction form of response. A major effect is the so-called P-delta, discussed in Sec. 11.5. Other problems include inelastic behavior, effects of lateral stability, and general effects of the geometry of member cross sections.

11.4. COMBINED STRESS: COMPRESSION PLUS BENDING

Combined actions of compression and bending produce various effects on structures. The general interaction of the two separate phenomena and possible concern for P-delta effects are discussed in earlier sections of this chapter. We will now consider the condition of the actual stress combinations that occur when axial compression is added to bending moment at some cross section. One common example of this is the development of stress on the bottom of a bearing footing; the "section" in this case being the bearing contact face of the footing bottom.

Figure 11.5 shows a situation in which a simple rectangular footing is subjected to a combination of forces that require the resistance of vertical force, horizontal sliding, and overturning moment. The development of resistance to horizontal movement is produced by some combination of friction on the bottom of the footing and horizontal earth pressure on the face of the footing. Our concern here is for the investigation of the vertical force and the overturning moment and the resulting combination of vertical soil pressures that they develop.

Figure 11.5 illustrates our usual approach to the combined direct force and moment on a cross section. In this case the "cross section" is the contact face of the footing with the soil. However, the combined force and moment may originate, we make a transformation into an equivalent eccentric force that produces the same effects on the cross section. The direction and magnitude of this mythical equivalent e are related to properties of the cross section in order to qualify the nature of the stress combination. The value of e is established simply by dividing the moment by the force normal to the cross section, as shown in the figure. The net, or combined, stress distribution on the section is visualized as the sum of the separate stresses due to the normal force and the moment. For the stresses on the two extreme edges of the footing the general formula for the combined stress is

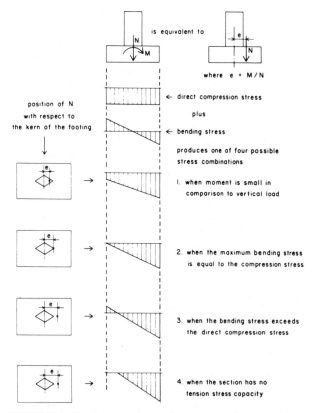

FIGURE 11.5. Development of stress due to combined compression and bending at a section.

$$p = \frac{N}{A} \pm \frac{Nec}{I}$$

We observe three cases for the stress combination obtained from this formula, as shown in the figure. The first case occurs when e is small, resulting in very little bending stress. The section is thus subjected to all compressive stress, varying from a maximum value on one edge to a minimum on the opposite edge.

The second case occurs when the two stress components are equal, so that the minimum stress becomes zero. This is the boundary condition between the first and third cases, since any increase in the eccentricity will tend to produce some tension stress on the section. This is a significant limit for the footing since tension stress is not possible for the soil-to-footing contact face. Thus case 3 is possible only in a beam or column where tension stress can be developed. The value of e that corresponds to case 2 can be derived by equating the two components of the stress formula as follows:

$$\frac{N}{A} = \frac{Nec}{I}, \qquad e = \frac{I}{Ac}$$

This value for e establishes what is called the *kern limit* of the section. The kern is a zone around the centroid of the section within which an eccentric force will not cause tension on the section. The form of this zone may be established for any shape of cross section by application of

the formula derived for the kern limit. The forms of the kern zones for three common shapes of section are shown in Fig. 11.6.

When tension stress is not possible, eccentricities beyond the kern limit will produce a *cracked section*, which is shown as case 4 in Fig. 11.5. In this situation some portion of the section becomes unstressed or cracked, and the com-

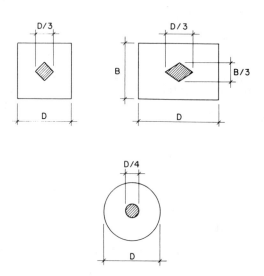

FIGURE 11.6. Kern limits for common shapes.

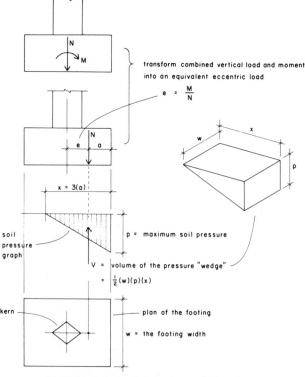

FIGURE 11.7. Analysis by the cracked section.

pressive stress on the remainder of the section must develop the entire resistance to the force and moment.

Figure 11.7 shows a technique for the analysis of the cracked section, called the *pressure wedge method*. The pressure wedge represents the total compressive force developed by the soil pressure. Analysis of the static equilibrium of this wedge and the force and moment on the section produces two relationships that may be utilized to establish the dimensions of the stress wedge. These relationships are as follows.

1. The total volume of the wedge is equal to the vertical force on the section. (Sum of the vertical forces equals zero.)
2. The centroid of the wedge is located on a vertical line with the force on the section. (Sum of the moments on the section equals zero.)

Referring to Fig. 11.7, the three dimensions of the stress wedge are w, the width of the footing, p, the maximum soil pressure, and x, the limit of the uncracked portion of the section. With w known, the solution of the wedge analysis consists of determining values for p and x. For the rectangular footing, the simple triangular stress wedge will have its centroid at the third point of the triangle. As shown in the figure, this means that x will be three times the dimension a. With the value for e determined, a may be found and the value of x established.

The volume of the stress wedge may be expressed in terms of its three dimensions as follows:

$$V = \tfrac{1}{2}wpx$$

Using the static equilibrium relationship stated previously, this volume may be equated to the force on the section. Then, with the values of w and x established, the value for p may be found as follows:

$$N = V = \tfrac{1}{2}wpx$$

$$p = \frac{2N}{wx}$$

Example 1.   Find the maximum value of the soil pressure for a square footing. The axial compression force at the bottom of the footing N is 100 k [450 kN], and the moment is 100 kip-ft [135 kN-m]. Find the pressure for footing widths of (a) 8 ft, (b) 6 ft, and (c) 5 ft.
Solution.   The first step is to determine the equivalent eccentricity and compare it to the kern limit for the footing to establish which of the cases shown in Fig. 11.5 applies.
(a) We thus compute for 8 ft:

$$e = \frac{M}{N} = \frac{100}{100} = 1 \text{ ft } [0.3 \text{ m}]$$

kern for the 8-ft wide footing $= \dfrac{8}{6} = 1.33$ ft [0.41 m]

and it is established that Case 1 applies.

We next determine the soil pressure, using the formula for the combined stress as previously derived.

$$p = \frac{N}{A} + \frac{Mc}{I} = \frac{100}{64} + \frac{100 \times 4}{341.3}$$
$$= 1.56 + 1.17 = 2.73 \text{ ksf}$$
$$[75.6 + 56.1 = 131.7 \text{ kPa}]$$

in which

$$A = (8)^2 = 64 \text{ ft}^2 \ [5.95 \text{ m}^2]$$

$$I = \frac{bd^3}{12} = \frac{(8)^4}{12} = 341.3 \text{ ft}^4 \ [2.95 \text{ m}^4]$$

(b) It may be observed that the kern limit is 6/6 = 1, which is equal to the eccentricity. Thus the situation is that shown as case 2 in Fig. 11.5, and the pressure is such that $N/A = Mc/I$. Thus

$$p = 2\left(\frac{N}{A}\right) = 2\left(\frac{100}{6(6)}\right) = 5.56 \text{ ksf } [266 \text{ kPa}]$$

(c) The eccentricity exceeds the kern limit, and the investigation must be done as illustrated in Fig. 11.7.

$$a = \frac{5}{2} - e = 2.5 - 1 = 1.5 \text{ ft } [0.76 - 0.3 = 0.46 \text{ m}]$$

$$x = 3a = 3(1.5) = 4.5 \text{ ft } [1.38 \text{ m}]$$

$$p = \frac{2N}{wx} = \frac{2(100)}{5(4.5)} = 8.89 \text{ ksf } [429 \text{ kPa}]$$

11.5. THE *P*-DELTA EFFECT

Bending moments can be developed in structural members in a number of ways. When the member is subjected to an axial compression force, there are various ways in which the bending effect and the compression effect can relate to each other. Figure 11.8a shows a very common situation that occurs in building structures when an exterior wall functions as a bearing wall or contains a column. The combination of gravity load plus lateral load due to wind or seismic action can result in the loading shown. If the member is quite flexible, and the actual deflection from the unloaded position is significant, an additional moment is developed as the axis of the member deviates from the action line of the compression force. This moment is the simple product of the load and the member deflection at any point; that is P times Δ, as shown in Fig. 11.8d. It is thus referred to as the *P*-delta effect.

There are various other situations that can result in this effect. Figure 11.8b shows an end column in a rigid frame structure, where moment is induced at the top of the column by the moment-resistive connection to the beam. Although slightly different in its geometric profile, the basic column response is similar to that in Fig. 11.8a—unless, of course, the frame is also subjected to a sideways deflection due to a lateral load or some lack of symmetry in the frame. Figure 11.8c shows the effect of a combination of gravity and lateral loads on a vertically cantilevered structure that supports a sign or a tank at its top. The two effects shown in Fig. 11.8b and c will be combined when a rigid

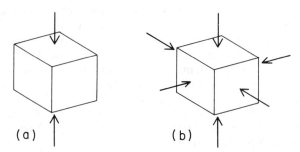

FIGURE 11.9. Effect of constraint on compression stress.

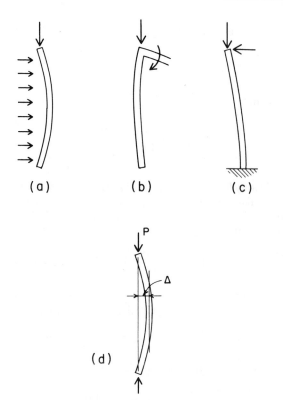

FIGURE 11.8. The *P*-delta effect.

frame is also subjected to the combined vertical and lateral loads.

In all of these—and various other—situations, the *P*-delta effect may or may not be critical. The major factor that determines its seriousness is the relative flexibility of the structure in general, but particularly the stiffness of the member that directly sustains the effect. In a worst-case scenario, the *P*-delta effect may be an accelerating one, in which the added moment due to the *P*-delta effect causes additional deflection, which in turn results in additional *P*-delta effect, which then causes more deflection, and so on.

The *P*-delta effect is seldom a solitary condition, except for the case of the member with a significant lack of initial straightness. In most cases, the effect will be combined with other conditions. For the slender element, the *P*-delta effect may work to precipitate a buckling failure. In other cases the moment due to the *P*-delta effect may simply combine with other moments, and the resulting action becomes one of interaction or combined stress as a critical condition.

There are actually not very many real situations in which the *P*-delta effect is critical. Although it occurs and adds to other conditions in many situations, it is most often only a minor effect. The potentially critical situations are usually those involving super-slender structural members, particularly very tall or very skinny columns. In the latter case, the *P*-delta effect should be carefully considered, regardless of the magnitude of other effects.

11.6. COMPRESSION OF CONFINED MATERIALS

Solid materials have capability to resist linear compression effects. Fluids can resist compression only if they are in a confined situation, such as air in an auto tire or oil in a hydraulic jack. Compression of a confined material results in a three-dimensional compressive stress condition, visualized as a triaxial condition, as shown in Fig. 11.9.

A major occurrence of the triaxial stress condition is that which exists in loose soil materials. Supporting materials for foundations are typically buried below some amount of soil overburden. The upper mass of soil—plus the general effect of the surrounding soil—creates a considerable confinement. This confinement permits some otherwise compression-weak materials to accept some amount of compression force. Loose sands and soft, wet clays must have this confinement, although they are still not very desirable bearing materials, even with the confinement. The confinement is a necessary continuing condition for their stability, and removal or reduction due to effects such as excavating near the soil mass, or a drop in the groundwater level, may cause problems.

While confinement is mandatory for fluid or loose, granular materials, it can also enhance the compression resistance of solids. An example of this is the concrete in the center of a reinforced concrete spiral column. The spiral column utilizes a continuous helix (in the general shape of a coiled spring) which is near the column perimeter. At loads near the ultimate capacity of the column, the spiral will develop tension and act to confine the concrete in the center mass of the column. This confined concrete will develop a level of compression higher than that of which it is capable in a simple linear stress situation.

Confinement is a basic technique used in air-inflated structures. There are several ways in which air can be used as a structural device, all of which essentially involve some means of confinement. The simplest case is that in which an enclosing surface is held rigid by the maintaining of an air pressure differential from inside to outside. In this case the entire building volume is confined. Another possibility is to inflate some object to make it reasonably rigid (like an air mattress or life raft) and then use it as a structural component (see Fig. 5.20).

CHAPTER TWELVE

Special Topics

Chapter 12 contains discussions of a number of special topics pertaining to various aspects of structural behavior.

12.1. THERMAL EFFECTS

A special case of strain is that which occurs when a material undergoes a change of temperature. In general, materials tend to expand when heated and contract when cooled. Although this is actually a volumetric change and does not occur at a constant rate for all ranges of temperature, it is possible to generalize with reasonably approximate accuracy for the case of a simple linear element at the temperature range of temperate climates from about $-30°$ to $+120°$ Fahrenheit. Assuming a constant rate of expansion for this range (see Table 12.1) we may determine the accompanying linear strain (unit deformation) as follows:

$$\varepsilon = C\ \Delta T$$

where

$$C = \text{linear coefficient of expansion}$$

$$\Delta T = \text{change in temperature}$$

When thermal expansion or contraction is prevented, stresses are developed in the material proportional to those that would occur if the member were free to move. If elastic conditions are assumed, stress due to thermal change may be determined as follows:

$$f = \varepsilon E = C\ \Delta T\ E$$

in which E is the direct stress modulus of elasticity.

TABLE 12.1. Coefficient of Linear Expansion per Degree

Material	Coefficient	
	Fahrenheit	Celsius
Aluminum	128×10^{-7}	231×10^{-7}
Copper	93×10^{-7}	168×10^{-7}
Steel (structural)	65×10^{-7}	117×10^{-7}
Concrete	55×10^{-7}	99×10^{-7}
Masonry (brick)	34×10^{-7}	61×10^{-7}
Wood (fir)	32×10^{-6}	58×10^{-6}

Example 1. A steel beam is subjected to a temperature range of $120°F$ from summer to winter. The beam is 60 ft long. Find (a) the length change if movement is not prevented, and (b) the stress developed if movement is completely prevented.

Solution. From Table 12.1 we find the coefficient of expansion of steel to be 65×10^{-7}. Thus the free length change is

$$e = \varepsilon L = C\ \Delta T\ L = [65(10^{-7})](120)[60(12)]$$
$$= 0.562 \text{ in.}$$

(b) If length change is prevented,

$$f = \varepsilon E = C\ \Delta T\ E = [65(10^{-7})](120)(29{,}000)$$
$$= 22.6 \text{ ksi}$$

The force exerted by the beam on its constraints would be the product of this stress times the cross-sectional area of the beam.

12.2. COMPOSITE ELEMENTS

A special stress condition occurs when two or more materials are assembled in a single unit so that when load is applied they strain as a single mass. An example of this is a column of reinforced concrete. In an idealized condition we assume both materials to be elastic and make the following derivation for the distribution of stresses between the two materials.

If the two materials deform the same total amount (see Fig. 12.1), we may express the total length change as

$$e = e_1 = e_2$$

where

$$e_1 = \text{length change of material 1}$$

$$e_2 = \text{length change of material 2}$$

Because both materials have the same original length and the same total deformation, the unit strains in the two materials are the same; thus

$$\varepsilon_1 = \varepsilon_2$$

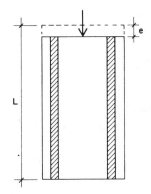

FIGURE 12.1. Deformation of the composite structure.

Assuming elastic conditions, these strains may be related to the stresses and moduli of elasticity for the two materials; thus

$$\varepsilon_1 = \frac{f_1}{E_1} = \varepsilon_2 = \frac{f_2}{E_2}$$

The relationship of the stresses in the two materials may thus be stated as

$$\frac{f_1}{f_2} = \frac{E_1}{E_2} \quad \text{or} \quad f_1 = f_2 \frac{E_1}{E_2}$$

Expressed in various ways, the relationship is simply that the stresses in the materials are proportional to their moduli of elasticity.

Example 1. A reinforced concrete column consists of a 12-in. square concrete section with four 0.75-in.-diameter round steel rods. The column sustains a compression load of 100 kips. Find the stresses in the concrete and steel. (Assume an E of 4000 ksi for concrete and 29,000 ksi for steel.)

Solution. Consider the load to be resisted by two internal forces: P_s and P_c. That is, the load resisted by the steel and the load resisted by the concrete. Then

$$\text{total } P = P_s + P_c = f_s A_s + f_c A_c$$

Using the previously derived relationship for the two stresses yields

$$f_s = \frac{E_s}{E_c} f_c = \frac{29,000}{4000} f_c = 7.25 f_c$$

Substituting this in the expression for P yields

$$\begin{aligned} P = 100 &= f_s A_s + f_c A_c \\ &= 7.25 f_c A_s + f_c A_c \\ &= f_c (7.25 A_s + A_c) \end{aligned}$$

for one steel bar

$$A = \pi R^2 = 3.14(0.375)^2 = 0.44 \text{ in.}^2$$

and for all four bars

$$A_s = 4(0.44) = 1.76 \text{ in.}^2$$

Then the concrete area is

$$A_c = (12)^2 - 1.76 = 142.24 \text{ in.}^2$$

Substituting values yields

$$100 = f_c\{7.25(1.76) + 142.24\} = f_c(155.0)$$

Thus

$$f_c = \frac{100}{155.0} = 0.645 \text{ ksi}$$

and

$$f_x = 7.25 f_c = 7.25(0.629) = 4.68 \text{ ksi}$$

12.3. LATERAL DEFORMATION

We discussed previously the case of material deformation in the direction of an applied stress (see Sec. 8.8). For most materials there is also a dimensional change in the lateral direction, that is, perpendicular to the direction of the stress (see Fig. 12.2). The ratio of these two deformations is called *Poisson's ratio*, after the French mathematician who studied the phenomenon. The symbol used for this property is the Greek lowercase letter nu (ν), the value being defined as

$$\nu = \frac{\text{unit lateral deformation}}{\text{unit axial deformation}}$$

The theoretical value for Poisson's ratio for an isotropic material (the same stress–strain properties in all directions) as derived from molecular theory is 0.25. This indicates some volumetric change under stress. Values for various materials, as derived from tests, are given in Table 12.2. Behaviors vary from that of rubber, which undergoes virtually no volumetric change, to that of cork, which has practically no lateral deformation.

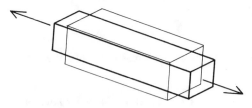

FIGURE 12.2. Longitudinal and lateral deformations.

TABLE 12.2. Poisson's Ratio

Material	Ratio
Steel	0.28
Aluminum	0.33
Concrete	0.10
Rubber	0.50−
Cork	0.00+

Poisson's ratio is of interest in several problems in structural investigation, principally those involving three-dimensional behavior. Shear deformation—as discussed in the Sec. 12.4—is one such problem.

12.4. SHEAR EFFECTS

Simple shear consists of a slicing kind of action, producing an internal force in the plane of a cut section (see the discussion in Sec. 8.4). For an idealized situation, the stress is assumed to be uniformly distributed over the area of the section and is expressed as

$$f = \frac{V}{A}$$

where

V = total shear force at the section

A = area of the section

Horizontal and Vertical Shear

Consider the particle of material shown in Fig. 12.3, having dimensions of a and b and a unit dimension as the third dimension. The particle is subjected to a direct shear stress of f_1 on the top face, the total force effect of which is expressed as

$$f_1 a(1) = f_1 a$$

For equilibrium of the particle, there must be an equal force of opposite direction on the bottom face. This force is expressed as

$$f_2 a(1) = f_2 a$$

By equating these, it may be observed that $f_1 = f_2$.

These opposed efforts will induce a rotation that must be resisted by the development of shear stresses f_3 and f_4 on the end faces of the particle. We may show that these two stresses will be equal; that is, $f_3 = f_4$ in the same manner that $f_1 = f_2$. Thus the force on the particle from these stresses is

$$f_3 b(1) = f_3 b$$

Finally, if we now consider an equilibrium of moments, we observe that

$$(f_1 a)b = (f_3 b)a \quad \text{or} \quad f_1 = f_3$$

Or, to summarize;

$$f_1 = f_2 = f_3 = f_4$$

We use this derivation to support the following statement. When a shear stress exists in a plane, there will be an equal shear stress in a mutually perpendicular plane.

Shear Deformation

With a particle under the action of the coexisting shear stresses on mutually perpendicular planes, we visualize the type of deformation shown in Fig. 12.4. The unit strain due to shear is expressed as the angle, indicated as γ (lowercase Greek gamma) in the figure. The relation of shear stress to shear strain is expressed as the shear modulus of elasticity; thus

$$G = \frac{f}{\gamma}$$

The shear modulus of elasticity is related to the direct stress modulus of elasticity (E). Since three-dimensional behavior is involved, one of the factors in the relationship is Poisson's ratio. The relationship is expressed as

$$G = \frac{E}{2(1 + \nu)}$$

Diagonal Stress

If we consider a particle that is subjected to shear to be cut along a diagonal (as shown in Fig. 12.5) it may be observed that tension stress exists on one diagonal, while compression stress exists on the other diagonal. It may be shown that the magnitude of these diagonal stresses is equal to that of the shear stress that generates them.

Referring to the free-body diagram in Fig. 12.5*b*, we note that the combined effects of the shear stresses on the two

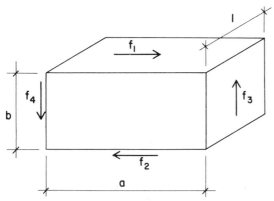

FIGURE 12.3. Three-dimensional development of shear.

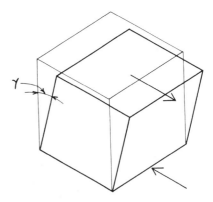

FIGURE 12.4. Shear deformation.

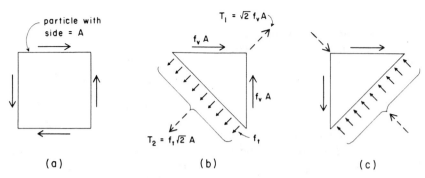

FIGURE 12.5. Direct stresses due to shear.

sides of the particle must be resisted by the total effect of the diagonal tension stress acting on the diagonal cut area. Thus

$$T_1 = T_2 \quad \text{or} \quad f_v A\sqrt{2} = f_t\sqrt{2}A$$

from which $f_v = f_t$.

The shear-developed diagonal stresses are additive to any other direct stresses that may exist due to other actions. Thus in a member that is subjected to several combined actions, the net effects and the true magnitudes and directions of critical stresses may be quite difficult to determine. See Sec. 12.5 for a discussion of the combined stress effects in a beam.

Stress on an Oblique Cross Section

Just as shear was shown to produce direct stresses, so we may show that direct force produces shear stress. Consider the element shown in Fig. 12.6, subjected to a tension force. If a section is cut that is not at a right angle to the force, but rather is at some angle θ to it, there may be seen to exist two components of the internal force P. One component is at a right angle to the cut section surface plane and the other is in the surface plane. These two components produce, respectively, direct tension stress and shear stress at the cut section. We may express these stresses as follows:

$$f = \frac{P\cos\theta}{A/\cos\theta} = \frac{P}{A}\cos^2\theta$$

$$v = \frac{P\sin\theta}{A/\cos\theta} = \frac{P}{A}\sin\theta\cos\theta$$

We note that when $\theta = 0$, $\cos\theta = 1$ and $\sin\theta = 0$, and therefore $f = P/A$ and $v = 0$. Also, when $\theta = 45°$, $\cos\theta = \sin\theta = (2)^{1/2}/2$, and therefore $f = \frac{1}{2}(P/A)$ and $v = \frac{1}{2}(P/A)$.

Example 1. The wood block shown in Fig. 12.7a has its grain at an angle of 30° to the direction of force. Find the compression and shear stresses on a plane parallel to the grain.
Solution. Note that as used in Fig. 12.6, $\theta = 60°$. Then for the free-body diagram shown in Fig. 12.7b.

$$N = P\cos 60°, \quad V = P\sin 60°, \quad A = \frac{4(3)}{\cos 60°}$$

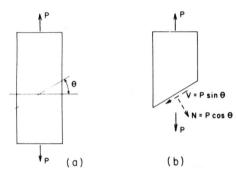

FIGURE 12.6. Development of stresses on an oblique section.

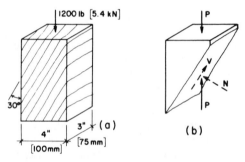

FIGURE 12.7.

The forces and the area may now be used directly to find the stresses. Or we may use the formulas derived in the preceding work.

$$f = \frac{P}{A}\cos^2\theta = \frac{1200}{12}(0.5)^2 = 25 \text{ psi } [180 \text{ kPa}]$$

$$v = \frac{P}{A}\sin\theta\cos\theta = \frac{1200}{12}$$
$$(0.5)(0.866) = 43.3 \text{ psi } [312 \text{ kPa}]$$

Shear Stress in Beams

Development of shear stress in a beam tends to produce lateral deformation. This may be visualized by considering the beam to consist of a layer of loose boards. Under the beam loading on a simple span, the boards tend to slide over each other, taking the form shown in Fig. 12.8. This type of deformation also tends to occur in a solid beam, but is resisted by the development of horizontal shearing stresses.

Shear stresses in beams are not distributed evenly over cross sections of the beam as was assumed for the case of simple direct shear (see Sec. 8.4). From observations of tested beams and derivations considering the equilibrium of beam segments under combined shear and moment actions, the following expression has been obtained for shear stress in a beam:

$$f_v = \frac{VQ}{Ib}$$

where

V = shear force at the beam section

Q = moment about the neutral axis of the area of the section between the point of stress and the edge of the section

I = moment of inertia of the section with respect to the neutral (centroidal) axis

b = width of the section at the point of stress

It may be observed from this formula that the maximum value for Q will be obtained at the neutral axis of the section, and that the stress will be zero at the edges of the section farthest from the neutral axis. The form of shear stress distribution for various geometric shapes of beam sections is shown in Fig. 12.9.

The following examples illustrate the use of this stress relationship.

Example 1. A beam section with depth of 8 in. and width of 4 in. [200 and 100 mm] sustains a shear force of 4 kips [18 kN]. Find the maximum shear stress.
Solution. For the rectangular section the moment of inertia about the centroidal axis is

$$I = \frac{bd^3}{12} = \frac{4(8)^3}{12} = 170.7 \text{ in.}^4 \ [67 \times 10^6 \text{ mm}^4]$$

The static moment (Q) is the product of the area a' and its centroidal distance from the axis of the section ($\bar{y}$) as shown in Fig. 12.10b. We thus compute Q as

$$Q = a'\bar{y} = [4(4)]2 = 32 \text{ in.}^3 \ [500 \times 10^3 \text{ mm}^3$$

The maximum shear stress at the neutral axis is thus

$$f_\theta = \frac{VQ}{Ib} = \frac{4000(32)}{170.7(4)} = 187.5 \text{ psi } [1.34 \text{ MPa}]$$

Example 2. A beam with the T-section shown in Fig. 12.10d is subjected to a shear of 8 kips [36 kN]. Find the

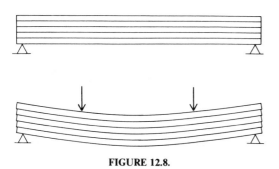

FIGURE 12.8.

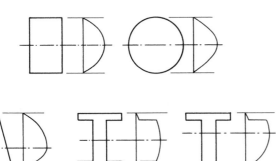

FIGURE 12.9. Development of shear in beams.

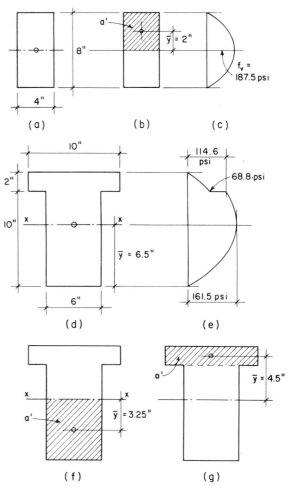

FIGURE 12.10. Form of shear stress distribution in beams with various cross sections.

maximum shear stress and the shear stress at the location of the juncture of the web and the flange of the T.

Solution. For this section the location of the centroid and the determination of the moment of inertia about the centroidal axis must be accomplished using processes explained in Appendix A. This work is summarized in Table 12.3. For determination of the maximum shear stress at the neutral axis (centroidal axis *x–x*, as shown in Fig. 12.10*d*) we find *Q* using the bottom portion of the web, as shown in Fig. 12.10*f*. Thus

$$Q = a'\bar{y} = [6.5(6)]3.25 = 126.75 \text{ in.}^3$$
$$[1.98 \times 10^6 \text{ mm}^3]$$

and the maximum stress at the neutral axis is thus

$$f_v = \frac{VQ}{Ib} = \frac{8000(126.75)}{1046.7(6)} = 161.5 \text{ psi } [1160 \text{ kPa}]$$

For the stress at the juncture of the web and flange we use the area shown in Fig. 12.10*g* for *Q*; thus

$$Q = [2(10)]4.5 = 90 \text{ in.}^3 [1.41 \times 10^6 \text{ mm}^3]$$

and the two shear stresses at this location, as displayed in Fig. 12.10*e*, are

$$f_{v1} = \frac{8000(90)}{1046.7(6)} = 114.6 \text{ psi } [828 \text{ kPa}]$$

$$f_{v2} = \frac{8000(90)}{1046.7(10)} = 68.8 \text{ psi } [497 \text{ kPa}]$$

In most design situations it is not necessary to use the complex form of the general expression for beam shear. In wood structures the beam sections are almost always of simple rectangular shape. For this shape we can make the following simplification.

$$I = \frac{bd^3}{12}, \quad Q = (b)\frac{d}{2}\frac{d}{4} = \frac{bd^2}{8}$$

$$f_v = \frac{VQ}{Ib} = \frac{V(bd^2/8)}{(bd^3/12)b} = \frac{3}{2}\frac{V}{bd}$$

For steel beams—which are mostly I-shaped cross sections—the shear is taken almost entirely by the web. (See

shear stress distribution for the I-shape in Fig. 12.9.) Since the stress distribution in the web is so close to uniform, it is considered adequate to use a simplified computation of the form

$$f_v = \frac{V}{dt_w}$$

in which *d* is the overall beam depth, and t_w is the thickness of the beam web.

For beams of reinforced concrete the shear mechanisms of the composite section are so complex that it is customary to use a highly simplified form for the shear computation. Various limits for this simplified stress and its use in design situations are then established by numerous requirements. This process is explained in Part 5 in discussions of the investigation and design for shear conditions of various types.

There are situations in which the general form of the shear stress formula must be used. Most of these involve the use of complex geometries for beam cross sections, which commonly occur in beams of prestressed concrete and the built-up, compound sections of steel or timber + plywood.

12.5. COMBINED DIRECT AND SHEAR STRESSES

The stress actions shown in Fig. 12.5 represent the conditions that occur when shear alone is considered. When shear occurs simultaneously with other effects, the various resulting stress conditions must be combined to produce the net effect. Figure 12.11 shows the result of combining a shear stress effect with a direct tension stress effect. For shear alone, the critical tension stress plane is at 45°, as shown in Fig. 12.11*a*. For tension alone, the critical tension stress plane is at 90° (Fig. 12.11*b*). For the combined stress condition, the unit stress will be some magnitude higher than either the shear or direct tension stress, and the critical tension stress plane will be at an angle somewhere between 45 and 90° (Fig. 12.11*c*).

Consider the beam shown in Fig. 12.12. Various combinations of direct and shear stresses may be visualized in terms of the conditions at the cross section labeled *S–S* in

TABLE 12.3. Computation of Properties for the Section in Example 2

Part	Area (in.²)	y from Bottom	Ay_1	I_0	Ay_2^2	I_x (in.⁴)
1	6(12) = 72	6	432	$\frac{6(12)^3}{12} = 864$	$72(0.5)^2 = 18$	882
2	2(2)(2) = 8	11	88	$2(2)(2)^3 = 2.7$	$8(4.5)^2 = 162$	164.7
Σ	80 in.² [50 × 10³ mm²]		520 in.³ [8.125 × 10⁶ mm³]			1046.7 in.⁴ [4.088 × 10⁸ mm⁴]

$$y_x = \frac{520}{80} = 6.5 \text{ in. } [162.5 \text{ mm}] \text{ (see Fig. 12.10}d).$$

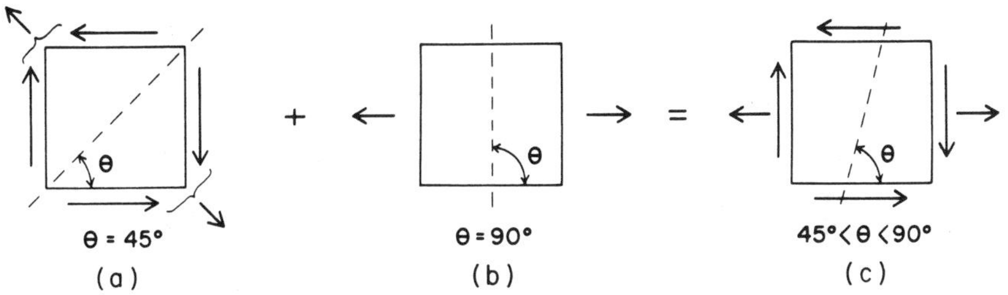

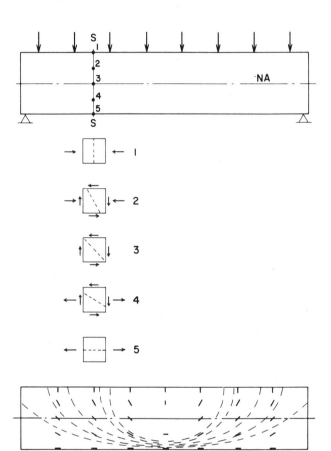

FIGURE 12.11. Shear stress examples.

the figure. With reference to the points on the section labeled 1 through 5, we observe the following:

At point 1 the beam shear stress is zero, and the dominant stress is the compressive bending stress in a horizontal direction.

At point 3 the shear stress is maximum, bending stress is zero, and the tension stress due to shear is at an angle of 45°.

At point 5 the shear stress is zero; bending stress in tension predominates, acting in a horizontal direction.

At point 4 the net tension stress due to the combination of shear and direct tension force will operate in a direction somewhere between the horizontal and 45°.

At point 2 the net tension will operate at an angle somewhat larger than 45°.

At point 1 the tension goes to zero as the angle of its direction approaches 90°.

The direction of the net tension stress is indicated for various points in the beam by the short dark bars on the beam elevation at the bottom of Fig. 12.12. The dashed lines indicate the flow of tension. If this figure were inverted, it would indicate the action of the net compressive stresses in the beam.

12.6. TORSION

Torsional shear stress and torsional rotation are discussed in Sec. 8.6, using the simplest case, that of an element with a round or cylindrical cross section. Most of the structural members in buildings are not of those shapes and their behavior in torsional actions is considerably more complex. The material in this section is presented for the purpose of developing some understanding of the problems of torsion in noncircular members. However, there are few situations in which actual design for torsion is performed in the development of building structural systems. In most cases the torsional effects are either quite minor, or provision of structural response characteristics for other actions results in more than adequate capacity for torsion. Actually, in many cases the design for torsion consists in preventing its occurrence by use of bracing. It is important, however, for the designer to have an awareness of the situations in which torsion is likely to occur and to have

FIGURE 12.12.

some appreciation of the potential for torsional resistance on the part of typical structural elements.

When a rectangular bar is twisted, it may be observed that lines on the bar surface do not remain straight and the bar cross section becomes distorted (see Fig. 12.13). If a rectangular grid is marked on the bar surface before it is twisted, investigation will show that the maximum distortions in the grid will occur at the middle of the long side of the bar. This indicates that the maximum shear distortion (strain) and a maximum shear stress will occur at this location. From theory and experiments, the maximum shear stress is found to be

$$f = \frac{T}{abc^2}$$

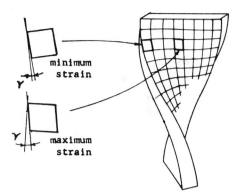

FIGURE 12.13. Torsion of a rectangular bar.

and the angle of twist of the bar per unit length is

$$\theta = \frac{T}{\beta b c^3 G}$$

In these formulas:

The values of b and c are the dimensions of the long and short sides of the bar, respectively.

The values of α and β are taken from Table 12.4. These are factors that vary with the ratio of b to c.

G is the modulus of elasticity for shear, as discussed in Sec. 12.4.

For a very thin element, such as a strip of metal, α and β both approach a value of $\frac{1}{3}$, and the expressions for the shear stress and rotation become

$$f = \frac{3T}{bc^2} \quad \text{and} \quad \theta = \frac{3T}{bc^3 G}$$

Torsion of noncircular sections is very complex. Theories and tests have produced a simple relationship for approximate evaluation of the torsional resistance of a sec-

TABLE 12.4. Factors for Investigation of Torsion in a Flat Bar[a]

Ratio, b/c	Factors	
	α	β
1.00	0.208	0.141
1.50	0.231	0.196
1.75	0.239	0.214
2.00	0.246	0.229
2.50	0.258	0.249
3.00	0.267	0.263
4.00	0.282	0.281
6.00	0.299	0.299
8.00	0.307	0.307
10.00	0.313	0.313
∞	0.333	0.333

[a] Dimension b is the width of the bar and dimension c is the thickness of the bar.

tion. The concept is that of an analogy between the torsional moment resistance of a cross section and the volume under a surface produced by inflating a thin membrane, which has been stretched across an opening of the shape of the cross section. The theory states that the volume under this surface is proportional to the torsional moment resistance of the section. By observing variations of this volume with changing shapes of the same magnitude of total area, certain general conclusions can be made. The theory is directly useful only for solid sections, but with a modification it can also be applied to hollow sections. (Similar experiments have been made using the volumes of mounded sand on the cross-sectional areas, producing approximately the same results.)

From consideration of the membrane (or mounded sand) theory, we observe the following (see Fig. 12.14):

1. The most efficient section is the solid circular one. This is usually used as the basis of comparison for evaluating the relative efficiency of other sections.
2. Rectangular sections are not as resistive as square ones; the longer and narrower the rectangle, the less the resistance.
3. Any addition to the section increases its resistance.
4. Notches, cuts, and holes reduce the torsional resistance more than they reduce the area.
5. The resistance of a very thin rectangular section is approximately the same as that of a section having a C, L, T, H, U, or X shape, providing that the thickness of the parts is the same as that of the rectangle, and the total length of the parts is the same as the

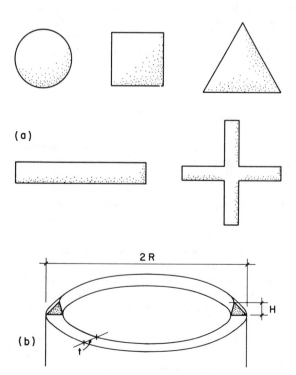

FIGURE 12.14. Evaluation of torsional resistance: membrane theory.

long side of the rectangle. This is useful in approximate evaluation of these shapes, which is otherwise a quite complex investigation.

The inflated membrane theory is true for closed, hollow sections if an adjustment is made. The correct volume for this type of section has the form shown in Fig. 12.14b. It is approximated by raising the inside edge of the membrane by the height H, producing a volume with a semitriangular section. For a circular tube, the value of H is $t/2\pi R^2$.

In general, the same observations hold for hollow sections of other than circular shape as were made for solid sections. One must be careful to consider the possibility of buckling of the walls of a hollow section, which will limit the torsional resistance. If buckling is not critical, however, the hollow sections are very efficient.

If a longitudinal slit is made in a member with a hollow section, resulting in an open, rather than a closed section, torsional resistance is greatly reduced. The resistance then drops to that of a rectangular section with width equal to the wall thickness and length equal to the perimeter of the hollow section. The resistance of a member with an open section can be increased by the use of stiffeners, such as the bulkhead type shown in Fig. 12.15. These may double or triple the resistance of the unbraced section, but will still leave a great gap with comparison to the resistance of the closed section.

12.7. SHEAR CENTER

Loadings on beams may produce other actions in addition to the normal bending stresses. Potential lateral or torsional buckling is one such type of action and is considered in Sec. 9.11. Another type of action is a torsional effect, which can occur if the plane of the bending moment does not coincide with the *shear center* of the beam cross section.

In Fig. 12.16a a concentrated load is shown on the end of a cantilevered beam. The location of the load results in a moment that coincides with the vertical centroidal axis of the beam's simple rectangular section. This will produce the beam deformation shown in Fig. 12.16c with a simple distribution of bending stress, the neutral axis occurring as the horizontal centroidal axis of the beam section. If, however, the load is moved off center, as shown in Fig. 12.16b, the beam is also subjected to a torsional twist and assumes the deformed shape shown in 12.16d.

For beams with biaxial symmetry (symmetrical about both centroidal axes) the shear center will coincide with the centroid of the section. Thus for the beam in Fig. 12.16 the twisting effect is avoided if the plane of the bending moment coincides with the centroidal axis. The relationship holds for other doubly symmetrical shapes, such as the steel I-shaped section shown in Fig. 12.17a.

For sections that have no axis of symmetry parallel to the plane of the bending moment, such as the C shape in Fig. 12.17c, the shear center is at a location separate from the centroid of the section, although it is always on any

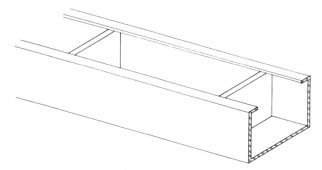

FIGURE 12.15. Increase of torsional resistance by use of stiffeners for open sections.

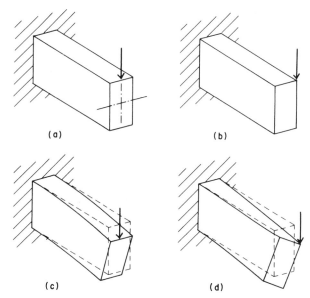

FIGURE 12.16. Relative stiffness of different cross sections with the same total area; arranged in general order of magnitude resistance (1 is highest).

other symmetrical axis that may exist (such as the horizontal axis for the C-shape). For the C-shape beam section, loading and bending moment that coincides with the vertical centroidal axis (Fig. 12.17c) will produce a torsional twist. If used as a beam, a C-shape member must be braced against the torsional rotation or have the loading made to coincide with the shear center, as shown in Fig. 12.17d, if torsional stress is to be avoided.

Figure 12.18 shows the relation between the centroid and the shear center for various sections consisting of rolled steel shapes. For the I-shape and pipe (Fig. 12.18a) the centroid and shear center coincide. For the C, T, and L shapes (Fig. 12.18b) the two are at different locations. It is possible, however, to use the C, T, or L shapes in ways that permit loading on a centroidal axis, as shown in Fig 12.18c and d, by orienting the loading on the one existing symmetrical axis, or by producing a compound section with a symmetrical axis.

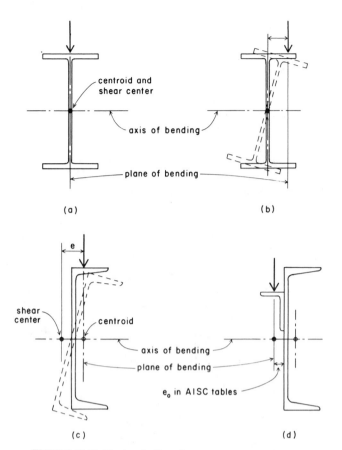

FIGURE 12.17. Torsional effect of load not through shear center.

12.8. UNSYMMETRICAL BENDING

Beams in flat-spanning floor or roof systems are ordinarily positioned so that the loads and the plane of the bending moment are perpendicular to one of the principal centroidal axes of the beam section. Thus the bending stresses are symmetrically distributed and the principal axis about which the bending occurs is also the neutral axis for the bending stress (see the discussion in Sec. 8.5).

There are various situations in which a structural member is subjected to bending in a manner that results in simultaneous bending about more than one axis. If the member is braced against torsion, the result may simply be a case of what is called *biaxial bending* or *unsymmetrical bending*. Figure 12.19 shows a common situation in which a roof beam is used for a sloping roof, spanning between trusses or other beams that generate the sloping profile. With respect to gravity loads, the beam will experience bending in a plane that is rotated with respect to its major axes, resulting in components of bending about both its axes, as shown in Fig. 12.19*b*. The bending stresses about the two axes of the beam section will produce the following stresses:

$$f_x = \frac{M_x}{S_x} \quad \text{and} \quad f_y = \frac{M_y}{S_y}$$

These are maximum stresses that occur at the edges of the section. Their actual distribution is of a form such as that shown in Fig. 12.19*c*, which can be described by determining the stresses at the four corners of the section. Not-

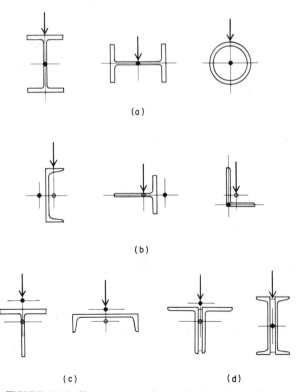

FIGURE 12.18. Shear centers and centroids of various sections.

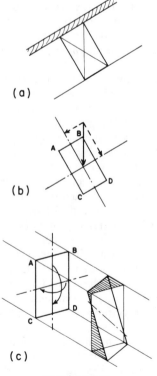

FIGURE 12.19.

ing the sense of the moments with respect to the two axes, and using plus for compression and minus for tension, the net stress conditions at the corners are as follows:

$$\text{At } A: \quad -f_x + f_y.$$
$$\text{At } B: \quad -f_x - f_y.$$
$$\text{At } C: \quad +f_x - f_y.$$
$$\text{At } D: \quad +f_x + f_y.$$

This is a somewhat idealized condition that ignores problems of torsional effects and potential lateral or torsional buckling. If a member is used in the situation shown in Fig. 12.19, it should be one with low susceptibility to these actions (such as the almost square one shown), or the construction should be carefully developed so as to provide bracing and other controls to prevent actions other than the simple bending.

Beams that occur in exterior walls, although having an orientation that is vertical, may experience biaxial bending under combinations of gravity and lateral loads. Columns subjected to bending are another case for possible multiple bending actions.

Example 1. Figure 12.20a shows the use of a beam in a sloping roof. The beam section is rotated to a position corresponding to the roof slope of 30° and consists of a wood member with a nominal size of 8 × 10. Find the net bending stress condition for the beam if the gravity load generates a moment of 10 kip-ft [14 kN-m] in a vertical plane.

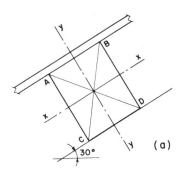

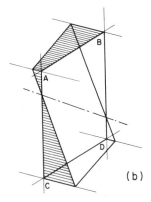

FIGURE 12.20. Development of biaxial bending.

Solution. From Table A.2 we find the properties of the beam to be: $S_x = 112.8$ in.^{3} [1.85×10^6 mm^3] and $S_y = 89.1$ in.^{3} [1.46×10^6 mm^3]. The components of the moment with respect to the major and minor axes of the section are

$$M_x = 10 \cos 30° = 10(0.866)$$
$$= 8.66 \text{ kip-ft [12.12 kN-m]}$$
$$M_y = 10 \sin 30° = 10(0.5) = 5 \text{ kip-ft [7 kN-m]}$$

and the corresponding maximum bending stresses are

$$f_x = \frac{M_x}{S_x} = \frac{8.66(12)}{112.8} = 0.921 \text{ ksi [6.55 MPa]}$$

$$f_y = \frac{M_y}{S_y} = \frac{5(12)}{89.1} = 0.532 \text{ ksi [4.79 MPa]}$$

The net stress conditions at the four corners of the section—lettered A through D in Fig. 12.20a—are then determined as follows, using plus for compression and minus for tension.

At A: $\quad +0.921 - 0.532 = +0.389$ ksi [$+1.76$ MPa].

At B: $\quad +0.921 + 0.532 = +1.453$ ksi [$+11.34$ MPa].

At C: $\quad -0.921 + 0.532 = +0.389$ ksi [-1.76 MPa].

At D: $\quad -0.921 - 0.532 = -1.453$ ksi [-11.34 MPa].

The form of the distribution of bending stresses is as shown in Fig. 12.20b.

In some situations bending stresses may be combined with stresses produced by direct force actions (see the discussions in Sections 10.5 and 11.4). The following example illustrates the situation that occurs when a column is subjected to biaxial bending together with the compressive loading.

Example 2. Figure 12.21 shows a situation in which a masonry column is subjected to a compressive load that is placed with eccentricities from both of the principal axes of the column. Find the distribution of the net direct stresses due to this loading.

Solution. We note that the eccentric load will generate bending moments equal to the product of the load times the eccentricity from the centroidal axis. Thus, calling the load N and the eccentricities e_x and e_y, the general expression for stress at a corner of the section is

$$f = \frac{N}{A} \pm \frac{N e_x c_x}{I_x} \pm \frac{N e_y c_y}{I_y}$$

where

N = axial load

A = area of the section

e_x = eccentricity from the x-axis

c_x = extreme distance from the x-axis

I_x = moment of inertia about the x-axis

e_y, c_y, I_y = corresponding values for the y-axis

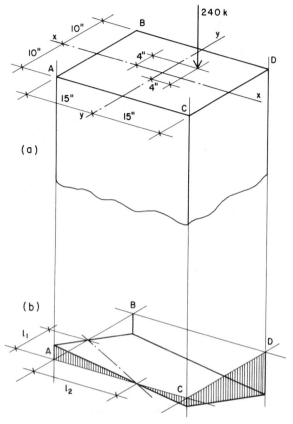

FIGURE 12.21.

Using the data given in Fig. 12.21a, we thus determine

$$I_x = \frac{bd^3}{12} = \frac{30(20)^3}{12}$$
$$= 20,000 \text{ in.}^4 \; [7.8 \times 10^9 \text{ mm}^4]$$

$$I_y = \frac{bd^3}{12} = \frac{20(30)^3}{12}$$
$$= 45,000 \text{ in.}^4 \; [1.76 \times 10^{10} \text{ mm}^4]$$

$$A = 20(30) = 600 \text{ in.}^2 \; [375 \times 10^3 \text{ mm}^2]$$

$$\frac{N}{A} = \frac{240,000}{600} = 400 \text{ psi} \; [2.85 \text{ MPa}]$$

$$\frac{Ne_x c_x}{I_x} = \frac{240,000(4)(10)}{20,000} = 480 \text{ psi} \; [3.43 \text{ MPa}]$$

$$\frac{Ne_y c_y}{I_y} = \frac{240,000(4)(5)}{45,000} = 320 \text{ psi} \; [2.28 \text{ MPa}]$$

The stresses at the four corners, as indicated in Fig. 12.21b, are:

At A: $+400 - 480 - 320 = -400$ psi $[-2.86$ MPa$]$.

At B: $+400 + 480 - 320 = +560$ psi $[+4.0$ MPa$]$.

At C: $+400 - 480 + 320 = +240$ psi $[+1.70$ MPa$]$.

At D: $+400 + 480 + 320 = +1200$ psi $[+8.56$ MPa$]$.

The two distances that locate the position of the neutral axis—l_1 and l_2 in Fig. 12.21b—may be found by consideration of similar triangles, as follows:

$$l_1 = \frac{400}{400 + 560} (20) = 8.33 \text{ in.} \; [208 \text{ mm}]$$

$$l_2 = \frac{400}{400 + 240} (30) = 18.75 \text{ in.} \; [470 \text{ mm}]$$

12.9. TIME-DEPENDENT BEHAVIOR

Building structures are intended to last for some time, in most situations. Loads that must be resisted by the structure occur in various time frames over the life of the structure. There are various considerations that must be made with regard to time-dependent behavior. These considerations are generally divided between those that relate to the structure and those that relate to the nature of the loads, although they are sometimes not separable.

The Response of the Structure

The structure may change over time. Its material may age, be chemically altered, biodegrade, corrode, and so on. It may also shrink, oxidize, evaporate, crumble, progressively crack, or otherwise change due simply to passage of time or to climate effects. Connections may deteriorate as well due to aging or to continuous actions. Usage of the building may result in wearing away, corrosion, or other effects. Choice of materials, details of the construction, use of safety factors, and the general conservatism of the design may be affected by some of these considerations. Building codes and industry design standards incorporate some of these concerns, but the judgment of designers is still required.

Time Character of Loads

Loads vary with respect to time. The extreme cases are the dead load (gravity weight), which essentially endures for the life of the structure, and the sudden, short-time shock of a major jolt from a gust of wind or an earthquake. In between these are the effects of people, furniture, snow, slow-moving vehicles, and sustained winds. Codes and design standards make various adjustments for the different effects of common loads. Considerations for possible combinations of load relate to some of the time factors for the individual loadings. Time may also relate to the likelihood of occurrence, as in the case of violent winds or major earthquakes.

Structure–Load Interaction

In many cases loads generate different responses from different materials or types of structures. Dead load has major effect on structures of wood or concrete; if the dead-load stresses are high, long-term, permanent deformations will occur. Long-term heavy load will also produce continuous settlement in some types of soil. On the other hand, impulse loads from wind or earthquakes will have effects that depend to a degree on the time-related dynamic responses of the structure, having magnified effects in some and reduced effects in others.

Various aspects of time-related structural behavior are discussed with respect to structural materials in the parts of the book that deal with them. Use of various code and industry standards are discussed in those parts and more generally in Part 9 with regard to the general problems of system design.

12.10. INELASTIC BEHAVIOR

There are various situations in which the structural behavior of elements deviates sufficiently from the idealized form of response visualized as pure elastic behavior that some more accurate investigation is required. There are actually very few structural materials that conform to a straight stress–strain relationship all the way to ultimate failure of the materials. The need to acknowledge a more realistic response depends on a number of considerations.

If the working stress (stress at service load conditions) method is used, the maximum stress permitted to occur in the structure will fall some distance below the ultimate failure limit for the material. Thus for the full range of working (service) load development, stress–strain responses may be close to those for an idealized material. Stress and deformation computations may thus be reasonably adequate within the usual safety margins. What may be critical in these situations is the establishment of the working stress limits, since these must be based on some realistic evaluation of the structure's true limiting capacity at failure. It is common, therefore, to use ultimate limit evaluation—with usually inelastic behaviors—in establishing limiting stresses and some realistic safety factor.

Although ultimate, inelastic responses must be used to establish limits for the working stress method, these investigations are the central theme of the strength method of design. For load capacity in terms of strength, this method does not concern itself with responses at the service load situation. Service loads are used only to derive necessary safety factors, which is done by multiplying service loads by some number. This *factored load* (increased load) is used for design. The design is then a limit design—literally a design for failure. If the failure at the factored load is accurately predicted, the response at the service load will be well within the limit of the structure. Use of this method requires a clear understanding of the full range of material response from zero stress to failure.

Ultimate responses are best dealt with in terms of the responses of specific materials and their common situations of usage. Discussion of these issues is therefore presented in Parts 3 through 6 for the four most common structural materials: wood, steel, concrete, and masonry. For all situations, the classic, idealized form of stress and strain development is used for a reference, and its understanding is essential as a starting point.

12.11. FRICTION

When forces act on objects in such a way as to tend to cause one object to slide on the surface of another object, a resistance to the sliding motion is often developed at the contact face between the objects. This resistance is called *friction*, and it constitutes a special kind of force.

For the object shown in Fig. 12.22*a*, being acted on by its own weight and the inclined force *F*, we may observe that the motion that is impending is one of sliding of the block toward the right along the surface of the plane. The force that tends to cause this motion is the component of *F* that is parallel to the plane. The component of *F* that is vertical works with the weight of the block to press the block against the plane. The sum of the weight plus the vertical component of *F* ($F \sin \theta$) is called the *pressure* on the plane or the *force normal* (perpendicular) to the plane.

A free-body diagram of the block is shown in Fig. 12.22*b*. For equilibrium of the block, two components of resistance must be developed. For equilibrium in a direction normal to the plane, the reactive force *N* is required, its magnitude being equal to $W + F \sin \theta$. For equilibrium in a horizontal direction, along the surface of the plane, a frictional resistance must be developed with a magnitude equal to $F \cos \theta$.

The situation just described may result in one of three possibilities, as follows:

1. The block does not move because the potential frictional resistance is more than adequate, that is,

$$F' > F \cos \theta$$

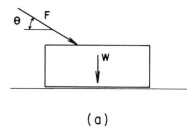

(a)

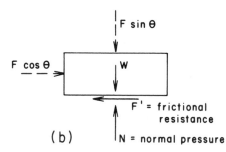

(b)

FIGURE 12.22. Development of net tension in a beam.

2. The block moves because the friction is not of sufficient magnitude, that is,

$$F' < F \cos \theta$$

3. The block is just on the verge of moving because the potential friction is exactly equal to the force tending to induce sliding, that is,

$$F' = F \cos \theta$$

From observations and experimentation the following deductions have been made about friction.

1. The frictional resisting force (F' in Fig. 12.22b) always opposes motion; that is, it acts opposite to the slide-inducing force.
2. For dry, smooth surfaces, the frictional resistance developed up to the point of motion is directly proportional to the normal pressure between the surfaces. We thus express the maximum value for the frictional resistance as

$$F' = \mu N$$

in which μ (lowercase Greek mu) is called the *coefficient of friction*.
3. The frictional resistance is independent of the area of contact.
4. The coefficient of static friction (before motion occurs) is greater than the coefficient of kinetic friction (during actual sliding). That is, for the same amount of normal pressure, the frictional resistance is reduced once motion actually occurs.

Frictional resistance is ordinarily expressed in terms of its maximum potential value. Coefficients for static friction are determined by finding the ratio between the sliding force and the normal pressure at the moment motion just occurs. Simple experiments consist of using a block on an inclined plane with the angle of the plane's slope slowly increased until sliding occurs (see Fig. 12.23a). Referring to the free-body diagram for the block (Fig. 12.23b) we note that

$$F' = \mu N = W \sin \phi$$
$$N = W \cos \phi$$

and as noted previously, the coefficient of friction is expressed as the ratio of F' to N, or

$$\mu = \frac{F'}{N} = \frac{W \sin \phi}{W \cos \phi} = \tan \phi$$

Approximate values of the coefficient of static friction for various combinations of joined objects are given in Table 12.5.

Problems involving friction are usually one of two types. The first involves situations in which friction is one of the forces in a system, and the problem is to determine whether the frictional resistance is sufficient to maintain the equilibrium of the system. For this type of problem the solution consists of writing the equations for equilibrium,

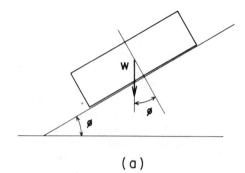

(a)

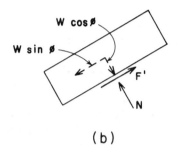

(b)

FIGURE 12.23. Development of sliding friction.

TABLE 12.5. Range of Values for Coefficient of Static Friction

Contact Surfaces	Coefficient, μ
Wood on wood	0.40–0.70
Metal on wood	0.20–0.65
Metal on metal	0.15–0.30
Metal on stone. masonry, concrete	0.30–0.70

including the maximum potential friction $F' = \mu N$, and interpreting the results. If the frictional resistance is not large enough, sliding will occur; if it is just large enough or excessive, sliding will not occur.

The second type of problem involves situations in which the force required to overcome friction is to be found. In this case the slide-inducing force is simply equated to the frictional resistance, and the required force is determined.

Note that in these problems sliding motions are usually of the form of pure translation (no rotation), and we may therefore treat the force systems as simple concurrent ones, ignoring moment effects. An exception is the problem in which tipping is possible, shown in Example 3.

Example 1. A block is placed on an inclined plane whose angle is slowly increased until sliding occurs. If the angle of the plane with the horizontal is 35° when sliding occurs, what is the coefficient of static friction between the block and the plane?

Solution. As previously derived, the coefficient of friction may simply be stated as the tangent of the angle of the plane. Thus

$$\mu = \tan\phi = \tan 35° = 0.70$$

Example 2. Find the horizontal force P required to slide a block weighing 100 lb if the coefficient of static friction is 0.30. (see Fig. 12.24).

Solution. For sliding to occur, the slide-inducing force P must be slightly larger than the frictional resistance F'. Equating P to F' gives us

$$P = F' = \mu N = 0.30(100) = 30 \text{ lb } \pm$$

The force must be slightly larger than 30 lb.

Example 3. Will the block shown in Fig. 12.25a tip or slide?

Solution. We first determine whether the block will slide by computing the maximum frictional resistance and comparing it to the slide-inducing force.

$$\text{Maximum } F' = 0.60(8) = 4.8 \text{ lb}$$

and therefore the block will not slide.

We then evaluate the tipping potential by considering the block to be restrained horizontally at the base by the frictional resistance. Tipping will be considered in terms of overturning about the lower corner (*a* in Fig. 12.25b). The overturning moment is

$$M = 4(6) = 24 \text{ in.-lb}$$

Resisting this is the moment of the block's weight, which is

$$M = 8(2) = 16 \text{ in.-lb}$$

Since the stabilizing moment offered by the weight is not sufficient, the block will tip.

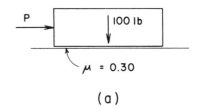

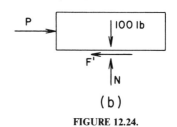

FIGURE 12.24.

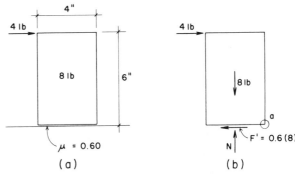

FIGURE 12.25.

CHAPTER THIRTEEN

Trusses

Chapter 13 presents material relating to the general investigation of structures that utilize trussing (or triangulation) as a basic structural technique. The basic form of analysis of behavior of simple planar trusses is illustrated in Chapter 7. Considerations of the use of trussing for three-dimensional systems are discussed in Sec. 14.6. Use of wood and steel trusses is discussed in Parts 3 and 4.

13.1. USE OF TRUSSES

Trussing is essentially a means of stabilizing a framework of linear elements by arranging them in a certain geometric pattern. Figure 13.1a shows an arrangement of four linear members in a simple rectangular pattern with the member ends connected by joints characterized as *pinned*. This type of joint is quite common in wood and steel frameworks, being one that can transmit a reasonable amount of direct or shear force, but an insignificant amount of moment or torsion. Under the action of the lateral force, the frame in Fig. 13.1a will clearly collapse sideways. The addition of the diagonal member to the frame, as shown in Fig. 13.1b, is a means of stabilizing the frame, the basic device being to reduce the rectangle to two triangles. In this case the frame is converted to a simple, vertically cantilevered truss.

In the pure truss the members are all considered to be connected by end pins, thus being subject to only one of two possible internal force actions: axial tension or axial compression. In a truss of slender elements, with loads applied only at the joints, this condition is usually relatively true. However, in some situations joints may be quite rigid, members may be relatively short and stout, or loads may be applied directly to members. For any of the latter conditions, the truss members may have some significant additional structural actions, which must be added to those of the simple truss functions.

Interior triangulation is a basic necessity for a truss. In addition, the external supports must have certain qualifications. There must be sufficient reaction components for the stability of the truss, but the truss must also be allowed to deform freely under the actions of the loads. Consider the truss in Fig. 13.1c, which is supported at its ends and subjected to the loads shown. The two vertical reaction components are sufficient for resistance of the loads shown. The horizontal reaction components are

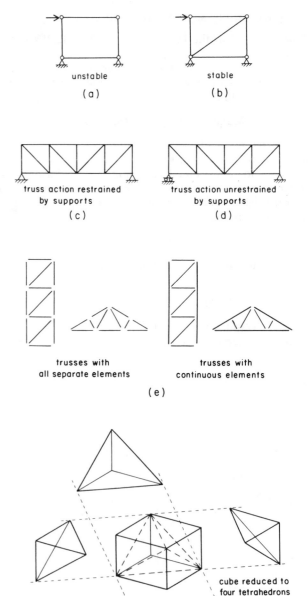

FIGURE 13.1. Aspects of trusses.

zero for this loading, but are potentially useful for resistance of any horizontal load. For the loading shown, the bottom chord is subjected to tension and tends to elongate. This is not possible, however, if both of the reactions resist horizontal motion; thus the truss as supported is not able to function properly. If one of the reactions has its horizontal resistance capacity removed, as shown in Fig. 13.1*d*, the truss is still stable but is capable of the necessary free deformation.

A construction simplification often used for small trusses is that of allowing some members to be continuous through the joints (see Fig. 13.1*e*). If members remain reasonably flexible, this does not usually affect the basic truss action, except to reduce slightly the deflections of the truss.

Trussing may also be used in three dimensions, that is, as a spatial framing system. Whereas the triangle is the basic unit of the planar truss, the tetrahedron (four-sided solid) is the basic unit of the spatial truss. Although other geometries are possible, three-dimensional trussing is often achieved by trussing of the three basic, mutually perpendicular planes of the orthogonal system (*x-y-z*); each plane being separately developed as a planar trussed system (see Fig. 13.1*f*).

Truss forms derive from many considerations of usage, efficiency, construction simplicity, and economics. Some traditional forms have been named for the designers who first developed them. Figure 13.2 shows a number of common truss forms for both simple flat-spanning elements and planar bents. As the size of the truss increases, the amount of interior triangulation increases in order to keep individual members relatively short.

13.2. STABILITY AND DETERMINANCY OF PLANAR TRUSSES

Any particular arrangement of truss elements and supports is subject to classification into one of the following cases with regard to its behavior.

1. *Unstable.* This may result from a lack of triangulation, too many joints, or insufficient reaction components.
2. *Stable and Determinate.* This is the result of an exact balance of elements, joints, and reaction components.
3. *Stable but Indeterminate.* This may result from too many elements, too few joints, or an excess of reaction components.

Obviously, a truss must be stable. If it is statically determinate, it offers the advantage of a simple analysis by the use of static equilibrium conditions alone. However, the determinate truss cannot spare any members or joints; thus failure (collapse due to instability) will occur if a single joint or member is lost. The indeterminate truss, although somewhat more complex to analyze, is sometimes considered an advantage because its redundancy of mem-

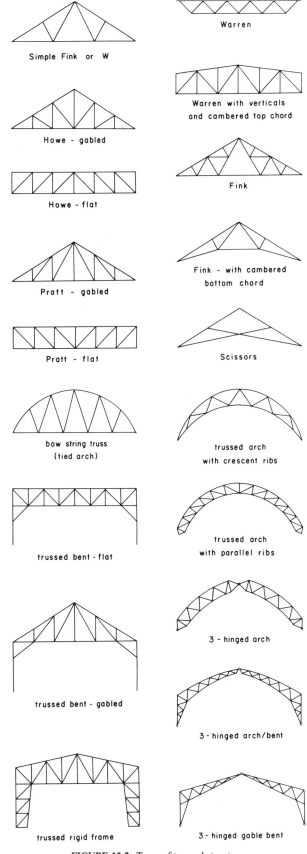

FIGURE 13.2. Types of trussed structures.

bers enhances safety and usually also reduces deflection and increases energy resistance for dynamic loadings.

A simple method exists for determining which of the three categories a particular truss falls into. The method is as follows. Consider each element between joints as a bar. Count the number of bars, the number of joints, and the number of reaction components. For case 2 these totals must satisfy the expression

$$B = 2J - R$$

where B = number of bars

J = number of truss joints

R = number of reaction components

If B is less than $2J - R$, the truss is unstable (case 1). If B is greater than $2J - R$, the truss is indeterminate (case 3). The following examples, shown in Fig. 13.3, illustrate the use of this expression.

Figure 13.3a. $B = 4$, $2J - R = 2(4) - 3 = 5$, the truss is unstable.

Figure 13.3b. $B = 5$, $2J - R = 2(4) - 3 = 5$, the truss is stable and determinate.

Figure 13.3c. $B = 6$ (assuming that the diagonals cross without attachment), $2J - R = 2(4) - 3 = 5$, the truss is stable but indeterminate.

Figure 13.3d. $B = 5$, $2J - R = 2(4) - 4 = 4$, the truss is stable but externally indeterminate.

Figure 13.3e. $B = 6$, $2J - R = 2(4) - 2 = 6$, which says that the truss is stable and determinate. However, the truss is actually internally indeterminate and externally unstable with respect to lateral force.

Figure 13.3f. $B = 33$, $2J - R = 2(18) - 3 = 33$, the truss is stable and determinate.

Figure *13.3g.* $B = 33$, $2J - R = 2(18) - 4 = 32$, the truss is stable but indeterminate (analogous to a two-span beam).

Figure 13.3h. The internal pin acts to separate the truss, so we consider the two halves independently. For the left half, $B = 19$, $2J - R = 2(11) - 3 = 19$, and for the right half, $B = 11$, $2J - R = 2(7) - 3 = 1$; both halves are stable and determinate.

Note that the truss in Fig. 13.3*e* has a basic flaw in the form of the supports. At least one support must be horizontally anchored or the structure is dynamically unstable; that is, it is easily collapsed by a slight misalignment of the load. The application of the formula does not work because the supports are inadequate.

For the right half in Fig. 13.3*h*, the internal pin represents a support with two components of resistance, since the other half is horizontally restrained.

Correct application of the expression will usually properly qualify a truss, but when the truss form is complex and the truss has several supports and internal pins, it is difficult to apply.

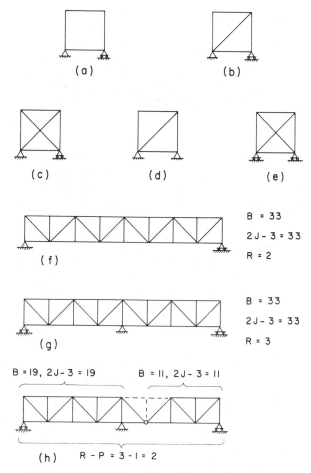

FIGURE 13.3. Stability and determinacy of planar trusses.

13.3. TRUSS LOADING

For truss behavior—involving only tension or compression forces in the members—we assume all loads to be applied only at joints of the truss. This may be literally true for some loads when they are applied through framing that occurs at the location of truss joints. However, loads are also frequently applied directly to the truss members; thus the members affected have two functions: as truss members (taking tension or compression) and as directly loaded members (taking bending, shear, etc.).

For investigation of the behavior of the whole truss, we consider only joint loadings. Loads that are actually distributed otherwise are collected as joint loads for the truss analysis. Then in the design of individual members the true loading is considered and the combined actions required of individual members are recognized.

Because different load sources (dead load, live load, wind, etc.) may have different effects on a truss, it is common to consider the load actions separately and then to consider their potential combinations. This is usually done in the process of designing the individual members, as discussed in Sec. 13.4.

13.4. DESIGN FORCES FOR TRUSS MEMBERS

The primary concern in analysis of trusses is the determination of the critical forces for which each member of the truss must be designed. The first step in this process is the decision about which combinations of loading must be considered. In some cases the potential combinations may be quite numerous. Where both wind and seismic actions are potentially critical, and more than one type of live loading occurs (e.g., roof loads plus hanging loads), the theoretically possible combinations of loadings can be overwhelming. However, designers are usually able to exercise judgment in reducing the sensible combinations to a reasonable number. For example, it is statistically improbable that a violent windstorm will occur simultaneously with a major earthquake shock.

Once the required design loading conditions are established, the usual procedure is to perform separate analyses for each of the loadings. The values obtained can then be combined at will for each individual member to ascertain the particular combination that establishes the critical result for the member. This means that in some cases certain members will be designed for one combination and others for different combinations.

In most cases design codes permit an increase in allowable stress for design of members when the critical loading includes forces due to wind or seismic loads. For wood trusses the codes also permit an increase in allowable stress for roof live loads. On the other hand, when the load is permanent (all dead load) the codes require a *decrease* of 10% for wood structures. These factors must be taken into account when considering load combinations. One procedure is to use adjusted values for the various combinations, as illustrated in the following example.

Table 13.1 illustrates a process for summarizing the results of load analysis on a truss. In this case the three basic design loads are dead load, live load, and wind load. Four loadings are produced since a separate analysis must be done for the wind from the two opposite directions, referred to as wind left and wind right. The results of the load analyses for these four conditions are shown for three members of the truss, with force magnitudes given in pounds and sense indicated using plus for tension and minus for compression.

The four load combinations considered in this example are dead load alone, dead load plus live load, dead load plus wind left, and dead load plus wind right. These are typical design combinations, but may not be the only ones required in all cases. Some codes require that some or all of the live load be included with the wind load. In this example we assume the truss to be of wood and the structure to be a roof. Thus the following adjustments must be made for all of the combinations.

For Dead Load Only. Allowable stresses must be reduced by 10%. We thus adjust the load by a factor of $1/0.9 = 1.11$. With this adjusted load the member can be designed for the full allowable stress, since the reduction has already been performed.

TABLE 13.1. Examples of Combination of Load Analysis for Critical Design Values

	Truss Members		
	AB	BC	CD
Loadings			
Dead load only	+2478	−1862	+3847
Live load	+3684	−2768	+5719
Wind left	−2862	+894	−2643
Wind right	+2074	+1046	+3427
Combinations			
DL only at $\frac{1}{0.9} = 1.11$	+2751	−2067	+4270
DL + LL at $\frac{1}{1.25} = 0.80$	+4930	−3704	+7653
DL + WL$_L$ at $\frac{1}{1.33} = 0.75$	−288	−726	+903
DL + WL$_R$ at $\frac{1}{1.33} = 0.75$	+3414	−2181	+5456
Design forces			
Maximum force	+4930	−3704	+7653
Reversal force	−288	—	—

For Dead Load plus Live Load. Allowable stresses may be increased by 25%, assuming the roof live load to be not more than 7 days in duration (see Ref. 7). We thus adjust the load by a factor of $1/1.25 = 0.8$ to obtain the adjusted load that may be used with the full allowable stresses.

For Dead Load plus Wind Load. Allowable stresses may be increased by 33%, and the adjustment factor becomes $1/1.33 = 0.75$.

With the four combinations determined, together with their adjustments, we next scan the list of combinations for the critical design values to be used for the actual design of the members. Of first concern is simply the largest number, which is the maximum force in the member. However, of possible equal concern, or in some situations even greater concern, is the case of a reversal of sign in some combination. In Table 13.1 it may be observed that the maximum force in member *AB* is 4930 lb in tension. However, the combination of dead load plus wind left produces a compression force, albeit of small magnitude. If the member is long, it is possible that the slenderness limitations for compression members may prove to be more critical in the selection of the member, even though the tension force is much larger.

Figure 13.4 shows a typical roof truss in which the actual loading consists of the roof load distributed continuously along the top chords and a ceiling loading distributed continuously along the bottom chords. The top chords are thus designed for a combination of axial compression and bending and the bottom chords for a combination of axial

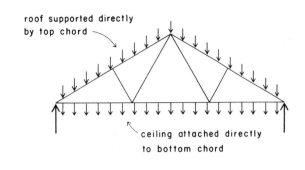

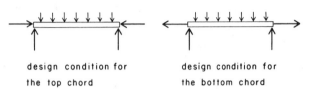

FIGURE 13.4. Effects of loads applied directly to truss chords.

tension plus bending. This will, of course, result in somewhat larger members being required for both chords, and any estimate of the truss weight should account for this anticipated additional requirement.

13.5.   METHODS OF INVESTIGATION

Simple, statically determinate, planar trusses are quite easily analyzed for the effects of ordinary loadings. Depending on the complexity of the truss form, its lack of symmetry, and the diversity of loadings, the analysis may more easily be done by one of several methods. These are as follows:

1. *Graphical Analysis.*   This method is explained in Sec. 7.5; it consists of successive applications of the use of force polygons for the joints. If done at a sufficient size and with a reasonable degree of accuracy, it can produce data adequate for most design work.
2. *Algebraic Method of Joints.*   This consists of solving the simple concurrent force systems at the truss joints, proceeding from joint to joint until the forces in all truss members are determined. It is illustrated in Sec. 7.6.
3. *Method of Sections.*   This consists of cutting sections through the entire truss and considering the action of the free body on either side of the cut section. The method is explained in Sec. 13.7.
4. *Beam Analogy Method.*   For flat-spanning, parallel-chord trusses a shortcut application of the method of sections is the beam analogy method, in which the variations of beam shear and moment along the span are used for rapid determination of member forces.
5. *Computer-Aided Methods.*   Programs are readily available for both investigation and design of ordinary trusses.

The method to be used in a particular situation may depend on many factors. Computer-aided methods offer many advantages to the professional designer. However, truss design is not usually an everyday occurrence, and once-in-a-great-while design may not justify the investment in the computer software or the time to learn how to use it. Unless an exceptionally complex form is required, or a highly optimized design is desired, simpler hand methods may be preferable. For trusses that are statically indeterminate, are three-dimensional in character, or sustain dynamic loading, there will be real benefits in use of computer-aided methods. For a simple, modest-sized truss with ordinary static loads, hand analysis will probably be quick and easy.

For those familiar with its application, the graphical method offers a quick analysis, useful for actual design or as a check on algebraic solutions. It is especially useful when geometry of the truss arrangement is of some irregular form, making algebraic computations clumsy.

The method of joints is probably the most common hand-analysis method. However, for a particular truss or a particular loading, any of the methods may offer some advantage. It is not unusual to make use of several methods in the course of a single truss design.

The usual reason for performing an analysis of a truss is to obtain data for its design. If that is the case, the method of investigation may be linked in some ways to the design process. Thus input data for the analysis may include not only truss form, support conditions, and loadings, but also member material and configuration, connection methods for the joints, cost of materials and fabrication, and so on.

13.6.   THE METHOD OF JOINTS

Both the graphical method and the algebraic method of joints use the same basic technique: the analysis of individual joints as simple concurrent force systems. Because the concurrent force system has only two conditions for static equilibrium, only two unknowns (the forces in two truss members) can be found at a single joint. The joints must therefore be investigated in a sequence that proceeds from joint to joint. In some cases the joint configuration may require a solution of two simultaneous equations, but in many situations this is not required. The mathematical work is thus quite simple, although it can be very laborious if the truss has many members, lacks symmetry, has members at many different angles of slope, and so on.

Some designers use the graphical method to find the forces in truss members, then use the algebraic method of joints or the method of sections to check a few members to verify the answers. In any process it is useful to have more than one method of finding answers, using one for a general solution and the other for a partial check.

The method of joints is explained in Sec. 7.5 (by graphics) and Sec. 7.6 (by algebra).

13.7. THE METHOD OF SECTIONS

The analysis for internal forces in the members of a truss by the method of sections consists of dealing with the truss in a manner similar to that used for beam analysis. The cut section utilized in the explanation of internal forces in a beam, as illustrated in Fig. 9.6, is utilized here to externalize the forces in the cut members of the truss. The following example will illustrate the technique.

Figure 13.5 shows a simple span, flat chorded truss with a vertical loading on the top chord joints. The Maxwell diagram for this loading and the answers for the internal forces are also shown in the figure. This solution is provided as a reference for comparison with the results that will be obtained by the method of sections.

In Fig. 13.5 the truss is shown with a cut plane passing vertically through the third panel. The free-body diagram of the portion of the truss to the left of this cut plane is shown in Fig. 13.5a. The internal forces in the three cut members become external forces on this free-body diagram, and their values may be found using the following analysis of the static equilibrium of the free body.

In Fig. 13.5b we observe the condition for vertical equilibrium. Because ON is the only cut member with a vertical force component, it must be used to balance the forces, resulting in the value for ON_v, of 500 lb acting downward. We may then establish the value for the horizontal component and the actual force in ON.

We next consider a moment equilibrium condition, picking a point for moment reference that will eliminate

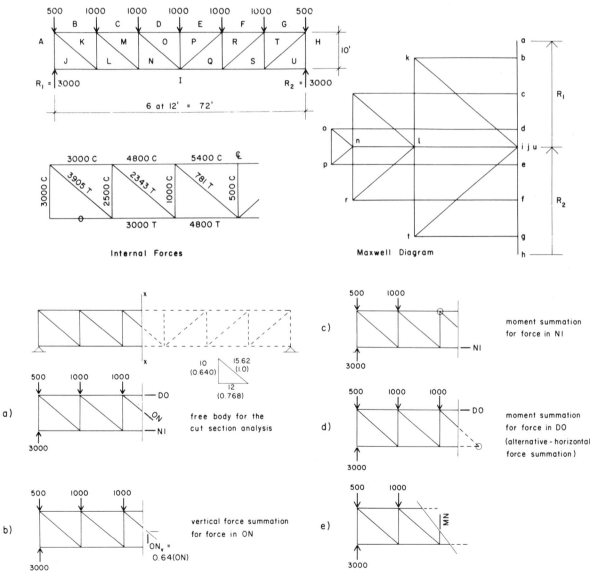

FIGURE 13.5. Investigation of a flat-chorded truss.

all but one of the unknown forces. If we select the top chord joint as shown in Fig. 13.5c, both the force in the top chord and the force in the diagonal member ON will be eliminated. The only remaining unknown force is then that in the bottom chord, and the summation is as follows.

$$M_1 = 0 = +3000(24) = +72,000$$
$$-500(24) = -12,000$$
$$-1000(12) = -12,000$$
$$-NI(10) = -10(NI)$$

Thus

$$10(NI) = +72,000 - 12,000 - 12,000 = 48,000$$
$$NI = \frac{48,000}{10} = 4800 \text{ lb}$$

Note that the sense of the force in NI was assumed to be tension, and the sign used for NI in the moment summation was based on this assumption.

One way to find the force in the top chord would be to do a summation of horizontal forces, since the horizontal component of ON and the force in NI are now known. An alternative method would be to use another moment summation, this time selecting the bottom chord joint shown in Fig. 13.5d in order to eliminate IN and ON from the summation

$$M_2 = 0 = +3000(36) = +108,000$$
$$-500(36) = -18,000$$
$$-1000(24) = -24,000$$
$$-1000(12) = -12,000$$
$$-DO(10) = -10(DO)$$

Thus

$$10(DO) = +54,000$$
$$DO = \frac{54,000}{10} = 5400 \text{ lb}$$

The forces in all of the diagonal and horizontal members in the truss may be found by cutting the truss with a series of vertical planes and doing static equilibrium analyses similar to that just illustrated. In order to find the forces in the vertical members of the truss, it is possible to cut the truss with an angled plane, as shown in Fig. 13.5e. A summation of vertical forces on this free body will yield the internal force of 1500 lb in compression in member MN.

The method of sections is sometimes useful when it is desired to find the forces in some members of the truss without performing a complete analysis for the forces in all members. By the method of joints it is necessary to work from one end of a truss all the way to the desired location, while the single cut plane may be used anywhere on the truss.

13.8. TRUSS DEFLECTIONS

When used in situations where they are most capable of being utilized to the best of their potential, trusses will seldom experience critical deflections. In general, trusses possess great stiffness in proportion to their mass. When the deflection of a truss is significant, it is usually the result of one of two causes. The first of these is the ratio of the truss span to the depth. This ratio is ordinarily quite low when compared to the normal ratio for a beam, but when it becomes as high as that for a beam, considerable deflection may be expected. The second principal cause of truss deflections is excessive deformation in the truss joints. A particular problem is that experienced with trusses that are fabricated with ordinary bolts. Because the bolt holes must be somewhat larger than the bolts to facilitate the assembly, considerable slippage is accumulated when the joints are loaded. This is a reason for favoring joints made with welds, split-ring connectors, or high-strength bolts, the latter functioning in friction resistance.

For most trusses deflection is essentially due to the lengthening and shortening of the members caused by the interior forces of tension and compression. Figure 13.6a shows a simple W truss in which the original, unloaded position is shown as a solid line and the deflected shape as a dashed line. In this example, with the left end held horizontally, there are two deflections of concern. The first is the vertical movement, which is measured as the sag of the bottom chord or as the drop of the peak of the truss. The second is the horizontal movement, which occurs at the right end, where horizontal restraint is not provided.

A relatively simple procedure for determining the deflected shape of a truss due to the length change of the members is to plot the deformations graphically. This consists simply of constructing the individual triangles of the truss with the sides equal to the deformed lengths. The procedure for this is illustrated in Fig. 13.6. We begin with one truss joint as a fixed reference. In this example it is logical to use the joint at the left support, since it truly remains fixed in location, both vertically and horizontally. We then assume that one of the sides of this first triangle remains in its angular position. Although this is probably not true, the result can be adjusted for when the construction is completed. The deformed location of joint C (shown as C') is now found simply as a point along the line of the fixed member by determining the length change of the member due to the internal force. For practical purposes this deformation is exaggerated by simply multiplying it by some factor. With the same factor used for all the deformations, the end produced deflections are simply divided by the same factor to find the true dimensions.

The other truss joint B', which defines the first deformed triangle, is now found by using the deformed lengths of the other two members, as shown in the figure. With this triangle ($AB'C'$) constructed, we next use the known locations of joints B' and C' as a reference for the construction of the next triangle, $B'C'D'$. This procedure continues until we have produced the completed figure shown in dashed lines

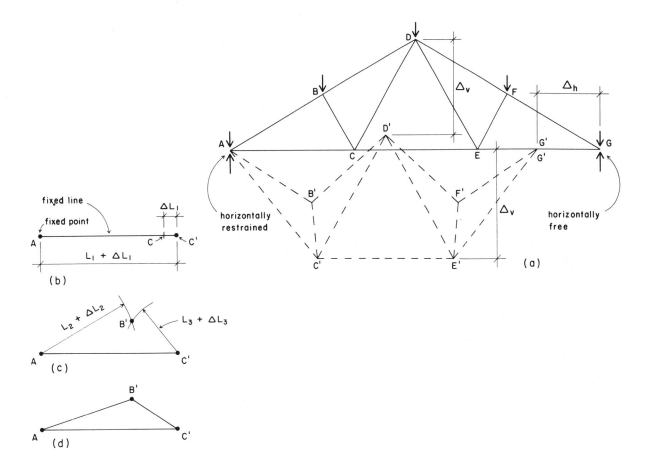

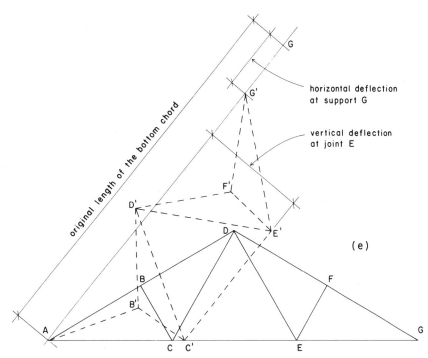

FIGURE 13.6. Investigation of truss deflections.

in Fig. 13.6e. We then simply superimpose the original figure of the truss on the deflected figure by matching the location of joint A in both figures and by aligning the line between joints A and G' with the original position of the bottom chord. The resulting figure will be that shown in Fig. 13.6a.

If the construction of the deformed truss is performed only to determine the values of specific deflections, it is not necessary to rotate the original truss form, as shown in Fig. 13.6a. In the construction in Fig. 13.6e, a line has been drawn from the fixed point A to the deformed location of point G, noted as point G' on the figure. As shown, the horizontal deflection at the right support and the vertical deflection of the truss joints may be determined with reference to this line.

The procedure for determining the deformed lengths of the individual members of the truss is shown in Table 13.2. The actual changes in the lengths of the members are found as follows.

1. Determine the magnitudes of the internal forces, the original lengths of the members, the gross cross-sectional area of the members, and the modulus of elasticity of the members.
2. Find the true change in the length of the members as follows.

$$\Delta L = \frac{FL}{AE}$$

where

ΔL = actual change in length; shortening when the member is in compression and lengthening when it is in tension

F = axial force in the member

A = gross cross-sectional area of the member

E = modulus of elasticity of the material

[Note: Be sure that A and L are in the same units (in.) and that F and E are in the same units (lb/in.² or kips/in.²).]

3. Multiply the actual length changes by some factor in order to produce dimensions that are between 5 and 10% of the original lengths. This is necessary for reasonable construction of the graphic figure. When the construction is completed, the deflections found using the figure are simply divided by this factor to find their true values.

In real situations there are always effects that produce some modification of the deflections caused by the axial forces in the members. The following effects are the most common.

Deformation of the Joints. These can be of considerable magnitude, especially with bolted joints not utilizing high-strength bolts for steel trusses or split-ring connectors for wood trusses. Nailed joints will also experience considerable deformation owing to the bending of the nails and bearing stress in the wood.

Continuous Chord Members. Observation of the deflected truss form in Fig. 13.6a will indicate that the top and bottom chords cannot take this form, unless they are discontinuous at joints B, C, E, and F. If the chords are continuous through any of these joints, there will be considerable reduction in the truss deflection owing to the bending resistance of the chords. If the chords are actually designed for bending, as the result of a directly applied loading, they will have considerable bending stiffness and will substantially reduce the net deflection of the truss.

Rigidity of the Joints. The typical simple analysis for the internal forces in the truss assumes a pure "pin" function of the truss joints. Except for a joint employing a single bolt, where the bolt is only moderately tightened, this is never the true condition. To the extent that the end connections of the members and the general stiffness of the joint facilitate transfer of moments between the members, the truss will actually function as a rigid frame in resisting deformations. If the joints are quite rigid (as with the all-welded truss) and the members are relatively short and

TABLE 13.2. Example of the Form of Length Change of the Truss Members

Truss Member	Axial Force		Area of Cross Section (in.²)	Length (in.)	Length Change (in.)	
	Type	Magnitude (lb)			True[a]	For Plot[b]
1	Tension	20,000	1.6	100	0.0431	10.78
2	Compression	26,000	2.8	87	0.0279	6.96

[a] ΔL = (force)(length)/(area)(29,000,000 lb/in.²) (actual change).
[b] $250(\Delta L)$ (for the graphic plot).

stiff, this effect will greatly reduce the truss deflections. The extreme case is the *Vierendeel truss*, which is without triangulation and functions entirely on the basis of the joint stiffnesses and the shear and bending resistance of the members.

Because of these possible effects, the true deflection of trusses is quite complex and can only be determined by using considerable judgment on the results of any calculations. To be honest, the only truly reliable method is full-scale test loading of the actual structure.

13.9. SECONDARY STRESSES

Secondary stresses are those induced in the truss members by effects other than the axial forces determined from the analysis of the pure pin jointed truss actions. The principal causes of such stresses are essentially the same effects described in Sec. 13.8 as modifying factors for truss deflections: joint deformation, joint rigidity, member stiffness, and continuous members.

In the pin joint analysis it is assumed that all of the members meet as concurrent, axial force-carrying elements at a single point, the truss joint. If joint deformations result in the misalignment of some members, the forces applied at the ends of the members may become other than axial; that is, they may not be aligned with the centroidal axes of the members. If this is the case, bending moments will be developed as the internal forces are eccentric from the axes of the members. Details for the truss joints should be carefully developed to assure that the members are propperly aligned and that the deformations that occur do not produce twisting or other unbalancing conditions in the joints.

When truss joints are quite rigid, or members are continuous through the joints, bending will be induced in the members as the truss performs partially as a rigid frame with moment-resistive joints. In addition to the bending stresses thus produced, there will also be some modification of the axial forces. The degree to which this occurs will depend on the relative stiffness of the members and the relative rigidity of the joints. When the members are quite short and stout and the joints are all welded, secondary stress effects may be considerable. When the truss members are relatively slender and the joints are capable of only moderate moment transfer to the ends of the members, secondary stresses are usually quite minor. For ordinary, light trusses for buildings, the latter is most often the case. In general, some investigation of secondary stress effects should be made when:

1. Joints are quite rigid, owing to all welding or to use of large gusset plates and many fasteners in the ends of the members.
2. Members are quite stiff, as indicated by the approximate limits L/r less than 50, or I/L greater than 0.5.

CHAPTER FOURTEEN

Special Structures

Elementary structural elements constitute the basic forms from which we derive most of the structures that are used for buildings. Walls, columns, beams, and trusses are enduring in the construction inventory. However, designers and builders over the years have used their imaginations and ingenuity to derive special forms of structures. Chapter 14 presents some variations on the ordinary structures, involving the use of moment-resistive joints to produce rigid frames, the use of internal zero moment joints (pins), the implementation of two-way spanning systems, and the Vierendeel truss.

14.1. RIGID FRAMES

Frames in which members are connected in a manner that permits the transfer of end moments from member-to-member are commonly called *rigid frames*. Rigid in this case refers to the character of the joints, not necessarily to the deformation character of the whole frame. In fact, many rigid frames have critical deflection problems, and the control of movements—especially sideways movements—is often an important design factor. Most rigid frames are highly indeterminate; however, for common situations it is often acceptable to use approximate methods for analysis. Chapter 14 deals with some simple frames that yield to reasonable analysis by statics alone or by simple methods of approximate analysis. Examples of some frame designs are given in Part 9.

Aspects of Rigid Frames

When members are connected to each other by joints that act essentially as pinned (moment-free) connections, the members are free to deflect and to rotate at the joints without affecting the deformation of the other connected members. When members are rigidly connected, they tend to offer restraint to each other's movements. This can be a positive effect, producing stability of the frame and reducing deflections of spanning members. It can also cause problems in some situations, such as the following.

1. *Highly Unbalanced Loading.* When live load is high in comparison to dead load, random loading can cause some members to deform excessively, resulting in transfer of major effects to attached members.

2. *Mismatched Member Sizes.* Long-span beams attached to small columns will transfer considerable twist to the columns; similarly, alternate long and short spans will result in serious deformation of the shortspan members. Frame layout and sizes of members must be matched in ways that are not critical in pinned structures.

3. *Restrained Deformations.* Member deflections and joint rotations are natural and necessary to the functioning of a rigid frame. If infilling construction (principally walls) restrict the frame deformations, loads will be transferred to the stiffer restraining construction. In this event, damage can occur if the restraining construction does not have adequate structural capacity.

Because of the interactions of members in rigid frames, each frame member is usually subjected to a combination of internal actions. This is made more complex when the frame is subjected to a number of different loading combinations, is three-dimensional in character, has a large number of members, or lacks symmetry. When all of these conditions are present, the investigation of frame behavior and the determination of critical design values for the individual members becomes an arduous chore.

Simple Determinate Frames

Consider the frame shown in Fig. 14.1*a*, consisting of two members rigidly joined at their intersection. The vertical member is fixed at its base, providing the necessary support condition for stability of the frame. The horizontal member is loaded with a uniformly distributed loading and functions as a simple cantilever beam. The frame is described as a cantilever frame because of the single fixed support. The five sets of figures shown in Fig. 14.1*b* through *f* are useful elements for the investigation of the behavior of the frame:

1. The free-body diagram of the entire frame, showing the loads and the components of the reactions (Fig. 14.1*b*). Study of this figure will help in establishing the nature of the reactions and in the determination of the conditions necessary for stability of the frame.

2. The free-body diagrams of the individual elements (Fig. 14.1*c*). These are of great value in visualizing

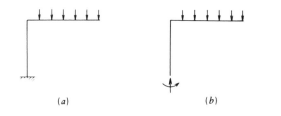

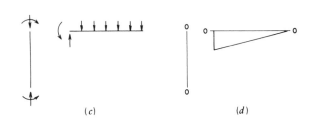

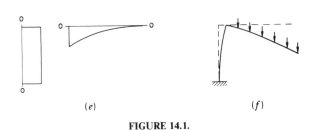

is generally recommended that the deformed shape be sketched first so that its correlation with other factors in the investigation may be used as a check on the work. The following examples illustrate the process of investigation for simple cantilever frames.

Example 1. Find the components of the reactions and draw the free-body diagrams, shear and moment diagrams, and the deformed shape of the frame shown in Fig. 14.2*a*.

Solution. The first step is the determination of the reactions. Considering the free-body diagram of the whole frame (Fig. 14.2*b*), we compute the reactions as follows:

$$\Sigma F = 0 = +8 - R_V, \quad R_V = 8 \text{ kips (up)}$$

and, with respect to the support

$$\Sigma M_0 = 0 = M_R - 8(4),$$
$$M_R = 32 \text{ kip-ft (counterclockwise)}$$

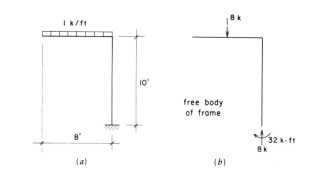

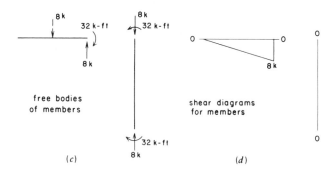

FIGURE 14.1.

the interaction of the parts of the frame. They are also useful in the computations for the internal forces in the frame.

3. The shear diagrams of the individual elements (Fig. 14.1*d*). These are sometimes useful for visualizing, or for actually computing, the variations of moment in the individual elements. No particular sign convention is necessary unless in conformity with the sign used for moment (see the discussion of relation of shear and moment in Chapter 9). Although good as visualization exercises, the shear diagrams have limited value in the investigation in most cases.

4. The moment diagrams for the individual elements (Fig. 14.1*e*). These are very useful, especially in determination of the deformation of the frame. The sign convention used is that of plotting the moment on the compression side of the element.

5. The deformed shape of the loaded frame (Fig. 14.1*f*). This is the exaggerated profile of the bent frame, usually superimposed on an outline of the unloaded frame for reference. This is very useful for the general visualization of the frame behavior. It is particularly useful for determination of the character of the external reactions and the form of interaction between the parts of the frame.

When performing investigations, these elements are not usually produced in the sequence just described. In fact, it

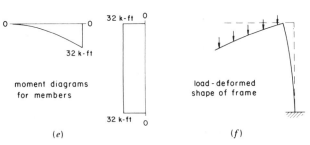

FIGURE 14.2.

Note that the sense, or sign, of the reaction components is visualized from the logical development of the free-body diagram.

Consideration of the free-body diagrams of the individual members will yield the actions required to be transmitted by the moment connection. These may be computed by application of the conditions for equilibrium for either of the members of the frame. Note that the sense of the force and moment is opposite for the two members, simply indicating that what one does to the other is the opposite of what is done to it.

In this example there is no shear in the vertical member. As a result, there is no variation in the moment from the top to the bottom of the member. The free-body diagram of the member, the shear and moment diagrams, and the deformed shape should all corroborate this fact.

The shear and moment diagrams for the horizontal member are simply those for a cantilever beam (see Fig. B.1, Case 4).

It is possible with this example, as with many simple frames, to visualize the nature of the deformed shape without recourse to any mathematical computations. It is advisable to do so and to continually check during the work that individual computations are logical with regard to the nature of the deformed structure.

Example 2. Find the components of the reactions and draw the shear and moment diagrams and the deformed shape of the frame in Fig. 14.3*a*.

Solution. In this frame there are three reaction components required for stability, since the loads and reactions constitute a general coplanar force system. Using the free-body diagram of the whole frame (Fig. 14.3*b*), the three conditions for equilibrium for a coplanar system are used to find the horizontal and vertical reaction components and the moment component. If necessary the reaction force components could be combined into a single-force vector, although this is seldom required for design purposes.

Note that the inflection occurs in the larger vertical member because the moment of the horizontal load about the support is greater than that of the vertical load. In this case this computation must be done before the deformed shape can be accurately drawn.

The reader should verify that the free-body diagrams of the individual members are truly in equilibrium and that there is the required correlation between all the diagrams.

The following examples of single-bent frames consist of frames with combinations of support and internal conditions that make the frames statically determinate. These conditions are technically achievable, but a bit on the weird side for practical use. We offer them here simply as exercises within the scope of our readers so that some experience in investigation may be gained.

Example 3. Investigate the frame shown in Fig. 14.4*a* for the reactions and internal conditions.

Solution. The typical elements of investigation, as illustrated for the Examples 1 and 2, are shown in the figure. The suggested procedure for the work is as follows:

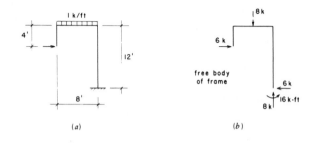

(a)   (b)

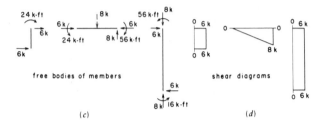

free bodies of members   shear diagrams

(c)   (d)

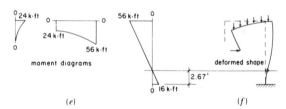

moment diagrams   deformed shape

(e)   (f)

FIGURE 14.3.

1. Sketch the deflected shape (a little tricky in this case, but a good exercise).
2. Consider the equilibrium of the free-body diagram for the whole frame to find the reactions.
3. Consider the equilibrium of the left-hand vertical member to find the internal actions at its top.
4. Proceed to the equilibrium of the horizontal member.
5. Finally, consider the equilibrium of the right-hand vertical member.
6. Draw the shear and moment diagrams and check for correlation of all the work.

Note that the right-hand support allows for an upward vertical reaction only, whereas the left-hand support allows for both vertical and horizontal components. Neither support provides moment resistance.

Before attempting the exercise problems, the reader is advised to attempt to produce the results shown in Fig. 14.4 independently.

Indeterminate Rigid Frames

There are many possibilities for the development of rigid frames for building structures. Two common types of frames are the single-span bent and the vertical, planar bent, con-

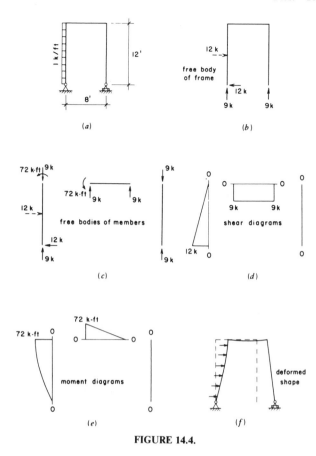

FIGURE 14.4.

highly flexible frame, however, the effects may be serious. In this case the actual lateral movements of the joints must be computed to obtain the eccentricities used for determination of the *P*-delta effect.

Lateral deflection of a rigid frame is related to the general stiffness of the frame. When several frames share a loading, as in the case of a multistory building with several bents, the relative stiffnesses of the frames must be determined. This is done by considering their relative deflection resistances.

The Single Span Bent

Figure 14.5 shows two possibilities for a rigid frame for a single-span bent. In Fig. 14.5a the frame has pinned bases for the columns, resulting in the load deformed shape shown in Fig. 14.5c, and the reaction components as shown in the free-body diagram for the whole frame in Fig. 14.5e. The frame in Fig. 14.5b has fixed bases for the columns, resulting in the slightly modified behavior indicated. These are common situations, the base condition depending on the supporting structure as well as the frame itself.

The frames in Fig. 14.5 are both technically not statically determinate and require analysis by something more than statics. However, if the frame is symmetrical and the loading is uniform, the upper joints do not move sideways and the behavior is of a classic form. For this condition, analysis by moment area, three-moment equation, or

sisting of the multistory columns and multispan beams in a single plane in a multistory building.

As with other structures of a complex nature, the highly indeterminate rigid frame presents a good case for use of computer-aided methods. Programs utilizing the finite element method are available and are used frequently by professional designers. So-called shortcut hand computation methods such as the moment distribution method were popular in the past. They are "shortcut" only in reference to more laborious hand computation methods; applied to a complex frame, they constitute a considerable effort—and then produce answers for only one loading condition.

Rigid-frame behavior is much simplified when the joints of the frame are not displaced; that is, they move only by rotating. This is usually only true for the case of gravity loading on a symmetrical frame—and only with a symmetrical gravity load. If the frame is not symmetrical, or the load is nonuniformly distributed, or lateral loads are applied, frame joints will move sideways (called *sidesway* of the frame) and additional forces will be generated by the joint displacements.

If joint displacement is considerable, there may be significant increases of force effects in vertical members due to the *P*-delta effect (see Sec. 11.5). In relatively stiff frames, with quite heavy members, this is usually not critical. In a

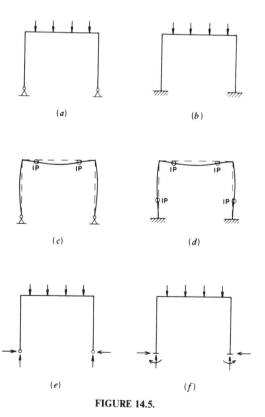

FIGURE 14.5.

moment distribution may be performed, although tabulated values for behaviors can also be obtained for this common form of structure.

Figure 14.6 shows the single-span bent under a lateral load applied at the upper joint. In this case the upper joints move sideways, the frame taking the shape indicated with reaction components as shown. This also presents a statically indeterminate situation, although some aspects of the solution may be evident. For the pinned base frame in Fig. 14.6, for example, a moment equation about one column base will cancel out the vertical reaction at that location, plus the two horizontal reactions, leaving a single equation for finding the value of the other vertical reaction. Then if the bases are considered to have equal resistance, the horizontal reactions will each simply be equal to one-half of the load. The behavior of the frame is thus completely determined, even though it is technically indeterminate.

For the frame with fixed column bases in Fig. 14.6b, we may use a similar procedure to find the value of the direct force components of the reactions. However, the value of the moment at the fixed base is not subject to such simplified procedures. For this investigation—as well as that of the frames in Fig. 14.5—it is necessary to consider the relative stiffness of the members, as is done in the moment distribution method or in any method for solution of the indeterminate structure.

A popular variation of the single-span bent is the gabled bent, having a form as shown in Fig. 14.7a. The significant angle of the slope provides both additional interior height and the possibility of using a fast-draining roof covering, which is usually much less expensive than that required for a flat roof. Although the bent may be as shown, with a moment-resistive joint at the peak, it is more often constructed with a pinned joint at the peak and pinned col-

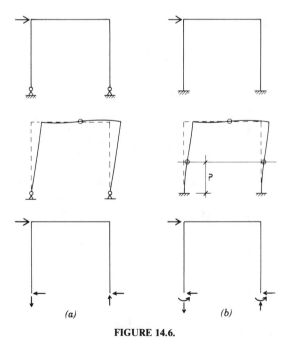

FIGURE 14.7.

umn bases, constituting a three-hinged structure, an example of which is shown in Fig. 14.14.

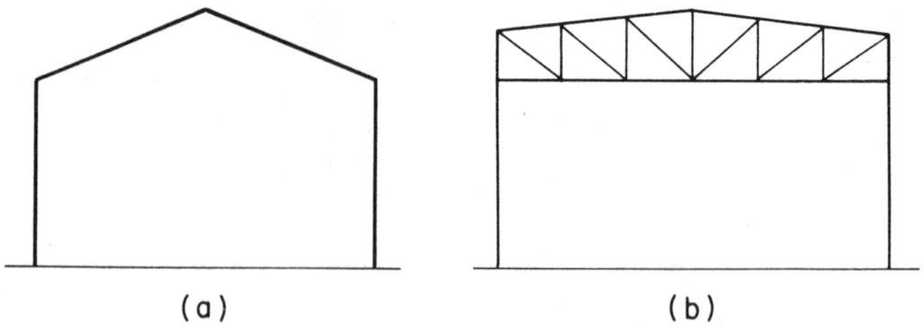

Another usage of the rigid bent is in the mixed type structure shown in Fig. 14.7b, called a trussed bent. In this case the horizontal member is a truss and the column continues upward to become part of the truss. Because of the typical extreme stiffness of the truss (very little vertical deflection or end rotation), it is usually assumed that the column is held fixed at the bottom of the truss. With this assumption, the column action is ignored for gravity loading, and the lateral force action is as for the simple bent in Example 1, without the component of lateral deflection due to rotation at the top of the column.

14.2. APPROXIMATE ANALYSIS OF MULTISTORY BENTS

The rigid-frame structure occurs quite frequently as a multiple-level, multiple-span bent, constituting part of the structure for a multistory building. In most cases, such a bent is used as a lateral bracing element; although once it is formed as a moment-resistive framework, it will respond characteristically for all types of loads.

The multistory rigid bent is quite indeterminate, and its investigation is complex—requiring considerations of several different loading combinations. When loaded or formed unsymmetrically, it will experience sideways movements that further complicate the analysis for internal forces. Except for very early design approximations, the analysis is now sure to be done with a computer-aided system. The software for such a system is quite readily available.

For preliminary design purposes, it is sometimes possible to use approximate analysis methods to obtain member sizes of reasonable accuracy. Actually, many of the older high-rise buildings still standing were completely designed with these techniques—a reasonable testimonial to their effectiveness. Demonstrations of these approximate methods are given in Chapter 58.

A general discussion of the use of the rigid frame for lateral bracing is given in Sec. 47.4.

14.3. STRUCTURES WITH INTERNAL PINS

In many structures qualifying conditions exist at supports or within the structure that modify the behavior of the

FIGURE 14.6.

structure, often eliminating some potential components of force actions. Qualification of supports as fixed or pinned has been a situation in most of the structures presented in this work. We now consider some qualification of conditions *within* the structure that modify its behavior.

Consider the structure shown in Fig. 14.8*a*. It may be observed that there are four potential components of the reaction forces: A_x, A_y, B_x, and B_y. These are all required for the stability of the structure for the loading shown, and there are thus four unknowns in the investigation of the external forces. Because the loads and reactions constitute a general planar force system, there are three conditions of equilibrium; for example: $\Sigma F_x = 0, \Sigma F_y = 0, \Sigma M_p = 0$). As shown in Fig. 14.8*a*, therefore, the structure is statically indeterminate, not yielding to complete investigation by use of static equilibrium conditions alone.

If the two members of the structure in Fig. 14.8 are connected to each other by a pinned joint, as shown in Fig. 14.8*b*, the number of reaction components is not reduced, and the structure is still stable. However, the internal pin establishes a fourth condition, which may be added to the three equilibrium conditions. There are then four conditions that may be used to find the four reaction components. The method of solution for the reactions of this type of structure is illustrated in the following example.

Example 1.   Find the components of the reactions for the structure shown in Fig. 14.9*a*.
Solution.   It is possible to write four equilibrium equations and to solve them simultaneously for the four unknown forces. However, it is always easier to solve these problems if a few tricks are used to simplify the equations. One trick is to write moment equations about points that

eliminate some of the unknowns, thus reducing the number of unknowns in a single equation. This was used in the finding of beam reactions in Chapter 9. Consider the free body of the entire structure, as shown in Fig. 14.9*b*.

$$\Sigma M_A = 0 = +400(5) + B_x(2) - B_y(24)$$
$$24B_y - 2B_x = 2000 \quad \text{or} \quad 12B_y - B_x = 1000 \quad (1)$$

Now consider the free-body diagram of the right member, as shown in Fig. 14.9*c*.

$$\Sigma M_C = 0 = +B_x(12) - B_y(9)$$

Thus

$$B_x = \frac{9}{12} B_y = 0.75 B_y \quad (2)$$

Substituting equation (2) in equation (1), we obtain

$$12B_y - 0.75B_y = 1000, \quad B_y = \frac{1000}{11.25} = 88.89 \text{ lb}$$

Then, from equation (2),

$$B_x = 0.75B_y = 0.75(88.89) = 66.67 \text{ lb}$$

Refer again to Fig. 14.9*b*:

$$\Sigma F_x = 0 = A_x + B_x - 400$$
$$0 = A_x + 66.67 - 400$$
$$A_x = 333.33 \text{ lb}$$
$$\Sigma F_y = 0 = A_y + B_y = A_y + 88.89, \quad A_y = 88.89 \text{ lb}$$

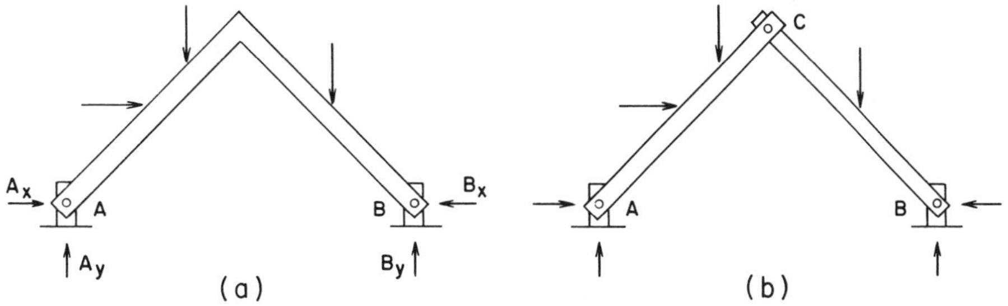

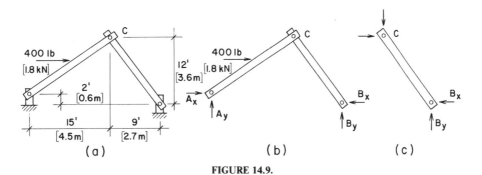

FIGURE 14.8.

FIGURE 14.9.

Note that the condition stated in equation (2) is true in this case because the right member behaves as a two-force member. This is not the case if load is directly applied to the member, and the solution of simultaneous equations would be necessary.

14.4. CONTINUOUS BEAMS WITH INTERNAL PINS

The actions of continuous beams are discussed in Chapter 9. It is observed that a beam such as that shown in Fig. 14.10*a* is statically indeterminate, having a number of reaction components (3) in excess of the conditions of equilibrium for a parallel force system (2). The continuity of such a beam results in the deflected shape and variation of moment as shown beneath the beam in Fig. 14.10*a*. If the beam is made discontinuous at the middle support, as shown in Fig. 14.10*b*, the two spans each behave independently as simple beams, with the deflected shapes and moment as shown.

If a multiple-span beam is made internally discontinuous at some point, its behavior may emulate that of a truly continuous beam. For the beam shown in Fig. 14.10*c* the internal pin is located at the point where the continuous beam inflects. Inflection of the deflected shape is an indication of zero moment, and thus the pin does not actually change the continuous nature of the structure. The deflected shape and moment variation for the beam in Fig. 14.10*c* is therefore the same as for the beam in Fig. 14.10*a*. This is true, of course, only for the single loading pattern that results in the inflection point at the same location as the internal pin.

In the first of the following examples the internal pin is deliberately placed at the point where the beam would inflect if it was continuous. In the second example the pins are placed slightly closer to the support than the location of the natural inflection points. The modification in the second example results in slightly increasing the positive moment in the outer spans, while reducing the negative moments at the supports; thus the values of maximum moment are made closer. If it is desired to use a single-size beam for the entire length, the modification in Example 2 permits design selection of a slightly smaller size.

Example 1. Investigate the beam shown in Fig. 14.11*a*. Find the reactions, draw the shear and moment diagrams, and sketch the deflected shape.
Solution. Because of the internal pin, the first 12 ft of the left-hand span acts as a simple beam. Its two reactions are therefore equal, being one-half the total load, and its shear, moment, and deflected shape diagrams are those for a simple beam with a uniformly distributed load (see Fig. B.1,

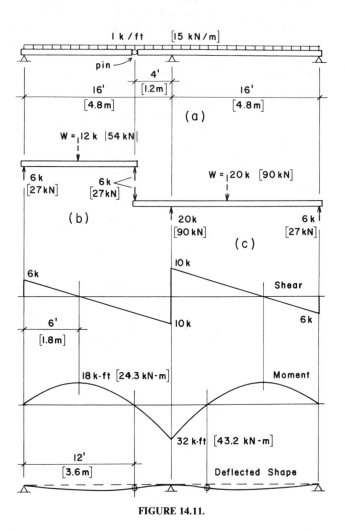

FIGURE 14.11.

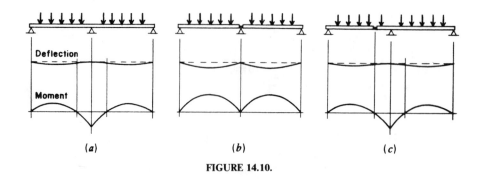

FIGURE 14.10.

Case 2.) As shown in Fig. 14.11*b* and *c*, the simple beam reaction at the right end of the 12-ft portion of the left span becomes a 6-kip concentrated load at the left end of the remainder of the beam. This beam (Fig. 14.11*c*) is then investigated as a beam with one overhanging end, carrying a single concentrated load at the cantilevered end and the total distributed load of 20 kips. (Note that on the diagram we indicate the total uniformly distributed load in the form of a single force, representing its resultant.) The second portion of the beam is statically determinate, and we can proceed to find its reactions.

With the reactions known, the shear diagram can be completed. We note the relation between the point of zero shear in the span and the location of maximum positive moment. For this loading the positive moment curve is symmetrical, and thus the location of the zero moment (and beam inflection) is at twice the distance from the end as the point of zero shear. As noted previously, the pin in this example is located exactly at the inflection point of the continuous beam (for comparison, see Fig. B.2).

Example 2. Investigate the beam shown in Figure 14.12. *Solution.* The procedure is essentially the same as for Example 1. Note that this beam with four supports requires two internal pins to become statically determinate. As

before, the investigation begins with the consideration of the end portion acting as a simple beam. The second step is to consider the center portion as a beam with two overhanging ends.

14.5. FRAMES WITH INTERNAL PINS

There are many situations in which framed structures are built with internal pins. The typical post-and-beam system is essentially of this general form, with all joints being pinned, except for the possibility of some continuous horizontal members. There are other cases, however, in which a mixture of pinned and rigid frame elements may constitute a frame.

Figure 14.13*a* shows a single-story frame in which the middle bay of the frame is constituted as a rigid frame, while the two end bays are completely pinned. This is a common situation that occurs in the development of lateral bracing; in this case the rigid framed bay is used to brace the entire plane of framing against lateral loading. Since the pinned frame construction is usually much less expensive in steel or wood, a savings can thus be made if the single bay is adequate for the bracing.

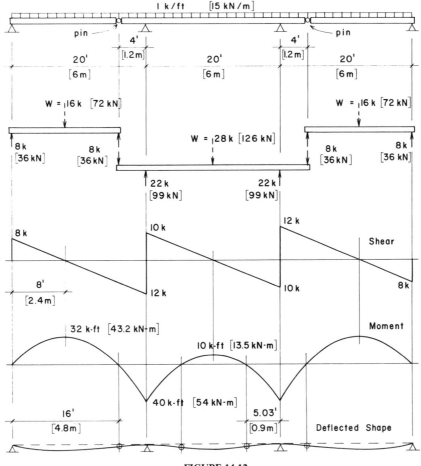

FIGURE 14.12.

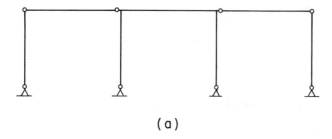

(a)

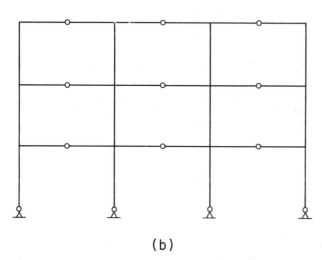

(b)

FIGURE 14.13.

The multistory bent shown in Fig. 14.13b utilizes pins at the midspan points of the beams. This is a form that is used in precast concrete construction, with the multibranched, treelike column units cast in a single piece. In this construction the assembled joint is one that is much simplified if it can function only for shear and direct force. The structure is stable for both gravity and lateral loads and is easy to form and erect.

Another framed structure that sometimes uses an internal pin is the gabled bent. If constituted without the pin, the frame takes the form shown in Fig. 14.7a. However, the frame is often fabricated in two symmetrical parts, of laminated timber, welded steel, or precast concrete. For simplicity of the field connections, the two-part fabrication is usually designed with pinned joints at the gable peak and at the column bases. The following example illustrates a procedure for analysis of such a frame for gravity loading.

Example 1. A gabled frame is composed of two parts with pinned connections, as shown in Fig. 14.14a. Investigate the actions of the frame members for the uniformly distributed gravity load shown.
Solution. For external equilibrium, the frame is of the form illustrated in Fig. 14.8. Using the free body of half of the symmetrical frame, shown in Fig. 14.14b, an equation for moment about the column base will yield a value for the horizontal force component; thus

$$H = \frac{32(16)}{32} = 16 \text{ kips}$$

With this value determined, the two elements of the half frame can now be dealt with as two spans of a continuous beam, as shown in Fig. 14.14c. Although this beam is technically statically indeterminate, the left end reaction (actually the horizontal support reaction component for the frame) is a known value. Thus the other two support reactions for the analogous two-span beam can be found by statics. Note that the load on the two-span beam is a modified value; actually, it is the component of the vertical load that is perpendicular to the sloping member of the frame. With the reactions determined, the shear and moment diagrams can be completed.

Both members of the frame are subjected to combinations of bending, shear, and axial compression, and their designs would need to include consideration for the combined effects. The column is subjected to the vertical support reaction force of 32 kips and the sloping member sustains a compression due to the component of the vertical load that is parallel to the member.

A complete investigation for design of the frame in the preceding example would need to include consideration for lateral load effects, although the usual allowable design

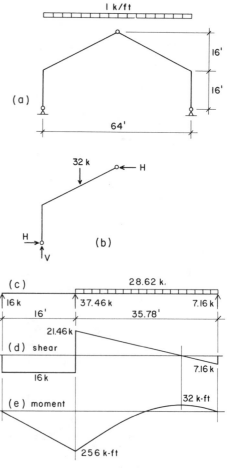

FIGURE 14.14.

stress increase (or lower load factors) may result in the combined gravity plus lateral loading not being a critical concern. The effects of wind and seismic loads are discussed in Part 8. Although somewhat more complex due to the unsymmetrical response of the frame, analysis of this example frame for lateral loads is also statically determinate.

14.6. TWO-WAY SPANNING STRUCTURES

Spanning structures often consist of assemblages of single, linear elements (trusses, beams, decks) that have some interaction in working with other elements to define a surface (roof or floor), but have individual structural actions that are essentially simple. However, the surface may also be defined by a system that has a more complex behavior if the structural functioning of individual elements is essentially interdependent. In the latter type of system it is not possible to consider the action of a single element without considering the action of the entire interacting system. Some structures that have this character are the following.

Two-Way Spanning Trusses. Trusses may function as linear elements in a system in which trusses spanning in one direction provide simple support for trusses spanning at right angles to them. However, such a system may also be developed as a two-way spanning system if the truss configurations, jointing, and support conditions are developed to achieve such action. An example of a two-way spanning truss system is shown in Fig. 5.14.

Cable Nets. Spanning cables may function individually or in simple tandem sets, such as the cables of a suspension bridge. However, they may also be used to define a surface in the form of laced sets of intersecting cables, draped in suspended form, or arched to restrain a pneumatically developed surface.

Concrete Systems. The essentially nonlinear nature of cast concrete can be exploited to produce two-way spanning solid slabs or intersecting beam systems. Continuity of both sets of intersecting beams is a natural feature of a cast system, although it is not generally feasible in assemblages of separate linear elements of steel or wood.

The two-way spanning structure is in general quite indeterminate, and its investigation is very complex. Inves-

tigation methods and computer-aided techniques exist and are frequently used, but design is considerably more laborious than that for simpler systems. Approximation techniques may be used for early stages of design in order to produce a model that is reasonable to use in the more exact and time-consuming investigations.

An advantage of the two-way spanning structure is the increased efficiency resulting from the mutually supporting nature of the member interactions. If the problems of more costly structural investigation and design and usually more complex construction can be overcome, these systems may actually result in reduced cost for the structure.

In planning of structures that have two-way spanning actions, if optimal utilization is to occur, some attention should be given to the nature of the system. A major consideration is that of the supports for the system. If the system is used to define a single square plan, there are several options for the location of supports. While architectural considerations must be made, the effects on the structure should be considered if the system is to achieve some reasonable efficiency. In Fig. 14.15a a square system is shown with supports at the four corneers. While the interior portion of the spanning system will function as a two-way spanning structure, the edges must function essentially as linear elements spanning directly between the supports and providing support for the interior system. The edges will therefore be quite heavy.

Figure 14.15b and c show supports for a square system that eliminate the heavy edge structure, replacing it with a bearing wall or closely spaced columns. This is better for the spanning structure if the supports are acceptable. Figure 14.15d is a variation on the corner-supported scheme in Fig. 14.15a, resulting in a reduction of the span effect of the interior portion, but requiring relatively heavy edge cantilevers to achieve the corners.

An optimal situation for the spanning structure is shown in Fig. 14.15e, in which the four corner columns are pulled in to allow the spanning structure to project as a cantilever on all sides. This system may be further improved if some widening of the effective perimeter is provided at the columns. The widened column reduces the punching shear effect and slightly reduces the span effect (see Fig. 5.14).

For a single unit of a two-way spanning system, the plan should describe a square as closely as possible. If the system has the same stiffness in both directions, this will normally result in equal sharing of the two-way action. As the plan becomes oblong rather than square (see Fig. 14.16), the increased stiffness of the shorter span tends to reduce the contribution of the span effect in the longer direction.

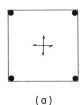 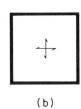  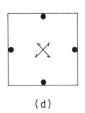 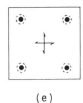

(a) (b) (c) (d) (e)

FIGURE 14.5. Alternative supports for a single two-way spanning truss.

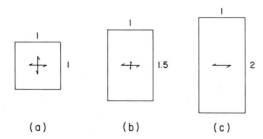

FIGURE 14.16. Effect of disymmetry on a two-way spanning element.

At a ratio of spans of as much as 1.5:1, the short span will effectively carry 80% or more of the load, making the two-way action of questionable use. If the ratio is as high as 2:1, the effect in the long direction is negligible.

Some additional considerations of two-way spanning concrete systems are discussed in Chapter 30.

14.7. THE VIERENDEEL TRUSS

The Vierendeel truss is not really a truss at all. It is essentially a rigid frame. In fact, the typical rigid-frame building structure designed for lateral loads on a multistory building is a sort of cantilevered Vierendeel truss (see Fig. 14.17a). The term is usually applied, however, to the spanning structure, such as that shown in Fig. 14.17b. What characterizes the Vierendeel truss is the absence of the diagonal members that produce the necessary triangulation that provides what we normally understand as truss action. Stability of the Vierendeel truss requires the development of major shear and bending effects in the truss members and the transfer of moments through the truss joints.

The major advantage of the Vierendeel truss is its ability to accommodate penetration without the interference represented by the usual truss diagonal members. This is a significant advantage to architectural planning in many situations, permitting placement of windows, doors, or corridors. In some circumstances these advantages may be worth the premium cost represented by the much heavier construction required for the Vierendeel truss.

The Vierendeel truss is statically indeterminate, and investigation of structural behavior is essentially similar to that required for ordinary rigid frames. In some cases a simplified investigation may be performed using some of the same techniques employed for approximate investigation of rigid frames. As with all rigid frames, behavior is affected by the relative stiffness of the frame members, as well as by the general form of the frame layout.

The Vierendeel principle is sometimes used for structures that are actually of a composite nature—blending the characteristics of more than one type of system. In the truss shown in Fig. 14.17c, for example, triangulation is provided for all but the center panel of the truss. If the truss is used to span across a building, the open space in the center panel may accommodate a corridor. If the truss is formed as shown in Fig. 14.17d, it will behave essentially as a simple truss at its ends (offering the advantages of simpler and

lighter construction) and as a Vierendeel only in the center portion.

Another example of a composite structure is that shown in Fig. 14.17e, in which a rigid-frame tower is used to support an ordinary spanning truss. The combined actions of such a structure for both gravity and lateral loadings may be quite complex, but the use of computer-aided design methods makes investigation quite feasible.

Investigation and design of Vierendeel trusses requires a background preparation in the general analysis of the indeterminate structure, a subject not developed in depth in this book.

14.8. THREE-DIMENSIONAL FRAMES

Buildings are generally three-dimensional, but are most often provided with structures that do not have a true three-dimensional character. That is, they may have parts that are arranged in a three-dimensional form, but the nature of individual parts is linear or planar for the purpose of consideration of structural behavior. However, this is not always true. Two particular types of structures with essen-

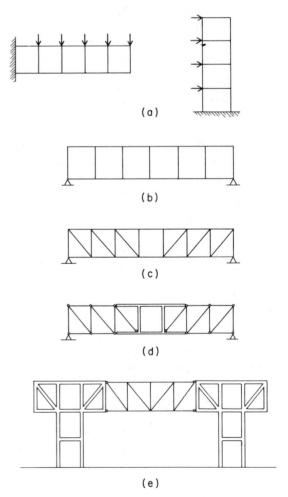

FIGURE 14.17. Vierendeel trusses.

tial three-dimensional character are the trussed tower and the three-dimensional rigid frame.

Trussed towers occur most often as open frames, used for signs, broadcast transmission, large power lines, or support towers for suspension bridges. The tower legs essentially constitute the vertical supports and serve major functions for overturn forces. Trussing is usually essentially effective for lateral load resistance and for reduction of slenderness of the individual legs. For resistance of gravity loads the tower legs may function primarily as simple vertical columns. For lateral loads, however, it is necessary to consider the whole frame. If the tower is square in plan, it may be considered to consist of individual vertical planar bents. If it is triangular in plan, or is subjected to torsional effects, it must be considered in its full three-dimensional form.

For buildings, trussed towers are often used as the stiffened elements that provide lateral bracing for the rest of the building. Towers of this type most often occur at the building "core," where the trussing may be incorporated into permanent walls around elevators and stairs. However, the towers may also occur at the building corners or at other locations where placement is both architecturally acceptable and structurally of strategic significance. Although it is possible to form bracing towers as simple triangulated trusses, it is becoming increasingly frequent to use a trussing arrangement that combines truss actions

with rigid-frame actions, so that plastic deformations of the frame can be used for their energy-absorbing contributions (see the discussion of eccentric bracing in Chapter 47).

When rigid-frame bracing is used for resistance to lateral loads in multistory buildings, it is usually arranged in sets that describe vertical planar bents, consisting of all of the columns in one row together with the beams that connect them at the various levels. These bents may actually occur in somewhat detached form in some cases, but when *all* of the columns and their connecting beams are so constituted, the net effect is that of a truly three-dimensional frame.

In the three-dimensional frame, the beams will tend to respond essentially in planar bent behavior; the columns, however, will be generally subjected to two-way bending and shear actions. The most notable unique action of the three-dimensional frame occurs when the frame is subjected to torsion (twisting in a horizontal plane). This is actually the general form of response under lateral load because perfect symmetry of the construction and application of the load is not really achievable. It is quite essential that the behavior of the fully three-dimensional rigid frame be studied for this form of response. The effects of such action are especially critical on the columns at greatest distance from the center of the plan, an effect not evident in investigations of simple planar bent actions.

CHAPTER FIFTEEN

Aspects of Dynamic Behavior

A good lab course in physics should provide a reasonable understanding of the basic ideas and relationships involved in dynamic behavior. A better preparation is a course in engineering dynamics that focuses on the topics in applied fashion, dealing directly with their applications in various engineering problems. The material in this section consists of a brief summary of basic concepts in dynamics that will be useful to those with a limited background and that will serve as a refresher for those who have studied the topics before.

The general field of dynamics may be divided into the areas of *kinetics* and *kinematics*. Kinematics deals exclusively with motion, that is, with time–displacement relationships and the geometry of movements. Kinetics adds the consideration of the forces that produce or resist motion.

15.1. KINEMATICS

Motion can be visualized in terms of a moving point, or in terms of the motion of a related set of points that constitute a body. The motion can be qualified geometrically and quantified dimensionally. In Fig. 15.1a the point is seen to move along a path (its geometric character) a particular distance. The distance traveled by the point between any two separate locations on its path is called *displacement*. The idea of motion is that this displacement occurs over time, and the general mathematical expression for the time–displacement function is

$$s = f(t)$$

Velocity is defined as the rate of change of the displacement with respect to time. As an instantaneous value, the velocity is expressed as the ratio of an increment of displacement (ds) divided by the increment of time (dt) elapsed during the displacement. Using the calculus, the velocity is thus defined as

$$v = \frac{ds}{dt}$$

That is, the velocity is the first derivative of the displacement.

If the displacement occurs at a constant rate with respect to time, it is said to have *constant velocity*. In this case the velocity may be expressed more simply without the calculus as

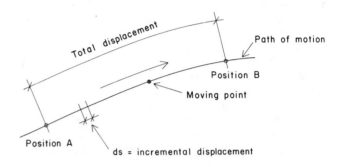

(a) Motion of a Point

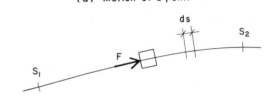

$$\text{Work} = \int_{S_1}^{S_2} F_t \ ds \quad \text{(F variable with time)}$$

$$= F(S_2 - S_1) \quad \text{(F constant with time)}$$

(b) Kinetics of a Moving Object

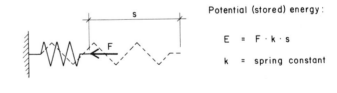

Potential (stored) energy:

$$E = F \cdot k \cdot s$$

$$k = \text{spring constant}$$

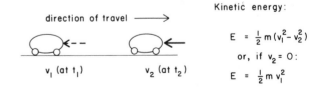

Kinetic energy:

$$E = \frac{1}{2} m (v_1^2 - v_2^2)$$

or, if $v_2 = 0$:

$$E = \frac{1}{2} m v_1^2$$

(c) Forms of Mechanical Energy

FIGURE 15.1. Moment of a point.

$$v = \frac{\text{total displacement}}{\text{total elapsed time}}$$

When the velocity changes over time, its rate of change is called the *acceleration (a)*. Thus, as an instantaneous change

$$a = \frac{dv}{dt} = \frac{d^2s}{dt^2}$$

That is, the acceleration is the first derivative of the velocity or the second derivative of the displacement with respect to time.

Except for the simplest cases, the derivation of the equations of motion for an object generally require the use of the calculus in the operation of these basic relationships. Once derived, however, motion equations are generally in algebraic form and can be used without the calculus for application to problems. An example is the set of equations that describe the motion of a free-falling object acted on by the earth's gravity field. Under idealized conditions (ignoring air friction, etc.) the distance of fall from a rest position will be

$$s = f(t) = 16.1t^2 \qquad (s \text{ in ft}, t \text{ in sec})$$

This equation indicates that the rate of fall (the velocity) is not a constant but increases with the elapsed time, so that the velocity at any instant of time may be expressed as

$$v = \frac{ds}{dt} = \frac{d(16.1t^2)}{dt} = 32.2t \qquad (v \text{ in ft/sec})$$

and the acceleration as

$$a = \frac{dv}{dt} = \frac{d(32.2t)}{dt} = 32.2 \text{ ft/sec}^2$$

which is the acceleration of gravity.

The following examples illustrate some additional displacement–velocity–acceleration relationships. We first consider the simplest case: a straight-line motion at constant velocity.

Example 1. A point moves along a straight line described by $y = \frac{3}{4}x$ at a constant velocity of 10 ft/sec. Express the relation of displacement along the line with time and also of displacement in the *x* direction with time. What is the *x*-direction displacement in 4 sec?

Solution. The relation of displacement along the line (Fig. 15.2) is expressed as

$$s = 10t$$

and the velocity along the line as

$$v = \frac{ds}{dt} = \frac{d(10t)}{dt} = 10 \text{ ft/sec}$$

which simply confirms the relationship, as the velocity was given.

The relation of displacement in the *x* direction is found by first noting that

$$s = \sqrt{x^2 + y^2} = \sqrt{x^2 + (\tfrac{3}{4}x)^2} = \sqrt{\tfrac{25}{16}x^2} = \tfrac{5}{4}x$$

Then

$$s = \tfrac{5}{4}x = 10t$$

and

$$x = \tfrac{4}{5}(10t) = 8t$$

and in 4 sec

$$x = 8(4) = 32 \text{ ft}$$

We next consider an example in which the velocity is a function of time.

Example 2. Data similar to Example 1, except that the velocity varies from zero when *s* is zero and increases with time such that $v = 2t$.
Solution. A change in velocity implies an acceleration, thus

$$a = \frac{dv}{dt} = \frac{d(2t)}{dt} = 2 \text{ ft/sec}^2$$

For the displacement, we note that

$$v = \frac{ds}{dt}$$

and thus

$$s = \int v \, dt = \int (2t) \, dt = t^2$$

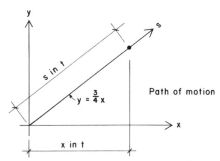

Path of motion

s (in ft)

s = 10t

40

32

x = 8t

t (in sec.)

4

Displacement – time graph

FIGURE 15.2.

Then, as before,

$$s = \tfrac{5}{4} x = t^2, \quad x = \tfrac{4}{5} t^2$$

and in 4 sec

$$x = \tfrac{4}{5} (4)^2 = 12.8 \text{ ft}$$

Example 3. A car accelerates from rest with a constant acceleration to a speed (velocity) of 60 mph in 30 sec (1/120 hour). Derive the expressions for s, v, and a and find the distance traveled.

Solution.   Using the given value for v at the time of the elapsed 30 sec, we note that

$$a = C, \quad v = a\,dt = \int_0^{1/120} C\,dt = 60$$

from which

$$(1/120)C = 60, \quad C = 7200 \text{ miles/hr}^2$$

Then

$$a = 7200 \text{ miles/hr}^2$$

$$v = \int_0^t a\,dt = \int_0^t 7200\,dt = 7200t$$

$$s = \int_0^t v\,dt = \int_0^t (7200t)\,dt = 3600t^2$$

and in the elapsed 1/120 hour,

$$s = 3600(1/120)^2 = 0.25 \text{ mile}$$

15.2. MOTION

A major aspect of consideration in dynamics is the nature of motion. While building structures are not really supposed to move (like machine parts), their responses to force actions involve consideration of motions. These motions may actually occur in the form of very small deformations, or merely be the failure response that the designer must visualize resisting. The following are some basic forms of motion.

Translation.   This occurs when an object moves in simple linear displacement, with the displacement measured as simple change of distance from some reference point.

Rotation.   This occurs when the motion can be measured in the form of angular displacement; that is, in the form of revolving about a fixed reference point.

Rigid-Body Motion.   A rigid body is one in which no internal deformation occurs and all particles of the body remain in fixed relation to each other. Three types of motion of such a body are possible. Translation occurs when all the particles of the body move in the same direction at the same time. Rotation occurs when all points in the body

describe circular paths about some common fixed line in space, called the *axis of rotation*. Plane motion occurs when all the points in the body move in planes that are parallel. Motion within the planes may be any combination of translation or rotation.

Motion of Deformable Bodies.   In this case motion occurs for the body as a whole, as well as for the particles of the body with respect to each other. This is generally of more complex form than rigid-body motion, although it may be broken down into simpler component motions in many cases. This is the nature of motion of fluids and of elastic solids. The deformation of elastic structures under load is of this form, involving both the movement of elements from their original positions and changes in their shapes.

15.3. KINETICS

As stated previously, kinetics includes the additional consideration of the forces that cause motion. This means that in addition to the variables of displacement and time, we must consider the mass of the moving objects. From Newtonian physics the simple definition of mechanical force is

$$F = ma = \text{mass} \times \text{acceleration}$$

Mass is the measure of the property of inertia, which is what causes an object to resist change in its state of motion. The more common term for dealing with mass is *weight*, which is a force defined as

$$W = mg$$

where g is the constant acceleration of gravity (32.2 ft/ sec^2).

Weight is literally a dynamic force, although it is the standard means of measurement of force in statics, when the velocity is assumed to be zero. Thus in static analysis we express forces simply as

$$F = W$$

and in dynamic analysis, when using weight as the measure of mass, we express force as

$$F = ma = \frac{W}{g} a$$

15.4. WORK, POWER, ENERGY, AND MOMENTUM

If a force moves an object, work is done. *Work* is defined as the product of the force multiplied by the displacement (distance traveled). If the force is constant during the displacement, work may be simply expressed as

$$w = Fs = \text{force} \times \text{total distance traveled}$$

If the force varies with time, the relationship is more generally expressed with the calculus as

$$w = \int_{s_2}^{s_1} F_t \, ds$$

indicating that the displacement is from position s_1 to position s_2, and the force varies in some manner with respect to time.

Figure 15.1*b* illustrates these basic relationships. In dynamic analysis of structures the dynamic "load" is often translated into work units in which the distance traveled is actually the deformation of the structure.

Energy may be defined as the capacity to do work. Energy exists in various forms: heat, mechanical, chemical, and so on. For structural analysis the concern is with mechanical energy, which occurs in one of two forms. *Potential energy* is stored energy, such as that in a compressed spring or an elevated weight. Work is done when the spring is released or the weight is dropped. *Kinetic energy* is possessed by bodies in motion; work is required to change their state of motion, that is, to slow them down or speed them up (see Fig. 15.1*c*).

In structural analysis energy is considered to be indestructible, that is, it cannot be destroyed, although it can be transferred or transformed. The potential energy in the compressed spring can be transferred into kinetic energy if the spring is used to propel an object. In a steam engine the chemical energy in the fuel is transformed into heat and then into pressure of the steam and finally into mechanical energy delivered as the engine's output.

An essential idea is that of the conservation of energy, which is a statement of its indestructibility in terms of input and output. This idea can be stated in terms of work by saying that the work done on an object is totally used and that it should therefore be equal to the work accomplished plus any losses due to heat, air friction, and so on. In structural analysis we make use of this concept by using a "work equilibrium" relationship similar to the static force equilibrium relationship. Just as all the forces must be in balance for static equilibrium, so the work input must equal the work output (plus losses) for "work equilibrium."

15.5. HARMONIC MOTION

A special problem of major concern in structural analysis for dynamic effects is that of *harmonic motion*. The two elements generally used to illustrate this type of motion are the swinging pendulum and the bouncing spring. Both the pendulum and the spring have a neutral position where they will remain at rest in static equilibrium. If one displaces either of them from this neutral position, by pulling the pendulum sideways or compressing or stretching the spring, they will tend to move back to the neutral position. Instead of stopping at the neutral position, however, they will be carried past it by their momentum to a position of displacement in the opposite direction. This sets up a cyclic form of motion (swinging of the pendulum; bouncing of the spring) that has some basic characteristics.

Figure 15.3 illustrates the typical motion of a bouncing spring. Using the calculus and the basic motion and force

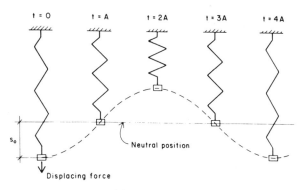

(a) The Moving Spring

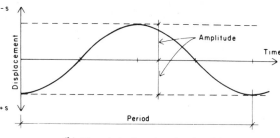

(b) Plot of the Equation: S = A cos BT

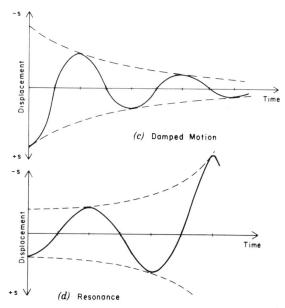

(c) Damped Motion

(d) Resonance

FIGURE 15.3. Kinetics of a moving object.

equations, the displacement—time relationship may be derived as

$$s = A \cos Bt$$

The cosine function produces the basic form of the graph, as shown in Fig. 15.3*b*. The maximum displacement from the neutral position is called the *amplitude*. The time elapsed for one full cycle is called the *period*. The number of full cycles in a given unit of time is called the *frequency* (usually expressed in cycles per second) and is equal to the inverse of the period. Every object subject to harmonic motion has a fundamental period (also called natural period), which is determined by its weight, stiffness, size, and so on.

Any influence that tends to reduce the amplitude in successive cycles is called a *damping effect*. Heat loss in friction, air resistance, and so on are natural damping effects. Shock absorbers, counterbalances, cushioning materials, and other devices can also be used to damp the amplitude. Figure 15.3c shows the form of a damped harmonic motion, which is the normal form of most such motions, because perpetual motion is not possible without a continuous reapplication of the original displacing force.

Resonance is the effect produced when the displacing effort is itself harmonic with a cyclic nature that corresponds with the period of the impelled object. An example is someone bouncing on a diving board in rhythm with the board's fundamental period, thus causing a reinforcement, or amplification, of the board's free motion. This form of motion is illustrated in Fig. 15.3d. Unrestrained resonance effects can result in intolerable amplitudes, producing destruction or damage of the moving object or its supports. A balance of damping and resonant effects can sometimes produce a constant motion with a flat profile of the amplitude peaks.

Loaded structures tend to act like springs. Within the elastic stress range of the materials, they can be displaced from a neutral (unloaded) position and, when released, will go into a form of harmonic motion. The fundamental period of the structure as a whole, as well as the periods of its parts, are major properties that affect responses to dynamic loads.

15.6.   DYNAMIC EFFECTS IN BUILDINGS

Load sources that involve motion, such as wind, earthquakes, walking people, moving vehicles, and vibrating heavy machinery, have the potential to cause dynamic effects on structures. Analyzing for their effects requires consideration of essential dynamic properties of the structure. These properties are determined by the size, weight, relative stiffness, fundamental period, type of support, and degree of elasticity of the materials of the structure and by various damping influences that may be present.

Dynamic load sources deliver an energy load to the structure that may be in the form of an impact, such as that caused by the moving air bumping into the stationary building. In this case the energy load is derived from the kinetic energy of the moving air, which is a product of its mass and velocity. In the case of an earthquake, or the vibration of heavy machinery, the load source is not a force as such but, rather, something that induces motion of the structure, in which case the mass of the building is actually the load source.

An important point to note is that the effects of a dynamic load on a structure are determined by the structure's response as well as by the nature of the load. Thus the same dynamic load can produce different effects in different structures. Two buildings standing side by side can have significantly different responses to the same earthquake shock if they have major differences in their dynamic properties.

Dynamic effects on structures may be of several types. Some of the principal effects are the following.

Total energy load is the balance between the peak magnitude of the load and the maximum work required by the structure and is known as the *work equilibrium concept.* Work done to the structure by the load equals the work done by the structure in resisting the load.

Unstabilizing effects occur if the dynamic load produces a stability failure of the structure. Thus a free-standing wall may topple over, an unbraced post and beam system may collapse sideways, and so on, because of the combined effects of gravity and the dynamic load.

Harmonic effects of various types may be set up in the structure, especially if the load source is cyclic in nature, such as the footsteps of marching troops. Earthquake motions are basically cyclic, in the form of vibration or shaking of the surface of the ground. Relations between these motions and the harmonic properties of the structure can result in various effects, such as the flutter of objects at a particular wind velocity, the resonant bouncing of floors, and the resonant reinforcing of the swaying of buildings during an earthquake.

Using up of the structure's energy capacity can occur if the energy of the load exceeds the limit for the structure. Actually, there are several degrees, or stages, of energy capacity, instead of a single limit.

Four significant stages are the following:

1. *The resilience limit*, or the limit beyond which some form of permanent damage—however slight—may occur.
2. *The minor damage limit*, the damage being either relatively insignificant or easily repairable.
3. *The major damage limit*, short of total destruction, but with loss of some minor elements. The structure as a whole remains intact, but some major repairs may be required to restore it to its original level of capacity.
4. *The toughness limit*, or the maximum, ultimate capacity represented by the destruction of the structure.

Failure under repeated loadings can result in some cases when structures progressively use up their dynamic resistance. The structures may successfully resist a single peak load of some dynamic effort, only to fail later under a similar, or even smaller, loading. This failure is usually due to the fact that the first loading used up some degree of structural failure, such as ductile yielding or brittle cracking, which absorbed enough energy to prevent total failure but was a one-time usable strength only.

A major consideration in design for dynamic loads is what the response of the structure means to the building as a whole. Thus, although the structure may remain intact, that may be only a minor accomplishment if there is significant damage to the building as a whole. A high-rise building may swing and sway in an earthquake without there being any significant damage to the structure, but if the occupants are tossed about, the ceilings fall, the windows shatter, the partitions and curtain walls collapse, the plumbing bursts, and the elevators derail, it can hardly be said that the building was adequately designed.

In many cases analysis and design for dynamic effects are not done by working directly with the dynamic relationships, but simply by using recommendations and rules of thumb that have been established by experience. Some testing or theoretical analysis may have helped in deriving ideas or data, but much of what is used is based on the observations and records from previous disasters. Even when actual calculations are performed, they are mostly done with data and relationships that have been translated into simpler static terms—so-called equivalent static analysis and design. The reasons for this practice have to do frankly with the degree of complexity of dynamic analysis. Even with the use of programmable calculators or computers, the work is quite laborious in all but the simplest of situations.

15.7. EQUIVALENT STATIC EFFECTS

Use of equivalent static effects essentially permits simpler analysis and design by eliminating the complex procedures of dynamic analysis. To make this possible the load effects and the structure's responses must be translated into static terms.

For wind load the primary translation consists of converting the kinetic energy of the wind into an equivalent static pressure, which is then treated in a manner similar to that for a distributed gravity load. Additional consid-erations are made for various aerodynamic effects, such as ground surface drag, building shape, and suction, but these do not change the basic static nature of the work.

For earthquake effects the primary translation consists of establishing a hypothetical horizontal static force that is applied to the structure to simulate the effects of sideward motions during ground movements. This force is calculated as some percentage of the dead weight of the building, which is the actual source of the kinetic energy loading once the building is in motion—just as the weight of the pendulum and the spring keep them moving after the initial displacement and release. The specific percentage used is determined by a number of factors, including some of the dynamic response characteristics of the structure.

An apparently lower safety factor is used when designing for the effects of wind and earthquake because an increase of one-third is permitted in allowable stresses. This is actually not a matter of a less-safe design but is merely a way of compensating for the fact that one is actually adding static (gravity) effects and *equivalent* static effects. The total stresses thus calculated are really quite hypothetical because in reality one is adding static strength effects to dynamic strength effects, in which case 2 + 2 does not necessarily make 4.

Regardless of the number of modifying factors and translations, there are some limits to the ability of an equivalent static analysis to account for dynamic behavior. Many effects of damping and resonance cannot be accounted for. The true energy capacity of the structure cannot be accurately measured in terms of the magnitudes of stresses and strains. There are some situations, therefore, in which a true dynamic analysis is desirable, whether it is performed by mathematics or by physical testing. These situations are actually quite rare, however. The vast majority of building designs present situations for which a great deal of experience exists. This experience permits generalizations on most occasions that the potential dynamic effects are really insignificant or that they will be adequately accounted for by design for gravity alone or with use of the equivalent static techniques.

PART 3

WOOD STRUCTURES

On the basis of cost and availability, wood is by far the most popular choice for structures for buildings in the United States. Unless usage requirements prevent its application, it is almost always the first-choice material. This is less so in other countries where the forest resources are less extensive, the lumber industry less developed, or the use of wood as a fuel source more prevalent. The material in this part presents considerations involving the use of wood as a structural material in the various forms of products produced by the wood industry.

CHAPTER SIXTEEN

General Concerns for Wood

Wood is both an ancient and modern material. In many places wood is still obtained directly from standing trees and worked by hand with methods that are thousands of years old. For building structures in the United States today, however, we know wood primarily in the form of industrial products. Chapter 16 deals with some basic properties and characteristics of wood and with the considerations of usage of the various wood products and accessory devices utilized to produce wood structures for buildings.

16.1. TYPES OF STRUCTURAL PRODUCTS

Wood, primarily from softwood or coniferous trees, is used for a wide range of products. The principal ones used for building structures are the following.

Solid-Sawn Lumber. These are the boards, planks, and timbers cut directly from logs and finished to produce a number of standard sizes.

Glued-Laminated Products. This includes a number of elements, major ones being plywood sheets and large beams laminated from boards and planks.

Reconstituted Fiber-Based Products. These range from paper to structural hardboard and consist of sheet-form materials produced by compressing and bonding wood fibers.

Built-Up and Fabricated Products. These are products produced by various manufacturers, including prefabricated trusses and composite beams of plywood and lumber.

Most of the material in this chapter deals with the use of solid sawn lumber. Fabricated products, including plywood and glued laminated beams, are largely "designed" by their producers with properties and code-accepted ratings preestablished. Designing structures with fabricated products begins with obtaining design data and recommendations from manufacturers or from the industry-wide organizations that set standards and provide coordinated references.

16.2. STRUCTURAL LUMBER

Structural lumber consists of standard-size solid-sawn wood pieces that are identified by various characteristics so that properties can be established for design use. The identification of a particular individual piece of lumber ordinarily includes considerations of its species, grade, usage, size, and moisture content. On the basis of these considerations, design values for allowable stresses and modulus of elasticity are defined by codes or industry standards.

Species refers to the type of tree from which the wood is obtained. Although many types of trees could be used—and are for special purposes—most of the lumber used in the United States comes from a few trees. Two of the principal sources are Douglas fir–larch and southern pine. Douglas fir–larch comes largely from stands of trees in the Pacific northwest and southern pine comes largely from the southeast. Lumber from both of these sources is shipped throughout the country, although each is dominant in its own region. Other species are also competitive in the regions where the trees grow.

Grade refers to named classifications (select, No. 1, utility, etc.) based on inspection of the integrity and density of the wood, the nature of the grain pattern, and the presence or absence of flaws. Grading rules are established by industry-wide standards and lumber that is graded is permanently marked to indicate the species, grade, and standards used in the grading.

Size and usage considerations are incorporated with grading to further identify a piece of lumber for a particular task. The smaller a piece of lumber, the more it may be affected by a knot or other flaw, but the more likely it may be that a reasonably flawless piece can actually be obtained. In extracting a reasonable amount of lumber from a log, the likelihood of obtaining a flawless piece of large size is very low. Therefore, the highest grades are generally obtainable only in the thinner and overall smaller sizes.

The size issue just raised also applies to moisture content. When wood is rushed fairly quickly from the felled tree to the lumberyard, the amount of original (living tree) moisture retained will generally be in proportion to the thickness of the piece. For this reason, it is assumed that large timbers (generally 6 in. or more in thickness) will be used in what is called a *green condition*. As the wood tends to stiffen up as it dries, both its strength in stress resistance and its stiffness in deformation resistance (modulus of elasticity) will be higher the more it cures from the green condition.

Industry standards, building codes, and general local practices in design and construction will usually combine to establish a common match between various construction uses and particular rated lumber. Thus the lumber used commonly for wall studs, wall sills, roof rafters, floor joists, and other typical construction elements is mostly in conformance with accepted rules of practice. In some cases this relates to specific structural tasks, but it is mostly a matter of experience deriving from the attempt to use the lowest grades (of lowest cost) that will do the job.

Structural lumber is produced in a range of standard sizes. Size is specified as the two dimensions *b* (thickness) and *d* (width) in full-inch increments. This is called the *nominal size*, because the actual dimensions are somewhat smaller. This practice goes back to earlier times when lumber was commonly produced as either rough sawn or finished (sanded smooth), and the even-inch dimensions were the rough-sawn ones. Sanding slightly reduced the surface faces, so some dimension was lost. Now, almost all lumber is finished (called dressed) but the size custom remains. Current standard sizes of structural lumber are given in Table A.2, together with various properties of the sections. Smaller sizes are quite widely available, but larger timbers are somewhat more difficult to obtain.

16.3.   DATA FOR STRUCTURAL DESIGN

Design Values for Structural Lumber

Recommended limits for structural properties of solid-sawn structural lumber, called design values, are established by building codes or by industry standards. Table 16.1 is an adaptation of a portion of a document prepared by the National Forest Products Association (Ref. 7). To obtain values from the table the following data must be determined for a particular piece of lumber.

1. *Species.* Table 16.1 shows only the values for the wood that comes from Douglas fir trees that grow in northwestern United States and western Canada, commonly called *Douglas fir-larch* or *Douglas fir north.*

2. *Moisture Content at Time of Construction (Use).* The table is based on a relatively high moisture content of 19%. Table footnotes (given in the reference document) allow for adjustments if the wood is in a better-cured (dried) condition or is used in a situation where it is exposed to the weather.

3. *Grade.* This is indicated in the first column of the table as dense select structural, No. 1, and so on.

4. *Size or Use.* The second column of the table identifies size ranges or usage (beams and stringers, for

example) of the lumber. Thus the grade "select structural" appears four times in the table for various sizes and uses.

The document from which Table 16.1 is adapted contains values for some 30 different species of wood. There are 19 footnotes to the table, extending over two full pages. Data from Table 16.1 will be used in example problems in this book and some of the issues treated in the document footnotes will be explained. The document (Ref. 7) is a standard reference and working designers should be aware of its full contents.

Example 1.   A wood beam used in the interior of a building consists of a 6 × 12 of Douglas fir–larch, No. 1 grade. What are the design values for bending, shear, and axial tension?
Solution.   Scanning the second column of Table 16.1, we note that this member falls in the classification "beams and stringers." The values in the table for the No. 1 grade are thus: F_b = 1300 psi (bending), F_v = 85 psi (shear), and F_t = 675 psi (tension).

Example 2.   Wood floor joists, to be used at 16 in. on center, are to be of Douglas fir–larch, No. 2 grade. What are the design values for the bending stress and the modulus of elasticity? Joist size is 2 × 10.
Solution.   In Table 16.1 this member falls in the size classification described as "2 in. to 4 in. thick, 5 in. and wider." For this group it is noted that two values are given for the bending stress limit. In this case the joist falls under the classification of "repetitive member uses," and the allowable bending stress is thus 1450 psi. This classification is discussed later in this section. The table value for the modulus of elasticity is 1700 ksi.

Obtaining the correct design values from the table is only the first step in determining the limiting values to be used in a design computation. In many cases there are modifications to the table values, as described in the following discussion.

It should be noted here that the data in Table 16.1 and the data used in general for the work in this book is taken from an earlier edition of the referenced document: the *National Design Specification* (NDS). The most recent edition contains some increases in design values, reflecting research and experience in the wood industry. However—for many reasons—the building codes and various agencies are slow to accept these changes, and other referenced materials are still based on the lower values in former editions of the specification. Use of any materials from this book for actual design work—as in all cases—should be evaluated for specific code requirements applicable to the work.

Modified Design Values

Basic allowable values as given in Table 16.1 may be used without modification in some instances. However, in many situations there are additional considerations that require

TABLE 16.1. Design Values for Structural Lumber of Douglas Fir–Larch[a]

		Design Values (psi)						
		Extreme Fiber in Bending. F_b		Tension Parallel to Grain, F_t	Horizontal Shear, F_v	Compression Perpendicular to Grain, $F_{c\perp}$	Compression Parallel to Grain, F_c	Modulus of Elasticity, E
Species and Commercial Grade	Size Classification	Single Member Uses	Repetitive Member Uses					
Dense select structural	2 to 4 in. thick.	2450	2800	1400	95	730	1850	1,900,000
Select structural	2 to 4 in. wide	2100	2400	1200	95	625	1600	1,800,000
Dense No. 1		2050	2400	1200	95	730	1450	1,900,000
No. 1		1750	2050	1050	95	625	1250	1,800,000
Dense No. 2		1700	1950	1000	95	730	1150	1,700,000
No. 2		1450	1650	850	95	625	1000	1,700,000
No. 3		800	925	475	95	625	600	1,500,000
Appearance		1750	2050	1050	95	625	1500	1,800,000
Stud		800	925	475	95	625	600	1,500,000
Construction	2 to 4 in. thick.	1050	1200	625	95	625	1150	1,500,000
Standard	4 in. wide	600	675	350	95	625	925	1,500,000
Utility		275	325	175	95	625	600	1,500,000
Dense select structural	2 to 4 in. thick.	2100	2400	1400	95	730	1650	1,900,000
Select structural	5 in. and wider	1800	2050	1200	95	625	1400	1,800,000
Dense No. 1		1800	2050	1200	95	730	1450	1,900,000
No. 1		1500	1750	1000	95	625	1250	1,800,000
Dense No. 2		1450	1700	775	95	730	1250	1,700,000
No. 2		1250	1450	650	95	625	1050	1,700,000
No. 3		725	850	375	95	625	675	1,500,000
Appearance		1500	1750	1000	95	625	1500	1,800,000
Stud		725	850	375	95	625	675	1,500,000
Dense select structural	Beams and stringers	1900	—	1100	85	730	1300	1,700,000
Select structural		1600	—	950	85	625	1100	1,600,000
Dense No. 1		1550	—	775	85	730	1100	1,700,000
No. 1		1300	—	675	85	625	925	1,600,000
Dense select structural	Posts and timbers	1750	—	1150	85	730	1350	1,700,000
Select structural		1500	—	1000	85	625	1150	1,600,000
Dense No. 1		1400	—	950	85	730	1200	1,700,000
No. 1		1200	—	825	85	625	1000	1,600,000
Select Dex	Decking	1750	2000	—	—	625	—	1,800,000
Commercial Dex		1450	1650	—	—	625	—	1,700,000

Source: Data adapted from *National Design Specification for Wood Construction.* 1982 ed. (Ref. 7). with permission of the publishers, National Forest Products Association. The table in the reference document lists values for 29 other wood species and has 19 footnotes.

[a] Values listed are for normal duration loading with wood that is surfaced dry or green and used at 19% maximum moisture content.

some adjustment of these values. Some of these are described in the footnotes to the table in the reference document (Ref. 7) from which Table 16.1 is adapted. Others are described in the body of the standard of which the table for design values is a part. Some of the considerations are the following.

1. *Moisture.* Stress values may be increased in some cases if the wood is of a better-cured condition (less retained moisture) than that assumed in the table. On the other hand, stresses may be reduced for some usage conditions; notably that of full exterior exposure to the weather.

2. *Repetitive Use.* This increase in the allowable bending stress is permitted when the individual member is one of a closely spaced set, as in the case of rafters, joists, and studs. The qualification is limited to members at least three in number and not spaced more than 24 in. center to center.

3. *Duration of Load.* These adjustments are made on the basis of the time character of the load, as described in Table 16.2. Although they are actually load adjustments and are also used in the form of direct adjustment of loads in the strength design method (sometimes called the *factored load method*), the usual method with wood is to modify the design stress values.

4. *Other Modifications.* Modification is also made for wood that is treated with preservatives or fire retardants, for slenderness effects in columns and thin joists or beams, and for size of large beams.

For real design situations, designers must be aware of the usage conditions and the various modifications that are applicable to specific structural members. Some of the modifications are incorporated in the design examples in this part as well as the building case studies in Part 9.

TABLE 16.2. Modification Factors for Design Values for Structural Lumber

Duration of Load and General Use	Multiply Design Values by:
10 years or more at the full load limit of a member (as for members carrying only dead load, such as headers in walls)	0.90
Two months' duration (as for snow)	1.15
7 days' duration (as for roof load where no snow pack is incurred)	1.25
Maximum force of wind or earthquake	1.33
Impact (such as wheel bumps, braking of moving equipment, or slamming of heavy doors)	2.00

Source: Adapted from specifications in *National Design Specification for Wood Construction*, 1982 ed. (Ref. 7), with permission of the publishers, National Forest Products Association.

Fasteners and Connectors

Wood structures are assembled from a large number of separate pieces, involving a great deal of fastening. Requirements for connections between elements vary with the size, shape, and functional relationships of the elements being connected. Most connections employ some fastening element of steel: nails, screws, lag bolts, machine bolts, and various special fabricated devices. Choice of the type of fastener and the manner of its use is mostly the responsibility of builders. Only if the appearance of the connection or the need for some specific structural capacity is of concern is it usually necessary for the structural designer to designate the details of connections. When it *is* necessary, the designer must have some reasonable understanding of the current technology and of common construction practices.

For some types of structures, such as trusses or exposed heavy timber structures, considerations of connection design may strongly influence the choice of form and dimensions of the assembled wood members. This may be a simple, practical matter of facilitating the fitting together of parts or may involve efforts to make a neater and generally better-looking structure when the construction is exposed to view.

The use of various common fastening elements and devices is discussed in Chapter 19. General considerations of various connection problems for wood structures are presented in the example designs in Part 9.

CHAPTER SEVENTEEN

Wood Beams, Joists, and Decks

Chapter 17 deals with the typical elements used to develop horizontal-spanning systems for roofs and floors. Concentration is on the use of solid-sawn lumber. Other systems for spanning are discussed in Chapter 20. Examples of complete system design are more fully presented in the building cases studies in Part 9.

17.1. GENERAL CONSIDERATIONS

Beam–joist–deck systems of wood are widely used for floors and for both flat and sloping roofs. Common usage has evolved an interactive relationship of design and construction practices, code regulations, and standardization of various industrial products. Standard lumber sizes and the old standard 48×96-in. sheet size for plywood conspire to create a dominace of the 12–16–24–32–48–96-in. modular system.

Industry standards and general economic pressures work to define minimal quality, which is reflected in the common minimum grades of materials used for ordinary construction situations. Local building codes and construction practices, as well as regional availability of products, may result in minor variations; but what constitutes minimally adequate construction is usually rather tightly defined at any given time. This "minimum" output may not be so great in all regards, and designers seeking some higher quality of performance may modify the usual construction, but they must do so from a position of thorough understanding of the prevailing norm.

Solid-sawn lumber has distinct limitations and problems that must be appreciated. First is the practical limit of availability of lumber sizes, which is a somewhat local market situation. Even in the big city, however, one should not expect to find all of the sizes listed in Table A.2 on the shelf in the neighborhood lumberyard. Furthermore, even where odd sizes such as 5×5 or 22×22 are available, they may be premium priced in comparison with more commonly used sizes.

A major problem with solid-sawn lumber, especially in the larger timber sizes, is that of dimensional stability. This generally refers to the likelihood that some change in shape (shrinkage, warp, twist, splitting, etc.) will occur after the piece is cut from the log. If the changes occur before the piece is used for construction, it may be rejected. However, large timbers are not likely to be fully dried when used; thus some shape change must be expected. If this has the potential for creating some objectionable appearance or causing some difficulty with the general construction, it must be considered.

Solid-sawn lumber has certain structural limitations, a notable one being its low shear capacity. In many structures consisting essentially of wood construction, it is fairly common to use some major structural elements of glued-laminated construction or rolled steel sections. This may be done to obtain more structural capacity, to reduce the size of element required, or to eliminate the concern for problems of dimensional stability.

17.2. BEAMS

Beam design in general involves a number of considerations, the principal ones being effects of flexure, shear, bearing, deflection, lateral stability, and connection to supports. We will consider these individually.

Flexure

Flexure, or bending, is generally considered in terms of the maximum fiber stress, basically determined from the formula

$$f_b = \frac{Mc}{I} \text{ or } f_b = \frac{M}{S}$$

Design for flexure consists of determining the maximum moment, establishing the limit for f_b, and finding the section modulus required for the beam. The section modulus thus determined is a minimum value required; however, there is no particular economy in holding to this minimum. The cost of a beam is mostly simply based on the amount of material, or the area of its cross section.

The bending moment resistance, or the allowable stress for design, may be modified by a number of factors. General modifications include those affecting all structures, as described in Sec. 16.3. Additional concerns for beams include the following.

1. *Repetitive Use.* This affects only rafters and joists. It involves using the increased allowable stress listed in Table 16.1 under the heading "repetitive member uses."

157

TABLE 17.1. Size Factors for Wood Beams[a]

Beam Depth (in.)	Form Factor,[b] C_F
13.5	0.987
15.5	0.972
17.5	0.959
19.5	0.947
21.5	0.937
23.5	0.928

[a] Reduce allowable bending stress or moment capacity by the factor C_F.

[b] $C_F = (12/d)^{1/9}$.

2. *Beam Shape.* Most beams are rectangular in section and sustain bending on one of the principal centroidal axes. For this there is no adjustment, except that noted for bending on the minor axis in some cases. For bending of round shapes or on the diagonal of a rectangular shape, adjustment is made using a form factor, as described in NDS Sec. 3.3.4.

3. *Size.* Beams greater than 12 in. in depth have reduced values for the maximum flexural stress. This is done by using a size factor (from NDS Sec. 4.3.4) defined as

$$C_F = \left(\frac{12}{d}\right)^{1/9}$$

Values for the size factor for standard lumber sizes are given in Table 17.1.

4. *Stability.* Specifications provide for the adjustment of bending capacity or allowable stress when the beam is vulnerable to a compression buckling condition. When beams are incorporated in framing systems, they will mostly be provided with bracing that is adequate to resist these effects, allowing use of the full value for flexural stress. Requirements for the type of bracing required to prevent both lateral and torsional buckling are given in NDS Sec. 4.4.1 and are summarized in Table 17.2. If details of the construction cannot provide adequate bracing, the rules of NDS Sec. 3.3.3 must be used to reduce bending capacity.

Shear

The principal concern for shear in a beam is for the horizontal splitting effect, usually occurring at the beam ends, unless the ends overhang the supports. Assuming a rectangular beam section, the unit shear stress is found as

$$f_v = \frac{3}{2}\frac{V}{bd} \quad \text{or simply} \quad f_v = \frac{3}{2}\frac{V}{A}$$

Allowable stress is quite low (see Table 16.1), so that heavily loaded beams are often quite critical in shear effect limits.

This is a case where glued-laminated or steel beams are often more feasible choices.

If a beam is supported so that the bottom of the beam bears directly on the support with full contact of the beam bottom surface, the code permits the decrease of the maximum shear for design to that occurring in the beam span a distance from the support equal to the beam depth. The vertical compression effect in this case tends to reduce the end-splitting effect due to shear.

Shear capacity may be affected by notches, holes, or other details of the end connections. Considerations for various situations are described in NDS Sec. 3.4.

Although shear is often critical in heavily loaded beams, it is almost never critical in rafters or joists, which tend to be relatively lightly loaded and long in span with respect to their depths. This is even more true for members used for spanning in a flat orientation (as planks or decking), to the extent that Table 16.1 does not even bother to give an allowable shear stress for decking.

Bearing

For most beams, bearing consists of compression stress perpendicular to the grain. The allowable stress for this is that given in Table 16.1, and is used simply by multiplying it by the bearing contact area. This is done for beam end bearing or for bearing under concentrated loads where the length of bearing along the beam is 6 in. or more. For shorter lengths an increase in allowable stress is permitted if the bearing does not occur closer than 3 in. from the end of the beam. The increase consists of multiplying the table value by a factor equal to

TABLE 17.2. Lateral Support Requirements for Wood Beams

Ratio of Depth to Thickness[a]	Required Conditions
2 : 1 or less	No support required
3 : 1, 4 : 1	Ends held in position to resist rotation
5 : 1	One edge held in position for entire span
6 : 1	Bridging or blocking at maximum spacing of 8 ft; or both edges held in position for entire span; or one edge held in position for entire span (compression edge) and ends held against rotation
7 : 1	Both edges held in position for entire span

Source: Adapted from data in Sec. 4.4.1 of *National Design Specification for Wood Construction*, 1982 ed. (Ref. 7), with permission of the publishers, National Forest Products Association.

[a] Ratio of nominal dimensions for standard sections.

$$\frac{(\text{bearing length}) + 0.375 \text{ in.}}{\text{bearing length}}$$

When a beam is not horizontal, as in the case of some roof members, there may be a bearing that occurs at some angle to the grain other than parallel or perpendicular. Allowable stress in this case is adjusted using the Hankinson formula. The same type of adjustment is made for bolted joints when the load is at an angle to the grain. The use of the graph that is a plot of the Hankinson formula is discussed in Sec. 19.3. For bearing at intermediate angles, the stress ranges between the two limiting values in Table 16.1.

Deflection

Deflections in wood structures tend to be most critical for rafters and joists, where span-to-depth ratios are often pushed to the limit. Maximum permitted spans for specific rafter or joist arrangements are usually limited by deflection. Because rafters and joists are usually of a simple span form with uniformly distributed loading, the deflection takes the form of the equation

$$\Delta = \frac{5WL^3}{384EI}$$

Substitutions of relations between W, M, and flexural stress in this equation can result in the form

$$\Delta = \frac{5L^2 f_b}{24Ed}$$

Using average values of $f_b = 1500$ psi and $E = 1500$ ksi, the expression reduces to

$$\Delta = \frac{0.03L^2}{d}$$

where Δ = deflection, inches

L = span, feet

d = beam depth, inches

Figure 17.1 is a plot of this expression with curves for standard lumber dimensions. The curves are labeled by the nominal dimensions, but the computations were done with the actual dimensions, as given in Table A.2. For reference the lines on the graph corresponding to ratios of deflection of $L/240$ and $L/360$ are shown. These are commonly used design limitations for total-load and live-load deflections, respectively. Also shown for reference is the limiting ratio of beam span to depth of 25 to 1, which is an approximate practical limit, even if deflection is not especially important. For beams with values of f_b and E other than those used for the graphs, actual deflections can be obtained as follows

$$\text{true } \Delta = \frac{\text{true } f_b}{1500} \times \frac{1,500,000}{\text{true } E \text{ (psi)}} \times \Delta \text{ from graph}$$

The following examples illustrate typical problems involving computation of deflections.

Example 1. An 8 x 12 wood beam with $E = 1,600,000$ psi is used to carry a total uniformly distributed load of 10 kips on a simple span of 16 ft. Find the maximum deflection of the beam.
Solution. Using the deflection formula for this loading (Case 2 in Fig. B.1) and the value of $I = 950$ in.[4] for the section (obtained from Table A.2), we compute

$$\Delta = \frac{5WL^3}{384EI} = \frac{5(10,000)(16 \times 12)^3}{384(1,600,000)(950)} = 0.61 \text{ in.}$$

Or, using the graphs in Fig. 17.1

$$M = \frac{WL}{8} = \frac{10,000(16)}{8} = 20,000 \text{ ft-lb}$$

$$f_b = \frac{M}{S} = \frac{20,000(12)}{165} = 1455 \text{ psi}$$

From Fig. 17.1, Δ = approximately 0.66 in. Then

$$\text{true } \Delta = \frac{1455}{1500}\left(\frac{1,500,000}{1,600,000}\right)0.66 = 0.60 \text{ in.}$$

which shows reasonable accuracy in comparison with the computed value.

Example 2. A beam consisting of a 6×10 section with $E = 1,400,000$ psi spans 18 ft and carries two concentrated loads. One load is 1800 lb, placed at 3 ft from one end of the beam, and the other load is 1200 lb, placed at 6 ft from the opposite end of the beam. Find the maximum deflection due only to the concentrated loads.
Solution. For an approximate computation, we use the equivalent uniform load method, consisting of finding the hypothetical total uniform load that will produce a moment equal to the actual maximum moment in the beam. For this loading the maximum moment is 6600 ft-lb (the reader should verify this by the usual procedures), and the equivalent uniform load is thus

$$W = \frac{8M}{L} = \frac{8(6600)}{18} = 2933 \text{ lb}$$

and the approximate deflection is

$$\Delta = \frac{5WL^3}{384EI} = \frac{5(2933)[(18)(12)]^3}{384(1,400,000)(393)} = 0.70 \text{ in.}$$

As in the previous example, the deflection could also be found by computing the maximum bending stress and using Fig. 17.1.

17.3. JOISTS AND RAFTERS

The most common wood construction is light wood framing of the type shown in Fig. 17.2, with walls framed with closely spaced studs and floors and roofs framed with

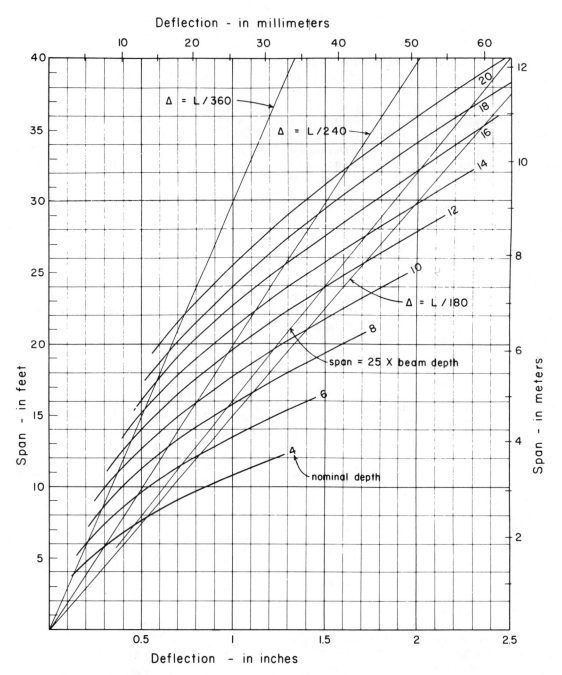

FIGURE 17.1. Deflection of wood beams. Assumed conditions: max imum f_b = 1500 psi [10.3 MPa], E = 1500 ksi [10.3 GPa].

joists and rafters. Wall covering varies, but roof and floor surfaces are mostly covered by plywood sheets, although board sheathing is also possible. Studs, rafters, and joists are ordinarily either 16 or 24 in. on center, but may be closer or farther apart for various reasons. Typical spacings ordinarily derive from the use of 48 × 96-in. standard sheet material for decks, walls, and ceilings. Efficient spacings are 12, 16, 24, 32, and 48 in.

Joists and rafters may be designed using the usual requirements for beams. However, for common loading and

span conditions, sizes and spacing of joists and rafters may usually be obtained from tabulations. Reproductions of five such tables from the *Uniform Building Code* (UBC) (Ref. 1) are presented in Appendix D. If pushed to the limit, the maximum spans permitted by these tables may result in some objectionable roof sag or bounciness of floors. The author recommends some conservative judgment in using these tables.

Roof and floor decks are most often utilized as horizontal diaphragms in developing lateral bracing of the build-

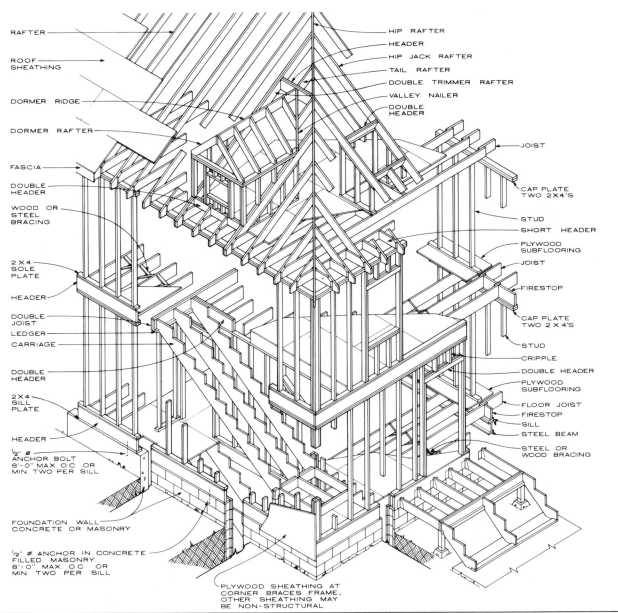

RAFTER

ROOF SHEATHING

DORMER RIDGE

DORMER RAFTER

FASCIA

DOUBLE HEADER

WOOD OR STEEL BRACING

2 X 4 SOLE PLATE

HEADER

DOUBLE JOIST

LEDGER

CARRIAGE

DOUBLE HEADER

2 X 4 SILL PLATE

HEADER

½" Ø ANCHOR BOLT 8'-0" MAX. O.C. OR MIN. TWO PER SILL

FOUNDATION WALL CONCRETE OR MASONRY

½" Ø ANCHOR IN CONCRETE FILLED MASONRY 8'-0" MAX. O.C. OR MIN. TWO PER SILL

HIP RAFTER

HEADER

HIP JACK RAFTER

TAIL RAFTER

DOUBLE TRIMMER RAFTER

VALLEY NAILER

DOUBLE HEADER

JOIST

CAP PLATE TWO 2X4'S

STUD

SHORT HEADER

PLYWOOD SUBFLOORING

JOIST

FIRESTOP

CAP PLATE TWO 2 X 4'S

STUD

CRIPPLE

DOUBLE HEADER

PLYWOOD SUBFLOORING

FLOOR JOIST

FIRESTOP

SILL

STEEL BEAM

STEEL OR WOOD BRACING

PLYWOOD SHEATHING AT CORNER BRACES FRAME, OTHER SHEATHING MAY BE NON-STRUCTURAL

PLATFORM FRAMING

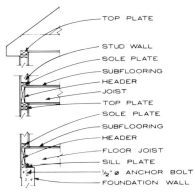

TOP PLATE

STUD WALL

SOLE PLATE

SUBFLOORING

HEADER

JOIST

TOP PLATE

SOLE PLATE

SUBFLOORING

HEADER

FLOOR JOIST

SILL PLATE

½" Ø ANCHOR BOLT

FOUNDATION WALL

Timothy B. McDonald; Washington, D.C.

NOTES

WESTERN OR PLATFORM FRAMING

Before any of the superstructure is erected, the first floor subflooring is put down making a platform on which the walls and partitions can be assembled and tilted into place. The process is repeated for each story of the building. This framing system is used frequently .

FIRESTOPPING

All concealed spaces in framing, with the exception of areas around flues and chimneys, are to be fitted with 2 in. blocking arranged to prevent drafts between spaces.

EXTERIOR WALL FRAMING

One story buildings: 2 x 4's, 16 in. or 24 in. o.c.;
2 x 6's, 24 in. o.c.
Two and three stories: 2 x 4's, 16 in. o.c.;
2 x 6's, 24 in. o.c.

BRACING EXTERIOR WALLS

Because floor framing and wall frames do not interlock, adequate sheathing must act as bracing and provide the necessary lateral resistance. Where required for additional stiffness or bracing, 1 x 4's may be let into outer face of studs at 45° angle secured at top, bottom, and to studs.

BRIDGING FOR FLOOR JOISTS

May be omitted when flooring is nailed adequately to joist; however, where nominal depth-to-thickness ratio of joists exceeds 6, bridging would be installed at 8 ft. 0 in. intervals. Building codes may allow omission of bridging under certain conditions.

Steel bridging is available. Some types do not require nails.

FIGURE 17.2. Typical light wood frame construction: western or platform type. Reproduced from *Architectural Graphic Standards*, 7th ed. (Ref. 13), with permission of John Wiley & Sons, Inc., New York.

ing. Required diaphragm nailing (as discussed in Part 8) and the need for blocked edges of the plywood sheets will usually satisfy the needs for lateral stability of the rafters and joists where the diaphragm is required to take significant forces.

For sloping rafters it is common practice to consider the span to be the horizontal projection, as shown in Fig. 17.3. This applies only to gravity loads. If rafters must be designed for wind, the wind forces are considered as applied perpendicular to the roof surface, and the span must be the actual rafter length.

Floor joists often serve dual functions, providing for attachment of the floor deck on top and attachment of the ceiling directly to the bottom of the joists. Note that the tables for rafters and joists always mention the ceiling construction (or lack of it) and some value assumed for total dead load. Ceilings may also be separately framed, using the framing obtained from Table 25-U-J-6 of the *UBC* (see Appendix D).

The following examples illustrate the use of the load-span tables from the *UBC*.

Example 1. Using Table 25-U-J-1 from the *UBC* (Ref. 1) in Appendix D, select joists to carry a live load of 40 psf on a span of 15 ft 6 in. if the spacing is 16 in. on center.
Solution. From the table we determine that 2 × 10 joists with E of 1,400,000 psi and F_b of 1150 psi may be used on a span of 15 ft 8 in. From Table 16.1 it may be noted that Douglas fir–larch No. 2 joists can be used, having table values of F_b = 1450 psi and E = 1,700,000 psi.

Example 2. Rafters are to be used on 24-in. centers for a roof span of 16 ft. Live load is 20 psf, dead load is 15 psf, and live-load deflection is limited to $L/240$. Find the rafter size required for No. 1 Douglas fir.
Solution. Table 16.1 gives values of F_b = 1750 psi and E = 1,800,000 psi for No. 1 joists 5 in. or wider. The stress value of 1750 psi does not appear in Table 25-U-R-1, but a stress of 1700 psi yields a permissible span for 2 × 10 rafters of 18 ft 7 in. while requiring an E of only 1,170,000 psi for the limiting deflection. Further inspection of the table will show that 2 × 8 rafters cannot be used unless the spacing is reduced. It may also be observed that the rafters could be of No. 2 grade, which would be even more true if modification is made for load duration (probably a 15% increase in the allowable stress for this case).

17.4. BOARD AND PLANK DECKS

Before plywood established its dominance as a decking material, most roof and floor decks were made with $\frac{3}{4}$-in. (nominal-1-in.) boards with interlocking edges of the type shown in Fig. 17.4—called *tongue-and-groove joints*. Today, this type of deck is mostly used only in regions where labor cost is relatively low and the boards are locally competitive in availability and cost in comparison to plywood.

When installed in a position with the boards perpendicular to the joists, board decks produce rather poor horizontal diaphragms. It is common, therefore, when significant diaphragm action for lateral loads is required, to install the deck at an angle of 45° to the joists, creating a trussed structure.

The $\frac{3}{4}$-in.-thick board deck is usually adequate for the spanning tasks of roof and floor decks where spacing of joists or rafters does not exceed 24 in. The type of roofing or the type of finish floor to be used must be considered, however. Roofing of all types must be anchored to the deck with nails of some kind, for which the board deck is quite adequate, possibly better than the thinner plywoods. Membrane roofing for flat roofs usually requires a minimum of $\frac{1}{2}$-in. plywood, making the board deck more competitive than in the case of sloping roofs with shingles, which might be achieved with thinner plywood.

For floors it is common to use some additional material on top of the structural deck, such as a layer of concrete fill or particleboard sheets. These will add considerable stiffness to the floor, so when they are *not* to be used, some conservative judgment should be exercised in choice of deck thickness and support spacings.

If a deck of the board type is thicker than $\frac{3}{4}$ in., it is generally referred to as planking or plank deck. The most widely used form is that made with 2-in.-nominal-thickness units, approximately 1.5 in. in actual dimension. There are usually special reasons for using such a deck, including one or all of the following.

1. The deck is to be exposed on the underside, and the appearance of the plank deck is hands down more handsome than board or plywood decks.
2. Exposed or not, the deck may require a fire rating, which is highly limited for other decks.

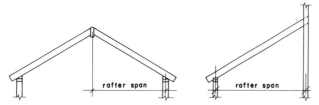

FIGURE 17.3. Design assumption for span of sloping rafters; gravity loading.

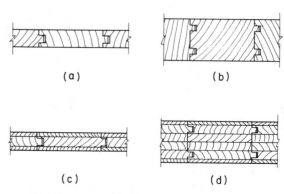

FIGURE 17.4. Typical forms of plank or timber decking.

3. It may be desired to have supporting members with spacing exceeding that feasible for board or plywood decks.
4. Concentrated loadings from vehicles or equipment may be too high for the thinner decks.

Nominal-2-in.-plank deck may be of the same form as board deck (Fig. 17.4a), but is also often made with laminated units as shown in Fig. 17.4c. Plank deck is also available in thicknesses greater than 2 in. nominal. When the thickness exceeds 2.5 in. or so, the units usually have a double tongue and groove on each face, as shown in Fig. 17.4b for a solid-sawn unit and in Fig. 17.4d for a laminated unit. The thicker plank deck units are capable of achieving considerable span distances and may be used in structures that are essentially without the usual rafters or joists, having clear deck spans from wall to wall or from beam to beam.

One problem with plank decks, as with board decks, is the low diaphragm capacity achieved when units are in a perpendicular orientation to supports. Diagonal placement, while common with 1-in.-thick-board decks, is not so common with the plank decks. When diaphragm capacity of some significant level is required, the usual solution is to use plywood nailed to the top of the deck units. This is quite commonly done and is frankly the main reason that diaphragm capacities are rated for decks of plywood as thin as $\frac{5}{16}$ in., a thickness not usually usable for a structural spanning deck (see *UBC* Tables 25-J-1 and 25-K-1 in Appendix D).

Plank deck units are essentially fabricated products and information about them should be obtained from the manufacturers or suppliers of the products. Type of units, finishes of exposed units, installation specifications, and structural properties vary greatly and the type of unit available regionally should be used for design.

17.5. PLYWOOD DECKS

Structural grades of plywood, mostly of Douglas fir, are the most widely used materials for roof and floor decking. Grading of plywood is discussed in Sec. 20.2. When used for finished surfaces or cabinet work, a important grading consideration is the quality of the face laminates. For structural purposes, however, this is not the only concern.

Plywood decks quite frequently serve the dual purposes of spanning support surfaces for gravity roof or floor loads and shear-resisting diaphragms for lateral forces on the building. Choice of the plywood quality and thickness must relate to both of these concerns, as well as to any additional functional problems, such as attachment of roofing materials. Nailing of the plywood sheets to supports is done minimally for gravity loads, usually with nails at 6-in. centers at all edges and at 12-in. centers in the middle of the sheets, called *field* or *intermediate nailing*. Nailing for diaphragm action is the basic means of making the separate sheets act together as a single large unit and transfer loads to chords, shear walls, and collectors. Diaphragm nailing is discussed in Part 8.

For gravity-load spanning functions, plywood is strongest when the face grain is perpendicular to the supports. However, for various reasons, it is sometimes desired to turn the sheets the other way, which results in considerably less spanning capacity in thinner sheets that usually have only three plys, but has decreasing influence as the number of plys and overall thickness increases. Thus $\frac{3}{8}$-in.-thick sheets are strongly affected by orientation, while $\frac{3}{4}$-in.-thick sheets are only minimally affected.

One reason for the orientation with face grain parallel to supports is that it reduces the amount of total unsupported edge length in the deck. This refers to the edges of the sheets that do not fall over a supporting rafter or joist. If left substantially unsupported (and not much nailed down), these edges do not contribute to the deck diaphragm capacity and offer some problems for membrane roofing of flat roofs and some types of floor finishes. If support is considered to be necessary, it is usually achieved by one of the following means.

1. Using tongue-and-groove joints between sheets. This is generally limited to sheets of $\frac{3}{4}$-in. or greater thickness.
2. Using a clipping device—usually a short metal H-shaped element between the sheet edges.
3. Providing nailable supports at the edges not having support by the structure beneath the deck. This is typically provided by blocking when the supports are solid-sawn rafters or joists. The blocks consist of short pieces of the same-size members as the rafters or joists, fitted between and nailed to the structural members.

Tongue-and-groove or clipped edges do not change the deck with regard to diaphragm action; thus the deck in these cases must usually need to function adequately as a so-called *unblocked diaphragm*. Tongue-and-groove decks are available in thicknesses up to 1.25 in. and are most often used for floors (with heavier loadings) or for roofs of longer spans where structural members may have spacings greater than 24 in.

The All-American sheet size is, of course, 48 in. × 96 in. If ordered in large enough batches, however, larger sheets can be obtained when the logical structural module simply does not fit in the 2–4–8-ft system.

For decks, ordinary structural plywood is produced in thicknesses from $\frac{5}{16}$ to $\frac{7}{8}$ in. Deck capacities and span limits are rated in industry standards or building codes. The following tables from the UBC are presented in Appendix D.

Table 25-J-1, giving capacities for shear in horizontal diaphragm action.

Table 25-S-1, giving capacities or span limits for gravity load spanning decks for roofs or floors with the face grain perpendicular to supports.

Table 25-S-2, giving capacities or span limits for roof decks with the face grain parallel to supports.

Common practice often establishes some minimum usages. The following are some widely used minimum choices for plywood decks in common situations.

1. For roofs of significant slope (usually 3 : 12 or greater) with shingle roofing—$\frac{3}{8}$ in.
2. For flat roofs with conventional membrane (three-ply, etc.) roofing and rafters not over 24 in. on center—$\frac{1}{2}$ in.
3. For floors with some permanent structural material between the deck and the finish (concrete fill, $\frac{1}{2}$-in.-or-thicker particleboard, etc.)—$\frac{5}{8}$ in.
4. For other floors—$\frac{3}{4}$ in.

17.6. TIMBER SYSTEMS

The light frame structure is essentially a wall bearing system, using the stud wall as a gravity load-carrying element. Where more open space is required, some parts of such a building may use freestanding columns, instead of solid walls to support the roof or upper floors. The light frame typically uses only stud walls and joist or rafter systems as the basic elements, with beams occurring only over large openings, at edges of stairs, and so on. If freestanding columns (or posts) are used extensively, they usually work directly with a beam system, the final extension of the system being one of all posts and beams with walls of a general nonstructural nature (called *partitions* for interior walls or *curtain walls* for the exterior).

A special structure that uses solid-sawn timbers of relatively large dimension (6-in.-nominal thickness or larger) and plank decking is called *heavy timber* construction. This system may be used for its enhanced fire resistance, or for its appearance as a rustic exposed structure. It is generally quite expensive in comparison to other construction alternatives, but may be justified by the increased fire resistance and/or the architectural character of the exposed appearance. In some cases the usual thin (2-in. nominal) rafters or joists and plywood decks may be used with heavy posts and beams, but the exposed condition will be considerably less charming and the fire rating will be lost.

Some aspects of design and construction detailing for wood post-and-beam construction are discussed in the considerations of alternative systems for the design case studies in Part 9.

CHAPTER EIGHTEEN

Wood Columns

The wood column most widely used is the 2×4 stud in the wall systems of light frame construction. Other solid sawn sections are used only where there is some need for the heavier timber element, as in the post-and-beam system. Round columns are sometimes used in pole-type structures or for their desired shape in an exposed situation. Special columns built up of multiple elements may be used with certain types of structures or may exist in the form of compression members in a wood truss. Chapter 18 presents material relating to the design of compression members in general with emphasis on the most common uses for building columns.

18.1. SOLID-SAWN COLUMNS

The simple solid column usually consists of a single piece of wood, square or rectangular in cross section. Round sections may also be used, usually as peeled logs of tapered form; these are discussed in Sec. 18.5. This section deals only with the square and rectangular elements in the standard sizes given in Table A.2.

In wood construction the slenderness ratio of a freestanding simple solid column is the ratio of the unbraced (laterally unsupported) length to the dimension of its least side, or L/d. (Fig. 18.1a.) When members are braced so that the unsupported length with respect to one face is less than that with respect to the other, L is the distance between the points of support that prevent lateral movement in the direction along which the dimension of the section is measured. This is illustrated in Fig. 18.1b. If the section is not square or round, it may be necessary to investigate two L/d conditions for such a column, to determine which is the limiting one. The slenderness ratio for simple solid columns is limited to $L/d = 50$; for spaced columns the limiting ratio is $L/d = 80$.

Figure 18.2 illustrates the typical form of the relationship between axial compression capacity and slenderness for a linear compression member (column). The two limiting conditions are those of the very short member and the very long member. The short member (such as a block of wood) fails in crushing, which is limited by the mass of material and the stress limit in compression. The very long member (such as a yardstick) fails in elastic buckling, which is determined by the stiffness of the member; stiff-

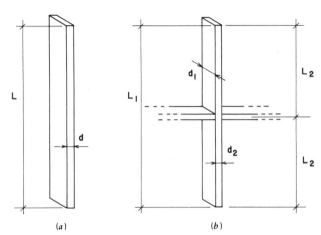

FIGURE 18.1. Slenderness ratio for columns.

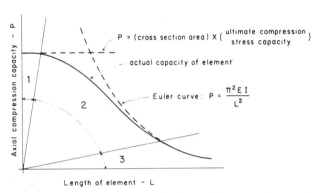

FIGURE 18.2. Relation of member length to axial compression capacity.

ness is determined by a combination of geometric property (shape of the cross section) and material stiffness property (modulus of elasticity). Between these two extremes—which is where most wood compression members fall—the behavior is indeterminate as the transition is made between the two distinctly different modes of behavior.

The National Design Specification currently provides for three separate compression stress calculations, corresponding to the three zones of behavior described in Fig. 18.2. The plot of these three stress formulas, for a specific

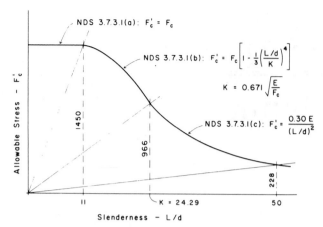

FIGURE 18.3. Allowable axial compression as a function of the slenderness ratio L/d. National Design Specification (NDS) requirements for Douglas fir-larch, dense No. 1 grade.

example wood, is shown in Fig. 18.3. Typical analysis and design procedures for simple solid wood columns are illustrated in the following examples.

Example 1.   A wood compression member consists of a 3×6 of Douglas fir-larch, dense No. 1 grade. Find the allowable axial compression force for unbraced lengths of (a) 2 ft [0.61 m], (b) 4 ft [1.22 m], and (c) 8 ft [2.44 m].
Solution.   We find from Table 16.1: $F_c = 1450$ psi [10.0 MPa] and $E = 1,900,000$ psi [13.1 GPa]. To establish the zone limits we compute the following:

$$11(d) = 11(2.5) = 27.5 \text{ in.}$$
$$50(d) = 50(2.5) = 125 \text{ in.}$$

and

$$K = 0.671 \sqrt{\frac{E}{F_c}} = 0.671 \sqrt{\frac{1,900,000}{1450}} = 24.29$$

(a) Thus for 2 ft, $L = 24$ in., which is in zone 1; $F_c' = F_c = 1450$ psi; allowable $C = F_c' \times$ gross area $= 1450 \times 13.75 = 19,938$ lb [88.7 kN].
   (b) $L = 48$ in.; $L/d = 48/2.5 = 19.2$, which is in zone 2.

$$F_c' = F_c \left\{ 1 - \frac{1}{3} \left(\frac{L/d}{K} \right)^4 \right\}$$

$$= 1450 \left\{ 1 - \frac{1}{3} \left(\frac{19.2}{24.29} \right)^4 \right\}$$

$$= 1262 \text{ psi}$$

Allowable $C = 1262 \times (13.75) = 17,353$ lb [77.2 kN].
   (c) $L = 96$ in.; $L/d = 96/2.5 = 38.4$, which is in zone 3.

$$F_c' = \frac{0.3(E)}{(L/d)^2} = \frac{0.3(1,900,000)}{(38.4)^2} = 387 \text{ psi}$$

Allowable compression $= 387(13.75) = 5321$ lb [23.7 kN].

Example 2.   Wood 2×4 elements are to be used as vertical compression members to form a wall (ordinary stud wall construction). If the wood is Douglas fir-larch, No. 3 grade, and the wall is 8.5 ft high, what is the column load capacity of a single stud?
Solution.   In this case it must be assumed that the surfacing materials used for the wall (plywood, drywall, plaster, etc.) will provide adequate bracing for the studs on their weak axis (the 1.5-in. [38-mm] direction). If not, the studs cannot be used, since the specified height of the wall is considerably in excess of the limit for L/d for a solid column (50). We therefore assume the direction of potential buckling to be that of the 3.5-in. [89-mm] dimension. Thus

$$\frac{L}{d} = \frac{8.5 \times 12}{3.5} = 29.1$$

In order to determine which column load formula must be used, we must find the value of K for this wood. From Table 16.1 we find $F_c = 675$ psi [4.65 MPa] and $E = 1,500,000$ psi [10.3 GPa]. Then

$$K = 0.671 \sqrt{\frac{E}{F_c}} = 0.671 \sqrt{\frac{1,900,000}{675}} = 31.63$$

We thus establish the condition for the stud as zone 2 (Fig. 18.2), and the allowable compression stress is computed as

$$F_c' = F_c \left\{ 1 - \frac{1}{3} \left(\frac{L/d}{K} \right)^4 \right\}$$

$$= 675 \left\{ 1 - \frac{1}{3} \left(\frac{29.1}{31.63} \right)^4 \right\} = 514 \text{ psi} [3.54 \text{ MPa}]$$

The allowable load for the stud is

$$P = F_c'(\text{gross area}) = 514(5.25) = 2699 \text{ lb} [12.0 \text{ kN}]$$

Example 3.   A wood column of Douglas fir-larch, dense No. 1 grade, must carry an axial load of 40 kips [178 kN]. Find the smallest section for unbraced lengths of: (a) 4 ft [1.22 m], (b) 8 ft [2.44 m], and (c) 16 ft [4.88 m].
Solution.   Since the size of the column is unknown, the values of L/d, F_c, and E cannot be predetermined. Therefore, without design aids (tables, graphs, or computer programs), the process becomes a cut-and-try approach, in which a specific value is assumed for d and the resulting values for L/d, F_c, E, and F_c' are determined. A required area is then determined and the sections with the assumed d compared with the requirement. If an acceptable member cannot be found, another try must be made with a different d Although somewhat clumsy, the process is usually not all that laborious, since a limited number of available sizes are involved.

   (a) We first consider the possibility of a zone 1 stress condition (Fig. 18.2), since this calculation is quite simple. If the maximum $L = 11(d)$, then the minimum $d = (4 \times 12)/11 = 4.36$ in. [111 mm]. This requires a nominal thickness of 6 in., which puts the size range into the "posts and timbers" category in Table 16.1, for which the allowable stress F_c is 1200 psi. The required area is thus

$$A = \frac{\text{load}}{F_c'} = \frac{40,000}{1200} = 33.3 \text{ in.}^2 [21,485 \text{ mm}^2]$$

The smallest section is thus a 6×8, with an area of 41.25 in.^{2}, since a 6×6 with 30.25 in.^{2} is not sufficient (see Table A.2). If the rectangular-shape column is acceptable, this becomes the smallest member usable. If a square shape is desired, the smallest size would be an 8×8.

(b) If the 6-in. nominal thickness is used for the 8-ft column, we determine that

$$\frac{L}{d} = \frac{8(12)}{5.5} = 17.45$$

Since this is greater than 11, the allowable stress is in the next zone, for which

$$F_c' = F_c \left\{ 1 - \frac{1}{3} \left(\frac{L/d}{K} \right)^4 \right\}$$

$$= 1200 \left\{ 1 - \frac{1}{3} \left(\frac{17.45}{25.26} \right)^4 \right\}$$

$$= 1109 \text{ psi } [7.65 \text{ MPa}]$$

in which

$F_c = 1200$ psi   and $E = 1,700,000$ (from Table 16.1)

and

$$K = 0.671 \sqrt{\frac{E}{F_c}} = 0.671 \sqrt{\frac{1,700,000}{1200}} = 25.26$$

The required area is thus

$$A = \frac{\text{load}}{F_c'} = \frac{40,000}{1109} = 36.07 \text{ in.}^3 \ [23,272 \text{ mm}^2]$$

and the choices remain the same as for the 4-ft column.

(c) If the 6-in. nominal thickness is used for the 16-ft column, we determine that

$$\frac{L}{d} = \frac{16 \times 12}{5.5} = 34.9$$

Since this is greater than the value of K, the stress condition is that of zone 3 (Fig. 18.2), and the allowable stress is

$$F_c' = \frac{0.30E}{(L/d)^2} = \frac{0.30(1,700,000)}{34.9^2} = 419 \text{ psi } [2.89 \text{ MPa}]$$

which requires an area for the column of

$$A = \frac{\text{load}}{F_c'} = \frac{40,000}{419} = 95.5 \text{ in.}^3 \ [61,617 \text{ mm}^2]$$

This is greater than the area for the largest section with a nominal thickness of 6 in., as listed in Table A.2. Although larger sections may be available in some areas, it is highly questionable to use a member with these proportions as a column. Therefore, we consider the next larger nominal thickness of 8 in. Then if

$$\frac{L}{d} = \frac{16(12)}{7.5} = 25.6$$

we are still in the zone 3 condition, and the allowable stress is

$$F_c' = \frac{0.30E}{(L/d)^2} = \frac{0.30(1,700,000)}{25.6^2} = 778 \text{ psi } [5.36 \text{ MPa}]$$

which requires an area of

$$A = \frac{\text{load}}{F_c'} = \frac{40,000}{778} = 51.4 \text{ in.}^3 \ [33,163 \text{ mm}^2]$$

The smallest member usable is thus an 8×8. It is interesting to note that the required square column remains the same for all the column lengths, even though the stress varies from 1200 psi to 778 psi. This is not uncommon and is simply due to the limited number of sizes available for the square column section.

18.2. USE OF DESIGN AIDS FOR WOOD COLUMNS

It should be apparent from the examples in Sec. 18.1 that the design of wood columns by these procedures is a laborious task. The working designer, therefore, typically utilizes some design aids in the form of graphs, tables, or computer-aided processes. One should exercise care in using such aids, however, to be sure that any specific values for E or F_c that are used correspond to the true conditions of the design work and that the aids are developed from criteria identical to those in any applicable code for the work.

Figure 18.4 consists of a graph from which the axial compression capacity of solid, square wood columns may be determined. Note that the graph curves are based on a specific species and grade of wood (Douglas fir–larch, dense No. 1 grade). The three circled points on the graph correspond to the design examples in Example 3 of Sec. 18.1.

Table 18.1 gives the axial compression capacity for a range of sizes and unbraced lengths of solid, rectangular wood sections. Note that the design values for elements with nominal thickness of 4 in. and less are different from those with nominal thickness of 6 in. and over, owing to the different size classifications as given in Table 16.1.

18.3. SPACED COLUMNS

A spaced column consists of two or more identical wood lumber elements that are fastened to each other in a manner that permits them to act in unison as a compression member. The most common occurrence of this is in the heavy timber trusses that use overlapping elements for the individual truss members. (See Fig. 20.3b and the discussion in Sec. 20.1.)

As in the case of the wood stud in a wall, the relatively thin members used become somewhat less critical with regard to buckling on their weak axes due to the bracing effect on the weak axis. This is achieved in the stud wall by

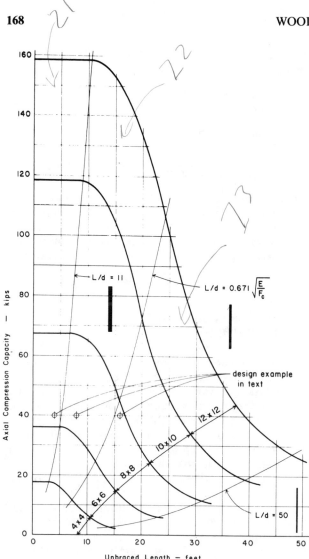

FIGURE 18.4. Axial compression load capacity for wood members of square cross section. Derived from *National Design Specification (NDS)* requirements for Douglas fir-larch, dense No. 1 grade.

18.4. STUDS

Studs are the vertical elements used for wall framing in the light wood frame system, as shown in Fig. 17.2. They serve utilitarian purposes of providing for attachment of the wall surfacing materials but also serve as vertical columns for support of roof or floor systems for which the wall may serve bearing support functions. The most common stud is the 2 × 4, spaced at 16- or 24-in. centers.

Studs of nominal 2-in.-thick lumber must be braced on their weak axis if used for story-high walls, a simple requirement deriving from the limit for slenderness of the solid sawn column. Wall finish materials will in most cases serve this bracing function, although some horizontal blocking at midheight can also provide for a reduction of unbraced length on the weak axis, as shown in Fig. 18.1b. Where walls are braced by finish on only one side, the blocking is essential.

Studs may also serve other functions, as in the case of exterior walls, where the studs must usually work as vertically spanning elements for lateral loads. For this situation the studs must be designed for the combined actions of axial compression plus bending, as discussed in Sec. 18.6.

In colder climates it is now common to use studs of greater width than the nominal 4 in. to create a larger void space for installation of insulation. This generally results in studs that are quite redundantly strong for the structural

TABLE 18.1. Axial Compression Capacity of Wood Elements (kips)

Element Size		Unbraced Length (ft)							
Designation	Area of Section (in²)	6	8	10	12	14	16	18	20
2 × 3	3.375	0.8			*L/d* greater than 50				
2 × 4	5.25	1.3							
3 × 4	8.75	6.0	3.4	2.2					
3 × 6	13.75	9.4	5.3	3.4					
4 × 4	12.25	14.7	9.3	5.9	4.1	3.0			
4 × 6	19.25	23.1	14.6	9.3	6.5	4.8			
4 × 8	25.375	30.4	19.2	12.3	8.5	6.3			
6 × 6	30.25	35.4	33.5	29.6	22.5	16.6	12.7	10.0	8.1
6 × 8	41.25	48.3	45.7	40.3	30.6	22.6	17.3	13.6	11.0
6 × 10	52.25	61.2	57.9	51.1	38.8	28.6	21.9	17.2	14.0
6 × 12	63.25	74.1	70.1	61.8	47.0	34.7	26.5	20.9	17.0
8 × 8	56.25	67.5	66.0	63.9	60.0	53.6	43.8	34.6	28.0
8 × 10	71.25	85.5	83.6	80.9	75.9	67.9	55.4	46.9	35.5
8 × 12	86.25	103.5	101.2	98.0	91.9	82.2	67.1	53.0	42.9
8 × 14	101.25	121.5	118.9	115.0	107.9	96.5	78.8	62.3	50.4
10 × 10	90.25	108.3	108.3	106.0	103.6	99.6	93.5	84.5	72.1
10 × 12	109.25	131.1	131.1	128.4	125.4	120.6	113.2	102.4	87.3
10 × 14	128.25	153.9	153.9	150.7	147.2	141.6	132.9	120.2	102.5
10 × 16	147.25	176.7	176.7	173.0	169.0	162.5	152.5	138.0	117.7
12 × 12	132.25	158.7	158.7	158.7	155.5	152.7	148.6	142.5	134.1

Note: Wood used is dense No. 1 Douglas fir–larch under normal moisture and load conditions.

the bracing provided by the structural surfacing materials on the wall. In the spaced column, the action involves the combined effect of the interacting members due to the fact that, although they are "spaced," they are bolted together at their ends and midpoints with spacer blocks, as shown in Fig. 18.5.

Design of a spaced column involves considerations for the axial compression load as limited by separate concerns on its two axes. For buckling in the y–y direction (see Fig. 18.5), an ordinary investigation is made on the basis of simple solid-sawn member action, using the combined areas of the members, using the slenderness determined by dimension d_2 in the figure. The true spaced column function involves a modified investigation using the d_1 dimension in the figure.

Requirements for structural investigation and specifications for bolting of spaced columns are provided in the *NDS* (see Ref. 7).

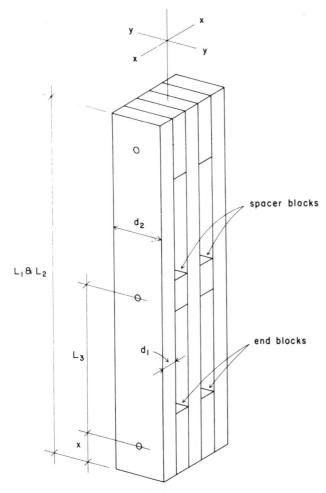

FIGURE 18.5. Spaced column.

18.5. POLES

Poles are round timbers consisting of the peeled logs of coniferous trees. In short lengths they may be of relatively constant diameter, but have a typically tapered profile when long—the natural form of the tree trunk. As columns, poles are designed with the same basic criteria used for rectangular-sawn sections. For slenderness considerations the d used for L/d computations is taken as that of a square section's side dimension, the square having an area equal to that of the round section. Thus, calling the pole diameter D,

$$d^2 = \frac{\pi D^2}{4}, \qquad d = 0.886D$$

For the tapered column, a conservative assumption for design is that the critical column diameter is the least diameter at the narrow end. However, for a slender column, with buckling occurring in the midheight of the column, this is very conservative and the code provides for some adjustment. Nevertheless, because of a typical lack of initial straightness and presence of numerous flaws, many designers prefer to use the unadjusted small end diameter for design computations.

Poles are used as timber piles, as discussed in Part 7. They are also used as buried-end posts for fences, signs, and utility transmission lines. Buildings of *pole construction* typically use buried-end posts as building columns. For lateral forces, buried-end posts must be designed for bending, and if sustaining significant vertical loads, as combined actions members with bending plus axial compression. For the latter situations it is common to consider the round pole to be equivalent to a square-sawn section with the same cross-sectional area and to use the code criteria for solid-sawn sections. The special case of foundation pressures on a laterally loaded buried-end post are considered in Part 7.

The wood-framed structure utilizing poles has a long history of use and continues to be a practical solution for utilitarian buildings where good poles are readily available. Accepted practices of construction for these buildings are based mostly on experience and do not always yield to highly rational analysis. If it works and many long-standing examples have endured the ravages of time and climate, it is hard to make a case on theory alone.

18.6. COLUMNS WITH BENDING

In wood structures columns with bending occur most frequently as shown in Fig. 18.6. Studs in exterior walls represent the situation shown in Fig. 18.6*a*, when considered for the case of vertical gravity plus horizontal wind loadings. In various situations, due to framing details, a column carrying only vertical loads may sustain bending if the load is not axial to the column, as shown in Fig. 18.6*b*. Investigation of columns with bending may be quite simple or very complex, depending on a number of qualifying conditions.

tasks of one- or two-story construction. Wall heights may also be increased with the wider studs, where the 2 × 4 limits free-standing unbraced height to about 14 ft.

If loads are high from gravity forces, as with multistory construction or long spans of roofs or floors, it may be necessary to strengthen the stud wall. This can be achieved in a number of ways, such as:

Decreasing the stud spacing to 12-in. centers.

Increasing the stud thickness to 3 in. nominal.

Increasing the stud width to a nominal size greater than 4 in.

Using doubled or tripled studs (or larger-size members) as posts directly under heavily loaded beams.

It is also sometimes necessary to use thicker studs or to restrict stud spacing when required nailing for plywood shear walls is excessive. This is discussed in Part 8.

In general, studs are columns and must comply to the various requirements for design of solid-sawn sections. Any appropriate grade of wood may be used, although special stud grades are commonly used for 2 × 4 members.

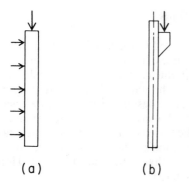

FIGURE 18.6. Columns with combined axial compression plus bending.

The general consideration for a member subjected to column and beam actions simultaneously is on the basis of interaction, as discussed in Sec. 11.3. The simplest form of this relationship is expressed by the equation for the straight line of the interaction graph:

$$\frac{f_a}{F_a} + \frac{f_b}{F_b} = 1$$

The form of this expression permits consideration of the combined actions while allowing for individual treatment of the two separate phenomena of column action and beam action. Current criteria for wood column design uses the simple interaction relationship as a fundamental reference with various adjustments for special cases. The basis of these adjustments is intended primarily to account for two potential complications. The first of these is the *P*-delta effect, described in Sec. 11.5. This effect occurs when the deflection due to bending moves the centroid of the column at midheight away from the line of action of the axial compression force. This results in some additional bending moment of a magnitude equal to the product of the compression force (*P*) times the deflection (delta). This is in general only critical if there is considerable deflection due to a very high span-to-depth ratio of the member or a high magnitude of the lateral loading. In the case of the loading in Fig. 18.6*b*, the entire moment is a *P*-delta effect.

The second adjustment to the simple interaction relationship involves the potential for lateral buckling due to bending. This is usually accounted for by using a value for F_b that is adjusted in the usual manner for a bending member with a critical laterally unsupported length condition.

As presented in the NDS, the general form of the interaction relationship is

$$\frac{f_c}{F_c'} + \frac{f_b}{F_b' - Jf_c} \leq 1$$

in which f_c and f_b are the computed axial compression and bending stresses, respectively.

F_c' is the usual allowable column compression stress as adjusted for the condition of slenderness, as discussed in Sec. 18.1. F_b' is the allowable flexural stress, adjusted if necessary for any stability considerations. *J* is a factor generally computed by the expression

$$J = \frac{(L_e/d) - 11}{K - 11}$$

in which *K* is the factor used in determination of F_c', as discussed in Sec. 18.1.

For the three zones of relative stiffness shown in Fig. 18.2, the use of *J* is as follows:

Zone 1, $L/d \leq 11$: $J = 0$.

Zone 2, $11 \leq L/d \leq K$: $J =$ the value from the expression.

Zone 3, $L/d \geq K$: $J = 1$.

For slender, unbraced columns it may be necessary to use an adjusted value for F_b. However, based on code qualifications, we note the following two exemptions.

1. For square sections used as beams no lateral support is required and no stress adjustment is made.
2. When the compression edge is continuously supported (as for a typical wall stud) the unsupported length may be considered as zero; thus no stress adjustment is required.

18.7. BUILT-UP COLUMNS

In various situations columns may consist of multiple elements of solid-sawn sections. Although the description includes glued-laminated and spaced columns, the term is generally used for multiple-element columns that do not qualify for those conditions. Glued-laminated columns are essentially designed as solid sections. Spaced columns that qualify as such are designed by the criteria discussed in Sec. 18.3.

Built-up columns generally have the elements attached to each other by mechanical devices, such as nails, lag screws, or machine bolts. Unless the particular assembly has been load tested for code approval, it is usually designed on the basis of the single element capacity. That is, the least load capacity is the sum of the capacities of the individually considered parts.

The most common built-up column is the multiple stud assembly that occurs at corners, wall intersections, and edges of door or window openings in the light wood-framed structure shown in Fig. 17.2. Since these are braced by wall surfacing in most cases, the aggregate capacity is simply the sum of the individual stud capacities.

When built-up columns occur as freestanding columns, it may be difficult to make a case for rational determina-

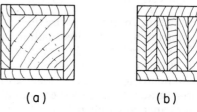

FIGURE 18.7. Sections of built-up columns.

tion of their capacities, unless single elements have low enough individual slendernesses to qualify for significant capacities. Two types of assembly that have some proven capacity as built-up columns are those shown in Fig. 18.7. In Fig. 18.1*a* a solid-core column is wrapped on all sides by thinner elements. The thinner elements may be assumed to borrow stiffness from the core and an allowable stress for the whole section might be that based on the core's slenderness. In Fig. 18.7*b* a series of thin elements is held together by two cover plates, which tend to restrict the buckling of the core elements on their weak axes. A reasonable assumption for the latter section is to assume the cover plates to brace the core elements so that their slenderness may be considered by using their larger dimension.

CHAPTER NINETEEN

Fasteners and Connections

Assemblage of wood structures involves the making of large numbers of connections. Various fasteners (nails, screws, etc.) are used for this purpose, along with an increasing array of connecting devices, mostly of steel. Chapter 19 deals with some of the common hardware methods for achieving the assemblage of wood structures for buildings.

19.1. NAILS AND SPIKES

Nails are made in a wide range of sizes and forms for many purposes. They range in size from tiny tacks to huge spikes. Most nails are driven by someone pounding a hammer—more or less as people have done for thousands of years. For situations where many nails must be driven, however, we now have a number of mechanical or automated driving devices. The use of driving equipment and industrialized processes of assemblage has resulted in ordinary nails being replaced by other fasteners in some situations. Staples are easier to drive in many cases and have replaced nails in many situations where mechanical driving equipment is used. Wide use of power-driven screwdrivers has resulted in the use of screws in place of nails, especially where a more positive resistance to withdrawal is desired.

For structural fastening in light wood framing, the nail most commonly used is called—appropriately—the *common wire nail*. Basic concerns for the use of the common nail, as shown in Fig. 19.1, are the following.

1. *Nail Size.* Critical dimensions are the diameter and length. Sizes are specified in pennyweight units, designated as 4d, 6d, and so on, and referred to as four penny, six penny, and so on.
2. *Load Direction.* Pull-out loading in the direction of the nail shaft is called *withdrawal*; shear loading perpendicular to the nail shaft is called *lateral load.*
3. *Penetration.* Nailing is typically done through one element and into another, and the load capacity is limited by the amount of length of embedment of

the nail in the second member. The length of embedment is called the penetration.

4. *Species and Grade of Wood.* The harder, tougher, and heavier the wood, the more the load resistance capability.

Design of good nail joints requires a little engineering and a lot of good carpentry. Some obvious situations to avoid are those shown in Fig 19.2. A little actual carpentry experience is highly desirable for anyone who designs nailed joints.

Withdrawal load capacities of common wire nails are given in Table 19.1. The capacities are given for both Douglas fir–larch and southern pine. Note that the table values are given in units of capacity per inch of penetration and must be multiplied by the actual penetration length to obtain the load capacity in pounds. In general, it is best not to use structural joints that rely on withdrawal loading resistance.

Lateral load capacities for common wire nails are given in Table 19.2. These values also apply to Douglas fir–larch and southern pine. Note that a penetration of at least 11 times the nail diameter is required for the development of the full capacity of the nail. A value of one-third of that in the table is permitted with a penetration of one-third of this length, which is the minimum penetration permitted. For actual penetration lengths between these limits, the load capacity may be determined by direct proportion. Orientation of the load to the direction of grain in the wood is not a concern when considering nails in terms of lateral loading.

The following example illustrates the design of a typical nailed joint for a wood truss.

Example 1. The truss heel joint shown in Fig. 19.3 is made with 2-in.-nominal wood elements of Douglas fir–larch, dense No. 1 grade, and gusset plates of $\frac{1}{2}$-in. plywood. Nails are 6d common, with the nail layout shown occurring on both sides of the joint. Find the tension force limit for the bottom chord (load 3 in the illustration).

Solution. The two primary concerns are for the lateral capacity of the nails and the tension, tearing stress in the gussets. For the nails, we observe from Table 19.2 that.

19.1. NAILS AND SPIKES

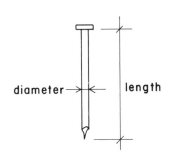

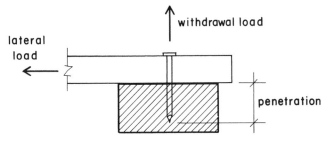

FIGURE 19.1. Typical common wire nail and loading conditions.

which is probably not a critical magnitude for the plywood.

A problem that must be considered in this type of joint is that of the pattern of placement of the nails (commonly called the *layout* of the nails). To accommodate the large number of nails required, they must be quite closely spaced, and since they are close to the ends of the wood pieces, the possibility of splitting the wood is a critical concern. The factors that determine this possibility include the size of the nail (essentially its diameter), the spacing of the nails, the distance of the nails from the end of the piece, and the tendency of the particular wood species to be split. There are no formal guidelines for this problem; it is largely a matter of good carpentry or some experimentation to establish the feasibility of a given layout.

One technique that can be used to reduce the possibility of splitting is to stagger the nails rather than to arrange them in single rows. Another technique is to use a single set of nails for both gusset plates, rather than to nail the plates independently, as shown in Fig. 19.4. The latter procedure consists simply of driving a nail of sufficient length so that its end protrudes from the gusset on the opposite side and

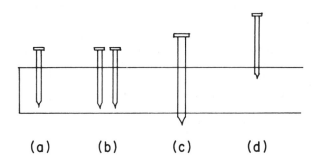

(a)   (b)   (c)   (d)

Nail length is 2 in. [5. mm].

Minimum penetration for full capacity is 1.24 in. [31.5 mm].

Maximum capacity is 63 lb/nail [0.28 kN].

From inspection of the joint layout, we see that

actual penetration = nail length − plywood thickness

= 2.0 − 0.5 = 1.5 in. [38 mm]

Therefore, we may use the full table value for the nails. With 12 nails on each side of the member, the total capacity is thus

$$F_3 = 24(63) = 1512 \text{ lb } [6.73 \text{ kN}]$$

If we consider the cross section of the plywood gussets only in the zone of the bottom chord member, the tension stress in the plywood will be approximately

$$f_t = \frac{1512}{0.5(2)(5 \text{ in. of width})} = 302 \text{ psi } [2.08 \text{ MPa}]$$

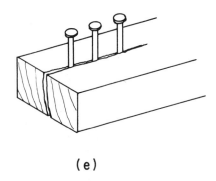

(e)

FIGURE 19.2. Poor nailing practices: (a) too close to edge; (b) nails too closely spaced; (c) nail too large for wood piece; (d) too little penetration into holding piece; (e) too many nails in a single row parallel to the wood grain.

TABLE 19.1. Withdrawal Load Capacity of Common Wire Nails (lb/in.)

	Size of Nail				
Pennyweight	6	8	10	12	16
Diameter (in.)	0.113	0.131	0.148	0.148	0.162
Douglas fir–larch	29	34	38	38	42
Southern pine	35	41	46	46	50

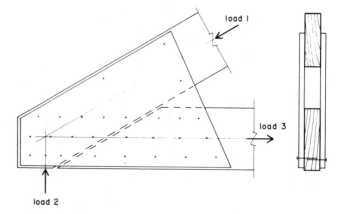

FIGURE 19.3. Truss joint with nails and plywood gusset plates.

then bending the end over—called *clinching*—so that the nail is anchored on both ends. A single nail may thus be utilized for twice its rated capacity for lateral load. This is similar to the development of a single bolt in double shear in a three-member joint.

It is also possible to glue the gusset plates to the wood pieces and to use the nails essentially to hold the plates in place only until the glue has set and hardened. The adequacy of such joints should be verified by load testing, and the nails should be capable of developing some significant percentage of the design load as a safety backup for the glue.

Box nails are slightly thinner versions of the common nail. They have lower structural load capacities, but have the advantage of easier driving and less tendency to split the wood. For connections involving low magnitudes of force transfer or with materials subject to splitting, box nails may be preferred to common nails.

Spikes are essentially sturdier versions of the common nail and are used with heavier members. Large spikes may be alternates to lag screws or bolts for some joints. Driving of large spikes requires real brute force, and the finished form of connections does not have a very precise or neat appearance.

TABLE 19.2. Lateral Load Capacity of Common Wire Nails (lb/nail)

	Size of Nail				
Pennyweight	6	8	10	12	16
Diameter (in.)	0.113	0.131	0.148	0.148	0.162
Length (in.)	2.0	2.5	3.0	3.25	3.5
Douglas fir–larch and southern pine	63	78	94	94	108
Penetration required for 100% of table value[a] (in.)	1.24	1.44	1.63	1.63	1.78
Minimum penetration[b] (in.)	0.42	0.48	0.54	0.54	0.59

[a] Eleven diameters; reduce by straight-line proportion for less penetration.
[b] One-third of that for full value; 11/3 diameters.

gussets nailed
independently

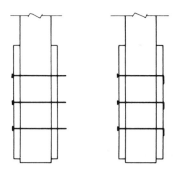

single set of nails driven
through and clinched

FIGURE 19.4. Nailing techniques for joints with plywood gusset plates.

19.2. SCREWS AND LAG BOLTS

Screws are used for wood connections in building structures whenever a positive anchoring is required or a real tightening of the connection is desired. When loosening or popping of nails is a problem, a more positive grabbing of the nail by the surrounding wood can be achieved by having some nonsmooth surface on the nail. Special nails are produced with such surfaces, but another simple means for achieving the same effect is to use threaded screws in place of the nails. Screws can also be tightened down to squeeze connected members together in a manner that is not quite possible with nails. For dynamic loading, such as shaking by an earthquake, the tight, positively anchored connection is a real advantage.

Three commonly used types of wood screw for building structures are shown in Fig. 19.5. Figure 19.5a shows a countersunk-head, tapered screw used widely for general wood work. Where protrusion of the screw head above the wood surface is not desired, this is the form of choice. For structural connections, the form of screw shown in Fig. 19.5b may be more useful. This screw is used where some thinner element, such as a sheet metal part or metal fastener device, is to be attached; the screw head functions like a bolt head and can be used with a washer if necessary.

The fastener shown in Fig. 19.5c is called a *lag screw* or *lag bolt*. Its form is in general the same as the screw in Fig. 19.5b, except that the head is designed to be turned by a wrench rather than a screw driver. While the screws shown in Fig. 19.5a and b can be obtained in large sizes, driving of the bolt-type head of the lag screw is generally more functional as the screw size increases.

Screws function essentially the same as nails, being used to resist either withdrawal or lateral, shear-type loading. Screws must usually be inserted into predrilled holes of a diameter slightly smaller than that of the screw. This is the more necessary when the wood is of harder species and grades or the length of the screw is great. Criteria for the sizes of guide holes and for the capacities of screws in withdrawal or lateral loading is provided in the NDS.

As with nailed joints, the use of screws involves much good judgment that is more craft than science. Screw type, size, spacing, length, and other details of a good joint may depend on some specifications, but are usually also a matter of some experience and care.

19.3. BOLTS

Steel bolts are used to connect wood structural members in many different situations. In general, a bolted joint is achieved by drilling holes slightly larger than the bolt diameter (usually $\frac{1}{16}$ in. larger) through the wood members and inserting the bolt like a shear pin. The end of the bolt is threaded and the joint is completed by placing a nut on the threaded end and tightening it down until the parts are squeezed together. The bolts may be used to resist tension or shear actions by the bolts, relating to the separation or slipping of the wood members.

Bolting may be done to achieve direct connection of wood pieces, usually in a lapped joint. Bolts are also used to connect various steel devices to the wood members. Column bases, column-to-beam connections, and splices of long members are frequently done with elements made of steel plate when the wood members are large. Heavier connection elements will usually be attached to the wood pieces by bolts or lag screws. Lighter elements used for smaller wood members may consist of sheet metal devices with connections achieved with nails or wood screws.

FIGURE 19.5. Wood screws: (a) flat or countersunk head; (b) round head; (c) lag screw (also called *lag bolt*) with square or hexagonal bolt-type head.

Where deformation or stress reversal in a joint is critical, it is sometimes necessary to use some element to reduce the tendency to slip in bolted joints. Some of the means for achieving this are discussed in Sec. 19.4.

There are many considerations to be made in the design of bolted joints. We will consider here only the limited situation of joints of modest size, using a relatively few number of bolts and wood members of smaller dimension. The principal concerns in this situation are the following:

1. *Net Stress in Member.* Holes drilled for the placing of bolts reduce the member cross section. For this analysis the hole is assumed to have a diameter $\frac{1}{16}$ in. larger than that of the bolt. The most common situations are those shown in Fig. 19.6. When bolts are staggered, it may be necessary to make two investigations, as shown in the illustration.

2. *Bearing of the Bolt on the Wood and Bending in the Bolt.* When the members are thick and the bolt thin and long, the bending of the bolt will cause a concentration of stress at the edges of the members. The bearing on the wood is further limited by the angle of the load to the grain, since wood is much stronger in compression in the grain direction.

3. *Number of Members Bolted.* The worst case, as shown in Fig. 19.7, is that of the two-member joint. In this case the lack of symmetry in the joint produces considerable twisting. This situation is referred to as *single shear*, since the bolt is subjected to shear on a single plane. When more members are joined, this twisting effect is reduced.

4. *Ripping Out the Bolt When Too Close to an Edge.* This problem, together with that of the minimum spacing of the bolts in multiple-bolt-joints, is dealt with by using the criteria given in Fig. 19.8. Note that the limiting dimensions involve the consideration of: the bolt diameter D; the bolt length L; the

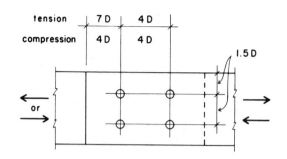

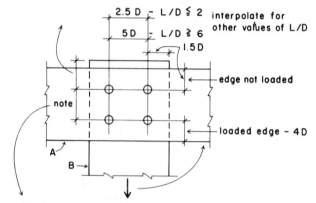

4D if member B is critical limit.
Straight line proportion if design load is less than limit.
1.5D minimum.

FIGURE 19.8. Edge, end, and spacing distances for bolts in wood structures.

type of force—tension or compression; and the angle of load to the grain of the wood.

The bolt design length is established on the basis of the number of members in the joint and the thickness of the wood pieces. There are many possible cases, but the most common are those shown in Fig. 19.9. The critical lengths for these cases are given in Table 19.3. Also given in the table is the factor for determining the allowable load on the bolt. Allowable loads ordinarily are tabulated for the three-member joint (Case 1 in Table 19.3), and the factors represent adjustments for other conditions.

Table 19.4 gives allowable loads for bolts with wood members of dense grades of Douglas fir–larch and southern pine. The two loads given are that for a load parallel to the grain (P load) and that for a load perpendicular to the grain (Q load). Figure 19.10 illustrates these two loading conditions, together with the case of a load at some other angle (θ). For such cases it is necessary to find the allowable load for the specific angle. Figure 19.10d is an adaptation of the Hankinson graph, which may be used to find values for loads at some angle to the grain.

The following examples illustrate the use of the data presented for the design of bolted joints.

Example 1. A three-member bolted joint is made with members of Douglas fir–larch, dense No. 1 grade. The members are loaded in a direction parallel to the grain

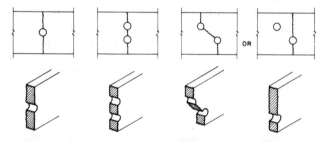

FIGURE 19.6. Effect of bolt holes on reduction of cross section for tension members.

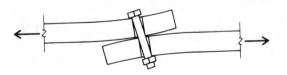

FIGURE 19.7. Behavior of the single-lapped joint with the bolt in single shear.

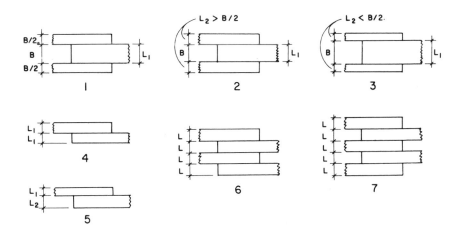

FIGURE 19.9. Various cases of lapped joints with relation to the determination of the critical bolt length (see Table 22.3).

TABLE 19.3. Design Length for Bolts

Case[a]	Critical Length	Modification Factor
1	L_1	1.0
2	L_1	1.0
3	$2L_2$	1.0
4	L_1	0.5
5	Lesser of L_2 or $2L_1$	0.5
6	L	1.5
7	L	2.0

[a] See Fig. 19.9.

TABLE 19.4. Bolt Design Values for Joints with Douglas Fir–Larch and Southern Pine

		Design Values for One Bolt in Double Shear[a] (lb)			
		Parallel to Grain Load (P)		Perpendicular to Grain Load (Q)	
Design Length of Bolt (in.)	Diameter of Bolt (in.)	Dense Grades	Ordinary Grades	Dense Grades	Ordinary Grades
1.5	1/2	1100	940	500	430
	5/8	1380	1180	570	490
	3/4	1660	1420	630	540
	7/8	1940	1660	700	600
	1	2220	1890	760	650
2.5	1/2	1480	1260	840	720
	5/8	2140	1820	950	810
	3/4	2710	2310	1060	900
	7/8	3210	2740	1160	990
	1	3680	3150	1270	1080
3.0	1/2	1490	1270	1010	860
	5/8	2290	1960	1140	970
	3/4	3080	2630	1270	1080
	7/8	3770	3220	1390	1190
	1	4390	3750	1520	1300
3.5	1/2	1490	1270	1140	980
	5/8	2320	1980	1330	1130
	3/4	3280	2800	1480	1260
	7/8	4190	3580	1630	1390
	1	5000	4270	1770	1520
5.5	5/8	2330	1990	1650	1410
	3/4	3350	2860	2200	1880
	7/8	4570	3900	2550	2180
	1	5930	5070	2790	2380
	1 1/4	8940	7640	3260	2790
7.5	5/8	2330	1990	1480	1260
	3/4	3350	2860	2130	1820
	7/8	4560	3890	2840	2430
	1	5950	5080	3550	3030
	1 1/4	9310	7950	4450	3800

[a] See Table 19.3 for modification factors for other conditions.

(Fig. 19.10a), with a tension force of 10 kips. The middle member is a 3 × 12 and the outer members are each 2 × 12. If four $\frac{3}{4}$-in. bolts are used for the joint, is the joint capable of carrying the tension load?

Solution. The first step is to identify the critical both length (Fig. 19.9) and the load factor (Table 19.3). Since the outer members are greater than one-half the thickness of the middle member, the condition is that of Case 2 in Fig. 19.9, and the effective length is the thickness of the middle member: 2.5 in. The load factor from Table 19.3 is 1.0, indicating that the tabulated load may be used with no adjustment.

From Table 19.4 the allowable load per bolt is 2710 lb (bolt design length of 2.5 in.; bolt diameter of $\frac{3}{4}$ in.; P load; dense grade.) With the four bolts, the total capacity of the bolts is thus 4 × 2710 = 10,840 lb.

For tension stress in the wood, the critical condition is for the middle member with the net section through the bolts being that shown in Fig. 19.6. With the holes being considered as $\frac{1}{16}$ in. larger than the bolts, the net area for tension stress is thus

$$A = 2.5\left[11.25 - \left(2 \times \frac{13}{16}\right)\right] = 24.1 \text{ in.}^2$$

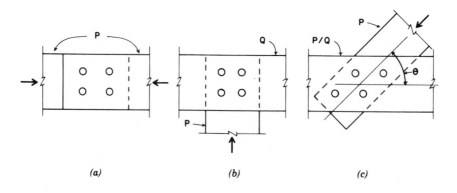

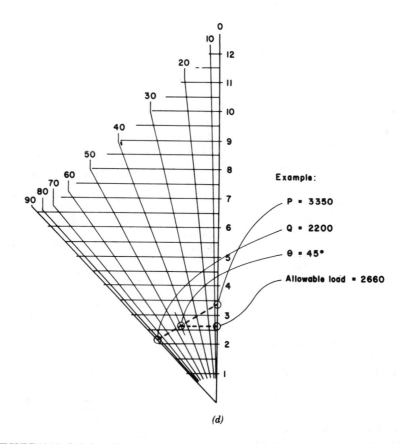

FIGURE 19.10. Relation of load to grain direction in bolted joints: (a) members parallel; (b) members perpendicular; (c) members at less than 90°; (d) Hankinson graph for determination of load values with the loading at an angle to the grain, as in (c). See Table 22.4 for values of P and Q.

From Table 16.1 the allowable tension stress is 1200 psi. The maximum tension capacity of the member in tension at the net section is thus

T = allowable stress × net area
$$= 1200 \times 24.1 = 28{,}920 \text{ lb}$$

and the joint is adequate for the required load.

Example 2. A bolted two-member joint consists of two 2 × 10 members of Douglas fir–larch, dense No. 2 grade, attached at right angles to each other, as shown in Fig. 19.10b. If the joint is made with two $\frac{7}{8}$-in. bolts, what is the maximum compression capacity of the joint?

Solution. This is Case 4 in Fig. 19.9, and the effective length is the member thickness of 1.5 in. The modification factor from Table 19.3 is 0.5, and the bolt capacity from Table 19.4 is 700 lb per bolt (bolt design length of 1.5 in.; bolt diameter of $\frac{7}{8}$ in.; Q load; dense grade). The total capacity of the bolts is thus

$$C = 2(700)(0.5) = 700 \text{ lb}$$

The net section is not a concern for the compression force, as the capacity of the members would be based on an analysis for the slenderness condition based on the L/d of the members.

Example 3. A three-member bolted joint consists of two outer members, each 2 × 10, and a middle member that is a 4 × 12. The outer members are arranged at an angle to the middle member, as shown in Fig. 19.10c, such that $\theta = 60°$. Find the maximum compression force that can be transmitted through the joint by the outer members. Wood is southern pine, No. 1 grade. The joint is made with two $\frac{3}{4}$-in. bolts.

Solution. In this case we must investigate both the outer and middle members. For the outer members the effective length is 2 × 1.5 = 3.0 in., and the modification factor is 1.0 (Case 3, Fig. 19.9 and Table 19.3). From Table 19.4 the bolt capacity based on the outer members is 2630 lb per bolt (Bolt design length of 3.0 in.; bolt diameter of $\frac{3}{4}$ in.; P load; ordinary grade).

For the middle member the effective bolt length is the member thickness of 3.5 in., and the unadjusted load per bolt from Table 19.4 is 2800 lb for the P condition and 1260 lb for the Q condition. If these values are used on the Hankinson graph in Fig. 19.10d, the load per bolt for the 60° angle is found to be approximately 1700 lb. Since this value

is lower than that found for the outer members, it represents the limit for the joint. The joint capacity based on the bolts is thus 2 × 1700 = 3400 lb.

19.4. SHEAR DEVELOPERS

When wood members are lapped at a joint and bolted (as shown in Fig. 19.9) it is difficult to maintain a tight joint that has no slip between the lapped members. If force reversals, such as those that occur with wind or seismic effects, cause back-and-forth stress on the joint, this looseness may be objectionable. Various devices are used between the lapped members to enhance resistance to slipping of the members. The type of device used is related to the size and form of the members and the magnitudes of the loads.

One device that is used for the shear-enhanced joint is the steel split-ring connector, shown in Fig. 19.11. Design considerations for the split ring include the following:

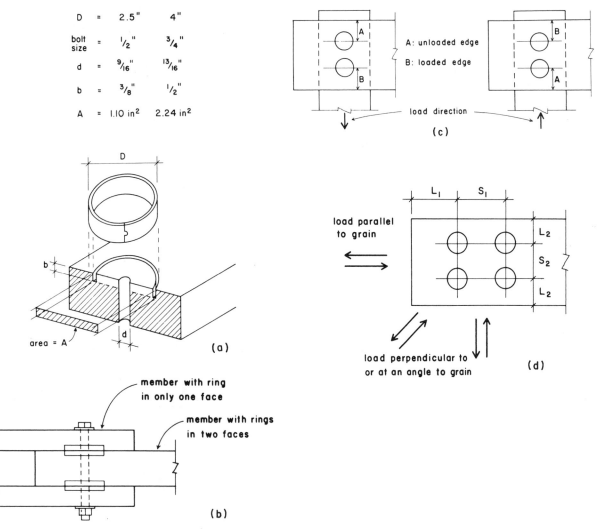

FIGURE 19.11. Split-ring connectors for bolted wood joints.

1. *Size of the Ring.* Rings are available in the two sizes shown in the figure, with nominal diameters of 2.5 and 4 in.
2. *Stress on the Net Section of the Wood Member.* As shown in Fig. 19.11*a*, the cross section of the wood piece is reduced by the ring profile (*A* in the figure) and the bolt hole. If rings are placed on both sides of a wood piece, there will be two reductions for the ring profile.
3. *Thickness of the Wood Piece.* If the wood piece is too thin, the cut for the ring will bite excessively into the cross section. Rated load values reflect concern for this.
4. *Number of Faces of the Wood Piece Having Rings.* As shown in Fig. 19.11*b*, the outside members in a joint will have rings on only one face, while the inside members will have rings on both faces. Thickness considerations, therefore, are more critical for the inside members.
5. *Edge and End Distances.* These must be sufficient to permit the placing of the rings and to prevent splitting out from the side of the wood piece when the joint is loaded. Concern is greatest for the edge in the direction of loading—called the *loaded edge* (see Fig. 19.11*c*).
6. *Spacing of Rings.* Spacing must be sufficient to permit the placing of the rings and the full development of the ring capacity of the wood piece. Figure 19.11*d* shows the four placement dimensions that must be considered.
7. *Load-to-Grain Orientation.* As with bolts, values for split rings are given for load directions both parallel to and perpendicular to the grain of the wood. Values for loadings at some other angle to the grain can be determined with the use of the Hankinson graph in Fig. 19.10*d*.

19.5. METAL FRAMING DEVICES

Metal devices of various kinds have been used for many centuries for the assembly of structures of heavy timber. Some of the steel devices commonly used today (see Fig. 19.12*b* and *c*) are essentially the same in function and general form to ones made of cast iron or bronze in ancient times. The ordinary tasks of attaching beams to columns and columns to footings continue to be required and the simple means of achieving the tasks evolve from practical concerns.

For connection of larger wood elements steel devices are mostly made from bent and welded steel plate. These are attached to the wood members by lag screws or bolts. For resistance to gravity loads, connections such as that shown in Fig. 19.12*b* have no essential purpose. It would, in theory, be possible to simply rest the beam on top of the column—as it is done in rustic construction in some cases. However, for resistance to lateral loads from wind or earthquakes, the tying functions of these connection devices are quite essential. Furthermore, the relative lack of depen-

(a)

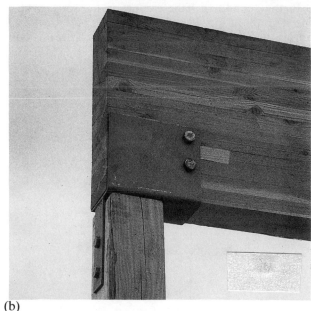

(b)

(c)

FIGURE 19.12. Metal connecting devices for wood structures.

dency on the craft of the field workers with the use of these devices is of worthy note as the craft level continuously declines.

An innovation of more recent times is the extension of the use of metal devices for assembly of light frame construction. Devices of thin sheet metal, such as those shown in Fig. 19.12*a* are now quite commonly used. These are mostly attached by nails; in some cases using special short nails that will not pass through the nominal 2-in.-thick members (see Fig. 19.12*a*). As with the heavier devices for large timber structures, these lighter devices serve well the function of tying the structure together to increase considerably resistance to lateral load effects.

Commonly used connection devices of both the light sheet metal type and the heavier steel plate type are available from neighborhood building supply stores. Many of these devices are rated by their makers for specific load resistances. If these ratings are approved by local building codes, the devices can be used for computed structural load transfers.

For unique situations it may be necessary to design a special framing device. However, the catalogs of manufacturers of these devices are filled with an amazing array of devices for all kinds of situations, and the designer should first assure that the required device is not already produced. Custom made and one-time-designed devices are sure to be more expensive.

19.6.   ADHESIVES

Adhesive compounds (good old glue, mostly) are used in wood structures for various purposes. The most extensive use is in the production of plywood and glued-laminated timber elements as discussed in Secs. 20.2 and 20.3. With the development of new products and the steady increase of their availability and reliability, however, new uses for adhesives continue to grow for structural components of buildings. Some common current uses are the following:

1. *Trusses.*   Light wood trusses are mostly fabricated with metal connection devices. However, for special situations, simple trusses are sometimes produced by using single 2-in.-nominal lumber elements in a single plane with joints consisting of nailed plywood gussets (see Fig. 19.3). These gusset plates are often glued to the truss members, although the gluing is usually not totally relied on for force transfer, and nails or screws are used. Nevertheless, the glue significantly stiffens the joints and reduces general deformation of the truss.

2. *Sandwich Panels.*   For wall units—or in some cases, for spanning roof or floor units—a form of construction sometimes used is that of a light wood frame with two faces of plywood, commonly called a *sandwich panel*. The ordinary hollow-core door is such an element. In most cases wall or spanning units will utilize nails or screws for attachment of the plywood to the frame. However, as with the truss gusset plates, glue may be used to enhance the stiffness of the panel. Also, loosening of screws or "popping" of nailheads is less likely.

3. *Floor Deck.*   A type of construction sometimes used is one in which a plywood floor deck is glued to the supporting joists with the intent of producing a composite action, similar to that with a concrete deck and steel beams. Whether the true composite action is accepted or not, the resulting floor structure is usually undeniably stiffer and there is less potential for squeaking due to the popping of nailheads. Some builders routinely glue the decks to joists, even though the composite action is not considered in the joist design.

4. *I-beams.*   Beams made with top and bottom flange elements of solid wood and webs of plywood or particleboard are being increasingly used. These are discussed in Sec. 20.5.

19.7.   DESIGN INFORMATION

Information about wood elements and the numerous devices used in association with them is available from many sources. A starting point for any design work should be the building code of jurisdiction for the work. The code itself may contain considerable information, including specifications for structural computations and tabulations of design data. Other sources, such as textbooks and handbooks, or industry-wide publications, may have data or specifications that differ from the building code. It is common for codes to lag somewhat behind the pace of change in industry and design practice. The code should be considered to be a prime source, as it will be the major reference for obtaining a permit for construction.

In addition to being dated, information may be somewhat limited regionally. There are several separate organizations in the wood industry, duplicating each other's efforts. Some are regionally based and have a relatively strong hold on some territory. Woods commonly used are also a regional matter, which adds to the provincial character of design work. Model building codes are also used predominantly in some regions. Finally, a real issue is the consideration of local climate and probability of windstorms or earthquakes.

Many of the elements used for wood structures (as well as structures of other materials) are produced as manufactured products. For these the designer must determine the particular details, materials, sizes, and other data from the manufacturer or local distributer of the products. Local availability should be a major consideration in selecting products. Just because a manufacturer's catalog is in *Sweet's Catalog File* does not mean that the product is available or competitive on the market everywhere. Specific data for the structural capacity of products must usually be obtained from manufacturers or suppliers. The acceptability of such data by local building codes should be established before the data are used.

CHAPTER TWENTY

Special Wood Structures

Ordinary solid-sawn structural lumber, as discussed in the preceding chapters of Part 3, is still widely used for buildings. However, considerable use is also made of various special structural elements and systems that utilize wood as a basic material. Chapter 20 deals with many of these in common use.

20.1. WOOD TRUSSES

Wood trusses are widely used—most often for roof structures, owing to the lighter loads, less concern for deflection, and ability of the truss to accommodate many forms. The feasible range of span for trusses generally begins at the point where ordinary rafters or joists end—at about 25 ft for roofs and 20 ft for floors. Chapter 20 discusses some of the general issues regarding the use of wood trusses. Design considerations of a more specific nature are presented in the examples of building design in Part 9.

Truss Form

As a structural type, the truss can be used to achieve a great range of possible forms. For building structures, some common truss profiles are those shown in Fig. 20.1. For floor structures, the form most used is the simple parallel-chorded truss with a flat top and flat bottom. The floor deck is directly attached to the top, and in most cases the ceiling for a lower space is attached directly to the bottom.

For roofs, the parallel-chorded truss may also be used when a flat roof surface is desired. However, there is really no such thing as a true flat roof, there always being a need for some drainage, with a minimum pitch of at least $\frac{1}{4}$ in. to a foot (1:48, or about 2%). Thus the top chord may be minimally pitched; all in one direction, or both ways from the center, as shown in Fig. 20.1. However, the truss is still essentially of the parallel type. Many manufactured trusses are of this form.

For steeper roof slopes, the simple gable or compound gable form may be used. Even the arched roof surface may be produced by a truss. If the truss is not exposed to view, it is common to use a horizontal bottom chord for direct support of a ceiling.

Other possible forms and uses are discussed in Chapters 5 and 13.

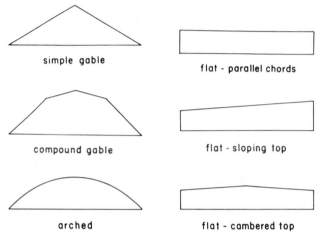

FIGURE 20.1. Typical profiles for wood trusses.

Truss Pattern

Pattern refers to the layout of the truss members. The starting point for this is the pattern of the chords, which defines the truss profile. While the simple gable form may consist only of the two top chords and a single bottom chord, most trusses have some number of interior members, called *web members*. As discussed in Chapter 13, the truss pattern must be one that produces a stable structure, essentially by creating a continuous series of connected triangles.

Some common patterns used to achieve the profiles shown in Fig. 20.1 are shown in Fig. 20.2. The two most widely used are the simple W pattern, used for the roof and ceiling of small buildings, and the Warren pattern, used for most manufactured trusses. The choice of a particular pattern depends on the size and profile of the truss, the types of members, and the method of making joints between members. It may also depend on the method of support, the manner of loading of the truss, the means for achieving lateral stability, and the magnitude of loads to be carried.

Truss Members and Joints

Three common forms of truss construction are those shown in Fig. 20.3. The single-member type, with all members in one plane, as shown in Fig. 20.3a, is that used most often to produce the simple W-form truss, with members usually of 2 in. nominal thickness. Joints may be made with plywood gussets, as shown in the figure, but are most often made with metal connecting devices when the trusses are produced as standard products by a manufacturer. In the latter case, the joint performance will be certified by load testing of prototypes.

For larger trusses the configuration shown in Fig. 20.3*b* may be used, with members consisting of multiples of standard lumber elements. If the member carries compression, it will usually be designed as a spaced column, as described in Sec. 18.3. For modest spans, members are usually of two elements of nominal 2-in. thickness. However, for large

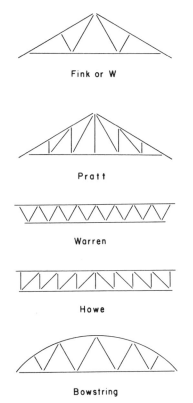

Fink or W

Pratt

Warren

Howe

Bowstring

FIGURE 20.2. Typical patterns utilized for wood trusses.

spans or very heavy loads, individual members may consist of several pieces with thicknesses greater than 2 in. nominal. Joints are usually achieved with bolts.

The so-called *heavy timber truss* is one in which the individual members are of large timber elements, usually occurring in a single plane as shown in Fig. 20.3*c*. A common type of joint in this case is one using steel side plates as gussets with the plates attached by lag screws or through bolts. Depending on the truss pattern and loading, it may also be possible to make some joints without the gussets, as shown for the connection of the interior compression member and the bottom chord in Fig. 20.4*a*. This was common in times past, but requires carpentry work not easily obtained today.

Although wood members have considerable capability to resist tension, the achieving of tension connections is not so easy, especially in the heavy timber truss. Thus in some trusses tension members are made with steel elements, such as the steel rod shown in Fig. 20.4*a*. Another example of the mixture of wood and steel members is the composite truss, for which a common form is that shown in Fig. 20.4*b*.

Both member selection and jointing method depend on the size and loading conditions for the truss. Unless the truss is exposed to view and its details are considered to be important architectural design features, the specific choice of members and connection details is most often left to the fabricator for determination.

Manufactured Trusses

The majority of wood trusses used for building structures in the United States at present are produced as manufactured products and their detailed, engineering design is done largely by engineers in the employ of the manufacturers. Since shipping of large trusses over great distances is not generally feasible, the use of these products is mostly limited to some region within a reasonable distance of a particular manufacturer. If you want to use these products, you must first establish what suppliers there are within range of the building site.

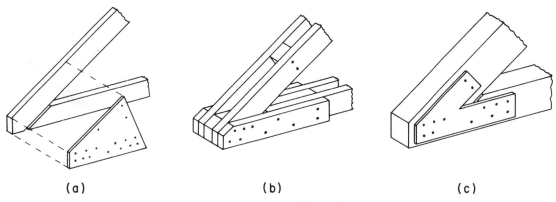

(a)   (b)   (c)

FIGURE 20.3. Typical details for wood trusses.

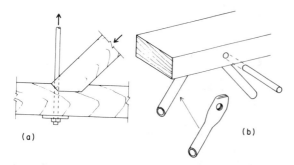

FIGURE 20.4. Composite trusses with wood and steel members.

There are three principal types of manufactured truss. Manufacturers vary in size of operation, some producing only one type and others having a range of products. The three basic types are:

1. Simple gable-form, W-pattern trusses with truss members of single-piece, 2-in.-nominal-thickness structural lumber. These are quite simple to produce and can be turned out in small quantities by neighborhood building suppliers. Larger companies use automated processes and more sophisticated handling in general, but the range of availability is considerably extended by the simplicity of the product.
2. Light, prefabricated trusses, usually consisting of some type of composite wood and steel elements (see Fig. 20.4). These are more sophisticated in detail than the simple W trusses and are produced primarily by larger companies. They compete in general with equivalent trusses of steel (called *open-web joists*). The choice of the composite wood + steel or the all-steel product is largely a matter of local situation, depending on the marketing practices and the location of producers. The fairly recent development of lamination of wood chord members has both improved the quality of the composite product and made larger, longer-spanning trusses possible.
3. Large long-span trusses, usually using multiple-element members, as shown in Fig. 20.3b. These may be produced with some standardization by a specific manufacturer, but are typically customized to some degree for a particular building. One form used for very long spans is the bowstring type, which is in reality essentially a tied arch (see Fig. 20.2).

Suppliers of manufactured trusses usually have some fairly standardized models, but almost always have some degree of variability to accommodate the specific usage considerations for a particular building. The possible range of these variations can be established only by working with the suppliers.

As in many other similar situations, using a product such as a manufactured truss system requires some management of the relationships between the architect, the architect's consulting engineer, and the engineers who do the actual design work for the product's manufacturer. Who does what, who is responsible for what, who gets paid for what, and who gets sued when the structure does not work should all be carefully considered.

Heavy Timber Trusses

Single members of large-dimension timber were used for trusses hundreds of years ago. Cast iron and steel elements were steadily employed for various tasks in this construction as the iron and steel technology developed in the nineteenth century. Figure 20.5 shows the details of this type of construction as it was developed early in this century. This type of construction is seldom used today, except for restoration of historic buildings or nostalgic imitations. The special parts, such as the cast-iron shoe, and the highly crafted notched joints are generally unobtainable.

Trusses of the size and general form of that shown in Fig. 20.5 are now mostly made with multiple-element members and bolted joints, as shown in Fig. 20.3b. Steel elements may be used for some tension members, but this is done less often today since joints using shear developers are capable of sufficient load resistance to permit use of wood tension members. It was essentially the connection problem that inspired the use of the steel rods in the timber truss. The one vestige of this is the use of the steel web members in the manufactured trussed joists (see Fig. 20.4).

The heavy timber structure is a rustically handsome exposed architectural feature and is still often used in buildings. It may be combined with other rough-textured natural materials, such as board deck and plank floors, or may be made a counterpoint in combination with slicker, more modern materials. As with other imitations of past methods of construction, it is often difficult to achieve in a true manner because the craft and supporting details are no longer obtainable. Some compromise with modern materials and processes is usually necessary.

Reference

F. Kidder and H. Parker, *Kidder-Parker Architects' and Builders' Handbook*, 18th ed., Wiley, New York, 1931.

Bracing for Trusses

Single planar trusses are very thin structures that require some form of lateral bracing. The compression chord of the truss must be designed for its laterally unbraced length. In the plane of the truss, the chord is braced by other truss members at each joint. However, if there is no lateral bracing, the unbraced length of the chord in a direction perpendicular to the plane of the truss becomes the full length of the truss. Obviously, it is not feasible to design a slender compression member for this unbraced length.

In most buildings other elements of the construction ordinarily provide some or all of the necessary bracing for the trusses. In the structural system shown in Fig. 20.6a, the

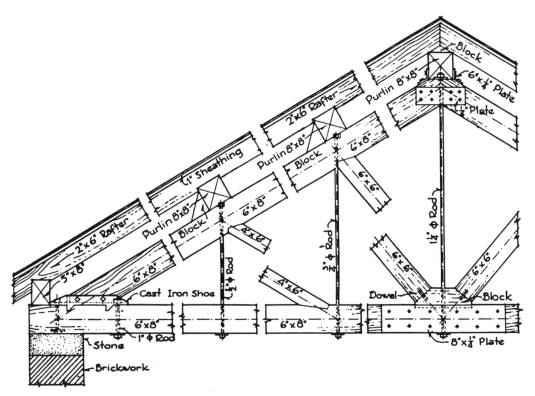

FIGURE 20.5. Typical form of old heavy timber truss. Reproduced from *Materials and Methods of Architectural Construction* by C. M. Gay and H. Parker, 1932, with permission of John Wiley & Sons, Inc., New York.

top chord of the truss is braced at each truss joint by the purlins. If the roof deck is a reasonably rigid planar structural element and is adequately attached to the purlins, this constitutes a very adequate bracing of the compression chord—which is the main problem for the truss. However, it is also necessary to brace the truss generally for out-of-plane movement throughout its height. In Fig. 20.6a this is done by providing a vertical plane of X-bracing at every other panel point of the truss. The purlin does an additional service by serving as part of this vertical plane of trussed bracing. One panel of this bracing is actually capable of bracing a pair of trusses, so that it would be possible to place it only in alternate bays between the trusses. However, the bracing may be part of the general bracing system for the building, as well as providing for the bracing of the individual trusses. In the latter case, it would probably be continuous.

Light truss joists that directly support a deck, as shown in Fig. 20.6b, usually are adequately braced by the deck. This constitutes continuous bracing, so that the unbraced length of the chord in this case is actually zero. Additional bracing in this situation often is limited to a series of continuous steel rods or single small angles that are attached to the bottom chords as shown in the illustration.

Another form of bracing that is used is that shown in Fig. 20.6c. In this case a horizontal plane of X-bracing is placed between two trusses at the level of the bottom chords. This single braced bay may be used to brace several other

bays of trusses by connecting them to the X-braced trusses with horizontal struts. As in the previous example, with vertical planes of bracing, the top chord is braced by the roof construction. It is likely that bracing of this form is also part of the general lateral bracing system for the building so that its use, location, and details are not developed strictly for the bracing of the trusses.

When bracing between trusses is not desired, it is sometimes possible to use a structure that is self-bracing. One form of self-bracing truss is the delta truss, which is discussed in Sec. 25.1.

Trussed Bracing for Frames

Trussing is sometimes used as the technique for developing resistance to the lateral forces due to wind or earthquakes on a framed structure. A type of construction long in use for the bracing of light wood frames with wall studs is that shown in Fig. 20.7a. This consists of the use of diagonal elements of 1-in.-nominal thickness, which are notched in (commonly called *let in*) the faces of the studs and are nailed at each stud. Although still used sometimes, this practice has largely been abandoned with the advent of rated shear design capacities for a wide range of wall surfacing materials, including plaster, drywall gypsum wallboard, and wood particleboard or fiberboard.

Where trussing is utilized as bracing for frames consisting of heavy timber elements, a form often used is that

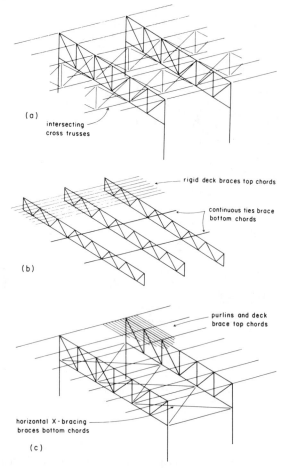

(a)

intersecting
cross trusses

rigid deck braces top chords

continuous ties brace
bottom chords

(b)

purlins and deck
brace top chords

horizontal X - bracing
braces bottom chords

(c)

FIGURE 20.6. Forms of lateral bracing for trusses.

shown in Fig. 20.7b. The trussing is formed with diagonal members consisting of steel rods, placed in X forms so that the diagonals work only in tension. Horizontal elements of the wood frame are usually utilized in combination with the vertical columns and X-bracing to form the complete trussing pattern.

20.2. PLYWOOD

Plywood is the term used to designate wood panels made by gluing together multiple layers of thin wood veneer (called *plies*) with alternate layers having their grain direction at right angles. The outside layers are called the *faces* and the others, *inner plies*. Inner plies with the grain direction perpendicular to the faces are called *crossbands*. There is usually an odd number of plies so that the faces have the grain in the same direction. For structural use as wall sheathing or roof and floor decks, the common range of panel thickness is $\frac{1}{4}$ to $1\frac{1}{8}$ in. and the usual panel size is 4×8 ft.

The alternating grain direction of the plies gives the panels considerable resistance to splitting, and as the number of plies is increased the panels become essentially

equally strong in both directions. Thinner panels will be most effective when spanning with the face plies having their grain direction perpendicular to the supporting studs or joists. For $\frac{3}{4}$-in. and larger thicknesses this becomes less critical. The most widely used references for plywood are those published by the American Plywood Association.

Types and Grades of Plywood

Structural plywood consists primarily of that made with all plies of Douglas fir. Many different kinds of panels are produced; the principal distinctions other than the panel thickness are the following:

1. *Glue Type.*   Panels are identified for exterior (exposed to the weather) or interior use, based on the type of glue used. Exterior type should also be used for any interior conditions involving high moisture.
2. *Grade of Plies.*   Individual plies are rated—generally as A, B, C, or D, with A best—on the basis of the presence of knots, splits, patches, or plugs. The most critical ratings are those of the face plies; thus a panel is typically designated by the ratings of the front and back plies. For example, C-C indicates both faces are of C grade; C-D indicates the front is C grade and the back is D grade.
3. *Structural Classification.*   In some cases panels are identified as structural I or structural II. This is mostly of concern when the panels are used for shear walls or horizontal roof or floor diaphragms. For this rating the grade of the interior plies is also a concern.
4. *Special Faces.*   Plywood with special facing, usually only on one side, is produced for a variety of uses. Special surfaces for use as exposed siding are one such example. These are usually produced as some particular manufacturer's special product and any structural properties or other usage considerations should be obtained from the supplier.

Some ratings and classifications are industry-wide and some are local variations due to the use of a particular building code or the general use of particular products. Designers should be aware of the general industry standards but also of what products are generally available and widely used in a given locality.

References

American Plywood Association publications:
    Plywood Design Specification
    Plywood Floor Systems
    Plywood Roof Systems
Uniform Building Code, 1991 ed., International Conference of Building Officials, 5360 South Workmanmill Road, Whittier, CA 90601, Chap. 25.

Panel Identification Index

Structural grades of plywood usually have a designation called the *Identification Index*, which is stamped on the

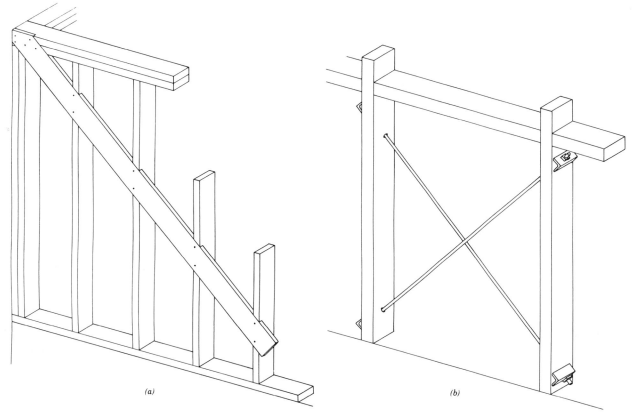

FIGURE 20.7. Trussing used for lateral bracing of wood structures: (a) diagonal let-in member of 1-in.-nominal thickness with light wood frame; (b) steel tie rods for X-bracing in heavy timber frame.

panel back as part of the grade trademark. This index is a measure of the strength and stiffness of the panel and consists of two numbers separated by a slash (/). The first number indicates the maximum center-to-center spacing of supports for a roof deck under average loading conditions and the second number indicates the maximum spacing for a floor deck for average residential loading. There are various conditions on the use of these numbers, but they generally permit the selection of panels for a specific ordinary situation without the need for further structural computations.

Usage Data

Data for structural design with plywood may be obtained from industry publications or from individual plywood producers. This type of data will usually be incorporated in a sales document with some other useful information about the product use and a lot of promotional material. Some building codes have data for plywood design for ordinary situations, usually adapted from industry-wide publications. Three tables from the *Uniform Building Code* consist of adaptations from materials produced by the American Plywood Association. The tables and their applications are as follows:

1. *UBC Table No. 25-N-1.* Plywood Wall Sheathing. This table yields requirements for ordinary use of panels for structural wall covering (called *sheathing*). If the plywood is also used for shear wall actions, it is also necessary to use data from other tables, as described in Part 8.
2. *UBC Table No. 25-S-1.* Allowable Spans for Plywood Subfloor and Roof Sheathing Continuous over Two or More Spans and Face Grain Perpendicular to Supports. This long title pretty well tells the story. As with wall sheathing, diaphragm action must be considered by using other data, although the basic deck functions for gravity loads must first be satisfied. (See Appendix D.)
3. *UBC Table No. 25-S-2.* Allowable Loads for Plywood Roof Sheathing Continuous Over Two or More Spans and Face Grain Parallel to Supports. This table is a complement to Table 25-S-1, yielding data relating to the placement of panels with the long panel dimension parallel to the rafters—an advantage in some situations as discussed in Sec. 20.3. (See Appendix D.)

Footnotes to these tables present various qualifications including some of the loading and deflection criteria. There are other panel thicknesses and types of panels not covered by these tables that may be designed from data available from manufacturers or suppliers. Code acceptability of such data should always be assured before use for any design work.

Usage Considerations

There are many uses for plywood in building construction and a broad array of products available. For ordinary structural applications the following are some of the principal usage considerations.

1. *Choice of Type and Grade.* This is essentially a matter of common usage and code acceptability. For economy the thinnest, lowest-grade panels will always be used for structural applications, unless other, non-structural concerns require better face grades or some special need. Roof decks may be required to be a minimum thickness to hold shingle nails or to support membrane roofing adequately.

2. *Modular Supports.* With the usual common panel size of 4 × 8 ft, logical spacing of studs, rafters, and joists become full number divisions of the 48- or 96-in. panel sides: 12, 16, 24, 32, 48. However, spacing of supports often also relates to what is attached to the other side, which may require closer spacing than that strictly required for structural considerations.

3. *Blocking.* Panel edges not falling on the supports may need support, especially for roofs and floors. This may be required for gravity load actions or for development of adequate diaphragm action and is discussed in detail in Sec. 17.5.

4. *Attachment.* Structural attachment of plywood is still largely made with common wire nails. Nail size and spacing relates to panel thickness and minimum code requirements and, where applicable, to requirements for diaphragm action. Structural gluing of panels to supports is done in some situations, but is seldom relied on exclusively. Nailing is now often done with mechanical equipment and the actual driven fastener is not always the usual common nail, although its rating is usually based on an equivalency to the rated capacities for common nails.

5. *Fabricated Units.* Various types of construction components are produced for use as wall, roof, or floor units, consisting of assemblages of plywood and solid wood elements. Some types of units presently used are discussed in Sec. 20.5.

20.3. GLUED-LAMINATED PRODUCTS

The basic process of gluing together of pieces of wood that is used to produce plywood panels can be extended to other applications. This section deals with some of the other products that are produced by glue laminating for use in building structures.

Types and Usage of Products

For structural applications the principal types of products and their usual usages are the following (see Fig. 20.8):

1. *Multiple Laminated Structural Lumber.* These elements are produced by laminating multiples of

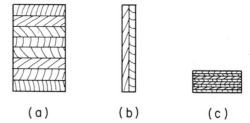

FIGURE 20.8. Forms of glued-laminated wood products.

standard 2X or 1X members. The most widely used product is the beam or girder produced from multiple 2X members; however, elements may also be curved and the 1X members are used when a shorter radius of curvature is desired. Size and length are virtually unlimited and many huge elements have been produced as girders, arches, and gabled bents.

2. *Vertical Laminated Joists.* These generally consist of $\frac{3}{4}$-in.-thick (nominal 1X) boards laminated in multiples of two or three for use as joists to compete with the high end of the range of standard nominal 2X and 3X standard lumber widths. For the spans at which they are competitive, they must also vie with the built-up elements described in Sec. 20.4.

3. *Thin Lamination Elements.* These consist of small cross sections produced with laminations of about 0.1-in. thickness. Their principal use is for the chords of manufactured trusses, where their high strength and unlimited lengths are major advantages.

Large Beams and Girders

Lamination of standard 2-in.-nominal-thickness lumber has been used for many years to produce large beams and girders. This is really the only option for a wood member beyond the feasible range of size of solid-sawn lumber. The limit of size range for the solid-sawn member is a local matter of availability. However, there are other reasons for using the laminated member which include the following:

1. *Higher Strength.* Lumber used for laminating is of a moisture condition described as kiln dried, which is the opposite end of the quality range from that of the green condition usually assumed for large solid-sawn sections. This, plus the minimizing of effects of flaws due to lamination, permits use of stresses for flexure and shear that are as much as twice those allowed for ordinary solid-sawn sections. The result is that much smaller sections can be used, which may be enough to justify the choice of the expensive laminated product in some situations.

2. *Better Dimensional Stability.* This refers to the tendency to warp, shrink, and so on. Both the use of the kiln-dried materials and the laminating process itself tend to create a very stable product. This is often a major consideration where the shape changes can affect the general building construction.

3. *Shape Variability.* Lamination permits the production of curved, tapered, and other profile forms for a beam, as shown in Fig. 20.9. Cambering for dead-load deflections, sloping of tops for roof drainage, and other desired customized profiling can be done with relative ease. This is otherwise possible only with a truss or a built-up section of the kind described in Sec. 20.4.

This type of member has seen wide use for may years and industry-wide standards are well-established [see the *Timber Construction Manual* (Ref. 8) or Chapter 25 of the *UBC* (Ref 1)]. Sizes used are standardized by the process and the standard lumber elements used for lamination. Thus depths are simply multiples of the 1.5-in. thicknesses, and widths are slightly less than the lumber widths as a result of the finishing of the product. Minor misalignment of laminations and the sloppiness of the gluing produce a generally unattractive face. Finishing may simply be a smoothing off, although special finishes such as rough-sawn ones can be produced.

These elements are manufactured products and information about them should be obtained from suppliers that serve the area in the region of a building site. There is a lot of industry standardization, but it is still best to know the real products available from a supplier.

Building codes and timber industry organizations provide data and design standards for the design of glued-laminated products. It is thus theoretically possible for designers to produce customized designs for individual members of this type. However, the finished design work will almost always be done by experienced engineers in the employ of the company that actually fabricates—and in many cases, actually erects—the finished products. Data for reasonably precise preliminary design of members is typically provided by the same manufacturers, or by industry publications. As a result of this situation, hardly any building designers bother to design such elements from scratch, except for very unusual forms or loading conditions.

Arches and Bents

Individual elements of glue-laminated lumber can be custom profiled to produce a wide variety of sizes and shapes for structures. The two most commonly produced are the three-hinged arch and the gabled bent. A critical consideration is the radius of curvature of the member, which must be limited to that which the wood species and the lamination thickness can tolerate. For very large elements this is not a problem, but for smaller structures the limits of the standard 2-in.-nominal lumber may be critical. The latter situation sometimes results in the choice of 1-in.-nominal ($\frac{3}{4}$-in.-actual)-thickness laminates; which is usually the only time they are used.

Manufacturers of laminated products usually produce the arch and gabled bent elements as more-or-less standard items. Actual structural design of the products is most often done by the manufacturers' engineers. Limitations, size ranges, connection details, and other considerations should be investigated with individual manufacturers.

Other shapes, of course, are possible, such as that of doubly curved members. Many highly imaginative structures have been designed using the form variation potentials of this process of wood construction (see Fig. 4.1).

Columns

Columns consisting of multiples of glued-laminated 2-in.-nominal-thickness lumber are sometimes used. The advantages of the higher-strength material may be significant—more so if combined compression and flexure must be developed. However, as in other situations, the much higher quality of dimensional stability may be a major factor.

The large glued-laminated column section offers the same general advantages of fire resistance that are assigned to heavy timber construction with solid-sawn sections. This does not make it more competitive in comparison to the solid section, but does in comparison to an exposed steel column.

It is also possible to produce columns of great length, tapered form, considerable width, and so on—in other words, all of the potentials offered for other laminated items. In general, laminated columns are used less frequently than laminated beams, and are generally chosen only when some of the limitations of other options for the column are restrictive.

A special application of the laminated column is in built-up sections, such as that shown in Fig. 18.7*b*. In these situations the laminated portion is the functioning structural member and the added solid-sawn elements are limited to decorative functions or use for other construction reasons.

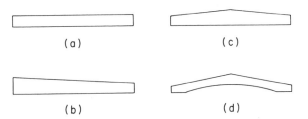

FIGURE 20.9. Laminated beam profiles: (*a*) straight; (*b*) single tapered-straight; (*c*) double tapered-curved; (*d*) double tapered-curved.

20.4. BUILT-UP PLYWOOD AND LUMBER ELEMENTS

Various types of structural components can be produced with combinations of plywood and pieces of standard structural lumber. Figure 20.10 shows some commonly used elements that can serve as structural components for buildings.

Stressed-Skin Panels

This type of unit consists of two sheets of plywood attached to a core frame of solid-sawn lumber elements (Fig. 20.11*a*). This is generally described as a plywood *sandwich panel*. However, when it is used in a manner that involves the development of spanning functions (as for a roof deck) and a true box-beam type of action is assumed, it is called a *stressed-skin panel*. These are mostly used for structures that have some modular, prefabricated system that may also involve the need for demountability for reuse. For ordinary situations, however, they are usually not competitive with typical construction with plywood panels, joists, and the usual ceiling surfacing.

For the stressed-skin panel the plywood is usually glued to the frame, although if appearance is not a concern, nails or screws may also be used. If gluing only is done, it must be done in a factory with real quality control of the process. The basic technique of construction for the simple unit shown in Fig. 20.10*a* may be extended to more complex forms, such as curved panels or nonrectangular shapes, as shown in Fig. 20.11. For special situations it may be possible to obtain plywood panels in larger sizes than the usual 4 × 8 ft to produce large stressed-skin units with no joints in the faces.

Some manufacturers produce these panels as more-or-less standard products. Plywood facings may vary for various applications and the unit voids may be filled with insulation for use as wall or roof construction.

Built-Up Beams

In the stressed skin panel tension and compression developed by bending are essentially resisted by the plywood top and bottom panels, while shear is resisted by the lumber frame elements. These roles are reversed in the built-up beams shown in Fig. 20.11*b* and *c*. Top and bottom lumber elements function like the chords of a truss or the flanges of a steel section, while single or multiple webs of plywood develop shear resistance. These members are highly variable in response to different load and span conditions. For short spans with heavy loads, shear will be critical and the multiple web member a likely choice. For long spans with relatively light loads, the single web element with a heavier flange member will be more effective.

This type of unit is also capable of variations in profile, such as that where the bottom is flat and the top is sloped, as shown for glued-laminated elements in Fig. 20.9. Webs may be penetrated for passage of wiring or piping in regions of low shear (not near the ends).

A great deal of customizing is possible with this type of unit. However, the widest usage is probably in the simple form shown in Fig. 20.10*c*, where a single web of plywood is glued into a groove cut in the single-piece flanges. This type of unit is used extensively for commercial buildings with roofs or floors of medium-span range (see Fig. 20.12). Flanges may be single standard lumber elements: 2 × 4, 3 × 4, and so on. Webs are usually of plywood, although beams with webs of particleboard are also produced.

Specifications and guidelines for design and construction with these units are provided by the American Plywood Association and the American Institute of Timber Construction. Many units are available as manufactured products and can be selected directly from suppliers' catalogs.

20.5. POLE STRUCTURES

The straight, slender trunks of various species of coniferous trees have been used for many structural purposes

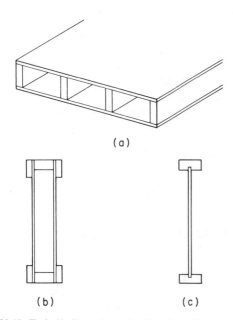

FIGURE 20.10. Typical built-up plywood and lumber elements: (*a*) stressed-skin panel; (*b*) box beams and I-beams; (*c*) I-beam with let-in web.

FIGURE 20.11. Prefabricated units of plywood-lumber construction used to form the folded plate roof of a church in St. Paul, Minnesota. Architects: Buetow Associates. Photo: American Plywood Association.

FIGURE 20.12. Manufactured plywood and lumber I-beams used as floor joists for a low-rise office building.

from ancient times to the present. Log cabins, pole stockades, fences, waterfront piers, and various utilitarian buildings have used these in ways largely unchanged for many centuries. Modern developments consist primarily of more sophisticated connection devices and pressure treatment for weather and insect resistance.

Poles may be used as beams or rafters, but are used more often as columns or for foundations. Timber piles are poles driven into the ground like large spikes by a pile driver (see the discussion in Part 7). Another way of using poles for foundations, however, is simply to dig a hole, insert the pole partway into the hole and partway above ground, and then fill the hole around the pole with soil or concrete. This is, of course, how poles are used for fence posts or utility transmission-line supports. For a building the buried pole may be cut off a short distance above the ground and support a structure of ordinary wood frame construction; or the poles may extend up to become columns for a frame structure.

Poles are usually produced simply by peeling the log of the bark and soft inner layers on the outside of the tree trunk. This leaves a member that may be essentially straight but typically has a tapered form and various surface irregularities, such as knots, splits, and pitch pockets. Precision of form and perfect straightness are not realistically

achievable. Poles can be shaved to a truer round cross section and short, untapered lengths are possible, but this adds greatly to the cost, and the surface will be less natural in appearance. One way to reduce the taper is to cut poles from a longer log, rather than use a single tree trunk of the necessary length.

Poles tend to be used somewhat regionally, where the climate, soil conditions, and availability of good, cheap poles favor their selection. They can be used for simple utilitarian structures, or for imaginative, sculptural building structures developed with a rough-textured and rustic timber style.

Design of timber piles, buried pole foundations, and round tapered wood columns can be done with criteria from industry specifications and guidelines. These may be used to develop a basis for structural computations, but much of the design for these structures is based on experience and simple recognition of the fact that the structures have endured successfully for so long that it must be okay to build them that way.

20.6.   WOOD FIBER PRODUCTS

Wood fiber products range from paper and cardboard to very dense and strong pressed hardboards (Masonite and the like). Actually, the major amount of commercially forested wood goes into these products, mostly for the production of newsprint, tissues, and container items. For building structures, some current uses are the following:

1. *Paper.*   Paper itself is not used much, but it is utilized in various composite products or forms of construction. *Plasterboard* is actually a sandwich with a core of gypsum plaster and two faces of paper. Stucco (exterior cement plaster) is sometimes applied directly to a wall frame using a backup of only a wire mesh bonded to paper. Paper may be coated, impregnated, or reinforced with various materials to improve resistance to moisture or other effects.

2. *Cardboard.*   Cardboard is in essence simply thick, stiff paper. Special cardboard is the familiar *corrugated cardboard* consisting of a formed (corrugated) core of paper glued in a sandwich with layers of flat cardboard or paper. This material is, of course, used primarily in the container industry, but sees some use in forming of special constructions of poured concrete.

3. *Pressed Wood Fiber Panels.*   Rigid sheet-form panels of pressed wood fiber have been used for many years. These have seen wider use in nonstructural applications, but have recently been steadily encroaching in areas of use traditionally led by plywood. Examples are for structural wall sheathing, roof decks, and the webs of built-up wood I-beams (see Sec. 20.4).

4. *Composite Panel Products.*   Various commercial products are made using wood fibers (or other vegetable fibers) in combination with cement, asbestos

fibers, asphalt, and so on. These may be used for essentially nonstructural purposes as insulation or filler (such as Celotex) or for structural decking units (such as Timdeck).

As in other situations, information for the use of manufactured products should be obtained from the suppliers of the products. Building codes have some data and industry-wide specifications exist for some basic types of products, such as particleboard and hardboard. Partly because they permit use of wood materials not usable for structural lumber and partly because there is already a large wood fiber industry, it is likely that wood fiber products will see increasing use. The trees from which fiber is obtained are very fast growing, making the raw material one of the most renewable resources.

20.7. FLITCHED BEAMS

Before the advent of glued-laminated wood beams, it was common practice to increase the strength of timber beams by the addition of steel plates. Two means of achieving such a built-up beam section are shown in Fig. 20.13. This composite steel and wood member is known as a *flitched beam.* With little increase in the beam size, a considerable increase in strength is achieved by this technique. Of greater significance, however, is the increase in stiffness, particularly with regard to long-time load effects. The quite common sag experienced with timber beams under large, long-time dead load can be considerably reduced through use of a flitched beam. Of course, the same can be accomplished by use of a steel beam or one of glued-laminated construction. Where the latter are prohibitively expensive or not easily available, however, the flitched beam is still in use.

The components of a flitched beam are securely held together with through bolts so that the elements act as a single unit. The computations for determining the strength of such a beam illustrate the phenomenon of two different materials in a beam acting as a unit. The computations are based on the premise that the two materials deform equally. Let

Δ_1, Δ_2 = deformations per unit length of the outermost fibers of the two materials, respectively

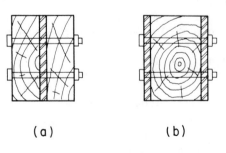

f_1, f_2 = unit bending stresses in the outermost fibers of the two materials, respectively

E_1, E_2 = moduli of elasticity of the two materials, respectively

Since, by definition, the modulus of elasticity of a material is equal to the unit stress divided by the unit deformation, then

$$E_1 = \frac{f_1}{\Delta_1} \quad \text{and} \quad E_2 = \frac{f_2}{\Delta_2}$$

and transposing gives

$$\Delta_1 = \frac{f_1}{E_1} \quad \text{and} \quad \Delta_2 = \frac{f_2}{E_2}$$

Since the two deformations must be equal,

$$\frac{f_1}{E_1} = \frac{f_2}{E_2} \quad \text{and} \quad f_2 = f_1 \frac{E_2}{E_1}$$

This simple equation for the relationship between the stresses in the two materials of a composite beam may be used as the basis for investigation or design of a flitched beam, as demonstrated in the following example.

Example 1. A flitched beam is formed as shown in Fig. 20.13a, consisting of two 2 × 12 planks of Douglas fir-larch, No. 1 grade, and a 0.5 × 11.25 in. [13 × 285 mm] plate of A36 steel. Compute the allowable uniformly distributed load this beam will carry on a simple span of 14 ft [4.2 m].
Solution. 1. We first apply the formula just derived to determine which of the two materials limits the beam action. For this we obtain the following data.

For the Steel. E = 29,000,000 psi [200 GPa], and the maximum allowable bending stress F_b is 22 ksi [150 MPa] (Table 21.1).

For the Wood. E = 1,800,000 psi [12.4 GPA], and the maximum allowable bending stress for single-member use is 1500 psi [10.3 MPa] (Table 16.1).

For a trial we assume the stress in the steel plate to be the limiting value and use the formula to find the maximum useable stress in the wood. Thus

$$f_w = f_s \frac{E_w}{E_s} = (22,000) \frac{1,800,000}{29,000,000}$$
$$= 1366 \text{ psi } [9.3 \text{ MPa}]$$

As this produces a stress lower than that of the table limit for the wood, our assumption is correct. That is, if we permit a stress higher than 1366 psi in the wood, the steel stress will exceed its limit of 22 ksi.
2. Using the stress limit just determined for the wood, we now find the capacity of the wood members. Calling the load capacity of the wood W_w, we find

FIGURE 20.13.

$$M = \frac{W_w L}{8} = \frac{W_w(14)(12)}{8} = 21 W_w$$

Then, using the S of 31.6 in.^{3} for the 2 × 12 (Table A.2), we find

$$M = 21 W_w = f_w S_w = 1366[2(31.6)]$$
$$W_w = 4111 \text{ lb } [18.35 \text{ kN}]$$

3. For the plate we first must find the section modulus as follows (see Fig. A.1):

$$S_s = \frac{bd^2}{6} = \frac{0.5(11.25)^2}{6}$$
$$= 10.55 \text{ in.}^3 \, [176 \times 10^3 \text{ mm}^3]$$

Then

$$M = 21 W_s = f_s S_s = 22,000(10.55)$$
$$W_s = 11,052 \text{ lb } [50.29 \text{ kN}]$$

and the total capacity of the combined section is

$$W = W_w + W_s = 4111 + 11,052$$
$$= 15,163 \text{ lb } [68.64 \text{ kN}]$$

Although the load-carrying capacity of the wood elements is actually reduced in the flitched beam, the resulting total capacity is substantially greater than that of the wood members alone. This significant increase in strength achieved with small increase in size is a principal reason for popularity of the flitched beam. In addition, there is a significant reduction in deflection in most applications, and—most noteworthy—a reduction in sag over time.

PART 4

STEEL STRUCTURES

Steel is used in some form in every type of construction. Steel nails for wood and steel reinforcing for concrete are indispensible items. This part deals essentially with the use of steel as a structural material in its own right. As such, steel is used in the form of industrial products of various types. A principal usage dealt with in this part is that of so-called *structural steel*, a term used for products produced by rolling to form a semimolten steel ingot into a linear member with some formed cross section; the product being described as a *rolled section*. Data and design specifications for rolled sections are published in the *Manual of Steel Construction* (Ref. 5) by the American Institute for Steel Construction (hereinafter referred to as the AISC). Many other types of products are also used for building structures and are also described and illustrated in this part. Other industry-wide organizations that deal with specific types of products are discussed in the appropriate sections in Part 4.

CHAPTER TWENTY-ONE

General Concerns for Steel

Steel is a highly variable material and is used for a wide range of products that serve many purposes for building construction. Chapter 21 deals with some of the general concerns for use of the basic material and with the use of the various common forms of products now being produced for structural applications.

21.1 TYPES OF STEEL PRODUCTS

Steel itself is formless, coming basically in the form of a molten material or a softened lump. The structural products produced derive their basic forms from the general potentialities and limitations of the industrial processes of forming and fabricating. Standard raw stock elements—deriving from the various production processes—are the following:

1. *Rolled Shapes.* These are formed by squeezing the heat-softened steel repeatedly through a set of rollers that shape it into a linear element with a constant cross section. Simple forms of round rods and flat bars, strips, plates, and sheets are formed, as well as more complex shapes of I, H, T, L, U, C, and Z. Special shapes, such as rails or sheet piling, can also be formed in this manner.
2. *Wire.* This is formed by pulling (called *drawing*) the steel through a small opening.
3. *Extrusion.* This is similar to drawing, although the sections produced are other than simple round shapes. This process is not frequently used for steel products of the sizes used for buildings.
4. *Casting.* This is done by pouring the molten steel into a form (mold). This is limited to objects of a three-dimensional form.
5. *Forging.* This consists of pounding the softened steel into a mold until it takes the shape of the mold. This is preferred to casting because of the effects of the working on the properties of the finished material.

Stock elements produced by the basic forming processes may be reworked by various means, such as the following:

1. *Cutting.* Shearing, sawing, punching, or flame cutting can be used to trim and shape specific forms.
2. *Machining.* This may consist of drilling, planing, grinding, routing, or turning on a lathe.
3. *Bending.* Sheets, plates, or linear elements may be bent if made from steel with a ductile character (see the discussion in Sec. 21.2).
4. *Stamping.* This is similar to forging; in this case sheet steel is punched into a mold that forms it into some three-dimensional shape, such as a hemisphere.
5. *Rerolling.* This consists of reworking a linear element into a curved form (arched) or of forming a sheet or flat strip into a formed cross section.

Finally, raw stock or reformed elements can be assembled by various means into objects of multiple parts, such as a manufactured truss or a prefabricated wall panel. Basic means of assemblage include the following:

1. *Fitting.* Threaded parts may be screwed together or various interlocking techniques may be used, such as the tongue-and-groove joint or the bayonet twist lock.
2. *Friction.* Clamping, wedging, or squeezing with high-tensile bolts may be used to resist the sliding of parts in surface contact.
3. *Pinning.* Overlapping flat elements may have matching holes through which a pin-type device (bolt, rivet, or actual pin) is placed to prevent slipping of the parts at the contact face.
4. *Nailing, Screwing.* Thin elements—mostly with some preformed holes—may be attached by nails or screws.
5. *Welding.* Gas or electric arc welding may be used to produce a bonded connection, achieved partly by melting the contacting elements together at the contact point.
6. *Adhesive Bonding.* This usually consists of some form of chemical bonding that results in some fusion of the materials of the connected parts.

We are dealing here with industrial processes, which at any given time relate to the state of development of the technology, the availability of facilities, the existence of the necessary craft, and competition with other materials and products.

21.2. PROPERTIES OF STEEL

The strength, hardness, corrosion resistance, and some other properties of steel can be varied through a considerable range by changes in the production processes. Hundreds of different steels are produced, although only a few standard products are used for the majority of the elements of building structures. Working and forming processes, such as rolling, drawing, machining, and forging may also alter some properties. Certain properties, such as density (unit weight), stiffness (modulus of elasticity), thermal expansion, and fire resistance tend to remain constant for all steels.

Basic structural properties, such as strength, stiffness, ductility, and brittleness, can be interpreted from load tests. Figure 21.1 displays typical forms of curves that are obtained by plotting stress and strain values from such tests. An important characteristic of many structural steels is the plastic deformation (yield) phenomenon. This is demonstrated by curve 1 in Fig. 21.1. For steels with this character there are two different stress values of significance: the yield limit and the ultimate failure limit.

Generally, the higher the yield limit, the less the degree of ductility. Curve 1 in Fig. 21.1 is representative of ordinary structural steel (ASTM A36), and curve 2 indicates the typical effect as the yield strength is raised a significant amount. Eventually, the significance of the yield phenomenon becomes negligible when the yield strength approaches as much as three times the yield of ordinary steel.

Some of the highest-strength steels are produced only in thin sheet or drawn wire forms. Bridge strand is made from wire with strength as high as 300,000 psi. At this level yield is almost nonexistent and the wires approach the brittleness of glass rods.

For economical use of the expensive material, steel structures are generally composed of elements with quite relatively thin parts. This results in many situations in which the limiting strength of elements in bending, compression, and shear is determined by buckling, rather than the stress limits of the material. Since buckling is a function of stiffness (modulus of elasticity) of the material, and since this property remains the same for all steels, there is limited opportunity to make effective use of higher-strength steels in many situations. The grade of steel most commonly used is to some extent one that has the optimal effective strength for most typical tasks.

For various applications, other properties will be significant. Hardness affects the ease with which cutting, drilling, planing, and other working can be done. For welded connections there is a property of weldability of the base material that must be considered. Resistance to rusting is normally low, but can be enhanced by various materials added to the steel, producing various types of special steels, such as stainless steel and the so-called rusting steel that rusts at a very slow rate.

Since many structural elements are produced as some manufacturer's product line, choices of materials are often mostly out of the hands of individual building designers. The proper steel for the task—on the basis of many properties—is determined as part of the product design.

Steel that meets the requirements of the American Society for Testing and Materials (ASTM) Specification A36 is the grade of structural steel commonly used to produce rolled steel elements for building construction. It must have an ultimate tensile strength of 58–80 ksi and a minimum yield point of 36 ksi. It may be used for bolted, riveted, or welded fabrication.

Allowable Stresses for Structural Steel

For structural steel the AISC Specification expresses the allowable unit stresses in terms of some percent of the yield stress F_y or the ultimate stress F_u. Selected allowable unit stresses used in design are listed in Table 21.1. Specific values are given for ASTM A36 steel, with values of 36 ksi for F_y and 58 ksi for F_u. This is not a complete list, but it generally includes the stresses used in the examples in this book. Reference is made to the more complete descriptions

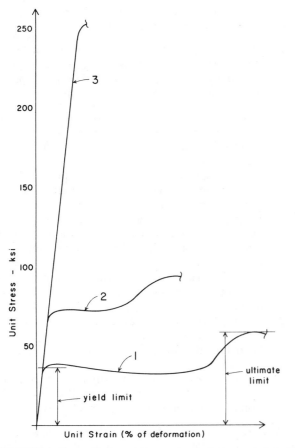

FIGURE 21.1. Form of stress–strain response of structural steel: (1) ordinary structural grade—A36; (2) higher grades used for rolled shapes; (3) high strength strand (cable, rope).

TABLE 21.1. Allowable Unit Stresses for Structural Steel: ASTM A36[a]

Type of Stress and Conditions	See Discussion in This Book in:	Stress Designation	AISC Specification	Allowable Stress ksi	Allowable Stress MPa
Tension					
1. On the gross (unreduced) area	Sec. 24.2	F_t	$0.60F_y$	22	150
2. On the effective net area, except at pin-holes			$0.50F_u$	29	125
3. Threaded rods on net area at thread			$0.33F_u$	19	80
Compression	Chapter 23	F_a	See discussion		
Shear					
1. Except at reduced sections	Sec. 22.4	F_v	$0.40F_y$	14.5	100
2. At reduced sections	Sec. 24.2		$0.30F_u$	17.4	120
Bending					
1. Tension and compression on extreme fibers of compact members braced laterally, symmetrical about and loaded in the plane of their minor axis	Sec. 22.2	F_b	$0.66F_y$	24	165
2. Tension and compression on extreme fibers of other rolled shapes braced laterally			$0.60F_y$	22	150
3. Tension and compression on extreme fibers of solid round and square bars, on solid rectangular sections bent on their weak axis, on qualified doubly symmetrical I and H shapes bent about their minor axis			$0.75F_y$	27	188
Bearing					
1. On contact area of milled surfaces		F_p	$0.90F_y$	32.4	225
2. On projected area of bolts and rivets in shear connections	Chapter 24		$1.50F_u$	87	600

[a] F_y = 36 ksi; assume that F_u = 58 ksi; some table values rounded off as permitted in the *AISC Manual* (Ref. 5). For SI units F_y = 250 MPa, F_u = 400 MPa.

in the AISC Specification, which is included in the *AISC Manual* (Ref. 5). There are in many cases a number of qualifying conditions, some of which are discussed in other portions of this book. Table 21.1 gives the location of some of these discussions.

21.3. ROLLED STRUCTURAL SHAPES

The products of the steel rolling mills used as beams, columns, and other structural members are known as *sections* or *shapes*, and their designations are related to the profiles of their cross sections. American standard I-beams (Fig. 21.2*a*) were the first beam sections rolled in the United States and are currently produced in sizes of 3–24 in. in depth. The wide-range shapes (Fig. 21.2*b*) are a modification of the I cross section and are characterized by parallel flange surfaces as contrasted with the tapered inside flange surfaces of standard I-beams; they are available in depths of 4–36 in. In addition to the standard I and wide-flange sections, the structural steel shapes most commonly used

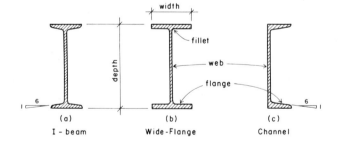

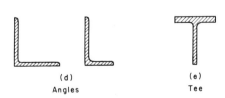

FIGURE 21.2. Rolled structural shapes.

in building construction are channels, angles, tees, plates, and bars. The tables in Appendix A list the dimensions and weights of some of these shapes with other properties, which are identified and discussed in Chapters 22 through 24. Complete tables of structural shapes are given in the AISC *Manual of Steel Construction* (Ref. 5).

Wide-Flange Shapes

In general, wide-flange shapes have greater flange widths and relatively thinner webs than standard I-beams; and, as noted above, the inner faces of the flanges are parallel to the outer faces. These sections are identified by the alphabetical symbol W, followed by the *nominal* depth in inches and the weight in pounds per linear foot. Thus the designation W 12 × 26 indicates a wide-flange shape of nominal-12-in. depth, weighing 26 lb per linear ft.

The actual depths of wide-flange shapes vary within the nominal depth groupings. By reference to Appendix A, it is found that a W 12 × 26 has an actual depth of 12.22 in., whereas the depth of a W 12 × 35 is 12.50 in. This is a result of the rolling process during manufacture in which the cross-sectional areas of wide-flange shapes are increased by spreading the rolls both vertically and horizontally. The additional material is thereby added to the cross section by increasing flange and web thickness as well as flange width (Fig. 21.2b). The resulting higher percentage of material in the flanges makes wide-flange shapes more efficient structurally than standard I-beams. A wide variety of weights is available within each nominal depth group.

In addition to shapes with profiles similar to the W 12 × 26, which has a flange width of 6.490 in., many wide-flange shapes are rolled with flange widths approximately equal to their depths. The resulting H configurations of these cross sections are much more suitable for use as columns than the I profiles. By reference to Appendix A it is found that the following shapes, among others, fall into this category: W 14 × 90, W 12 × 65, W 10 × 60, and W 8 × 40.

Standard I-Beams

American standard I-beams are identified by the alphabetical symbol S, the designation S 12 × 35 indicating a standard shape 12 in. deep, weighing 35 lb per linear ft. In Appendix A it is shown that this section has an *actual* depth of 12 in., a flange width of 5.078 in., and a cross-sectional area of 10.3 in.². Unlike wide-flange sections, standard I-beams in a given depth group have uniform depths, and shapes of greater cross-sectional area are made by spreading the rolls in one direction only. Thus the depth remains constant, whereas the width of flange and thickness of web are increased. A comparison of Sections S 12 × 35 and S 12 × 50 in Appendix A will clarify these relationships. Since a bar of steel 1 sq in. in cross section and 1-ft long weighs approximately 3.4 lb, the weight per linear ft of any structural shape is 3.4 times the cross-sectional area.

All standard I-beams have a slope on the inside faces of the flanges of $16\frac{2}{3}$%, or 1 in 6. In general, standard I-beams are not so efficient structurally as wide-flange sections and consequently are not so widely used. Also, the variety available is not nearly so large as that for wide-flange shapes. Characteristics that may favor the use of American standard I-beams in any particular situation are constant depth, narrow flanges, and thicker webs.

Standard Channels

The profile of an American standard channel is shown in Fig. 21.2c. These shapes are identified by the alphabetical symbol C. The designation C 10 × 20 indicates a standard channel 10 in. deep and weighing 20 lb per linear ft. Appendix A shows that this section has an area of 5.88 in.², a flange width of 2.739 in., and a web thickness of 0.379 in. Like the standard I-beams, the depth of a particular group remains constant and the cross-sectional area is increased by spreading the rolls to increase flange width and web thickness. Because of their tendency to buckle when used independently as beams or columns, channels require lateral support or bracing. They are generally used as elements of built-up sections such as columns and lintels. However, the absence of a flange on one side makes channels particularly suitable for framing around floor openings.

Angles

Structural angles are rolled sections in the shape of the letter L. Appendix A gives dimensions, weights, and other properties of equal and unequal leg angles. Both legs of an angle have the same thickness.

Angles are designated by the alphabetical symbol L, followed by the dimensions of the legs and their thickness. Thus the designation L 4 × 4 × $\frac{1}{2}$ indicates an equal leg angle with 4-in. legs, $\frac{1}{2}$ in. thick. By reference to Appendix A it is found that this section weighs 12.8 lb per linear ft and has an area of 3.75 in.². Similarly, the designation L 5 × $3\frac{1}{2}$ × $\frac{1}{2}$ indicates an unequal leg angle with one 5-in. and one $3\frac{1}{2}$-in. leg, both $\frac{1}{2}$ in. thick. Appendix A shows that this angle weighs 13.6 lb per linear ft and has an area of 4 in.². To change the weight and area of an angle of a given leg length the thickness of each leg is increased the same amount. Thus if the leg thickness of the L 5 × $3\frac{1}{2}$ × $\frac{1}{2}$ is increased to $\frac{5}{8}$ in., Appendix A shows that the resulting L 5 × $3\frac{1}{2}$ × $\frac{5}{8}$ has a weight of 16.8 lb per linear ft and an area of 4.92 in.². It should be noted that this method of spreading the rolls changes the leg lengths slightly.

Single angles are often used as lintels and pairs of angles, as members of light steel trusses. Angles were formerly used as elements of built-up sections such as plate girders and heavy columns, but the advent of the heavier wide-flange shapes has largely eliminated their usefulness for this purpose. Short lengths of angles are common connecting members for beams and columns.

Structural Tees

A structural tee is made by splitting the web of a wide-flange shape (Fig. 21.2*e*) or a standard I-beam. The cut, normally made along the center of the web, produces tees with a stem depth equal to half the depth of the original section. Structural tees cut from wide-flange shapes are identified by the symbol WT; those cut from standard I shapes, by ST. The designation WT 6 × 53 indicates a structural tee with a 6-in. depth and a weight of 53 lb per linear ft. This shape is produced by splitting a W 12 × 106 shape. Similarly, ST 9 × 35 designates a structural tee 9 in. deep, weighing 35 lb per linear ft and cut from a S 18 × 70. Tables of properties of these shapes appear in Appendix A. Structural tees are used for the chord members of welded steel trusses and for the flanges in certain types of plate girder.

Plates and Bars

Plates and bars are made in many different sizes and are available in all the structural steel specifications listed in Table 21.1.

Flat steel for structural use is generally classified as follows:

Bars. 6 in. or less in width, 0.203 in. and more in thickness.

6–8 in. in width, 0.230 in. and more in thickness.

Plates. More than 8 in. in width, 0.230 in. and more in thickness.

More than 48 in. in width, 0.180 in. and more in thickness.

Bars are available in varying widths and in virtually any required thickness and length. The usual practice is to specify bars in increments of $\frac{1}{4}$ in. for widths and $\frac{1}{8}$ in. in thickness.

For plates the preferred increments for width and thickness are the following:

Widths. Vary by even inches, although smaller increments are obtainable.

Thickness. $\frac{1}{32}$-in. increments up to $\frac{1}{2}$ in.

$\frac{1}{32}$-in. increments of more than $\frac{1}{2}$–2 in.

$\frac{1}{8}$-in. increments of more than 2–6 in.

$\frac{1}{4}$-in. increments of more than 6 in.

The standard dimensional sequence when describing steel plate is

thickness × width × length

All dimensions are given in inches, fractions of an inch, or decimals of an inch.

Column base plates and beam bearing plates may be obtained in the widths and thicknesses noted. For the design of column base plates and beam bearing plates, see Secs. 22.7 and 23.7, respectively.

Designations for Structural Steel Elements

As noted earlier, wide-flange shapes are identified by the symbol W, American standard beam shapes, by S. It was also pointed out that W shapes have essentially parallel flange surfaces, whereas S shapes have a slope of approximately $16\frac{2}{3}\%$ on the inner flange faces. A third designation, M shapes, covers miscellaneous shapes that cannot be classified as W or S: these shapes have various slopes on their inner flange surfaces and many of them are of only limited availability. Similarly, some rolled channels cannot be classified as C shapes. These are designated by the symbol MC.

Table 21.2 lists the standard designations used for rolled shapes, formed rectangular tubing, and round pipe.

21.4. TABULATED DATA FOR STEEL PRODUCTS

Information in general regarding steel products used for building structures must be obtained from steel industry publications. The AISC is the primary source of design information regarding structural rolled products, which are the principal elements used for major structural components: columns, beams, large trusses, and so on. Several other industry-wide organizations also publish documents that provide information about particular products, such as manufactured trusses (open-web joists), cold-formed sections, and formed sheet-steel decks. Many of these organizations and their publications are described in the appropriate chapters and sections of Part 4.

Individual manufacturers of steel products usually conform to some industry-wide standards in the design and fabrication of their particular products. Still, there is often some room for variation of products, so that it is advised that the manufacturers' own publications be used for specific data and details of the actual products. As in other

TABLE 21.2. Standard Designations for Structural Steel Elements

Type of Element	Designation
Wide-flange shapes	W 12 × 27
American standard beams	S 12 × 35
Miscellaneous shapes	M 8 × 18.5
American standard channels	C 10 × 20
Miscellaneous channels	MC 12 × 45
Angles	
Equal legs	L 4 × 4 × $\frac{1}{2}$
Unequal legs	L 5 × 3$\frac{1}{2}$ × $\frac{1}{2}$
Structural tees	
Cut from wide flange shapes	WT 6 × 53
Cut from American standard beams	ST 9 × 35
Cut from miscellaneous shapes	MT 4 × 9.25
Plate	PL $\frac{1}{2}$ × 12
Structural tubing: square	TS 4 × 4 × 0.375
Pipe	Pipe 4 std.

similar situations in building design, the designer should strive to design and specify components of the building construction so that only those controls that are critical are predetermined; leaving flexibility in the choice of a particular manufactured product.

Some of the tabulated data presented in this book have been reproduced or abstracted from industry publications. In many cases the data presented here are abbreviated and limited to uses pertinent to the work displayed in the text example computations and are necessary for the exercise problems. The reference sources cited should be consulted for more complete information, the more so since change is the name of the game due to growth of the technology, advances in research, and modifications of codes and industry standards.

21.5. USAGE CONSIDERATIONS

The following are some problems that must often be considered in the use of structural components of steel for buildings.

Rust

Exposed to air and moisture, most steels will rust at the surface of the steel mass. Rusting will generally continue at some rate until the entire steel mass is eventually rusted away. Response to this problem may involve one or more of the following actions.

1. Do nothing, if there is essentially no exposure, as when the steel element is encased in poured concrete or other encasing construction.
2. Paint the steel surface with rust-inhibiting material.
3. Coat the surface with nonrusting metal, such as zinc or aluminum.
4. Use a steel that contains ingredients in the basic material that prevent or retard the rusting action (see the discussion in Sec. 21.2 on corrosion-resistant steels).

Rusting is generally of greater concern when exposure conditions are more severe. It is also of greatest concern for the thinner elements, especially those formed of thin sheet steel, such as formed roof decks.

Fire

As with all materials, the stress and strain response of steel varies with its temperature. The rapid loss of strength (and stiffness, which may be more important when buckling is possible) at high temperatures, coupled with rapid heat gain due to the high conductivity of the material and the common use of thin parts, makes steel structures highly susceptible to fire. On the other hand, the material is noncombustible and less critical for some considerations in comparison to constructions with thin elements of wood.

The chief strategy for improving fire safety with steel structures is to prevent the fire (and the heat buildup) from getting to the steel by providing some coating or encasement with fire-resistant, insulative materials. Ordinary means for this include use of concrete, masonry, plaster, asbestos fiber, or gypsum plasterboard elements. The general problem and some specific design situations are presented in Part 9.

Cost

Steel is relatively expensive, on a volume basis. The real cost of concern, however, is the final *installed cost*, that is, the total cost for the erected structure. Economy concerns begin with attempts to use the least volume of the material, but this is applicable only within the design of a single type of item. Rolled structural shapes do not cost the same per pound as fabricated open web joists. Furthermore, each must be transported to the site and erected, using various auxiliary devices to complete the structure, such as framing elements for structural components and bridging for joists. Cost concerns in general are discussed in Part 9. In this part, cost considerations are limited to attempts to find the least-weight member of a specific type in a single design task.

21.6. NOMENCLATURE

The standard symbols that are used in steel design work (called the *general nomenclature*) are primarily those established by the AISC and are contained in its standard specifications. Although these are in part special to the area of steel, the attempt is being made increasingly to standardize the nomenclature among all fields. The following list is an abridged version of a more extensive list in the *AISC Manual* (Ref. 5); it contains all of the symbols used in this book for steel design work.

A = Cross-sectional area (in.^{2}).
Gross area of an axially loaded compression member (in.^{2}).

A_e = Effective net area of an axially loaded tension member (in.^{2}).

A_n = Net area of an axially loaded tension member (in.^{2}).

C_c = Column slenderness ratio separating elastic and inelastic buckling.

D = Beam deflection (in.).

E = Modulus of elasticity of steel (29,000 ksi).

F_a = Axial compressive stress permitted in a prismatic member in the absence of a bending moment (ksi).

F_b = Bending stress permitted in a prismatic member in the absence of axial force (ksi).

F_e' = Euler stress for a prismatic member divided by the factor of safety (ksi).

F_p = Allowable bearing stress (ksi).

F_t = Allowable axial tensile stress (ksi).

F_u = Specified minimum tensile strength of the type of steel or fastener being used (ksi).

F_v = Allowable shear stress (ksi).

F_y = Specified minimum yield stress of the type of steel being used (ksi); as used in this book, *yield stress* denotes either the specified minimum yield point (for those steels that have a yield point) or specified minimum yield strength (for those steels that have no yield point).

I = Moment of inertia of a section (in.⁴).

I_x = Moment of inertia of a section about the X–X axis (in.⁴).

I_y = Moment of inertia of a section about the Y–Y axis (in.⁴).

J = Torsional constant of a cross section (in.⁴).

K = Effective length factor for a prismatic member.

L = Span length (ft).
Length of connection angles (in.).

L_c = Maximum unbraced length of the compression flange at which the allowable bending stress may be taken at $0.66F_y$ or as determined by AISC Specification Formula (1.5-5a) or Formula (1.5-5b), when applicable (ft).
Unsupported length of a column section (ft).

L_u = Maximum unbraced length of the compression flange at which the allowable bending stress may be taken at $0.6F_y$ (ft).

M = Moment (kip-ft).

M_D = Moment produced by a dead load.

M_L = Moment produced by a live load.

M_p = Plastic moment (kip-ft).

M_R = Beam resisting moment (kip-ft).

N = Length of base plate (in.).
Length of bearing of applied load (in.).

P = Applied load (kips).
Force transmitted by a fastener (kips).

S = Elastic section modules (in.³).

S_x = Elastic section modulus about the X–X (major) axis (in.³).

S_y = Elastic section modulus about the Y–Y (minor) axis (in.³).

V = Maximum permissible web shear (kips).
Statical shear on a beam (kips).

Z = Plastic section modulus (in.³).

Z_x = Plastic section modulus with respect to the major (X–X) axis (in.³).

Z_y = Plastic section modulus with respect to the minor (Y–Y) axis (in.³).

b_f = Flange width of a rolled beam or plate girder (in.).

d = Depth of a column, beam, or girder (in.).
Nominal diameter of a fastener (in.).

e_o = Distance from the outside face of a web to the shear center of a channel section (in.).

f_a = Computed axial stress (ksi).

f_b = Computed bending stress (ksi).

f_c' = Specified compression strength of concrete at 28 days (ksi).

f_p = Actual bearing pressure on a support (ksi).

f_t = Computed tensile stress (ksi).

f_v = Computed shear stress (ksi).

g = Transverse spacing locating fastener gage lines (in.).

k = Distance from the outer face of a flange to the web toe of a fillet of rolled shape or equivalent distance on a welded section (in.).

l = For beams, the distance between cross sections braced against twist or lateral displacement of the compression flange (in.).
For columns, the actual unbraced length of a member (in.).
Length of a weld (in.).

m = Cantilever dimension of a base plate (in.).

n = Number of fasteners in one vertical row.
Cantilever dimension of a base plate (in.).

r = Governing radius of gyration (in.).

r_x = Radius of gyration with respect to the X–X axis (in.).

r_y = Radius of gyration with respect to the Y–Y axis (in.).

s = Longitudinal center-to-center spacing (pitch) of any two consecutive holes (in.).

t = Girder, beam, or column web thickness (in.).
Thickness of a connected part (in.).
Wall thickness of a tubular member (in.).
Angle thickness (in.).

t_f = Flange thickness (in.).

t_w = Web thickness (in.).

x = Subscript relating a symbol to strong-axis bending.

y = Subscript relating a symbol to weak-axis bending.

CHAPTER TWENTY-TWO

Steel Beams, Joists, and Decks

There are many steel elements that can be used for the basic function of spanning, including rolled sections, cold-formed sections, and fabricated beams and trusses. Chapter 22 deals with some of the fundamental considerations of use of steel elements for beams, with an emphasis on the rolled shapes. Use of some other types of members and the special considerations for their design are discussed in other appropriate portions of Part 4. For simplicity, it is assumed that all the rolled shapes used for the work in Chapter 22 are of ASTM A36 steel with F_y = 36 ksi [250 MPa].

22.1. SECTIONS AND USAGE

All of the rolled structural shapes shown in Fig. 21.2 are used as flexural members in various situations. The shape used most often for beams in framing systems is the wide-flange section (Fig. 21.2b). Its biaxial symmetry and various attributes of its geometry make it well suited for such tasks. The other sections shown in Fig. 21.2 are used for special tasks, such as the following:

1. The S section, or I-beam as it is commonly called, is used where the relatively high stiffness of its flanges or the relatively thick web is of some special advantage. This was the original shape of the basic I-shaped section and was largely abandoned once the wide-flange sections were developed.
2. The channel has the major disadvantage of lack of symmetry on its minor axis, which causes critical concern for torsional stability. However, its flat side is an advantage in some framing situations, such as those occurring at openings or at the building edge.
3. Angles also have the problem of torsion when used singly. Their form makes them particularly handy for use as lintels over door or window openings.
4. The tee section is symmetrical about its minor axis, but its thin web is quite unstable against lateral buckling. It is sometimes used in an inverted position as a lintel, but is most popular for use as a chord member in a welded truss.

Rolled steel elements may be combined in various ways to form built-up sections. Some examples of these are shown in Fig. 22.1. Very large structural members can be

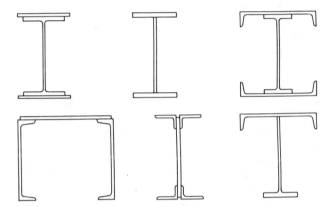

FIGURE 22.1. Common built-up sections, formed with rolled elements.

formed in this manner, for use in situations where requirements exceed the range of available rolled shapes. However, the combinations are also done to produce a section with particular capabilities when there is a need for resistance of torsion, two-way bending, high local shear, and so on.

22.2. DESIGN FOR FLEXURE

For the simplest case, design for bending moment consists primarily of simply determining that the limiting value for bending stress is not exceeded. In the design procedure this usually means solving a computation to determine the basic required property for the beam cross section: the section modulus (S). To do this we first find the maximum bending moment required of the beam, then establish the limiting flexural stress, and finally use the formula for maximum bending stress in the form

$$\text{required } S = \frac{M}{F_b}$$

With this desired property determined, we then proceed to find a beam shape that is acceptable. If no other considerations must be made in the selection (hardly ever the case), we typically choose the shape with the desired S value and the least weight. The following example demonstrates this process.

Example 1. Design a simply supported beam to carry a superimposed load of 2 kips/ft [29.2 kN/m] on a span of 24 ft [7.3 m]. Allowable bending stress is 24 ksi [165 MPa].
Solution. The bending moment due to the superimposed load is

$$M = \frac{wL^2}{8} = \frac{2(24)^2}{8} = 144 \text{ kip-ft [195 kN-m]}$$

The required section modulus for this moment is

$$S = \frac{M}{F_b} = \frac{144(12)}{24} = 72.0 \text{ in.}^3 \ [1182 \times 10^3 \text{ mm}^3]$$

Scanning of the tables of properties for wide-flange shapes in Appendix A (Table A. 3) will reveal a W 16 × 45 with a section modulus of 72.7 in.³. However, this leaves very little excess over the requirement for the superimposed load; some margin being required for the additional moment due to the weight of the beam. Further searching of the table will reveal a W 18 × 46 with $S = 78.8$ in.³. With no other criteria for selection, we may try the 18-in. member, for which the moment due to the beam weight is

$$M = \frac{wL^2}{8} = \frac{46(24)^2}{8} = 3312 \text{ ft-lb}$$
$$\text{or } 3.3 \text{ kip-ft [4.46 kN-m]}$$

Thus the total maximum moment is 144 + 3.3 = 147.3 kip-ft [199.5 kN-m] and the required S for this moment is

$$S = \frac{M}{F_b} = \frac{147.3(12)}{24} = 73.7 \text{ in.}^3 \ [1209 \times 10^3 \text{ mm}^3]$$

and the W 18 × 46 is adequate.

Use of Section Modulus Tables

Selection of rolled shapes on the basis of required section modulus may be achieved by the use of the tables in the *AISC Manual* (Ref. 5) in which beam shapes are listed in descending order of their section modulus values. Material from these tables is presented in Table 22.1. Note that certain shapes have their designations listed in boldface type. These are sections that have an especially efficient bending moment resistance, indicated by the fact that there are other sections of greater weight but the same or smaller section modulus. Thus for a savings of material cost these *least-weight* sections offer an advantage. Consideration of other beam design factors, however, may sometimes make this a less important concern.

Data are also supplied in Table 22.1 for the consideration of compact sections and lateral support for beams of A36 steel with $F_y = 36$ ksi [246 MPa]. Shapes that are noncompact for A36 steel are noted by a mark in the listings of designations. For consideration of lateral support the values are given for the two limiting lengths L_c and L_u. If a calculation has been made by assuming the maximum allowable stress of 24 ksi [165 MPa], the required section modulus obtained will be proper only for sections not indicated as noncompact and beams in which the lateral unsupported length is equal to or less than L_c.

A second method of using Table 22.1 for beams of A36 steel omits the calculation of a required section modulus and refers directly to the listed values for the maximum bending resistance of the sections, given as M_R in the tables. Although the condition of the noncompact section may be noted, in this case the M_R values have taken the reduced values for bending stress into account.

Example 2. Rework the problem in Example 1 using Table 22.1.
Solution. As before, we determine that the bending moment due to the superimposed loading is 144 kip-ft [195 kN-m]. Noting that some additional M_R capacity will be required because of the beam's own weight, we scan the tables for shapes with an M_R of slightly more than 144 kip-ft [195 kN-m]. Thus we find

Shape	M_R (kip-ft)	M_R (kN-m)
W 21 × 44	163	221
W 16 × 50	162	220
W 18 × 46	158	214
W 12 × 58	156	212
W 14 × 53	156	212

Although the W 21 × 44 is the least-weight section, other design considerations, such as restricted depth, may make any of the other shapes the appropriate choice.

Examples 1 and 2 demonstrate the simple case that occurs when the beam is loaded so that bending moment occurs in the plane of its minor axis (*Y–Y* axis). It is also necessary that the load plane coincide with the shear center of the cross section (see Sec. 12.7) and that lateral and torsional buckling are not critical.

The allowable bending stress of 24 ksi that was used in Example 1 is the value permitted for rolled sections of A36 steel with several qualifications regarding the form of the cross section. One qualification is that the section must be symmetrical about both major axes. In A36 steel, all wide-flange sections and S and M sections qualify for this allowable stress. Most other sections must use some lower value for the limiting flexural stress.

Another qualification for the allowable flexural stress is based on conditions of lateral support. This is dealt with in Sec. 22.3.

22.3. STABILITY CONSIDERATIONS

The ability of a beam to develop the full resisting moment implied by its cross section may be limited by considerations of the beam's stability. The two basic potential modes of stability failure are by sideways buckling (called *lateral buckling*) of the compression side of the beam and twisting (called *torsional rotational buckling*).

TABLE 22.1. Section Modulus and Resisting Moment Values for Selected Rolled Structural Shapes ~ F.S. IS CALCED IN ↖ *(handwritten annotation)*

All columns are for $F_y = 36$ ksi.

S_x (In.³)	Shape	L_c (ft)	L_u (ft)	M_R (Kip-ft)
1110	W 36x300	17.6	35.3	2220
1030	W 36x280	17.5	33.1	2060
953	W 36x260	17.5	30.5	1910
895	W 36x245	17.4	28.6	1790
837	W 36x230	17.4	26.8	1670
829	W 33x241	16.7	30.1	1660
757	W 33x221	16.7	27.6	1510
719	W 36x210	12.9	20.9	1440
684	W 33x201	16.6	24.9	1370
664	W 36x194	12.8	19.4	1330
663	W 30x211	15.9	29.7	1330
623	W 36x182	12.7	18.2	1250
598	W 30x191	15.9	26.9	1200
580	W 36x170	12.7	17.0	1160
542	W 36x160	12.7	15.7	1080
539	W 30x173	15.8	24.2	1080
504	W 36x150	12.6	14.6	1010
502	W 27x178	14.9	27.9	1000
487	W 33x152	12.2	16.9	974
455	W 27x161	14.8	25.4	910
448	W 33x141	12.2	15.4	896
439	W 36x135	12.3	13.0	878
414	W 24x162	13.7	29.3	828
411	W 27x146	14.7	23.0	822
406	W 33x130	12.1	13.8	812
380	W 30x132	11.1	16.1	760
371	W 24x146	13.6	26.3	742
359	W 33x118	12.0	12.6	718
355	W 30x124	11.1	15.0	710
329	W 30x116	11.1	13.8	658
329	W 24x131	13.6	23.4	658
329	W 21x147	13.2	30.3	658
299	W 30x108	11.1	12.3	598
299	W 27x114	10.6	15.9	598
295	W 21x132	13.1	27.2	590
291	W 24x117	13.5	20.8	582
273	W 21x122	13.1	25.4	546
269	W 30x99	10.9	11.4	538
267	W 27x102	10.6	14.2	534
258	W 24x104	13.5	18.4	516
249	W 21x111	13.0	23.3	498
243	W 27x94	10.5	12.8	486
231	W 18x119	11.9	29.1	462
227	W 21x101	13.0	21.3	454
222	W 24x94	9.6	15.1	444
213	W 27x84	10.5	11.0	426
204	W 18x106	11.8	26.0	408
196	W 24x84	9.5	13.3	392
192	W 21x93	8.9	16.8	384
190	W 14x120	15.5	44.1	380
188	W 18x97	11.8	24.1	376
176	W 24x76	9.5	11.8	352
175	W 16x100	11.0	28.1	350
173	W 14x109	15.4	40.6	346
171	W 21x83	8.8	15.1	342
166	W 18x86	11.7	21.5	332
157	W14x99	15.4	37.0	314
155	W 16x89	10.9	25.0	310
154	W 24x68	9.5	10.2	308
151	W 21x73	8.8	13.4	302
146	W 18x76	11.6	19.1	292
143	W 14x90	15.3	34.0	286
140	W 21x68	8.7	12.4	280
134	W 16x77	10.9	21.9	268
131	W 24x62	7.4	8.1	262
127	W 21x62	8.7	11.2	254
127	W 18x71	8.1	15.5	254
123	W 14x82	10.7	28.1	246
118	W 12x87	12.8	36.2	236
117	W 18x65	8.0	14.4	234
117	W 16x67	10.8	19.3	234
114	W 24x55	7.0	7.5	228
112	W 14x74	10.6	25.9	224
111	W 21x57	6.9	9.4	222
108	W 18x60	8.0	13.3	216
107	W 12x79	12.8	33.3	214
103	W 14x68	10.6	23.9	206
98.3	W 18x55	7.9	12.1	197
97.4	W 12x72	12.7	30.5	195
94.5	W 21x50	6.9	7.8	189
92.2	W 16x57	7.5	14.3	184
92.2	W 14x61	10.6	21.5	184
88.9	W 18x50	7.9	11.0	178
87.9	W 12x65	12.7	27.7	176
81.6	W 21x44	6.6	7.0	163
81.0	W 16x50	7.5	12.7	162
78.8	W 18x46	6.4	9.4	158
78.0	W 12x58	10.6	24.4	156
77.8	W 14x53	8.5	17.7	156
72.7	W 16x45	7.4	11.4	145
70.6	W 12x53	10.6	22.0	141
70.3	W 14x48	8.5	16.0	141
64.4	W 18x40	6.3	8.2	137
66.7	W 10x60	10.6	31.1	133
64.7	W 16x40	7.4	10.2	129
64.7	W 12x50	8.5	19.6	129
62.7	W 14x43	8.4	14.4	125
60.0	W 10x54	10.6	28.2	120
58.1	W 12x45	8.5	17.7	116
57.6	W 18x35	6.3	6.7	115
56.5	W 16x36	7.4	8.8	113
54.6	W 14x38	7.1	11.5	109
54.6	W 10x49	10.6	26.0	109
51.9	W 12x40	8.4	16.0	104
49.1	W 10x45	8.5	22.8	98
48.6	W 14x34	7.1	10.2	97
47.2	W 16x31	5.8	7.1	94
45.6	W 12x35	6.9	12.6	91
42.1	W 10x39	8.4	19.8	84
42.0	W 14x30	7.1	8.7	84
38.6	W 12x30	6.9	10.8	77
38.4	W 16x26	5.6	6.0	77
35.3	W 14x26	5.3	7.0	71
35.0	W 10x33	8.4	16.5	70
33.4	W 12x26	6.9	9.4	67
32.4	W 10x30	6.1	13.1	65
31.2	W 8x35	8.5	22.6	62
29.0	W 14x22	5.3	5.6	58
27.9	W 10x26	6.1	11.4	56
27.5	W 8x31	8.4	20.1	55
25.4	W 12x22	4.3	6.4	51
24.3	W 8x28	6.9	17.5	49
23.2	W 10x22	6.1	9.4	46
21.3	W 12x19	4.2	5.3	43
21.1	M 14x18	3.6	4.0	42
20.9	W 8x24	6.9	15.2	42
18.8	W 10x19	4.2	7.2	38
18.2	W 8x21	5.6	11.8	36
17.1	W 12x16	4.1	4.3	34
16.7	W 6x25	6.4	20.0	33
16.2	W 10x17	4.2	6.1	32
15.2	W 8x18	5.5	9.9	30
14.9	W 12x14	3.5	4.2	30
13.8	W 10x15	4.2	5.0	28
13.4	W 6x20	6.4	16.4	27
13.0	M 6x20	6.3	17.4	26
12.0	M 12x11.8	2.7	3.0	24
11.8	W 8x15	4.2	7.2	24
10.9	W 10x12	3.9	4.3	22
10.2	W 6x16	4.3	12.0	20
10.2	W 5x19	5.3	19.5	20
9.91	W 8x13	4.2	5.9	20
9.72	W 6x15	6.3	12.0	19
9.63	M 5x18.9	5.3	19.3	19
8.51	W 5x16	5.3	16.7	17
7.81	W 8x10	4.2	4.7	16
7.76	M 10x 9	2.6	2.7	16
7.31	W 6x12	4.2	8.6	15
5.56	W 6x 9	4.2	6.7	11
5.46	W 4x13	4.3	15.6	11
5.24	M 4x13	4.2	16.9	10
4.62	M 8x 6.5	2.4	2.5	9
2.40	M 6x 4.4	1.9	2.4	5

Source: Reprinted from the *Manual of Steel Construction*, 8th ed. (Ref. 5), with permission of the publishers, American Institute of Steel Construction.

206

Lateral Buckling

A beam may fail by sideways buckling of the top (compression) flange when lateral deflection is not prevented. The tendency to buckle increases as the compressive bending stress in the flange increases and as the unbraced length of the span increases. Consequently the full value of the allowable extreme fiber stress $F_b = 0.66F_y$ can be used only when the compression flange is adequately braced. As for the compact section, the value of this required length includes the variable of the F_y of the beam steel.

When the compression flanges of compact beams are supported laterally at intervals not greater than L_c the full allowable stress of $0.66F_y$ may be used. For lateral unsupported lengths greater than L_c, but not greater than L_u, the allowable bending stress is reduced to $0.60F_y$. (In certain instances the AISC Specification permits a proportionate reduction in the allowable stress between the two limits; however, in the examples in this book we assume that the drop occurs totally as L_c is exceeded.)

When the laterally unsupported length exceeds L_u the specifications provide a formula for the determination of the allowable stress, based on the specific value of the unsupported length. The design of beams based on these requirements is not a simple matter, and the *AISC Manual* (Ref. 5) contains supplementary charts to aid the designer. These beam charts include "allowable moments in beams" as a function of the laterally unsupported length and provide a workable approach to this otherwise rather cumbersome problem. Reproductions of the charts for beams of A36 steel appear in Fig. 22.2.

After determining the maximum bending moment in kip-ft and noting the longest unbraced length of the compression flange, these two coordinates are located on the sides of the chart and are projected to their intersection. Any beam whose curve lies above and to the right of this intersection point satisfies the bending stress requirement. The nearest curve that is a solid (versus dashed) line represents the most economical, or least-weight, section in terms of beam weight, a relationship similar to that discussed in Sec 22.2 with regard to selection from Table 22.1. It should be noted that selection from the charts incorporates the considerations of bending stress, compact sections, and lateral support: however, deflection, shear, and other factors may also have to be considered.

Example 1. A simple beam carries a total uniform load of 19.6 kips [87.2 kN], including its own weight, over a span of 20 ft [6.10 m]. It has no lateral support except at the ends of the span. Assuming A36 steel, select from the charts of Fig. 22.2 a beam that will meet bending strength requirements and not deflect more than 0.75 in. [19 mm] under the full load.
Solution. For this loading the maximum bending moment is

$$M = \frac{WL}{8} = \frac{19.6(20)}{8} = 49 \text{ kip-ft [66.5 kN-m]}$$

Entering the appropriate chart with the values for the moment and the unsupported length, we locate the critical intersection point. The nearest solid line curve above and to the right of this point is that for a W 8 × 31.

For consideration of the deflection limit we may now find the actual deflection for the W 8 × 31 for the given span and load. If this is excessive, we then return to the chart to read the other shapes whose curves are above and to the right of the critical intersection point and examine the corresponding deflections until we can make an acceptable choice. An alternative to this pick-and-try method is a separate calculation for the required moment of inertia of the beam with a transformed version of the formula for maximum deflection for the beam (see Appendix B, Fig. B.1, Case 2). Thus

$$I = \frac{5}{384} \times \frac{Wl^3}{E\Delta} = \frac{5(19.6)(20 \times 12)^3}{384(29,000)(0.75)}$$
$$= 162 \text{ in.}^4 \ [67.82 \times 10^6 \text{ mm}^4]$$

For the W 8 × 31, from Table A.3, $I = 110$ in.^{4}. Because this is less than that required, we return to the chart and select the W 10 × 33 for which $I = 170$ in.^{4}. The W 10 × 33 is an acceptable selection for the criteria established.

It is not always a simple matter to decide that a beam is laterally supported. In cases such as that shown in Fig. 22.3a lateral support is supplied to beams by the floor construction; it is evident from the figure that lateral deflection of the top flange is prevented by the concrete slab. On the other hand, the type of floor construction shown in Fig. 22.3b, where wood joists simply rest on the top flange of a steel beam, offers no resistance to sideways buckling. Floor systems of the types indicated in Fig. 22.3c–f usually furnish adequate lateral support of the top flange. However, metal or precast floor systems held in place by clips generally have insufficiently rigid connections to the flange to provide adequate lateral bracing. If a beam acts as a girder and supports other beams with connections like those shown in Fig. 22.3g, lateral bracing is provided at the connections, and the laterally unsupported length of the top flange to be checked against L_c or L_u becomes the distance between the supported beams.

Torsional Buckling

In various situations steel beams may be subjected to torsional twisting effects in addition to the primary conditions of shear and bending. These effects may occur when the beam is loaded in a plane that does not coincide with the shear center of the section. This problem, of special concern for channel and angle shapes, is discussed in Sec. 12.7. Even for the doubly symmetrical W, M, and S shapes, loadings may produce torsion when applied off center, as shown in Fig. 22.4.

A special torsional effect is that of the rotational effect known as *torsional buckling*. Beams that are weak on their minor axes are subject to this effect, which occurs at the points of support or at the location of concentrated loads (Fig. 22.5a).

When potential torsional effects threaten they may be dealt with in one of two ways. In the first we simply com-

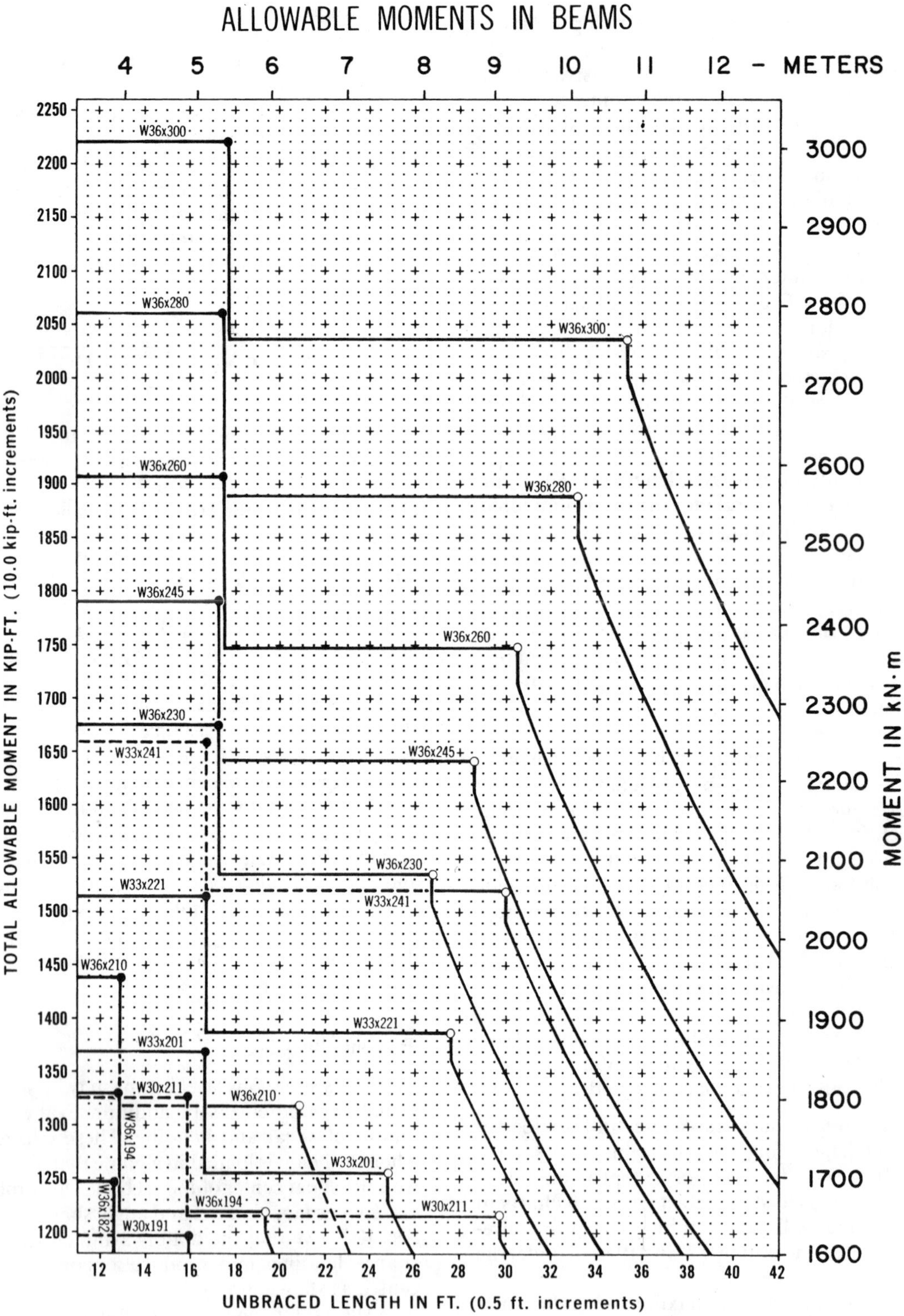

ALLOWABLE MOMENTS IN BEAMS

FIGURE 22.2. Allowable bending moment in beams with various unbraced lengths. F_y = 36 ksi [250 MPa]. Adapted from the *AISC Manual, 8th ed.* (Ref. 5), with permission of the publishers, American Institute of Steel Construction.

ALLOWABLE MOMENTS IN BEAMS

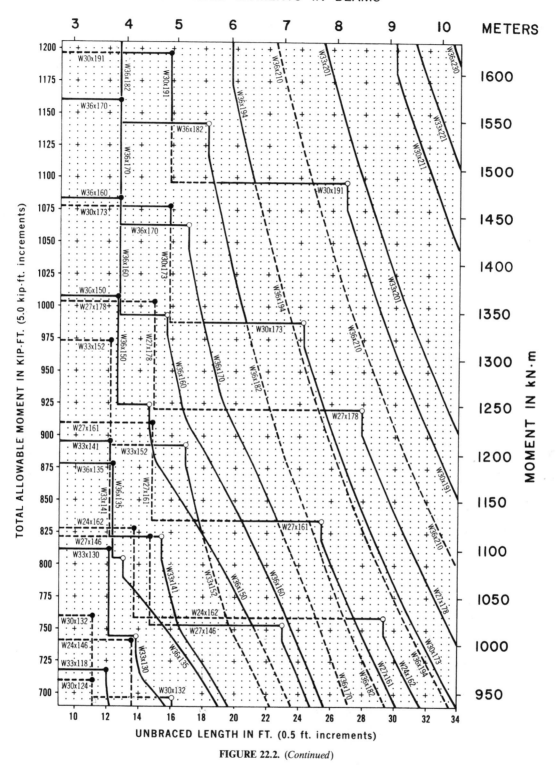

FIGURE 22.2. (*Continued*)

ALLOWABLE MOMENTS IN BEAMS

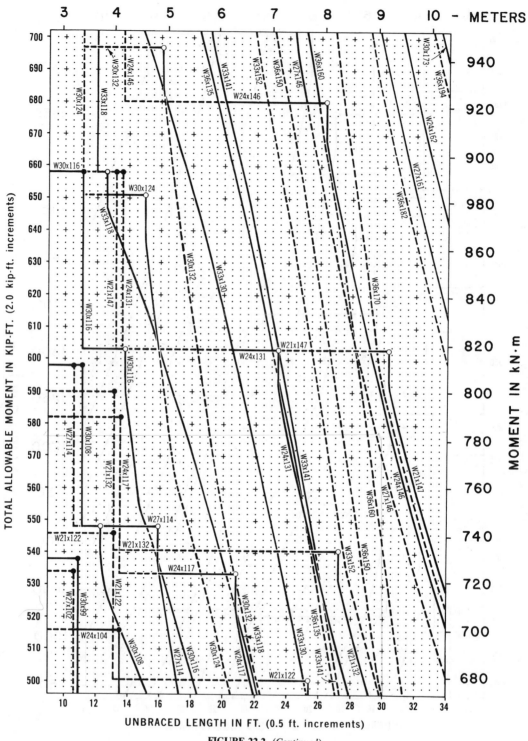

FIGURE 22.2. (*Continued*)

ALLOWABLE MOMENTS IN BEAMS

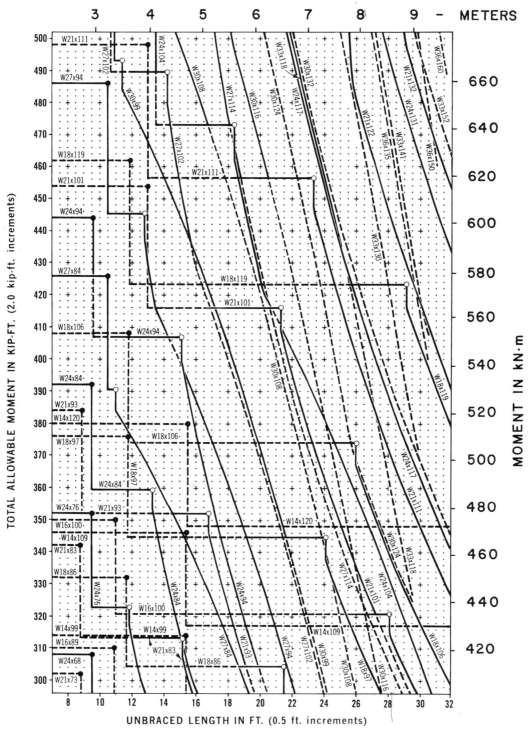

FIGURE 22.2. (*Continued*)

ALLOWABLE MOMENTS IN BEAMS

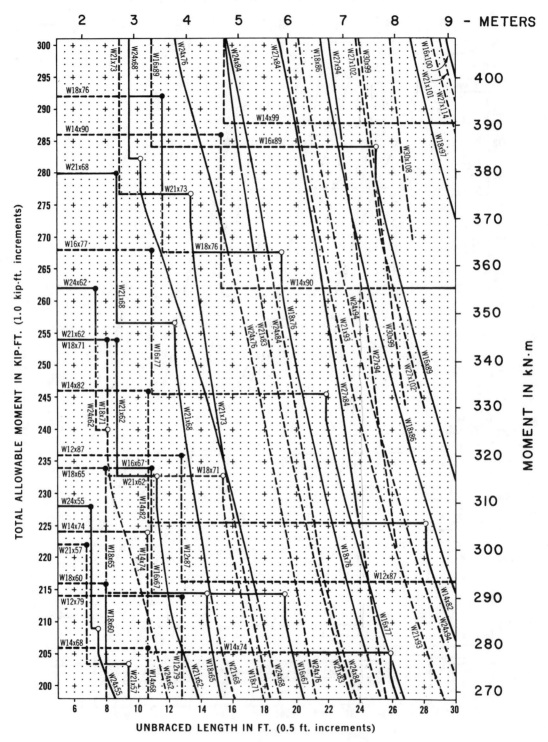

FIGURE 22.2. (*Continued*)

ALLOWABLE MOMENTS IN BEAMS

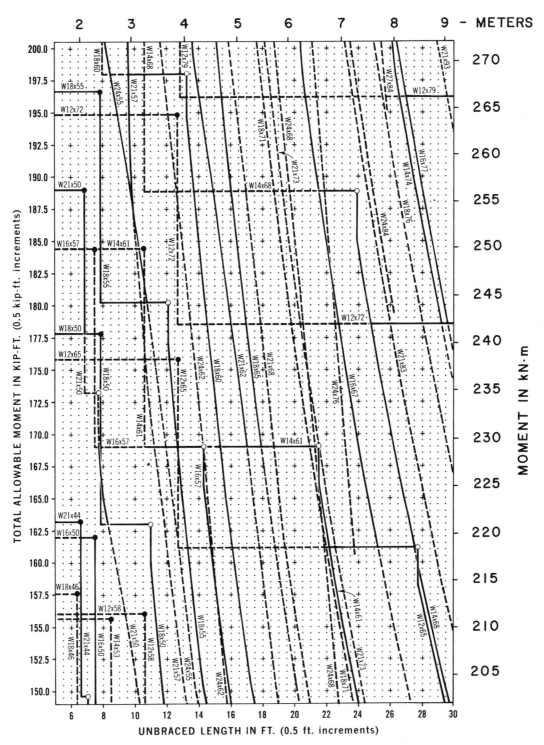

FIGURE 22.2. (*Continued*)

ALLOWABLE MOMENTS IN BEAMS

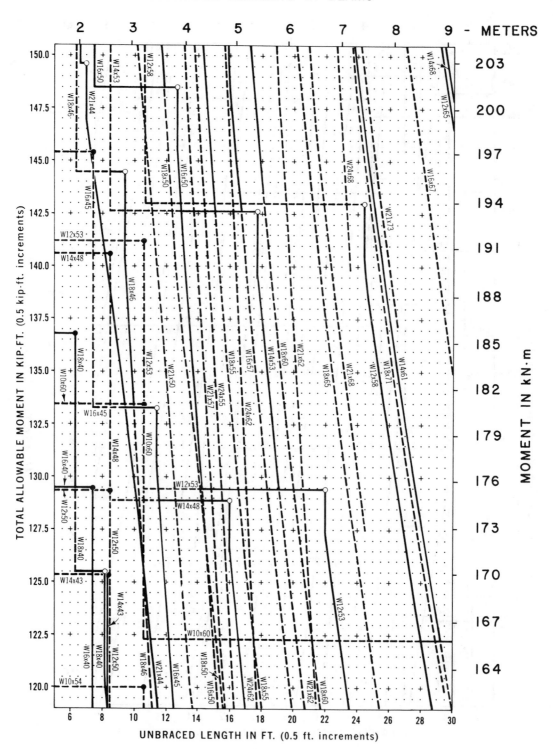

FIGURE 22.2. (*Continued*)

ALLOWABLE MOMENTS IN BEAMS

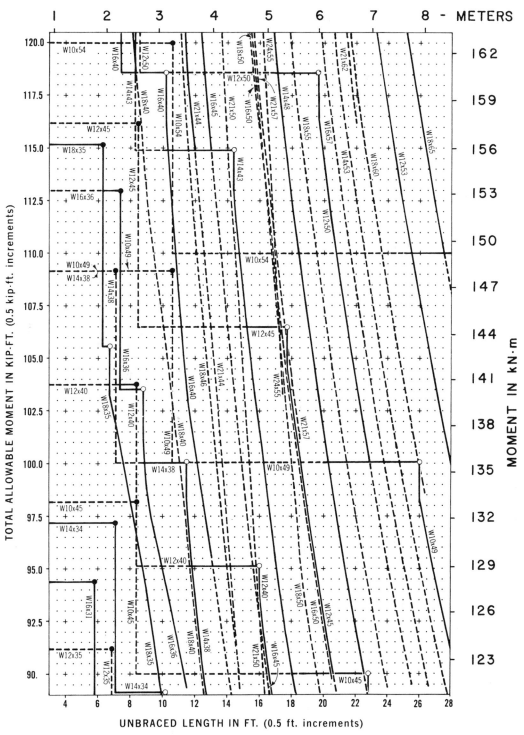

FIGURE 22.2. (*Continued*)

ALLOWABLE MOMENTS IN BEAMS

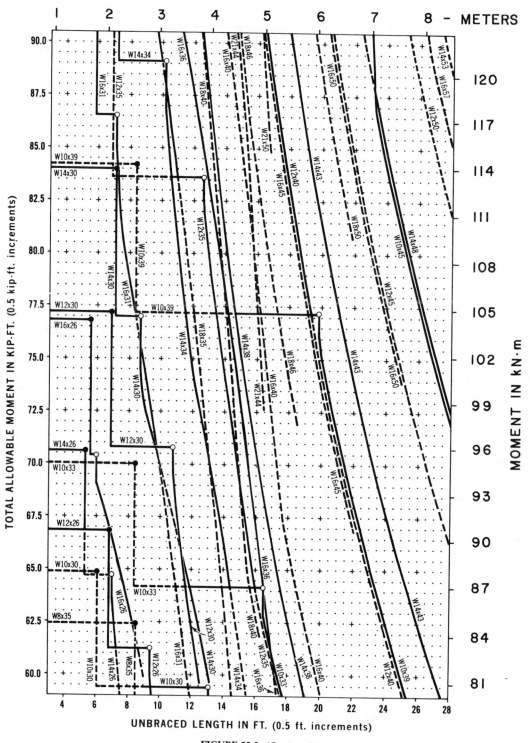

FIGURE 22.2. (*Continued*)

ALLOWABLE MOMENTS IN BEAMS

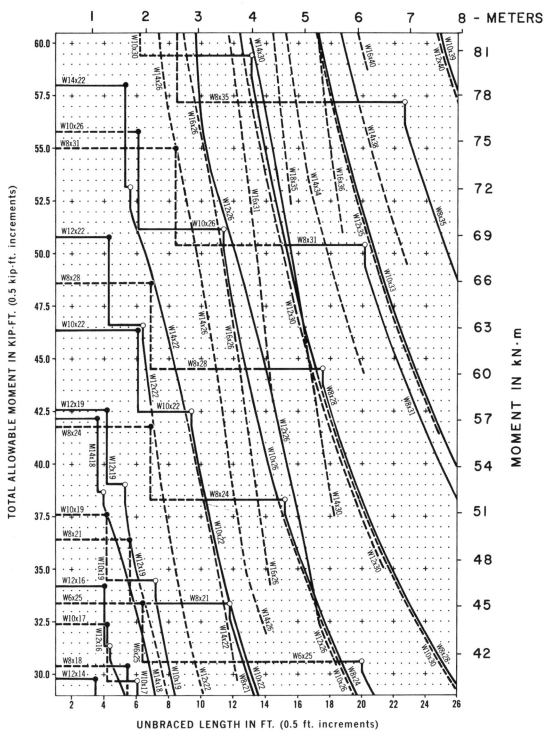

FIGURE 22.2. (*Continued*)

ALLOWABLE MOMENTS IN BEAMS

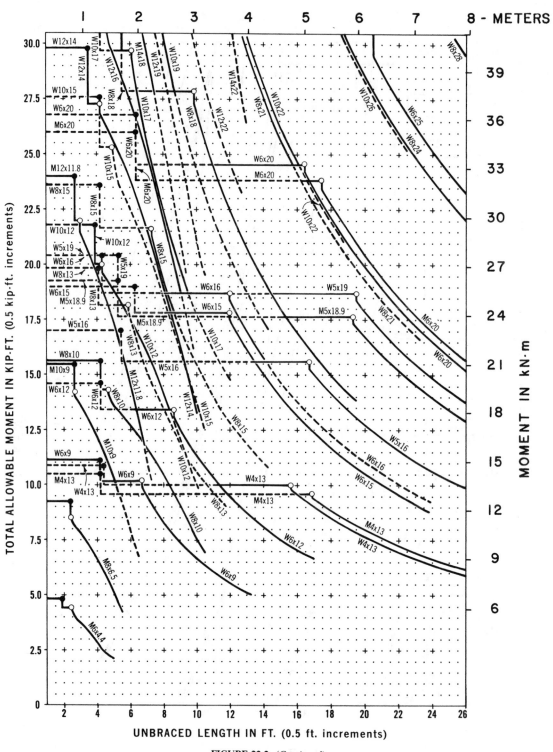

FIGURE 22.2. (*Continued*)

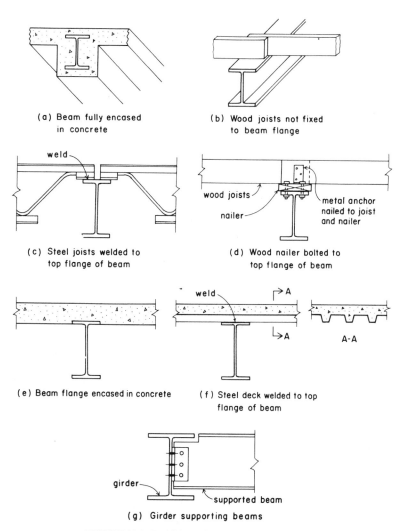

FIGURE 22.3. Provision of lateral bracing for beams.

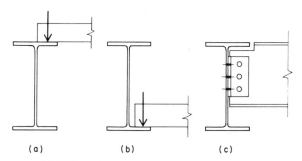

FIGURE 22.4. Torsion-inducing loads on beams.

pute the torsional moments and design the beam to resist them, or we determine the torsional buckling and reduce the allowable stress accordingly. The analysis required, however, is complex and beyond the scope of this book. In the second method, which is usually the preferred, adequate bracing is provided for the beam to prevent the potential torsional rotation, in which case the torsional effect is essentially avoided.

In many cases the ordinary details of construction will result in adequate bracing against torsion. In the detail shown in Fig. 22.4c, for example, although the supported beam applies a slightly off-center load, the connection between the beam and girder may be adequately stiff to prevent torsional twisting of the girder. Some judgment must be exercised, of course, with regard to the relative sizes of girder, beam, and the connection itself.

In Fig. 22.5b the torsional buckling of the beam shown in Fig. 22.5a is prevented by the framing that is attached at right angles to the beam.

22.4. SHEAR

Shear stress in a steel beam is seldom a factor in determining its size. It is customary to determine first the size of the beam to resist bending stresses. Having done this, the beam is then investigated for shear, which means that we compute the actual maximum unit shear stress to see that it does not exceed the allowable stress. The AISC Specifica-

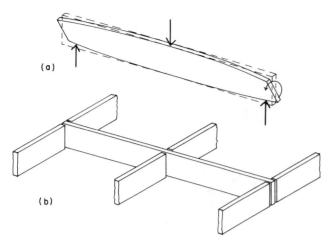

FIGURE 22.5. Bracing for torsional buckling.

tion gives F_v, the allowable shear stress in beam webs, as $0.40F_y$ on the gross section of the web, which is computed as the product of the web thickness and the overall beam depth. For A36 steel $F_v = 0.40 \times 36$ or 14.4 ksi. This value is rounded off at $F_v = 14.5$ ksi [100 MPa], as shown in Table 21.1.

As noted in Sec. 12.4, shearing stresses in beams are not distributed uniformly over the cross section but are zero at the extreme fibers, with the maximum value occurring at the neutral surface. Consequently, the material in the flanges of wide flange sections, I-beams, and channels has little influence on shearing resistance, and the working formula for determining shearing stress is taken as

$$f_v = \frac{V}{A_w}$$

where

f_v = unit shearing stress
V = maximum vertical shear
A_w = gross area of the web (actual depth of section times the web thickness, or $d \times t_w$)

The shearing stresses in beams are seldom excessive. If however, the beam has a relatively short span with a large load placed near one of the supports, the bending moment is relatively small and the shearing stress becomes comparatively high. This situation is demonstrated in the following example:

Example 1. A simple beam of A36 steel is 6 ft [1.83 m] long and has a concentrated load of 36 kips [160 kN] applied 1 ft [0.3 m] from the left end. It is found that a W 8 × 24 is large enough to support this load with respect to bending stresses (required S = 15 in.³ [246 × 10³ mm³]). Investigate the beam for shear, neglecting the weight of the beam.
Solution. The two reactions are computed by the methods explained in Sec. 9.4; we find that the left reaction is 30 kips [133 kN] and the right reaction is 6 kips [27 kN]. The maximum vertical shear is thus 30 kips [133kN].

From Table A.3 we find that $d = 7.93$ in. [201.4 mm] and $t_w = 0.245$ in. [6.22 mm] for the W 8 × 24. Then

$$A_w = dt_w = 7.93(0.245) = 1.94 \text{ in.}^2 [1253 \text{ mm}^2]$$

and

$$f_v = \frac{V}{A_w} = \frac{30}{1.94} = 15.4 \text{ ksi} [106 \text{ MPa}]$$

Because this exceeds the allowable value of 14.5 ksi [100 MPa], the W 8 × 24 is not acceptable.

Recalling from Sec. 27.3 that S shapes have somewhat thicker webs than W shapes, we find in Table A.4 that an S 8 × 23 has a depth of 8 in. [203.2 mm] and a web thickness of 0.441 in. [11.2 mm]. Then

$$A_w = dt_w = 8(0.441) = 3.53 \text{ in.}^2 [2276 \text{ mm}^3]$$

and

$$f_v = \frac{V}{A_w} = \frac{30}{3.53} = 8.50 \text{ ksi} [58.4 \text{ MPa}]$$

which is less than the allowable stress. It should be observed that both W and S shapes are adequate for bending stress. Thus both sections are acceptable for bending resistance, but only the S 8 × 23 will provide adequate shearing resistance.

22.5. DEFLECTION

For steel beams we present three means for determining deflection.

1. *By Formula for the Loading Condition.* This involves use of the formulas given for the various cases in Appendix B. The most common case being the simple beam with uniformly distributed load, for which the deflection is determined as

$$D = \frac{5 WL^3}{384EI}$$

2. *By Use of Deflection Factors.* This is limited to the case of the simple beam with uniformly distributed load. The process involves use of the factors given in Table 22.2.
3. *By Use of the Deflection Graphs in Fig. 22.6.* This is essentially the same operation as that using the deflection factors, presented in a different format.

The several cases of loading and support conditions presented in Appendix B cover most of the common situations occurring in building structures. Where only the maximum deflection is required, the formulas given can be used directly. The other two methods presented here permit a somewhat simplified procedure for the special case of the simple beam with uniformly distributed load. Since this is the singly most common case for beams, the shortcut methods are worth considering.

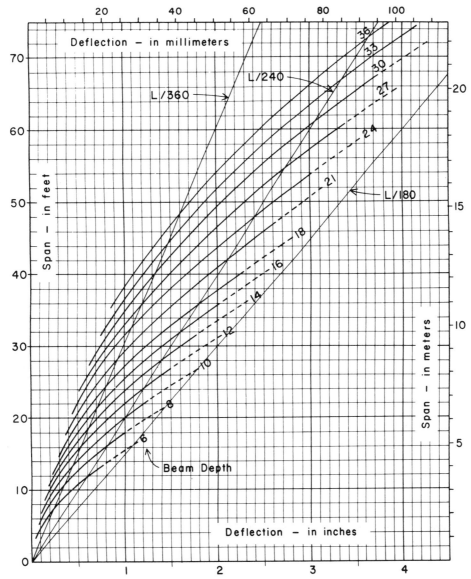

FIGURE 22.6. Deflection of steel beams with bending stress of 24 ksi [165 MPa].

The relationship used for the deflection factor and graph methods is derived as follows. We begin with the formula for deflection of the simple beam, as stated above. For the same beam the maximum bending moment is expressed as

$$M = \frac{WL}{8} \quad (W = \text{total load}; L \text{ in ft})$$

and if the beam is symmetrical about its bending axis,

$$\text{maximum } f = \frac{Mc}{I} = \frac{M(d/2)}{I} = \frac{Md}{2I}$$

or

$$M = \frac{2If}{d}$$

Equating the two expressions for M gives us

$$\frac{WL}{8} = \frac{2If}{d}$$

or

$$W = \frac{16If}{Ld}$$

Substituting this expression for W in the deflection formula, we obtain

$$D = \frac{5}{384}\left(\frac{L^3}{EI}\right)\frac{16If}{Ld} = \frac{5}{24}\frac{L^2f}{Ed}$$

This is the general form of the equation that may be used for any beam of any material, subject to the various qualifications mentioned (see the application to a wood beam in Sec. 17.2). For the A36 steel beam, if we assume the maximum bending stress is 24 ksi and the modulus of elasticity of the steel is 29,000 ksi, the expression can be simplified further to

$$D = \frac{5}{24}\left(\frac{24}{29,000}\right)\frac{(12L)^2}{d} = 0.02483\,\frac{L^2}{d}$$

This is the expression used to derive the deflection factors in Table 22.2 and to plot the graphs for Fig. 22.6. Note that the length of the span was converted to inches with the factor of 12, so that the value for L remains in feet in the formula. The following examples show some uses for the deflection computation shortcuts.

Note that for SI units $f = 165$ MPa and $E = 200$ GPa. Then, with D in millimeters and L in meters,

$$D = 0.1719\,\frac{L^2}{d}$$

Example 1. A simple beam has a span of 20 ft [6.10 m] and a total uniformly distributed load of 39 kips [173.5 kN]. The beam is a W 14 × 34. Find the maximum deflection.
Solution. The first step is to compute the actual value for the maximum bending stress; thus

$$M = \frac{WL}{8} = \frac{39(20)}{8} = 97.5\ \text{kip-ft}\ [132.3\ \text{kN-m}]$$

From Table A.3 we find the S for the beam to be 48.6 in.³ [797 x 10³ mm³]. Then the maximum stress is

$$f = \frac{M}{S} = \frac{97.5(12)}{48.6} = 24.07\ \text{ksi}\ [166\ \text{MPa}]$$

Since this is very close to the value of 24 ksi that was used in deriving the factors, they may be used without adjustment. Thus, from Table 22.2 we find the deflection factor for the 20-ft span to be 9.93, and the deflection is computed as

$$D = \frac{\text{deflection factor}}{\text{beam depth}} = \frac{9.93}{14} = 0.71\ \text{in.}\ [18\ \text{mm}]$$

(Note that the actual beam depth is not precisely 14 in., but it is usually not necessary to be highly accurate in deflection computations.)

Inspection of Fig. 22.6, using only the span of 20 ft and the curve for 14-in. beams, will yield a deflection value of just over 0.7 in., which is in reasonable agreement with the computations.

Example 2. A W 12 × 30 is used on a simple span to carry a total uniformly distributed load of 30 kips [133 kN]. Find the deflection if the span is 18 ft [5.5 m].
Solution. As before, we find the moment and stress as

$$M = \frac{WL}{8} = \frac{30(18)}{8} = 67.5\ \text{ksi}\ [91.4\ \text{kN-m}]$$

$$f = \frac{M}{S} = \frac{67.5(12)}{38.6} = 21.0\ \text{ksi}\ [145\ \text{MPa}]$$

From Table 22.2 we obtain the factor for the span as 8.05, with which the deflection is

$$D = \frac{8.05}{12} = 0.671\ \text{in.}\ [17\ \text{mm}]$$

This, however, is the deflection obtained when the stress in bending is 24 ksi. To find the true deflection, we simply multiply this value by the ratio of the actual deflection to the table value of 24. Thus

$$D = \frac{21}{24}\,(0.671) = 0.587\ \text{in.}\ [15\ \text{mm}]$$

Again, we note that Fig. 22.6 yields a deflection of approximately 0.67 in., which must also be adjusted for the true stress condition.

Some other uses for Fig. 22.6 are illustrated in the design examples in Part 9.

22.6. SAFE-LOAD TABLES

The simple span beam loaded entirely with uniformly distributed load occurs so frequently in steel structural systems that it is useful to have a rapid design method for quick selection of shapes for a given load and span condition. Use of Table 22.2 allows for a simple design procedure where design conditions permit its use. When beams of A36 steel are loaded in the plane of their minor axis (Y–Y) and have lateral bracing spaced not farther than L_c, they may be selected from Table 22.2 after determination of only the total load and the beam span.

For a check, the values of L_c are given for each shape in the table. If the actual distance between points of lateral support for the compression (top) flange exceeds L_c, the table values must be reduced, as discussed in Sec. 22.3.

Deflections may be determined by using the factors given in the table or the graphs in Fig. 22.6. Both of these are based on a maximum bending stress of 24 ksi [165 MPa], and deflections for other stresses may be proportioned from these values if necessary.

The loads in the tables will not result in excessive shear stress on the beam webs if the full beam depth is available for stress development. Where end framing details result in some reduction of the web area, investigation of the stress on the reduced section may be required.

The following examples illustrate the use of Table 22.2 for some common design situations.

Example 3. A simple span beam of A36 steel is required to carry a total uniformly distributed load of 40 kips [178 kN] on a span of 30 ft [9. 14 m]. Find the lightest shape permitted and the shallowest (least deep) shape permitted.
Solution. From Table 22.2 we find the following:

Shape	Allowable Load (kips)
W 21 × 44	43.5
W 18 × 46	42.0
W 16 × 50	43.2
W 14 × 53	41.5

TABLE 22.2. Load-Span Values for Beams

Shape	Span (ft) / L_c** (ft)	8	10	12	14	16	18	20	22	24	26	28	30
	Deflection Factor*	1.59	2.48	3.58	4.87	6.36	8.05	9.93	12.0	14.3	16.8	19.5	22.3
M 8 × 6.5	2.4	9.24	7.39	6.16	5.28	4.62	4.11						
M 10 × 9	2.6	15.5	12.4	10.3	8.87	7.76	6.90	6.21	5.64				
W 8 × 10	4.2	15.6	12.5	10.4	8.92	7.81	6.94						
W 8 × 13	4.2	19.8	15.9	13.2	11.3	9.91	8.81						
W 10 × 12	3.9	21.8	17.4	14.5	12.5	10.9	9.69	8.72	7.93				
W 8 × 15	4.2	23.6	18.9	15.7	13.5	11.8	10.5						
M 12 × 11.8	2.7	24.0	19.2	16.0	13.7	12.0	10.7	9.60	8.73	8.00	7.38	6.86	
W 10 × 15	4.2	27.6	22.1	18.4	15.8	13.8	12.3	11.0	10.0				
W 12 × 14	3.5	29.8	23.8	19.9	17.0	14.9	13.2	11.9	10.8	9.93	9.17	8.51	
W 8 × 18	5.5	30.4	24.3	20.3	17.4	15.2	13.5						
W 10 × 17	4.2	32.4	25.9	21.6	18.5	16.2	14.4	13.0	11.8				
W 12 × 16	4.1	34.2	27.4	22.8	19.5	17.1	15.2	13.7	12.4	11.4	10.5	9.77	
W 8 × 21	5.6	36.4	29.1	24.3	20.8	18.2	16.2						
W 10 × 19	4.2	37.6	30.1	25.1	21.5	18.8	16.7	15.0	13.7				
W 8 × 24	6.9	41.8	33.4	27.9	23.9	20.9	18.6						
M 14 × 18	3.6	42.2	33.8	28.1	24.1	21.1	18.7	16.9	15.3	14.1	13.0	12.0	11.2
W 12 × 19	4.2	42.6	34.1	28.4	24.3	21.3	18.9	17.0	15.5	14.2	13.1	12.2	
W 10 × 22	6.1	46.4	37.1	30.9	26.5	23.2	20.6	18.5	16.9				
W 8 × 28	6.9	48.6	38.9	32.4	27.8	24.3	21.6						

Shape	Span (ft) / L_c** (ft)	12	14	16	18	20	22	24	26	28	30	32	34
	Deflection Factor*	3.58	4.87	6.36	8.05	9.93	12.0	14.3	16.8	19.5	22.3	25.4	28.7
W 12 × 22	4.3	33.9	29.0	25.4	22.6	20.3	18.5	16.9	15.6	14.5			
W 10 × 26	6.1	37.2	31.9	27.9	24.8	22.3	20.3						
W 14 × 22	5.3	38.7	33.1	29.0	25.8	23.2	21.1	19.3	17.8	16.6	15.5	14.5	
W 10 × 30	6.1	43.2	37.0	32.4	28.8	25.9	23.6						
W 12 × 26	6.9	44.5	38.2	33.4	29.7	26.7	24.3	22.3	20.5	19.1			
W 10 × 33	8.4	46.7	40.0	35.0	31.0	28.0	25.4						
W 14 × 26	5.3	47.1	40.3	35.3	31.4	28.2	25.7	23.5	21.7	20.2	18.8	17.6	
W 16 × 26	5.6	51.2	43.9	38.4	34.1	30.7	27.9	25.6	23.6	21.9	20.5	19.2	18.1
W 12 × 30	6.9	51.5	44.1	38.6	34.3	30.9	28.1	25.7	23.8	22.0			
W 14 × 30	7.1	56.0	48.0	42.0	37.3	33.6	30.5	28.0	25.8	24.0	22.4	21.0	
W 10 × 39	8.4	56.1	48.1	42.1	37.4	33.7	30.6						
W 12 × 35	6.9	60.8	52.1	45.6	40.5	36.5	33.2	30.4	28.1	26.0			
W 16 × 31	5.8	62.9	53.9	47.2	41.9	37.8	34.3	31.5	29.0	27.0	25.2	23.6	22.2
W 14 × 34	7.1	64.8	55.5	48.6	43.2	38.9	35.3	32.4	29.9	27.8	25.9	24.3	
W 10 × 45	8.5	65.5	56.1	49.1	43.6	39.3	35.7						

Shape	Span (ft) / L_c** (ft)	16	18	20	22	24	26	28	30	32	34	36	38
	Deflection Factor*	6.36	8.05	9.93	12.0	14.3	16.8	19.5	22.3	25.4	28.7	32.2	35.9
W 12 × 40	8.4	51.9	46.1	41.5	37.7	34.6	31.9	29.6					
W 14 × 38	7.1	54.6	48.5	43.7	39.7	36.4	33.6	31.2	29.1	27.3			
W 16 × 36	7.4	56.5	50.2	45.2	41.1	37.7	34.8	32.3	30.1	28.2	26.6	25.1	
W 18 × 35	6.3	57.8	51.4	46.2	42.0	38.5	35.6	33.0	30.8	28.9	27.2	25.7	24.3
W 12 × 45	8.5	58.1	51.6	46.5	42.2	38.7	35.7	33.2					
W 14 × 43	8.4	62.7	55.7	50.1	45.6	41.8	38.6	35.8	33.4	31.3			
W 12 × 50	8.5	64.7	57.5	51.7	47.0	43.1	39.8	37.0					
W 16 × 40	7.4	64.7	57.5	51.7	47.0	43.1	39.8	37.0	34.5	32.3	30.4	28.7	
W 18 × 40	6.3	68.4	60.8	54.7	49.7	45.6	42.1	39.1	36.5	34.2	32.2	30.4	28.8
W 14 × 48	8.5	70.3	62.5	56.2	51.1	46.9	43.3	40.2	37.5	35.1			
W 12 × 53	10.6	70.6	62.7	56.5	51.3	47.1	43.4	40.3					
W 16 × 45	7.4	72.7	64.6	58.2	52.9	48.5	44.7	41.5	38.8	36.3	34.2	32.3	
W 14 × 53	8.5	77.8	69.1	62.2	56.6	51.9	47.9	44.4	41.5	38.9			
W 18 × 46	6.4	78.8	70.0	63.0	57.3	52.5	48.5	45.0	42.0	39.4	37.1	35.0	33.2
W 16 × 50	7.5	81.0	72.0	64.8	58.9	54.0	49.8	46.3	43.2	40.5	38.1	36.0	

Note: Total allowable uniformly distributed load is in kips for simple span beams of A36 steel with yield stress of 36 ksi [250 MPa].

* Maximum deflection in inches at the center of the span may be obtained by dividing this factor by the depth of the beam in inches. This is based on a maximum bending stress of 24 ksi [165 MPa].

** Maximum permitted distance in feet between points of lateral support. If distance exceeds this, use the charts in Fig. 22.2.

TABLE 22.2 (*continued*)

Shape	L_c** (ft) / Deflection Factor*	16 / 6.36	18 / 8.05	20 / 9.93	22 / 12.0	24 / 14.3	27 / 18.1	30 / 22.3	33 / 27.0	36 / 32.2	39 / 37.8	42 / 43.8	45 / 50.3
W 21 × 44	6.6	81.6	72.5	65.3	59.3	54.4	48.3	43.5	39.6	36.3	33.5	31.1	29.0
W 18 × 50	7.9	88.9	79.0	71.1	64.6	59.3	52.7	47.4	43.1	39.5	36.5		
W 14 × 61	10.6	92.2	81.9	73.8	67.0	61.5	54.6	49.2	44.7				
W 16 × 57	7.5	92.2	81.9	73.8	67.0	61.5	54.6	49.2	44.7	41.0			
W 21 × 50	6.9	94.5	84.0	75.6	68.7	63.0	56.0	50.4	45.8	42.0	38.8	36.0	33.6
W 18 × 55	7.9	98.3	87.4	78.6	71.5	65.5	58.2	52.4	47.7	43.7	40.3		
W 18 × 60	8.0	108	96.0	86.4	78.5	72.0	64.0	57.6	52.4	48.0	44.3		
W 21 × 57	6.9	111	98.7	88.6	80.7	74.0	65.8	59.2	53.8	49.3	45.5	42.3	39.5
W 24 × 55	7.0	114	101	91.2	82.9	76.0	67.5	60.8	55.3	50.7	46.8	43.4	40.5
W 16 × 67	10.8	117	104	93.6	85.1	78.0	69.3	62.4	56.7	52.0			
W 18 × 65	8.0	117	104	93.6	85.1	78.0	69.3	62.4	56.7	52.0	48.0		
W 18 × 71	8.1	127	113	102	92.4	84.7	72.2	67.7	61.5	56.4	52.1		
W 21 × 62	8.7	127	113	102	92.4	84.7	72.2	67.7	61.5	56.4	52.1	48.4	45.1
W 24 × 62	7.4	131	116	105	95.3	87.3	77.6	69.9	63.5	58.2	53.7	49.9	46.6
W 16 × 77	10.9	134	119	107	97.4	89.3	79.4	71.5	65.0	59.5			
W 21 × 68	8.7	140	124	112	102	93.3	83.0	74.7	67.9	62.2	57.4	53.3	49.8
W 18 × 76	11.6	146	130	117	106	97.3	86.5	77.9	70.8	64.9	59.9		
W 21 × 73	8.8	151	134	121	110	101	89.5	80.5	73.2	67.1	61.9	57.5	53.7
W 24 × 68	9.5	154	137	123	112	103	91.2	82.1	74.7	68.4	63.2	58.7	54.7
W 18 × 86	11.7	166	147	133	121	111	98.4	88.5	80.5	73.8	68.1		
W 21 × 83	8.8	171	152	137	124	114	101	91.2	82.9	76.0	70.1	65.1	60.8

Shape	L_c** (ft) / Deflection Factor*	24 / 14.3	27 / 18.1	30 / 22.3	33 / 27.0	36 / 32.2	39 / 37.8	42 / 43.8	45 / 50.3	48 / 57.2	52 / 67.1	56 / 77.9	60 / 89.4
W 24 × 76	9.5	117	104	93.9	85.3	78.2	72.2	67.0	62.6	58.7			
W 21 × 93	8.9	128	114	102	93.1	85.3	78.8	73.1	68.3				
W 24 × 84	9.5	131	116	104	95.0	87.1	80.4	74.7	69.7	65.3			
W 27 × 84	10.5	142	126	114	103	94.7	87.4	81.1	75.7	71.0	65.5	60.8	
W 24 × 94	9.6	148	131	118	108	98.7	91.1	84.6	78.9	74.0			
W 21 × 101	13.0	151	134	121	110	101	93.1	86.5	80.7				
W 27 × 94	10.5	162	144	130	118	108	99.7	92.6	86.4	81.0	74.8	69.4	
W 24 × 104	13.5	172	153	138	125	115	106	98.3	91.7	86.0			
W 27 ×·102	10.6	178	158	142	129	119	109	102	94.9	89.0	82.1	76.3	
W 30 × 99	10.9	179	159	143	130	120	110	102	95.6	89.7	82.8	76.9	71.7
W 24 × 117	13.5	194	172	155	141	129	119	111	103	97.0			
W 27 × 114	10.6	199	177	159	145	133	123	114	106	99.7	92.0	85.4	
W 30 × 108	11.1	199	177	159	145	133	123	114	106	99.7	92.0	85.4	79.7
W 30 × 116	11.1	219	195	175	159	146	135	125	117	110	101	94.0	87.7
W 30 × 124	11.1	237	210	189	172	158	146	135	126	118	109	101	94.7

Shape	L_c** (ft) / Deflection Factor*	30 / 22.3	33 / 27.0	36 / 32.2	39 / 37.8	42 / 43.8	45 / 50.3	48 / 57.2	52 / 67.1	56 / 77.9	60 / 89.4	65 / 105	70 / 122
W 33 × 118	12.0	191	174	159	147	137	128	120	110	103	95.7	88.4	
W 30 × 132	11.1	203	184	169	156	145	135	127	117	109	101		
W 33 × 130	12.1	216	197	180	166	155	144	135	125	116	108	99.9	
W 27 × 146	14.7	219	199	183	169	156	146	137	126	117			
W 36 × 135	12.3	234	213	195	180	167	156	146	135	125	117	108	100
W 33 × 141	12.2	239	217	199	184	171	159	149	138	128	119	110	
W 33 × 152	12.2	260	236	216	200	185	173	162	150	139	130	120	
W 36 × 150	12.6	269	244	224	207	192	179	168	155	144	134	124	115
W 30 × 173	15.8	287	261	239	221	205	192	180	166	154	144		
W 36 × 160	12.7	289	263	241	222	206	193	181	167	155	144	133	124
W 36 × 170	12.7	309	281	258	238	221	206	193	178	166	155	143	132
W 30 × 191	15.9	319	290	268	245	228	213	199	184	171	159		
W 36 × 182	12.7	332	302	277	256	237	221	208	192	178	166	153	142
W 36 × 194	12.8	354	322	295	272	253	236	221	204	190	177	163	152
W 33 × 201	16.6	365	332	304	281	260	243	228	210	195	182	168	
W 36 × 210	12.9	383	349	319	295	274	256	240	221	205	192	177	164
W 33 × 221	16.7	404	367	336	310	288	269	252	233	216	202	186	
W 33 × 241	16.7	442	402	368	340	316	295	276	255	237	221	204	
W 36 × 230	17.4	446	406	372	343	319	298	279	257	239	223	206	191
W 36 × 245	17.4	477	434	398	367	341	318	298	275	256	239	220	204
W 36 × 260	17.5	508	462	423	391	363	339	318	293	272	254	234	218
W 36 × 280	17.5	549	499	458	422	392	366	343	317	294	275	253	235
W 36 × 300	17.6	592	538	493	455	423	395	370	341	317	296	273	254

Thus the lightest shape is the W 21 × 44, and the shallowest is the W 14 × 53.

Example 4. A simple span beam of A36 steel is required to carry a total uniformly distributed load of 25 kips [111 kN] on a span of 24 ft [7.32 m] while sustaining a maximum deflection of not more than 1/360 of the span. Find the lightest shape permitted.
Solution. From Table 22.2 we find the lightest shape that will carry this load to be the W 16 × 26. For this beam the deflection will be

$$D = \frac{25}{25.6}\left(\frac{14.3}{16}\right) = \frac{\text{actual load}}{\text{table load}} \times \frac{\text{deflection factor}}{\text{beam depth}}$$
$$= 0.873 \text{ in.}$$

which exceeds the allowable of $[24(12)]/360 = 0.80$ in.

The next heaviest shape from Table 22.2 is a W 16 × 31, for which the deflection will be

$$D = \frac{25}{31.5}\left(\frac{14.3}{16}\right) = 0.709 \text{ in.}$$

which is less than the limit, so this shape is the lightest choice.

22.7. MISCELLANEOUS BEAM PROBLEMS

The preceding sections in this chapter have presented the routine considerations for steel beams. This section presents some special problems that may need consideration in particular situations.

Compact Sections

To qualify for use of the maximum allowable bending stress of $0.66 F_y$, a beam consisting of a rolled section must satisfy several qualifications. The principal ones are the following.

1. The beam section must be symmetrical about its minor $(Y–Y)$ axis, and the plane of the loading must coincide with the plane of this axis; otherwise, a torsional twist will be developed along with the bending.
2. The web and flanges of the section must have width–thickness ratios that qualify the section as *compact.*
3. The compression flange of the beam must be adequately braced against lateral buckling.

The criteria for establishing whether or not a section is compact include as a variable the F_y of the steel. It is therefore not possible to identify sections for this condition strictly on the basis of their geometric properties. The yield stress limits for the qualification of sections as compact for bending and those for combined actions of compression and bending are given, together with other properties, in the tables in the *AISC Manual* (Ref. 5). For beams of A36 steel, all S and C sections and all wide flange sec-

tions, except the W 6 × 15, qualify as compact when used for bending alone. When sections do not qualify as compact, the allowable bending stress must be reduced with the use of formulas given in the AISC Specification.

Bearing and Web Crippling

An excessive end reaction on a beam or an excessive concentrated load at some point along the interior of the span may cause crippling, or localized yielding, of the beam web. The AISC Specification requires that end reactions or concentrated loads for beams without stiffeners or other web reinforcement will not exceed the following (Fig. 22.7):

$$\text{maximum end reaction} = 0.75F_y t(N + k)$$
$$\text{maximum interior load} = 0.75F_y t(N + 2k)$$

where

t = thickness of the beam *web*, inches
N = length of the bearing or length of the concentrated load (not less than k for end reactions), inches
k = distance from the outer face of the flange to the web toe of the fillet, inches
$0.75F_y$ = 27 ksi for A36 steel [186 MPa]

When these values are exceeded, the webs of the beams should be reinforced with stiffeners, the length of bearing increased, or a beam with a thicker web selected.

Example 1. A W 21 x 57 beam of A36 steel has an end reaction that is developed in bearing over a length of $N = 10$ in. [254 mm]. Check the beam for web crippling if the reaction is 44 kips [196 kN].
Solution. In Table A.3 we find that $k = 1.38$ in. [35 mm] and the web thickness is 0.405 in. [10 mm]. To check for web crippling, we find the maximum end reaction permitted and compare it with the actual value for the reaction.

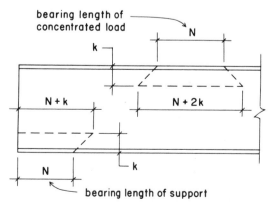

FIGURE 22.7. Bearing length considerations.

Thus

$$R = F_p t(N + k) = 27(0.405)(10 + 1.38)$$

$$= 124 \text{ kips (the allowable reaction)}$$

$$\left[R = \frac{186(10)(254 + 35)}{10^3} = 538 \text{ kN} \right]$$

Because this is greater than the actual reaction, the beam is not critical with regard to web crippling.

Example 2. A W 14 × 22 of A36 steel supports a column load of 70 kips [311 kN] at the center of the span. The bearing length of the column on the beam is 10 in. [254 mm]. Investigate the beam for web crippling under this concentrated load.

Solution. In Table A.3 we find that $k = 0.88$ in. [22 mm] and the web thickness is 0.230 in. [5.84 mm]. The allowable load that can be supported on the given bearing length is

$$P = F_p t(N + 2k) = 27(0.230)\{10 + (2 \times 0.88)\}$$

$$= 73 \text{ kips}$$

$$\left[P = \frac{186(5.84)}{10^3} \{254 + [2(22)]\} = 324 \text{ kN} \right]$$

which exceeds the required load. Because the column load is less than this value, the beam web is safe from web crippling.

Beam Bearing Plates

Beams that are supported on walls or piers of masonry or concrete usually rest on steel bearing plates. The purpose of the plate is to provide an ample bearing area. The plate also helps to seat the beam at its proper elevation. Bearing plates provide a level surface for a support and, when properly placed, afford a uniform distribution of the beam reaction over the area of contact with the supporting material.

By reference to Fig. 22.8, the area of the bearing plate is *BN*. It is found by dividing the beam reaction by F_p, the allowable bearing value of the supporting material. Then

$$A = \frac{R}{F_p}$$

where

$A = BN$, the area of the plate, in.²

R = reaction of the beam, pounds or kips

R_p = allowable bearing pressure on the supporting material, psi or ksi (see Table 22.3)

The thickness of the wall generally determines N, the dimension of the plate parallel to the length of the beam. If the load from the beam is unusually large, the dimension *B* may become excessive. For such a condition, one or more shallow-depth I-beams, placed parallel to the wall length, may be used instead of a plate. The dimensions *B* and *N* are

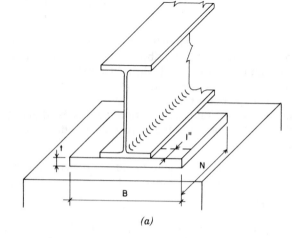

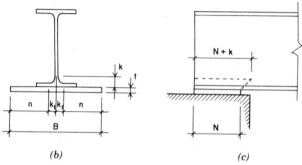

FIGURE 22.8. Beam end bearing plates.

TABLE 22.3. Allowable Bearing Pressure on Masonry and Concrete

Type of Material and Conditions	Allowable Unit Stress in Bearing, F_p	
	psi	kPa
Solid brick, unreinforced, type S mortar		
$f'_m = 1500$ psi	170	1200
$f'_m = 4500$ psi	338	2300
Hollow unit masonry, unreinforced, type S mortar, $f'_m = 1500$ psi (on net area of masonry)	225	1500
Concrete[a]		
(1) Bearing on full area of support		
$f'_c = 2000$ psi	500	3500
$f'_c = 3000$ psi	750	5000
(2) Bearing on 1/3 or less of support area		
$f'_c = 2000$ psi	750	5000
$f'_c = 3000$ psi	1125	7500

[a] Stresses for areas between these limits may be determined by direct proportion.

usually in even inches, and a great variety of thicknesses is available.

The thickness of the plate is determined by considering the projection n (Fig. 22.8b) as an inverted cantilever; the uniform bearing pressure on the bottom of the plate tends to curl it upward about the beam flange. The required thickness may be computed readily by the following formula, which does not involve direct computation of bending moment and section modulus:

$$t = \sqrt{\frac{3f_p n^2}{F_b}}$$

where

t = thickness of the plate, inches

F_p = *actual* bearing pressure of the plate on the masonry, psi or ksi

F_b = allowable bending stress in the plate (the ASIC Specification gives the value of F_b as $0.75F_y$; for A36 steel F_y = 36 ksi; therefore, F_b = 0.75(36) = 27 ksi)

n = $(B/2) - k_1$, inches (see Fig. 22.8b)

k_1 = distance from the center of the web to the toe of the fillet; values of k_1 for various beam sizes may be found in the tables in the *AISC Manual*

When the dimensions of the bearing plate are determined, the beam should be investigated for web crippling on the length $(N + k)$ shown in Fig. 22.8c.

Example 3. A W 21 × 57 of A36 steel transfers an end reaction of 44 kips [196 kN] to a wall built of solid brick by means of a bearing plate of A36 steel. Assume type S mortar and a brick with f'_m = 1500 psi. The N dimension of the plate (see Fig. 22.8) is 10 in. [254 mm]. Design the bearing plate.

Solution. In the *AISC Manual* we find that k_1 for the beam is 0.875 in. [22 mm]. From Table 22.3 the allowable bearing pressure F_p for this wall is 170 psi [1200 kPa]. The required area of the plate is then

$$A = \frac{R}{F_p} = \frac{44,000}{170} = 259 \text{ in.}^2 \text{ [163,333 mm}^2\text{]}$$

Then, because N = 10 in. [254 mm],

$$B = \frac{259}{10} = 25.9 \text{ in. [643 mm]}$$

which is rounded off to 26 in. [650 mm].

With the true dimensions of the plate, we now compute the true bearing pressure.

$$f_p = \frac{R}{A} = \frac{44,000}{10(26)} = 169 \text{ psi [1187 kPa]}$$

To find the thickness, we first determine the value of n.

$$n = \frac{B}{2} - k_1 = \frac{26}{2} - 0.875 = 12.125 \text{ in. [303 mm]}$$

Then

$$t = \sqrt{\frac{3f_p n^2}{F_p}} = \sqrt{\frac{3(169)(12.125)^2}{27,000}}$$

$$= \sqrt{2.760} = 1.66 \text{ in. [42 mm]}$$

The complete design for this problem would include a check of the web crippling in the beam. This has been done previously as Example 1.

22.8. ROOF AND FLOOR SYSTEMS

Steel elements may be used to produce a variety of horizontal-spanning floor structures and horizontal or sloping roof structures. Choice of steel may be related to a need for exceptional spans, but is most often due to a requirement for noncombustible construction in order to classify the structure for a higher fire resistance rating. Some of the issues involved in the general development of steel roof and floor systems are discussed in this section. Design examples are generally presented in the building design studies in Part 9.

Decks

Steel decks consist of formed sheet steel, produced in a variety of configurations, as shown in Fig. 22.9. The simplest is the corrugated sheet, shown in Fig. 22.9a. This may be used as the single, total surface for walls or roofs of utilitarian buildings (tin shacks?). For more serious buildings it is mostly used only in pairing with a light steel joist system where the joist spacing is quite close and a poured concrete fill is used.

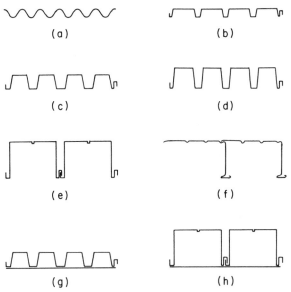

(a) (b)

(c) (d)

(e) (f)

(g) (h)

FIGURE 22.9. Formed sheet steel deck units.

A widely used product is that shown in three variations in Fig. 22.9b–d. When used for roof deck, where loads are light and a need exists for a flat top surface to support insulation and roofing, the deck unit shown in Fig. 22.9b is popular. For heavier floor loads and use with a structural-grade concrete fill, the units shown in Fig. 22.9c and d are used. These come in different sizes as measured by the overall height of the deck pleats; common sizes being 1.5, 3, and 4.5 in. The units shown in Fig. 22.9b–d are produced by many different manufacturers.

There are also steel deck units produced as proprietary items by individual manufacturers. Figure 22.9e and f show two units of considerable depth, capable of achieving quite long spans. Use of these special units is somewhat a matter of regional marketing.

Although less used now, with the advent of other wiring products and techniques, a possible use for the steel deck is as a conduit for power or communication wiring. This is accomplished by closing the deck cells with a flat sheet of steel, as shown in Fig. 22.9g and h. This provides for wiring in one direction in a wiring grid; the perpendicular wiring being achieved in conduits buried in the concrete fill on top of the deck.

Decks vary in form, including ones not shown in Fig. 22.9, and in the thickness (gage) of the sheet steel. Choice relates to form and to the load and span conditions for the deck. Units are typically available in single pieces of 30 or more ft length. For the shallower decks, this usually assures a multiple-span support condition. Required fire ratings for the construction and use of the deck for diaphragm action in resisting lateral forces must also be considered.

Roof decks are most often used with either a very low density poured concrete fill or preformed rigid foam plastic insulation units on top of the deck. The deck itself is the functioning structure in these cases. For floor decks, however, the concrete fill is usually of sufficient strength to perform structural tasks, so the steel unit + concrete fill combination is often designed as a composite structural element.

Table 22.4 presents data relating to the use of the type of deck unit shown in Fig. 22.9b for roof structures. These data are adapted from a publication distributed by an industry-wide organization referred to in the table notes. In general, data for these manufactured products is best obtained from the manufacturers of the products, especially for the design of the composite floor units.

In general, economy is obtained with the use of the thinnest sheet steel (highest gauge number) that can be utilized for a particular application. Where lateral forces are critical, this choice may be determined on the basis of consideration of diaphragm actions.

Rusting is a critical problem with the very thin deck elements, so that some rust-inhibiting finish is commonly used, except for surfaces that will be covered with a concrete fill. The deck weights given in Table 22.4 are based on painted surfaces, which is the least expensive and most common finish.

Figure 22.10 shows four possibilities for a floor deck used in conjunction with a framing system of rolled steel sections. When a wood deck is used, it is usually nailed to a series of wood joists or trusses, which are supported by the steel beams. However, in some cases the deck may be nailed to wood strips bolted to the top of the steel framing, as shown in the illustration.

A site-cast concrete slab may be used, with the slab forms placed on the underside of the top flanges of the steel beams, producing the detail shown in Fig. 22.10b. It is common in this case to use steel devices welded to the top of the beams to develop composite action of the beams and the slab.

Concrete may also be used in the form of precast units that are welded to the beams using steel devices cast into the units. A concrete fill is ordinarily used with such units in order to develop a smooth surface for flooring (see Fig. 22.10c).

As described in Sec. 22.9, formed sheet steel units may be used with a concrete fill, the steel units being welded to the tops of the beams. It is also now common to use a concrete fill on top of a plywood deck. For both cases, the fill adds stiffness, mass to resist bouncing, enhanced acoustic properties, and some possibility for buried wiring.

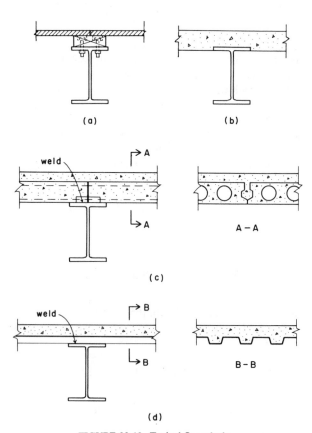

FIGURE 22.10. Typical floor decks.

TABLE 22.4. Load Capacity of Steel Roof Deck

Deck[a] Type	Span Condition	Weight[b] (psf)	4-0	4-6	5-0	5-6	6-0	6-6	7-0	7-6	8-0	8-6	9-0	9-6	10-0
			Total (Dead & Live) Safe Load[c] for Spans Indicated in ft-in.												
NR22	Simple	1.6	73	58	47										
NR20		2.0	91	72	58	48	40								
NR18		2.7	121	95	77	64	54	46							
NR22	Two	1.6	80	63	51	42									
NR20		2.0	96	76	61	51	43								
NR18		2.7	124	98	79	66	55	47	41						
NR22	Three or More	1.6	100	79	64	53	44								
NR20		2.0	120	95	77	63	53	45							
NR18		2.7	155	123	99	82	69	59	51	44					
IR22	Simple	1.6	86	68	55	45									
IR20		2.0	106	84	68	56	47	40							
IR18		2.7	142	112	91	75	63	54	46	40					
IR22	Two	1.6	93	74	60	49	41								
IR20		2.0	112	88	71	59	50	42							
IR18		2.7	145	115	93	77	64	55	47	41					
IR22	Three or More	1.6	117	92	75	62	52	44							
IR20		2.0	140	110	89	74	62	53	46	40					
IR18		2.7	181	143	116	96	81	69	59	52	45	40			
WR22	Simple	1.6			(89)	(70)	(56)	(46)							
WR20		2.0			(112)	(87)	(69)	(57)	(47)	(40)					
WR18		2.7			(154)	(119)	(94)	(76)	(63)	(53)	(45)				
WR22	Two	1.6				98	81	68	58	50	43				
WR20		2.0			125	103	87	74	64	55	49	43			
WR18		2.7			165	137	115	98	84	73	65	57	51	46	41
WR22	Three or More	1.6			122	101	85	72	62	54	(46)	(40)			
WR20		2.0			156	129	108	92	80	(67)	(57)	(49)	(43)		
WR18		2.7			207	171	144	122	105	(91)	(76)	(65)	(57)	(50)	(44)

approx. 6"
1" max. 1.5"
0.375" min.
Narrow Rib Deck – NR

1.75" max.
0.5" min.
Intermediate Rib Deck – IR

2.5" max.
1.75" min.
Wide Rib Deck – WR

Source: Adapted from the *Steel Deck Institute Design Manual for Composite Decks, Form Decks, and Roof Decks,* 1981–82 issue, with permission of the publishers, the Steel Deck Institute. May not be reproduced without express permission of the publishers.

[a] Letters refer to rib type (see the key). Numbers indicate gage (thickness) of steel.

[b] Approximate weight with paint finish; also available.

[c] Total safe allowable load in lb/ft². Loads in parentheses are governed by live-load deflection not in excess of 1/240 of the span, assuming a dead load of 10 psf.

Figure 22.11 shows three possibilities for a roof deck used in conjunction with a framing system of steel. A fourth possibility is that of the plywood deck shown in Fig. 22.10a. Many of the issues discussed for the floor decks also apply here. However, roof loads are usually lighter, and there is less concern for bounciness due to people walking; thus some lighter systems are feasible here and not for floors.

Formed steel units, such as those presented in Table 22.4, are normally used with a rigid insulation (as shown in Fig. 22.11a) or a poured concrete fill. If the concrete fill is of the very low density type, it does not have significant strength and the deck is the functioning spanning structure. However, a slightly stronger fill may be used in some cases with a light steel deck, the steel units functioning primarily only to form the concrete deck.

To facilitate roof drainage, concrete fill may be varied in thickness, allowing the deck units to remain flat. Rigid insulation units can also be obtained in modular packages with tapered pieces. This is only possible for a few inches of elevation change; the structure must be tilted for substantial slopes.

A special deck is that shown in Fig. 22.11c, consisting of a low-density concrete fill poured on top of a forming system of inverted steel tees and lay-in rigid panel units. The panel units can be used to form the finished ceiling surface where it is possible to leave the steel framing elements exposed.

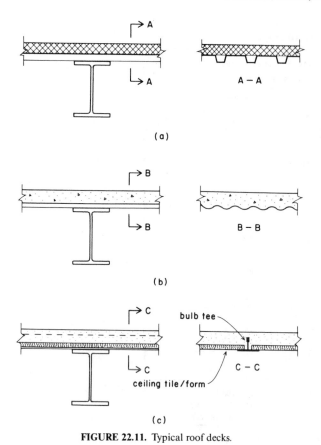

FIGURE 22.11. Typical roof decks.

For a system such as that shown in Fig. 22.12a, the basic planning begins with the location of the system supports, usually columns or bearing walls. The character of the spanning system is closely related to the magnitude of the spans it must achieve. Decks are mostly quite short in span, requiring relatively close spacing of the elements that provide their direct support. Joists and beams may be small or large, depending mostly on their spans. The larger they are, the less likely they will be very closely spaced. Thus very long-span systems may have several levels of components before ending with the elements that directly support the deck.

Figure 22.12b shows a plan and elevation of a system that uses trusses for the major span. If the trusses are very large and the purlin spans quite long, the purlins may have to be quite widely spaced. A constraint on the purlin locations is usually that they coincide with the joints in the top of the truss, so as to avoid high shear and bending in the truss top chord. In the latter case, it may be advisable to use joists between the purlins to provide support for the deck. On the other hand, if the truss spacing is a modest distance, it may be possible to use a long-span deck with no purlins. The basic nature of the system can thus be seen to change

System Planning

The planning of the layout of the structural framing system for a roof or floor must respond to many concerns. In this section we discuss some typical problems that relate to the individual functioning of the system components and to their interrelationships with other components in the system.

Figure 22.12a shows a framing plan for a system that has four basic components. In developing the system layout and choosing the components, considerations such as the following must be made.

1. *Deck Span.* The type of deck as well as its specific variation (thickness of plywood, gage of steel sheet, etc.) will relate to the deck span.
2. *Joist Spacing.* This determines the deck span and the magnitude of load on the joist. The type of joist selected may limit the spacing, based on the joist capacity. The type and spacing of joists must be coordinated with the selection of the deck.
3. *Beam Span.* For systems with some plan regularity, the joist spacing should be some full number division of the beam span.
4. *Column Spacing.* The spacing of the columns determines the spans for the beams and joists, and is thus related to the planning modules for all the other components.

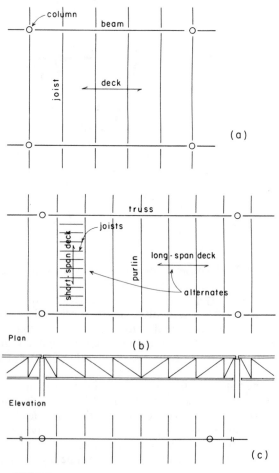

FIGURE 22.12. Beam framing—planning considerations.

with different positioning of the system supports. In any event, the truss span and panel module, the column spacing, the purlin span and spacing, the joist span and spacing, and the deck span are interrelated and the selection of the components is a highly interactive exercise.

For systems with multiple elements, some consideration must be given to the various intersections and connections of the components. For the framing plan shown in Fig. 22.12a there is a five-member intersection at the column, involving the column, the two beams, and the two joists (plus an upper column, if the building is multistory). Depending on the materials and forms of the members, the forms of connections, and the types of force transfer at the joint, this may be a routine matter of construction or a real mess. Some relief of the traffic congestion may be achieved by the plan layout shown in Fig. 22.12b, in which the module of the joist spacing is off-set at the columns, leaving only the column and beam connections. A further reduction possible is that shown in Fig. 22.12c, where the beam is made continuous through the column, with the beam splice occurring off the column. In the plan in Fig. 22.12c the connections are all only two-member relationships: column to beam, beam to beam, and beam to joist.

Bridging, blocking, and cross-bracing for trusses must also be planned with care. These members may interfere with ducting or other building elements as well as create connection problems similar to those just discussed. Use of these extra required elements for multiple purposes should be considered. Blocking required for plywood nailing may also function as edge nailing for ceiling panels and lateral bracing for slender joists or rafters. The cross-bracing required to brace tall trusses may be used to support ceilings, ducts, building equipment, catwalks, and so on.

In the end, structural planning must be carefully coordinated with the general planning of the building and its various subsystems. Real optimization of the structure may need to yield to other, more pragmatic concerns.

Cantilevered Edges

A problem that occurs frequently is that of providing for the extension of the horizontal structure beyond the plane of the building's exterior walls. This most often occurs with an overhanging roof, but can also be required to provide for balconies or exterior walkways for a floor. Figure 22.13a shows one possibility for achieving this by simply extending the ends of joists or rafters that are perpendicular to the wall. With steel framing, this type of cantilever is most easily achieved if the extended members simply rest on top of the supporting beam or bearing wall at the wall plane. With an exterior column system, an alternate is shown in Fig. 22.13b, in which the column line members are extended to support a member at the cantilevered edge, which in turn supports simple span members between the columns. Loading, member size and type, and the magnitude of the cantilever would all affect the choice of one of these schemes over the other.

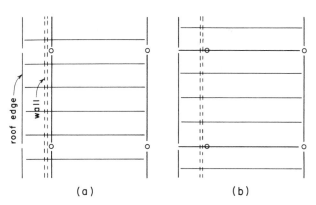

FIGURE 22.13. Framing of cantilevered edges.

A special problem with the cantilevered edge is that occurring at a building corner, when both sides of the building have the cantilever condition, as shown in Fig. 22.14a. With the framing system shown in Fig. 22.13a, a possibility for the corner is that shown in Fig. 22.14b, where the supporting beam is cantilevered to support edge member 1, and the joists are cantilevered to support edge member 2.

For the system shown in Fig. 22.13b, a way of achieving the corner is that shown in Fig. 22.14c. In this case the column-line member is cantilevered as usual to support edge member 1, which in turn cantilevers to the corner to support edge member 2.

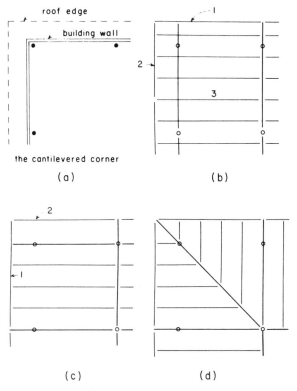

FIGURE 22.14. Framing of cantilevered corners.

A third possibility for the corner is the use of a diagonal member, as shown in Fig. 22.14*d*. A feature of this solution is the achieving of a reorientation of the framing system as the corner is turned. This layout is more often utilized in wood than in steel, and is a common one used for sloping roofs when the diagonal member defines a ridge as the roof slopes to both edges. It may be noted that there is a rather busy intersection at the interior column in the plan in Fig. 22.14*d*.

Nonstructural Concerns

General planning concerns for structures are discussed in Part 9. The following are some particular issues that relate to design of steel framing systems.

Ceilings

Where ceilings exist, they are generally provided for in one of three ways: by direct attachment to the overhead structure (underside of the roof or floor above), by some independent structure, or by suspension from the overhead structure. Suspended ceilings are quite common, as the space created between the ceiling and the structure is used for ducting and registers of the HVAC system, wiring and recessed fixtures of the lighting system, and the installation of some items of building equipment. If a joist or rafter system is used with closely spaced members (4 ft center-to-center or less), the structure for the ceiling is usually suspended from these members. The other means for suspension is to use hangers attached to the deck, an advantage being the freeing of the modules of the structure and the ceiling framing from each other. The suspended ceiling is also used when the form of the ceiling does not correspond to that of the overhead structure.

Roof Drainage

Providing for the minimum slopes required for drainage of flat roofs is always a problem for a roof framing system. The most direct means is simply to tilt the framing to provide the slope patterns required. For a complex roof this gets to be quite complicated with regard to the specification of the levels of the various framing members. The desired patterns of slopes and the locations of drains may not relate well to the layout of the roof framing members. Another possibility is to keep the framing flat but vary the thickness of the deck (applicable only to poured concrete decks) or the insulation fill on top of the deck. The latter technique simplifies the framing details, but is usually capable of developing only a few inches of slope differential. If a flat ceiling is required and is to be attached directly to the roof structure, this must be considered in facilitating the drainage.

For some types of structural members—most notably the manufactured trusses—it is possible to slope the top of the member while keeping the bottom flat. This makes it possible to have a sloping roof surface and a flat ceiling, with the ceiling directly attached.

Dynamic Behavior

Roof structures may usually be optimized for light weight without major restriction, the weight reduction providing a benefit of load reduction for both the spanning structure and its supports. Lightweight floor structures, on the other hand, tend to be bouncy, which is generally not a desirable characteristic. Bounciness can also be a result of an excessive span-to-depth ratio for the spanning elements. Experience is the primary guide in this matter, but some general rules are the following:

1. Restrict live-load deflections to a conservative ratio of the span (usually not greater than 1/360 of the span for any floor).
2. Limit span-to-depth ratios well below those of the maximum permitted. Suggestions: maximum of 20 for solid members, 15 for trussed joists.
3. Use a very stiff deck for its load-distributing function (the repetitive member effect, as described for wood joists).
4. Even if load distributing is not significant, do not use decks for the longest spans listed in design data references.

A major factor in reducing bounciness is the presence of the concrete fill on top of steel decks. This fill is now commonly also used on top of wood decks.

Holes

Both floor and roof surfaces are commonly pierced by a number of passages for various items. Large openings are required for stairs and elevators, medium-sized ones for ducts and chimneys, and small ones for piping and wiring. The structure must be planned and detailed to accommodate these openings, which entails some of the following considerations.

1. *Location of Openings.* Openings may often occur at locations not convenient for the framing. This may indicate some poor planning of the framing, or may be essentially unavoidable. For structures that utilize column line rigid frame bents for lateral bracing, the integrity of the bents generally requires that openings be kept off of the column lines. For regularly spaced systems in general, the layout of the framing and locations of required openings should be coordinated to maintain a maximum regularity of the system. Openings should not interrupt the major elements of the system (large trusses or girders).
2. *Size of Openings.* Large openings must have some framing around their perimeters, which are also likely to be locations of heavy wall construction. Small openings (for single pipes, for example) may simply pierce the deck with no special provision. For sizes of openings between these extremes, the accommodation requirements depend on the form and size of the elements of the structure. For closely

spaced joists, provision for openings of a size that fits between the joists is usually quite simple; when the size requires the interruption of one or more joists, it entails some more difficult measures, such as doubling the joists on each side of the opening.

3. *Openings Near Columns.* For efficiency in architectural planning, it is sometimes convenient to locate duct shafts or chases for piping or wiring next to a structural column. If this can be done without interrupting a major spanning member, it may not present a problem. If the opening must be on the column line, it may require straddling of the opening with a double framing member of two spaced elements.

4. *Loss of Diaphragm Effectiveness.* Presence of large openings must be considered with regard to effects on the effectiveness of the floor or roof system as a horizontal diaphragm for lateral bracing. It may be necessary to provide special framing or connections to develop collector functions, drag struts, or the subdivision of the diaphragm, as described in Part 8.

CHAPTER TWENTY-THREE

Steel Columns

Steel columns in buildings vary from small, single-piece members (single pipe, tube, wide-flange shape, etc) to gigantic assemblages of many elements for large towers. The basic column function is the development of axial compression force, but many columns must also sustain some amount of bending, shear, or torsion. This chapter presents a general discussion of column design issues, with an emphasis on the simple, single-piece column.

23.1. COLUMN SHAPES AND USAGE

The most common building columns are the round, cylindrical pipe, the rectangular tubular element, and the wide-flange shape. Pipes and tubes are useful for one-story buildings in which attachment to supported framing is achieved by setting the spanning members on top of the column. Framing that requires spanning members to be attached to the side of a column is more easily achieved with wide-flange members. The most commonly used columns for multistory buildings are the approximately square wide-flange shapes of nominal 10-, 12-, and 14-in. size. These are available in a wide range of flange and web thicknesses, up to the heaviest rolled column section: the W 14 × 730.

Pipes and square tubes are ideally suited by the geometry of their cross sections for resistance of axial compression. The wide flange section has a strong axis (the X–X axis) and a weak axis (the Y–Y axis), although the close-to-square column shapes have the least difference in stiffness and radius of gyration on the two principal axes. The wide-flange section is well-suited to the development of bending in combination with the axial force, if it is bent about its strong axis.

For various reasons, it is sometimes necessary to make a column section by assembling two or more individual steel elements. Figure 23.1 shows some commonly used assemblages, which are used for special purposes. These are often required where a particular size or shape is not available from the inventory of stock rolled sections, where a particular structural task is required, or where details of the framing require a particular form. These are somewhat less used now, as the range of size and shape of stock sections has steadily increased. They are now mostly used for exceptionally large columns or for special framing problems. The customized fabrication of built-up sections is

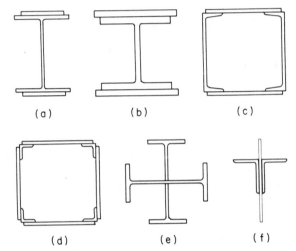

FIGURE 23.1. Built-up column sections.

usually costly, so a single piece is typically favored if one is available.

One widely used built-up section is that of the double angle, shown in Fig. 23.1*f*. This occurs most often as a member of a truss or as a bracing member in a frame, the general stability of the paired members being much better than that of the single angle. This section is not much used as a building column.

23.2. STABILITY CONCERNS

Axially loaded columns are designed for a uniformly distributed compressive stress. The value of the stress is determined from formulas in the AISC Specification that include variables of the yield strength, modulus of elasticity, relative stiffness, and special conditions of restraint of the column. As a measure of resistance to buckling, the basic property of the column is its slenderness, computed as L/r, in which L is the unbraced length and r the radius of gyration of the column cross section. Effect of various end conditions, as shown in Fig. 23.2, is considered by use of a modifying factor (K) resulting in some reduced or magnified value for L.

Figure 23.3 is a graph of the allowable axial compressive stress for a column of A36 steel (F_y = 36 ksi [165 MPa]) as determined from the AISC formulas. Values for full number increments of KL/r are also given in Table 23.1. Note that the table has two parts for values of KL/r over 120. Building columns would be considered as "main members" in this classification.

It is generally recommended that building columns not have an L/r in excess of 120. This is a matter of some concern for efficient use of the material, but is also a desire not to have excessively skinny columns.

23.3. DESIGN FOR AXIAL LOAD

The allowable axial compression load for a column is computed by multiplying the allowable stress (F_a) by the cross-sectional area of the column. The following examples demonstrate the process. For single-piece columns, the most direct method is through the use of column load tables, as discussed in the next section. For built-up sec-

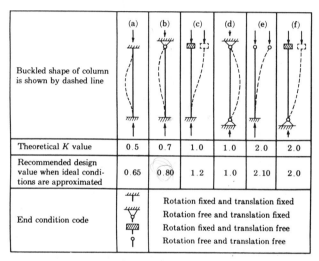

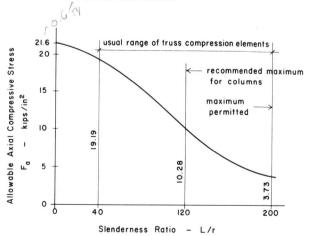

FIGURE 23.2. Determination of effective column length. Reproduced from the *AISC Manual*, 8th ed. (Ref. 5), with permission of the publishers, American Institute of Steel Construction.

FIGURE 23.3. Allowable compression stress as a function of slenderness. AISC requirements for A36 steel.

tions, it is necessary to compute the properties of the section and to use the process demonstrated in Example 3.

Example 1. A W 12×58 is used as a column with an unbraced length of 16 ft [4.99 m]. Compute the allowable load.

Solution. Referring to Table A.3, we find that $A = 17.0$ in.^{2} [10,968 mm^2], $r_x = 5.28$ in. [134 mm], and $r_y = 2.51$ in. [63.8 mm]. If the column is unbraced on both axes, it is limited by the lower r value for the weak axis. With no qualifying end conditions, we assume an end condition, as shown in Fig. 23.2d, for which $K = 1$. Thus the relative slenderness is computed as

$$\frac{Kl}{r} = \frac{1(16)(12)}{2.51} = 76.5$$

In design work it is usually considered acceptable to round the value for the slenderness ratio off to the next highest figure in front of the decimal point. Therefore we consider the Kl/r ratio to be 77 and find from Table 23.1 that the allowable stress (F_a) is 15.69 ksi [108.2 MPa]. The allowable load for the column is thus

$$P = AF_a = 17.0(15.69) = 266.7 \text{ kips [1187 kN]}$$

Example 2. Compute the allowable load for the column in Example 1 if the top is pinned and the bottom is fixed.
Solution. Referring to Fig. 23.2, the recommended K factor for this condition is 0.8. The modified stiffness is thus

$$\frac{KL}{r} = \frac{0.8[16(12)]}{2.51} = 61.2 \text{ say } 61$$

and the allowable stress from Table 23.1 is 17.33 ksi [119.5 MPa]. The allowable load is thus

$$P = AF_a = 17.0(17.33) = 294.6 \text{ kips [1311 kN]}$$

Conditions frequently exist to cause different modification of the basic column action on the two axes of a wide-flange column. In this event it may be necessary to investigate the conditions relating to the separate axes in order to determine the limiting condition. Example 3 illustrates such a case.

Example 3. Figure 23.4a shows an elevation of the steel framing at the location of an exterior wall. The column is laterally restrained but rotation free at both its top and bottom. With respect to the X-axis of the section, the column is laterally unbraced for its full height. However, horizontal framing in the wall plane provides lateral bracing at a point between the top and bottom with respect to the Y-axis of the section. If the column is a W 12×58 of A36 steel, what is the allowable compression load? $L_1 = 30$ ft and $L_2 = 18$ ft.
Solution. With respect to the X-axis, the column functions as a pin-ended member for its full height (Fig. 23.2d). However, with respect to the Y-axis, the form of buckling is as shown in Fig. 23.4b, and the laterally unsupported height is 18 ft [5.49 m]. For both conditions the K factor is 1, and the investigation is as follows (see Example 1 for data for the section):

TABLE 23.1. Allowable Unit Stresses for Columns of A36 Steel (in ksi)

Main and Secondary Members Kl/r not over 120						Main Members Kl/r 121 to 200				Secondary Members* l/r 121 to 200			
$\dfrac{Kl}{r}$	F_a (ksi)	$\dfrac{Kl}{r}$	F_a (ksi)	$\dfrac{Kl}{r}$	F_a (ksi)	$\dfrac{Kl}{r}$	F_a (ksi)	$\dfrac{Kl}{r}$	F_a (ksi)	$\dfrac{l}{r}$	F_{as} (ksi)	$\dfrac{l}{r}$	F_{as} (ksi)
1	21.56	41	19.11	81	15.24	121	10.14	161	5.76	121	10.19	161	7.25
2	21.52	42	19.03	82	15.13	122	9.99	162	5.69	122	10.09	162	7.20
3	21.48	43	18.95	83	15.02	123	9.85	163	5.62	123	10.00	163	7.16
4	21.44	44	18.86	84	14.90	124	9.70	164	5.55	124	9.90	164	7.12
5	21.39	45	18.78	85	14.79	125	9.55	165	5.49	125	9.80	165	7.08
6	21.35	46	18.70	86	14.67	126	9.41	166	5.42	126	9.70	166	7.04
7	21.30	47	18.61	87	14.56	127	9.26	167	5.35	127	9.59	167	7.00
8	21.25	48	18.53	88	14.44	128	9.11	168	5.29	128	9.49	168	6.96
9	21.21	49	18.44	89	14.32	129	8.97	169	5.23	129	9.40	169	6.93
10	21.16	50	18.35	90	14.20	130	8.84	170	5.17	130	9.30	170	6.89
11	21.10	51	18.26	91	14.09	131	8.70	171	5.11	131	9.21	171	6.85
12	21.05	52	18.17	92	13.97	132	8.57	172	5.05	132	9.12	172	6.82
13	21.00	53	18.08	93	13.84	133	8.44	173	4.99	133	9.03	173	6.79
14	20.95	54	17.99	94	13.72	134	8.32	174	4.93	134	8.94	174	6.76
15	20.89	55	17.90	95	13.60	135	8.19	175	4.88	135	8.86	175	6.73
16	20.83	56	17.81	96	13.48	136	8.07	176	4.82	136	8.78	176	6.70
17	20.78	57	17.71	97	13.35	137	7.96	177	4.77	137	8.70	177	6.67
18	20.72	58	17.62	98	13.23	138	7.84	178	4.71	138	8.62	178	6.64
19	20.66	59	17.53	99	13.10	139	7.73	179	4.66	139	8.54	179	6.61
20	20.60	60	17.43	100	12.98	140	7.62	180	4.61	140	8.47	180	6.58
21	20.54	61	17.33	101	12.85	141	7.51	181	4.56	141	8.39	181	6.56
22	20.48	62	17.24	102	12.72	142	7.41	182	4.51	142	8.32	182	6.53
23	20.41	63	17.14	103	12.59	143	7.30	183	4.46	143	8.25	183	6.51
24	20.35	64	17.04	104	12.47	144	7.20	184	4.41	144	8.18	184	6.49
25	20.28	65	16.94	105	12.33	145	7.10	185	4.36	145	8.12	185	6.46
26	20.22	66	16.84	106	12.20	146	7.01	186	4.32	146	8.05	186	6.44
27	20.15	67	16.74	107	12.07	147	6.91	187	4.27	147	7.99	187	6.42
28	20.08	68	16.64	108	11.94	148	6.82	188	4.23	148	7.93	188	6.40
29	20.01	69	16.53	109	11.81	149	6.73	189	4.18	149	7.87	189	6.38
30	19.94	70	16.43	110	11.67	150	6.64	190	4.14	150	7.81	190	6.36
31	19.87	71	16.33	111	11.54	151	6.55	191	4.09	151	7.75	191	6.35
32	19.80	72	16.22	112	11.40	152	6.46	192	4.05	152	7.69	192	6.33
33	19.73	73	16.12	113	11.26	153	6.38	193	4.01	153	7.64	193	6.31
34	19.65	74	16.01	114	11.13	154	6.30	194	3.97	154	7.59	194	6.30
35	19.58	75	15.90	115	10.99	155	6.22	195	3.93	155	7.53	195	6.28
36	19.50	76	15.79	116	10.85	156	6.14	196	3.89	156	7.48	196	6.27
37	19.42	77	15.69	117	10.71	157	6.06	197	3.85	157	7.43	197	6.26
38	19.35	78	15.58	118	10.57	158	5.98	198	3.81	158	7.39	198	6.24
39	19.27	79	15.47	119	10.43	159	5.91	199	3.77	159	7.34	199	6.23
40	19.19	80	15.36	120	10.28	160	5.83	200	3.73	160	7.29	200	6.22

Source: Reprinted from *Manual of Steel Construction* (Ref. 5), with permission of the publishers, American Institute of Steel Construction.

* K is taken as 1.0 for secondary members.

X-axis: $\dfrac{KL}{r} = \dfrac{30(12)}{5.28} = 68.2.$

Y-axis: $\dfrac{KL}{r} = \dfrac{18(12)}{2.51} = 86.1,$ say 86.

Despite the bracing, the column is still critical on its Y-axis. From Table 23.1 we obtain $F_a = 14.67$ ksi [101 MPa], and the allowable load is thus

$$P = AF_a = 17.0(14.67) = 249.4 \text{ kips [1108 kN]}$$

Column Load Tabulations

As demonstrated in the examples in Sec. 23.3 the determination of the allowable compression load for a column is quite simple once the necessary data are known. When designing columns, however, the process is complicated by the fact that the allowable stress is unknown if the section is undefined. This results in a pick-and-try method of design, unless some aids are used. Since there are a fixed number of stock sections for ordinary use as single-piece columns, it is possible to determine their allowable loads

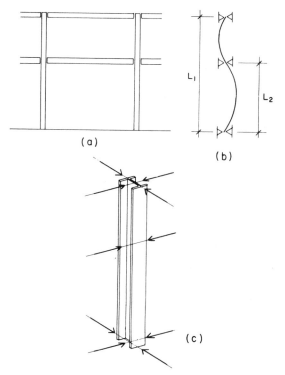

FIGURE 23.4. Biaxial bracing of columns.

for various unbraced lengths and record the data in tabular form. Extensive tables exist in the *AISC Manual* (Ref. 5) for single-piece columns as well as for commonly used combinations of double angles. Selected data from the AISC tables is displayed in Tables 23.2 through 23.5. The following examples illustrate the use of tabular data for the design of simple, axially loaded columns.

Example 4. A column is pinned at top and bottom and has a laterally unsupported height of 20 ft [6.10 m]. Select a wide-flange shape for an axial load of 200 kips [890 kN].
Solution. Scanning down the column for 20 ft in Table 23.2, we find the following:

Shape	Table Load (kips)	Actual L/r_y
W 8 × 67	221	113
W 10 × 54	217	94
W 12 × 53	209	97
W 14 × 61	237	98

Although all of these shapes are adequate for the load, other considerations in the development of the structure and building construction would probably point to the preference for one size over others. The L/r values are not taken from the table, but were determined from the r_y properties of the shapes. The table cuts off at the maximum value of $L/r = 200$, but most designers would prefer to stay below a value of 120 for major building columns.

Example 5. Same data as for Example 4; select a pipe of A36 steel.

Solution. From Table 23.3 the minimum size is the 10-in.-nominal-diameter pipe. It should be noted that Table A.9 gives the actual outside diameter as 10.75 in. Table A.9 also gives properties for the two heavier classes of pipe, although Table 23.3 gives loads only for standard pipe. When size is critical, it is possible to use the heavier pipe, for which loads are given in the AISC Manual.

Example 6. Same data as for Example 4; select a square tubular section.
Solution. In this case the table (Table 23.4) yields a single answer: the 8-in. square with wall thickness of $\frac{3}{8}$ in. Again, the tables presented here are not as complete as those in the reference, and other sizes may also be used. As with the wide-flange column, there will usually be other considerations for the selection of the column dimensions. It should also be noted that the tubular sections are available in oblong cross sections, for which properties and column load values are given in the reference.

Further examples of uses of these tables are given in other sections in this part and in the design examples in Part 9.

Double-Angle Members

Pairs of angles, separated by the thickness of some connecting element (usually a plate), are commonly used for truss members or for bracing elements in truss-braced frames. The *AISC Manual* contains tables for several different combinations of angles. Tables of properties may be used for design of both tension and compression members. Column load tables, of which Table 23.5 is a sampling, provide data for design of compression members.

When the angles used do not have legs of equal length, there are two possibilities for matched sets: those with the long legs back to back and those with the short legs back to back. In general, the set with long legs back to back will have a better balance of properties with respect to its two axes (notably r_x and r_y). However, in some cases there may be special circumstances that indicate the use of the other combination. The sets listed in Table 23.5 all consist of unequal leg angles with long legs back to back, a combination used quite frequently.

Because various situations can occur with the use of these members, it is useful to have the loads tabulated for column action with respect to both axes, as is done in Table 23.5. If fully unbraced on both axes, the least value obtained must be used. If different unbraced lengths occur with respect to the two axes, or the member is fully braced on one axis, the table values may be used appropriately.

Example 7 illustrates the use of the data from Table 23.5 for a simple design selection.

Example 7. A double-angle member is to be used for a 12-ft-long compression member carrying 60 kips. The member is unbraced and has a K factor of 1.0 on both axes. Select an appropriate angle size for the member.
Solution. In this case it is necessary to read the table in both portions at the same time. We thus obtain the following possible choices from those listed.

TABLE 23.2. Allowable Column Loads for Selected W and M Shapes (kips)[a]

[handwritten marginal note: KL/r ≥ 200]

(Left panel)

Shape	Effective length (KL) in feet										Bending factor	
	8	9	10	11	12	14	16	18	20	22	B_x	B_y
M 4 × 13	48	42	35	29	24	18					0.727	2.228
W 4 × 13	52	46	39	33	28	20	16				0.701	2.016
W 5 × 16	74	69	64	58	52	40	31	24	20		0.550	1.560
M 5 × 18.9	85	78	71	64	56	42	32	25			0.576	1.768
W 5 × 19	88	82	76	70	63	48	37	29	24		0.543	1.526
W 6 × 9	33	28	23	19	16	12					0.482	2.414
W 6 × 12	44	38	31	26	22	16					0.486	2.367
W 6 × 16	62	54	46	38	32	23	18				0.465	2.155
W 6 × 15	75	71	67	62	58	48	38	30	24	20	0.456	1.424
M 6 × 20	98	92	87	81	74	61	47	37	30	25	0.453	1.510
W 6 × 20	100	95	90	85	79	67	54	42	34	28	0.438	1.331
W 6 × 25	126	120	114	107	100	85	69	54	44	36	0.440	1.308
W 8 × 24	124	118	113	107	101	88	74	59	48	39	0.339	1.258
W 8 × 28	144	138	132	125	118	103	87	69	56	46	0.340	1.244
W 8 × 31	170	165	160	154	149	137	124	110	95	80	0.332	0.985
W 8 × 35	191	186	180	174	168	155	141	125	109	91	0.330	0.972
W 8 × 40	218	212	205	199	192	168	160	143	124	104	0.330	0.959
W 8 × 48	263	256	249	241	233	215	196	176	154	131	0.326	0.940
W 8 × 58	320	312	303	293	283	263	240	216	190	162	0.329	0.934
W 8 × 67	370	360	350	339	328	304	279	251	221	190	0.326	0.921
W 10 × 33	179	173	167	161	155	142	127	112	95	78	0.277	1.055
W 10 × 39	213	206	200	193	186	170	154	136	116	97	0.273	1.018
W 10 × 45	247	240	232	224	216	199	180	160	138	115	0.271	1.000
W 10 × 49	279	273	268	262	256	242	228	213	197	180	0.264	0.770
W 10 × 54	306	300	294	288	281	267	251	235	217	199	0.263	0.767
W 10 × 60	341	335	328	321	313	297	280	262	243	222	0.264	0.765
W 10 × 68	388	381	373	365	357	339	320	299	278	255	0.264	0.758
W 10 × 77	439	431	422	413	404	384	362	339	315	289	0.263	0.751
W 10 × 88	504	495	485	475	464	442	417	392	364	335	0.263	0.744
W 10 × 100	573	562	551	540	528	503	476	446	416	383	0.263	0.735
W 10 × 112	642	631	619	606	593	565	535	503	469	433	0.261	0.726
W 12 × 40	217	210	203	196	188	172	154	135	114	94	0.227	1.073
W 12 × 45	243	235	228	220	211	193	173	152	129	106	0.227	1.065
W 12 × 50	271	263	254	246	236	216	195	171	146	121	0.227	1.058
W 12 × 53	301	295	288	282	275	260	244	227	209	189	0.221	0.813
W 12 × 58	329	322	315	308	301	285	268	249	230	209	0.218	0.794
W 12 × 65	378	373	367	361	354	341	326	311	294	277	0.217	0.656
W 12 × 72	418	412	406	399	392	377	361	344	326	308	0.217	0.651
W 12 × 79	460	453	446	439	431	415	398	380	360	339	0.217	0.648
W 12 × 87	508	501	493	485	477	459	440	420	398	376	0.217	0.645
W 12 × 96	560	552	544	535	526	506	486	464	440	416	0.215	0.635
W 12 × 106	620	611	602	593	583	561	539	514	489	462	0.215	0.633
W 12 × 120	702	692	660	636	611	584	555	525	493	460	0.217	0.630
W 12 × 136	795	772	747	721	693	662	630	597	561	524	0.215	0.621
W 12 × 152	891	866	839	810	778	745	710	673	633	592	0.214	0.614
W 12 × 170	998	970	940	908	873	837	798	757	714	668	0.213	0.608

(Right panel)

Shape	Effective length (KL) in feet										Bending factor	
	8	10	12	14	16	18	20	22	24	26	B_x	B_y
W 12 × 190	1115	1084	1051	1016	978	937	894	849	802	752	0.212	0.600
W 12 × 210	1236	1202	1166	1127	1086	1042	995	946	894	840	0.212	0.594
W 12 × 230	1355	1319	1280	1238	1193	1145	1095	1041	985	927	0.211	0.589
W 12 × 252	1484	1445	1403	1358	1309	1258	1203	1146	1085	1022	0.210	0.583
W 12 × 279	1642	1600	1554	1505	1452	1396	1337	1275	1209	1141	0.208	0.573
W 12 × 305	1799	1753	1704	1651	1594	1534	1471	1404	1333	1260	0.206	0.564
W 12 × 336	1986	1937	1884	1827	1766	1701	1632	1560	1484	1404	0.205	0.558
W 14 × 43	230	215	199	181	161	140	117	96	81	69	0.201	1.115
W 14 × 48	258	242	224	204	182	159	133	110	93	79	0.201	1.102
W 14 × 53	286	268	248	226	202	177	149	123	104	88	0.201	1.091
W 14 × 61	345	330	314	297	278	258	237	214	190	165	0.194	0.833
W 14 × 68	385	369	351	332	311	289	266	241	214	186	0.194	0.826
W 14 × 74	421	403	384	363	341	317	292	265	236	206	0.195	0.820
W 14 × 82	465	446	425	402	377	351	323	293	261	227	0.196	0.823
W 14 × 90	536	524	511	497	482	466	449	432	413	394	0.185	0.531
W 14 × 99	589	575	561	546	529	512	494	475	454	433	0.185	0.527
W 14 × 109	647	633	618	601	583	564	544	523	501	478	0.185	0.523
W 14 × 120	714	699	682	663	644	623	601	578	554	528	0.186	0.523
W 14 × 132	786	768	750	730	708	686	662	637	610	583	0.186	0.521
W 14 × 145	869	851	832	812	790	767	743	718	691	663	0.184	0.489
W 14 × 159	950	931	911	889	865	840	814	786	758	727	0.184	0.485
W 14 × 176	1054	1034	1011	987	961	933	904	874	842	809	0.184	0.484
W 14 × 193	1157	1134	1110	1083	1055	1025	994	961	927	891	0.183	0.477
W 14 × 211	1263	1239	1212	1183	1153	1121	1087	1051	1014	975	0.183	0.477
W 14 × 233	1396	1370	1340	1309	1276	1241	1204	1165	1124	1081	0.183	0.472
W 14 × 257	1542	1513	1481	1447	1410	1372	1331	1289	1244	1198	0.182	0.470
W 14 × 283	1700	1668	1634	1597	1557	1515	1471	1425	1377	1326	0.181	0.465
W 14 × 311	1867	1832	1794	1754	1711	1666	1618	1568	1515	1460	0.181	0.459
W 14 × 342		2022	1985	1941	1894	1845	1793	1738	1681	1621	0.181	0.457
W 14 × 370		2181	2144	2097	2047	1995	1939	1881	1820	1756	0.180	0.452
W 14 × 398		2356	2304	2255	2202	2146	2087	2025	1961	1893	0.183	0.447
W 14 × 426		2515	2464	2411	2356	2296	2234	2169	2100	2029	0.178	0.442
W 14 × 455		2694	2644	2589	2530	2467	2401	2332	2260	2184	0.177	0.441
W 14 × 500		2952	2905	2845	2781	2714	2642	2568	2490	2409	0.175	0.434
W 14 × 550		3272	3206	3142	3073	3000	2923	2842	2758	2670	0.174	0.429
W 14 × 605		3591	3529	3459	3384	3306	3223	3136	3045	2951	0.171	0.421
W 14 × 665		3974	3892	3817	3737	3652	3563	3469	3372	3270	0.170	0.415
W 14 × 730		4355	4277	4196	4100	4019	3923	3823	3718	3609	0.168	0.408

Source: Adapted from data in the *Manual of Steel Construction*, 8th ed. (Ref. 5), with permission of the publishers, American Institute of Steel Construction.

[a] Loads in kips for shapes of steel with yield stress of 36 ksi [250 MPa], based on buckling with respect to the Y-axis.

TABLE 23.3. Allowable Column Loads for Standard Steel Pipe with Yield Stress of 36 ksi (kips)

Nominal Dia		12	10	8	6	5	4	3½	3
Wall Thickness		0.375	0.365	0.322	0.280	0.258	0.237	0.226	0.216
Weight per Foot		49.56	40.48	28.55	18.97	14.62	10.79	9.11	7.58
F_y		36 ksi							
Effective length in feet KL with respect to radius of gyration	0	315	257	181	121	93	68	58	48
	6	303	246	171	110	83	59	48	38
	7	301	243	168	108	81	57	46	36
	8	299	241	166	106	78	54	44	34
	9	296	238	163	103	76	52	41	31
	10	293	235	161	101	73	49	38	28
	11	291	232	158	98	71	46	35	25
	12	288	229	155	95	68	43	32	22
	13	285	226	152	92	65	40	29	19
	14	282	223	149	89	61	36	25	16
	15	278	220	145	86	58	33	22	14
	16	275	216	142	82	55	29	19	12
	17	272	213	138	79	51	26	17	11
	18	268	209	135	75	47	23	15	10
	19	265	205	131	71	43	21	14	9
	20	261	201	127	67	39	19	12	
	22	254	193	119	59	32	15	10	
	24	246	185	111	51	27	13		
	25	242	180	106	47	25	12		
	26	238	176	102	43	23			
	28	229	167	93	37	20			
	30	220	158	83	32	17			
	31	216	152	78	30	16			
	32	211	148	73	29				
	34	201	137	65	25				
	36	192	127	58	23				
	37	186	120	55	21				
	38	181	115	52					
	40	171	104	47					
Properties									
Area A (in.^{2})		14.6	11.9	8.40	5.58	4.30	3.17	2.68	2.23
I (in.^{4})		279	161	72.5	28.1	15.2	7.23	4.79	3.02
r (in.)		4.38	3.67	2.94	2.25	1.88	1.51	1.34	1.16
B } Bending factor		0.333	0.398	0.500	0.657	0.789	0.987	1.12	1.29
* a		41.7	23.9	10.8	4.21	2.26	1.08	0.717	0.447

* Tabulated values of a must be multiplied by 10^6.
Note: Heavy line indicates Kl/r of 200.

Source: Adapted from data in the *Manual of Steel Construction*, 8th ed. (Ref. 5), with permission of the publishers, American Institute of Steel Construction.

Angle Size	Load for X-axis (kips)	Load for Y-axis (kips)	Weight (lb)
$4 \times 3 \times \frac{1}{2}$	71	77	22.2
$5 \times 3 \times \frac{3}{8}$	81	61	19.6

Actual choice would of course be influenced by other concerns, most notably those relating to the development of framing connections. It should also be noted that the data in Table 23.5 is a tiny sample of the material in the reference.

23.4. COLUMNS WITH BENDING

Many steel columns must sustain bending in addition to the usual axial compression. Figure 23.5 shows three of the most common situations that result in this combined effect. When framing members are supported at the column face or on a bracket, the compression load may actually occur with some eccentricity, as shown in Fig. 23.5a. When moment-resistive connections are used and the column becomes a member of a rigid frame bent, moments will be induced in the ends of the columns, as shown in Fig. 23.5b. Columns in exterior walls frequently function as part of the general wall framing; if vertical spanning for wind load is involved, the column may receive a direct beam loading, as shown in Fig. 23.5c.

Basic problems of columns with bending are discussed in Chapter 11. The fundamental relationship is one of interaction, the simplest form of which is expressed by the straight-line interaction formula:

$$\frac{f_a}{F_a} + \frac{f_b}{F_b} = 1$$

In reality, the problem has the potential for great complexity. The usual problems of dealing with buckling effects of the column must be combined with that of the laterally unsupported beam and some possible synergetic interactions, such as the *P*-delta effect (see Sec. 11.5). Although the basic form of the interaction formula is used for investigation, numerous adjustments must often be made for various situations. The AISC Specifications are quite extensive and complex and not very self-explanatory with regard to this problem. For a full treatment of the topic the reader is referred to one of the major textbooks on steel design, such as *Steel Buildings* by S. W. Crawley and R. M. Dillon (Ref. 21).

An approximation method for obtaining a trial section for a column with bending is discussed in Sec. 23.5. Some additional discussion and illustration of the problems of designing columns with bending in various situations is presented in the building design examples in Part 9.

23.5. APPROXIMATE DESIGN OF COLUMNS WITH BENDING

For use in preliminary design work, or to obtain a first trial section to be used in a more extensive design investigation, a procedure may be used that involves the determination of an equivalent total design load that incorporates the bending effects. This is done by using the bending factors, B_x and B_y, which are listed in Table 23.2. The equivalent design load is determined as

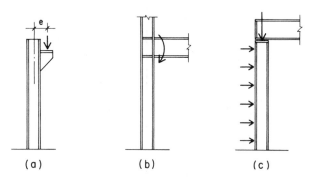

FIGURE 23.5. Columns with bending.

TABLE 23.4. Allowable Column Loads for Square Standard Steel Tubing with Yield Stress of 46 ksi (kips)

Nominal Size: 8 × 8 and 7 × 7

Nominal Size	8 × 8					7 × 7				
Thickness	5/8	1/2	3/8	5/16	1/4	1/2	3/8	5/16	1/4	3/16
Wt./ft.	59.32	48.85	37.60	31.84	25.82	42.05	32.59	27.59	22.42	17.08
F_y	46 ksi									
Effective length in feet KL with respect to radius of gyration										
0	480	397	306	258	209	342	264	224	182	139
6	448	370	286	242	196	314	244	207	168	128
7	441	364	281	238	193	308	239	203	165	126
8	433	358	277	234	190	301	234	199	162	123
9	425	352	272	230	187	294	229	195	158	121
10	417	345	267	226	184	287	224	190	155	118
11	408	338	262	222	180	280	218	185	151	115
12	399	330	256	217	176	272	212	180	147	113
13	389	323	251	212	173	264	206	175	143	109
14	379	315	245	207	169	255	200	170	139	106
15	369	307	238	202	165	246	193	165	135	103
16	358	298	232	197	160	237	186	159	130	100
17	347	289	225	191	156	228	179	153	125	96
18	336	280	219	186	151	218	172	147	120	92
19	324	271	212	180	147	208	165	141	115	89
20	312	261	205	174	142	197	157	134	110	85
21	300	251	197	168	137	187	149	128	105	81
22	287	241	190	162	132	176	140	121	99	77
23	274	231	182	155	127	164	132	114	94	73
24	261	220	174	148	122	152	123	106	88	68
25	247	209	165	141	116	140	114	99	82	64
26	232	198	157	134	110	130	105	91	76	59
27	218	186	148	127	105	120	98	85	70	55
28	203	174	139	120	99	112	91	79	65	51
29	189	162	130	112	93	104	85	73	61	47
30	177	151	122	105	87	98	79	69	57	44
32	155	133	107	92	76	86	70	60	50	39
34	137	118	95	82	67	76	62	53	44	35
36	123	105	84	73	60	68	55	48	40	31
38	110	94	76	65	54	61	49	43	35	28
40	99	85	68	59	49	55	45	39	32	25
Properties										
A (in.^{2})	17.40	14.40	11.10	9.36	7.59	12.40	9.58	8.11	6.59	5.02
I (in.^{4})	153	131	106	90.9	75.1	84.6	68.7	59.5	49.4	38.5
r (in.)	2.96	3.03	3.09	3.12	3.15	2.62	2.68	2.71	2.74	2.77
B Bending factor	0.455	0.437	0.420	0.412	0.404	0.511	0.488	0.477	0.467	0.457

Nominal Size: 6 × 6 and 5 × 5

Nominal Size	6 × 6					5 × 5				
Thickness	1/2	3/8	5/16	1/4	3/16	1/2	3/8	5/16	1/4	3/16
Wt./ft.	35.24	27.48	23.34	19.02	14.53	28.43	22.37	19.08	15.62	11.97
F_y	46 ksi									
Effective length in feet KL with respect to radius of gyration										
0	287	223	189	154	118	231	182	155	127	97
6	257	201	171	140	107	200	159	136	111	86
7	251	196	167	137	105	193	153	131	108	83
8	244	191	163	133	102	186	148	127	104	80
9	237	186	158	130	99	178	142	122	100	77
10	229	180	154	126	96	169	135	116	96	74
11	221	174	149	122	93	160	129	111	92	71
12	212	168	143	117	90	151	122	105	87	67
13	203	161	138	113	87	141	115	99	82	64
14	194	154	132	108	83	131	107	93	77	60
15	185	147	126	104	80	120	99	86	72	56
16	175	140	120	99	76	109	90	79	66	52
17	164	132	113	94	72	97	82	72	60	47
18	153	124	107	88	68	87	73	64	54	43
19	142	115	100	83	64	78	65	58	49	39
20	131	107	93	77	60	70	59	52	44	35
21	119	98	85	71	56	64	54	47	40	32
22	108	89	78	65	51	58	49	43	36	29
24	91	75	65	55	43	49	41	36	31	24
26	77	64	56	47	36	41	35	31	26	21
28	67	55	48	40	31	36	30	27	22	18
30	58	48	42	35	27	31	26	23	20	15
31	54	45	39	33	26		25	22	18	14
32	51	42	37	31	24				17	14
34	45	37	33	27	21					
36	40	33	29	24	19					
37		32	27	23	18					
38			26	22	17					
39					16					
Properties										
A (in.^{2})	10.40	8.08	6.86	5.59	4.27	8.36	6.58	5.61	4.59	3.52
I (in.^{4})	50.5	41.6	36.3	30.3	23.8	27.0	22.8	20.1	16.9	13.4
r (in.)	2.21	2.27	2.30	2.33	2.36	1.80	1.86	1.89	1.92	1.95
B Bending factor	0.615	0.583	0.567	0.553	0.539	0.773	0.722	0.699	0.677	0.656

Nominal Size: 4 × 4 and 3 × 3

Nominal Size	4 × 4					3 × 3		
Thickness	1/2	3/8	5/16	1/4	3/16	5/16	1/4	3/16
Wt./ft.	21.63	17.27	14.83	12.21	9.42	10.58	8.81	6.87
F_y	46 ksi							
Effective length in feet KL with respect to radius of gyration								
0	176	140	120	99	76	86	71	56
2	168	134	115	95	73	80	67	53
3	162	130	112	92	71	77	64	50
4	156	126	108	89	69	73	61	48
5	150	121	104	86	67	68	57	45
6	143	115	100	83	64	63	53	42
7	135	110	95	79	61	57	49	39
8	126	103	90	75	58	51	44	35
9	117	97	84	70	55	44	38	31
10	108	89	78	65	51	37	33	27
11	98	82	72	60	47	31	27	22
12	87	74	65	55	43	26	23	19
13	75	65	58	49	39	22	19	16
14	65	57	51	43	35	19	17	14
15	57	49	44	38	30	16	15	12
16	50	43	39	33	27	14	13	11
17	44	38	34	29	24	13	11	9
18	39	34	31	26	21		10	8
19	35	31	28	24	19			
20	32	28	25	21	17			
21	29	25	23	19	16			
22	26	23	21	18	14			
23	24	21	19	16	13			
24		19	17	15	12			
25				14	11			
Properties								
A (in.^{2})	6.36	5.08	4.36	3.59	2.77	3.11	2.59	2.02
I (in.^{4})	12.3	10.7	9.58	8.22	6.59	3.58	3.16	2.60
r (in.)	1.39	1.45	1.48	1.51	1.54	1.07	1.10	1.13
B Bending factor	1.04	0.949	0.910	0.874	0.840	1.30	1.23	1.17

Source: Adapted from data in the *Manual of Steel Construction*, 8th ed. (Ref. 5), with permission of the publishers, American Institute of Steel Construction.

Note: Heavy line indicates KL/r of 200.

TABLE 23.5. Allowable Axial Compression for Double Angle Struts[a]

Left portion

Size (in.)	8 × 6		6 × 4			5 × 3 1/2		5 × 3		
Thickness (in.)	3/4	1/2	5/8	1/2	3/8	1/2	3/8	1/2	3/8	5/16
Weight (lb/ft)	67.6	46.0	40.0	32.4	24.6	27.2	20.8	25.6	19.6	16.4
Area (in²)	19.9	13.5	11.7	9.50	7.22	8.00	6.09	7.50	5.72	4.80
r_x (in.)	2.53	2.56	1.90	1.91	1.93	1.58	1.60	1.59	1.61	1.61
r_y (in.)	2.48	2.44	1.67	1.64	1.62	1.49	1.46	1.25	1.23	1.22

Effective Length (KL) with Respect to Indicated Axis

X–X Axis

KL	8×6 (3/4)	8×6 (1/2)	KL	6×4 (5/8)	6×4 (1/2)	6×4 (3/8)	KL	5×3½ (1/2)	5×3½ (3/8)	KL	5×3 (1/2)	5×3 (3/8)	5×3 (5/16)
0	430	266	0	253	205	142	0	173	129	0	162	121	94
10	370	231	8	214	174	122	4	159	119	4	149	112	88
12	353	222	10	200	163	115	6	150	113	6	141	106	83
14	334	211	12	185	151	107	8	139	105	8	130	98	77
16	315	200	14	168	137	99	10	126	96	10	119	90	71
20	271	175	16	150	123	89	12	113	86	12	106	81	64
24	222	148	20	110	90	69	14	97	75	14	92	70	57
28	168	117	24	76	62	48	16	81	63	16	76	59	49
32	129	90	28	56	46	36	20	52	40	20	49	38	32
36	102	71											

Y–Y Axis

KL	8×6 (3/4)	8×6 (1/2)	KL	6×4 (5/8)	6×4 (1/2)	6×4 (3/8)	KL	5×3½ (1/2)	5×3½ (3/8)	KL	5×3 (1/2)	5×3 (3/8)	5×3 (5/16)
0	430	266	0	253	205	142	0	173	129	0	162	121	94
10	368	229	6	222	179	125	4	158	118	4	145	108	85
12	351	219	8	207	167	117	6	148	110	6	132	99	78
14	332	207	10	190	153	108	8	136	101	8	118	88	69
16	311	195	12	171	137	97	10	122	91	10	101	75	60
20	266	169	14	151	120	86	12	107	79	12	82	61	49
24	216	139	16	129	102	74	14	90	67	14	62	46	38
28	162	106	20	85	66	49	16	72	53	16	47	35	29
32	124	81	24	59	46	34	20	46	34	20	30	22	19
36	98	64											

Right portion

Size (in.)	4 × 3			3 1/2 × 2 1/2			3 × 2		
Thickness (in.)	1/2	3/8	5/16	3/8	5/16	1/4	3/8	5/16	1/4
Weight (lb/ft)	22.2	17.0	14.4	14.4	12.2	9.8	11.8	10.0	8.2
Area (in²)	6.50	4.97	4.18	4.22	3.55	2.88	3.47	2.93	2.38
r_x (in.)	1.25	1.26	1.27	1.10	1.11	1.12	0.940	0.948	0.957
r_y (in.)	1.33	1.31	1.30	1.11	1.10	1.09	0.917	0.903	0.891

X–X Axis

KL	4×3 (1/2)	4×3 (3/8)	4×3 (5/16)	KL	3½×2½ (3/8)	3½×2½ (5/16)	3½×2½ (1/4)	KL	3×2 (3/8)	3×2 (5/16)	3×2 (1/4)
0	140	107	90	0	91	77	60	0	75	63	51
2	134	103	86	2	86	73	57	2	70	59	48
4	126	96	81	4	80	67	53	3	67	57	46
6	115	88	74	6	71	60	48	4	63	54	44
8	102	78	66	8	61	52	41	6	55	46	38
10	88	67	57	10	50	42	34	8	44	38	31
12	71	55	47	12	37	31	26	10	32	27	23
14	54	42	36	14	27	23	19	12	22	19	16
16	41	32	27	16	21	18	15	14	16	14	12
18	33	25	22	18	16	14	12				
20	26	20	17								

Y–Y Axis

KL	4×3 (1/2)	4×3 (3/8)	4×3 (5/16)	KL	3½×2½ (3/8)	3½×2½ (5/16)	3½×2½ (1/4)	KL	3×2 (3/8)	3×2 (5/16)	3×2 (1/4)
0	140	107	90	0	91	77	60	0	75	63	51
2	135	103	86	2	87	73	57	2	70	59	48
4	127	97	81	4	80	67	53	3	67	56	46
6	117	89	74	6	72	60	47	4	63	53	43
8	105	80	67	8	62	52	41	6	54	45	36
10	92	70	58	10	50	42	33	8	43	36	28
12	77	58	48	12	37	31	25	10	30	25	20
14	61	45	37	14	28	23	18	12	21	17	14
16	47	35	29	16	21	17	14	14	15	13	10
18	37	27	23	18	17	14	11				
20	30	22	18								

Source: Adapted from data in the *Manual of Steel Construction* (Ref. 5), with permission of the publishers, American Institute of Steel Construction.
[a] Loads in kips; F_y = 36 ksi [250 MPa]; long legs back-to-back; 3/8-in. separation between angles.

$$P' = P + B_x M_x + \dot{B_y} M_y$$

where

P' = equivalent axial load

P = actual axial load

B_x = bending factor for the section's x-axis

M_x = bending moment about the x-axis

B_y = bending factor for the section's y-axis

M_y = bending moment about the y-axis

Example 1 illustrates the use of the method.

Example 1. It is desired to use a 10-in. W shape for a column in a situation such as that shown in Fig. 23.6. The axial compression load from above is 120 kips and the beam load is 24 kips, with the beam attached at the column face. The column is 16 ft-high and has a K factor of 1.0. Select a trial section for the column.

Solution. Since bending occurs only about the x-axis, we use only the B_x factor for this case. Scanning the column of B_x factors in Table 23.2, we observe that the factor for 10 W sections varies from 0.261 to 0.277. As we have not yet determined the section to be used, it is necessary to make an assumption for the factor and to verify the assumption after the selection is made. Let us assume a B_x of 0.27, with which we find

$$P' = P + B_x M_x = (120 + 24) + 0.27(24)(5)$$

$$= 144 + 32.4 = 176.4 \text{ kips}$$

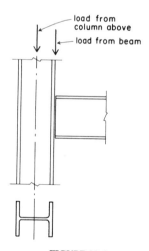

load from column above
load from beam

FIGURE 23.6.

From Table 23.2, for a KL of 16 ft, we obtain a W 10×45. For this shape the B_x factor is 0.271, which is very close to our assumption.

For most situations use of the bending factors in the manner just demonstrated will result in conservative selections. If the designer intends to use the section thus obtained in a more thorough investigation, it is probably wise to reduce the size slightly before proceeding with the work.

23.6. COLUMNS IN FRAMED BENTS

A major occurrence of the condition of a column with bending is that of the case of a column in a rigid-frame bent. For steel structures, this most often occurs when a steel frame is developed as a moment-resisting space frame in three dimensions with the rigid-frame action being used for lateral bracing. If the frame is made rigid, both vertical gravity loads and horizontal wind or earthquake loads will result in bending and shear in the columns. Frames of more than a single bay in size are statically indeterminate and require investigation for various load combinations for determination of critical internal forces and deformations.

Consideration of the general actions of rigid frames is treated in Sec. 14.1. A discussion of rigid frames as lateral bracing elements is given in Sec. 47.4. Approximate investigation and design techniques for multistory rigid frames are illustrated in Chapter 58.

23.7. COLUMN CONNECTIONS

Connection details for columns must be developed with considerations of the column shape and size, the shape and orientation of other framing, and the particular structural functions of the joints. Some common connections are shown in Fig. 23.7. When beams sit directly on the top of a column, the usual solution is to fasten a bearing plate on top of the column, with provision for the attachment of the beam to the plate. The plate serves no specific structural purpose in this case, functioning essentially only as an attachment device, assuming that the load transfer is one of simple vertical bearing (see Fig. 23.7a).

In many situations beams must frame into the side of a column. If simple transfer of vertical load is all that is required, a common solution is the connection shown in Fig. 23.7b, in which a pair of steel angles are used to connect the beam web to the column flange or the column web. This type of connection is discussed in Sec. 24.4. If moment must be transferred between the columns and beams, as in the development of rigid framed bents, the most common solution involves the use of welding to achieve a direct connection between the members, as shown in Fig. 23.7c. The details of rigid connections vary considerably, mostly on the basis of the size of the members and the magnitude of forces.

Another common connection is that of the bottom of a column to its supports. Figure 23.7d shows a typical solution for transfer of bearing to a concrete footing, consisting of a steel plate welded to the column bottom and bolted to preset anchor bolts. In this case the plate has a definite structural function, serving to transform the highly concentrated, punching effect of the column force into a low-valued bearing stress on the much softer concrete. Thus, while the general form of the connection is similar, the role of the plates in Fig. 23.7a and d are considerably different.

Part 4 of the *AISC Manual* provides extensive data relating to the development of connections for structural steel members. Some of the material is presented in the discussions in this book in Chapter 24. Issues of specific concern are also discussed in some of the design examples in Part 9. In this section we limit the presentation to consideration of the development of column base plates.

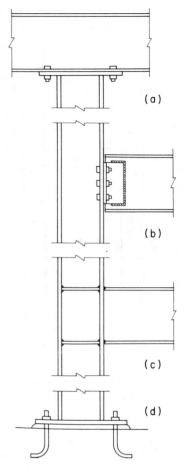

FIGURE 23.7. Typical column connections; lightly loaded frames.

Design of Column Base Plates

Column base plates vary from relatively modest ones for small, lightly loaded columns to huge thick ones for the heavy 14 W shapes in high-strength steel grades. The following procedure is based on the specifications and design illustrations in the *AISC Manual* and the recommendations in the *ACI Code* (Ref. 6).

The plan area required for the base plate is determined as

$$A_1 = \frac{P}{F_p}$$

where

A_1 = plan area of the bearing plate
P = compression load from the column
F_p = allowable bearing stress on the concrete

Allowable bearing is based on the concrete design strength, f_c'. It is limited to a value of $0.3f_c'$ when the plate covers the entire area of the support member, which is seldom the case. If the support member has a larger plan area, A_2, the allowable bearing stress may be increased by a factor of $\sqrt{A_2/A_1}$, but not greater than 2. For modest-size columns supported on footings, the maximum factor is most likely to be used. For large columns or those supported on pedestals or piers, the adjustment may be less.

For a W-shape column, the basis for determination of the thickness of the plate due to bending is shown in Fig. 23.8. Once the required value for A_1 is found, the dimensions B and N are established so that the projections m and n are approximately equal. Choice of dimensions must also relate to the locations of anchor bolts and to any details for development of the attachment of the plate to the column. The required plate thickness is determined with the formula

$$t = \sqrt{\frac{3f_p m^2}{F_b}} \quad \text{or} \quad t = \sqrt{\frac{3f_p n^2}{F_b}}$$

where

t = thickness of the bearing plate, inches
f_p = actual bearing pressure: P/A_1
F_b = allowable bending stress in the plate: $0.75F_y$

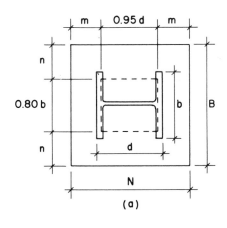

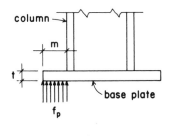

FIGURE 23.8. Reference dimensions for column base plates.

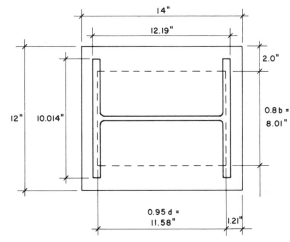

FIGURE 23.9.

The following example illustrates the process for a column with a relatively light load.

Example 1. Design a base plate of A36 steel for a W 12×58 column with a load of 250 kips. The column bears on a concrete footing with $f_c' = 3$ ksi.
Solution. We assume the footing area to be considerably larger than the plate area; thus $F_p = 0.6f_c' = 1.8$ ksi. Then

$$A_1 = \frac{P}{F_p} = \frac{250}{1.8} = 138.9 \text{ in.}^2$$

If the plate is square,

$$B = N = \sqrt{138.9} = 11.8 \text{ in.}$$

Since this is almost the same size as the column, we will assume the plan size layout shown in Fig. 23.9, which allows the welding of the plate to the column and the placing of the anchor bolts. It may be observed that the dimension labeled n in Fig. 23.8 is critical in this case, and the plate thickness is thus found as

$$t = \sqrt{\frac{3f_p n^2}{F_b}}$$

for which

$$f_p = \frac{P}{A_1} = \frac{250}{12(14)} = 1.49 \text{ ksi}$$

and

$$t = \sqrt{\frac{3(1.49)(2)^2}{0.75(36)}} = \sqrt{0.662} = 0.814 \text{ in.}$$

Plates are usually specified in thickness increments of $\frac{1}{8}$ in., so the minimum thickness would be $\frac{7}{8}$ in. (0.875 in.).

CHAPTER TWENTY-FOUR

Steel Connections

At one time a major technique for attaching steel elements at the scale of building structures was by riveting. Bolting was done mostly for temporary connections and for special situations. Riveting is now hardly seen in building construction, with connections achieved with structural bolts or by welding. This chapter deals with the use of bolts for various types of structural joints.

24.1. BOLTED CONNECTIONS

The diagrams in Fig. 24.1 show a simple connection between two steel bars that functions to transfer a tension force from one bar to another. Although this is a tension-transfer connection, it is also referred to as a shear connection because of the manner in which the connecting device (the bolt) works in the connection (see Fig. 24.1*b*). If the bolt tension (due to tightening of the nut) is relatively low, the bolt serves primarily as a pin in the matched holes, bearing against the sides of the holes as shown in Fig. 24.1*d*. In addition to these functions, the bars develop tension stress that will be a maximum at the section through the bolt holes.

In the connection shown in Fig. 24.1, the failure of the bolt involves a slicing (shear) failure that is developed as a shear stress on the bolt cross section. The resistance of the bolt can be expressed as an allowable shear stress F_v times the area of the bolt cross section, or

$$R = F_v \times A$$

With the size of the bolt and the grade of steel known, it is a simple matter to establish this limit. In some types of connections, it may be necessary to slice the same bolt more than once to separate the connected parts. This is the case in the connection shown in Fig. 24.2, in which it may be observed that the bolt must be sliced twice to make the joint fail. When the bolt develops shear on only one section (Fig. 24.1), it is said to be in *single shear*; when it develops shear on two sections (Fig. 24.2), it is said to be in *double shear*.

When the bolt diameter is large or the bolt is made of strong steel, the connected parts must be sufficiently thick if they are to develop the full capacity of the bolts. The maximum bearing stress permitted for this situation by the AISC Specification is $F_p = 1.5F_u$, where F_u is the ultimate

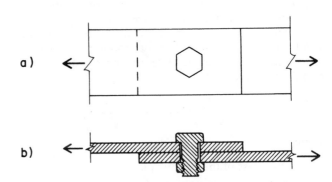

a)

b)

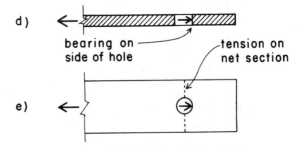

c)

d)

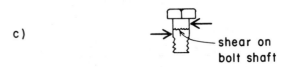

e)

FIGURE 24.1. Action of bolted joints.

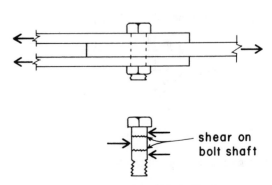

FIGURE 24.2. Bolted joint with double shear.

tensile strength of the steel in the part in which the hole occurs.

Bolts used for the connection of structural steel members come in two types. Bolts designated A307 and called *unfinished* have the lowest load capacity of the structural bolts. The nuts for these bolts are tightened just enough to secure a snug fit of the attached parts; because of this, plus the oversizing of the holes, there is some movement in the development of full resistance. These bolts are generally not used for major connections, especially when joint movement or loosening under vibration or repeated loading may be a problem.

Bolts designated A325 or A490 are called *high-strength bolts*. The nuts of these bolts are tightened to produce a considerable tension force, which results in a high degree of friction resistance between the attached parts. High-strength bolts are further designated as F, N, or X. The F designation denotes bolts for which the limiting resistance is that of friction. The N designation denotes bolts that function ultimately in bearing and shear but for which the threads are not excluded from the bolt shear planes. The X designation denotes bolts that function like the N bolts but for which the threads are excluded from the shear planes.

When bolts are loaded in tension, their capacities are based on the development of the ultimate resistance in tension stress at the reduced section through the threads.

When loaded in shear, bolt capacities are based on the development of shear stress in the bolt shaft. The shear capacity of a single bolt is further designated as S for single shear (Fig. 24.1) or D for double shear (Fig. 24.2). The capacities of structural bolts in both tension and shear are given in Table 24.1. The size range given in the table ($\frac{5}{8}$–$1\frac{1}{2}$ in.) is that listed in the *AISC Manual*. However, the most commonly used sizes for structural steel framing are $\frac{3}{4}$ and $\frac{7}{8}$ in.

Bolts are ordinarily installed with a washer under both head and nut. Some manufactured high-strength bolts have specially formed heads or nuts that in effect have self-forming washers, eliminating the need for a separate, loose washer. When a washer is used, it is sometimes the limiting dimensional factor in detailing for bolt placement in tight locations, such as close to the fillet (inside radius) of angles or other rolled shapes.

For a given diameter of bolt, there is a minimum thickness required for the bolted parts in order to develop the full shear capacity of the bolt. This thickness is based in the bearing stress between the bolt and the side of the hole, which is limited to a maximum of $F_p = 1.5F_u$. The stress limit may be established by either the bolt steel or the steel of the bolted parts.

Steel rods are sometimes threaded for use as anchor bolts or tie rods. When they are loaded in tension, their

TABLE 24.1. Capacity of Structural Bolts (in kips)

ASTM Designation	Connection Type*	Loading Condition**	Nominal Diameter (in.)							
			5/8	3/4	7/8	1	1-1/8	1-1/4	1-3/8	1-1/2
			Area, Based on Nominal Diameter (in²)							
			0.3068	0.4418	0.6013	0.7854	0.9940	1.227	1.485	1.767
A307		S	3.1	4.4	6.0	7.9	9.9	12.3	14.8	17.7
		D	6.1	8.8	12.0	15.7	19.9	24.5	29.7	35.3
		T	6.1	8.8	12.0	15.7	19.9	24.5	29.7	35.3
A325	F	S	5.4	7.7	10.5	13.7	17.4	21.5	26.0	30.9
		D	10.7	15.5	21.0	27.5	34.8	42.9	52.0	61.8
	N	S	6.4	9.3	12.6	16.5	20.9	25.8	31.2	37.1
		D	12.9	18.6	25.3	33.0	41.7	51.5	62.4	74.2
	X	S	9.2	13.3	18.0	23.6	29.8	36.8	44.5	53.0
		D	18.4	26.5	36.1	47.1	59.6	73.6	89.1	106.0
	All	T	13.5	19.4	26.5	34.6	43.7	54.0	65.3	77.7
A490	F	S	6.7	9.7	13.2	17.3	21.9	27.0	32.7	38.9
		D	13.5	19.4	26.5	34.6	43.7	54.0	65.3	77.7
	N	S	8.6	12.4	16.8	22.0	27.8	34.4	41.6	49.5
		D	17.2	24.7	33.7	44.0	55.7	68.7	83.2	99.0
	X	S	12.3	17.7	24.1	31.4	39.8	49.1	59.4	70.7
		D	24.5	35.3	48.1	62.8	79.5	98.2	119.0	141.0
	All	T	16.6	23.9	32.5	42.4	53.7	66.3	80.2	95.4

* F = friction; N = bearing, threads not excluded; X = bearing, threads excluded.

** S = single shear; D = double shear; T = tension.

Source: Adapted from data in the *Manual of Steel Construction* (Ref. 5), with permission of the publishers, American Institute of Steel Construction.

capacities are usually limited by the stress on the reduced section at the threads. Tie rods are sometimes made with *upset ends*, which consist of larger diameter portions at the ends. When these enlarged ends are threaded, the net section at the thread is the same as the gross section in the remainder of the rods; the result is no loss of capacity for the rod.

Layout of Bolted Connections

Design of bolted connections generally involves a number of considerations in the dimensioned layout of the bolt-hole patterns for the attached structural members. Although we cannot develop all the points necessary for the production of structural steel construction and fabrication details, the material in this section presents basic factors that often must be included in the structural calculations.

Figure 24.3 shows the layout of a bolt pattern with bolts placed in two parallel rows. Two basic dimensions for this layout are limited by the size (nominal diameter) of the bolt. The first is the center-to-center spacing of the bolts, usually called the *pitch*. The AISC Specification limits this dimension to an absolute minimum of $2\frac{2}{3}$ times the bolt diameter. The preferred minimum, however, which is used in this book, is 3 times the diameter.

The second critical layout dimension is the *edge distance*, which is the distance from the center line of the bolt to the nearest edge. There is also a specified limit for this as a function of bolt size. This dimension may also be limited by edge tearing, which is discussed later in this section.

Table 24.2 gives the recommended limits for pitch and edge distance for the bolt sizes used in ordinary steel construction.

In some cases bolts are staggered in parallel rows (Fig. 24.4). In this case the diagonal distance, labeled *m* in the illustration, must also be considered.

Location of bolt lines is often related to the size and type of structural members being attached. This is especially

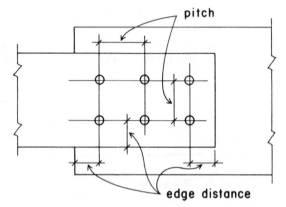

FIGURE 24.3. Pitch and edge distances for bolts.

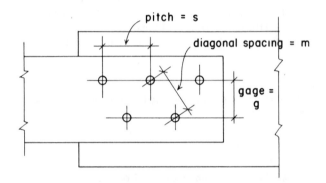

FIGURE 24.4. Reference dimensions for layout of bolted joints.

TABLE 24.2. Pitch and Edge Distances for Bolts

Rivet or Bolt Diameter, d (in.)	Minimum Edge Distance for Punched, Reamed, or Drilled Holes (in.)		Pitch, Center to Center (in.)	
	At Sheared Edges	At Rolled Edges of Plates, Shapes, or Bars, or Gas-Cut Edges[a]	Minimum Recommended 2–2/3d	3d
$\frac{5}{8}$	1.125	0.875	1.67	1.875
$\frac{3}{4}$	1.25	1	2	2.25
$\frac{7}{8}$	1.5[b]	1.125	2.33	2.625
1	1.75[b]	1.25	2.67	3

Source: Adapted from data in the *Manual of Steel Construction*, 8th ed. (Ref. 5), with permission of the publishers, American Institute of Steel Construction.

[a] May be reduced $\frac{1}{8}$ in. when the hole is at a point where stress does not exceed 25% of the maximum allowed in the connected element.

[b] May be $1\frac{1}{4}$ in. at the ends of beam connection angles.

true of bolts placed in the legs of angles or in the flanges of W, M, S, C, and structural tee shapes. Figure 24.5 shows the placement of bolts in the legs of angles. When a single row is placed in a leg, its recommended location is at the distance labeled g from the back of the angle. When two rows are used, the first row is placed at the distance g_1, and the second row is spaced a distance g_2 from the first. Table 24.3 gives the recommended values for these distances.

When placed at the recommended locations in rolled shapes, bolts will end up a certain distance from the edge of the part. Based on the recommended edge distance for rolled edges given in Table 24.2, it is thus possible to determine the maximum size of bolt that can be accommodated. For angles, the maximum fastener may be limited by the edge distance, especially when two rows are used; however, other factors may in some cases be more critical. The distance from the center of the bolts to the inside fillet of the angle may limit the use of a large washer where one is required. Another consideration may be the stress on the net section of the angle, especially if the member load is taken entirely by the attached leg. These problems are given some discussion in Sec. 24.4.

Sections 1-16-4 and 1-16-5 of the AISC Specification provide additional criteria for minimum spacing and edge distances for fasteners as a function of the load per fastener and the thickness and the ultimate stress capacity of the connected parts.

Tension Connections with Bolts

When tension members have reduced cross sections, two stress investigations must be considered. This is the case for members with holes for bolts or for bolts or rods with cut threads. For the member with a hole (Fig. 24.1d) the

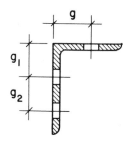

FIGURE 24.5. Gage distances for angles.

TABLE 24.3. Usual Gage Dimensions for Angles (in.)

Gage Dimension	Width of Angle Leg								
	8	7	6	5	4	$3\frac{1}{2}$	3	$2\frac{1}{2}$	2
g	$4\frac{1}{2}$	4	$3\frac{1}{2}$	3	$2\frac{1}{2}$	2	$1\frac{3}{4}$	$1\frac{3}{8}$	$1\frac{1}{8}$
g_1	3	$2\frac{1}{2}$	$2\frac{1}{4}$	2					
g_2	3	3	$2\frac{1}{2}$	$1\frac{3}{4}$					

Source: Adapted from data in the *Manual of Steel Construction*, 8th ed. (Ref. 5), with permission of the publishers, American Institute of Steel Construction.

allowable tension stress at the reduced cross section through the hole is $0.50F_u$, where F_u is the ultimate tensile strength of the steel. The total resistance at this reduced section (also called the *net section*) must be compared with the resistance at other, unreduced sections at which the allowable stress is $0.60F_y$.

For threaded steel rods the maximum allowable tension stress at the threads is $0.33F_u$. For steel bolts the allowable stress is specified as a value based on the type of bolt. The load capacity of various types and sizes of bolt is given in Table 24.1.

When tension elements consist of W, M, S, and tee shapes, the tension connection is usually not made in a manner that results in the attachment of all the parts of the section (e.g., both flanges plus the web for a W). In such cases the AISC Specification requires the determination of a reduced effective net area, A_e, that consists of

$$A_e = C_t A_n$$

where

A_n = actual net area of the member
C_t = reduction coefficient

Unless a larger coefficient can be justified by tests, the following values are specified.

1. For W, M, or S shapes with flange widths not less than two-thirds the depth and structural tees cut from such shapes, when the connection is to the flanges and has at least three fasteners per line in the direction of stress, $C_t = 0.90$.
2. For W, M, or S shapes not meeting the conditions above and for tees cut from such shapes, provided the connection has not fewer than three fasteners per line in the direction of stress, $C_t = 0.85$.
3. For all members with connections that have only two fasteners per line in the direction of stress, $C_t = 0.75$.

Angles used as tension members are often connected by only one leg. In a conservative design, the effective net area is only that of the connected leg, less the reduction caused by bolt holes. Rivet and bolt holes are punched larger in diameter than the nominal diameter of the fastener. The punching damages a small amount of the steel around the perimeter of the hole; consequently the diameter of the hole to be deducted in determining the net section is $\frac{1}{8}$ in. greater than the nominal diameter of the rivet.

When only one hole is involved, as in Fig. 24.1 or in a similar connection with a single row of fasteners along the line of stress, the net area of the cross section of one of the plates is found by multiplying the plate thickness by its net width (width of member minus diameter of hole).

When holes are staggered in two rows along the line of stress (Fig. 24.6), the net section is determined somewhat differently. The AISC Specification reads:

In the case of a chain of holes extending across a part in any diagonal or zigzag line, the net width of the part shall be obtained by deducting from the gross width the sum of

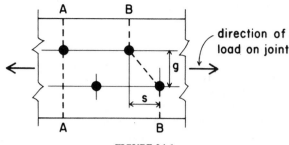

FIGURE 24.6.

the diameters of all the holes in the chain and adding, for each gage space in the chain, the quantity $s^2/4g$, where

  s = longitudinal spacing (pitch) in inches of any two successive holes

and

  g = transverse spacing (gage) in inches for the same two holes

The critical net section of the part is obtained from the chain that gives the least net width.

The AISC Specification also provides that in no case will the net section through a hole be considered as more than 85% of the corresponding gross section.

One possible form of failure in a bolted connection is that of tearing out the edge of one of the attached members. The diagrams in Fig. 24.7 show this potentiality in a connection between two plates. The failure in this case involves a combination of shear and tension to produce the torn-out form shown. The total tearing force is computed as the sum required to cause both forms of failure. The allowable stress on the net tension area is specified as $0.50F_u$, where F_u is the maximum tensile strength of the steel. The allowable stress on the shear areas is specified as $0.30F_u$. With the edge distance, hole spacing, and diameter of the holes known, the net widths for tension and shear are determined and multiplied by the thickness of the part in which the tearing occurs. These areas are then multiplied by the appropriate stresses to find the total tearing force that can be resisted. If this force is greater than the connection design load, the tearing problem is not critical.

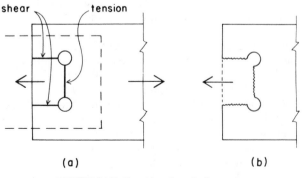

FIGURE 24.7. Consideration of edge tearing.

Another case of potential tearing is shown in Fig. 24.8. This is the common situation for the end framing of a beam in which support is provided by another beam, whose top is aligned with that of the supported beam. The end portion of the top flange of the supported beam must be cut back to allow the beam web to extend to the side of the supporting beam. With the use of a bolted connection, the tearing condition shown is developed.

24.2. DESIGN OF BOLTED CONNECTIONS

The issues raised in several of the preceding sections are illustrated in the following design example. Before proceeding with the problem data, we should consider some of the general requirements for this joint.

If friction-type bolts are used, the surfaces of the connected parts must be cleaned and made reasonably true. If high-strength bolts are used, the determination to exclude threads from the shear failure planes must be established.

The AISC Specification has a number of general requirements for connections:

1. Need for a minimum of two bolts per connection.
2. Need for a minimum connection capacity of 6 kips.
3. Need for the connection to develop at least 50% of the full potential capacity of the member (for trusses only).

Although a part of the design problem may be the selection of the type of fastener or the required strength of steel for the attached parts, we provide this as given data in the example problem.

Example 1.   The connection shown in Fig. 24.9 consists of a pair of narrow plates that transfer a load of 100 kips [445 kN] in tension to a single 10-in. [254 mm]-wide plate. The plates are A36 steel with F_u = 58 ksi [400 MPa] and are attached with $\frac{3}{4}$-in. A325F bolts placed in two rows. Determine the number of bolts required, the width and thickness of the narrow plates, the thickness of the wide plate, and the layout of the bolts.

Solution.   From Table 24.1 we find the double shear (D) capacity for one bolt is 15.5 kips [69 kN]. The required number of bolts is thus

$$n = \frac{\text{connection load}}{\text{bolt capacity}} = \frac{100}{15.5} = 6.45$$

and the minimum number for a symmetrical connection is eight.

With eight bolts used, the load on one bolt is

$$P = \frac{100}{8} = 12.5 \text{ kips } [55.6 \text{ kN}]$$

According to Table 24.2, the $\frac{3}{4}$-in. bolts require a minimum edge distance of 1.25 in. (at a sheared edge) and a recommended pitch of 2.25 in. The minimum width for the narrow plates is therefore (see Fig. 24.9)

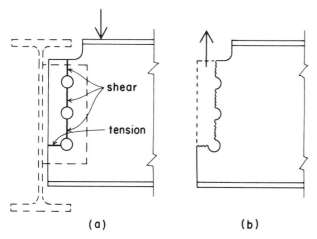

FIGURE 24.8. Tearing in a beam connection.

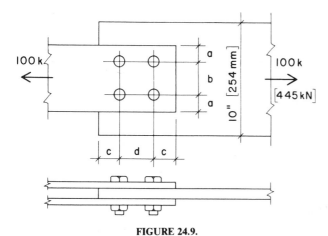

FIGURE 24.9.

$$w = b + 2(a)$$

$$w = 2.25 + 2(1.25) = 4.75 \text{ in. [121 mm]}$$

With no other constraining conditions given, we arbitrarily select a width of 6 in. [152.4 mm] for the narrow plates. Checking first for the requirement of a maximum tension stress of $0.60F_y$ on the gross area, we find

$$F_t = 0.60F_y = 0.60(36) = 21.6 \text{ ksi [149 MPa]}$$

$$A_{\text{req}} = \frac{100}{21.6} = 4.63 \text{ in.}^2 \text{ [2987 mm}^2\text{]}$$

and the required thickness with the width selected is

$$t = \frac{4.63}{2(6)} = 0.386 \text{ in. [9.80 mm]}$$

We therefore select a minimum thickness of $\frac{7}{16}$ in. (0.4375 in.) [11 mm]. The next step is to check the stress condition on the net section through the holes, for which the allowable stress is $0.50F_u$. For the computations, we assume a hole diameter $\frac{1}{8}$ in. [3.18 mm] larger than the bolt. Thus

hole size $= 0.875$ in. [22.22 mm]

net width $= 2\{6 - [2(0.875)]\} = 8.5$ in. [215.9 mm]

aand the stress on the net section of the two plates is

$$f_t = \frac{100}{0.4375(8.5)} = 26.89 \text{ ksi [187 MPa]}$$

This computed stress is compared with the specified allowable stress of

$$F_t = 0.50F_u = 0.50(58) = 29 \text{ ksi [200 MPa]}$$

Bearing stress is computed by dividing the load on a single bolt by the product of the bolt diameter and the plate thickness. Thus

$$f_p = \frac{12.5}{2(0.75)(0.4375)} = 19.05 \text{ ksi [146 MPa]}$$

This is compared with the allowable stress of

$$F_p = 1.5F_u = 1.5(58) = 87 \text{ ksi [600 MPa]}$$

For the middle plate the procedure is essentially the same except that, in this case, the plate width is given. As before, on the basis of stress on the unreduced section, we determine that the total area required is 4.63 in.² [2987 mm²]. Thus the thickness required is

$$t = \frac{4.63}{10} = 0.463 \text{ in. [11.76 mm]}$$

We therefore select a minimum thickness of $\frac{1}{2}$ in. (0.50 in.) [13 mm]. We then proceed as before to check the stress on the net width. The net width through the two holes is

net width $= 10 - 2(0.875) = 8.25$ in. [209.6 mm]

and the tension stress on this net cross section is

$$f_t = \frac{100}{8.25(0.5)} = 24.24 \text{ ksi [177 MPa]}$$

which is less than the allowable stress of 29 ksi [200 MPa] determined previously.

The computed bearing stress on the wide plate is

$$f_p = \frac{12.5}{0.75(0.50)} = 33.3 \text{ ksi [243 MPa]}$$

which is considerably less than the allowable determined before: $F_p = 87$ ksi [600 MPa].

In addition to the layout restrictions given in Sec. 24.1, the AISC Specification requires that the minimum spacing in the direction of the load be

$$\frac{2P}{F_u t} + \frac{d}{2} \qquad \text{(dimension } d \text{ in Fig. 24.9)}$$

and that the minimum edge distance in the direction of the load be

$$\frac{2P}{F_u t} \qquad \text{(dimension } c \text{ in Fig. 24.9)}$$

where

P = force transmitted by one fastener to the critical connected part

F_u = specified minimum (ultimate) tensile strength of the connected part

t = thickness of the critical connected part

For our case

$$\frac{2P}{F_u t} = \frac{2(12.5)}{58(0.5)} = 0.862 \text{ in.}$$

which is considerably less than the specified edge distance listed in Table 24.2 for a $\frac{3}{4}$-in. bolt at a sheared edge: 1.25 in.

For the spacing

$$\frac{2P}{F_u t} + \frac{d}{2} = 0.862 + 0.375 = 1.237 \text{ in.}$$

which is also not critical.

A final problem that must be considered is the potential of tearing out the two bolts at the ends of the plates. Because the combined thickness of the two outer plates is greater than that of the middle plate, the critical case in this connection is that of the middle plate. Figure 24.10 shows the condition for the tearing, which involves tension on the section labeled "1" and shear on the two sections labeled "2."

For the tension section

$$\text{net width} = 3 - 0.875 = 2.125 \text{ in. [54 mm]}$$

$$F_t = 0.50F_u = 29 \text{ ksi [200 MPa]}$$

For the shear sections

$$\text{net width} = 2\left(1.25 - \frac{0.875}{2}\right) = 1.625 \text{ in. [41.3 mm]}$$

$$F_v = 0.30F_u = 17.4 \text{ ksi [120 MPa]}$$

The total resistance to tearing is thus

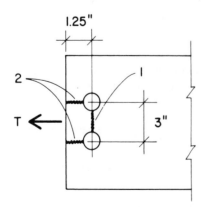

FIGURE 24.10.

$$T = 2.125(0.5)(29) + 1.625(0.5)(17.4)$$

$$= 30.8 + 14.1 = 44.9 \text{ kips [205 kN]}$$

Because this is greater than the combined load of 25 kips [111.2 kN] on the two bolts, the problem is not critical.

Connections that transfer compression between the joined parts are essentially the same with regard to the bolt stresses and bearing on the parts. Stress on the net section is less likely to be critical because the compression members will usually be designed for column action, with a considerably reduced value for the allowable compression stress.

24.3. BOLTED FRAMING CONNECTIONS

The joining of structural steel members in a structural system generates a wide variety of situations, depending on the form of the connected parts, the type of connecting device used, and the nature and magnitude of the forces that must be transferred between the members. Figure 24.11 shows a number of common connections that are used to join steel columns and beams consisting of rolled shapes.

In the joint shown in Fig. 24.11*a* a steel beam is connected to a supporting column by the simple means of resting it on top of a steel plate that is welded to the top of the column. The bolts in this case carry no computed loads if the force transfer is limited to that of the vertical end reaction of the beam. The only computed stress condition that is likely to be of concern in this situation is that of crippling the beam web (Sec. 22.6). This is a situation in which the use of unfinished bolts is indicated.

The remaining details in Fig. 24.11 illustrate situations in which the beam end reactions are transferred to the supports by attachment to the beam web. This is, in general, an appropriate form of force transfer because the vertical shear at the end of the beam is resisted primarily by the beam web. The most common form of connection is that which uses a pair of angles (Fig. 24.11*b*). The two most frequent examples of this type of connection are the joining of a steel beam to the side of a column (Fig. 24.11*b*) or to the side of another beam (Fig. 24.11*d*). A beam may also be joined to the web of a W-shape column in this manner if the column depth provides enough space for the angles.

An alternative to this type of connection is shown in Fig. 24.11*c*, where a single angle is welded to the side of a column, and the beam web is bolted to one side of the angle. This is generally acceptable only when the magnitude of the load on the beam is low because the one-sided connection experiences some torsion.

When the two intersecting beams must have their tops at the same level, the supported beam must have its top flange cut back, as shown at Fig. 24.11*e*. This is to be avoided, if possible, because it represents an additional cost in the fabrication and also reduces the shear capacity of the beam. Even worse is the situation in which the two beams have the same depth and which requires cutting

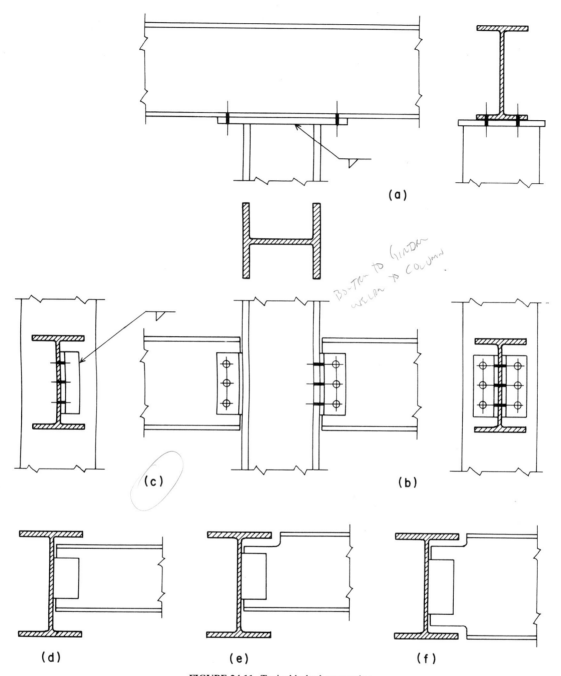

FIGURE 24.11. Typical bolted connection.

both flanges of the supported beam. When these conditions produce critical shear in the beam web it will be necessary to reinforce the beam end. The problem of tearing the beam web in these situations is discussed in Sec. 24.2.

Alignment of the tops of beams is usually done to simplify the installation of decks on top of the framing. When steel deck is used it may be possible to adopt some form of the detail shown in Fig. 24.12, which permits the beam tops to be offset by the depth of the deck ribs. Unless the flange of the supporting beam is quite thick, it will probably pro-

vide sufficient space to permit the connection shown, which does not require cutting the flange of the supported beam.

Figure 24.13 shows additional framing details that may be used in special situations. The technique described in Fig. 24.13a is sometimes used when the supported beam is shallow. The vertical load in this case is transferred through the seat angle, which may be bolted or welded to the carrying beam. The connection to the web of the supported beam merely provides additional resistance to rollover, or torsional rotation, on the part of the beam. Another reason for favoring this detail is the possibility that the seat angle

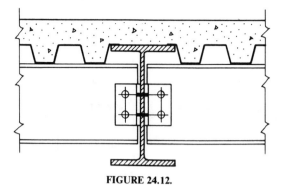

FIGURE 24.12.

may be welded in the shop and the web connection made with small unfinished bolts in the field, which greatly simplifies the field work.

Figure 24.13b shows the use of a similar connection for joining a beam and column. For heavy beam loads the seat angle may be braced with a stiffening plate. Another variation of this detail involves the use of two plates rather than the angle, which may be used if more than four bolts are required for attachment to the column.

Figure 24.13c and d show connections commonly used when pipe or tube columns carry the beams. Because the one-sided connection in Fig. 24.13c produces some torsion in the beam, the seat connection is favored when the beam load is high.

Framing connections quite commonly involve the use of welding and bolting in a single connection, as illustrated in the figures. In general, welding is favored for fabrication in the shop and bolting for erection in the field. If this practice is recognized, the connections must be developed with a view to the overall fabrication and erection process, and some decision must be made regarding what is to be done where. With the best of designs, however, the contractor who is awarded the job may have some of his own ideas about these procedures and may suggest alterations in the details.

Development of connection details is particularly critical for structures in which a great number of connections occur. The truss is one such structure. Some of the problems of truss connections are discussed in Sec. 24.4.

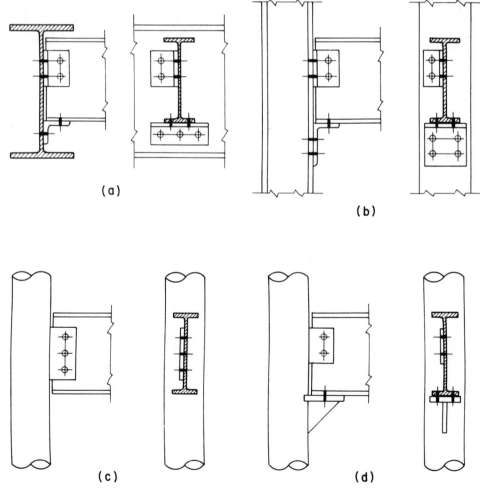

FIGURE 24.13. Bolted connections for special situations.

Framed Beam Connections

The connection shown in Fig. 24.11b is the type used most frequently in the development of structures that consist of I-shaped beams and H-shaped columns. This device is referred to as a *framed beam connection*, for which there are several design considerations:

1. *Type of Fastening.*   This may be accomplished with rivets or with any of the several types of structural bolt. The angles may also be welded in place, as described in Chapter 32. The most common practice is to weld the angles to the beam web in the fabricating shop and to bolt them to the supports in the field.
2. *Number of Fasteners.*   This refers to the number of bolts used on the beam web; there are twice this number in the outstanding legs of the angles. The capacities are matched, however, because the web bolts are in double shear, the others in single shear.
3. *Size of the Angles.*   This depends on the size of the fasteners, the magnitude of the loads, and the size of the support, if it is a column with a particular limiting dimension. Two sizes used frequently are 4 × 3

in. and 5 × 3½ in. Thickness of the angle legs is usually based on the size and type of the fastener.
4. *Length of the Angles.*   This is primarily a function of the size of the fasteners. As shown in Fig. 24.14, typical dimensions are an end distance of 1.25 in. and a pitch of 3 in. In special situations, however, smaller dimensions may be used with bolts of 1 in. or smaller diameter.

The *AISC Manual* (Ref. 5) provides considerable information to assist in the design of this type of connection in both the bolted and welded versions. A sample for bolted connections that use A325F bolts and angles of A36 steel is given in Table 24.4. The angle lengths in the table are based on the standard dimensions, as shown in Fig. 24.14. For a given beam shape the maximum size of connection (designated by the number of bolts) is limited by the dimension of the flat portion of the beam web. By referring to the tables in Appendix A we can determine this dimension for any beam designation.

Although there is no specified limit for the minimum size of a framed connection to be used with a beam, the general rule is to choose one with an angle length of at least

TABLE 24.4.   Framed Beam Connections[a]

No. of bolts n (Fig. 31.14)	Angle length L (in.)	Total shear capacity of bolts (kips)			Use with the following rolled shapes
		Bolt diameter, d (in.)			
		¾	⅞	1	
		Usual angle thickness, t (in.)			
		¼	⁵⁄₁₆	⅜	
10	29½	155	210	275	W 36
9	26½	139	189	247	W 36, 33
8	23½	124	168	220	W 36, 33, 30
7	20½	108	147	192	W 36, 33, 30, 27, 24, S 24
6	17½	92.8	126	165	W 36, 33, 30, 27, 24, 21, S 24
5	14½	77.3	105	137	W 30, 27, 24, 21, 18, S 24, 20, 18, C 18
4	11½	61.9	84.2	110	W 24, 21, 18, 16, S24, 20, 18, 15, C 18, 15
3	8½	46.4	61.9[b]	82.5	W 18, 16, 14, 12, 10, S 18, 15, 12, 10, C 18, 15, 12, 10
2	5½	30.9	39.4[b]	55.0	W 12, 10, 8, S12, 10, 8, C 12, 10, 9, 8
1	2½	15.4	21.0	27.5	W 6, 5, M 6, 5, C 7, 6, 5

Source: Adapted from data in the *Manual of Steel Construction*, 8th ed. (Ref. 5), with permission of the publishers, American Institute of Steel Construction.

[a] Connections made with A325F bolts and angles of A36 steel.
[b] Capacity limited by shear on the angles.

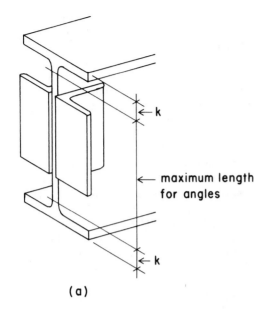

(a)

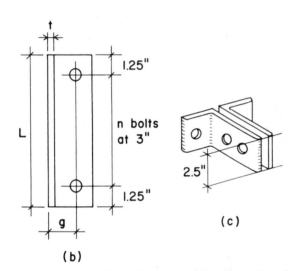

(b)

(c)

FIGURE 24.14. Ordinary framed beam connection with double angles.

one-half the beam depth. This is intended in the most part to ensure some rotational stability for the beam end.

The one-bolt connection with an angle length of only 2.5 in. (Fig. 24.14c) is the shortest. This special connection has double-gage spacing of bolts in the beam web to ensure its stability.

The following example illustrates the general design procedure for a framed beam connection. In practice, this process can be shortened because experience permits judgments that will eventually make some of the steps unnecessary. Other design aids in the *AISC Manual* (Ref. 5) will shorten the work for some computations.

Example 1. A beam consists of a W 27 × 94 of A36 steel with F_u of 58 ksi [400 MPa], which is needed to develop an end reaction of 80 kips [356 kN]. Design a standard framed

beam connection with A325F bolts and angles of A36 steel.

Solution. A scan of Table 24.4 reveals that the range of possible connections for a W 27 is $n = 5$ to $n = 7$. For the required load, possible choices are

$n = 6$, $\frac{3}{4}$-in. bolts, angle $t = \frac{1}{4}$ in., load
$$= 92.8 \text{ kips } [413 \text{ kN}]$$

$n = 5$, $\frac{7}{8}$-in. bolts, angle $t = \frac{5}{16}$ in., load
$$= 105 \text{ kips } [467 \text{ kN}]$$

Bolt size is ordinarily established for a series of framing rather than for each element. Having no other criterion, we make an arbitrary choice of the connection with $\frac{7}{8}$-in. bolts.

The bolt capacity in double shear is the primary consideration in the development of data in Table 24.4. We must make a separate investigation of the bearing on the beam web because it is not incorporated in the table data. It is actually seldom a problem except in heavily loaded beams, but the following procedure should be used:

From Table A.3 the thickness of the beam web is 0.490 in. [12.5 mm]. The total bearing capacity of the five bolts is

$$V = n(\text{bolt diameter})(\text{web } t)1.5F_u$$

$$= 5(0.875)(0.490)(87) = 186.5 \text{ kips } [832 \text{ kN}]$$

which is considerably in excess of the required load of 80 kips [356 kN].

Another concern in the typical situation is that for the shear stress through the net section of the web, reduced by the chain of bolt holes. If the connection is made as shown in Fig 24.11b or d, this section is determined as the full web width (beam depth), less the sum of the hole diameters, times the web thickness, and the allowable stress is specified as $0.40F_y$. From Table A.3 for the W 27, $d = 26.92$ in. [684 mm]. The net shear width through the bolt holes is thus

$$w = 26.92 - 5(1.0) = 21.92 \text{ in. } [557 \text{ mm}]$$

and the computed stress due to the load is

$$f_v = \frac{80}{21.92(0.490)} = 7.45 \text{ ksi } [51.4 \text{ MPa}]$$

which is less than the allowable of $0.40 \times 36 = 14.4$ ksi [100 MPa].

If the top flange of the beam is cut back to form the type of connection shown in Fig. 24.11e, a critical condition that must be investigated is that of tearing out the end portion of the beam web, as discussed in Sec. 24.4. This is also called *block shear*, which refers to the form of the failed portion (Fig. 24.15). If the angles are placed with the edge distances shown in Fig. 24.15, this failure block will have the dimensions of 14 × 2.25 in. [356 × 57 mm]. The tearing force V is resisted by a combination of tension stress on the section labeled 1 and shearing stress on the combined sections labeled 2. The allowable stresses for this situation are $0.30F_u$ for shear and $0.50F_u$ for tension. Next find the net widths of the sections, multiply them by the web thickness to obtain the areas, and multiply by the allowable stresses to obtain the total resisting forces.

For the tension resistance

$$w = 2.25 - \frac{1}{2} = 1.75 \text{ in. [44.5 mm]}$$

For the shear resistance

$$w = 14 - 4\frac{1}{2}(1) = 9.5 \text{ in. [241 mm]}$$

For the total resisting force

$$V = (\text{tension } w \times t_w)(0.50F_u)$$
$$+ (\text{shear } w \times t_w)(0.30F_u)$$
$$= 1.75(0.49)(29) + 9.5(0.49)(17.4)$$
$$= 24.9 + 81.0 = 105.9 \text{ kips [471 kN]}$$

Because this potential total resistance exceeds the load required, the tearing is not critical.

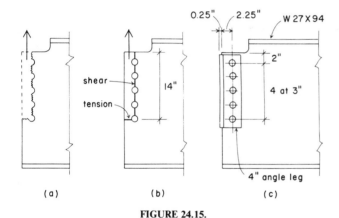

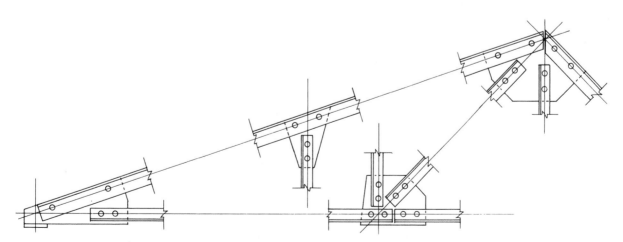

FIGURE 24.15.

24.4. BOLTED TRUSS CONNECTIONS

A major factor in the design of trusses is the development of the truss joints. Because a single truss typically has several joints, the joints must be relatively easy to produce and economical, especially if there are many trusses of a single type in the building structural system. Considerations involved in the design of connections for the joints include the truss configuration, member shapes and sizes, and the fastening method—usually welding or high-strength bolts.

In most cases the preferred method of fastening for connections made in the fabricating shop is welding. In most cases trusses will be shop-fabricated in the largest units possible, which means the whole truss for modest span trusses or the maximum-size unit that can be transported for large trusses. Bolting is mostly used for connections made at the building site. For the small truss, the only bolting is usually done for the connections to supports and to supported elements or bracing. For the large truss, bolting may also be done at splice points between shop-fabricated units. All of this is subject to many considerations relating to the nature of the rest of the building structure, the particular location of the site, and the practices of local fabricators and erectors.

Figure 24.16 shows the joint layouts for a small truss. Truss members consist of pairs of angles, and joints are formed with steel plates (called *gusset plates*). Joint connections are made by bolting the angle legs through the gusset plates. This is a form of construction that was highly developed more than 100 years ago, although the fasteners at that time were rivets instead of bolts.

Development of the joint designs for the truss shown in Fig. 24.16 would involve many considerations, including the following.

1. *Member Size and Load Magnitude.* This determines primarily the size and type of connector (bolt) required, based on individual connector capacity.

FIGURE 24.16. Typical light steel truss with bolted joints.

2. *Angle Leg Size.* This relates to the maximum diameter of bolt that can be used, based on minimum edge distances.

3. *Thickness and Profile Size of Gusset Plates.* The preference is to have the lightest weight added to the structure (primarily for the cost per pound of the steel), which is achieved by reducing the plates to a minimum thickness and general minimum size.

4. *Layout of Members at Joints.* The general attempt is to have the action lines of the forces (vested in the rows of bolts) all meet at a single point, thus avoiding twisting in the joint.

Many of the points mentioned are determined by data. Minimum edge distances for bolts (Table 24.2) can be matched to usual gage dimensions for angles (Table 24.3). Forces in members can be related to bolt capacities in Table 24.1, the general intent being to keep the number of bolts to a minimum in order to make the required size of the gusset plate smaller.

Other issues involve some judgment or skill in the manipulation of the joint details. For really tight or complex joints, it is often necessary to study the form of the joint with carefully drawn large-scale layouts. Actual dimensions and form of the member ends and the gusset plates may be derived from these drawings.

The truss shown in Fig. 24.16 has some features that are quite common for small trusses. All member ends are connected by only two bolts, the minimum required by the specifications. This simply indicates that the minimum-size bolt chosen has sufficient capacity to develop the forces in all members with only two bolts. At the top chord joint between the support and the peak, the top chord member is shown as being continuous (uncut) at the joint. This is quite common where the lengths of members available are greater than the joint-to-joint distances in the truss, a cost savings in member fabrication as well as connection.

If there are only one or a few of the trusses as shown in Fig. 24.16 to be used in a building, the fabrication may indeed be as shown in the illustration. However, if there are many such trusses, or the truss is actually a manufactured, standardized product, it is much more likely to be fabricated with joints as described previously, employing welding for shop work and bolting only for field connections. Possible forms of joints for such trusses are discussed in Sec. 24.5.

24.5. WELDED CONNECTIONS

Welding is in some instances an alternative means of making connections in a structural joint, the other principal option being structural bolts. A common situation is that of a connecting device (bearing plate, framing angles, etc.) that is welded to one member in the shop and fastened by bolting to a connecting member in the shop. However, there are also many instances of joints that are fully welded, whether done in the shop or at the site of the building construction. For some situations the use of welding may be the only reasonable means of making attachment for a joint. As in many other situations, the design of welded joints requires considerable awareness of the problems of achieving the work on the parts of the welder and the fabricator of the welded parts. This chapter presents some of the problems and potential uses of welding in building structures.

Types and Usage of Welds

One advantage of welding is that it offers the possibility for direct connection of members, often eliminating the need for intermediate devices, such as gusset plates or framing angles. Another advantage is the lack of need for holes (required for bolts), which permits development of the capacity of the unreduced cross section of tension members. Welding also offers the possibility of developing exceptionally rigid joints, an advantage in moment-resistive connections or generally nondeforming connections.

Although there are many welding processes, electric arc welding is the one most used for connections of structural steel elements. In arc welding an electric current is used to form an arc in the gap between the end of an electrode (welding rod) and the joint between two members to be connected The intense heat of the arc melts the end of the rod and portions of the two joined members, resulting in a fusion of all three when the molten materials solidify.

Figure 24.17 shows the three general classifications of joints: butt joints, tee joints, and lap joints. There are various ways of achieving these joints with different types of welds. For butt and tee joints members may be shaped to facilitate the making of the welds, especially when the connected elements are quite thick. The choice of type and size of weld depends on the form of the attached members, the magnitude of forces to be transferred, and the general structural actions of the joints and the members.

The groove weld, such as that shown in Fig. 24.17a, is usually made to develop the full capacity of the member being connected. This type of joint is used for full splicing of members, for the full development of beam flanges in moment-resistive joints, and so on. There is no specified size for such a weld; the specifications deal only with the preparation of the joint (with ends beveled, etc.) and the manner of building up of the weld. For exposed structures (in view after construction) there may also be specifications for grinding down and smoothing of the joint, as the typical weld is quite roughly formed.

The butt weld is the most effective means for achieving full member development and is usually made with a groove weld. The groove weld may also be used for a tee joint where full development of the member is critical, although the tee is also formed with fillet welds, which are described in the next section.

Fillet Welds

The weld most commonly used for structural steel in building construction is the *fillet weld*. It is approximately tri-

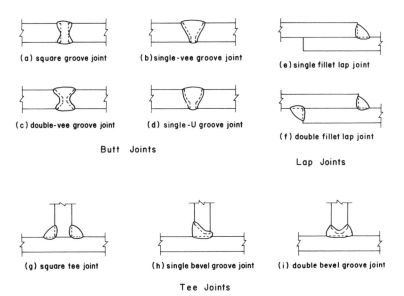

FIGURE 24.17. Typical welded joints.

angular in cross section and is formed between the two intersecting surfaces of the joined members (see Fig. 24.18a and b). The *size* of a fillet weld is the leg length of the largest inscribed isosceles right triangle, *AB* or *BC* (see Fig. 24.18a). The root of the weld is the point at the bottom of the weld, point *B* in Fig. 24.18a. The *throat* of a fillet weld is the distance from the root to the hypotenuse of the largest isosceles right triangle, which can be inscribed within the weld cross section, distance *BD* in Fig. 24.18a. The exposed surface of a weld is not the plane surface indicated in Fig. 24.18a but is usually somewhat convex, as shown in Fig. 24.18b. Therefore the actual throat may be greater than that shown in Fig. 24.18a. This additional material is called *reinforcement*. It is not included in determining the strength of a weld.

A single-vee groove weld between two members of unequal thickness is shown in Fig. 24.18c. The *size* of a butt weld is the thickness of the thinner part joined, with no allowance made for the weld reinforcement. If the dimension (size) of *AB* in Fig. 24.18a is one unit in length, $(AD)^2 + (BD)^2 = 1^2$. Because *AD* and *BD* are equal, $2(BD)^2 = 1^2$, and $BD = \sqrt{0.5}$, or 0.707. Therefore, the throat of a fillet weld

is equal to the *size* of the weld multiplied by 0.707. As an example, consider a $\frac{1}{2}$-in. fillet weld. This would be a weld with dimensions *AB* or *BC* equal to $\frac{1}{2}$ in. In accordance with the above, the throat would be 0.5×0.707, or 0.3535 in. Then, if the allowable unit shearing stress on the throat is 21 ksi, the allowable working strength of a $\frac{1}{2}$-in. fillet weld is $0.3535 \times 21 = 7.42$ kips per lineal inch of weld. If the allowable unit stress is 18 ksi, the allowable working strength is $0.3535 \times 18 = 6.36$ kips per lineal inch of weld.

The permissible unit stresses used in the preceding paragraph are for welds made with E 70 XX- and E 60 XX-type electrodes on A36 steel. Particular attention is called to the fact that the stress in a fillet weld is considered as shear on the throat, regardless of the direction of the applied load. The allowable working strengths of fillet welds of various sizes are given in Table 24.5.

The stresses allowed for the metal of the connected parts (known as the *base metal*) apply to complete penetration groove welds that are stressed in tension or compression parallel to the axis of the weld or are stressed in tension perpendicular to the effective throat. They apply also to complete or partial penetration groove welds stressed in

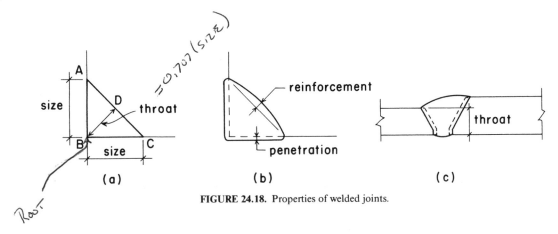

FIGURE 24.18. Properties of welded joints.

TABLE 24.5. Allowable Working Strength of Fillet Welds

Size of Weld (in.)	Allowable Load (kips/in.)		Allowable Load (kN/mm)		Size of Weld (mm)
	E 60 XX Electrodes $F_{vw} = 18$ (ksi)	E 70 XX Electrodes $F_{vw} = 21$ (ksi)	E 60 XX Electrodes $F_{vw} = 124$ (MPa)	E 70 XX Electrodes $F_{vw} = 145$ (MPa)	
$\frac{3}{16}$	2.4	2.8	0.42	0.49	4.76
$\frac{1}{4}$	3.2	3.7	0.56	0.65	6.35
$\frac{5}{16}$	4.0	4.6	0.70	0.81	7.94
$\frac{3}{8}$	4.8	5.6	0.84	0.98	9.52
$\frac{1}{2}$	6.4	7.4	1.12	1.30	12.7
$\frac{5}{8}$	8.0	9.3	1.40	1.63	15.9
$\frac{3}{4}$	9.5	11.1	1.66	1.94	19.1

compression normal to the effective throat and in shear on the effective throat. Consequently, allowable stresses for butt welds are the same as for the base metal.

The relation between the weld size and the maximum thickness of material in joints connected only by fillet welds is shown in Table 24.6. The maximum size of a fillet weld applied to the square edge of a plate or section that is $\frac{1}{16}$ in. or more in thickness should be $\frac{1}{16}$-in. less than the nominal thickness of the edge. Along edges of material less than $\frac{1}{4}$ in. thick, the maximum size may be equal to the thickness of the material.

The effective area of butt and fillet welds is considered to be the effective length of the weld multiplied by the effective throat thickness. The minimum effective length of a fillet weld should not be less than four times the weld size. For starting and stopping the arc, approximately $\frac{1}{4}$ in. should be added to the design length of fillet welds.

Figure 24.19a represents two plates connected by fillet welds. The welds marked A are longitudinal; B indicates a transverse weld. If a load is applied in the direction shown by the arrow, the stress distribution in the longitudinal weld is not uniform, and the stress in the transverse weld is approximately 30% higher per unit of length.

Added strength is given to a transverse fillet weld that terminates at the end of a member, as shown in Fig. 24.19b, if the weld is returned around the corner for a distance not less than twice the weld size. These end returns, sometimes

called *boxing*, afford considerable resistance to the tendency of tearing action on the weld.

The $\frac{1}{4}$-in. fillet weld is considered to be the minimum practical size, and a $\frac{5}{16}$-in. weld is probably the most economical size that can be obtained by one pass of the electrode. A small continuous weld is generally more economical than a larger discontinuous weld if both are made in one pass. Some specifications limit the singlepass fillet weld to $\frac{5}{16}$ in. Large fillet welds require two or more passes (multipass welds) of the electrode, as shown in Fig. 24.19c.

24.6. DESIGN OF WELDED CONNECTIONS

The appropriate weld to use for a given condition depends on several factors. It should be borne in mind that members to be connected by welding must be firmly clamped or held rigidly in position during the welding process. The designer must consider the actual conditions during erection and must provide for economy and ease in working the welds. Seat angles or similar members used to facilitate erection are *shop-welded* before the material is sent to the site. The welding done during erection is called *field-welding*.

The following examples illustrate the basic principles on which welded connections are designed.

Example 1.   A bar of A36 steel, $3 \times \frac{7}{16}$ in. [76.2 × 11 mm] in cross section, is to be welded with E 70 XX electrodes to the back of a channel so that the full tensile strength of the bar may be developed. What is the size of the weld? (See Fig. 24.20).

Solution.   The area of the bar is $3 \times 0.4375 = 1.313$ in² [76.2 × 11 = 838.2 mm²]. Because the allowable unit tensile stress of the steel is 22 ksi (Table 21.1), the tensile strength of the bar is $F_t \times A = 22 \times 1.313 = 28.9$ kips [152 × 838.2/10³ = 127 kN]. The weld must be of ample dimensions to resist a force of this magnitude.

A $\frac{3}{8}$-in. [9.52-mm] fillet weld will be used. Table 24.5 gives the allowable working strength as 5.6 kips/in. [0.98 kN/mm]. Hence the required length of weld to develop the strength of the bar is 28.9 ÷ 5.6 = 5.16 in. [127 ÷ 0.98 = 130

TABLE 24.6. Relation between Material Thickness and Minimum Size of Fillet Welds

Material Thickness of the Thicker Part Joined		Minimum Size of Fillet Weld	
in.	mm	in.	mm
To $\frac{1}{4}$ inclusive	To 6.35 inclusive	$\frac{1}{8}$	3.18
Over $\frac{1}{4}$ to $\frac{1}{2}$	Over 6.35 to 12.7	$\frac{3}{16}$	4.76
Over $\frac{1}{2}$ to $\frac{3}{4}$	Over 12.7 to 19.1	$\frac{1}{4}$	6.35
Over $\frac{3}{4}$	Over 19.1	$\frac{5}{16}$	7.94

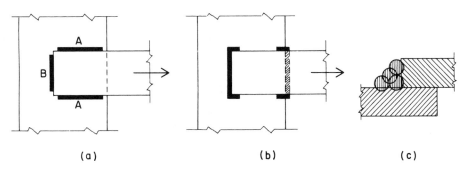

(a)   (b)   (c)

FIGURE 24.19. Welding of lapped plates.

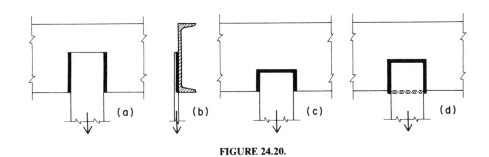

(a)   (b)   (c)   (d)

FIGURE 24.20.

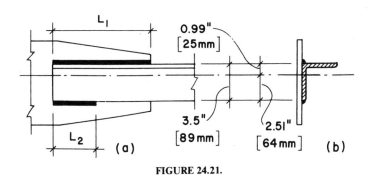

0.99"
[25 mm]

3.5"
[89 mm]

2.51"
[64 mm]

L_1 L_2 (a)   (b)

FIGURE 24.21.

D: minimum = t + $^5/_{16}$ in.   L: maximum = 10 X t_1
    maximum = 2 $^1/_4$ X t_1

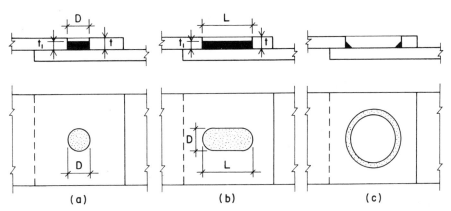

(a)   (b)   (c)

FIGURE 24.22. Welds in holes: (a) plug weld; (b) slot weld; (c) fillet weld in large hole.

mm]. The position of the weld with respect to the bar has several options, three of which are shown in Fig. 24.20*a*, *c*, and *d*.

Example 2.   A $3\frac{1}{2} \times 3\frac{1}{2} \times \frac{5}{16}$-in. [89 × 89 × 7.94-mm] angle of A36 steel subjected to a tensile load is to be connected to a plate by fillet welds, using E 70 XX electrodes. What should the dimensions of the welds be to develop the full tensile strength of the angle?

Solution.   We shall use a $\frac{1}{4}$-in. fillet weld, which has an allowable working strength of 3.7 kips/in. [0.65 kN/mm] (Table 24.5). From Table A.6 the cross-sectional area of the angle is 2.09 in² [1348 mm²]. By using the allowable tension stress of 22 ksi [152 MPa] for A36 steel (Table 21.1), the tensile strength of the angle is 22 × 2.09 = 46 kips [152 × 1348/10³ = 205 kN]. Therefore, the required total length of weld to develop the full strength of the angle is 46 ÷ 3.7 = 12.4 in. [205 ÷ 0.65 = 315 mm].

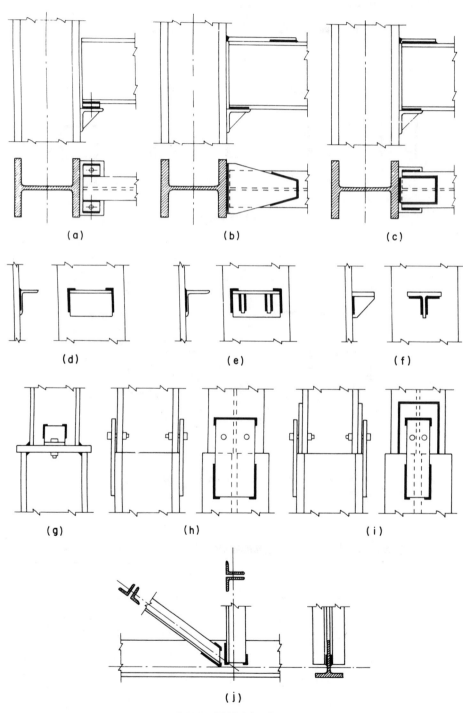

(a)   (b)   (c)

(d)   (e)   (f)

(g)   (h)   (i)

(j)

FIGURE 24.23. Welded framing connections.

An angle is an unsymmetrical cross section, and the welds marked L_1 and L_2 in Fig. 24.21 are made unequal in length so that their individual resistance will be proportioned in accordance to the distributed area of the angle. From Table A.6 we find that the centroid of the angle section is 0.99 in. [25 mm] from the back of the angle; hence the two welds are 0.99 in. [25 mm] and 2.51 in. [64 mm] from the centroidal axis, as shown in Fig. 24.5. The lengths of welds L_1 and L_2 are made inversely proportional to their distances from the axis, but the sum of their lengths is 12.4 in. [315 mm]. Therefore,

$$L_1 \;=\; \frac{2.51}{3.5}\,(12.4) \;=\; 8.9 \text{ in. } [227 \text{ mm}]$$

and

$$L_2 \;=\; \frac{0.99}{3.5}\,(12.4) \;=\; 3.5 \text{ in. } [88 \text{ mm}]$$

These are the design lengths required, and as noted earlier, each weld would actually be made $\frac{1}{4}$-in. [6.4 mm] longer than its computed length.

When angle shapes are used as tension members and connected by fastening only one leg, it is questionable to assume a stress distribution of equal magnitude on the entire cross section. Some designers therefore prefer to ignore the stress in the unconnected leg and to limit the capacity of the member in tension to the force obtained by multiplying the allowable stress by the area of the connected leg only. If this is done, it is logical to use welds of equal length on each side of the leg, as in Example 1.

Part 4 of the *AISC Manual* contains a series of tables that pertain to the design of welded connections. The tables cover free-end as well as moment-resisting connections. In addition, suggested framing details are shown for various situations.

24.7.   MISCELLANEOUS WELDED CONNECTIONS

Plug and Slot Welds

One method of connecting two overlapping plates uses a weld in a hole made in one of the two plates (see Fig. 24.22).

Plug and slot welds are those in which the entire area of the hole or slot receives weld metal. The maximum and minimum diameters of plug and slot welds and the maximum length of slot welds are shown in Fig. 24.22. If the plate containing the hole is not more than $\frac{5}{8}$-in. thick, the hole should be filled with weld metal. If the plate is more than $\frac{5}{8}$-in. thick, the weld metal should be at least one-half the thickness of the material but not less than $\frac{5}{8}$ in.

The stress in a plug or slot weld is considered to be shear on the area of the weld at the plane of contact of the two plates being connected. The allowable unit shearing stress, when E 70 XX electrodes are used, is 21 ksi [145 MPa].

A somewhat similar weld consists of a continuous fillet weld at the circumference of a hole, as shown in Fig. 24.22c. This is not a plug or slot weld and is subject to the usual requirements for fillet welds discussed in Sec. 24.5.

Miscellaneous Framing Connections

A few common connections are shown in Fig. 24.23. As an aid to erection, certain parts are welded together in the shop before being sent to the site. Connection angles may be shop-welded to beams and the angles field-welded or field-bolted to girders or columns. The beam connection in Fig. 24.23a shows a beam supported on a seat that has been shop-welded to the column. A small connection plate is shop-welded to the lower flange of the beam, and the plate is bolted to the beam seat. After the beams have been erected and the frame plumbed, the beams are field-welded to the seat angles. This type of connection provides no degree of continuity in the form of moment transfer between the beam and column.

The connections shown in Fig. 24.23b and c are designed to develop some moment transfer between the beam and its supporting column. Auxiliary plates are used to make the connection at the upper flanges.

Beam seats shop-welded to columns are shown in Fig. 24.23d–f. A short length of angle welded to the column with no stiffeners is shown in Fig. 24.23d. Stiffeners consisting of triangular plates are welded to the legs of the angles shown in Fig. 24.23e and add materially to the strength of

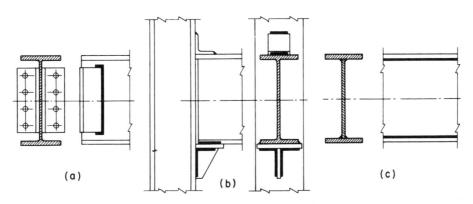

FIGURE 24.24. Additional uses of welding.

the seat. Another method of forming a seat, using a short piece of structural tee, is shown in Fig. 24.23*f*.

Various types of column splice are shown in Fig. 24.23*g–i*. The auxiliary plates and angles are shop-welded to the columns and provide for bolted connections in the field before the permanent welds are made.

Figure 24.23*j* shows a type of welded construction used in light trusses in which the lower chord consists of a structural tee. Truss web members consisting of pairs of angles are welded to the stem of the tee chord.

Some additional connection details are given in Fig. 24.24. The detail in Fig 24.24*a* is an arrangement for framing a beam to a girder, in which welds are substituted for bolts or rivets. In this figure welds replace the fasteners that secure the connection angles to the web of the supported beam.

A welded connection for a stiffened seated beam connection to a column is shown in Fig. 24.24*b*. Figure 24.24*c* shows the simplicity of welding in connecting the upper and lower flanges of a plate girder to the web plate.

Special Steel Structures

Chapter 25 deals with a number of special structures that use steel as a basic structural material.

25.1. STEEL TRUSSES

Steel is used for trusses of a great range of sizes, from 8-in.-deep prefabricated open-web joists to huge long-span bridge trusses. This chapter discusses the use of trusses for building roofs and floors of medium span.

Truss Form and Usage

Some of the commonly used forms for steel trusses are those shown in Fig. 25.1 Some of the considerations of their use are as follows.

Gable Truss. The W-form truss—popular in wood—is less used in steel, primarily because of the desire to reduce the length of the truss members in steel trusses. Because of the high stress capacity of steel, it is often possible to use very small elements for the truss members. This results in increased concern for the problems of slenderness of compression members, of span length for members subjected to bending, and of the sag of horizontal tension members. This often makes it more practical to use some pattern that

produces shorter members, such as the compound Fink shown in Fig. 25.1.

Warren Truss. The Warren is the form used for light open-web joists, in which the entire web sometimes consists of a single round steel rod with multiple bends. When a large scale is used, the Warren offers the advantage of providing a maximum of clear open space for the inclusion of building service elements that must pass through the trusses (ducts, piping, catwalks, etc.).

Flat Pratt Truss. For the parallel-chord truss, the Pratt offers the advantage of having the longest web members in tension and the shorter vertical members in compression.

Delta Truss. Another popular form of truss is the three-dimensional arrangement referred to as a delta truss. This truss derives its name from the form of its cross section—an equilateral triangle resembling the capital Greek letter delta (Δ). Where lateral bracing is not possible—or is not desired—for ordinary planar trusses, it may be possible to use the delta truss, which offers resistance to both vertical and horizontal loading. The delta form is also one used for trussed columns.

Members and Joints for Steel Trusses

For the small- to medium-size truss (up to a 200-ft span or so), truss members typically consist of single-piece rolled shapes (T, I, C, etc.), pairs of angles, round pipe or structural tubing, or cold-formed shapes of heavy-gage sheet metal. For heavily loaded or very long-span trusses, members may consist of the heavier wide-flange column shapes or of built-up sections. For economic reasons, member selection often depends on the preferences of fabricators who bid for the work.

Joints are most often achieved by welding, the much preferred method for work done in the fabricating shop. Bolting is generally used only for attachment of bracing, attachment of the truss to the supports, and for splicing of individual shop-fabricated units when the truss is too large to transport to the site in one piece.

A common form for small trusses, which was used in the past, is that shown in Fig. 24.16, employing members of pairs of angles with joints achieved with gusset plates be-

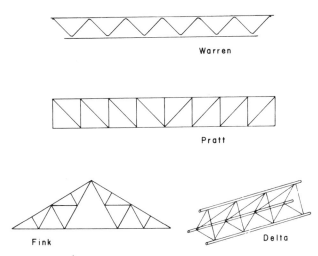

FIGURE 25.1. Common forms of light steel trusses.

tween the angles. When riveting was used extensively for shop fabrication, the joints shown in Fig. 24.16 would have been made with rivets, except possibly for any field splice joints and for attachment to bracing and supports. If this form of truss is used today, joints are more likely to be made with welds, as shown in Fig. 25.2a. However, a more popular form at present is that shown in Fig. 25.2b, in which double-angle web members are directly welded to single T-shaped chords.

When trusses are exposed to view, popular forms are those using members of round pipe, solid round rods, or structural steel tubing, as shown in Fig. 25.2c and d. The complex end cuts and welds for these trusses were once quite expensive, but now often yield to automated processes in the shop, resulting in a feasibility that was elusive in the past.

As mentioned previously, unless member and joint design is affected by building construction details or es-

thetics, it should be based on consideration of the current preferences of fabricators who will bid for the work. This situation varies as the technology changes and is also a matter of locally available facilities, as transportation may be a major factor for larger trusses.

25.2. MANUFACTURED TRUSSES

Shop-fabricated parallel-chord trusses are produced in a wide range of sizes with various fabrication details by a number of manufacturers. Most producers comply with the regulations developed by the major industry organization in this area: the Steel Joist Institute (SJI). Publications of the institute are a chief source of design information (see Ref. 11), although the products of individual manufacturers vary some, so that more information may be provided by suppliers or the manufacturers themselves.

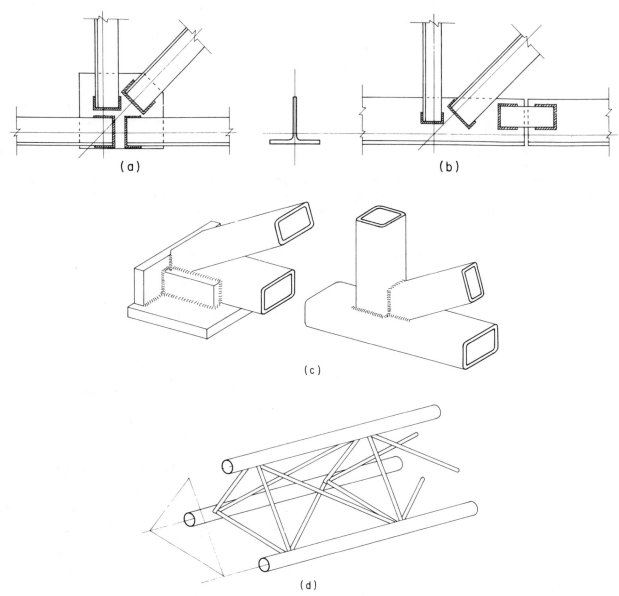

FIGURE 25.2. Members and joints for light steel trusses.

The smallest and lightest members produced, called *open-web joists*, are used for the direct support of roof and floor decks, sustaining essentially only uniformly distributed loads on their top chords. A popular form for these is that shown in Fig. 25.3, with chords of cold-formed sheet steel and webs of steel rods. Chords may also be double angles, with the rods sandwiched between the angles at joints (see also Fig. 4.2).

Table 25.1 is adapted from the standard tables of the SJI. This table lists the range of joist sizes available in the basic K series. (*Note:* A few of the heavier sizes have been omitted to shorten the table.) Joists are identified by a three-unit designation. The first number indicates the overall depth of the joist, the letter tells the series, and the second number gives the class of size of the members used; the higher the number, the heavier the joist.

Table 25.1 can be used to select the proper joist for a determined load and span condition. There are usually two entries in the table for each span; the first number represents the total load capacity of the joist, and the number in parentheses identifies the load that will produce a deflection of 1/360 of the span. The following examples illustrate the use of the table data for some typical design situations.

Note that the specific identification of what constitutes the span dimension varies with different forms of support. The common situations are indicated in Fig. 25.4, which is taken from the reference publication for Table 25.1.

Example 1. Open-web steel joists are to be used to support a roof with a unit live load of 20 psf and a unit dead load of 15 psf (not including the weight of the joists) on a span of 40 ft. Joists are spaced at 6 ft center to center. Select the lightest joist if deflection under live load is limited to 1/360 of the span.
Solution. We first determine the load per ft on the joist:

Live load:	6 × 20 =	120 lb/ft
Dead load:	6 × 15 =	90 lb/ft
Total load:	=	210 lb/ft

We then scan the entries in Table 25.1 for the joists that will just carry these loads, noting that the joist weight must be deducted from the entry for total capacity. The possible choices for this example are summarized in Table 25.2. Although the joist weights are all very close, the 24K6 is the lightest choice.

Example 2. Open-web steel joists are to be used for a floor with a unit live load of 75 psf and a unit dead load of 40 psf (not including the joists) on a span of 30 ft. Joists are 2 ft center to center, and deflection is limited to 1/360 of the span under live load only and to 1/240 of the span under

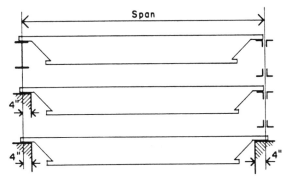

FIGURE 25.4. Definition of span for open-web steel joists, as given in Ref. 11. Reprinted by permission of the Steel Joist Institute.

total load. Determine the lightest joist possible and the joist with the least depth possible.
Solution. As in Example 1, we first find the loads on the joist:

Live load:	2 × 75 =	150 lb/ft
Dead load:	2 × 40 =	80 lb/ft
Total load:	=	230 lb/ft

To satisfy the deflection criteria, we must find a table entry in parentheses of 150 lb/ft (for live load only) or 240/360 × 230 = 153 lb/ft (for total load). The possible choices obtained from scanning Table 25.1 are summarized in Table 25.3, from which we observe

The lightest joist is the 20K4.

The shallowest joist is the 18K5.

In real situations there may be compelling reasons for selection of a deeper joist, even though its load capacity may be redundant.

For heavier loads and longer spans, trusses are produced in series described as long span and deep long span, the latter achieving depths of 7 ft and spans approaching 150 ft. In some situations the particular loading and span may clearly indicate the choice of the series, as well as the specific size of member. In many cases, however, the separate series overlap in capabilities, making the choice dependent on the product costs.

Open-web, long-span, and deep long-span trusses are all essentially designed for the uniformly loaded condition. This load may be due only to a roof or floor deck on the top, or may include a ceiling attached to the bottom. For roofs, an often used potential is that for sloping the top chord to facilitate drainage, while maintaining the flat bottom for a ceiling. Relatively small concentrated loads may be tolerated, the more so if they are applied close to the truss joints.

Load tables for all the standard products, as well as specifications for lateral bracing, support details, and so on, are available from the SJI (see Ref. 11). Planar trusses, like very thin beams, need considerable lateral support. Attached roof or floor decks and even ceilings may provide the necessary support if the attachment and the stiffness of the bracing construction are adequate. For trusses where no ceiling exists, bridging or other forms of lateral bracing must be used.

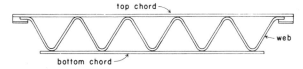

FIGURE 25.3. Open-web steel joist.

TABLE 25.1. Allowable Loads for Open-Web Steel Joists[a]

Joist Designation	12K1	12K3	12K5	14K1	14K3	14K4	14K6	16K2	16K3	16K4	16K6	18K3	18K4	18K5	18K7	20K3	20K4	20K5	20K7
Weight (lb/ft)	5.0	5.7	7.1	5.2	6.0	6.7	7.7	5.5	6.3	7.0	8.1	6.6	7.2	7.7	9.0	6.7	7.6	8.2	9.3
Span (ft)																			
20	241 (142)	302 (177)	409 (230)	284 (197)	356 (246)	428 (287)	525 (347)	368 (297)	410 (330)	493 (386)	550 (426)	463 (423)	550 (490)	550 (490)	550 (490)	517 (517)	550 (550)	550 (550)	550 (550)
22	199 (106)	249 (132)	337 (172)	234 (147)	293 (184)	353 (215)	432 (259)	303 (222)	337 (247)	406 (289)	498 (351)	382 (316)	460 (370)	518 (414)	550 (438)	426 (393)	514 (461)	550 (490)	550 (490)
24	166 (81)	208 (101)	282 (132)	196 (113)	245 (141)	295 (165)	362 (199)	254 (170)	283 (189)	340 (221)	418 (269)	320 (242)	385 (284)	434 (318)	526 (382)	357 (302)	430 (353)	485 (396)	550 (448)
26				166 (88)	209 (110)	251 (129)	308 (156)	216 (133)	240 (148)	289 (173)	355 (211)	272 (190)	328 (222)	369 (249)	448 (299)	304 (236)	366 (277)	412 (310)	500 (373)
28				143 (70)	180 (88)	216 (103)	265 (124)	186 (106)	207 (118)	249 (138)	306 (168)	234 (151)	282 (177)	318 (199)	385 (239)	261 (189)	315 (221)	355 (248)	430 (298)
30								161 (86)	180 (96)	216 (112)	266 (137)	203 (123)	245 (144)	276 (161)	335 (194)	227 (153)	274 (179)	308 (201)	374 (242)
32								142 (71)	158 (79)	190 (92)	233 (112)	178 (101)	215 (118)	242 (132)	294 (159)	199 (126)	240 (147)	271 (165)	328 (199)
36												141 (70)	169 (82)	191 (92)	232 (111)	157 (88)	189 (103)	213 (115)	259 (139)
40																127 (64)	153 (75)	172 (84)	209 (101)

Joist Designation	22K4	22K5	22K6	22K9	24K4	24K5	24K6	24K9	26K5	26K6	26K9	28K6	28K7	28K8	28K10	30K7	30K8	30K9	30K12
Weight (lb/ft)	8.0	8.8	9.2	11.3	8.4	9.3	9.7	12.0	9.8	10.6	12.2	11.4	11.8	12.7	14.3	12.3	13.2	13.4	17.6
Span (ft)																			
28	348 (270)	392 (302)	427 (328)	550 (413)	381 (323)	429 (362)	467 (393)	550 (456)	466 (427)	508 (464)	550 (501)	548 (541)	550 (543)	550 (543)	550 (543)				
30	302 (219)	341 (245)	371 (266)	497 (349)	331 (262)	373 (293)	406 (319)	544 (419)	405 (346)	441 (377)	550 (459)	477 (439)	531 (486)	550 (500)	550 (500)	550 (543)	550 (543)	550 (543)	550 (543)
32	265 (180)	299 (201)	326 (219)	436 (287)	290 (215)	327 (241)	357 (262)	478 (344)	356 (285)	387 (309)	519 (407)	418 (361)	466 (400)	515 (438)	549 (463)	501 (461)	549 (500)	549 (500)	549 (500)
36	209 (126)	236 (141)	257 (153)	344 (201)	229 (150)	258 (169)	281 (183)	377 (241)	280 (199)	305 (216)	409 (284)	330 (252)	367 (280)	406 (306)	487 (366)	395 (323)	436 (353)	475 (383)	487 (392)
40	169 (91)	190 (102)	207 (111)	278 (146)	185 (109)	208 (122)	227 (133)	304 (175)	227 (145)	247 (157)	331 (207)	266 (183)	297 (203)	328 (222)	424 (284)	319 (234)	353 (256)	384 (278)	438 (315)
44	139 (68)	157 (76)	171 (83)	229 (109)	153 (82)	172 (92)	187 (100)	251 (131)	187 (108)	204 (118)	273 (155)	220 (137)	245 (152)	271 (167)	350 (212)	263 (176)	291 (192)	317 (208)	398 (258)
48					128 (63)	144 (70)	157 (77)	211 (101)	157 (83)	171 (90)	229 (119)	184 (105)	206 (117)	227 (128)	294 (163)	221 (135)	244 (148)	266 (160)	365 (216)
52									133 (65)	145 (71)	195 (93)	157 (83)	175 (92)	193 (100)	250 (128)	188 (106)	208 (116)	226 (126)	336 (184)
56												135 (66)	151 (73)	166 (80)	215 (102)	162 (84)	179 (92)	195 (100)	301 (153)
60																141 (69)	156 (75)	169 (81)	262 (124)

Source: Data adapted from more extensive tables in the *Standard Specifications, Load Tables, and Weight Tables for Steel Joists and Joist Girders*, 1986 ed. (Ref. 11), with permission of the publishers, Steel Joist Institute. The Steel Joist Institute publishes both specifications and load tables; each of these contain standards that are to be used in conjunction with one another.

[a] Loads in pounds per foot of joist span; first entry represents the total joist capacity; entry in parentheses is the load that produces a maximum deflection of 1/360 of the span.

For development of a complete truss system, a special type of truss available is that described as a *joist girder*. This truss is specifically designed to carry the spaced, concentrated loads consisting of the end supports of joists. The general form of joist girders is shown in Fig. 25.5. Also shown in Fig. 25.5 is the form of standard designation for a joist girder, which includes indications of the nominal girder depth, weight, spacing of members, and the unit load to be carried.

Predesigned joist girders may be selected from the catalogs of manufacturers in a manner similar to that for open-web joists. The procedure is usually as follows:

1. The designer first determines the desired spacing of joists, loads to be carried, and span of the girder. The joist spacing must be an even-number division nof the girder span.
2. The total design load for one supported joist is determined. This is the unit concentrated load on the girder.
3. With the girder span, joist spacing, and joist load, the weight and depth for the girder is found from design tables in the manufacturer's catalog. Specification of the girder is as shown in Fig. 25.5.

An illustration of the use of a complete truss system using open-web joists and joist girders is shown in the design example in Chapter 57.

25.3. BRACING FOR TRUSSES

Planar trusses, like very thin beams, need considerable lateral support. Attached roof or floor decks and even ceilings may provide the necessary support if the attachment and the stiffness of the bracing construction are adequate. For trusses where no ceiling exists, bridging or other forms of lateral bracing must be used. General means for achieving bracing are discussed in Sec. 20.1 for wood trusses; the basic techniques are the same for steel trusses.

For trusses exposed to view, it may be possible to use the delta form or other type of truss that is in effect self-bracing (see Figs. 25.1 and 25.2d). In some cases such a truss may not only brace itself, but provide some of the lateral bracing for the building in general.

Development of bracing must be carefully studied in many cases. This is a three-dimensional problem, involving not only the stability of the trusses, but the general development of the building construction. Elements used for bracing can often do service as supports for ducts, lighting, building equipment, catwalks, and so on.

25.4. TRUSSED BRACING FOR FRAMES

Single steel diagonals, X-braces, or knee braces may be used to provide stability for frames of wood or steel. The general use of such bracing is discussed in Part 8 under the topic "braced frames," which is the building code term for the lateral bracing achieved with trussing. In general, the stability produced by the two-dimensional (planar) triangle or the three-dimensional tetrahedron can be used extensively as a bracing device. This potential can be observed in the designs of bicycle frames, transmission towers, bridges, freeway sign supports, and so on. Very large structural members of quite light weight can be produced as individually trussed elements: for example, the legs of the Eiffel Tower and the chords of large bridges.

For the open, exposed-to-view structure, trussing offers the possibilities for lightness in a literal, weight-reduction sense, as well as a visual sense. It also, however, has the potential for keeping entire armies of painters permanently employed to keep the exposed structure from rusting away.

For very tall columns, the trussed form offers the possibility of a relatively light, stiff element, in comparison to standard solutions with single-piece rolled shapes or heavy, built-up members.

TABLE 25.2. Possible Choices for the Roof Joist

Load Condition	Load per Foot for the Indicated Joists			
	22K9	24K6	26K5	28K6
Total capacity (from Table 25.1)	278	227	227	266
Joist weight (from Table 25.1)	11.3	9.7	9.8	11.4
Net usable capacity	266.7	217.3	217.2	254.6
Load for 1/360 deflection (from Table 25.1)	146	133	145	183

TABLE 25.3. Possible Choices for the Floor Joist

Load Condition	Load per Foot for the Joists Indicated		
	18K5	20K4	22K4
Total capacity (from Table 25.1)	276	274	302
Joist weight (from Table 25.1)	7.7	7.6	8.0
Net usable capacity	268.3	266.4	294
Load for 1/360 deflection (from Table 25.1)	161	179	219

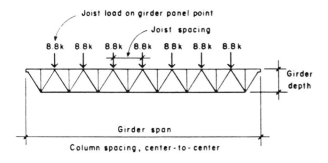

Specify: 48G8N8.8K

FIGURE 25.5. Considerations for joist girders, From Ref. 11.

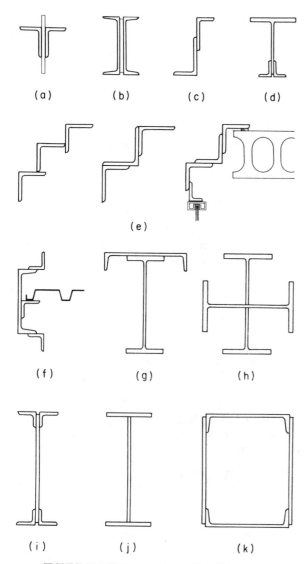

FIGURE 25.6. Built-up sections with rolled elements.

25.5. BUILT-UP MEMBERS

Primary structural elements of steel most often consist of single-piece rolled shapes. The variety of shapes and the range of size variations available generally reflects the extent of usage demand. If usage trends dictate the need for some particular shape or a size variation, the steel industry responds by adding the new items to its inventory of standard products. Each new edition of the *AISC Manual* generally contains some new shapes and drops some old ones, reflecting usage as well as refinement based on new developments in materials and structural theory. However, it is also possible to make virtually endless combinations of the standard items into built-up members.

Built-up members may be developed in response to some particular structural task, to facilitate some details of the building construction, or just as a sculptural exercise. Figure 25.6 shows a number of built-up sections that may

be used for any of the foregoing reasons. One of the most common built-up members is the double angle (Fig. 25.6a), used in this form to facilitate connections as well as to improve its bending, column action, and torsional resistance. In some situations the double channel (Fig. 25.6b) is used for purposes similar to those of the double angle. Another use of the double channel is where some vertical penetration of the structure must be made (to accommodate piping or wiring, for example) on the column-beam line; the channels being spread the necessary distance, possibly straddling the column.

Figure 25.6c shows the use of two angles to produce a Z-shape. Although one piece Z-shapes are among the standard rolled products, a wider range of variations of dimensions can be obtained by the two-angle combination. Similarly, a customized I-shape can be obtained by using a T-shape in combination with a pair of angles (Fig. 25.6d) or a plate.

In exposed steel structures it is common to have some customized built-up sections to facilitate various details of the building construction. Figure 25.6e shows a building exterior corner developed with a combination of steel shapes. The corner member may be a part of the steel structure, or may simply be used to facilitate the construction. Figure 25.6f shows a similar situation, in which a roof edge is developed with an exposed channel section, in this case utilizing the channel to support the edge of the roof deck.

Members such as those shown in Fig. 25.6g and h may be used for decorative reasons, but offer certain enhancements in the form of structural properties that may be significant. The section in Fig. 25.6g is commonly used for a crane rail support in mill buildings, the channel being used to absorb the horizontal thrust caused by the braking action of the moving crane. The star-shaped column section in Fig. 25.6h may be cute, but is also a means for giving two-way bending or biaxial stiffness enhancement to the column; however, it does not much improve torsional resistance, if that is a concern.

When a beam is required with a depth greater than that of the largest rolled shape, a common response is the so-called plate girder, two typical forms being those shown in Fig. 25.6i and j. When riveting was the predominant means of attaching the parts in built-up members, the form shown in Fig. 25.6i was widely used. Now, with welding the major means of attachment in shop fabrication, the form shown in Fig. 25.6j is more typical.

A column section used often in the past is that shown in Fig. 25.6k, consisting of two channels spaced to form a square and connected by other elements. To form a closed section, the connecting elements may be plates, in which case the structural properties of all elements would contribute to the composite properties of the built-up section. However, the connecting elements may also be bars or short pieces of plate; resulting in an open, laced form, with the two channels alone forming the structural member. While the laced form may still be made with channel sections, the closed form is more likely to be made with plates if the plates are of sufficient thickness to facilitate the corner welds.

There is virtually no limit to the form or the size of built-up members. The columns of high-rise buildings, chord members of large trusses, girder sections for long-span structures, and legs of towers have been made in this manner, producing some really gigantic sections.

25.6. COLD-FORMED STEEL PRODUCTS

Steel elements formed by the rolling process (angles, wide-flange shapes, etc.) must be heat-softened, whereas those formed from sheet steel are ordinarily made by bending or punching the cold steel; thus the common description of these products is *cold-formed.* Because they are usually formed from relatively thin sheet steel, they are also referred to as *light-gage* steel products. The design of cold-formed structural elements is described in the *Cold-Formed Steel Design Manual*, published by the American Iron and Steel Institute.

Figure 25.7 shows the typical cross sections of some ordinary structural products formed of sheet steel. The corrugated or fluted deck units shown in Fig. 25.7a are widely used for roof and floor decks. They are also used on occasion to develop wall panels, forming the surfaces of a sandwich-type unit with an insulative core. Use of these elements for structural decks is described in Sec. 22.9.

Cold-formed structural shapes range from the simple L, C, U, and so on (Fig. 25.7b), to the various special forms produced for particular purposes. Sections such as those shown in Fig. 25.7c are commonly used to form door and window frames, combining trim and structural functions in a single piece. The section in Fig. 25.7d is typical of that produced as a built-up member; in this case overcoming the limitations of the bent sheet element, which cannot easily produce a single-piece I-shape.

Elements such as those shown in Fig. 25.7b and d are widely used to develop framed wall and ceiling constructions where fire regulations prevent the use of combustible materials (e.g., wood). There is thus a need for a steel equivalent of the ubiquitous 2 × 4, and the necessary parts to form the usual sills, plates, lintels, furring, and so on.

The structures of modest size buildings may be almost completely formed with sheet metal products. Indeed, several manufacturers produce patented kits of such products for the formation of predesigned, packaged buildings. However, the widest use of cold-formed products is for direct use for utilitarian functions in ordinary buildings, as described in this section.

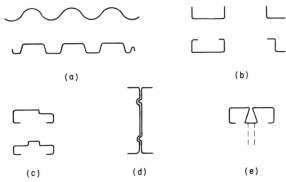

FIGURE 25.7. Cold-formed sheet steel elements.

PART 5

CONCRETE STRUCTURES

The process of generating a solid substance by using a binder to adhere a mass of loose material is applied in various ways to produce filled plastics, asphalt pavement, particleboard, and plaster, as well as what we generally call concrete. Although the term has broader generic meaning, we usually apply the word *concrete* to the material that is produced in rocklike form with a binder of water and portland cement and a loose filler consisting of sand and gravel. Forms of concrete, made with natural binders, were used by many ancient builders, but modern concrete, as we use it today, dates primarily from the development of calcined portland cement in the early part of the nineteenth century. The potential for this highly improved material was at first not fully recognized, and concrete continued to be used mostly in the old ways—for crude, filler functions in massive construction until late in the nineteenth century, when various builders and designers began to experiment with new ways to utilize the truly new material. Basic systems and construction techniques developed in a few decades around the beginning of the twentieth century continue largely unchanged in form as major uses for building structures.

CHAPTER TWENTY-SIX

General Concerns for Concrete

Concrete itself is a somewhat complex material, and its use involves many concerns, such as those for forming, reinforcing, finishing, and curing of the cast material. Chapter 26 deals with some of the critical concerns for the material and its production in the form of a structure. Considerations of its structural functions and the process of design must be built on some understanding of these general concerns. As compared to structures of wood, steel, or masonry, concrete structures both offer a higher degree of freedom of variability and require a greater responsibility on the parts of the designer and builder in terms of control of the finished product.

26.1 USAGE CONSIDERATIONS

Most of the concrete produced does not go into buildings but rather, into pavements, bridges, dams, retaining walls, waterways, and other types of structures. Furthermore, most of the concrete used for buildings goes into foundations, basement walls and grade beams, and grade-level floor paving slabs. Almost every building utilizes these elements, while only a relatively few have a structure aboveground made of concrete. This is said only so that it may be appreciated that the concrete industry is not oriented principally to the production of buildings.

On the other hand, concrete does lend itself to the possible production of all the basic building components: foundations, floors, roofs, walls, and frames, as well as a great range of the various types of structural systems, including arches, domes, shells, trusses, and space frames. It is also generally the most inert and durable construction material, resisting aging, weather effects, rot, insects, fire, and chemical change or decomposition. Given the right circumstances, it is a very usable material.

Most concrete is produced by pouring the semifluid mixed material into a hole or a forming mold at the building site. For aboveground construction, forming costs often exceed those of the basic material itself. This and other factors have led to the increasing use of *precast* construction, in which a cast unit is formed elsewhere and then transported and erected into its final position. This operation has major implications in terms of the character of the structure, making it more like those of steel or wood than that of the typical sitecast construction. Design of precast structures is discussed in Chapter 34.

A major structural limitation for concrete is its low resistance to tension. For major structural tasks that involve bending and shear, it is necessary to overcome this limitation by using either inert reinforcing or some method of prestressing. Proper design of the reinforcing or prestressing is often a major part of the design effort for concrete structures. The objective in design of reinforced concrete is to place the tension-strong steel rods in locations where tension must be developed, generally ignoring the tensile capacity of the concrete. However, deformations under loading will result in tension in the concrete as well as the steel; thus cracking is an inevitable consequence in ordinary reinforced concrete structures. Reduction of tension cracking is a major plus for prestressing, with which the effort is made to eliminate tension stress by producing canceling compressive stresses (the prestress). Most of the work in Part 5 deals with structures of reinforced concrete, which is the method still most widely used. Some considerations in design of prestressed structures are presented in Chapter 34.

Producing the finished concrete structure as a sitecast object involves attention to several concerns. If the structure is to be exposed to view, its appearance will be vitally dependent on careful consideration of these matters. Even for the covered structure, however, the integrity of the finished construction may be critical to structural behavior. Principal factors that must be considered are the following.

1. *Forming.* The shapeless, fluid-mixed concrete must be poured into a mold (called the *forms*) of the desired shape and held until it is adequately hardened. For footings and paving slabs the mold is mostly the dug and leveled ground surfaces. For walls, columns, beams, and spanning slabs, the molds must be built and removed after the concrete is hardened. Built forms must be reasonably watertight and strong enough to support the wet concrete without bulging or sagging. It is not unusual for forming costs to exceed the cost of the concrete material itself, especially when the forms must be hand built, used once, and discarded. Use of reusable forms, repetitive shapes, and precast units are means of overcoming forming cost.

2. *Installation of Reinforcing.* Steel reinforcing must be placed in the forms and held firmly in place—

literally in midair—so that the concrete can be poured around the bars. Detailing of the reinforcing must be done with the placement and pouring operations in mind, as well as the structural functions required.

3. *Pouring.* Getting the concrete into the forms and working it into place is a challenge when the forms are intricate and filled with reinforcing and the various accessories required to hold the reinforcing and brace the forms. Distances between bars (called *spacing*) and distances between the bars and the face of the forms (called *cover*) must be controlled to allow pouring as well as to satisfy structural or fire protection considerations.

4. *Finishing (Wet).* The top surface of the concrete in a single pour must be finished. This may consist essentially of simply leveling it to a desired flat surface. For floor slabs the surface may be made quite smooth if it constitutes the finished floor, or deliberately roughened if bonding to a filler of some kind is required.

5. *Finishing (Dry).* Surfaces may be reworked after the concrete is hardened. If this is a reduction type of working (chipping, sand-blasting, etc.) the surface dimension lost should be considered in establishing the required cover for the reinforcing.

6. *Curing.* Ordinary concrete hardens within a few hours after being mixed, but does not attain significant strength for several days or weeks. Various means can be used to slow down or speed up these actions, depending on the circumstances. What is critical is that the concrete be kept moist and within some range of temperature during the curing period. If allowed to dry out, to freeze, or to heat excessively from its own chemical reactions or the sun, it may become damaged or at the least not attain its potential of quality.

Much of what must be done to achieve good concrete falls on the suppliers and builders and the craft capabilities of the workers. However, the designer must exercise some decision-making and specifying actions and generally set up a situation that favors good construction. Choice of shapes and finishes and detailing of the construction can best be done by someone who is sensitive to the realities of the construction work.

Ingredients

Figure 26.1 shows the composition of ordinary concrete. The binder consists of the water and cement, whose chemical reaction results in the hardening of the mass. The binder is mixed with some aggregate (loose, inert particles) so that the binder coats the surfaces and fills the voids between the particles of the aggregate. For materials such as grout, plaster, and stucco, the aggregate consists of sand of reasonably fine grain size. For concrete the grain size is extended into the category of gravel, with the maximum particle size limited only by the size of the structure.

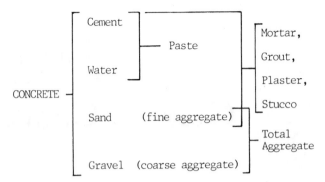
FIGURE 26.1. Composition of ordinary concrete.

Cement

Cement is commonly calcined portland cement, industrially produced in a dry powdery form and bagged in units of 1 ft³ (94 lb per bag). The two most used types are type 1, which produces structural concrete in approximately a month of ordinary curing, and type 3, which cures in about a week. Special cements can be obtained for slow curing (to retard heat buildup in large masses), to reduce shrinkage, or to control color.

Water

Water must be reasonably clean, free of oil, organic matter, and any substances that may affect the actions of hardening, curing, or general finish quality of the concrete. In general, drinking-quality (potable) water is usually adequate. Salt-bearing seawater may be used for plain concrete (without reinforcing), but may cause corrosion of steel bars in reinforced concrete.

Aggregate

The most common aggregates are sand, crushed stone, and pebbles. Particles smaller than $\frac{3}{8}$ in. in diameter constitute the *fine aggregate*. There should be only a very small amount of very fine materials, to allow for the free flow of the water—cement mixture between the aggregate particles. Material larger than $\frac{3}{8}$ in. is called the *coarse aggregate*. The maximum size of aggregate particle is limited by specification, based on the thickness of poured elements, spacing and cover of the reinforcing, and some consideration of finishing methods. In general, the aggregate should be well graded, with some portion of large to small particles over a range to permit the smaller particles to fill the spaces between the larger ones. The volume of the concrete is thus virtually determined by the volume of the total aggregate, the water and cement going into the spaces remaining between the smallest aggregate particles. The weight of the concrete is determined largely by the weight of the coarse aggregate. Strength is also dependent to some degree on the integrity of the large aggregate particles.

Admixtures

Substances added to concrete to improve its workability, accelerate its set, harden its surface, and increase its water-proof qualities are known as *admixtures*. The term embraces all materials other than the cement, water, and aggregates that are added just before or during mixing. Many of the proprietary compounds contain hydrated lime, calcium chloride, and kaolin. Calcium chloride is the most commonly used admixture for accelerating the set of concrete, but corrosion of steel reinforcement may be the consequence of its excessive use. Caution should be exercised in the use of admixtures, especially those of unknown composition.

Air-entrained concrete is produced by using an air-entraining portland cement (ASTM C175) or by introducing an air-entraining admixture as the concrete is mixed. In addition to improving workability, entrained air permits lower water–cement ratios and significantly improves the durability of hardened concrete. Air-entraining agents produce billions of microscopic air cells per cubic foot; they are distributed uniformly throughout the mass. These minute voids prevent the accumulation of water, which, on freezing, would expand and result in spalling of the exposed surface under frost action.

Structural Properties

The primary index of strength of concrete is the specified compressive strength, designated f_c'. This is the unit compressive stress used for structural design and for a target for the mix design. It is usually given in units of psi, and it is common to refer to the structural quality of the concrete simply by calling it by this number: 3000-lb concrete, for example. For strength design, this value is used to represent the ultimate compressive strength of the concrete. For working stress design, allowable maximum stresses are based on this limit; specified as some fraction of f_c'.

Table 26.1 is reproduced from the 1963 *ACI Code* and indicates the various allowable stresses used in the working stress method in that code. The 1983 *ACI Code* (Ref. 6) contains vestiges of these standards in the alternate design method described in Appendix B. Use of these references is explained in the various portions of this part.

The value for the modulus of elasticity of concrete is established by a formula that incorporates variables of the weight (density) of the concrete and its strength. Distribution of stresses and strains in reinforced concrete is dependent on the concrete modulus, the steel modulus being a constant. This is discussed in Chapter 27.

When subjected to long-duration stress at a high level, concrete has a tendency to *creep*, a phenomenon in which strain increases over time under constant stress. This has effects on deflections and on the distributions of stresses between the concrete and reinforcing. Some of the implications of this for design are discussed in the chapters dealing with design of beams and columns.

Hardness of concrete refers essentially to its surface density. This is dependent primarily on the basic strength, as indicated by the value for compressive stress. However, surfaces may be somewhat softer than the central mass of concrete, owing to early drying at the surface. Some techniques are used to deliberately harden surfaces, especially those of the tops of slabs. Fine troweling will tend to draw a very cement-rich material to the surface, resulting in an enhanced density. Chemical hardeners can also be used, as well as sealing compounds that trap surface water.

Nonstructural Properties

In addition to the basic structural properties, there are various properties of concrete that bear on its use as a construction material and in some cases on its structural integrity.

Workability. This term generally refers to the ability of the wet mixed concrete to be handled, placed in the forms, and finished while still fluid. A certain degree of workability is

TABLE 26.1. Allowable Stresses in Concrete

Description		For any strength of concrete in accordance with Section 502	For strength of concrete shown below			
			$f_c' = 2500$ psi	$f_c' = 3000$ psi	$f_c' = 4000$ psi	$f_c' = 5000$ psi
Modulus of elasticity ratio: n		29,000,000 $w^{1.5} 33\sqrt{f_c'}$				
For concrete weighing 145 lb per cu ft (see Section 1102)	n		10	9	8	7
Flexure: f_c						
Extreme fiber stress in compression	f_c	$0.45 f_c'$	1125	1350	1800	2250
Extreme fiber stress in tension in plain concrete footings and walls	f_t	$1.6\sqrt{f_c'}$	80	88	102	113
Shear: v (as a measure of diagonal tension at a distance d from the face of the support)						
Beams with no web reinforcement*	v_c	$1.1\sqrt{f_c'}$	55*	60*	70*	78*
Joists with no web reinforcement	v_c	$1.2\sqrt{f_c'}$	61	66	77	86
Members with vertical or inclined web reinforcement or properly combined bent bars and vertical stirrups	v	$5\sqrt{f_c'}$	250	274	316	354
Slabs and footings (peripheral shear, Section 1207)*	v_c	$2\sqrt{f_c'}$	100*	110*	126*	141*
Bearing: f_c						
On full area			625	750	1000	1250
On one-third area or less†		$0.25 f_c'$ $0.375 f_c'$	938	1125	1500	1875

*For shear values for lightweight aggregate concrete see Section 1208.
†This increase shall be permitted only when the least distance between the edges of the loaded and unloaded areas is a minimum of one-fourth of the parallel side dimension of the loaded area. The allowable bearing stress on a reasonably concentric area greater than one-third but less than the full area shall be interpolated between the values given.

Source: Table 1002(a) from *Building Code Requirements for Reinforced Concrete* ACI 318-63; reproduced with permission of the publishers, American Concrete Institute.

essential to the proper forming and finishing of the material. However, the fluid nature of the mix is largely determined by the amount of water present, and the easiest way to make it more workable is to add water. Up to a point this may be acceptable, but the extra water usually means less strength, greater porosity, and more shrinkage; all generally undesirable properties. Use of vibration, admixtures, and other techniques to facilitate handling without increasing the water content are often used to obtain the best-quality concrete.

Watertightness.   It is usually desirable to have a generally nonporous concrete. This may be quite essential for walls or for floors consisting of paving slabs, but is good in general for protection of reinforcing from corrosion. Watertightness is obtained by having a well-mixed, high-quality concrete (low water content, etc.), that is worked well into the forms and has dense surfaces with little cracking or voids. Subject to the continuous presence of water, however, concrete is absorptive and will become saturated. Moisture or waterproof barriers must be used where water penetration must be positively prevented.

Density.   Concrete unit weight is essentially determined by the density of the coarse aggregate (ordinarily two-thirds or more of the total volume) and the amount of air in the mass of the finished concrete. With ordinary gravel aggregate and air limited to not more than 4% of the total volume, air dry concrete weighs around 145 lb/ft^3. Use of strong but lightweight aggregates can result in weight reduction to close to 100 lb/ft^3 with strengths generally competitive with that obtained with gravel. Lower densities are achieved by entraining air up to 20% of the volume and using very light aggregates, but strength and other properties are quickly reduced.

Fire Resistance.   Concrete is noncombustible and its insulative, fire protection character is used to protect the steel reinforcing. However, under long exposure to fire, popping and cracking of the material will occur, resulting in actual structural collapse or a diminished capacity that requires replacement or repair after a fire. Design for fire resistance involves the following basic concerns:

1. *Thickness of Parts.*   Thin slabs or walls will crack quickly, permitting penetration of fire or gases.
2. *Cover of Reinforcing.*   More is required for higher fire rating of the construction.
3. *Character of the Aggregate.*   Some are more vulnerable than others to fire actions.

Design specifications and building code regulations deal with these issues, some of which are discussed in the development of the building design illustrations in Part 9.

Shrinkage.   Water-mixed materials, such as plaster, mortar, and concrete, tend to shrink in volume during the hardening process. For ordinary concrete, the shrinkage averages about 2% of the volume. The actual dimensional change of structural members is usually less, due to the presence of the steel bars; however, some consideration must be given to the shrinkage effects. Stresses caused by shrinkage are in some ways similar to those caused by thermal change; the combination resulting in specifications for minimum two-way reinforcing in walls and slabs. For the structure in general, shrinkage is usually dealt with by limiting the size of individual pours of concrete, as the major shrinkage ordinarily occurs quite rapidly in the fresh concrete. For special situations it is possible to modify the concrete with admixtures or special cements that cause a slight expansion to compensate for the normal shrinkage.

26.2. CONTROL OF CONCRETE PROPERTIES AND FINISHED QUALITY

From a structural design point of view, mix design basically means dealing with the considerations involved in achieving a particular design strength, as measured by the value of the fundamental property: f_c'. For this consideration alone, principal factors are the following:

1. *Cement Content.* The amount of cement per unit volume is a major factor determining the richness of the cement–water paste and its ability to fully coat all of the aggregate particles and fill the voids between them. Cement content is normally measured in terms of the number of sacks of cement (1 ft^3) per cubic yard of concrete mixed. The average for structural concrete is about 5 sacks per yd. If tests show that the mix exceeds or falls short of the desired results, the cement content is increased or decreased. The cement is by far the costliest ingredient, so its volume is critical in cost control.
2. *Water–Cement Ratio.*   This is expressed in terms of gallons of water per sack of cement or gallons of water per yard of mixed concrete. The latter is usually held very close to an average of about 35 gal/yd; less and workability is questionable; more and strength becomes difficult to obtain. Attaining of very strong concrete usually means employing various means to reduce water content and improve the ratio of cement to water without losing workability.
3. *Fineness Modulus of the Sand.*   If too coarse, the wet mix will be grainy and finishing of surfaces will be difficult. If too fine, an excess of water will be required, resulting in high shrinkage and loss of strength. Grain size is controlled by specification.
4. *Character of the Coarse Aggregate.*   Shape, size limits, and type of material must be considered. Since this represents the major portion of the concrete volume, its properties are quite important to strength, weight, fire performance, and so on.

Concrete is obtained primarily from plants that mix the materials and deliver them in mixer trucks. The mix design is developed cooperatively with the structural designer and the management of the mixing plant. Local materials must be used, and experience with them is an important consideration.

Testing

Various types of tests are performed on concrete. The two most common are the slump test for wetness and the compression test on cured samples. Both of these tests are forms of quality control, one being made on the wet concrete mix and the other on the finished, hardened material.

The slump test is a means of determining the consistency of the mix, generally indicating the amount of water in the mix and the degree of mixing, and is a reasonably reliable prediction of its potential for workability in the pouring and placing processes. It is performed at the building site on samples taken from the mixed concrete before it is placed in the forms. The desired degree of workability depends in some degree on the type of construction, being different for large, massive elements, such as piers and footings, and for small, intricate forms with numerous reinforcing.

The equipment for making a slump test consists of a sheet metal truncated cone 12 in. high with a base diameter of 8 in. and a top diameter of 4 in. Both top and bottom are open. Handles are attached to the outside of the mold. When a test is made, freshly mixed concrete is placed in the mold in a stipulated number of layers and each is rodded separately a specified number of times with a steel rod. When the mold is filled, the top is leveled off and the mold lifted at once. The slumping action of the concrete is measured by taking the difference in height between the top of the mold and the top of the slumped mass of concrete (Fig. 26.2).

If the concrete settles 3 in., we say that the particular sample has a 3-in. slump. This measured value is compared with some preestablished range of permitted slumps appropriate to the situation. Table 26.2 gives a sample of recommended slump values for various types of construction. In the table values in the example is a consideration for the use of vibration to assist in the consolidation of the

TABLE 26.2. Recommended Slumps for Various Types

Type of Construction	Slump (in.)	
	Maximum[a]	Minimum
Reinforced foundation walls and footings	3	1
Plain footings, caissons, and substructure walls	3	1
Beams and reinforced walls	4	1
Building columns	4	1
Pavements and slabs	3	1
Mass concrete	2	1

Source: Data abstracted from *Recommended Practice for Selecting Proportions for Normal and Heavy Weight Concrete* (ACI 211.1-74) with permission of the American Concrete Institute.

[a] May be increased 1 in. for methods of consolidation other than vibration.

wet concrete in the forms. Adjustments of various kinds may be made to establish desired values for slump, including consideration of the desired strength (f_c'), use of admixtures that affect workability, use or ommission of vibration, and so on.

Compression tests are performed for the primary reason of establishing that the finished concrete has attained an ultimate resistance to compression within an acceptable range of the value assumed (the specified f_c') in the design work. Tests of compressive strength are made at periods of 7 and 28 days on specimens prepared and cured in accordance with prescribed ASTM testing procedures. The specimen to be tested is cylindrical in shape and has a height twice its diameter. The standard cylinder is 6 in. in diameter and 12 in. high when the maximum size of the coarse aggregate does not exceed 2 in. For larger aggregates, the cylinder should have a diameter at least three times the maximum size of the aggregate and its height should be twice the diameter.

Because the strength of a specimen is greatly affected by temperature changes, exposure to drying, and disturbances due to movement, it is customary to keep it at the construction site for 24 hr. It is then taken to the laboratory and cured under controlled conditions. At the end of the curing period, each specimen is placed in the testing machine and a gradually increasing compressive load is applied until the specimen fails. The load causing failure is divided by the cross-sectional area of the cylinder, indicating the ultimate compressive unit stress of the specimen. The same test is made on other specimens taken at the same time and cured under similar conditions, which, of course, results in a range of values for the compressive strength.

There are other types of tests that can be performed, including tests made on samples cut from the finished construction. Some of these tests are alternatives to the slump or compression tests, being more economical or logical for

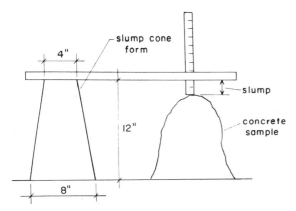

FIGURE 26.2. Concrete slump test.

particular situations. The need for testing, as well as directions for testing procedures, are often established by building code regulations, following standard procedures specified in ASTM regulations or the *ACI Code*. Testing is sometimes a costly and job-interrupting procedure, and may be undesirable and of questionable need, especially for small projects or very minor types of construction. For this reason, building codes and regulatory agencies usually provide for the ommission of testing under certain conditions. One such condition may be when the design work is done with the assumption of a very low value for f_c', such as 2000 psi. It may generally be assumed that virtually any concrete that will not crumble away when the forms are removed or crack off from a swift kick will attain such a low strength.

26.3. REINFORCEMENT

Reinforcement of concrete is done most often with *deformed bars*. These are steel bars of approximately round cross section, with patterned ridges projecting from their surfaces, the projections being used to increase the bonding between the bars and the surrounding concrete. Cross-section properties used for design are those of equivalent round sections. Bar sizes are designated by number, from No. 2 to No. 18. For sizes up to No. 8 the equivalent round shape is one with a diameter equal to the bar number in eighths of an inch (No. 5 = $\frac{5}{8}$ in. diameter). Table 26.3 gives properties of standard steel reinforcing bars.

Deformed bars are available in various steels. The steel is usually specified by a grade number, indicating the basic structural property of concern: the yield strength (f_y). Smaller bar sizes (up to No. 11) are generally available in grade 40 (f_y = 40 ksi [275 MPa]), grade 50 (f_y = 50 ksi [345 MPa]), or grade 60 (f_y = 60 ksi [414 MPa]). No. 14 and No. 18 bars are used mostly for compressive reinforcing in large columns and piers and are usually rolled in higher-strength steels, with f_y up to 90 ksi [620 MPa] or more.

Another type of reinforcement is *welded wire fabric*, which consists of a series of parallel longitudinal wires welded at regular intervals to transverse wires. It is available in sheets or rolls and is widely used as reinforcement in floor slabs and walls. The configuration of welded wire fabric achieves a more uniform distribution of steel by the use of smaller members more closely spaced than that provided by larger bars spaced more widely. Also, improved bond between the concrete and steel is obtained by the mechanical anchorage that results from the continuously welded transverse wires.

Welded wire fabric is made from cold-drawn wire, either smooth or deformed; the former has a yield strength of 65,000 or 56,000 psi, depending on the wire size, whereas the yield strength of the deformed wire is 70,000 psi.

A new method of designating wire sizes has superseded the *steel wire gage* system once employed. The cross-sectional area of the wire is the basic element in the new system. Deformed wire sizes are specified by the letter D, followed by a number indicating hundredths of a square inch. Thus D20 designates a deformed wire with a cross-sectional area of 0.20 in.^{2}. Smooth wire sizes are similarly indicated by substituting W for D. The term *style* is used to identify the spacings and sizes of the wires in welded wire fabric. A typical style designation is 4 × 8—W16 × W10, which denotes a welded smooth wire fabric in which

$$\text{spacing of longitudinal wires} = 4 \text{ in.}$$

$$\text{spacing of transverse wires} = 8 \text{ in.}$$

$$\text{size of longitudinal wires} = \text{W16 (0.16 in.}^2)$$

$$\text{size of transverse wires} = \text{W10 (0.10 in.}^2)$$

A welded deformed-wire fabric style would be indicated in the same manner by substituting D for W.

To facilitate the shop fabrication and field installation of reinforcing bars, the bar supplier usually prepares a set of drawings—commonly called the *shop drawings*. These drawings consist of the supplier's interpretation of the engineering contract drawings, with the information necessary for the workers who fabricate the bars in the shop and those who install the bars in the field prior to pouring of the concrete. The exact cut lengths of bars, the location of all bends, the number of each type of bar, and so on will

TABLE 26.3. Properties of Standard Deformed Steel Reinforcing Bars

Size	Nominal diameter (in.)	Nominal diameter (mm)	Nominal area (in.^{2})	Nominal area (mm^2)	Nominal perimeter (in.)	Nominal perimeter (mm)	Weight (lb/ft)	Weight (kg/m)
3	0.375	9.52	0.11	71	1.178	29.92	0.376	0.560
4	0.500	12.70	0.20	129	1.571	39.90	0.668	0.994
5	0.625	15.88	0.31	200	1.963	49.86	1.043	1.552
6	0.750	19.05	0.44	284	2.356	59.84	1.502	2.235
7	0.875	22.22	0.60	387	2.749	69.82	2.044	3.042
8	1.000	25.40	0.79	510	3.142	79.81	2.670	3.973
9	1.128	28.65	1.00	645	3.544	90.02	3.400	5.060
10	1.270	32.26	1.27	819	3.990	101.35	4.303	6.404
11	1.410	35.81	1.56	1006	4.430	112.52	5.313	7.907
14	1.693	43.00	2.25	1452	5.320	135.13	7.650	11.380
18	2.257	57.33	4.00	2581	7.090	180.09	13.600	20.240

be indicated on these drawings. While the correctness of these drawings is the responsibility of the supplier, it is usually a good idea for the designer to verify the drawings in order to reduce mistakes in the construction.

Reinforcing bars must be held firmly in place during the pouring of the concrete. Horizontal bars must be held up above the forms; vertical bars must be braced from swaying against the forms. The positioning and holding of bars is done through the use of various accessories and a lot of light-gage tie wire. When concrete surfaces are to be exposed to view after being poured, it behooves the designer to be aware of the various problems of holding bars and bracing forms, since many of the accessories used ordinarily will be partly in view on the surface of the finished concrete.

Installation of reinforcing may be relatively simple and easy to achieve, as in the case of a simple footing or a single beam. In other cases, where the reinforcing is extensive or complex, the problems of installation may require consideration during the design of the members. When beams intersect each other, or when beams intersect columns, the extended bars from the separate members must pass each other at the joint. Consideration of the "traffic" of the intersecting bars at such joints may affect the positioning of bars in the individual members.

26.4. GENERAL REQUIREMENTS FOR REINFORCED CONCRETE STRUCTURES

This section discusses some of the general requirements that apply to all concrete structures. Only those that apply to the structure as a whole are presented here. Requirements for particular types of members are given in the chapters that deal with those members.

Cover of Reinforcement

Steel bars are usually placed as close as possible to the outside surface of concrete members in order to be most effective in resisting flexure or to help in the reduction of surface cracking. The distance between the edge of the bars and the outside surface of the concrete is called the *cover*. General requirements for cover are as follows:

1. 3 in. for the sides of members cast directly against soil.
2. For concrete exposed to soil or the weather and cast in forms: 2 in. for No. 6 bars and larger; 1.5 in. for No. 5 bars and smaller.
3. For concrete not exposed to weather or in contact with soil: 0.75 in. for No. 11 bars or smaller in slabs, walls, and joists; 1.5 in. in beams and columns.

Spacing of Reinforcement

Where multiple bars are used in members (which is the common situation), there are both upper and lower limits for the spacing of the bars. Lower limits are intended to permit adequate development of the concrete-to-steel

stress transfers and to facilitate the flow of the wet concrete during pouring. For columns, the minimum clear distance between bars is specified as 1.5 times the bar diameter or a minimum of 1.5 in. For other situations, the minimum is one bar diameter or a minimum of 1 in.

For walls and slabs, maximum center-to-center bar spacing is specified as three times the wall or slab thickness or a maximum of 18 in. This applies to reinforcement required for computed stresses. For reinforcement that is required for control of cracking due to shrinkage or temperature change, the maximum spacing is five times the wall or slab thickness or a maximum of 18 in.

For adequate placement of the concrete, the largest size of the coarse aggregate should be not greater than three-fourths of the clear distance between bars.

Bending of Reinforcement

In various situations, it is sometimes necessary to bend reinforcing bars. Bending is done preferably in the fabricating shop instead of at the job site, and the bend diameter (see Fig. 26.3) should be adequate to avoid cracking the bar.

Bending of bars is sometimes done in order to provide anchorage for the bars. The code defines such a bend as a "standard hook," and the requirements for the details of this type of bend are given in Fig. 26.3b.

As the yield stress of the steel is raised, bending becomes increasingly difficult. Bending of bars should be avoided when the yield stress exceeds 60 ksi [414 MPa]; and where it is necessary, should be done with bend diameters slightly greater than those given in Fig. 26.3

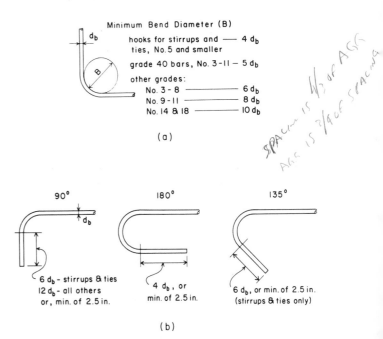

FIGURE 26.3. Bend requirements for reinforcing bars.

Minimum Dimensions for Concrete Members

For practical reasons, as well as the satisfying of various requirements for cover and bar spacing, there are minimum usable dimensions for various reinforced concrete members. When flexural reinforcement is required in slabs, walls, or beams, its effectiveness will be determined in part by the distance between the tension-carrying steel and the far edge of the compression-carrying concrete. Thus extremely shallow beams and thin slabs or walls will have reduced efficiency for flexure.

In slabs and walls, it is usually necessary to provide two-way reinforcing. Even where the bending actions occur in only one direction, the code requires a minimum amount of reinforcing in the other direction for control of cracking due to shrinkage and temperature changes. As shown in Fig. 26.2a, even with minimum cover and small bars, a minimum slab thickness is approximately 2 in. Except for joist or waffle construction, however, slab thicknesses are usually greater, for reasons of development of practical levels of flexural resistance. Thus reinforcing is more often as shown in Fig. 26.4b, with the bars closer to the top or bottom, depending on whether the moment is positive or negative.

In many cases it is desirable for the slab to have a significant fire rating. Building codes often require additional cover for this purpose, and typically specify minimum slab thicknesses of 4 in. or more. Slab thicknesses required for this purpose also depend on the type of aggregate that is used for the concrete.

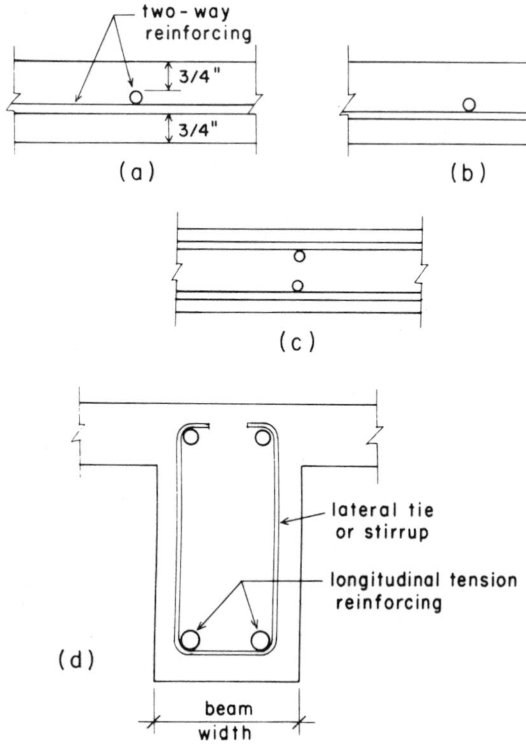

Walls of 10-in. or greater thickness often have two separate layers of reinforcing, as shown in Fig. 26.4c. Each layer is placed as close as the requirements for cover permit to the outside wall surface. Walls with crisscrossed reinforcing (both vertical and horizontal bars) are seldom made less than 6 in. thick.

As shown in Fig. 26.4d, concrete beams usually have a minimum of two reinforcing bars and a stirrup or tie of at least No. 2 or No. 3 size. Even with small bars, the minimum beam width in this situation is at least 8 in., with 10 in. being much more practical.

For rectangular columns with ties, a limit of 8 in. is usual for one side of an oblong cross section and 10 in. for a square section. Round columns may be either tied or spiral wrapped. A 10-in. diameter may be possible for a round tied column, but 12 in. is more practical and is the usual minimum for a spiral column, with larger sizes required where more cover is necessary.

Shrinkage and Temperature Reinforcement

The essential purpose of steel reinforcing is to prevent the cracking of the concrete due to tension stresses. In the design of concrete structures, investigation is made for the anticipated structural actions that will produce tensile stress: primarily the actions of bending, shear, and torsion. However, tension can also be induced by the shrinkage of the concrete during its drying out after the initial pour. Temperature variations may also induce tension in various situations. To provide for these latter actions, the *ACI Code* requires a minimum amount of reinforcing in members such as walls and slabs even when structural actions do not indicate any need. These requirements are discussed in the sections that deal with the design of these members.

Minimum Reinforcement

In the design of most reinforced concrete members the amount of steel reinforcing required is determined from computations and represents the amount determined to be necessary to resist the required tensile force in the member. In various situations, however, there is a minimum amount of reinforcing that is desirable, which may on occasion exceed that determined by the computations. The *ACI Code* makes provisions for such minimum reinforcing in columns, beams, slabs, and walls. The minimum reinforcing may be specified as a minimum percentage of the member cross-sectional area, as a minimum number of bars, or as a minimum bar size. These requirements are discussed in the sections that deal with the design of the various types of members.

26.5. DESIGN METHODS

There are two basic methods that may be used for the design of structural members of reinforced concrete. *Working stress design* consists of the general application of the

FIGURE 26.4.

traditional method of designing for service load conditions using stress computations based on allowable unit stresses. Safety is established by the setting of allowable stress values at some fraction of the test-determined ultimate stress capacity of the materials. *Strength design* is based on determination of the ultimate capacity of members, as demonstrated in tests and expressed in largely empirical formulas. For strength design computations, the only values of stress used are the limits for the materials: f_c' for the concrete and f_y for the steel reinforcement. Safety in strength design is established by using a load that is some multiple of the true service load, called the *factored load.*

Strength design is now predominantly favored by the codes and design standards and is generally used in the work in professional design offices. Some design work by the working stress method is tolerated (covered by a terse appendix section in the current *ACI Code*), but its use is largely ignored by engineers. Still, the working stress method is outstandingly simpler to explain and has some arguably valid applications. Although some work is presented in this part using the strength method, it is done primarily to demonstrate some of the aspects of the basic concepts and procedures. Design work is mostly performed with the working stress method, because the procedures are frankly simpler and allow more carry over from similar methods used for wood and steel design. A full presentation of current strength design is an exhaustive undertaking—if you do not think so, take a look at any current engineering textbook on the subject.

The last edition of the *ACI Code* to develop the working stress method in a full manner was the 1963 edition (ACI Publication 318–63). Some of the material presented here is based on that publication, although the general reference is Appendix B of the current edition (ACI Publication 318–83, Ref. 6), which limits the method considerably.

It is not our purpose to advocate a return to the "good old days" of the working stress method, except as a learning experience. Persons expecting to become professional designers must become well versed in the theories and procedures of the strength design method. For practical purposes, however, some of the simple, direct procedures of the working stress method can be used for preliminary determination of approximate member properties or even for finished design of members with low amounts of reinforcing and low concrete strength.

26.6. NOMENCLATURE

Notation used in this book complies with that used in the 1989 *ACI Code*. The following list includes all of the notation used in this book and is compiled and adapted from a more extensive list given in Appendixes B and C of the code.

A_c = Area of concrete; gross area minus area of reinforcing.

A_g = Gross area of section $(A_c + A_s)$.

A_s = Area of reinforcing.

A_s' = Area of compressive reinforcement in a doubly reinforced section.

A_v = Area of shear reinforcing.

A_1 = Loaded area in bearing.

A_2 = Gross area of bearing support member.

E_c = Modulus of elasticity of concrete.

E_s = Modulus of elasticity of steel.

M = Design moment.

N = Design axial load.

V = Design shear force.

a = Depth of equivalent rectangular stress block (strength design).

b = Width of compression face of member.

b_w = Width of stem in a T-beam.

c = Distance from extreme compression fiber to the neutral axis (strength design).

d = Effective depth, from extreme compression fiber to centroid of tensile reinforcing.

e = Eccentricity of a nonaxial load, from the centroid of the section to the point of application of the load.

f_c = Unit compressive stress in concrete.

f_c' = Specified compressive strength of concrete.

f_s = Stress in reinforcement.

f_y' = Specified yield stress of steel.

h = Overall thickness of member; unbraced height of a wall.

jd = Length of internal moment arm.

kd = Distance from extreme compression fiber to the neutral axis (working stress).

n = Modular ratio of elasticity: E_s/E_c.

p = Percent of reinforcing with working strength design, expressed as a ratio: A_s/A_g.

s = Spacing of stirrups.

t = Thickness of a solid slab.

ρ = Percent of reinforcing with ultimate strength design expressed as a ratio: A_s/A_g

ϕ = Strength reduction factor (strength design).

CHAPTER TWENTY-SEVEN

Flexure

The principal use of reinforcing in concrete structures is to develop internal tension resistance associated with bending. Chapter 27 considers the actions of bending in simple beams and slabs and presents the design methods based on consideration of the interactions of the concrete and steel in the composite reinforced sections.

27.1. DEVELOPMENT OF BENDING RESISTANCE

When a member is subjected to bending, such as the beam shown in Fig. 27.1a, internal resistances of two basic kinds are generally required. Internal actions are "seen" by visualizing a cut section, such as that taken at X–X in Fig. 27.1a. Removing the portion of the beam to the left of the cut section, we visualize its free-body actions as shown in Fig. 27.1b. At the cut section, consideration of static equilibrium requires the development of the internal shear force (V in the figure) and the internal resisting moment (represented by the force couple: C and T in the figure).

If the beam consists of a simple rectangular concrete section with tension reinforcing only, as shown in Fig.

27.1c, the force C is considered to be developed by compressive stresses in the concrete—indicated by the shaded area above the neutral axis. The tension force, however, is considered to be developed by the steel alone, ignoring the tensile resistance of the concrete. For low-stress conditions the latter is not true, but at a serious level of stress the tension-weak concrete will indeed crack, virtually leaving the steel unassisted, as assumed.

At moderate levels of stress, the resisting moment is visualized as shown in Fig. 27.2a, with a linear variation of compressive stress from zero at the neutral axis to a maximum value of f_c at the edge of the section. As stress levels increase, however, the nonlinear stress–strain character of the concrete becomes more significant, and it becomes necessary to acknowledge a more realistic form for the compressive stress variation, such as that shown in Fig. 27.2b. As stress levels approach the limit of the concrete, the compression becomes vested in an almost constant magnitude of unit stress, concentrated near the top of the section. For strength design, in which the moment capacity is expressed as the ultimate limit, it is common to assume the form of stress distribution shown in Fig. 27.2c, with the

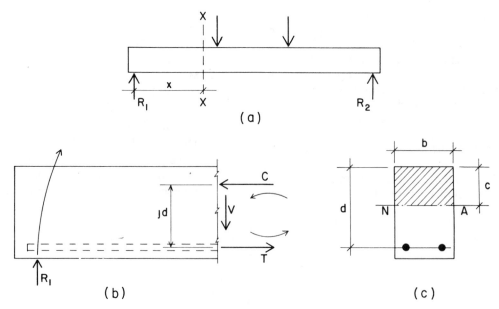

FIGURE 27.1. Bending in a reinforced beam.

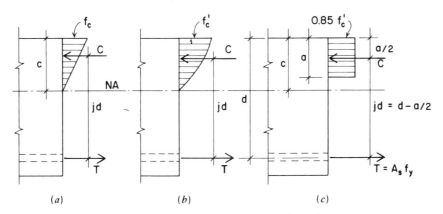

FIGURE 27.2. Development of internal bending resistance.

limit for the concrete stress set at 0.85 times f_c'. Expressions for the moment capacity derived from this assumed distribution have been shown to compare reasonably with the response of beams tested to failure in laboratory experiments.

Response of the steel reinforcing is more simply visualized and expressed. Since the steel area in tension is concentrated at a small location with respect to the size of the beam, the stress in the bars is considered to be a constant. Thus at any level of stress the total value of the internal tension force may be expressed as

$$T = A_s f_s$$

and for the practical limit of T,

$$T = A_s f_y$$

27.2. WORKING STRESS METHOD: INVESTIGATION AND DESIGN

In working stress design a maximum allowable (working) value for the extreme fiber stress is established (Table 26.1), and the formulas are predicated on elastic behavior of the reinforced concrete member under service load. The straight-line distribution of compressive stress is valid at working stress levels because the stresses developed vary approximately with the distance from the neutral axis, in accordance with elastic theory.

The following is a presentation of the formulas and procedures used in the working stress method. The discussion is limited to a rectangular beam section with tension reinforcing only.

Referring to Figure 27.3, the following are defined:

b Width of the concrete compression zone.

d Effective depth of the section for stress analysis; from the centroid of the steel to the edge of the compression zone.

A_s Cross-sectional area of the reinforcing.

p Percentage of reinforcing, defined as $p = A_s/bd$.

n Elastic ratio $= \dfrac{E \text{ of the steel reinforcing}}{E \text{ of the concrete}}$

kd Height of the compression stress zone; used to locate the neutral axis of the stressed section; expressed as a percentage (k) of d.

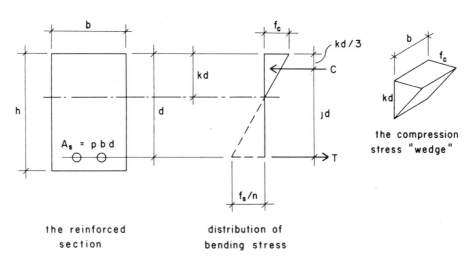

FIGURE 27.3. Bending resistance of a rectangular section with tension reinforcement only.

jd Internal moment arm, between the net tension force and the net compression force; expressed as a percentage (j) of d.

f_c Maximum compressive stress in the concrete.

f_s Tensile stress in the reinforcing.

The compression force C may be expressed as the volume of the compression stress "wedge," as shown in the figure:

$$C = \tfrac{1}{2}(kd)(b)(f_c) = \tfrac{1}{2}kf_cbd$$

Using the compression force, the moment resistance of the section may be expressed as

$$M = Cjd = (\tfrac{1}{2}kf_cbd)(jd) = \tfrac{1}{2}kjf_cbd^2 \qquad (1)$$

This may be used to derive an expression for the concrete stress:

$$f_c = \frac{2M}{kjbd^2} \qquad (2)$$

The resisting moment may also be expressed in terms of the steel and the steel stress as

$$M = Tjd = (A_2)(f_s)(jd)$$

This may be used for determination of the steel stress or for finding the required area of steel.

$$f_s = \frac{M}{A_s jd} \qquad (3)$$

$$A_s = \frac{M}{f_s jd} \qquad (4)$$

A useful reference is the *balanced section*, which occurs when the exact amount of reinforcing used results in the simultaneous limiting stresses in the concrete and steel. The properties that establish this relationship may be expressed as follows:

$$\text{balanced } k = \frac{1}{1 + f_s/nf_c} \qquad (5)$$

$$j = 1 - \frac{k}{3} \qquad (6)$$

$$p = \frac{f_ck}{2f_s} \qquad (7)$$

$$M = Rbd^2 \qquad (8)$$

in which

$$R = \tfrac{1}{2}kjf_c \qquad (9)$$

[derived from formula (1)]. If the limiting compression stress in the concrete ($f_c = 0.45f_c'$) and the limiting stress in the steel are entered in formula (5), the balanced section value for k may be found. Then the corresponding values for j, p, and R may be found. The balanced p may be used to determine the maximum amount of tensile reinforcing that may be used in a section without the addition of compressive reinforcing. If less tensile reinforcing is used, the moment will be limited by the steel stress, the maximum stress in the concrete will be below the limit of $0.45f_c'$, the value of k will be slightly lower than the balanced value, and the value of j slightly higher than the balanced value. These relationships are useful in design for the determination of approximate requirements for cross sections.

Table 27.1 gives the balanced properties for various combinations of concrete strength and limiting steel stress. The values of n, k, j, and p are all without units. However, R must be expressed in particular units; the unit used in the table is kip-inches (kip-in.).

When the area of steel used is less than the balanced p, the true value of k may be determined by the following formula:

$$k = \sqrt{2np - (np)^2} - np \qquad (10)$$

Figure 27.4 may be used to find approximate k values for various combinations of p and n.

TABLE 27.1. Balanced Section Properties for Rectangular Concrete Sections with Tension Reinforcing Only

| f_s | | f_c'' | | | | | | R | |
ksi	MPa	ksi	MPa	n	k	j	p	kip-in.	kN-m
16	110	2.0	13.79	11.3	0.389	0.870	0.0109	0.152	1045
		2.5	17.24	10.1	0.415	0.862	0.0146	0.201	1382
		3.0	20.68	9.2	0.437	0.854	0.0184	0.252	1733
		4.0	27.58	8.0	0.474	0.842	0.0266	0.359	2468
20	138	2.0	13.79	11.3	0.337	0.888	0.0076	0.135	928
		2.5	17.24	10.1	0.362	0.879	0.0102	0.179	1231
		3.0	20.68	9.2	0.383	0.872	0.0129	0.226	1554
		4.0	27.58	8.0	0.419	0.860	0.0188	0.324	2228
24	165	2.0	13.79	11.3	0.298	0.901	0.0056	0.121	832
		2.5	17.24	10.1	0.321	0.893	0.0075	0.161	1107
		3.0	20.68	9.2	0.341	0.886	0.0096	0.204	1403
		4.0	27.58	8.0	0.375	0.875	0.0141	0.295	2028

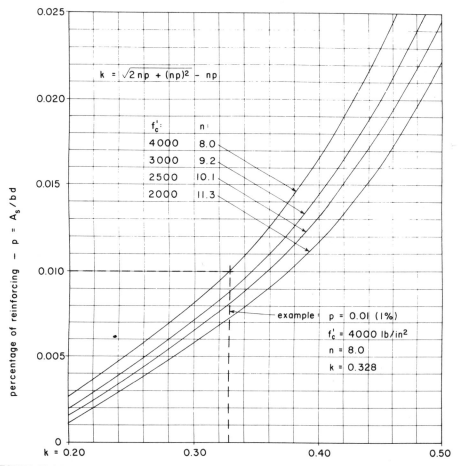

FIGURE 27.4. Flexural k factors for rectangular sections with tension reinforcement only—as a function of p and n.

Design of beams is usually done in the course of development of a complete framing system. The interactive decisions and numerous considerations that affect choice of the beam dimensions and the form of reinforcing are best demonstrated in the system design examples in Part 9. The following examples demonstrate the use of the formulas for solution of two of the most common design situations for a rectangular beam section with tension reinforcing only.

Example 1.   A rectangular concrete beam of concrete with f_c' of 3000 psi [20.7 MPa] and steel reinforcing with f_s = 20 ksi [138 MPa] must sustain a bending moment of 200 kip-ft [271 kN-m]. Select the beam dimensions and the reinforcing for a section with tension reinforcing only.
Solution.   1. With tension reinforcing only, the minimum size beam will be a balanced section, since a smaller beam would have to be stressed beyond the capacity of the concrete to develop the required moment. Using formula (8),

$$M = Rbd^2 = 200 \text{ kip-ft [271 kN-m]}$$

Then from Table 27.1, for f_c' of 3000 psi and f_s of 20 ksi,

$$R = 0.226 \text{ (in units of kip-in.)[1554 in units of kN-m]}$$

Therefore,

$$M = 200 \times 12 = 0.226bd^2, \text{ and } bd^2 = 10{,}619$$
$$[M = 271 = 1554bd^2, \text{ and } bd^2 = 0.1744]$$

2. Various combinations of b and d may be found; for example,

$$b = 10 \text{ in., } d = \sqrt{\frac{10{,}619}{10}} = 32.6 \text{ in. [0.829 m]}$$

$$b = 15 \text{ in., } d = \sqrt{\frac{10{,}619}{10}} = 26.6 \text{ in. [0.677 m]}$$

Although they are not given in this example, there are often some considerations other than flexural behavior alone that influence the choice of specific dimensions for a beam. If the beam is of the ordinary form shown in Fig. 27.5 the specified dimension is usually that given as h. Assuming the use of a No. 3 U-stirrup, a cover of 1.5 in. [38 mm], and an average-size reinforcing bar of 1 in. [25-mm] diameter (No. 8 bar), the design dimension d will be less than h by 2.375 in. [60 mm]. Lacking other considerations, we will assume a b of 15 in. [380 mm] and an h of 29 in. [740

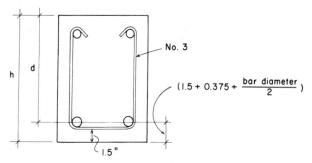

FIGURE 27.5.

mm], with the resulting d of $29 - 2.375 = 26.625$ in. [680 mm].

3. We next use the specific value for d with formula (4) to find the required area of steel A_s. Since our selection is very close to the balanced section, we may use the value of j from Table 27.1. Thus

$$A_s = \frac{M}{f_s jd} = \frac{200(12)}{20(0.872)(26.625)} = 5.17 \text{ in.}^2 \text{ [3312 mm}^2\text{]}$$

Or using the formula for the definition of p and the balanced p value from Table 27.1,

$$A_s = pbd = 0.0129[15(26.625)] = 5.15 \text{ in.}^2 \text{ [3333 mm}^2\text{]}$$

4. We next select a set of reinforcing bars to obtain this area. As with the beam dimensions, there are other concerns. For the purpose of our example, if we select bars all of a single size (see Table 26.3), the number required will be:

For No. 6 bars: $\dfrac{5.17}{0.44} = 11.75$, or 12

For No. 7 bars: $\dfrac{5.17}{0.60} = 8.62$, or 9

For No. 8 bars: $\dfrac{5.17}{0.79} = 6.54$, or 7

For No. 9 bars: $\dfrac{5.17}{1.00} = 5.17$, or 6

For No. 10 bars: $\dfrac{5.17}{1.27} = 4.07$, or 5

For No. 11 bars: $\dfrac{5.17}{1.56} = 3.31$, or 4

For all except the No. 11 bars, the requirements for bar spacing (as discussed in Sec. 26.4) would result in the need to place the bars in stacked layers in the 15-in.-wide beam. While this is possible, it would require some increase in the dimension h in order to maintain the effective depth of approximately 26.6 in., since the centroid of the steel bar areas would move farther away from the edge of the concrete.

Example 2. A rectangular concrete beam of concrete with f_c' of 3000 psi [20.7 MPa] and steel with f_s of 20 ksi [138 MPa]

has dimensions of $b = 15$ in. [380 mm] and $h = 36$ in. [910 mm]. Find the area required for the steel reinforcing for a moment of 200 kip-ft [271 kN-m].

Solution. The first step in this case is to determine the balanced moment capacity of the beam with the given dimensions. If we assume the section to be as shown in Fig. 27.5, we may assume an approximate value for d to be h minus 2.5 in. [64 mm], or 33.5 in. [851 mm]. Then with the value for R from Table 27.1,

$$M = Rbd^2 = 0.226(15)(33.5)^2 = 3804 \text{ kip-in.}$$

or

$$M = \frac{3804}{12} = 317 \text{ kip-ft [427 kN-m]}$$

Since this value is considerably larger than the required moment, it is thus established that the given section is larger than that required for a balanced stress condition. As a result, the concrete flexural stress will be lower than the limit of $0.45f_c'$, and the section is qualified as being under-reinforced; which is to say that the reinforcing required will be less than that required to produce a balanced section (with moment capacity of 317 k-ft). In order to find the required area of steel, we use formula (4) just as we did in the preceding example. However, the true value for j in the formula will be something greater than that for the balanced section (0.872 from Table 27.1).

As the amount of reinforcing in the section decreases below the full amount required for a balanced section, the value of k decreases and the value of j increases. However, the range for j is small: from 0.872 up to something less than 1.0. A reasonable procedure is to assume a value for j, find the corresponding required area, and then perform an investigation to verify the assumed value for j, as follows. Assume that $j = 0.90$. Then

$$A_s = \frac{M}{f_s jd} = \frac{200(12)}{20(0.90)(33.5)} = 3.98 \text{ in.}^2$$

and

$$p = \frac{A_s}{bd} = \frac{3.98}{15(33.5)} = 0.00792$$

Using this value for p in Fig. 27.4, we find $k = 0.313$. Using formula (6), we then determine j to be

$$j = 1 - \frac{k}{3} = 1 - \frac{0.313}{3} = 0.896$$

which is reasonably close to our assumption, so the computed area is adequate for design.

27.3. STRENGTH METHOD: INVESTIGATION AND DESIGN

Figure 27.6 shows the equivalent rectangular compressive stress distribution in the concrete permitted by the *ACI Code* for use in the strength design method. As stated in Sec. 27.1, the rectangular stress block is based on the as-

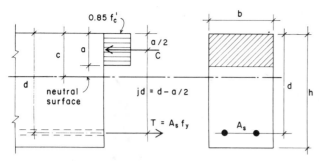

FIGURE 27.6.

sumption that, at ultimate load, a concrete stress of $0.85 f_c'$ is uniformly distributed over the compression zone. The dimensions of this zone are the beam width b and the distance a, which locates a line parallel to and above the neutral axis. Although we have not yet considered how the value of a is determined, let us turn our attention to Fig. 27.6 and develop equations for the theoretical resisting moment M_t.

We observe that the resultant (sum) of the compressive stresses is

$$C = 0.85 f_c' ba$$

and that it acts at a distance of $a/2$ from the top of the beam. The arm of the resisting moment couple jd then becomes $d - a/2$ and the theoretical resisting moment as governed by the concrete is

$$M_t = C \left(d - \frac{a}{2} \right) = (0.85 f_c' ba) \left(d - \frac{a}{2} \right) \quad (1)$$

Similarly, the theoretical moment strength as controlled by the steel reinforcement is

$$M_t = T \left(d - \frac{a}{2} \right) = A_s f_y \left(d - \frac{a}{2} \right) \quad (2)$$

If *balanced* conditions exist, that is, if the concrete reaches its full compressive strength when the steel reaches its yield strength, the two equations will be equal to each other, or

$$0.85 f_c' ba = A_s f_y = \rho b d f_y \quad (3)$$

where

$$\rho = \frac{A_s}{bd}$$

(Note: The *ACI* Code uses ρ to indicate steel percent with the strength method, whereas ρ is used with the working stress method.) Then, from formula (3),

$$a = \frac{\rho b d f_y}{0.85 f_c' b} = \frac{\rho f_y d}{0.85 f_c'} \quad (4)$$

and

$$\rho = \frac{a}{d} \left(\frac{0.85 f_c'}{f_y} \right) \quad (5)$$

The symbol ρ_b is used to denote the balancing ratio of reinforcement and a_b, the depth of the stress block under balanced conditions.

Using this expression for a and considering the strain relationship between concrete and steel, the following formula for the balancing ratio of reinforcement may be derived:

$$\rho_b = \frac{0.85 f_c' \beta_1}{f_y} \times \frac{87,000}{87,000 + f_y} \quad (6)$$

in which β_1 is a coefficient relating the depth of the rectangular stress block to the depth from the compression face to the neutral axis, or $a = \beta_1 \times c$ (see Fig. 27.6). The value of β_1 varies with the strength of the concrete. The *ACI Code* prescribes a value of 0.85 for concrete strengths up to 4000 psi [27.6 MPa] and a reduction of 0.05 for each 1000 psi [6.895 MPa] of strength in excess of 4000 psi, with a minimum value of 0.65. (For example, if f_c' = 5000 psi [34.5 MPa], β_1 = 0.80, and so on.)

By equating formulas (5) and (6), we can derive an expression for the balanced value of a/d. Thus

$$\frac{a}{d} \left(\frac{0.85 f_c'}{f_y} \right) = \frac{0.85 f_c' \beta_1}{f_y} \times \frac{87,000}{87,000 + f_y}$$

and

$$\frac{a}{d} = \beta_1 \frac{87,000}{87,000 + f_y} \quad (7)$$

$$\left[\begin{array}{l} \text{For SI units:} \\[4pt] \dfrac{a}{d} = \beta_1 \dfrac{600,000}{600,000 + f_y} \\[4pt] \text{when stress is in MPa} \end{array} \right]$$

Referring to formula (1), we may derive another form for this equation as follows:

$$M_t = 0.85 f_c' ba \left(d - \frac{a}{2} \right)$$

$$= 0.85 f_c' b \frac{a}{d} d \left(1 - \frac{1}{2} \frac{a}{d} \right)$$

$$= bd^2 \left[0.85 f_c' \left\{ \frac{a}{d} - \frac{1}{2} \left(\frac{a}{d} \right)^2 \right\} \right]$$

$$= Rbd^2 \quad (8)$$

where

$$R = 0.85 f_c' \left[\frac{a}{d} - \frac{1}{2} \left(\frac{a}{d} \right)^2 \right] \cdot \quad (9)$$

Formulas (5), (7), and (9) may be used to derive factors for balanced sections, which can be used in the design of beams. Table 27.2 contains a compilation of these factors for five values of concrete strength and two values of steel yield strength. The use of this material is demonstrated in the following section.

TABLE 27.2. Balanced Section Properties for Rectangular Concrete Sections with Tension Reinforcing Only: Strength Design[a]

f_c'		$f_y = 40$ ksi [276 MPa]					$f_y = 60$ ksi [414 MPa]				
		Balanced	Usable a/d	Usable	Usable R		Balanced	Usable a/d	Usable	Usable R	
psi	MPa	a/d	(75% Balance)	ρ	k-in.	kN-m	a/d	(75% Balance)	ρ	k-in.	kN-m
2000	13.79	0.5823	0.4367	0.0186	0.580	4000	0.5031	0.3773	0.0107	0.520	3600
2500	17.24	0.5823	0.4367	0.0232	0.725	5000	0.5031	0.3773	0.0137	0.650	4500
3000	20.69	0.5823	0.4367	0.0278	0.870	6000	0.5031	0.3773	0.0160	0.781	5400
4000	27.58	0.5823	0.4367	0.0371	1.161	8000	0.5031	0.3773	0.0214	1.041	7200
5000	34.48	0.5480	0.4110	0.0437	1.388	9600	0.4735	0.3551	0.0252	1.241	8600

[a] See Sec. 27.3 for derivation of formulas used to obtain table values.

Application of the working stress method consists of designing members to *work* in an adequate manner (without exceeding established stress limits) under actual service load conditions. The basic procedure in strength design is to design members to *fail*; thus the ultimate strength of the member at failure (called its *design strength*) is the only type of resistance considered. Safety in strength design is not provided by limiting stresses, as in the working stress method, but by using a factored design load (called the *required strength*) that is greater than the service load. The code establishes the value of the required strength, called U, as not less than

$$U = 1.4D + 1.7L \tag{1}$$

where D = effect of dead load
L = effect of live load

Other adjustment factors are provided when design conditions involve consideration of the effects of wind, earth pressure, differential settlement, creep, shrinkage, or temperature change.

The design strength of structural members (i.e., their *usable* ultimate strength) is determined by the application of assumptions and requirements given in the code and is further modified by the use of a *strength reduction factor* ϕ as follows:

ϕ = 0.90 for flexure, axial tension, and combinations of flexure and tension
= 0.75 for columns with spirals
= 0.70 for columns with ties
= 0.85 for shear and torsion
= 0.70 for compressive bearing
= 0.65 for flexure in plain (not reinforced) concrete

Thus while formula (1) may imply a relatively low safety factor, an additional margin of safety is provided by the stress reduction factors.

Use of the strength design formulas is illustrated in the following examples.

Example 1. The service load bending moments on a rectangular beam 10 in. [254 mm] wide are 58 kip-ft [78.6 kN-m] for dead load and 38 kip-ft [51.5 kN-m] for live load. If $f_c' = 4000$ psi [27.6 MPa] and $f_y = 60$ ksi [414 MPa], deter-

mine the depth of the beam and the required area of tension reinforcing.
Solution. 1. The required ultimate moment strength (M_u) is first determined as

$$U = 1.4D + 1.7L$$
$$M_u = 1.4M_{DL} + 1.7M_{LL}$$
$$= 1.4(58) + 1.7(38) = 146 \text{ kip-ft}$$
$$[M_u = 1.4(78.6) + 1.7(51.5) = 198 \text{ kN-m}]$$

2. To find the required design moment strength (M_t) we apply the capacity reduction factor $\phi = 0.90$ and the relationship $M_u = \phi M_t$; thus

$$M_t = \frac{M_u}{\phi} = \frac{146}{0.90} = 162 \text{ kip-ft} \quad \text{or} \quad 1944 \text{ kip-in.}$$
$$\left[M_t = \frac{198}{0.90} = 220 \text{ kN-m} \right]$$

3. The maximum usable reinforcement ratio as given in Table 27.2 is $\rho = 0.0214$. If a balanced section is used, we may thus determine the required area of reinforcement from the relationship

$$A_s = \rho bd$$

While there is nothing especially desirable about a balanced section, it does represent the beam section with least depth if tension reinforcing only is used. We will therefore proceed to find the required balanced section for this example.

4. To determine the required effective depth d, we use formula (8); thus

$$M_t = Rbd^2$$

With the value of $R = 1.041$ from Table 27.2,

$$M_t = 1944 = 1.041(10)(d)^2$$

and

$$d = \sqrt{\frac{1944}{1.041(10)}} = \sqrt{186.7} = 13.66 \text{ in. } [0.347 \text{ m}]$$

5. If this value is used for d, the required steel area may be found as

$$A_s = \rho b d = 0.0214(10)(13.66) = 2.92 \text{ in.}^2$$

The *ACI Code* requires a minimum ratio of reinforcing as follows:

$$\rho_{min} = \frac{200}{f_y} = \frac{200}{60,000} = 0.0033$$

which is clearly not critical for this example.

Selection of the actual beam dimensions and the actual number and size of reinforcing bars would involve various considerations, as discussed in Sec. 27.2 and illustrated in Fig. 27.5. We will not complete the example in this case, since more complete design situations will be illustrated in the examples in the succeeding chapters.

If there are reasons, as there often are, for not selecting the least deep section with the greatest amount of reinforcing, a slightly different procedure must be used, as illustrated in the following example.

Example 2. Using the same data as in Example 1, find the reinforcing required if the desired beam section has $b = 10$ in. [254 mm] and $d = 18$ in. [457 mm].

Solution. The first two steps in this situation would be the same as in Example 1—to determine M_u and M_t. The next step would be to determine whether the given section is larger than, smaller than, or equal to a balanced section. Since this investigation has already been done in Example 1, we may observe that the 10×18-in. section is larger than a balanced section. Thus the actual value of a/d will be less than the balanced section value of 0.3773. The next step would then be as follows:

4. Estimate a value for a/d—something smaller than the balanced value. For example, try $a/d = 0.25$. Then

$$a = 0.25d = 0.25(18) = 4.5 \text{ in. [114 mm]}$$

With this assumed value for a, we may use formula (2) to find a value for A_s.

5. Referring to Fig. 27.6,

$$A_s = \frac{M_t}{f_y(d - a/2)} = \frac{1944}{60(15.75)} = 2.057 \text{ in.}^2$$

6. We next test to see if the estimate for a/d was close by finding a/d using formula (5). Thus

$$\rho = \frac{A_s}{bd} = \frac{2.057}{10(18)} = 0.0114$$

and

$$\frac{a}{d} = \rho \frac{f_y}{0.85 f'_c} = 0.0114 \frac{60}{0.85(4)} = 0.202$$

$$a = 0.202(18) = 3.63 \text{ in.}, \quad d - \frac{a}{2} = 16.2 \text{ in.}$$

If we replace the value for $d = a/2$ that was used earlier with this new value, the required value of A_s will be slightly reduced. In this example, the correction will be only a few percent. If the first guess for a/d had been way off, it may justify a second run through steps 4, 5, and 6 to get closer to an exact answer.

27.4. SLABS

Concrete slabs are frequently used as spanning roof or floor decks, often occurring in monolithic, cast-in-place slab and beam framing systems. There are generally two basic types of slabs: one-way spanning and two-way spanning slabs. The spanning condition is not so much determined by the slab as by its support conditions. The two-way spanning slab is discussed in Sec. 30.5. As part of a general framing system, the one-way spanning slab is discussed in Sec. 30.1. The following discussion relates to the design of one-way solid slabs using procedures developed for the design of rectangular beams.

Solid slabs are usually designed by considering the slab to consist of a series of 12-in.-wide planks. Thus the procedure consists of simply designing a beam section with a predetermined width of 12 in. Once the depth of the slab is established, the required area of steel is determined, specified as the number of square inches of steel required per foot of slab width.

Reinforcing bars are selected from a limited range of sizes, appropriate to the slab thickness. For thin slabs (4–6 in. thick) bars may be of a size from No. 3 to No. 6 or so (nominal diameters from $\frac{3}{8}$–$\frac{3}{4}$ in.). The bar size selection is related to the bar spacing, the combination resulting in the amount of reinforcing in terms of so many square inches per one foot unit of slab width. Spacing is limited by code regulation to a maximum of three times the slab thickness. There is no minimum spacing, other than that required for proper placing of the concrete; however, a very close spacing indicates a very large number of bars, making for laborious installation.

Every slab must be provided with two-way reinforcement, regardless of its structural functions. This is to satisfy requirements for shrinkage and temperature effects, as discussed in Sec. 26.4. The amount of this minimum reinforcement is specified as a percentage p of the gross cross-sectional area of the concrete, as follows:

1. For slabs reinforced with grade 40 or grade 50 bars:

$$p = \frac{A_s}{bt} = 0.0020 \text{ or } 0.2\%$$

2. For slabs reinforced with grade 60 bars:

$$p = \frac{A_s}{bt} = 0.0018 \text{ or } 0.18\%$$

Center-to-center spacing of this minimum reinforcement must not be greater than five times the slab thickness or 18 in.

Minimum cover for slab reinforcement is normally 0.75 in., although exposure conditions or need for a high fire rating may require additional cover. For a thin slab reinforced with large bars, there will be a considerable difference between the slab thickness and the effective depth—t and d, as shown in Fig. 27.7. Thus the practical efficiency of the slab in flexural resistance decreases rapidly as the slab thickness is decreased. For this and other

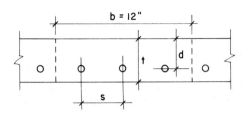

FIGURE 27.7. Reference for slab design.

reasons, very thin slabs (less than 4 in. thick) are often reinforced with wire fabric rather than sets of loose bars.

Shear reinforcement is seldom used in one-way slabs, and consequently the maximum unit shear stress in the concrete must be kept within the limit for the concrete without reinforcement. This is usually not a concern, as unit shear is usually low in one-way slabs, except for exceptionally high loadings.

Table 27.3 gives data that are useful in slab design, as demonstrated in the following example. Table values indicate the average amount of steel area per foot of slab width provided by various combinations of bar size and spacing. Table entries are determined as follows:

$$A_s/\text{ft} = (\text{bar area})\ \frac{12}{\text{bar spacing}}$$

Thus for No. 5 bars at 8-in. centers,

$$A_s/\text{ft} = (0.31)\ \frac{12}{8} = 0.465\ \text{in.}^2/\text{ft}$$

It may be observed that the table entry for this combination is rounded off to a value of 0.46 in.²/ft.

Example 1. A one-way solid concrete slab is to be used for a simple span of 14 ft [4.27 m]. In addition to its own weight, the slab carries a superimposed dead load of 30 psf [1.44 kN/m²] and a live load of 100 psf [4.79 kN/m²]. Using $f_c' = 3$

ksi [20.7 MPa], $f_y = 40$ ksi [276 MPa], and $f_s = 20$ ksi [138 MPa], design the slab for minimum overall thickness.
Solution.   Working Stress Method
Using the general procedure for design of a beam with rectangular section (Sec. 27.2), we first determine the required slab thickness. Thus

For deflection, from Table 30.2

$$\text{minimum } t = \frac{L}{25} = \frac{14(12)}{25} = 6.72\ \text{in. [171 mm]}$$

For flexure we first determine the maximum bending moment. The loading must include the weight of the slab, for which we use the thickness required for deflection as a first estimate. Assuming a 7-in. [178-mm]-thick slab, then slab weight is $\frac{7}{12}(150\ \text{pcf}) = 87.5$ psf, say 88 psf and total load is 100 psf LL + 118 psf DL = 218 psf.

The maximum bending moment for a 12-in.-wide design strip of the slab thus becomes

$$M = \frac{wL^2}{8} = \frac{218(14)^2}{8} = 5341\ \text{ft-lb [7.24 kN-m]}$$

For minimum slab thickness, we consider the use of a balanced section, for which Table 27.1 yields the following properties.

$$j = 0.872, \qquad p = 0.0129, \qquad R = 0.226$$

Then

$$bd^2 = \frac{M}{R} = \frac{5.341(12)}{0.226} = 284\ \text{in.}^3$$

And since b is the 12-in. design strip width,

$$d = \sqrt{\frac{284}{12}} = \sqrt{23.7} = 4.86\ \text{in. [123 mm]}$$

TABLE 27.3. Average Areas of Bars in Slabs per Foot of Width

Spacing (in.)	Areas of bars (in.²)									
	No. 2	No. 3	No. 4	No. 5	No. 6	No. 7	No. 8	No. 9	No. 10	No. 11
3	0.20	0.44	0.79	1.23	1.77	2.41	3.14	4.00		
3½	0.17	0.38	0.67	1.05	1.51	2.06	2.69	3.43	4.36	
4	0.15	0.33	0.59	0.92	1.33	1.80	2.36	3.00	3.81	4.68
4½	0.13	0.29	0.52	0.82	1.18	1.60	2.09	2.67	3.39	4.16
5	0.12	0.26	0.47	0.74	1.06	1.44	1.88	2.40	3.05	3.74
5½	0.11	0.24	0.43	0.67	0.96	1.31	1.71	2.18	2.77	3.40
6	0.10	0.22	0.40	0.61	0.88	1.20	1.57	2.00	2.54	3.12
6½	0.09	0.20	0.36	0.57	0.82	1.11	1.45	1.85	2.35	2.88
7	0.08	0.19	0.34	0.53	0.76	1.03	1.35	1.71	2.18	2.67
7½	0.08	0.18	0.31	0.49	0.71	0.96	1.26	1.60	2.03	2.50
8	0.07	0.17	0.29	0.46	0.66	0.90	1.18	1.50	1.91	2.34
8½	0.07	0.16	0.28	0.43	0.62	0.85	1.11	1.41	1.79	2.20
9	0.07	0.15	0.26	0.41	0.59	0.80	1.05	1.33	1.69	2.08
9½	0.06	0.14	0.25	0.39	0.56	0.76	0.99	1.26	1.60	1.97
10	0.06	0.13	0.24	0.37	0.53	0.72	0.94	1.20	1.52	1.87
11	0.05	0.12	0.21	0.33	0.48	0.66	0.86	1.09	1.39	1.70
12	0.05	0.11	0.20	0.31	0.44	0.60	0.79	1.00	1.27	1.56

Assuming an average bar size of a No. 6 ($\frac{3}{4}$-in.-nominal diameter) and cover of $\frac{3}{4}$ in., the minimum required slab thickness based on flexure becomes

$$t = 4.86 + \frac{0.75}{2} + 0.75 = 5.985 \text{ in. [152 mm]}$$

We thus observe that the deflection limitation controls in this situation, and the minimum overall thickness is the 6.72-in. dimension. If we continue to use the 7-in. overall thickness, the actual effective depth with a No. 6 bar will be

$$d = 7.0 - 1.125 = 5.875 \text{ in.}$$

Since this d is larger than that required for a balanced section, the value for j will be slightly larger than 0.872, as found from Table 27.1. Let us assume a value of 0.9 for j and determine the required area of reinforcement as

$$A_s = \frac{M}{f_s j d} = \frac{5.341(12)}{20(0.9)(5.875)} = 0.606 \text{ in.}^2$$

From Table 27.3, we find that the following bar combinations will satisfy this requirement:

Bar Size	Spacing from Center to Center (in.)	Average A_s in a 12-in. Width
5	6	0.61
6	8.5	0.62
7	12	0.60
8	15	0.63

The *ACI Code* requires a maximum spacing of three times the slab thickness (21 in. in this case). Minimum spacing is largely a matter of the designer's judgment. Many designers consider a minimum practical spacing to be one approximately equal to the slab thickness. Within these limits, any of the bar size and spacing combinations listed are adequate.

As described previously, the *ACI Code* requires a minimum reinforcement for shrinkage and temperature effects to be placed in the direction perpendicular to the flexural reinforcement. With the grade 40 bars in this example, the minimum percentage of this steel is 0.0020, and the steel area required for a 12-in. strip thus becomes

$$A_s = p(bt) = 0.0020[12(7)] = 0.168 \text{ in.}^2$$

From Table 27.3, we find that this requirement can be satisfied with No. 3 bars at 8-in. centers or No. 4 bars at 14-in. centers. Both of these spacings are well below the maximum of five times the slab thickness.

Although simply supported single slabs are sometimes encountered, the majority of slabs used in building construction are continuous through multiple spans. An example of the design of such a slab is given in Chapter 30.

Strength Design Method

Strength design procedures for the slab are essentially the same as for the rectangular beam, as described in Sec. 27.3. In most cases, slab sections will be reinforced with steel areas well below those for a balanced section, so the procedure for a so-called under-reinforced section should be used. If the procedure illustrated in Sec. 27.3 is used for this example, it will be found that the required steel area is approximately 15% less than that required from the working stress method computations.

27.5. T-BEAMS

When a floor slab and its supporting beams are poured at the same time, the result is a monolithic construction in which a portion of the slab on each side of the beam serves as the flange of a T-beam. The part of the section that projects below the slab is called the *web* or *stem* of the T-beam. This type of beam is shown in Fig. 27.8a. For positive moment, the flange is in compression and there is ample concrete to resist compressive stresses, as shown in Fig. 27.8b or c. However, in a continuous beam, there are negative bending moments over the supports, and the flange here is in the tension stress zone with compression in the web.

It is important to remember that only the area formed by the width of the web b_w and the effective depth d is to be considered in computing resistance to shear and to bending moment over the supports. This is the hatched area shown in Fig. 27.8d.

The effective flange width to be used in the design of symmetrical T-beams is limited to one-fourth the span length of the beam. In addition, the overhanging width of the flange on either side of the web is limited to eight times the thickness of the slab or one-half the clear distance to the next beam.

In monolithic construction with beams and one-way solid slabs, the effective flange area of the T-beams is usually quite capable of resisting the compressive stresses caused by positive bending moments. With a large flange area, as shown in Fig. 27.9, the neutral axis of the section usually occurs quite high in the beam web, resulting in only minor compressive stresses in the web. If the compression developed in the web is ignored, the net compression force may be considered to be located at the centroid of the trapezoidal stress zone that represents the stress distribution in the flange. On this basis, the compression force is located at something less than $t/2$ from the top of the beam.

An approximate analysis of the T-section by the working stress method, which avoids the need to find the location of the neutral axis and the centroid of the trapezoidal stress zone, consists of the following steps.

1. Ignore compression in the web and assume a constant value for compressive stress in the flange (see Fig. 27.9). Thus

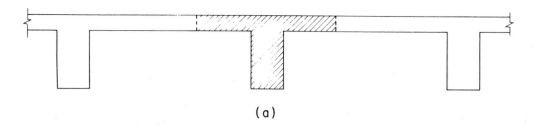

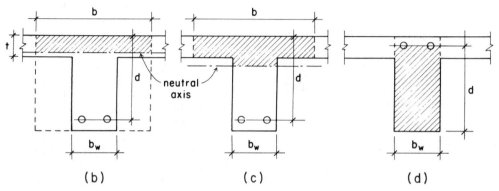

FIGURE 27.8. Considerations for T-beams.

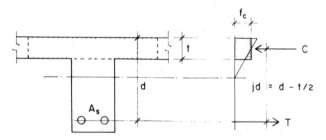

FIGURE 27.9. Basis for simplified analysis of a T-beam.

$$jd = d - \frac{t}{2}$$

2. Find the required steel area as

$$A_s = \frac{M}{f_s jd} = \frac{M}{f_s (d - t/2)}$$

3. Check the compressive stress in the concrete as

$$f_c = \frac{C}{b_f t}, \quad \text{where } C = \frac{M}{jd} = \frac{M}{d - t/2}$$

The actual value of maximum compressive stress will be slightly higher, but will not be critical if this computed value is significantly less than the limit of $0.45 f_c'$.

The following example illustrates the use of this procedure. It assumes a typical design situation in which the dimensions of the section (b_f, b_w, d, and t) are all predetermined by other design considerations and the design of the

T-section is reduced to the requirement to determine the area of tension reinforcing.

Example 1. A T-section is to be used for a beam to resist positive moment. The following data is given: beam span = 18 ft [5.49 m], beams are 9 ft [2.74 m] center to center, slab thickness is 4 in. [0.102 m], beam stem dimensions are b_w. = 15 in. [0.381 m] and d = 22 in. [0.559 m], f_c' = 4 ksi [27.6 MPa], f_y = 60 ksi [414 MPa], f_s = 24 ksi [165 MPa]. Find the required area and pick reinforcing bars for a dead load moment of 100 kip-ft [136 kN-m] plus a live load moment of 100 kip-ft [136 kN-m].

Solution. Using working stress design with the approximate method described previously:

1. Determine the effective flange width (necessary only for a check on the concrete stress). The maximum value for the flange width is

$$b_f = \frac{\text{span}}{4} = \frac{18(12)}{4} = 54 \text{ in. [1.37 m]}$$

or

$$b_f = \text{center-to-center beam spacing}$$
$$= 9(12) = 108 \text{ in. [2.74 m]}$$

or

$$b_f = \text{beam stem width plus 16 times the slab thickness}$$
$$= 15 + 16(4) = 79 \text{ in. [2.01 m]}$$

The limiting value is therefore 54 in. [1.37 m].

2. Find the required steel area

$$A_s = \frac{M}{f_s(d - t/2)} = \frac{200(12)}{24(22 - 4/2)}$$

$$= 5.00 \text{ in.}^2 \text{ [3364 mm}^2\text{]}$$

3. Pick bars using Table 26.3 and check the adequacy of the stem width using Table 30.1. From the properties table: Choose five No. 9 bars, actual A_s = 5.00 in.² From Table 30.1 required width for five No. 9 bars is 14 in., less than the 15 in. provided.

4. Check the concrete stress.

$$C = \frac{M}{jd} = \frac{200(12)}{20} = 120 \text{ kips [353 kN]}$$

$$f_c = \frac{C}{b_f t} = \frac{120}{54(4)} = 0.556 \text{ ksi [3.83 MPa]}$$

limiting stress $0.45 f_c' = 0.45(4)$

$$= 1.8 \text{ ksi [12.4 MPa]}$$

Thus compressive stress in the flange is clearly not critical.

In a real design situation, of course, consideration would have to be given to problems of shear and possibly to problems of development lengths for the bars.

When using strength design methods for T-sections, we recommend a procedure similar to that described for the working stress method. This method and procedure assumes that the flange area of the T is so large that the concrete stress never gets up to its ultimate limit before the yield stress develops in the reinforcing. The following example illustrates the procedure.

Example 2. Perform the design for the beam described in Example 1 using strength design methods.
Solution.

1. As in Example 1, effective b_f = 54 in. [1.37 m].
2. Design moment is found as

$$M_u = 1.4 M_d + 1.7 M_l$$

$$= 1.4(100 + 1.7(100)$$

$$= 310 \text{ kip-ft [420 kN-m]}$$

3. Required design strength is found as

$$M_t = \frac{M_u}{0.90} = \frac{310}{0.90} = 345 \text{ kip-ft [467 kN-m]}$$

4. Assuming the location of the net compression force to be at the center of the flange area, as described for the working stress method, the steel area is found as

$$A_s = \frac{M}{f_y(d - t/2)} = \frac{345(12)}{60(20)}$$

$$= 3.45 \text{ in.}^2 \text{ [2221 mm}^2\text{]}$$

5. A possible choice for the reinforcing is two No. 8 bars plus two No. 9 bars, providing an actual area of

3.58 in.² [2310 mm²]. Table 30.1 shows that a minimum width for four No. 9 bars is only 12 in., so the stem width is more than adequate for bar spacing.

6. Assuming the steel bars to be stressed to the yield point, the average stress in the flange would be as follows.

$$C = T = f_y A_s = 60(3.58)$$

$$= 215 \text{ kips [956 kN]}$$

$$f_c = \frac{C}{b_f t} = \frac{215}{54(4)} = 0.995 \text{ ksi [6.86 MPa]}$$

which is considerably lower than f_c'.

The examples in this section illustrate procedures that are reasonably adequate for beams that occur in ordinary beam and slab construction. When special T-sections occur with thin flanges (t *less than* $d/8$ or so) or narrow effective flange widths (b_f less than three times b_w or so), these methods may not be valid. In such cases more accurate investigation should be performed, using the requirements of the *ACI Code*.

27.6. BEAMS WITH COMPRESSIVE REINFORCEMENT

There are many situations in which steel reinforcing is used on both sides of the neutral axis in a beam. When this occurs, the steel on one side of the axis will be in tension and that on the other side in compression. Such a beam is referred to as a doubly reinforced beam or simply as a beam with compressive reinforcing (it being naturally assumed that there is also tensile reinforcing). Various situations involving such reinforcing have been discussed in the preceding sections. In summary, the most common occasions for such reinforcing include:

1. The desired resisting moment for the beam exceeds that for which the concrete alone is capable of developing the necessary compressive force.
2. Other functions of the section require the use of reinforcing on both sides of the beam. These include the need for bars to support U-stirrups and situations when torsion is a major concern.
3. It is desired to reduce deflections by increasing the stiffness of the compressive side of the beam. This is most significant for reduction of long-term creep deflections.
4. The combination of loading conditions on the structure result in reversal moments on the section. That is, the section must sometimes resist positive moment, and other times resist negative moment.
5. Anchorage requirements (for development of reinforcing) require that the bottom bars in a continuous beam be extended a significant distance into the supports.

The precise investigation and accurate design of doubly reinforced sections, whether performed by the working stress or by strength design methods, is quite complex and is beyond the scope of work in this book. The following discussion presents an approximation method that is adequate for preliminary design of a doubly reinforced section. For real design situations, this method may be used to establish a first trial design, which may then be more precisely investigated using more rigorous methods.

For the beam with double reinforcing, as shown in Fig. 27.10, we consider the total resisting moment for the section to be the sum of the following two component moments.

M_1 (Fig. 27.10b) is comprised of a section with tension reinforcing only (A_{s1}). This section is subject to the usual procedures for design and investigation, as discussed in Sec. 27.2.

M_2 (Fig. 27.10c) is comprised of two opposed steel areas (A_{s2} and A_s') that function in simple moment couple action, similar to the flanges of a steel beam or the top and bottom chords of a truss.

Ordinarily, we expect that $A_{s2} = A_s'$, since the same grade of steel is usually used for both. However, there are two special considerations that must be made. The first involves the fact that A_{s2} is in tension, while A_s' is in compression. A_s' must therefore be dealt with in a manner similar to that for

column reinforcing, as discussed in Chapter 31. This requires, among other things, that the compressive reinforcing be braced against buckling, using ties similar to those in a tied column.

The second consideration involves the distribution of stress and strain on the section. Referring to Fig. 27.10, it may be observed that, under normal circumstances (kd less than $0.5d$), A_s' will be closer to the neutral axis than A_{s2}. Thus the stress in A_s' will be lower than that in A_{s2} if pure elastic conditions are assumed. However, it is common practice to assume steel to be doubly stiff when sharing stress with concrete in compression, due to shrinkage and creep effects. Thus, in translating from linear strain conditions to stress distribution, we use the relation $f_s'/2n$ (where $n = E_s/E_c$, as discussed in Sec. 27.1. Utilization of this relationship is illustrated in the following examples.

Example 1. A concrete section with $b = 18$ in. [0.457 m] and $d = 21.5$ in. [0.546 m] is required to resist service load moments as follows: dead-load moment = 150 kip-ft [203.4 kN-m], live-load moment = 150 kip-ft [203.4 kN-m]. Using working stress methods, find the required reinforcing. Use $f_c' = 4$ ksi [27.6 MPa] and $f_y = 60$ ksi [414 MPa].
Solution. For the grade 60 reinforcing, we use an allowable stress of $f_s = 24$ ksi [165 MPa]. Then, using Table 27.1, find

$$n = 8, \quad k = 0.375, \quad j = 0.875, \quad p = 0.0141$$

$$R = 0.295 \text{ in kip-in. units [2028 in kN-m units]}$$

Using the R value for the balanced section, the maximum resisting moment of the section is

$$M = Rbd^2 = \frac{0.295}{12}(18)(21.5)^2$$

$$= 205 \text{ kip-ft [278 kN-m]}$$

This is M_1, as shown in Fig. 27.10. Thus

$$M_2 = \text{total } M - M_1 = 300 - 205 = 95 \text{ kip-ft}$$

$$[129 \text{ kN-m}]$$

For M_1 the required reinforcing (A_{s1} in Fig. 27.10) may be found as

$$A_{s1} = pbd = 0.0141(18)(21.5)$$

$$= 5.46 \text{ in.}^2 \text{ [3523 mm}^2\text{]}$$

And assuming that $f_s' = f_s$, we find A_s' and A_{s2} as follows:

$$M_2 = A_s'(d - d') = A_{s2}(d - d')$$

$$A_s' = A_{s2} = \frac{M_2}{f_s(d - d')} = \frac{95(12)}{24(19)}$$

$$= 250 \text{ in.}^2 \text{ [1613 mm}^2\text{]}$$

The total tension reinforcing is thus

$$A_s = A_{s1} + A_{s2} = 5.46 + 2.50$$

$$= 7.96 \text{ in.}^2 \text{ [5136 mm}^2\text{]}$$

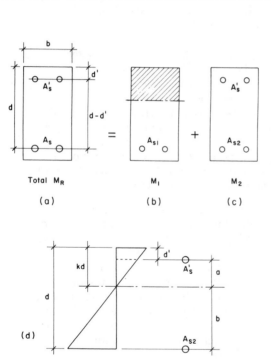

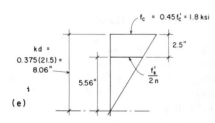

FIGURE 27.10. Basis for simplified analysis of a doubly reinforced section.

For the compressive reinforcing, we must find the proper limit for f_s'. To do this, we assume the neutral axis of the section to be that for the balanced section, producing the situation that is shown in Fig. 27.10. Based on this assumption, the limit for f_s' is found as follows:

$$\frac{f_s'}{2n} = \frac{5.56}{8.06}\,[0.45(4)] = 1.24 \text{ ksi}$$

$$f_s' = 2n(1.24) = 2(8)(1.24) = 19.84 \text{ ksi [137 MPa]}$$

Since this is less than the limit of 24 ksi, we must use it to find A_s'; thus

$$A' = \frac{M_2}{f_s(d-d')} = \frac{95(12)}{19.84(19)} = 3.02 \text{ in.}^2 \text{ [1948 mm}^2]$$

In practice, compressive reinforcing is often used even when the section is theoretically capable of developing the necessary resisting moment with tension reinforcing only. The following two examples illustrate procedures that are applicable in this situation.

Example 2. Design the beam in Example 1 using strength design methods.
Solution. We first find the design moment in the usual manner.

$$M_u = 1.4M_d + 1.7M_l = 1.4(150) + 1.7(150)$$

$$= 465 \text{ kip-ft [631 } kN\text{-}m]$$

$$M_t = \frac{M_u}{\phi} = \frac{465}{0.9} = 517 \text{ kip-ft [701 kN-m]}$$

As a point of reference, we next determine the maximum resisting moment for the section with tension reinforcing only. Thus, using Table 27.2, we find

$$R = 1.041, \quad p = 0.0214, \quad \frac{a}{d} = 0.3773$$

and

$$M = Rbd^2 = \frac{1.041}{12}\,(18)(21.5)^2$$

$$= 722 \text{ kip-ft [1979 kN-m]}$$

This indicates that the section could actually function without compressive reinforcing. However, we will assume that there are compelling reasons for having some compressive reinforcing although its *amount* (magnitude of A_s') becomes somewhat arbitrary. As a rough guide, we suggest a trial design with A_s' approximately one-third of A_s. On the basis of the previous computation, we know that the value for A_s will be less than that required for the full maximum resisting moment. That is, A_s will be less than

$$A_s = pbd = 0.0214(18)(21.5) = 8.28 \text{ in.}^2 \text{ [5342 mm}^2]$$

For a trial design, we choose compressive reinforcing consisting of two No. 9 bars, with $A_s' = 2.0$ in.^{2} [1290 mm^2]. With this reinforcing we may now compute a value for M_2, but to do so we must first establish a value for f_s', the usable stress in the compressive reinforcing. For an approximate de-

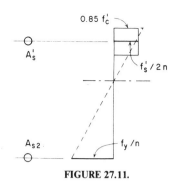

FIGURE 27.11.

sign, we may use the relationship shown in Fig. 27.11, in which we visualize the limit for f_s' to be $2n$ times the maximum stress of $0.85f_c'$ in the concrete. Thus

$$f_s' = 2n(0.85f_c') = 2(8)[0.85(4)]$$

$$= 54.4 \text{ ksi [375 MPa]}$$

Since this value is less than the limiting yield strength, we use it to find M_2, thus

$$M_2 = A_s'f_s'(d-d')$$

$$= 2.0(54.4)(19)(\tfrac{1}{12}) = 172 \text{ kip-ft [233 kN-m]}$$

A_{s2} will have a value different from A_s', since the value for stress for A_{s2} will be the full yield stress of 60 ksi. Thus

$$M_2 = A_{s2}f_y(d-d') = 172 \text{ kip-ft}$$

$$A_{s2} = \frac{172(12)}{60(19)} = 1.81 \text{ in.}^2 \text{ [1168 mm}^2]$$

With the value of M_2 established, we now find the required value for M_1. Thus

$$M_1 = M_t - M_2 = 517 - 172 = 345 \text{ kip-ft}$$

$$[701 - 233 = 468 \text{ kN-m}]$$

To find the required value for A_{s1}, we use the usual procedure for a section with tension reinforcing only, as described in Sec. 27.2. Since the required value for M_1 is almost half of the maximum resisting moment (722 kip-ft, as previously computed), we may assume that a/d will be considerably smaller than the table value of 0.377. For a first guess try

$$\frac{a}{d} = 0.20, \quad a = 0.20(21.5) = 4.3 \text{ in.}$$

Rounding this off to 4 in., we find

$$A_{s1} = \frac{M_1}{f_y(d-a/2)} = \frac{345(12)}{60(19.5)} = 3.53 \text{ in.}^2 \text{ [2277 mm}^2]$$

With this area of steel, $p = 3.53/[18(21.5)] = 0.00912$ and

$$a = pd \frac{f_y}{0.85 f_c'} = 0.00912(21.5) \frac{60}{0.85(4)}$$

$$= 3.46 \text{ in.}$$

For a second try, guess $a = 3.4$ in. Then

$$A_{s1} = \frac{345(12)}{60(19.8)} = 3.48 \text{ in.}^2$$

and

$$\text{total } A_s = A_{s1} + A_{s2} = 3.48 + 1.81 = 5.29 \text{ in.}^2$$

With these computations completed, we now make a choice of reinforcing for the section as follows.

Compressive reinforcing: two No. 9 bars, $A_s' = 2.0$ in.^{2} [1290 mm^2].

Tensile reinforcing: two No. 10 + two No. 11 bars, $A_s = 5.66$ in.^{2} [3652 mm^2].

The following example illustrates a procedure that may be used with the working stress method when the required resisting moment is less than the balanced section limiting moment. It is generally similar to the procedure used with strength design in Example 2.

Example 3. Design a section by the working stress method for a moment of 180 kip-ft [244 kN-m]. Use the section dimensions and data given in Example 1.
Solution. The first step is to investigate the section for its balanced stress limiting moment, as was done in Example 1. This will show that the required moment is less than the balanced moment limit, and that the section could function without compressive reinforcing. Again, we assume that compressive reinforcing is desired, so we assume an arbitrary amount for A_s' and proceed as in Example 2. We make a first guess for the total tension reinforcing as

$$A_s = \frac{M}{f_s(0.9d)} = \frac{180(12)}{24[0.9(21.5)]} = 4.65 \text{ in.}^2 \text{ [3000 mm}^2\text{]}$$

Try

$$A_s' = \tfrac{1}{3} A_s = \tfrac{1}{3}(4.65) = 1.55 \text{ in.}^2 \text{ [1000 mm}^2\text{]}$$

Choose two No. 8 bars,

$$\text{actual } A_s' = 1.58 \text{ in.}^2 \text{ [1019 mm}^2\text{]}$$

Thus

$$A_{s1} = A_s - A_s' = 4.65 - 1.58 = 3.07 \text{ in.}^2 \text{ [1981 mm}^2\text{]}$$

Using A_{s1} for a rectangular section with tension reinforcing only (see Sec. 27.3).

$$p = \frac{3.07}{18(21.5)} = 0.0079$$

Then, from Fig 27.4, we find $k = 0.30$, $j = 0.90$.

Using these values for the section, and the formula involving the concrete stress in compression from Sec. 27.2, we find

$$f_c = \frac{2M_1}{kjbd^2} = \frac{2(120)(12)}{0.3(0.9)(18)(21.5)^2}$$

$$= 1.28 \text{ ksi [8.83 MPa]}$$

With this value for the maximum concrete stress and the value of 0.30 for k, the distribution of compressive stress will be as shown in Fig. 27.12. From this, we determine the limiting value for f_s' as follows.

$$\frac{f_s'}{2n} = \frac{3.95}{6.45} (1.28) = 0.784 \text{ ksi}$$

$$f_s' = 2n(0.784) = 2(8)(0.784) = 12.5 \text{ ksi [86.2 MPa]}$$

Since this is lower than f_s, we use it to find the limiting value for M_2. Thus

$$M_2 = A_s' f_s' (d - d')$$

$$= 1.58(12.5)(19) \left(\frac{1}{12}\right) = 31 \text{ kip-ft [42 kN-m]}$$

To find A_{s2} we use this moment with the full value of $f_s = 24$ ksi. Thus

$$A_{s2} = \frac{M_2}{f_s(d - d')} = \frac{31(12)}{24(19.0)} = 0.82 \text{ in.}^2 \text{ [529 mm}^2\text{]}$$

To find A_{s1}, we determine that

$$M_1 = \text{total } M - M_2 = 180 - 31$$

$$= 149 \text{ kip-ft [202 kN-m]}$$

$$A_{s1} = \frac{M_1}{f_s jd} = \frac{149(12)}{24(0.9)(21.5)} = 3.85 \text{ in.}^2 \text{ [2484 MPa]}$$

Then the total tension reinforcing is found as

$$A_s = A_{s1} + A_{s2} = 3.85 + 0.82$$

$$= 4.67 \text{ in.}^2 \text{ [3013 mm}^2\text{]}$$

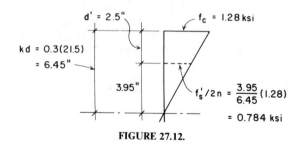

FIGURE 27.12.

CHAPTER TWENTY-EIGHT

Shear

There are many situations in concrete structures that involve the development of shear. In most cases the shear effect itself is not the major concern but rather, the diagonal tension that accompanies the shear action. The material in this chapter deals primarily with the situation of shear in slabs and beams. Shear conditions in footings, walls, and columns are dealt with in other chapters that deal generally with those elements.

28.1. SHEAR SITUATIONS IN CONCRETE STRUCTURES

The most common situations involving shear in concrete structures are shown in Fig 28.1. Shear in beams (Fig. 28.1a) is ordinarily critical near the supports, where the shear force is greatest. In short brackets (Fig. 28.1b) and keys (Fig. 28.1c) the shear action is essentially a direct slicing effect. Punching shear, also called *peripheral shear*, occurs in column footings and in slabs that are directly supported on columns (Fig. 28.1e). When walls are used as bracing elements for shear forces that are parallel to the wall surface (called *shear walls*), they must develop resistance to the direct shear effect that is similar to that in a bracket.

In all of these situations consideration must be given to the shear effect and the resulting shear stresses. Both the magnitude and the direction of the shear stresses must be considered. In many cases, however, the shear effect occurs in combination with other effects, such as bending moment, axial tension, or axial compression. In combined force situations the resulting net combined stress situations must be considered.

28.2. DEVELOPMENT OF SHEAR AND DIAGONAL STRESSES

The general consideration of shear effects is developed in several sections in Part 2, including Sec. 8.4, 9.5, and 12.4. From that development we may note the following:

1. Shear is an ever-present phenomenon, produced directly by slicing actions, by lateral loading in

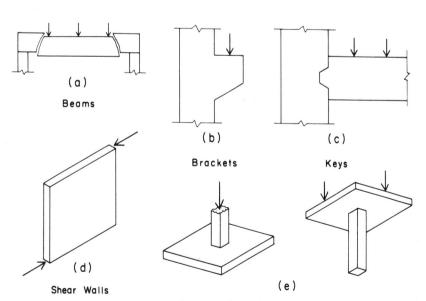

(a)
Beams

(b)
Brackets

(c)
Keys

(d)
Shear Walls

(e)
Punching Shear in Footings and Slabs

FIGURE 28.1. Situations involving shear in concrete structures.

beams, and on oblique sections in tension and compression members.

2. Shear forces produce shear stress in the plane of the force and equal unit shear stresses in planes that are perpendicular to the shear force.

3. Diagonal stresses of tension and compression, having magnitudes equal to that of the shear stress, are produced in directions of 45° from the plane of the shear force.

4. Direct slicing shear force produces a constant magnitude shear stress on affected sections, but beam shear action produces shear stress that varies on the affected sections, having magnitude of zero at the edges of the section and a maximum value at the centroidal neutral axis of the section.

In the discussions that follow, it is assumed that the reader has a general familiarity with these relationships. If not, it would be best to review the appropriate sections in Part 2 before proceeding.

28.3. SHEAR IN BEAMS

Let us consider the case of a simple beam with uniformly distributed load and end supports that provide only vertical resistance (no moment restraint). The distribution of internal shear and bending moment are as shown in Fig. 28.2a. For flexural resistance, it is necessary to provide longitudinal reinforcing bars near the bottom of the beam. These bars are oriented for primary effectiveness in resistance to tension stresses that develop on a vertical (90°) plane (which is the case at the center of the span, where the bending moment is maximum and the shear approaches zero).

Under the combined effects of shear and bending, the beam tends to develop tension cracks as shown in Fig. 28.2b. Near the center of the span, where the bending is predominant and the shear approaches zero, these cracks approach 90°. Near the support, however, where the shear predominates and bending approaches zero, the critical tension stress plane approaches 45°, and the horizontal bars are only partly effective in resisting the cracking.

For beams, the most common form of shear reinforcement consists of a series of U-shaped bent bars (Fig. 28.2d), placed vertically and spaced along the beam span, as shown in Fig. 28.2c. These bars are intended to provide a vertical component of resistance, working in conjunction with the horizontal resistance provided by the flexural reinforcement. In order to develop tension near the support face, the horizontal bars must be bonded to the concrete beyond the point where the stress is developed. Where the beam ends extend only a short distance over the support (a common situation), it is often necessary to bend or hook the bars, as shown in Fig. 28.2.

The simple span beam and the rectangular section shown in Fig. 28.2 occur only infrequently in building structures. The most common case is that of the beam section shown in Fig. 28.3a, which occurs when a beam is poured monolithically with a supported concrete slab. In

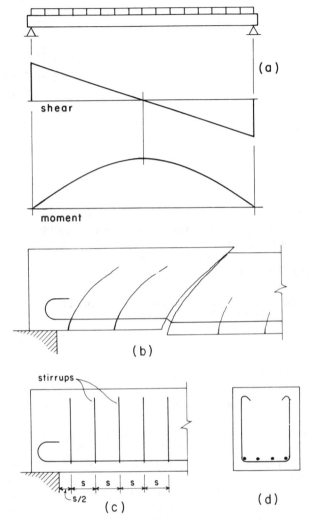

FIGURE 28.2. Considerations of shear in beams.

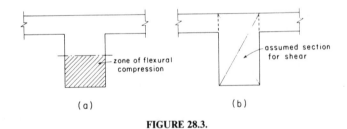

FIGURE 28.3.

addition, these beams normally occur in continuous spans with negative moments at the supports. Thus the stress in the beam near the support is as shown in Fig. 28.3a, with the negative moment producing compressive flexural stress in the bottom of the beam stem. This is substantially different from the case of the simple beam, where the moment approaches zero near the support.

For the purpose of shear resistance, the continuous, T-shaped beam is considered to consist of the section indicated in Fig. 28.3b. The effect of the slab is ignored, and the section is considered to be a simple, rectangular one. Thus for shear design, there is little difference between the simple span beam and the continuous beam, except for the effect of the continuity on the distribution of shear along the beam span. It is important, however, to understand the relationships between shear and moment in the continuous beam.

Figure 28.4 illustrates the typical condition for an interior span of a continuous beam with uniformly distributed load. Referring to the portions of the beam span numbered 1, 2, and 3, we note:

1. In this zone the high negative moment requires major flexural reinforcing consisting of horizontal bars near the top of the beam.
2. In this zone, the moment reverses sign; moment magnitudes are low; and, if shear stress is high, the design for shear is a predominant concern.
3. In this zone, shear consideration is minor and the predominant concern is for positive moment requiring major flexural reinforcing in the bottom of the beam.

Vertical U-shaped stirrups, similar to those shown in Fig. 28.5a. may be used in the T-shaped beam. An alternate detail for the U-shaped stirrup is shown in Fig. 28.5b, in which the top hooks are turned outward; this makes it possible to spread the negative moment reinforcing bars to make placing of the concrete somewhat easier. Figure 28.5c and d show possibilities for stirrups in beams that occur at the edges of large openings or at the outside edge of the structure. This form of stirrup is used to enhance the torsional resistance of the section and also assists in developing the negative moment resistance in the slab at the edge of the beam.

Closed stir-ups, similar to ties in columns, are sometimes used for T-shaped beams, as shown in Fig. 28.5c and f. These are generally used to improve the torsional resistance of the beam section.

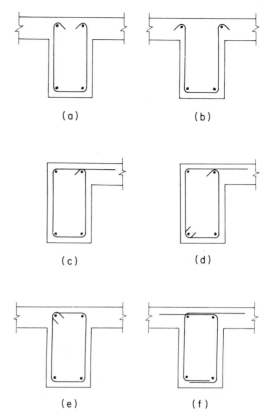

FIGURE 28.5. Various forms for vertical stirrups.

Stirrup forms are often modified by designers or by the reinforcing fabricator's detailers to simplify the fabrication and/or the field installation. The stirrups shown in Fig. 28.5d and f are two such modifications of the basic details in Fig. 28.5c and e, respectively.

Beam Shear: General Considerations

The following are some of the general considerations and code requirements that apply to current practices of design for beam shear.

Concrete Capacity. Whereas the tensile strength of the concrete is ignored in design for flexure, the concrete is assumed to take some portion of the shear in beams. If the capacity of the concrete is not exceeded—as it sometimes is for lightly loaded beams—there may be no need for reinforcing. The typical case, however, is as shown in Fig. 28.6, where the maximum shear V exceeds the capacity of the concrete alone (V_c), and the steel reinforcing is required to absorb the excess, which is indicated as the shaded portion in the shear diagram.

Minimum Shear Reinforcing. Even when the maximum computed shear stress falls below the capacity of the concrete, the present code requires the use of some minimum amount of shear reinforcing. Exceptions are made in some situations, such as for slabs and very shallow beams. The

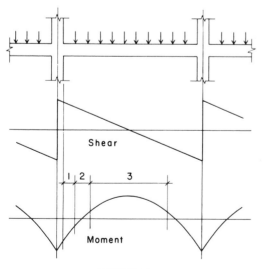

FIGURE 28.4.

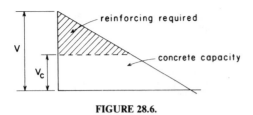

FIGURE 28.6.

objective is essentially to toughen the structure with a small investment in additional reinforcing.

Type of Stirrup. The most common stirrups are the simple U shape or closed forms shown in Fig. 28.5, placed in a vertical position at intervals along the beam. It is also possible to place stirrups at an incline (usually 45°), which makes them somewhat more effective in direct resistance to the potential shear cracking near the beam ends (see Fig. 28.2). In large beams with excessively high unit shear stress, both vertical and inclined stirrups are sometimes used at the location of the greatest shear.

Size of Stirrups. For beams of moderate size, the most common size for U-stirrups is a No. 3 bar. These bars can be bent relatively tightly at the corners (small radius of bend) in order to fit within the beam section. For larger beams, a No. 4 bar is sometimes used, its strength (as a function of its cross-sectional area) being almost twice that of a No. 3 bar.

Reinforcing for Narrow Beams. When beams are less than about 10 in. wide, it is not possible to bend a U-shaped stirrup to fit within the beam profile. If shear reinforcing is required, one form that is used is the *ladder* stirrup, shown in Fig. 28.7. This consists of a series of single vertical bars welded to horizontal bars at the top and bottom. A variation on this type of reinforcing consists of using a portion of heavy-gage welded wire fabric.

Spacing of Stirrups. Stirrup spacings are computed (as discussed in the following sections) on the basis of the amount of reinforcing required for the unit shear stress at the location of the stirrups. A maximum spacing of $d/2$ (i.e., one-half the effective beam depth d) is specified in order to assure that at least one stirrup occurs at the location of any potential diagonal crack (see Fig. 28.2). When shear stress is excessive, the maximum spacing is limited to $d/4$.

Critical Maximum Design Shear. Although the actual maximum shear value occurs at the end of the beam, the code permits the use of the shear stress at a distance of d (effective beam depth) from the beam end as the critical maximum for stirrup design. Thus, as shown in Fig. 28.8, the shear requiring reinforcing is slightly different from that shown in Fig. 28.6.

Total Length for Shear Reinforcing. On the basis of computed shear stresses, reinforcing must be provided along the beam length for the distance defined by the shaded portion of the shear stress diagram shown in Fig. 28.8. For the center portion of the span, the concrete is theoretically capable of the necessary shear resistance without the assistance of reinforcing. However, the code requires that some reinforcing be provided for a distance beyond this computed cutoff point. The 1963 *ACI Code* required that stirrups be provided for a distance equal to the effective depth of the beam beyond the cutoff point. The 1989 *ACI Code* requires that minimum shear reinforcing be provided as long as the computed shear stress exceeds one-half of the capacity of the concrete. However it is established, the total extended range over which reinforcing must be provided is indicated as R on Fig. 28.8.

28.4. DESIGN FOR BEAM SHEAR

Working Stress Method

The following is a description of the procedure for design of shear reinforcing for beams, which is in compliance with Appendix B of the 1989 *ACI Code* (Ref. 6).

Shear stress is computed as

$$v = \frac{V}{bd}$$

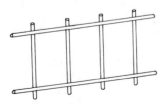

FIGURE 28.7. "Ladder" shear reinforcement for a narrow beam.

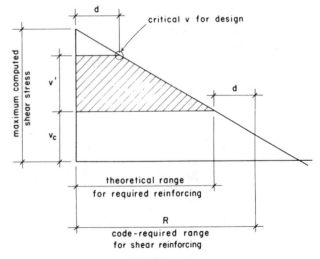

FIGURE 28.8.

where V = total shear force at the section

b = beam width (of the stem for T-shapes)

d = effective depth of the section

For beams of normal weight concrete, subjected only to flexure and shear, shear stress in the concrete is limited to

$$v_c = 1.1 \sqrt{f_c'}$$

When v exceeds the limit for v_c, reinforcing must be provided, complying with the general requirements discussed in Sec. 28.2. Although the code does not use the term, we coin the notation of v' for the excess unit shear for which reinforcing is required. Thus

$$v' = v - v_c$$

Required spacing of shear reinforcement is determined as follows. Referring to Fig. 28.9, we note that the capacity in tensile resistance of a single, two-legged stirrup is equal to the product of the total steel cross-sectional area times the allowable steel stress. Thus

$$T = A_v f_s$$

This resisting force opposes the development of shear stress on the area s times b, where b is the width of the beam and s is the spacing (half the distance to the next stirrup on each side). Equating the stirrup tension to this force, we obtain the equilibrium equation

$$A_v f_f = bsv'$$

From this equation, we can derive an expression for the required spacing; thus

$$s = \frac{A_v f_s}{v' b}$$

The following example illustrates the procedure for a simple beam.

Example 1.   Using the working stress method, design the required shear reinforcing for the simple beam shown in Fig 28.10. Use f_c' = 3 ksi [20.7 MPa] and f_s = 20 ksi [138 MPa] and single U-shaped stirrups.

Solution.   The maximum value for the shear is 40 kips [178 kN], and the maximum value for shear stress is computed as

$$v = \frac{V}{bd} = \frac{40,000}{12(24)} = 139 \text{ psi [957 kPa]}$$

We now construct the shear stress diagram for one-half of the beam, as shown in Fig. 28.10c. For the shear design, we determine the critical shear stress at 24 in. (the effective depth of the beam) from the support. Using proportionate triangles, this value is

$$\frac{72}{96}(139) = 104 \text{ psi [718 kPa]}$$

The capacity of the concrete without reinforcing is

$$v_c = 1.1\sqrt{f_c'} = 1.1\sqrt{3000} = 60 \text{ psi [414 kPa]}$$

At the point of critical stress, therefore, there is an excess shear stress of $104 - 60 = 44$ psi [$718 - 414 = 304$ kPa] that must be carried by reinforcing. We next complete the construction of the diagram in Fig. 28.10c to define the shaded portion, which indicates the extent of the required reinforcing. We thus observe that the excess shear condition extends to 54.4 in. [1.382 m] from the support.

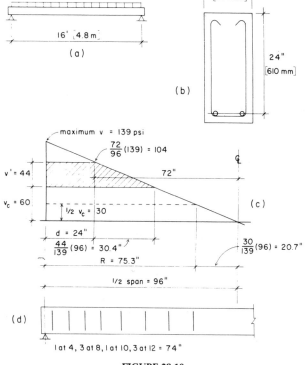

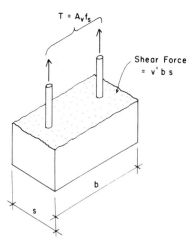

FIGURE 28.9.   **FIGURE 28.10.**

In order to satisfy the requirements of the 1983 *ACI Code*, shear reinforcing must be used wherever the computed unit stress exceeds one-half of v_c. As shown in Fig. 28.10c, this is a distance of 75.3 in. from the support. The code further stipulates that the minimum cross-sectional area of this reinforcing be

$$A_v = 50 \frac{bs}{f_y}$$

If we assume an f_y value of 50 ksi [345 MPa] and use the maximum allowable spacing of one-half the effective depth, the required area is

$$A_v = (50) \frac{12(12)}{50,000} = 0.144 \text{ in.}^2$$

which is less than the area of $2 \times 0.11 = 0.22$ in.² provided by the two legs of the No. 3 stirrup.

For the maximum v' value of 44 ksi, the maximum spacing required is determined as

$$s = \frac{A_v f_s}{v'b} = \frac{(0.22 \text{ in.}^2)(20,000 \text{ psi})}{(44 \text{ psi})(12 \text{ in.})} = 8.3 \text{ in.}$$

Since this is less than the maximum allowable of 12 in., it is best to calculate at least one more spacing at a short distance beyond the critical point. We thus determine that the unit stress at 36 in. from the support is

$$v = \frac{60}{96}(139) = 87 \text{ psi}$$

and the value of v' at this point is $87 - 60 = 27$ psi. The spacing required at this point is thus

$$s = \frac{0.22(20,000)}{27(12)} = 13.6 \text{ in.}$$

which indicates that the required spacing drops to the maximum allowed at less than 12 in. from the critical point. A possible choice for the stirrup spacings is shown in Fig. 28.10d, with a total of eight stirrups that extend over a range of 74 in. from the support. There are thus a total of 16 stirrups in the beam, 8 at each end.

Example 2. Determine the required number and spacings for No. 3 U-stirrups for the beam shown in Fig. 28.11. Use $f_c = 3$ ksi [20.7 MPa] and $f_s = 20$ ksi [138 MPa].
Solution. As in Example 1, the shear values and corresponding stresses are determined, and the diagram in Fig. 28.11c is constructed. In this case, the maximum critical shear stress of 89 psi results in a maximum v' value of 29 psi, for which the required spacing is

$$s = \frac{0.22(20,000)}{29(10)} = 15.2 \text{ in.}$$

Since this value exceeds the maximum limit of $d/2 = 10$ in., the stirrups may all be placed at the limiting spacing, and a possible arrangement is as shown in Fig. 28.11d.

Note that in both Examples 1 and 2 the first stirrup is placed at one-half the required distance from the support.

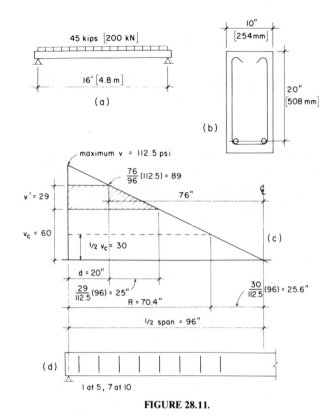

FIGURE 28.11.

Example 3. Determine the required number and spacings for No. 3 U-stirrups for the beam shown in Fig. 28.12. Use $f_c' = 3$ ksi [20.7 MPa] and $f_s = 20$ ksi [138MPa].
Solution. In this case, the maximum critical design shear stress is found to be less than v, which in theory indicates that reinforcing is not required. To comply with the code requirement for minimum reinforcing, however, we provide stirrups at the maximum permitted spacing out to the point where the shear stress drops to 30 psi (one-half of v_c). To verify that the No. 3 stirrup is adequate, we compute

$$A_v = (50) \frac{10(10)}{50,000} = 0.10 \text{ in.}^2 \quad \text{(see Example 1)}$$

which is less than the area of 0.22 in. provided, so the No. 3 stirrup at 10-in. spacing is adequate.

Examples 1 through 3 have illustrated what is generally the simplest case for beam shear design—that of a beam with uniformly distributed load and with sections subjected only to flexure and shear. When concentrated loads or unsymmetrical loadings produce other forms for the shear diagram, these must be used for design of the shear reinforcing. In addition, where axial forces of tension or compression exist in the concrete frame, consideration must be given to the combined effects when designing for shear.

When torsional moments exist (twisting moments at right angles to the beam), their effects must be combined with beam shear.

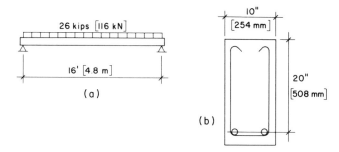

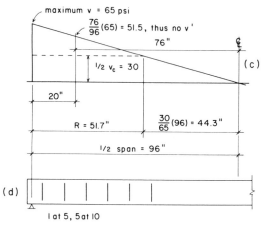

FIGURE 28.12.

Strength Method

The requirements and procedures for strength design are essentially similar to those for working stress design. The principal difference is in the use of ultimate resistance as opposed to working stresses at service loads. The basic requirement in strength design is that the modified ultimate resistance of the section be equal to or greater than the factored load. This condition is stated as

$$V_u \leq \phi V_n$$

where V_u = factored shear force at the section
        V_n = nominal shear strength of the section

The nominal strength is defined as

$$V_n = V_c + V_s$$

where V_c = nominal strength provided by concrete
        V_s = nominal strength provided by reinforcing

The term *nominal strength* is used to differentiate between the computed resistances and the usable value of total resistance, which is reduced for design by the strength reduction factor ϕ.

For members subjected to shear and flexure only, the nominal concrete strength is defined as

$$V_c = 2 \sqrt{f_c'} \; bd$$

Translated into unit stress terms, this means that the limiting nominal shear stress in the concrete is $2 \sqrt{f_c'}$, and when reduced by ϕ, the limiting *working* ultimate strength is $0.85 \times 2 \sqrt{f_c'} = 1.7 \sqrt{f_c'}$.

When shear reinforcing consists of vertical stirrups, the nominal reinforcing strength is defined as

$$V_s = \frac{A_v \, f_y d}{s}$$

with a limiting value for V_s established as

$$V_s = 8 \sqrt{f_c'} \; bd$$

The following example illustrates the use of strength design methods for shear reinforcing. The problem data are essentially the same as for Example 1, so that a comparison of the design results can be made.

Example 4. Using strength design methods, determine the spacing required for No. 3 U-stirrups for the beam shown in Fig. 28.13. Use $f_c' = 3$ ksi [20.7 MPa] and $f_y = 50$ ksi [345 Mpa].
Solution. The loads shown in Fig. 28.13*a* are service loads. These must be converted to *factored loads* for strength design, as discussed in Sec. 26.5. We thus determine the factored load to be

$$
\begin{aligned}
W_u &= 1.4(\text{dead load}) + 1.7(\text{live load}) \\
&= 1.4(40) + 1.7(40) \\
&= 124 \text{ kips}
\end{aligned}
$$

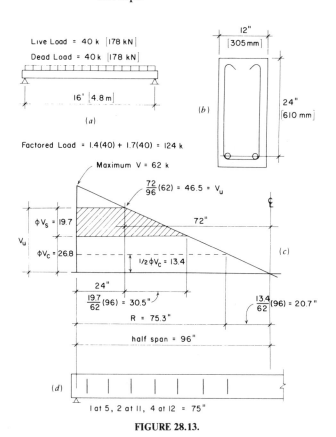

FIGURE 28.13.

The maximum shear force is thus 62 kips, and the shear diagram for one-half the beam is as shown in Fig. 28.13c. The critical value for V_u at 24 in. (effective beam depth) from the support is determined from proportionate triangles to be 46.5 kips. The usable capacity of the concrete is determined as

$$\phi V_c = \phi 2 \sqrt{f_c'} \ bd$$
$$= 0.85(2 \sqrt{3000})(12)(24)$$
$$= 26,816 \ lb \quad or \quad approximately \ 26.8 \ kips$$

and for the reinforcing

$$\phi V_s = 46.5 - 26.8 = 19.7 \ kips$$

Therefore,

$$V_s = \frac{19.7}{\phi} = \frac{19.7}{0.85} = 23.18 \ kips$$

and the required spacing is determined from

$$V_s = \frac{A_v f_y d}{s}$$
$$s = \frac{A_v f_y d}{V_s} = \frac{0.22(50)(24)}{23.18} = 11.4 \ in.$$

Referring to Example 1, we may see that this value is larger than that computed by the working stress method; thus the strength design is somewhat less conservative for this example.

A possible choice of stirrup spacings is that shown in Fig. 28.12d, using seven stirrups at each end of the beam.

To verify that the value for V_s is within the limit previously given, we compute the maximum value of

$$V_s = 8 \sqrt{f_c'} \ bd = 8 \sqrt{3000}(12)(24) = 126 \ kips$$

which is far from critical.

CHAPTER TWENTY-NINE

Bond and Development

Bond stresses develop on the surfaces of reinforcing bars whenever some structural action requires the steel and concrete to interact. In times past, working stress procedures included the establishment of allowable stresses for bond and the computation of bond stresses for various situations. At present, however, the codes deal with this problem as one of development length. Chapter 29 presents a discussion of bond stress situations and the current practices in establishing required lengths for the development of reinforcement.

29.1. BOND STRESS AND BAR DEVELOPMENT

The basic concept of bond stress development is illustrated by the example shown in Fig. 29.1, in which a steel bar is embedded in a block of concrete and is required to resist a pull-out tension force. Figure 29.1b shows the static equilibrium relationship for the steel bar, with the pull-out force developed as the product of a tensile stress × the area of the bar cross section $[f_s(\pi D^2/4)]$ and the resisting force developed by a bond stress (u) operating on the surface of the bar $[u(\pi D)L]$. By equating these two forces, we can

derive an expression either for the unit bond stress or the required embedment length for a limiting bond stress.

Bond stress development is affected by a number of considerations; some of the major ones are the following:

1. *Grade of Steel.* As the f_y of the steel is increased, the allowable f_s value will also increase, requiring the development of higher bond stresses or the need for greater embedment lengths.
2. *Strength of Concrete.* In general, as f_c' is increased, the capability for development of bond stress is also increased.
3. *Bar Size.* Consideration of the expression for the tension force in the bar in Fig. 29.1 will indicate that the force capability of the bar increases with the square of the diameter. On the other hand, the resistance developed by bond stress increases only linearly with increase of the bar diameter. Thus bond stresses tend to be more critical on bars of large diameter.
4. *Concrete Encasement.* The bonding force must be developed in the concrete mass around each bar. This development is limited when this mass is constrained due to closely spaced groups of bars or where bars are placed close to the edge of the concrete member.
5. *Location of Bars.* When concrete is poured into forms and cured into its hardened state, the concrete near the bottom of the member tends to develop slightly higher quality than that near the top. The weight of the concrete mass above produces a denser material in the lower concrete, and the exposed top surface tends to dry more rapidly, resulting in less well-cured concrete near the top. This difference in quality affects the potential for bond resistance, so some adjustment is made for bars placed near the top (such as reinforcement for negative moment in beams).

Development of Reinforcement

The *ACI Code* defines *development length* as the length of embedment required to develop the design strength of the reinforcing at a critical section. For beams, critical sections occur at points of maximum stress and at points within the

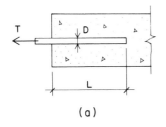

(a)

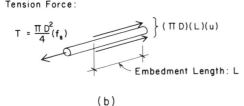

Tension Force:

$$T = \frac{\pi D^2}{4}(f_s)$$

Bond Force:

$(\pi D)(L)(u)$

Embedment Length: L

(b)

FIGURE 29.1. Development of bond stress.

span where some of the reinforcement terminates or is bent up or down. For a uniformly loaded simple span beam, one critical section is at midspan, where the bending moment is a maximum. The tensile reinforcing required for flexure at this point must extend on both sides a sufficient distance to develop the stress in the bars; however, except for very short spans with large bars, the bar lengths will ordinarily be more than sufficient.

In the simple beam, the bottom reinforcing required for the maximum moment at midspan is not entirely required as the moment decreases toward the end of the span. It is thus sometimes the practice to make only part of the midspan reinforcing continuous for the whole beam length. In this case it may be necessary to assure that the bars that are of partial length are extended sufficiently from the midspan point and that the bars remaining beyond the cutoff point can develop the stress required at that point.

When beams are continuous through the supports, top reinforcing is required for the negative moments at the supports. These top bars must be investigated for the development lengths in terms of the distance they extend from the supports.

For tension reinforcing consisting of bars of No. 11 size and smaller, the code specifies a minimum length for development (l_d) as follows:

$$l_d = 0.04 A_b \frac{f_y}{\sqrt{f_c'}}$$

but not less than $0.0004 d_b f_y$ or 12 in. In these formulas A_b is the cross-sectional area of the bar and d_b is the bar diameter.

Modification factors for l_d are given for various situations, as follows:

For top bars in horizontal members with at least 12 in. of concrete below the bars — 1.4

For sets of bars where the bars are 6 in. or more on center — 0.8

For flexural reinforcement that is in excess of that required by computations — $\dfrac{A_s \text{ required}}{A_s \text{ provided}}$

Additional modification factors are given for lightweight concrete, for bars encased in spirals, and for bars with f_y in excess of 60 ksi.

Table 29.1 gives values for minimum development lengths for tensile reinforcing, based on the requirements of the *ACI Code*. The values listed under "other bars" are the unmodified length requirements; those listed under "top bars" are increased by the modification factor for this situation. Values are given for two concrete strengths and for the two most commonly used grades of tensile reinforcing.

29.2. HOOKS

When details of the construction restrict the ability to extend bars sufficiently to produce required development lengths, partial development can sometimes be achieved by use of a hooked end. Section 12.5 of the *ACI Code* provides a means by which a so-called standard hook may be evaluated in terms of an equivalent development length. Detailed requirements for standard hooks are given in Chapter 7 of the *ACI Code*. Bar ends may be bent at 90, 135, or 180° to produce a hook. The 135° bend is used only for

TABLE 29.1. Minimum Development Length of Tensile Reinforcement (in.)[a]

Bar Size	$f_y = 40$ ksi [276 MPa]				$f_y = 60$ ksi [414 MPa]			
	$f_c' = 3$ ksi [20.7 MPa]		$f_c' = 4$ ksi [27.6 MPa]		$f_c' = 3$ ksi [20.7 MPa]		$f_c' = 4$ ksi [27.6 MPa]	
	Top Bars[b]	Other Bars	Top Bars[b]	Other Bars	Top Bars[b]	Other Bars	Top Bars[b]	Other Bars
3	12	12	12	12	13	12	13	12
4	12	12	12	12	17	12	17	12
5	14	12	12	12	21	15	21	15
6	18	13	16	12	27	19	15	18
7	25	18	21	15	37	26	32	23
8	32	23	28	20	48	35	42	30
9	41	29	36	25	61	44	53	38
10	52	37	45	32	78	56	68	48
11	64	46	55	40	96	68	83	59
14	87	62	75	54	130	93	113	81
18	113	80	98	70	169	120	146	104

[a] Lengths are based on requirements of the *ACI Code* (Ref. 6).
[b] Horizontal bars so placed that more than 12 in. [305 mm] of concrete is cast in the member below the reinforcement.

ties and stirrups, which normally consist of relatively small diameter bars (see Fig. 26.3).

Table 29.2 gives values for standard hooks, using the same variables for f_c' and f_y that are used in Table 29.1. The table values given are in terms of the equivalent development length provided by the hook. Comparison of the values in Table 29.2 with those given for the unmodified lengths ("other") in Table 29.1, will show that the hooks are mostly capable of only partial development. The development length provided by a hook may be added to whatever development length is provided by extension of the bar, so that the total development may provide for full utilization of the bar tension capacity (at f_y) in many cases. The following example illustrates the use of the data from Tables 29.1 and 29.2 for a simple situation.

Example 1. The negative moment in the short cantilever shown in Fig. 29.2 is resisted by the steel bar in the top of the beam. Determine whether the development of the reinforcing is adequate. Use $f_c' = 3$ ksi [20.7 MPa] and $f_y = 60$ ksi [414 MPa].

Solution. The maximum moment in the cantilever is produced at the face of the support; thus the full tensile capacity of the bar should be developed on both sides of this section. In the beam itself the condition is assumed to be that of a "top bar," for which Table 29.1 yields a required minimum development length of 27 in., indicating that the length of 46 in. provided is more than adequate. Within the support, the condition is unmodified, and the requirement is for a length of 19 in. The actual extended development length provided within the support is 14 in., which is measured as the distance to the end of the hooked bar end, as shown in the figure. If the hooked end qualifies as a *standard hook* (in accordance with the requirements of Chapter 7 of the *ACI Code*), the equivalent development length provided (from Table 29.2) is 9.5 in. Thus the total development provided by the combination of extension and hooking is 14 + 9.5 = 23.5 in., which exceeds the requirement of 19 in., so the development is adequate.

In a real situation, it is probably not necessary to achieve the full development lengths given in Table 29.1,

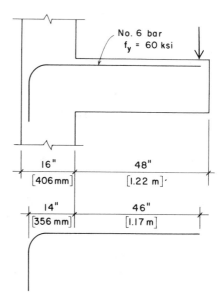

FIGURE 29.2.

since bar selection often results in some slight excess in the actual steel cross-sectional area provided. In such a case, the required development length can be reduced by the modification factor given in Sec. 29.2.

29.3. DEVELOPMENT OF COMPRESSIVE REINFORCEMENT

The discussion of development length so far has dealt with tension bars only. Development length in compression is, of course, a factor in column design and in the design of beams reinforced for compression.

The absence of flexural tension cracks in the portions of beams where compression reinforcement is employed, plus the beneficial effect of the end bearing of the bars on the concrete, permit shorter development lengths in compres-

TABLE 29.2. Equivalent Embedment Lengths of Standard Hooks (in.)

| Bar Size | $f_y = 40$ ksi [276 MPa] | | $f_y = 60$ ksi [414 MPa] | |
	$f_c' = 3$ ksi [20.7 MPa]	$f_c' = 4$ ksi [27.6 MPa]	$f_c' = 3$ ksi [20.7 MPa]	$f_c' = 4$ ksi [27.6 MPa]
3	3.0	3.4	4.4	5.1
4	3.9	4.5	5.9	6.8
5	4.9	5.7	7.4	8.5
6	6.3	6.8	9.5	10.2
7	8.6	8.6	12.9	12.9
8	11.4	11.4	17.1	17.1
9	14.4	14.4	21.6	21.6
10	18.3	18.3	24.4	24.4
11	22.5	22.5	26.2	26.2

sion than in tension. The *ACI Code* prescribes that l_d for bars in compression shall be computed by the formula

$$l_d = \frac{0.02 f_y d_b}{\sqrt{f_c'}}$$

but shall not be less than $0.0003 f_y d_b$ or 8 in., whichever is greater. Table 29.3 lists compression bar development lengths for a few combinations of specification data.

In reinforced concrete columns both the concrete and the steel bars share the compression force. Ordinary construction practices require the consideration of various situations for development of the stress in the reinforcing bars. Figure 29.3 shows a multistory concrete column with its base supported on a concrete footing. With reference to the illustration, we note the following.

1. The concrete construction is ordinarily produced in multiple, separate pours, with construction joints between the separate pours occurring as shown in the illustration.

2. In the lower column, the load from the concrete is transferred to the footing in direct compressive bearing at the joint between the column and footing. The load from the reinforcing must be developed by extension of the reinforcing into the footing: distance L_1 in the illustration. Although it may be possible to place the column bars in position during pouring of the footing to achieve this, the common practice is to use dowels, as shown in the illustration. These dowels must be developed on both sides of the joint: L_1 in the footing and L_2 in the column. If the f_c value for both the footing and the column are the same, these two required lengths will be the same.

3. The lower column will ordinarily be cast together with the supported concrete framing above it, with a construction joint occurring at the top level of the framing (bottom of the upper column), as shown in the illustration. The distance L_3 is that required to develop the reinforcing in the lower column—bars a1 in the illustration. As for the condition at the top of the footing, the distance L_4 is re-

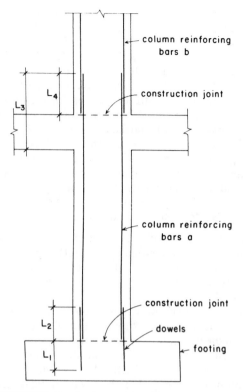

FIGURE 29.3. Bar development considerations for columns.

quired to develop the reinforcing in bars b in the upper column. L_4 is more likely to be the critical consideration for the determination of the extension required for bars *a*.

29.4. BAR DEVELOPMENT IN BEAMS

The *ACI Code* defines *development length* as the length of embedded reinforcement required to develop the design

TABLE 29.3. Minimum Development Length for Compressive Reinforcement (in.)

Bar Size	f_y = 40 ksi [276 MPa]		f_y = 60 ksi [414 MPa]	
	f_c' = 3 ksi [20.7 MPa]	f_c' = 4 ksi [27.6 MPa]	f_c' = 3 ksi [20.7 MPa]	f_c' = 4 ksi [27.6 MPa]
3	8.0	8.0	8.0	8.0
4	8.0	8.0	11.0	9.5
5	9.2	8.0	13.7	11.9
6	10.9	9.5	16.4	14.2
7	12.8	11.1	19.2	16.6
8	14.6	12.7	21.9	19.0
9	16.5	14.3	24.8	21.5
10	18.5	16.1	27.8	24.1
11	20.6	17.9	31.0	26.8
14	—	—	37.1	32.1
18	—	—	49.5	42.8

strength of the reinforcement at a critical section. Critical sections occur at points of maximum stress and at points within the span at which adjacent reinforcement terminates or is bent up into the top of the beam. For a uniformly loaded simple beam, one critical section is at midspan where the bending moment is maximum. This is a point of maximum tensile stress in the reinforcement (peak bar stress), and some length of bar is required over which the stress can be developed. Other critical sections occur between midspan and the reactions at points where some bars are cut off because they are no longer needed to resist the bending moment; such terminations create peak stress in the remaining bars that extend the full length of the beam.

When beams are continuous through their supports, the negative moments at the supports will require that bars be placed in the top of the beams. Within the span, bars will be required in the bottom of the beam for the positive moments. While the positive moment will go to zero at some distance from the supports, the codes require that some of the positive moment reinforcing be extended for the full length of the span and a short distance into the support.

Figure 29.4 shows a possible layout for reinforcing in a beam with continuous spans and a cantilevered end at the first support. Referring to the notation in the illustration, we make the following observations.

1. Bars a and b are provided for the maximum moment of positive sign that occurs somewhere near the beam midspan. If all these bars are made full length (as shown for bars a), the length L_1 must be sufficient for development (this situation is seldom critical). If bars b are partial length, as shown in the illustration, then length L_2 must be sufficient to develop bars b and length L_3 must be sufficient to develop bars a. As was discussed for the simple beam, the partial length bars must actually extend beyond the theoretical cutoff point (B in the illustration) and the true length must include the dashed portions indicated for bars b.

2. For the bars at the cantilevered end, the distances L_4 and L_5 must be sufficient for development of bars c. L_4 is required to extend beyond the actual cutoff point of the negative moment by the extra length described for the partial length bottom bars. If L_5 is not adequate, the bar ends may

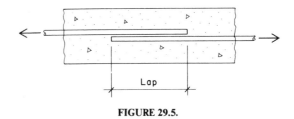

FIGURE 29.5.

be bent into the 90° hook as shown or the 180° hook shown by the dashed line.

3. If the combination of bars shown in the illustration is used at the interior support, L_6 must be adequate for the development of bars d and L_7 adequate for the development of bars e.

For a single loading condition on a continuous beam it is possible to determine specific values of moment and their location along the span, including the locations of points of zero moment. In practice, however, most continuous beams are designed for more than a single loading condition, which further complicates the problems of determining development lengths required.

29.5. SPLICES

In various situations in reinforced concrete structures it becomes necessary to transfer stress between steel bars in the same direction. Continuity of force in the bars is achieved by splicing, which may be affected by welding, by mechanical means, or by the so-called lapped splice. Figure 29.5 illustrates the concept of the lapped splice, which consists essentially of the development of both bars within the concrete. The length of the lap becomes the development length for both bars. Because a lapped splice is usually made with the two bars in contact, the lapped length must usually be somewhat greater than the simple development length required in Table 29.1.

Sections 12.15 to 12.20 of the *ACI Code* give requirements for various types of splices. For a simple tension lap splice, the full development of the bars requires a lap length of 1.7 times that required for simple development of the bars. Lap splices are generally limited to bars of No. 11 size and smaller.

For pure tension members, lapped splicing is not permitted, and splicing must be achieved by welding the bars or by some mechanical connection. End-to-end butt welding of bars is usually limited to compression splicing of large diameter bars with high f_y for which lapping is not feasible.

When members have several reinforcing bars that must be spliced, the splicing must be staggered. In general, splicing is not desirable, and is to be avoided where possible. Because bars are obtainable only in limited lengths, however, some situations unavoidably involve splicing. Horizontal reinforcing in long walls is one such case.

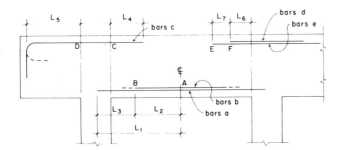

FIGURE 29.4. Various situations for consideration of development length.

CHAPTER THIRTY

Concrete Framing Systems

There are many different reinforced concrete floor systems, both cast in place and precast. The cast-in-place systems are generally of one of the following types:

1. One-way solid slab and beam.
2. Two-way solid slab and beam.
3. One-way concrete joist construction.
4. Two-way flat slab or flat plate without beams.
5. Two-way joist construction, called *waffle construction*.

Each system has its distinct advantages and limitations, depending on the spacing of supports, magnitude of loads, required fire rating, and cost of construction. The floor plan of the building and the purpose for which the building is to be used determine loading conditions and the layout of supports. Whenever possible, columns should be aligned in rows and spaced at regular intervals in order to simplify and lower the cost of the building construction.

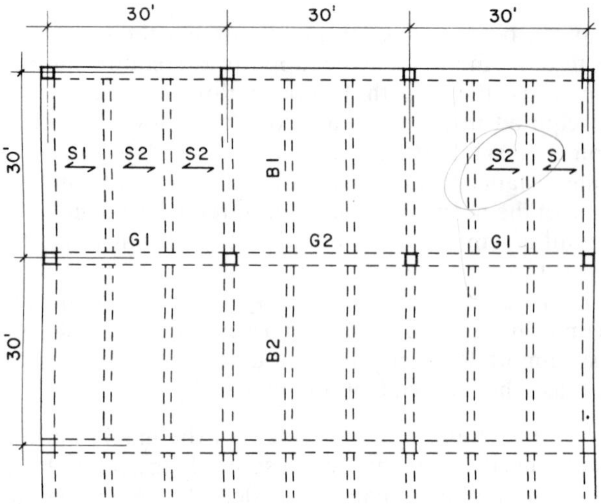

FIGURE 30.1. Plan of a typical slab-beam-girder framing system.

30.1. SLAB AND BEAM SYSTEMS

The most widely used and most adaptable poured-in-place concrete floor system is that which utilizes one-way solid slabs supported by one-way spanning beams. This system may be used for single spans, but occurs more frequently with multiple-span slabs and beams in a system such as that shown in Fig. 30.1. In the example shown, the continuous slabs are supported by a series of beams that are spaced at 10 ft center to center. The beams, in turn, are supported by a girder and column system with columns at 30-ft centers, every third beam being supported directly by the columns and the remaining beams being supported by the girders.

Because of the regularity and symmetry of the system shown in Fig. 30.1, there are relatively few different elements in the system, each being repeated several times. While special members must be designed for conditions that occur at the outside edge of the system and at the location of any openings for stairs, elevators, and so on, the general interior portions of the structure may be determined by designing only six basic elements: S1, S2, B1, B2, G1, and G2, as shown in the framing plan. The design of these typical elements is illustrated in Chapter 58.

In computations for reinforced concrete, the span length of freely supported beams (simple beams) is generally taken as the distance between centers of supports or bearing areas; it should not exceed the clear span plus the depth of beam or slab. The span length for continuous or restrained beams is taken as the clear distance between faces of supports. For a simple beam, that is, a single span having no restraint at the supports, the maximum bending moment for a uniformly distributed load is at the center of the span, and its magnitude is $M = WL/8$. The moment is zero at the supports and is positive over the entire span length. In continuous beams, however, negative bending moments are developed at the supports and positive moments at or near midspan. This may be readily observed from the exaggerated deformation curve of Fig. 30.2a. The exact values of the bending moments depend on several factors, but in the case of approximately equal spans supporting uniform loads, when the live load does not exceed three times the dead load, the bending moment values given in Fig. 30.2b and c may be used for design.

The values given in Fig. 30.2 are in general agreement with the recommendations of Chapter 8 of the *ACI Code*. These values have been adjusted to account for partial live loading of multiple-span beams. Note that these values apply only to uniformly loaded beams.

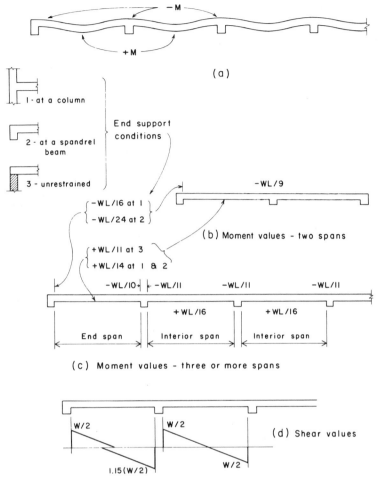

FIGURE 30.2. Approximate design factors for continuous structures.

Design moments for continuous-span slabs are given in Fig. 30.2c. Where beams are relatively large, and the slab spans are small, the rotational (torsional) stiffness of the beam tends to minimize the effect of individual slab spans on the bending in adjacent spans. Thus most slab spans in the slab-beam systems tend to function much like individual spans with fixed ends. For slabs with spans of 10 ft or less, the *ACI Code* permits the consideration of individual spans as isolated fixed-ended spans, with a maximum negative moment at all supports of $wL^2/12$.

Continuous One-Way Slabs

There are a great number of considerations to be made in the design of a spanning concrete slab. The basic functioning of this structural element is discussed in Sec. 27.4. That discussion also treats various concerns, such as those for reinforcing spacing, temperature and shrinkage reinforcing, concrete cover of reinforcement, and selection of bars. An example of the design of a continuous slab is presented with the design of the concrete floor structure in Chapter 58.

A critical factor in design is the selection of the slab thickness. Various concerns for this include the following.

1. *Flexural Stress.* Compressive steel is not used, so the limit is for a balanced section, although economy is usually achieved with lower percentages of reinforcement.
2. *Shear Stress.* Shear reinforcement is not used, so the limit is the stress permitted without reinforcement.
3. *Deflection.* It is wise to be conservative on this score; code-specified minimum recommendations are given in Table 30.2.
4. *Fire Resistance.* For fire separation, code-required thicknesses must be used for the rating required, usually at least about 5 in. for a 2-hour rating.

Beam Design: General Considerations

The design of a single beam involves a large number of pieces of data, most of which are established for the system as a whole, rather than individually for each beam. System-wide decisions usually include those for the type of concrete and its design strength (f_c'), the type of reinforcing steel (f_y), the cover required for the necessary fire rating, and various generally used details of forming and reinforcement. Most beams occur in conjunction with solid slabs that are poured monolithically with the beams. Slab thickness is established by the structural requirements of the spanning action between beams and by various concerns, such as those for fire rating, thermal and acoustic separation, type of reinforcement, and so on. Design of a single beam is usually limited to determination of the following.

1. Choice of shape and dimensions of the beam cross section.
2. Selection of the type, size, and spacing of shear reinforcement.
3. Selection of the flexural reinforcement to satisfy requirements based on the variation of moment along the beam span.

The following are some factors that must be considered in effecting these decisions.

Beam Shape

Figure 30.3 shows the most common shapes used for beams in poured-in-place construction. The single, simple rectangular section is actually uncommon, but does occur in some situations. Design of the concrete section consists of selecting the two dimensions: the width b and the overall height or depth h.

As mentioned previously, beams occur most often in conjunction with monolithic slabs, resulting in the typical T shape shown in Fig. 30.3b or the L shape shown in Fig.

30.3c. The full T shape occurs at the interior portions of the system, while the L shape occurs at the outside edge of the system or at the side of large openings. As shown in the illustration, there are four basic dimensions for the T and L that must be established in order to fully define the beam section.

t is the slab thickness; it is ordinarily established on its own, rather than as a part of the single-beam design.

h is the overall beam stem depth, corresponding to the same dimension for the rectangular section.

b_w is the beam stem width, which is critical for consideration of shear and for problems of fitting reinforcing into the section.

b_f is the *effective width* of the flange, which is the portion of the slab assumed to work with the beam.

A special beam shape is that shown in Fig. 30.3d. This occurs in concrete joist and waffle construction when "pans" of steel or reinforced plastic are used to form the concrete, the taper of the beam stem being required for easy removal of the forms. The smallest width dimension of the beam stem is ordinarily used for the beam design in this situation.

Beam Width

The width of a beam will affect its resistance to bending. Consideration of the flexure formulas given in Secs. 27.2 and 27.3 will show that the width dimension affects the bending resistance in a linear relationship (double the width and you double the resisting moment, etc.). On the other hand, the resisting moment is affected by the *square* of the effective beam depth. Thus efficiency—in terms of beam weight or concrete volume—will be obtained by striving for deep, narrow beams, instead of shallow, wide ones. (Just as a 2 × 8 is more efficient than a 4 × 4 in wood.)

Beam width also relates to various other factors, however, and these are often critical in establishing the minimum width for a given beam. The formula for shear stress ($v = V/bd$) indicates that the depth is less effective in shear resistance than in moment resistance. Placement of reinforcing bars is also a problem in narrow beams. Table 30.1 gives minimum beam widths required for various bar combinations, based on considerations of bar spacing (Sec. 26.4), minimum concrete cover of 1.5 in., and use of a No. 3 stirrup. Situations requiring additional concrete cover, use of larger stirrups, or the intersection of beams with columns, may necessitate widths greater than those given in Table 30.1.

Beam Depth

For specification of the construction, the beam depth is defined by the overall concrete dimension: h in Fig. 30.4. For structural design, however, the critical depth dimen-

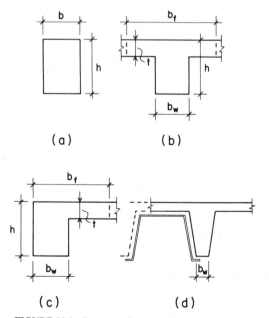

FIGURE 30.3. Common shapes for beam cross sections.

TABLE 30.1. Minimum Beam Widths[a]

Number of Bars	Bar Size								
	3	4	5	6	7	8	9	10	11
2	10	10	10	10	10	10	10	10	10
3	10	10	10	10	10	10	10	11	11
4	10	10	10	10	11	11	12	13	14
5	10	11	11	12	12	13	14	16	17
6	11	12	13	14	14	15	17	18	20

[a] Minimum width in inches for beams with 1.5-in cover, No. 3 U-stirrups, clear spacing between bars of one bar diameter or minimum of 1.0 in. Minimum practical width for beam with No. 3 U-stirrups: 10 in.

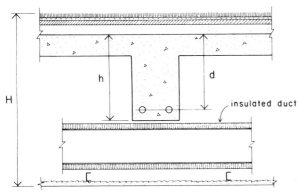

FIGURE 30.4.

sion is that from the center of the tension reinforcing to the far side of the concrete: d in Fig. 30.4. While the selection of the depth is partly a matter of satisfying structural requirements, it is typically constrained by other considerations in the building design.

Figure 30.4 shows a section through a typical building floor/ceiling with a concrete beam-slab structure. In this situation the critical depth from a general building design point of view is the overall thickness of the construction, shown as H in the illustration. In addition to the concrete structure, this includes allowances for the floor finish, the ceiling framing, and the passage of an insulated air duct. The net usable portion of H for the structure is shown as the dimension h, with the effective structural depth d being something less than h. Since the space defined by H is not highly usable for the building occupancy, there is a tendency to constrain it which works to limit the extravagant use of d.

Most concrete beams tend to fall within a limited range in terms of the width-to-depth ratio. The typical range is for a width-to-depth ratio between 1:1.5 and 1:2.5, with an average of 1:2. This is not a code requirement or a magic rule; it is merely the result of satisfying typical requirements for flexure, shear, bar spacing, and deflection.

Deflection Control

Deflection of spanning slabs and beams of poured-in-place concrete is controlled primarily by using recommended minimum thicknesses (overall height) expressed as a percentage of the span. Table 30.2 is adapted from a similar table given in Section 9.5 of the *ACI Code* and yields minimum thicknesses as a fraction of the span. Table values apply only for concrete of normal weight (made with ordinary sand and gravel) and for reinforcing with f_y of 60 ksi [414 MPa]. The *Code* supplies correction factors for other concrete weights and reinforcing grades. The *Code* further stipulates that these recommendations apply only where beam deflections are not critical for other elements of the building construction, such as supported partitions subject to cracking caused by beam deflections.

Table 30.3 yields maximum spans for beams with various overall depths. These are based on the requirements given in Table 30.2. It should be noted that these are *limits* and are not necessarily practical or efficient values. Use of these limits will usually result in beams having a great amount of reinforcing, whereas economy is generally achieved by using minimum amounts of reinforcing.

Deflection of concrete structures presents a number of special problems. For concrete with ordinary reinforcing (not prestressed), flexural action normally results in tension cracking of the concrete at points of maximum bending. Thus the presence of cracks in the bottom of a beam at midspan points and in the top over supports is to be expected. In general, the size (and visibility) of these cracks will be proportionate to the amount of beam curvature produced by deflection. Crack size will also be greater for long spans and for deep beams. If visible cracking is considered objectionable, more conservative depth-to-span ratios should be used, especially for spans over 30 ft and beam depths over 30 in.

Creep of concrete (see Sec. 26.1) results in additional deflections over time. This is caused by the sustained loads—essentially the dead load of the construction. Deflection controls reflect concern for this as well as for the instantaneous deflection under live load, the latter being the major concern in structures of wood and steel.

TABLE 30.2. Minimum Thickness of One-Way Slabs or Beams Unless Deflections Are Computed

Type of Member	End Conditions	Minimum Thickness of Slab or Height of Beam	
		$f_y = 40$ ksi [276 MPa]	$f_y = 60$ ksi [414 MPa]
Solid one-way slabs[a]	Simple support	$L/25$	$L/20$
	One end continuous	$L/30$	$L/24$
	Both ends continuous	$L/35$	$L/28$
	Cantilever	$L/12.5$	$L/10$
Beams or joists	Simple support	$L/20$	$L/16$
	One end continuous	$L/23$	$L/18.5$
	Both ends continuous	$L/26$	$L/21$
	Cantilever	$L/10$	$L/8$

Source: Data adapted from *Building Code Requirements for Reinforced Concrete* (ACI 318-89), 1989 ed., with permission of the publishers, American Concrete Institute.

[a] Valid only for members not supporting or attached to partitions or other construction likely to be damaged by large deflections.

TABLE 30.3. Maximum Span for Beams[a]

Overall Beam Depth, h (in.)	Maximum Permissible Span (ft)			
	Simply Supported	One End Continuous	Both Ends Continuous	Cantilever
10	13.3	15.4	17.5	6.7
12	16	18.5	21	8
14	18.7	21.6	24.5	9.3
16	21.3	24.7	28	10.7
18	24	27.7	31.5	12
20	26.7	30.8	35	13.3
24	32	37.0	42	16
30	40	46.2	52.5	20
36	48	55.5	63	24

[a] Based on requirements of Table 30.2. For normal-weight concrete and reinforcing with $f_y = 60$ ksi. For $f_y = 40$ ksi, multiply table values by 1.25.

In beams, deflections, especially creep deflections, may be reduced by the use of some compressive reinforcing. Where deflections are of concern, or where depth-to-span ratios are pushed to their limits, it is advisable to use some compressive reinforcing, consisting of continuous top bars.

When, for whatever reasons, deflections are deemed to be critical, computations of actual values of deflection may be necessary. Section 9.5 of the *ACI Code* provides directions for such computations; they are quite complex in most cases, and beyond the scope of this work. In actual design work, however, they are required very infrequently.

Continuous Beams

Continuous beams are typically indeterminate and must be investigated for the bending moments and shears that are critical for the various loading conditions. When the beams are not involved in rigid-frame actions (as when they occur on column lines in multistory buildings), it may be possible to use approximate analysis methods, as described in Section 8.3 of the *ACI Code* (Ref. 6). An illustration of such a procedure is shown in the design of the concrete floor structure in Chapter 58.

30.2. ONE-WAY JOIST CONSTRUCTION

Figure 30.5 shows a partial framing plan and some details for a type of construction that utilizes a series of very closely spaced beams and a relatively thin solid slab. Because of its resemblance to ordinary wood joist construction, this is called *concrete joist* construction. This system is generally the lightest (in dead weight) of any type of flat-spanning, poured-in-place concrete construction and is structurally well suited to the light loads and medium spans of office buildings and commercial retail buildings.

Slabs as thin as 2 in. and joists as narrow as 4 in. are used with this construction. Because of the thinness of the parts and the small amount of cover provided for reinforcement (typically $\frac{3}{4}$–1 in. for joists vs 1.5 in. for ordinary beams), the construction has very low resistance to fire, especially when exposed from the underside. It is therefore necessary to provide some form of fire protection, as for steel construction, or to restrict its use to situations where high fire ratings are not required.

The relatively thin, short span slabs are typically reinforced with welded wire mesh rather than ordinary deformed bars. Joists are often tapered at their ends, as shown in the framing plan in Fig. 30.5. This is done to provide a larger cross section for increased resistance to shear and negative moment at the supports. Shear reinforcement in the form of single vertical bars may be provided, but is not frequently used.

Early joist construction was produced by using lightweight hollow clay tile blocks to form the voids between joists. These blocks were simply arranged in spaced rows on top of the forms, the joists being formed by the spaces between the rows. The resulting construction provided a flat underside to which a plastered ceiling surface could be directly applied. Hollow, lightweight concrete blocks later replaced the clay tile blocks (see Fig. 30.6). Other forming systems have utilized plastic-coated cardboard boxes, fiber glass reinforced pans, and formed sheet metal pans. The latter method was very widely used, the metal pans being pried off after the pouring of the concrete and reused for several additional pours. The tapered joist cross section shown in Fig. 30.5 is typical of this construction, since the removal of the metal pans requires it.

In contrast to beams of wood and steel, those of concrete must be designed for the changing internal force conditions along their length. The single, maximum values for bending moment and shear may be critical in establishing

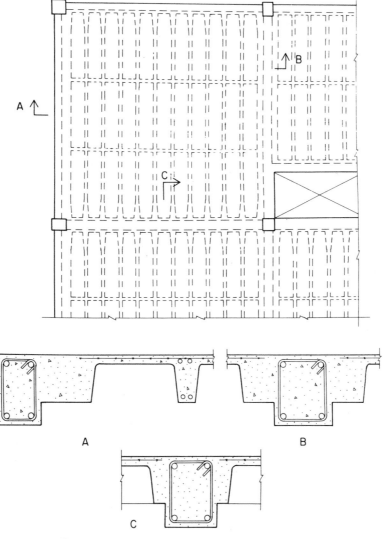

FIGURE 30.5. Typical concrete one-way joist construction.

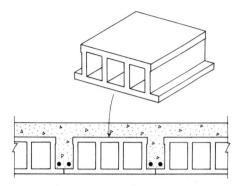

FIGURE 30.6. Joist construction formed with tile or concrete units.

the required beam size, but requirements for reinforcement must be investigated at all supports and midspan locations. A procedure for this is illustrated in the design examples in Chapter 58.

Wider joists can be formed by simply increasing the space between forms, with large beams being formed in a similar manner or by the usual method of extending a beam stem below the construction, as shown for the beams in Fig. 30.5. Because of the narrow joist forms, cross bridging is usually required, just as with wood joist construction. The framing plan in Fig. 30.5 shows the use of two bridging strips in the typical bay of the framing.

Design of joist construction is essentially the same as for ordinary slab and beam construction. Some special regulations are given in the *ACI Code* for this construction, such as the reduced cover mentioned previously. Because joists are so commonly formed with standard-sized metal forms, there are tabulated designs for typical systems in various handbooks. The *CRSI Handbook* (Ref. 9) has extensive tables offering complete designs for various spans, loadings, pan sizes, and so on. Whether for final design or simply for a quick preliminary design, the use of such tables is quite efficient.

One-way joist construction was highly popular in earlier times, but has become less utilized, due to its lack of fire resistance and the emergence of other systems. The popularity of lighter, less fire resistive ceiling construction has been a contributing factor. In the right situation, however, it is still a highly efficient type of construction.

30.3. WAFFLE CONSTRUCTION

Waffle construction consists of two-way spanning joists that are formed in a manner similar to that for one-way spanning joists, using forming units of metal, plastic, or cardboard to produce the void spaces between the joists. The most widely used type of waffle construction is the waffle flat slab, in which solid portions around column supports are produced by omitting the void-making forms. An example of a portion of such a system is shown in Fig. 30.7. This type of system is analogous to the solid flat slab, which will be discussed in Sec. 30.4. At points of discontinuity in the plan—such as at large openings or at edges of

the building—it is usually necessary to form beams. These beams may be produced as projections below the waffle, as shown in Fig. 30.7, or may be created within the waffle depth by omitting a row of the void-making forms, as shown in Fig. 30.8.

If beams are provided on all of the column lines, as shown in Fig. 30.8, the construction is analogous to the two-way solid slab with edge supports, as discussed in Sec. 30.4. With this system, the solid portions around the column are not required, since the waffle itself does not achieve the transfer of high shear or development of the high negative moments at the columns.

As with the one-way joist construction, fire ratings are low for ordinary waffle construction. The system is best suited for situations involving relatively light loads, medium-to-long spans, approximately square column bays, and a reasonable number of multiple bays in each direction.

For the waffle construction shown in Fig. 30.7, the edge of the structure represents a major discontinuity when the column supports occur immediately at the edge, as shown. Where planning permits, a more efficient use of the system is represented by the partial framing plan shown in Fig. 30.9, in which the edge occurs some distance past the columns. This projected edge provides a greater shear periphery around the column and helps to generate a negative moment, preserving the continuous character of the spanning structure. With the use of the projected edge, it may be possible to eliminate the edge beams shown in Fig. 30.7, thus preserving the waffle depth as a constant.

Another variation for the waffle is the blending of some one way joist construction with the two-way waffle joists. This may be achieved by keeping the forming the same as for the rest of the waffle construction and merely using the ribs in one direction to crate the spanning structure. One reason for doing this would be a situation similar to that shown in Fig. 30.8, where the large opening for a stair or elevator results in a portion of the waffle (the remainder of the bay containing the opening) being considerably out of square, that is, having one span considerably greater than the other. The joists in the short direction in this case will tend to carry most of the load due to their greater stiffness (less deflection than the longer spanning joists that intersect them). Thus the short joists would be designed as one-way spanning members and the longer joists would have only minimum reinforcing and serve as bridging elements.

The two-way spanning waffle systems are quite complex in structural behavior and their investigation and design is beyond the scope of this book. Some aspects of this work are discussed in the next article, since there are many similarities between the two-way spanning waffle systems and the two-way spanning solid slab systems. As with the one-way joist system, there are some tabulated designs in various handbooks that may be useful for either final or preliminary design. The *CRSI Handbook* (Ref. 9) mentioned previously has some such tables.

For all two-way construction, such as the waffle system, real feasible use of the system depends on some logic in terms of development of the general building plans regard-

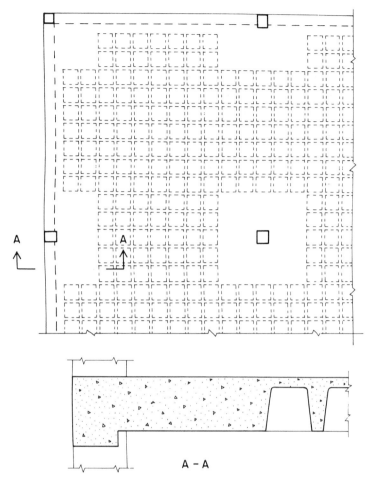

FIGURE 30.7. Typical concrete waffle construction.

ing the arrangements of structural supports, locations of openings, length of spans, and so on. In the right situation, these systems may be able to realize their full potential, but if lack of order, symmetry, and other factors result in major adjustments away from the simple two-way functioning of the system, it may be very unreasonable to select such a structure. In some cases the waffle has been chosen strictly for its underside appearance, and has been pushed into use in situations not fitted to its nature. There may be some justifications for such a case, but the resulting structure is likely to be quite awkward.

30.4. TWO-WAY SOLID SLABS

If reinforced in both directions, the solid concrete slab may span two ways as well as one. The widest use of such a slab is in flat slab or flat plate construction. In flat slab construction, beams are used only at points of discontinuity, with the typical system consisting only of the slab and the strengthening elements used at column supports. Typical details for a flat slab system are shown in Fig. 30.10. Drop panels consisting of thickened portions square in plan are used to give additional resistance to the high shear and negative moment that develops at the column supports. Enlarged portions are also sometimes provided at the tops of the columns (called *column capitals*) to further reduce the stresses in the slab.

Two-way slab construction consists of multiple bays of solid two-way-spanning slabs with edge supports consisting of bearing walls of concrete or masonry or of column-line beams formed in the usual manner. Typical details for such a system are shown in Fig 30.11.

Two-way solid slab construction is generally favored over waffle construction where higher fire rating is required for the unprotected structure or where spans arc short and loadings high. As with all types of two-way spanning systems, they function most efficiently where the spans in each direction are approximately the same.

For investigation and design, the flat slab (Fig. 30.10) is considered to consist of a series of one-way-spanning solid slab strips. Each of these strips spans through multiple bays in the manner of a continuous beam and is supported either by columns or by the strips that span in a direction

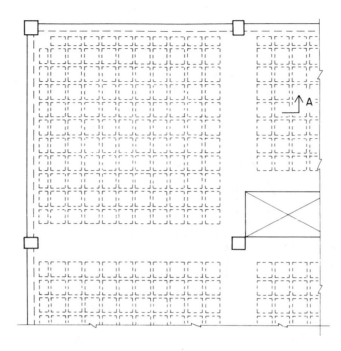

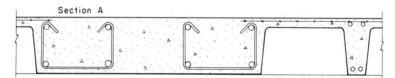

FIGURE 30.8. Waffle construction with column-line beams within the waffle depth.

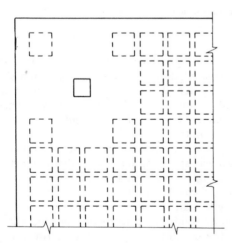

FIGURE 30.9. Waffle construction with cantilevered edge and no edge beam.

perpendicular to it. The analogy for this is shown in Fig. 30.12.

As shown in Fig. 30.12*b*, the slab strips are divided into two types: those passing over the columns, and those passing between columns—called *middle strips*. The complete structure consists of the intersecting series of these strips, as shown in Fig. 30.12*c*. For the flexural action of the sys-

tem there is two-way reinforcing in the slab at each of the boxes defined by the intersections of the strips. In box 1 in Fig. 30.12*c*, both sets of bars are in the bottom portion of the slab, due to the positive moment in both intersecting strips. In box 2, the middle-strip bars are in the top (for negative moment) while the column-strip bars are in the bottom (for positive moment). And in box 3, the bars are in the top in both directions.

30.5. COMPOSITE CONSTRUCTION: CONCRETE PLUS STRUCTURAL STEEL

Figure 30.13 shows a section detail of a type of construction generally referred to as *composite construction*. This consists of a poured concrete spanning slab supported by structural steel beams, the two being made to interact by the use of shear developers welded to the top of the beams and embedded in the cast slab. The concrete slab may be formed by use of plywood sheets, resulting in the detail as shown in Fig. 30.13. However, a more popular form of construction is that in which a formed steel deck is used in the usual manner, welded to the top of the beams. The shear developers are then site-welded through the deck to the top of the beam. The steel deck may function essentially only

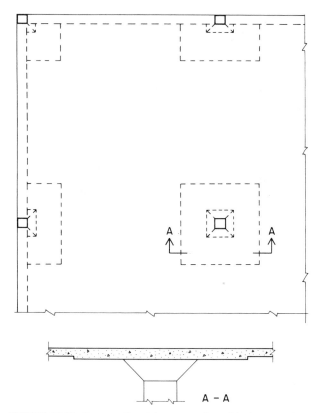

FIGURE 30.10. Concrete flat slab construction with drop panels and column caps.

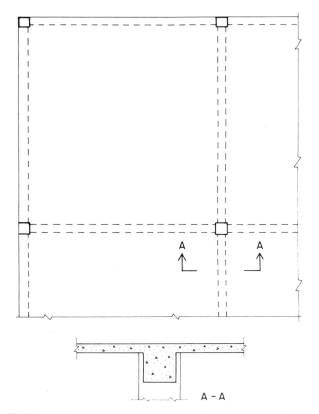

FIGURE 30.11. Two-way spanning concrete slab construction with edge supports.

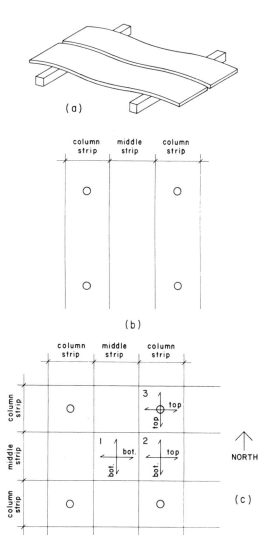

FIGURE 30.12. Development of the two-way concrete flat slab.

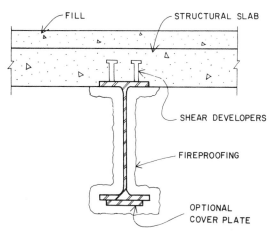

FIGURE 30.13. Steel frame with poured-in-place concrete slab.

to form the concrete, or may itself develop a composite action with the poured slab.

Part 2 of the *AISC Manual* (Ref. 5) contains data and design examples of this type of construction.

30.6. USE OF DESIGN AIDS

The design of various elements of reinforced concrete can be aided—or in many cases totally achieved—by the use of various prepared materials. Handbooks (see Refs. 9, 13, and 14) present complete data for various elements, such as footings, columns, one-way slabs, joist construction, waffle systems, and two-way slab systems. For the design of a single footing or a one-way slab, the handbook merely represents a convenience, or a shortcut, to a final design. For columns subjected to bending, for waffle construction, and for two-way slab systems, "longhand" design (without aid other than a pocket calculator) is really not feasible. In the latter cases, handbook data may be used to establish a reasonable preliminary design, which may then be custom fit to the specific conditions by some investigation and computations. Even the largest of handbooks cannot present all possible combinations of values of f_c', grade of reinforcing bars, value of superimposed loads, and so on. Thus only coincidentally will handbook data be exactly correct for any specific design job.

CHAPTER THIRTY-ONE

Concrete Columns

Concrete columns occur most often as the vertical support elements in a structure generally built of sitecast concrete. This is the situation discussed in Chapter 31. Columns may also occur as precast elements; the general concerns of the precast structure are discussed in Chapter 34. Very short columns, called *pedestals*, are sometimes used in the support system for columns or other structures. The ordinary pedestal is discussed as a foundation transitional device in Chapter 33. The special case of an abutment that resolves both horizontal and vertical support forces is discussed in Part 7. Walls that serve as vertical compression supports (called *bearing walls*) are discussed in Chapter 32.

31.1. GENERAL CONSIDERATIONS

The sitecast concrete column usually falls into one of the following categories:

1. Square columns with tied reinforcing.
2. Oblong columns with tied reinforcing.
3. Round columns with tied reinforcing.
4. Round columns with spiral-bound reinforcing.
5. Square columns with spiral-bound reinforcing.
6. Columns of other geometries (L- or T-shaped, octagonal, etc.) with either tied or spiral-bound reinforcing.

Obviously, the choice of column cross-sectional shape is an architectural as well as a structural decision. However, forming methods and costs, arrangement and installation of reinforcing, and relations of the column form and dimensions to other parts of the structural system must also be dealt with.

In tied columns the longitudinal reinforcing is held in place by loop ties made of small-diameter reinforcing bars, commonly No. 3 or No. 4. Such a column is represented by the square section shown in Fig. 31.1a. This type of reinforcing can quite readily accommodate other geometries as well as the square.

Spiral columns are those in which the longitudinal reinforcing is placed in a circle, with the whole group of bars enclosed by a continuous cylindrical spiral made from steel rod or large-diameter steel wire. Although this reinforcing system obviously works best with a round col-

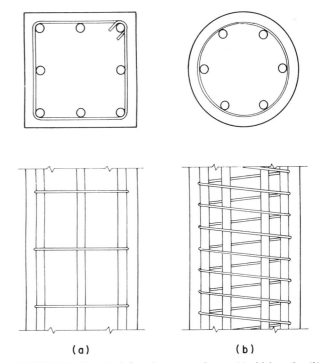

FIGURE 31.1. Typical reinforced concrete columns: (a) with loop ties; (b) with a spiral wrap.

umn section, it can be used also with other geometries. A round column of this type is shown in Fig. 31.1b.

Experience has shown the spiral column to be slightly stronger than an equivalent tied column with the same amount of concrete and reinforcing. For this reason, code provisions allow slightly more load on spiral columns. Spiral reinforcing tends to be expensive, however, and the round bar pattern does not always mesh well with other construction details in buildings. Thus tied columns are often favored where restrictions on the outer dimensions of the sections are not severe.

General Requirements

Code provisions and practical construction considerations place a number of restrictions on column dimensions and choice of reinforcing.

Column Size.   The current code does not contain limits for column dimensions. For practical reasons, the following limits are recommended. Rectangular tied columns should be limited to a minimum area of 100 in.² and a side dimension of 10 in. if square and 8 in. if oblong. Spiral columns should be limited to a minimum size of 12 in. if either round or square.

Reinforcing.   Minimum bar size is No. 5. The minimum number of bars is four for tied columns, five for spiral columns. The minimum amount of area of steel is 1% of the gross column area. A maximum area of steel of 8% of the gross area is permitted, but bar spacing limitations makes this difficult to achieve; 4% is a more practical limit. The 1989 *ACI Code* stipulates that for a compression member with a larger cross section than required by considerations of loading, a reduced effective area not less than one-half the total area may be used to determine minimum reinforcement and design strength.

Ties.   Ties shall be at least No. 3 for bars that are No. 10 and smaller. No. 4 ties should be used for bars that are No. 11 and larger. Vertical spacing of ties shall be not more than 16 times the bar diameter, 48 times the tie diameter, or the least dimension of the column. Ties shall be arranged so that every corner and alternate longitudinal bar is held by the corner of a tie with an included angle of not greater than 135°, and no bar shall be farther than 6 in. clear from such a supported bar. Complete circular ties may be used for bars placed in a circular pattern.

Concrete Cover.   A minimum of 1.5 in. is needed when the column surface is not exposed to weather or in contact with the ground; 2 in. should be used for formed surfaces exposed to the weather or in contact with ground; 3 in. are necessary if the concrete is cast against earth.

Spacing of Bars.   Clear distance between bars will not be less than 1.5 times the bar diameter, 1.33 times the maximum specified size for the coarse aggregate, or 1.5 in.

Column Load Capacity

Due to the nature of most concrete structures, current design practices generally do not consider the possibility of a concrete column with axial compression alone. That is, the existence of some bending moment is always considered together with the axial force. Figure 31.2 illustrates the nature of the so-called *interaction response* for a concrete column, with a range of combinations of axial load plus bending moment. In general, there are three basic ranges of this behavior, as follows:

1. *Large Axial Force, Minor Moment.*   For this case the moment has little effect, and the resistance to pure axial force is only negligibly reduced.
2. *Significant Values for Both Axial Force and Moment.*   For this case the analysis for design must include the full combined force effects, that is, the

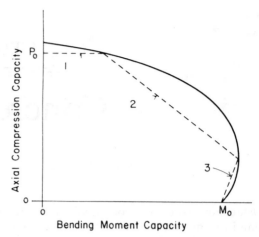

FIGURE 31.2. Interaction of axial compression and bending for a reinforced column.

interaction of the axial force and the bending moment.

3. *Large Bending Moment, Minor Axial Force.*   For this case, the column behaves essentially as a doubly reinforced (tension and compression reinforced) member, with its capacity for moment resistance affected only slightly by the axial force.

In Fig. 31.2 the solid line on the graph represents the true response of the column—a form of behavior verified by many load tests on laboratory specimens. The dashed line figure on the graph represents the generalization of the three types of response just described.

Design Methods

At present, design of concrete columns is mostly achieved by using either tabulations from handbooks or computer-aided procedures. The present *ACI Code* (ACI 318-89) does not permit design of columns by working stress methods in a direct manner; stipulating instead that a capacity of 40% of that determined by strength methods be used if working stress procedures are used in design. Using the code formulas and requirements to design by "hand operation" with both axial compression and bending present at all times is prohibitively laborious. The number of variables present (column shape and size, f_c', f_y, number and size of bars, arrangement of bars, etc.) adds to the usual problems of column design to make for a situation much worse than those for wood or steel columns.

The large number of variables also works against the efficiency of handbook tables. Even if a single concrete strength (f_c') and a single steel yield strength (f_y) are used, tables would be very extensive if all sizes, shapes, and types (tied and spiral) of columns were included. Even with a very limited range of variables, handbook tables are much larger than those for wood or steel columns. They are, nevertheless, often quite useful for preliminary design estimation of column sizes. A computer-aided system is the obvious preference when relationships are complex, requirements are tedious and extensive, and there are a large number of variables.

As in other situations, the common practices at any given time tend to narrow down to a limited usage of any type of construction, even though the potential for variation is extensive. It is thus possible to use some very limited but easy-to-use design aids to make early selections for design. These approximations may be adequate for preliminary building planning, cost estimates, and some preliminary structural analyses.

One highly useful reference is the *CRSI Handbook* (Ref. 9), which contains quite extensive tables for design of both tied and spiral columns. Square, round, and some oblong shapes, plus some range of concrete and steel strengths are included. Table format uses the equivalent eccentric load technique which is discussed in Sec. 31.2.

31.2. APPROXIMATE DESIGN OF TIED COLUMNS

Tied columns are much preferred due to the relative simplicity and usually lower cost of their construction, plus their adaptability to various column shapes (T, L, etc.) Even round columns—most naturally formed with spiral-bound reinforcing—are often made with ties instead, when the structural demands are modest. An exception to this is the situation of columns in rigid-frame structures in zones of high seismic risk, where the toughness of spiral columns is much preferred.

The column with moment is often designed using the equivalent eccentric load method. The general case for this is discussed in Sec. 11.4. The method consists of translating a compression plus bending situation into an equivalent one with an eccentric load, the moment becoming the product of the load and the eccentricity, as shown in Fig. 31.3. This method is often used in presentation of tabular data for column capacities. It is also used in the development of the graphs in Figs. 31.4 and 31.5, which yield safe service load capacities for simple square and round tied columns of concrete with $f_c' = 4$ ksi [27.6 MPa] and reinforcing with $f_y = 60$ ksi [414 MPa].

Figure 31.4 gives safe loads for a selected number of sizes of square tied columns. Loads are given for various degrees of eccentricity, which is a means for expressing axial load and bending moment combinations. The computed moment on the column is translated into an equivalent eccentric loading, as shown in Fig. 31.3. Data for the curves were computed by using 40% of the load determined by strength design methods, as required by the 1983 *ACI Code*.

The following examples illustrate the use of Fig. 31.4 for the design of tied columns.

Example 1.   A column with $f_c' = 4$ ksi and steel with $f_y = 60$ ksi sustains an axial compression load of 400 kips. Find the minimum practical column size if reinforcing is a maximum of 4% and the maximum size if reinforcing is a minimum of 1%.
Solution.   Using Fig. 31.4, we find from the sizes given:

Minimum column is 20 in.² with 8 No. 9 (curve 14).
Maximum capacity is 410 kips, $p_g = 2.0\%$
Maximum size is 24 in.² with 4 No. 11 (curve 17).
Maximum capacity is 510 kips, $p_g = 1.08\%$.

It should be apparent that it is possible to use an 18-in. or 19-in. column as the minimum size and to use a 22-in. or 23-in. column as the maximum size. Since these sizes are not given in Fig. 31.4, we cannot verify them for certain without using strength design procedures.

Example 2.   A square tied column with $f_c' = 4$ ksi and steel with $f_y = 60$ ksi sustains an axial load of 400 kips and a bending moment of 200 kip-ft. Determine the minimum size column and its reinforcing.
Solution.   We first determine the equivalent eccentricity, as shown in Fig. 31.3. Thus

$$e = \frac{M}{P} = \frac{200\ (12)}{400} = 6 \text{ in.}$$

Then, from Fig. 31.4, we find:

Minimum size is 24 in. square with 16 No. 10 bars.
Capacity at 6 in. eccentricity is 410 kips.

Round columns may be designed and built as spiral columns, or they may be developed as tied columns with the bars placed in a circle and held by a series of round circumferential ties. Because of the cost of spirals, it is often more economical to use the tied columns, so they are often used unless the additional strength or other behavioral characteristics of the spiral column are required. In such cases, the column is usually designed as a square column using the square shape that can be included within the round form. It is thus possible to use a four-bar column for small-diameter, round column forms.

Figure 31.5 gives safe loads for round columns that are designed as tied columns. Load values have been adapted from values determined by strength design methods. The curves in Fig. 31.5 are similar to those for the square columns in Fig. 31.4. and their use is similar to that demonstrated in Examples 1 and 2.

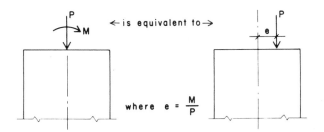

FIGURE 31.3. Moment plus axial compression on a column visualized as an equivalent eccentric load.

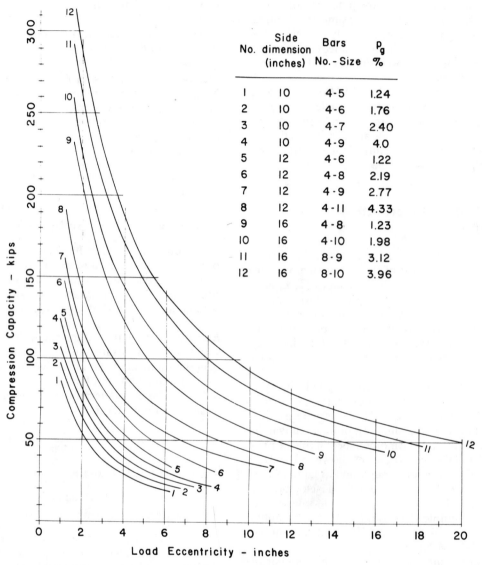

No.	Side dimension (inches)	Bars No. - Size	p_g %
1	10	4 - 5	1.24
2	10	4 - 6	1.76
3	10	4 - 7	2.40
4	10	4 - 9	4.0
5	12	4 - 6	1.22
6	12	4 - 8	2.19
7	12	4 - 9	2.77
8	12	4 - 11	4.33
9	16	4 - 8	1.23
10	16	4 - 10	1.98
11	16	8 - 9	3.12
12	16	8 - 10	3.96

FIGURE 31.4. Safe service loads for square tied columns with $f_c' = 4$ ksi and $f_y = 60$ ksi.

31.3.   DETAIL CONSIDERATIONS FOR COLUMNS

Usually, a number of possible combinations of reinforcing bars may be assembled to satisfy the steel area requirement for a given column. Aside from providing for the area, the number of bars must also work reasonably in the layout of the column. Figure 31.6 shows a number of tied columns with various number of bars. When a column is small, the preferred choice is usually that of the simple four-bar layout, with one bar in each corner and a single peripheral

tie. As the column gets larger, the distance between the corner bars gets larger, and it is best to use more bars so that the reinforcing is spread out around the column periphery. For a symmetrical layout and the simplest of tie layouts, the best choice is for numbers that are multiples of four, as shown in Fig. 31.6a. The number of additional ties required for these layouts depends on the size of the column and the considerations discussed in Sec. 31.1.

An unsymmetrical bar arrangement is not necessarily bad, even though the column and its construction details

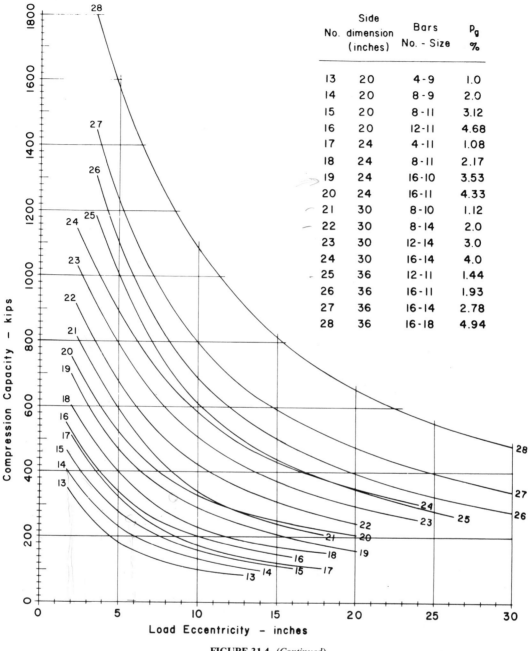

No.	Side dimension (inches)	Bars No. - Size	P_g %
13	20	4-9	1.0
14	20	8-9	2.0
15	20	8-11	3.12
16	20	12-11	4.68
17	24	4-11	1.08
18	24	8-11	2.17
19	24	16-10	3.53
20	24	16-11	4.33
21	30	8-10	1.12
22	30	8-14	2.0
23	30	12-14	3.0
24	30	16-14	4.0
25	36	12-11	1.44
26	36	16-11	1.93
27	36	16-14	2.78
28	36	16-18	4.94

FIGURE 31.4. (*Continued*)

are otherwise not oriented differently on the two axes. In situations where moments may be greater on one axis, the unsymmetrical layout is actually preferred; in fact, the column shape will also be more effective if it is unsymmetrical, as shown for the oblong shapes in Fig. 31.6c.

Figure 31.6 also shows a number of special column shapes developed as tied columns. Although spirals could be used in some cases for such shapes, the use of ties allows much greater flexibility and simplicity of construction. One reason for using ties may be the column dimensions;

there being a practical limit of about 12 in. in width for a spiral-bound column.

Round columns are most often formed as shown in Fig. 31.6e, if built as tied columns. This allows for a minimum reinforcing with four bars. If a round pattern is used (as it must be for a spiral-bound column), the usual minimum number recommended is six bars. Spacing of bars is much more critical in spiral-bound circular arrangements, making it very difficult to use high percentages of steel in the column section. For very large diameter columns it is pos-

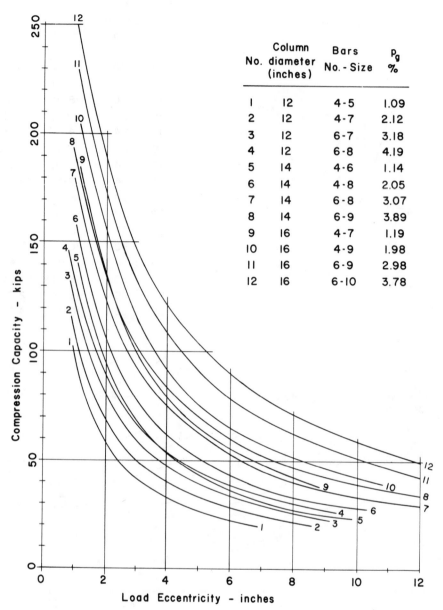

No.	Column diameter (inches)	Bars No. - Size	p_g %
1	12	4-5	1.09
2	12	4-7	2.12
3	12	6-7	3.18
4	12	6-8	4.19
5	14	4-6	1.14
6	14	4-8	2.05
7	14	6-8	3.07
8	14	6-9	3.89
9	16	4-7	1.19
10	16	4-9	1.98
11	16	6-9	2.98
12	16	6-10	3.78

FIGURE 31.5. Safe service loads for round tied columns with f'_c = 4 ksi and f_y = 60 ksi.

sible to use sets of concentric spirals, as shown in Fig. 31.6e.

For columns a concern that must be dealt with is that for vertical splicing of the steel bars. As shown in Fig. 29.3, the two places where this commonly occurs are at the top of the foundation and at floors where a multistory column continues upward. At these points there are three ways to achieve the vertical continuity (splicing) of the steel bars, any of which may be appropriate for a given situation.

1. Bars may be lapped the required distance for development of the compression splice, as shown in

Fig. 29.3. For bars of smaller dimension and of lower yield strengths, this is usually the desired method.

2. Bars may have milled square-cut ends butted together with a grasping device to prevent separation in a horizontal direction.

3. Bars may be welded with full-penetration butt welds or by welding of the grasping device described for method 2.

The choice of splicing methods is basically a matter of cost comparison, but is also affected by the size of the bars,

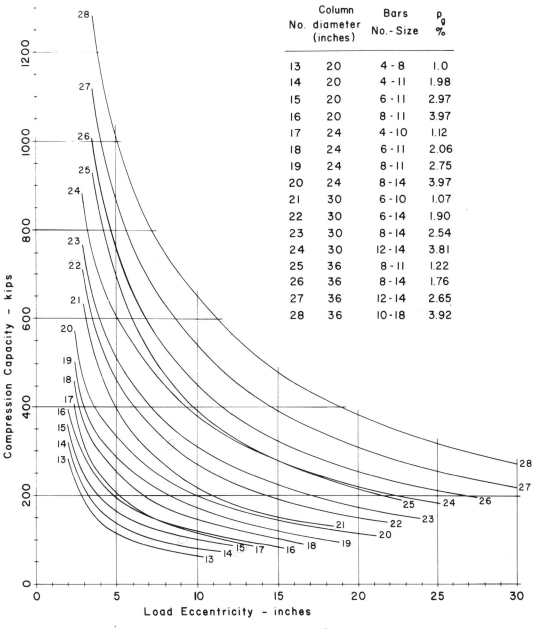

No.	Column diameter (inches)	Bars No.-Size	p_g %
13	20	4 - 8	1.0
14	20	4 - 11	1.98
15	20	6 - 11	2.97
16	20	8 - 11	3.97
17	24	4 - 10	1.12
18	24	6 - 11	2.06
19	24	8 - 11	2.75
20	24	8 - 14	3.97
21	30	6 - 10	1.07
22	30	6 - 14	1.90
23	30	8 - 14	2.54
24	30	12 - 14	3.81
25	36	8 - 11	1.22
26	36	8 - 14	1.76
27	36	12 - 14	2.65
28	36	10 - 18	3.92

FIGURE 31.5. (*Continued*)

the degree of concern for bar spacing in the column arrangement, and possibly for a need for some development of tension through the splice if uplift or high magnitudes of moments exist. If lapped splicing is used, a problem that must be considered is the bar layout at the location of the splice, at which point there will be twice the usual number of bars. The lapped bars may be adjacent to each other, but the usual considerations for space between bars must be made. If spacing is not critical, the arrangement shown in Fig. 31.7a is usually chosen, with the spliced sets of bars next to each other at the tie perimeter. If spacing does not permit the arrangement in Fig. 31.7a, that shown in Fig.

31.7b may be used, with the lapped sets in concentric patterns. The latter arrangement is commonly used for spiral-bound columns, where spacing is often critical.

31.4. SLENDERNESS

Poured-in-place concrete columns tend to be quite stout in profile, so that slenderness is much less often a critical concern than with columns of wood or steel. The code provides for consideration of slenderness, but permits the issue to be

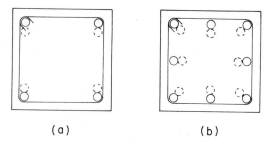

FIGURE 31.7. Bar placement at splice locations.

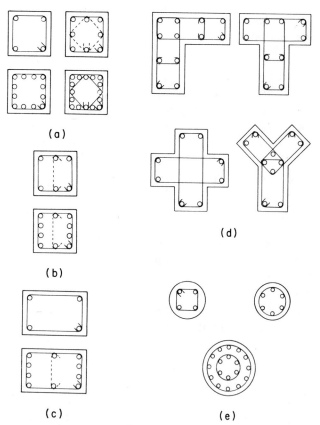

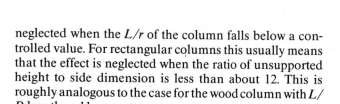

FIGURE 31.6. Typical shapes and reinforcement for concrete columns: (*a*) symmetrically reinforced square columns; (*b*) square columns with a strong axis for bending; (*c*) oblong columns; (*d*) columns of miscellaneous shapes; (*e*) round columns.

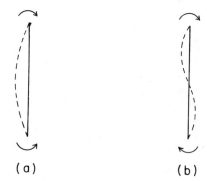

FIGURE 31.8. Bar placement at splices.

neglected when the L/r of the column falls below a controlled value. For rectangular columns this usually means that the effect is neglected when the ratio of unsupported height to side dimension is less than about 12. This is roughly analogous to the case for the wood column with L/D less than 11.

Slenderness effects must also be related to the conditions of bending for the column. Since bending is usually induced at the column ends, the two typical cases are those shown in Fig. 31.8. If two equal end moments as shown in Fig. 31.8*a* exist, the buckling effect is magnified, the *P*-delta

effect is maximum, and the code limits slenderness without reduction to L/d ratios of 6.6 or less. The condition in Fig. 31.8*a* is not the common case, however, the more typical condition in framed structures being that shown in Fig. 31.8*b*, for which the L/d limit for equal end moments jumps to 13.8 before reduction for slenderness is required.

When slenderness must be considered, the complex procedures required are simply built into your friendly neighborhood software program. One should be aware, however, that reduction for slenderness is not considered in the usual design aids, such as tables or graphs.

CHAPTER THIRTY-TWO

Concrete Walls

Concrete walls are by nature quite strong and in most situations serve some structural purpose—often multiple structural purposes. Chapter 32 discusses the various structural functions of walls and some of the general problems of their design and construction.

32.1. TYPES OF STRUCTURAL WALLS

The slablike concrete element, reinforced in both directions, has considerable application for various structural tasks. Concrete walls are all basically the same in character, the differences occurring in how they are used. The following are some of the common structural uses for concrete walls.

1. *Bearing Walls, Uniformly Loaded.* These may be single story or multistory, carrying loads from floors, roofs, and/or walls above.
2. *Bearing Walls with Concentrated Loads.* These are walls that provide support for beams or columns. In most cases they also support uniformly distributed loads.
3. *Basement Walls, Earth Retaining.* These are walls that occur at the exterior boundary between interior sublevel spaces and the surrounding earth. In addition to functioning as bearing walls (in most cases) they also span either vertically or horizontally as slabs to resist horizontal earth pressures.
4. *Retaining Walls.* This term is usually used to refer to walls that function to achieve grade-level changes, working essentially as vertical cantilevers to resist the horizontal earth pressures from the high side.
5. *Shear Walls.* These are walls that are used to brace the building against horizontal (lateral) forces due to wind or earthquakes. The shear referred to is generated in the plane of the wall, as opposed to shear generated in slab-spanning action.
6. *Freestanding Walls.* These are walls used as fences or partitions, being supported only at their bases.
7. *Grade Walls.* These are walls that occur in buildings without basements; they function to support walls above grade and grade-level floor slabs. They may also function as grade beams or ties in build-

ings with isolated foundations consisting of column footings, piles, or piers.

It is possible, of course, for walls to serve more than one of these functions. Concrete walls are quite expensive when compared to other types of wall construction and, when used, are usually exploited for all their potential value for structural purposes.

32.2. GENERAL REQUIREMENTS FOR REINFORCED CONCRETE WALLS

Regardless of their structural functions, a number of basic considerations apply to all walls. Some considerations of major concern are:

1. *Wall Thickness.* Nonstructural walls may be as thin as 4 in.; structural walls must be at least 6 in. thick. In general, slenderness ratio (unsupported height divided by thickness) should not exceed 25. A practical limit for a single pour (total height achieved in one continuous casting) is 15 times the wall thickness; taller walls will require multiple pours. Walls 10 in. or more in thickness should have two layers of reinforcing, one near each wall surface. Basement walls, foundation walls, and party walls must be at least 8 in. thick. Of course, the thickness must also be appropriate to the structural tasks.
2. *Reinforcement.* A minimum area of reinforcing equal to 0.0025 times the wall cross section must be provided in a horizontal direction; 0.0015 in a vertical direction. A reduction is possible if bars No. 5 or smaller of grade 60 or higher steel are used. As noted previously, two layers are required for walls 10 in. or more in thickness. The distribution of the total area required between the two layers depends on the wall functions.
3. *Special Reinforcing Requirements.* General practice is to provide extra reinforcing at the top, bottom, ends, corners, intersections, and at openings in the wall. Suggested details for placement of reinforcing at these locations is given in the *ACI Detailing Manual* and various requirements are given in the *ACI Code*.

32.3. BEARING WALLS

Bearing walls have two principal structural concerns. The first is for the compressive bearing stress that occurs beneath the applied loads. When the full wall cross section is utilized for bearing, the bearing strength P is limited as follows: by working stress,

$$P = 0.30f_c'A_1$$

and by strength methods,

$$P_u = 0.7(0.85f_c'A_1)$$

When the supporting surface is wider on all sides than the loaded area, design bearing strength may be increased by multiplying by $\sqrt{A_2/A_1}$, but not more than 2. A_1 is the actual contact bearing area, and A_2 is the full area of the supporting surface—in this case, the full horizontal cross section of the wall.

Bearing must usually be considered only when concentrated loads such as columns or the ends of steel or precast concrete beams are placed on the wall. In these cases the limits for bearing usually become the basis for determining the required contact bearing area.

The bearing wall also functions as a vertical compression member and may be designed as a column or by empirical means provided by the code. The present code does not permit direct design by working stress methods, so the following discussion is limited to the application of the empirical strength design method presented in Sec. 14.5 of the 1989 *ACI Code* (Ref. 6).

When the resultant vertical compression force on a wall falls within the middle third of the wall thickness, the wall may be signed as an axial loaded column, using the following empirical formula with the strength method.

$$\phi P_{nw} = 0.55\phi f_c'A_g \left[1 - \left(\frac{l_c}{40h} \right)^2 \right]$$

where $\phi = 0.70$
      P_{nw} = nominal axial load strength of wall
      A_g = the effective area of the wall cross section
      l_c = vertical distance between lateral supports
      h = overall thickness of the wall

If the wall carries concentrated loads, the length of the wall to be considered as effective for each load will not exceed the center-to-center distance between the loads nor the actual width of bearing plus four times the wall thickness.

The following example illustrates the procedure for design of a wall with concentrated loads. Design for a uniformly distributed load is essentially the same except that bearing stress and reduced effective area considerations need not be made.

Example 1. A reinforced concrete bearing wall supports a floor system consisting of precast concrete T-units spaced 8 ft [2.4 m] on centers. The bottom of the stem of the T-unit is 7 in. [180 mm] wide and the units bear on the full thickness of the wall. The height of the wall is 11.5 ft [3.5 m], and

the end reaction of each T-unit due to service loads is 22 kips [98 kN] for dead load and 12 kips [53 kN] for live load. Design the wall for the following data: $f_c' = 4$ ksi [27.6 MPa], and $f_y = 40$ ksi [276 MPa].

Solution. The factored load is

$$P_U = 1.4P_D + 1.7P_L = 1.4(22) + 1.7(12)$$

$$= 51.2 \text{ kips [228 kN]}$$

Assuming a poured-in-place wall, recommended minimum thickness is 15 times the wall thickness or

$$h = \frac{11.5(12)}{15} = 9.2 \text{ in.} \quad \text{say, 10 in. [250 mm]}$$

1. For bearing stress:

$$P_U = 0.7(0.85f_c'A_1) = 0.7[0.85(4)(7)(10)]$$

$$= 166.6 \text{ kips} > 51.2 \text{ kips [739 kN} > 228 \text{ kN]}$$

Thus bearing stress is not critical.

2. For column action:

effective wall length = bearing width
                               + 4(wall thickness)

$$l_e = 7 + 4(10) = 47 \text{ in. [1.18 m]}$$

Then

$$A_g = 10(47) = 470 \text{ in.}^2 \text{ [0.295 m}^2\text{]}$$

and

$$\phi P_{nw} = 0.55\phi f_c'A_g \left[1 - \left(\frac{l_c}{40h} \right)^2 \right]$$

$$= 0.55(0.7)(14)(470) \left[1 - \left(\frac{11.5(12)}{40(10)} \right)^2 \right]$$

$$= 638 \text{ kips} > 51.2 \text{ kips [2817 kN} > 228 \text{ kN]}$$

For the minimum wall reinforcing (see Sec. 32.2) we determine the following average required steel areas.

Horizontal A_s: $0.0025(10)(12) = 0.30$ in.^{2}/ft.
Vertical A_s: $0.0015(10)(12) = 0.18$ in.^{2}/ft.

From Table 27.3 select

Horizontal bars: No. 5 at 12, $A_s = 0.31$ in^2/ft.
Vertical bars: No. 4 at 13, $A_s = 0.18$ in^2/ft.

32.4. BASEMENT WALLS

Basement walls usually perform an earth-retaining function. They also often function as vertical load-carrying bearing walls or spanning grade beams or as the base for building shear walls. A complete design must include consideration for all of the load combinations resulting from any of these multiple functions.

For their earth-retaining function, basement walls ordinarily span vertically between levels of support. For a single-story basement the support at the bottom of the wall is provided by the edge of the concrete basement floor slab, and the support at the top of the wall is provided by the first-floor structure of the building. If an active soil pressure of the fluid type is assumed, the load and structural actions for the wall will be as shown in Fig. 32.1*a* when the ground level is at the top of the wall. When the ground level is below the top of the wall, the pressure is as shown in Fig. 32.1*b*. A surcharge load at the edge of the building will increase the pressure as shown in Fig. 32.1*c*. If the basement is multilevel, the wall will usually function as a multiple span element for the pressure as shown in Fig. 32.1*d*.

The following example illustrates the design of a simple vertically spanning basement wall of reinforced concrete. The design conforms to general requirements of the *ACI Code* for cover of reinforcing and minimum reinforcing in both vertical and horizontal directions. The example illustrates the basis for determination of the entries in Table 32.1. Figure 32.2 illustrates the data given in Table 32.1.

Example 1: *Design of a Basement Wall.* Design data and criteria:

Concrete design strength: $f_c' = 3000$ psi.

Reinforcing: $f_s = 20,000$ psi.

Active soil pressure: 30 lb/ft² per foot of depth below surface.

Surcharge: 300 lb/ft² on ground surface (equivalent to 3 ft of additional soil).

Solution. The wall is shown in Fig. 32.3. The wall spans vertically between the lateral supports provided by the basement floor and the structure supported on top of the

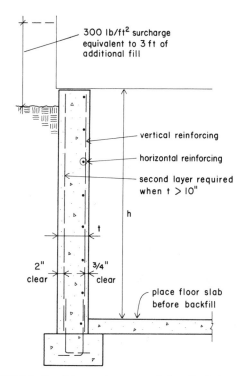

FIGURE 32.2. Conditions and recommendations for the basement walls in Table 32.1.

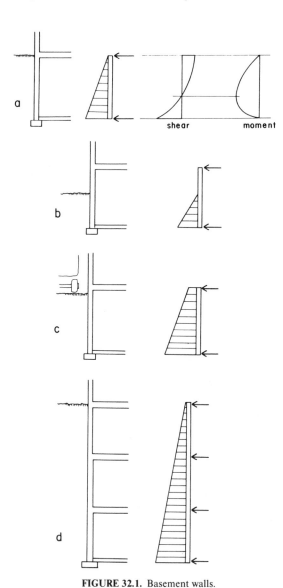

FIGURE 32.1. Basement walls.

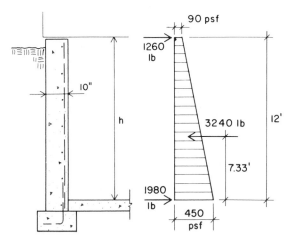

FIGURE 32.3.

TABLE 32.1. Reinforced Concrete Basement Walls[a]

	Vertical reinforcement (bar size – spacing in in./mm)			
Thickness of wall t (in./mm)	One layer of reinforcement		Two layers of reinforcement	
	8 / 200	10 / 250	12 / 300	
Height of wall h (ft/m)	¾ in. [20mm] clear of inside of wall	¾ in. [20mm] clear of inside of wall	¾ in. [20mm] clear of inside of wall	2 in. [50mm] clear of outside of wall
8 / 2.4	No. 4 at 14 / 350	No. 4 at 13 / 325	No. 4 at 18 / 450	No. 4 at 18 / 450
9 / 2.7	No. 4 at 11 / 275	No. 4 at 13 / 325	No. 4 at 16 / 400	No. 4 at 18 / 450
10 / 3.0	No. 5 at 12 / 300	No. 5 at 16 / 400	No. 5 at 18 / 450	No. 4 at 18 / 450
11 / 3.3		No. 5 at 13 / 325	No. 5 at 15 / 375	No. 4 at 18 / 450
12 / 3.6		No. 5 at 10 / 250	No. 5 at 12 / 300	No. 4 at 18 / 450
13 / 3.9	Not recommended		No. 5 at 10 / 250	No. 4 at 18 / 450
14 / 4.2	h > 15 t		No. 6 at 11 / 275	No. 4 at 18 / 450
Horizontal reinforcement	No. 5 at 15 / 375	No. 5 at 12 / 300	No. 5 at 10 / 250 if in one layer No. 4 at 13 / 325 if in two layers	

[a] See Figure 32.2.

wall. We assume the span to be the clear height of the wall. For ease of pouring the concrete we recommend limiting the wall height to approximately 15 times the thickness. For the 10-in.-thick wall this limits the height to $12\frac{1}{2}$ ft. Subtracting for the basement floor, we will therefore consider the maximum clear height to be 12 ft, and will design the wall for this span.

The *ACI Code* (Ref. 6) recommends the following minimum reinforcing for the wall:

Horizontal:

$$A_s = 0.0025A_g = 0.0025(120)$$
$$= 0.30 \text{ in.}^2/\text{ft}$$

Vertical:

$$A_s = 0.0015A_g = 0.0015(120)$$
$$= 0.18 \text{ in.}^2/\text{ft}$$

Thus we would use the following as minimum reinforcing for the wall, unless structural calculations indicate larger areas:

Horizontal: No. 5 bars at 12 in., $A_s = 0.31$ in.²/ft.
Vertical: No. 4 bars at 13 in., $A_s = 0.185$ in.²/ft.

The beam action of the wall is shown in Fig. 32.3. Because of the surcharge, the pressure at the top of the wall is $3(30) = 90$ lb/ft² using the equivalent fluid pressure method. This pressure increases at the rate of 30 lb/ft² per addi-

tional ft of depth to the maximum value of 450 lb/ft² at the bottom of the wall. In terms of the span and the unit of the pressure variation, the maximum moment produced for this loading will be approximately

$$M = 0.064 \, ph^3 + 0.375 \, ph^2$$

and for this example,

$$M = 0.064(30)(12)^3$$
$$+ 0.375(30)(12)^2$$
$$= 3318 + 1620 = 4938 \text{ lb-ft}$$

This moment may be compared to the balanced moment capacity in order to consider the concrete bending stress and the relative values to be used for k and j:

$$\text{balanced } M = Rbd^2$$
$$= 226(12)(9)^2(\tfrac{1}{12})$$
$$= 18,306 \text{ lb-ft}$$

which indicates that concrete stress is not critical and the section will be considerably underreinforced, permitting a conservative guess of 0.90 or higher for j.

We have used the value of 9 in. for the effective depth of the reinforced concrete section, which assumes the reinforcing to be placed with the minimum clearance of $\frac{3}{4}$ in. on the inside face of the wall. With these approximations for j and d, the area of steel required is determined as follows:

$$A_s = \frac{M}{f_s jd} = \frac{4938(12)}{20,000(0.9)(9)}$$

$$= 0.366 \text{ in.}^2/\text{ft}$$

which could be furnished with

No. 5 bars at 10 in., $A_2 = 0.37$ in.^{2}/ft.
No. 6 bars at 14 in., $A_2 = 0.38$ in.^{2}/ft.

Note from the Fig. 32.3 that the required support force at the bottom of the wall is 1980 lb. Unless there is considerable dead load on the top of the wall, it will probably be necessary to require that the basement floor slab be placed before the backfill is deposited against the wall.

It is possible, of course, that with a considerable vertical load the wall may have a critical combined stress or load/moment interaction. It is also possible that the vertical load may be sufficiently off center of the wall to produce significant moment, which may add to the bending due to the soil pressure. These conditions, plus any others due to grade beam action, load distributing, and so forth, should be considered in the full design of a basement wall.

Table 32.1 gives reinforcing recommendations for some concrete walls as determined by the procedures illustrated in Example 1.

32.5. RETAINING STRUCTURES

Strictly speaking, any wall that sustains significant lateral soil pressure is a retaining wall. However, the term is usually used with reference to a *cantilever retaining wall*, which is a freestanding wall without lateral support at its top. For such a wall the major design consideration is for the actual dimension of the ground-level difference that the wall serves to facilitate. The range of this dimension establishes some different categories for the retaining structure as follows:

Curbs. Curbs are the shortest freestanding retaining structures. The two most common forms are as shown in Fig. 32.4a, the selection being made on the basis of whether or not it is necessary to have a gutter on the low side of the curb. Use of these structures is typically limited to grade level changes of about 2 ft or less.

Short Retaining Walls. Vertical walls up to about 10 ft in height are usually built as shown in Fig. 32.4b. These consist of a concrete or masonry wall of uniform thickness. The wall thickness, footing width and thickness, vertical wall reinforcing, and transverse footing reinforcing are all designed for the lateral shear and cantilever bending moment plus the vertical weights of the wall, footing, and earth fill.

When the bottom of the footing is a short distance below grade on the low side of the wall and/or the lateral passive resistance of the soil is low, it may be necessary to use an extension below the footing—called a *shear key*—to increase the resistance to sliding. The form of such a key is shown in Fig. 32.4c.

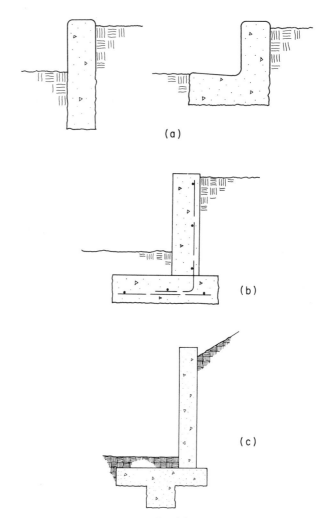

FIGURE 32.4. Retaining structures.

Tall Retaining Walls. As the wall height increases it becomes less feasible to use the simple construction shown in Fig. 32.4b or c. The overturning moment increases sharply with the increase in height of the wall. For very tall walls one modification used is to taper the wall thickness. This permits the development of a reasonable cross section for the high bending stress at the base without an excessive amount of concrete. However, as the wall becomes really tall, it is often necessary to consider the use of various bracing techniques, as shown in the other illustrations in Fig. 32.5.

Under most circumstances it is reasonable to design short retaining walls (up to about 10 ft high) by the equivalent fluid method. The following example illustrates this simplified method of design, using the working stress method for the investigation of the concrete elements.

Example 1. Short Concrete Retaining Wall. The wall is to be of reinforced concrete with the proposed profile shown in Fig. 32.6a. Design data and criteria are as follows:

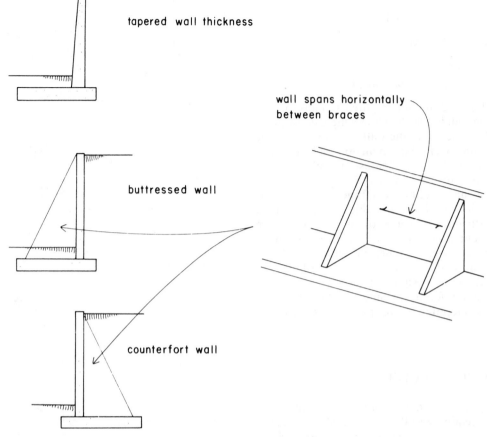

tapered wall thickness

wall spans horizontally
between braces

buttressed wall

counterfort wall

FIGURE 32.5. Tall retaining walls.

Active soil pressure: 30 lb/ft^2 per foot of height.
Soil weight: assumed to be 100 lb/ft^3.
Maximum allowable soil pressure: 1500 lb/ft^2.
Concrete strength: $f_c' = 3000$ psi.
Allowable tension on reinforcing: 20,000 psi.

Solution. The loading condition used to analyze the stress conditions in the wall is shown in Fig. 32.6*b*.

Maximum lateral pressure:

$$p = 30(4.667 \text{ ft}) = 140 \text{ lb/ft}^2$$

Total horizontal force:

$$H_1 = \frac{140(4.667)}{2} = 327 \text{ lb}$$

Moment at base of wall:

$$M = 327(\tfrac{56}{3}) = 6104 \text{ lb-in.}$$

For the wall we assume an approximate effective *d* of 5.5 in. The tension reinforcing required for the wall is thus

$$A_s = \frac{M}{f_s jd} = \frac{6104}{20,000(0.9)(5.5)}$$
$$= 0.061 \text{ in.}^2/\text{ft}$$

This may be provided by using No. 3 bars at 20-in. centers, which gives an actual A_s of 0.066 in.^{2}/ft. Since the embedment length of these bars in the footing is quite short, they should be selected conservatively and should have hooks at their ends for additional anchorage.

The loading condition used to investigate the soil stresses and the stress conditions in the footing is shown in Fig. 32.6*c*. In addition to the limit of the maximum allowable soil-bearing pressure, it is usually required that the resultant vertical force be kept within the kern limit of the footing. The location of the resultant force is therefore usually determined by a moment summation about the centroid of the footing plan area, and the location is found as an eccentricity from this centroid.

Table 32.2 contains the data and calculations for determining the location of the resultant force that acts at the bottom of the footing. The position of this resultant is found by dividing the net moment by the sum of the vertical forces as follows:

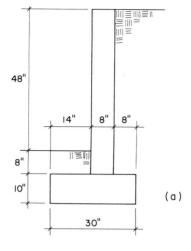

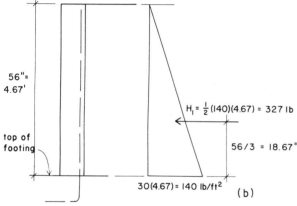

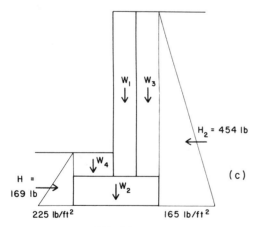

FIGURE 32.6.

$$e = \frac{5793}{1167} = 4.96 \text{ in.}$$

For the rectangular footing plan area the kern limit will be one-sixth of the footing width, or 5 in. The resultant is thus within the kern, and the combined soil stress may be determined by the stress formula

TABLE 32.2. Determination of the Eccentricity of the Resultant Force

Force (lb)		Moment Arm (in.)	Moment (lb-in.)
H_2	454	22	+9988
w_1	466	3	−1398
w_2	312	0	0
w_3	311	11	−3421
w_4	78	8	+624
Σ_w = 1167 lb		Net moment = +5793 lb-in.	

$$p = \frac{N}{A} \pm \frac{M}{S}$$

where N = total vertical force
A = plan area of the footing
M = net moment about the footing centroid
S = section modulus of the rectangular footing plan area, which is determined as

$$S = \frac{bh^2}{6} = \frac{1(2.5)^2}{6} = 1.042 \text{ ft}^3$$

The limiting maximum and minimum soil pressures are thus determined as

$$p = \frac{N}{A} \pm \frac{M}{S} = \frac{1167}{2.5} \pm \frac{5793/12}{1.042}$$

$$= 467 \pm 463$$

$$= 930 \text{ lb//ft}^2 \text{ maximum}$$

$$\text{and } 4 \text{ lb/ft}^2 \text{ minimum}$$

Since the maximum stress is less than the established limit of 1500 lb/ft², vertical soil pressure is not critical for the wall. For the horizontal force analysis the procedure varies with different building codes. The criteria given in this example for soil friction and passive resistance are those in the *UBC* (Appendix D) for ordinary sandy soils. This code permits the addition of these two resistances without modification. Using these data and technique, the analysis is as follows:

Total active force: 454 lb, as shown in Fig. 32.6c.
Friction resistance [(friction factor) (total vertical dead load)]:

$$0.25(1167) = 292 \text{ lb}$$

Passive resistance: 169 lb, as shown in Fig. 32.6c.
Total potential resistance:

$$292 + 169 = 461 \text{ lb}$$

(*Note*: Many designers reduce the passive resistance by one-third when combining it with sliding.) Since the total potential resistance is greater than the active force, the wall is not critical in horizontal sliding.

As with most wall footings, it is usually desirable to select the footing thickness to minimize the need for tension reinforcing due to bending. Thus shear and bending stresses are seldom critical, and the only footing stress concern is for the tension reinforcing. The critical section for bending is at the face of the wall, and the loading condition is as shown in Fig. 32.7. The trapezoidal stress distribution produces the resultant force of 833 lb, which acts at the centroid of the trapezoid, as shown in the illustration. Assuming an approximate depth of 6.5 in. for the section, the analysis is as follows:

Moment:

$$M = 833(7.706) = 6419 \text{ lb-in.}$$

Required area:

$$A_s = \frac{M}{f_s jd} = \frac{6149}{20,000(0.9)(6.5)}$$

$$= 0.055 \text{ in.}^2/\text{ft}$$

This requirement may be satisfied by using No. 3 bars at 24-in. centers. For ease of construction it is usually desirable to have the same spacing for the vertical bars in the wall and the transverse bars in the footing. Thus in this example the No. 3 bars at 20-in. centers previously selected for the wall would probably also be used for the footing bars. The vertical bars can then be held in position by wiring the hooked ends to the transverse footing bars.

Although bond stress is also a potential concern for the footing bars, it is not likely to be critical as long as the bar size is relatively small (less than a No. 6 bar or so).

Reinforcing in the long direction of the footing should be determined in the same manner as for ordinary wall footings. We recommend a minimum of 0.15% of the cross section. For the 10-in.-thick and 30-in.-wide footing this requires

$$A_s = 0.0015(300) = 0.45 \text{ in.}^2$$

We would therefore use three No. 4 bars with a total area of $3(0.2) = 0.6 \text{ in.}^2$.

In most cases designers consider the stability of a short cantilever wall to be adequate if the potential horizontal resistance exceeds the active soil pressure and the resultant of the vertical forces is within the kern of the footing. However, the stability of the wall is also potentially questionable with regard to the usual overturn effect. If this investigation is considered to be necessary, the procedure is as follows.

The loading condition is the same as that used for the soil stress analysis and shown in Fig. 32.6. As with the vertical soil stress analysis, the force due to passive soil resistance is not used in the moment calculation since it is only a potential force. For the overturn investigation the moments are taken with respect to the toe of the footing. The

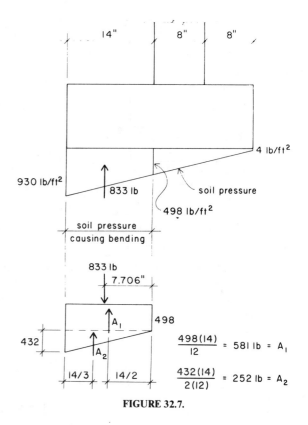

FIGURE 32.7.

TABLE 32.3. Analysis for Overturning Effect

Force (lb)	Moment Arm (in.)	Moment (lb-in.)
Overturn		
H_2 454	22	9988
Restoring moment		
w_1 466	18	8388
w_2 312	15	4680
w_3 311	26	8086
w_3 78	7	546
	Total	20,686 *lb-in.*

calculation of the overturning and the dead load restoring moments is shown in Table 32.3. The safety factor against overturn is determined as

$$\text{SF} = \frac{\text{restoring moment}}{\text{overturning moment}}$$

$$= \frac{20,686}{9988} = 2.07$$

The overturning effect is usually not considered to be critical as long as the safety factor is at least 1.5.

CHAPTER THIRTY-THREE

Concrete Foundations

Concrete construction—usually of poured-in-place, reinforced concrete—is used extensively in the development of building foundations. Common foundation elements include basement walls, basement floor slabs, wall footings, column footings, pedestals, abutments, pile caps, poured concrete deep foundations (piles and piers), and special bases for elevators and other equipment. General problems of foundations are discussed in Part 7. Chapter 33 deals with the design of three widely used, simple elements: wall footings, column footings, and pedestals. Design of some other foundation elements of reinforced concrete is presented in Part 7 and in the building design examples in Part 9.

33.1. COLUMN FOOTINGS

The great majority of independent or isolated column footings are square in plan, with reinforcing consisting of two sets of bars at right angles to each other. This is known as *two-way reinforcement*. The column may be placed directly on the footing block, or it may be supported by a pedestal. A pedestal, or pier, is a short, wide compression block that serves to reduce the punching effect on the footing. For steel columns a pier may also serve to raise the bottom of the steel column above ground level.

The design of a column footing is usually based on the following considerations:

1. *Maximum Soil Pressure.* The sum of the superimposed load on the footing and the weight of the footing must not exceed the limit for bearing pressure on the supporting material. The required total plan area of the footing is determined on this basis.

2. *Control of Settlement.* Where buildings rest on highly compressible soil, it may be necessary to select footing areas that assure a uniform settlement of all the building columns rather than to strive for a maximum use of the allowable soil pressure.

3. *Size of the Column.* The larger the column, the less will be the shear, flexural, and bond stresses in the footing, since these are developed by the cantilever effect of the footing projection beyond the edges of the column.

4. *Shear Stress Limit for the Concrete.* For square-plan footings this is usually the only critical stress condition for the concrete. In order to reduce the required amount of reinforcing, the footing depth is usually established well above that required by the flexural stress limit for the concrete.

5. *Flexural Stress and Development Length Limits for the Bars.* These are considered on the basis of the moment developed in the cantilevered footing at the face of the column.

6. *Footing Thickness for Development of Column Reinforcing.* When a footing supports a reinforced concrete column, the compressing force in the column bars must be transferred to the footing by bond stress—called *doweling* of the bars. The thickness of the footing must be sufficient for the necessary development length of the column bars.

The following example illustrates the design process for a simple, square column footing.

Example 1. A 16-in. [406-mm] square concrete column exerts a load of 240 [1068 kN] on a square column footing. Determine the footing dimensions and the necessary reinforcing using the following data: $f_c' = 3$ ksi [20.7 MPa], grade 40 bars with $f_y = 40$ ksi [276 MPa], $f_s = 20$ ksi [138 MPa], and maximum permissible soil pressure = 4000 psf [192 kN/m²].

Solution. The first decision to be made is that of the height, or thickness, of the footing. This has to be a raw first guess unless the dimensions of similar footings are known. In practice this knowledge is generally available from previous design work or from handbook tables. In lieu of this, a reasonable guess is made, the design work is performed, and an adjustment is made if the assumed thickness proves inadequate. We will assume a footing thickness of 20 in. [508 mm] for a first try for this example.

The footing thickness establishes the weight of the footing on a per-square-foot basis. This weight is then subtracted from the maximum permissible soil pressure, and the net value is then usable for the superimposed load on the footing. Thus

Footing weight: $\frac{20}{12}$ (150 psf) = 250 psf [12 kN/m²].

Net usable pressure: $4000 - 250 = 3750$ psf [180 kN/m²].

Required footing plan area: $\dfrac{240,000}{3750} = 64$ ft² [59.3 m²].

Length of the side of the square footing (L): $\sqrt{64}$ = 8 ft [2.44 m].

Two shear stress situations must be considered for the concrete. The first occurs as ordinary beam shear in the cantilevered portion and is computed at a critical section at a distance d (effective depth of the beam) from the face of the column as shown in Fig. 33.1a. The shear stress at this section is computed in the same manner as for a beam, as discussed in Chapter 27, and the stress limit is $v_c = 1.1\sqrt{f_c'}$. The second shear stress condition is that of peripheral shear, or "punching" shear, and is investigated at a circumferential section around the column at a distance of $d/2$ from the column face as shown in Fig. 33.1b. For this condition the allowable stress is $v_c = 2.0\sqrt{f_c'}$.

With two-way reinforcing, it is necessary to place the bars in one direction on top of the bars in the other direction. Thus, although the footing is supposed to be the same in both directions, there are actually two different d distances—one for each layer of bars. It is common practice to use the average of these two distances for the design value of d; that is, d = the footing thickness less the sum of the concrete cover and the bar diameter. With the bar diameter as yet undetermined, we will assume an approximate d of the footing thickness less 4 in. [102 mm] (a concrete cover of 3 in. plus a No. 8 bar). For the example this becomes

$$d = t - 4 = 20 - 4 = 16 \text{ in. [306 mm]}$$

It should be noted that it is the *net* soil pressure that causes stresses in the footing, since there will be no bending or shear in the footing when it rests alone on the soil. We thus use the net soil pressure of 3750 psf [180 kN/m²] to determine the shear and bending effects for the footing.

For the beam shear investigation, we determine the shear force generated by the net soil pressure acting on the shaded portion of the footing plan area shown in Fig. 33.1a. Thus

$$V = 3750(8)\left(\frac{24}{12}\right) = 60,000 \text{ lb [267.5 kN]}$$

and, using the formula for shear stress in a beam

$$v = \frac{V}{bd} = \frac{60,000}{96(16)} = 39.1 \text{ psi [0.270 MPa]}$$

which is compared to the allowable stress of

$$v_c = 1.1\sqrt{f_c'} = 1.1\sqrt{3000} = 60 \text{psi [0.414 MPa]}$$

indicating that this condition is not critical.

For the peripheral shear investigation, we determine the shear force generated by the net soil pressure acting on the shaded portion of the footing area shown in Fig. 33.1b.

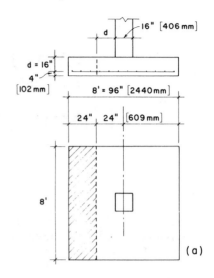

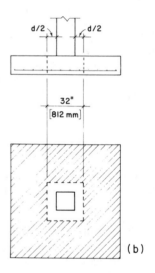

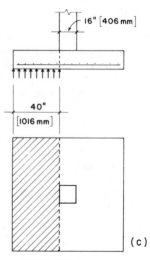

FIGURE 33.1.

Thus

$$V = 3750\left[(8)^2 - \left(\frac{32}{12}\right)^2 \right] = 213,333 \text{ lb } [953 \text{ kN}]$$

Shear stress for this case is determined with the same formula as for beam shear, with the dimension b being the total peripheral circumference. Thus

$$v = \frac{V}{bd} = \frac{213,333}{[4(32)]16} = 104.2 \text{ psi } [0.723 \text{ MPa}]$$

which is compared to the allowable stress of

$$v_c = 2\sqrt{f_c'} = 2\sqrt{3000} = 109.5 \text{ psi } [0.755 \text{ MPa}]$$

This computation indicates that the peripheral shear stress is not critical, but since the actual stress is quite close to the limit, the assumed thickness of 20 in. is probably the least full-inch value that can be used. Flexural stress in the concrete should also be considered, although it is seldom critical for a square footing. One way to verify this is to compute the balanced moment capacity of the section with $b = 96$ in. and $d = 16$ in. Using the factor for a balanced section from Table 27.1, we find

$$M_R = Rbd^2 = 0.226(96)(16)^2$$
$$= 5554 \text{ kip-in. or } 463 \text{ kip-ft}$$

which may be compared with the actual moment computed in the next step.

For the reinforcing we consider the stresses developed at a section at the edge of the column as shown in Fig. 33.1c. The cantilever moment for the 40-in. [1016-mm] projection of the footing beyond the column is

$$M = 3750(8)\left(\frac{40}{12}\right)\frac{1}{2}\left(\frac{40}{12}\right) = 166,667 \text{ lb-ft}$$
$$[227 \text{ kN/m}]$$

Using the formula for required steel area in a beam, with a conservative guess of 0.9 for j, we find (see Sec. 27.2)

$$A_s = \frac{M}{f_s jd} = \frac{166,667(12)}{20(0.9)(16)(10^3)}$$
$$= 6.95 \text{ in.}^2 \ [4502 \text{ mm}^2]$$

This requirement may be met by various combinations of bars, such as those in Table 33.1. Data for consideration of the development length and the center-to-center bar spacing is also given in the table. The flexural stress in the bars must be developed by the embedment length equal to the projection of the bars beyond the column edge, as discussed in Sec. 29.1. With a minimum of 2 in. [51 mm] of concrete cover at the edge of the footing, this length is 38 in. [965 mm]. The required development lengths indicated in the table are taken from Table 29.1. It may be noted that all of the combinations in the table are adequate in this regard.

If the distance from the edge of the footing to the first bar at each side is approximately 3 in. [76 mm], the center-to-

TABLE 33.1. Reinforcing Alternatives for the Column Footing

Number and Size of Bars	Area of Steel Provided		Required Development Length[a]		Center-to-Center Spacing	
	in.^{2}	mm^2	in.	mm	in.	mm
12 No. 7	7.20	4645	18	457	8.2	208
9 No. 8	7.11	4687	23	584	11.3	286
7 No. 9	7.00	4516	29	737	15	381
6 No. 10	7.62	4916	37	940	18	458

[a] From Table 29.1; values for "other bars," $f_y = 40$ ksi, $f_c' = 3$ ksi.

center distance for the two outside bars will be $96 - 2(3) = 90$ in. [2286 mm], and the rest of the bars evenly spaced, the spacing will be 90 divided by the number of total bars less one. This value is shown in the table for each set of bars. The maximum permitted spacing is 18 in. [457 mm], and the minimum should be a distance that is adequate to permit good flow of the wet concrete between the two-way grid of bars—say 4 in. [102 mm] or more.

All of the bar combinations in Table 33.1 are adequate for the footing. Many designers prefer to use the largest possible bar, as this reduces the number of bars that must be handled and supported during construction. On this basis, the footing will be the following:

8 ft^2 by 20 in. thick with six No. 10 bars each way.

For ordinary situations we often design square column footings by using data from tables in various references. Even when special circumstances make it necessary to perform the type of design illustrated in Example 1, such tables will assist in making a first guess for the footing dimensions.

Table 33.2 gives the allowable superimposed load for a range of footings and soil pressures. This material has been adapted from a more extensive table in *Simplified Design of Building Foundations* (Ref. 16). Designs are given for concrete strengths of 2000 and 3000 psi. The low strength of 2000 psi is sometimes used for small buildings, since many building codes permit the omission of testing of the concrete if this value is used for design.

33.2. WALL FOOTINGS

Wall footings consist of concrete strips placed under walls. The most common type of wall footing is that shown in Fig. 33.3, consisting of a strip with a rectangular cross section placed in a symmetrical position with respect to the wall and projecting an equal distance as a cantilever from both faces of the wall. For soil stress the critical dimension of the footing is the width of the footing bottom measured perpendicular to the wall face.

In most situations, the wall footing is utilized as a platform upon which the wall is constructed. Thus a minimum

TABLE 33.2. Allowable Loads on Square Column Footings (See Fig. 33.2)

Maximum soil pressure (lb/ft²)	Minimum column width t (in.)	$f'_c = 2000$ psi				$f'_c = 3000$ psi			
		Allowable load on footing[a] (k)	Footing dimensions h (in.)	w (ft)	Reinforcing each way	Allowable load on footing (k)	Footing dimensions h (in.)	w (ft)	Reinforcing each way
1000	8	7.9	10	3.0	2 No. 3	7.9	10	3.0	2 No. 3
	8	10.7	10	3.5	3 No. 3	10.7	10	3.5	3 No. 3
	8	14.0	10	4.0	3 No. 4	14.0	10	4.0	3 No. 4
	8	17.7	10	4.5	4 No. 4	17.7	10	4.5	4 No. 4
	8	22	10	5.0	4 No. 5	22	10	5.0	4 No. 5
	8	31	10	6.0	5 No. 6	31	10	6.0	5 No. 6
1500	8	12.4	10	3.0	3 No. 3	12.4	10	3.0	3 No. 3
	8	16.8	10	3.5	3 No. 4	16.8	10	3.5	3 No. 4
	8	22	10	4.0	4 No. 4	22	10	4.0	4 No. 4
	8	34	11	5.0	5 No. 5	34	10	5.0	6 No. 5
	8	48	12	6.0	6 No. 6	49	11	6.0	6 No. 6
	8	65	14	7.0	7 No. 6	65	13	7.0	6 No. 7
	8	83	16	8.0	7 No. 7	84	15	8.0	7 No. 7
2000	8	17	10	3.0	4 No. 3	17	10	3.0	4 No. 3
	8	23	10	3.5	4 No. 4	23	10	3.5	4 No. 4
	8	30	10	4.0	6 No. 4	30	10	4.0	6 No. 4
	8	46	12	5.0	6 No. 5	46	11	5.0	5 No. 6
	8	65	14	6.0	6 No. 6	66	13	6.0	7 No. 6
	8	88	16	7.0	8 No. 6	89	15	7.0	7 No. 7
	8	113	18	8.0	8 No. 7	114	17	8.0	9 No. 7
	8	142	20	9.0	8 No. 8	143	19	9.0	8 No. 8
	10	174	21	10.0	9 No. 8	175	20	10.0	10 No. 8
3000	8	26	10	3.0	3 No. 4	26	10	3.0	3 No. 4
	8	35	10	3.5	4 No. 5	35	10	3.5	4 No. 5
	8	45	12	4.0	4 No. 5	46	11	4.0	5 No. 5
	8	70	14	5.0	5 No. 6	71	13	5.0	6 No. 6
	8	100	17	6.0	7 No. 6	101	15	6.0	8 No. 6
	10	135	19	7.0	7 No. 7	136	18	7.0	8 No. 7
	10	175	21	8.0	10 No. 7	177	19	8.0	8 No. 8
	12	219	23	9.0	9 No. 8	221	21	9.0	10 No. 8
	12	269	25	10.0	11 No. 8	271	23	10.0	10 No. 9
	12	320	28	11.0	11 No. 9	323	26	11.0	12 No. 9
	14	378	30	12.0	12 No. 9	381	28	12.0	11 No. 10
4000	8	35	10	3.0	4 No. 4	35	10	3.0	4 No. 4
	8	47	12	3.5	4 No. 5	47	11	3.5	4 No. 5
	8	61	13	4.0	5 No. 5	61	12	4.0	6 No. 5
	8	95	16	5.0	6 No. 6	95	15	5.0	6 No. 6
	8	135	19	6.0	8 No. 6	136	18	6.0	7 No. 7
	10	182	22	7.0	8 No. 7	184	20	7.0	9 No. 7
	10	237	24	8.0	9 No. 8	238	22	8.0	9 No. 8
	12	297	26	9.0	10 No. 8	299	24	9.0	9 No. 9
	12	364	29	10.0	13 No. 8	366	27	10.0	11 No. 9
	14	435	32	11.0	12 No. 9	440	29	11.0	11 No. 10
	14	515	34	12.0	14 No. 9	520	31	12.0	13 No. 10
	16	600	36	13.0	17 No. 9	606	33	13.0	15 No. 10
	16	688	39	14.0	15 No. 10	696	36	14.0	14 No. 11
	18	784	41	15.0	17 No. 10	793	38	15.0	16 No. 11

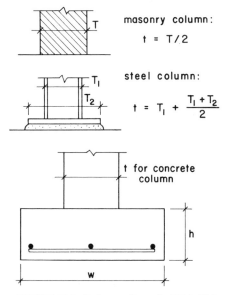

FIGURE 33.2. Reference figure for Table 33.2.

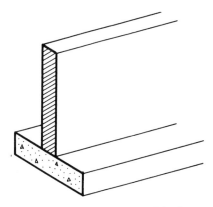

FIGURE 33.3. Continuous wall footing.

width for the footing is established by the wall thickness, the footing usually being made somewhat wider than the wall. With a concrete wall this additional width is used to support the wall forms while the concrete is poured. For masonry walls this added width assures an adequate base for the mortar bed for the first course of the masonry units. The exact additional width required for these purposes is a matter of judgment. For support of concrete forms, it is usually desirable to have at least a 3-in. projection; for masonry, the usual minimum is 2 in.

With relatively lightly loaded walls, the minimum width required for platform considerations may be more than adequate in terms of the allowable bearing stress on the soil. If this is the case, the short projection of the footing from the wall face will produce relatively insignificant transverse bending and shear stresses, permitting a minimal thickness for the footing and the omission of transverse reinforcing. Most designers prefer, however, to provide some continuous reinforcing in the long direction of the footing, even when none is used in the transverse direction. The purpose is to reduce shrinkage cracking and also to give some enhanced beamlike capabilities for spanning over soft spots in the supporting soil.

As the wall load increases, the increased width of the footing required to control soil stress eventually produces significant transverse bending and shear in the footing. At some point this determines the required thickness for the footing and for required reinforcing in the transverse direction. If the footing is not reinforced in the transverse direction, the controlling stress is usually the transverse tensile bending stress in the concrete. If the footing has transverse reinforcing, the controlling concrete stress is usually the shear stress. The following example illustrates the procedure for design of a wall footing with transverse reinforcing.

Example 1. Using concrete with $f_c' = 2$ ksi and grade 40 reinforcing, design a wall footing for the following data: wall thickness = 6 in., load on footing = 8750 lb/ft of wall length, and maximum allowable soil pressure = 2000 psf.
Solution. For the wall footing, the only concrete stress of concern is that of shear. Compression stress in flexure is seldom critical due to design for shear and the desire for minimal transverse reinforcing. The usual design procedure consists of making a guess for the footing thickness and determining conditions to verify the guess. Code restrictions establish a minimum thickness of 10 in. Try

$$h = 12 \text{ in.}$$

Then

Footing weight is 150 psf.
Usable soil pressure is: $2000 - 150 = 1850$ psf.
Required width: $8750/1850 = 4.73$ ft or 56.8 in.

Try

$$\text{width} = 57 \text{ in.} \quad \text{or} \quad 4 \text{ ft } 9 \text{ in.}$$

Then

$$\text{design soil pressure} = \frac{8750}{4.75} = 1842 \text{ psf}$$

With a 3-in. cover and a No. 6 bar (a guess), the effective depth will be 8.625 in., say 8.6 in. approximately.

The *ACI Code* requires that shear be investigated as a beam shear condition, with the critical section at a distance d (effective depth) from the face of the wall. This condition is shown in Fig. 33.4a. This is reasonably valid when the cantilever distance is larger than the footing thickness by a significant amount, but is questionable for short cantilevers. In fact, the code recommends that this shortened span not be used for brackets and short cantilevers. Therefore, we recommend that the critical section for shear be taken at the face of the wall unless the cantilever exceeds three times the overall footing thickness. However, if the latter assumption is made (short cantilever analysis), it is

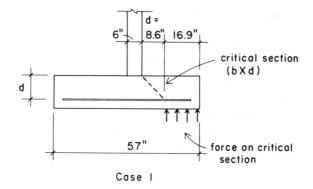

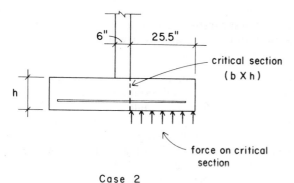

FIGURE 33.4.

amount of reinforcing, and any reduction in the footing thickness will shorten the moment arm for the tension reinforcing, requiring an increase in steel area. It therefore becomes a matter of judgment about the ideal value for the footing thickness.

If we reduce the footing thickness to 11 in., a second try would proceed as follows:

New footing weight: $\frac{11}{12}$ (150) = 137.5, say 138 lb/ft².

Usable soil pressure (p): 2000 − 138 = 1862 lb/ft².

Required width (w): $\frac{8750}{1862}$ = 4.70 ft or 56.4 in.

which does not change the footing width or design soil pressure.

$$\text{new } d = h - 3 - \frac{D}{2} = 11 - 3.375 = 7.625,$$

$$\text{say } 7.6 \text{ in.}$$

For the case 2 shear stress, the shear force is the same as for the first try, and the new shear stress is

$$v_c = \frac{V}{bh} = \frac{3914}{12(11)} = 30 \text{ psi}$$

For the case 1 shear stress, the shear section is now an inch closer to the wall and the shear force becomes

$$V = 1842 \left(\frac{17.9}{12} \right) = 2748 \text{ lb}$$

$$v_c = \frac{V}{bd} = \frac{2748}{12(7.6)} = 30 \text{ psi}$$

The bending moment to be used for concrete stress and determination of the steel area is

$$M = 3914 \left(\frac{25.5}{2} \right) = 49,903 \text{ lb-in.}$$

and the required steel area per ft of wall length is

$$A_s = \frac{M}{f_s jd} = \frac{49,903}{20(0.9)(7.6)} = 0.365 \text{ in.}^2$$

Since the steel area requirement has been determined in the same manner as for a slab, Table 27.3 may be used to select the bars and their spacing. The following should be considered in making the selection.

1. Maximum recommended spacing is 18 in.
2. Minimum recommended spacing is 6 in. to minimize the number of bars and allow for easy placing of the concrete during construction.
3. For proper development of the bars smaller bar sizes are usually preferable.

Table 33.3 presents a summary of the possible alternatives for reinforcing in the transverse direction, as determined from the data in Table 27.3. Our preference would be for the No. 5 bars at 10 in. center to center. Reference to

reasonable to use the full thickness of the footing for the stress computation rather than the effective depth of the cross section. Both cases are shown in Fig. 33.4, and we will show the computations for both.

Case 1. Shear at the d distance from the wall (see Fig. 33.4a).

Shear force:

$$V = 1842 \left(\frac{16.9}{12} \right) = 2594 \text{ lb}$$

Stress:

$$v_c = \frac{V}{bd} = \frac{2594}{12(8.6)} = 25 \text{ psi}$$

Case 2. Shear at the wall face (see Fig. 33.4b).

$$V = 1842 \left(\frac{25.5}{12} \right) = 3914 \text{ lb}$$

Stress:

$$v_c = \frac{V}{bh} = \frac{3914}{12(12)} = 27 \text{ psi}$$

Both of these are well below the allowable stress of $1.1 \sqrt{f_c'}$, which is the same as in Example 1 of Sec. 33.1: 49 psi. It is possible, therefore, to reduce the footing thickness if the shear stress is considered to be an important criterion. However, as has been discussed previously, reduction of cost in construction is usually obtained by minimizing the

Table 29.1 will show that development is more than adequate for these bars. (*Note*: Data are not given in Table 29.1 for $f_c' = 2000$ psi; however, the table shows only 12 in. required for $f_c' = 3000$ psi, whereas over 23 in. is available with our footing.)

Whether transverse reinforcing is provided or not, we recommend a minimum reinforcing for shrinkage stresses

in the long direction of the footing consisting of 0.0015 times the gross area of the cross section. Thus

$$A_s = 0.0015(11)(57) = 0.94 \text{ in.}^2$$

This area can be supplied by using three No. 5 bars with a total area of 0.93 in.^{2}.

Table 33.4 gives values for wall footings for four different soil pressures. Table data were derived using the procedures illustrated in the example. Figure 33.5 shows the dimensions referred to in the table.

TABLE 33.3. Selection of Reinforcing for Example 1

Bar Size	Area of Bar (in.^{2})	Area Required for Flexure (in.^{2})	Spacing Required (in.)	Selected Spacing (in.)
3	0.11	0.365	3.6	3.5
4	0.20	0.365	6.6	6.5
5	0.31	0.365	10.2	10
6	0.44	0.365	14.5	14.5
7	0.60	0.365	19.7	19.5

33.3. PEDESTALS

A pedestal (also called a *pier*) is defined by the *ACI Code* as a short compression member whose height does not exceed three times its width. Pedestals are frequently used as transitional elements between columns and the bearing footings that support them. Figure 33.6 shows the use of

TABLE 33.4. Allowable Loads on Wall Footings (See Fig. 33.5)

Maximum Soil Pressure (lb ft^2)	Minimum Wall Thickness, t		Allowable Load on Footing[a] (lb/ft)	Footing Dimensions		Reinforcing	
	Concrete (in.)	Masonry (in.)		h (in.)	w (in.)	Long Direction	Short Direction
1000	4	8	2625	10	36	3 No. 4	No. 3 at 16
	4	8	3062	10	42	2 No. 5	No. 3 at 12
	6	12	3500	10	48	4 No. 4	No. 4 at 16
	6	12	3938	10	54	3 No. 5	No. 4 at 13
	6	12	4375	10	60	3 No. 5	No. 4 at 10
	6	12	4812	10	66	5 No. 4	No. 5 at 13
	6	12	5250	10	72	4 No. 5	No. 5 at 11
1500	4	8	4125	10	36	3 No. 4	No. 3 at 10
	4	8	4812	10	42	2 No. 5	No. 4 at 13
	6	12	5500	10	48	4 No. 4	No. 4 at 11
	6	12	6131	11	54	3 No. 5	No. 5 at 15
	6	12	6812	11	60	5 No. 4	No. 5 at 12
	6	12	7425	12	66	4 No. 5	No. 5 at 11
	8	16	8100	12	72	5 No. 5	No. 5 at 10
2000	4	8	5625	10	36	3 No. 4	No. 4 at 14
	6	12	6562	10	42	2 No. 5	No. 4 at 11
	6	12	7500	10	48	4 No. 4	No. 5 at 12
	6	12	8381	11	54	3 No. 5	No. 5 at 11
	6	12	9250	12	60	4 No. 5	No. 5 at 10
	8	16	10106	13	66	4 No. 5	No. 5 at 9
	8	16	10875	15	72	6 No. 5	No. 5 at 9
3000	6	12	8625	10	36	3 No. 4	No. 4 at 10
	6	12	10019	11	42	4 No. 4	No. 5 at 13
	6	12	11400	12	48	3 No. 5	No. 5 at 10
	6	12	12712	14	54	6 No. 4	No. 5 at 10
	8	16	14062	15	60	5 No. 5	No. 5 at 9
	8	16	15400	16	66	5 No. 5	No. 6 at 12
	8	16	16725	17	72	6 No. 5	No. 6 at 10

[a] Allowable loads do not include the weight of the footing, which has been deducted from the total bearing capacity. Criteria $f_c' = 2000$ psi, grade 40 reinforcing, $v_c = 1.1 \sqrt{f_c'}$.

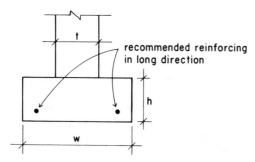

FIGURE 33.5. Reference figure for Table 33.4.

pedestals with both steel and reinforced concrete columns. The most common reasons for use of pedestals are:

1. To spread the load on top of the footing. This may relieve the intensity of direct bearing pressure on the footing or may simply permit a thinner footing with less reinforcing due to the wider column.
2. To permit the column to terminate at a higher elevation where footings must be placed at depths con-

siderably below the lowest parts of the building. This is generally most significant for steel columns.
3. To provide for the required development length of reinforcing in reinforced concrete columns, where footing thickness is not adequate for development within the footing.

Figure 33.6c illustrates the third situation described. Referring to Table 29.3, we may observe that a considerable development length is required for large diameter bars made from high grades of steel. If the minimum required footing does not have a thickness that permits this development, a pedestal may offer a reasonable solution. However, there are many other considerations to be made in the decision, and the column reinforcing problem is not the only factor in this situation.

If a pedestal is quite short with respect to its width (see Fig. 33.6e), it may function essentially the same as a column footing, with significant values for shear and bending stresses. This condition is likely to occur if the pedestal width exceeds twice the column width and the pedestal height is less than one-half of the pedestal width. In such cases, the pedestal must be designed by the same procedures used for an ordinary column footing.

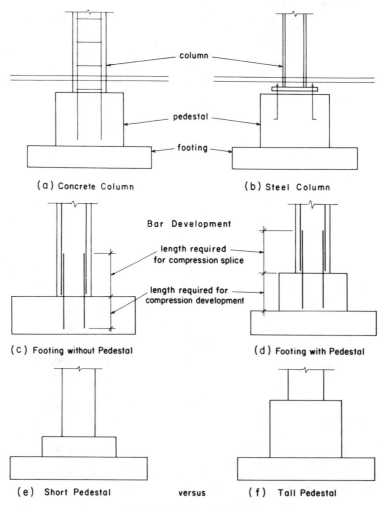

FIGURE 33.6. Use of pedestals.

CHAPTER THIRTY-FOUR

Special Structures

Despite the extensive treatment in this part, the work presented is only a sampling of the most basic issues and the most ordinary uses of structural concrete. A full treatment of all problems of investigation and design of all possible concrete structures would comprise a total presentation several times the size of this entire book. Concrete is a highly versatile material, lending itself to highly mundane and practical uses as well as to some of the most imaginative and sophisticated constructions produced by architects and engineers. In Chapter 34 we deal briefly with some of the uses of concrete for building structures—beyond the simple wall and frame constructions thus far presented in Part 5.

34.1. PRESTRESSED CONCRETE

Prestressing consists of the deliberate inducing of some internal stress condition in a structure prior to its sustaining of service loads. The purpose is to compensate in advance for some anticipated service load stress, which for concrete means some high level of tension stress. The "pre-" or "before" stress is therefore usually a compressive or reversal bending stress. This chapter discusses some uses of prestressing and some of the problems encountered in utilizing it for building structures.

Use of Prestressing

The principal use of prestressing is for spanning elements, in which the major stress conditions to be counteracted are tension due to bending and diagonal tension due to shear. A principal advantage of prestressing is that when properly achieved, it does not result in the natural tension cracking associated with ordinary reinforced concrete. Since flexural cracking is proportionate to the depth of the member, which in turn is proportionate to the span, the use of prestressing frees spanning concrete members from the span limits associated with ordinary reinforcing. Thus gigantic beam cross sections and phenomenal spans are possible—and indeed have been achieved, although mostly in bridge construction.

The cracking problem also limits the effective use of very high concrete strengths with ordinary reinforcing. Free of this limit, the prestressed structure can utilize effectively the highest strengths of concrete achievable, and thus weight saving is possible, resulting in a span-to-weight ratio that partly overcomes the usual massiveness of spanning concrete structures.

The advantages just described have their greatest benefit in the development of long, flat-spanning roof structures. Thus a major use of prestressing has been in the development of precast, prestressed units for roof structures. The hollow-cored slab, single-tee, and double-tee sections shown in Fig. 34.1 are the most common forms of such units—now a standard part of our structural inventory. These units can also be used for floor structures, having a major advantage when span requirements are at the upper limits of feasibility for ordinary reinforced construction.

Some other uses of prestressing in building structures are the following:

1. *For Columns.* Concrete shafts may be prestressed for use as building columns, precast piles, or posts for street lights or signs. In this case the prestressing compensates for bending, shear, and torsion associated with service use and handling during production, transportation, and installation. The ability to use exceptionally high strength concrete is often quite significant in these applications.

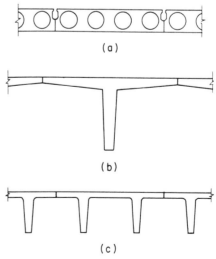

(a)

(b)

(c)

FIGURE 34.1. Typical precast concrete spanning units.

2. *For Two-Way Spanning Slabs.* Two-way, continuous prestressing can be used to provide for the complex deformations and stress conditions in concrete slabs with two-way spanning actions. A special usage is that for a paving slab designed as a spanning structure where ground settlement is anticipated. Crack reduction may be a significant advantage in these applications.

3. *Tiedown Anchors.* Where exceptionally high anchorage forces must be developed, and development of ordinary tension reinforcing may be difficult or impossible, it is sometimes possible to use the tension strands employed for prestressing. Large abutments, counterforts for large retaining walls, and other elements requiring considerable tension anchorage are sometimes built as prestressed elements.

4. *Horizontal Ties.* Single-span arches and rigid frames that develop outward thrusts on their supports are sometimes tied with prestressing strands.

For any structure it is necessary to consider various loading conditions that occur during construction and over a lifetime of use. For the prestressed structure this is a quite complex issue, and design must incorporate many different events over the life of the structure. For common usages, experience has produced various empirical adjustments (educated fudge factors) that account for the usual occurrences. For unique applications there must be some reasonable tolerance for errors in assumptions or some provision for tuning up the finished structure. The prestressed structure is a complex object, and design of other than very routine elements should be done by persons with considerable training and experience.

Pretensioned Structures

Prestressing is generally achieved by stretching high-strength steel strands (bunched wires) inside the concrete element. The stretching force is eventually transferred to the concrete, producing the desired compression in the concrete. There are two common procedures for achieving the stretching of the strands: pretensioning and post-tensioning.

Pretensioning consists of stretching the strands prior to pouring the concrete. The strands are left exposed, and as the concrete hardens it bonds to the strands. When the concrete is sufficiently hardened, the external stretching force is released and the strand tension is transferred to the concrete through the bond action on the strand surfaces. This procedure requires some substantial element to develop the necessary resistance to the jacking force used to stretch the strands before the concrete is poured. Pretensioning is used mostly for factory precast units, for which the stretch force resisting element is the casting form, sturdily built of steel and designed for continuous, multiple use.

Pretensioning is done primarily for cost-saving reasons. There is one particular disadvantage to pretensioning: It does not allow for any adjustment and the precise stress and deformation conditions of the finished product are only approximately predictable. The exact amount of the strand bonding and the exact properties of the finished concrete are somewhat variable. Good quality control in production can keep the range of variability within some bounds, but the lack of precision must be allowed for in the design and construction. A particular problem is that of control of deflection of adjacent units in systems consisting of side-by-side units.

Post-Tensioned Structures

In post-tensioning the prestressing strands are installed in a slack condition, typically wrapped with a loose sleeve or conduit. The concrete is poured and allowed to harden around the sleeves and the end anchorage devices for the strand. When the concrete has attained sufficient strength, the strand is anchored at one end and stretched at the other end by jacking against the concrete. When the calibrated jacking force is observed to be sufficient, the jacked end of the strand is locked into the anchorage device, pressurized grout is injected to bond the strand inside the sleeve, and the jack is released.

Post-tensioning is generally used for elements that are cast in place, since the forms need not resist the jacking forces. However, it may also be used for precast elements when jacking forces are considerable and/or a higher control of the net existing force is desired.

Until the strands are grouted inside the sleeves, they may be rejacked to a higher stretching force condition repeatedly. In some situations this is done as the construction proceeds, permitting the structure to be adjusted to changing load conditions.

Post-tensioning is usually more difficult and more costly, but there are some situations where it is the only alternative for achieving the prestressed structure.

34.2. PRECAST CONCRETE

Precasting refers to the process of construction in which a concrete element is cast somewhere other than where it is to be used. The other place may be somewhere else on the building site or away from the site, probably in a casting yard or factory. The precast element may be prestressed, may be of ordinary reinforced construction, or may even be without reinforcement. The single precast element may be a component of a general precast concrete system, or may serve a singular purpose in a construction system of mixed materials or types of elements. This chapter gives some discussion of the uses of precasting and the problems encountered in designing precast elements and systems.

Use of Precasting

The technique of precasting is utilized in a variety of ways. Undoubtedly, the most widely used precast element is the ordinary concrete block—called a *CMU*, short for *concrete masonry unit.* Most structural masonry is made from these units. In ordinary construction the block form shown in Fig. 34.2*a* is commonly used; for reinforced masonry construction (used exclusively in zones of high seismic risk) a different form is used, as shown in Fig. 4.6. Construction with CMUs is discussed in Chapter 36.

Another widely used precast element is the tilt-up wall unit, shown in Fig. 34.2*b*. This element is site-cast in a horizontal position, then tilted up and moved by a crane to its desired location. The casting bed usually consists of the building floor slab on grade, resulting in a considerable reduction in forming cost. This type of construction is widely used for one-story and low-rise commercial structures in the southern and western regions of the United States.

As discussed in Sec. 34.1, some of the most widely used prestressed elements are the flat-spanning units of hollow core or tee form used for roof and floor construction. These units are produced in casting factories in continuous production processes.

Structural systems consisting of connected components of precast concrete have been produced in great variety. Some of these have been produced as patented, manufactured systems; but mostly they have been the single, innovative products of individual designers.

Design and construction of precast concrete is strongly influenced by the standards and publications of the Precast Concrete Institute (PCI). Industry products are largely developed in conformance with these standards. Anyone contemplating the design of a unique system or element of precast concrete is advised to investigate the information available from PCI.

Advantages of Precast Concrete

There are various reasons for considering the use of precast concrete construction. In some cases the choice is between precasting and ordinary construction of cast-in-place concrete, with elements formed and cast at the location where they are to be used. In other cases the choice may be between using precast elements or some other material or type of construction. The following are some advantages offered by the precasting process, generally in comparison to cast-in-place construction.

Faster Site Work. Cast-in-place concrete construction usually proceeds quite slowly, requiring construction of forms, installation of reinforcing, pouring of concrete, hardening to sufficient strength to permit removal of forms, and so on. Erection of precast elements is more akin to construction with steel or timber structures, and the faster site work may be an advantage where construction time is highly constrained. However, it is the total building construction time that is significant, not just the time to get up the structure. If time cannot also be gained in other parallel construction activities, the rapidly erected structure may just sit there and wait for the other project work to catch up.

Forming Economies. For the ordinary cast-in-place concrete structure, a major portion of the total cost is represented by the forming. This includes the cost of the construction, bracing and support, and removal of the forming. Some reuse of items may be possible, but the process tends to use materials up rapidly. Precasting offers more potential for reuse of forms, even with site-cast construction. Factory processes involve extensive reuse of forms or production by forming processes such as extrusion. Reduction of on-site labor costs may often be the major gain in this area.

Quality Control. Precision of detail, quality, and uniformity of finishes, and uniformity of concrete properties (color, density, compaction, etc.) may be assured in factory production to a degree not possible with site casting. Here it is not just precasting but factory conditions that are the issue. All of this is more true if the element is a standard manufactured product subject to ongoing quality control in its production. This is obviously of greatest concern for construction elements exposed to view, especially wall components.

Use of Predesigned Elements and Systems. Face it—design effort means time and money. Use of a standard predesigned building construction component means a shortcut in design development effort. If the component is part of a system, with system-wide concerns for total building

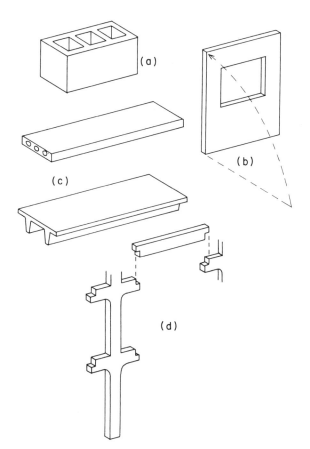

FIGURE 34.2. Structural elements of precast concrete.

utilization carefully preconsidered and standardized, the savings in design effort and the relative assurance of end results may be quite significant. This is treacherous ground for designers, with the possible end result being a direct connection between supplier and building owner, leaving the designer out in the cold.

Utilization of High-Quality Concrete and Prestressing. The potential for utilization of the structural properties of very high strength concrete usually occurs in association with prestressing (versus ordinary reinforcing). Factory-produced concrete is routinely of higher quality than the concrete produced by site casting. In some situations, denser surfaces, lower permeability, and reduced shrinkage may also be significant. This may be a factor in the use of mixed systems of precast and cast-in-place elements.

Problems with Precast Structures

As with any form of construction, there are some particular problems associated with precast concrete. These are not necessarily unsurmountable, but designers should be aware of them and of the considerations that may be required in using the construction method. The following are some major concerns that may be of significance in various situations.

Handling and Transporting. Concrete construction is usually heavy—precast elements included. Precast units are usually of considerable size, and the combination presents a major problem of handling and transporting the heavy and relatively fragile units. Stresses induced during handling and erecting of units may be significant structural design concerns. Use of factory-cast units is usually feasible only within some reasonable distance from the factory.

Cost. The cost of production, handling, and transporting of precast units is considerable. There needs to be some considerable list of other advantages to make this form of construction generally competitive—not so much with alternative cast-in-place concrete, but with other materials and systems. In the end, the total building construction cost is most significant, not just structural cost.

Connections. From a general construction development point of view, the single biggest problem in design with precast elements is usually the connection of elements. From a structural response consideration, it is here that the major difference occurs between precast and ordinary cast-in-place concrete construction. The completely precast structure has more in common with structures of steel and timber than with ordinary concrete structures. The adequate development of individual connections is a problem, but almost more significant is the overall loss of natural continuity and inherent stability of the fully cast-in-place system. A major effort by the PCI is in the development of recommended practices for design and construction of connections of precast elements.

Integration. This refers to the general problem of incorporating the precast concrete elements into the general building construction. A major problem to be dealt with in this regard is the loss of some opportunities that are present with other types of construction. Installation of hidden items such as wiring, piping, ducts, and housings for light switches, power outlets, recessed lighting fixtures, bathroom medicine cabinets, fire hose cabinets, and exit signs is made somewhat more difficult. With light wood frames, ordinary cast-in-place concrete, and most other types of construction, precise locations of these items and the actual installation can be done during site construction work. With precast concrete construction, provisions must be made in advance—a procedure that is not impossible, but that does not fit with the routine operations with which most designers and builders are familiar.

Seismic-Effects. Usage of precast concrete has received some setbacks in response to the performance of some structures during major seismic events. Promotional literature of the wood and steel industry frequently contain some dramatic pictures of collapsed structures of precast concrete. This is indeed a major problem to be dealt with in design, especially of connections of structural components. The loss of continuity in changing from cast-in-place to precast construction is a major concern. Learning from failures (a basic process in all structural design areas), the precast industry has developed more stringent criteria and techniques. Nevertheless, the seismic response of heavy, individually stiff, and brittle elements, with many joints in the assembled system, makes the precast structural system less than ideal in seismic response. In spite of this, some systems—such as the tilt-up wall—see ongoing, extensive use in regions of high seismic risk, attesting to the fact that design can be effectively achieved when other factors are sufficiently persuasive in the overall decision of system selection.

Mixed Sitecast and Precast Construction

The completely precast concrete structure does not represent the widest usage of precasting. Precast components are used in many situations in conjunction with a building structure and general construction with other materials and systems. This is, of course, the *usual* situation; few buildings are all steel, all wood, all concrete, or all masonry.

Precast concrete decks, especially of the hollow-cored form, are used with frames of steel or poured-in-place concrete and with bearing walls of masonry or poured concrete. Precast concrete wall panels are used with structures of steel, wood, and poured-in-place concrete. The separate component/separate material situation is a common one in building construction. Of course, the individual functions of the components and the interfacing of components for structural interaction must be dealt with in design and in the development of proper construction details.

The blending of components of precast and poured-in-place concrete offers some opportunities for complementary enhancement of the two methods. Figure 34.3*a* shows a joint detail commonly used with tilt-up wall panels. In this case a formed and site-poured concrete column is used to effect the structural connection between two panels, as well as to achieve a positive interaction of the panels (as shear walls) with the continuous poured concrete frame. A similar situation is shown in Fig. 34.3*b*, in which two levels of a structure, consisting of slab-and-beam concrete systems, are connected to a precast concrete wall panel.

A slightly different situation is shown in Fig. 34.3*c*. In this case precast concrete units are used to form poured-in-place concrete columns and the edges of a poured-in-place framing system. In this case the precast units do not serve significant structural tasks in their own right, in comparison to the situations in Fig. 34.3*a* and *b*. They do, however, eliminate the need for other, temporary forming. More important probably, they provide the possibility for a quality of detail and textural control of the exposed concrete, which cannot be achieved with sitecast concrete. An example of this usage is shown in Fig. 4.4.

34.3. ARCHES, VAULTS, AND DOMES

The fluid, take-any-form character of concrete, plus its basically compressive-resistant nature, make it a natural choice for the forming of these types of structures. It can mimic the forms developed in masonry by the Romans and builders of the gothic cathedrals, but can also be made much more structurally sophisticated by reinforcing or prestressing. Its resistance to weather and fire in an exposed condition give it an advantage over construction with wood or steel.

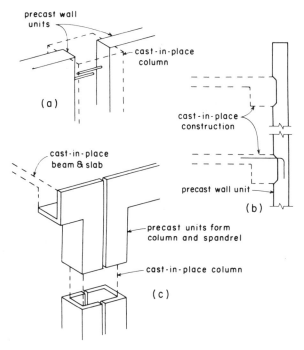

FIGURE 34.3. Mixing of precast and poured-in-place concrete.

As with other concrete construction, weight is a problem—becoming increasingly so with the very large, long-spanning structure. Use of higher-strength concrete, low-density aggregates, prestressing, and ribbed forms instead of solid ones are all potential weight-reducing techniques. And, of course, design of structural forms that are efficient in force resistance (sufficient span-to-rise ratios, etc.) will result in lower stresses and less need for structural mass.

Forming is a major problem, and as with many other complex constructions of concrete, its cost may exceed that of the concrete itself. Precasting may be used to some advantage, but only if it uses simple, repetitive, easily supported and assembled units. If the structure is large, complex, and lacking in repetitive form, forming costs are likely to be considerable.

As with all large concrete structures, considerable attention must be given to problems of shrinkage, thermal movement and deformations, and potential heat buildup in very massive elements during the curing of the concrete. Lessons learned from the construction of large dams and bridges must be applied to any massive concrete construction. These concerns plus the practical ones of the construction process require careful study of the structure for the possibility of locating control joints. These joints may also be significant in terms of control of structural actions, such as seismic separation.

Recent technological developments, such as use of shrinkage-compensating cement, prestressing, super-high-strength concrete, and fibrous concrete, have potentialities that have yet to be fully realized by designers. Concrete is basically a synthetic material, offering possibilities of infinite variation and constant challenge to both the technical experimenter and the imaginative designer.

Because of the relative market competitiveness of wood and steel and the high cost of labor in the United States, concrete has been used somewhat less for the more complex constructions here. In regions or times that favor the use of concrete, innovation and sophisticated usages tend to flourish.

34.4. SHELLS AND FOLDED PLATES

A few years ago—inspired by the works of Torroja, Candella, and others—designers became fascinated by the richness and gracefulness of form that could be achieved with shell and folded plate construction. Many buildings resulted that resembled large exercises in Japanese paper folding. Some were well done, but many were badly conceived.

A principal potential advantage of this construction is that of weight reduction, achievable when the forms are carefully developed in response to the development of forces. While single, simple geometries are possible (see Fig. 2.2), rich, complex, multiunit or multifaceted forms are also possible (see Fig. 5.21). For the multiunit structure, precasting or modular, repetitive-use forming may offer considerable economy. Although weight reduction may be

a technical achievement, however, the material saved and
the dead load eliminated may not be all that significant.

Involvement with these structures requires first, a lot of
learning of three-dimensional geometry, a largely lost art
in current education. It also requires a real understanding
of force operations in three dimensions. This is excellent
exercise for anyone, but a truly challenging field of endeavor. Finally, it requires a thorough knowledge of the
problems of dealing with concrete construction in its most
skill-demanding usages. It probably also works best when
the designer is able to carry the work all the way from form
conception to final construction with command control.

Forming of curved shell surfaces—especially those with
double curvature—is challenging. Some special geometries, such as hyperbaloids, can be generated with families
of straight lines, actually permitting simple board forming.
Folded plate surfaces can, of course, be flat formed, offering at least that simplification in the construction process.

34.5. PRECAST BENTS AND TRUSSES

Most precast elements for structures consist of single wall
panels, columns, or spanning elements such as deck units
or beams. However, the advantages of the casting operation may be utilized to produce single-piece units that
comprise planar rigid-frame bents or trusses. In this case
some of the problems of effecting connections that are
associated with wood and steel structures are eliminated.
For the single-span bent shown in Fig. 34.4a, for example,
the three elements of the bent are monolithically formed
with the cast material.

The single-span gabled bent shown in Fig. 34.4b can be
produced as indicated with two halves cast in single pieces.
This is a commonly used form of structure, particularly for
small utilitarian buildings where the durability of the exposed concrete structure is an advantage.

For multispan or multistory buildings, it is sometimes
possible to subdivide the entire structure into units that
provide for a simpler form of assemblage. For the bents in
Fig. 34.4c, the assemblage connections are reduced to simple pin-type connections, preserving both the continuity of
the horizontal spans and the stability of the system for
lateral loads. For the multistory bents in Fig. 34.4d, the
need for moment connections and column splices is
eliminated.

The concept of trussing seems somewhat inappropriate
for concrete, since the idea of a concrete tension member is
unreasonable. Nevertheless, trussing has been utilized in

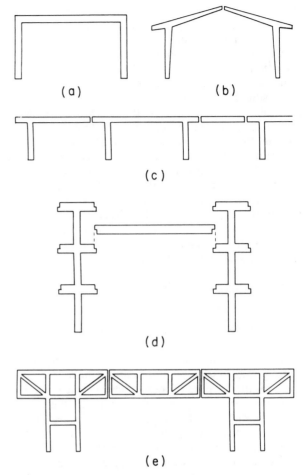

FIGURE 34.4. Use of precast concrete for structural elements.

various ways for some structures. One possibility is the use
of crisscrossed members, or X-bracing. In steel structures,
this often means double tensioning, that is, providing two
tension members, each responding to a different load direction. In concrete, the corresponding usage involves
double compression, with each diagonal functioning in
compression for a different load direction.

Tension is possible, of course, with either steel reinforcing or prestressing. Figure 34.4e shows a structure consisting of precast units in which the units have a composite
trussed and rigid flame form. At a relatively large scale,
where high-strength concrete and prestressing are feasible,
this type of structure is quite reasonable.

PART 6

MASONRY

Part 6 deals with some aspects of the use of masonry construction for structures. The brevity of treatment does not reflect any lack of concern for the subject, but rather a response to current practices. Most of the masonry seen as finished surfaces in new construction these days is either nonstructural or of a few limited types of structural masonry. Most of what must be dealt with in using masonry falls in the general category of building construction rather than with strictly structural design considerations. One

must learn a great deal about the materials and processes of masonry construction to become generally capable of using it. For that large body of information we refer the reader to the various sources, including textbooks, handbooks, and industry standards. The discussion in Part 6 is limited to the most common uses of structural masonry and the general design considerations relating to building structures.

CHAPTER THIRTY-FIVE

General Concerns for Masonry

There are many types of masonry and many factors that must be dealt with in producing a good masonry structure. Chapter 35 presents a general discussion of basic issues and a description of the various types of masonry.

35.1. MASONRY CONSTRUCTION

Units

Masonry consists generally of a solid mass produced by bonding separate units. The traditional bonding material is mortar. The units include a range of materials, the common ones being the following:

Stone. These may be in essentially natural form (called *rubble* or *field stone*) or may be cut to specified shape.
Brick. These vary from unfired, dried mud (adobe) to fired clay (kiln-baked) products. Form, color, and structural properties vary considerably.
Concrete Blocks (CMUs). Called *concrete masonry units*, these are produced from a range of types of material in a large number of form variations.
Clay Tile Blocks. Used widely in the past, these are hollow units similar to concrete blocks in form. They were used for many of the functions now performed by concrete blocks.
Gypsum Blocks. These are precast units of gypsum concrete, used mostly for nonstructural partitions.

The potential structural character of masonry depends greatly on the material and form of the units. From a material point of view, the high-fired clay products (brick and tile) are the strongest, producing very strong construction with proper mortar, a good arrangement of the units, and good construction craft and work in general. This is particularly important if the general class of the masonry construction is the traditional, unreinforced variety. Although some joint reinforcing is typical in all structural masonry these days, the term *reinforced masonry* is reserved for a class of construction in which major vertical and horizontal reinforcing is used, quite analogous to reinforced concrete construction.

Mortar

Mortar is usually composed of water, cement, and sand, with some other materials added to make it stickier (so that it adheres to the units during laying up of the masonry), faster setting, and generally more workable during the construction. Building codes establish various requirements for the mortar, including the classification of the mortar and details of its use during construction. The quality of the mortar is obviously important to the structural integrity of the masonry, both as a structural material in its own right and as a bonding agent that holds the units together. While the integrity of the units is dependent primarily on the manufacturer, the quality of the finished mortar work is dependent primarily on the skill of the mason who lays up the units.

There are several classes of mortar established by codes, with the higher grades being required for uses involving major structural functions (bearing walls, shear walls, etc.) Specifications for the materials and required properties determined by tests are spelled out in detail. Still—the major ingredient in producing good mortar is always the skill of the mason. This is a dependency that grows increasingly critical as the general level of craft in construction erodes with the passage of time.

Basic Construction and Terminology

Figure 35.1 shows some of the common elements of masonry construction. The terminology and details shown apply mostly to construction with bricks or concrete blocks.

Units are usually laid up in horizontal rows, called *courses*, and in vertical planes, called *wythes*. Very thick walls may have several wythes, but most often walls of brick have two wythes and walls of concrete block are single wythe. If wythes are connected directly, the construction is called *solid.* If a space is left between wythes, as shown in the illustration, the wall is called a *cavity wall.* If the cavity is filled with concrete, it is called a *grouted cavity wall.*

The multiple-wythe wall must have the separate wythes bonded together in some fashion. If this is done with the masonry units, the overlapping unit is called a *header.* Various patterns of headers have produced some classic

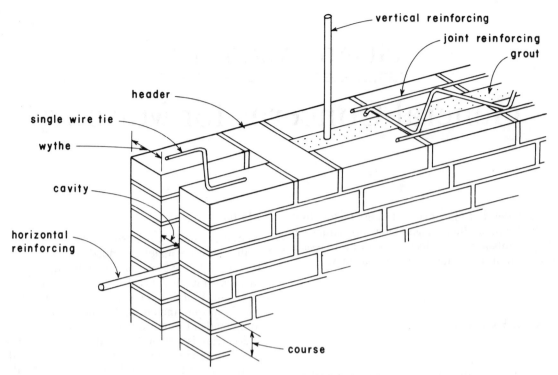

FIGURE 35.1. Elements of masonry construction.

forms of arrangement of bricks in traditional masonry construction. For cavity walls, bonding is often done with metal ties, using single ties at intervals, or a continuous wire trussed element that provides both the tying of the wythes and some minimal horizontal reinforcing.

The continuous element labeled *joint reinforcing* in Fig. 35.1 is now commonly used in both brick and concrete block construction that is code-classified as unreinforced. For seriously reinforced masonry, the reinforcement consists of steel rods (the same as those used for reinforced concrete) that are placed at intervals both vertically and horizontally, and are encased in concrete poured into the wall cavities.

Structural Masonry

Masonry intended for serious structural purpose includes that used for bearing walls, shear walls, retaining walls, and spanning walls of various type. The types of masonry most used for these purposes include the following:

Solid Brick Masonry. This is unreinforced masonry, usually of two or more wythes, with the wythes directly connected (no cavities) and the whole consisting of a solid mass of bricks and mortar. This is one of the strongest forms of unreinforced masonry if the bricks and the mortar are of reasonably good quality.

Grouted Brick Masonry. This is usually a two-wythe wall with the cavity filled completely with lean concrete (grout). If unreinforced, it will usually have continuous joint reinforcing (see Fig. 35.1). If reinforced, the steel rods are placed in the cavity. For the reinforced wall, strength derives considerably from the two-way reinforced, concrete-filled cavity, so that considerable strength of the construction may be obtained, even with a relatively low-quality brick or mortar.

Unreinforced Concrete Block Masonry. This is usually of the single-wythe form shown in Fig. 35.2a. The faces of blocks as well as cross parts are usually quite thick, although thinner units of lightweight concrete are also produced for less serious structural uses. While it is possible to place vertical reinforcement and grout in cavities, this is not the form of block used generally for reinforced construction. Structural integrity of the construction derives basically from unit strength and quality of the mortar. Staggered vertical joints are used to increase the bonding of units.

Reinforced Concrete Block Masonry. This is usually produced with the type of unit shown in Fig. 35.2b. This unit has relatively large individual cavities, so that filled vertical cavities become small reinforced concrete columns. Horizontal reinforcing is placed in courses with the modified blocks shown in Fig. 35.2c or d. The block shown in Fig. 35.2d is also used to form lintels over openings.

The various requirements for all of these types of masonry are described in building codes and industry standards. Regional concerns for weather and critical design loads make for some variation in these requirements. Designers should be careful to determine the particular code requirements and general construction practices for any specific building location.

Walls that serve structural purposes and have a finished surface of masonry may take various forms. All of the walls just described use the structural masonry unit as the finished wall face, although architecturally exposed surfaces may be specially treated in various ways. For the brick walls, the single brick face that is exposed may have a special treatment (textured, glazed, etc.). It is also possible to use the masonry strictly for structural purposes and to finish the wall surface with some other material, such as stucco or tile.

Another use of masonry consists of providing a finish of masonry on the surface of a structural wall. The finish masonry may merely be a single wythe bonded to other backup masonry construction, or it may be a nonstructural veneer tied to a separate structure. For the veneered wall, the essential separation of the elements makes it possible for the structural wall to be something other than masonry construction. Indeed, many brick walls are really single-wythe brick veneers tied to wood or steel stud structural walls—still a structural wall, just not a masonry one.

Finally, a masonry-appearing wall may not be masonry at all, but rather, a surface of thin tiles adhesively bonded to some structural wall surface. Entire "brick" buildings are now being produced with this construction.

35.2.   UNREINFORCED MASONRY

This section deals with masonry construction that is not designed and built in a manner analogous to that for rein-

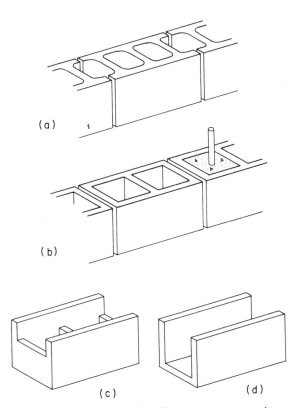

FIGURE 35.2. Construction with concrete masonry units.

forced concrete. The term *unreinforced* thus relates to the usual building code designation of such construction. As discussed in Sec. 35.1, however, there are many means for enhancing the basic masonry construction that can be considered forms of reinforcement, and every masonry structure usually contains some such elements.

Nonstructural Masonry

Masonry materials can be used for a variety of functions in building construction. Units of fired clay, precast concrete, cut stone, or field stone can be used to form floor surfaces, wall finishes, or nonload-bearing walls (partitions or curtain walls). Indeed, most walls that appear to be made of brick, cut stone, or field stone in present-day construction are likely to be of veneered construction, with the surface of masonry units attached to some backup structure.

Although structural utilization of the masonry in these situations may be minor or nonexistent, it is still necessary to develop the construction with attention to many of the concerns given to the production of structural masonry for bearing walls, shear walls, and so on. The quality of the masonry units and the quality and workmanship of the mortar joints is often just as critical for these uses, although structural behavior or safety is not at issue. This also extends to concerns for shrinkage, thermal change, stress concentrations at discontinuities, and other aspects of the general behavior of the materials. From an appearance point of view, cracking is just as objectionable in nonstructural masonry facing or paving as it is in a masonry bearing wall.

Masonry veneer facings and nonload-bearing partitions must be provided with control joints and various forms of anchorage and support. Since nonstructural masonry is used extensively, there are many situations for these concerns and a large inventory of recommended construction details. Since these fall mostly in the category of general building construction issues, they are not treated generally in this book.

Masonry units used for nonstructural applications may be of the same structural character as those used for construction of serious masonry structures. Some of the strongest bricks are those that are high fired to produce great hardness and color intensity for use in veneer construction. However, it is also possible to use some lower structural grades of material, or even some materials which are not usable for structural masonry. Gypsum tile, lightweight concrete blocks, and some very soft bricks may be limited to use in situations not involving structural demands—and probably also not involving exposure to weather or other hazardous situations.

A problem with some nonstructural masonry construction is that of unintended structural response. If the masonry is indeed real masonry (with real masonry units and joints of mortar or grout), it will usually have considerable stiffness. Thus, just as with a plaster surface on a light wood-framed wall structure, it may tend to absorb load due to deformation of the supporting structure. Much of the cracking of nonstructural masonry (and plaster) is due to this phenomenon. The whole construction must be carefully studied for potential problems of this kind. Flexible attachments, control joints, and possibly some reinforcement can be used to alleviate many of these situations.

A particular problem of the type just mentioned is that of the unintended bracing effect of nonstructural partitions during seismic activity. Actually, there are probably many buildings, which have structures that are not adequate for significant lateral loads, but are being effectively braced by supposedly nonstructural walls. However, in some cases the general response of the building may be significantly altered by the stiffening effect of rigid partitions. This issue is discussed with regard to the design of rigid-frame structures for seismic effects in Part 8.

Structural Masonry

Many structures of unreinforced masonry have endured for centuries, and this form of construction is still widely used. Although it is generally held in low regard in regions that have frequent earthquakes, it is still approved by most building codes for use within code-defined limits. With good design and good-quality construction, it is possible to have structures that are more than adequate by present standards.

If masonry is essentially unreinforced, the character and structural integrity of the construction are highly dependent on the details and the quality of the masonry work. Strength and form of units, arrangements of units, general quality of the mortar, and the form and details of the general construction are all important. Thus the degree of attention paid to design, to writing the specifications, to detailing the construction, and to careful inspection during the work must be adequate to ensure good finished construction.

There are a limited number of structural applications for unreinforced masonry, and structural computations for design of common elements are in general quite simple. We hesitate to show examples of such work, since the data and general procedures are considerably variable from one region to another, depending on local materials and construction practices as well as variations in building code requirements. In many instances forms of construction not subject to satisfactory structural investigation are tolerated simply because they have been used with success for many years on a local basis—a hard case to argue against. Nevertheless, there are some general principles and typical situations that produce some common problems and procedures. The following discussion deals with some major concerns of design of the unreinforced masonry structure.

Minimal Construction

As in other types of construction, there is a minimal form of construction that results from the satisfaction of various general requirements. Industry standards result in some standardization and classification of products, which is usually reflected in building code designations. Structural usage is usually tied to specified minimum grades of units, mortar, construction practices, and in some cases to need for reinforcement or other enhancement. This results in most cases in a basic minimum form of construction that is adequate for many ordinary functions; which is in fact usually the intent of the codes. Thus there are many instances in which buildings of a minor nature are built without benefit of structural computations, simply being produced in response to code-specified minimum requirements.

Design Strength of Masonry

As with concrete, the basic strength of the masonry is measured as its resistive compressive strength. This is established in the form of the *specified compressive strength*, designated f_m'. The value for f_m' is usually taken from code specifications, based on the strength of the units and the class of the mortar.

Allowable Stresses

Allowable stresses are directly specified for some cases (such as tension and shear) or are determined by code formulas that usually include the variable value of f_m'. There are usually two values for any situation: that to be used when special inspection of the work is provided (as specified by the code), and that to be used when it is not. For minor construction projects it is usually desirable to avoid the need for the special inspection. For construction with hollow units that are not fully grouted, stress computations are based on the net cross section of the hollow construction.

Avoiding Tension

While codes ordinarily permit some low stress values for flexural tension, many designers prefer to avoid tension in unreinforced masonry. An old engineering definition of mortar is "the material used to keep masonry units *apart*," reflecting a lack of faith in the bonding action of mortar.

Reinforcement or Enhancement

The strength of a masonry structure can be improved by various means, including the insertion of steel reinforcing rods as is done to produce reinforced masonry. Vertical rods are sometimes used with construction that is essentially classified as unreinforced, usually to enhance bending resistance or to absorb localized stress conditions. Horizontal wire type reinforcing is commonly used to reduce stress effects due to shrinkage and thermal change.

Another type of structural reinforcement is achieved through the use of form variation, examples of which are shown in Fig. 35.3. Turning a corner at the end of a wall (Fig. 35.3a) adds stability and strength to the discontinuous edge of the structure. An enlargement in the form of a pilaster column (Fig. 35.3b) can also be used to improve the end of a wall or to add bracing or concentrated

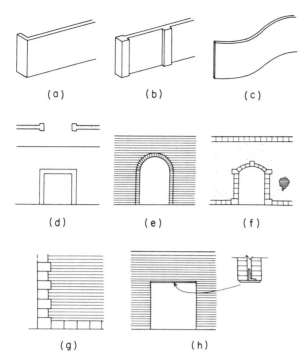

FIGURE 35.3. Reinforcement in masonry construction.

design procedures used for reinforced masonry are in general similar to those used for concrete structures. Until recently, the general methods of the working stress design procedure were used for masonry. However, recent industry standards and some building codes (including the 1991 edition of the *UBC*) have promoted the use of strength methods for investigation and design. As with concrete, the use of strength methods is quite complex and abstract. We have therefore chosen to use the simpler methods of the working stress design procedures for most of the example computations to permit a briefer treatment; our interest being more in demonstrating the problem and the basic concerns for investigation rather than particular means for investigation.

Reinforced Brick Masonry

Reinforced brick masonry typically consists of the type of construction shown in Fig. 35.1. The wall shown in the illustration consists of two wythes of bricks with a cavity space between them. The cavity space is filled completely with grout so that the construction qualifies firstly as *grouted masonry*, for which various requirements are stipulated by the code. One requirement is for the bonding of the wythes, which can be accomplished with the masonry units, but is most often done by using the steel wire joint reinforcement shown in the illustration in Fig. 35.1. Added to this basic construction are the vertical and horizontal reinforcing rods in the grouted cavity space, making the resulting construction qualify as *reinforced grouted masonry*.

General requirements and design procedures for the reinforced brick masonry wall are similar to those for concrete walls. There are stipulations for minimum reinforcement and provisions for stress limits for the various structural actions of walls in vertical compression, bending, and shear wall functions. Structural investigation is essentially similar to that for the hollow unit masonry wall, which is discussed in the next section.

Despite the presence of the reinforcement, the type of construction shown in Fig. 35.1 is still essentially a masonry structure, highly dependent on the quality and structural integrity of the masonry itself—particularly the skill and care exercised in laying up the units and handling of the construction process in general. The grouted, reinforced cavity structure is in itself often considered to be a third wythe, constituted as a very thin reinforced concrete wall panel. The enhancement of the construction represented by the cavity wythe is considerable, but the major bulk of the construction is still basically just solid brick masonry.

strength at some intermediate point along the wall. Heavy concentrated loads are ordinarily accommodated by using pilaster columns when wall thickness is otherwise minimal. Curving a wall in plan (Fig. 35.3c) is another means of improving the stability of the wall.

Building corners and openings for doors or windows are other locations where enhancement is often required. Figure 35.3d shows the use of an enlargement of the wall around the perimeter of a door opening. If the top of the opening is of arched form, the enlarged edge may continue as an arch to span the opening, as shown in Fig. 35.3e, or a change may be made to a stronger material to effect the edge and arch enhancement, as shown in Fig. 35.3f. Building corners in historic buildings were often strengthened by using large cut stones to form the corner, as shown in Fig. 35.3g.

While form variations or changes of the masonry units can be used to affect spans over openings, the more usual means of achieving this—especially for flat spans—is by using a steel beam (called a *lintel*, as shown in Fig. 35.31h.

35.3. REINFORCED MASONRY

As we use it here, the term *reinforced masonry* designates a type of masonry construction specifically classified by building code definitions. Essential to this definition are the assumptions that the steel reinforcement is designed to carry forces and the masonry does not develop tensile stresses. This makes the design basically analogous to that for reinforced concrete, and indeed the present data and

Reinforced Hollow Unit Masonry

This type of construction most often consists of single-wythe walls formed as shown in Fig. 35.2b–d (see also Fig. 4.6 for a construction site view). Cavities are vertically aligned so that small reinforced concrete columns can be formed within them. At some interval, horizontal courses are also used to form reinforced concrete members. The intersecting vertical and horizontal concrete members thus constitute a rigid frame bent inside the wall. This reinforced concrete frame is the major structural component of the construction. Besides providing forming, the concrete blocks serve to brace the frame, provide protection for the reinforcement, and interact in composite action with the rigid frame. Nevertheless, the structural character of the construction derives largely from the concrete frame created in the void spaces in the wall.

The code requires that reinforcement be a maximum of 48 in. on center; thus the maximum spacing of the concrete members inside the wall is 48 in., both vertically and horizontally. With 16 in.-long blocks this means that every sixth vertical void space is grouted. With blocks having a net section of approximately 50% (half solid, half void), this means that the minimum construction is an average of approximately 60% solid. For computations based on the net cross section of the wall, the wall may thus usually be considered to be a minimum of 60% solid.

If all void spaces are grouted, the construction is fully solid. This is usually required for structures such as retaining walls and basement walls, but may also be done simply to increase the wall section for the reduction of stress levels. Finally, if reinforcing is placed in all of the vertical voids (instead of every sixth one), the contained reinforced concrete structure is considerably increased and both vertical bearing capacity and lateral bending capacity are significantly increased. Heavily loaded shear walls are developed in this manner.

For shear wall actions, there are two conditions defined by the code. The first case involves a wall with minimum reinforcing, in which the shear is assumed to be taken by the masonry. The second case is one in which the reinforcing is designed to take all of the shear. Allowable stresses are given for both cases, even though the reinforcement must be designed for the full shear force in the second case.

Design of typical elements of reinforced masonry is discussed in Chapter 36.

35.4. USAGE CONSIDERATIONS

Utilization of masonry construction for structural functions requires the consideration of a number of factors that relate to the structural design and to the proper details and specifications for construction. The following are some major concerns that must ordinarily be dealt with.

Units

The material, form, and specific dimensions of the units must be established. Where code classifications exist, the specific grade or type must be defined. Type and grade of unit, as well as usage conditions, usually set the requirements for type of mortar required.

Unit dimensions may be set by the designer, but the sizes of industrially produced products such as bricks and concrete blocks are often controlled by industry standard practices. As shown in Fig. 35.4a, the three dimensions of a brick are the height and length of the exposed face and the width that produces the thickness of a single wythe. There is no single standard size brick, but most fall in a range close to that shown in the illustration.

Concrete blocks are produced in families of modular sizes. The size of block shown in Fig. 35.4b is one that is equivalent to the 2 × 4 in wood—not the only size, but the most common. Concrete block have both nominal and actual dimensions. The nominal dimensions are used for designating the blocks and relate to modular layouts of building dimensions. Actual dimensions are based on the assumption of a mortar joint thickness of $\frac{3}{8}$–$\frac{1}{2}$ in. (the sizes shown in Fig. 35.4b reflect the use of $\frac{3}{8}$-in. joints).

A construction frequently used in unreinforced masonry is that of a single wythe of brick bonded to a single wythe of concrete block, as shown in Fig. 35.4c. In order to install the metal ties that affect the bonding, as well as to have the bricks and blocks come out even at the top of a wall, a special-height brick is sometimes used, based on either two or three bricks to one block.

Since transportation of large quantities of bricks or concrete blocks is difficult and costly, units used for a building are usually those obtainable on a local basis. Although some industry standardization exists, the type of locally produced products should be investigated for any design work.

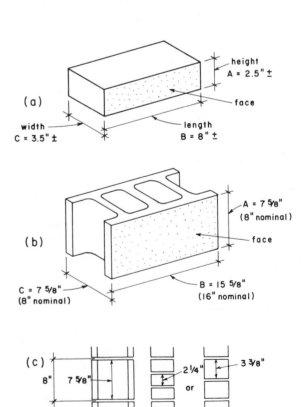

FIGURE 35.4. Dimensional considerations with masonry units.

Unit Layout Pattern

When exposed to view, masonry units present two concerns: that for the face of the unit and for the pattern of layout of the units. Patterns derive from unit shape and the need for unit bonding, if unit-bonded construction is used. Classic patterns were developed from these concerns, but other forms of construction, now more widely used, free the unit pattern somewhat. Nevertheless, classic patterns such as running bond, English bond, and so on, are still widely used.

Patterns also have some structural implications; indeed, the need for unit bonding was such a concern originally. For reinforced construction with concrete blocks, a major constraint is the need to align the voids in a vertical arrangement to facilitate installation of vertical bars. Generally, however, pattern as a structural issue is more critical for unreinforced masonry.

Structural Functions

Masonry walls vary from those that are essentially of a nonstructural character to those that serve major and often multiple structural tasks. The type of unit, grade of mortar, amount and details of reinforcement, and so on, may depend on the degree of structural demands. Wall thickness may relate to stress levels as well as to construction considerations. Most structural tasks involve force transfers: from supported structures, from other walls, and to supporting foundations. Need for brackets, pilaster columns, vertically tapered or stepped form, or other form variations may relate to force transfers, wall stability, or other structural concerns.

Reinforcement

In the broad sense, reinforcement means anything that is added to help. Structural reinforcement thus includes the use of pilasters, buttresses, tapered form, and other devices, as well as the usual added steel reinforcement. Reinforcement may be generally dispersed or may be provided at critical points, such as at wall ends, tops, edges of openings, and locations of concentrated loads. Both form variation and steel rods are used in both unreinforced masonry and in what is technically referred to as *reinforced masonry*.

Control Joints

Shrinkage of mortar, temperature variation, and movements due to seismic actions or settlement of foundations are all sources of concern for cracking failures in masonry. Stress concentrations and cracking can be controlled to some extent by reinforcement. However, it is also common to provide some control joints (literally, preestablished cracks) to alleviate these effects. Planning and detailing of control joints is a complex problem and must be studied carefully as a structural and architectural design issue. Code requirements, industry recommendations, and common construction practices on a local basis will provide guides for this work.

Attachment

Attachment of elements of the construction to masonry is somewhat similar to that required with concrete. Where the nature and exact location of attached items can be predicted, it is usually best to provide some built-in device, such as an anchor bolt, threaded sleeve, and so on. Adjustment of such attachments must be considered, as precision of the construction is limited. Attachment can also be effected with drilled-in anchors or adhesives. These tend to be less constrained by the problem of precise location, although the exact nature of the masonry at the point of attachment may be a concern. This is largely a matter of visualization of the complete building construction and of the general problem of integrating the structure into the whole building. It is simply somewhat more critical with masonry structures, as the simple use of nails, screws, and welding is not possible in the direct way that it is with structures of wood and steel.

CHAPTER THIRTY-SIX

Masonry with Concrete Units

Most structural masonry is now achieved with precast concrete units, principally because of the low cost of the units and the lower labor costs. Chapter 36 deals with the use of concrete units for common structural elements. Many of the structural elements discussed here could also be achieved in either brick or stone, but are more likely to be done with CMUs.

36.1. UNREINFORCED CONSTRUCTION

The standard form of unit used most often for unreinforced forms of construction is that shown in Fig. 36.1a. This is called a *three-cell unit*, although it actually has two half-cells at each end; thus the module of repetition is four cells per block. The actual dimensions shown here—$\frac{3}{8}$ in. less than the modular dimensions—are frequently used, anticipating slightly thinner mortar joints. This is done partly to allow for interfacing with brick construction, which sometimes occurs in multiwythe work with bricks of modular height.

Staying within this standard unit module, there are typically some special units provided for achieving common elements of the construction. As shown in Fig. 36.1, these include the following:

End and Corner Units (Figs. 36.1b and c). These are used at the ends and corners of walls, where the ends of

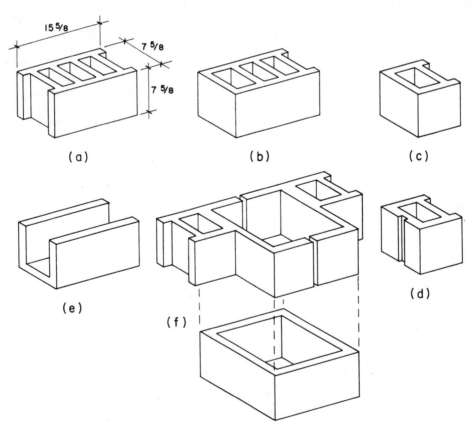

FIGURE 36.1. Forms of CMUs for unreinforced construction.

TABLE 36.1. Average Properties of Hollow Concrete Block Construction

| Nominal Block Thickness (in.) | Net Area of Block (in.² per ft of wall length) | Wall Weight in psf of Wall Surface, Where Density of Concrete in Block in lb/ft³ Is | | | | | Moment of Inertia, I, in in.⁴ and Section Modulus, S, in in.³, Where Mortar Is | | | |
| | | | | | | | Face Shell Bedded[a] | | Fully Bedded[a] | |
		60	80	100	120	140	I	S	I	S
4	28	14	18	22	27	31	38	21	45	25
6	37	20	26	33	40	46	130	46	139	50
8	48	24	32	40	47	55	309	81	334	88
10	60	28	37	47	56	65	567	118	634	132
12	68	34	45	55	67	78	929	160	1063	183

[a] Horizontal mortar joint only at face shells.
[b] Horizontal mortar joint on full unit cross section.

Source: Adapted from table in NCMA-TEK Publication 2A, *Concrete units*, with permission of the publisher, National Concrete Masonry Association.

the units are exposed. The half-blocks (one 8-in. module) are used to develop the typical running-bond wall face pattern of units.

Sash Units (Fig. 36.1*d*). These are one (as shown) or two module units with squared ends and a groove to facilitate the anchoring of window or door frames at the vertical edges of openings.

Lintel or Bond Beam Units (Fig. 36.1*e*). These are used over openings or at the tops of walls to create cast-in-place concrete with steel reinforcement in the form of a reinforced concrete beam internal to the masonry construction.

Pilaster Units (Fig. 36.1*f*). These are used to form pilaster columns within the modular system of the wall construction. The two types of units shown are alternated in vertical courses to maintain the running-bond pattern on the flat face of the wall. Units of various size are available to create different size pilasters. The units shown here would normally be filled with concrete and steel reinforcement to create a cast-in-place concrete column inside the masonry.

For simple structural functions, resisting compression, bending or shear, the hollow construction with concrete units is treated much the same as the solid brick construction, as illustrated in the examples in Chapter 37. The exception here is that the construction is *not* solid, and the wall cross section used for the computations must be the net section, with the voids from the full area of the wall's outer dimensions subtracted. This net section must be determined specifically from the suppliers of units, although approximate investigations can be made using the average values provided in Table 36.1. The following example demonstrates some simple cases.

Example. CMU construction is to be used for a single-wythe wall of nominal 8-in. thickness. The wall is 10 ft high and sustains on its top an axially applied, uniformly distributed load of 2000 lb/ft of wall length. Type N, ASTM C-90 units with f_m' = 1350 psi are to be used. Investigate the

wall for average maximum compressive stress. (Note: Type N is required for weather exposure; f_m' of 1350 psi is the typical assumed maximum code limit in the absence of tests on units.)

Solution. From Table 36.1, we note that the nominal 8-in.-thick wall has an average net cross section of 48 in.²/ft of wall length. Thus the applied load produces a stress of

$$f_a = \frac{P}{A} = \frac{2000}{48} = 41.7 \text{ psi}$$

To this must be added the stress due to the weight of the wall, which has a maximum value at the base of the wall. Without the unit density of the concrete in the CMUs, we will assume the maximum value in Table 36.1 (140 lb/ft³), yielding a weight of 55 lb/ft of wall height. Thus the 10-ft wall weights 10(55) = 550 lb/ft of wall length, the stress due to its own weight is

$$f_a = \frac{W}{A} = \frac{550}{48} = 11.5 \text{ psi}$$

and the total stress is 41.7 + 11.5 = 53.2 psi.

From Table 36.2, we find the maximum allowable stress for hollow-unit masonry with Type S mortar (the minimum mortar for weather-exposed structural work) to be 150 psi. The construction is therefore quite adequate.

If bending occurs in combination with axial compression, the combined stresses must be investigated for both the maximum net compression and any possible net tension. For flexural stresses we may use the section properties given in Table 36.1 (*S* for flexural stress). Note that Table 36.2 provides very low allowable tension stress in flexure, generally discouraging any actual net tension.

Walls must also be investigated for slenderness effects and concentrated loadings, following the procedures illustrated in the examples in Chapter 37. Unreinforced piers, columns, and pedestals can also be designed with these procedures.

TABLE 36.2. Allowable Working Stresses in Unreinforced Unit Masonry

Material	Type M Compression[1]	Type S Compression[1]	Type M or Type S Mortar Shear or Tension in Flexure[2,3]		Tension in Flexure[4]		Type N Compression[1]	Type N Shear or Tension in Flexure[2,3]	
1. Special inspection required	No	No	Yes	No	Yes	No	No	Yes	No
2. Solid brick masonry									
4500 plus psi	250	225	20	10	40	20	200	15	7.25
2500–4500 psi	175	160	20	10	40	20	140	15	7.5
1500–2500 psi	125	115	20	10	40	20	100	15	7.5
3. Solid concrete unit masonry									
Grade N	175	160	12	6	24	12	140	12	6
Grade S	125	115	12	6	24	12	100	12	6
4. Grouted masonry									
4500 plus psi	350	275	25	12.5	50	25			
2500–4500 psi	275	215	25	12.5	50	25			
1500–2500 psi	225	175	25	12.5	50	25			
5. Hollow unit masonry[5]	170	150	12	6	24	12	140	10	5
6. Cavity wall masonry solid units[5]									
Grade N or 2500 psi plus	140	130	12	6	30	15	110	10	5
Grade or 1500–2500 psi	100	90	12	6	30	15	80	10	5
Hollow units[5]	70	60	12	6	30	15	50	10	5
7. Stone masonry									
Cast stone	400	360	8	4	—	—	320	8	4
Natural stone	140	120	8	4	—	—	100	8	4
8. Unburned clay masonry	30	30	8	4	—	—	—	—	—

[1] Allowable axial or flexural compressive stresses in pounds per square inch gross cross-sectional area (except as noted). The allowable working stresses in bearing directly under concentrated loads may be 50 percent greater than these values.

[2] This value of tension is based on tension across a bed joint, i.e., vertically in the normal masonry work.

[3] No tension allowed in stack bond across head joints.

[4] The values shown here are for tension in masonry in the direction of running bond, i.e., horizontally between supports.

[5] Net area in contact with mortar or net cross-sectional area.

Source: Table 24-H in the *Uniform Code*, 1988 ed., reproduced with permission of the publishers, International Conference of Building Officials.

Although the typical unreinforced class of construction is considered as a simple homogeneous material, permitting simple stress calculations for most loadings, investigation is somewhat more complex when some voids are grouted solid and possibly have steel reinforcement. This is often done partially, even in what is generally qualified as unreinforced construction. Procedures for these situations are described more fully in the next section.

Some procedures are generally followed in all construction with hollow concrete units, whether the work is intended for structural purposes or not. Horizontal joint reinforcement is commonly used to reduce cracking and absorb the considerable stress due to shrinkage of the mortar. Bond beams (grout-filled with steel rods) are often used for the top course. These practices simply ensure better construction. Other enhancements may relate to direct concerns for structural actions.

Reinforcement for Unreinforced Construction

Taken in its broader context, reinforcement means anything that adds strength to the otherwise unadorned construction. As discussed in other situations in this book, this includes the use of steel elements in mortar joints or in grout-filled cavities, but it also refers to the use of pilasters and stronger masonry units at strategic locations.

Without fully transforming construction with hollow masonry units into the reinforced class, it is possible to use either stronger units or some reinforcement to give significant improvement in strength at various critical points. The general means for improving strength of hollow-unit masonry are the following:

1. Use denser concrete. This generally results in increased unit strength in the material. Some strong

light-weight concretes can be achieved, but usually the lower the density, the weaker the concrete.

2. Use units with thicker walls. Typically, units made with dense concrete are intended for major structural use and sometimes have thicker walls than those made with significantly less dense material and generally intended for less demanding structural applications. The combination of denser material and thicker walls can make a considerable difference in strength.

3. Fill voids with grout (generally a slightly runny concrete). This can be done with or without adding steel bars; the concrete in the voids generally produces equivalent solid construction, which constitutes a considerable gain in wall strength by itself.

4. Use larger and/or stronger units at strategic locations. Pilasters are one example of this and are often used at points of concentrated loads. The increased local size alone may be significant, but the units used to form pilasters are usually either thicker and stronger themselves or are formed to provide for a cast-in-place reinforced concrete column of significant size.

5. Add steel elements (bars or heavy wire) to mortar joints or grout-filled voids. This amounts to building the equivalent of reinforced concrete beams or columns into the masonry construction. This may be done to add local required strength or simply to tie the construction together. Reinforced bond beams are ordinarily used along the tops of walls for the latter purpose, even in construction that is essentially nonstructural.

Reinforcement may have a specific purpose in many cases, but is also frequently done simply to improve the general integrity of the construction. Many of the classic, ornamental details of masonry construction, handed down from ancient times, were initially developed from experience and the pragmatic concerns for producing better construction.

36.2. REINFORCED CONSTRUCTION

Hollow-unit masonry that qualifies for the building codes' definition of reinforced is typically made with standard units that provide larger voids. Thus the vertical alignment of voids provides for the forming of a reinforced concrete column of some greater dimension. Instead of the three-core unit typically used for unreinforced forms of construction (see Fig. 36.1), the standard unit used for reinforced construction is the two-core unit, as shown in Fig. 35.2*b*. Actually, as discussed in Sec. 36.1, the units shown in Fig. 36.1 have a four-core module, due to the half core at each end. Thus the units used for reinforced construction actually have voids that are close to twice in size to those in unreinforced construction.

With fewer voids, there are fewer cross walls in the units, and the general strength of the masonry is slightly less in reinforced forms of construction. Compensating significantly for this is the existence of the cast-in-place reinforced concrete column and beam rigid frame that is developed inside the masonry, as shown in Fig. 36.2. This frame does the double job of tying the masonry together into an integral mass and developing its own independent strength as a reinforced concrete structure.

Building code requirements ensure the presence of the vertical and horizontal reinforced members at least every 4 ft. Additional elements are specifically required at the tops

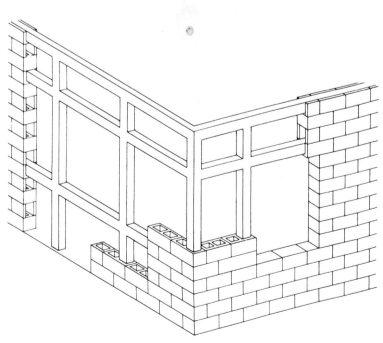

FIGURE 36.2. Form of the internal reinforced concrete structure in reinforced construction with CMUs.

and ends of walls and around all sides of openings. The result, shown in Fig. 36.2, is a rather extensive framework of reinforced concrete. The overall strength of the construction derives significantly from this encased structure and often somewhat minimally from the encasing masonry.

There are, in fact, proprietary systems that use units placed without mortar, which derive their structural integrity almost entirely from the encased structure. However, most reinforced construction is done with structurally rated units, laid with mortar of high grade and generally producing a masonry construction with some significant integrity. For some minor structural tasks, such as may occur with low-rise buildings with short-span roof systems, the masonry itself may be sufficient for general structural purposes; in which case the encased reinforced concrete structure is a bonus.

To a large extent, the fully reinforced forms of hollow-unit masonry now in use were developed in response to needs for structures with enhanced resistance to the effects of earthquakes and windstorms. The poor response of unreinforced construction in these situations is well documented and is, unfortunately, repeatedly demonstrated in almost every major earthquake and windstorm. Building codes in regions at high risk to these disasters now generally prevent use of unreinforced construction.

Structural investigation and design of reinforced masonry is typically achieved with theories and procedures adapted from those used for reinforced concrete. Still in use considerably are methods deriving from the working-stress techniques used extensively in the past. Because of its relatively simpler procedures and formulas, the computational work in this book generally uses these methods.

As in most areas of structural design today, strength methods using load and resistance factors and based on ultimate failures are used for major structural design work in professional design offices. In most situations these latest techniques are the basis for various computer-aided procedures, and building code requirements tend toward an acknowledgment of this situation. Still, for design of minor work and for simple structures in general, considerable use is made of tabulations in various references and some of the very simple, approximate procedures of the working-stress method.

Minimum Construction

As in other types of construction, building codes and industry standards result in a certain minimum form of construction. The structural capacity of this minimum construction represents a threshold that may be sufficiently high to provide for many situations of structural demand. Such is frequently the case with reinforced construction, so quite frequently, for buildings of modest size, the construction developed in simple response to code minimum requirements is more than adequate. In any event, the minimum construction represents a take-off point for consideration of any modifications that may be required for special structural tasks.

For example, the construction shown in Fig. 36.2 represents the general form of minimal construction. With vertical reinforced members at a minimum spacing of 4 ft on center, this means that only every sixth void is filled (voids are 8 in. on center). From this minimum, increased strength can be produced by filling more voids, all the way up to a completely filled wall if necessary. In addition, the amount of reinforcing is also stipulated as a minimum, so this alone may be increased for additional strength.

Design of various types of structural elements of reinforced masonry with hollow concrete units is discussed in the rest of this chapter. Utilization of many of these elements is illustrated in the building design examples in Part 9.

36.3. BEARING WALLS

Walls made with CMUs are often used to support gravity loads from roofs, floors, or walls in upper stories. Walls serving such functions are called *bearing walls*. The principal structural task is resistance to vertical compression force, which may involve considerations for one or more of the following:

Average Compressive Stress. This is simply the total compression load (including the wall weight) divided by the area of the wall cross section. Care should be taken to indicate whether this is the average stress for the gross section, defined by the wall's outer dimensions, or the net section, which is the actual solid portion of the hollow units.

Bearing Stress. This is the actual contact pressure under a load that is concentrated, such as that under the end of a supported beam.

Effective Column Stress. This is the stress considered when the wall supports widely spaced, concentrated loads, and only a limited part of the wall is considered effective.

Bending Stress. This may be produced when the vertical load is not applied in an axial manner to the wall, such as when a beam is not placed on top of a wall, but is supported by a bracket or ledger on the face of the wall, resulting in bending plus compression loadings.

Investigations for stress conditions in solid brick walls for all of these conditions are demonstrated in the examples in Chapter 37. For construction with unreinforced hollow units, the procedures are essentially the same, except for consideration of the voided cross section of the wall. See Example 1 in Sec. 36.1.

Walls may also sustain concentrated loads, as illustrated in Example 2 in Sec. 37.3 for a brick wall. The following example demonstrates a similar procedure for a CMU wall.

Example 1. A CMU wall sustains concentrated loads of 12,000 lb at 8-ft centers. The wall is of 8 in. nominal (7.625 in. actual) thickness, with blocks of concrete with density

of 100 lb/ft³, producing a wall with approximate weight of 40 psf of wall surface. Blocks are approximately 50% solid, f'_m for the construction is 1350 psi, and the wall has an unbraced height of 10 ft. The load is placed axially on the wall through a steel bearing plate that is 6 in. by 16 in. Investigate the wall for bearing and compression stresses.
Solution. For the calculated bearing stress we find

$$f_{br} = \frac{P}{A} = \frac{12,000}{6 \times 16} = 125 \text{ psi}$$

which is compared to the allowable bearing of

$$F_{br} = 0.26 f'_m = 0.26(1350) = 351 \text{ psi}$$

The bearing stress was calculated without reduction for the hollow-wall cross section, since it is assumed that the bearing will be on a fully grouted bond beam at the top of the wall. However, in this example, the stress would not be critical even with the reduced effective area.

For the maximum compressive stress we consider the situation at the bottom of the wall, for which the wall weight must be added to the applied load. We find the wall weight to be

$$W = 40(10)$$
$$= 400 \text{ lb/ft of wall length in plan}$$

Using an effective wall pier of six times the wall thickness or four times the wall thickness plus the bearing width, we use 6(7.625) = 45.75 in. or 4(7.625) + 16 = 46.5 in., say 46 in. The total weight of this pier is therefore

$$W = \frac{46}{12} (400) = 1533 \text{ lb}$$

Thus the total load is 12,000 + 1533 = 13,533 lb, and the average compressive stress in the pier is

$$f_a = \frac{P}{A_e} = \frac{13,533}{(7.625)(46)(0.50)} = 77.2 \text{ psi}$$

This must be compared with the allowable stress, which is typically taken from Table 36.2 for the unreinforced construction. For this case, the limiting value is 150 psi, and the construction is adequate.

Unreinforced walls may sustain reasonably heavy loads, as long as no bending is present. Bending may be added from the effects of eccentricity of the vertical load or from lateral loads due to wind, earthquakes, or soil pressures. This results in a combined stress condition that may be critical for either the net total compression or the net tension, where such is present.

Details of the construction often make it difficult to place supported loads axially on walls (load center over the wall center), especially with multistory or parapet walls. The following example illustrates the effect of such a condition on an unreinforced wall.

Example 2. Suppose that the load in Example 1 is placed on the wall face, as shown in Fig. 36.3, so that an eccentricity of 6.8125 in. occurs between the load center and the wall center. Investigate the wall for the combined stress condition.
Solution.   With the eccentricity as shown in Fig. 36.3, the load develops a bending moment of

$$M = Pe = (12,000)(6.8125)$$
$$= 81,750 \text{ in.-lb}$$

For the bending stress, we find from Table 36.1 that the wall has a section modulus (S) of 88 in.³/ft of wall. For the 46-in. pier the total value is

$$S = \frac{46}{12} (88) = 337 \text{ in.}^3$$

and the maximum bending stress is

$$f_b = \frac{M}{S} = \frac{81,750}{337}$$
$$= 243 \text{ psi}$$

Since this alone is in excess of the allowable stress of 150 psi from Table 36.2, there is no need to add it to the axial compression. Although the allowable compression is exceeded, the more critical concern is for the net tension, which will exceed 170 psi near the top of the wall (where the wall weight will not reduce it).

Clearly, an unreinforced wall cannot develop resistance to this magnitude of bending. Efforts must be made to reduce the bending effects or to make other provisions for it. Reduction can be made by moving the load closer to the wall center. Other provisions may consist of adding steel reinforcement or a pilaster column.

Reinforced walls can sustain both greater axial compression and higher ranges of bending. Grouting of some void spaces to install the reinforcement adds to the wall cross section. For concrete block walls, the grouted spaces can usually be assumed to be at least as strong as the concrete units. Thus a fully grouted wall may be considerably stronger for direct compression.

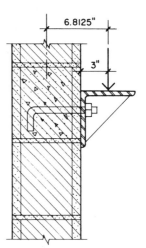

FIGURE 36.3.

Various methods can be used for design of reinforced walls. We use the working-stress method here because it is simple, but it will often produce quite conservative results.

Example 3. Suppose the wall in Example 2 is developed with reinforced construction. Using steel with $F_y = 40$ ksi, investigate the wall.

Solution. The general condition that must be satisfied for the reinforced wall is

$$\frac{f_a}{F_a} + \frac{f_b}{F_b} \lessgtr 1$$

In this formula, f_a and f_b are the calculated axial and bending stresses, and F_a and F_b are the corresponding allowable stresses. Each value can be determined, and the combination can be evaluated. However, a first condition worth investigating is whether the wall can sustain the moment at all. This is determined by calculating the maximum resisting moment as Kbd^2. (See Appendix E.) For the wall with $f_m' = 1350$ psi and steel with $F_y = 40$ ksi, Table E.2 yields $K = 66.6$ (in in.-lb units). Considering the "beam" width to be the full 46-in. width of the pier, and $d = 7.625/2 = 3.8125$ in. with reinforcing in the center of the wall, the maximum bending moment without any axial compression is

$$M_R = Kbd^2 = (66.6)(46)(3.8125)^2$$
$$= 44{,}530 \text{ in.-lb.}$$

which indicates that the moment in Example 2 cannot be sustained, even without consideration for the combined stresses.

Adding grouted void spaces and some vertical steel reinforcement does indeed increase wall strength, but not without limits. In many situations the major improvement to the general construction when reinforced masonry is used is the overall toughness gained by the structure.

Example 4. Consider the wall in the preceding examples to be subjected to a uniform load of 2500 lb/ft and a lateral wind force of 20 lb/ft² on the wall surface. Investigate the wall (a) as an unreinforced wall and (b) as a reinforced wall of minimum construction.

Solution. For the wind pressure, the wall spans vertically as a simple beam on the 10-ft span, producing of maximum bending moment of

$$M = \frac{wL^2}{8} = \frac{(20)(10)^2}{8}$$
$$= 250 \text{ ft-lb, or } 250(12)$$
$$= 3000 \text{ in.-lb}$$

(a) Unreinforced Wall

Investigating for combined compression and bending stresses, we will consider the situation of the wall at approximately midheight, where the bending moment is maximum. From Table 36.1, assuming a density of 100 lb/

ft³ for the concrete and fully bedded mortar joints, the properties of the wall are

Average weight = 40 psf of wall surface

Average net cross section of wall = 48 in.²/ft of length

Average section modulus (S) = 88 in.³/ft of wall length

At midheight we add to the applied load a wall weight of

$$W = \text{(unit weight)(half the wall height)}$$
$$= (40)\left(\frac{10}{2}\right) = 200 \text{ lb}$$

The total compression load at midheight is thus $2500 + 200 = 2700$ lb, and the average net compression stress is

$$f_a = \frac{P}{A_e} = \frac{2700}{48} = 56.25 \text{ psi}$$

The maximum bending stress is

$$f_b = \frac{M}{S} = \frac{3000}{88} = 34.09 \text{ psi}$$

and the combined stresses are

Maximum $f = 56.25 + 34.09$
$$= 90.34 \text{ psi}$$

Minimum $f = 56.25 - 34.09$
$$= 22.16 \text{ psi (compression)}$$

This indicates that there is no net tension stress, and the maximum value for compression is well below the limit of $150(1.33) = 200$ psi as given in Table 36.2 and is increased for wind loading.

The allowable stress based on slenderness (h/t) of the wall should also be investigated, and the possible effect of interaction should be considered ($f_a/F_a + f_b/F_b \lessgtr 1$). This will not be critical in this case. The procedure for this is demonstrated in the following investigation of the reinforced wall.

(b) Reinforced Wall

To qualify as reinforced, the wall must satisfy various criteria, including the following:

Horizontal and vertical reinforcement in grouted voids, maximum of 48 in. spacing

Minimum bar size, No. 4

Minimum percentage of A_s, sum of both directions, $0.002A_g$ of wall

Minimum percentage of A_s, either direction, $0.0007A_g$ of wall

The minimum total reinforcement (sum of both ways) is thus

$$A_s = (0.002)(7.625)(12) = 0.183 \text{ in.}^2/\text{ft}$$

Favoring the horizontal direction due to increased shrinkage effects, a possible choice is therefore (see Table E.1)

Horizontal: No. 6 at 48 in., $A_s = 0.110 \text{ in.}^2/\text{ft}$
Vertical: No. 5 at 48 in., $A_s = 0.077 \text{ in.}^2/\text{ft}$

Based on the minimum vertical reinforcement, the total moment resistance can be expressed as (See Appendix E.)

$$\begin{aligned} M_R &= A_s f_s jd \\ &= (0.007)(20{,}000 \times 1.33)(0.9)(3.8125) \\ &= 7028 \text{ in.-lb} \end{aligned}$$

For the masonry, flexural compression is limited to

$$\begin{aligned} F_b &= 0.333 f'_m = (0.333)(1350)(1.333) \\ &= 600 \text{ psi} \end{aligned}$$

which must be reduced by 50% if special inspection and testing are not provided.

Using the lower value, we can express the total moment resistance as

$$\begin{aligned} M_R &= \tfrac{1}{2}(F_b kjbd^2) \\ &= \tfrac{1}{2}[(300)(0.33)(0.9)(12)(3.8125)^2] \\ &= 7770 \text{ in.-lb} \end{aligned}$$

Axial compression will be approximately the same as for the unreinforced wall. The grouted construction weighs slightly more, but the net wall cross section is slightly increased. Referring to the calculation for the unreinforced wall, we will assume an average value of $f_a = 60$ psi.

For the allowable axial compression stress, we use the formula based on wall slenderness:

$$\begin{aligned} F_a &= 0.20 f'_m \left[1 - \left(\frac{h'}{42t} \right)^3 \right] \\ &= 0.20(1350) \left[1 - \left(\frac{120}{42 \times 7.625} \right)^3 \right] (1.333) \\ &= 340 \text{ psi (or 170 psi without inspection)} \end{aligned}$$

The investigation for the interaction condition is thus as follows:

$$\frac{f_a}{F_a} + \frac{f_b}{F_b} = \frac{60}{170} + \frac{3000}{7028} = 0.353 + 0.426 = 0.779$$

which indicates a safe condition.

Note that we have substituted a ratio of actual bending moment to limiting bending moment for the f_b/F_b term. This is possible because f_b is produced by the actual moment and F_b by the limiting moment; thus the ratio of the moments is the same as the ratio of the stresses.

These calculations indicate that the wall is adequate with or without reinforcement. In fact, there seems to be no real gain in strength by addition of the reinforcement. This is due partly to the limitations of the working-stress method; use of strength methods will usually indicate more capacity for the reinforced construction. However, the real gains achieved by the two-way reinforcement are truly more in the form of the enhanced toughness of the general construction. This character of toughness is particularly significant when walls are called upon to resist the dynamic effects of windstorms or earthquakes. These are the load conditions associated with use as shear walls, as discussed in Sec. 36.4.

36.4. SHEAR WALLS

Structural masonry walls are quite often used as shear walls, forming part of the lateral-load-resisting system for a building. Various systems can be used for lateral bracing, but the most common one for low-rise construction is the box system, typically formed with a combination of horizontal and vertical diaphragms.

Use of Stiffness Factors

A common situation that occurs is that in which a number of individual shear walls or piers share some total lateral force, requiring that the distribution of force to the individual bracing elements be determined. Figure 36.4 shows three situations in which this can occur for a single wall.

In Fig. 36.4a a wall is formed by a series of separated, but linked, masonry walls, with lighter construction forming the wall portions between the masonry. If the individual masonry piers are all of the same size and similarly constructed, the total lateral force in the wall will be simply divided equally between the piers. If they have different dimensions, however, their relative stiffnesses must be used to apportion the load to the piers.

For the wall in Fig. 36.4b a similar situation occurs if the distribution of shear on a horizontal plane through the window openings is considered. In this case the individual piers of masonry between the window openings act as fully fixed elements.

For the piers between the door openings in Fig. 36.4c, the condition may be one of full fixity (36.4b) or simple cantilever (Fig. 5.15a), depending on the nature of the anchorage and support at the base of the piers.

Assuming that the piers are all similarly constructed, the basic factor that distinguishes them from each other in any of the walls in Fig. 36.4 is their aspect ratio (vertical dimension to horizontal dimension). Strictly on the basis of this ratio, plus their qualification of single fixity (simple cantilever) or double fixity, their relative stiffnesses can be established and used as a basis for distribution of lateral loads. Factors for this purpose are given in the tables in

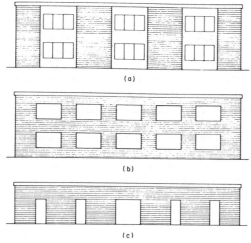

FIGURE 36.4. Form variations for masonry shear walls: (*a*) individual, isolated, linked piers (vertical cantilevers with fixed bases; (*b*) continuous wall with fully fixed, individual piers; (*c*) continuous wall with individual cantilever piers, fixed at their tops.

Appendix E, and their use is demonstrated in the following example.

Example. A lateral force is delivered to the wall shown in Fig. 36.5. Find the percentage of the total load (*H*) resisted by each of the individual piers at the level of the window openings.

Solution. In this case the piers are considered to be fixed at their tops and bottoms. Stiffness factors, R_c, for the individual piers are thus obtained from Table E.3, on the basis of the h/d ratios for the piers. The load distribution to

each individual pier is then determined by multiplying the total load by a distribution factor, DF:

$$\text{DF} = \frac{\text{factor for the individual pier}}{\text{sum of the factors for all the piers}}$$

The computations for the distribution are summarized in Table 36.3.

Construction

Just about any form of structural masonry offers some potential for development of a shear wall. Wind and earthquake forces acted on ancient buildings just as they now do on modern ones. Ancient walls still standing testify to their adequacy in this regard.

However, we now generally build much lighter and thinner masonry structures and have enhancements at our disposal to create better structures. We also have many ways to attach the separate elements of building systems to each other. And, of major significance, we generally document and disseminate data regarding our collective experiences for all to share. Although we inherited many valuable lessons from the masons and builders of the past, the art of design for lateral force effects is a fairly recent development.

Where wind forces are moderate and earthquakes virtually unknown to have occurred, it is possible to develop shear walls with unreinforced construction. Where windstorms are prevalent or major earthquakes are a high risk, or simply where major loads must be resisted, it is now preferred to use reinforced construction. Codes provide data and procedures for both situations, and typically require

TABLE 36.3. Load Distribution to the Masonry Piers

Pier	h (ft)	d (ft)	h/d	R_c	DF[a]	Share of Lateral Load (%)
1	8	4	2.0	0.1786	0.087	8.7
2	8	8	1.0	0.6250	0.304	30.4
3	8	10	0.8	0.8585	0.417	41.7
4	8	6	1.33	0.3942	0.192	19.2
			Sum =	2.0563		

[a] $\text{DF} = \dfrac{R_c \text{ for pier}}{\text{sum of } R_c}$.

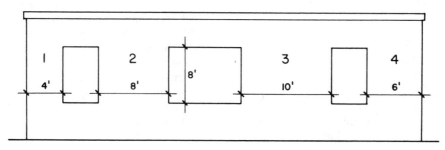

FIGURE 36.5. Multiple-pier wall for the example problem.

only reinforced construction where severe wind or seismic effects are present.

Usage of masonry structures for lateral load resistance is illustrated in the building design cases in Part 9. These are simple cases using common construction elements. Structural design and development of construction details for complex structures or for very severe loading conditions must be done with the highest state of the art in terms of engineering and construction and are beyond the scope of this book.

Some general considerations for shear wall design are the following:

1. Anchorage of walls to supports for resistance to sliding or overturn is generally achieved by doweling the reinforcement in reinforced construction. Special anchors or keys may be necessary with unreinforced construction, although the sheer weight of the wall is often sufficient for its stability.
2. Walls must be adequately supported, and bearing foundations should be very conservatively designed to minimize any soil deformations.
3. Connections of supported structures that deliver lateral loads to walls should be very carefully designed for the combined gravity and lateral loads. Positive resistance to lateral loads is critical, especially for seismic loads.
4. Special attention should be given to stress conditions at the wall discontinuities at openings, wall corners, and intersecting walls. These should be reinforced, even in otherwise unreinforced construction.

36.5. PEDESTALS

There are many instances in which a short compression element, called a *pier* or *pedestal*, is used as a transition between a footing and some supported element. Some of the purposes for pedestals are the following.

To Permit Thinner Footings. By widening the bearing area on the top of the footing, the pedestal will achieve a reduction in the shear and bending stresses in the footing, permitting the use of a thinner footing. This may be a critical issue where allowable soil bearing pressures are quite low.

To Keep Wood or Metal Elements above the Ground. When the bottom of the footing must be some distance below the ground surface, a pedestal may be used to keep vulnerable elements above ground.

To Provide Support for Elements at Some Distance above the Footing. In addition to the previous situation, there are others in which elements to be supported may be some distance above the footing. This may occur in tall crawl spaces, basements, or where the footing elevation must be dropped a considerable distance to obtain adequate bearing.

Pedestals of masonry or concrete are essentially short columns. When carrying major loads, they should be designed as reinforced columns with appropriate vertical reinforcement, ties, and dowels. When loads are light and the pedestal height is less than three times the thickness, however, they may be designed as unreinforced elements. If built of hollow masonry units (concrete blocks, etc.) they should have the voids completely filled with concrete. Table 36.4 gives data for short, unreinforced masonry pedestals. The table entries represent only a sampling of the potential sizes for these pedestals and are intended to give a general idea of the load range, not to be construed as standard or recommended sizes.

There are minimum as well as maximum heights for pedestals. If the pedestal is very short and the area of the bearing contact with the object supported by the pedestal is small, there will be considerable bending and shear in the pedestal, similar to that in a footing. This can generally be avoided if the pedestal is at least as tall as it is wide. For a pedestal of concrete it is theoretically possible to reinforce the pedestal in the same manner as a footing, although this is usually not feasible.

Masonry pedestals are subject to considerable variation. Primary concerns are for the type of masonry unit, the class of mortar, and the general type of masonry construction. For the pedestals in Table 36.4 we have assumed a widely used type of construction, utilizing hollow concrete masonry units (concrete blocks) with all voids filled with concrete. For this construction we have used the design criteria in the *UBC* (Ref. 1) as given in Table 36.2, assuming Type S mortar. The allowable stress in compression on the gross cross-sectional area of the pedestal is 150 psi [1.03 MPa]. According to a footnote to the table, an increase of 50% is permitted for the stress at the bearing contact with the object supported by the pedestal. Based on these requirements, it is possible to establish a maximum allowable load for a given size pedestal, and to derive the minimum column size on the basis of the ratio of the two allowable stresses. These two items are given in Table 36.4.

For supported objects with contact area smaller than the minimum size listed, the pedestal load will be limited by the contact area rather than by the potential pedestal capacity. Thus the table includes pedestal capacities for a range of column sizes, based on the stress limit for the contact area.

Although the pedestals in Table 36.4 are designed using criteria for unreinforced masonry, we recommend the use of vertical reinforcing in the four corners of the pedestal when the pedestal height exceeds twice the pedestal thickness. This reinforcing is quite minimal in cost and need not be doweled into the footing, but it adds some degree of toughness to the pedestal against the ravages of shrinkage, temperature changes, and possible damage during construction of the supported structure. As stated previously, if the pedestal height exceeds three times its thickness, the pedestal should be designed as a reinforced masonry column with appropriate vertical reinforcement, horizontal ties, and footing dowels.

TABLE 36.4. Unreinforced Masonry Pedestals

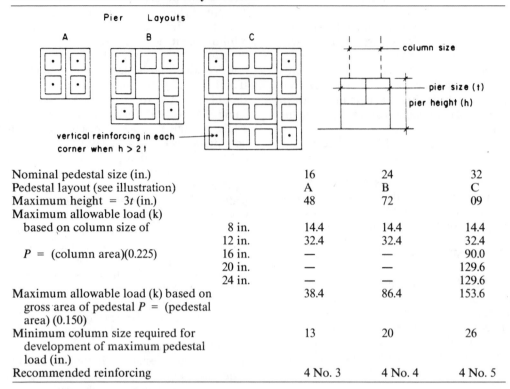

		16	24	32
Nominal pedestal size (in.)		16	24	32
Pedestal layout (see illustration)		A	B	C
Maximum height = 3t (in.)		48	72	09
Maximum allowable load (k) based on column size of	8 in.	14.4	14.4	14.4
	12 in.	32.4	32.4	32.4
P = (column area)(0.225)	16 in.	—	—	90.0
	20 in.	—	—	129.6
	24 in.	—	—	129.6
Maximum allowable load (k) based on gross area of pedestal P = (pedestal area) (0.150)		38.4	86.4	153.6
Minimum column size required for development of maximum pedestal load (in.)		13	20	26
Recommended reinforcing		4 No. 3	4 No. 4	4 No. 5

Pedestals may also be constructed with bricks, one possible advantage being the potential for increased bearing stress. However, because pedestals are mostly not exposed to view, the construction with CMUs, as shown in Table 36.4, is likely to be more economical in most situations.

Forms of construction with CMUs, other than those shown in Table 36.4 are also possible. Special units ordinarily used to form masonry columns, as discussed in Sec. 36.6, can be used with or without reinforcement.

36.6. COLUMNS

Masonry columns may take many forms, the most common being a simple square or oblong rectangular cross section. The general definition of a column is a member with a cross section having one dimension not less than one-third the other and a height of at least three or more times its lateral dimension. Shorter columns are called *pedestals* (see Sec. 36.5), and elements with longer, thinner plan dimensions are considered as wall piers.

The three most common forms of construction for structural columns are the following (see Fig. 36.6):

Unreinforced Masonry. These may be brick (Fig. 36.6a), CMU construction (Fig. 36.6b) or stone (Fig. 36.6c), and are generally limited to forms bordering on the pedestal category—that is, very stout. Slender columns, or any column required to develop significant bending or shear, should be reinforced.

Reinforced Masonry. These may be any recognized form of reinforced construction, but are mostly either fully grouted and reinforced brick construction or CMU construction with all voids filled (Fig. 36.6a, b and d). Large columns formed with shells of CMUs are more likely in the next category.

Masonry-Faced Concrete. These are essentially reinforced concrete columns with masonry shells. The shells may be veneers (applied after the concrete is cast) or laid up to form the cast concrete. In the latter case, the structure may be considered as a composite one, but is often more conservatively designed, ignoring the capacity of the masonry facing.

Short columns of unreinforced construction are generally designed for simple compression resistance, using the appropriate limit for compressive stress from Table 36.2. Bearing stresses are generally allowed to be higher, but the general design of the column will not be based on bearing, unless the contact bearing area is close to the column gross cross-sectional area.

If simple compression is the major concern, the design of an unreinforced column is similar to that for a vertical load-bearing wall. (See Sec. 36.3.) If the compression load is applied with some eccentricity, it is necessary to consider the combined compression and bending stresses, as discussed in Appendix E.

Example 1. A short column is to be formed of CMU units as shown in Fig. 36.2b, with 8 × 8 × 16-in.-nominal blocks

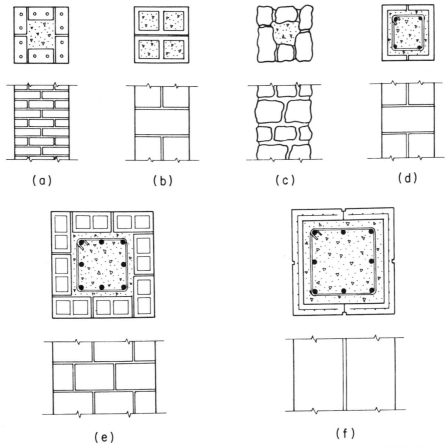

FIGURE 36.6. Forms of masonry columns: (*a*) brick or low-void CMUs, unreinforced; (*b*) CMUs with or without reinforcement; (*c*) stone, fully grouted; (*d*) CMU shell with cast concrete column; (*e*) large concrete column with CMU shell; (*f*) sitecast concrete column formed with precast concrete units (larger scale version of *d*).

having $f_m' = 1500$ psi. What is the maximum axial compression load for this column, if it is unreinforced and the units are laid with Type S mortar?

Solution. Assuming bearing is not critical, the maximum stress for the cross section from Table 36.2 is 150 psi for hollow-unit masonry. With the voids unfilled and assuming a 50%-solid block, the total load capacity is

$$P = F_a(\text{net area}) = (150)(15.5)^2(0.50)$$

$$= 18{,}019 \text{ lb}$$

Reinforced masonry columns are designed as described in Appendix E. As with present design practice for reinforced concrete columns, it is generally accepted that all columns should be designed for some bending, with a minimum bend assumed to be that created by a load eccentricity equal to one-tenth of the column side dimension. The following example demonstrates the process for a simple column.

Example 2. Assume the column in Example 1 to have its voids fully grouted and to be reinforced with four No. 5 bars with $F_y = 40$ ksi. Investigate the column for a compression load of 20 kips placed so that it is 2 in. off of the column center. Unbraced height is 16 ft.

Solution. If bending is ignored, the axial load capacity is limited to

$$P_a = (0.20 f_m' A_e + 0.65 A_s F_{sc}) \times$$

$$\left[1 - \left(\frac{h}{42t} \right)^3 \right]$$

where

$$A_e = \text{effective net area} = (15.5)^2$$
$$= 155 \text{ in.}^2 \text{ (fully grouted)}$$
$$A_s = 4(0.31) = 1.24 \text{ in.}^2$$
$$F_{sc} = \text{allowable steel stress} = 0.4 F_y$$
$$= 16 \text{ ksi}$$

Thus

$$P_a = [(0.20)(1500)(155) + (0.65)(1.24)(16{,}000)] \times$$

$$\left[1 - \left(\frac{16(12)}{42(15.5)} \right)^3 \right]$$

$$= 53{,}474 \text{ lb}$$

For the interaction relationship, therefore, $f_a/F_a = 40/53.5$ = 0.748, which indicates some margin for bending. For an approximate estimate of the bending capacity, we may consider the section as one with tension reinforcement only. Thus, with two No. 5 bars in the centers of the voids, the approximate effective depth is 11.6 in., and the moment limited by the steel is

$$M_R = A_s F_s jd = (0.62)(20,000)(0.9)(11.6)$$

$$= 129,456 \text{ in.-lb}$$

Comparing this with the actual moment of $40(2) = 80$ kip-in., or 80,000 in.-lb, we find the ratio of f_b/F_b to be

$$\frac{80,000}{129,456} = 0.618$$

and the interaction response is thus

$$\frac{f_a}{F_a} + \frac{f_b}{F_b} = 0.748 + 0.618 = 1.366$$

Since this exceeds 1, the column is not adequate.

This is a conservative, approximate analysis, and a more accurate one using strength methods and including the effect of the compression in the other bars would probably show less overstress.

Masonry columns are also frequently developed as pilasters—that is, columns built monolithically with a wall. These may take various cross-sectional shapes. Pilasters typically serve multiple functions—reinforcing the wall for concentrated loads, excessive bending, or spanning, and bracing the wall to reduce its slenderness.

Except for the bracing afforded by the wall, pilasters are usually designed structurally as freestanding columns. This may in some cases reduce the required wall functions to that of infill between the columns.

Pilasters may be used at the ends of walls or at the edges of large openings for reinforcement. The chords for shear walls may be developed as pilasters where significant overturn is present.

CHAPTER THIRTY-SEVEN

Brick Masonry

The noble brick has a long history and endures steadily as a popular visual element in architecture. Structural use of bricks has decreased for reasons of cost, but the potential for structural brick masonry still exists and is currently experiencing some expanded application. Chapter 37 deals with some of the possible uses of bricks for structures; not merely for decoration.

37.1. TYPICAL ELEMENTS OF BRICK CONSTRUCTION

For structural purposes, brick is used primarily to produce walls, piers, columns, and pedestals. In masonry terminology, *piers* are segments of walls, short in plan length, while *pedestals* are very short columns. There is thus a sort of geometric order of progression for a singular element that is basically a compression member oriented for vertical loading. (See Fig. 37.1.)

Structural masonry, in the form of walls or columns, is often used for support of horizontal-spanning structures and possibly for support of other walls or columns above in multistory construction. A major consideration, therefore, is that for resistance to vertical gravity loads that induce compression stress in the walls or columns. In ancient, predominantly very heavy, masonry, a significant amount of the total vertical gravity load often came from the weight of the masonry itself.

Required thicknesses for walls and horizontal cross sections for columns were developed from experience. As larger and taller buildings were produced, various rules of thumb were developed and passed on, becoming part of the traditions and craft. This tradition-based design was dominant until a relatively short time ago. Figure 37.2 shows the requirements for the graduated thickness of a 12-story high bearing wall of brick, as defined by the Chicago Building Code some 60 years ago. This criteria was largely judgmental and based primarily on demonstrated success with years of construction.

Structural walls and columns in present-day masonry construction tend to be considerably slimmer and lighter than those in ancient buildings. Economic pressures and the availability of better, stronger materials partly account for this. A major factor, however, is the considerable development of the fields of materials testing and structural engineering, permitting reasonably precise predictions for

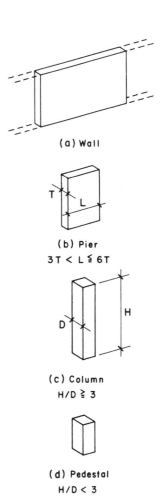

(a) Wall

(b) Pier
$3T < L \leqq 6T$

(c) Column
$H/D \geqq 3$

(d) Pedestal
$H/D < 3$

FIGURE 37.1. Classification of vertical compression members.

structural behaviors and, as a result, some heightened confidence in predictions of safety against failure. This has somewhat reduced the dependency on experience alone as a guarantee of success, although confidence is still boosted significantly when experience is there to back up speculations about safety.

Actually, in spite of the lack of sophisticated structural analysis and design theories and procedures, ancient builders produced many technological feats of daring with masonry construction that generally cannot be duplicated today, even though some of their achievements are still

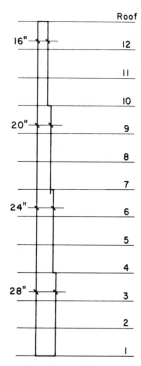

FIGURE 37.2. Required thickness of brick walls for a 12-story mercantile building, Chicago Building Code, 1932.

standing as testimonials to the builders' successes. Structures like the Roman aqueducts and the Gothic cathedrals would never be executed in masonry today, despite our considerable advances in technology of materials and the science of materials behaviors. Structures can be built to look like the ancient masonry ones, but will certainly be actually achieved with steel or reinforced concrete.

37.2. GENERAL CONCERNS FOR BRICK WALLS

Structural brick walls can be used for many purposes, and basic design concerns derive initially from the specific structural requirements. Compression stress from vertical loading is a predominant condition, deriving—if from nothing else—from the weight of the masonry itself. For simple bearing walls, this may be the only real concern, and a simple axial compression stress investigation may suffice for the structural design. Thus the total, critical compression load is determined, and the simple stress calculation

$$f = \frac{P}{A}$$

is made, in which

f = average compression stress
P = total vertical compression force
A = cross-sectional area of the wall

This calculated stress is compared to an allowable stress for the construction, based on code requirements and the specifications for the masonry materials.

For solid masonry construction, the wall cross section is simply the product of the wall thickness and some defined, design increment of the wall length (typically a 1-ft or 12-in. length in U.S. units). For cavity walls, or walls made with hollow units, the actual net solid cross section must be used.

Additional structural functions for walls may include any of those illustrated in Fig. 37.3. Exterior walls in buildings frequently must serve multiple purposes: bearing walls, spanning walls resisting direct wind pressures, and shear walls. These situations produce multiple concerns for stress combinations, as well as for the logical development of wall form, construction details, attachment of any supported elements, and so on. Structural walls are also themselves supported by other structures, either directly by foundation elements or by other supporting walls or frameworks.

In the past, brick construction was frequently used as a "cheap" infill material to achieve solid masonry with higher-quality surface materials. Now brick has graduated mostly to the higher class itself, and its use must usually be justified by exposure of the expensive construction. Thus, appearance considerations become major concerns, even when serious structural tasks are required. Brick materials, mortar color and joint form, and wall face patterns are typically major architectural design concerns, as are various construction details if exposed to view. Wall ends, supports, and tops and edges of openings for windows and doors must be sensitively detailed, both for the appearance and for the special structural tasks at these locations.

For various reasons, brick walls are mostly used in only a few common ways, especially for structural purposes. The following sections describe some specific uses and some of the additional concerns that derive from them. Usage considerations in a broader context are also discussed in the building design cases in Part 9.

37.3. BRICK BEARING WALLS

Brick bearing walls often occur as exterior walls. They thus quite frequently have other structural functions, in addition to that of resisting simple vertical compression. Three common additional functions are those shown in Fig. 37.3b (spanning for wind), 37.3c (shear wall), and 37.3f (basement wall). For complete design it thus becomes necessary to consider these combined actions. Some combined effects will be discussed later in this section, but we will first consider the simple problem of vertical compression.

Readers with less experience in structural investigation should review Chapter 11 before pursuing the following discussions. The general concerns for compression members are presented there, and the various methods for investigation are discussed. Additionally, the general basis

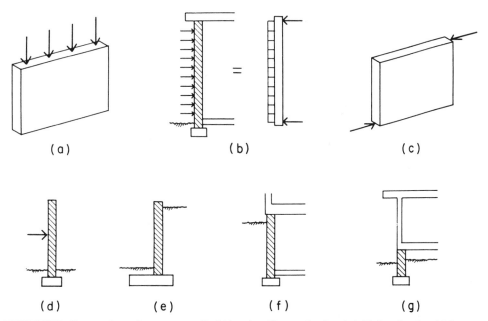

FIGURE 37.3. Structural uses for masonry walls; (*a*) bearing; (*b*) spanning for wind; (*c*) shear bracing; (*d*) free-standing; (*e*) cantilever retaining; (*f*) foundation—basement; (*g*) foundation—grade.

for working-stress investigation of reinforced members is presented in Part 5.

Investigation of compression in unreinforced construction may be done by using maximum stress limits, such as those in Table 36.2, which is a reprint of Table 24-H from the 1988 *UBC*. This limiting stress may be used as long as the wall slenderness does not exceed limits established by the code. When slenderness is excessive, a general formula for allowable compression stress is

$$F_a = 0.20 f_m' \left[1 - \left(\frac{h'}{42t} \right)^3 \right]$$

in which

F_a = allowable average compressive stress
h' = effective unbraced height
t = effective thickness

Example 1.   A brick wall of solid construction is 9 in. thick and has an unbraced height of 10 ft. The construction has a specified value of $f_m' = 2500$ psi and is unreinforced. The wall sustains a bearing load on its top of 2000 lb/ft. The brick construction weighs approximately 140 lb/ft³. Investigate the condition for average compressive stress.
Solution.   The total load on the wall includes the applied loading on the top plus the weight of the masonry. This total gravity load is greatest at the bottom of the wall. We thus determine the total wall weight for a 1-ft (12-in.)-wide strip to be

$$\text{Weight} = \frac{9 \times 12}{144} (10)(140) = 1050 \text{ lb}$$

The total load is then 1050 + 2000 = 3050 lb, and the maximum compressive stress at the bottom of the wall is

$$f_a = \frac{P}{A} = \frac{3050}{9 \times 12} = 28.2 \text{ psi } [194 \text{ kPa}]$$

Since the allowable-stress formula is based on consideration for the wall slenderness, it is technically critical near the wall midheight (where buckling occurs). However, for a conservative investigation we can compare the calculated stress just determined with that found from the formula for allowable stress, as follows:

$$F_a = 0.20 f_m' \left[1 - \left(\frac{h'}{42t} \right)^3 \right]$$

$$= 0.20(2500) \left[1 - \left(\frac{120}{42 \times 9} \right)^3 \right]$$

$$= 484 \text{ psi}$$

This value is compared with the maximum limit given in Table 36.2. For a minimum Type S mortar, the table yields a maximum value of 160 psi. Although this is lower than that determined from the formula, it is still greater than the computed average stress, so the wall is still safe.

Allowable stresses for design may be modified by various conditions, one of which relates to the type of inspection and testing provided for the construction. The design code used must be carefully studied for these qualifications.

When walls sustain concentrated loads, rather than distributed loads, it is usually necessary to investigate for two stress conditions. The first involves the concentrated bearing stress directly beneath the applied load. The second investigation involves a consideration of the effective portion of the wall that serves to resist the load. The length of wall for the latter situation is usually limited to six times the wall thickness, four times the wall thickness plus the actual bearing width, or the center-to-center spacing of the loads. The following example demonstrates this type of investigation.

Example 2.   Assume that the wall in Example 1 sustains a concentrated load of 12,000 lb from the end of a truss. The truss load is transferred to the wall through a steel bearing plate that is 2 in. narrower than the 9-in.-thick wall and 16 in. long along the wall length. Investigate for bearing and average compression stresses in the wall.

Solution.   Using the given data, we first determine the actual values for the bearing and average compression stresses. For the bearing,

$$f_{br} = \frac{P}{A} = \frac{12,000}{7 \times 16} = 107 \text{ psi [739 kPa]}$$

For the compression, we assume an effective wall unit that is $6 \times 9 = 54$ in. or $(4 \times 9) + 16 = 52$ in., the latter being critical. Using the unit density given in Example 1, we find that this portion of the wall weighs

$$(140) \frac{9 \times 52}{144} (10 \text{ ft}) = 4550 \text{ lb}$$

When this is added to the applied load, the total vertical compression at the base of the wall is thus $12,000 + 4550 = 16,550$ lb, and the average compressive stress is

$$f_a = \frac{P}{A} = \frac{16,550}{9 \times 54} = 34.0 \text{ psi [234 kPa]}$$

The allowable compression stress for this case is the same as that found in Example 1, 160 psi. The actual stress is thus well below the limit for safety.

For the bearing stress, the code provides two values, one based on "full bearing," and the other on a defined condition where the bearing contact area is a fraction of the full cross section of the masonry. Although the actual bearing area in this case is indeed a fraction of the full wall cross section, the short distance from the edge of the bearing area to the edge of the masonry qualifies this situation as full bearing. We therefore use the code limit for allowable bearing as

$$F_{br} = 0.26f'_m = (0.26)(2500)$$
$$= 650 \text{ psi [4482 kPa]}$$

which indicates a safe condition for bearing for the example.

Solid brick masonry is usually the strongest form of unreinforced masonry, and thus is capable of considerable force resistance for situations involving only simple compression and bearing stresses.

Design of unreinforced masonry columns generally follows the same procedures just demonstrated for the bearing wall. From the data of Example 2, it may be demonstrated that a 9-in.-square column could sustain the load, using the allowable stress limit of 160 psi. While this may satisfy code limits, a 10-ft-high column only 9 in. across would appear quite slender. The author does not recommend the use of unreinforced columns in general, except for very low magnitudes of load and columns in the pedestal class, where the column height is less than three times its side dimension.

Compression Plus Bending

From eccentrically placed compression loads or some form of direct bending action, compression members frequently must resist a combination of direct compression plus flexure. The general case for this may be visualized as that occurring with the eccentric compression force, which induces a bending moment equal to the product of the force and its eccentricity from the compression member's centroidal axis. See discussions in Appendix E and Sec. 11.4.

The condition of net stress caused by a combination of compression and bending can be visualized as an addition of the separate stresses of uniformly distributed compressive stress plus the varying stresses caused by bending, ranging from a maximum tension to a maximum compression across the stressed section. In a member capable of resisting both compression and tension, this will result in some limiting, boundary values of the stress variation on the section. A special case occurs when the section is actually not capable of tension resistance—a conservative assumption for masonry, which has a very low tensile resistance. An analogy can be made with bearing foundations on soil, where the foundation—soil contact face can resist only compression.

The discussion in Sec. 11.4 develops some relationships for investigation of compression plus bending occurring with a simple bearing footing, a condition we will use to represent the tension-weak, unreinforced masonry.

When tension stress is not possible, eccentricities beyond the kern limit will produce a *cracked section*, which is shown as Case 4 in Fig. 11.5. In this situation some portion of the section becomes unstressed, or cracked, and the compressive stress on the remainder of the section must develop the entire resistance to the force and moment.

The cracked-section stress condition is really not desirable for either soil bearing or masonry. The kern limit, with a zero stress at one edge, should be used in most cases as a design limit. A possible exception may be for stress combinations that include wind or seismic effects, which are extremely short in duration. Masonry is now generally permitted to sustain a very low tensile stress, but it is still better to design without it.

The following example illustrates an application of the combined stress relationships derived in Sec. 11.4.

Example 3.   Investigate the wall in Example 1 for a combined loading that includes the gravity loads as given plus a wind pressure of 20 psf on the wall surface.

Solution.   The stress due to the vertical gravity load was found in Example 1. That is actually the value of the compressive stress at the base of the wall, since the load includes the entire wall weight. Bending stress will have its greatest value at midheight of the wall, as the following discussion indicates.

The wall spans vertically a distance of 10 ft (its unbraced height) and sustains a uniformly distributed load on a 1-ft-wide strip of 20 lb/ft. Using the formula for maximum moment on a simple beam, we find the maximum bending moment at midheight in the wall:

$$M = \frac{wL^2}{8} = \frac{(20)(10)^2}{8} = 250 \text{ ft-lb}$$

For the solid rectangular section, 9 in. × 12 in., the section modulus is determined to be

$$S = \frac{bd^2}{6} = \frac{(12)(9)^2}{6} = 162 \text{ in.}^3$$

and the maximum bending stress is found to be

$$f_b = \frac{M}{S} = \frac{250 \times 12}{162} = 18.5 \text{ psi}$$

Adding this stress to the axial compression stress produces

$$f = f_a + f_b = 28.2 \pm 18.5$$

$$= 46.7 \text{ psi in net compression (maximum)}$$

$$= 9.7 \text{ psi in net compression (minimum)}$$

(See Case 1 in Fig. 11.5.)

This is obviously not a critical stress condition, especially since the load combination including wind permits the allowable stress of 160 psi (see Example 1) to be increased by one-third.

As previously described, the wind load produces a simple beam-bending action in the wall with a maximum bending moment at midheight, as shown in Fig. 37.4a. The

stress combination just determined is thus conservative, since it uses the maximum compression at the base of the wall. However, the low-stress condition does not justify a more accurate investigation.

Bending moments in walls can also be induced by other loading conditions. A common situation is that shown in Fig. 37.4b, in which a supported gravity load is placed off the wall centroid (in this case, the center of the wall). This may occur when a wall is continuous past a floor level in multistory construction or when a wall forms a parapet by extending above a roof level. In these cases, a bending moment is created with a magnitude equal to the product of the load and its distance of eccentricity from the centroid of the wall cross section. The following example illustrates such a situation.

Example 4. Assume that the supported load of 12,000 lb in Example 2 is placed on a bracket attached to the wall so that the center of the loading is 8 in. from the center of the wall. Investigate the condition for combined bending and axial stresses.
Solution. The average unit compressive stress was determined in Example 2 to be 34.0 psi, assuming the effective wall segment of 9 in. × 54 in. The bending moment is

$$M = Pe = (12,000)(8) = 96,000 \text{ in.-lb}$$

The section modulus for the 9-in. × 54-in. wall segment is

$$S = \frac{bd^2}{6} = \frac{(54)(9)^2}{6} = 729 \text{ in.}^3$$

and the maximum bending stress is

$$f_b = \frac{M}{S} = \frac{96,000}{729} = 131.7 \text{ psi}$$

The limiting combined stresses are thus

$$f = f_a + f_b = 34.0 \pm 131.7$$

$$= 165.7 \text{ psi in net compression}$$

$$= 97.7 \text{ psi in net tension}$$

(See Case 3 in Fig. 11.4.)

The investigation indicates that the allowable compression stress of 160 psi is slightly exceeded. More critical, however, is the tension stress. Table 36.2 yields allowable tension stress in flexure of 20 psi or 40 psi, depending on the job conditions regarding inspection (as described in the code). Even if the higher value is permitted, the eccentric loading produces a situation of considerable overstress. Remedial design considerations in this case include the possibilities for modifying the support details to reduce the amount of eccentricity of the load, using a pilaster at the concentrated load (as described in Sec. 37.5), or development of the construction with steel reinforcement, as described in Sec. 37.4.

The load eccentricity described in Example 4 is typically produced by the use of supporting devices attached to

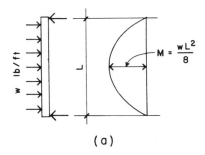

(a)

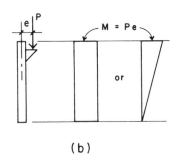

(b)

FIGURE 37.4. Bending moment in walls: (*a*) from wind or soil pressure; (*b*) from eccentric compression load.

the face of a wall, such as that shown in Fig. 37.5a. The advantage of this type of construction is that it generally permits an undisturbed, continuous development of the wall, with only minor accommodation for anchor bolts or other anchoring devices. The disadvantage is in the degree of bending induced when the supported load is of considerable magnitude.

In earlier times, with thick masonry walls, it was common to use a recessed pocket in the wall to permit supported beams or ends of trusses to slightly penetrate the wall in order to get closer to the wall center and avoid the bending moments due to an eccentric load. This technique is not used as often today due to increased use of reinforced or veneered forms of construction.

Figure 37.5b shows a possibility for creating a pocket in one wythe of the two-wythe brick wall, producing a bearing area about 5 in. deep in the 9-in.-thick wall. This may be sufficient for light loads, such as those from closely spaced floor joists. For heavier loads, it may be possible to use a steel unit built into the wall (called a *beam box*) to spread

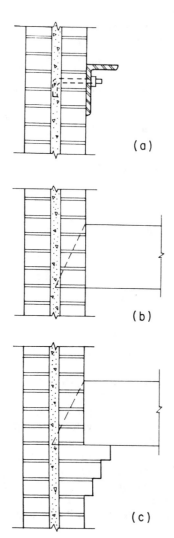

FIGURE 37.5. Beam supports in brick walls.

the bearing load. It may also be possible to corbel (progressively cantilever) the face of the masonry, as shown in Fig. 37.5c, to create a wider bearing, although this increases the eccentricity of the load.

A standard detail for pocketed beams is the fire-cut end, which is intended to permit the beam to fall without tearing the wall by its end rotation.

37.4. REINFORCED BRICK MASONRY

Brick masonry is sometimes used with steel reinforcement, emulating the nature of reinforced concrete or the more frequently used reinforced masonry with CMUs. Reinforced walls typically consist of two-wythe construction with an internal cavity of sufficient width to permit the insertion of two-way steel rods in the cavity space. The cavity is then filled with concrete as the construction proceeds.

A minimum wall is usually approximately 9 in. thick, assuming a wythe thickness of 3.75 in. and a cavity width of 1.5 in. The cavity dimension is a bare minimum to permit the steel bars to pass each other, if they are as large as No. 5 bars (approximately 0.625 in. diameter).

Various code requirements must be satisfied if the construction is to qualify as reinforced in the codes' definitions of the class of construction. However, reinforcement can also be used simply for enhancement for local stress conditions, even with what is otherwise "unreinforced" construction.

Consider the situation of Example 4 in the preceding section, in which an excess of bending was determined to be created by the eccentric load. If some steel reinforcement is inserted in the grouted cavity at the location of the load, as shown in the wall plan section in Fig. 37.6, it is possible to consider the resistance to flexural tension as being developed solely by the steel, which is traditionally assumed in design of reinforced concrete. The following example demonstrates such a design process.

Example. Redesign the wall for the conditions described in Example 4 in Sec. 37.3, using vertical steel reinforcement in the wall cavity to develop resistance to the flexural tension. (See Fig. 37.6.)

Solution. The vertical steel rods will tend to reinforce the wall for both the bending and the vertical compression. Since the wall without the steel was found to be almost adequate for the combined compression stresses in Example 4, the assistance provided by the steel for column action will certainly suffice in this regard. What remains is to consider the tension. Actually, the steel will be slightly precompressed by the vertical compression, but for a conservative design we will ignore this effect.

Using the bending moment determined in the previous example, a value of 20,000 psi for allowable tension stress in the steel, an effective depth of 4.5 in. (assuming the rods in the center of the wall), and the basic working-stress formula for flexural tension in reinforced concrete, we find

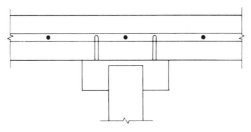

FIGURE 37.6. Plan of the reinforced wall at the beam.

$$A_s' = \frac{M}{f_s jd}$$

$$= \frac{96,000}{(20,000)(0.9)(4.5)} = 1.185 \text{ in.}^2$$

A conservative choice would be three No. 6 (approximately 0.75 in. diameter; see Table 26.3) rods, providing a total area of

$$A_s = (3)(0.44) = 1.32 \text{ in.}^2$$

Reinforced masonry is developed more frequently with CMUs. Design of reinforced construction is demonstrated more extensively in Chapter 36, although many of the procedures used there are generally applicable to brick construction as well.

37.5. MISCELLANEOUS BRICK CONSTRUCTION

Brick can be used for many purposes, other than for walls of buildings. Past and current uses include foundation walls, short retaining walls, freestanding fences, columns, pedestals, fireplaces, and chimneys. Construction may be all brick, or, when only some faces are exposed to view, a composite of brick and other materials: CMUs, concrete, or stone. This section considers some of these uses for brick masonry.

Columns

Built-up brick columns usually begin with the smallest size, which consists of two bricks per layer, as shown in Fig. 37.7a. The column side dimension in this case will be determined by the brick width dimension plus the mortar joint thickness. There is no real opportunity for installation of vertical steel rods, so this is strictly an unreinforced member; add the small size, and it is not likely to be used for major loads.

A pinwheel arrangement of the courses will produce the column shown in Fig. 37.7b, resulting in a small center cavity. A single steel rod might be placed in this cavity, but to create a real reinforced member it is usually necessary to use the arrangement shown in Fig. 37.7c. The minimum, officially reinforced brick column is therefore about 16 in. wide.

A slightly narrower reinforced column can be produced if the bricks are laid with the narrow side (face) down, as shown in Fig. 37.8. This form may be structurally acceptable, but appearance may be questionable, since bricks are usually produced with the intent of having their faces exposed, and the flat sides may have a significantly cruder finished surface.

Unreinforced columns are generally questionable, unless of very stout proportions (height-to-width ratio of, say, 10 or less). A significant usage, however, may be for pedestals, which are actually very short columns (usually de-

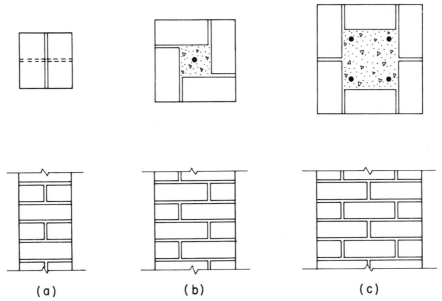

(a)   (b)   (c)

FIGURE 37.7. Forms of brick columns.

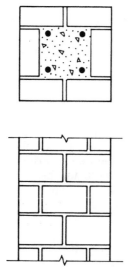

FIGURE 37.8. Column formed with bricks face down.

fined by codes as having a height of less than three times their width).

Design of unreinforced brick columns and pedestals is generally accomplished with the same procedures used for walls, as demonstrated in the examples in Sec. 37.3. Design of reinforced columns is done with code-specified procedures, generally adapted from those used with reinforced concrete.

A special column is the pilaster, consisting essentially of a bulged-out portion of a continuous wall. It may occur at a corner, at the edge of an opening, at the end of a wall, or at some intermediate point in a wall. The functional purposes for pilasters are usually the provision of some reinforcement for the otherwise relatively thin wall. They may serve to receive some large, concentrated load or simply to brace the wall when slenderness is a problem. Very tall walls may be braced with closely spaced pilasters, so that the unbraced distance for the wall becomes the pilaster spacing, rather than the wall height.

A frequent use of pilasters is that of receiving a load at the wall face, such as that described in Example 4 in Sec. 37.3. The portion of the pilaster extending beyond the wall face can be used to support the load, thus eliminating the need for penetration of the supported member into the wall. The pilaster also reinforces the wall for the concentrated load and any bending effects it produces.

When built monolithically with the wall, the pilaster column consists of the extended portion plus some segment of the wall. It may be possible to consider a T-shaped effective column cross section, although for a conservative design it is customary to consider only the portion of the wall equal to the pilaster width. General design of pilasters is the same as for freestanding columns, except for the consideration of the lateral bracing afforded by the wall in the direction of the wall plane (even as the wall sheathing braces a 2 x 4 stud on its weak axis).

Large brick columns, as well as large piers and abutments, are most likely to be built with only a facing of bricks. The major interior portion of the mass is most likely to be cast-in-place concrete, although CMUs or rubble stone may be options for some situations.

Arches

The arch was undoubtedly invented in rough stone and eventually refined in cut stone and brick construction. Massive arches, vaults, and domes were eventually achieved in many cultures. The brick arch saw major use in the spanning of wall openings for doors, windows, and arcades. Many basic forms were developed, a sampling of which is shown in Fig. 37.9, which is reproduced from an early edition of *Architectural Graphic Standards* (Ref. 13.).

Long-span arches, vaults, and domes are now achieved almost entirely with other forms of construction. Short-span arches for windows, doors, or arcades can be developed in brick masonry, but are most likely to be done so only for decorative purposes. The flat span is now most common, and although a form is used in masonry (see Fig. 37.9) a reinforcement of some kind is usually built into the construction.

Arches can still be built of masonry, and when the construction is otherwise generally of structural masonry with bricks or stone, short-span arches may be feasible. Two major concerns must be noted for arches. The first has to do with the horizontal, outward thrust at the base. This may not be a problem when the arch occurs within a large continuous wall, but must be considered for other situations, or when an opening is quite close to the end of a wall.

The other primary concern for arches is for the rise-to-span ratio. This should be as high as possible, preferably approaching that for a semicircle, as opposed to a flat profile.

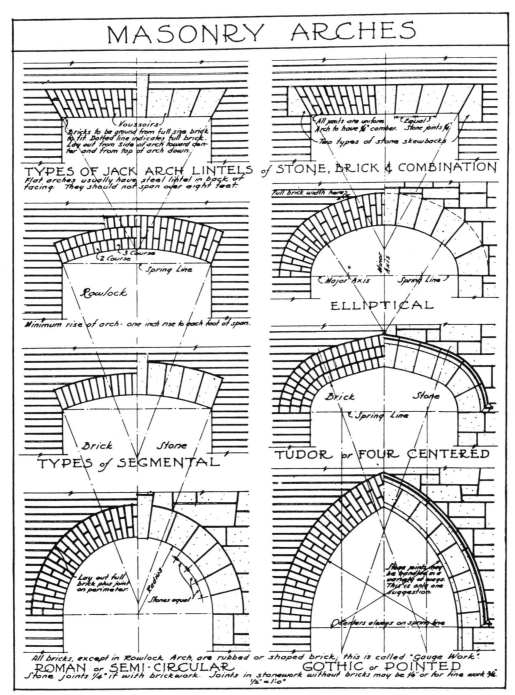

FIGURE 37.9. Forms of masonry arches in walls. Reprinted from *Architectural Graphic Standards,* 3rd edition, 1941, with permission of John Wiley & Sons, Inc., New York.

CHAPTER THIRTY-EIGHT

Miscellaneous Masonry Construction

Masonry construction has been, and still can be, produced with many different materials in various forms. The principal topic of this book is structural masonry as presently used in the United States, which is represented primarily in the treatments in Chapters 36 and 37. Nevertheless, there are other forms of masonry in use for special purposes or simply because of enduring attachments to historic procedures and symbols. Chapter 38 presents some of the other forms of masonry construction that have endured.

38.1. STONE MASONRY

Prior to this century stone was used as a major structural building material. Now it is mostly either too expensive, too scarce, or too difficult to work with in comparison to alternative structural materials. However, real stone remains a popular material for building exteriors, so extensive use is made in the form of surfacing materials. Chapter 38 deals primarily with current structural uses for stone; which is no longer the major use of the material.

Rubble and Fieldstone Construction

The earliest uses of stone were undoubtedly in the form of rock piles made with stones as found in nature. Fitting the stones together to form a stable pile surely required a long patient search for the right size and shape of stone as the pile grew. Even today, stone structures are best achieved with some attention to fundamental principles slowly developed by skillful pilers of rocks. Some of these basic tricks, as shown in Fig. 38.1, are the following:

1. Most stones should be of a flat or angular form to minimize the tendency for upper stones to roll off of lower ones.
2. Vertical joints in successive layers (courses) should be offset rather than aligned. This helps develop a certain horizontal continuity of the structure.
3. As with a pile of any generally loose material (soil, sand, pebbles), a tapered profile with a wider base is preferred for stability. If a single-direction lateral force must be resisted, the pile may be leaned in opposition to the force. (See Fig. 38.1e.)

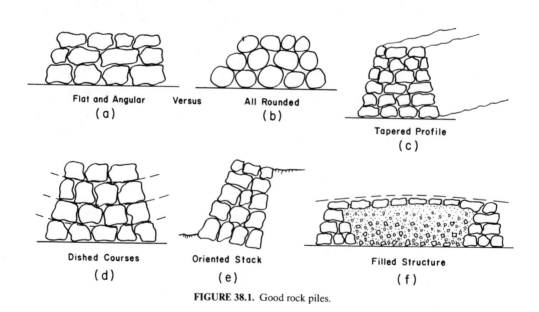

FIGURE 38.1. Good rock piles.

4. If the stack is tall, the horizontal layers should be dished slightly (Fig. 38.1*d*) so that the whole stack leans slightly to the inside.

5. Very wide stacks may be filled (Fig. 38.1*f*), with larger stones reserved for sides and topping. Most of the filler should be coarse, granular material (broken rock, gravel, coarse sand), but top and sides may be filled with some clay materials to seal the stack (ancient form of crude mortar).

As the ancient builders learned, the best stone structures are those that maintain a stable equilibrium without assistance. Mortar should be used primarily to fill voids after the rocks are settled in place. Once hardened, the mortar may further stabilize the pile, but it should not be used initially to prop up the stones.

Stones may be used as found, or they may be shaped. Depending on the source, natural forms may be rounded, angular, or flat. All forms may be used, but the angular and flat shapes will usually produce more stable structures. Rounded shapes may be used to fill spaces between some stones, but should not be used extensively.

Shaping may be minor and crude (just breaking or chipping), or it may be done with great precision and accuracy. Construction with natural or minor shaped stones is called rubble. Work done with stones shaped to reasonably accurate rectangular forms is called *ashlar.*

If stones are laid in precise horizontal layers, the work is described as coursed. If there is no particular attempt to achieve layers within the mass of the pile, the work is described as random. The general combinations of rubble, ashlar, coursed, and random are shown in Fig. 38.2. Variations and combinations are possible, and complex structures may use many forms and types of stone and great variety in arrangements.

Unreinforced Construction

Stone structures for buildings must generally be built with the same care and requirements as required for construction with bricks or CMUs. Code specifications and data can be used for work that follows good construction practices. The resulting construction can be as structurally sound as other forms of unreinforced masonry.

Stone structures will generally be somewhat thicker and heavier than those of bricks or CMUs. Joining of the stone work with other elements of the building construction must be achieved with details that allow for the somewhat rough dimensional accuracy of the stone construction. Installation of anchor bolts or other attachment devices may require some greater effort in random rubble construction.

For long walls some horizontal reinforcement should be used. If horizontal courses are achieved, ordinary joint reinforcement may be used. Otherwise, or as an alternative, bond beams may be developed as with CMU construction.

A bond beam should be used at the top of a wall. If a stone top is desired, the bond beam may be in the second course. If the wall provides bearing for a supported structure, however, the top should be developed as a cast concrete member for this purpose; serving to provide better bearing and to distribute the loads to the stone work.

Reinforced Construction

Stone walls that carry major loads are best developed as reinforced structures. These may take various forms, but the two shown in Fig. 38.3 are frequently used. The wall shown in Fig. 38.3*a* is developed in the same general way as a fully grouted, reinforced brick wall. A significant cavity is

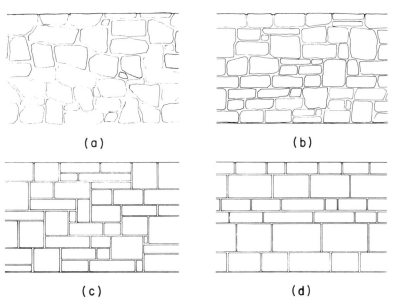

(a)

(b)

(c)

(d)

FIGURE 38.2. Patterns of stone masonry: (*a*) random rubble; (*b*) coursed rubble; (*c*) random ashlar; (*d*) coursed ashlar. Reproduced from *Fundamentals of Building Construction* (Ref. 16), with permission of John Wiley & Sons, Inc., New York.

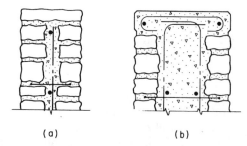

FIGURE 38.3. Reinforced stone walls: (*a*) as fully grouted (solid) masonry; (*b*) as a faced concrete wall.

formed in the center of the wall and regularly spaced vertical and horizontal bars are grouted into the cavity.

The wall in Fig. 38.3*b* forms a significantly wider cavity that is essentially filled with a sitecast reinforced concrete wall. The stone in this case is developed as faced construction, working as a composite structure with the reinforced concrete in the cavity. For a conservative design, such a structure may be designed simply as a concrete wall, ignoring the structural capacity of the stone.

For any form of construction, including that which is basically unreinforced, it is advisable to provide some steel reinforcement around openings and at wall ends, corners, and intersections.

Cut Stone Construction

Construction of masonry structures with large cut stones (quarried granite, etc.) is now rarely done, except in restoration of old buildings. The source material is generally too expensive to use in this bulk form, and other means of creating structures are much more economical and less craft-intensive. Cut stone is almost entirely used for veneers, often with supporting structures of steel or reinforced concrete.

Where stone is plentiful and other technologies in limited use, it may be possible to use cut stone to generate structures, but they will generally be most feasible in the form shown in Fig. 38.3*b*. If the stone indeed forms a mold for the cast concrete, anchorage of the stone to the concrete will most likely be achieved with metal ties set into slots in the stone or into joints between the stone units. Because of the potential for damage to the stone during the pouring of the concrete, however, it is still probably more feasible to develop such a structure as a veneered one, with the concrete wall cast first and the stone anchored to it later.

Construction Without Mortar

Low retaining structures for site construction are sometimes built of stone without mortar. These may be of a general rubble form in random pattern or achieved with flat stones in coursed layers. Lack of need to relate to supported elements or other construction in general makes the general dimensional stability of the construction much less critical.

The unmortared construction may actually be a better choice in many cases, since it can adjust in minor ways as it settles into place without development of the cracking that often occurs in similar structures of brick or CMUs. The structure must nevertheless keep some level of stability to remain functional, and the same amount of care is required in fitting the units as with mortared stone work.

Another advantage of the stone structure without mortar is its ability to drain water through the structure, preventing the buildup behind the retaining structure, which is a concern for solid walls in the same situation.

Although we do not now consider using stone construction without mortar for buildings, this was the basic form of the construction in ancient times.

38.2. ADOBE

Adobe is a term that refers to a masonry unit (a sun-dried unburned brick), the construction process in which it is used, and the building produced with the process. It is a very ancient form of construction, probably the antecedent of all other forms of brick construction.

Adobe bricks are simply produced with soil materials that somewhat emulate a good concrete mix. There must be a binder (cement–clay) and preferably a graded aggregate (for adobe a mixture of silt and fine-to-coarse sand). This just happens to be the normal constituency of the surface soils in most temperate, arid regions, exactly where abode construction thrives. So, if you live in Arizona and want to make adobe bricks, just go out in your backyard and scoop up some dirt, mix a little water with it, and make bricks.

To make a wall, you lay out a row of bricks, fill the cracks between them with the same mud you used to make the bricks, spread a layer of mud on top of them, and lay another row of bricks on top of the mud, continuing the process until the wall is as high as you want it. Normal time required for the hand labor (including siestas and festivals) will ensure that lower courses dry well, and that shrinkage in the slowly drying mass is steadily and incrementally accumulative, resulting in a very stable structure.

Figure 38.4 is a reproduction of a page from the third edition (1941) of *Architectural Graphic Standards*, presenting a modern version of the ages-old process with a few high-tech touches, such as steel industrial windows, flashing, anchor bolts, and strengthening with reinforced concrete members. Also indicated are the basic techniques of protecting the soft, moisture-susceptible adobe with an exterior coating of stucco (cement plaster) and reinforcing the stucco with good old chicken wire. Pragmatic references to "cheap" and "good" construction are also used for communication with the practical builder.

It is interesting to note that the latest (eighth) edition of *Architectural Graphic Standards* (Ref. 13) has three pages of adobe construction.

A variation of adobe brick construction is the use of rammed earth, in which the same basic material (sand–

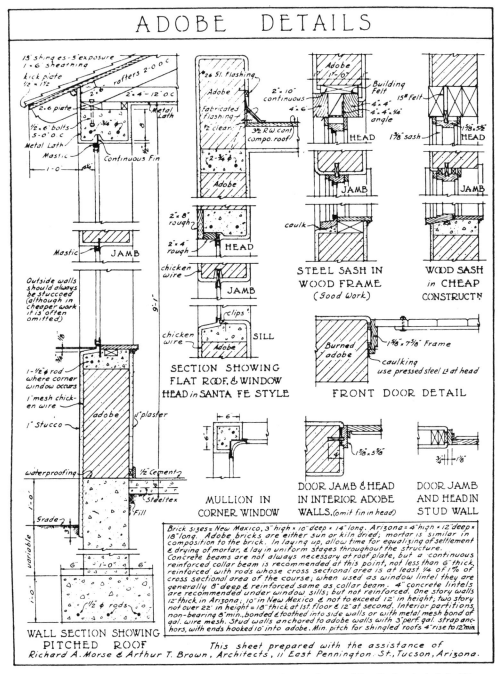

FIGURE 38.4. Details for adobe construction. Reproduced from *Architectural Graphic Standards*, 3rd edition, 1941, with permission of John Wiley & Sons, Inc., New York.

silt–clay) is used in a manner similar to sitecast concrete. Forms are erected, and shallow layers of the soil are tamped (rammed) into the forms to form a dense mass. A variation on this is the use of soil–cement, consisting of the addition of a small quantity of portland cement to the soil mixture. Soil–cement construction may also be used to produce foundations for ordinary adobe buildings.

Adobe is fundamentally brick construction, and all of the ideas and tricks for improving and strengthening ma-

sonry in general can be used and in fact were probably largely learned originally in building with adobe. Use of pilasters, lintels, general strengthening of openings, corners, and wall intersections, and bonding of multiple wythes apply equally with adobe construction.

Bricks for adobe construction are ordinarily made for use in single-wythe walls. The standard brick is therefore quite wide, usually 10–12 in. It is also usually made as large as possible, to reduce the number required and increase the speed of the construction. The unit size therefore more closely relates to that of current CMUs, and a popular form of CMU construction today uses 4-in. or 6-in.-high units with a bulging side form (called *slump block*) that consciously imitates the appearance of ancient construction with adobe.

Because of its enduring popularity, modern building codes usually give some acknowledgment to adobe and provide some criteria for its design and construction. (See Sec. 2407(i)9 in the 1991 *UBC*.)

38.3.   HOLLOW CLAY TILE

Man-made masonry units are generally formed of fired clay or cast concrete. Small units are generally solid, while large units are made with significant voids. In earlier times, fired clay was used to produce large, voided units, called *tiles* or *tile blocks*. Although now largely displaced by CMUs, the clay units were forerunners of present CMUs, and much of the detail of present CMU construction was originally developed with clay tile units.

A special form of clay tile unit is referred to as *architectural terra-cotta*. These consist of units with glazed surfaces, intended for exposure to view. They were developed as economical imitations of cut stone and widely used for elaborate cornices and other decorative features on buildings in the form of architecture that prevailed in the nineteenth century and early part of the twentieth century. Since this coincided with a major use of masonry, in general, the use of clay tile was perpetuated until other materials were developed to imitate it.

Fired clay can quite easily attain strength superior to most cast concrete, so clay tile units were characteristically quite thin and light despite the relatively large size of units. The large units, however, resulted in a rather inflexible modular system, so finished wall dimensions were often achieved by patching in with bricks at ends, tops, and openings.

PART 7

BUILDING FOUNDATIONS

Almost all buildings, regardless of their size, shape, intended purpose, type of construction, or geographic location, share a common problem: They must rest on the ground. Thus the design of adequate foundations is a general problem in building design. Since each building site is unique in terms of specific geological conditions, each building foundation presents unique design problems.

CHAPTER THIRTY-NINE

General Considerations

Chapter 39 summarizes the general issues involved in foundation design, the properties and behavioral characteristics of foundation materials of significance for design work, and the problems of establishing useful design data and criteria.

39.1. BASIC PROBLEMS IN FOUNDATION DESIGN

The design of the foundation for a building cannot be separated from the overall problems of the building structure and the building and site designs in general. Nevertheless, it is useful to consider the specific aspects of the foundation design that must be dealt with.

Site Exploration

For purposes of the foundation design, as well as for the building and site development in general, it is necessary to know the actual site conditions. This investigation usually consists of two parts: determination of the ground surface conditions, and of the subsurface conditions. The surface conditions are determined by a site survey that establishes the three-dimensional geometry of the surface and the location of various objects and features on the site. Where they exist, the location of buried objects such as sewer lines, underground power and telephone lines, and so on, may also be shown on the site survey.

Unless they are known from previous explorations, the subsurface conditions must be determined by penetrating the surface to obtain samples of materials at various levels below the surface. Inspection and testing of these samples in the field, and possibly in a testing lab, is used to identify the materials and to establish a general description of the subsurface conditions.

Site Design

Site design consists of positioning the building on the site and the general development, or redevelopment, of the site contours and features. The building must be both horizontally and vertically located. Recontouring the site may involve both taking away existing materials (called *cutting*) and building up to a new surface with materials brought in or borrowed from other locations on the site (called *filling*).

Development of controlled site drainage for water runoff is an important part of the site design.

Selection of Foundation Type

The first formal part of the foundation design is the determination of the type of foundation system to be used. This decision cannot normally be made until the surface and subsurface conditions are known in some detail and the general size, shape, and location of the building are determined. In some cases it may be necessary to proceed with an approximate design of several possible foundation schemes so that the results can be compared.

Design of Foundation Elements

With the building and site designs reasonably established, the site conditions known, and the type of foundation determined, work can proceed to the detailed design of individual structural elements of the foundation system.

Construction Planning

In many cases the construction of the foundation requires a lot of careful planning. Some of the possible problems include conditions requiring dewatering the site during construction, bracing the sides of the excavation, underpinning adjacent properties or buildings, excavating difficult objects such as large tree roots or existing constructions, and working with difficult soils such as wet clays, quick sands or silts, soils with many large boulders, and so on. The feasibility of dealing with these problems, primarily in terms of cost and delays, may influence the foundation design as well as the positioning of the building on the site and the general site development.

Inspection and Testing

During the design and construction of the foundation there are several times when it may be necessary to perform inspection or testing. Whether done by the designer or by others, the results of the inspections and tests will be used to influence design decisions or to verify the adequacy of the completed designs or construction. The need for this work will depend on the size of the building, the type of construction, the specific subsurface conditions, the type

389

of foundation system, and the various problems encountered during construction. Some of the ordinary inspections or tests are as follows:

Preliminary Site Investigation. The preliminary investigation usually consists of a site survey and some minimal subsurface investigation prior to the construction and often prior to the final design of the foundation. For major projects or difficult subsurface conditions it is usually necessary to have this information even before the preliminary site design and building design can be done.

Detailed Site Design. In some cases it is necessary to have additional information prior to the final design or the construction of the foundation. In some instances it is possible to incorporate this investigation with the early stages of the foundation work, with any necessary design adjustments made as the work progresses.

Inspection and Testing during Construction. At a bare minimum the completed excavation should be visually inspected prior to any construction to verify that the actual conditions encountered are those assumed for the design. In some cases the site conditions, the type of foundation, or the nature of the building may require extensive and continuous inspection and testing throughout the foundation construction process. Inspections by both the designer and the permit-granting agency may be required.

Inspection and Testing after Construction. In some cases it may be necessary to perform inspection and testing after the foundation construction is complete. This is usually required where progressive soil deformation is anticipated over time or with seasonal changes.

Remedial Alterations

For various reasons it is often necessary to modify the foundation in some way from the original design. This is best done prior to construction, of course, but must sometimes be done as repair or renovation. The remedial measures may be obvious and simple to accomplish, or may require the best efforts of the most-qualified experts. Some of the situations that may require remedial alterations are:

Unanticipated Subsurface Conditions. Where the site conditions are very nonuniform or the preliminary investigations sketchy, or for other reasons, it may be necessary to modify the design due to actual encountered conditions.

Unanticipated Construction Problems. Weather conditions, unusual excavation problems, unavoidable delays, and a host of other possibilities may necessitate expedient change of the design.

Construction Errors. Foundation construction is usually done under the crudest and sloppiest of working conditions. Great accuracy and perfection is not to be expected. Overexcavation, mislocation of elements, errors in dimensions, omission of details, and so on, are common.

Inadequate Performance of the Foundation. During construction, or even at some time after completion of the building, there may be evidence of excessive settlement, uneven settlement, horizontal shifting, tilting, or other forms of foundation failure.

39.2. SOIL CONSIDERATIONS RELATED TO FOUNDATION DESIGN

The principal properties and behavior characteristics of soils that are of direct concern in foundation design are the following:

Strength. For bearing-type foundations the main concern is resistance to vertical compression. Resistance to horizontal pressure and to friction are of concern when foundations must resist the force of wind, earthquakes, or retained earth.

Strain Resistance. Deformation of soil under stress is of concern in designing for limitations of the movements of foundations, such as the vertical settlement of bearing foundations.

Stability. Frost action, fluctuations in water content, seismic shock, organic decomposition, and disturbance during construction are some of the things that may produce changes in physical properties of soils. The degree of sensitivity of the soil to these actions is called its *relative stability*.

Properties Affecting Construction Activity

A number of possible factors may affect construction activity, including the following:

The relative ease of excavation.

Ease of and possible effects of site dewatering during construction.

Feasibility of using excavated materials as fill material.

Ability of the soil to stand on a vertical side of an excavation.

Effects of construction activity—notably the movement of workers and equipment—on unstable soils.

Miscellaneous Conditions

In specific situations various factors may affect the foundation design or the problems to be dealt with during construction. Some examples are the following:

Location of the water table, affecting soil strength or stability, need for waterproofing basements, requirement for dewatering during construction, and so on.

Nonuniform soil conditions on the site, such as soil strata that are not horizontal, strips or pockets of poor soil, and so on.

Local frost conditions, affecting the depth required for bearing foundations and possible heave and settlement of exterior pavements.

Deep excavation or dewatering operations, possibly affecting the stability of adjacent properties, buildings, streets, and so on.

All of these concerns must be anticipated and dealt with in designing buildings and in planning and estimating construction costs. Persons charged with responsibility for design and planning foundation construction must have some understanding of the characteristics of ordinary soils so that they can translate information about site conditions into usable data. The discussions that follow deal with the basic nature of soils of various types, the behavior and design considerations of various foundation elements and systems, and the means for obtaining and using information about specific site conditions.

39.3. FOUNDATION DESIGN CRITERIA

For the design of ordinary bearing-type foundations several structural properties of a soil must be established. The principal values are the following:

Allowable Bearing Pressure. This is the maximum permissible value for vertical compression stress at the contact surface of bearing elements. It is typically quoted in units of pounds or kips per square foot of contact surface.

Compressibility. This is the predicted amount of volumetric consolidation that determines the amount of settlement of the foundation. Quantification is usually done in terms of the actual dimension of vertical settlement predicted for the foundation.

Active Lateral Pressure. This is the horizontal pressure exerted against retaining structures, visualized in its simplest form as an equivalent fluid pressure. Quantification is in terms of a density for the equivalent fluid given in actual unit weight value or as a percentage of the soil unit weight.

Passive Lateral Pressure. This is the horizontal resistance offered by the soil to forces against the soil mass. It is also visualized as varying linearly with depth in the manner of a fluid pressure. Quantification is usually in terms of a specific pressure increase per unit of depth.

Friction Resistance. This is the resistance to sliding along the contact bearing face of a footing. For cohesionless soils it is usually given as a friction coefficient to be multiplied by the compression force. For clays it is given as a specific value in pounds per square foot to be multiplied by the contact area.

Whenever possible, stress limits should be established as the result of a thorough investigation and the recommendations of a qualified soils engineer. Most building codes allow for the use of *presumptive* values for design. These are average values, on the conservative side usually, that may be used for soils identified by groupings used by the codes. Reprints of portions of the *UBC*, 1991 edition, and the *Building Code of the City of Los Angeles*, 1976 edition, are given in Appendix D; both contain presumptive values for design. Soil types are identified only rather broadly in the *UBC*, whereas the Los Angeles code uses what is essentially the unified system in establishing allowable bearing pressures.

CHAPTER FORTY

Soil Properties and Foundation Behavior

Information about the materials that constitute the earth's surface is forthcoming from a number of sources. Persons and agencies involved in fields such as agriculture, landscaping, highway and airport paving, waterway and dam construction, and the basic earth sciences such as geology, mineralogy, and hydrology have generated research and experience that is useful to the field of foundation engineering. Chapter 40 consists of a brief summary of the issues and data regarding soil materials and behavior that directly concern the designer of building foundations. It will provide the reader with a general understanding of the problems of soil identification and the means for establishing criteria for foundation design.

40.1. SOIL PROPERTIES AND IDENTIFICATION

A general distinction can be made between two basic materials: *soil* and *rock*. At the extreme the distinction is clear, loose sand versus solid granite, for example. A precise distinction is somewhat more difficult, however, since some highly compressed soils may be quite hard, while some types of rock are quite soft or contain many fractures, making them relatively susceptible to disintegration. For practical use in engineering, soil is generally defined as material consisting of discrete particles that are relatively easy to separate, while rock is any material that requires considerable brute force for excavation.

A typical soil mass is visualized as consisting of three parts, as shown in Figure 40.1. The total soil volume is taken up partly by the solid particles and partly by the open spaces between the particles, called the *void*. The void is typically filled partly by liquid (usually water) and partly by gas (usually air). There are several soil properties that can be expressed in terms of this composition, such as the following:

Soil Weight (γ). Most of the materials that constitute the solid particles in ordinary soils have a unit density that falls within a narrow range; expressed as specific gravity, the range is from 2.60 to 2.75. Sands typically average about 2.65; clays about 2.70. Notable exceptions are soils containing large amounts of organic materials. Specific gravity refers to a comparison of the density to that of water, usually considered to weigh 62.4 lb/ft³.

Thus for a dry soil sample the soil weight may be determined as follows:

soil unit weight = (% of solids)(specific gravity)
(unit weight of water)

Thus for a sandy soil with a void of 30%, the weight may be approximated as follows:

$$\text{soil weight} = \gamma = \frac{70}{100}(2.65)(62.4) = 116 \text{ lb/ft}^3$$

Void Ratio (e). Instead of expressing the void as a percentage, as was done in the preceding example, the term generally used is the void ratio, *e*, which is defined as follows:

$$e = \frac{\text{volume of the void}}{\text{volume of the solid}}$$

In practice the void ratio is often determined by using the relationship between soil weight, specific gravity of the solids, and percentage of the void, as follows: If

$$\gamma = \frac{\text{measured dry weight of the sample}}{\text{measured volume of the sample}} = 116 \text{ lb/ft}^3$$

then, assuming a specific gravity (G_s) of 2.65,

$$\gamma = \frac{\% \text{ of solids}}{100}(2.65)(62.4) = 116 \text{ lb/ft}^3$$

$$\% \text{ of solids} = \frac{116}{2.65(62.4)}(100) = 70\%$$

$$\% \text{ of void} = 100 - 70 = 30\%$$

and with the volume expressed as a percentage,

$$e = \frac{\text{volume of the void}}{\text{volume of the solid}} = \frac{30}{70} = 0.43$$

Porosity (n). The actual percentage of the void is expressed as the porosity of the soil, which in coarse-grained soils (sands and gravels) is generally an indication of the rate at which water flows through or drains from the soil. The actual water flow is deter-

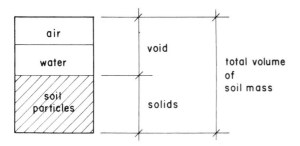

FIGURE 40.1. Three-part composition of a soil mass.

mined by standard tests, however, and is described as the relative *permeability* of the soil. Porosity, when used, is simply determined as

$$n(\text{percent}) = \frac{\text{volume of the void}}{\text{total soil volume}}(100)$$

Thus for the preceding example, $n = 30\%$.

Water Content (w).   The amount of water in a soil sample can be expressed in two ways: by the water content (w) and by the saturation (S). They are defined as follows:

$$w(\text{in percent}) = \frac{\text{weight of water in the sample}}{\text{weight of solids in the sample}}(100)$$

The weight of the water is simply determined by weighing the wet sample and then drying it to find the dry weight.

The saturation is expressed in a ratio, similar to the void ratio, as follows:

$$S = \frac{\text{volume of water}}{\text{volume of void}}$$

Full saturation ($S = 1.0$) thus occurs when the void is totally filled with water. Oversaturation ($S > 1.0$) is possible in some soils when the water literally floats some of the solid particles, increasing the void above that in the partly saturated soil mass.

In the preceding example, if the soil weight of the sample as taken at the site was found to be 125 lb/ft³, the water content and saturation would be as follows:

weight of water = (original sample weight)

$$- \text{(dry weight)}$$

$$= 125 - 116 = 9 \text{ lb/ft}^3$$

Then

$$w = \frac{9}{116}(100) = 7.76\%$$

The volume of water may be found as

$$V_w = \frac{\text{weight of water in sample}}{\text{unit weight of water}} = \frac{9}{62.4}$$

$$= 0.144 \quad \text{or} \quad 14.5\%$$

Then

$$S = \frac{14.4}{30} = 0.48$$

The size of the discrete particles that constitute the solids in a soil is significant with regard to the identification of the soil and the evaluation of many of its physical characteristics. Most soils have a range of particles of various sizes, so the full identification typically consists of determining the percentage of particles of particular size categories.

The two common means for measuring grain size are by sieve and sedimentation. The sieve method consists of passing the pulverized dry soil sample through a series of sieves with increasingly smaller openings. The percentage of the total original sample retained on each sieve is recorded. The finest sieve is a No. 200, with openings of approximately 0.003 in. A broad distinction is made between the total amount of solid particles that pass the No. 200 sieve and those retained on all the sieves. Those passing are called the *fines* and the total retained is called the *coarse fraction*.

The fine-grained soil particles are subjected to a sedimentation test. This consists of placing the dry soil in a sealed container with water, shaking the container, and measuring the rate of settlement of the particles. The coarser particles will settle in a few minutes; the finest will take several days.

Figure 40.2 shows a graph that is commonly used to record the grain size characteristics for soils. A log scale is used for the grain size, since the range is quite large. The common soil names, based on grain size, are given at the top of the graph. These are approximations, since some overlap occurs at the boundaries, particularly for the fines. The distinction between sand and gravel is specifically established by the No. 4 sieve, although the actual materials that constitute the coarse fraction are sometimes the same across the grain size range. The curves shown on the graph are representative of some particularly characteristic soils, described as follows:

A *well-graded* soil consists of some significant percentages of a wide range of soil particles.

A *uniform* soil has a major portion of the particles grouped in a small size range.

A *gap-graded* soil has a wide range of sizes, but with some concentrations of single sizes and small percentages over some ranges.

These size-range characteristics are specifically established by using some actual numeric values from the size-range graph. The three size values used are points at which the curve crosses the percent lines for 10, 30, and 60%. The values are interpreted as follows:

Major Size Range.   This is established by the value of the grain size in mm at the 10% line, called D_{10}. This expresses the fact that 90% of the solids are above a cer-

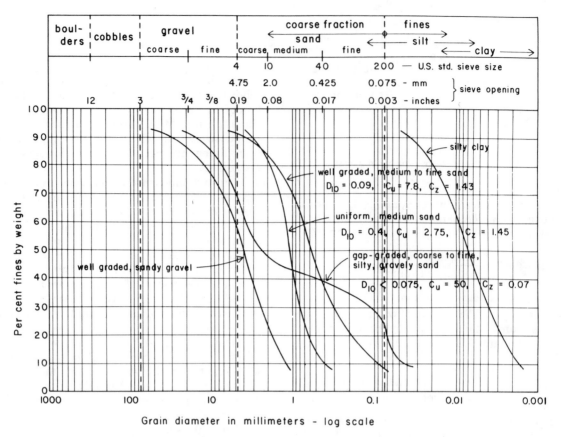

FIGURE 40.2. Grain size range for typical soils.

tain grain size. The D_{10} value is specifically defined as the *effective grain size*.

Degree of Size Gradation. The distinction between uniform and well-graded has to do with the slope of the major portion of the size-range curve. This is established by comparing the size value at the 10% line, D_{10}, with the size value at the 60% line, D_{60}. The relationship is expressed by the *uniformity coefficient* (C_u), which is defined as

$$C_u = \frac{D_{60}}{D_{10}}$$

The higher this number, the greater the degree of size gradation.

Continuity of Gradation. The value of C_u does not express the character of the graph between D_{60} and D_{10}; that is, it does not establish whether the soil is gap graded or well graded, only that it is graded. To establish this, another property is defined, called the *coefficient of curvature* (C_z), which uses all three size values, D_{60} D_{30}, and D_{10}, as follows:

$$C_z = \frac{(D_{30})^2}{(D_{10})(D_{60})}$$

These coefficients are used only for classification of the coarse-grained soils: sand and gravel. For a well-graded

gravel, C_u should be greater than 4, and C_z between 1 and 3. For a well-graded sand, C_u should be greater than 6, and C_z between 1 and 3.

The shape of soil particles is also significant for some soil properties. The three major classes of shape are bulky, flaky, and needlelike, the latter being quite rare. Sand and gravel are typically bulky; further distinction is made with regard to the degree of roundedness of the particle form. Bulky-grained soils are usually quite strong in resisting static loads, especially when the grain shape is quite angular, as opposed to well rounded. Unless a bulky-grained soil is well graded or contains some significant amount of fine-grained material, however, it tends to be subject to displacement and consolidation due to vibration or shock.

Flaky-grained soils tend to be easily deformable and highly compressible, similar to the action of randomly thrown loose sheets of paper or dry leaves in a container. A small percentage of flaky-grained particles can impart the character of a flaky soil to an entire soil mass.

Water has various effects on soils, depending on the proportion of water and on the particle shape, size, and chemical properties. A small amount of water tends to make sand particles stick together somewhat. As a result, the sand behaves differently than usual, no longer acting as a loose, flowing mass. When saturated, however, most sands behave like viscous fluids, moving easily under stress due to gravity or other sources. The effect of the variation of water content is generally more dramatic in fine-grained soils. These will change from rocklike solids when totally dry to virtual fluids when supersaturated.

Table 40.1 describes for fine-grained soils the Atterberg limits, which are the water content limits, or boundaries, between four stages of structural character of the soil. An important property of such soils is the *plasticity index*, I_p, which is the numeric difference between the liquid limit and plastic limit. A major physical distinction between clays and silts is the range of the plastic state, referred to as the relative plasticity of the soil. Clays have a considerable plastic range, and silts generally have practically none, going almost directly from the semisolid to the liquid state. The plasticity chart, shown in Fig. 40.3, is used to classify clays and silts on the basis of two properties, liquid limit and plasticity. The line on the chart is the classification boundary between the two soil types.

Another water-related property is the relative ease with which water flows through, or can be extracted from, the soil mass. Coarse-grained soils tend to be rapid draining, or permeable. Fine-grained soils tend to be nondraining, or impervious, and may literally seal out flowing water.

Soil structure may be classified in many ways. A major distinction is that made between soils that are considered to be *cohesive* and those considered *cohesionless*. Cohesionless soils are those consisting predominantly of sand and gravel with no significant bonding of the discrete soil particles. The addition of a small amount of fine-grained material will cause the cohesionless soil to form a weakly

bonded mass when dry, but the bonding will virtually disappear with a small percentage of moisture. As the percentage of fine materials is increased, the soil mass becomes progressively more cohesive, tending to retain some defined shape right up to the fully saturated, liquid consistency.

The extreme cases of cohesive and cohesionless soils are typically personified by a pure clay and a pure, or clean, sand respectively. Typical soil mixtures will range between

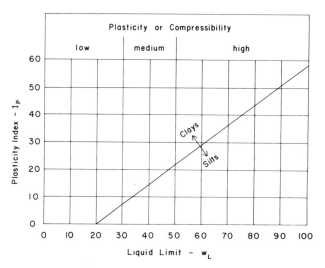

FIGURE 40.3. Plasticity chart using Atterberg limits.

TABLE 40.1. Atterberg Limits for Water Content in Fine-Grained Soils

Description of Structural Character of the Soil Mass	Analagous Material and Behavior	Water Content Limit
Liquid	Thick soup; flows or is very easily deformed	
		------ Liquid limit: w_L
Plastic	Thick frosting or toothpaste; retains shape, but is easily deformed without cracking	— Magnitude of range is *plasticity index:* I_p
		------ Plastic limit: w_p
Semisolid	Cheddar cheese or hard caramel candy; takes permanent deformation but cracks	
		---- Shrinkage limit: w_S
Solid	Hard cookie; crumbles up if deformed	(Least volume attained upon drying out)

these two extremes, so they are useful in establishing the boundaries for classification. For a clean sand the structural nature of the soil mass will be largely determined by three properties: the particle shape (well rounded versus angular), the nature of size gradation (well graded, gapgraded, or uniform), and the density or degree of compaction of the soil mass.

The density of a sand deposit is related to how closely the particles are fit together and is essentially measured by the void ratio. The actions of water, vibration and shock, and compressive force will tend to pack the particles into tighter arrangements. Thus the same sand particles may produce strikingly different soil deposits as a result of density variation.

Table 40.2 gives the range of density classifications that are commonly used in describing sand deposits, varying from very loose to very dense. The general character of the deposit and the typical range of usable bearing strength are shown as they relate to the density. As mentioned previously, however, the effective nature of the soil depends on additional considerations, principally the particle shape and the size gradation. Also of concern are the absolute

particle size, generally established by the D_{10} property, and the amount of water present, as measured by w or S. A minor amount of water will often tend to give a slight cohesiveness to the sand, as the surface tension in the water partially bonds the discrete sand particles. When fully saturated, however, the sand particles are subject to a buoyancy effect that can work to substantially reduce the stability of the soil.

Many physical and chemical properties affect the structural character of clays. Major considerations are the particle size, the particle shape, and whether the particles are inorganic or organic. The percentage of water in a clay has a very significant effect on its structural nature, changing it from a rocklike material when dry to a viscous fluid when saturated. The property of a clay corresponding to the density of sand is its consistency, varying from very soft to very hard. The general nature of clays and their typical usable bearing strengths as they relate to consistency are shown in Table 40.3.

Another major structural property of fine-grained soils is relative plasticity. This was discussed in terms of the Atterberg limits and the classification was made using the

TABLE 40.2. Average Properties of Cohesionless Soils

Relative Density	Blow Count, N (blows/ft)	Void Ratio, e	Simple Field Test with $\frac{1}{2}$-in.- Diameter Rod	Usable Bearing Strength (kips/ft^2)
Loose	<10	0.65–0.85	Easily pushed in by hand	0–1.0
Medium	10–30	0.35–0.65	Easily driven in by hammer	1.0–2.0
Dense	30–50	0.25–0.50	Driven in by repeated hammer blows	1.5–3.0
Very dense	>50	0.20–0.35	Barely penetrated by repeated hammer blows	2.5–4.0

TABLE 40.3. Average Properties of Cohesive Soils

Consistency	Unconfined Compressive Strength (kips/ft^2)	Simple Field Test by Handling of an Undisturbed Sample	Usable Bearing Strength (kips/ft^2)
Very soft	<0.5	Oozes between fingers when squeezed	0
Soft	0.5–1.0	Easily molded by fingers	0.5–1.0
Medium	1.0–2.0	Molded by moderately hard squeezing	1.0–1.5
Stiff	2.0–3.0	Barely molded by strong squeezing	1.0–2.0
Very stiff	3.0–4.0	Barely dented by very hard squeezing	1.5–3.0
Hard	4.0 or more	Dented only with a sharp instrument	3.0+

plasticity chart shown in Fig. 40.3. Most fine-grained soils contain both silt and clay, and the predominant character of the soil is evaluated in terms of various measured properties, most significant of which is the plasticity index. Thus identification as "silty" usually indicates a lack of plasticity (crumbly, friable, etc.) while that of "claylike" or "clayey" usually indicates some significant degree of plasticity (moldable even when only partly wet).

Various special soil structures are formed by actions that help produce the original soil deposit or work on the deposit after it is in place. Coarse-grained soils with a small percentage of fine-grained material may develop arched arrangements of the cemented coarse particles resulting in a soil structure that is called *honeycombed*. Organic decomposition, electrolytic action, or other factors can cause soils consisting of mixtures of bulky and flaky particles to form highly voided soils that are called *flocculent*. The nature of formation of these soils is shown in Fig. 40.4a and b. Water-deposited silts and sands, such as those found at the bottom of dry streams or ponds, should be suspected of this condition if the tested void ratio is found to be quite high.

Honeycombed and flocculent soils may have considerable static strength and be quite adequate for foundation purposes as long as no unstabilizing effects are anticipated. A sudden, unnatural increase in the water content or significant vibration or shock may disturb the fragile bonding, however, resulting in major consolidation of the soil. This can produce major settlement of ground surfaces or foundations if the affected soil mass is large.

Behavior under stress is usually quite different for the two basic soil types: sand and clay. Sand has little resistance to stress unless it is confined. Consider the difference in behavior of a handful of dry sand and sand rammed into a strong container. Clay, on the other hand, has resistance to tension in its natural state all the way up to its liquid consistency. If a hard, dry clay is pulverized, however, it becomes similar to loose sand until some water is added.

In summary, the basic nature of structural behavior and the significant properties that affect it for the two soil types are as follows:

Sand. Little compression resistance without some confinement; principal stress mechanism is shear resistance (interlocking particles grinding together); important properties are angle of internal friction (ϕ), penetration resistance (N) to a driven object such as a soil sampler, unit density in terms of weight or void ratio, grain shape, predominant grain size, and nature of size gradation. Some reduction in capacity with high water content.

Clay. Principal stress resistance in tension; confinement generally of concern only in soft, wet clays (to prevent flowing or oozing of the mass); important properties are the unconfined compressive strength (q_u), liquid limit (w_L), plastic index (I_p), and relative consistency (soft to hard).

We must remind the reader that these represent the cases for pure clay and clean sand, which generally represent the outer limits for the range of soil types. Soil deposits typically contain some percentage of all three basic soil ingredients: sand, silt, and clay. Thus most soils are neither totally cohesive nor totally cohesionless and possess some of the characteristics of both of the basic extremes.

Soil classification or identification must deal with a number of properties for precise categorization of a particular soil sample. Many systems exist and are used by various groups with different concerns. The three most widely used systems are the triangular textural system used by the U. S. Department of Agriculture; the AASHO system, named for its developer, the American Association of State Highway Officials; and the so-called unified system, which is primarily used in foundation engineering. Each of these systems reflects some of the primary concerns of the developers of the system.

The unified system relates to properties of major concern in stress and deformation behavior, excavation and dewatering problems, stability under load, and other issues of concern to foundation designers. The unified system is shown in Fig. 40.5. It consists of categorizing the soil

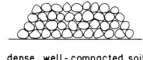

dense, well-compacted soil

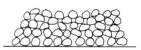

loose, compactible soil

honeycombed soil

(a) Cohesionless Soils

oriented, well dispersed
soil formation

partly flocculent
soil formation

highly flocculent
soil formation

(b) Mixed-grain Soils

FIGURE 40.4. Typical soil structures.

into one of 15 groups, each identified by a two-letter symbol. The primary data used are the grain size analysis, the liquid limit, and the plasticity index. It is not significantly superior to other systems in terms of its data base, but it provides more distinct identification of the soil pertaining to significant considerations of structural behavior.

Building codes and engineering handbooks often use some simplified system of grouping soil types for the purpose of regulating or recommending foundation design criteria and some construction details. This topic is discussed in Sec. 40.3 with regard to the establishment of foundation design criteria.

40.2. BEHAVIOR OF FOUNDATIONS

Most foundations consist of some elements of concrete, primarily because of the relative cost of the material and its high resistance to water, rot, insects, and various effects resulting from being buried in the ground. The two fundamental types of foundations are shallow bearing foundations and deep foundations. This distinction has to do primarily with where the load transfer to the ground occurs. With shallow foundations it occurs near the bottom of the building; with deep foundations the load transfer involves soil strata at some distance below the building.

The most common types of shallow foundations are wall and column footings, consisting of concrete strips and pads poured directly on the ground and directly supporting structural elements of the building. The basic stress transfer between the footing and the ground is by direct contact bearing pressure, inducing general mechanisms of soil behavior. Occasionally several structural elements of the building may be supported by a single large footing. The ultimate extension of this is to turn the entire underside of the building into one large footing, simulating the

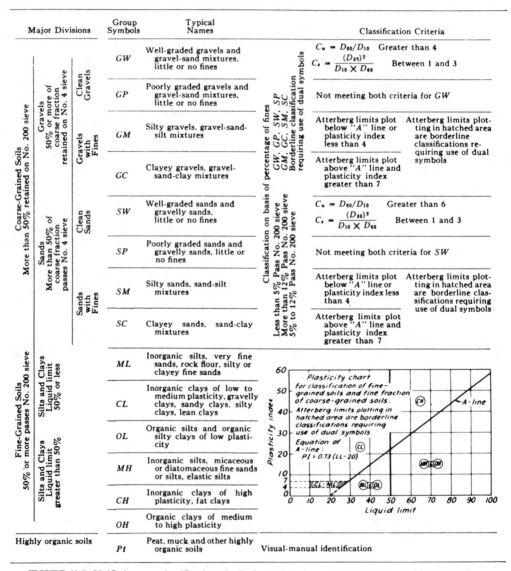

FIGURE 40.5. Unified system classification of soils for engineering purposes. Reproduced from *Foundation Engineering* (Ref. 26), with permission of John Wiley & Sons, Inc., New York.

action of the hull of a ship. This is actually done in rare cases and, indeed, such a foundation does literally float on the soil and is called a *raft*.

If the soil at the bottom of the building is not adequate for the load transfers, or possibly is underlain by weak materials, it becomes necessary to utilize the resistance of lower soil strata. This may require going all the way down to bedrock, or merely to some more desirable soil layer. The technique for accomplishing this is simply to place the building on stilts, or tall legs, in the ground. The two basic types of elements used to do this are piles and piers. Piles are elements that are driven into the ground, much the same as a nail is driven into wood. Piers are shafts that are excavated and then filled with concrete. Some of the principles of their behavior, examples of typical elements used, and various problems encountered in design and construction of buildings on deep foundations are discussed and illustrated in Chapter 42.

When loads are applied to a bearing foundation, stresses are generated in the soil mass. In order to visualize these stresses and the accompanying strains it is necessary to consider the nature of movement of the foundation and the soil mass. Figure 40.6 shows the typical failure mechanism for a simple bearing footing as it is pushed into a soil mass. Part of the vertical movement of the footing is accounted for by the consolidation, or squeezing, of the soil immediately beneath the footing. If any additional movement of the footing is to occur, it must be accomplished by pushing some of the soil out from under the footing, which then involves stresses and movements in the soil mass adjacent to and even above the footing.

Figure 40.7 shows the so-called bulbs of pressure that occur in a typical compression soil beneath a bearing footing. The contour lines of pressure indicate both the net direction of the pressure and the location of equal points of pressure magnitude in terms of percentages of the contact pressure, q, at the bottom of the footing. Although the foundation load is directed vertically downward, the net pressure is vertically downward only in the soil mass immediately beneath the center of the footing. As we move away from the center, the net pressure becomes increasingly horizontal. Adjacent to the footing and above the level of its bottom the net pressure will actually be upward, if the footing is pushed into the soil mass.

Figure 40.8 shows a typical subsurface profile with stratified layers of soils with different properties. If the footings placed at the same level are considerably different in width, there will be significant pressure at some depth below the wider footings. Thus the existence of the highly compressible soil in stratum 4 will have a negligible effect on the narrower footings, but may produce some significant settlement of the wider footings.

40.3. SPECIAL SOIL PROBLEMS

Expansive Soils

In climates with long dry periods, fine-grained soils often shrink to a minimum volume, sometimes producing vertical cracking in the soil masses that extends to considerable

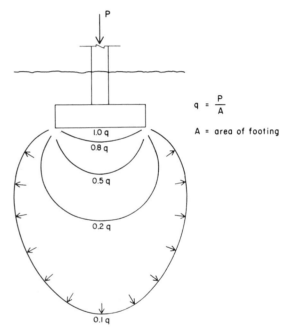

$$q = \frac{P}{A}$$

A = area of footing

FIGURE 40.7. Locations of equal pressure under a bearing footing.

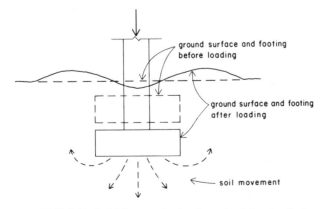

FIGURE 40.6. Typical failure mechanism for a simple bearing footing.

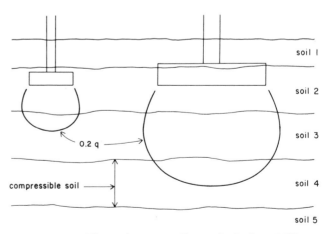

FIGURE 40.8. Difference in pressure effects under footings of different widths.

depths. When significant rainfall occurs, two phenomena occur that can produce problems for structures. The first is the swelling of the ground mass as water is absorbed, which can produce considerable upward or sideways pressures on structures. The second is the rapid seepage of water into lower soil strata through the vertical cracks.

The soil swelling can produce major stresses in foundations, especially when it occurs nonuniformly, which is the general case because of paving, landscaping, and so on. Compensation for these stresses depends on the details of the building construction, the type of foundation system, and the relative degree of expansive character in the soil. Local building codes usually have provisions for design with these soils in regions where they are common. If this property is to be expected, it is highly advised that the tests necessary to establish the expansive property be performed and the results together with necessary considerations for foundation design be reviewed by an experienced foundation engineer.

Collapsing Soils

Collapsing soils are in general soils with large voids. The collapse mechanism is essentially one of rapid consolidation as whatever tends to maintain the soil structure in the large void condition is removed or altered. Very loose sands may display such behavior when they experience drastic changes in water content or are subjected to shock or vibration. The more common cases, however, are those involving soil structures in which fine-grained materials achieve a bonding or molding of cellular voids. These soil structures may be relatively strong when dry but may literally dissolve when the water content is significantly raised. The bonded structures may also be destroyed by shock or simply by excessive compression stress.

This behavior is generally limited to a few types of soil and can usually be anticipated when such soils display large void ratios. Again, the phenomenon of collapse is often a local condition and is given special consideration in local building codes and practices. The two ordinary methods of dealing with collapsing soils are to stabilize the soil by introducing materials to partly fill the void and reduce the potential degree of collapse, or to use vibration, saturation, or other means to cause the collapse to occur prior to construction. When considerable site grading is to be done, it is sometimes possible to temporarily place soil to some depth on the site, providing sufficient compression to cause significant deformation of the potential foundation-bearing materials. The latter method is effective only for soils in which the static pressure thus produced is truly capable of causing significant consolidation.

Differential Settlements

It is generally desirable that the foundation of any building settles uniformly. If separate elements of the foundation settle significantly different amounts, there will be some distortion of the building structure. The seriousness of this situation depends upon the materials and type of the construction; the most critical cases are ones involving masonry, concrete, and plaster constructions that tend to be quite rigid and subject to brittle cracking.

A number of situations can result in differential settlements. Some of these can be adjusted for by careful design of the foundation, whereas others are more difficult to compensate for. The following are some of the situations that can cause this problem.

Nonuniform Subgrade Conditions. Pockets of poor soil and soil strata that are not horizontal can result in different settlement conditions for footings at different locations on the site. Any attempts to equalize settlements in this case require extensive information about the subsurface soil conditions at all points on the site where foundations occur. In addition, settlement calculations must be carefully done for each foundation, which can constitute considerable work in the structural design.

Footings of Significantly Different Size. As discussed before, the vertical stresses under large footings can reach great depths. This can produce larger settlements of the large footings, even though the contact bearing pressure is the same for all footings. Precise compensation in design for this situation also requires extensive knowledge of subsurface conditions and entails laborious settlement calculations.

Footings Placed at Different Elevations. Footings placed at different elevations may bear on different soil strata with significantly different settlement resistance, may have considerable difference in the containment due to overlying soil, or may result in some footings being below the water table and some above it. All of these conditions may result in the development of different settlements under the same soil pressure.

Varying Ratios of Live and Dead Loads. Although the foundations must carry all the building loads, the effect of dead load is often more critical. One reason for this is that the dead load tends to be more "real," while the live load is often quite vaguely established. Another reason is that the dead load is permanent and may thus have more influence on settlements that are progressive with time, such as those caused by clay soils or soils subject to repetitive effects such as fluctuations of water content or frost actions. Thus if footing bearing pressures are equal for the total load, the settlements due to dead load alone will be equal only when the percentage of dead load is the same on all footings. Design for equalized dead-load pressures is commonly done when conditions require it.

40.4. SOIL INVESTIGATION

The amount of information necessary for a good foundation design depends on a number of factors. For a small building located on a flat site with good soil conditions the necessary information may be minimal. For a large building on a difficult site considerable site exploration and extensive field or laboratory testing may be necessary. In any case, the investigation of site and subsurface conditions and the interpretation of information obtained from them should be done by persons with experience and competence in this work.

It is not our purpose here to explain how to do such investigations, but rather to describe the usual processes, the form of information generally provided, and the relation in general of soil investigation to foundation design. We will also discuss the various types of information that may be useful and some of the possible sources of information other than formal soil testing service programs.

Visual inspection of the building site by the foundation designer is highly desirable. Although a site survey will give the actual ground surface contours and location of items such as existing constructions, streets, and so on, a visit to the site may provide much more information.

Persons with some experience in soil investigation may be able to determine some major aspects of the subsurface conditions without elaborate equipment or tests. Soil samples to some depth may be obtained with a posthole digger or a hand auger, and the general character as well as some fairly significant properties may be reasonably determined from handling these samples. Color, odor, general texture, moisture content, density, and ease of excavation of the samples can be correlated to give a fair approximation of the classification and average structural properties of the soil. This type of investigation may be quite adequate for small projects where no unusual conditions exist. It must be emphasized that while the investigation may be simple, the person performing it should have considerable knowledge about soils and their structural properties, and in addition should preferably be familiar with the local climate conditions and regional geology as well as foundation design and construction practices.

A soil investigation performed by a soil testing service can be quite expensive if explorations must be made to a great depth and at several locations on the site. Before such an investigation is performed, it is highly desirable that there be some information about the building size, location, and type of construction, and preferably some idea about the general subsurface conditions that can be anticipated. In such a case the desired location for the soil borings, the necessary depth to which they must be carried, and the extent of testing of samples required can be more intelligently planned. If soil exploration is done on a large site with virtually unknown subsurface conditions and little idea about the location or type of construction that is planned, a second exploration may be necessary. This may not be a serious economic problem with a large building project, but on a small project can result in the total soil investigation cost being as much as the engineering fee for the entire foundation design. It is thus desirable to explore all possible sources of information prior to ordering formal testing by a professional organization.

A formal soil testing program ordinarily consists of the following:

Exploration of the subsurface conditions by some means to obtain samples of soil at various levels below the ground surface.

Field tests consisting of some observations made during the exploration as well as observations and tests on the samples obtained.

Laboratory tests on some samples, such as determination of dry weight, particle size gradation, and so on.

Interpretation of the information obtained and recommendations for criteria and some details of the foundation design.

The extent of exploration, the type of equipment used, the type of tests performed, and the soil properties determined are all subject to considerable variation and require some judgment on the part of those conducting the investigation. The actual type of soil encountered has a lot to do with this. In many cases the information recorded and the terminology of identification of samples is done in relation to local practices and the requirements of local codes and building regulatory agencies.

When the need for deep foundations is anticipated, the exploration will generally be carried to a significant depth. If unusual soil conditions are discovered, more extensive testing may be performed. If separate explorations indicate nonuniform soil strata, many explorations may be necessary in order to obtain a clear picture of the contour of subsurface soil strata. On the other hand, if building loads are modest and the soil encountered is of generally good character for bearing, the necessary exploration and testing may be quite minimal.

40.5. PROPERTIES FOR FOUNDATION DESIGN

For foundation design purposes, soil properties may be broadly separated into two groups. The first consists of those properties that are significant to the identity of the soil type. With reference to classification by the unified system, these are the properties necessary to establish the soil identity as one of the 15 types in the unified system chart, or in some cases, as a soil with marginal properties between two closely related types. For sand, the significant properties are the amount of fines and the size gradation characteristics of the coarse fraction. For silt and clay, the presence of organic material, the liquid limit, and the plasticity index are the major factors affecting identity.

The second group of soil properties includes those that relate directly to the structural character or stress and deformation behavior of the soil. While some of these properties can be presumed in a general way on the basis of the soil identity, there are some specific tests that provide more information, permitting more accurate predictions of structural performance.

For a sand, there are typically four items of information not included in the data used for classification by the unified system that are significant to structural behavior. These are the following:

Grain Shape. The shape ranges from well rounded to angular.

Water Content. Water content is expressed as measured in the natural state, but also as anticipated on the basis of annual climate variations.

Density. Density ranges from loose to dense and indicates the potential degree of consolidation and the settlement that will result from it.

Penetration Resistance. Penetration resistance is usually quoted as the N value, which is the number of blows required to advance a particular type of soil sampler into the soil deposit.

For a clay, the principal tested structural property is the unconfined compression strength, q_u. This may be approximated by some simple field tests, but is most accurately established by a laboratory test on a carefully excavated, so-called undisturbed sample of the soil.

Many soil properties are interrelated or derivative; thus a crosscheck is possible when considerable information is available. Thus unit dry weight and penetration resistance are both related to the relative density of sand. Similarly, unconfined compression strength, relative consistency, and plasticity are interrelated for clay.

The structural character of silts ranges considerably from that resembling a low-plasticity clay to that resembling a fine sand. Thus it is sometimes necessary to use tested properties pertinent to both cohesive and cohesionless soils for complete evaluation of the structural character of silty soil.

40.6. FOUNDATION DESIGN DATA

For the foundation designer, information about soils and ground conditions in general has as a primary purpose the establishment of data useful for design applications. The type of data usually required is described in Sec. 40.3. Standard notation, classification, and general format for presentation of such data usually follows the forms used in building codes or by local engineering and testing organizations.

Table 40.4 presents a summary of information for the basic soil types classified by the unified system. Information is grouped as follows:

Significant Properties. These are the properties that relate directly to the identity or stress-and-strain evaluation of the soil.

Simple Field Identification. These are the various characteristics of the soil that may be used for identification in the absence of testing. They may also be used to verify that the soil encountered during construction is that indicated by the soil exploration and assumed for the foundation design.

Average Properties. These are approximate values and are best verified by exploration and testing, but should be adequate for preliminary design. Stress values are generally those in the approximate range recommended by building codes.

TABLE 40.4. Summary of Properties and Recommended Design Values for Typical Soils

Description	Gravel, well-graded; little or no fines			Gravel, poorly graded, little or no fines			Silty gravel and gravel-sand-silt mixes		
ASTM classification	GW			GP			GM		
Significant properties	$\geq$95% retained on No. 200 sieve (0.003 in.) $\geq$50% of coarse fraction retained on No. 4 sieve ($\frac{3}{16}$ in.) $C_u > 4$ $1 < C_z < 3$			$\geq$95% retained on No. 200 sieve (0.003 in.) $\geq$50% of coarse fraction retained on No. 4 sieve ($\frac{3}{16}$ in.) Does not meet C_u and/or C_z requirements for well-graded (GW)			50–88% retained on No. 200 sieve (0.003 in.) $\geq$50% of coarse fraction retained on No. 4 sieve ($\frac{3}{16}$ in.) Atterberg plot below A line or $I_p < 4$		
Field identification	Significant amounts of coarse rock fragments; easily pulverized; fast draining; wide range of grain sizes			Significant amounts of coarse rock fragments; easily pulverized; fast draining; has narrow range of sizes or is gap-graded			Gravely but forms clumps that pulverize with moderate effort; wet sample takes little or no remolding before disintegrating; slow draining		
Average properties	Loose	Medium	Dense	Loose	Medium	Dense	Loose	Medium	Dense
Allowable bearing (lb/ft^2) with minimum of one ft	($N < 10$)	($10 < N < 30$)	($N > 30$)	($N < 10$)	($10 < N < 30$)	($N > 30$)	($N < 10$)	($10 < N < 30$)	($N > 30$)
surcharge	1300	1500	2000	1300	1500	2000	1000	1500	2000
increase for surcharge (%/ft)	20	20	20	20	20	20	20	20	20
maximum total	8000	8000	8000	8000	8000	8000	8000	8000	8000
Lateral pressure									
active coefficient	0.25	0.25	0.25	0.25	0.25	0.25	0.30	0.30	0.30
passive (lb/ft^2 per ft depth)	200	300	400	200	300	400	167	250	333
Friction (coefficient or lb/ft^2)	0.50	0.60	0.60	0.50	0.60	0.60	0.40	0.50	0.50
Weight (lb/ft^3)									
dry	100	110	115	90	100	110	100	115	130
saturated	125	130	135	120	125	130	125	135	145
Compressibility	Medium	Low	Very low	Medium	Low	Very low	Medium	Low	Very low

TABLE 40.4. (*Continued*)

Description	Clayey gravel and gravel-sand-clay mixes			Sand, well-graded; gravely sand; little or no fines			Sand, poorly graded; gravely sand; little or no fines		
ASTM classification	GC			SW			SP		
Significant properties	50–88% retained on No. 200 sieve (0.003 in.) $\geq$50% of coarse fraction retained on No. 4 sieve ($\frac{3}{16}$ in.) Atterberg plot above A line or $I_p > 7$			$\geq$95% retained on No. 200 sieve (0.003 in.) >50% of coarse fraction passes No. 4 sieve ($\frac{3}{16}$ in.) $C_u > 6$ C_z 1–3			$\geq$95% retained on No. 200 sieve (0.003 in.) >50% of coarse fraction passes No. 4 sieve ($\frac{3}{16}$ in.) Does not meet C_u and/or C_z requirements for well graded (SW)		
Field identification	Gravely but forms hard clumps that require considerable effort to pulverize; wet sample takes some remolding before disintegrating; very slow draining			Relatively clean sand with wide size range; easily pulverized; fast draining			Relatively clean sand with narrow size range or gaps in grading; easily pulverized; fast draining		
Average properties	Loose ($N < 10$)	Medium ($10 < N < 30$)	Dense ($N < 30$)	Loose ($N < 10$)	Medium ($10 < N < 30$)	Dense ($N > 30$)	Loose ($N < 10$)	Medium ($10 < N < 30$)	Dense ($N > 30$)
Allowable bearing (lb/ft^2) with minimum of one ft									
surcharge	1000	1500	2000	1000	1500	2000	1000	1500	2000
increase for surcharge (%/ft)	20	20	20	20	20	20	20	20	20
maximum total	8000	8000	8000	6000	6000	6000	6000	6000	6000
Lateral pressure									
active coefficient	0.30	0.30	0.30	0.25	0.25	0.25	0.25	0.25	0.25
passive (lb/ft^2 per ft depth)	167	250	333	183	275	367	75	150	200
Friction (coefficient or lb/ft^2)	0.40	0.50	0.50	0.35	0.40	0.40	0.35	0.40	0.40
Weight (lb/ft^3)									
dry	110	120	125	100	110	115	90	100	110
saturated	125	130	135	125	130	135	120	125	130
Compressibility	Medium	Low	Very low	Medium high	Medium	Low	Medium high	Medium	Low

Description	Silty sand and sand-silt mixes			Clayey sand and sand-clay mixes			Inorganic silt, very find sand, rock flour, silty or clayey fine sand		
ASTM classification	SM			SC			ML		
Significant properties	50–80% retained on No. 200 sieve (0.003 in.) >50% of coarse fraction passes No. 4 sieve ($\frac{3}{16}$ in.) Atterberg plot below A line, or $I_p < 4$			50–80% retained on No. 200 sieve (0.003 in.) >50% passes No. 4 sieve ($\frac{3}{16}$ in.) Atterberg plot above A line or $I_p > 7$			$\geq$50% passes No. 200 sieve (0.003 in.) $w_L \leq 50\%$ Atterberg plot below A line $I_p < 20$		
Field identification	Sandy soil; forms clumps that can be pulverized with moderate effort; wet sample takes little remolding before disintegrating; slow draining			Sandy soil; forms clumps that offer some resistance to being pulverized; wet sample takes some remolding before disintegrating; very slow draining			Fine-grained soils of low plasticity; slow draining; dry clumps easily pulverized; won't form thin thread when molded		
Average properties	Loose ($N < 10$)	Medium ($10 < N < 30$)	Dense ($N > 30$)	Loose or soft ($N < 10$)	Medium ($10 < N < 30$)	Dense or stiff ($N > 30$)	Loose or soft ($N < 10$)	Medium ($10 < N < 30$)	Dense or stiff ($N < 30$)
Allowable bearing (lb/ft^2) with minimum of one ft									
surcharge	500	1000	1500	1000	1500	2000	500	750	1000
increase for surcharge (%/ft)	20	20	20	20	20	20	20	20	20
maximum total	4000	4000	4000	4000	4000	4000	3000	3000	3000
Lateral pressure									
active coefficient	0.30	0.30	0.30	0.30	0.30	0.30	0.35	0.35	0.35
passive (lb/ft^2 per ft depth)	100	167	233	133	217	300	67	100	133
Friction (coefficient or lb/ft^2)	0.35	0.40	0.40	0.35	0.40	0.40	0.35 or 250	0.40 or 375	0.40 or 500
Weight (lb/ft^3)									
dry	105	115	120	105	115	120	105	115	120
saturated	125	130	135	125	130	135	125	130	135
Compressibility	Medium	Low	Low	Medium	Low	Low	Medium high	Medium	Low

TABLE 40.4. (*Continued*)

Description	Lean clay; inorganic clay of low to medium plasticity; gravely clay; sandy clay; silty clay			Organic silt and organic silty clay of low plasticity			Inorganic silt; micaceous or diatomaceous fine sands or silt; elastic silt		
ASTM classification	CL			OL			MH		
Significant properties	$\geqq$50% passes No. 200 sieve (0.003 in.) $w_L \leqq 50\%$ Atterberg plot above A line $I_p > 7$			$\geqq$50% passes No. 200 sieve (0.003 in.) $w_L \leqq 50\%$ Atterberg plot below A line $I_p < 20$			$\geqq$50% passes No. 200 sieve (0.003 in.) $w_L > 50\%$ Atterberg plot below A line $I_p < 20$		
Field identification	Fine-grained soil of low plasticity; slow draining; dry clumps quite hard, but not very difficult to pulverize			Fine-grained soil of low plasticity; slow draining; dry clumps quite hard, but not very difficult to pulverize; typical slight musty, rotting odor			Fine-grained soils of low plasticity; slow draining; dry clumps quite hard, but not very difficult to pulverize; spongy; compressible wet or dry		
Average properties	Soft	Medium	Stiff	Loose or soft ($N < 10$)	Medium ($10 < N < 30$)	Dense or stiff ($N > 30$)	Loose or soft ($N < 10$)	Medium ($10 < N < 30$)	Dense or stiff ($N > 30$)
Allowable bearing (lb/ft^2) with minimum of one ft surcharge	1000	1500	2000	500	750	1000	500	750	1000
increase for surcharge (%/ft)	20	20	20	10	10	10	10	10	10
maximum total	3000	3000	3000	2000	2000	2000	1500	1500	1500
Lateral pressure active coefficient	0.40	0.50	0.75	0.75	0.85	0.95	0.50	0.60	0.75
passive (lb/ft^2 per ft depth)	267	467	667	33	50	67	33	50	67
Friction (coefficient or lb/ft^2)	500	750	1000	250	375	500	200	300	400
Weight (lb/ft^2) dry	80	95	105	75	90	100	70	85	100
saturated	110	120	130	95	105	115	100	110	120
Compressibility	High	Medium high	Low	High	Medium high	Medium	Very high	High	Medium high

Description	Fat clay; inorganic clay of high plasticity			Organic clay of medium to high plasticity			Peat, muck, topsoil
ASTM classification	CH			OH			Pt
Significant properties	$\geqq$50% passes No. 200 sieve (0.003 in.) $w_L > 50\%$ Atterberg plot above A line $I_p > 20$			$\geqq$50% passes No. 200 sieve (0.003 in.) $w_L > 50\%$ Atterberg plot below A line			Highly organic Low density
Field identification	Fine-grained soil of high plasticity; sticky and highly moldable without fracture when wet; non-draining; impervious; dry clumps very hard and very difficult to pulverize; highly compressible			Fine-grained soil of medium to high plasticity; sticky and moderately moldable without fracture when wet; non-draining; impervious; dry clumps hard; moderately difficult to pulverize; highly compressible; typical slight musty, rotting odor			Contains large amounts of partially decomposed plant or animal material strong rotting odor; slow draining highly compressible
Average properties Allowable bearing (lb/ft^2) with minimum of one ft	Soft	Medium	Stiff	Soft	Medium	Stiff	
surcharge	500	750	1000	500	500	500	Not usable
increase for surcharge (%/ft)	10	10	10	0	0	0	
maximum total	1500	1500	1500	500	500	500	
Lateral pressure active coefficient	0.75	0.85	0.95	0.75	0.85	0.95	0.30
passive (lb/ft^2 per ft depth)	33	100	167	33	33	33	Not usable
Friction (coefficient or lb/ft^2)	200	300	400	150	200	200	Not usable
Weight (lb/ft^3) dry	75	90	105	65	85	100	70–90
saturated	95	110	130	90	110	125	90–110
Compressibility	Very high	High	Medium high	Very high	High	Medium high	Very high

CHAPTER FORTY-ONE

Shallow Bearing Foundations

Shallow foundation is the term usually used to describe the type of foundation that transfers vertical loads by direct bearing on soil strata close to the bottom of the building and a relatively short distance below the ground surface. The form of construction used mostly for these structures is reinforced concrete and some material relating to the topics in Chapter 41 is presented in Part 5, notably in Chapter 33.

Figure 41.1 illustrates a variety of elements ordinarily used in bearing foundation systems.

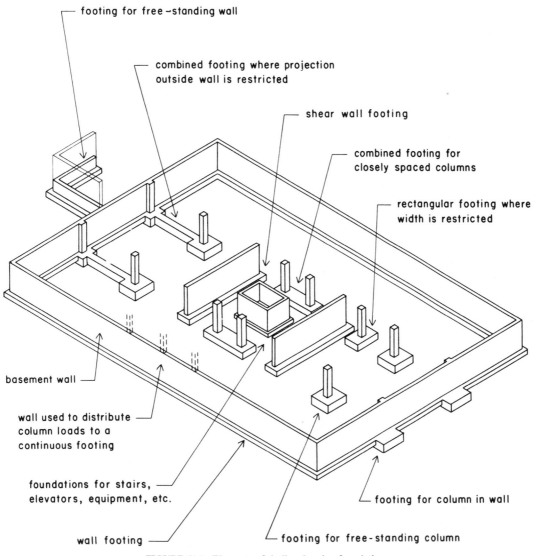

footing for free-standing wall

combined footing where projection outside wall is restricted

shear wall footing

combined footing for closely spaced columns

rectangular footing where width is restricted

basement wall

wall used to distribute column loads to a continuous footing

foundations for stairs, elevators, equipment, etc.

wall footing

footing for free-standing column

footing for column in wall

FIGURE 41.1. Elements of shallow bearing foundations.

41.1. WALL FOOTINGS

Wall footings usually serve to transfer loads from the walls by bearing, but often also serve a primary function as a platform for the beginning of the wall construction. If the supported wall is of concrete or reinforced masonry, dowels for the wall vertical reinforcing will be anchored in the footing. If the wall is constructed with wood or metal, and bearing is at some distance below grade, there will often be a foundation wall that serves to keep the vulnerable wall materials above the soil.

Design of ordinary wall footings is discussed in Sec. 33.2. The various functions of foundation walls are discussed in Sec. 41.6. The special situation of the wall footing subjected to moment is discussed in Sec. 41.3.

41.2. COLUMN FOOTINGS

Footings for columns must relate to the structural actions of the columns they support. Action is often limited to simple, direct, vertical compression, but may be combined with bending moment and lateral shear. For simple compression, the most common footing is the square concrete pad with reinforcing in two layers at right angles to each other. The design of this simple footing is described in Sec. 33.1. The remaining sections in this chapter deal with the design of special column footings that are required when the simple square footing cannot be used.

As with wall footings, it is often necessary to consider factors other than that of the simple bearing function of column footings. Provisions for dowels or anchor bolts, or other details related to the supported structure, may affect the design in a critical way. Pedestals (also called *piers*) are often used with column footings, serving as transitional devices for various purposes. The design of pedestals of reinforced concrete is presented in Sec. 33.3, and the design of pedestals of masonry is presented in Sec. 36.5.

Various special concerns for the supported structure are also discussed in Sec. 41.6.

Oblong Column Footings

When constraints on the size of a column footing prevent the use of the required square footing, it is sometimes possible to use a rectangular footing that is oblong. Although a square footing is actually also rectangular, the term *rectangular* is traditionally used to describe a footing with one plan dimension larger than the other. The close proximity of other elements of the building construction or property lines are the usual reasons for using such a footing.

An oblong footing must be designed separately for bending in two directions, called the *long* and *short* directions. The greater the difference between these two dimensions, the more the footing tends to act primarily in one-way bending. Regardless of the relative proportions, there tends to be a concentration of the bending in the short direction in the vicinity of the column. The *ACI Code* provides for this latter phenomenon by requiring that the reinforcing found to be required for bending in the short direction must be proportioned to provide a specific percentage of the total within a zone equal to the width in the short direction. There is no particular limit on the proportions of the two plan dimensions, but general feasibility for such a footing usually limits the ratio of the two dimensions to approximately 2:1.

Combined Column Footings

When two or more columns are quite close together, it is sometimes desirable or necessary to use a single footing, as shown in Fig. 41.2. If the columns and their loads are symmetrical, the footing is usually designed to function in its long direction as a uniformly loaded double cantilever beam, as shown in Fig. 41.2. If the column loads are not equal, the usual procedure is to find the location of the centroid of the column loads and to design a footing whose plan centroid coincides with that of the column loads. This will assure a uniform soil pressure on the bottom of the

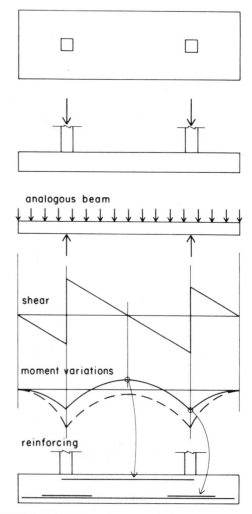

FIGURE 41.2. Actions of a symmetrical combined footing.

footing. There are many possible variations for such a combined footing.

Occasionally, because of property line location, adjacent construction, or excavation difficulties, it is not possible to allow a footing to extend beyond a column on one side. A possible solution to this problem is to use a *strap*, or *cantilever*, footing. This technique consists of developing a combined footing for the restricted column and an adjacent column. This situation usually occurs because of columns at the building edge with restrictions because of the property line or an adjacent building.

Figure 41.3 shows the two most common forms for such a footing shared by an edge column and an interior column. In the upper illustration a simple rectangular footing is provided with its centroid developed to coincide with

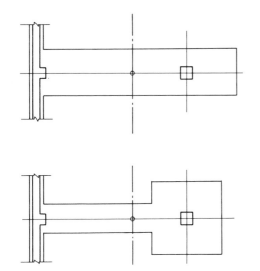

FIGURE 41.3. Common plan forms for the cantilever footing.

that of the total load on the footing. In the lower illustration the two columns are supported by individual square footings and the eccentric load effect on the exterior footing is resisted by a strap between the two footings. In either of these cases there is likely to be considerable bending and shear in the footing between the two columns unless the columns are very closely spaced.

41.3.   MOMENT-RESISTIVE FOOTINGS

Bearing footings must occasionally resist moments, in addition to some combination of vertical and horizontal forces. Some of the situations that produce this effect, as shown in Fig. 41.4, are the following:

Freestanding Walls.   When a wall is supported only at its base and must resist horizontal forces on the wall, it requires a moment-resistive foundation. Examples are exterior walls used as fences and interior walls that are not full story in height. The horizontal forces are usually due to wind or seismic effects.

Cantilever Retaining Walls.   Cantilever retaining walls are essentially freestanding walls that must sustain horizontal earth pressures. The various aspects of behavior and the problems of design of such walls are discussed in Sec. 32.5.

Bases for Shear Walls and Trussed Frames.   The overturning effect at the bottom of a shear wall must be resisted by the foundation. When the wall is relatively isolated in plan, as in the case of some interior walls, the foundation for the wall may be developed in a manner similar to that for a freestanding tower.

Supports for Rigid Frames, Arches, Cable Structures, etc. The foundations for these types of structures most often

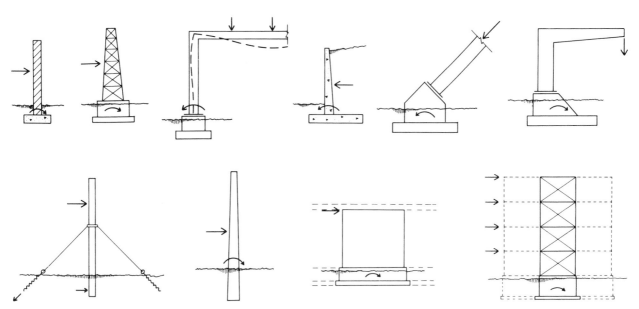

FIGURE 41.4. Structures with moment-resistive foundations.

sustain horizontal forces and moments, even for vertical gravity loading. The special problems of abutments are discussed in Sec. 43.2.

Bases for Chimneys, Signs, Towers, Flagpoles, etc. Any freestanding vertical element supported only at its base must have a moment-resistive foundation. Such a foundation may be quite simple and modest when the element is small, or may be a major engineering undertaking when the element is very tall and the horizontal forces are very high.

Moment-resistive footings are typically subjected to a combination of vertical compression and overturning moment. The general combined stress condition that this produces is discussed in Sec. 11.4. Of special note is the case in which maximum tension stress due to bending exceeds the compression stress due to the vertical load, resulting in the *cracked section*. This case is analyzed by the *pressure wedge method* in Sec. 11.4 and illustrated in Fig. 11.7.

All four cases of combined stress shown in Fig. 11.5 will cause rotation of the footing due to deformation of the soil. The extent of this rotation and the concern for its effect on the supported structure must be considered carefully in the design of the footing. It is generally desirable that long-term loads (such as dead loads) not develop uneven stress on the footing. This is especially true when the soil is highly deformable or is subject to long-term continued deformation, as is the case with soft, wet clay. Thus it is preferred that stress conditions as shown for Cases 2 or 4 in Fig. 41.5 be developed only with short-term live loads.

When foundations have significant depth below the ground surface, other forces will develop to resist moment, in addition to the vertical pressure on the bottom of the footing. Figure 41.5a shows the general case for such a

foundation. The moment effect of the horizontal force is assumed to develop a rotation of the foundation at some point between the ground surface and the bottom of the footing. The position of the rotated structure is shown by the dashed outline. Resistance to this movement is visualized in terms of the three major soil pressure effects (A, B, C) plus the friction on the bottom of the footing (D).

When the foundation is quite shallow, as shown in Fig. 41.5b, the rotation point for the foundation moves down and toward the toe of the footing. It is common in this case to assume the rotation point to be at the toe, and the overturning effect to be resisted only by the weights of the structure, the foundation, and the soil on top of the footing. Resisting force A in this case is considered to function only in assisting the friction to develop resistance to the horizontal force in direct force action.

When a foundation is very deep and is essentially without a footing, as in the situation of a pole, resistance to moment must be developed entirely by the forces A and B, as shown in Fig. 41.5c. If the structure is quite flexible, its bending will cause the two forces to develop quite close to the ground surface, making the extension of the element into the ground beyond this point of little use in developing resistance to moment.

41.4. FOUNDATIONS FOR SHEAR WALLS

When shear walls rest on bearing foundations the situation is usually one of the following:

1. The shear wall is part of a continuous wall and is supported by a foundation that extends beyond the shear wall ends.

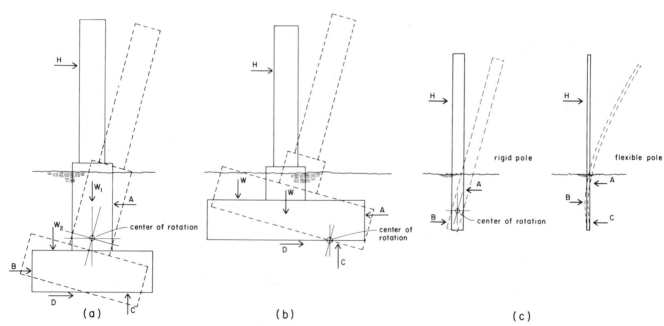

FIGURE 41.5. Moment resistance of foundations: (a) deep-bearing type; (b) shallow-bearing type; (c) pole type.

2. The shear wall is a separate wall and is supported by its own foundation in the manner of a freestanding tower.

We will consider item 2 first. The basic problems to be solved in the design of such a foundation are the following:

Anchorage of the Shear Wall. The shear wall anchorage consists of the attachment of the shear wall to the foundation to resist the sliding and the overturning effects due to the lateral loads on the wall. This involves a considerable range of possible situations, depending on the construction of the wall and the magnitude of forces.

Overturning Effects. The overturning effect is taken into consideration by performing the usual analysis for the overturning moment due to the lateral loads and the determination of the safety factor resulting from the resistance offered by the dead loads and the passive soil pressure.

Horizontal Sliding. Horizontal sliding is the direct, horizontal force resistance in opposition to the lateral loads. It may be developed by some combination of soil friction and passive soil pressure or may be transferred to other parts of the building structure.

Maximum Soil Pressure and Its Distribution. The magnitude and form of distribution of the vertical soil pressure on the foundation caused by the combination of vertical load and moment must be compared with the established design limits.

Example 1. *Independent Shear Wall Footing.* The wall and proposed foundation are shown in Fig. 41.6a. The wall is assumed to function as a bearing wall as well as a shear wall, and the vertical loads applied to the top of the foundation are the sum of the wall weight and the support loads on the wall. The following are design data and criteria:

Allowable soil pressure: 1500 lb/ft².
Soil type: group 4, *UBC* Table 29-B, (Appendix D).
Concrete design strength: $f_c' = 2000$ psi.
Allowable tension on reinforcing: 20,000 psi.

Solution. The various forces acting on the foundation are shown in Fig. 41.6b. For the overturning analysis the usual procedure is to assume a rotation about the toe of the footing and to include only the gravity loads in determining the resistive moment. With these assumptions the analysis is as follows:

Overturning moment:

$$M = 3000(11.83) = 35,490 \text{ lb/ft}$$

Weight of foundation wall:

$$2\left(\frac{10}{12}\right)(10.5)(150) = 2625 \text{ lb}$$

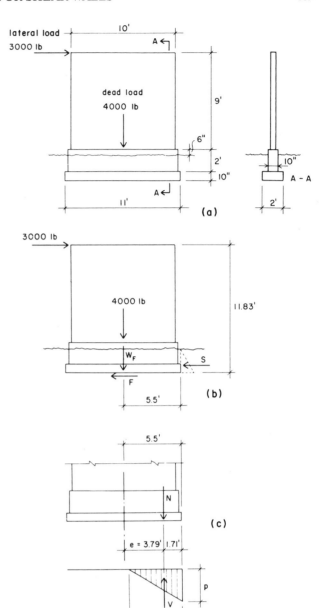

FIGURE 41.6. Shear wall foundation design example.

Weight of footing:

$$\frac{10}{12}(11)(2)(150) = 2750 \text{ lb}$$

Weight of soil over footing:

$$1.17(1.5)(11)(80) = 1544 \text{ lb}$$

Total vertical load:

$$4000 + 2625 + 2750 + 1544 = 10,919 \text{ lb}$$

Resisting moment:

$$10,919(5.5) = 60,055 \text{ lb-ft}$$

Safety factor:

$$SF = \frac{60,055}{35,490} = 1.69$$

Since this safety factor is greater than the usual requirement of 1.5, the foundation is not critical for overturning effect.

For the soil group given the soil friction coefficient is 0.25, and the total-sliding resistance offered by friction is thus

$$F = 0.25(10,919) = 2730 \text{ lb}$$

Since we assume that the lateral load on the shear wall is due to either wind or seismic force, this resistance may be increased by one-third. Thus, although there is some additional resistance developed by the passive soil pressure on the face of the foundation wall and the footing, it is not necessary to consider it in this example.

For the soil stress analysis we combine the overturning moment as calculated previously with the total vertical load to find the equivalent eccentricity as follows, deducting soil weight for N:

$$e = \frac{M}{N} = \frac{35,490}{9375} = 3.79 \text{ ft}$$

This eccentricity is considerably outside the kern limit for the 11-ft-long footing ($\frac{11}{6}$ or 1.83 ft) so that the stress analysis must be done by the pressure wedge method, as discussed in Sec. 11.4. As illustrated in Fig. 41.6c, the analysis follows.

Distance of the eccentric load from the footing end is

$$5.5 - 3.79 = 1.71 \text{ ft}$$

Therefore,

$$x = 3(1.71) = 5.13 \text{ ft}$$

$$p = \frac{2N}{wx} = \frac{2(9375)}{2(5.13)} = 1827 \text{ lb/ft}^2$$

Since this is less than the allowable design pressure with the permissible increase of one-third [$p = 1.33(1500) = 2000$], the condition is not critical as long as this type of soil pressure distribution is acceptable. This acceptance is a matter of judgment, based on concern for the rocking effect, as discussed in the example of the freestanding wall. In this case, with the wall relatively short with respect to the footing length, we would judge the concern to be minor and would therefore accept the foundation as adequate.

The design considerations remaining for this example are concerned with the structural adequacy of the foundation wall and footing. The short wall in this case is probably adequate without any vertical reinforcing, although it would be advisable to provide at least one vertical dowel at each end of the wall, extended with a hook into the footing. The 2-ft-wide footing is adequate without lateral reinforcing. Both the wall and footing, however, should be provided with some minimal longitudinal reinforcing for shrinkage and temperature stresses.

In this example we have assumed the shear wall and its foundation to be completely independent of the building structure and have dealt with it as a freestanding tower. This is sometimes virtually the true situation, and the design approach that we have used is a valid one for such cases. However, various relations may occur between the shear wall structure and the rest of the building. One of these possibilities is shown in Fig. 41.7. Here the shear wall and its foundation extend some distance below a point at which the building structure offers a bracing force in terms of horizontal constraint to the shear wall. This situation may occur when there is a basement and the first-floor structure is a heavy, rigid concrete system. If the floor structure is capable of transferring the necessary horizontal force directly to the outside basement walls, the shear wall foundation may be relieved of the usual sliding resistance function.

As shown in Fig. 41.7, when the upper level constraint is present, the rotation point for overturn moves to this point. The forces that contribute to the resisting moment become the gravity load W, the sliding friction F, and the passive soil pressure S.

Another relationship that may occur between the shear wall structure and the rest of the building is that of some connection between the shear wall footing and other adjacent foundations. This occurs commonly in buildings designed for high seismic risk because it is usually desirable to assure that the foundation system moves in unison during seismic shocks. This may be a useful relationship for the shear wall in that additional horizontal resistance may be developed to add to that produced by the friction and passive soil pressure on the shear wall foundation itself. Thus, if the elements to which the shear wall foundation is tied do not have lateral load requirements, their potential friction and passive pressures may be enlisted to share the loads on the shear wall.

Shear walls on the building exterior often occur as individual wall segments, consisting of solid portions of the wall between openings or other discontinuities in the wall construction. In these situations the foundation often consists of a continuous wall and footing or a grade beam

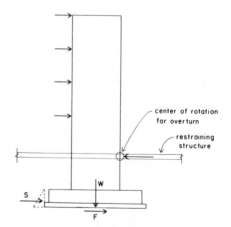

FIGURE 41.7. Tall shear wall with upper-level constraint.

that extends along the entire wall. The effect of the overturning moment on such a foundation is shown in Fig. 41.8. The loading tends to develop a shear force and moment in the foundation wall, both of which are one-half of the forces in the wall. If the foundation wall is capable of developing this shear and bending, it functions as a distributing member, spreading the overturning effect along an extended length of the foundation.

The overturning effect just described must be added to other loadings on the wall for a complete investigation of the foundation wall and footing stresses. It is likely that the continuous foundation wall also functions as a distributing member for the gravity loads.

41.5. MISCELLANEOUS PROBLEMS OF SHALLOW FOUNDATIONS

A number of problems occur in the general design of foundation systems that utilize shallow bearing footings. While the individual problems of various types of foundation elements are discussed in various other parts of this book, the following are some problems that are often shared by the several elements that constitute the complete foundation system for a building.

Equalizing of Settlements

It is usually desired that all of the elements of a foundation system settle the same amount. If part of a wall settles more than another, or if columns settle more than walls, there are a number of problems that can result, such as:

Cracking of walls, especially those consisting of rigid materials such as masonry, plaster, and concrete.

Jamming of doors and operable windows.

Fracturing of plumbing or electrical conduits incorporated into the structure.

Producing of undesirable stress or stability conditions in structures that have some degree of continuity, such as multispan beams or rigid frames of steel or reinforced concrete.

A technique sometimes used to reduce these problems is to design for so-called equalized settlements. This is a process in which the sizes of bearing elements are determined on the basis of developing equal vertical settlement, rather than producing a common maximum soil pressure. The ease of accomplishing this depends on several factors, including the specific soil conditions. The simplest case occurs when all the footings are at the same approximate level and all bear on the same type of soil. In this case the technique most often used is to design for a relatively constant pressure under the dead load, since this most often represents the critical loading condition for settlements. The process is as follows:

1. Design loads are established for each bearing element, separating the live load from the dead load.

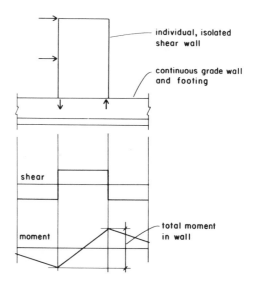

FIGURE 41.8. Isolated shear wall on a continuous footing.

2. The element with the highest ratio of live load to dead load is selected and designed for a total soil pressure using the limit of allowable bearing pressure.

3. The pressure under the dead load only is determined for the selected control element. This represents the desired constant pressure for design.

4. The required plan dimensions for all other bearing elements are determined, using only their dead loads and the control pressure determined in step 3.

5. With their plan sizes established, the other elements are then designed structurally for their own individual design soil pressures under the total loading of the dead and live loads.

The effectiveness of this technique is limited. Some of the conditions that may require a different design approach are the following:

Nonuniform Soil Conditions. Where the soil conditions vary for separate footings, maintaining a constant soil pressure may not assure equal settlements. Nonhorizontal soil strata, pockets of poor soil, or weak underlying strata affected by large footings are some situations of this type. These conditions may require actual settlement calculations for each individual footing.

Time-Dependent Settlements. Granular soils, consisting of clean sand and gravel, tend to settle instantly under the application of the maximum load. In this case the live load may have a major effect. Soft clays, on the other hand, tend to produce long-time, continuous settlements, for which the dead load is more truly critical.

Significant Range in Load Magnitudes. In some buildings, parts of the foundation system may be quite lightly loaded, while others sustain very heavy loads. An example of this is a building with adjacent high-rise and low-rise portions. This situation often makes the equalizing of pressures unfeasible. If significant differences in settlements are involved, it may be necessary to provide some structural separation between the adjacent parts of the building. If settlements can be reasonably accurately predicted, it may be possible to provide for independent vertical settlements during the building construction, with the adjacent parts arriving at some alignment when construction is complete.

Proximity of Foundation Elements

Building planning often results in situations in which separate parts of the building structure are located so that their foundations are close together. Some of the situations of this type and the design problems that result are the following:

Closely Spaced Columns. Columns are occasionally so closely spaced that it is not possible to use the usual square column footing under each column. When this occurs, the three main options are to place an oblong footing under one column with one dimension restricted to that required to clear the adjacent column, to place oblong footings under both columns, or to use a single footing for both columns.

Closely Spaced Buildings. When new construction must be placed very close to an existing building, the excavation and foundation construction problems must be studied very closely. Excavation must be performed in a manner that does not cause settlement or collapse of the adjacent, existing construction. If the new footings must be lower than the existing ones, it will most likely be necessary to underpin the adjacent building.

Footings Adjacent to Pits, Tunnels, etc. Columns or bearing walls must sometimes be located near steam tunnels, elevator pits, underground utility vaults, or other structures that must be placed at an elevation near or below the normal elevation for the footings. When loads on the walls or columns are relatively light, it may be possible to combine the structures and their foundations. When the structural loads are high, however, it is usually necessary to drop the footings for the walls or columns below the adjacent construction. If this elevation change is considerable, it may create a problem with regard to other footings in the vicinity. This is treated in the discussion that follows.

Adjacent Footings at Different Elevations

When individual footings are horizontally closely spaced but occur at different elevations, a number of potential problems may be created. As shown in Fig. 41.9, some of these are as follows:

Disturbance of the Upper Footing. Excavation for the lower footing may result in settlement or collapse of the upper footing, if the upper footing is already in place

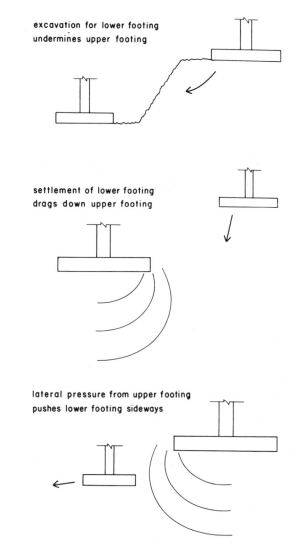

FIGURE 41.9. Problems with adjacent footings at different elevations.

and carrying a significant load. Even when the two excavations are performed at the same time, however, the deep cut required for the lower footing may significantly disturb the soil under the upper footing. There are a number of variables in this situation, so that each case must be carefully studied. The relative size and load on the two footings must be considered. Also of major concern is the effect of a steep cut on the soil. In highly cohesionless material, such as clean sand or gravel, this will surely cause considerable disturbance of the soil structure. In a soft clay there will be an oozing due to the pressure relief.

Additional Settlement of the Upper Footing. If the lower footing is large, its pressure may spread sufficiently through the soil mass to cause some additional settlement of the upper footing, as it is dragged down with the lower one.

Lateral Pressure Effect on the Lower Footing. If the upper footing is large, the horizontal spread of pressure beneath it may produce some lateral movement of the lower footing.

The critical design limit in these situations is the relation of the vertical separation to the horizontal distance between the footings, as shown in Fig. 41.10. This limit is very much a matter of judgment and cannot be generalized for all situations. We hesitate to recommend any single number for this ratio, since it would be highly conservative for some situations and potentially critical for others. Some building codes or building regulatory agencies set limits for this ratio, although sometimes it is possible to make a case for an exception where good information about soil conditions is available and a careful engineering analysis can be performed.

Although the ratio of the dimensions h and L is a critical concern with regard to the various effects illustrated in Fig. 41.9, of equal concern is the actual value of L. When this distance is close to or less than the dimension of the larger footing, the soil stress may cause difficulties, even though the footings are at the same elevation. Thus a value of h/L of one half or less does not mean the design is conservative. Conversely, with the same soil conditions, when L is several times the dimension of the larger footing, a considerable elevation difference may be tolerated.

Another problem involving differences in footing elevations occurs when the bottom of a continuous foundation wall must be lowered at some point. A common situation of this type occurs when a building has only a partial basement, requiring the foundation wall to be considerably lower in the area of the basement. As shown in Fig. 41.11, the solution to this problem is to either slope or step the wall footing between the two levels. Unless the angle of the required slope is quite low (h/L of one-fifth or less) the usual preference is for a stepped footing.

There are three critical considerations for the design of the stepped footing.

1. *The Length of the Step.* If the step length is too short, the individual steps will have questionable validity as individual footings, and the footing is effectively the same as a sloped one. As shown in Fig. 41.12, the toe portion of the step is essentially unusable for bearing. Thus if the step is very short, the length remaining for use as a bearing footing may be quite minor.

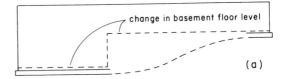
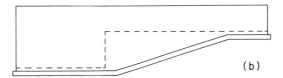
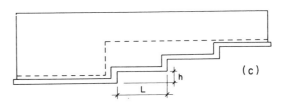

FIGURE 41.11. Change in elevation with a continuous wall.

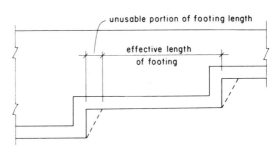
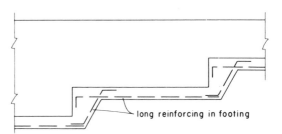

FIGURE 41.12. Considerations for a stepped footing.

2. *The Height of the Step.* The higher the step, the longer the unusable toe portion of the flat step. This has something to do with the soil type and is related to excavation problems. In soils that are unfeasible to excavate with a vertical cut it may be necessary to slope the stepped cut, as shown in the lower part of Fig. 41.12.

3. *The Angle of the Step.* The step angle—the h/L ratio as shown in the figures—is essentially the

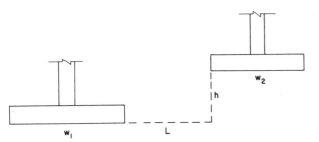

FIGURE 41.10. Dimensional relationships for footings at different elevations.

same problem as was discussed for the individual footings and illustrated in Fig. 41.10.

As in the case of individual footings, the details of stepped footings may be regulated by building codes or regulatory agencies. For a generally conservative design, and in the absence of other design limitations, we recommend the following.

Limit the step length to not less than three times the footing width (footing dimension perpendicular to the wall plane).

Limit the step height to not more than $1\frac{1}{2}$ times the footing width, or 2 ft, whichever is smaller.

Limit the h/L ratio to one-third or less.

Footings on Fill

In general it is desirable that bearing footings be placed on undisturbed soil, undisturbed meaning that the soil has not been previously excavated. If this is not feasible, the choices are limited to the use of a deep foundation or footings placed on fill materials. In some situations, when footing loads are relatively light, it may be reasonable to consider the latter option. Some such situations are the following:

1. When a relatively thin layer of undesirable material is encountered at the level desired for the footings. In this event it may be possible to deposit a thin layer of highly compacted material to replace the poor soil, and to end up with a much improved bearing condition.

2. When the soil at the desired bearing level is highly sensitive to disturbance by the excavation and foundation construction activities. This may be dealt with in a manner similar to the previous situation. If the footings are small, the pressures developed below the level of the fill may be able to be sustained by the weaker materials.

3. When the excavation of some unanticipated buried object, such as a large tree root, old sewer line, or large boulder, leaves the usable level for bearing a single footing considerably below that of adjacent footings. As in the previous cases, if a small amount of fill can be used, it may be a better choice than dropping all the adjacent footings.

41.6. ELEMENTS OF FOUNDATION SYSTEMS

The complete foundation system for most buildings includes various elements in addition to the basic supporting objects (footings or deep foundations). This chapter discusses various ordinary components of foundation systems that serve transitional functions between the building and the major foundation elements.

Foundation Walls

Foundation walls are walls that extend below the ground surface and perform some sort of transition between the foundation and the portions of the building that are above-ground. They are typically built of concrete or masonry. The structural and architectural functions of foundation walls vary considerably, depending on the type of foundation, the size of the building, climate and soil conditions, and whether or not they serve to form a basement.

Figure 41.13 shows some typical situations for the use of foundation walls for buildings without basements. In these cases the walls are not actually walls in the usual architectural sense. A principal difference relates to the construction of the building floor, whether it consists of a framed structure elevated above the ground, or concrete placed directly on the ground. Another major difference has to do with the distance of the foundation below the ground surface. If this distance is great, because of the need for frost protection or the need to reach adequate soil for bearing, the walls may be quite high. If these problems do not exist, the walls may be quite short, and scarcely constitute walls at all in the usual sense, especially when there is no crawl space, as shown in Fig. 41.13a.

When the wall is very short and building loads are low, it is sometimes possible to use a construction known as a grade beam. This consists of combining the functions of

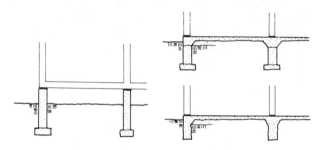

framed floor over a crawl space   concrete floor poured on the ground
(a)

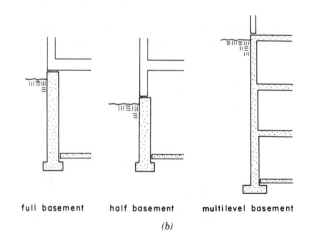

full basement   half basement   multilevel basement
(b)

FIGURE 41.13. Foundation walls.

foundation wall and wall footing into a single element that provides continuous support for elements of the building construction.

When a basement is required foundation walls are usually quite high. An exception is the case of a half-basement in which the basement floor is only a short distance below grade and the portion of the basement wall aboveground may be of different construction than that extending into the ground. If the basement floor is a significant distance below grade, major soil pressure will be exerted horizontally against the outside of the wall. In such a case the wall will function as a spanning element supported laterally by the footing or the basement floor at its lower end and by the building floor at its top. The design of basement walls is discussed in Sec. 32.4.

Figure 41.13*b* shows a number of situations in which foundation walls function to provide basement spaces. For buildings with multilevel basements, walls may become quite massive due to the accumulation of vertical load and the potential for considerable lateral soil pressure.

In addition to their usual functions of providing a ground-level edge for the building and support for elements of the building, walls often serve a variety of functions for the building foundation system. Some of these are as follows:

Load Distributing or Equalizing. Walls of some length and height typically constitute rather stiff, beamlike elements. Their structural potential in these cases is often utilized for load distributing or equalizing, as shown in Fig. 41.14. Even the shallow grade beam is typically designed with continuous top and bottom reinforcing to serve as a continuous beam for these purposes; hence the derivation of its name.

Spanning as a Load-Carrying Beam. Walls may be used as spanning members, carrying their own weight as well as some supported loads, as shown in Fig. 41.14. This is often the case in buildings with deep foundations and a column structure. Deep foundation elements are placed under the columns, and walls are used to span from column to column. This can also be the situation with bearing foundations if the column footings are quite large and reasonably closely spaced; rather than bearing on its own small footing, the stiff wall tends to span between the larger footings.

Distribution of Column Loads. When columns occur in the same plane as a foundation wall, many different relationships for the structural action of the walls, columns, and column foundations are possible. If the walls and columns are monolithically constructed, some load sharing is generally unavoidable. The various possibilities are discussed in the following section.

Transfer of Building Lateral Loads to the Ground. Although there are many possible situations for the development of lateral resistive structural systems, the total lateral load on the building must ultimately be transferred to the ground. The horizontal force com-

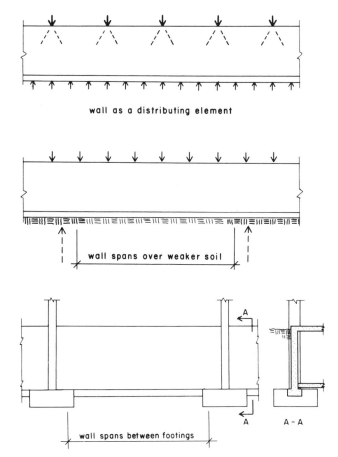

wall as a distributing element

wall spans over weaker soil

wall spans between footings

FIGURE 41.14. Spanning action of walls.

ponent is usually transferred through some combination of soil friction on the bottoms of footings and the development of passive lateral soil pressure against the sides of the footings and foundation walls. For large buildings with shallow below-grade structures this may be a major task for the foundations walls.

Ties, Struts, Collectors, etc. Foundation walls may be used to push or pull forces between separate elements of the below-grade structure. For seismic design it is usually required that the separate elements of the foundation be adequately tied to permit them to move as a single mass; where they exist, foundation walls may help to serve this purpose.

If a foundation wall is sufficiently tall, it should be treated as a wall and reinforced in both directions with minimal reinforcing for shrinkage and temperature stresses, as recommended by the *ACI Code* (see Sec. 32.2).

Columns in Walls

A common problem in the design of building foundations is a foundation wall that must share its location with a row of columns. This happens quite frequently along the exterior walls of buildings with frame structures. In the process of transferring loads to the ground there are various possibilities for the relationships between the columns, the wall, and the foundation elements. Some of the variables to consider in this situation are the following.

Magnitude of the Column Loads. It is one thing if the column loads are light, as in the case of a one-story building with light construction and short spans. It is quite another thing if the column loads are large, as in the case of a multistory building.

Spacing of the Columns. If columns are closely spaced, the idea of using the wall as a spanning or distributing element will have more merit. If columns are quite far apart, it becomes less reasonable to expect the wall to achieve these functions. There is no cutoff dimension for this effect by itself; it must be considered along with the column load, wall height, type of foundation, and so on.

Height of the Wall. For spanning or distributing functions the most critical dimension for the wall is its height. This relates to the relative efficiency of the wall as a beam and the classification of the wall as a spanning member. For the latter relationship, the wall may fall into three categories of behavior as a function of its span-to-depth ratio. As shown in Fig. 41.15, these are:

1. An ordinary flexural member (beam) with shear, moment, and deflection as we ordinarily consider it to develop for a beam.
2. A deep beam with virtually no deflection, and with stiffness in proportion to its span that significantly affects the nature of stress and strain on a vertical section in the member.
3. A member so stiff with respect to its span that there is essentially no flexure involved in its action. Instead, it functions in arching, or corbeling, action to bridge the space between supports.

The numerical values shown in Fig. 41.15 are approximate limits for identifying these behavioral differences. As the wall changes in depth-to-span ratio there is no sudden switch from one form of action to another, but rather a gradual shift.

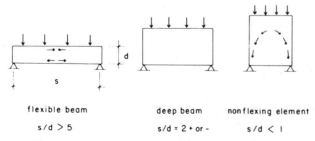

FIGURE 41.15. Effects of the span-to-depth ratio on the action of spanning elements.

flexible beam   deep beam   nonflexing element

$s/d > 5$ $s/d = 2 + or -$ $s/d < 1$

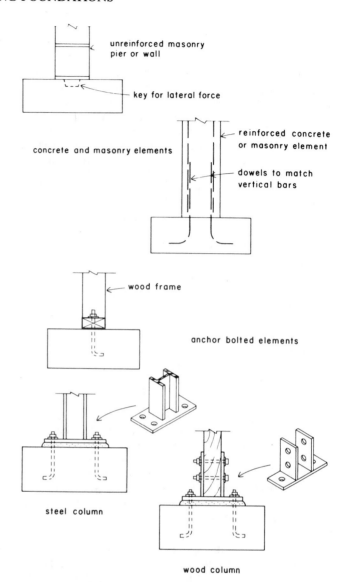

unreinforced masonry pier or wall

key for lateral force

concrete and masonry elements

reinforced concrete or masonry element

dowels to match vertical bars

wood frame

anchor bolted elements

steel column

wood column

FIGURE 41.16. Common attachments to footings.

Type of Foundation. If the foundation consists of deep elements, either piers or groups of piles, the wall is most likely to be designed as a spanning element of one type or another. When the foundation is of the shallow bearing type, there may be several options for the column/wall/footing relationships, as illustrated in the following example.

Support Considerations

Objects that sit on footings may be attached to the footings in a number of ways. As shown in Fig. 41.16, the basic types of attachment are as follows:

Direct Bearing without Anchorage. Direct bearing is the usual case with small, unreinforced masonry piers. No

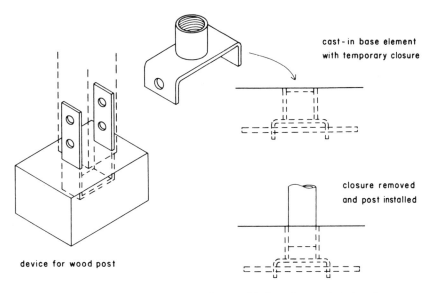

cast-in base element
with temporary closure

device for wood post

closure removed
and post installed

demountable base for round steel post

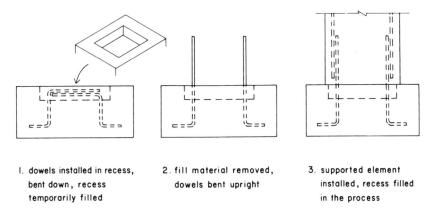

1. dowels installed in recess,
   bent down, recess
   temporarily filled

2. fill material removed,
   dowels bent upright

3. supported element
   installed, recess filled
   in the process

provision for future concrete or masonry element

FIGURE 41.17. Special attachment devices for footings.

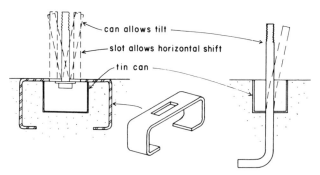

can allows tilt

slot allows horizontal shift

tin can

FIGURE 41.18. Means for providing adjustment for location of anchor bolts.

uplift resistance and little lateral load transfer are possible with this detail. Adhesion of mortar is the only actual attachment mechanism, although friction due to dead load will develop a lateral load transfer potential.

Doweling of Reinforcing Bars. Dowels provided for vertical reinforcement in concrete or masonry elements have the potential for providing uplift resistance and some lateral resistance.

Anchor Bolts. Bolts are commonly used to attach elements of wood, metal, and precast concrete. In most cases the principal function of these bolts is simply to hold the supported elements in place during construc-

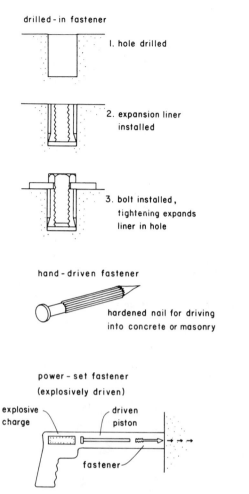

drilled - in fastener

1. hole drilled

2. expansion liner installed

3. bolt installed, tightening expands liner in hole

hand - driven fastener

hardened nail for driving into concrete or masonry

power - set fastener (explosively driven)

explosive charge

driven piston

fastener

FIGURE 41.19. Attachment devices installed after casting of concrete.

tion. However, as with dowels, there is a potential for development of resistance to uplift or lateral forces.

Special Embedded Anchors. Embedded anchors may be patented devices or custom-designed elements for various purposes. A variety of elements are available for attachment of wood columns. One special situation involves attaching elements to provide for subsequent removal, and possibly for repeated removal and reattachment. Another involves providing for future permanent attachment where addition to the structure is anticipated. Figure 41.17 shows a number of these types of attachment.

When a reinforced concrete column rests on a footing it is usually necessary to develop the vertical compressive stress in the column reinforcing by doweling action into the supporting concrete structure. For reinforcing bars of large size and high yield strength, the distance required for this development length can become considerable and may be a major consideration for determining the footing thickness. The various requirements for this situation are given in Section 15.8 of the *ACI Code* (Ref. 6).

Tension anchorage of reinforcing bars is commonly developed by a combination of dowel embedment action and hooking of the end of the bar. The *ACI Code* also specifies the details of hooks.

A major construction detail problem with using anchor bolts and other embedded anchoring elements is the accuracy of their placement. Foundation construction in general is typically quite crude and not capable of high precision. It therefore becomes necessary to make some provision for this potential inaccuracy if elements to be attached later require relatively precise positioning. Leveling beds of grout are used to develop precise vertical positioning. Some of the techniques for providing horizontal adjustment are shown in Fig. 41.18. It is possible to provide for adjustment of the anchor device itself, the object to be anchored, or a combination of both means.

It is also possible to attach elements to foundations after the foundation concrete is hardened through the use of drilled-in, driven, or explosively set anchoring devices. Although not desirable for major structural elements, this is quite commonly done for items such as partitions, window and door frames, signs, and equipment bases. Figure 41.19 shows some of the types of devices used for these purposes.

CHAPTER FORTY-TWO

Deep Foundations

In most instances deep foundations are utilized only where it is not possible to have shallow foundations. The decision to use deep foundations, the selection of the type of system, and the design of the elements of the system should all be done by persons with considerable knowledge and experience in foundation design. The material in Chapter 42 is provided to help the reader gain familiarity with the types of systems, their capabilities and limitations, and some of the problems involved in utilizing such systems for building foundations. This familiarity is important for all building designers, even though they may need expert advice and assistance for completion of the foundation design.

42.1. NEED FOR DEEP FOUNDATIONS

The most common reasons for using deep foundations are the following:

1. *Lack of Adequate Soil Conditions for Bearing Footings.* As shown in Figure 42.1, there are a number of soil conditions that may make it impossible to place the usual bearing foundation elements near the bottom of the building. The deep foundation thus becomes essentially a means of reaching a desirable bearing level at some distance from the bottom of the building.

2. *Heavy Loads on the Foundations.* In some cases the soil at upper levels may be sufficient for the use of bearing elements for relatively light loads, but the size of footings required, the need for limited settlement, or other factors occurring with excessively heavy loads, may require a deep foundation. Highrise buildings, long-span structures, and construction of massive elements of concrete or masonry are cases in which loads may become considerable and may exceed the simple bearing capabilities of ordinary soils.

3. *Potential Instability of Ground-Level Soil.* Use of deep foundations may be necessary where soil at the level of the bottom of the building is subject to erosion, subsidence, slippage, decomposition, or other forms of change in the soil structure or the general state of the soil mass. This situation some-

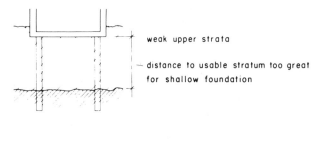

weak upper strata

distance to usable stratum too great for shallow foundation

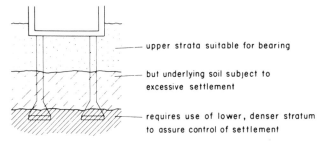

upper strata suitable for bearing

but underlying soil subject to excessive settlement

requires use of lower, denser stratum to assure control of settlement

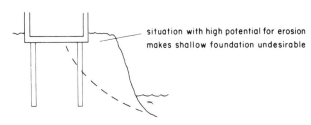

situation with high potential for erosion makes shallow foundation undesirable

FIGURE 42.1. Situations requiring deep foundations.

times occurs at waterfront and hillside locations. In these cases the use of deep foundations is a means for anchoring the building to a more reliable, stable ground mass.

4. *Support of Structures Highly Sensitive to Settlement.* In some situations settlement of foundations is highly critical. Examples are buildings with stiff, rigid frame structures and buildings housing equipment that requires precise and continuous alignment. Bearing foundations will settle on almost any soil other than solid bedrock. Deep foundations tend to have little settlement, espe-

cially when the elements are carried down to rock or to a highly consolidated soil stratum.

As stated previously, deep foundations are usually used only in situations in which shallow footings are not possible. The primary reason is cost. Where ordinary footings can be used and their size is not excessive, the cost of deep foundations will seldom be competitive. When footings cannot be used, this cost must be borne. For marginal situations—where footings are possible but their size is excessive—it may be necessary to perform a cost analysis of alternatives. Such an analysis will usually require relatively complete designs of the alternate systems and can be further complicated when the alternate foundation systems require significant differences in the building structure.

Foundation design and construction practices are often strongly influenced by the regional location of a building. This may be partly because of natural phenomena such as the local climate and soil conditions. Where there is some history of construction in the area, however, it may also be due to local experience with particular foundation systems or construction procedures. In the case of deep foundations, an important consideration is often the availability of local contractors with the necessary equipment and expertise for particular types of construction.

Most deep foundation elements—piles or piers—are installed by special foundation contractors, often using particular types of equipment or construction elements. Thus while the design of a foundation may be developed with a basic type of deep foundation in mind, the final design and detailing may have to be delayed until a particular contractor is selected.

In areas of considerable construction activity, especially near to large metropolitan areas, there may be competition between several companies specializing in deep foundations. In more remote areas there may be only a few companies, or even a single company, available for such work. Since the equipment necessary for this work is often quite large and not easily transportable, the distance from a contractor's home base to the building site is a critical cost factor.

Before beginning the design of any building foundation it is advisable to learn as much as possible about local experiences and practices. Evolution of technology and advances in design techniques may result in more intelligent designs, but it is still wise to proceed from a base of knowledge of previous construction experience. The proven need for and general feasibility of deep foundation construction is an area particularly sensitive to these considerations.

42.2. TYPES OF DEEP FOUNDATIONS

As shown in Fig. 42.2, the common basic types of deep foundations are the following:

Friction Piles. These usually consist of timber, steel, or precast concrete shafts that are forcibly inserted into the

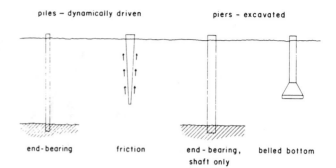

FIGURE 42.2. Types of deep foundations.

ground, most often by dynamic driving similar to a hammer pounding a nail into a piece of wood. Vertical load-carrying capability is developed by surface friction between the pile and the ground. Specific rating of load capacity is commonly established on the basis of the measured effort required for advancing the pile the last few feet of penetration.

End-Bearing Piles. These are usually elements similar to those used for friction piles, although in this case they are driven a specific distance in order to lodge their ends in some highly resistive soil stratum or in rock. While considerable skin friction may be developed during the advancement of the pile, the major load capacity is developed at the point of the pile. The chief function of the upper soil strata in this case is to maintain the lateral stability of the long compression member.

Piers or Caissons. For various reasons it may be better to place an end-bearing element by excavating the soil down to the level at which bearing is desired and then backfilling the excavation with concrete. The structural function of such an element is essentially the same as that of an endbearing pile. An advantage of this technique is that the material encountered at the bottom of the shaft can be inspected before pouring the concrete. The capacity of many endbearing piles, on the other hand, can only be inferred on the basis of the length driven, the difficulty of driving, or by load testing the driven pile.

Belled Piers. When piers bear on solid rock they usually have an end bearing capacity approximately equal to that of the concrete shaft in column action. When they bear on soil, however, they must usually be enlarged at their ends in order to increase the bearing area. The usual conical form of this end enlargement yields the term *belled* to describe such an element.

Pile foundations are discussed in Sec. 42.3 and piers in Sec. 42.4. The following are some of the problems encountered with both types of elements and some of their relative merits.

Piles must generally be driven in groups. One reason for this is their limited individual capacities, another is the problem of precisely controlling their locations during the placing process. Even when loads are small, it is generally not feasible to place a concentrated column load on top of a single pile since precise alignment of the column and pile is not feasible. The preferred minimum group for a concentrated load is three piles, typically arranged in a symmetrical, triangular pattern, as shown in Fig. 42.3. With piles spread a minimum distance, usually at least 30–36 in. center to center, a small mislocation of the column and pile group centroids can usually be tolerated. If one of the piles is slightly mislocated during driving it may be possible to alter the location of the rest of the group to compensate for this, as shown in the illustration. However, if the mislocation cannot be adjusted for in this manner, it becomes necessary to add more piles to the group in order to regain a reasonable alignment of the column and the group centroid.

When piles are driven in tight clusters and closely spaced groups, the driving of each successive pile often causes lateral movement of the top of previously driven piles. Thus it is necessary to check the final location of all piles before proceeding to the completion of the foundation construction. If substantial movements have occurred it may be necessary to add piles to some groups to once

again align the group centroids with the required column locations.

In order to achieve the transfer of load from a column to the several piles in a group it is necessary to use a stiff reinforced concrete cap on top of the piles. This element functions like an isolated column footing and also accommodates anchor bolts or reinforcing dowels.

Piers can be produced in a large range of sizes. Small piers of 12- to 36-in. diameter can be drilled with an auger-type rig, similar to the simple truck-mounted equipment used for well drilling or posthole drilling. This equipment is usually relatively light and portable compared to that required for pile driving. Belled ends can be excavated with the drilling rig using a spreader device once the pier shaft has been drilled.

For both piles and drilled piers limiting soil conditions are required. The presence of many large boulders may preclude either operation. Loose sand or a high water table may make it difficult to maintain the excavation while placing the concrete. Wet clay may offer resistance to pile driving or prove impossible to drill out so that excavation may have to be hand advanced. The limiting conditions for each type of system and each individual type of element and process must be carefully analyzed when establishing the feasibility of alternatives.

Large-diameter piers are usually excavated by direct digging, with the shaft walls lined and braced as the excavation proceeds. Once the pier excavation is completed, the lining is usually removed as the shaft is filled with concrete.

Placing of piers generally permits more precise control over the location of the shaft than that possible with piles. For this reason piers are usually not placed in clusters, or groups, but are merely increased or decreased in size as the load requirement varies. For drilled piers the shaft diameter may be held constant while the bells of individual piers are varied on the basis of the loads to be carried and/or the soil actually encountered upon excavation of the shaft to its final depth. Pier groups may be used when the structure to be carried is not a single column. Thus bearing walls, stair and elevator towers, and large items of equipment may be supported on several piers.

Although caps are not basically required for single piers, they are sometimes used. Caps may be used to eliminate the need for high accuracy of placement of anchor bolts or column reinforcing dowels at the time of the relatively rough foundation construction work. Caps are also used where a transition must be made between a very high strength concrete column and the usual low-strength pier shaft; thus the cap becomes similar in function to a pier for a footing.

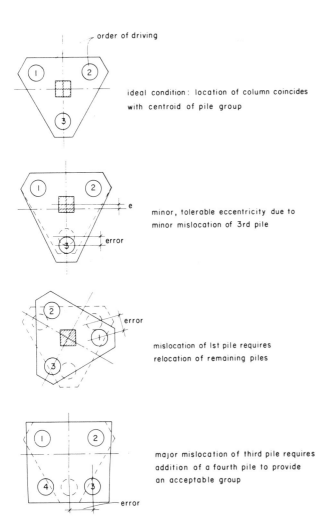

FIGURE 42.3. Adjusting for mislocation of piles.

42.3. PILES

Timber piles were used by ancient builders and are still in wide use today, primarily because of their low cost and widespread availability. However, industrialized countries also make use of steel and concrete for pile construction, especially where loads are high or rotting of timber piles is a potential problem. The major types of piles, in addition to timber, are steel, cast-in-place concrete, and precast concrete.

Timber Piles

Timber piles consist of straight tree trunks, similar to those used for utility poles, that are driven with the small end down, primarily as friction piles. Their length is limited to that obtainable from the species of tree available. In most areas where timber is plentiful, lengths up to 50 or 60 ft are obtainable, whereas piles up to 80 or 90 ft may be obtained in some areas. The maximum driving force, and consequently the usable load, is limited by the problems of shattering either the leading point or the driven end. It is generally not possible to drive timber piles through very hard soil strata or through soil containing large rocks. Usable design working loads are typically limited to 50–60 kips.

Decay of the wood is a major problem, especially where the tops of piles are above the groundwater line. Treatment with creosote will prolong the pile life, but is only a delaying measure, not one of permanent protection. One technique is to drive the wood piles below the water line and then build concrete piers on top of them up to the desired support level for the building.

For driving through difficult soils or to end bearing, wood piles are sometimes fitted with steel points. This reduces the problem of damage at the leading point, but does not increase resistance to shattering at the driven end.

Because of their relative flexibility, long timber piles may be relatively easily diverted during driving, with the pile ending up in something other than a straight, vertical position. The smaller the pile group, the more this effect can produce an unstable structural condition. Where this is considered to be a strong possibility, piles are sometimes deliberately driven at an angle, with the outer piles in a group splayed out for increased lateral stability of the group. While not often utilized in buildings, this splaying out, called *battering*, of the outer piles is done routinely for foundations for isolated towers and bridge piers in order to develop resistance to lateral forces.

Timber piles are somewhat limited in their ability to accommodate to variations in driven length. In some situations the finished length of piles can only be approximated, as the actual driving resistance encountered establishes the required length for an individual pile. Thus the specific length of the pile to be driven may be either too long or too short. If too long, the timber pile can easily be cut off. However, if it is too short, it is not so easy to splice on additional length. Typically, the lengths chosen for the piles are quite conservatively long, with considerable cutting off tolerated in order to avoid the need for splicing.

Cast-in-Place Concrete Piles

Figure 42.4 shows various methods of installing concrete piles in which the shaft of the pile is cast in place in the ground. Most of these systems utilize materials or equipment produced by a particular manufacturer, who in some cases is also the installation contractor. As shown in Fig. 42.4, the systems are as follows:

Armco System. In this system a thin-walled steel cylinder is driven by inserting a heavy steel driving core, called a *mandrel*, inside the cylinder. The cylinder is then dragged into the ground as the mandrel is driven. Once in place, the mandrel is removed for reuse, and the hollow cylinder is filled with concrete.

Raymond Step-Taper Pile. This is similar to the Armco system in that a heavy core is used to insert a thin-walled cylinder into the ground. In this case the cylinder is made of spirally corrugated sheet steel and has a tapered vertical profile, both of which tend to increase the skin friction.

Union Metal Monotube Pile. With this system the hollow cylinder is fluted longitudinally to increase its stiffness, permitting it to be driven without the mandrel. The fluting also increases the surface area, which tends to add to the friction resistance.

Franki Pile with Permanent Steel Shell. The Franki pile is created by depositing a mass of concrete into a shallow hole and then driving this concrete "plug" into the ground. Where a permanent liner is desired for the pile shaft, a spirally corrugated steel shell is engaged to the concrete plug and is dragged down with the driven plug. When the plug has arrived at the desired depth, the steel shell is then filled with concrete.

Franki Pile without Permanent Shell. In this case the plug is driven without the permanent shell. If conditions require it, a smooth shell is used and is withdrawn as the concrete is deposited. The concrete fill is additionally rammed into the hole as it is deposited, which assures a tight fit for better friction between the concrete and the soil.

Both length and load range is limited for these systems, based on the size of elements, the strength of materials, and the driving techniques. The load range generally extends from timber piles at the lower end up to as much as 400 kips for some systems.

Precast Concrete Piles

Some of the largest and highest load capacity piles have been built of precast concrete. In larger sizes these are usually hollow cylinders in order to reduce both the amount of material used and the weight for handling. These are more generally used for bridges and waterfront construction. A problem with these piles is establishing their precise in-place length. They are usually difficult to cut off as well as to splice. One solution is to produce them in modular lengths with a typical splice joint, which permits some degree of adjustment. The final finished top is then produced as a cast-in-place concrete cap.

In smaller sizes these piles are competitive in load capacity with those of cast-in-place concrete and steel. For

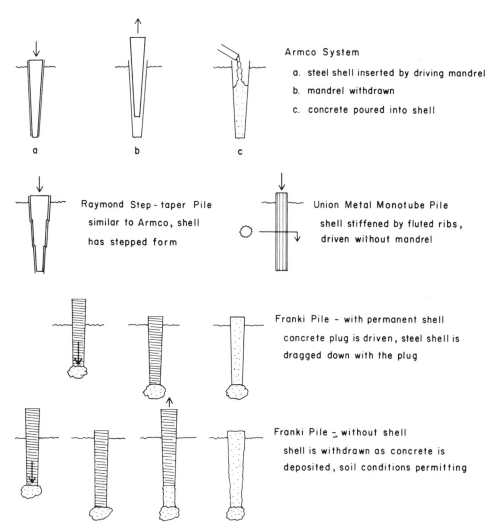

FIGURE 42.4. Systems for producing cast-in-place concrete piles.

deep-water installations huge piles several hundred feet in length have been produced. These are floated into place and then dropped into position with their own dead weight ramming them home, since driving such a large element is not possible.

Steel Piles

Steel pipes and H-sections are widely used for piles, especially where great length or load capacity is required or where driving is difficult and requires excessive driving force. Although the piles themselves are quite expensive, their ability to achieve great length, their higher load capacity, and the relative ease of cutting or splicing them may be sufficient advantages to offset their price. As with timber piles of great length, their relative flexibility presents the problems of assuring exact straightness during driving.

As with timber piles, deterioration above the ground-water level can be a problem. One solution is to cast a concrete jacket around the steel pile down to a point below the water level.

Table 42.1 summarizes information about some of the common types of piles for building construction. The data given are approximate and specific information about any type of pile should be obtained from local contractors or manufacturers.

When placed in groups, piles are ordinarily driven as close to each other as possible, primarily to reduce the structural requirements for the pile cap construction. The exact spacing allowable is related to the pile size and the driving technique. Ordinary spacings are 2 ft 6 in. for small timber piles and 3 ft for most other piles of the size range ordinarily used in building foundations.

Pile caps function much like column footings, and will generally be of a size close to that of a column footing for the same total load with a relatively high soil pressure. Pile layouts typically follow classical patterns, based on the number of piles in the group. Typical layouts are shown in Fig. 42.5. Special layouts, of course, may be used for groups carrying bearing walls, shear walls, elevator towers, combined foundations for closely spaced columns, and other special situations.

TABLE 42.1. Information for Commonly Used Piles

Type of Pile	Usual Size Range		Usual Load Capacity (kips)
	Diameter at Top (in.)	Length (ft)	
Wood	12 and up	50–80	60–100
Concrete, precast			
Solid	8–12	30–60	60–100
Hollow	No limit, 18–36 common	Up to 200	80–200
Concrete, cast-in-place	15–36	40–100	60–150
Steel			
H-section	8–14 (nominal)	Up to 300	80–400
Pipe	10–36	Up to 200	100–400

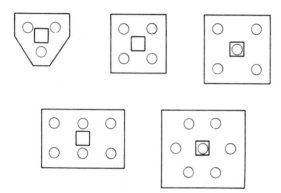

FIGURE 42.5. Typical pile group arrangements and caps.

FIGURE 42.6. Typical form of the drilled pier with a belled bottom.

Although the three-pile group is ordinarily preferred as the minimum for a column, the use of lateral bracing between groups may offer a degree of additional stability permitting the possibility of using a two-pile group, or even a single pile, for lightly loaded columns. This may extend the feasibility of using piles for a given situation, especially where column loads are less than that developed by even a single pile, which is not uncommon for single-story buildings of light construction and a low roof live load. Lateral bracing may be provided by foundation walls or grade beams or by the addition of ties between pile caps.

42.4. PIERS

When loads are relatively light, the most common form of pier is the drilled-in pier consisting of a vertical round shaft and a bell-shaped bottom, as shown in Fig 42.6. When soil conditions permit, the pier shaft is excavated with a large auger-type drill similar to that used for large post holes and water wells. When the shaft has reached the desired bearing soil strata, the auger is withdrawn and an expansion element is inserted to form the bell. The deci-

sion to use such a foundation, the determination of the necessary sizes and details for the piers, and the development of any necessary inspection or testing during the construction should all be done by persons with experience in this type of construction.

This type of foundation is usually feasible only when a reasonably strong soil can be reached with a minimum-length pier. The pier shaft is usually designed as an unreinforced concrete column, although the upper part of the shaft is often provided with some reinforcement. This is done to give the upper part of the pier some additional resistance to bending caused by lateral forces or column loads that are slightly eccentric from the pier centroid.

The usual limit for the bell diameter is three times the shaft diameter. With this as an upper limit, actual bell diameters are sometimes determined at the time of drilling on the basis of field tests performed on the soil actually encountered at the bottom of the shaft.

Table 42.2 gives the capacities for various-size belled piers for a range of allowable soil pressures. Loads are given for bell diameters of two and three times the shaft diameter. In addition to the load capacity based on the area of the bell, the table gives the corresponding minimum value for the concrete in the shaft. This minimum value for F_c' is based on a matching of the pier strength to the limit for the bell bearing capacity and the use of an allowable stress in the shaft of $f_c = 0.3f_c'$. While this is a stress limit that is commonly used, individual building codes may establish lower limits.

One of the advantages of drilled piers is that they may usually be installed with a higher degree of control on the final position of the pier tops than is possible with driven piles. It thus becomes more feasible to consider the use of a single pier for the support of a column load. For the support of walls, shear walls, elevator pits, or groups of closely spaced columns, however, it may be necessary to use clusters or rows of piers. The minimum spacing for such groups of piers is essentially limited by the bell diameters, with some spacing between edges of bells required to assure that the drilling of one pier does not disturb perviously installed piers.

TABLE 42.2. Load Capacity of Drilled Piers

Shaft Diameter (ft)	Bell Diameter (ft)	Bell Bearing Capacity (kips) and Minimum f_c' Required for Shaft[a] (kips/in²)			
		Allowable Soil Pressure for Bell (lb/ft²)			
		10,000	15,000	20,000	30,000
1.5	3	71	106	141	212
		2	2	2	3
	4.5	159	239	318	477
		2.5	3.5	4.5	6.5
2	4	126	188	251	377
		2	2	2	3
	6	283	424	565	848
		2.5	3.5	4.5	6.5
3	6	283	424	565	848
		2	2	2	3
	9	636	954	1272	1909
		2.5	3.5	4.5	6.5
4	8	503	754	1005	1508
		2	2	2	3
	12	1131	1696	2262	3393
		2.5	3.5	4.5	6.5
5	10	785	1178	1571	2356
		2	2	2	3
	15	1767	2651	3534	5301
		2.5	3.5	4.5	6.5
6	12	1131	1696	2262	3393
		2	2	2	3
	18	2545	3817	5089	7634
		2.5	3.5	4.5	6.5
7	14	1539	2309	3079	4618
		2	2	2	3
	21	3464	5195	6927	10391
		2.5	3.5	4.5	6.5

[a] Based on allowable $f_c = 0.30f_c'$.

42.5.   LATERAL, UPLIFT, AND MOMENT EFFECTS ON DEEP FOUNDATIONS

The resistance to horizontal forces, vertically directed upward forces, and moments presents special problems for deep foundation elements. Whereas a bearing footing has no potential for the development of tension between the structure and the ground, both piles and piers have considerable capacity for uplift resistance. On the other hand, the sliding friction that constitutes a major resistance to horizontal force by a bearing footing is absent with deep foundation elements. The following discussion deals with some of the problems of designing deep foundations for force effects other than the primary one of vertically directed downward load.

Lateral Force Resistance

Resistance to horizontal force at the top of both piles and piers is very poor in most cases. The relatively narrow profile offers little contact surface for the development of passive soil pressure. In addition, the process of installation generally causes considerable disturbance of the soil around the top of the foundation elements. Although there are procedures for determination of the resistance that can be developed by the passive pressure, this resistance is seldom utilized as the major force-resolving effect in building design. In addition to the low magnitude of the force that can be developed, there is the problem of considerable movement due to the soil deformation, which constitutes a dimensional distortion that few buildings can tolerate.

The usual method for resolving horizontal forces with deep foundation systems is to transfer the forces from the piles or piers to other parts of the building construction. In most cases this means the use of ties and struts to transfer the forces to grade walls or basement walls that offer considerable surface area for the development of passive soil pressures. This procedure is also used with bearing footings when the footings themselves are not capable of the total force resistance required.

For large freestanding structures without grade walls, the horizontal force resistance of pile groups is usually developed through the use of some piles driven at an angle. These *battered piles* are capable of considerable horizontal force resistance, in both compression and tension. This practice is common in the construction of foundations for large towers, bridge abutments, and so on, but is rarely utilized in building construction, where the load sharing described previously is usually the more feasible design option.

Uplift Resistance

Friction piles and large piers have considerable resistance to upward forces. If skin friction is truly the main resistive force that constitutes the pile capacity to sustain downward load, then it should also resist force in the opposite direction. An exception is the pile with a tapered form, which will have slightly higher resistance in one direction. Another exception is the unreinforced concrete pile, which has considerably more resistance to compressive stresses than it has to tensile stresses on the pile shaft.

The combined weight of the shaft and bell of a drilled pier offers a considerable potential force for resistance to upward loads. In addition, if the bell is large in diameter, considerable soil pressure can be developed against the upward withdrawal of the pier from the ground. Finally, if the shaft is long and the hole is without a permanent steel casing, some skin friction will be developed. These effects can be combined into a major resistance to upward force. However, the concrete shaft must be heavily reinforced, or prestressed, in order to develop the full potential resistance of the pier.

Endbearing piles usually have much less resistance to uplift in comparison to their compressive load capacities. However, in spite of the existence of relatively weak upper soil strata, there is usually some potential for skin friction resistance to withdrawal of the pile.

Uplift resistance of piles must usually be established by field load tests if they are to be relied on for major design loads. For large piers, and also for very large individual piles, this may not be feasible, because of the load magnitudes involved. However, in many design situations the uplift forces required are less than the limiting capacity of the deep elements, in which case it is sometimes acceptable to rely on a very conservative calculation of the uplift capacity.

The potential for uplift resistance on the part of deep foundation elements is an element of design that is not present with bearing foundations. Thus design for moments or for actual tension anchorage may be approached differently with deep foundations. In the case of the shear wall examples illustrated in Sec. 41.4, for example, the concerns for overturn and maximum soil pressure must be resolved without any reliance on tensile resistance between the footing and the soil. Thus the footing length and the length of the large grade beam must both be increased until the necessary relationships are developed through the use of weight and compression stress on the soil. With deep foundations this structure could possibly be reduced by shortening the grade beam and relying on the tensile capacity of the deep foundation elements.

Moment Resistance

Piles and piers are seldom deliberately designed to develop moments. Although any element strong enough to function as a pile or pier inevitably possesses some bending strength, it is generally considered desirable to design for an ideal condition of axial load only. However, the unavoidable inaccuracies inherent in the construction processes sometimes make it impractical to assure the perfect alignment inferred by the assumption of axial loading. Thus the true evaluation of any structural element that is primarily intended to carry axial load usually includes some consideration of the probability of accidental moments produced by eccentricity of the load.

The larger the diameter of a pile or pier, the larger the potential eccentricity that can be tolerated with a significant load magnitude. If the construction process can reasonably assure a dimensional control of placement error within this tolerable eccentricity, then potential accidental misalignment may not be a problem. This is the primary judgment to be made in justifying the use of a single pile or pier for a single concentrated load, such as that from an individual column.

42.6. POLE FOUNDATIONS

When a slender, vertically cantilevered element develops resistance to lateral force by its simple embedment in the ground, it is described as a pole or a polelike structure. Wood fence posts and utility poles for power lines are examples of these structures. The general nature of such a structure is discussed in Sec. 41.4 and illustrated in Fig. 41.5c.

Building code requirements generally deal with the pole structure as being one of two situations, as illustrated in Fig. 42.7. If construction exists at grade level, the lateral movement of the pole may be constrained at this location, so that a rotation of the pole occurs at grade level, and the development of resistance to lateral load is as shown in Fig. 42.7a. If grade-level constraint is minor or nonexistent, lateral resistance will be developed by opposing soil pressures on the buried portion of the pole, as shown in Fig. 42.7b.

The following example illustrates the use of design criteria for the unconstrained pole taken from the *UBC* and the *City of Los Angeles Building Code*.

Example 1. A 12-in.-diameter round wood pole is used as shown in Fig. 42.8. The soil around the buried pole is generally a medium compacted silty sand. Investigate the adequacy of the 10-ft embedment.

Solution. Using criteria from Sec. 91.2311 of the *City of Los Angeles Building Code* (see Appendix D), a determination is made of the two critical soil stresses, f_1 and f_2, as shown in Fig. 42.7b. These computed stresses are then compared to the allowable pressures. Using the formulas from the code, we find

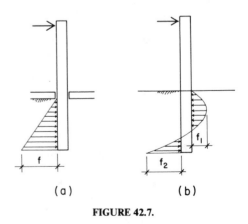

FIGURE 42.7.

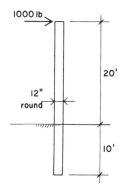

FIGURE 42.8.

$$f_2 = \frac{7.62P(2h + d)}{bd^2} = \frac{7.62(1000)(50)}{1(10)^2}$$

$$= 3810 \text{ psf}$$

$$f_1 = \frac{2.85P}{bd} + \frac{f_2}{4} = \frac{2.85(1000)}{1(10)} + \frac{3810}{4}$$

$$= 285 + 953 = 1238 \text{ psf}$$

From Table 28-B of the *Los Angeles Code* the allowable lateral-bearing pressure for the compact silty sand is 233 psf/ft of depth. For f_1 the depth is taken as one-third the total; thus

$$\text{allowable } p = \frac{10}{3}(233)\left(\frac{4}{3}\right) = 1036 \text{ psf}$$

(assuming that the lateral force is due to wind or seismic load, and that the increase of one-third is permitted).

For f_2 the allowable pressure is

$$p = 10(233)\left(\frac{4}{3}\right) = 3107 \text{ psf}$$

As both of the computed pressures exceed the allowable values, it is observed that the embedment is not adequate.

Solving for the required depth of embedment is not very direct with the formulas from the *Los Angeles Code*. On the other hand, the UBC provides a formula for the direct determination of the depth of embedment as follows (see Appendix D):

From *UBC*, Sec. 2907(g)2,

$$d = \frac{A}{2}\left(1 + \sqrt{1 + \frac{4.36h}{A}}\right)$$

$$A = \frac{2.34P}{S_1 b}$$

From *UBC* Table 29-B, p for silty sand = 150 psf/ft of depth; thus

$$S_1 = 150\left(\frac{10}{3}\right)\left(\frac{4}{3}\right) = 667 \text{ psf}$$

$$A = \frac{2.34(1000)}{667(1)} = 3.51$$

$$d = \frac{3.51}{2}\left(1 + \sqrt{1 + \frac{4.36(20)}{3.51}}\right)$$

$$= 14.24 \text{ ft}$$

which also indicates that the proposed embedment is not adequate.

CHAPTER FORTY-THREE

Special Foundation Problems

43.1. HORIZONTAL FORCES IN SOILS

There are a number of situations involving horizontal forces in soils. The three major ones of concern in foundation design are the following:

Active Soil Pressure. Active soil pressure originates with the soil mass; that is, it is pressure exerted *by* the soil on something, such as the outside surface of a basement wall.

Passive Soil Pressure. Passive soil pressure is exerted *on* the soil, for example, that developed on the side of a footing when horizontal forces push on the footing.

Friction. Friction is the sliding effect developed between the soil and the surface of some object in contact with the soil. To develop friction there must be some pressure between the soil and the contact face of the object.

The development of all these effects involves a number of different stress mechanisms and structural behaviors in the various types of soils. A complete treatment of these topics is beyond the scope of this book, and the reader is referred to other references for such a discussion (see *Foundation Engineering*, Ref. 20). The discussion that follows will explain the basic phenomena and illustrate the use of some of the simple techniques for design utilizing data and procedures from existing codes.

Active Soil Pressure

The nature of active soil pressure can be visualized by considering the situation of an unrestrained vertical cut in a soil mass, as shown in Fig. 43.1. In most soils, such a cut, as shown in Fig. 43.1a, will not stand for long. Under the action of various influences, primarily gravity, the soil mass will tend to move to a form as shown in Fig. 43.1b, producing an angled profile rather than the vertical face.

There are two general forces involved in the change from the vertical to the sloped cut profile. The soil near the top of the cut tends to simply drop vertically under its own weight. The soil near the bottom of the cut tends to bulge out horizontally from the cut face, being squeezed by the soil mass above it. Another way to visualize this movement is to consider that the whole moving soil mass tends to

rotate with respect to a slip plane such as that indicated by the heavy dashed line in Fig. 43.1c.

If a restraining structure of some type is introduced at the vertical face, the forces exerted on it by the soil will tend to be those involved in the actions illustrated in Fig. 43.1. As shown in Fig. 43.1d, the soil mass near the top of the cut will have combined vertical and horizontal effects. The horizontal component of this action will usually be minor, since the mass at this location will move primarily downward, and may develop significant friction on the face of the restraining structure. The soil mass near the bottom of the cut will exert primarily a horizontal force, very similar to that developed by a liquid in a tank. In fact, the most common approach to the design for this situation is to assume that the soil acts as a fluid with a unit density of some percentage of the soil weight and to consider a horizontal pressure that varies with the height of the cut, as shown in Fig. 43.1d.

The simplified equivalent fluid pressure assumption is in general most valid when the retained soil is a well-drained sandy soil. If the soil becomes saturated, the water itself will increase the horizontal pressure, and the buoyancy effect will tend to reduce the resistance to the slip-plane rotation of the soil mass as illustrated in Fig. 43.1c. If the soil contains a high percentage of silt or clay, the simple linear pressure variation as a function of the height is quite unrealistic.

In addition to considerations of the soil type and the water content, it is sometimes necessary to deal with a surcharge on the retained soil mass. As shown in Fig. 43.1e, the two common situations involving a surcharge occur when the soil mass is loaded by some added vertical force, such as the wheel load of a vehicle, and when the ground profile is not flat, producing the effect of raising its level with respect to the top of the restraining structure. The surcharge tends to increase the pressure near the top of the wall. When the equivalent fluid pressure method is used, the usual procedure is to consider the top of the soil mass (and the zero stress point for the fluid pressure) to be above the top of the structure. This results in a general overall increase in the fluid pressure, which is somewhat conservative in the case of the wheel load whose effect tends to diminish with distance below the contact point. When handled as fluid pressure, the surcharge is sometimes sim-

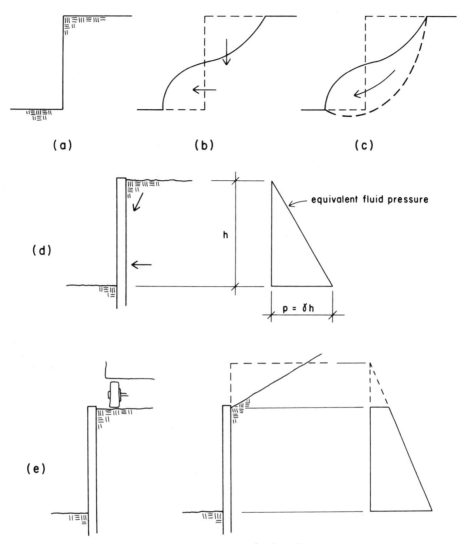

FIGURE 43.1. Development of active soil pressure.

ply visualized either as an increase in the assumed density of the equivalent fluid or as the addition of a certain height of soil mass above the top of the retaining structure.

Passive Soil Pressure

Passive soil pressure is visualized by considering the effect of pushing some object through the soil mass. If this is done in relation to a vertical cut, as shown in Fig. 43.2a, the soil mass will tend to move inward and upward, causing a bulging of the ground surface behind the cut. If the slip-plane type of movement is assumed, the action is similar to that of active soil pressure, with the directions of the soil forces simply reversed. Since the gravity load of the upper soil mass is a useful force in this case, passive soil resistance will generally exceed active pressure for the same conditions.

If the analogy is made to the equivalent fluid pressure, the magnitude of the passive pressure is assumed to vary with depth below the ground surface. Thus for structures

whose tops are at ground level, the pressure variation is the usual simple triangular form as shown in the left-hand illustration in Fig. 43.2b. If the structure is buried below the ground surface, as is the typical case with footings, the surcharge effect is assumed and the passive pressures are correspondingly increased.

As with active soil pressure, the type of soil and the water content will have some bearing on development of stresses. This is usually accounted for by giving values for specific soils to be used in the equivalent fluid pressure analysis.

Soil Friction

The potential force in resisting the slipping between some object and the soil depends on a number of factors, including the following principal ones:

Form of the Contact Surface. If a smooth object is placed on the soil, there will be a considerable tendency for it to slip. Our usual concern is for a contact surface

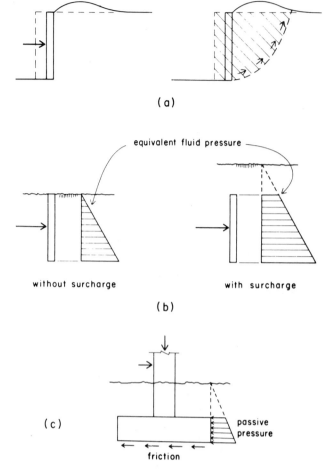

FIGURE 43.2. Development of passive soil pressure.

created by pouring concrete directly onto the soil, which tends to create a very nonsmooth, intimately bonded surface.

Type of Soil. The grain size, grain shape, relative density, and water content of the soil are all factors that will affect the development of soil friction. Well-graded dense, angular sands and gravels will develop considerable friction. Loose, rounded, saturated, fine sand and soft clays will have relatively low friction resistance. For sand and gravel the friction stress will be reasonably proportional to the compressive pressure on the surface, up to a considerable force. For clays, the friction tends to be independent of the normal pressure, except for the minimum pressure required to develop any friction force.

Pressure Distribution on the Contact Surface. When the normal surface pressure is not constant, the friction will also tend to be nonuniform over the surface. Thus, instead of an actual stress calculation, the friction is usually evaluated as a total force in relation to the total load generating the normal stress.

Friction seldom exists alone as a horizontal resistive force. Foundations are ordinarily buried with their bot-

toms some distance below the ground surface. Thus pushing the foundation horizontally will also usually result in the development of some passive soil pressure, as shown in Fig. 43.2c. Since these are two totally different stress mechanisms, they will actually not develop simultaneously. Nevertheless, the usual practice is to assume both forces to be developed in opposition to the total horizontal force on the structure.

In situations where simple sliding friction is not reliable or the total resistance offered by the combination of sliding and passive pressure is not adequate for total force resistance, a device called a *shear key* is used. Utilizing such a device is discussed in Sec. 41.4 in connection with the design of retaining walls. The enhancement of force resistance offered by a shear key is particularly desirable when the soil at the footing bottom is quite slippery (wet clay, etc.) or the footing bottom is a very short distance below grade.

43.2. ABUTMENTS

The support of some types of structures, such as arches, gables, and shells, often requires the resolution of both horizontal and vertical forces. When this resolution is accomplished entirely by the supporting foundation element, the element is described as an abutment. Figure 43.3a shows a simple abutment for an arch consisting of a rectangular footing and an inclined pier. The design of such a foundation has three primary concerns as follows:

Resolution of the Vertical Force. This consists of assuring that the vertical soil pressure does not exceed the maximum allowable value for the soil.

Resolution of the Horizontal Force. If the abutment is freestanding, resolution of the horizontal force means the development of sufficient soil friction and passive horizontal pressure.

Resolution of the Moment Effect. In this case the aim is usually to keep the resultant force as close as possible to the centroid of the footing plan area. If this is truly accomplished, that is, $e = 0$, there will literally be no moment effect on the footing itself.

Figure 43.3b shows the various forces that act on an abutment such as that shown for the arch in Fig. 43.3a. The active forces consist of the load and the weights of the pier, the footing, and the soil above the footing. The reactive forces consist of the vertical soil pressure, the horizontal friction on the bottom of the footing, and the passive horizontal soil pressure against the sides of the footing and pier. The dashed line in the illustration indicates the path of the resultant of the active forces; the condition shown is the ideal one, with the path coinciding with the centroid of the footing plan area at the bottom of the footing.

If the passive horizontal pressure is ignored, the condition shown in Fig. 43.3b will result in no moment effect on the bottom of the footing and a uniform distribution of the vertical soil pressure. If the passive horizontal pressure is included in the force summation, the resultant path will move slightly to the right of the footing centroid. However, for the abutment as shown, the resultant of the passive pressure will be quite close to the bottom of the footing, so that the error is relatively small.

If the pier is tall and the load is large with respect to the pier weight or is inclined at a considerable angle from the vertical, it may be necessary to locate the footing centroid at a considerable distance horizontally from the load point at the top of the pier. This could result in a footing of greatly extended length if a rectangular plan form is used. One device that is sometimes used to avoid this is to use a T-shape, or other form, that results in a relocation of the centroid without excessive extension of the footing. Figure 43.3c shows the use of a T-shape footing for such a condition.

When the structure being supported is symmetrical, such as an arch with its supports at the same elevation, it may be possible to resolve the horizontal force component at the support without relying on soil stresses. The basic technique for accomplishing this is to tie the two opposite

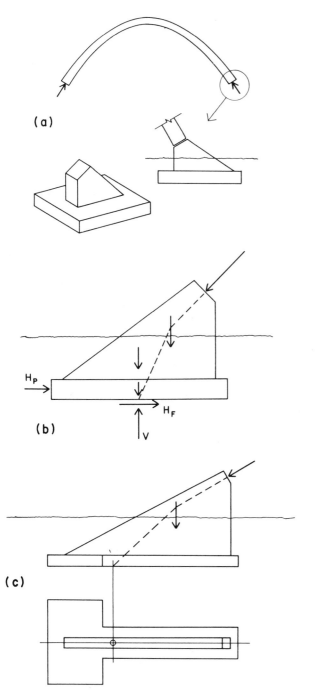

FIGURE 43.3. Abutments for arches.

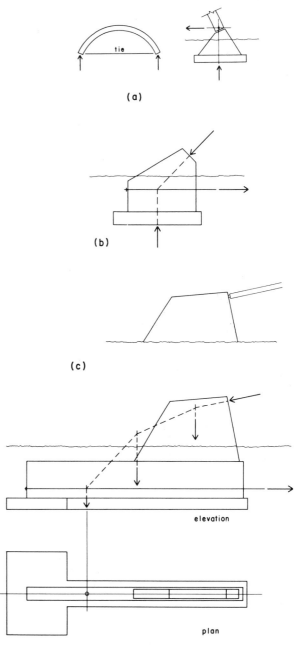

FIGURE 43.4. Abutments for tied arches.

supports together, as shown in Fig. 43.4a, so that the horizontal force is resolved internally (within the structure) instead of externally (by the ground). If this tie is attached at the point of contact between the structure and the pier, as shown in Fig. 43.4a the net load delivered to the pier is simply a vertical force, and the pier and footing could theoretically be developed in the same manner as that for a truss or beam without the horizontal force effect. However, since either wind or seismic loading will produce some horizontal force on the supports, the inclined pier is still the normal form for the supporting structure. The position of the footing, however, would usually be established by locating its centroid directly below the support point, as shown in the illustration.

For practical reasons it is often necessary to locate the tie, if one is used, below the support point for the structure. If this support point is above ground, as it usually is, the existence of the tie above ground is quite likely to interfere with the use of the structure. A possible solution in this problem is to move the tie down to the pier, as shown in Fig. 43.4b. In this case the pier weight is added to the load to find the proper location for the footing centroid.

When the footing centroid must be moved a considerable distance from the load point, it is sometimes necessary to add another element to the abutment system. Figure 43.4c shows a structure in which a large grade beam has been inserted between the pier and the footing. The main purpose of this element is to develop the large shear and bending resistance required by the long cantilever distance between the ends of the pier and footing. In the example, however, it also serves to provide for the anchorage of the tie. Because of this location of the tie, the weights of both the pier and grade beam would be added to the load to find the proper location of the footing centroid. In this way the heavy grade beam further assists the footing by helping to move the centroid closer to the pier and reducing the cantilever distance.

43.3. PAVING SLABS

Sidewalks, driveways, and basement floors are typically produced by depositing a relatively thin coating of concrete directly on the ground surface. While the basic construction process is simple, a number of factors must be considered in developing details and specifications for a paving slab.

Thickness of the Slab

Pavings vary in thickness from a few inches (for residential basement floors) to several feet (for airport landing strips). Although more strength is implied by a thicker slab, thickness alone does not guarantee a strong pavement. Of equal concern is the reinforcement provided and the character of the subbase on which the concrete is poured. The minimum slab thickness commonly used in building floor slabs is $3\frac{1}{2}$ in. This relates specifically to the actual dimension of a nominal wood 2×4 and simplifies forming of the

edges of a slab pour. Following the same logic, the next-size jump would be to a $5\frac{1}{2}$-in.-thickness, which is the dimension of a nominal 2×6.

The $3\frac{1}{2}$-in.-thick slab is usually considered adequate for interior floors not subjected to wheel loadings or other heavy structural demand. At this thickness, usually provided with very minimal reinforcing, the slab has relatively low resistance to bending and shear effects of concentrated loads. Thus walls, columns, and heavy items of equipment should be provided with separate footings.

The $5\frac{1}{2}$-in.-thick slab is adequate for heavier live loads and for light vehicular wheel loads. It is also strong enough to provide support for light partitions, so that some of the extra footing construction can be eliminated.

For heavy truck loadings, for storage warehouses, and for other situations involving very heavy loads—especially concentrated ones—thicker pavements should be used, although thickness alone is not sufficient, as mentioned previously.

Reinforcing

Thin slabs are ordinarily reinforced with welded wire mesh. The most commonly used meshes are those with a square pattern of wires—typically 4- or 6-in. spacings—with the same wire size in both directions. This reinforcing is generally considered to provide only for shrinkage and temperature effects and to add little to the flexural strength of the slab. The minimum mesh, commonly used with the $3\frac{1}{2}$-in. slab, is a 6×6-10/10, which denotes a mesh with No. 10 wires at 6 in. on center in each direction. For thicker slabs the wire gage should be increased or two layers of mesh should be used.

Small-diameter reinforcing bars are also used for slab reinforcing, especially with thicker slabs. These are generally spaced at greater distances than the mesh wires and must be supported during the pouring operation. Unless the slab is actually designed to span, this reinforcing is still considered to function primarily for shrinkage and temperature stress resistance. However, since cracking in the exposed top surface of the slab is usually the most objectionable, specifications usually require the reinforcing to be kept some minimum distance from the top of the slab.

Subbase

The ideal subbase for floor slabs is a well-graded soil, ranging from fine gravel to coarse sand with a minimum of fine materials. This material can usually be compacted to a reasonable density to provide a good structural support, while retaining good drainage properties to avoid moisture concentrations beneath the slab. Where ground water conditions are not critical, this base is usually simply wetted down before pouring the concrete and the concrete is deposited directly on the subbase. The wetting serves somewhat to consolidate the subbase and to reduce the bleeding out of the water and cement from the bottom of the concrete mass.

To reduce further the bleeding-out effect, or where moisture penetration is more critical, a lining membrane is often used between the slab and the subgrade or base. This usually consists of a 6-mil plastic sheet or a laminated paper—plastic—fiberglass product that possesses considerably more tear resistance and that may be desirable where the construction activity is expected to increase this likelihood.

Joints

Building floor slabs are usually poured in relatively small units, in terms of the horizontal dimension of the slab. The main reason is to control shrinkage cracking. Thus a full break in the slab, formed as a joint between successive pours, provides for the incremental accumulation of the shrinkage effects. Where larger pours are possible or more desirable, control joints are used. These consist of tooled or sawed joints that penetrate some distance down from the finished top surface.

Surface Treatment

Where the slab surface is to serve as the actual wearing surface, the concrete is usually formed to a highly smooth surface by troweling. This surface may then be treated in a number of ways, such as brooming it to make it less slippery, or applying a hardening compound to further toughen the wearing surface. When a separate material—such as tile or a separate concrete fill—is to be applied as the wearing surface, the surface is usually kept deliberately rough. This may be achieved by simply reducing the degree of finished troweling.

Weather Exposure

Once the building is enclosed, interior floor slabs are not ordinarily exposed to exterior weather conditions. In cold climates, however, freezing and extreme temperature ranges should be considered if slabs are exposed to the weather. This may indicate the need for more temperature reinforcing, less distance between control joints, or the use of materials added to the concrete mix to enhance resistance to freezing.

43.4. FRAMED FLOORS ON GRADE

It is sometimes necessary to provide a concrete floor poured directly on the ground in a situation that precludes the use of a simple paving slab and requires a real structural spanning capability of the floor structure. One of these situations is where a deep foundation is provided for support of walls and columns and the potential settlement of upper ground masses may result in a breaking up and subsidence of the paving. Another situation is where considerable fill must be placed beneath the floor, and it is not feasible to produce a compaction of this amount of fill to assure a steady support for the floor.

Figure 43.5 illustrates two techniques that may be used to provide what amounts to a framed concrete slab and beam system poured directly on the ground. Where spans are modest and beam sizes not excessive it may be possible to produce the system in a single pour by simply trenching for the beam forms, as shown in the upper illustration in Fig. 43.5. When large beams must be provided, it may be more feasible to use the system shown in the lower part of Fig. 43.5. In this case the stems of the large beams are formed and poured, and the slab is poured separately on the fill placed between the beam stems. These two techniques can be blended, of course, with smaller beams trenched in the fill between the large formed beams.

If the system with separately poured beams and slabs is used, it is necessary to provide for the development of shear between the slab and the top of the beam stems. Depending on the actual magnitude of the shear stresses involved, this may be done by various means. If stress is low, it may be sufficient to require a roughening of the surface of the top of the beam stems. If stress is of significant magnitude, shear keys, similar to those used for shear walls, may be used. If stirrups or ties are used in the beam stems, these will extend across the joint and assist in the development of shear.

43.5. TENSION-RESISTIVE FOUNDATIONS

Tension-resistive foundations are a special, although not unique, problem. Some of the situations that require this type of foundation are the following:

Anchorage of very lightweight structures, such as tents, air-inflated structures, light metal buildings, and so on.

Anchorage of cables for tension structures or for guyed towers.

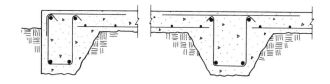

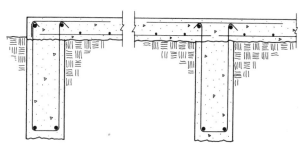

FIGURE 43.5. Details of concrete framed systems poured on grade.

Anchorage for uplift resistance as part of the development of overturn resistance for the lateral bracing system for a building.

Figure 43.6 illustrates a number of elements that may be used for tension anchorage. The simple tent stake is probably the most widely used temporary tension anchor. It has been used in sizes ranging from large nails up to the huge stakes used for large circus tents. Also commonly used is

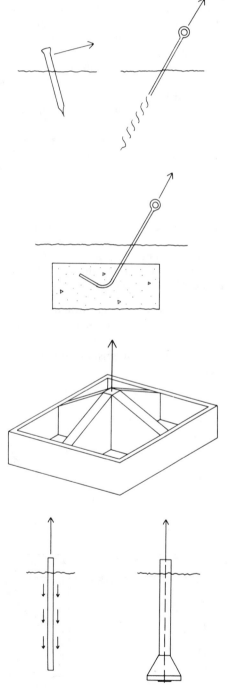

FIGURE 43.6. Forms of tension anchors.

the screw-ended stake, which offers the advantages of being somewhat more easily inserted and withdrawn, and having less tendency to loosen.

Ordinary concrete bearing foundations offer resistance to tension in the form of their own dead weight. The *dead man anchor* consists simply of a buried block of concrete similar to a simple footing. Column and wall footings, foundation walls, concrete and masonry piers, and other such heavy elements may be utilized for this type of anchorage. Many lightweight building structures are essentially anchored by being fastened to their heavy foundations.

Where resistance to exceptionally high uplift force is required, special anchoring foundations may be used that develop resistance through a combination of their own dead weight plus the ballast effect of earth fill placed in or on them. Friction piles and piers may also be used for major uplift resistance, although if their shafts are of concrete, care should be taken to reinforce them adequately for the tension force. A special technique is to use a belled pier to resist force by its own dead weight plus that of the soil above it, since the soil must be pushed up by the bell in order to extract the pier. One method for the development of the tension force through the bell end is to anchor a cable to a large plate, which is cast into the bottom of the bell, as shown in Fig. 43.6.

The nature of tension forces must be considered as well as their magnitude. Forces caused by wind or seismic shock will have a jarring effect that can loosen or progressively weaken anchorage elements. If the surrounding soil is soft and easily compressed, the effectiveness of the anchor may be reduced.

43.6. HILLSIDE FOUNDATIONS

Figure 43.7a illustrates the situation of a footing in a hillside location that could be a wall footing or an isolated column footing. The dimension in Fig. 43.7a, called the *daylight dimension* must be sufficient to provide resistance to failure of the soil. Lateral load on the footing will further aggravate failure if the daylight dimension is too small. The preferred solution, if the construction permits it, is to use a tension tie to transfer the lateral effect to some other part of the structure. Otherwise, the level of the footing should be lowered until a conservative distance is developed for the daylight dimension.

Many building codes require a minimum distance for the daylight dimension. Logically, the limit for this distance should depend on the type of soil and the angle of the slope of the ground surface. For low slope angles a reasonable daylight dimension is assured simply because of the usual requirements for a minimum depth of the bottom of the footing below the surface in a vertical direction. As the slope increases, the daylight dimension is less assured and should be limited to some minimum distance.

A common problem with hillside construction is that shown in Fig. 43.7b, where recontouring of the construction site results in a portion of the building being placed on

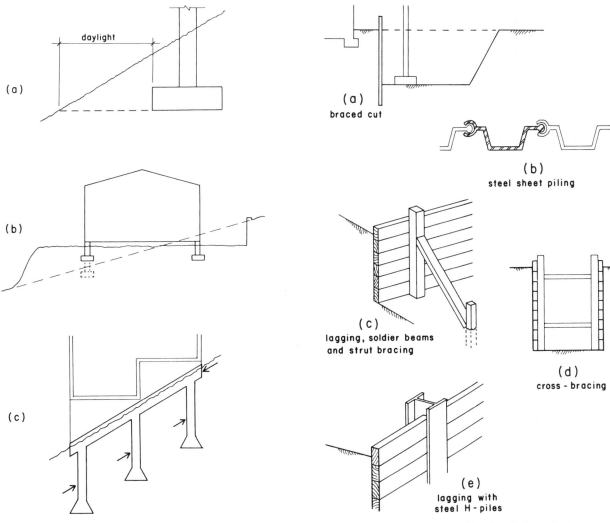

FIGURE 43.7. Hillside foundations.

FIGURE 43.8. Considerations for braced cuts.

some significant depth of fill. If possible, all of the footings should be carried down to a depth below the fill. Where this is not feasible it may be necessary to use deep foundation elements of piles or piers. If the structure is relatively light and the fill is of proper materials placed with reasonable compaction, it may be possible to rest footings in the fill, although such work should be done only with consultation and inspection by an experienced soils engineer.

A special structure sometimes used for hillside and beach locations is shown in Fig. 43.7c. This consists of a set of deep foundation elements—usually driven or cast-in-place piles—that are imbedded in a reasonably dense lower soil stratum and utilized as vertical cantilevers for lateral resistance. A grade level concrete frame is used to tie the piles together and provide direct support for the building. The general stiffness and overall effectiveness of this system are greatly increased if the tops of the piles can be made to form a rigid frame with the grade level structure. The term *downhill frame* is sometimes used to describe this

type of structure. It is frequently used where beach front or hillside surface erosion is a potential hazard.

43.7. BRACED CUTS

Shallow excavations can usually be made with no provision for the bracing of the sides of the excavation. However, when the cut is quite deep, the soil has little cohesive character, or undercutting of adjacent construction or property is a concern, it may be necessary to provide some form of bracing for the vertical face of the cut.

One means for bracing is through the use of sheet piling (Fig. 43.8b), which consists of pleated interlocking units of steel. These units are driven individually prior to the excavation work and may be withdrawn for reuse or become part of the permanent construction. If the cut is quite deep, it may be necessary to brace the top of the piling as the excavation proceeds.

A simpler form of bracing, used for shallow cuts and cuts where some minor movement is not critical, consists of stacked horizontal planks—called *lagging*—that are installed as the excavation proceeds. These must be braced by spaced vertical members—called *soldier beams*—which in turn require some bracing element, such as the diagonal struts shown in Fig. 43.8c. If the excavation is narrow, bracing may be achieved with cross members that use the opposed actions of the facing cuts (Fig. 43.8d).

For relatively deep cuts a form of bracing used consists of soldier beams of steel H-shaped sections driven like piles, with lagging installed between them as the excavation proceeds (Fig. 43.8e). As with sheet piling, the steel sections may be withdrawn when the bracing is no longer required.

Bracing is quite commonly required for construction on urban sites where undermining of sidewalks, streets, or adjacent property is critical. The design of such bracing must be carefully coordinated with the design of the building substructure. Elements of the bracing system may become parts of the permanent construction in some cases, or the permanent construction may merely replace the temporary bracing progressively as the construction work proceeds.

Existence of a high water level in the ground or the presence of troublesome soils, such as soft clay or fine sand, can seriously complicate excavation and the use of bracing.

Cut faces of soil intended as permanent conditions may, of course, be braced by various forms of permanent construction, such as cantilevered retaining walls or foundation walls of supported structures. Another device used in some situations, for both permanent and temporary cuts, is to anchor a wall facing back into the soil mass with drilled-in anchors.

PART 8

LATERAL FORCE EFFECTS

Part 8 deals generally with the topic of horizontal force effects in buildings. The term used for these effects is *lateral*, meaning sideways, which identifies them in relation to the major orientation of the effect of gravity as a vertical force. Conceptually, therefore, designing for lateral forces is typically viewed in terms of bracing a building against sideways collapse. In truth, most load sources that produce lateral forces also generate some vertical effects and so it is of limited use to treat the horizontal force effects in isolation. Even where this may be valid as an investigative technique, it should always be borne in mind that lateral effects always occur in combination with some vertical gravity effects. In the end it is the full combined effects that must be understood and dealt with.

While the issues of lateral forces are generally treated in this part, lateral effects are included in the work in several other sections in this book. The problem of horizontal forces on foundations is developed in Chapter 43. Some design examples of lateral resistive structures are presented in Part 8, but the building design examples in Part 9 present lateral force design in the broader context of the whole building structure.

CHAPTER FORTY-FOUR

General Considerations for
Lateral Effects

Chapter 44 describes the various sources of lateral loads and the effects that they produce.

44.1. SOURCES OF LATERAL LOADS

Wind

Wind is moving air. Air is a fluid, and some general knowledge of fluid mechanics is helpful for fully understanding the various effects of wind on buildings. Our primary concern here is for the effect of wind on the lateral bracing system for the building. As a net effect this force is an aggregate of the various aspects of the fluid flow of the air around the stationary object (the building) on the ground surface.

Earthquakes

Earthquakes—or *seismic activity* as it is called in engineering circles—produce various disastrous effects, including tidal waves, massive ruptures along earth faults, and violent vibratory motions. It is the last effect for which we design the lateral bracing systems for buildings, dealing mostly with the horizontal aspect of the ground motion. In a static equivalent sense, the force applied to the building structure is actually generated by the momentum of the building mass as it is impelled and rapidly reversed in direction. This activity cannot be fully understood in terms of static force alone, however, as dynamic aspects of both the ground motion and the building response must be considered. For readers with a limited background in dynamics the discussion presented in Chapter 15 may prove useful.

Soil Pressure

The problems of soil-retaining structures are discussed in Chapter 57. Horizontal forces on foundations are discussed with regard to various load sources and the problems of various types of lateral resistive structures.

Structural Actions

The natural action of various structures in resisting gravity loads may result in some horizontal forces on the supports of the structure even though the direction of the gravity load is vertical. Common examples of such structures are arches, gable roofs, cable structures, rigid frames, and pneumatic structures sustained by internal pressure.

Volume Change: Thermal, Moisture, and Shrinkage

Thermal expansion and contraction, moisture swelling and shrinkage, and the shrinkage of concrete, mortar, and plaster are all sources of dimensional change in the volume of building materials. Ideally, these effects are controlled through use of expansion joints, deformable joint materials, or other means. However, the potential forces that they represent must be understood and must actually be provided for in terms of structural resistance in some instances.

Relation of Lateral to Gravity Effects

Eventually, the structural designer must deal with the potential net critical effects of the various load combinations on a structure. Many of the examples of investigation in Part 8 are limited to the determination of the lateral effects alone. This is simply because this is the principal purpose of this part. The reader should be cautioned, however, that concentration on the resistance of a single loading condition may obscure the overall intelligent design of a structure or the building for which it exists.

Problems of Quantification

A major problem in designing for lateral effects is simply that of determining the magnitudes of the loads. This is probably easiest for structurally induced effects, although approximations in determining weights of the construction and complex behaviors of highly indeterminate structures may make quantification suspect in these situations as well. Translating the fluid flow effects of the wind into so many pounds of force on a structure is a convoluted exercise in fantasy. Precise prediction of potential ground movements caused by some hypothetical earthquake at a specific site and estimating the effects of the building's response and any site/structure interaction are conjectures of ethereal proportions. Determining or controlling the conditions of a specific soil mass is a highly approximate exercise. Precisely determining the dimensional changes of complex masses of construction due to thermal or moisture variation is not possible.

This is not intended as an argument for not being serious in designing for these effects. It is merely to make the point that what we do chiefly is provide the *kind* of structure that is likely to have the character suited for its task; its *exact* capacity is a very soft target. We do our best, but cannot pretend to claim great precision.

44.2.   APPLICATION OF WIND AND SEISMIC FORCES

To understand how a building resists the lateral load effects of wind and seismic force it is necessary to consider the manner of application of the forces and then to visualize how these forces are transferred through the lateral resistive structural system and into the ground.

Wind Forces

The application of wind forces to a closed building is in the form of pressures applied normal to the exterior surfaces of the building. In one design method the total effect on the building is determined by considering the vertical profile, or silhouette, of the building as a single vertical plane surface at right angles to the wind direction. A direct horizontal pressure is assumed to act on this plane.

Figure 44.1 shows a simple rectangular building under the effect of wind normal to one of its flat sides. The lateral resistive structure that responds to this loading consists of the following:

Wall surface elements on the windward side are assumed to take the total wind pressure and are typically designed to span vertically between the roof and floor structures.

Roof and floor decks, considered as rigid planes (called diaphragms), receive the edge loading from the windward wall and distribute the load to the vertical bracing elements.

Vertical frames or shear walls, acting as vertical cantilevers, receive the loads from the horizontal diaphragms and transfer them to the building foundations.

The *foundations* must anchor the vertical bracing elements and transfer the loads to the ground.

The propagation of the loads through the structure is shown in the left part of Fig. 44.1, and the functions of the major elements of the lateral resistive system are shown in the right part of the figure. The exterior wall functions as a simple spanning element loaded by a uniformly distributed pressure normal to its surface and delivering a reaction force to its supports. In most cases, even though the wall may be continuous through several stories, it is considered as a simple span at each story level, thus delivering half of its load to each support. Referring to Fig. 44.1, this means that the upper wall delivers half of its load to the roof edge and half to the edge of the second floor. The lower wall delivers half of its load to the second floor and half to the first floor.

This may be a somewhat simplistic view of the function of the walls themselves, depending on their construction. If they are framed walls with windows or doors, there may be many internal load transfers within the wall. Usually, however, the external load delivery to the horizontal structure will be as described.

The roof and second-floor diaphragms function as spanning elements loaded by the edge forces from the exterior wall and spanning between the end shear walls, thus producing a bending that develops tension on the leeward edge and compression on the windward edge. It also produces shear in the plane of the diaphragm that becomes a maximum at the end shear walls. In most cases the shear is assumed to be taken by the diaphragm, but the tension and compression forces due to bending are transferred to framing at the diaphragm edges. The means of achieving this transfer depends on the materials and details of the construction.

The end shear walls act as vertical cantilevers that also develop shear and bending. The total shear in the upper story is equal to the edge load from the roof. The total shear in the lower story is the combination of the edge loads from the roof and second floor. The total shear force in the wall is delivered at its base in the form of a sliding friction between the wall and its support. The bending caused by the lateral load produces an overturning effect at the base of the wall as well as the tension and compression forces at the edges of the wall. The overturning effect is resisted by the stabilizing effect of the dead load on the wall. If this stabilizing moment is not sufficient, a tension tie must be made between the wall and its support.

If the first floor is attached directly to the foundations, it may not actually function as a spanning diaphragm, but rather will push its edge load directly to the leeward foundation wall. In any event, it may be seen in this example that only three-fourths of the total wind load on the building is delivered through the upper diaphragms to the end shear walls.

This simple example illustrates the basic nature of the propagation of wind forces through the building structure, but there are many other possible variations with more complex building forms or with other types of lateral resistive structural systems.

Seismic Forces

Seismic loads are actually generated by the dead weight of the building construction. In visualizing the application of

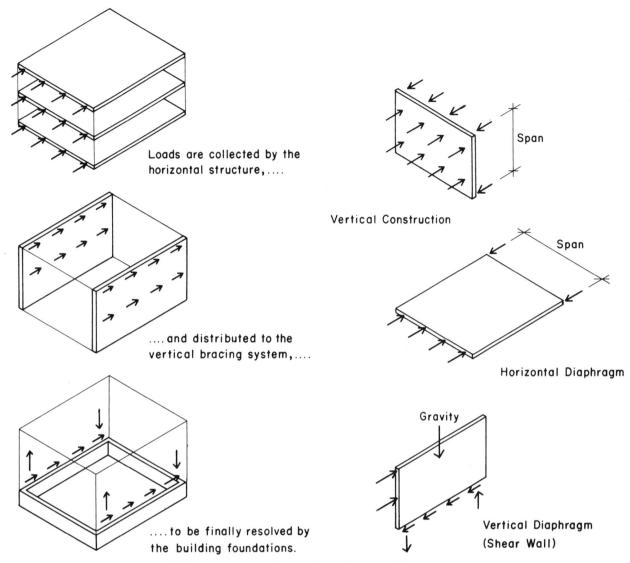

FIGURE 44.1. Propagation of wind force and basic functions of elements in a box building.

seismic forces, we look at each part of the building and consider its weight as a horizontal force. The weight of the horizontal structure, although actually distributed throughout its plane, may usually be dealt with in a manner similar to the edge loading caused by wind. In the direction normal to their planes, vertical walls will be loaded and will function structurally in a manner similar to that for direct wind pressure. The load propagation for the box-shaped building in Fig 44.1 will be quite similar for both wind and seismic forces.

If a wall is reasonably rigid in its own plane, it tends to act as a vertical cantilever for the seismic load in the direction parallel to its surface. Thus, in the example building, the seismic load for the roof diaphragm would usually be considered to be caused by the weight of the roof and ceiling construction plus only those walls whose planes are normal to the direction being considered. These different functions of the walls are illustrated in Fig. 44.2. If this

assumption is made, it will be necessary to calculate a separate seismic load in each direction for the building.

For determination of the seismic load, it is necessary to consider all elements that are permanently attached to the structure. Ductwork, lighting and plumbing fixtures, supported equipment, signs, and so on, will add to the total dead weight for the seismic load. In buildings such as storage warehouses and parking garages it is also advisable to add some load for the building contents.

44.3. TYPES OF LATERAL RESISTIVE SYSTEMS

The building in the previous example illustrates one type of lateral resistive system: the box or panelized system. As shown in Fig. 44.3, the general types of systems are those discussed in the following paragraphs.

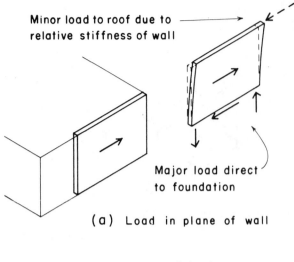

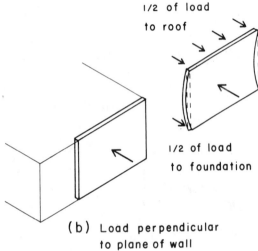

FIGURE 44.2. Seismic loads caused by wall weight.

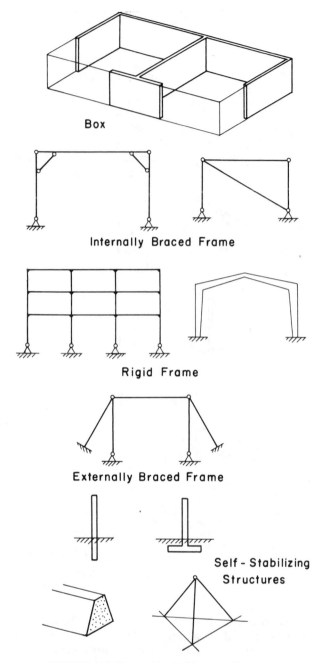

FIGURE 44.3. Types of lateral load resistive systems.

The Box or Panelized System

The box or panelized system is usually of the type shown in the previous example, consisting of some combination of horizontal and vertical planar elements. Actually, most buildings use horizontal diaphragms simply because the existence of roof and floor construction provides them as a matter of course. The other types of systems usually consist of variations of the vertical bracing elements. An occasional exception is a roof structure that must be braced by trussing or other means when there are a large number of roof openings or a roof deck with little or no diaphragm strength.

Internally Braced Frames

The typical assemblage of post and beam elements is not inherently stable under lateral loading unless the frame is braced in some manner. Shear wall panels may be used to achieve this bracing, in which case the system functions as a box even though there is a frame structure. It is also pos-

sible, however, to use diagonal members, X-bracing, knee braces, struts, and so on, to achieve the necessary stability of the rectangular frame. The term *braced frame* usually refers to these techniques.

Rigid Frames

Although the term *rigid frame* is a misnomer since this technique usually produces the most *flexible* lateral resistive system, the term refers to the use of moment-resistive joints between the elements of the frame.

Externally Braced Frames

The use of guys, struts, buttresses, and so on, that are applied externally to the structure or the building results in externally braced frames.

Self-Stabilizing Elements and Systems

Retaining walls, flagpoles, pyramids, tripods, and so on, in which stability is achieved by the basic form of the structure, are examples of self-stabilizing elements and systems.

Each of these systems has variations in terms of materials, form of the parts, details of construction, and so on. These variations may result in different behavior characteristics, although each of the basic types has some particular properties. An important property is the relative stiffness or resistance to deformation, which is of particular concern in evaluating energy effects, especially for response to seismic loads. A box system with diaphragms of poured-in-place concrete is usually very rigid, having little deformation and a short fundamental period. A multistory rigid frame of steel, on the other hand, is usually quite flexible and will experience considerable deformation and have a relatively long fundamental period. In seismic analysis these properties are used to modify the percentage of the dead weight that is used as the equivalent static load to simulate the seismic effect.

Elements of the building construction developed for the gravity load design, or for the general architectural design, may become natural elements of the lateral resistive system. Walls of the proper size and in appropriate locations may be theoretically functional as shear walls. Whether they can actually serve as such will depend on their construction details, on the materials used, on their length-to-height ratio, and on the manner in which they are attached to the other elements of the system for load transfer. It is also possible, of course, that the building construction developed only for gravity load resistance and architectural planning considerations may *not* have the necessary attributes for lateral load resistance, thus requiring some replanning or the addition of structural elements.

Many buildings consist of mixtures of the basic types of lateral resistive systems. Walls existing with a frame structure, although possibly not used for gravity loads, can still be used to brace the frame for lateral loads. Shear walls may be used to brace a building in one direction whereas a braced frame or rigid frame is used in the perpendicular direction. Multistory buildings occasionally have one type of system, such as a rigid frame, for the upper stories and a different system, such as a box system or braced frame, for the lower stories to reduce deformation and take the greater loads in the lower portion of the structure.

In many cases it is neither necessary nor desirable to use every wall as a shear wall or to brace every bay of the building frame. The illustrations in Fig. 44.4 show various situations in which the lateral bracing of the building is achieved by partial bracing of the system. This procedure does require that there be some load-distributing elements, such as the roof and floor diaphragms, horizontal struts,

and so on, that serve to tie the unstabilized portions of the building to the lateral resistive elements.

There is a possibility that some of the elements of the building construction that are not intended to function as bracing elements may actually end up taking some of the lateral load. In frame construction, surfacing materials of plaster, drywall, wood paneling, masonry veneer, and so on, may take some lateral load even though the frame is braced by other means. This is essentially a matter of relative stiffness, although connection for load transfer is also a consideration. What can happen in these cases is that the stiffer finish materials take the load first, and if they are not strong enough, they fail and the intended bracing system then goes to work. Although collapse may not occur, there can be considerable damage to the building construction as a result of the failure of the supposed nonstructural elements.

The choice of the type of lateral resistive system must be related to the loading conditions and to the behavior characteristics required. It must also, however, be coordinated with the design for gravity loads and with the architectural planning considerations. Many design situations allow for alternatives, although the choice may be limited by the size of the building, by code restrictions, by the magnitude of lateral loads, by the desire for limited deformation, and so on.

44.4. LATERAL RESISTANCE OF ORDINARY CONSTRUCTION

Even when buildings are built with no consideration given to design for lateral forces, they will have some natural capacity for lateral force resistance. It is useful to understand the limits and capabilities of ordinary construction as a starting point for the consideration of designing for enhanced levels of lateral force resistance.

Wood Frame Construction

Wood structures can be categorized broadly as light frame or heavy timber. Light frames—using mostly 2 × dimension lumber for wall studs, floor joists, and roof rafters—account for the vast majority of small, low-rise buildings in the United States. In most cases the frames are covered on both surfaces by some type of surfacing material. Many of these surfacing systems have a usable, quantifiable capacity for diaphragm resistance. Thus, without any major alteration of the basic structure, most light wood frames can be made resistive to lateral forces through the use of a combination of horizontal and vertical diaphragms.

Many of the ordinary elements of light wood frames can be utilized as parts of the lateral resistive system, serving as diaphragm chords, collectors, and edge transfer members. Wall studs, posts, sills and plates, and roof and floor framing members, occurring naturally in the structure, are often able to be utilized for these functions. Alterations necessary to make them more functional are often limited to moderate increases in sizes or to the use of some ad-

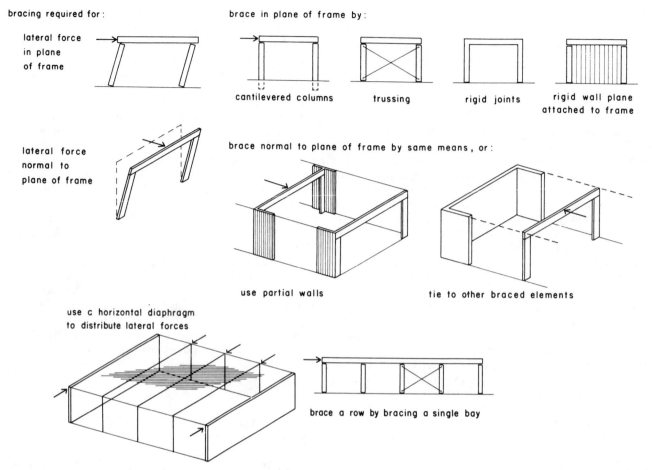

FIGURE 44.4. Bracing of framed structures for lateral loads.

ditional fastening or anchoring. When members are long and not able to be installed as a single piece (as with the top plate on a long wall), it may be necessary to use stronger splicing than is ordinarily required for gravity resistance alone.

Common practices of carpentry result in a considerable amount of fastening between members of a light wood frame. Building codes often specify minimum requirements for such fastening. (See reprint of Table 25—Q from the *UBC* in Appendix D.) Before undertaking to design such a structure for seismic loads, it well behooves the designer to become thoroughly familiar with these requirements as well as current local practices of contractors and workmen. References with suggested details for construction become rapidly out of date as codes and construction practices and the availability of materials and equipment change.

In recent times there has been a trend toward the extensive use of sheet metal fastening devices for the assemblage of light wood structures. In general, these tend to increase resistances to lateral loads because the continuity of the frame is greater and the anchorage of members is more positive.

There is a considerable range in the diaphragm shear capability of various surfacing materials. The following are some widely used products:

Plywood. This may be used as the structural backup for a variety of finishes or may be used with a special facing as the complete surfacing. Most plywoods offer considerable potential for shear resistance. With an increase in structural quality, greater thickness than the minimum required for other functions, and a greater number of nails, it is possible to develop considerable strength for either horizontal or vertical diaphragm use.

Board Sheathing. Boards of 1-in.-nominal thickness, with shiplap or tongue-and-groove edges, were once quite commonly used for sheathing. When applied in a position diagonal to the frame, they can produce some diaphragm action. Capacities are nowhere near that of plywood, however, so that for this and other reasons, this type of structural covering is not much used at present.

Plaster. Portland cement plaster, applied over wire-reinforced backing and adequately secured to the framing, produces a very stiff diaphragm with load capacities equal to that of the thinner plywoods. On the exterior it is called *stucco*, which due to its popularity, is the definitive vertical bracing material for light structures in southern and western United States.

Miscellaneous Surfacing. Gypsum drywall, gypsum plaster, nonstructural plywood, and particleboard can develop some diaphragm capacity that may be sufficient for low-stress situations. Stiffness is minimal, thus narrow diaphragms should be avoided.

When the same material is applied to both sides of a wall, the code permits the use of the sum of the resistances of the two surfaces. For interior walls this is quite common and permits the utilization of low-strength surfacing. However, for exterior walls the two surfaces are seldom the same, thus the stronger (usually the exterior) must be used alone.

In the past a widely used method for bracing light wood frames consisted of diagonal bracing, typically in the form of 1-in.-nominal-thickness boards with their outside surfaces made flush with the face of the framing by cutting notches in the framing members (called *let-in bracing*). The acceptance of rated load capacities for a wider range of popular surfacing materials has made this practice largely redundant. When diagonal bracing is used today, it often consists of thin steel straps applied to the faces of the framing members.

Some of the problems encountered in developing seismic resistance with light wood frame construction follow.

1. *Lack of Adequate Solid Walls to Serve as Shear Walls.* This may be due to the building planning, with walls insufficient at certain locations or in a particular direction. Walls may also simply be too broken up in short lengths by doors and windows. For multistory buildings there may be a problem where upper level walls do not occur above walls in lower levels.

2. *Lack of Adequate Diaphragm Surfacing.* Many types of surfacing have rated capacities for shear. Each, however, has its limits and some materials are not rated for code-acceptable loadings.

3. *Lack of Continuity of Framing.* Because it is often not required for gravity load conditions, members that could function as chords or collectors may consist of separate pieces that are not spliced for tension or compression continuity.

4. *Lack of Adequate Connections.* Load transfers—most notably those from horizontal to vertical diaphragms—may not be possible without modification of the construction details, involving additional framing, increased nailing, or use of special anchorage devices. Ordinary code-required minimum wall sill bolting is not acceptable for overturn resistance and is often not adequate for sliding resistance.

Wood post-and-beam structures can sometimes be made to function as braced frames or moment-resistive frames, the latter usually being somewhat more difficult to achieve. Often, however, these structures occur in combination with wall construction of reinforced masonry or wood frame plus surfacing so that the post-and-beam frame need not function for lateral bracing. Floor and roof decks are often the same as for light frames and thus are equally functional as horizontal diaphragms.

A problem with post-and-beam construction is often the lack of ability for load transfers between members required for resistance to lateral loads. At present this is less the case due to the increasing use of metal framing devices for beam seats, post caps and bases, and so on.

When heavy timber frames are exposed to view, a popular choice for roof or floor decks is a timber deck of 1.5-in. or greater thickness. Although such a deck has a minimal shear capacity, it is usually not adequate for diaphragm development except for small buildings. The most common and economical means for providing the necessary diaphragm resistance is to simply nail a continuous plywood surface to the top of the timber deck.

Structural Masonry

For seismic zones 3 and 4 the only masonry structural construction permitted is *reinforced masonry.* These comments are confined to structural masonry walls constructed with hollow concrete units (concrete blocks), with the voids partly or wholly reinforced and filled with grout (see Fig. 4.6).

Structural masonry walls have considerable potential for utilization as shear walls. There are, however, a number of problems that must be considered.

1. *Increased Load.* Due to their weight, stiffness, and brittleness, masonry walls must be designed for higher lateral seismic forces. The *UBC* requires an additional increase of 50% in the load for stress analysis of the wall.

2. *Limited Stress Capacity.* The unit strength and the mortar strength must be adequate for the required stress resistances. In addition, both vertical and horizontal reinforcing is required for major shear wall functions.

3. *Cracks and Bonding Failures.* Walls not built to the specifications usually used for seismic-resistive construction often have weakened mortar joints and cracking. These reduce seismic resistance, especially in walls with minimal reinforcing.

Code specifications for concrete block walls result in a typical minimum construction that has a particular limit of shear wall capacity. This limit is beyond the limit for the strongest of the wood-framed walls with plywood on a single side, and so the change to a masonry wall is a significant step. Beyond the minimum value the load capacity is increased by adding additional reinforcing and filling more of the block voids. At its upper limits the reinforced masonry wall approaches the capacity of a reinforced concrete wall.

Anchorage of masonry walls to their supports is usually simply achieved. Resistance to vertical uplift and horizontal sliding can typically be developed by the usual doweling of the vertical wall reinforcing. The anchorage of horizontal diaphragms to masonry walls is another matter and typically requires the use of more "positive" anchoring methods than are ordinarily used when seismic risk is low.

Reinforced Concrete Construction

Poured concrete elements for most structures are ordinarily quite extensively reinforced, thus providing significant compensation for the vulnerability of the tension-weak material. Even where structural demands are not severe, minimum two-way reinforcing is required for walls and slabs to absorb effects of shrinkage and fluctuation of temperature. This form of construction has considerable natural potential for lateral force resistance.

Subgrade building construction most often consists of thick concrete walls, in many cases joined to horizontal concrete structures with solid poured-in-place slabs. The typical result is a highly rigid, strong boxlike structure. Shears in the planes of the walls and slabs can be developed to considerable stress levels with minimum required reinforcing. Special attention must be given to the maintaining of continuity through pour joints and control joints and to the anchorage of reinforcing at wall corners and intersections and at the joints between slabs and walls. This does not always result in an increase in the amount of reinforcing, but may alter some details of its installation.

Structures consisting of poured concrete columns used in combination with various concrete spanning systems require careful study for the development of seismic resistance as rigid frame structures. The following are some potential problems:

1. *Weight of the Structure.* This is ordinarily considerably greater than that of wood or steel construction, with the resulting increase in the total seismic force.
2. *Adequate Reinforcing for Seismic Effects.* Of particular concern are the shears and torsions developed in framing elements and the need for continuity of the reinforcing or anchorage at the intersections of elements. A special problem is that of vertical shears developed by vertical accelerations, most notably punching shear in slab structures.
3. *Ductile Yielding of Reinforcing.* This is the desirable first mode of failure, even for gravity load resistance. With proper design it is a means for developing a yield character in the otherwise brittle, tension-weak structure.
4. *Detailing of Reinforcing.* Continuity at splices and adequate anchorage at member intersections must be assured by careful layout of reinforcing installation.
5. *Tying of Compression Bars.* Column and beam bars should be adequately tied in the region of the column-beam joint.

When poured concrete walls are used in conjunction with concrete frames, the result is often similar to that of the plywood-braced wood frame, the walls functioning to absorb the major portion of the lateral loads due to their relative stiffness. At the least, however, the frame members

function as chords, collectors, drag struts, and so on. The forces at the intersections of the walls and the frame members must be carefully studied to assure proper development of necessary force transfers. This is even more critical if there is an actual interaction of the wall and frame systems.

As with masonry structures, considerable cracking is normal in poured concrete structures; much of it is due to shrinkage, temperature expansion and contraction, settlement or deflection of supports, and the normal development of internal tension forces. In addition, built-in cracks of a sort are created at the cold joints that are unavoidable between successive, separate pours. Under the back-and-fourth swaying actions of an earthquake, these cracks will be magnified, and a grinding action may occur as stresses reverse. The grinding action can be a major source of energy absorption but can also result in progressive failures or simply a lot of pulverizing and flaking off of the concrete. If reinforcing is adequate, the structure may remain safe, but the appearance is sure to be affected.

It is virtually impossible to completely eliminate cracking from masonry and poured-in-place concrete buildings. Good design, careful construction detailing, and quality construction can reduce the amount of cracking and possibly eliminate some types of cracking. However, the combination of shrinkage, temperature expansion, settlement of supports, creep, and flexural stress is a formidable foe.

Steel Frame Construction

Structures with frames of steel can often quite readily be made resistive to lateral loads, usually by producing either a braced (trussed) frame or a moment-resistive frame. Steel has the advantage of having a high level of resistance to all types of stress and is thus not often sensitive to multidirectional stresses or to rapid stress reversals. In addition, the ductility of ordinary structural steel provides a toughness and a high level of energy absorption in the plastic behavior mode of failure.

The high levels of stress obtained in steel structures are accompanied by high levels of strain, resulting often in considerable deformation. The actual magnitudes of the deformations may affect the building occupants or contents or may have undesirable results in terms of damage to nonstructural elements of the building construction. Deformation analysis is often a critical part of the design of steel structures, especially for moment-resistive frames.

The ordinary post and beam steel frame is essentially unstable under lateral loading. Typical framing connections have some minor stiffness and moment resistance but are not effective for development of the rigid joints required for a moment-resistive frame. Frames must therefore either be made self-stable with diagonal bracing or with specially designed moment-resistive connections or be braced by shear walls.

Steel frames in low-rise buildings are often braced by walls, with the steel structure serving only as the horizontal spanning structure and vertical gravity load resisting structure. Walls may consist of masonry or of wood or metal frames with various shear-resisting surfacing. For the wall-braced structure, building planning must incorporate the necessary solid wall construction for the usual shear wall braced building. In addition, the frame will usually be used for chord and collector actions; therefore, the connections between the decks, the walls, and the frame must be designed for the lateral load transfers.

The trussed steel structure is typically quite stiff, in a class with the wall-braced structure. This is an advantage in terms of reduction of building movements under load, but it does mean that the structure must be designed for as much as twice the total lateral force as a moment-resistive frame. Incorporating the diagonal members in vertical planes of the frame is often a problem for architectural planning, essentially similar to that of incorporating the necessary solid walls for a shear wall-braced structure.

In the past, steel frames were mostly used in combination with decks of concrete or formed sheet steel. A popular construction for low-rise buildings at present—where fire-resistance requirements permit its use—is one that utilizes a wood infill structure of joists or trusses with a plywood deck. A critical concern for all decks is the adequate attachment of the deck to the steel beams for load transfers to the vertical bracing system. Where seismic design has not been a factor, typical attachments are often not sufficient for these load transfers.

Buildings of complex unsymmetrical form sometimes present problems for the development of braced or moment-resistive frames. Of particular concern is the alignment of the framing to produce the necessary vertical planar bents. Randomly arranged columns and discontinuities due to openings or voids can make bent alignment or continuity a difficult problem.

Precast Concrete Construction

Precast concrete structures present unique problems in terms of lateral bracing. Although they share many characteristics with poured-in-place concrete structures, they lack the natural member-to-member continuity that provides considerable lateral stability. Precast structures must therefore be dealt with in a manner similar to that for post and beam structures of wood or steel. This problem is further magnified by the increased dead weight of the structure, which results in additional lateral force.

Separate precast concrete members are usually attached to each other by means of steel devices that are cast into the members. The assemblage of the structure thus becomes a steel-to-steel connection problem. Where load transfer for gravity resistance is limited to simple bearing, connections may have no real stress functions, serving primarily to hold the members in position. Under lateral load, however, all connections will likely be required to transfer shear, tension, bending, and torsion. Thus for seismic resistance many of the typical connections used for gravity resistance alone will be inadequate.

Because of their weight, precast concrete spanning members may experience special problems due to vertical accelerations. When not sufficiently held down against upward movement, members may be bounced off their supports (a failure described as *dancing*).

Precast concrete spanning members are often also prestressed rather than simply utilizing ordinary steel reinforcing. This presents a possible concern for the effects of the combined loading of gravity and lateral forces or for upward movements due to vertical acceleration. Multiple loading conditions and stress reversals tend to greatly complicate the design of prestressing.

As with frames of wood or steel, those of precast concrete must be made stable with trussing, moment connections, or infill walls. If walls are used, they must be limited to masonry or concrete. Connections between the frame and any bracing walls must be carefully developed to assure the proper load transfers.

Foundations

Where considerable below-grade construction occurs—with heavy basement walls, large bearing footings, basement or sublevel floor construction of reinforced concrete, and so on—the below-grade structure as a whole usually furnishes a solid base for the above-grade building. Not many extra details or elements are required to provide for seismic actions. Of principal concern is the tying together of the base of the building, which is where the seismic movements are transmitted to the building. If the base does not hold together as a monolithic unit, the result will be disastrous for the supported building. Buildings without basements or those supported on piles or piers may not ordinarily be sufficiently tied together for this purpose, thus requiring some additional construction. Foundation design problems are discussed more thoroughly in Chapter 43.

Freestanding Structures

Freestanding structures include exterior walls used as fences as well as large signs, water towers, and detached stair towers. The principal problem is usually the large overturning effect. Rocking and permanent soil deformations that result in vertical tilting must be considered. It is generally advisable to be quite conservative in the design for soil pressure due to the overturning effects. When weight is concentrated at the top—as in the case of signs or water towers—the dynamic rotational effect is further increased. These concerns apply also to the elements that may be placed on the roof of a building.

44.5. DESIGN PROBLEMS

When the need to develop resistance to seismic forces is kept in mind throughout the entire process of the building design, it will have bearing on many areas of the design development. This chapter deals with various considerations that may influence the general planning as well as choices for materials, systems, and construction details.

When lateral design is dealt with as an afterthought rather than being borne in mind in the earliest decisions on form and planning of the building, it is quite likely that optimal conditions will not be developed. Some of the major issues that should be kept in mind in the early planning stages follow.

1. *The Need for Some Kind of Lateral Bracing System.* In some cases, because of the building form or size or the decision to use a particular structural material or system, the choice may be highly limited. In other situations there may be several options, with each having different required features (alignment of columns, incorporation of solid walls, etc.). The particular system to be used should be established early, although it may require considerable exploration and development of the options in order to make an informed decision.

2. *Implications of Architectural Design Decisions.* When certain features are desired, it should be clearly understood that there are consequences in the form of problems with regard to lateral design. Some typical situations that commonly cause problems are the following:

General complexity and lack of symmetry in the building form.

Random arrangement of vertical elements (walls and columns), resulting in a haphazard framing system in general.

Lack of continuity in the horizontal structure due to openings, multiplane roofs, split-level floors, or open spaces within the building.

Building consisting of aggregates of multiple, semidetached units, requiring considerations for linking or separation for seismic interaction.

Special forms (curved walls, sloping floors, etc.) that limit the performance of the structure.

Large spans, heights, or wall openings that limit placement of structural elements and result in high concentrations of load.

Use of nonstructural materials and construction details that result in high vulnerability to damage caused by seismic movements.

3. *Allowance for Lateral Design Work.* Consideration should be given to the time, cost, and scheduling for the lateral investigation and design development. This is most critical when the building is complex or when an extensive dynamic analysis is required. Sufficient time should be allowed for a preliminary investigation of possible alternatives for the lateral bracing system, since a shift to another system at late stages of the architectural design work will undoubtedly cause problems.

4. *Design Styles Not Developed with Seismic Effects in Mind.* In many situations popular architectural design styles or features are initially developed in areas where seismic effects are not of concern. When these are imported to regions with high risk of seismic activity, a mismatch often occurs. Early European colonizers of Central and South America and the west coast of North America learned this the hard way. The learning goes on. The form of a building has a great deal to do with the determination of the effects of seismic activity on the building. This chapter discusses various aspects of building form and the types of problem commonly experienced.

Most buildings are complex in form. They have plans defined by walls that are arranged in complex patterns. They have wings, porches, balconies, towers, and roof overhangs. They are divided vertically by multilevel floors. They have sloping roofs, arched roofs, and multiplane roofs. Walls are pierced by openings for doors and windows. Floors are pierced by stairways, elevators, ducts, and piping. Roofs are pierced by skylights, vent shafts, and chimneys. The dispersion of the building mass and the overall response of the building to seismic effects can thus be complicated; difficult to visualize, let alone to quantitatively evaluate.

Despite this typical complexity, investigation for seismic response may often be simplified by the fact that we deal mostly with those elements of the building that are directly involved in the resistance of lateral forces, what we refer to as the *lateral resistive system.* Thus most of the building construction, including parts of the structure that function strictly for resistance of gravity loads, may have only minimal involvement in seismic response. These nonstructural elements contribute to the load (generated by the building mass) and may offer damping effects to the structure's motion, but may not significantly contribute to the development of resistance to lateral force.

A discussion of the issues relating to building form must include consideration of two separate situations: the form of the building as a whole and the form of the lateral resistive system. Figure 44.5 shows a simple one-story building,

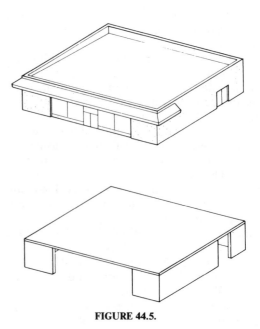

FIGURE 44.5.

with the general exterior form illustrated in the upper figure. The lower figure shows the same building with the parapet, canopy, window wall, and other elements removed, leaving the essential parts of the lateral resistive system. This system consists primarily of the horizontal roof surface and the portions of the vertical walls that function as shear walls. The whole building must be considered in determining the building mass for the lateral load, but the stripped down structure must be visualized in order to investigate the effects of lateral forces.

In developing building plans and the building form in general, architectural designers must give consideration to many issues. Seismic response has to take its place in line with the needs for functional interior spaces, control of traffic, creation of acoustic privacy, separation for security, energy efficiency, and general economic and technical feasibility. In this section we dwell primarily on the problems of lateral response, but it must be kept in mind that the architect must deal with all of these other concerns.

Development of a reasonable lateral resistive structural system within a building may be easy or difficult and for some proposed plan arrangements may be next to impossible. Figure 44.6 shows a building plan in the upper figure for which the potentiality for development of shear walls in the north–south direction is quite reasonable but in the east–west direction is not so good as there is no possibility for shear walls on the south side. If the modification shown in the middle figure is acceptable, the building can be adequately braced by shear walls in both directions. If the open south wall is really essential, it may be possible to brace this wall by using a column and beam structure that is braced by trussing or by rigid connections, as shown in the lower figure.

In the plan shown in Fig. 44.7a the column layout results in a limited number of possible bents that may be developed as moment-resistive frames. In the north–south direction the interior columns are either offset from the exterior columns or the bent is interrupted by the floor opening; thus the two end bents are the only ones usable. In the east–west direction the large opening interrupts two of the three interior bents, leaving only three usable bents that are not disposed symmetrically in the plan. The modification shown in Fig. 44.7b represents an improvement in lateral response, with six usable bents in the north–south direction and four symmetrically placed bents in the east–west direction. This plan, however, has more interior columns, smaller open spaces, and a reduced size for the opening—all of which may present some drawbacks for architectural reasons.

In addition to planning concerns the vertical massing of the building has various implications on its seismic response. The three building profiles shown in Fig. 44.8a–c represent a range of potential response with regard to the fundamental period of the building and the concerns for lateral deflection. The short, stiff building shown in Fig. 44.8a tends to absorb a larger jolt from an earthquake because of its quick response (short period of natural vibration). The tall, slender building, on the other hand, responds slowly, dissipating some of the energy of the

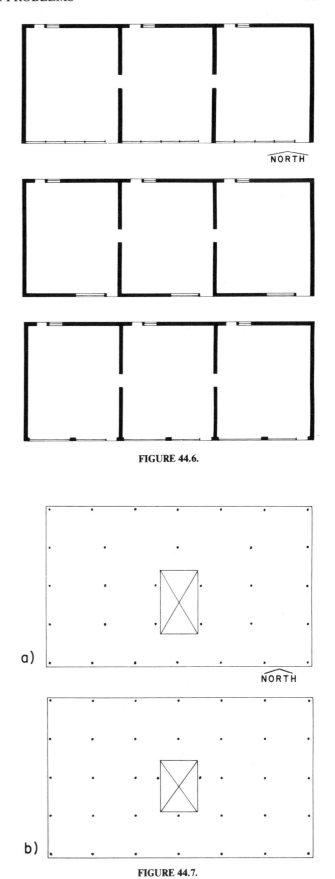

FIGURE 44.6.

a)

NORTH

b)

FIGURE 44.7.

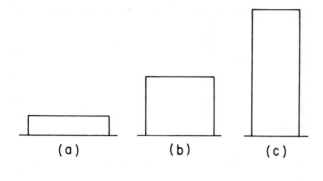

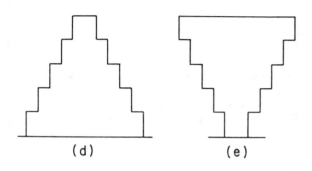

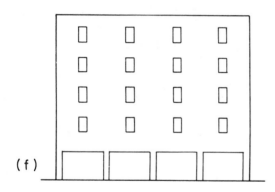

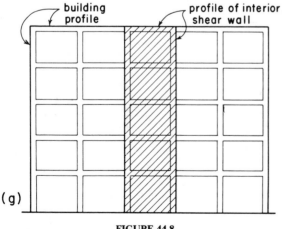

FIGURE 44.8.

seismic action in its motion. However, the tall building may develop some multimodal response, a whiplash effect, or simply so much actual deflection that it may have problems of its own.

The overall inherent stability of a building may be implicit in its vertical massing or profile. The structure shown in Fig. 44.8*d* has considerable potential for stability with regard to lateral forces, whereas that shown in Fig. 44.8*e* is highly questionable. Of special concern is the situation in which abrupt change in stiffness occurs in the vertical massing. The structure shown in Fig. 44.8*f* has an open form at its base, resulting in the so-called soft story. While this type of system may be designed adequately by the general requirements of the equivalent static force method, a true dynamic analysis will indicate serious problems, as borne out by some recent serious failures.

As with the building plan, consideration of the vertical massing must include concerns for the form of the lateral resistive system as well as the form of the whole building. The illustration in Fig. 44.8*g* shows a building whose overall profile is quite stout. However, if the building is braced by a set of interior shear walls, as shown in the section, it is the profile of the shear walls that must be considered. In this case the shear wall is quite slender in profile.

Investigation of the seismic response of a complex building is, in the best of circumstances, a difficult problem. Anything done to simplify the investigation will not only make the analysis easier to perform but will tend to make the reliability of the results more certain. Thus, from a seismic design point of view, there is an advantage in obtaining some degree of symmetry in the building massing and in the disposition of the elements of the lateral resistive structure.

When symmetry does not exist, a building tends to experience severe twisting as well as the usual rocking back and forth. The twisting action often has its greatest effects on the joints between elements of the bracing system. Thorough investigation and careful detailing of these joints for construction are necessary for a successful design. The more complex the seismic response and the more complicated and unusual the details of the construction, the more difficult it becomes to assure a thorough and careful design.

Most buildings are not symmetrical, being sometimes on one axis, often not on any axis. However, real architectural symmetry is not necessarily the true issue in seismic response. Of critical concern is the alignment of the net effect of the building mass (or the centroid of the lateral force) with the center of stiffness of the lateral resistive system—most notably the center of stiffness of the vertical elements of the system. The more the eccentricity of the centroid of the lateral force from the center of stiffness of the lateral bracing system, the greater the twisting effect on the building.

Figure 44.9 shows an extreme example—the *three-sided building*. In this situation the lack of resistive vertical elements on one side of the building requires that the opposite wall take all of the direct effect of the lateral force that is parallel to it. Assuming the centroid of the building mass to

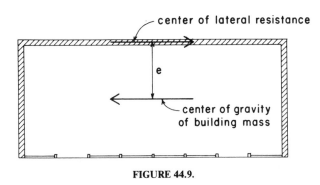

FIGURE 44.9.

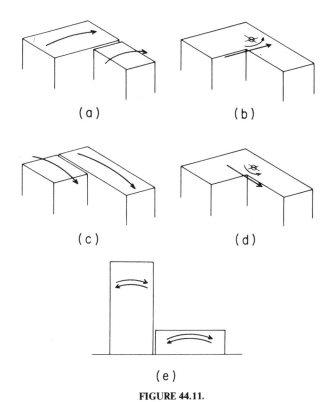

FIGURE 44.11.

be approximately at the center of the plan, this results in a large eccentricity between the load and the resisting wall. The twisting action that results will be partly resisted by the two end walls that are at right angles to the load, but the general effect on the building is highly undesirable. This type of structure is presently highly restricted for use in regions of high seismic risk.

When a building is not architecturally symmetrical, the lateral bracing system must either be adjusted so that its center of stiffness is close to the centroid of the mass or it must be designed for major twisting effects on the building. As the complexity of the building form increases, it may be necessary to consider the building to be multimassed.

Many buildings are multimassed rather than consisting of a single geometric form. The building shown in Fig. 44.10 is multimassed, consisting of an L-shaped tower that is joined to an extended lower portion. Under lateral seismic movement the various parts of this building will have different responses. If the building structure is developed as a single system, the building movements will be very complex, with extreme twisting effects and considerable strain at the points of connection of the discrete parts of the mass.

If the elements of the tower of the building in Fig. 44.10 are actually separated, as shown in Fig. 44.11a or c, the in-

dependent movements of the separated elements will be different due to their difference in stiffness. It may be possible to permit these independent movements by providing structural connections that are detailed to tolerate the type and magnitude of the actual deformations. Thus the twisting effects on the building and the strain at the joints between the elements of the mass may be avoided.

There is also the potential for difference in response movements of the tower and the lower portion of the building, as shown in Fig. 44.11e. Actual separation may be created at this connection of the masses to eliminate the need for investigation of the dynamic interaction of the separate parts. However, it may not be feasible or architecturally desirable to make the provisions necessary to achieve either of the types of separation described. The only other option is therefore to design for the twisting effects and the dynamic interactions that result from having a continuous, single structural system for the entire building. The advisability or feasibility of one option over the other is often difficult to establish and may require considerable study of alternate designs.

Figure 44.12a shows an L-shaped building in which the architectural separation of the masses is accentuated. The linking element, although contiguous with the two parts, is unlikely to be capable of holding them together under seismic movements. If it cannot, there are two forms of differential movement that must be provided for, as shown in Fig. 44.12b and c. In addition to providing for these movements, it is also necessary to consider the bracing of the linkage element. If it is not capable of being independently

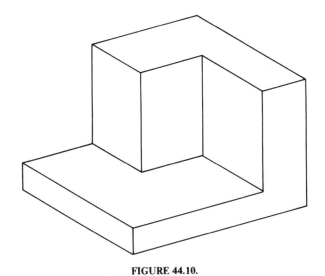

FIGURE 44.10.

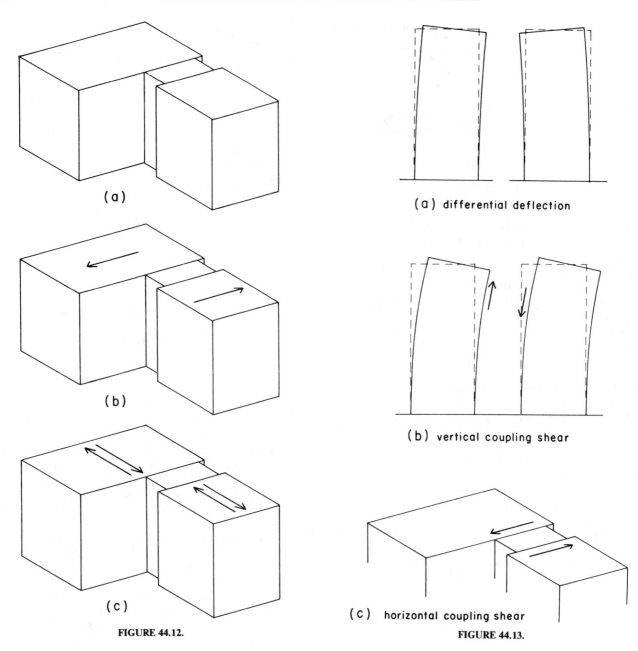

FIGURE 44.12.

(a) differential deflection

(b) vertical coupling shear

(c) horizontal coupling shear

FIGURE 44.13.

braced, it must be attached to one or the other of the larger elements for support, making for a quite complex study of actions at the connection of the masses.

When individual parts of multimassed buildings are joined, there are many potential problems, some of which were described in the preceding section. Figure 44.13 shows three types of action that must often be considered for such structures. When moving at the same time, as shown in Fig. 44.13a, a problem for the separate masses becomes the actual dimension of the separation that must be provided to prevent them from bumping each other (called *battering* or *hammering*). If they are not actually separated, their independent deflections may be a basis for

consideration of the forces that must be considered in preventing them from being torn apart.

Another potential action of separately moving parts is that shown in Fig. 44.13b. This involves a shearing action on the joint similar to that which occurs in laminated elements subjected to bending. For the vertically cantilevered elements shown, both the shear and lateral deflection effects vary from zero at the base to a maximum at the top. The taller the structure, the greater the actual dimension of the critical movements near the top.

A third type of action is the horizontal shearing effect illustrated in Fig. 44.13c. This is probably the most common type of problem that must be dealt with, as it occurs frequently in one story structures, whereas the vertical shear and lateral deflection problems are usually severe only in taller structures.

Individual joined masses are sometimes so different in size or stiffness that the indicated solution is to simply attach the smaller part to the larger and let it tag along. Such is the case for the buildings shown in Fig. 44.14a and b in which the smaller lower portion and the narrow stair tower would be treated as attachments.

In some instances the tag along relationship may be a conditional one, as shown in Fig. 44.14c, where the smaller element extends a considerable distance from the larger mass. In this situation the movement of the smaller part to and away from the larger may be adequately resisted by the attachment. However, some bracing would probably be required at the far end of the smaller part to assist resistance to movements parallel to the connection of the two parts.

The tag-along technique is often used for stairs, chimneys, entries, and other elements that are part of a building, but are generally outside the main mass. It is also possible, of course, to consider the total structural separation of such elements in some cases.

Another classic problem of joined elements is that of coupled shear walls. These are shear walls that occur in sets in a single wall plane and are connected by the continuous construction of the wall. Figure 44.15 illustrates such a situation in a multistory building. The elements that serve to link such walls—in this example the spandrel panels beneath the windows—are wracked by the vertical shearing effect illustrated in Fig. 44.13b. As the building rocks back and forth, this effect is rapidly reversed, developing the diagonal cracking shown in Fig. 44.15b and c. This results in the X-shaped crack patterns shown in Fig. 44.15d, which may be observed on the walls of many masonry, concrete, and stucco-surfaced buildings in regions of frequent seismic activity.

Forces applied to buildings must flow with some direct continuity through the elements of the structure, be transferred effectively from element to element, and eventually be resolved into the ground. Where there are interruptions in the normal flow of the forces, problems will occur. For example, in a multistory building the resolution of gravity forces requires a smooth, vertical path; thus columns and bearing walls must be stacked on top of each other. If a column is removed in a lower story, a major problem is created, requiring the use of a heavy transfer girder or other device to deal with the discontinuity.

A common type of discontinuity is that of openings in horizontal and vertical diaphragms. These can be a problem as a result of their location, size, or even shape. Figure 44.16 shows a horizontal diaphragm with an opening. The diaphragm is braced by four shear walls, and if it is considered to be uninterrupted it will distribute its load to the walls in the manner of a continuous beam (see the discussion of flexibility of horizontal diaphragms in Sec. 47.1). If the relative size of the opening is as shown in Fig. 44.16a, this assumption is a reasonable one. What must be done to assure the integrity of the continuous diaphragm is to reinforce the edges and corners of the opening and to be sure that the net diaphragm width at the opening is adequate for the shear force.

If the opening in a horizontal diaphragm is as large as that shown in Fig. 44.16b, it is generally not possible to maintain the continuity of the whole diaphragm. In the example the best solution would be to consider the diaphragm as consisting of four individual parts, each resisting some portion of the total lateral force. For openings of sizes between the ones shown in Fig. 44.16 judgment must be exercised as to the best course.

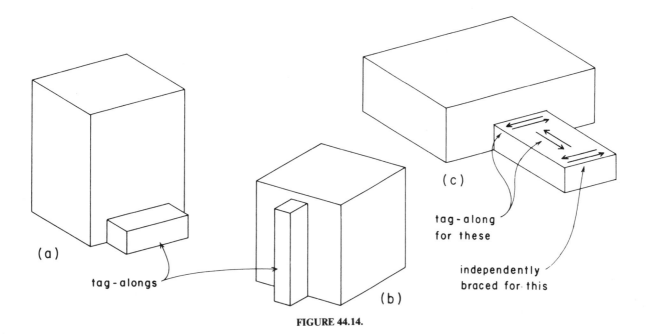

FIGURE 44.14.

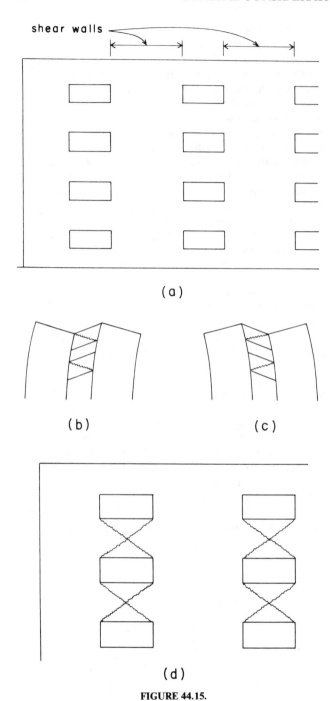

FIGURE 44.15.

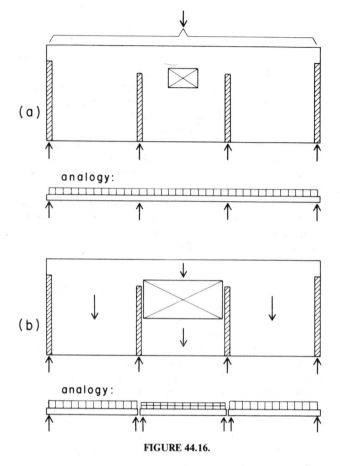

FIGURE 44.16.

Another discontinuity that must sometimes be dealt with is that of the interrupted multistory shear wall. Figure 44.17 shows such a situation, with a wall that is not continuous down to its foundation. In this example it may be possible to utilize the horizontal structure at the second level to redistribute the horizontal shear force to other shear walls in the same plane. The overturn effect on the upper shear wall, however, cannot be so redirected, thus requiring that the columns at the ends of the shear wall continue down to the foundation.

It is sometimes possible to redistribute the shear force from an interrupted wall, as shown in Fig. 44.17, with walls that are sidestepped rather than in the same vertical plane of the upper wall. Again, however, the overturn on the upper wall must be accommodated by continuing the structure at the ends of the wall down to the foundation.

Figure 44.18 shows an X-braced frame structure with a situation similar to that of the shear wall in Fig. 44.17a. The individual panels of X-bracing are sufficiently similar in function to the individual panels of the shear wall to make the situation have the same general options and requirements for a solution.

Discontinuities are usually inevitable in multistory and multimassed buildings. They add to the usual problems of dissymmetry to create many difficult situations for analysis and design and require careful study for the proper assumptions of behavior and the special needs of the construction.

44.7. SPECIAL PROBLEMS

Vulnerable Elements

There are many commonly used elements of buildings that are especially vulnerable to damage due to earthquakes.

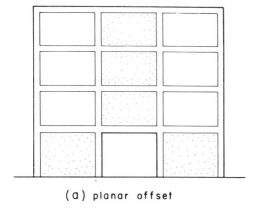

(a) planar offset

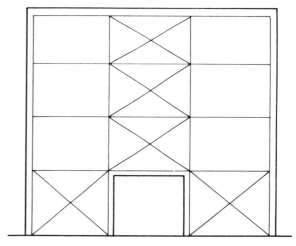

FIGURE 44.18. Offset trussed bracing.

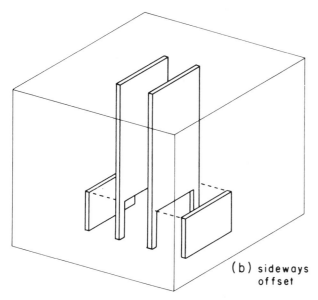

(b) sideways offset

FIGURE 44.17. Offset multistory shear walls.

Most of these are nonstructural, that is, not parts of the structural system for resistance of gravity or lateral loads. Because of their nonstructural character, they do not routinely receive thorough design study by the structural designer; thus in earthquake country they constitute major areas of vulnerability. Some typical situations are the following:

1. *Suspended Ceilings.* These are subject to horizontal movement. If not restrained at their edges, or hung with elements that resist horizontal movement, they will swing and bump other parts of the construction. Another common failure consists of the dropping of the ceiling due to downward acceleration if the supports are not resistive to a jolting action.

2. *Cantilevered Elements.* Balconies, canopies, parapets, and cornices should be designed for significant seismic force in a direction perpendicular to the cantilever. In most cases, codes provide criteria for consideration of these forces.

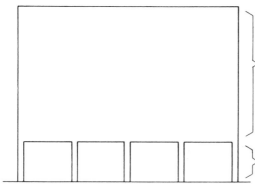

May be
· Solid wall
· Wall with few small windows
· Rigid frame with shorter story height or closer column spacing

Relatively flexible story

FIGURE 44.19. The soft story.

3. *Miscellaneous Suspended Objects.* Lighting fixtures, signs, HVAC equipment, loudspeakers, catwalks, and other items that are supported by hanging should be studied for the effects of pendulumlike movements. Supports should tolerate the movement or should be designed to restrain it.

4. *Piping.* Building movements during seismic activity can cause the rupture of piping that is installed in a conventional manner. In addition to the usual allowances for thermal expansion, provisions should be made for the flexing of the piping or for isolation from the structure that is sufficient to prevent any damage. This is obviously most critical for piping that is pressurized.

5. *Stiff Weak Elements.* Any parts of the building construction that are stiff but not strong are usually vulnerable to damage. This includes window glazing, plastered surfaces, wall and floor tile (especially of ceramic or cast materials), and any masonry (especially veneers of brick, tile, precast concrete, or stone). Reduction of damage and of hazard to occupants requires careful study of installation details for attachment. In most cases extensive use of control joints is advised to permit movements without fractures.

The Soft Story

Any discontinuity that constitutes an abrupt change in the structure is usually a source of some exceptional distress. This is true for static load conditions as well, but is especially critical for dynamic loading conditions. Any abrupt increase or decrease in stiffness will result in some magnification of deformation and stress in a structure subjected to energy loading. Openings, notches, necking-down points, and other form variations produce these abrupt changes in either the horizontal or vertical structure. An especially critical situation is the so-called soft story, as shown in Fig. 44.19.

The soft story could—and indeed sometimes does—occur at an upper level. However, it is more common at the ground-floor level between a rigid foundation system and some relatively much stiffer upper level system. The tall, open ground floor has both historical precedent and current popularity as an architectural feature. This is not always strictly a matter of design style, as it is often required for functional reasons.

Several failures of such structures in recent years have focused attention on their vulnerability and on the inadequacy of the equivalent static method for their design. The *UBC* presently disclaims the use of static methods for buildings with this sort of discontinuity. The soft story presents a strong case for a true dynamic analysis, or at the least a very conservative design with the static load method.

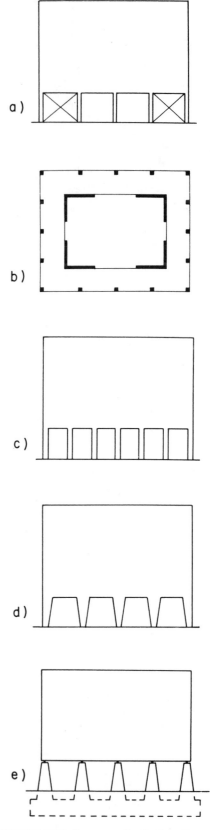

FIGURE 44.20. Some remedies for the soft story.

If the tall, relatively open ground floor is necessary, Fig. 44.20 presents some possibilities for having this feature with a reduction of the soft-story effect. The methods shown consist of the following:

1. Bracing some of the open bays (Fig. 44.20a). If designed adequately for the forces, the braced frame (truss) should have a class of stiffness closer to a rigid shear wall, which is the usual upper structure in these situations. However, the soft story effect can also occur in rigid frames where the soft story is simply significantly less stiff.

2. Keeping the building plan periphery open while providing a rigidly braced interior (Fig. 44.20b).

3. Increasing the number and/or stiffness of the ground-floor columns for an all-rigid frame structure (Fig. 44.20c).

4. Using tapered or arched forms for the ground-floor columns to increase their stiffness (Fig. 44.20d).

5. Developing a rigid first story as an upward extension of a heavy foundation structure (Fig. 44.20e).

The soft story is actually a method for providing critical damping or major energy absorption, which could be a *positive* factor in some situations. However, the major stress concentrations and deformations must be carefully provided for, and a true dynamic analysis is certainly indicated.

CHAPTER FORTY-FIVE

Wind Effects on Buildings

Wind is moving air. The air has a particular mass (density or weight) and moves in a particular direction at a particular velocity. It thus has kinetic energy of the form expressed as

$$E = \tfrac{1}{2} m v^2$$

When the moving fluid air encounters a stationary object, there are several effects that combine to exert a force on the object. The nature of this force, the many variables that affect it, and the translation of the effects into criteria for structural design are dealt with in Chapter 45.

45.1. WIND CONDITIONS

Of primary concern in wind evaluation is the maximum velocity that is achieved by the wind. Maximum velocity usually refers to a sustained velocity and not to gust effects. A gust is essentially a pocket of higher velocity wind within the general moving fluid air mass. The resulting effect of a gust is that of a brief increase, or surge, in the wind velocity, usually of not more than 15% of the sustained velocity and for only a fraction of a second in duration. Because of both its higher velocity and its slamming effect, the gust actually represents the most critical effect of the wind in most cases.

Winds are measured regularly at a large number of locations. The standard measurement is at 10 m (approximately 33 ft) above the surrounding terrain, which provides a fixed reference with regard to the drag effects of the ground surface. The graph in Fig. 45.1 shows the correlation between wind velocity and various wind conditions. The curve on the graph is a plot of the general equation used to relate wind velocity to equivalent static pressure on buildings, as discussed in Sec. 45.2.

Although wind conditions are usually generalized for a given geographic area, they can vary considerably for specific sites because of the nature of the surrounding terrain, of landscaping, or of nearby structures. Each individual building design should consider the possibilities of these localized site conditions.

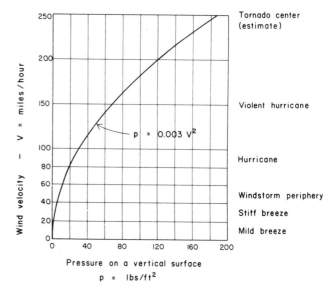

FIGURE 45.1. Relation of wind velocity to pressure.

45.2. WIND EFFECTS

The effects of wind on stationary objects in its path can be generalized as in the following discussions (see Fig. 45.2).

Direct Positive Pressure. Surfaces facing the wind and perpendicular to its path receive a direct impact effect from the moving mass of air, which generally produces the major portion of force on the object unless it is highly streamlined in form.

Aerodynamic Drag. Because the wind does not stop upon striking the object but flows around it like a fluid, there is a drag effect on surfaces that are parallel to the direction of the wind. These surfaces may also have inward or outward pressures exerted on them, but it is the drag effect that adds to the general force on the object in the direction of the wind path.

Negative Pressure. On the leeward side of the object (opposite from the wind direction) there is usually a suction effect, consisting of pressure outward on the sur-

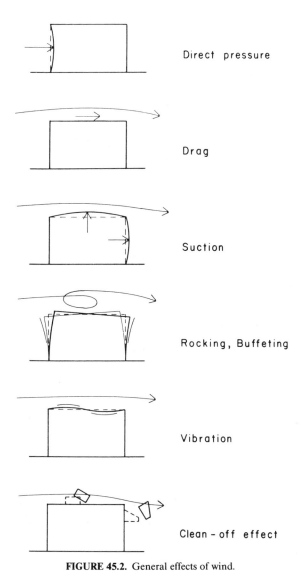

Direct pressure

Drag

Suction

Rocking, Buffeting

Vibration

Clean - off effect

FIGURE 45.2. General effects of wind.

ric surfaces that are not taut) are most susceptible to these effects.

Harmonic Effects. Anyone who plays a wind instrument appreciates that wind can produce vibration, whistling, flutter, and so on. These effects can occur at low velocities as well as with wind storm conditions. This is a matter of some match between the velocity of the wind and the natural period of vibration of the object or of its parts.

Clean-Off Effect. The friction effect of the flowing air mass tends to smooth off the objects in its path. This fact is of particular concern to objects that protrude from the general mass of the building, such as canopies, parapets, chimneys, and signs.

The critical condition of individual parts or surfaces of an object may be caused by any one, or some combination, of the above effects. Damage can occur locally or be total with regard to the object. If the object is resting on the ground, it may be collapsed or may be slid, rolled over, or lifted from its position. Various aspects of the wind, of the object in the path of the wind, or of the surrounding environment determine the critical wind effects. With regard to the wind itself some considerations are the following:

The magnitude of sustained velocities.

The duration of high-level velocities.

The presence of gust effects, swirling, and so on.

The prevailing direction of the wind (if any).

With regard to objects in the path of the wind some considerations are the following:

The size of the object (relates to the relative effect of gusts, to variations of pressure above ground level, etc.).

The aerodynamic shape of the object (determines the critical nature of drag, suction, uplift, etc.).

The fundamental period of vibration of the object or of its parts.

The relative stiffness of surfaces, tightness of connections, and so on.

With regard to the environment, possible effects may result from the sheltering or funneling caused by ground forms, landscaping, or adjacent structures. These effects may result in an increase or reduction of the general wind effects or in turbulence to produce a very unsteady wind condition.

The actual behavior of an object during wind storm conditions can be found only by subjecting it to a real wind situation. Wind tunnel tests in the laboratory are also useful, and because we can create the tests more practically on demand, they have provided much of the background for data and procedures used in design.

The major effects of wind on buildings can be generalized to some degree because we know a bracketed range of characteristics that cover the most common con-

face of the object. By comparison to the direction of pressure on the windward side, this is called *negative pressure*.

These three effects combine to produce a net force on the object in the direction of the wind that tends to move the object along with the wind. In addition to these there are other possible effects on the object that can occur due to the turbulence of the air or to the nature of the object. Some of them are as follows.

Rocking Effects. During wind storms, the wind velocity and its direction are seldom constant. Gusts and swirling winds are ordinary, so that an object in the wind path tends to be buffeted, rocked, flapped, and so on. Objects with loose parts, or with connections having some slack, or with highly flexible surfaces (such as fab-

ditions. Some of the general assumptions made are as follows:

> Most buildings are boxy or bulky in shape, resulting in typical aerodynamic response.
>
> Most buildings present closed, fairly smooth surfaces to the wind.
>
> Most buildings are fit snugly to the ground, presenting a particular situation for the drag effects of the ground surface.
>
> Most buildings have relatively stiff structures, resulting in a fairly limited range of variation of the natural period of vibration of the structure.

These and other considerations allow for the simplification of wind investigation by permitting a number of variables to be eliminated or to be lumped into a few modifying constants. For unusual situations, such as elevated buildings, open structures, highly flexible structures, and unusual aerodynamic shapes, it may be advisable to do more thorough investigation, including the possible use of wind tunnel tests.

The primary effect of wind is visualized in the form of pressures normal to the building's exterior surfaces. The basis for this pressure begins with a conversion of the kinetic energy of the moving air mass into an equivalent static pressure using the basic formula

$$p = Cv^2$$

in which C is a constant accounting for the air mass, the units used, and a number of the assumptions previously described. With the wind in miles per hour (mph) and the pressure in pounds per square foot (psf), the C value for the total wind effect on a simple box-shaped building is approximately 0.003, which is the value used in deriving the graph in Fig. 45.1. It should be noted that this pressure does not represent the actual effect on a single building surface, but rather the *entire* effect of all surface pressures visualized as a single pressure on the windward side of the building.

Inward Pressure on Exterior Walls

Surfaces directly facing the wind are usually required to be designed for the full base pressure, although this is somewhat conservative, because the windward force usually accounts for only about 60% of the total force on the building. Designing for only part of the total force is, however, partly compensated for by the fact that the base pressures are not generally related to gust effects, which tend to have less effect on the building as a whole and more effect on parts of the building.

Suction on Exterior Walls

Most codes also require suction on exterior walls to be the full base pressure, although the preceding comments about inward pressure apply here as well.

Pressure on Roof Surfaces

Depending on their actual form, as well as that of the building as a whole, nonvertical surfaces may be subjected to either inward or suction pressures because of wind. Actually such surfaces may experience both types of pressure as the wind shifts direction. Most codes require an uplift (suction) pressure equal to the full design pressure at the elevation of the roof level. Inward pressure is usually related to the actual angle of the surface as an inclination from the horizontal.

Overall Horizontal Force on the Building

Overall horizontal force is calculated as a horizontal pressure on the building silhouette, as previously described, with adjustments made for height above the ground. The lateral resistive structural system of the building is designed for this force.

Horizontal Sliding of the Building

In addition to the possible collapse of the lateral resistive system, there is the chance that the total horizontal force may slide the building off its foundations. For a tall building with fairly shallow foundations, this may also be a problem for the force transfer between the foundation and the ground. In both cases, the dead weight of the building generates a friction that helps to resist this force.

Overturn Effect

As with horizontal sliding, the dead weight tends to resist the overturn, or toppling, effect. In practice, the overturn effect is usually analyzed in terms of the overturn of individual vertical elements of the lateral resistive system rather than for the building as a whole.

Wind on Building Parts

The previously discussed clean-off effect is critical for elements that project from the general mass of the building. In some cases codes require for such elements a design pressure higher than the base pressure, so that gust effects as well as the clean-off problem are allowed for in the design.

Harmonic Effects

Design for vibration, flutter, whipping, multinodal swaying, and so on requires a dynamic analysis and cannot be accounted for when using the equivalent static load method. Stiffening, bracing, and tightening of elements in general may minimize the possibilities for such effects, but only a true dynamic analysis or a wind tunnel test can assure the adequacy of the structure to withstand these harmonic effects.

Effect of Openings

If the surface of a building is closed and reasonably smooth, the wind will slip around it in a fluid flow. Openings or building forms that tend to cup the wind can greatly affect the total wind force on the building. It is difficult to account for these effects in a mathematical analysis, except in a very empirical manner. Cupping of the wind can be a major effect when the entire side of a building is open, for example. Garages, hangars, band shells, and other buildings of similar form must be designed for an increased force that can only be estimated unless a wind tunnel test is performed.

Torsional Effect

If a building is not symmetrical in terms of its wind silhouette, or if the lateral resistive system is not symmetrical within the building, the wind force may produce a twisting effect. This effect is the result of a misalignment of the centroid of the wind force and the centroid (called *center of stiffness*) of the lateral resistive system and will produce an added force on some of the elements of the structure.

Although there may be typical prevailing directions of wind in an area, the wind must be considered to be capable of blowing in any direction. Depending on the building shape and the arrangement of its structure, an analysis for wind from several possible directions may be required.

45.3. BUILDING CODE REQUIREMENTS FOR WIND

Model building codes such as the *UBC* (Ref. 1) and the *BOCA Basic National Building Code* (Ref. 3) are not legally binding unless they are adopted by ordinances by some state, country, or city. Although smaller communities usually adopt one of the model codes, states, counties, and cities with large populations usually develop their own codes using one of the model codes as a basic reference. In the continental United States the *UBC* is generally used in the west, the Southern Building Code in the southeast, and *BOCA* in the rest of the country.

Where wind is a major local problem, local codes are usually more extensive with regard to design requirements for wind. However, many codes still contain relatively simple criteria for wind design. One of the most up-to-date and complex standards for wind design is contained in the *American National Standard Minimum Design Loads for Buildings and Other Structures*, ANSI A58.1-1982, published by the American National Standards Institute in 1982 (Ref. 2).

Complete design for wind effects on buildings includes a large number of both architectural and structural concerns. Of primary concern for the work in this book are those requirements that directly affect the design of the lateral bracing system. The following is a discussion of some of the requirements for wind as taken from the 1991

edition of the *UBC* (Ref. 1), which is in general conformance with the material presented in the ANSI standard just mentioned. Reprints of some of the materials referred to are given in Appendix D.

Basic Wind Speed

This is the maximum wind speed (or velocity) to be used for specific locations. It is based on recorded wind histories and adjusted for some statistical likelihood of occurrence. For the continental United States the wind speeds are taken from *UBC*, Fig. 23-1 (see Appendix D). As a reference point, the speeds are those recorded at the standard measuring position of 10 m (approximately 33 ft) above the ground surface.

Exposure

This refers to the conditions of the terrain surrounding the building site. The ANSI standard (Ref. 2) describes four conditions (A, B, C, and D), although the *UBC* uses only three (B, C, and D). Condition C refers to sites surrounded for a distance of one half-mile or more by flat, open terrain. Condition B has buildings, forests, or ground surface irregularities 20 ft or more in height covering at least 20% of the area for a distance of 1 mi or more around the site.

Exposure D represents the most severe exposure in aeras with basic wind speed of 80 mph or greater and with terrain that is flat and unobstructed facing large bodies of water over one mile or more in width relative to any quadrant of the building site. Exposure D extends inland from the shoreline one quarter mile or ten times the building height, whichever is greater.

Wind Stagnation Pressure (q_s)

This is the basic reference equivalent static pressure based on the critical local wind speed. It is given in *UBC* Table 23-F (see Appendix D) and is based on the following formula as given in the ANSI standard

$$q_s = 0.00256V^2$$

Example: For a wind speed of 100 mph,

$$q_s = 0.00256V^2 = 0.00256(100)^2$$
$$= 25.6 \text{ psf } [1.23 \text{ kPa}]$$

which is rounded off to 26 psf in the *UBC* table.

Design Wind Pressure

This is the equivalent static pressure to be applied normal to the exterior surfaces of the building and is determined from the formula

$$p = C_eC_qq_sI$$

(*UBC* formula 16-1, Section 2316)

where p = design wind pressure, psf

C_e = combined height, exposure, and gust factor coefficient as given in *UBC* Table 23-G (see Appendix D)

C_q = pressure coefficient for the structure or portion of structure under consideration as given in *UBC* Table 23-H (see Appendix D)

q_s = wind stagnation pressure at 30 ft given in *UBC* Table 23-F (see Appendix D)

I = importance factor

The importance factor is 1.15 for facilities considered to be essential for public health and safety (such as hospitals and government buildings) and buildings with 300 or more occupants. For all other buildings the factor is 1.0.

The design wind pressure may be positive (inward) or negative (outward, suction) on any given surface. Both the sign and the value for the pressure are given in the *UBC* table. Individual building surfaces, or parts thereof, must be designed for these pressures.

Design Methods

Two methods are described in the code for the application of the design wind pressures in the design of structures. For design of individual elements particular values are given in *UBC* Table 23-H for the C_q coefficient to be used in determining p. For the primary bracing system the C_q values and their use is to be as follows:

Method I (Normal Force Method). In this method wind pressures are assumed to act simultaneously normal to all exterior surfaces. This method is required to be used for gabled rigid frames and may be used for any structure.

Method 2 (Projected Area Method). In this method the total wind effect on the building is considered to be a combination of a single inward (positive) horizontal pressure acting on a vertical surface consisting of the projected building profile and an outward (negative, upward) pressure acting on the full projected area of the building in plan. This method may be used for any structure less than 200 ft high, except for gabled rigid frames. This is the method generally employed by building codes in the past.

Uplift

Uplift may occur as a general effect, involving the entire roof or even the whole building. It may also occur as a local phenomenon such as that generated by the overturning moment on a single shear wall. In general, use of either design method will account for uplift concerns.

Overturning Moment

Most codes require that the ratio of the dead load resisting moment (called the *restoring moment, stabilizing moment,*

etc.) to the overturning moment be 1.5 or greater. When this is not the case, uplift effects must be resisted by anchorage capable of developing the excess overturning moment. Overturning may be a critical problem for the whole building, as in the case of relatively tall and slender tower structures. For buildings braced by individual shear walls, trussed bents, and rigid frame bents, overturning is investigated for the individual bracing units. Method 2 is usually used for this investigation, except for very tall buildings and gabled rigid frames.

Drift

Drift refers to the horizontal deflection of the structure due to lateral loads. Code criteria for drift are usually limited to requirements for the drift of a single story (horizontal movement of one level with respect to the next above or below). The *UBC* does not provide limits for wind drift. Other standards give various recommendations, a common one being a limit of story drift to 0.005 times the story height (which is the *UBC* limit for seismic drift). For masonry structures wind drift is sometimes limited to 0.0025 times the story height. As in other situations involving structural deformations, effects on the building construction must be considered; thus the detailing of curtain walls or interior partitions may affect limits on drift. See further discussion of lateral drift in Sec. 49.2.

Combined Loads

Although wind effects are investigated as isolated phenomena, the actions of the structure must be considered simultaneously with other phenomena. The requirements for load combinations are given by most codes, although common sense will indicate the critical combinations in most cases. With the increasing use of load factors the combinations are further modified by applying different factors for the various types of loading, thus permitting individual control based on the reliability of data and investigation procedures and the relative significance to safety of the different load sources and effects. Required load combinations are described in Sec. 2303 of the *UBC*.

Special Problems

The general design criteria given in most codes are applicable to ordinary buildings. More thorough investigation is recommended (and sometimes required) for special circumstances such as the following:

Tall Buildings. These are critical with regard to their height dimension as well as the overall size and number of occupants inferred. Local wind speeds and unusual wind phenomena at upper elevations must be considered.

Flexible Structures. These may be affected in a variety of ways, including vibration or flutter as well as the simple magnitude of movements.

Unusual Shapes. Open structures, structures with large overhangs or other projections, and any building with a complex shape should be carefully studied for the special wind effects that may occur. Wind tunnel testing may be advised or even required by some codes.

Use of code criteria for various ordinary buildings is illustrated in the design examples in Part 9.

45.4. GENERAL DESIGN CONSIDERATIONS FOR WIND

The relative importance of design for wind as an influence on the general building design varies greatly among buildings. The location of the building is a major consideration, the basic design pressure varying by a factor of 2.4 from the lowest wind area to the highest on the *UBC* map. Other important variations include the dead weight of the construction, the height of the building, the type of structural system (especially for lateral load resistance), the aerodynamic shape of the building and its exposed parts, and the existence of large openings, recessed portions of the surface, and so on.

The following is a discussion of some general considerations of design of buildings for wind effects. Any of these factors may be more or less critical in specific situations.

Influence of Dead Load

Dead load of the building construction is generally an advantage in wind design, because it is a stabilizing factor in resisting uplift, overturn, and sliding and tends to reduce the incidence of vibration and flutter. However, the stresses that result from various load combinations, all of which include dead load, may offset these gains when the dead load is excessive.

Anchorage for Uplift, Sliding, and Overturn

Ordinary connections between parts of the building may provide adequately for various transfers of wind force. In some cases, such as with lightweight elements, wind anchorage may be a major consideration. In most design cases the adequacy of ordinary construction details is considered first, and extraordinary measures are used only when required. Various situations of anchorage are illustrated in the examples in Part 9.

Critical Shape Considerations

Various aspects of the building form can cause increase or reduction in wind effects. Although it is seldom as critical in building design as it is for racing cars or aircraft, streamlining can improve the relative efficiency of the building in wind resistance. Some potential critical situations, as shown in Fig. 45.3, are as follows:

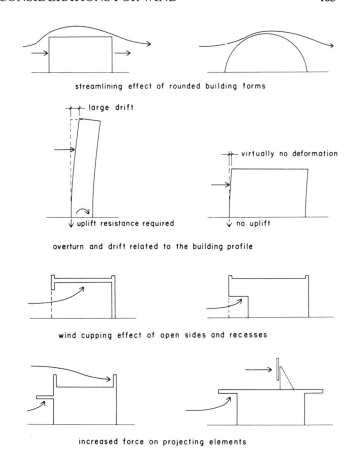

FIGURE 45.3. Wind effects related to building form.

1. Flat versus curved forms. Buildings with rounded forms, rather than rectangular forms with flat surfaces, offer less wind resistance.
2. Tall buildings that are short in horizontal dimension are more critical for overturn and possibly for the total horizontal deflection at their tops.
3. Open-sided buildings or buildings with forms that cup the wind tend to catch the wind, resulting in more wind force than that assumed for the general design pressures. Open structures must also be investigated for major outward force on internal surfaces.
4. Projections from the building. Tall parapets, solid railings, cantilevered balconies and canopies, wide overhangs, and freestanding exterior walls catch considerable wind and add to the overall drag effect on the building. Signs, chimneys, antennae, penthouses, and equipment on the roof of a building are also critical for the clean-off effect discussed previously.

Relative Stiffness of Structural Elements

In most buildings the lateral resistive structure consists of two basic elements: the horizontal distributing elements and the vertical cantilevered or braced frame elements. The manner in which the horizontal elements distribute forces and the manner in which the vertical elements share forces are critical considerations in wind analysis. The relative stiffness of individual elements is the major property that affects these relationships. The various situations that occur are discussed in Chapter 47 and illustrated in the examples in Part 9.

Stiffness of Nonstructural Elements

When the vertical elements of the lateral resistive system are relatively flexible, as with rigid frames and wood shear walls that are short in plan length, there may be considerable lateral force transferred to nonstructural elements of the building construction. Wall finishes of masonry veneer, plaster, or drywall can produce relatively rigid planes whose stiffnesses exceed those of the structures over which they are placed. If this is the case, the finish material may take the load initially, with the structure going to work only when the finish fails. This result is not entirely a matter of relative stiffness, however, because the load propagation through the building also depends on the attachments between elements of the construction. This problem should be considered carefully when developing the details of the building construction.

Allowance for Movement of the Structure

All structures deform when loaded. The actual dimension of movement may be insignificant, as in the case of a poured-concrete shear wall, or it may be considerable, as in the case of a slender steel rigid frame. The effect of these movements on other elements of the building construction must be considered. The case of transfer of load to nonstructural finish elements, as just discussed, is one example of this problem. Another critical example is that of windows and doors. Glazing must be installed so as to allow for some movement of the glass with respect to the frame. The frame must be installed so as to allow for some movement of the structure of the building without load being transferred to the window frame.

All these considerations should be kept in mind in developing the general design of the building. If the building form and detail are determined and the choice of materials made before any thought is given to structural problems, it is not likely that an intelligent design will result. This is not to suggest that structural concerns are the most important concerns in building design, but merely that they should not be relegated to afterthoughts.

CHAPTER FORTY-SIX

Earthquake Effects on Buildings

Earthquakes are essentially vibrations of the earth's crust caused by subterranean ground faults. They occur several times a day in various parts of the world, although only a few each year are of sufficient magnitude to cause significant damage to buildings. Major earthquakes occur most frequently in particular areas of the earth's surface that are called *zones of high probability*. However, it is theoretically possible to have a major earthquake anywhere on the earth at some time.

During an earthquake the ground surface moves in all directions. The most damaging effects on structures are generally the movements in a direction parallel to the ground surface (that is, horizontally) because of the fact that structures are routinely designed for vertical gravity loads. Thus, for design purposes the major effect of an earthquake is usually considered in terms of horizontal force similar to the effect of wind.

46.1. CHARACTERISTICS OF EARTHQUAKES

Following a major earthquake, it is usually possible to retrace its complete history through the recorded seismic shocks over an extended period. This period may cover several weeks, or even years, and the record will usually show several shocks preceding and following the major one. Some of the minor shocks may be of significant magnitude themselves, as well as being the foreshocks and aftershocks of the major quake.

A major earthquake is usually rather short in duration, often lasting only a few seconds and seldom more than a minute or so. During the general quake, there are usually one or more major peaks of magnitude of motion. These peaks represent the maximum effect of the quake. Although the intensity of the quake is measured in terms of the energy release at the location of the ground fault, called the *epicenter*, the critical effect on a given structure is determined by the ground movements at the location of the structure. The extent of these movements is affected mostly by the distance of the structure from the epicenter, but they are also influenced by the geological conditions directly beneath the structure and by the nature of the entire earth mass between the epicenter and the structure.

Modern recording equipment and practices provide us with representations of the ground movements at various locations, thus allowing us to simulate the effects of major earthquakes. Figure 46.1 shows the typical form of the graphic representation of one particular aspect of motion of the ground as recorded or as interpreted from the recordings for an earthquake. In this example the graph is plotted in terms of the acceleration of the ground in one horizontal direction as a function of elapsed time. For use in physical tests in laboratories or in computer modeling, records of actual quakes may be "played back" on structures in order to analyze their responses.

These playbacks are used in research and in the design of some major structures to develop criteria for design of lateral resistive systems. Most building design work, however, is done with criteria and procedures that have been evolved through a combination of practical experience, theoretical studies, and some empirical relationships derived from research and testing. The results of the current collective knowledge are put forth in the form of recommended design procedures and criteria that are incorporated into the building codes.

Although it may seem like a gruesome way to achieve it, we advance our level of competency in design every time there is a major earthquake that results in some major structural damage to buildings. Engineering societies and other groups routinely send investigating teams to the sites of major quakes to report on the effects on buildings in the area. Of particular interest are the effects on recently built structures, because these buildings are in effect full-scale tests of the validity of our most recent design techniques. Each new edition of the building codes usually reflects some of the results of this cumulative growth of knowledge culled from the latest disasters.

46.2. EFFECTS OF EARTHQUAKES

The ground movements caused by earthquakes can have several types of damaging effects. Some of the major effects are:

Direct Movement of Structures. Direct movement is the motion of the structure caused by its attachment to the ground. The two primary effects of this motion are a general destabilizing effect due to the shaking and to the impelling force caused by the inertia of the structure's mass.

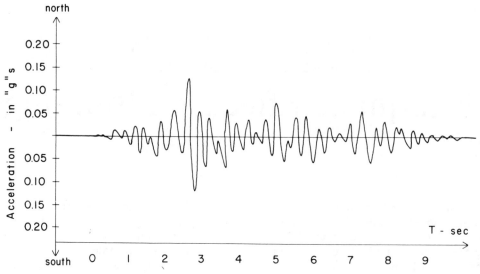

FIGURE 46.1. Characteristic form of ground acceleration graph for an earthquake.

Ground Surface Faults. Surface faults may consist of cracks, vertical shifts, general settlement of an area, landslides, and so on.

Tidal Waves. The ground movements can set up large waves on the surface of bodies of water that can cause major damage to shoreline areas.

Flooding, Fires, Gas Explosions, and so on. Ground faults or movements may cause damage to dams, reservoirs, river banks, buried pipelines, and so on, which may result in various forms of disaster.

Although all these possible effects are of concern, we deal in this book only with the first effect—the direct motion of structures. Concern for this effect motivates us to provide for some degree of dynamic stability (general resistance to shaking) and some quantified resistance to energy loading of the structure.

The force effect caused by motion is generally directly proportional to the dead weight of the structure—or, more precisely, to the dead weight borne by the structure. This weight also partly determines the character of dynamic response of the structure. The other major influences on the structure's response are its fundamental period of vibration and its efficiency in energy absorption. The vibration period is basically determined by the mass, the stiffness, and the size of the structure. Energy efficiency is determined by the elasticity of the structure and by various factors such as the stiffness of supports, the number of independently moving parts, and the rigidity of connections.

A relationship of major concern is that which occurs between the period of the structure and that of the earthquake. Figure 46.2 shows a set of curves, called *spectrum curves*, that represent this relationship as derived from a large number of earthquake "playbacks" on structures with different periods. The upper curve represents the major effect on a structure with no damping. Damping

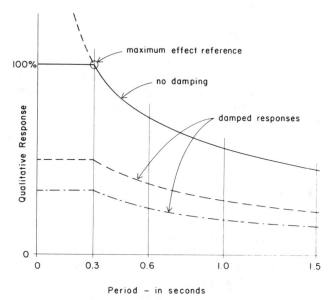

FIGURE 46.2. Spectrum response graph.

results in a lowering of the magnitude of the effects, but a general adherence to the basic form of the response remains.

The general interpretation of the spectrum effect is that the earthquake has its major direct force effect on buildings with short periods. These tend to be buildings with stiff lateral resistive systems, such as shear walls and X-braced frames, and buildings that are small in size and/or squat in profile.

For very large, flexible structures, such as tall towers and high-rise buildings, the fundamental period may be so long that the structure develops a whiplash effect, with different parts of the structure moving in opposite directions at the same time, as shown in Fig. 46.3. Analysis for this

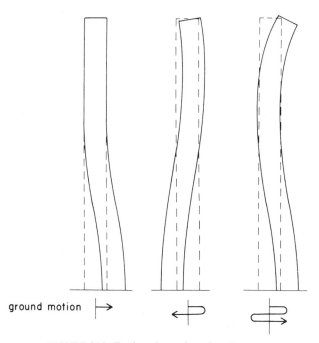

FIGURE 46.3. Earthquake motion of a tall building.

In addition to the movement of the structure as a whole, there are independent movements of individual parts. These each have their own periods of vibration, and the total motion occurring in the structure can thus be quite complex if it is composed of a number of relatively flexible parts.

Earthquake Effects on Buildings

The principal concern in structural design for earthquake forces is for the laterally resistive system of the building. In most buildings this system consists of some combination of horizontally distributing elements (usually roof and floor diaphragms) and vertical bracing elements (shear walls, rigid frames, trussed bents, etc.). Failure of any part of this system, or of connections between the parts, can result in major damage to the building, including the possibility of total collapse.

It is well to remember, however, that an earthquake shakes the whole building. If the building is to remain completely intact, the potential movement of all its parts must be considered. The survival of the structural system is a limited accomplishment if suspended ceilings fall, windows shatter, plumbing pipes burst, and elevators are derailed.

A major design consideration is that of tying the building together so that it is quite literally not shaken apart. With regard to the structure, this means that the various separate elements must be positively secured to one another. The detailing of construction connections is a major part of the structural design for earthquake resistance.

In some cases it is desirable to allow for some degree of independent motion of parts of the building. This is especially critical in situations where a secure attachment between the structure and various nonstructural elements, such as window glazing, can result in undesired transfer of force to the nonstructural elements. In these cases use must be made of connecting materials and details that allow for

behavior requires the use of dynamic methods that are beyond the scope of this book. The three general cases of structural response are illustrated by those shown in Fig. 46.4. Referring to the spectrum curves, for buildings with a period below that representing the upper cutoff of the curves (approximately 0.3 sec), the response is that of a rigid structure with virtually no flexing. For buildings with a period slightly higher, there is some reduction in the force effect caused by the slight "giving" of the building and its using up some of the energy of the motion-induced force in its own motion. As the building period increases, the behavior approaches that of the slender tower, as shown in Fig. 46.3.

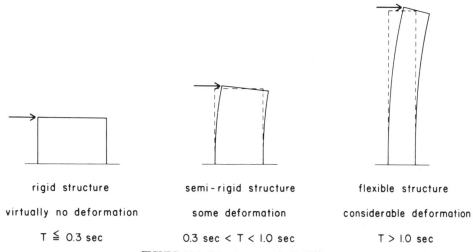

rigid structure

virtually no deformation

$T \lesssim 0.3$ sec

semi-rigid structure

some deformation

0.3 sec $< T < 1.0$ sec

flexible structure

considerable deformation

$T > 1.0$ sec

FIGURE 46.4. Seismic response of buildings.

the holding of the elements in place while still permitting relative independence of motion.

When the building form is complex, various parts of the building may tend to move differently, which can produce critical stresses at the points of connection between the parts of the building. The best solution to this sometimes is to provide connections (or actually in some cases nonconnections) that allow for some degree of independent movement of the parts. This type of connection is called a *seismic separation joint*, and its various problems are discussed in Sec. 47.8.

Except for the calculation and distribution of the loads, the design for lateral loads from earthquakes is generally similar to that for the horizontal forces that result from wind. In some cases the code requirements are the same for the two loading conditions. There are many special requirements for seismic design in the *UBC*, however, and the discussion in the next section, together with the examples in Part 9, deal with the use of the code for analysis and design for earthquake effects.

46.3. BUILDING CODE REQUIREMENTS FOR EARTHQUAKE EFFECTS

The model building code that generally presents the most up-to-date, complete guidelines for design for earthquake effects is the *UBC*. This code is reissued every three years and its section on seismic design requirements is continually revised to reflect current developments in the field.

The 1988 edition of the *UBC* contained a considerable revision of the seismic design requirements. While it represents a major revision in comparison to the previous codes, it actually merely reflects developments that have been in preparation for some time, as reflected in various research reports and recommendations by engineering groups and government-sponsored agencies.

The following is a brief digest of the current *UBC* requirements. Applications of much of this material is illustrated in the building design examples in Part 9.

Considerations for seismic effects are included in several chapters of the code—such as requirements for foundations, reinforced concrete, and so on. The basic requirements for investigation of seismic effects, however, are presented in Sec. 2312. This section begins with a list of definitions of many of the special terms that are used in the section, and it is recommended that this list be studied for the purpose of a clear understanding of the terminology. Some selected parts of the code are reproduced here in Appendix D.

A critical determination for seismic design is what is called the *base shear*, which is essentially the total lateral force assumed to be delivered to the building at its base. For structural design, the force is actually assumed to be distributed vertically in the building, but the total force thus distributed is visualized as the base shear.

As an equivalent static force effect, the base shear is expressed as some percent of the building weight. In the earliest codes this took the simple form of

$$V = 0.1W$$

or simply 10% of the building weight. This simple formula was embellished over the years to incorporate variables that reflected such issues as potential risk, dynamic response of the building, potential building/site interaction, and relative importance of the building's safety.

The changes in the 1988 *UBC* include considerations for an increased number of variations of the building form and the dynamic characteristics of the building. The formula for base shear now takes the form

$$V = \frac{ZIC}{R_W} W$$

The terms in this formula are as follows:

Z is a factor that adjusts for probability of risk; values are given in *UBC* Table 23-I for zones established by a map in *UBC* Fig. 2. (Note: Values for *Z* have been changed in the 1988 *UBC*.)

I is an importance factor; a token acknowledgment of the fact that in the event of a major disaster, the destruction of some buildings is more critical than that of others. In this view, three such buildings are those that house essential emergency facilities (police and fire stations, hospitals, etc.), potentially hazardous materials (toxic, explosive, etc.), or simply a lot of people in one place.

I factors are given in *UBC* Table 23-L and the categories of the occupancies on which they are based are described in detail in *UBC* Table 23-K.

C is a general factor that accounts for the specific, fundamental character of the building's dynamic response, as related to the general dynamic nature of major, recorded seismic events (translation: big earthquakes). Determination of *C* and the various factors that affect it are described in the following section.

R_W is the principal new factor in the 1988 *UBC*, although it relates to some issues dealt with previously by use of other factors. This factor generally accounts for considerations of the building's materials, type of construction, and type of lateral bracing system: a lot for one factor to deal with, but a considerable amount of code data goes into the establishing of this factor. Section 46.4 discusses the considerations for the determination of this factor.

W is the weight (read: *mass*) of the building. This is simply, basically, the dead load of the building construction, but may also include some considerations for the weight of some contents of the building: notably heavy equipment, furnishings, stored materials, and so on. The basic consideration is simply: What is the mass that is impelled by the seismic motion?

It should be noted that building code criteria in general are developed with a particular concern in mind. This concern is not for the preservation of the building's appearance, protection of the general security of the property as a financial investment, or the assurance that the building will remain functional after the big earthquake—or, at the least, be feasibly repairable for continued use. The building codes are concerned essentially—and pretty much only—with life safety: the protection of the public from injury or death. Designers or building owners with concerns beyond this basic one should consider the building code criteria to be really minimal and generally not sufficient to assure the other protections mentioned previously, relating to the security of the property.

Although it is of major importance, the determination of the base shear is only the first step in the process of structural design or investigation for seismic effects. Many other factors must be dealt with, some of which are discussed in later sections of this chapter. However, the process of design goes beyond merely a response to commands of the codes, and many issues relating to a general consideration of earthquake-resistive construction are discussed in Chapters 44 and 47. These include concerns for the building's lateral bracing system as well as general concern for the integrity of the whole of the building construction.

Although the basic considerations for seismic effects are dealt with in *UBC* Sec. 2312, there are numerous requirements scattered throughout the code, many of which will be discussed in relation to the design of specific types of structures, use of particular materials, and development of various forms of construction.

Specific Dynamic Properties

The predictable response of an individual building to an earthquake can be generalized on the basis of many factors having to do with the type of building. These general considerations are largely dealt with in deriving the R_W factor for the base shear. Each building also has various specific properties, however, which must be determined individually for each case. The actual building weight, for example, is such a property, and the value for W in the equation for base shear is specific to a single building.

Additional specific properties include the following:

Fundamental period (T) of the building, as described in Sec. 46.2

Fundamental period of the building site

Interaction of the building and site, due to the matching of their individual periods and their linkage during seismic movements

The general incorporation of these considerations is through the use of the C factor, which is expressed as follows:

$$C = \frac{1.25S}{T^{2/3}}$$

where T is the fundamental period of the building and S is the site coefficient. Provisions are made in the code for limiting the value of C and the value of the C/R_W combination. Provisions are also made for the determination of both T and S on either an approximate basis, using generalized data, or an analytical basis, using more precise dynamic investigative methods.

Although the derivation of the C and R_W factors incorporates many considerations for dynamic effects, it must be noted that their use is still limited to producing shear force (V) for the equivalent static investigation. As much as this process may be refined, it is still not a real dynamic analysis using energy and work instead of static force.

One further adjustment required for the equivalent static method involves the distribution of the total lateral force effect to the various elements of the lateral bracing system. This issue is discussed further in Sec. 46.4.

Applications of much of the code criteria for the equivalent static method are illustrated in the examples in Part 9.

General Categories for Dynamic Response

Many of the aspects of a building's response to an earthquake can be predicted on the basis of the general character of the building in terms of form, materials, and general construction. In previous editions of the *UBC*, this was accounted for by use of a K factor, with data providing for the differentiation of six different basic forms of construction. A major change in the 1988 *UBC* is the use of the R_W factor, generally replacing the K factor and providing for 14 categories of the construction of the lateral load-resisting system.

Another major change in the 1988 *UBC* is the identification of several cases of *structural irregularity*, which can limit the use of the equivalent static method and require a more rigorous, dynamic investigation. Two types of irregularity are defined: those which are related to vertical form and relationships, and those that are related to plan (horizontal) form and relationships. These issues are discussed more fully in Sec. 46.4.

Previous editions of the *UBC* have been somewhat vague about what specifically constitutes a building that cannot be reliably investigated or designed by the equivalent static method. The 1988 edition has considerable definition of the categories of buildings that require a true dynamic investigation for design. These requirements include considerations for the risk zone, building size or number of stories, bracing system, and degree of either vertical or plan irregularity.

A general effect of the 1988 edition of the *UBC* has been to cause a much greater concern for general architectural features of the building. While the real concern in many cases is for the response of the lateral load-resisting system, the form of that system is often substantially defined by the form of the building in general. The various forms of irregularity are usually derived from architectural features of the building plans and vertical profile. Beyond this are many special requirements based on concerns for damage to various nonstructural elements, such as ceilings, parapets and cornices, signs, heavy suspended light fixtures or equipment, and freestanding partitions, shelving, or other items subject to lateral movement, overturning, or detachment during an earthquake.

As mentioned previously, the main concern of codes is for life safety. Building collapse is the principal worry, but flying or falling elements of the building construction or furnishings can also represent major hazards. Various factors used to modify or regulate the design of the structure may be partly derived from concerns for nonstructural damage. Restrictions on lateral deflection (drift) are largely based on this concern.

Distribution of Seismic Base Shear

The total horizontal force computed as the base shear (V) must be distributed both vertically and horizontally to the elements of the lateral load-resisting system. This begins with a consideration of the actual distribution of the building mass, which essentially develops the actual inertial forces. However, for various purposes in simulating dynamic response, the distribution of forces for the actual investigation of the structure may be modified.

Section 2312(i)4 of the *UBC* requires a redistribution of the lateral forces at the various levels of multistory buildings. As visualized vertically, these forces are assumed to be applied at the levels of the horizontal diaphragms, although the redistribution is intended to modify the loading to the vertical bracing system. The purpose of this modification is to move some of the lateral load to upper levels of the building, simulating more realistically the nature of the response of the vertical cantilever to the dynamic loads. Use of this criterion is illustrated in the multistory building examples in Chapter 48.

In a horizontal direction, the total shear at any level of the building is generally considered to be distributed to the vertical elements of the system in proportion to their stiffness (resistance to lateral deflection). If the lateral bracing elements are placed symmetrically, and their centroid corresponds to the center of gravity of the building mass, this simple assumption may be adequate. However, two considerations may alter this simple distribution. The first has to do with the coincidence of location of the centroid of the bracing system (usually called the *center of resistance*) and the center of gravity of the building mass. If there is a major discrepancy in the location of these, there will be a horizontal torsional effect, which will produce shears that must be added to those produced by the direct shear force. Even when no actual theoretical eccentricity of this type occurs, the code may require the inclusion of a so-called accidental eccentricity as a safety measure.

The second modification of horizontal distribution has to do with the relative horizontal stiffness of the horizontal diaphragm (in effect, its deflection resistance in the diaphragm/beam action). The two major considerations that affect this are the aspect ratio of the diaphragm in plan (length-to-width ratio), and the basic construction of the diaphragm. Wood and formed steel deck diaphragms tend to be flexible, while concrete decks are stiff. The actions of diaphragms in this respect are discussed in Sec. 47.1.

In some cases it may be possible to manipulate the distribution of forces by alterations of the construction. Seismic separation joints represent one such alteration (see discussion in Sec. 47.8). Another technique is to modify the stiffnesses of various vertical bracing elements to cause them to resist more or less of the total lateral load; thus a building may have several vertical bracing elements, but a few may take most of the load if they are made very stiff.

46.4. GENERAL DESIGN CONSIDERATIONS FOR EARTHQUAKES

The influence of earthquake considerations on the design of building structures tends to be the greatest in the zones of highest probability of quakes. This fact is directly reflected in the *UBC* by the Z factor, which varies from $\frac{3}{16}$ to 1, or by a ratio of more than 5:1. As a result, wind factors often dominate the design in the zones of lower seismic probability.

A number of general considerations in the design of lateral resistive systems were discussed in Chapter 45. Most of these also apply to seismic design. Some additional considerations are included in the following discussion.

Influence of Dead Load

Dead load is in general a disadvantage in earthquakes, because the lateral force is directly proportional to it. Care should be exercised in developing the construction details and in choosing materials for the building in order to avoid creating unnecessary dead load, especially at upper levels in the building. Dead load is useful for overturn resistance and is a necessity for the foundations that must anchor the building.

Advantage of Simple Form and Symmetry

Buildings with relatively simple forms and with some degree of symmetry usually have the lowest requirements for elaborate or extensive bracing or for complex connections for lateral loads. Design of plan layouts and of the building form in general should be done with a clear understanding of the ramifications in terms of structural requirements when wind or seismic forces are high. When complex form is deemed necessary, the structural cost must be acknowledged.

Following Through with Load Transfers

It is critical in design for lateral loads that the force paths be complete. Forces must travel from their points of origin through the whole system and into the ground. Design of the connections between elements and of the necessary drag struts, collectors, chords, blocking, hold-downs, and so on, is highly important to the integrity of the whole lateral resistive system. The ability to visualize the load paths and a reasonable understanding of building construction details and processes are prerequisites to this design work.

Use of Positive Connections

Earthquake forces often represent the most severe demands on connections because of their dynamic, shaking effects. Many means of connection that may be adequate for static force resistance fail under the jarring, loosening effects of earthquakes. Failures of a number of recently built buildings and other structures in earthquakes have been due to connection failures, even though the structures were designed in accordance with current code requirements and accepted practices. Increasing attention is being paid to this problem in the development of recommended details for building construction.

Determination of Lateral Loads

For the equivalent static load method, the determination of lateral loads consists of the following:

1. *Visualization of the Loading Pattern.* This consists of determining the manner in which the mass of various parts of the buildings become horizontal forces that are applied to the lateral resistive structure. This varies with the form of the building and with the type and disposition of the structure.
2. *Determination of the Building Mass.* This is a matter of figuring out how much the various parts weigh, and how they aggregate as combined quantities for the loading analysis. (*Example*: The total mass that constitutes the lateral load applied to the roof diaphragm.)
3. *Conversion of Mass to Force.* Whether for the load on a single element or the total base shear for the building, this is a matter of using the code formulas with the various appropriate modifying factors.
4. *Redistribution (if any) of Forces.* This refers primarily to the adjustments for multi-story buildings, as described in Sec. 46.3.
5. *Determination of Design Values for Individual Elements.* Once the load magnitudes and their patterns of application are found, this is a matter of analyzing the behavior of the particular structure—box, truss, frame, and so on.

Procedures for various types of structures are illustrated in the examples in Part 9. Data for the determination of the building weight are provided in Appendix C.

Calculation of Deformations

There are a few situations in seismic design in which determination of structural deformations must be made. In some cases the actual dimension of the movement is required, in other cases it is mostly a matter of establishing relative stiffness. Some common situations are the following.

1. *Lateral Drift of Frames.* This is quoted in terms of the relative horizontal movements of the levels of the frame. The code establishes limits as a percent of the story height. Of concern in some situations is the effect of the frame distortion on walls in the plane of the frame. The calculation of drift for braced frames is discussed in Sec. 47.3, where the primary contribution is the change in length of the diagonals. Drift of rigid frames is discussed in Sec. 47.4.
2. *Deflection of Shear Walls.* This is mostly of concern for wood-framed walls. Empirical formulas are used for plywood walls, incorporating the various contributions of chord length change, shear distortion of the panels, and bending of the nails.
3. *Deflection of Horizontal Diaphragms.* These are mostly of concern for diaphragms that are wood framed or those that use light steel deck. Empirical formulas and available data can be used for both situations.

Actual dimensions of movement may be of concern for various reasons. Satisfying code limits for story drift or determining effects on nonstructural elements—as mentioned previously—are examples. Another problem is that of providing a basis for the detailing and dimensioning of structural separation joints. In truth, however, movements calculated by the static load method are quite fictitious, and considerable judgment is needed if they are used for design.

Dynamic Analysis

For the equivalent static load method, the energy loads are translated into static forces, and the analysis then proceeds in the usual manner as for real static loads. Loads are quantified in static units (pounds or kips), and the response of the structure is evaluated in static terms (stress). A more realistic dynamic analysis requires that the load be dealt with in dynamic units (energy, not force) and the structure be evaluated in dynamic terms (work done or energy absorbed).

In fact, the current application of the static load method—as specified in the *UBC*—is considerably modified by dynamic considerations. Determination of the loading is modified by dynamic responses of the structure through use of the K, C, and S factors. Distribution of loads for tall structures is modified to approximate the true distribution of dynamic shear. Stress magnification is required for shear in masonry walls and for forces in the members and connections of braced frames.

Nevertheless, true dynamic responses can only be determined using dynamic relationships, dynamic properties of the structure, and dynamic units for loads. This type of analysis is being increasingly done in engineering practice as the operating facilities (mostly computer software) become more readily available and usable. In time, dynamic analysis will probably be routinely done for most buildings. At present, however, it is mostly done only as a check or design adjustment after a preliminary design has been done by conventional methods. And then it is mostly only for buildings of unusual configuration, complex structural form, considerable size and cost, or where there is a heightened concern for safety (e.g., nuclear power plants).

Load Transfers Between Elements

The most difficult aspect of visualizing the propagation of loads due to seismic activity is that it is essentially a three-dimensional problem. The seismic actions, the resulting forces, and the building structure are all three dimensional. Add this to the fact that the actions are dynamic and the problem is quite enough.

The difficulty of the design and investigative work may be a contributory reason for the number of failures of connections between elements of lateral resistive systems. Neither architects nor civil engineers are, by training, well prepared to deal with three-dimensional problems of force resolution.

Load propagation in general is discussed in Sec. 44.2, and the actions of various elements and systems are discussed and illustrated in Chapter 47. More direct and practical applications are presented in the building design examples in Part 9.

Interactions in the lateral resistive system are commonly illustrated as single static load conditions for individual elements (the classic free-body diagram). Since the separate elements are usually two dimensional in character (single shear wall, single rigid bent, etc.), illustrations are simplified as two-dimensional force resolutions. The concept of the three-dimensional actions fades away when one works continuously with the relatively simple, two-dimensional problem sets.

Individual elements are not loaded individually, but rather are loaded simultaneously with all other parts of the system. Load transfers are thus a matter of load distributions (or load sharing) in an interactive, multiunit system. This is similar to the action of a continuous rigid frame in which the action of any one individual member is conditioned by the response of all the members in the system.

Anchoring and Splicing

Two functions that occur repeatedly in lateral resistive systems are those of anchoring and splicing. Anchoring is the general term for the securing of some structural entity at its boundaries. Thus a horizontal diaphragm is anchored to the vertical system, shear walls are anchored at their bases, and so on. Splicing occurs in all structures in one form or another, being generally necessary whenever continuity is required in long elements that are assembled from short individual parts.

For either anchoring or splicing there is a need for so-called positive connection. This is a somewhat nebulous term but generally means a connecting method that is relatively secure against the jolting, repetitive, multidirectional actions during an earthquake. Many connecting devices and methods that are acceptable for resistance to static gravity forces are not so qualified. The seismic be-

havior of ordinary connection methods is discussed in Sec. 44.3 and specific issues relating to typical elements of lateral resistive systems are presented in Chapter 47.

The positive character of anchorage is often a matter of degree. In some cases it may be achieved by simple redundancy, much like wearing both belt and suspenders, or using a shotgun to kill a flea.

Figure 46.5 shows the situation of anchoring the top reinforcing for a beam in a supporting column or spandrel girder. The positive character of this connection may be increased by degrees as follows:

1. Anchorage can be achieved by simple embedment by providing a sufficient length (*L* in Fig. 46.5*a* and *b*) for development of bond between the bar surface and the surrounding concrete.

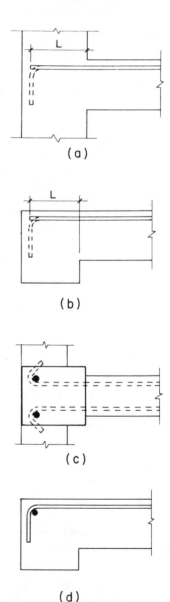

FIGURE 46.5. Progressive development of bar anchorage.

2. Additional anchorage is provided by adding a hook to the end of the bar, a common procedure when the length L is not sufficient. The hook substantially increases the positive character of the connection.

3. Finally, the hook may be made more effective if it is not merely made to grasp the concrete mass but is wrapped around a transverse reinforcing bar, as shown in Fig. 46.5c and d.

The following are some additional considerations regarding the positive character of connections:

Screws or bolts are more positive than nails.

Nails loaded laterally are more positive than those subjected to withdrawal.

Welding of steel is more positive than bolting; high-strength friction bolts are more positive than unfinished bolts.

Lapped splices of reinforcing bars that comply with the requirements of the *ACI Code* are usually adequate for either tension or compression anchorage. However, some designers prefer the more positive action of welded or mechanically achieved anchorage at splice points. The latter, however, must be done by methods approved by local codes and may require special inspection during construction.

Discontinuities

In general, seismic response will be better when the flow of forces in the structure is simple, direct, and uninterrupted. The existence of various discontinuities will work to reduce all of these qualities so that the force flow becomes complex and inefficient. This is not a blank indictment for complexity, as the primary concern is for the lateral resistive structure. It is possible to have richness of form and detail in the building while still maintaining a simple, continuous, well-ordered bracing system.

Certain discontinuities in the building plan or general form, however, will often be reflected in the form of the lateral bracing system. Some of these situations are as follows:

1. *Openings.* Windows, doors, skylights, stairs, ducts, and elevators require openings. These may result in holes in horizontal or vertical diaphragms or may produce misalignment or discontinuity of rigid frame bents.

2. *Multiplane Roofs.* The farther the roof is from a single, flat, continuous plane, the less effective it will be for use as a horizontal diaphragm.

3. *Split-Level Floors.* This is the same problem as for the roof. Where this is required, the locations of level changes should be carefully worked out with the layout of the vertical elements of the bracing system so that the horizontal diaphragm does not have to try to maintain its continuity across level changes.

4. *Interrupted Shear Walls or Offset Columns.* While not uncommon, these represent major discontinuities and should be avoided if possible.

5. *Nonlinear Edges.* If the building edge is curved, sawtoothed, zigzagged, or otherwise not straight, the development of diaphragm chords and collectors may be difficult. A useful technique may be to use a line somewhat inside the actual edge as the boundary of the diaphragm and simply cantilever the edge for horizontal force.

Site-Structure Interaction

Building site problems are usually of one of the following type:

1. Subsurface conditions that complicate the building foundation design problem: weak, compressible soil deposits; high groundwater levels; highly organic materials.

2. Topographic situations: hillsides, surface water.

3. Instability conditions: collapsing soils, highly erodable soils, soils subject to dynamic liquefaction.

4. Structure-site interaction during motion that increases seismic effects on the building. (See the *UBC* for use of the *S* factor.)

In effect, the ground must be considered to be an element of the lateral resistive system. Once the building is set in motion, the forces induced by its momentum must be resolved into the ground, just as must be done for gravity forces. Vertical soil pressure, horizontal sliding and lateral soil pressure, pressure concentrations due to overturning, and resistance to uplift must all be given consideration in the design of the foundations.

Concern for site–structure interaction in the form of vibratory relationships has only relatively recently been incorporated in code criteria. It is likely that this criterion will be further refined in future editions of the *UBC*. With more experience in the use of design criteria and the data in the form of geological studies required for its implementation, it is likely that this area of investigation will have more influence. To an extent this is a trend of concern for geological factors beyond the scale of the immediate site as traditionally reflected in ordinary shallow soil borings at the location of the building.

Load Sharing

A principal concern in the design of the vertical elements of the lateral bracing system is the distribution of loads to the individual elements of the system. The portion of the total horizontal force on the building that is carried by a single element will be affected by several factors. The principal considerations are as follows:

1. *Location of the Vertical Elements in the Building Plan.* This may be of significance primarily with respect to the locations of other vertical elements. If loads are distributed on a peripheral basis (also called *tributary load basis*), the periphery for a single element is defined by the location of adjacent elements. If the building is subject to significant torsion, the location of an element will determine its contribution to torsional resistance.

2. *Relative Stiffness.* Where elements exist in rows, such as a series of shear wall piers in a single wall surface, load sharing is on the basis of the relative stiffness of the elements in the linked series. If the distributing element (usually the horizontal diaphragm) is quite stiff, distribution to all members may be on a stiffness basis rather than on a peripheral basis. Illustrations and discussion of these situations are presented in the design examples in Part 9.

3. *Control Devices.* Load distribution may be controlled in part through the use of various devices such as expansion joints, seismic separation joints, pinned joints, collectors, drag struts, and ties. Natural force paths may thus be diverted or redirected to some extent.

Loads may be redistributed or shifted when staged responses occur. Initial failures, such as yielding of steel reinforcing and plastic hinging in frames, will change the relative resistance of elements so that loads may be shifted to other members in the linked system. From initial response to final failure of the entire system, load distributions will most likely shift considerably among the vertical elements.

In many cases determination of the design load for a single element of the vertical system requires some amount of judgment involving many qualifying assumptions. When a rational process seems elusive, it is not uncommon to consider a range of possibilities, to investigate each, and to provide for all the worst eventualities.

Coupling and Decoupling

It is generally desirable that the building move as a single unit in response to seismic actions. However, in the typical multimassed building, with many parts of different mass and stiffness, it is not reasonable to think of movements as simple as those of a tuning fork or a single bouncing spring. Nevertheless, it is usually a first preference to try to tie the separate elements of the building together for a single joined response if possible. This is particularly critical for the building base; thus other options are seldom considered for the foundations.

As in other situations elements of the construction may serve useful purposes for the tying together of the building.

Horizontal diaphragms do the most work in this regard. Also effective are continuous spandrel beams and the continuous top plates of stud walls. If tying is a specific need, these elements often need positive splicing, although they may already be developed for use as chords or collectors.

Building corners and wall intersections are critical locations for interactions. If the presence of framing does not offer potential for tying at these locations, the wall joints themselves should be studied for this purpose.

There are times, of course, when tying elements is not practical or not even desirable. When this is the case, the construction must be designed to permit independent actions by use of flexible attachments or structural separation joints. Detachment of rigid finishes from flexible supports, decoupling of tall shear walls, and separation of separate parts of a multimassed building are examples of these instances. The judgment as to the need for separation, and the development of the construction details to facilitate the degree of required independence, is one of the most difficult parts of seismic design of buildings.

Determination of Building Weight

Translation of seismic motions into force effects involves the use of many theoretical relationships, most of which are empirical, complex, judgmental, and to say the least, mystical. One evident piece of hard data used in the process is that of the building weight. Here is a touchpoint with reality—something believable and logically determinable. However, the physical fabric of most buildings is quite complex, and the computation of building weight is typically a process that employs considerable approximation. Some of the factors that contribute to this situation are as follows:

1. The densities of many building materials vary over some range, requiring the use of *average* values, unless precise material specifications are available. "Plaster" is a description that permits considerable range of interpretation—from the very low density of sprayed-on acoustic coating of ceilings to the high density of exterior stucco.

2. Installation details vary. Thicknesses vary; required auxiliary devices, such as attachments, reinforcing, bracing, and so on, may vary; designer judgment affects many relatively common details. A simple exterior wood stud wall with drywall on the inside and plywood plus exterior finish on the outside can range over a considerable spread in terms of weight per square foot of wall surface. Variables include the size, spacing, and type of wood of the studs, the need for blocking between studs, the thickness of the plywood, the thickness of the drywall, and the type of exterior finish—from light wood siding to real brick veneer.

3. Structural design calculations are often done with information of a vague nature regarding the materials and details of the building construction. This is simply because the definitive structural design

must often be done quite early in the design process since the architectural designer needs to know basic facts about the structure quite early. It is hoped that major changes in the process of development of the construction will be reflected in later adjustments to the structural design. Nevertheless, the preliminary structural design—which often affects major decisions—is usually done with unavoidably sketchy information.

4. Discontinuities in the construction make determination of material quantities difficult. Doors, windows, stairs, openings for ducts, and other building features make precise quantity takeoffs quite laborious.

In short, the determination of the building weight is generally accepted to be an imprecise process that involves considerable judgment and some sheer guesswork. This does not excuse the structural designer from trying to do the best job possible, but the pursuit of precision is doomed to frustration.

Data are provided in Appendix C to assist in the determination of the weight of building construction, once some amount of detail is known about the choices of materials and type of construction.

Distribution of Lateral Loads

To a degree the design requirements and the formulas and data presented in codes are based on buildings of simple shape with lateral resistive structures that are well ordered and free of major discontinuities. Most real buildings are not of this type, and the degree to which they do not conform to simple conditions determines the relative validity of the code criteria.

A major point of consideration in this regard is the relationship between the disposition of the building mass (as generator of lateral forces) and the arrangement of the lateral resistive system. The idea is to try to get the catcher positioned to catch the ball. This may be imaged as a relationship between the entire mass and the entire resistive system, as illustrated in Fig. 44.9. In most real situations the problem is both more complex in nature and less subject to simple computation.

Figure 46.6a shows an irregular building plan with an arrangement of shear walls that are disposed for resistance to lateral forces in one direction. The distribution of the lateral loads to these walls on a peripheral basis is illustrated by the zones defined by the dashed lines on the plan. On this basis the total building weight between the lines labeled 1 and 2 would be used to find the load on the horizontal diaphragm in this area, and this load would be distributed to the three shear walls labeled H, I, and J. Even though these three walls are not symmetrically disposed in the zone area, the stiffness of the diaphragm and the torsional stabilizing effect of the walls and horizontal diaphragm outside the zone would be considered adequate to prevent any significant torsion on the affected walls.

The situation described in Fig. 46.6 is one that is subject to considerable judgmental interpretation. Some factors that could allow for different interpretations are as follows:

1. Is this a one-story or a multistory building? If it is multistory, it may be advisable to use a seismic separation joint at one side or the other of the narrow, necked-down portion near the middle of the plan. This would alter the conditions for distribution to the shear walls near the middle of the building.

2. Are the shear walls of wood frame or of masonry or concrete construction? This relates primarily to the load distribution to individual walls. However, if the walls are of masonry or concrete, some designers would move the peripheral zone boundaries slightly to acknowledge the relative stiffness of the longer walls. Possibilities for such shifting are shown in Fig. 46.6b.

3. What is the construction of the horizontal diaphragm? If it is wood frame with plywood or a light-gage metal deck, its flexibility will affect some judgments. The separation joint would be less indicated in this case, and the shift in peripheral boundaries more questionable. If the diaphragm is a poured-in-place concrete deck, the situation is reversed.

4. Should the system be designed for torsion? If the deck is flexible, and the separation joint is used, probably not. If the deck is concrete, and the separation joint *is* used, each independent half of the building should probably be investigated for torsion.

The use of the peripheral load distribution method and the procedures for consideration of torsional effects are illustrated in the design examples in Part 9.

46.5. BUILDING FORM AND THE 1988 *UBC*

A major effect of the changes in seismic design criteria in the 1988 edition of the *UBC* is a significant advancement of seismic design of buildings as an architectural design concern. This is not entirely a new concern, but has been given considerably more attention in this edition of the code.

There is a steadily growing body of evidence—obtained mostly from postquake inspections of buildings following major seismic events (translation: big earthquakes), which indicates that building form and choice of materials has considerable influence on the response of buildings to earthquake effects. A major response to this is represented by the addition of the definition of the *regular* structure (and by corollary, the *irregular* structure). The qualifications of irregularity described in *UBC* Tables 23-M and 23-N—while describing structural properties—are generally created by architectural planning of the building. By general inference, the building form conditions described in

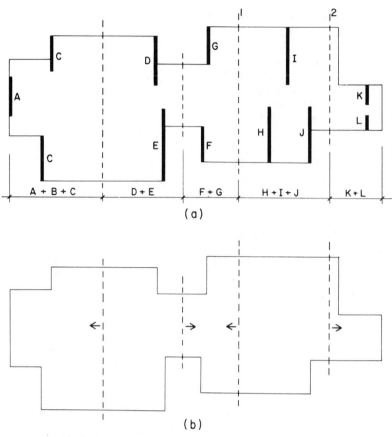

FIGURE 46.6. Load distribution in an irregular shear wall system.

these tables are all negative factors with regard to the seismic resistance of buildings.

There is really very little in the new seismic design requirements in the 1988 *UBC* that was unfamiliar to experienced structural engineers. The *UBC* essentially simply reflects what is well established in engineering and research circles, and quite frankly follows—rather than leads—what is common practice by the leading structural design practitioners. Publication of these materials in the 1988 *UBC* merely brings the architectural designer more specifically into the circle of participants in the development of intelligent designs for seismic-resistive buildings.

We do not advocate that all buildings should be "regular" and that any irregularity is intolerable. However, it behooves all architectural designers who work on buildings in zones of high seismic risk to understand the significance and the consequences of various types of irregularity and how structural irregularity relates to architectural design factors.

The following is a brief discussion of some of the building form considerations that relate to the requirements of the *UBC* Tables 23-M and 23-N.

UBC Table 23-M, Vertical Irregularities

There are five cases—labeled A through E—described in the table, as follows:

A: Soft Story. This situation is described in Sec. 44.7. The common ways in which this occurs are when an individual story (often the ground-level story) is made taller and/or more open in construction. This form is quite commonly used with commercial multistory buildings, so the condition must frequently be considered.

B: Weight Irregularity. This often occurs in conjunction with other irregularities—such as setbacks or interior open spaces. However, another common occurrence is a very heavy roof with rooftop HVAC equipment above a lightweight framed building.

C: Vertical Geometric Irregularity. This refers to setbacks of more than a specified amount in the lateral resistive structure. If core bracing or some other system that does not follow the building's exterior profile is used, this irregularity may be avoided, even when the building itself has major setbacks. If a general perimeter bracing system is used, however, this irregularity may be unavoidable.

D: In-Plane Discontinuity in Vertical Bracing. This refers to offsets of the form illustrated in Fig. 44.17a, limiting them to a specified distance. This is generally less of a problem on the building exterior than on the interior, where changes in building usage (or occupancy) on different levels requires the reorganization of spaces.

E: Weak Story. This is also described briefly in Sec. 44.7. One way that this occurs is when code requirements for minimum construction result in major redundancy of structural capacity: for example, a continuous, long wall with plywood sheathing with code-required minimum nailing, possibly not intended as a shear wall, but obviously having a great resistance to lateral force. If this construction occurs above a more open-walled story, both a soft story and a weak story may result. Minimum masonry or concrete wall construction are similar situations.

UBC Table 23-N, Plan Irregularities

Again, there are five cases, labeled A through E, as follows:

A: Torsional Irregularity. This occurs when there is a major discrepancy between the centroid of the building mass and the centroid of the whole lateral resistive system. The limits in the table refer to disproportionate deformations of opposite ends of the building. This occurs most commonly when the plan of the building is unsymmetrical—probably the general, rather than a special, case for all buildings. The problem is not essentially one of maintaining symmetry in the building, although that may be a simple way to deal with the problem. What is essential is to distribute bracing elements with stiffnesses that match the distribution of the building mass. The key word here is *stiffness*, not strength. A case of this irregularity may occur when mixed bracing systems are used (shear walls plus frames, etc.). This generally requires a very careful analysis of the relative deformations of the vertical bracing elements when plan dissymmetry or other complexity of the building form occurs.

B: Reentrant Corners. This has to do with situations such as that for the building with an L-shaped plan, whose actions are illustrated in Fig. 44.12. The table defines a limit for extension of the building on both sides of the corner, so that the irregular classification is not given for minor corners, such as those shown in Fig. 46.7c and d. It should also be noted that the table considers this case only when both the building and its bracing system have the reentrant corner condition. If perimeter bracing is used, this may be the case. However, it may be possible to develop a bracing system that does not follow the building plan in this manner; such a case is that for the system shown in Fig. 46.7e.

C: Diaphragm Discontinuity. In most cases we consider a roof or floor diaphragm to be a contiguous,

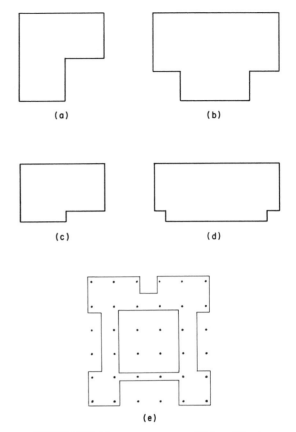

FIGURE 46.7. The reentrant corner; *UBC* qualification.

rigid planar element, capable of distributing lateral forces to the vertical elements without significant concern for its own deformation. However, either plan form discrepancies or changes in construction may modify the behavior of the diaphragm sufficiently to make this not the case. Large openings for skylights or enclosed atrium spaces may create this situation.

D: Out-of-Plane Offsets. This refers to the situation illustrated in Fig.44.17b, which is considered to be an irregularity without qualification.

E: Nonparallel Systems. This generally refers to shear walls or braced bents that are curved or angled in plan, with respect to a simple *x*- and *y*-axis system.

Again, we state that having an irregularity is not necessarily an irredeemable disgrace, but—at the least—calls some attention to special concerns for seismic response. And quite possibly, a collection of several irregularities in the same building may be cause for serious reflection on the quality of the building design.

CHAPTER FORTY-SEVEN

Elements of Lateral Resistive Systems

Chapter 47 presents a discussion of the ordinary structural elements used to develop lateral force resistance in buildings. The primary use of the elements described here is in the development of resistance to the effects of wind and earthquakes. The general problem of structurally induced lateral effects and the structural elements utilized for various situations of such effects are discussed in Chapter 49. The problems of lateral forces on foundations are discussed in Part 7. Design examples of building structures utilizing many of the elements discussed in this chapter are included in Part 9.

47.1. HORIZONTAL DIAPHRAGMS

Most lateral resistive structural systems for buildings consist of some combination of vertical elements and horizontal elements. The horizontal elements are most often the roof and floor framing and decks. When the deck is of sufficient strength and stiffness to be developed as a rigid plane, it is called a *horizontal diaphragm*.

General Behavior

A horizontal diaphragm typically functions by collecting the lateral forces at a particular level of the building and then distributing them to the vertical elements of the lateral resistive system. For wind forces the lateral loading of the horizontal diaphragm is usually through the attachment of the exterior walls to its edges. For seismic forces the loading is partly a result of the weight of the deck itself and partly a result of the weights of other parts of the building that are attached to it.

The particular structural behavior of the horizontal diaphragm and the manner in which loads are distributed to vertical elements depend on a number of considerations that are best illustrated by various example cases in Part 9. Some of the general issues of concern are discussed in the following sections.

Relative Stiffness of the Horizontal Diaphragm. If the horizontal diaphragm is relatively flexible, it may deflect so much that its continuity is negligible and the distribution of load to the relatively stiff vertical elements is essentially on a peripheral basis. If the deck is quite rigid, on the other hand, the distribution to vertical elements will be essentially in proportion to their relative stiffness with respect to each other. The possibility of these two situations is illustrated for a simple box system in Fig. 47.1.

Torsional Effects. If the centroid of the lateral forces in the horizontal diaphragm does not coincide with the centroid of the stiffness of the vertical elements, there will be a twisting action (called *rotation effect* or *torsional effect*) on the structure as well as the direct force effect. Figure 47.2 shows a structure in which this effect occurs because of a lack of symmetry of the structure. This effect is usually of significance only if the horizontal diaphragm is relatively stiff. This stiffness is a matter of the materials of the construction as well as the depth-to-span ratio of the horizontal diaphragm. In general, wood and metal decks are quite flexible, whereas poured concrete decks are very stiff.

Relative Stiffness of the Vertical Elements. When vertical elements share load from a rigid horizontal diaphragm, as shown in the lower figure in Fig. 47.1, their relative stiffness must usually be determined in order to establish the manner of the sharing. The determination is comparatively simple when the elements are similar in type and materials such as all plywood shear walls. When the vertical elements are different, such as a mix of plywood and masonry shear walls or of some shear walls and some braced frames, their actual deflections must be calculated in order to establish the distribution, and this may require laborious calculations.

Use of Control Joints. The general approach in design for lateral loads is to tie the whole structure together to assure its overall continuity of movement. Sometimes, however, because of the irregular form or large size of a building, it may be desirable to control its behavior under lateral loads by the use of structural separation joints. In some cases these joints function to create total separation, allowing for completely independent motion of the separate parts of the building. In other cases the joints may control movements in a single direction while achieving connection for load transfer in other directions. A general discussion of separation joints is given in Sec. 47.8.

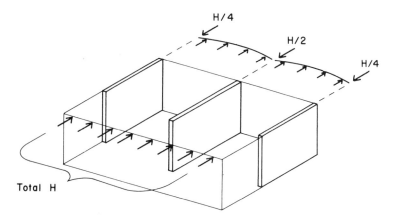

Peripheral distribution – flexible horizontal diaphragm

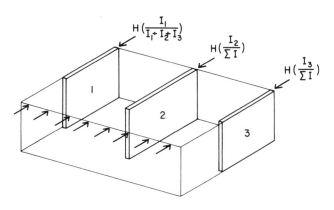

Proportionate stiffness distribution – rigid horizontal diaphragm

FIGURE 47.1. Peripheral distribution versus proportionate stiffness distribution.

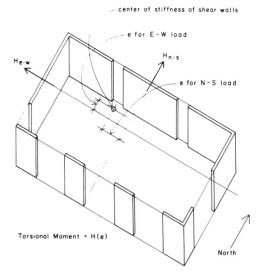

FIGURE 47.2. Torsional effect of lateral load.

Design and Usage Considerations

In performing their basic tasks, horizontal diaphragms have a number of potential stress problems. A major consideration is that of the shear stress in the plane of the diaphragm caused by the spanning action of the diaphragm as shown in Fig. 47.3. This spanning action results in shear stress in the material as well as a force that must be transferred across joints in the deck when the deck is composed of separate elements such as sheets of plywood or units of formed sheet metal. The sketch in Fig. 47.4 shows a typical plywood framing detail at the joint between two sheets. The stress in the deck at this location must be passed from one sheet through the edge nails to the framing member and then back out through the other nails to the adjacent sheet.

As is the usual case with shear stress, both diagonal tension and diagonal compression are induced simultaneously with the shear stress. The diagonal tension becomes critical in materials such as concrete. The diagonal compres-

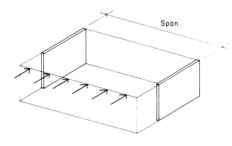

Beam Analogy

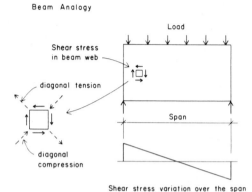

FIGURE 47.3. Function of a horizontal diaphragm.

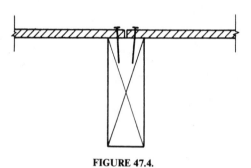

FIGURE 47.4.

sion is a potential source of buckling in decks composed of thin sheets of plywood or metal. In plywood decks the thickness of the plywood relative to the spacing of framing members must be considered, and it is also why the plywood must be nailed to intermediate framing members (not at edges of the sheets) as well as at edges. In metal decks the gauge of the sheet metal and the spacing of stiffening ribs must be considered. Tables of allowable loads for various deck elements usually incorporate some limits for these considerations.

Diaphragms with continuous deck surfaces are usually designed in a manner similar to that for webbed steel beams. The web (deck) is designed for the shear, and the flanges (edge-framing elements) are designed to take the moment, as shown in Fig. 47.5. The edge members are called *chords*, and they must be designed for the tension and compression forces at the edges. With diaphragm edges of some length, the latter function usually requires

that the edge members be spliced for some continuity of the forces. In many cases there are ordinary elements of the framing system, such as spandrel beams or top plates of stud walls, that have the potential to function as chords for the diaphragm.

The diaphragm shear capacities for commonly used decks of various materials are available from the codes or from load tables prepared by deck manufacturers. Loads for plywood decks are given in the *UBC* (see Appendix D). Other tabulations are available from product manufacturers, although care should be exercised in their use to be certain that they are acceptable to the building code of jurisdiction.

A special situation is a horizontal system that consists partly or wholly of a braced frame. Care may be required when there are a large number of openings in the roof deck, or when the diaphragm shear stress is simply beyond the capacity of the deck. In the event of a deck with no code-accepted rating for shear, the braced frame may have to be used for the entire horizontal system.

The horizontal deflection of flexible decks, especially those with high span-to-depth ratios, may be a critical factor in their design. Calculation of actual deflection dimensions may be required to determine the effect on vertical elements of the building construction or to establish positively whether the deck must be considered as essentially flexible or rigid, as discussed previously.

The use of subdiaphragms may also be required in some cases, necessitating the design of part of the whole system as a separate diaphragm, even though the deck may be continuous.

Typical Construction

The most common horizontal diaphragm is the plywood deck for the simple reason that wood frame construction is so popular, and plywood is mostly used for roof and floor decks. For roofs the deck may be as thin as $\frac{3}{8}$ in., but for flat roofs with waterproof membranes decks are usually $\frac{1}{2}$ in. or more. Attachment is typically by nailing, although glued floor decks are used for their added stiffness and to avoid nail popping and squeaking. Mechanical devices for nailing may eventually become so common that shear capaci-

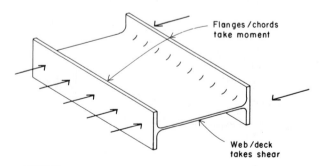

FIGURE 47.5. Horizontal diaphragm functions—beam analogy.

ties will be based on some other fastener; at present the common wire nail is still the basis for load rating.

Attachment of plywood to chords and collectors and load transfers to vertical shear walls are also mostly achieved by nailing. Code-acceptable shear ratings are based on the plywood type and thickness, the nail size and spacing, and features such as size and spacing of framing and use of blocking. Load capacities for plywood decks are given in *UBC* Table 25-J-1, which is reprinted in Appendix D.

In general, plywood decks are quite flexible and should be investigated for deflection when spans are large or span-to-depth ratios are high.

Decks of boards or timber, usually with tongue-and-groove joints, were once popular but are given low rating for shear capacity at present. Where the exposed plank-type deck is desired, it is not uncommon to use a thin plywood deck on top of it for lateral force development.

Steel decks offer possibilities for use as diaphragms for either roof or floors. Acceptable shear capacities should be obtained from the supplier for any particular product, as capacities vary considerably and code approval is not consistent. Stiffnesses are generally comparable with those of plywood decks. Floor decks receive concrete fill, which significantly increases the stiffness of the deck. In some situations it is practical to use the concrete as the basic shear-resisting element.

Poured-in-place concrete decks provide the strongest and stiffest diaphragms. Precast concrete deck units, as well as the slab portions of precast systems, can be used for diaphragms. Precast units must be adequately attached to each other and to supporting members for diaphragm actions; if designed only for gravity, ordinary attachments will usually not be adequate for lateral forces. As with steel deck, concrete fill is sometimes placed on top of precast units, which both stiffens and strengthens the system.

Many other types of roof deck construction may function adequately for diaphragm action, especially when the required unit shear resistance is low. Acceptability by local building code administration agencies should be determined if any construction other than those described is to be used.

Stiffness and Deflection. As spanning elements, the relative stiffness and actual dimensions of deformation of horizontal diaphragms depend on a number of factors such as:

The materials of the construction.

The continuity of the spanning diaphragm over a number of supports.

The span-to-depth ratio of the diaphragm.

The effect of various special conditions, such as chord length changes, yielding of connections, and influence of large openings.

In general, wood and light-gauge metal decks tend to produce quite flexible diaphragms, whereas poured concrete decks tend to produce the most rigid diaphragms. Ranging between these extremes are decks of lightweight concrete, gypsum concrete, and composite constructions of lightweight concrete fill on metal deck. For true dynamic analysis the variations are more complex because the weight and degree of elasticity of the materials must also be considered.

With respect to their span-to-depth ratios, most horizontal diaphragms approach the classification of deep beams. As shown in Fig. 47.6, even the shallowest of diaphragms, such as the maximum 4:1 case allowed for a plywood deck by the *UBC*, tends to present a fairly stiff flexural member. As the span-to-depth ratio falls below about 2, the deformation characteristic of the diaphragm approaches that of a deep beam in which the deflection is primarily caused by shear strain rather than by flexural strain. Thus the usual formulas for deflection caused by flexural strain become of limited use.

The following formula is used for the calculation of deflection of simple-span plywood diaphragms:

$$\Delta = \frac{5vL^3}{8EAb} + \frac{vL}{4Gt} + 0.094Le_n + \sum \frac{\Delta_c X}{2b}$$

in which the four terms account for four different contributions to the deflection, as follows:

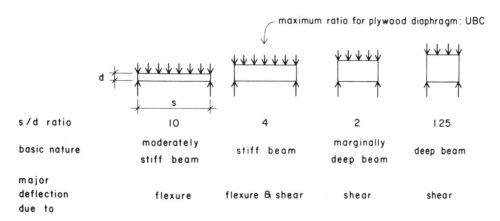

FIGURE 47.6. Behavior of horizontal diaphragms related to depth-to-span ratios.

Term 1 accounts for the length change of the chords.

Term 2 accounts for the shear strain in the plywood panels.

Term 3 accounts for the lateral bending of the nails.

Term 4 accounts for additional change in the chord lengths caused by slip in the chord splices.

The deflection of steel deck diaphragms is discussed and illustrated in the *Inryco Lateral Diaphragm Data Manual 20-2* published by Inryco, Inc. The formula used for calculating the deflection of a simple span deck is

$$\Delta_t = \frac{5WL_s^4(1728)}{384EI} + qF\frac{L_s}{2}(10^{-6})$$

in which the first term accounts for flexural deflection caused by the length change of the chords, and the second term for shear strain and panel distortion in the diaphragm web.

The quantities q and F vary as a function of the type and gauge of the deck, the fastening patterns and methods used, and the possible inclusion of concrete fill.

As with deck load capacities, deflections and the relative stiffness of deck systems are generally based on materials presented in the *Diaphragm Design Manual*, published by the Steel Deck Institute. For a specific product, however, designers should obtain information from the product supplier and verify that any data or procedures used are acceptable to the building permit-approving agency for the work.

47.2. VERTICAL DIAPHRAGMS

Vertical diaphragms are usually the walls of buildings. As such, in addition to their shear wall function, they must fulfill various architectural functions and may also be required to serve as bearing walls for the gravity loads. The location of walls, the materials used, and some of the details of their construction must be developed with all these functions in mind.

The most common shear wall constructions are those of poured concrete, masonry, and wood frames of studs with surfacing elements. Wood frames may be made rigid in the wall plane by the use of diagonal bracing or by the use of surfacing materials that have sufficient strength and stiffness. Choice of the type of construction may be limited by the magnitude of shear caused by the lateral loads, but will also be influenced by fire code requirements and the satisfaction of the various other wall functions, as described previously.

General Behavior

Some of the structural functions usually required of vertical diaphragms are the following (see Fig. 47.7):

1. *Direct Shear Resistance.* This usually consists of the transfer of a lateral force in the plane of the wall from some upper level of the wall to a lower level or

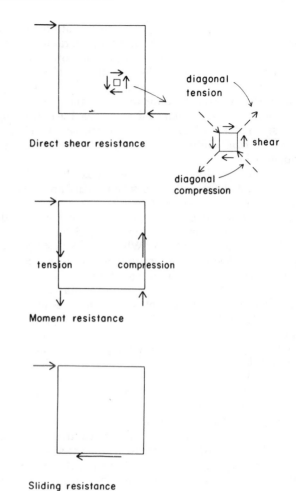

FIGURE 47.7. Functions of a shear wall.

to the bottom of the wall. This results in the typical situation of shear stress and the accompanying diagonal tension and compression stresses, as discussed for horizontal diaphragms.

2. *Cantilever Moment Resistance.* Shear walls generally work like vertical cantilevers, developing compression on one edge and tension on the opposite edge and transferring an overturning moment (M) to the base of the wall.

3. *Horizontal Sliding Resistance.* The direct transfer of the lateral load at the base of the wall produces the tendency for the wall to slip horizontally off its supports.

The shear stress function is usually considered independently of other structural functions of the wall. The maximum shear stress that derives from lateral loads is compared to some rated capacity of the wall construction, with the usual increase of one-third in allowable stresses because the lateral load is most often a result of wind or earthquake forces.

For concrete and masonry walls the actual stress in the material is calculated and compared with the allowable stress for the material. For structurally surfaced wood frames the construction as a whole is generally rated for its total resistance in pounds per foot of the wall length in plan. For a plywood-surfaced wall this capacity depends on the type and thickness of the plywood; the size, wood species, and spacing of the studs; the size and spacing of the plywood nails; and the inclusion or omission of blocking at horizontal plywood joints.

For wood stud walls the *UBC* provides tables of rated load capacities for several types of surfacing, including plywood, diagonal wood boards, plaster, gypsum drywall, and particleboard.

Although the possibility exists for the buckling of walls as a result of the diagonal compression effect, this is usually not critical because other limitations exist to constrain wall slenderness. The thickness of masonry walls is limited by maximum values for the ratio of unsupported wall height or length-to-wall thickness. Concrete thickness is usually limited by forming and pouring considerations, so that thin walls are not common except with precast construction. Slenderness of wood studs is limited by gravity design and by the code limits as a function of the stud size. Because stud walls are usually surfaced on both sides, the resulting sandwich–panel effect is usually sufficient to provide a reasonable stiffness.

As in the case of horizontal diaphragms, the moment effect on the wall is usually considered to be resisted by the two vertical edges of the wall acting as flanges or chords. In the concrete or masonry wall this results in a consideration of the ends of the wall as columns, sometimes actually produced as such by thickening of the wall at the ends. In wood-framed walls the end-framing members are considered to fulfill this function. These edge members must be investigated for possible critical combinations of loading because of gravity and the lateral effects.

The overturn effect of the lateral loads must be resisted with the safety factor of 1.5, which is required by the *UBC*. The form of the analysis for the overturn effect is as shown in Fig. 47.8. If the tiedown force is actually required, it is developed by the anchorage of the edge-framing elements of the wall.

For seismic effects, *UBC* Sec. 2312(h)1 specifies that only 85% of the dead load be used to resist uplift effects when using the working stress method for materials. This means that any anchorage elements that are required can be designed for their working stress resistance.

For an individual shear wall, the overturn investigation is summarized in Fig. 47.8. Specific applications are illustrated in the design examples in Chapter 48.

Resistance to horizontal sliding at the base of a shear wall is usually at least partly resisted by friction caused by the dead loads. For masonry and concrete walls with dead loads, which are usually quite high, the frictional resistance may be more than sufficient. If it is not, shear keys must be provided. For wood-framed walls the friction is usually ignored, and the sill bolts are designed for the entire load.

Design and Usage Considerations

An important judgment that must often be made in designing for lateral loads is that of the manner of distribution of lateral force between a number of shear walls that share the load from a single horizontal diaphragm. In some cases the existence of symmetry or of a flexible horizontal diaphragm may simplify this consideration. In many cases, however, the relative stiffnesses of the walls must be determined for this calculation.

If considered in terms of static force and elastic stress–strain conditions, the relative stiffness of a wall is inversely proportionate to its deflection under a unit load. Figure 47.9 shows the manner of deflection of a shear wall for two assumed conditions. In Fig. 47.9a the wall is considered to be fixed at its top and bottom, flexing in a double curve

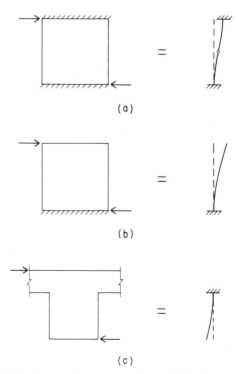

FIGURE 47.9. Shear wall support conditions: (a) fixed top and bottom; (b) and (c) cantilevered.

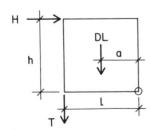

To determine T:

for wind – $DL(a) + T(l) = 1.5\,[H(h)]$

for seismic – $0.85\,[DL(a)] + T(l) = H(h)$

FIGURE 47.8. Determination of tiedown requirements for a shear wall.

with an inflection point at midheight. This is the case usually assumed for a continuous wall of concrete or masonry in which a series of individual wall portions (called *piers*) are connected by a continuous upper wall or other structure of considerable stiffness. In Fig. 47.9*b* the wall is considered to be fixed at its bottom only, functioning as a vertical cantilever. This is the case for independent, freestanding walls or for walls in which the continuous upper structure is relatively flexible. A third possibility is shown in Fig. 47.9*c* in which relatively short piers are assumed to be fixed at their tops only, which produces the same deflection condition as in Fig. 47.9*b*.

In some instances the deflection of the wall may result largely from shear distortion, rather than from flexural distortion, perhaps because of the wall materials and construction or the proportion of wall height to plan length. Furthermore, stiffness in resistance to dynamic loads is not quite the same as stiffness in resistance to static loads. The following recommendations are made for single-story shear walls:

1. For wood-framed walls with height-to-length ratios of 2 or less, assume the stiffness to be proportional to the plan length of the wall.
2. For wood-framed walls with height-to-length ratios over 2 and for concrete and masonry walls, assume the stiffness to be a function of the height-to-length ratio and the method of support (cantilevered or fixed top and bottom). Use the values for pier rigidity given in Appendix E.
3. Avoid situations in which walls of significantly great differences in stiffness share loads along a single row. The short walls will tend to receive a small share of the loads, especially if the stiffness is assumed to be a function of the height-to-length ratio.
4. Avoid mixing of shear walls of different construction when they share loads on a deflection basis.

In addition to the various considerations mentioned for the shear walls themselves, care must be taken to assure that they are properly anchored to the horizontal diaphragms. Problems of this sort are illustrated in the examples in Part 9.

A final consideration for shear walls is that they must be made an integral part of the whole building construction. In long building walls with large door or window openings or other gaps in the wall, shear walls are often considered as entities (isolated, independent piers) for their design. However, the behavior of the entire wall under lateral load should be studied to be sure that elements not considered to be parts of the lateral resistive system do not suffer damage because of the wall distortions.

Typical Construction

The various types of common construction for shear walls mentioned in the preceding section are wood frames with various surfacing, reinforced masonry, and concrete. The only wood frame wall used extensively in the past was the plywood-covered one. Experience and testing have established acceptable ratings for other surfacing, so that plywood is used somewhat less when shear loads are low.

For all types of walls there are various considerations (good carpentry, fire resistance, available products, etc.) that establish a certain minimum construction. In many situations this "minimum" is really adequate for low levels of shear loading, and the only additions are in the area of attachments and joint load transfers. Increasing wall strength beyond the minimum usually requires increasing the size or quality of units, adding or strengthening attachments, developing supporting elements to function as chords or collectors, and so on. It well behooves the designer to find out the standards for basic construction to know what the minimum consists of so that added strength can be developed when necessary—but using methods consistent with the ordinary types of construction.

Load Capacity. Load capacities for ordinary wood-framed shear walls are given in the load tables in the *UBC*. For seismic actions the only masonry construction ordinarily acceptable is that of reinforced masonry, of which the most common form is one using hollow precast concrete units (concrete blocks). The design of masonry construction should be done with the references that present material acceptable to the building code of jurisdiction. The reprints from the 1988 *UBC* in Appendix D give some criteria for masonry, but the complete design guide is in Chapter 24 of the code.

Most concrete design is based on the current edition of the *ACI Code* (*Building Code Requirements for Reinforced Concrete*, Ref. 6), although local codes are sometimes slow in accepting new changes in the *ACI Code* editions. The current edition of the *ACI Code* provides some criteria for seismic design, but recent developments—mostly in the form of suggested details for construction—are more stringent for areas of high seismic risk. Concrete design in general and seismic design in particular have become quite sophisticated and complex, and there are few simple guides. The shear wall is a relatively simple concrete element, but its design should be done in conformance with the latest codes and practices.

Stiffness and Deflection. As with the horizontal diaphragm, there are several potential factors to consider in the deflection of a shear wall. As shown in Fig. 47.10*a*, shear walls also tend to be relatively stiff in most cases, approaching deep beams instead of ordinary flexural members.

The two general cases for the vertical shear wall are the cantilever and the doubly fixed pier. The cantilever, fixed at its base, is the most commonly used. Fixity at both the top and bottom of the wall usually affects deflection only when the wall is relatively short in length with respect to its height. Walls with long lengths in proportion to their height fall into the deep beam category in which the predominant shear strain is not affected by the fixity of the support.

As shown in Fig. 47.10*b*, the doubly fixed pier is assumed to have an inflection point at its midheight, its

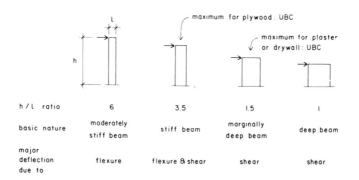

h / l ratio   6   3.5   1.5   1

basic nature   moderately   stiff beam   marginally   deep beam
          stiff beam   deep beam

major
deflection   flexure   flexure & shear   shear   shear
due to

(a)   Behavior of cantilevered elements related to height-to-length ratios

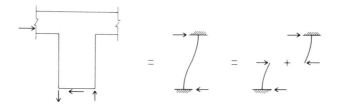

(b)   Deflection assumption for a fully fixed masonry pier

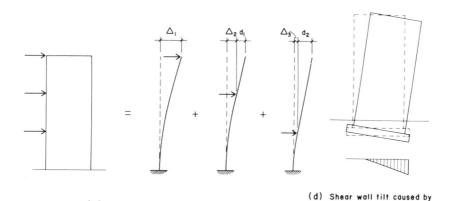

(c)   Deflection of a multistory shear wall   (d)   Shear wall tilt caused by
                                                uneven soil pressure

FIGURE 47.10. Aspects of deflection of shear walls.

deflection can be approximated by considering it to be the sum of the deflections of two half-height cantilevered piers. Yielding of the supports and flexure in the horizontal structure will produce some rotation of the assuredly fixed ends, which will result in some additional deflection.

The following formula is used for calculating deflection of cantilevered plywood shear walls similar to that for the plywood horizontal diaphragm:

$$\Delta = \frac{8vh^3}{EAb} + \frac{vh}{Gt} + 0.376he_n + d_a$$

in which the four terms account for the following:

Term 1 accounts for the change in length of the chords (wall end framing).
Term 2 accounts for the shear strain in the plywood panels.

Term 3 accounts for the nail deformation.
Term 4 is a general term for including the effects of yield of the anchorage.

The formula can also be used for calculating the deflection of a multistory wall, as shown in Fig. 47.10c. For the loading as shown in the illustration, a separate calculation would be made for each of the three loads (Δ_1, Δ_2, and Δ_3). To these would be added the deflection at the top of the wall caused by the rotation effects of the lower loads (d_1 and d_2). Thus the total deflection at the top of the wall would be the sum of the five increments of deflection.

Rotation caused by soil deformation at the base of the wall can also contribute to the deflection of shear walls (see Fig. 47.10d). This is especially critical for tall walls on isolated foundations placed on relatively compressible soils, such as loose sand and soft clay—a situation to be avoided if at all possible.

47.3. BRACED FRAMES

Although there are actually several ways to brace a frame against lateral loads, the term *braced frame* is used to refer to frames that utilize trussing as the primary bracing technique. In buildings, trussing is mostly used for the vertical bracing system in combination with the usual horizontal diaphragms. It is also possible, however, to use a trussed frame for a horizontal system, or to combine vertical and horizontal trussing in a truly three-dimensional trussed framework. The latter is more common for open tower structures, such as those used for large electrical transmission lines and radio and television transmitters.

Use of Trussing for Bracing

Post-and-beam systems, consisting of separate vertical and horizontal members, may be inherently stable for gravity loading, but they must be braced in some manner for lateral loads. The three basic ways of achieving this are through shear panels, moment-resistive joints between the members, or by trussing. The trussing, or triangulation, is usually formed by the insertion of diagonal members in the rectangular bays of the frame.

If single diagonals are used, they must serve a dual function: acting in tension for the lateral loads in one direction and in compression when the load direction is reversed (see Fig. 47.11a). Because long tension members are more efficient than long compression members, frames are often braced with a crisscrossed set of diagonals (called *X-bracing*) to eliminate the need for the compression members. In any event the trussing causes the lateral loads to induce only axial forces in the members of the frame, as compared to the behavior of the rigid frame. It also generally results in a frame that is stiffer for both static and dynamic loading, having less deformation than the rigid frame.

Single-story, single-bayed buildings may be braced as shown in Fig. 47.11a. Single-story, multibayed buildings may be braced by bracing less than all of the bays in a single plane of framing, as shown in Fig. 47.11b. The continuity of the horizontal framing is used in the latter situation to permit the rest of the bays to tag along. Similarly, a single-bayed, multistoried, towerlike structure, as shown in Fig. 47.11c, may have its frame fully braced, whereas the more common type of frame for the multistoried building, as shown in Fig. 47.11d, is usually only partly braced. Since either the single diagonal or the crisscrossed X-bracing causes obvious problems for interior circulation and for openings for doors and windows, building planning often makes the limited bracing a necessity.

Just about any type of floor construction used for multistoried buildings usually has sufficient capacity for diaphragm action in the lateral bracing system. Roofs, however, often utilize light construction or are extensively perforated by openings, so that the basic construction is not capable of the usual horizontal, planar diaphragm action. For such roofs or for floors with many openings, it may be necessary to use a trussed frame for the horizontal part of the lateral bracing system. Figure 47.11e shows a

roof for a single-story building in which trussing has been placed in all the edge bays of the roof framing in order to achieve the horizontal structure necessary. As with vertical trussed frames, the horizontal trussed frame may be partly trussed, as shown in Fig. 47.11e, rather than fully trussed.

For single-span structures, trussing may be utilized in a variety of ways for the combined gravity and lateral load resistive system. Figure 47.11f shows a typical gable roof with the rafters tied at their bottom ends by a horizontal member. The tie, in this case, serves the dual functions of resisting the outward thrust due to gravity loads and of one of the members of the single triangle, trussed structure that is rigidly resistive to lateral loads. Thus the wind force on the sloping roof surface, or the horizontal seismic force caused by the weight of the roof structure, is resisted by the triangular form of the rafter-tie combination.

The horizontal tie shown in Fig. 47.11f may not be architecturally desirable in all cases. Some other possibilities for the single-span structure—all producing more openness beneath the structure—are shown in Fig. 47.11g–i. Figure 47.11g shows the so-called scissors truss, which can be used to permit more openness on the inside or to permit a ceiling that has a form reflecting that of the gable roof. Figure 47.11h shows a trussed bent that is a variation on the three-hinged arch. The structure shown in Fig. 47.11i consists primarily of a single-span truss that rests on end columns. If the columns are pin-jointed at the bottom chord of the truss, the structure lacks basic resistance to lateral loads and must be separately braced. If the column in Fig. 47.11i is continuous to the top of the truss, it can be used in rigid frame action for resistance to lateral loads. Finally, if the knee-braces shown in the figure are added, the column is further stiffened, and the structure has more load resistance and less deflection under lateral loading.

The knee-brace (Fig. 47.11i) is one form of diagonal bracing described as *eccentric bracing*, a name deriving from the fact that one or more of the bracing connections is off the column-to-beam joint. Other forms of eccentric bracing are the K-brace, V-brace, and inverted V-brace (see Fig. 47.12). V-bracing is also sometimes described as *chevron bracing*.

Use of eccentric bracing results in some combined form of truss and rigid frame actions. The trussed form of the bracing produces the usual degree of stiffness associated with a braced frame, while the bending induced by the eccentricity of the bracing adds rigid frame deformations to the behavior. For ultimate load failure, a significant event is the development of plastic hinging in the members with the eccentric bracing joints.

Eccentric bracing was historically developed for use as core wind bracing in high-rise steel structures. It has more recently, however, become quite favored for bracing of steel frames for seismic resistance, where its combination of high stiffness and high energy capacity is advantageous. In zones of high seismic risk, braced frames are now most commonly achieved with either very heavy X-bracing or some form of eccentric bracing.

Planning of Bracing.   Some of the problems to be considered in using braced frames are the following:

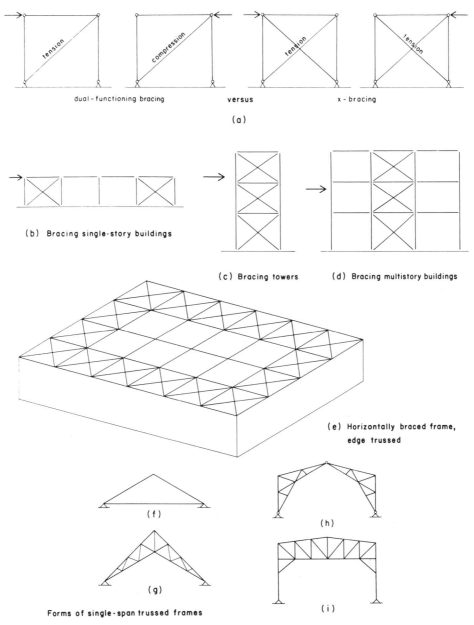

dual-functioning bracing   versus   x-bracing

(a)

(b) Bracing single-story buildings

(c) Bracing towers   (d) Bracing multistory buildings

(e) Horizontally braced frame,
edge trussed

(f)

(h)

(g)

(i)

Forms of single-span trussed frames

FIGURE 47.11. Considerations of braced frames.

1. Diagonal members must be placed so as not to interfere with the action of the gravity-resistive structure or with other building functions. If the bracing members are designed essentially as axial stress members, they must be located and attached so as to avoid loadings other than those required for their bracing functions. They must also be located so as not to interfere with door, window, or roof openings or with ducts, wiring, piping, light fixtures, and so on.

2. As mentioned previously, the reversibility of the lateral loads must be considered. As shown in Fig. 47.11a, such consideration requires that diagonal members be dual functioning (as single diagonals)

or redundant (as X-bracing) with one set of diagonals working for load from one direction and the other set working for the reversal loading.

3. Although the diagonal bracing elements usually function only for lateral loading, the vertical and horizontal elements must be considered for the various possible combinations of gravity and lateral load. Thus the total frame must be analyzed for all the possible loading conditions, and each member must be designed for the particular critical combinations that represent its peak response conditions.

4. Long, slender bracing members, especially in X-braced systems, may have considerable sag due to

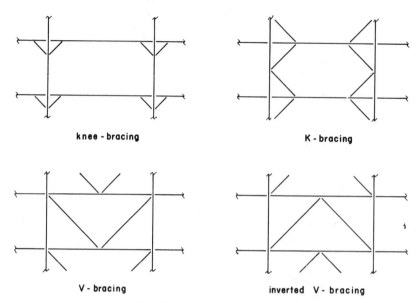

knee - bracing   K - bracing

V - bracing   inverted   V - bracing

FIGURE 47.12. Forms of eccentric bracing.

their own dead weight, which requires that they be supported by sag rods or other parts of the structure.

5. The trussed structure should be "tight." Connections should be made in a manner to assure that they will be initially free of slack and will not loosen under the load reversals or repeated loadings. This means generally avoiding connections that tend to loosen or progressively deform, such as those that use nails, loose pins, and unfinished bolts.

6. To avoid loading on the diagonals, the connections of the diagonals are sometimes made only after the gravity-resistive structure is fully assembled and at least partly loaded by the building dead loads.

7. The deformation of the trussed structure must be considered, and it may relate to its function as a distributing element, as in the case of a horizontal structure, or to the establishing of its relative stiffness, as in the case of a series of vertical elements that share loads. It may also relate to some effects on nonstructural parts of the building, as was discussed for shear walls.

8. In most cases it is not necessary to brace every individual bay of the rectangular frame system. In fact, this is often not possible for architectural reasons. As shown in Fig. 47.11*b*, walls consisting of several bays can be braced by trussing only a few bays, or even a single bay, with the rest of the structure tagging along like cars in a train.

The braced frame can be mixed with other bracing systems in some cases. Figure 47.13*a* shows the use of a braced frame for the vertical resistive structure in one direction and a set of shear walls in the other direction. In this example the two systems act independently, except for the possibility of torsion, and there is no need for a deflection analysis to determine the load sharing.

Figure 47.13*b* shows a structure in which the end bays of the roof framing are X-braced. For loading in the direction shown, these braced bays take the highest shear in the horizontal structure, allowing the deck to be designed for a lower shear stress.

Although buildings and their structures are often planned and constructed in two-dimensional components (horizontal floor and roof planes and vertical wall or framing bent planes), it must be noted that the building is truly three dimensional. Bracing against lateral forces is thus a three-dimensional problem, and although a single horizontal or vertical plane of the structure may be adequately stable and strong, the whole system must interact appropriately. While the single triangle is the basic unit for a planar truss, the three-dimensional truss may not be truly stable just because its component planes are braced.

In a purely geometric sense the basic unit for a three-dimensional truss is the four-sided figure called a *tetrahedron*. However, since most buildings consist of spaces that are rectangular boxes, the three-dimensional trussed building structure usually consists of rectangular units rather than multiples of the pyramidal tetrahedral form. When so used, the single planar truss unit is much the same as a solid planar wall or deck unit, and general reference to the box-type system typically includes both forms of construction.

Typical Construction

Development of the details of construction for trussed bracing is in many ways similar to the design of spanning trusses. The materials used (generally wood or steel), the form of individual truss members, the type of jointing (nails, bolts, welds, etc.), and the magnitudes of the forces are all major considerations. Since many of the members of the complete truss serve dual roles for gravity and lateral loads, member selection is seldom based on truss action alone. Quite often trussed bracing is produced by simply adding diagonals (or X-bracing) to a system already conceived for the gravity loads and for the general development of the desired architectural forms and spaces.

Figure 47.14 shows some details for wood framing with added diagonal members. Wood-framing members are most often rectangular in cross section and metal connecting devices of various form are used in the assembly of frameworks. Figure 47.14a shows a typical beam and column assembly with diagonals consisting of pairs of wood members bolted to the frame. When X-bracing is used, and the diagonal members need take only tension forces, slender steel rods may be used; a possible detail for this is shown in Fig 47.14b. For the wood diagonal an alternative to the bolted connection is the type of joint in Fig. 47.14c, employing a gusset plate to attach single members all in a single plane. If architectural detailing makes the protruding members shown in Fig 47.14a or even the protruding

gussets in Fig. 47.14c undesirable, a bolted connection like that shown in Fig.47.14d may be used.

As discussed in the next section, a contributing factor in the deformation of the bracing under loading may be movements within the connections. Bolted connections are especially vulnerable when used in shear resistance, since both oversizing of the holes and shrinkage of the wood contribute to a lack of tightness in the joints. In some cases it may be possible to increase the tightness of the joints by using some form of shear developer such as steel split rings.

Gusset plates ordinarily consist of plywood, sheet steel, or steel plate, depending mostly on the magnitude of the loads. Plywood joints should be glued or the nails should be ring or spiral shafted to increase the joint tightness. Steel plate gussets are usually attached by either lag screws or through bolts. Thin sheet metal gussets are either nailed or screwed in place, the latter being preferred for maximum tightness.

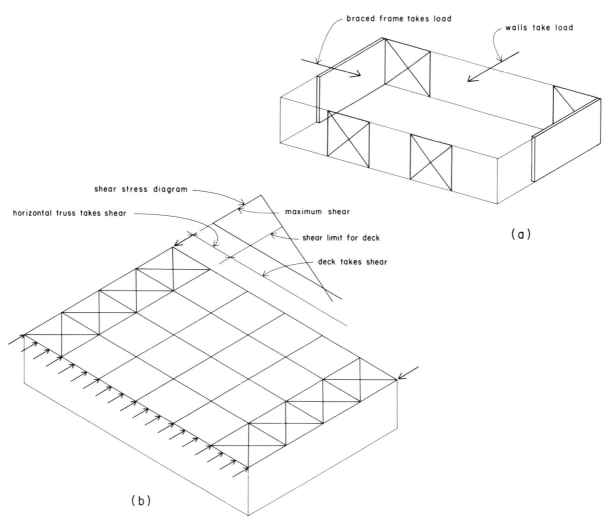

FIGURE 47.13. Use of braced systems: (a) mixed vertical elements; (b) mixed horizontal diaphragm and trussing.

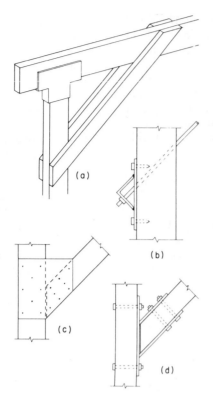

FIGURE 47.14. Framing of trussed bents in wood.

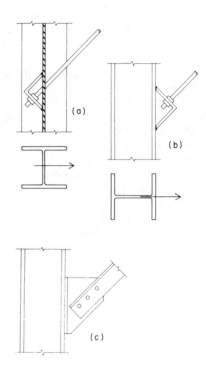

FIGURE 47.15. Details for trussed steel bents.

Figure 47.15 shows some details for the incorporation of diagonal bracing in steel frames. As with wood structures, bolt loosening is a potential problem. For bolts used in tension—or for the threaded ends of round steel rods—loosening of nuts can be prevented by welding them in place or by scarring the threads. For shear-type connections, highly tensioned, high-strength bolts are preferred over ordinary, unfinished bolts. A completely welded connection will produce the stiffest joint, but on-site bolting in the field is usually preferred over welding.

Various steel elements can be used for diagonal members, depending on the magnitude of loads, the problems of incorporating or exposing the members in the construction, and the requirements for attachment to the structural frame.

Stiffness and Deflection.   As has been stated previously, the braced frame is typically a relatively stiff structure. This is based on the assumption that the major contribution to the overall deformation of the structure is the shortening and lengthening of the members of the frame as they experience the tension and compression forces due to the truss action. However, the two other potentially significant contributions to the movement of the braced frame, with either or both of major concern, are:

1. *Movement of the Supports.*   This includes the possibilities of deformation of the foundations and yielding of the anchorage connections. If the foun-

dations rest on compressible soil, there will be some movement due to soil stress. Deformation of the anchorage may be due to a combination of lengthening of anchor bolts and bending of column base plates.
2. *Deformation in the Frame Connections.*   This is a complex problem having to do with the general nature of the connection (e.g., welds or glue versus bolts or nails) as well as its form and layout and the deformation of the parts being connected.

It is good design practice in general to study the connection details for braced frames with an eye toward reduction of deformation within the connections. As has been mentioned previously, this generally favors the choice of welding, gluing, high-strength bolts, wood screws, and other fastening techniques that tend to produce stiff, tight joints. It may also favor the choice of materials or form of the frame members as these choices may affect the deformations within the joints or the choice of connecting methods.

The deflection, or drift, of single-story X-braced frames is usually caused primarily by the tension elongation of the diagonal X members. As shown in Fig. 47.16a, the elongation of one diagonal moves the rectangular-framed bay into a parallelogram form. The approximate value of the deflection, *d* in Fig. 47.16a, can be derived as follows.

Assuming the change in the angle of the diagonal, $\Delta\theta$ in the figure, to be quite small, the change in length of the diagonal may be used to approximate one side of the triangle of which *d* is the hypotenuse. Thus

$$d = \frac{\Delta L}{\cos \theta} = \frac{TL/AE}{\cos \theta} = \frac{TL}{AE \cos \theta}$$

where T = tension in the X caused by the lateral load

   A = cross-sectional area of the X

   E = elastic modulus of the X

   θ = angle of the X from the horizontal

The deflection of multistory X-braced frames has two components, both of which may be significant. As shown in Fig. 47.16b, the first effect is caused by the change in length of the vertical members of the frame as a result of the overturning moment. The second effect is caused by the elongation of the diagonal X, as discussed for the single-story frame. These deflections occur in each level of the frame and can be calculated individually and summed up for the whole frame. Although this effect is also present in the single-story frame, it becomes more pronounced as the frame gets taller with respect to its width. These deflections of the cantilever beam can be calculated using standard

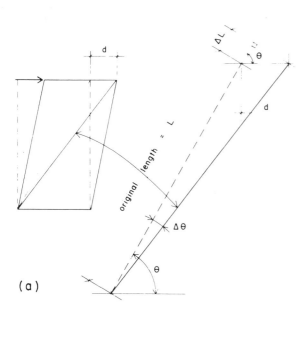

(a)

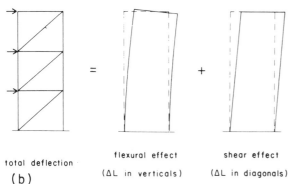

total deflection
(b)

FIGURE 47.16. Deflection of trussed bents.

formulas such as those given in the beam diagrams and formulas in Sec. 2 of the *Manual of Steel Construction* (Ref. 5).

47.4. MOMENT-RESISTIVE FRAMES

There is some confusion over the name to be used in referring to frames in which interactions between members of the frame include the transfer of moments through the connections. In years past the term most frequently used was *rigid frame*. This term came primarily from the classification of the connections or joints of the frame as *fixed* (or rigid) versus *pinned*, the latter term implying a lack of capability to transfer moment through the joint. As a general descriptive term, however, the name was badly conceived, since the frames of this type were generally the most deformable under lateral loading when compared to trussed frames or those braced by vertical diaphragms. The *UBC* (Ref. 1) uses the specific term *moment-resisting frame* and gives various qualifications for such a frame when it is used for seismic resistance. With apologies to the *UBC*, although we will assume the type of frames they thus define, we prefer not to use this rather cumbersome mouthful of a term, so will use the simpler term *rigid frame* in our discussions.

General Behavior

In rigid frames with moment-resistive connections, both gravity and lateral loads produce interactive moments between the members. The *UBC* requires that a rigid frame designed for seismic loading be classified as a "moment-resisting space frame." Generally, frames of steel possess this character, but frames of reinforced concrete require special consideration of the reinforcing in order to meet this qualification.

In most cases rigid frames are actually the most flexible of the basic types of lateral resistive systems. This deformation character, together with the required ductility, makes the rigid frame a structure that absorbs energy loading through deformation as well as through its sheer brute strength. The net effect is that the structure actually works less hard in force resistance because its deformation tends to soften the loading. This is somewhat like rolling with a punch instead of bracing oneself to take it head on.

Most moment-resistive frames consist of either steel or concrete. Steel frames have either welded or bolted connections between the linear members to develop the necessary moment transfers. Frames of concrete achieve moment connections through the monolithic concrete and the continuity and anchorage of the steel reinforcing. Because concrete is basically brittle and not ductile, the ductile character is essentially produced by the ductility of the reinforcing. The type and amount of reinforcing and the details of its placing become critical to the proper behavior of rigid frames of reinforced concrete.

A complete presentation of the design of moment-resistive ductile frames for seismic loads is beyond the scope of this book. Such design can be done only by using plastic design for steel and ultimate strength design for reinforced concrete. We present only a brief discussion of this type of structure. For wind loading the analysis and design may be somewhat more simplified. However, if the structure is considerably indeterminate, an accurate analysis requires a complex and laborious calculation. We show some examples in Part 9, but limit the analysis to approximate methods.

For lateral loads in general, the rigid frame offers the advantage of a high degree of freedom in architectural terms. Walls and interior spaces are freed of the necessity for solid diaphragms or diagonal members. For building planning as a whole, this is a principal asset. Walls, even where otherwise required to be solid, need not be of a construction qualifying them as shear walls.

When seismic load governs as the critical lateral load, the moment-resistive frame has the advantage of being required to carry less load than other types of lateral resistive systems. Table 23-O in the *UBC* gives the value of R_W for a special moment-resisting frame as 12 versus 4–8 for a box system with shear walls or braced frames.

Deformation-analysis is a critical part of the design of rigid frames because such frames tend to be relatively deformable when compared to other lateral resistive systems. The deformations have the potential of causing problems in terms of movements of a disturbing nature that can be sensed by the building occupants or of damage to nonstructural parts of the building, as previously discussed. The need to limit deformations often results in the size of vertical elements of the frame being determined by stiffness requirements rather than by stress limits.

Loading Conditions

Unlike shear walls or X-bracing, rigid frames are not generally able to be used for lateral bracing alone. Thus their structural actions induced by the lateral loads must always be combined with the effects of gravity loads. These combined loading conditions may be studied separately in order to simplify the work of visualizing and quantifying the structural behavior, but it should be borne in mind that they do not occur independently.

Figure 47.17a shows the form of deformation and the distribution of internal bending moments in a single-span rigid frame, as induced by vertical gravity loading. If the frame is not required to resist lateral loads, the singular forms of these responses may be assumed, and the various details of the structure may be developed in this context. Thus the direction of rotation at the column base, the sign of moment at the beam-to-column joint, the sign of the bending moment and nature of the corresponding stresses at midspan of the beam, and the location of inflection points in the beam may all relate to choices for the form and details of the members and development of any connection details. If the frame is reasonably symmetrical, the only concerns for deflection are the outward bulging of the columns and the vertical sag of the beam.

Under action of lateral loading due to wind or seismic force, the form of deformation and distribution of internal

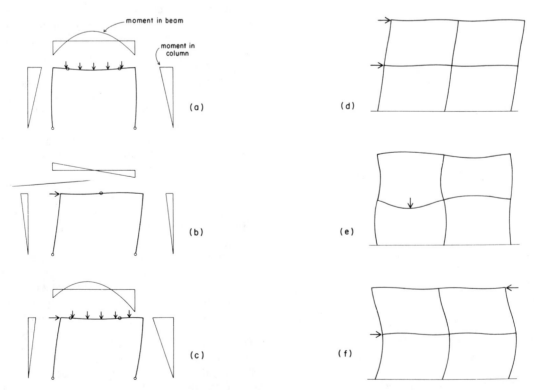

FIGURE 47.17. Behavior of rigid frames.

bending moment will be as shown in Fig. 47.17b. If the gravity and lateral loadings are combined, the net effect will be as shown in Fig. 47.17c. Observing the effects of the combined loading, we note the following:

1. Horizontal deflection at the top of the frame (called *drift*) must now be considered, in addition to the deflections mentioned previously for gravity load alone.

2. The maximum value for the moment at the beam-to-column connection is increased on one side and reduced on the other side of the bent. The increased moment requires that the beam, the column, and the connection at the joint must all be stronger for the combined loading.

3. If the lateral load is sufficient, the minimum value for the moment at the beam-to-column joint may be one of opposite sign from that produced by gravity loading alone. The form of the connection and possibly the design of the members may need to reflect this reversal of the sense of the moment.

4. The direction of the lateral load shown in Fig. 47.17b is reversible so that two combinations of load must be considered: gravity plus lateral load to the right and gravity plus lateral load to the left.

While single-span rigid frames are often used for buildings, the multispan or multistory frame is the more usual case. Figure 47.17d and e shows the response of a two-bay, two-story frame to lateral loads and to a gravity-type load applied to a single beam. The response to lateral loads is essentially similar to that for the single bent in Fig. 47.16. For gravity loads the multiunit frame must be analyzed for a more complex set of potential combinations, because the live load portion of the gravity loads must be considered to be random, and thus may or may not occur in any given beam span.

Lateral loads produced by winds will generally result in the loading condition shown in Fig. 47.17d. Because of its relative flexibility and size, however, a multistory building frame may quite likely respond so slowly to seismic motions that upper levels of the frame experience a whiplashlike effect; thus separate levels may be moving in opposite directions at a single moment. Figure 47.17f illustrates a type of response that may occur if the two-story frame experiences this action. Only a true dynamic analysis can ascertain whether this action occurs and is of critical concern for a particular structure.

Approximate Analysis for Gravity Loads.   Most rigid frames are statically indeterminate and require the use of some method beyond simple statics for their analysis. Simple frames of few members may be analyzed by some *hand* method using handbook coefficients, moment distribution, and so on. If the frame is complex, consisting of several bays and stories, or having a lack of symmetry, the analysis will be quite laborious unless performed with some computer-aided method.

For preliminary design it is often useful to have some approximate analysis, which can be fairly quickly performed. Internal forces, member sizes, and deflections thus determined may be used for a quick determination of the structural actions and the feasibility of some choices of systems and components. For the simple, single-bay bent shown in Fig. 47.18, the analysis for gravity loads is quite simple, since a single-loading condition exists (as shown) and the only necessary assumption is that of the relative stiffnesses of the beam and columns. For the frame with pin-based columns the analogy is made to a three-span beam on rotation-free supports. If the column bases are fixed, the end supports of the analogous beam are assumed fixed.

For multibayed, multistoried frames, an approximate analysis may be performed using techniques such as that described in Chapter 8 of the *ACI Code* (Ref. 6). This is more applicable to concrete frames, of course, but can be used for quick approximation of welded steel frames as well.

Even when using approximate methods, it is advisable to analyze separately for dead and live loads. The results can thus be combined as required for the various critical combinations of dead load, live load, wind load, and seismic load.

Approximate Analysis for Lateral Loads .   Various approximate methods may be used for the analysis of rigid frames under statically applied lateral loading. For ordinary frames, whether single-bay, multibay, or even multistory, approximate methods are commonly used for loading due to wind or as obtained from an equivalent static load analysis for seismic effects. More exact analyses are possible, of course, especially when performed by computer-aided methods.

For frames that are complex—due to irregularities, lack of symmetry, tapered members, and so on—analysis is hardly feasible without the computer. This is also true for analyses that attempt to deal with the true dynamic behavior of the structure under seismic load. With the increasing availability of the software, and the accumulation

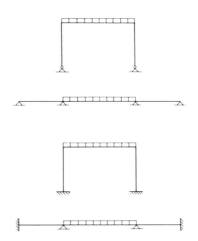

FIGURE 47.18. Continuous beam analogy for a single-span frame.

of experience with its use, this type of analysis is becoming more widespread in use. For quick approximations for preliminary design, however, approximation methods are likely to continue in use for some time.

For the simple bent shown in Fig. 47.19, the effects of the single lateral force may be quite simply visualized in terms of the deflected shape, the reaction forces, and the variation of moment in the members.

If the columns are assumed to be pin based and of equal stiffness, it is reasonable to assume that the horizontal reactions at the base of the columns are equal, thus permitting an analysis by statics alone. If the column bases are assumed to be fixed, the frame is truly able to be analyzed only by indeterminate methods, although an approximate analysis can be made with an assumed location of the inflection point in the column. In truth, the column bases will most likely be somewhere between these two idealized conditions. Approximate designs are sometimes done by combining the results from both idealized conditions (pinned and fixed bases) and designing for both. Adjustments are made when the more precise nature of the condition is established by the detailed development of the base construction.

For multibayed frames, such as those shown in Fig. 47.20, an approximate analysis may be done in a manner similar to that for the single-bay frame. If the columns are all of equal stiffness, the total load is simply divided by the number of columns. Assumptions about the column base condition would be the same as for the single-bay frame. If the columns are not all of the same stiffness, an approximate distribution can be made on the basis of relative stiffness.

Figure 47.20c illustrates the basis for an approximation of the horizontal shear forces in the columns of a multi-story building. As for the single-story frame, the individual column shears are distributed on the basis of assumed column stiffnesses. For the upper columns the inflection point is assumed to occur at midheight, unless column splice points are used to control its location.

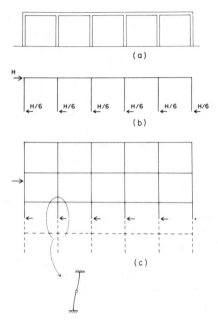

FIGURE 47.20. Distribution of lateral shear in a multiunit frame.

General Design Considerations

From a purely structural point of view, there are many advantages of the rigid frame as a lateral bracing system for resisting seismic effects. The current codes tend to favor it over shear walls or braced (trussed) frames by requiring less design load due to the higher assigned R_W value. For large frames the combination of slow reaction time (because of long fundamental period of vibration) and various damping effects means that forces tend to dissipate rapidly in the remote portions of the frame.

Architecturally, the rigid frame offers the least potential for interference with planning of open spaces within the building and in exterior walls. It is thus highly favored by architects whose design style includes the use of ordered rectangular grids and open spaces of rectangular or cubical form. Geometries other than rectangular ones are possible also; what remains as the primary advantage is the absence of the need for diagonal bracing or solid walls at determined locations.

Some of the disadvantages of rigid frames are the following:

1. Lateral deflection, or drift, of the structure is likely to be a problem. As a result, it is often necessary to stiffen the frame—mostly by increasing the number and/or size of the columns.
2. Connections must be stronger, especially in steel frames. Addition of heavily welded steel joints, of additional reinforcing for bar anchorage, shear, and torsion in concrete frames, and for cumbersome moment-resistive connections in wood frames can add appreciably to the construction cost and time.
3. For large frames dynamic behavior is often quite complex, with potential for whiplash, resonance,

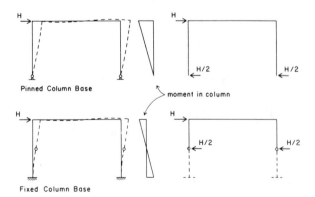

FIGURE 47.19. Effects of column base conditions on a single-span frame.

and so on. Dynamic analysis is more often required for acceptable design when a rigid frame is used. In time, dynamic analysis may be routine and easily performed with available aids on all structures; at the present it is quite expensive and time consuming.

When rigid frames are to be used for the lateral bracing system for a building, there must be a high degree of coordination between the planner of the building form and the structural designer. As illustrated in the examples in Part 9, the building plan must accommodate some regularity and alignment of the columns that constitute the vertical members of the rigid frame bents. This may not include all of the building columns, but those that function in the rigid frames must be aligned to form the planes of the bents. Openings in floors and roofs must be planned so as not to interrupt beams that occur as horizontal members of the bents.

There must, of course, be a coordination between the designs for lateral and gravity loads. If a rigid frame is used for bracing, it will in most cases be used in conjunction with some type of framed floor and roof systems. Horizontal members of the floor and roof systems designed for gravity loads will also be used as horizontal elements of the rigid frames. However, in some cases, much of the horizontal structure may be used only for gravity loads. In the system that utilizes only the bents that occur in exterior walls, none of the interior portion of the framing is involved in the rigid frame actions; thus its planning is free of concerns for the bent development.

Selection of members and construction details for rigid frames depends a great deal on the materials used. The following discusses separately the problems of bents of steel and reinforced concrete.

Steel Frames.   Steel frames with moment-resistive connections were used for early skyscrapers. Fasteners consisted of rivets, which were widely used until the development of high-strength bolts. Today, rigid frame construction in steel mostly utilizes welded joints, although bolting is sometimes used for temporary connection during erection. Figure 47.21 shows the erected frame for a low-rise office building, using a steel frame for both gravity and lateral load resistance. The principal members of the frame are wide-flange (I-shape) rolled steel sections, and moment connections are welded. This is the most common form of steel rigid frame for building construction.

Another form of steel rigid frame is the trussed bent. A single-span bent is illustrated in Fig. 47.11*i*, and the use of a two-span bent is described in the building design study in Fig. 48.17.

For tall buildings a currently popular system is one that uses a peripheral bracing arrangement with closely spaced, stiff steel columns and heavy spandrel beams. This type of structure offers a major advantage in its overall stiffness in resistance to lateral drift (horizontal deflection). Figure 47.22 shows the erected frame for such a structure. Note that the exterior rigid bents are discontinuous at the cor-

FIGURE 47.21. Steel frame structure with a peripheral bracing system of rigid bents.

ners, thus avoiding the high concentration of forces on the corner columns, especially those due to torsional action of the building.

Reinforced Concrete Frames.   Poured-in-place frames with monolithic columns and beams have a natural rigid frame

FIGURE 47.22. Use of closely spaced columns and stiff spandrels in a peripheral bent system.

action. For seismic resistance both columns and beams must be specially reinforced for the shears and torsions at the member ends. Beams in the column-line bents ordinarily use continuous top and bottom reinforcing with continuous loop ties that serve the triple functions of resisting shear, torsion, and compression bar buckling. It is possible, of course, to brace such a building with poured concrete shear walls and to use the frame strictly for gravity resistance, except for collector and chord functions.

Precast concrete structures are often difficult to develop as rigid frames, unless the precast elements are developed as individual bent units instead of the usual, single, linear members. Moment-resistive joints for these structures are usually quite difficult to develop, the more common solution being to use a shear wall system for lateral bracing.

47.5. INTERACTION OF FRAMES AND DIAPHRAGMS

Most buildings consist of combinations of walls and some framing of wood, steel, or concrete. The planning and design of the lateral-resistive structure require some judgments and decisions regarding the roles of the frame and the walls. This section discusses some of the issues relating to this aspect of design.

Coexisting, Independent Elements

Most buildings have some solid walls, that is, walls with continuous surfaces free of openings. When the gravity load-resistive structure of the building consists of a frame, the relationship between the walls and the frame has several possibilities with regard to action caused by lateral loads.

The frame may be a braced frame or a moment-resistive frame designed for the total resistance of the lateral loads, in which case the attachment of walls to the frame must be done in a manner that prevents the walls from absorbing

lateral loads. Because solid walls tend to be quite stiff in their own planes, such attachment often requires the use of separation joints or flexible connections that will allow the frame to deform as necessary under the lateral loads.

The frame may be essentially designed for gravity resistance only, with lateral load resistance provided by the walls acting as shear walls. This method requires that some of the elements of the frame function as collectors, stiffeners, shear wall end members, or diaphragm chords. If the walls are intended to be used strictly for lateral bracing, care must be exercised in the design of construction details to assure that beams that occur above the walls are allowed to deflect without transferring loads to the walls.

Load Sharing.   When walls are firmly attached to vertical elements of the frame, they usually provide continuous lateral bracing in the plane of the wall, thus permitting the vertical frame elements to be designed for column action using their stiffness in the direction perpendicular to the wall. Thus 2×4 studs may be designed as columns using $h{:}d$ ratios based on their larger dimension.

In some cases both walls and frames may be used for lateral load resistance at different locations or in different directions. Figure 47.23 shows two such situations. In Fig. 47.23a a shear wall is used at one end of the building and a frame at the other end for the wind from one direction. In Fig. 47.23b walls are used for the lateral loads from one direction and frames for the load from the other direction. In both cases the walls and frames do not actually interact, that is, they act independently with regard to load sharing.

Figure 47.23c shows a situation in which walls and frames interact to share a direct load. The interior walls and the end frames share the total load from a single direction. If the horizontal structure is a rigid diaphragm, the load sharing will be on the basis of the relative stiffness of the vertical elements. This relative stiffness must be established by the calculated deflection resistance of the elements, as previously discussed.

Figure 47.23d shows a situation in which the walls and frames interact to share a single direct load. This represents a highly indeterminate situation. Case D in Table 23-O of the *UBC* describes such a situation, called a *dual system*, which consists of shear walls or a braced frame and a moment-resistive frame.

Structures of the type illustrated in Fig. 47.23 should be analyzed using dynamic methods to determine the load distribution and ultimate or plastic strength analysis for the behavior of the elements of the system. If the simpler equivalent static load method is used and the elements are designed using working stress analysis, the stiffer elements of the system should be designed quite conservatively because they will tend to take more loading than the static analysis implies.

47.6. COLLECTORS AND TIES

Transfer of loads from horizontal to vertical elements in laterally resistive structural systems frequently involves the use of some structural members that serve the func-

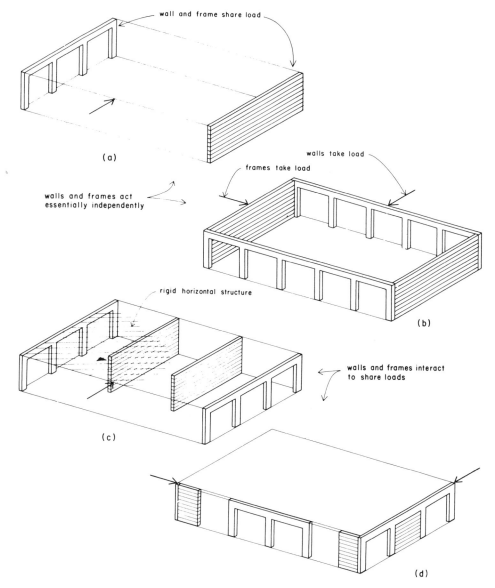

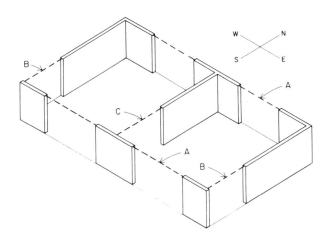

FIGURE 47.23. Mixed shear walls and rigid frames.

tions of struts, drags, ties, collectors, and so on. These members often serve two functions—as parts of the gravity-resistive system or for other functions in lateral load resistance.

Figure 47.24 shows a structure consisting of a horizontal diaphragm and a number of exterior shear walls. For loading in the north–south direction the framing members labeled A serve as chords for the roof diaphragm. In most cases they are also parts of the roof edge or top of the wall framing. For the lateral load in the east–west direction they serve as collectors. This latter function permits us to consider the shear stress in the roof diaphragm to be a constant along the entire length of the edge. The collector "collects" this constant stress from the roof and distributes it to the isolated shear walls, thus functioning as a tension/compression member in the gaps between the walls.

In the example in Fig. 47.24 the collector A must be attached to the roof edge to develop the transfer of the con-

FIGURE 47.24. Collector functions in a box system.

stant shear stress. The collector A must be attached to the individual shear walls for the transfer of the total load in each wall. In the gaps between walls the collector gathers the roof edge load and functions partly as a compression member, pushing some of the load to the forward wall, and partly as a tension member, dragging the remainder of the collected load into the rearward wall.

Collectors B and C in Fig. 47.24 gather the edge load from the roof deck under the north–south lateral loading. Their function over the gap reverses as the load switches direction. They work in compression for load in the northerly direction, pushing the load into the walls. When the load changes to the southerly direction, they work in tension, dragging the load into the walls.

The complete functioning of a lateral resistive structural system must be carefully studied to determine the need for such members. As mentioned previously, ordinary members of the building construction will often serve these functions: top plates of the stud walls, edge framing of roofs and floors, headers over openings, and so on. If so used, such members should be investigated for the combined stress effects involved in their multiple roles.

47.7. ANCHORAGE ELEMENTS

The attachment of elements of the lateral resistive structure to one another, to collectors, or to supports usually involves some type of anchorage element. There is a great variety of these, encompassing the range of situations with regard to load transfer conditions, magnitude of the forces, and various materials and details of the structural members and systems.

Tiedowns

Resistance to vertical uplift is sometimes required for elements of a braced- or moment-resistive frame for the ends of shear walls, or for light roof systems subject to the force of upward wind suction. For concrete and reinforced masonry structures, such resistance is most often achieved by doweling and/or hooking of reinforcing bars. Steel columns are usually anchored by the anchor bolts at their bases. The illustrations in Fig. 47.25 show some of the devices that are used for anchoring wood structural elements. In many cases these devices have been load tested and their capacities rated by their manufacturers. When using them, it is essential to determine whether the load ratings have been accepted by the building code agency with jurisdiction for a specific building design.

The term *tiedown* or *hold-down* is used mostly to describe the type of anchor shown in the lower right corner of Fig. 47.25.

Horizontal Anchors

In addition to the transfer of vertical gravity load and lateral shear load at the edges of horizontal diaphragms, there is usually a need for resistance to the horizontal pull-

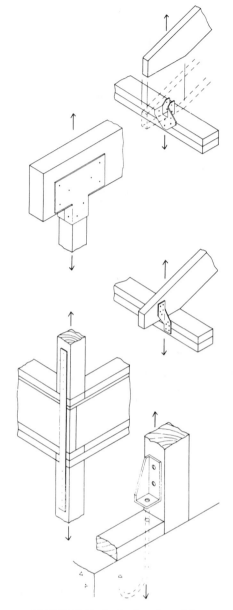

FIGURE 47.25. Anchoring of wood frames.

ing away of walls from the diaphragm edge. In many cases the connections that are provided for other functions also serve to resist this action. Codes usually require that this type of anchorage be a "positive" one, not relying on such things as the withdrawal of nails or lateral force on toe nails. Figure 47.26 shows some of the means used for achieving this type of anchorage.

Shear Anchors

The shear force at the edge of a horizontal diaphragm must be transferred from the diaphragm into a collector or some other intermediate member, or directly into a vertical diaphragm. Except for poured-in-place concrete structures,

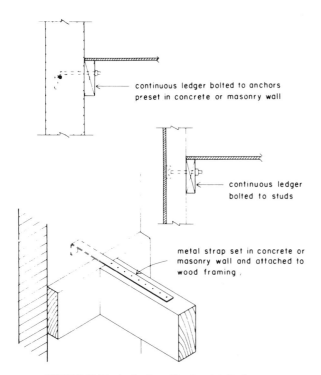

FIGURE 47.26. Anchoring of horizontal diaphragms.

some lateral shear resistance. If a more positive anchor is desired, or if the calculated load requires it, shear keys may be provided by inserting wood blocks in the concrete, as shown in Fig. 47.27.

Transfer of Forces

The complete transfer of force from the horizontal to the vertical elements of the lateral resistive system can be quite complex in some cases. Figure 47.27 shows a joint between a horizontal plywood diaphragm and a vertical plywood shear wall. For reasons other than lateral load resistance, it is desired that the studs in the wall run continuously past the level of the roof deck. This necessitates the use of a con-

this process usually involves some means of attachment. For wood structures the transfer is usually achieved through the lateral loading of nails, bolts, or lag screws for which the codes or industry specifications provide tabulated load capacities. For steel deck diaphragms the transfer is usually achieved by welding the deck to supporting steel framing.

If the vertical system is a steel frame, these members are usually parts of the frame system. If the vertical structure is concrete or masonry, the edge transfer members are usually attached to the walls with anchor bolts set in the concrete or in solid-filled horizontal courses of the masonry. As in other situations, the combined stresses on these connections must be carefully investigated to determine the critical load conditions.

Another shear transfer problem is that which occurs at the base of a shear wall in terms of a sliding effect. For a wood-framed wall some attachment of the wall sill member to its support must be made. If the support is wood, the attachment is usually achieved by using nails or lag screws. If the support is concrete or masonry, the sill is usually attached to preset anchor bolts. The lateral load capacity of the bolts is determined by their shear capacity in the concrete or the single shear limit in the wood sill. The *UBC* [Sec. 2907(f)] gives some minimum requirements for sill bolting, which should be used as a starting point in developing this type of connection.

For walls of concrete and masonry, in which there is often considerable dead load at the base of the wall, sliding resistance may be adequately developed by friction. Doweling provided for the vertical wall reinforcing also offers

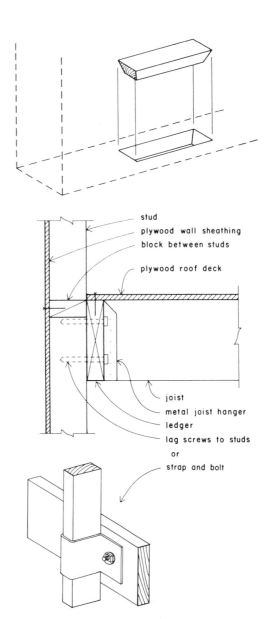

FIGURE 47.27. Transfer of loads between horizontal and vertical elements.

tinuous edge-framing member, called a *ledger*, which serves as the vertical support for the deck as well as the chord and edge collector for the lateral forces. This ledger is shown to be attached to the faces of the studs with two lag screws at each stud. The functioning of this joint involves the following:

1. The vertical gravity load is transferred from the ledger to the studs directly through lateral load on the lag screws.
2. The lateral shear stress in the roof deck is transferred to the ledger through lateral load on the edge nails of the deck. This stress is in turn transferred from the ledger to the studs by horizontal lateral load on the lag screws. The horizontal blocking is fit between the studs to provide for the transfer of the load to the wall plywood, which is nailed to the blocking.
3. Outward loading on the wall is resisted by the lag screws in withdrawal. This is generally not considered to be a good positive connection, although the load magnitude should be considered in making this evaluation. A more positive connection is achieved by using the bolts and straps shown in the lower sketch in Fig. 47.27.

47.8. SEPARATION JOINTS

During the swaying motions induced by earthquakes, different parts of a building tend to move independently because of the differences in their masses, their fundamental periods, and variations in damping, support constraint, and so on. With regard to the building structure, it is usually desirable to tie it together so that it moves as a whole as much as possible. Sometimes, however, it is better to separate parts from one another in a manner that permits them a reasonable freedom of motion with respect to one another.

Figure 47.28 shows some building forms in which the extreme difference of period of adjacent masses of the building makes it preferable to cause a separation. Designing the building connection at these intersections must be done with regard to the specific situation in each case. Some of the considerations to be made in this are the following.

The Specific Direction of Movements. In generally rectangular building forms, such as those shown in the examples in Fig. 47.28, the primary movements are in the direction of the major axes of the adjacent masses. Thus the joint between the masses has two principal forms of motion: a shear effect parallel to the joint, and a together-apart motion perpendicular to the joint. In building forms of greater geometric complexity, the motions of the respective masses are more random and the joint action is much more complex.

The Actual Dimensions of Movement at the Joint. If the joint is to be truly effective and if the adjacent parts are not

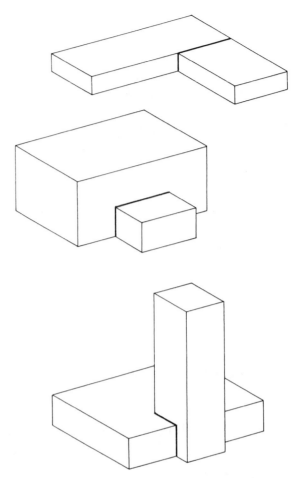

FIGURE 47.28. Potential situations requiring separation joints.

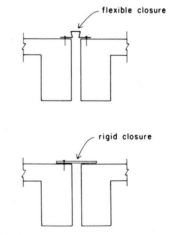

FIGURE 47.29. Closure of horizontal separation joints.

allowed to pound each other, the actual dimension of the movements must be safely tolerated by the separation. The more complex the motions of the separate masses, the more difficult it is to predict these dimensions accurately, calling for some conservative margin in the dimension of the separation provided.

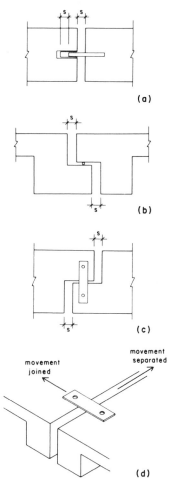

FIGURE 47.30. Means for achieving partial separation.

Detailing the Joint for Effective Separation. Because the idea of the joint is that structural separation is to be provided while still achieving the general connection of the adjacent parts, it is necessary to make a joint that performs both of

these seemingly contradictory functions. Various techniques are possible using connections that employ sliding, rolling, rotating, swinging, or flexible elements that permit one type of connection while having a freedom of movement for certain directions or types of motion. The possibilities are endless, and the specific situation must be carefully analyzed in order to develop an effective and logical joint detail. In some cases the complexity of the motions, the extreme dimension of movement to be facilitated, or other considerations may make it necessary to have complete separation—that is, literally to build two separate buildings very close together.

Facilitating Other Functions of the Joint. It is often necessary for the separation joint to provide for functions other than those of the seismic motions. Gravity load transfer may be required through the joint. Nonstructural functions, such as weather sealing, waterproofing, and the passage of wiring, piping, or ductwork through the joint may be required. Figure 47.29 shows two typical cases in which the joint achieves structural separation while providing for a closing of the joint. The upper drawing shows a flexible flashing or sealing strip used to achieve weather or water tightness of the joint. The lower drawing symbolizes the usual solution for a floor in which a flat element is attached to one side of the joint and is allowed to slip on the other side.

Figure 47.30 shows a number of situations in which partial structural separation is achieved. The details of such joints are often quite similar to those used for joints designed to provide separation for thermal expansion. Figure 47.30a shows a key slot, which is the type of connection usually used in walls where the separation is required only in a direction parallel to the wall plane. Figure 47.30b and c shows means for achieving the transfer of vertical gravity forces through the joint while permitting movement in a horizontal direction. Figure 47.30d shows a means for achieving a connection in one horizontal direction while permitting movement in the perpendicular direction.

CHAPTER FORTY-EIGHT

Design for Wind and Earthquake Effects

Chapter 48 considers the general process of design of whole systems for resistance to the lateral force effects of both wind and earthquakes. Some examples of building systems are presented here, but the major illustrations of system design are contained in the building case studies in Part 9. In the examples in Part 9 it is possible to consider the context for the lateral force problems in relation to the general design of the whole structure. The examples demonstrated here are intended to supplement the Part 9 illustrations by showing some buildings and forms of lateral bracing not covered by the example buildings in Part 9.

48.1. PROCESS OF DESIGN

Design is essentially a continuing task of inquiry and decision. The inquiry continues as long as potential questions can be brought up; decisions must be made with much judgment and with the weighing of the importance of many factors, some of which are usually in opposition. Final design solutions for complex systems, such as those for building structures, often contain many compromises and many relatively arbitrary choices. The most economical, most fire-resistive, most quickly erected, most handsome, or most architecturally accommodating structure is only accidentally likely to be the optimal choice for the resistance of lateral forces. We cannot pretend to show the complete process of structural design in these examples, although consideration of factors other than lateral force actions is frequently mentioned. The design solutions developed are thus unavoidably somewhat simplistic and myopic in their concentration on the lateral force problem.

In general, with regard to the lateral force resistive system, the design process incorporates the following:

1. *Determination of the Basic Scheme.* This includes the choice of type and layout of the basic elements of the system.
2. *Determination of Loads.* This involves the establishment of criteria and the choice of investigative methods.
3. *Determination of the Load Propagation.* This consists of the tracing of the load through the structure, from element to element of the system, until it is finally resolved into the supporting ground.

4. *Design of Individual Elements.* Based on their load-sharing roles, each separate element of the system must be investigated and designed.
5. *Design for Interactions.* Connections between elements of the structure, and between structural and nonstructural parts of the building, must be investigated and designed.
6. *Design Documentation.* Because the design as such is essentially only an imagined idea, all the information necessary to clearly and unequivocally communicate the idea must be documented.

In the discussions of the examples that follow, all of these aspects of the design process are given some treatment.

48.2. DESIGN EXAMPLES IN PART 9

The following is a list of the design examples in Part 9.

1. One-story box building with plywood roof deck and shear walls, Building One, Scheme 2, Sec. 51.5.
2. One-story masonry shear walls, Building One, Scheme 3, Sec. 52.4.
3. Steel roof deck diaphragm, Building One, Scheme 4, Sec. 53.4.
4. Three-story building with plywood shear walls, Building Two, Scheme 1, Sec. 55.4.
5. Three-story building with steel-braced frame, Building Two, Scheme 2, Sec. 56.4.
6. Three-story building with concrete moment-resisting frame, Building Two, Scheme 4, Sec. 58.4.
7. Three-story building with masonry shear walls, pierced wall system, Building Two, Scheme 6, Sec. 60.4.
8. Three-story building with masonry shear walls, individual masonry piers, Building Two, Scheme 8, Sec. 62.3.

The examples in the remaining sections of Chapter 48 illustrate some additional cases of building forms and types of lateral bracing systems.

48.3. BRACED FRAME, ONE-STORY BUILDING

In this example Building One is considered to have the same general form as indicated in Fig. 51.1 with the structure consisting of a light steel frame. Lateral bracing of the vertical structure consists of trussing developed with X-bracing between members of the perimeter steel frame. Figure 48.1 shows a possible development for such a bracing system. For resistance in the north–south direction the east and west walls are braced with the diagonals in the same walls that were used as shear walls in Scheme 2 (Sec. 51.5). In the other direction, however, only two walls on each side (north and south) are used so that there are a total of four walls with diagonals (see Fig. 48.1*b*). It is assumed that the diagonal members and their connections to the steel frame can be adequatedly developed with only four trussed bays in each direction.

Some considerations for the design of this system are as follows:

1. For lateral deflection and general dynamic response the trussed system is considered to be equivalent to the shear wall system. The codes recognize this by assigning the same general range of R_w factors for determination of the seismic load.
2. Connections stressed during lateral force actions should be tight and nonloosening (generally called *positive*), for the buffeting during wind storms and shaking during earthquakes will tend to shake them loose. Rigid frames will normally have such connections, but the simple pinned connections used routinely for trussing do not always have this character.
3. Although trussed structures are normally quite stiff, there are a number of things that may contribute to lateral deflection. These include the shortening and lengthening of the truss members, the deformation of connections, and the deformation of column anchorage. For the X-braced structure, a major contribution is that of the tension elongation of the X members, as these will be quite highly stressed and are usually the longest members of the truss system.
4. Planning of trussing is often a major architectural problem. Although trussing diagonals may be exposed if fire codes permit, they are frequently incorporated in wall constructions. Locating solid walls at points that are strategically useful to the lateral bracing system may be difficult.

The following computations demonstrate the process of investigation and design for the X-braced bents on the front of the building (south side). For this example we consider the flexible roof diaphragm to result in a simple peripheral distribution of equal portions of the total east–west seismic force to the two walls on the north and south sides. The total east–west load of 44.4 kips, as determined in Sec. 51.5, is thus divided in half to produce a load from the roof diaphragm to the south wall of 22.2 kips.

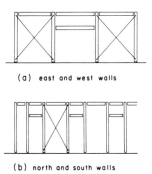

(a) east and west walls

(b) north and south walls

FIGURE 48.1. Braced frame scheme: Building One.

As determined in Sec. 51.5, this load tabulation does not include the seismic load due to the weight of the east–west walls (51 kips in Table 51.1). We will therefore add a force of $0.14(51) = 7$ kips to the total roof load; resulting in an additional 3.5 kips to the south wall. The total load to the wall thus becomes $22.2 + 3.5 = 25.7$ kips, and with two braced bents in the wall, the load per bent is $25.7 \div 2 = 12.85$ kips per bent.

Assuming the bent layout as shown in Fig. 48.2, the tension force in the diagonal is thus

$$T = \frac{1}{0.53} \times 12.85 = 24.2 \text{ kips [108 kN]}$$

The code requires that this force be increased by 25% for the design of the member and its connections [see *UBC*, Sec. 2312(j)1G]. We thus design for a force of $1.25(24.2) = 30.2$ kips [135 kN]. If a round rod of A36 steel is used and design is based on the maximum stress of 22 ksi in the rod, the area of rod required is

$$A = \frac{T}{1.33F_t} = \frac{30.2}{1.33 \times 22}$$

$$= 1.03 \text{ in.}^2 \text{ [666} \times 10^3 \text{ mm}^2\text{]}$$

It is possible therefore to use a 1.25-in.-diameter rod with a gross cross-sectional area of 1.23 in.². If this-size rod is used, the lateral deflection of the top of the bent due only to the tension stretching of the rod will be (see Sec. 47.3, Fig. 47.16)

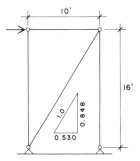

FIGURE 48.2. East–west bent.

$$d = \frac{TL}{AE \cos \theta}$$

$$= \frac{30.2[10(12)]/0.53}{1.23(29,000)(1/0.53)}$$

$$= 0.163 \text{ in.} \quad \text{or} \quad \text{about } \tfrac{3}{16} \text{ in. [3 mm]}$$

This indicates a very small deflection, even with the use of the relatively small diameter rod. The true deflection will be larger, as discussed previously, but should not be more than two or three times that computed. Most designers, however, would consider this rod to be too skinny, and would increase it for some more stiffness as well as to reduce stress and elongation.

It is also possible to use other types of steel members for the X-braces, although the simple round rod is often used when forces are low.

Overturning, sliding, and anchorage must also be considered, essentially in the same manner as for the shear walls in Sec. 51.5.

While the 1.25-in.-diameter rod is quite sturdy, it is nevertheless relatively slender with the overall length of about 18 ft in the bent. (Radius of gyration = D/4; L/r = 691.) The bent is thus quite likely to function as assumed, with the compression diagonal buckling at a quite low force. (Like a T-square, not a baseball bat.)

This form of bent may be acceptable for wind loading, but is now not considered desirable for seismic resistance in high-risk zones. Construction with stiffer diagonals is favored for the higher resulting energy capacity. The round rod is thus likely to be replaced by rolled angle or channel sections. Actually, although resulting in some additional total steel weight, the details for the construction with rolled shapes are usually simpler and less expensive, and the total cost for the bent may actually be reduced. See the details and discussion for the three-story bent in Sec. 56.4.

48.4. MOMENT-RESISTING FRAME, ONE-STORY BUILDING

Figure 48.3 shows a scheme for Building One with clear-span bents at 25-ft centers. These bents are to be achieved as moment-resisting frames and will be utilized for the lateral resistance of the central portion of the building. It is assumed that the end walls will also be braced with moment-resisting bents, developed with the wall construction.

Because of the higher R_w factor for the moment-resisting frame, the total seismic force will be reduced for this scheme. This may indeed result in a situation where the wind loading becomes critical, even in a high-risk seismic zone. Without showing the computations, we will assume the lateral and gravity loadings for the individual bents to be as shown in Figure 48.3c and d, respectively.

Figure 48.3c shows the free-body diagram of the bent with the loads and reactions. The deflected shape under load is shown by the dashed line. Although this problem is essentially indeterminate, if we assume the bent to be sym-

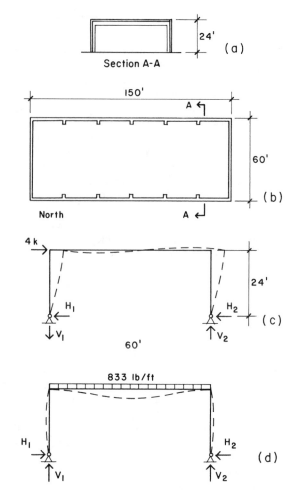

FIGURE 48.3. Building One, moment-resisting frame bents: (a) building section; (b) building plan; (c) lateral loading of the bents; (d) gravity loading of the bents.

metrical, we may reasonably assume the two horizontal reactions to be equal. In any event the vertical reactions are statically determinate and may be determined as follows:

$$V_1 = V_2 = \frac{4(24)}{60} = 1.6 \text{ kips}$$

On the basis of this analysis for the reactions, the distribution of internal forces is shown in the illustrations in Fig. 48.4. These forces must be combined with the forces caused by the gravity load to determine the critical design conditions for the bents.

Figure 48.3d shows the bent as loaded by a uniform load of 1000 lb/ft on top of the horizontal member. This effect is based on an assumption that the roof framing delivers an approximately uniform loading with a total dead plus live load of 40 psf as an average for the roof construction, including the weight of the horizontal bent member. As with the lateral load, the vertical reactions may be found on the basis of the bent symmetry to be

$$V_1 = V_2 = \frac{1(50)}{2} = 25 \text{ kips each}$$

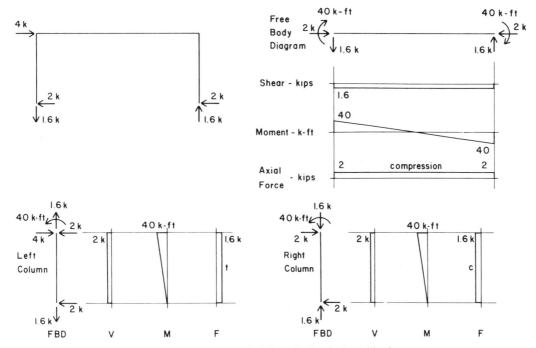

FIGURE 48.4. Example 4: investigation for lateral load.

Determination of the horizontal reactions in this case, however, is indeterminate and must consider the relative stiffness ($I:L$) of the bent members. For the pin-based columns it will be found that the horizontal reactions will each be

$$H = \frac{wL^3I_c}{8I_gh^2 + 12I_chL}$$

In this calculation we may use the relative, rather than actual, values of the column and girder stiffness (I_c and I_g in the formula). If we assume the girder to be approximately 1.5 times as stiff as the column, the horizontal reaction will be

$$H = \frac{1(50)^3(1)}{8(1.5)(20)^2 + 12(1)(20)(50)}$$

$$= \frac{125,000}{4800 + 12,000} = 7.44 \text{ kips}$$

With these values for the reactions the free-body diagrams and distribution of internal forces for the gravity loading are as shown in Fig. 48.5. These must next be combined with the previously determined lateral forces, as shown in Fig. 48.6. The design conditions for the individual bent members would be selected from the maximum values of Fig. 48.5 (gravity only) or three-quarters of the maximum values of the combined loading, as shown in Fig. 48.6. The adjustment of three-fourths for the comparison is based on the increased allowable stress for the combined forces that include the seismic load.

There are many possibilities for the construction of bents such as those shown for this building. Welded steel,

precast concrete, poured-in-place concrete, and built-up timber and plywood are possible choices for such a structure. Obviously, the general building construction and the exact form of the roof surface would be considered in selecting the materials for the structure.

A popular means for achieving the rigid frame and providing a drainage profile for the roof surface is through the use of a trussed bent, such as that shown in Fig. 47.11i. This type of structure allows for quite light construction, except for the columns which must be designed for considerable stiffness to resist lateral deflection.

If the horizontal portion of the frame is quite stiff, the lateral deflection of the bent may be approximated by considering the columns to be simple cantilevers. The deflection to be permitted for design is somewhat nebulous, and probably depends mostly on considerations of the effects of frame movements on the interior wall construction. Movement of anything more than a small fraction of an inch may become critical in this respect.

48.5.   MIXED BRACING SYSTEMS

The plan and section in Fig. 48.7 show a building with two symmetrical wings connected by a narrow central portion. The presence of solid walls permits the consideration of the development of a shear wall system for lateral resistance, except for the center portion with respect to north–south loading. A possible solution in this situation is the use of a rigid frame bent at the east and west sides of the center section, as shown in the plan in Fig. 48.7c.

Unless the roof deck diaphragms are exceptionally rigid (as in the case of a poured-in-place concrete struc-

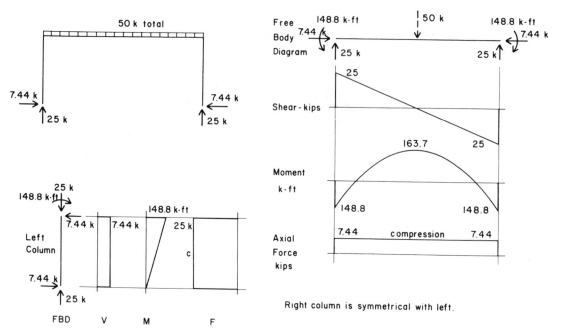

FIGURE 48.5. Example 4: investigation for gravity load.

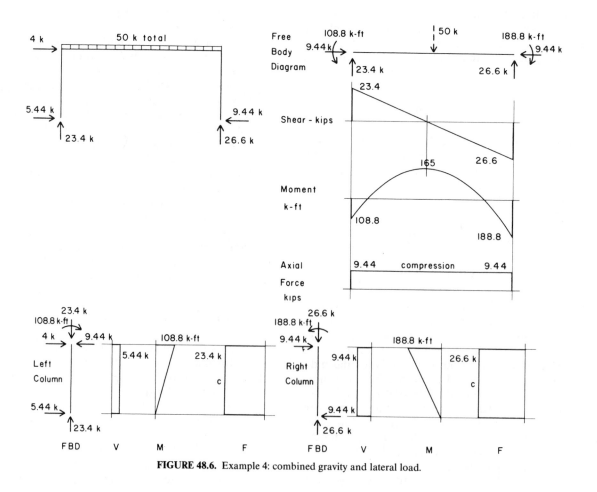

FIGURE 48.6. Example 4: combined gravity and lateral load.

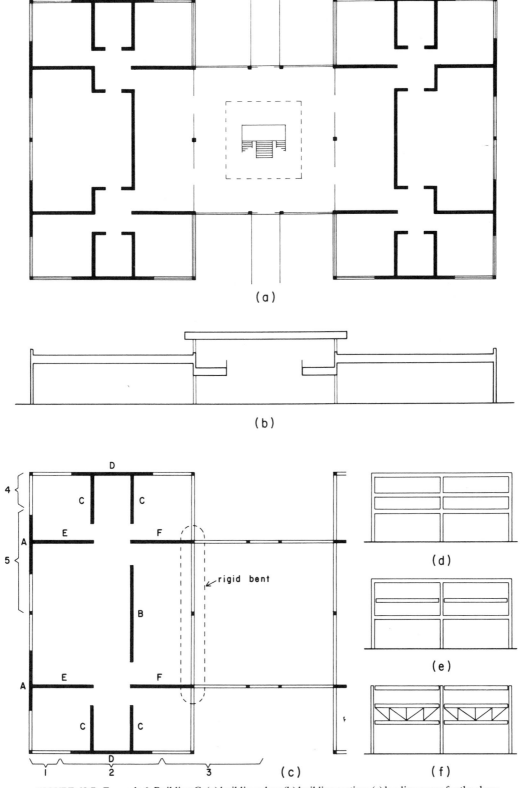

FIGURE 48.7. Example 6: Building C: (a) building plan; (b) building section; (c) loading zones for the shear walls and bents; (d) and (e) alternate forms for the rigid bents; (f) form of the trussed bent.

507

508   DESIGN FOR WIND AND EARTHQUAKE EFFECTS

tural deck), the usual method of determining the distribution to the mixed vertical elements in this situation would be on the basis of peripheral distribution. For this distribution, the zones are as shown in Fig. 48.7c.

As the section shows, there are three upper levels of framing in the bent. The bents may thus take one of the two forms shown in Fig. 48.7d and e. In (d) the frame is developed as a three-story rigid bent, whereas in (e) the roof of the wings is developed with shear-type connections only and the bent is only two stories.

Figure 48.7f illustrates an alternative means of achieving the braced bent, that is, through the use of a trussed bent, which may be simpler for fabrication and more economical in general if the truss depth is adequate. In this case, if the trussing is incorporated in a solid portion of the wall, it offers no intrusion in the architectural form or detail of the building.

As with other systems involving the mixing of shear walls and bents, it is necessary for the bents to be made quite stiff. Otherwise the peripheral distribution will not be valid. The more rigid the shear walls, the more this condition is critical. If the walls are of reinforced masonry or concrete and are quite long in plan with respect to their heights, the mixing for a peripheral distribution may be questionable.

We will not proceed with any computations or design of the elements for this example. Other examples show the methods for using the peripheral distribution and the design of shear walls and multistory rigid bents.

48.6. TALL BUILDING, SHEAR WALL BRACED

Figure 48.8a shows a typical upper-floor plan for a multistory apartment building that utilizes shear walls as the bracing system in both directions. In the north–south direction, the interior walls between the apartment units are used together with the two end walls at the stairs. There are thus a total of 14 walls, all approximately 20 ft long in plan.

In the east–west direction, the vertical bracing consists of the interior corridor walls. While there is usually a desire to have a minimum of permanent interior wall construction for some occupancies (such as the office building in the preceding section), it is reasonable to consider this structure for buildings such as apartments, hotels, dormitories, jails, and hospitals.

The exterior wall structure consists of a column and beam system. The ends of the shear walls are used as columns in this system. The typical floor structure could be a concrete slab or a deck and joist system. In either of these the shear walls would also be used as bearing walls. The solid concrete slab is a popular system for this situation, primarily because it permits a minimum floor-to-floor distance.

There is some reasonable limit for the height or number of stories for the shear wall-braced building. Shear walls have some limit for height-to-width ratio, lest overturn or lateral drift become unreasonable. In this example, the

north–south shear walls are relatively short in plan length, and the short slab span across the hall may be subject to major distortion if drift is significantly high. As shown in Fig. 48.8, the north–south walls have a height-to-width ratio of approximately 6:1, which may be a reasonable, practical limit.

The east–west corridor shear walls, on the other hand, would probably be built as continuous walls, with the doorways developed as reinforced openings in the continuous wall. In this event, these walls are actually longer in plan length than they are high, and the height limit would derive from the profile of the north–south walls.

The concrete shear wall structure is quite successful for resisting wind, but less successful for tall buildings in zones of high seismic risk. The combination of wall stiffness and heavy construction generates major lateral seismic force, whereas the weight is actually useful for overturn resistance against wind. For a building one-half this height, however, either concrete or reinforced masonry shear walls may be equally practical for wind and seismic resistance.

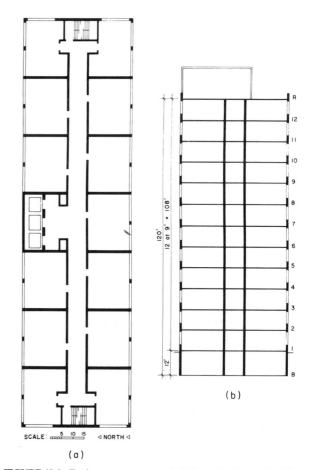

FIGURE 48.8. Twelve-story apartment building with shear walls; (a) typical upper-floor plan; (b) north–south section.

48.7. TALL BUILDING, BRACED FRAME

If steel construction is in general a feasible solution for the apartment building, lateral resistance may be developed with a braced frame. Some options for the development of trussing are shown in Fig. 48.9. The schemes shown in Fig. 48.9a, b, and c utilize various forms of bracing in the same walls that were used as shear walls in Example 13. In this case, however, it is unlikely that it would be necessary to develop the bracing in every wall or for the full height of the building. Figures 48.10 and 48.11 show the use of a staggered or stepped bracing system, in which many bays of the framing in the lower stories are braced, but the bracing is reduced in upper stories. This essentially reflects the actual variation of the magnitude of lateral shear and overturning moment.

Selection of the bracing form and the arrangement both in plan and vertically would depend on the relative magnitudes of both wind and seismic effects. For seismic resistance, the eccentric system with V-bracing, as shown in Fig. 48.9c, may be a better choice, since the code favors such a

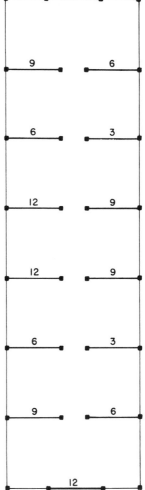

FIGURE 48.10. Stepped bracing system.

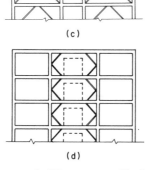

FIGURE 48.9. Apartment building, cross-corridor bracing: (a) X-braced bents; (b) single diagonal-braced bents; (c) inverted V-bracing; (d) K-bracing.

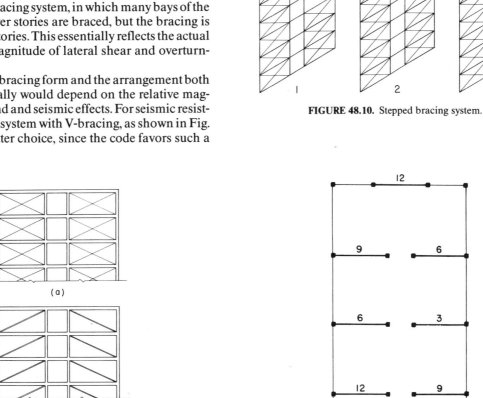

FIGURE 48.11. Plan layout for the stepped bents.

system with a much reduced magnitude for the lateral shear (see *UBC*, Table 23-O for R_W factors). Another option for the framing is shown in Fig. 48.9*d*, where the columns are shifted in position to permit K-bracing on both sides of the central hallway.

All of the bracing schemes shown for this example are more likely to be used for larger, taller buildings, and mostly when design for wind is critical.

48.8. TALL BUILDING, MOMENT-RESISTING FRAME

The plan in Fig. 48.12 shows a scheme for the development of a lateral bracing system for Building G with a moment-resisting frame of reinforced concrete. East–west bracing is limited to the two perimeter bents, with columns stiffened by use of oblong sections. These would be used in conjunction with quite deep spandrels to maintain the relative stiffness of the bent members.

For the north–south bracing, it is not feasible to expect the horizontal diaphragm to span from end to end for this example. In any event, since the bents are short in plan length, it is probably necessary to have more than two

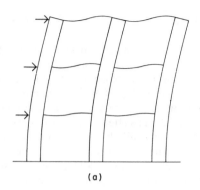

(a)

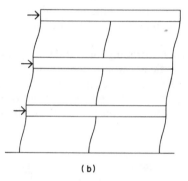

(b)

FIGURE 48.13. Lateral deformation of frames with members of disproportionate stiffnesses: (a) stiff columns and flexible beams; (b) stiff beams and flexible columns.

bents to share the load. The scheme shown therefore uses four bents—two at the center and one at each end. These bents are stiffened by using closely spaced columns.

There would, of course, be additional vertical structure for the support of the roof and floors. The plan in Fig. 48.12 shows only the elements considered for lateral bracing.

As with any building, the choice of the lateral bracing system has implications for architectural planning. In this example, if the scheme in Fig. 48.12 is used, the layout of rooms would be much less constrained than it would be with the shear wall system in Sec. 48.6. If some variety of interior spaces is desired, this may be a significant feature.

A special problem to consider with the rigid frame system is the potential for modification of the lateral response of the structure caused by nonstructural partitioning. This may cause quite critical damping or significantly change the fundamental period of the building for seismic response. Additionally, for either wind or seismic response, rigid partitioning may attract considerable lateral force with some unintended shear wall actions. Selection of materials and the detailing of attachments of the nonstructural partitioning and curtain walls must be done carefully to assure that the rigid (really probably flexible) frame is truly the lateral bracing element, and that its structural action is as assumed for rigid frame behavior.

A second consideration regarding stiffnesses has to do with the relative stiffnesses of the members in a single bent.

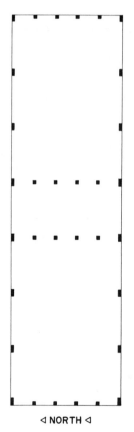

◁ NORTH ◁

FIGURE 48.12. Plan for the apartment building with moment-resisting bents.

It is generally desirable that all of the members of a bent that are connected at a single joint have close to the same relative stiffness This is to assure that the assumed bent deformations will actually occur. If columns are excessively stiff, the deformation will tend to be more of the character shown in Fig. 48.13a. If horizontal members (beams or trusses) are excessively stiff, deformation will tend to be more of the character shown in Fig. 48.13b. In this situation relative stiffness refers to the ratio of $I:L$ for individual members. In this example the columns are approximately half as long as the girders; thus equal stiffness will occur if the girder sections have twice the moment of inertia of the column sections. Exact equality is not to be expected, but when connected members have $I:L$ ratios of more than about 3 or 4:1 in difference, full flexing of the individual members may be doubtful.

CHAPTER FORTY-NINE

Special Lateral Effect Problems

49.1. STRUCTURALLY INDUCED LATERAL FORCES

There are many situations in which the vertical effect of gravity loads generates lateral (horizontally directed) forces due to the nature of the structure. This section discusses several types of these structures and the structural design and architectural planning concerns that they generate.

Single-Span Rigid Frames

Figure 49.1a shows a single-span structure consisting of two columns rigidly connected (for moment transfer) to a horizontal beam. If the column bases are free to move horizontally, the deflection of the beam under vertical loading will rotate the tops of the columns and cause the bases to move outward, as indicated by the dashed line figure. If horizontal restraint is provided with a pin-type connection at the column base, the rotation at the top of the column will result in bending and shear in the columns and an outward pushing force on the supports, as shown in Fig 49.1b. The distribution of moment in the beam and the columns for this condition is shown in Fig 49.1c.

If the column base is fixed against rotation as well as horizontal movement, the behavior of the frame under vertical loading will be as shown in Fig. 49.1d and e. Although full rotational restraint is not often feasible, many frames will have some degree of rotational fixity at their base, resulting in a true condition somewhere between those shown in Fig. 49.1b and d. It is not our purpose here to discuss the problems of designing moment-resistive column bases, so we limit the concern for resistance to horizontal thrust, as shown in Fig. 49.1b.

It is hard to imagine a building in which the outward movement of the column bases as shown in Fig. 49.1a could be accommodated. It seems realistic, therefore, to assume that some lateral restraining system must be provided. The three most common ways to achieve lateral restraint are shown in Fig. 49.2. Under some conditions it may be possible to develop resistance in the form of lateral movement restraint of the foundations alone. As shown in Fig. 49.2a, this requires the development of some combination of passive horizontal soil pressure on the side of the foundation and the sliding friction on the bottom of the footing. The latter is not possible, of course, if a deep foundation is used.

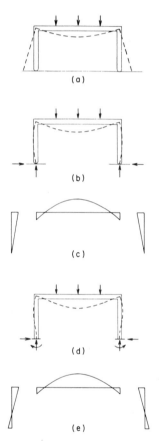

FIGURE 49.1. Functioning of a single rigid bent under gravity load.

If the footings are shallow (where frost is not a problem) and the soil is highly compressible, reliance on the passive soil pressure is highly questionable. For any significant lateral force, the foundations must be quite deep and probably must be designed in the form of abutments, as described elsewhere. In any event, whatever the actual capacity for resistance, some movement must be expected in the development of high levels of lateral soil pressure; this requires some design for adjustment during construction that permits the movement due to dead load to occur.

If movement must be restrained as fully as possible, or simply if the forces are large, it may be necessary to use the technique illustrated in Fig. 49.2b. In this case the two out-

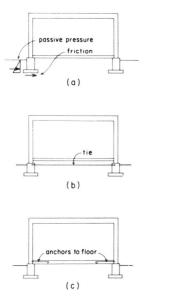

FIGURE 49.2. Development of the horizontal thrust in a single bent.

ward thrusts are balanced against each other by a tie member. This is often the most practical way of achieving a movement-free column base, although some problems must be considered. If the tie is quite long, its tension elongation under high stress may be considerable, requiring some adjustment during construction. If the location of the tie ends at the column bases is too high, it may be difficult to install the tie under the floor structure. The details of the construction of the floor, walls, frame base, and foundation must be worked out to provide for this.

Another way of achieving the tie for the column base is to tie each base separately to the floor structure, as shown in Fig. 49.2c. For relatively short-span steel frames this is sometimes accomplished by welding reinforcing bars to the column base plate and extending them into the poured concrete floor slab. This must be coordinated with the locations of poured or cut joints in the slab to be sure that a sufficiently large segment of slab is engaged by the anchoring reinforcing.

It is necessary, of course, to develop the design for gravity load effects with due consideration for the effects of wind or seismic loads. Overturning may require that the column bases resist uplift in addition to the effects just described. Lateral sliding may add to the forces, which must be resisted as shown in Fig. 49.2a and c, and may add more considerations for the design of the tie member and its connections.

Gabled Roofs and Three-Hinged Structures

Double-pitched roofs, commonly called *gabled roofs*, may be formed in a number of ways. When spans are short, the most simple and direct structure is usually a pair of inclined rafters, as shown in Fig. 49.3a. If the two rafters are supported only by each other at the top, the structure is of the three-hinged variety, as indicated in Fig. 49.3a, and the stability of the structure requires the development of both

vertical and horizontal support forces at the bottom end of the rafters. This must have been first discovered several thousand years ago when some early builders tried to put such a roof on top of two walls, only to have the walls topple outward as soon as any weight was placed on the rafters.

As with the rigid frame (Fig. 49.2b) a simple solution is to use a tie for balancing the two outward thrusts against each other. If it is desired to have a flat ceiling in the space below, the ceiling joists may often be used to achieve this tie. The rafters and tie will thus form an elementary truss; if the span is very long, this truss may be further subdivided. A simple truss form used commonly in wood-framed construction is that of the W-shape truss (named for the pattern of the interior members) shown in Fig. 49.3c.

When a ridge beam is used to support the upper ends of the rafters, as shown in Fig. 49.3d the outward thrust at the lower ends is essentially removed. This assumes, of course, that the ridge beam is supported by posts or walls and is not merely a member against which the rafters lean for framing purposes. Such a ridge member would ordinarily be used for the structure in Fig. 49.3a, but would be supported by the rafters rather than vice versa. With the ridge beam there is a possibility for a different problem due to the tendency for the rafter to rotate about its upper end, as shown in

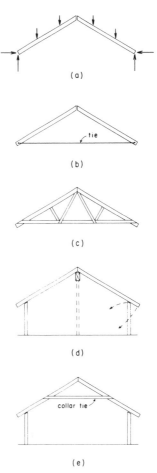

FIGURE 49.3. Actions of gabled structures under gravity loads.

Fig. 49.3d. This is more pronounced if the roof slope is high, and may cause inner movement at the top of the outside wall or post.

Another means for restraining inclined rafters is through the use of a collar tie, as shown in Fig. 49.3e. In effect, this turns the top joint into a rigid connection, and stability becomes dependent on the bending of the rafters. This solution is reasonably feasible only for very short spans (such as single car garages) and should use rafters of reasonable stiffness.

The three-hinged structure shown in Fig. 49.3a is a basic type of system that has various possible forms. One such structure is the two-part-gabled frame shown in Fig. 49.4a. This consists of two symmetrical rigid frames pinned at their bases and joined by a pin at the roof peak. With regard to forces at the supports, this structure behaves fundamentally the same as the gabled rafters in Fig. 49.3a. Options for the development of lateral resistance at the base are basically the same as those shown for the rigid frame in Fig. 49.2. Details for the attachment of the base of the frame to the support and the anchorage of the base to the floor structure may take various form, depending on the materials and size of the frame and the general construction of the walls and floor.

Figure 49.4b shows a typical building section with a modest span frame of welded steel. With a floor consisting of a poured concrete slab on grade, the simplest means for lateral restraint in this case is most likely a direct tie between the frame legs. Possible details for the frame base and tie are shown in Fig. 49.4c.

If the frame in Fig. 49.4b was made of glue-laminated wood, it would not be possible to simply bury the base in

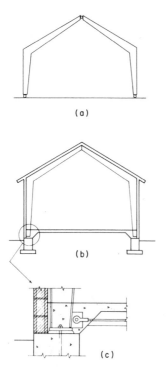

FIGURE 49.4. Form and details of a three-hinged bent.

the concrete to hide the tie and base plate. Thus the direct tying of the frame legs would probably not be a solution. In such a situation it would be necessary to design the connection of the frame base to the support to fully transmit the lateral kick to the support and then to tie, or otherwise brace, the supports.

Arches and Vaults

Arches and barrel vaults have support forces similar to those illustrated for the rigid frame and the three-hinged systems. One form of the arch is the three-hinged arch with a pin connection at the crown and an external stability condition essentially the same as the gabled rafter or the frame in Fig. 49.4. The arch or vault may also be continuous from spring point to spring point; in any case, the need for lateral as well as vertical support is present.

Figure 49.5a shows the general case for an arch under vertical gravity loading. If the arch springs from the ground, the support will be developed at that level as an abutment or a tied support; the same basic options as illustrated for the rigid frame in Fig. 49.2. In various building situations it is sometimes necessary to raise the arch or to create a higher space under it. Figure 49.5b illustrates the situation of an arch or vault that sits on top of supporting walls. The major problem to be resolved in this case is the outward thrust of the arch at the top of the walls. As in previous situations, a direct solution is the use of a tie, as shown in Fig. 49.5c. In many cases, however, the arch form will have been chosen for its interior architectural form, and the presence of the ties will be objectionable.

Where they can be accommodated at the building exterior, external braces may be used, as shown in Fig. 49.5d. These may take various form such as simple direct struts (as shown), counterfort walls (see discussion of tall retaining walls in Chapter 32), or as the famous flying buttresses of Gothic cathedrals. Two variations on the external buttress are shown in Fig. 49.5e and f. Where adjacent spaces occur in the building, it may be possible to use some of the crosswalls as shear walls. Even if these walls do not coincide with the location of the arches, it may be possible to use this scheme by utilizing the roof structure over the adjacent spaces as a horizontal distributing element, as is commonly done with the roof and floor diaphragms in box systems.

A second variation on the external brace is shown in Fig. 49.5f. In this case the arch is allowed to spring from low abutments at ground level, but the walls are placed some distance closer to the center. The building interior space—and even the exterior form—is almost the same as with the separate exterior braces.

Depending on site conditions, as well as building planning considerations, it may be possible to use the scheme shown in Fig. 49.5g, in which the building floor is lowered below the exterior grade. The external building form is thus simply that of a clean arch springing from the ground, whereas the taller interior space is achieved without ties or external elements cluttering the building exterior. Indeed, the external braces do occur, but in the form of counterfort

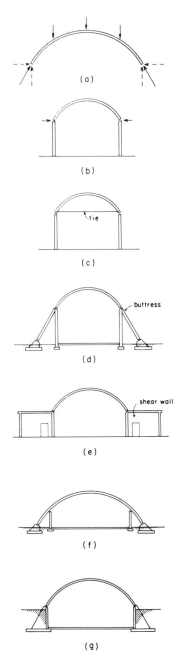

FIGURE 49.5. Resolution of arch thrust.

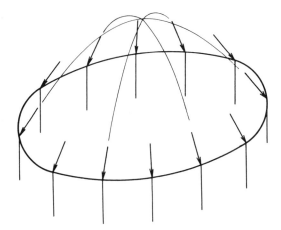

FIGURE 49.6. Tension ring for dome thrust.

retaining walls. This is a neat solution, but one that can be used only if windows in the side walls are not required and where siting and ground material conditions are favorable.

Domes present situations similar to arches and vaults, whether the dome is a shell or a framed structure. The round plan of the dome, however, offers a possibility that does not occur with the arch or vault structure. As shown in Fig. 49.6, it is often possible to resolve the outer thrusts at the base of the dome by developing a tension ring. This is quite commonly done, with the ring sometimes being part of the dome and other times being developed in the supporting structure.

Where the bottom edge of the dome is not continuous, due to openings or intersecting cross vaults, it is possible to use other methods of bracing. The techniques illustrated for the arch in Fig. 49.5d–g can be equally applied to the dome if the tension ring is not possible. However, the ring is usually the best solution when it is an option. In some situations, it may be possible to use one or more rings at points above the dome spring point to partly or totally restrain the outward thrusts. Even when the external bracing system is used, it is often possible to turn the base of the foundations into a ring to avoid dependence on soil pressure for lateral resistance.

Tension Structures

Tension can be used as the primary resolution in a number of ways for spanning structures. Figure 49.7a shows a simple draped tension element, which may be a cable or a membrane surface. This is basically the same as the gabled rafter or the arch with the load direction reversed; thus the lateral thrusts at the supports are inward instead of outward. In this case a direct solution might be a simple horizontal member between the supports, except that then it would not be a tension tie but a compression strut. If the span is long, the problem of slenderness for the strut becomes a major concern. The strut is therefore not a common solution, regardless of considerations of architectural design.

One method of resolving the lateral force at the supports for the draped structure is that shown in Fig. 49.7b. This consists of matching external braces similar to those shown for the arch in Fig. 49.5d; except that in this case they are tension guys instead of compression struts. If the span is long, the tension anchors required for these guys become major design elements. Furthermore, the guys exert a downward force on the vertical supports, so that the vertical walls or columns must carry considerably more than simply their share of the gravity load on the draped spanning structure. Add these problems to those of the intrusion on the building exterior and thus it is not a very popular solution.

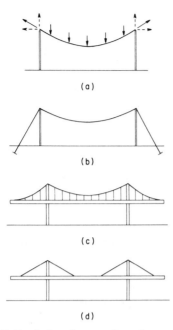

FIGURE 49.7. Resolution of support forces for spanning cables.

Figure 49.7c shows a slightly different use of the draped tension element. In this situation the roof itself is a separate, flat structure supported in the manner of the deck of a suspension bridge. Used as shown, the cable lateral force is resolved back into a compression in a horizontal direction in the roof structure. This may be possible, but the geometry of the cables and the design of the outer ends of the roof structure must be carefully developed. Rigid arches may be made into a predetermined profile, but flexible cables will take their own form in response to the loads.

A variation on the suspended "bridge" structure is shown in Fig. 49.7d. In this case the suspending cables are not draped, but rather are used for direct tension support. For the structure shown, the flat roof system is thus supported in the manner of a continuous beam with multiple supports provided by the mast/columns and the cables. Actually, to control deformations by using lower stresses, it is quite possible that the tension members may be steel bars or shapes rather than cables. As with the structure in Fig. 49.7c, the horizontal forces at the bottom end of the suspending members must be resisted by the roof structure. An advantage in this structure, however, is that the profile of the tension members is truly predetermined because their flexibility is not an issue.

As with the round dome, dished-shape suspended systems may be used with a round edge in plan. The same principle of the continuous ring edge may be used; in this situation it requires a compression ring. Although the compression strut is not as good for the single direction-draped structure (Fig. 49.7a), the compression ring is quite feasible and has been often used for the round suspended structure. The major problem for such a structure is usually water runoff and the potential for ponding in the center, rather than lateral forces at the edge.

Pneumatic Structures

Air pressure as a structural device is usually used in one of three ways.

Inflated Buildings. This consists of a whole enclosure with the enclosing surface held up by air pressure from the inside.

Inflated Buildings—Cable-Stayed. This is a variation on the inflated membrane, with the building form controlled by tension cables that wrap the membrane.

Inflated Structures. This consists of inflating some element to make it rigid (such as a giant air mattress) and using it as a roof or wall.

For the inflated building, the support condition is somewhat similar to that for the arch or dome except that the load direction is reversed. The inflating pressure causes an upward force, which must exceed that of gravity, or the structure will not stand. The condition for the supports is thus as shown in Fig. 49.8. If the floor level occurs at the point of support, as shown in the figure, the inward lateral force can usually be resisted by the floor. If there is no floor, as in the case of a tentlike structure, or the air-supported structure sits on top of a wall, other means must be used to resist the lateral effect.

For round, or close to round, structures, the support structure may be constituted as a compression ring, similar to that for the round dishlike suspended surface. In some instances this may be done with plans that are ovoid, polygonal, oblong, as well as perfectly circular in plan.

If foundations are shallow, the vertical support force may be a greater problem. The angle at the edge of an inflated membrane structure is usually quite steep, thus making the support force mostly vertical and the lateral component of negligible concern. With cable-stayed structures the situation is usually reversed, with the angle of slope of the surface quite shallow at the edge and the horizontal force component in the cables quite high.

Because of their usual light weight, wind forces on air-supported structures tend to be quite critical. Any development of support systems must be done with the consideration of the combined effects of wind and gravity. Arches and domes, on the other hand, tend to be quite heavy, and the use of systems that are logical for the heavier structures may not be appropriate for the air-supported systems.

Cantilevered Structures

Canopies, marquees, carport roofs, and similar elements are sometimes formed as cantilevers from the side of a building, as shown in Fig. 49.9a. The cantilever requires the development of a resisting moment, which often occurs as a separated pair of opposed forces—outward at the bottom and inward at the top. If these forces are generated at the point where roof or floor structures exist, they may be carried into these structures. If they occur at midheight of the walls, as shown in the figure, the wall structure must be made to resist them.

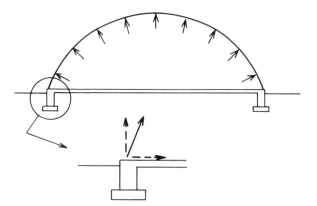

FIGURE 49.8. Support forces in pneumatic structures.

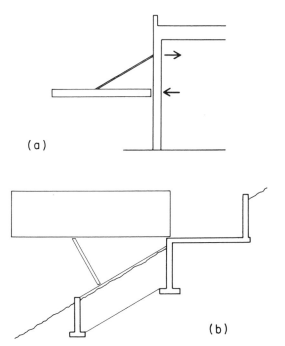

(a)

(b)

FIGURE 49.9. Cantilever structures.

In some situations entire buildings may be developed as cantilevered structures. These occur on steep hillsides, at edges of cliffs, and sometimes at waterfront sites. Development of supports for such a building is a serious affair. The cantilever may be achieved by using an anchoring structure as a counterweight; that is, by having something that is *not* cantilevered to extend from or grab onto. If this is not possible, the building may be perched on stilts that are either vertical or inclined as shown in Fig. 49.9b. The inclined stilt support requires the development of both vertical and lateral resistance at the base of the stilts. In addition, if the stilts occur well back of the cantilever edge, as shown in the illustration, there will be a significant component of lateral outward force at the upper support. This outward (and possibly upward) force is much more difficult to develop than the inward force. It may be necessary to

develop a heavy foundation structure at the uphill location as an anchor primarily for its simple weight.

A special structure sometimes used for hillside and beach locations is shown in Fig. 49.10. This consists of a set of deep foundation elements—usually driven or cast-in-place piles—that are embedded in a reasonably dense lower soil stratum and utilized as vertical cantilevers for lateral resistance. A grade-level concrete frame is used to tie the piles together and provide direct support for the building. The general stiffness and overall effectiveness of this system are greatly increased if the tops of the piles can be made to form a rigid frame with the grade-level structure. The term *downhill frame* is sometimes used to describe this type of structure. It is frequently used where beach front or hillside surface erosion is a potential hazard.

49.2. EFFECTS OF VOLUME CHANGE

Volume change resulting from fluctuation of temperature, moisture content, or other effects induces movements. Under various circumstances these movements may present difficulties, including the potential for development of lateral forces.

Thermal Change

Thermal change occurs primarily from daily and seasonal changes in the air temperature. This is generally most critical for the elements that constitute the building's exposed surfaces. Critical concerns include the following:

1. *Size of the Building.* The amount of volume change will generally be proportional to the size of the building. Dimensions of movements will be greater if the building is wider or higher. If the building is quite small, little provi-

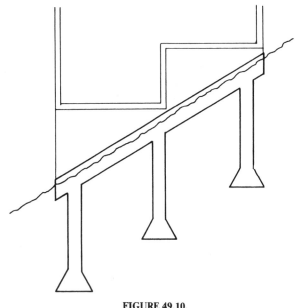

FIGURE 49.10.

sion may be required. If the building is very long or tall, extreme measures may be required.

2. *Continuous Structures or Surfaces.* If the building structure or the surface construction is made with a form of continuous construction (such as poured concrete or masonry), movements due to thermal change will be cumulative between breaks in the continuous elements. Assemblages with considerable jointing or other discontinuities will tend to absorb the small incremental changes. For the materials that are normally continuous, cumulative movement may be reduced or eliminated by the introduction of control joints that constitute breaks in the continuity.

3. *Differential Expansion.* Although large potential movements can be a major concern, small movements may also be critical, especially when attached elements of the construction move differently. Some situations where this may occur are the following:

a. When different materials are attached. Aluminum trim elements supported by steel structures are a common example of this. This is not often a source of distress to the structure, but must be considered for the effect of the movements on the attached elements or the jointing.

b. When an exposed skin is attached to a covered structure. Assuming a temperature-controlled interior, there will be a time lag of temperature change, with the surface elements changing more rapidly. This may be compounded if the materials are different, as discussed in item 3a. This is also seldom a source of concern for the structure but rather of concern for the surface elements and their attachment.

c. Exposed exterior structure and enclosed interior structure. When structural elements (usually columns or spandrels) are exposed, they will experience greater thermal change than the enclosed interior elements of the structure. This is a potential source of major stress if the building is long or tall, the climate temperature range is great, and the structure is relatively continuous in nature.

d. Structure partly aboveground and partly below ground. The building foundations normally experience little thermal change in mild climates, with the ground temperature a short distance below the surface experiencing little seasonal change. If there is considerable cumulative movement in the construction aboveground, there may be significant structural interaction—again, most critical for long buildings of continuous construction in cold climates.

4. *Long Construction Period.* When the time for the completion of construction stretches out (as it most often does), there are some special considerations that may be required. A common one occurs when the structure is erected during the coldest portion of the year. When the building is later enclosed and heated, or simply when summer comes, there may be thermal movements that were not considered in the design because the entire structure was expected to be enclosed. Since the construction schedule is seldom known at the time of the design of the construction,

consideration of this problem is often forgotten. A common example is a steel beam supported by relatively rigid masonry or concrete construction. A beam of even modest length, if erected in cold temperature and rigidly fastened in place, may likely cause failure of the supporting materials.

In some instances it is possible—and possibly the simplest solution—to simply design the structure for the stresses caused by thermal actions, and to add them to those for other loading combinations. Where the details of the construction necessary to alleviate the effects are costly, disruptive to the construction process, or disturbing to the appearance of the building, this is a reasonable approach. Often, however, it is necessary, or more desirable, to reduce or eliminate the thermal effects by one means or another. Some common techniques for achieving this are the following:

1. *Planned Discontinuity.* The general planning of the building may be developed with the deliberate intention of avoiding continuous construction. Thus the magnitudes of cumulative movement are kept small, even though the building is large.

2. *Provision of Deliberate Construction Breaks.* If continuous construction is unavoidable, it may be possible to use expansion joints at reasonable intervals to contain the expansion within some tolerable limit. The specific dimension between such joints will depend on the materials of the construction and the local climate. Joints so used may also function as seismic separation joints or controls for other types of structural movement.

3. *Use of Movement-Tolerating Joints.* Joints that permit controlled slipping, flexing stretching or other movement may be used to avoid the transfer of forces or the cumulative effect of thermal changes. These may be used between elements of the structure or between the structure and the attached parts of the construction.

4. *Staged Jointing during Construction.* Where the use of permanent movement-tolerating joints may not be desirable, it may be possible to use them on a temporary basis during construction. Thus the problem of the long construction period may be solved by leaving joints untightened until the structure is enclosed and heated, and then making the final connection.

Change of Moisture Content

Wood and products made from wood (paper, cardboard, and fiberboard) are subject to significant changes in volume due to fluctuations of moisture content. For elements of solid wood, there is a greater change in the direction perpendicular to the wood grain than along the grain. Although the greatest change occurs during the time of curing of the wood as it dries out from its condition in the live tree, changes may also occur with long duration shifts in the percent of moisture in the air around the wood. In northern climates interiors tend to be quite low in moisture during cold months, whereas humid summer conditions present the opposite situation. Elements of the building construction can experience long duration major change in moisture content with accompanying changes in volume.

In general, moisture change does not present a major source of lateral loads. The considerable jointing of wood structures allows for incremental development of dimensional changes, without major cumulative effects. In addition, any provisions made for thermal change will usually also provide for changes due to moisture content. However, thermal and moisture changes are not always linked, and provision for one may not be required where the other is. The interior moisture changes mentioned previously for northern climates can occur with little change in interior temperatures. Thus a condition not considered critical for thermal change should nevertheless be considered for expansion control caused by moisture change. Obviously, local climate conditions, as well as the materials and details of the building construction, must be carefully studied for good design.

Shrinkage

Shrinkage occurs in wood with the reduction of moisture content; the major effect occurs as the wood cures from its live condition. Shrinkage also occurs significantly during the drying of wet plaster, mortar, and concrete. For ordinary concrete not in a permanent submerged situation, the volume change is usually between 1 and 2%.

Good construction practices with plaster, masonry, and concrete require numerous provisions for the effects of shrinkage. The three basic techniques are to provide reinforcement (for the tension developed), to build sufficiently small increments so that large cumulative changes are avoided, and to provide control joints. As with moisture change in general, provisions made for thermal change will also work to reduce shrinkage effects. However, although thermal and air moisture content changes may vary with the local climate or the interior climate control, shrinkage is universal and must be provided for, even where other conditions are not critical.

49.3. LATERAL DRIFT

Lateral deflections are discussed for various types of structures and various loadings in other portions of this book. Lateral movements of the building structure may be critical for one or more of the following reasons.

Visible Movement. Exceptionally tall and slender structures may be seen to sway by persons inside or outside the building. While presenting a possibly disquieting experience, this is not necessarily a dangerous situation. Some movement is inevitable, but its theoretical computation as well as the setting of meaningful limitations are quite difficult for the average building. Once-in-a-lifetime movements due to major forces may be tolerable, but frequent occurrences due to average loadings are most likely objectionable. Experience is pretty much the only guide for design in this case.

Sensible Movements. Occupants of a building may not be able to see movement, but may experience the swaying effect. This is the way that most people know that an earthquake has occurred. Although this is also disquieting, it is inevitable and really objectionable only if it is of excessive magnitude or occurs frequently under ordinary loading conditions.

Distortions of Building Elements. Damage due to major movements is more often critical for other parts of the building than for the building structure. Plaster wall and ceiling surfaces, window glass, ceramic tile, and nonstructural masonry are all relatively intolerant of significant distortions. Although these rigid materials may suffer damage, other elements may also be affected. Rigid piping, operable windows, doors, and elevators are generally not able to remain functional with major distortions of the building. As in other situations where movements present problems, control jointing or other devices may be necessary if movement cannot be prevented sufficiently.

Objectionable Interactions. Structural movements are often considered with regard to the movement of some individual component of the structural system. Many problems, however, occur as interactions—between the structure and other parts of the building or between separate parts of the structure. Many of these situations are discussed in other sections of the book with regard to battering, multimassed buildings, unsymmetrical buildings, load sharing, interactive elements, and so on.

Codes are of limited help in design for lateral movements. (See discussion of story drift.) Experience and judgment are mostly required, although some general considerations may be made for techniques to reduce movements by a general stiffening of the lateral bracing structure as follows.

1. *Use Stiffer Materials.* This may involve a choice of basic materials (concrete or masonry versus wood) or of a stiffer grade of the same material (higher E value). Deflections of rigid frames, tall shear walls, and long horizontal diaphragms will be most affected.

2. *Increase Member Stiffness by Increasing I/L.* This may be done by increasing I or reducing L. For example, use a rectangular rather than a square shape for a wood or concrete column (increased I) or use closer spacing of columns in a bent (reduced L for the bent beams). This mostly affects rigid frames.

3. *Fix the Base of Frame Columns.* As discussed in Sec. 63.3, this greatly reduces story drift. The supports, of course, must be capable of developing the required fixity.

4. *Select a Stiffer Bracing System.* In general, shear walls and trusses are stiffer than rigid frames. However, the rigid frame with deep, short-span beams and many closely spaced columns may be quite stiff.

5. *Avoid Forms That Increase Movement.* Tall, narrow shear walls and long, narrow horizontal diaphragms are examples of poor form.

6. *Use Positive, Unyielding Jointing.* This most affects trussed bents, but may also be a factor for shear walls and horizontal diaphragms.

49.4. HORIZONTAL MOVEMENTS DUE TO VERTICAL LOADS

Lateral movements occur mostly under lateral loads. There are various situations, however, in which some horizontal movement occurs in structures that are subjected to vertical loads. An example is the movement of the columns of a rigid frame, as discussed in Sec 47.4. Figure 49.11a illustrates a common occurrence with tied rafters or the bottom chords of trusses. If the supports are as shown, outward movement at the supports is inevitable due to the tension elongation of the horizontal bottom member. This is reversed in the case of the parallel chord truss when support is provided for the top chords, as shown in Fig. 49.11b. In this case, inward movement is inevitable because of the compression shortening of the top chord.

Another situation of this type occurs with very long-span trusses or girders. As these members are generally quite deep, the end rotation causes an outward movement at the bottom and an inward movement at the top, as shown in Fig. 49.11c. For this situation one possible solution is to provide support that approaches that of a pin joint at the neutral axis where no length change occurs.

Obviously, the size of the structure has a great deal to do with the magnitude of the movements just illustrated. For modest size structures the movements may be absorbed in joint distortions. However, some detailing similar to that used for control of thermal expansion effects may be indicated. In some situations a technique that may be used is to leave the joint only partly fastened until all dead load movement has occurred. If the joint is then fully tightened, the movements to be considered will be only those due to live load—especially effective if live load is low in comparison with dead load.

49.5. LONG-TIME EFFECTS

Buildings exist for the most part over a long time, and various effects occur as a result of long time. Dead load is in effect permanent, so that structures that sustain high percentages of dead load are the most affected by the time duration of loads. Wind and earthquake effects are gen-

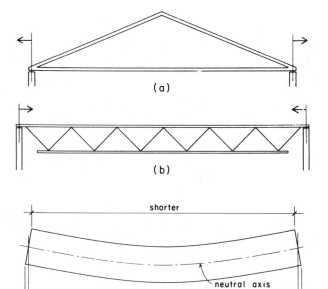

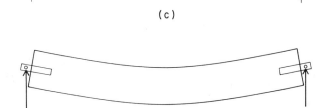

FIGURE 49.11. Horizontal support movements with deep spanning structures.

erally of quite short duration in their critical magnitudes, although repetitions of lower magnitude effects may also be critical. The following are some problems for concern with regard to time over the life of the building.

Soil Movements.   Soils with high clay content that are subject to high stress may deform on a more or less continuous basis. This is especially critical for structures in which the movements may result in tilting or horizontal displacement. Cantilever retaining walls and untied abutments must be carefully designed for these effects, and preferably not placed on soft clays or other continuously deformable materials. Movements may also occur due to seasonal soaking of the soil or heavy irrigation. Silty soils with high void ratios and very fine sands may be adversely affected in these situations.

Creep of Concrete.   Long-term creep effects in concrete may result in progressive movements over time. Where dead load is high or soil pressures constant over time, considerable deflection may occur. Tilt of tall cantilever retaining walls and inward bulging of concrete basement walls are examples of this phenomenon. Increased stiffness or use of bracing may be possible remedies.

Dimensional Change of Wood.   Moisture reduction due to curing, long-time fluctuations of moisture, and general aging of the material tend to result in some dimensional changes in solid wood elements: beams sag, columns twist, and walls and floors warp and curl. Where these can result in objectionable conditions—other than visual ones—measures should be taken to reduce the effects. Use of better cured wood for construction, bracing with steel elements, and substitution of glue-laminated products are some of the measures.

Progressive Loosening.   As discussed in Chapters 46 and 47, it is desirable to use connections that do not experience loosening over time. Connecting techniques vary in this aspect, and appropriate ones should be selected where their permanent tightness is critical to the maintaining of a general stiffness and tightness of the bracing system.

PART 9

DESIGN EXAMPLES

Work in Part 9 consists of the design of structural systems for example buildings. The principal purpose of this work is to illustrate the process of dealing with the design of the whole building structure; whereas work in other parts is focused on limited topics.

As discussed in the Introduction, the work here is linked to other parts by the device of using elements of the example buildings here for many of the example exercise problems in other parts. This helps to conserve on book space and reduce the size of the book, but also provides for two useful exchanges. When studying the more limited treatments in other parts, the whole building context of the isolated exercise problems may be viewed by reference to the presentations in this part. Conversely, fuller discussions of many problems encountered in the design work in this part can be quickly referenced by the notation here of the appropriate book sections that contain the computational work for elements of the structures developed here.

Buildings of similar size, shape, and purpose often have several alternatives for their basic construction, with each choice generally satisfying the basic design goals for the building. To illustrate that situation, several different schemes are presented for the two simple buildings used here for examples. For our purposes, this also allows the illustration of use of a wide range of the different structural materials and systems presented in earlier parts of the book. Thus while presenting only two different building examples, the materials in this part demonstrate the use of some 12 different structural schemes and systems. The buildings and the variations of the construction presented are as follows:

Building One: Single story, flat roofed box.
Scheme 1: Type V, light wood frame, with interior and exterior bearing walls.
Scheme 2: Same as 1, except clear-span roof trusses.
Scheme 3: CMU exterior bearing walls with a framed girder/purlin/rafter roof system and interior columns using wood members.
Scheme 4: Tilt-up concrete exterior walls with a roof frame using W shape steel girders, open web joists, and steel pipe columns.

Building Two: Five story office building.
Scheme 1: Type V all-wood system.
Scheme 2: Steel frame with W shape columns and beams, curtain walls, and a braced frame core for lateral resistance.
Scheme 3: Same as 2, except open web joists and joist girders for roof and floors.
Scheme 4: Sitecast concrete column, beam, slab system with curtain walls and a moment-resistive perimeter bent system for lateral resistance.
Scheme 5: Same as 4, except a two-way slab floor and roof system, a different curtain wall, and a core-braced shear wall system for lateral resistance.
Scheme 6: Exterior CMU walls providing both bearing wall and perimeter shear wall functions with continuous pierced walls; composite floor and roof structures with steel and wood elements.
Scheme 7: Isolated exterior CMU piers with combination sitecast and precast roof and floor systems.
Scheme 8: Same as 6, except reinforced brick masonry walls and an interior heavy timber structure.

CHAPTER FIFTY

General Concerns for Structural Design

Chapter 50 treats a number of issues that relate to the general work of designing building structures. Many of these issues are also mentioned in discussions in other parts of the book.

50.1. PROCESS AND METHODS

In general, the design of a building consists of the conceptualization and decision process by which the final form and fabric of the finished building is descriptively determined. The output of the design work is the recorded description of the desired object. The work of generating the ideas and recording them is called *designing*. The displayed collection of recorded ideas, usually in some combination of graphic and written documents, is called the *design*. The person who generates the ideas is the *designer*.

Design work may be viewed as the collection of decisions that determine the finished image of the design object. Designers exert judgment in making some of these decisions, although other sources strongly influence many decisions. How this all works for a specific design case depends on many factors. Some buildings are built using exclusively predesigned, prefabricated, off-the-shelf parts, which reduces the design work to selecting and arranging of cataloged items. Other buildings may use newly developed materials or old materials put together in new ways, requiring considerable imagination and innovation in the design work. A building may be simple in its behavior or may be complex and present problems that are difficult to analyze. What is exactly involved in design work varies considerably from case to case.

Figure 50.1 presents an image of the design process, viewed as a succession of activities. While a particular design is finished when the designer completes the design work and communicates the description to the persons who will create the actual object, the continuation of the process through final occupancy and use of the building has effects on final evaluation of the design and on the ongoing design activities of the working designer.

Designing of buildings is often thought of as being primarily the function of the architect—the master designer who maintains command of the design process and stamps the final design with his personal judgment, skill, style, and personality. While it is true that architects serve as prime designers for a large number of buildings, there is typically a long list of other participants in the design process.

The fact that innovation and creative design occurs to the extent that it does is the more impressive when the collective constraints of all of these influential parties is considered. Clearly, effective designers learn to deal with these realities, as well as with the creative process of design.

It is foolish to think of design as an activity that flows easily from one decision to the next, progressing smoothly toward a final statement. Indeed, a final, conclusive statement must be made. Getting there, however, usually involves some false starts, some steps backward, and a lot of dilemmas.

Most designers never really finish a design; they just quit. That is, there is almost always another idea to explore, another alternative to consider, or another approach to try. Given open-ended time and infinite resources, the design of an individual object may never be completed to the total satisfaction of the designer. In real situations, however, time is usually short and resources limited, and the designers are usually forced to quit at some specified time. The next idea or the alternative will have to wait for the next design project.

Whether organized with a computer program, or allowed to drift in a leisurely fashion, design work usually has to have a time schedule of some kind. Decisions are unavoidably sequential and interactive. Critical decisions must be frozen in stages, and the costs in time and resources involved in going back to reconsider decisions must be acknowledged.

Responsibility for various design decisions must be clearly established. This is especially important for two kinds of decisions: those that critically affect the time schedule, and those that have major effects on the final solution. In many regards, the person who accepts responsibility for the major decisions may be considered to be the prime designer because the elimination of alternatives and the major aspects of the final solution are in this person's control.

The time and money allotted for the performance of the design work has a strong influence on what kind of design is done. Whether the end product is a unique, innovative, pioneering effort or a carbon copy using predesigned pro-

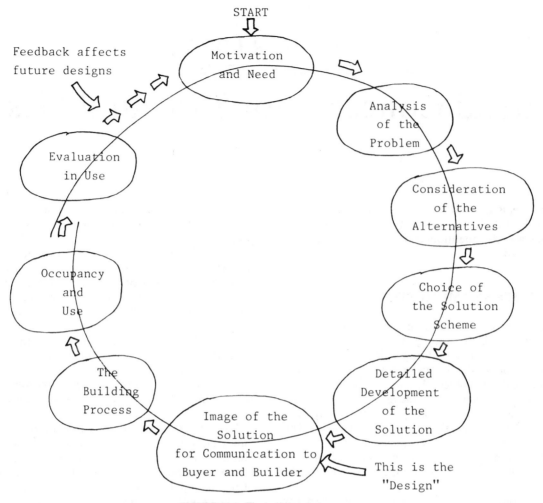

FIGURE 50.1. The building design process.

ducts and systems will often depend on the amount of design time and effort that money will permit. Design skills and creative imagination are also important, but they cannot be expected to flourish when the work required overwhelms the designer in terms of sheer time and energy.

Finally, how design gets done depends a great deal on the personal style and philosophy of the prime designer. Does the designer want testimonials to his or her creativity and skill? Does he or she want the cheapest product in the least time with a minimum effort? Does he or she want to achieve the whole design effort without anyone else's expertise or judgement getting in the way? Obviously, the obtaining of optimal fit, the degree of originality, the amount of experimentation, and the general care with which the work is done will depend on these considerations.

It is difficult to generalize about a specific level of mathematics or a set of ideas and tools that need some mastery in order to allow for professional work in the investigation and problem-solving activities related to building design work. Even if a person's specific role, area of involvement, and level of activity can be tightly defined,

it is hard to say just what is and what is not required in mathematical training. A great deal of the complex, serious engineering work for buildings is largely arithmetic, with some occasional simple trigonometry or algebra. With the advent of the computer and the growing stock of user-friendly programs, even much of that work is eliminated.

50.2. DESIGN STANDARDS AND AIDS

There are many sources for information, guidance, and assistance to support work in structural design for buildings. A bibliography of published materials is provided at the back of this book and selected references are listed at the ends of some of the sections of the book.

Industry Standards

There are many organizations that serve the general interests of various factions of the building industry. Groups are formed on the basis of common interest in a single material, in a type of product, in a method of building, and so on. In many cases the organizations perform promotional activities for the particular industry group, but also serve to police the group for some adherence to mutually accepted standards. These standards are in many cases published and widely distributed; they often become basic references for design practices and for the appropriate portions of building codes.

Table 50.1 indicates several areas of basic concern in building structures and the organizations that provide essential information relating to those areas. The publications of several of these organizations are listed in the bibliography. In some cases the publications provide general data about products, but in most situations it is advisable to obtain such information directly from manufacturers or suppliers of the products, as variations are sometimes possible and the actual products available in an area should be used.

There are also organizations, such as the American Society for Testing and Materials (ASTM) and the American National Standards Institute (ANSI), that deal with the industry as a whole, providing references used by many other groups.

For any actual design work, it is wise to determine what specific standards are used as references by the building code with jurisdiction for the proposed work. New industry standards are sometimes not readily accepted, so it is not uncommon for the legally enforced building codes to lag behind the latest industry standards. This is a general problem with any reference material, but is a chronic one with building codes.

Building Codes

Building codes are the legal ordinances enacted by some entity (city, county, state) for regulation of the construction of buildings in its jurisdiction. The code is the basis for granting or refusing to grant a permit for the construction. The ordinances are frequently revised, on the basis of recommendations by experts, or for various political reasons. Local builders, trade unions, real estate development interests, and others often exert pressure to influence the regulations in their interests.

Model building codes are prepared as recommendations by various organizations. The model standards of the AISC, ACI, NFPA, and others deal with limited concerns of a single material or other portion of the whole code. Some organizations prepare whole model codes, which have no legal jurisdiction, but may be accepted in whole or in part by a political entity, and thus become legal documents for that jurisdiction. Some of the model codes in wide use in the United States are the following:

1. *The Uniform Building Code (UBC)*, published by the International Conference of Building Officials; used extensively in western states.
2. *The BOCA National Building Code (BOCA)*, published by the Building Officials and Code Administrators International; used widely in the midwest and northeastern states.
3. *The Southern Standard Building Code (SSBC)*, published by the Southern Building Code Congress; used in the southeastern states.

TABLE 50.1. Industry Organizations

Acronym	Name of Organization	Areas of Concern
AISC	American Institute of Steel Construction	Steel construction, rolled products, steel connectors
AISI	American Iron and Steel Institute	Steel construction, steel products, cold-formed (light-gage) products
SJI	Steel Joist Institute	Prefabricated, light steel trusses (open-web joists, etc.)
SDI	Steel Deck Institute	Formed sheet steel products
AWS	American Welding Society	Welding
ACI	American Concrete Institute	Cement, concrete construction
PCA	Portland Cement Association	Cement, concrete construction
PCI	Precast Concrete Institute	Precast concrete products, construction
CRSI	Concrete Reinforcing Steel Institute	Steel reinforcing for concrete, concrete structural design
NFPA	National Forest Products Association	Wood, structural lumber and fasteners, wood structures
AITC	American Institute of Timber Construction	Wood, wood products and fasteners, wood construction
APA	American Plywood Association	Plywood, building construction with plywood
MIA	Masonry Institute of America	Masonry products, masonry structures

Most codes are also revised frequently. The *UBC* is generally reissued every 3 years. However, just as with the latest industry standards and other new recommendations, local building codes may be slow to accept the latest model codes in some cases. To get a building permit, of course, some respect must be paid to the local code, even though it may lag behind the prevailing practices and latest experience.

Model codes and legally enacted codes differ in various ways, but all deal with the same basic issues; with major concerns for public safety. Real differences are few in most cases, having mostly to do with some concentrations on local concerns (earthquakes in California, windstorms in Florida and Kansas, and heavy snow in Minnesota).

Professional Organizations

There are a number of professional organizations that produce journals and other publications with recommendations for aspects of building design. The two national groups most active in this context are the American Institute of Architects (AIA) and the American Society of Civil Engineers (ASCE). The Structural Division of the ASCE has both a journal and other publications that have some considerable influence, often being the first sounding place for ideas that eventually emerge in the industry standards and model codes.

Many other groups, with essentially local interests or more specialized involvements, also contribute efforts. These may be in the form of publications, or merely a concerted pressure to influence design practices or local building regulation. Collectively, these efforts are often very instrumental in evolving more responsible design, better construction practices, and more public concern and attention to improvement of the built environment.

Textbooks and Handbooks

There are textbooks of considerable variety available for every subject in the area of structural design. Most of these are of a form designed for use in courses in engineering schools, covering topics from an introductory level to highly advanced and specialized problems. The general purpose of these is to provide for investigation and design of situations relating to work of practicing structural engineers. Basic texts occur in multiples as several publishers compete for the market.

The level of mathematics in most structural engineering work is quite modest. Calculus, vector analysis, differential equations, and other forms of higher mathematics are generally limited to highly theoretical work and to the derivations of relationships. Specific relationships, as used for most structural computations, are usually in simple algebraic form, with some occasional simple trigonometry or very elementary calculus.

There are, of course, potentially complex situations, such as those occurring in dynamic behavior or highly indeterminate structures under multiple loadings. The vast majority of structural computations, however, use arithmetic and elementary high school algebra.

Current texts for engineering courses generally anticipate the usual working conditions for engineers; namely, that most computations, especially those that are complex and laborious, will be done with computer-aided routines. This does not free the engineer from the need to develop mathematics skills in order to understand the derivations of relationships and investigative procedures. It does, however, somewhat reduce the emphasis on learning of laborious "hand" computation methods, except for the experience of following the procedures.

For the person with limited mathematical training, using texts that are intended for engineering classes as self-study materials is difficult. The best of texts need some companion resource (teacher, tutor, etc.) in order to be effectively understood by inexperienced persons. Add any complex mathematics and the difficulty is compounded. Self-taught engineers are about as rare these days as self-taught doctors or lawyers.

In spite of all of this, it is possible to learn a great deal about the design of building structures without pursuing a degree in engineering. In the first place, computations are a small fraction of the work to be done in developing the complete design of a building structure. In addition, the vast majority of design problems are simple and repetitive. Thus it is possible to learn much of what is required to design many ordinary structures and to perform a great deal of the total work in design, without recourse to complex mathematical computations and highly sophisticated engineering investigative procedures.

Texts designed for use by architecture students, or for training of high-level technicians with limited engineering backgrounds, are available in the usual class text form, or sometimes are designed especially for self-study. The series of texts developed by the late Harry Parker, starting in the 1930s, are still in use and quite popular for their condensed, simplified procedures and direct solutions of practical problems.

Most handbooks, such as the manuals of the AISC and CRSI, are compilations of data and various aids for short-cut design work. They also usually have some amount of text, in a form designed to explain the use of the materials in the book. They are not usually self-standing as beginning texts, but have great practical value as learning sources, especially because they are usually quite up to date with industry practices.

Much of what goes into any current building construction is in the form of standardized products. The actual design and detailing of these products is preestablished, so that using them involves highly limited computations. This situation is generally increasing, so that much of structural design is similar to that for the building construction in general, consisting of selection and specification of products from manufacturers' catalogs. Catalogs of this type often virtually constitute handbooks of predesigned structures. Some materials abstracted from such catalogs are presented in Appendix F.

50.3. STRUCTURAL PLANNING

Planning a structure requires the ability to perform two major tasks. The first is the logical arranging of the structure itself, regarding form, dimensions, and proportions, and the ordering of the elements for basic stability and reasonable interaction. These issues must be faced, whether the building is simple or complex, small or large, of ordinary construction or totally unique. Spanning beams must be supported and have depths adequate for the spans; thrusts of arches must be resolved; columns above should be centered over columns below; and so on.

A second major task in structural planning is the development of the relationships between the structure and the building in general. The building plan must be "seen" as a structural plan. The two may not be quite the same, but they must fit together. "Seeing" the structural plan (or possibly alternative plans) inherent in a particular architectural plan is a major task for designers of building structures.

Consider the building plan shown in Fig. 50.2a. The general building form is defined by the combination of the solid portions of walls, the windows and doors, and the dashed lines outlining the edges and openings of the roof above. If a single ceiling height and a flat roof are assumed, the building form is easily visualized.

A possible alternative for the development of the roof structure is shown in Fig. 50.2b. This system uses some of the walls as bearing walls to support a roof framing system. Beams are placed at gaps between the walls with the beam supports developed in the walls (by the walls themselves or by posts buried in the walls). This is a reasonable solution, although one limitation may exist to question its logic. Commitment of a wall as a bearing support (and possibly as a shear wall) requires the wall to be permanent, which inhibits future rearrangement. If this is not seen as a problem, this solution may indeed be the simplest, least costly choice.

Figure 50.2c shows another possibility for the development of a support system for the roof structure, using columns only; the columns being located so as to be incorporated into the walls. A framing system based on this support is shown in Fig. 50.2d. The walls in this case may be developed completely free of building structural functions. However, lateral as well as gravity loads must be considered, and rigid-frame or trussed bracing must be developed if the walls are truly nonstructural.

Figure 50.2e shows a possibility for future rearrangement of some interior walls to create a larger open space. This is more easily achieved with the column support system, especially if lateral bracing uses only exterior walls.

Hopefully, architectural planning and structural planning are done interactively, not one after the other. The more the architect knows about structural problems and the structural designer (if another person) knows about architectural problems, the more likely it is possible that effective communication and an interactive design development may occur.

50.4. CHOICE OF STRUCTURE

Although each individual building is a unique situation if all of the variables are considered, the majority of building design problems are highly repetitive. Problems usually have many alternative solutions, each with its own pluses and minuses in terms of various points of comparison. Development of the final design involves the comparative evaluation of definable alternatives and the eventual selection of one.

When the problems are truly new in terms of a new building use, a scale jump, or a new performance situation, there is a real need for innovation. Usually, however, when new solutions to old problems are presented, their merits must be compared to established previous solutions in order to justify them. In its broadest context the selection process includes the consideration of all possible alternatives: those well known, those new and unproven, and those only imagined.

Selection may be done in one single stroke, similar to buying a new car off the dealer's showroom floor. For predesigned package building systems this may be the actual case. Usually, however, selection consists of a series of decisions, starting with broad ones of basic system type, form and materials, and progressing to increasingly detailed ones of shape of parts, connections, finishes, and so on.

Usually the broader the decision, the more difficult it is to make. Ideally, all of the detailed decisions should be anticipated when making the initial broad ones. However, the operational difficulty of this is immense. Quite often it is necessary to explore some alternatives in considerable detail before intelligent broad decisions can be made. The more innovative the solution or unique the problem, the more this is required.

It is not unusual to allow for some alternatives in the final design, that is, to present the final design with some parts of it stated in terms of a range of choice or in terms of general performance requirements rather than in the form of specific chosen elements. This is most feasible when the alternatives do not significantly affect the building design in general. Even when alternatives are not provided for, builders and suppliers frequently request them, and the designer must reconsider the choices and the decision criteria.

Simply knowing all the reasonable alternatives is in itself a considerable task. Information about building materials and products is not disseminated in a way that makes uniform, objective evaluations for comparison easy to perform. There is no single, well-organized information source that presents the various alternatives in a comprehensive, impartial way. Consequently, knowing alternatives is usually a fragmentary, imperfect thing, highly dependent on the personal experience and particular resources of the individual designer.

Assuming that a designer is able to know of a reasonable range of possible alternatives, the task of choosing between them must still be faced. Ideally, this calls for

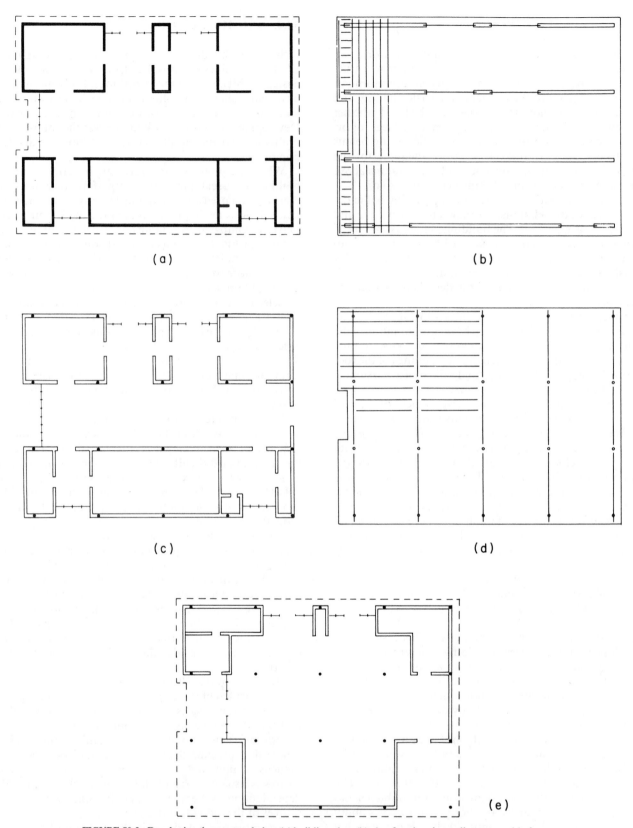

FIGURE 50.2. Developing the structural plan: (a) building plan; (b) plan for a bearing wall structure; (c) plan for a column and frame system; (d) roof-framing plan; (e) alternate architectural plan with the framed structure.

some system of evaluation of characteristics, including considerations of cost, time, fire behavior, energy use, installation problems, and so on. For all of the reasons discussed previously, this information is difficult to obtain.

A major aspect of this problem is simply that we have a highly dynamic society. We continually create new situations and problems for designers, produce new materials and products, shift our priorities (from dollar cost to pollution to energy use, etc), and generally keep two steps ahead of anyone trying to organize the design work. Any effort to deal with this problem requires the recognition of its essential dynamic nature.

As design work progresses from initial broad decisions to increasingly detailed ones, there is the everpresent possibility that previous decisions may need to be reconsidered. The farther back this reaches, the more it may disturb the progress of the work. This places a great pressure on the earliest, broadest decisions. For this reason, most people consider the preliminary design to be the most sensitive activity, requiring the most sound, experience-based judgments.

50.5. ECONOMICS

Dealing with dollar cost is a very difficult, but necessary, part of structural design. For the structure itself, the bottom-line cost is the delivered cost of the finished structure, usually measured in units of dollars per square foot of the building. For individual components, such as a single wall, units may be used in other forms. The individual cost factors or components, such as cost of materials, labor, transportation, installation, testing, and inspection, must be aggregated to produce a single unit cost for the entire structure.

Designing for control of the cost of the structure is only one aspect of the design problem, however. The more meaningful cost is that for the entire building construction. It is possible that certain cost-saving efforts applied to the structure may result in increases of cost for other parts of the construction. A common example is that of the floor structure for multistory buildings. Efficiency of floor beams occurs with the generous provision of beam depth in proportion to the span. However, adding inches to beam depths with the unchanging need for dimensions required for floor and ceiling construction and installation of ducts and lighting elements means increasing the floor-to-floor distance and the overall height of the building. The resulting increases in cost for the added building skin, interior walls, elevators, piping, ducts, stairs, and so on, may well offset the small savings in cost of the beams. The really effective cost-reducing structure is often one that produces major savings of nonstructural costs, in some cases at the expense of less structural efficiency.

Real-cost figures can only be determined by those who deliver the completed construction. Estimates of cost are most reliable in the form of actual offers or bids for the construction work. The farther the cost estimator is from the actual requirement to deliver the goods, the more speculative the estimate.

Designers, unless they are in the actual employ of the builder, must base any cost estimates on educated guesswork deriving from some comparison with similar work recently done in the same region. This kind of guessing must be adjusted for the most recent developments in terms of the local markets, competitiveness of builders and suppliers, and the general state of the economy. Then the four best guesses are placed in a hat, and one is drawn out.

Serious cost estimating requires a lot of training and experience and an ongoing source of reliable, timely information. For major projects various sources are available in the form of publications or computer data banks.

The following are some general rules for efforts that can be made in the structural design work in order to have an overall, general cost-saving attitude.

1. Reduction of material volume is usually a means of reducing cost. However, unit prices for different grades must be noted. Higher grades of steel or wood may be proportionally more expensive than the higher stress values they represent; more volume of cheaper material may be less expensive.

2. Use of standard, commonly stocked products is usually a cost savings, as special sizes or shapes may be premium prices. Wood 2×3 studs may be higher in price than 2×4s since the 2×4 is so widely used and bought in large quantities.

3. Reduction in the complexity of systems is usually a cost savings. Simplicity in purchasing, handling, managing of inventory, and so on, will be reflected in lower bids as builders anticipate simpler tasks. Use of the fewest number of different grades of materials, sizes of fasteners, and other such variables is as important as the fewest number of different parts. This is especially true for any assemblage done on the building site; large inventories may not be a problem in a factory, but usually are on the site.

4. Cost reduction is usually achieved when materials, products, and construction methods are highly familiar to local builders and construction workers. If real alternatives exist, choice of the "usual" one is the best course.

5. Do not guess at cost factors; use real experience, yours or others. Costs vary locally, by job size, and over time. Keep up to date with cost information.

6. In general, labor cost is greater than material cost. Labor for building forms, installing reinforcement, pouring the concrete, and finishing concrete surfaces is *the* major cost factor for sitecast concrete. Savings in these areas are much more significant than saving of material volume.

7. For buildings of an investment nature, time is money. Speed of construction may be a major advantage. However, getting the structure up fast is

not a true advantage unless the other aspects of the construction can take advantage of the time gained. Steel frames often go up quickly, only to stand around and rust while the rest of the work catches up.

50.6. COMPUTER-AIDED DESIGN

Two products of modern technology have had tremendous influence on the working style of building designers: the computer and the pocket calculator. In a short time, these have transformed the level of complexity and the accuracy of mathematical computation possible for even the most routine work. Through the use of prepared routines and computer programs, minimally trained designers can utilize highly sophisticated analyses and design procedures.

Computer Software

Computers—and various supportive equipment and materials that surround them—have been in a state of marketing hype for a long time. The frenzy has resulted in the development of some useful materials relating to work in various fields. Obviously, the larger the market (in terms of potential buyers), the more that market has been exploited and is now rich with available support equipment and materials.

Building design work does not generally represent a major market for just about anything—textbooks, computer equipment and software, or whatever—and thus the buildup of useful items has been slower. This is sure to change, as upcoming students and young working professionals are increasingly computer conversant. Some offices are already heavily involved, equipped, and operative with all that can be done with computers. However, of the total amount of work done in building design, only a small fraction is presently supported directly by computers. This is not true of all areas or stages of design; thus the engineering work in building design is generally far ahead of many other aspects of the whole building design field. This has significance to the engineering work itself, but the total impact on buildings will not be realized until the whole building design team is computer interactive.

For the field of structural engineering there is a considerable existing inventory of software available from various sources. Most of the major industry organizations and professional groups (AISC, ACI, AIA, ASCE, etc.) have either directly developed, or are somehow involved with the development of, computer data storage and computer-aided design software systems. Inquiry for information about currently available materials can be directed to those organizations. Private producers and marketers of software also have materials, but the limited market precludes a broad coverage. Individuals and small groups (schools, individual offices, small professional groups, etc.) have pet systems that they use and want to share. A lot is going on, but it is not easy to assess or to take full advantage of it.

In time (possibly by the time this book is published) it will become relatively easy to find out what can be done in terms of computer-aided design work to obtain the materials most useful for particular tasks and to learn how to use the materials productively. If the current level of competition continues, it is also likely that tooling up with the equipment, as well as the software, will become quite painless.

There is no shortcut to finding out about all of this, and plugging into it requires a commitment to doing whatever you can to get involved—learning about computers and how to use them.

50.7. DESIGN LOADS

Dead Load

Dead load consists of the weight of the materials of which the building is constructed such as walls, partitions, columns, framing, floors, roofs, and ceilings. In the design of a beam, the dead load must include an allowance for the weight of the beam itself. Table C.1. in Appendix C, which lists the weights of many construction elements, may be used in the computation of dead loads. Dead loads are due to gravity, and they result in downward vertical forces.

Dead load is generally a permanent load once the building construction is completed, unless frequent remodeling or rearrangement of the construction occurs. Because of this permanent, long-time character, the dead load requires certain considerations in design, such as the following:

1. It is always included in design loading combinations, except for investigations of singular effects, such as deflections due to only live load.
2. Its long-time character has some special effects causing sag and requiring reduction of design stresses in wood structures, producing creep effects in concrete structures, and so on.
3. It contributes some unique responses, such as the stabilizing effects that resist uplift and overturn due to wind forces.

Building Code Requirements

Structural design of buildings is most directly controlled by building codes, which are the general basis for the granting of building permits—the legal permission required for construction. Building codes (and permit-granting process) are administered by some unit of government: city, county, or state. Most building codes, however, are based on some model code, of which—as discussed in Sec. 50.2—there are three widely used in the United States: *UBC*, *BOCA* and *SSBC*.

Model codes are more similar than different, and are in turn largely derived from the same basic data and standard reference sources, including many industry standards. In the several model codes and many city, county, and state codes, however, there are some items that reflect particular regional concerns.

With respect to control of structures, all codes have materials (all essentially the same) that relate to the following issues:

1. *Minimum Required Live Loads.* Examples of these are Tables 23-A and 23-B in Appendix D, which are reproduced from the *UBC.*
2. *Wind Loads.* These are highly regional in character with respect to concern for local windstorm conditions. Model codes provide data with variability on the basis of geographic zones.
3. *Seismic (Earthquake) Effects.* These are also regional with predominant concerns in the western states.
4. *Load Duration.* Loads or design stresses are often modified on the basis of the time span of the load, varying from the life of the structure for dead load to a fraction of a second for a wind gust or a single major seismic shock. Some applications are illustrated in the work in the design examples in this part.
5. *Load Combinations.* These were formerly mostly left to the discretion of designers, but are now quite commonly stipulated in codes, mostly because of the increasing use of ultimate strength design and the use of factored loads.
6. *Design Data for Types of Structures.* These deal with basic materials (wood, steel, concrete, masonry, etc.), specific structures (towers, balconies, pole structures, etc.), and special problems (foundations, retaining walls, stairs, etc.). Industry-wide standards and common practices are generally recognized, but local codes may reflect particular local experience or attitudes. Minimal structural safety is the general basis, and some specified limits may result in questionably adequate performances (bouncy floors, cracked plaster, etc.).
7. *Fire Resistance.* For the structure, there are two basic concerns: structural collapse and the containment of the fire to control its spread. These concerns produce limits on the choice of materials (e.g., combustible or noncombustible) and some details of the construction (cover on reinforcement in concrete, fire insulation for steel beams, etc.).

The work in the design examples in this part is based largely on criteria from the *UBC.* The choice of this model code reflects only the fact of the degree of familiarity of the author with specific codes.

Live Loads

Live loads technically include all the nonpermanent loadings that can occur, in addition to the dead loads. However, the term as commonly used usually refers only to the vertical gravity loadings on roof and floor surfaces. These loads occur in combination with the dead loads, but are generally random in character and must be dealt with as potential contributors to various loading combinations.

Roof Loads

In addition to the dead loads they support, roofs are designed for a uniformly distributed live load that includes snow accumulation and the general loadings that occur during construction and maintenance of the roof. Snow loads are based on local snowfalls and are specified by local building codes.

Table 23-C from the *UBC* (see Appendix D) gives minimum roof liveload requirements. Note the adjustments for roof slope and for the total area or roof surface supported by a structural element. The latter accounts for the increase in probability of the lack of total surface loading as the size of the surface area increases.

Roof surfaces must also be designed for wind pressure, for which the magnitude and manner of application are specified by local building codes based on local wind histories. For very light roof construction, a critical problem is sometimes that of the upward (suction) effect of the wind, which may exceed the dead load and result in a net upward lifting force.

Although the term *flat roof* is often used, there is generally no such thing; all roofs must be designed for some water drainage. The minimum required pitch is usually $\frac{1}{4}$ in./ft, or a slope of approximately 1:50. With roof surfaces that are close to flat, a potential problem is that of *ponding*, a phenomenon in which the weight of the water on the surface causes deflection of the supporting structure, which in turn allows for more water accumulation (in a pond), causing more deflection, and so on, resulting in an accelerated collapse condition.

Floor Loads

The live load on a floor represents the probable effects created by the occupancy. It includes the weights of human occupants, furniture, equipment, stored materials, and so on. All building codes provide minimum live loads to be used in the design of buildings for various occupancies. Since there is a lack of uniformity among different codes in specifying live loads, the local code should always be used. Tables 23-A and 23-B from the *UBC* (see Appendix D) contain values for floor live loads.

Although expressed as uniform loads, code-required values are usually established large enough to account for ordinary concentrations that occur. For offices, parking garages, and some other occupancies, codes often require the consideration of a specified concentrated load as well as the distributed loading. Where buildings are to contain heavy machinery, stored materials, or other contents of unusual weight, these must be provided for individually in the design of the structure.

When structural framing members support large areas, most codes allow some reduction in the total live load to be used for design. These reductions, in the case of roof loads, are incorporated into the data in *UBC* Table 23-C. The following is the method given in the *UBC* for determining the reduction permitted for beams, trusses, or columns that support large floor areas.

Except for floors in places of assembly (theaters, etc.), and except for live loads greater than 100 psf [4.79 kN/m²], the design live load on a member may be reduced in accordance with the formula

$$R = 0.08 (A - 150)$$
$$[R = 0.86 (A - 14)]$$

The reduction shall not exceed 40% for horizontal members or for vertical members receiving load from one level only, 60% for other vertical members, nor *R* as determined by the formula

$$R = 23.1 \left(1 + \frac{D}{L}\right)$$

In these formulas

R = reduction, in percent

A = area of floor supported by a member

D = unit dead load/sq ft of supported area

L = unit live load/sq ft of supported area

In office buildings and certain other building types, partitions may not be permanently fixed in location but may be erected or moved from one position to another in accordance with the requirements of the occupants. In order to provide for this flexibility, it is customary to require an allowance of 15–20 psf [0.72–0.96 kN/m²], which is usually added to other dead loads.

CHAPTER FIFTY-ONE

Building One: Wood Structure

Building One is a simple box: a single-story, flat-roofed, rectangular building. The general configuration of the building and some architectural details for an all-wood structure are shown in Figures 51.1 and 51.2.

Two schemes for the structure are presented in Chapter 51, both using the same basic construction for the exterior walls. Scheme l uses the long interior walls between building units as bearing and shear walls; thus reducing the roof span required to the 25-ft distance of the wall spacings. This allows for the use of ordinary 2-by rafters and an ordinary plywood deck for the roof structure; by far, the simplest roof framing.

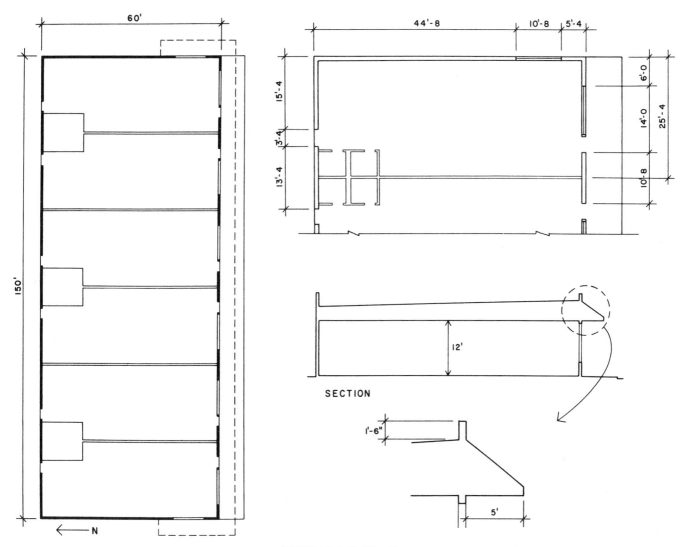

FIGURE 51.1. Building One.

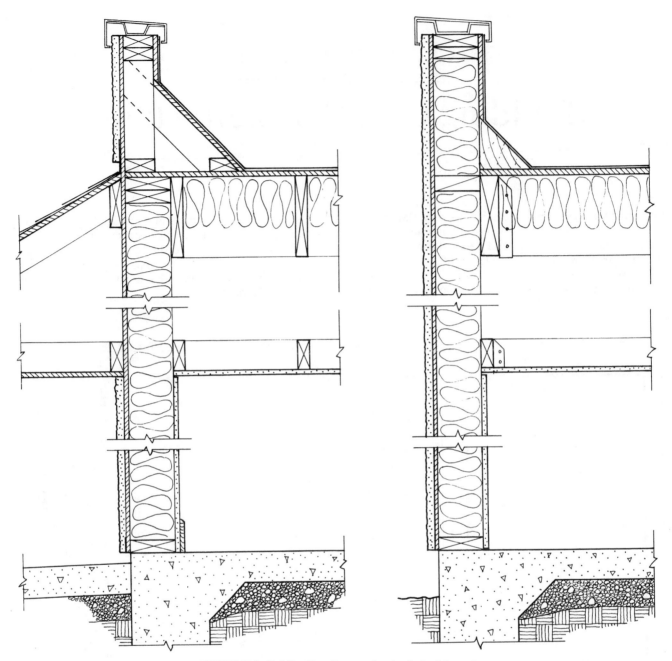

FIGURE 51.2. Building One: Construction details for Scheme 1.

Scheme 2 uses composite wood and steel trusses to develop a clear-spanning structure, which eliminates the need for any vertical structure on the building interior. Where this freedom of the interior space is significant to architectural planning, this is undoubtedly the easiest way to achieve the flat-topped, 60-ft clear span required. Composite trusses of this kind, or open web steel joists, as illustrated for the scheme in Chapter 53, are widely available from manufacturers and can achieve much greater spans than that shown here with relative ease. Still, it is generally *not* more economical to achieve longer spans versus shorter ones, and some real need for the clear-spanned space must usually justify such a choice.

51.1. GENERAL CONSIDERATIONS FOR BUILDING ONE

The following will be used for the structural design work.

Building code: 1991 *UBC* (Ref. 1).

Live loads:

Roof: 20 psf [0.96 KPa], *UBC* Table 23-C.

Wind: map speed of 80 mph, exposure B.

Seismic: zone 4, $Z = 0.4$.

Soil capacity: 2000 psf for shallow-bearing footings.

For the design of the roof we will assume construction loads as follows.

Roofing: built-up membrane at 6.5 psf [0.31 kPa].

Suspended ceiling: 10 psf [0.48 kPa].

Insulation, lights, ducts, etc.: 5 psf [0.24 kPa].

For the structure we will assume the use of the following materials.

For the light wood frame, structural lumber of Douglas fir–larch of the following grades:

2-by and 3-by, No. 2.

4-by and larger, No. 1.

Studs, Stud grade.

Wall sheathing and roof deck of Douglas fir plywood, C–D or Structural II grade.

Glue-laminated timbers of Douglas fir, 2400 psi stress grade.

Structural steel, A36 with F_y = 36 ksi [250 MPa].

Concrete, f_c' = 3 ksi [20.7 MPa].

For the development of the roof and ceiling construction, it is assumed that there is an all-air HVAC system, using ducts and registers incorporated in the ceiling space and some rooftop equipment. Roof drainage is to be accomplished by sloping the roof surface to the rear of the building and draining through scuppers in the parapet to vertical leaders on the rear wall.

51.2. SCHEME 1: DESIGN FOR GRAVITY LOADS

Roof Structure

For this scheme, the primary structural elements for the support of the roof are the roof deck and rafters, the wall studs, and the wall footings. Some special framing is required to achieve the wall openings, which probably involves the use of beams as headers above the openings and columns at the sides of the openings. The simplest form for headers is a structural lumber member with a width equal to the broad dimension of the wall studs and a depth as required for the span of the opening. Columns can most likely be achieved with doubled studs; a form used ordinarily at wall ends and edges of openings.

Selection of roof decking and wall sheathing involves many considerations for the full development of the constructions for the roof, exterior walls, and interior walls. All surfaces could be achieved with plywood, but many other surfacing materials are available and increasingly likely to be used. Some alternatives are:

Roof Deck. Wood fiber products of flake board or oriented strand board.

Exterior Wall Surface. Same as roof, plus possibly stucco without sheathing, applied directly to the studs.

Interior Walls. Gypsum drywall, plaster, or wood-fiber products.

For the roof deck the primary concerns are for the span distance between rafters, the resistance to the horizontal diaphragm shear stresses for lateral forces, and the attachments necessary to develop the roof covering and insulation. The latter usually requires a minimum of $\frac{1}{2}$-in.-thick plywood for the flat roof (now $\frac{15}{32}$ technically, although the difference is not measurable). If blocking is used for the plywood panel edges that do not fall on rafters, the 4 ft × 8 ft sheets are usually placed with the face grain parallel to the rafters to reduce the amount of blocking. In this case, with rafters not more than 24 in. on center, the $\frac{1}{2}$-in. plywood will be adequate, even in the cross-grain span of the panels.

A partial plan for the layout of the roof framing for Scheme 1 is shown in Fig. 51.3. With the $\frac{1}{2}$-in. plywood and the construction described earlier, the total roof dead load without the rafters becomes approximately 23 psf. For this situation, rafters could be selected from the tables in *UBC*, Chapter 25. A problem in this is that of finding the proper table, with the correct combination of the design live load (20 psf) and the actual dead load (23 psf). From the tables in Appendix D, it may be noted that neither Table 25-U-R-1 nor 25-U-R-2 quite matches up. However, a good guess can be made by comparing the answers from both tables. We thus observe:

Table 25-U-R-1 has a dead load 8 psf lower than ours, so the answer produced is on the skimpy edge.

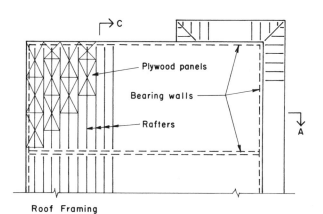

Roof Framing

Ground Floor and Foundations

FIGURE 51.3. Building One, Scheme 1: partial plan of roof framing.

Table 25-UR-2 has a total load (DL+LL) 2 psf larger than ours and a live load 50% greater. Since deflection under live load is often the limiting condition for maximum spans, the answer from this table is quite conservative.

For the No. 2 Douglas fir–larch rafters, the allowable bending stress for repetitive stress members is 1450 psi and the modulus of elasticity is 1,700,000 psi. (See Chapter 16, Table 16.1.) With the usual increase of 25% in the allowable stress for the roof live load condition, the design bending stress is increased to 1.25(1450) = 1812 psi. From *UBC* Table 25-U-R-1, we observe that 2 × 12s at 16 in. can span 28 ft 6 in.; and from Table 25-U-R-2, that they can span 25 ft 2 in. with a maximum stress of 1800 psi. In both cases, the required values for the modulus of elasticity indicated in the tables are less than that for the No. 2 rafters, so deflection is not a critical concern. We thus note that this is a reasonable choice for the rafters, although a routine check of bending stress and deflection could be performed to verify the choice, if somebody challenges it.

Lateral bracing must be provided for the 2 × 12 rafters in the form of bridging or blocking. If the blocking as described earlier for the nailing of the plywood is provided, this requirement will be quite satisfactorily fulfilled. If an unblocked decking is used, the requirement for lateral bracing must be otherwise considered.

This span is roughly at the cutoff point for the feasibility of using 2-by nominal lumber rafters. For any longer spans, it becomes necessary to use very closely spaced rafters, lumber deeper than the nominal 12 in., or something thicker than the nominal 2 in. At this point it is undoubtedly advisable to switch to other options, probably involving some form of special product, such as vertically laminated members, composite I-shaped members, or light prefabricated trusses with wood chords. All of the latter can be obtained in depths greater than 12 in., making design for reduced deflections more feasible.

Bearing Walls

With the 2 × 12 rafters at 16-in. centers, the total roof dead load including the rafters becomes approximately 27 psf, including a slight increase for assumed use of blocking. The total load on the interior walls thus becomes:

$$(25 \text{ ft}) + (27 + 20) = 1175 \text{ lb/ft of wall length}$$

This load can easily be borne by 2 × 4 studs at 16-in. centers, if the wall height is not too great. *UBC* Table 25-R-3 recommends a maximum height of 10 ft, although an investigation will probably show that the 2 × 4 studs could extend to 14 ft in this case. These walls, however, are major dividing walls between tenant spaces, and their general sturdiness and/or desired acoustic separation function may indicate some different choices. The minimum typical partition with 2 × 4 studs and gypsum drywall on both sides is really quite flimsy as a serious dividing wall.

Adding the roof load to the weight of the wall still produces a load of less than 2000 lb/ft at the bottom of the wall.

Although a wall footing should be provided for these bearing walls, its width will be quite minimal in this case, even with the very low allowable soil pressure of 2000 psf.

The exterior walls on the long sides of the building (parallel to the rafters) carry mostly only their own dead weight and that of the edge of the roof. Those on the short sides, however, carry these loads plus a half-span of the rafters. In addition to the gravity loads, the exterior stud walls must function to resist the direct wind pressure or lateral seismic force due to their own weights. The design of these studs therefore requires consideration for the combined axial compression and bending condition. This is taken up in Sec. 51.3, with the general discussion of design for wind forces.

51.3. SCHEME 1: DESIGN FOR WIND

Design for wind on Building One, Scheme 1, involves the following considerations.

Uplift on the roof: critical for the attachment of the deck, and for the attachment of the roof framing to supports. The latter is fundamentally a problem only if the combined dead loads on the framing is less than the wind uplift force. Use of ordinary sheet metal connectors for the framing will usually provide for sufficient anchorage for all but extreme wind conditions in high-risk windstorm areas.

Horizontal pressure on the vertical wall surfaces: critical for the selection of glazing, design of window framing, and design of the studs for the bending effect combined with the axial compression due to gravity loads.

Action of the roof as a horizontal diaphragm: not likely to be critical here if the interior walls are used as shear walls, cutting the diaphragm span to only 25 ft. The construction joint at the top of the interior walls must be studied for the necessary load transfer from the roof.

Action of individual wall sections as shear walls: involving shear in the wall sheathing, overturn, sliding, and chord forces in the framing.

Design of the Exterior Wall Studs

The end walls of the building carry the ends of the 25-ft span rafters or purlins. Because of the sloping roof surface, it is easier for construction to make these studs continuous to the top of the parapet and to carry the rafters or purlins on a ledger bolted to the face of the studs. The unbraced height of the studs is thus the distance from the floor to the bottom of the ledger. With the roof slope, this distance varies 15 in. from front to rear of the building. To attain the desired 12-ft clear ceiling height under the deep girder, the bottom of the rafters at the front of the building will be at approximately 15.5 ft above the floor. For this height the *UBC* requires that the studs be 2 × 6.

The studs should be checked for the combined compression plus bending due to the wind load and gravity load.

Gravity load equals:

> Roof:
>   12.5 ft (20 + 27) = 588 lb/ft of wall
> Canopy:
>   Assume   = 100
> Total DL + LL   = 688 lb/ft [10 kN/m]

or

> 688(1.33) = 915 lb/stud at 16-in. centers [4.07 kN]

For the 15.5-ft high stud, $h/t = 15.5(12)/5.5 = 33.8$,

$$F_c' = \frac{0.3E}{(h/t)^2} = \frac{0.3(1,700,000)}{(33.8)^2} = 446 \text{ psi } [3.08 \text{ MPa}]$$

Without the wind load, the allowable load per stud is thus

> load = 446(8.25) = 3680 lb [16.4 kN]

As discussed in Sec. 45.3, the design wind pressure on the wall is determined as

$$p = C_e C_q q_s$$

where

> C_e for the assumed exposure = 0.7
> C_q for the wall = 1.2
> q_s for the wind speed of 80 mph = 17 psf

Thus

$$p = 0.7(1.2)(17) = 14.28 \text{ psf} \qquad \text{say, 15 psf } [0.72 \text{ kPa}]$$

With the 16-in.-on-center studs spanning vertically as simple beams for their 15.5 ft height, the moment due to wind is thus

$$M = \frac{wL^2}{8} = \frac{15(16/12)(15.5)^2}{8} = 599 \text{ ft-lb } [0.812 \text{ kN-m}]$$

We thus consider the interaction of the column and beam effects as follows, noting that stresses may be increased by one third with the wind load included.

$$\frac{P/A}{F_c'} + \frac{M/S}{F_b - P/A}$$

where

> P/A = 915/8.25 = 111 psi [765 kPa]
> F_c' = 1.33(466) = 593 psi [4089 kPa]
> M/S = 599(12)/7.56 = 951 psi [6557 kPa]
> $F_b - (P/A)$ = 1.33(1450 − 111) = 1781 psi [12280 kPa]

and the interaction is considered as

$$\frac{111}{593} + \frac{951}{1781} = 0.187 + 0.534 = 0.721 < 1$$

Had the combined action values exceeded 1, it would be necessary to increase the size or stress grade of the studs or reduce the spacing to 12 in.

Note that the wall is also subjected to an outward suction pressure due to wind. However, the C_q factor is lower for this effect and the combined actions are similar. Thus the column action combined with inward pressure is the critical action for the stud. The suction effect should be considered, however, in developing the details for construction of the wall and for the attachment of windows and doors. Glazing in the windows should also be installed to resist both inward and outward forces.

Because the canopy is cantilevered from the wall, the studs should also be checked for this load unless the cantilever forces are carried directly back into the roof construction with struts and ties. Since we are not detailing the canopy construction, we will assume the 2 × 6s to be adequate for this condition.

We have checked the heaviest loaded stud so that we may safely use the 2 × 6s for the other walls. Actually, 2 × 4s can probably be used for the shorter rear wall.

51.4.   SCHEME 2: DESIGN FOR GRAVITY LOADS

A partial framing plan and some details for a modified wood roof structure are shown in Fig. 51.4. This structure uses light, composite wood and steel trusses to achieve a clear span of the 60-ft-wide building, eliminating all interior structural supports. Choice for such a structure would most likely be motivated by an architectural design decision relating to the desire for flexibility in future changes of interior space arrangements. For commercial buildings with investment concerns for speculative rental, this is more often than not a major goal.

The partial framing plan in Fig. 51.4 shows the layout for this system, indicating the spacing of the trusses and the need for lateral bracing of the laterally unstable trusses. For practical purposes, the lateral bracing also functions to assure the accurate spacing, vertical straightness, and longitudinal straightness of the quite flexible trusses. The

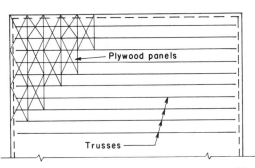

FIGURE 51.4. Building One, Scheme 2: partial plan of roof framing.

trussed system is quite sturdy when properly erected, but the correct bracing of the system is essential to its stability.

These trusses will undoubtedly be the proprietary products of some manufacturer, and such products are quite widely available. The trusses shown in the details in Fig. 51.4 are of a form of product made by the TrussJoist Co., a major distributor of these and other building products. Safe load tables, prepared by the manufacturer, may be used for the selection of the trusses. Major considerations are for the nominal depth of the trusses and the sizes of individual members—in particular, the top and bottom wood chords. The chords used by this manufacturer are laminated in continuous form from thin wood strips.

Selection of the truss spacing relates to not only the load capacity of the trusses, but also to the developments of the roof decking and ceiling construction. A 2-ft spacing would permit use of ordinary plywood panels for the roof deck and gypsum drywall for the ceiling; both surfaces being developed by direct nailing to the truss chords. As shown in the drawings in Figures 51.1 and 51.4, the truss chords may be fabricated in a nonparallel manner to achieve the sloping roof surface and the flat ceiling. This is undoubtedly the simplest and most economical form of construction.

However, the 2-ft spacing is a bit close for this span, so the scheme shown here uses a 32-in. spacing (still a module of the 96-in. standard panel) with a $\frac{7}{8}$-in.-thick plywood deck. These panels are available with tongue-and-groove edges, which permit the practical development of an unblocked deck. Need for some blocking, however, relates to the development of the roof deck as a horizontal diaphragm, which is taken up in Sec. 51.5.

The long-span trusses generate a significant vertical load for the stud walls on the long sides of the building. The combination of this vertical loading plus the direct wind load on the wall surface requires a design for bending plus axial compression for the studs. The conditions for this and the other major elements of the structure are summarized in Fig. 51.5. Example computations for these elements are developed in other parts of the book, as indicated in Fig. 51.5.

An alternative for this system would be to use slightly heavier trusses at even wider spacing with a plywood deck supported on small joists spanning between the trusses.

1. Typical roof deck, plywood panels, 32" span, face grain perpendicular to supports.

   UBC Table 25-S-1, No. 7, 40/20 panel, 7/8" thick, allowable span of 32" if unblocked

2. Joist (truss), at 32" centers, 60 ft span (+ or -)

   LL at 20 psf (32/12) = 53.3 lb/ft

   DL: roofing + insul. + deck + ceiling + equip., say 25 psf
   25(32/12) = 66.7 lb/ft + weight of joist

   Total load = 53.3 + 66.7 = 120 lb/ft

   Table 25.1: use 30K7, weight = 12.3 lb/ft
   capacity = 141 - 12.3 = 128.7 lb/ft
   LL capacity = 69 lb/ft
   actual DL = 66.7 + 12.3 = 79 lb/ft
   average DL of joists, 12.3(12/32) = 4.6 psf
   total roof DL = approximately 30 psf, including joists

3. Front bearing wall, 15 ft high, with 5 ft canopy

   DL = roof at (30 X 30) =   900 lb/ft   LL = 20(30 + 5) = 700 lb/ft
   wall at (20 X 15) =   300
   canopy, say   100   Total load: 2000 lb/ft
   Total DL =   1300 lb/ft

   See example calculations in Sec. 51.3. No. 2 grade 2 X 6 studs can carry 3680 lb/ft. Gravity thus uses only 2000/3680, or about 55% of the stud capacity. Adding the wind load (even with LL) and increasing allowable stresses by one third is probably still OK. If not, use No. 1 grade or place studs at 12" centers in solid portions of wall.

   Design individual wall posts and headers at openings for the wall loads plus wind loads.

4. Wall footing, front wall.

   Wall load of 2000 + weight of foundation wall & footing, say 3000 lb/ft

   Table 33.4: could safely use 36" wide by 10" thick footing with soil pressure of only 1500 psf. Or design a smaller footing, if allowable pressure is significantly higher.

FIGURE 51.5. Building One, Scheme 2: summary of design for gravity.

This system would more likely be used if no ceiling is required.

51.5. DESIGN FOR LATERAL FORCES

Lateral forces include those developed by wind and seismic effects. Design for wind is based on the requirements of the 1991 edition of the *UBC* (Ref. 1) using the basic wind speed of 80 mph. Seismic requirements are also taken from the *UBC* assuming a seismic risk zone 4 condition. It is quite common, when designing for both wind and seismic forces, to have some parts of the structure designed for wind and others for seismic effects. What is required is to investigate for both effects and to design each element of the structure for the condition that produces the greater effect. Thus the shear walls may be designed for seismic effects, the exterior wall framing and window glazing for wind, and so on.

Design for Wind

Primary design for wind consists of consideration of the wind acting on the whole building and developing forces in the general lateral bracing system. Wind must be investigated in terms of a horizontal force that may come from any direction. For a very complex building form, investigation may be required for several different directions of wind. For this building it should be sufficient to consider only the two major directions: north–south and east–west.

Using the projected area method, as described in Sec. 45.3, we first determine the design wind pressure, specified by the *UBC* as

$$p = C_e C_q q_s I$$

where

C_e is found from *UBC* Table 23-G to be 0.7 for the assumed exposure condition B and the vertical height zone from 0 to 20 ft above grade.

C_q is that for method 2 (projected area) and for the condition of investigation of the primary frame and the determination of horizontal pressure on the building visualized as a vertically projected area $C_q = 1.3$ from 0 to 40 ft above grade.

q_s is the wind stagnation pressure at the standard measuring height of 30 ft. From the *UBC* the q_s value for a wind speed of 80 mph is 17 psf.

I is the importance factor (*UBC* Table 23-L) for which we use a value of 1.0.

The design pressure is thus

$$p = C_e C_q q_s I = 0.7(1.3)(17)(1) = 15.47 \text{ psf}$$

We will round this off to 16 psf [0.77 kPa] for use in the wind analysis. This is the total wind force in a horizontal direction to be resisted by the building structure as a whole. It is visualized as applied to a single exterior wall, but the individual building surfaces (walls and roof) must be designed for other pressures.

We will assume the building exterior walls to be 18 ft high from grade to the top of the parapet. The roof deck, which horizontally braces the walls, is at a varying elevation due to the slope of the surface. We will assume the average height of the roof deck to be 15 ft. The wall is thus visualized, as shown in Fig. 51.6, as spanning between the floor and the roof deck. The wind pressure on the upper 10.5 ft of the wall is resisted by the roof deck and the pressure on the lower 7.5 ft is resisted by the concrete floor slab. The total wind load delivered to the roof deck with the wind blowing north is thus

$$F = (16 \text{ psf})(10.5 \text{ ft})(150 \text{ ft}) = 25,200 \text{ lb [112 kN]}$$

We will not proceed further with the design for wind, as it will be found later that the seismic load is greater. We will therefore illustrate the design of the roof deck diaphragm and shear walls using the seismic load.

In addition to the design for the total wind force on the building, individual elements of the building surface must be investigated for wind effects. The code gives criteria for this in terms of pressures for various surfaces. Some considerations to be made are the following.

1. Design of the stud walls for the combined effect of vertical gravity load and bending due to wind pressure on the wall.

2. Design of the roof for vertical uplift if it exceeds the gravity dead load.

3. Design of large panes of the window glazing for wind pressure.

Design for Seismic Force

Design for the seismic forces on the building includes the following considerations:

1. Design of the roof diaphragm for forces in both directions.

2. Design of the vertical shear walls.

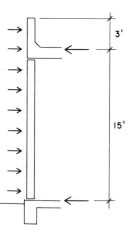

FIGURE 51.6. Assumed action of the exterior wall in resisting wind.

3. Development of the various construction details for transfer of the forces from the horizontal to the vertical diaphragms and for transfer of the forces from the shear walls to the foundations.

A critical preliminary decision is the identification of the walls to be used as shear walls. Some of the considerations in this decision are:

1. The actual magnitude of force that the walls in each direction must resist. A quick estimate of the load should be made to determine this.
2. Which walls lend themselves to being used. This has to do with their plan location, their length, and the type of bracing used. The code establishes maximum height-to-length ratios for various bracing: up to $3\frac{1}{2}$:1 for plywood, $1\frac{1}{2}$:1 for plaster or drywall.
3. What materials are planned for the wall surfaces, which may be used for their shear resistance, and what materials or bracing can be added where the surfacing materials are not adequate.

We assume that the ordinary construction to be used is drywall on the interior and cement plaster (stucco) on the exterior of the walls. This means that only one of these surfacings can be used for the exterior walls because the code does not permit addition of dissimilar materials [*UBC* Sec. 4714(a)]. Where the stucco is not sufficient, we add plywood to the wall, ignoring the surfacing materials. Where plywood is not required for the full length of a wall it sometimes simplifies the detailing if it is added to the interior, rather than to the exterior, of the wall.

Figure 51.7 shows the proposed layout of the shear wall system. For load in the short direction the roof will span from end to end of the building, transferring the shear to the two 45-ft-long end walls. For the load in the long direction the five 10-ft-long walls will be used on the front and the entire wall will be used on the rear, that consists of a net wall length of 130 ft. The latter will result in some eccentricity between the load and the centroid of the walls in the long direction, thus requiring an investigation of the torsional effect.

Design of the Roof Diaphragm

The dead loads to be used for the seismic force to the roof diaphragm are shown in Table 51.1. Wall loads are taken as the weight of the upper half of the walls, ignoring openings that are generally in the bottom portion. In each direction the wall loads considered are only those of the walls perpendicular to the load direction. The canopy load is assumed to be taken by the roof in both directions. A nominal load is assumed for rooftop HVAC units.

The total load in both directions is reasonably symmetrically placed. The canopy load on the front wall is offset by the toilet walls and the heavier rear wall in the long direction loading. The seismic design load is determined as (see *UBC* 2312)

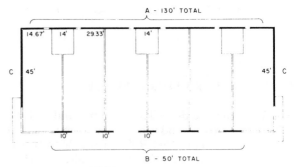

THE SHEAR WALL SYSTEM

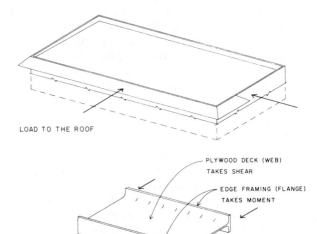

LOAD TO THE ROOF

PLYWOOD DECK (WEB) TAKES SHEAR

EDGE FRAMING (FLANGE) TAKES MOMENT

BEAM ANALOGY

FIGURE 51.7. The lateral resistive system.

$$V = \frac{ZIC}{R_W} W$$

where

Z = 0.40 (for zone 4, *UBC* Table 23-I)

I = 1 (*UBC* Table 23-L, no special qualification)

C = 2.75 [maximum required from *UBC* Sec. 2312(e)2A, with S not determined]

R_W = 8 (*UBC* Table 23-O, plywood shear wall)

Thus

$$V = \frac{(0.4)(1)(2.75)}{8} = 0.1375\,W, \quad \text{say } 0.14\,W$$

Using the loads from Table 51.1, we therefore determine the seismic loads to the roof diaphragm as

V = 0.14(329.4)

  = 46.1 kips in the N–S direction [205 kN]

V = 0.14(316.8)

  = 44.4 kips in the E–W direction [198 kN]

TABLE 51.1. Loads to the Roof Diaphragm

	Loads (kips)	
Load Source and Calculation	N–S	E–W
Roof dead load: 150 × 60 × 27 psf	243	243
East and west exterior walls: ½ × 60 × 17 × 20 psf × 2	0	20.4
North and south exterior walls: ½ × 150 × 17 × 20 psf × 2	51	0
Interior dividing walls: ½ × 60 × 15 × 8 psf × 5	0	18
Toilet walls: ½ × 190 × 15 × 8 psf	11.4	11.4
Canopy: 190 ft × 100 lb/ft	19	19
Rooftop HVAC units (estimate)	5	.5
Total load (W for seismic calculation)	329.4	316.8

If the load in the north–south direction is considered to be uniformly distributed and symmetrically placed, the maximum shear in the deck will be one-half of the total load, and the unit shear in the roof plywood is

$$v = \frac{\text{maximum shear}}{\text{diaphragm width}} = \frac{46,100/2}{60}$$

$$= 384 \text{ lb/ft } [5.6 \text{ kN/m}]$$

For the membrane roofing it is usually required to have a minimum plywood thickness of $\frac{1}{2}$ in. Assuming the use of the 2× rafters, structural II grade plywood, and 8d nails, *UBC* Table 25-J-1 (see Appendix D) yields the following requirements for the nail spacing.

Nails at 2.5 in. at the diaphragm boundary and at continuous panel edges parallel to the load.
Nails at 4 in. at other panel edges.

In the center of the plywood panels other code requirements usually ask for nails at a minimum of 12 in. along supports. This is usually called *field nailing* or *nailing at intermediate supports*.

We will assume the use of a rafter and purlin system for which the layout of the plywood panels will be Case 5 as shown in the footnotes for *UBC* Table 25-J-1 (see Appendix D). For this situation we would most likely specify the required boundary nailing for all panel edges.

If the seismic load is assumed to be uniformly distributed along the length of the single-span diaphragm, the magnitude of the shear will diminish continuously from the maximum value at the end to zero at the center of the span. It is therefore possible to consider a reduction in the amount of nailing for portions of the diaphragm nearer the center of the building. An example of this so-called zoned nailing is shown in Fig 51.8. We have used only the values for a blocked diaphragm in the example. It would be pos-

sible to consider the use of an unblocked diaphragm for the center portion of the building where the shear is lowest.

For the chord force (see Fig. 47.5) we first find the maximum moment for the spanning diaphragm. Then, assuming the chords to act with a moment arm equal to the width of the building, the analysis is as follows:

$$M = \frac{WL}{8}$$

chord force: $T = C = \dfrac{M}{d}$

where *d* is the building width. Combining these yields

$$T = C = \frac{WL}{8d} = \frac{46,100(150)}{8(60)}$$

$$= 14,400 \text{ lb } [65 \text{ kN}]$$

If the double top plate of the stud wall is used for the chord (see Fig. 51.11), an analysis will show that this is too much force for the member comprised of two 2 × 6s. Alternatives are to raise the usually low stress grade of the plates (Douglas fir–larch, No. 2) or to increase the size to 3 × 6.

Because it is not possible to use continuous members for the top plates (150 ft long), the splicing of the plates must be done in a manner that preserves the tension capacity of the chord. For low chord forces the usual practice of staggering the splices of the two plate members and the minimum required nailing between them (see *UBC* Table 25-P) may be sufficient. For a force of the magnitude of that

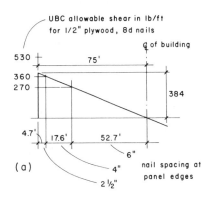

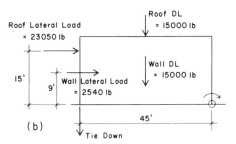

FIGURE 51.8. Investigation of the lateral resistive elements; (a) zoned nailing for the roof; (b) stability of the end shear wall.

required for our example, it will be necessary that the splicing use steel straps or steel bolts.

If the studs run full height to the top of the parapet, the edge of the roof deck would be nailed directly to a ledger fastened to the face of the studs (see the construction of the end walls in Sec. 51.6). If the ledger serves as the chord, it must be designed for the chord forces and its splicing must be developed to preserve the tension capacity required.

The load in the east–west direction will produce less stress in the deck and is therefore not critical for the selection of the deck or its nailing. The maximum stress at the deck boundary is

$$v = \frac{44,400/2}{150} = 148 \text{ lb/ft } [2.16 \text{ kN/m}]$$

This is less than the minimum capacity produced by using the minimum nailing (8d at 6 in.), but the load represented must be collected along the wall. If the deck is nailed to supporting header beams over large openings, the forces collected by these members must be transferred into the shear walls. Development of the construction detailing must include these considerations.

Design of the Shear Walls

The shear walls carry the edge shear forces from the roof diaphragm. In addition, they resist the full force of the wall weight in the plane of the shear wall, for this was not included in the load to the roof. The end walls thus resist the load of 23.05 kips from the roof plus the following:

Wall weight:

$$20 \text{ psf} \times 17 \text{ ft} \times 45 \text{ ft} = 15,300 \text{ lb}$$
$$20 \text{ psf} \times 7 \text{ ft} \times 15 \text{ ft} = 2,100$$
$$5 \text{ psf} \times 10 \text{ ft} \times 15 \text{ ft} = \underline{750}$$
$$\text{Total} = 18,150 \text{ lb } [76.3 \text{ kN}]$$

Lateral load:

$$0.14 \, W = 0.14 \times 18,150 = 2540 \text{ lb } [11.3 \text{ kN}]$$

The loading of the end shear wall is thus as shown in Fig. 51.8b. The total shear force is 25,590 lb [113.8 kN] and the unit shear in the wall is

$$v = \frac{25,590}{45} = 569 \text{ lb/ft } [8.30 \text{ kN/m}]$$

From UBC Table 25-K (see Appendix D) this requires $\frac{1}{2}$-in. structural II plywood with 10d nails at 3-in. centers at all panel edges. A footnote to the table requires the use of 3× studs and the staggering of the nails for this nailing.

The overturn analysis is:

Overturn M:

$$23.05 \times 15 = 346 \text{ kip-ft}$$
$$+ \; 2.54 \times 9 = \underline{23}$$
$$\text{Total} = 369 \text{ kip-ft } [500 \text{ kN-m}]$$

Dead load M:

$$30 \times 22.5 = 675 \text{ kip-ft } [915 \text{ kN-m}]$$

Safety factor:

$$\frac{\text{dead load } M}{\text{overturn } M} = \frac{675}{369} = 1.83$$

This would generally be considered to be adequate, for the usual requirement is for a minimum safety factor of 1.5. If the safety factor is less than 1.5, a tiedown anchor, as shown in Fig. 51.8b, must be used.

At the base of the wall the total horizontal sliding force must be transferred to the foundation through the wall sill anchor bolts. Using Table 25-F from the UBC and assuming a $2\frac{1}{2}$-in.-thick sill, we determine the following: For $\frac{1}{2}$-in. bolts:

$$\text{allowable load per bolt} = 1.33(630)$$
$$= 838 \text{ lb } [3.7 \text{ kN}]$$
$$\text{number required} = \frac{25,590}{838} = 30.5 \text{ or } 31$$

This requires bolts at approximately 18-in. centers, assuming the first bolt to be 12 in. from the end of the wall. For $\frac{3}{4}$-in. bolts:

$$\text{allowable load per bolt} = 1.33(1115)$$
$$= 1483 \text{ lb } [6.6 \text{ kN}]$$
$$\text{number required} = \frac{25,590}{1483} = 17.2$$

at approximately 2 ft 4 in. on center.

In most cases the builders would prefer the smaller number of bolts.

In the east–west direction there is a lack of symmetry in the disposition of the shear walls (the north and south building exterior walls). There are two approaches to the analysis for the seismic force on these walls, as follows:

1. *Analysis by Peripheral Distribution.* In this analysis it is assumed that the roof acts as a simple beam and one-half of the total lateral load is thus delivered to each wall. Thus the shear stresses in the walls would be

$$v = \frac{22,200}{130} = 171 \text{ lb/ft } [2.49 \text{ kN/m}] \text{ for the north wall}$$

$$v = \frac{22,200}{50} = 444 \text{ lb/ft } [6.48 \text{ kN/m}] \text{ for the south wall}$$

The length of wall used for each calculation is simply the total length of shear wall in each of the building walls.

2. *Analysis for Torsion.* In this analysis the north and south walls are considered in terms of their respective proportional stiffnesses, with stiffness of the plywood walls assumed to be proportionate to the wall plan lengths. Stress in the walls is considered as the sum of the direct stress plus the stress due to torsion as follows:

direct stress $= \dfrac{\text{total lateral force}}{\text{sum of all wall lengths}}$

$= \dfrac{44,400}{180}$

direct stress $= 247$ lb/ft [3.6 kN/m]

Referring to Fig. 51.9a, the torsional analysis follows. For the center of stiffness:

$y = \dfrac{50(60)}{180}$

$= 16.67$ ft from the rear wall

The torsional moment of inertia is as shown in Table 51.2, and the torsional stress on the front wall is

$v = \dfrac{Tc}{J} = \dfrac{[44,400(13.33)]43.33}{636,250}$

$= 40$ lb/ft [0.59 kN/m]

The total stress is thus

$v = 247 + 40 = 287$ lb/ft [4.19 kN/m]

In theory, the torsional shear is negative for the rear wall; that is, the true shear will be the direct shear minus the torsional shear. However, the usual practice is not to

TABLE 51.2. Torsional Moment of Inertia of the Shear Walls

Wall	Length (ft)	Distance from Center of Stiffness (ft)	$J = L(d)^2$
A	130	16.67	36,126
B	50	43.33	93,874
C	2(45)	75	506,250
Total J for the shear walls			636,250

deduct but only to add the torsional effect. Thus for the rear wall we use only the direct shared stress of 247 lb/ft.

For the most conservative design we may use the peripheral method for the lateral force in the front walls and the torsional method for the rear walls. If this is done, we consider

$v = 444$ lb/ft for the front wall

$v = 247$ lb/ft for the rear wall

As with the end shear walls, the weight of the walls in the plane of the shear walls should be added to the loads for the design of the individual piers. We have not done so, as these are quite minor loads in this structure. The procedure is the same as that illustrated for the end wall.

For the rear wall UBC Table 25-K permits the use of $\frac{3}{8}$-in. plywood with 8d nails at the maximum spacing of 6 in. Note that the footnote to Table 25-K permits an increase of 20% in the table values for $\frac{3}{8}$-in. plywood when studs are 16 in. on center.

For the overturn analysis of the rear wall we consider two approaches as shown in Fig. 51.9b. The first is to regard it as a series of independent piers linked together. For overturn analysis these piers would be considered to have a height from the sill to the roof deck level. The other option is to regard the wall as a continuous diaphragm with piers having a height equal to the door opening height. The latter option results in considerably less overturn, but requires some additional framing and tying to reinforce the wall at the openings.

In referring to Fig. 51.9b, the overturn analysis for the short end pier is:

Lateral load:
247 × 14 = 3,458 lb

Overturn M:
3458 × 14 × 1.5 SF = 72,618 ft-lb

DL M:
6300 × 7 = 44,100

Net M for hold-down = 28,518 ft-lb

Required T: $\dfrac{28,518}{14}$ = 2,037 lb.

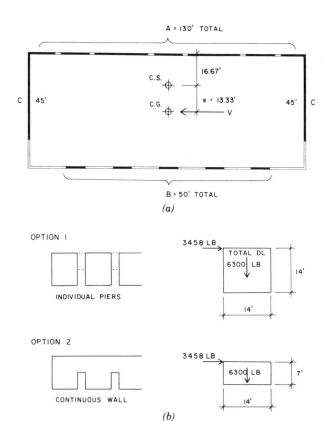

FIGURE 51.9. Torsional effect of the east–west load: (a) development of the torsional moment; (b) options for the rear shear walls.

This option would therefore require a hold-down device with an ultimate resistance of 2037 lb or more. For the other option, the analysis is

Overturn M: 3458(7)(1.5) = 36,309 ft-lb.

Because the DL moment is the same, observe that no hold-down is required; that is, the actual safety factor is greater than 1.5 (actually, 1.82 if analyzed as for the end wall).

For the front wall the *UBC* permits a $\frac{3}{8}$-in. plywood with 8d nails at 2.5 in. or $\frac{1}{2}$-in. with 10d nails at 4 in. Overturn for this wall will not be critical because the end reaction of the header delivers a considerable dead load at the ends of the walls. Since it is not possible to tilt the wall without lifting the header, the situation is quite safe.

The overturn forces on these walls must be transmitted to, and resisted by, the foundations. This requires that there be sufficient dead weight in the grade wall and footings and some bending and shear resistance by the grade wall. Assuming the depth of grade wall as shown in the construction drawings, this resistance can be developed with minimal top and bottom continuous reinforcing. If the grade wall is quite shallow, this problem should be carefully investigated.

Obviously, some reconsideration of the building details could reduce the requirements for lateral load resistance. Use of a lighter roofing and a lighter ceiling material would considerably reduce the roof dead load and consequently the lateral force. Use of one or more permanent cross walls in the interior would reduce the stresses in the roof deck and the end shear walls and eliminate some of the large girders.

51.6. CONSTRUCTION DETAILS FOR THE WOOD STRUCTURE

The drawings that follow show the basic construction details for the wood structure. The drawings are essentially intended to show the structural details and are not all fully complete as architectural details. In many instances there are equivalent alternatives for achieving the basic structural tasks, which could be more intelligently evaluated if all information about finish materials and architectural details were known. Some details may also be affected by considerations of the design of the lighting, electrical power, HVAC, and plumbing systems or by problems of security, acoustics, fire ratings, and so on.

Structural Plans

Note that the roof framing system used is Scheme 1. If Scheme 1 with only the rafters is used, some of the wall details would change.

Foundation and roof framing layouts are shown in the partial plans in Fig. 51.3. Detail sections shown on the plans are illustrated in Fig. 51.10 through 51.11. The following is a discussion of some of the considerations made in developing these details.

Construction Details

Detail A (Fig. 51.10).   Detail A shows the canopy, parapet, shear wall, and roof at the front of the building. Depending on the height of the parapet and the location of the top of the canopy, it may be advisable to run the wall studs continuously to the top of the parapet. This would permit the top of the canopy to be higher than the roof deck. In any event, if the top of the canopy is not exactly at the level of the roof deck, as shown in Fig.51.10, additional framing would be required for the anchoring of the tie straps.

In the detail shown, both the roof deck and wall sheathing are nailed directly to the top plate of the wall. This achieves a direct transfer of load from the horizontal to the vertical diaphragm. If the wall studs were continuous to the top of the parapet, a ledger would be provided at the face of the studs to support the rafters and provide for the edge nailing of the plywood. Transfer of the roof seismic load to the wall would then require the addition of blocking in the wall. This condition is illustrated in detail C, Fig. 51.10, in which the load transfer is from the roof plywood, through boundary nailing to the ledger, then from the ledger to the blocking, and finally from the blocking to the wall plywood.

At the point at which the bottom of the canopy kicks into the wall the strut shown may be required to brace the studs. This may not be required at the solid portion of the wall, but is most likely required at openings, unless the header is designed for the combined vertical and horizontal loads.

If the parapet is simply built on top of the roof deck, as shown, the diagonal struts may be used to brace the canopy and form the cant at the roof edge.

Detail B (Fig. 51.10).   Depending on the level of the exterior grade, the drainage situation and the wall finish materials, the sill plate may be simply put directly on the floor slab, as shown, or may be placed on top of a short curb to raise it above the floor level.

The details of the wall footing, grade wall, and floor slab are subject to considerable variation. Some considerations are:

1. *Depth of the Footing Below Grade.*   For frost protection or simply to reach adequate bearing soil it may be necessary to have a deep grade wall. At some point this requires that it be treated as a true vertical wall with vertical reinforcement, a separate footing with dowels, and so on. In this case there would likely be three concrete pours with a cold joint between the footing, wall, and slab.

2. *Need for Thermal Insulation for the Floor.*   In cold climates it is desirable to provide a thermal break between the slab and the cold ground at the building edge. This would be done by placing insulation on the inside of the

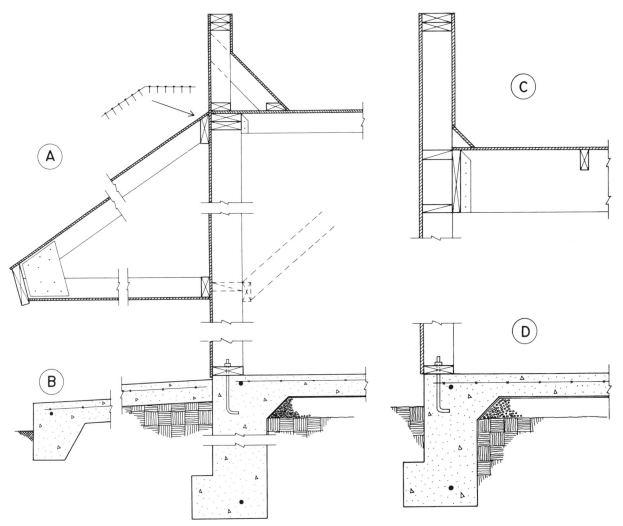

FIGURE 51.10. Construction details.

grade wall and/or by placing it under the slab along the wall.

3. *Cohesive Nature of the Sill.* If the soil at grade level is reasonably cohesive (just about anything but clean sand), and a shallow grade wall is possible, the footing, wall, and slab can be poured in a single pour. In this case the wall and footing are formed by simply trenching and providing a form for the outside wall surface and slab edge.

4. *Beam Action of the Grade Wall.* Because of expansive soil, highly varying soil bearing conditions, or the use of the grade wall for distribution of concentrated forces from posts, tiedowns, and so on, it may be necessary to provide top and bottom reinforcing. In any event, a minimum of one bar in the top and one in the bottom should be used for shrinkage and temperature stresses.

Detail C (Fig. 51.10). Detail C shows the end wall condition where the roof purlins are supported by the wall. In this detail the wall studs are continuous to the top of the parapet and a ledger is provided to which the joist hangers for the purlins are attached. Because of the roof slope, the elevation of the purlins varies. This is simply accommodated by sloping the ledger, whereas if the wall were as in detail A, all the studs would be different in length and the top plate of the wall would have to be sloped. Then the parapet studs would also be all different in length to achieve the level parapet top.

As discussed for detail A, the seismic load must be transferred from the roof to the wall by the circuitous route through the ledger and blocking. The vertical roof loading must also be transferred from the ledger to the studs. For a more positive tie it is recommended that the ledger be attached to the studs with lag screws. Two lag screws in

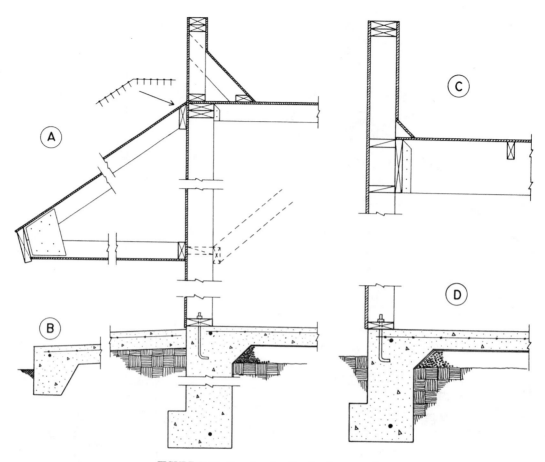

FIGURE 51.11. Framing elevation for the rear wall.

each stud and one in each block would probably provide all the load transfers necessary.

Detail D (Fig. 51.10). This is essentially the same condition as detail B except for the absence of the exterior paving slab and the posts. Depending on the level of the exterior grade and the exterior wall finish, it may be necessary to raise the sill on a curb, as previously discussed.

Detail E (Fig. 51.11). Detail E is a partial elevation of the framing of the front wall with the plywood removed. Shown are the following:

1. Support of the window header by a connection attached to the face of the wall end post, which permits the post to be continuous to the top of the parapet.
2. Strapping of the end of the header to blocking in the wall, which reinforces the opening corner and allows the shear wall action as shown in Fig. 51.8.
3. Splicing of the top plate with bolts as described in the calculations. For ease of construction it would probably be desirable to oversize the top plate member and recess the bolt heads to clear the roof plywood.

CHAPTER FIFTY-TWO

Building One, Scheme 3: Masonry and Timber Structure

Although the predetermination of the structure may well affect many details of the building form and general plan, it is usually possible to achieve a given building form with various construction systems. This scheme uses different materials to achieve the same basic building form that was developed as a light wood frame structure in Chapter 51.

52.1. GENERAL CONSIDERATIONS

Figure 52.1 shows the plan layouts for a structure consisting of exterior walls of structural masonry with CMUs (concrete blocks) supporting a wood roof with timber elements. To relate to the usual 16-in. length of the CMUs, the

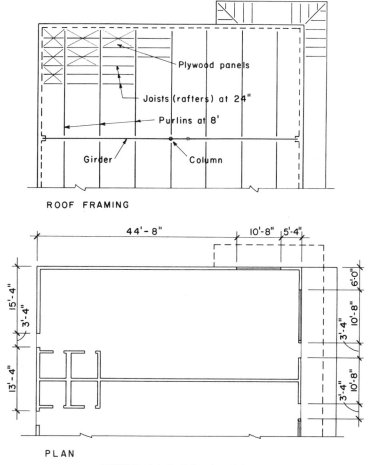

FIGURE 52.1. Building One, Scheme 3.

plan dimensions have been modified slightly, as shown in the figure.

The wood roof structure uses a single row of interior columns to support solid-sawn wood timber beams, which in turn support a panel system with timber purlins, light wood joists, and a plywood deck.

The following data and criteria are used.

Building Code: 1991 *UBC* (Ref. 1).

Design Loads:
  Roof: 20 psf [0.96 Kpa], *UBC* Table 23-C.
  Wind: map speed 80 mph [145 km/h], exposure B.
  Seismic: zone 4.

Soil capacity: 2000 psf [96 kPa].

Concrete: sand and gravel aggregate, $f_c' = 3$ ksi [20.7 MPa].

Masonry: reinforced CMU masonry, CMUs of medium weight, grade N, ASTM C90, $f_c' = 1350$ psi [9.3 MPa], mortar type S.

Reinforcement for concrete and masonry: grade 40, $f_y = 40$ ksi [276 MPa].

Timber of Douglas fir–larch, No. 1 grade.

Joists of Douglas fir–larch, No. 2 grade.

Plywood: Douglas fir, C-D face grade, structural type as required for gravity and lateral design.

52.2. DESIGN OF THE ROOF STRUCTURE

The roof is laid out to relate to the efficient use of 4 ft × 8 ft plywood panels; which also works for the 16-in. CMU module. The joists are thus 24 in. on center and the timber purlins are 8 ft on center. This generally provides for nailing at all plywood panel edges without the use of additional blocking.

Plywood Deck

Placement of the plywood panels involves their use for spanning in a direction perpendicular to the face grain. This is not optimal usage, but an inspection of *UBC* Table 25-S-2 (Appendix D) shows that $\frac{15}{32}$-in.-thick panels are adequate for the 24 in. span. This thickness is generally the minimum acceptable for application of roofing materials on a near-flat roof surface.

Further consideration of the plywood is deferred until the requirements for development of the roof as a horizontal diaphragm as part of the lateral resistive systems are studied.

Joists

The deck is assumed to carry only the roof live load plus its own dead weight and that of the roofing and insulation materials. The joists will probably carry the same load plus their own weight; however, they might also be used to support some suspended loads as well. On the latter basis, the joist load is considered to be as follows.

Live load: 20 psf
Dead load:

Roofing and insulation	= 6 psf
Ducts, lights, and ceiling	= 6.5
Plywood deck	= 1.5
Joists (assumed)	= 1.0
Total dead load	= 15 psf

Investigation of the joists for this span and loading condition can be achieved using Table 25-U-R-1 in the *UBC* (Appendix D). This will indicate that the smallest member in the table, a 2 × 6, is more than adequate. In fact, computations will show a 2 × 4 to be sufficient.

Purlins

The purlins span 24 ft and carry a total load that includes the joist load plus the purlin weight. At 8 ft on center, the purlins support a total peripheral area of $8 \times 24 = 192$ ft², and thus require a roof live load of 20 psf (see *UBC* Table 23-C, Appendix D). The superimposed uniformly distributed load on the beams is thus 8 times the unit live load plus dead load, or $8(20+15) = 280$ lb/ft.

Design of the beams is summarized in Fig. 52.2, using procedures described in Chapter 17.

Girder

The continuous girders carry their own weight as a uniformly distributed load plus the end reactions of the beams as concentrated loads at 8-ft centers. With the total girder length of 60 ft, it is necessary to have a splice point.

Figure 52.3 shows two possibilities for placement of splice points. The off-column placement is intended to retain some advantage of the potential for continuity for the multispan girder, as discussed in Sec. 14.4. Choice of

---

Purlins 8 ft on center, 24 ft clear span

LL = 8(20) = 160 lb/ft

DL = 8(15) = 120 + weight of purlin, say 140 lb/ft

Total load = 300 lb/ft

Maximum $M = wL^2/8 = (300)(24)^2/8 = 21{,}600$ ft-lb

UBC allowable stress (Table 16.1): 1300 psi
Times 1.25 for load duration (Table 16.2): 1.25(1300) = 1625 psi

Required $S = M/F = 21{,}600(12)/1625 = 160$ in³

From Appendix A, Table A.2: 6 × 14, S = 167 in³, A = 74.25², I = 1127 in⁴, weight = 18 lb/ft

Maximum shear = 300(24/2) = 3600 lb

Maximum shear stress = 1.5(V/A) = 1.5(3600/74.25) = 72.7 psi

Allowable shear = 1.25(85) = 106 psi; shear not critical

From Fig. 17.1, approx. LL deflection = 0.8 in.
This is L/360, so is probably not critical.

FIGURE 52.2. Building One, Scheme 3: summary of computations for the beams.

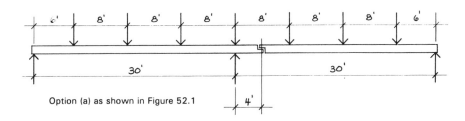

Option (a) as shown in Figure 52.1

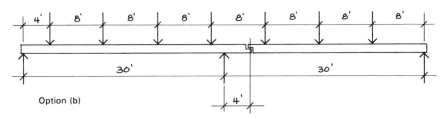

Option (b)

FIGURE 52.3. Options for the girder splices.

this scheme depends to some extent on the maximum feasible length of a single piece of the girder. This dimension is mostly up to the fabricators and erectors of the girder, but a reasonable length for this relatively modest-sized girder is probably around 40 ft. On this basis, the splice joint scheme in Fig. 52.3*b* seems acceptable.

Investigation of the bending moments and shears in the girder is shown in Fig. 52.4. Based on these values, the design of the girder is given in the summary in Fig. 52.5. For the concentrated loads, the beam end reactions are computed as follows.

> Live load: 12 psf, from *UBC* Table 23-C (Appendix D), based on the girder periphery of $25 \times 30 = 750$ ft².
> Purlin load = $8(12 + 15) = 216$ lb/ft + purlin, say 235 lb/ft
> Girder load = $24(235) = 5640$ lb, say 5600 lb at 8-ft centers.

Column

Assuming the girder to weight approximately 50 lb/ft, the load for the column and footing is

$4(5600) + 30(50) = 22,400 + 1500 = 23,900$ lb, say 24 kips

Column choices for this load and an unbraced height of 14 ft are:

Solid sawn wood column, Table 18.1, 8×8.
Steel pipe column, Table 23.3, standard 4-in. pipe.
Steel Tubular column, Table 23.4, $4 \times 4 \times \frac{3}{16}$ in.

Footing

Table 33.2 yields a load capacity of 30 kips for a 4-ft-square footing, 10 in. thick, reinforced with six No. 4 bars in each direction.

52.3.   DESIGN OF THE WALLS

We assume that the walls will consist of reinforced, hollow concrete blocks with finishes of stucco (cement plaster) on the exterior and gypsum drywall on wood furring strips on the interior. We assume this construction to weigh approximately 70 psf of wall surface.

The exterior walls must be designed for the following combinations of vertical gravity and lateral wind or seismic forces (see Fig. 52.6):

1. Gravity dead plus live loads.
2. Gravity vertical dead load plus bending due to lateral load.
3. Horizontal shear and overturn due to shear wall actions.

We first consider the long expanses of wall at the building ends and rear. For the end wall the laterally unsupported height varies because of roof slope. We assume it to be a maximum of 15 ft at the end of the solid wall portion nearest the front of the building. With an 8-in. block thickness, the maximum h/t of the wall is thus $[15(12)]/7.625 = 23.6$, which is just short of the usual limit of 27.

Assuming that code-required inspection is not provided during construction, the maximum stress for vertical compression is

$$F_a = 0.10f'_m \left[1 - \left(\frac{h}{42t} \right)^3 \right]$$

$$= 0.10(1350) \left[1 - \left(\frac{180}{42(7.625)} \right)^3 \right]$$

$$= 111 \text{ psi } [0.77 \text{ MPa}]$$

and the maximum allowable bending stress is

$$F_b = 0.166f'_m = 224 \text{ psi } [1.54 \text{ MPa}]$$

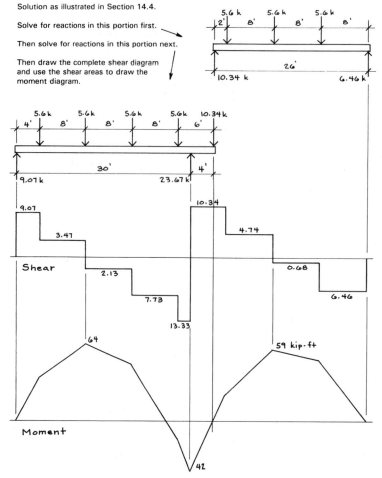

FIGURE 52.4. Investigation of the girder.

For a total wall height of 18 ft the wall dead load is $18 \times 70 = 1260$ lb/ft [18.4 kN/m]. Assuming the clear purlin span to be 24 ft, the loads from the purlins are

$$\text{dead load} = 12 \times 25 \text{ psf}$$
$$= 300 \text{ lb/ft } [4.38 \text{ kN/m}]$$
$$\text{live load} = 12 \times 20 \text{ psf}$$
$$= 240 \text{ lb/ft } [3.50 \text{ kN/m}]$$

The total gravity vertical load on the wall is thus 1800 lb/ft and the average net compression stress assuming the wall to be 65% solid is

$$f_a = \frac{P}{A} = \frac{1800}{0.65[12(7.625)]}$$
$$= 30.3 \text{ psi } [0.21 \text{ MPa}]$$

Assuming the purlins to be supported by a ledger that is bolted to the wall surface, the roof loading will cause a bending moment equal to the load times one half the wall thickness; thus

$$M = (540)\frac{7.625}{2}$$
$$= 2059 \text{ in.-lb per foot of wall length}$$

Using Fig. E.1, we find an approximate bending stress as follows.

Assume an average reinforcing with No. 5 bars at 40-in. centers. Thus

$$p = \frac{0.31(12/40)}{12(7.625)} = 0.001$$
$$np = 44(0.001) = 0.045$$
$$K = \frac{M}{bd^2} = \frac{2059}{12(3.813)^2} = 11.8$$

From the graph, $f_m = \pm 90$ psi $= f_b$ [0.62 MPa]. Then

$$\frac{f_a}{F_a} + \frac{f_b}{F_b} = \frac{30.2}{111} + \frac{90}{224} = 0.27 + 0.40 = 0.67$$

Because this is less than 1.0 the wall is adequate for the vertical gravity load alone. For the case of gravity plus lateral

From Figure 52.4:

Maximum moment = 64 kip-ft

Maximum shear = 13.33 kips

Then:

Required S = M/F = (64)(12,000)/(1.25)(1300) = 473 in³

Required A = 1.5(V/F) = (1.5)(13,333/1.25 × 85) = 188 in²

From Appendix A, Table A.2, area requirement is critical.

Choose: 10 X 22, A = 204 in²

Unless timber beams are cheap and readily available, it is probably smarter to use a steel or a glued laminated member for this girder.

FIGURE 52.5. Summary of the girder design.

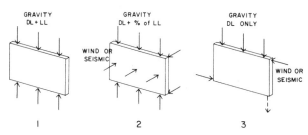

FIGURE 52.6. Load cases for the walls.

bending we must determine the maximum bending moment due to wind pressure or seismic load. For seismic action the code requires a force in the direction perpendicular to the wall surface equal to 30% of the wall weight, or 0.30 × 70 psf = 21 psf. Because this slightly exceeds the wind pressure of 15 psf, we use it for the bending. The wall spans the vertical distance of 15 ft from the floor to the roof. The dowelling of the reinforcing at the base plus the cantilever effect of the wall above the roof will reduce the positive moment at the wall midheight. We thus use an approximate moment for design of

$$M = \frac{qL^2}{10} = \frac{21(15)^2}{10} = 473 \text{ ft-lb } [0.64 \text{ kN-m}]$$

To this we add the moment due to the eccentricity of the roof dead load; thus

$$M = 300 \times \frac{7.625}{2(12)} = 95 \text{ ft-lb } [0.13 \text{ kN-m}]$$

and we now design for a total moment of 568 ft-lb [0.77 kN-m]. Assuming an approximate value of $j = 0.85$, we find that the required reinforcing is

$$A_s = \frac{M}{f_s jd} = \frac{0.568(12)}{[1.33(20)](0.85)(3.813)} = 0.079 \text{ in.}^2/\text{ft}$$

We try No. 5 at 32 in.

$$A_s = 0.31(12/32) = 0.116 \text{ in.}^2/\text{ft}$$

Then

$$p = \frac{A_s}{bd} = \frac{0.116}{12(3.813)} = 0.0025$$

$$np = 44(0.0025) = 0.112$$

$$K = \frac{M}{bd^2} = \frac{568(12)}{12(3.813)^2} = 39$$

From Fig. E.1, $f_m = \pm 240$ psi. For axial compression due to dead load only.

$$f_a = \frac{1560}{0.65[12(7.625)]} = 26.2 \text{ psi } [0.18 \text{ MPa}]$$

and

$$\frac{f_a}{F_a} + \frac{f_b}{F_b} = \frac{26.2}{111} + \frac{240}{224} = 0.24 + 1.07 = 1.31$$

This indicates a combination close to the limit of 1.33. However, the analysis is conservative because the axial stress used is actually that at the bottom of the wall and the tension resistance of the masonry is ignored.

The rear walls have less load from the roof and a slightly shorter unsupported height. For these walls it is possible that the minimum reinforcing required by the code is adequate. The code requirements are:

1. Minimum of 0.002 times the gross wall area in both directions (sum of the vertical and horizontal bars).
2. Minimum of 0.0007 times the gross area in either direction.
3. Maximum spacing of 48 in.
4. Minimum bar size of No. 3.
5. Minimum of one No. 4 or two No. 3 bars on all sides of openings.

With the No. 5 bars at 32 in., the gross percentage of vertical reinforcing is

$$p_g = \frac{0.116}{12(7.625)} = 0.00127$$

To satisfy the requirement for total reinforcing, it is thus necessary to have a minimum gross percentage for the horizontal reinforcing of

$$p_g = 0.002 - 0.00127 = 0.00073$$

which requires an area of

$$A_s = p_g \times A_g = 0.0073[12(7.625)] = 0.067 \text{ in.}^2/\text{ft}$$

This can be provided by

No. 4 at 32 in., $A_s = (0.20)\dfrac{12}{32} = 0.075 \text{ in.}^2/\text{ft}$.

No. 5 at 48 in., $A_s = (0.31)\dfrac{12}{48} = 0.0775 \text{ in.}^2/\text{ft}$.

Choice of this reinforcing must also satisfy the requirements for shear wall functions.

At the large wall openings the headers will transfer both vertical and horizontal loads to the ends of the supporting walls. The ends of these walls will be designed as reinforced masonry columns for this condition. Figure 52.7 shows the details and the loading condition for the header columns. In addition to this loading the columns are part of the wall and must carry some of the axial load and bending as previously determined for the typical wall.

Figure 52.8a shows a plan layout for the entire solid front wall section between the window openings. A pilaster column is provided for the support of the girder. Because of the stiffness of the column, it will tend to take a large share of the lateral load. We will thus assume the end column to take only a 2-ft strip of the wall lateral load. As shown in Fig. 52.8a, the end column is a doubly reinforced beam for the direct lateral load.

The gravity dead load on the header is:

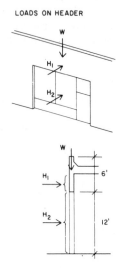

LOADS ON HEADER

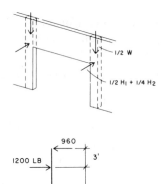

LOADS ON HEADER COLUMNS

FIGURE 52.7. Loads on the headers and columns.

Roof	= 100 lb/ft
Wall	= (70 psf)(6 ft)
	= 420 lb/ft
Canopy	= 100 lb/ft (assumed)
Total load	= 620 lb/ft [9.05 kN/m]

For the lateral load we will use the wind pressure of 15 psf because the weight of the window wall will produce a low seismic force. Assuming the window mullions span vertically, the wind loads are as shown in Fig. 52.7.

$$H_1 = (15\text{ psf})(2\text{ ft} \times 15\text{ ft}) = 450\text{ lb [2.0 kN]}$$

$$H_2 = (15\text{ psf})(12\text{ ft} \times 15\text{ ft}) = 2700\text{ lb [12.0 kN]}$$

Thus the column loads from the header are:

$$\text{vertical load} = (620\text{ plf})(15/2)$$
$$= 4650\text{ lb [20.7 kN]}$$

$$\text{horizontal load} = (\tfrac{1}{2}H_1 + \tfrac{1}{4}H_2)$$
$$= 225 + 675$$
$$= 900\text{ lb [4.0 kN]}$$

$$\text{moment} = 960(3)$$
$$= 2880\text{ ft-lb [3.9 kN-m] (see Fig. 52.7)}$$

For the direct wind load on the wall we assume a 15-ft vertical span and a 2-ft-wide strip of wall loading. Thus

$$M = \frac{wL^2}{8} = \frac{(15\text{ psf})(2)(15)^2}{8}$$

$$= 844\text{ ft-lb [1.15 kN-m]}$$

These two moments do not peak at the same point; thus without doing a more exact analysis we assume a maximum combined moment of 3600 ft-lb. Then, for the moment alone, assuming a j of 0.85,

$$\text{required } A_s = \frac{M}{f_s j d} = \frac{3.6(12)}{26.7(0.85)(5.9)}$$

$$= 0.32\text{ in.}^2 \text{ [208 mm}^2\text{]}$$

$$\text{approximate } f_m = \frac{M}{bd^2}\frac{2}{kj}$$

$$= \frac{3600(12)(2)}{(16)(5.9)^2(0.4)(0.85)}$$

$$= 457\text{ psi [3.15 MPa]}$$

Although f_m appears high, we have ignored the effect of the compressive reinforcing in the doubly reinforced member. The following is an approximate analysis based on the two-moment theory with two No. 5 bars on each side of the column.

For the front wall it is reasonable to consider the use of a fully grouted wall because the pilaster and the end columns already constitute a considerable solid mass. For the fully grouted wall we may use $f'_m = 1500$ psi, and the allowable bending stress thus increases to

$$F_b = 1.33 \times 0.166 \times 1500 = 331\text{ psi [2.28 MPa]}$$

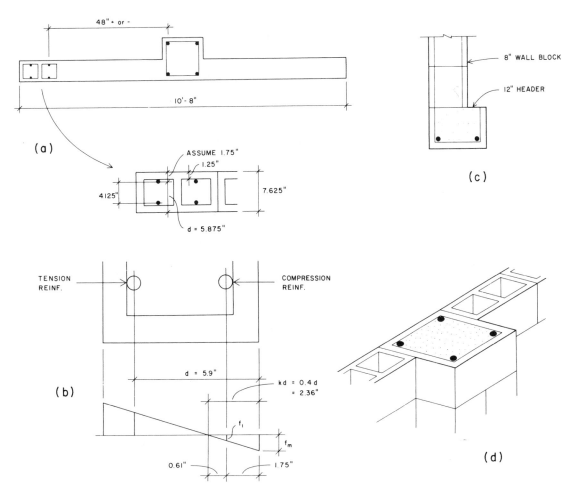

FIGURE 52.8. Details of the front wall.

Assuming the axial load to be almost negligible compared to the moment, we analyze for the full moment effect only. With a maximum stress of 331 psi we first determine the moment capacity with tension reinforcing only as

$$M_1 = \frac{f_m(bd^2)(k)(j)}{2}\left(\frac{1}{12}\right)$$

$$= \frac{331(16)(5.9)^2(0.4)(0.85)}{2(12)}$$

$$= 2612 \text{ ft-lb } [3.54 \text{ kN-m}]$$

This leaves a moment for the compressive reinforcing of

$$M_2 = 3600 - 2600$$

$$= 1000 \text{ ft-lb } [1.63 \text{ kN-m}]$$

If the compressive reinforcing is two No. 5 bars, then

$$f_s' = \frac{M_2}{A_s'(d - d')} = \frac{1000(12)}{0.62(4.125)}$$

$$= 4692 \text{ psi } [32.4 \text{ MPa}]$$

This is a reasonable stress even with the assumed low k value of 0.4. As shown in Fig. 52.8b, if k is 0.4 and f_m is 331, the compatible strain value for f_s' will be

$$f_s' = 2n(f_c) = 2(40)(3.31)\left(\frac{0.61}{2.36}\right)$$

$$= 6844 \text{ psi } [47.2 \text{ MPa}]$$

As shown by the preceding calculation, the stress in the tension reinforcing will not be critical. This approximate analysis indicates that the column is reasonably adequate for the moment. The axial load capacity should also be checked, using the procedure shown later for the pilaster design.

Window Header

As shown in Fig. 52.7, the header consists of a 6-ft-deep section of wall. This section will have continuous reinforcing at the top of the wall and at the bottom of the header. In addition there will be a continuous reinforced bond beam in the wall at the location of the steel ledger that supports the edge of the roof deck.

Using the loading previously determined and an approximate design moment of $wL^2/10$, the steel area required for gravity alone will be

$$A_s = \frac{M}{f_s(jd)}$$

where

$$M = wL^2/10 = 620(15)^2/10$$
$$= 13,950 \text{ ft-lb } [18.9 \text{ kN-m}]$$
$$d = \text{approximately 68 in. } [1.727 \text{ m}]$$

Then

$$A_s = \frac{13.95(12)}{20(0.85)(68)}$$
$$= 0.145 \text{ in.}^2 [94 \text{ mm}^2]$$

This indicates that the minimum reinforcing at the top of the wall may be two No. 3 bars or one No. 4 bar. This should be compared with the code requirement for minimum wall reinforcing. The *UBC* calls for a minimum of 0.0007 times the gross cross-sectional area of the wall in either direction and a sum of 0.002 times the gross cross-sectional area of the wall in both directions. Thus

$$\text{minimum } A_s = 0.0007(7.625)(12)$$
$$= 0.064 \text{ in.}^2/\text{ft of width or height}$$

with two No. 3 bars $A_s = 0.22$ in.^{2}

$$\text{required spacing} = \frac{0.22}{0.064}$$
$$= 3.44 \text{ ft} \quad \text{or} \quad 41.3 \text{ in.}$$

The minimum horizontal reinforcing would then be two No.3 bars at 40 in., or every fifth block course.

At the bottom of the header there is also a horizontal force consisting of the previously calculated wind load plus some force from the cantilevered canopy. Estimating this total horizontal force to be 250 lb/ft, we add a horizontal moment as

$$M = \frac{wL^2}{10} = \frac{0.25(15)^2}{10}$$
$$= 5.625 \text{ kip-ft } [7.63 \text{ kN-m}]$$

for which we require

$$A_s = \frac{M}{f_s(jd)} = \frac{5.625(12)}{26.7(0.85)(5.9)}$$
$$= 0.504 \text{ in.}^2 [325 \text{ mm}^2]$$

This must be added to the previous area required for the vertical gravity loads:

$$\text{total } A_s = 0.504 + \frac{\frac{1}{2}(0.145)}{1.33}$$
$$= 0.504 + 0.055$$
$$= 0.559 \text{ in.}^2 [361 \text{ mm}^2]$$

The requirement for vertical load is divided by two because it is shared by both bottom bars. It is divided by 1.33, since the previous calculation did not include the increase of allowable stresses for wind loading. If this total area is satisfied, the bottom bars in the header would have to be two No. 7s. An alternative would be to increase the width of the header at the bottom by using a 12-in.-wide block for the bottom course, as shown in Fig. 52.8d. This widened course would be made continuous in the wall.

The Pilaster-Column

To permit the wall construction to be continuous, the girder stops short of the inside of the wall and rests on the widened portion of the wall called a *pilaster*. As shown in Fig. 52.8d, the pilaster and wall together form a 16-in. square column. The principal gravity loading on the column is due to the end reaction of the girder. Since this load is eccentrically placed, it produces both axial force and bending on the column. The parapet, canopy, and column weight add to the axial compression.

Because of its increased stiffness, the column tends to take a considerable portion of the wind pressure on the solid portion of the wall. We will assume it to take a 6-ft-wide strip of this load. As shown in Fig. 52.9, the direct wind pressure on the wall (pushing inward on the outer surface) causes a bending moment of opposite sign from that due to the eccentric girder load. The critical wind load is therefore due to the outward wind pressure (suction force) on the wall. For a conservative design we will take this to be equal

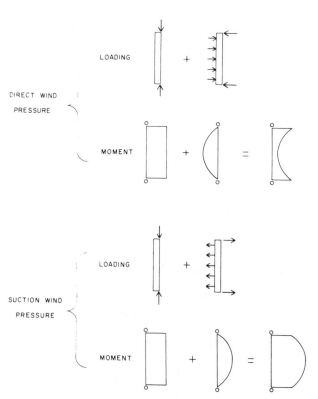

FIGURE 52.9. Load combinations for the pilaster column.

to the inward pressure of 15 psf. The combined moments are thus

$$\text{wind moment} = \frac{wL^2}{8} = \frac{15(6)(13.33)^2}{8}$$

$$= 2000 \text{ ft-lb } [2.71 \text{ kN-m}]$$

Assuming an eccentricity of 4 in. for the girder (see Fig. 53.3),

$$\text{girder moment} = \frac{23.5(4)}{12}$$

$$= 7.833 \text{ kip-ft} \quad \text{or} \quad 7833 \text{ ft-lb}$$
$$[10.62 \text{ kN-m}]$$

For the combined wind plus gravity loading we have used only half the live load. With the allowable stress increase, it should be apparent that this loading condition is not critical, so we will design for the gravity loads only. For this we will redetermine the girder-induced moment with full live load:

$$\text{girder } M = \frac{27.6(4)}{12}$$

$$= 9.2 \text{ kip-ft } [12.48 \text{ kN-m}]$$

The gravity loads of the canopy, parapet, roof edge, and column must be added. We will therefore assume a total vertical design load of approximately 35 kips. With this total load the equivalent eccentricity for design will be

$$e = \frac{M}{N} = \frac{9.2(12)}{35} = 3.15 \text{ in. } [80 \text{ mm}]$$

The *UBC* requires a minimum percentage of reinforcing of 0.005 of the gross column area. Thus

$$\text{minimum } A_s = 0.005(16)^2$$

$$= 1.28 \text{ in.}^2 \text{ } [826 \text{ mm}^2]$$

with four No. 7 bars $A_s = 2.40$ in^2 [1548 mm^2]. Then from the *UBC* the allowable axial load is

$$P = \frac{1}{2}(0.20f_m'A_e + 0.65A_sF_{sc}) \times \left[1 - \left(\frac{h'}{42t}\right)^3\right]$$

where

A_e = the total area of the fully grouted column
    $= (15.625)^2 = 244$ in.^{2}
$F_{sc} = 0.40F_y = 16$ ksi
h' = effective (unbraced) height of the column
    = 13.3 ft or 160 in.
t = column thickness

$$P = \frac{1}{2}[0.20(1.5)(244) + 0.65(2.40)(16)]$$

$$\times \left[1 - \left(\frac{160}{42(15.6)}\right)^3\right]$$

$$= 48.4 \text{ kips } [215 \text{ kN}]$$

Ignoring the compression steel, the approximate moment capacity is

$$M = A_sf_s(jd) = \frac{1.20(20)(0.85)(13.5)}{12}$$

$$= 22.95 \text{ kip-ft } [31.1 \text{ kN-m}]$$

and, for the combined effect

$$\frac{\text{actual } P}{\text{allowable } P} + \frac{\text{actual } M}{\text{allowable } M}$$

$$= \frac{27.6}{48.4} + \frac{9.2}{22.95} = 0.55 + 0.40 = 0.95$$

Note that the $\frac{1}{2}$ factor has been used with the axial load formula, assuming no special inspection during construction. Although a more exact analysis should be performed, this indicates generally that the column is reasonably adequate for the axial load and moment previously determined.

The Wall Foundations

The foundations for this structure will be essentially similar to those for the wood structure. Continuous wall footings will be provided under all the exterior walls except at the columns. The same options described for Scheme 2 are possible for the column footing.

For the end walls the load is:

Roof:
    45.5(12.5) = 569 plf
Wall:
    80(18) = 1440 plf
Grade wall and footing
    = 300 plf (estimate)
Total load = 2309 plf [33.7 kN/m]
Width required: 2309/2000 = 1.15 ft or
    14 in. [350 mm].

At the front wall the column load and header loads are carried by the solid wall portion. If the same scheme used in the previous structure is desired, we would provide a 12-ft-long footing for this total load.

Girder end reaction	= 27.6 kips
Roof edge load:	
3(25)(37.5 psf)	= 2.8 kips
Header dead load:	
80(6)(14)	= 6.7 kips
Wall dead load:	
80(18)(10.67)	= 15.4 kips
Pilaster	= 1.8 kips
Grade wall and footing:	
700 plf(12)	= 8.4 kips (estimate)
Total footing load	= 62.7 kips [279 kN]
Width required: 62.7/[2(12)]	= 2.61 ft [796 mm].

This is actually less than the width used for the other, lighter structure, because the design in that case was done for equalized dead load. With the higher proportion of dead to live load in this structure, this equalization is more questionable. If done, however, it would probably result in approximately the same footing as for the wood structure.

52.4.   DESIGN FOR LATERAL FORCE

The lateral load resistive system for this structure is basically the same as that for the wood structure in that it consists of the horizontal roof diaphragm and the exterior vertical shear walls. One significant difference is the increased load due to the heavier exterior walls. For determination of the seismic force the code requires an R_W factor of 6 for this building. This increases the design shear to $(8/6)(0.1375)W = 0.1833W$. (See previous discussion of seismic force for Scheme 2 in Sec. 51.5.)

Design of the Roof Diaphragm

The calculation of the loads for the horizontal seismic force (V) is shown in Table 52.1. In the north–south direction the load is symmetrically placed, the shear walls are symmetrical in plan, and the long diaphragm is reasonably flexible; all of which results in very little potential torsion. Although the code requires that a minimum torsional effect be considered by placing the load off-center by 5% of the building long dimension, the effect will be very little on the shear walls.

At the ends of the building the shear stress in the edge of the diaphragm will be

North–south force = 0.1833(448) = 82 kips

Maximum shear = 82/2 = 41 kips, or 41,000 lb

Maximum unit shear = 41,000/60 = 683 lb/ft

This is a very high stress for the plywood diaphragm, so some consideration should be given to a zoned roof deck, with the minimum construction used for the center of the building length and increased nailing toward the building ends. See discussion of the roof diaphragm for Scheme 2.

In the east–west direction the shear will be considerably less.

East–west force = (0.1833(323) = 59 kips

Maximum shear = 59/2 = 29.5 kips, or 29,500 lb

Maximum unit shear = 29,500/145 = 203 lb/ft

This is a very small stress, so the deck construction will be primarily determined by the north–south loading.

These calculations assume the deck to be nailed continuously at its boundaries, with collector elements of the framing serving to gather the deck load and push or drag it into the walls. Development of the framing supports for these elements should be done with consideration for the horizontal loads on the joints to achieve these collector functions. The elements must also be designed for the compression and/or tension effects, combined with other functions (bending in the headers, for example).

TABLE 52.1. Loads to the Roof Diaphragm (kips)

Load Source and Calculation	North–South Load	East–West Load
Roof dead load:	174	174
145 × 60 × 20 psf		
East and west exterior walls		
50 × 11 × 70 psf × 2	0	77
10 × 6 × 70 psf × 2	0	9
10 × 6 × 10 psf × 2	0	1
North wall:	126	0
150 × 12 × 70 psf		
South wall		
65.3 × 10 × 70 psf	46	0
84 × 6 × 70 psf	35	0
84 × 6 × 10 psf	5	0
Interior north–south partitions:	21	21
60 × 7 × 10 psf × 5		
Toilet walls: estimated	17	17
250 × 7 × 10 psf		
Canopy		
South: 150 × 100 plf	15	15
East and west: 40 × 100 plf	4	4
Rooftop HVAC units (estimate)	5	5
Total load	448 [1993 kN]	323 [1434 kN]

Design of the Masonry Shear Walls

In the north–south direction, with no added shear walls, the end shear forces will be taken almost entirely by the long solid walls because of their relative stiffness. The shear force will be the sum of the end shear from the roof and the force due to the weight of the end wall. For the latter we compute the following.

Wall weights:

18 ft × 50 ft × 70 psf	= 63,000 lb
6 ft × 10.67 ft × 70 psf	= 4,481 lb
12 ft × 10.67 ft × 5 psf	= 640
Total	= 68,121 lb [303 kN]

Lateral force: 0.1833 W = 0.1833 × 68

= 12.5 kips [55.6 kN]

The total force on the wall is thus 12.5 + 41 = 53.5 kips[238 kN] and the unit shear is

$$v = \frac{53,500}{44.67} = 1198 \text{ lb/ft } [20.3 \text{ kN/m}]$$

The code requires that this force be increased by 50% for shear investigation. Assuming a 60% solid wall with 8-in. blocks, the unit stress on the net area of the wall is thus

$$v = \frac{1198(1.5)}{12(7.625)(0.60)}$$

$$= 33 \text{ psi } [228 \text{ kP/a}]$$

From *UBC* Table 24-H (see Appendix D), with the reinforcing taking all shear and no special inspection, the allowable shear stress is dependent on the value of M/Vd for the wall. This is determined as

$$\frac{M}{Vd} = \frac{49.5(15) + 12.7(9)}{62.2(44.67)} = 0.308$$

Interpolating between the table values for M/Vd of 0 and 1.0 yields

$$\text{allowable } v = 35 + 0.69(25) = 52 \text{ psi}$$

This may be increased by the usual one third for seismic load to $1.33(52) = 69$ psi [476 kPa]. This indicates that the masonry stress is adequate, but we must check the wall reinforcing for its capacity as shear reinforcement. With the minimum horizontal reinforcing determined previously—No. 5 at 48 in.—the load on the bars is

$$V = 1198\left(\frac{48}{12}\right)(1.5)$$

$$= 7188 \text{ lb } [32 \text{ kN}] \text{ per bar}$$

and the required area for the bar is

$$A_s = \frac{V}{f_s} = \frac{8352}{26,667}$$

$$= 0.27 \text{ in.}^2 [174 \text{ mm}^2]$$

This indicates that the minimum reinforcing is adequate. Some additional stress will be placed on these walls by the effects of torsion, so that some increase in the horizontal reinforcing is probably advisable.

Overturn is not a problem for these walls because of their considerable dead weight and the natural tiedown provided by the dowelling of the vertical wall reinforcing into the foundations. These dowels also provide the necessary resistance to horizontal sliding.

In the east–west direction the shear walls are not symmetrical in plan, which requires that a calculation be made to determine the location of the center of rigidity so that the torsional moment may be determined. The total loading is reasonably centered in this direction, so we will assume the center of gravity to be in the center of the plan.

The following analysis is based on the examples in the *Masonry Design Manual* (Ref. 10). The individual piers are assumed to be fixed at top and bottom and their stiffnesses are found from Table E.1. The stiffness of the piers and the total wall stiffnesses are determined in Fig. 52.10. For the location of the center of stiffness we use the values determined for the north and south walls:

$$\bar{y} = \frac{(R \text{ for the S wall})(60 \text{ ft})}{(\text{sum of the } R \text{ values for the N and S walls})} = \frac{2.96(60)}{17.57}$$

$$= 10.11 \text{ ft } [3.08 \text{ m}]$$

The torsional resistance of the entire shear wall system is found as the sum of the products of the individual wall rigidities times the square of their distances from the center of stiffness. This summation is shown in Table 52.2. The torsional shear load for each wall is then found as

$$V_w = \frac{Tc}{J} = \frac{(V)(e)(c)(R \text{ for the wall})}{\text{sum of the } Rd^2 \text{ for all walls}}$$

In the north–south direction the *UBC* requires that the load be applied with a minimum eccentricity of 5% of the building length, or 7.5 ft. Although this produces less torsional moment than the east–west load, it is additive to the direct north–south shear and therefore critical for the end walls. The torsional load for the end walls is thus

$$V_w = \frac{82(7.5)(75)(3.17)}{44,524} = 3.28 \text{ kips } [14.6 \text{ kN}]$$

As mentioned previously, this should be added to the direct shear of 53,500 lb for the design of these walls.

For the north wall:

$$V_w = \frac{59(19.89)(10.11)(14.61)}{44,524} = 3.89 \text{ kips } [17.3 \text{ kN}]$$

This is actually opposite in direction to the direct shear, but the code does not allow the reduction and thus the direct shear only is used.

For the south wall:

$$V_w = \frac{59(19.89)(49.89)(2.96)}{44,524} = 3.89 \text{ kips } [17.3 \text{ kN}]$$

The total direct east–west shear will be distributed between the north and south walls in proportion to the wall stiffnesses.

For the north wall:

$$V_w = \frac{59(14.61)}{17.57} = 49.1 \text{ kips } [218 \text{ kN}]$$

For the south wall:

$$V_w = \frac{59(2.96)}{17.57} = 9.94 \text{ kips } [44.2 \text{ kN}]$$

The total shear loads on the walls are therefore

North: $V = 49.1$ kips [218 kN].
South: $V = 3.89 + 9.94 = 13.83$ kips [61.5 kN].

The loads on the individual piers are then distributed in proportion to the pier stiffnesses (R) as determined in Fig. 52.10. The calculation for this distribution and the determination of the unit shear stresses per foot of wall are shown in Table 52.3. A comparison with the previous calculations for the end walls will show that these stresses are not critical for the 8-in. block walls.

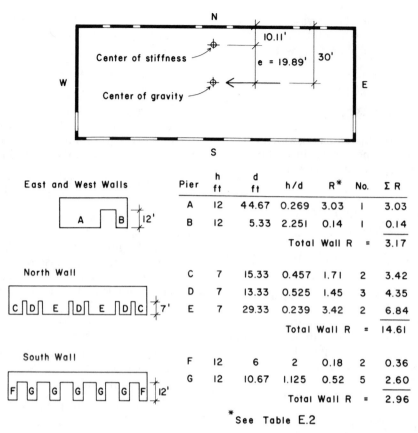

Pier	h ft	d ft	h/d	R*	No.	Σ R
A	12	44.67	0.269	3.03	1	3.03
B	12	5.33	2.251	0.14	1	0.14
				Total Wall R	=	3.17
C	7	15.33	0.457	1.71	2	3.42
D	7	13.33	0.525	1.45	3	4.35
E	7	29.33	0.239	3.42	2	6.84
				Total Wall R	=	14.61
F	12	6	2	0.18	2	0.36
G	12	10.67	1.125	0.52	5	2.60
				Total Wall R	=	2.96

*See Table E.2

FIGURE 52.10. Stiffness analysis of the masonry walls.

TABLE 52.2. Torsional Resistance of the Masonry Shear Walls

Wall	Total Wall R	Distance from Center of Stiffness (ft)	$R(d)^2$
South	2.96	49.89	7,367
North	14.61	10.11	1,495
East	3.17	75	17,831
West	3.17	75	17,831
Total torsional moment of inertia (J)			44,524

In most cases the stabilizing dead loads plus the doweling of the end reinforcing into the foundations will be sufficient to resist overturn effects. The heavy loading on the header columns and the pilasters will provide considerable resistance for most walls. The only wall not so loaded is wall C, for which the loading condition is shown in Fig. 52.11. The overturn analysis for this wall is as follows:

$$\text{overturn } M = 7440(7.0)(1.5)$$
$$= 78,120 \text{ ft-lb } [106 \text{ kN-m}]$$

$$\text{stabilizing } M = 23,000 \left(\frac{15.33}{2} \right)$$

$$= 176,295 \text{ ft-lb } [239 \text{ kN-m}]$$

This indicates that the wall is stable without any requirement for anchorage even though the wall weight in the plane of the shear wall was not included in computing the overturning moment.

52.5. CONSTRUCTION DETAILS

The drawings that follow illustrate the construction of the timber and masonry structure for Building One. There are numerous alternatives for many of the details shown, some of which are discussed in the design work in the preceding sections. The reader is reminded that many of the details shown are essentially limited to the illustration of the structure, and thus there are additional elements of the building construction required for finishes, waterproofing, and so on. Some details may also be affected by considerations for lighting, electrical power, HVAC, and plumbing systems, or by problems of building security, acoustics, fire resistance, and so on.

Structural Plans

Figure 52.12 shows partial plans for the roof structure and the foundation system. The basic components of the modular roof system are indicated. While odd-sized or cut plywood panels are possible, this system optimizes the use

TABLE 52.3. Shear Stresses in the Masonry Walls

Wall	Shear Force on Wall (kips)	Wall R	Pier	Pier R	Shear Force on Pier (kips)	Pier Length (ft)	Shear Stress in Pier (lb/ft)
North	49.1	14.61	C	1.71	5.75	15.33	375
			D	1.45	4.87	13.33	365
			E	3.42	11.49	29.33	392
South	13.83	2.96	F	0.18	0.84	6	140
			G	0.52	2.43	10.67	228

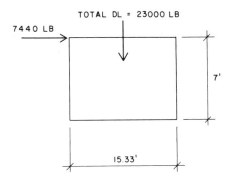

FIGURE 52.11. Stability of wall C.

of the standard 4 ft × 8 ft panels commonly available from material suppliers.

In addition to the plans, a nailing schedule or plan would be used to show the variation of nailing required for the roof diaphragm actions. For this system the joists and plywood are usually prefabricated in panel units, so some of the plywood nailing occurs in the fabricating plant and some in the field. For design, nailing is usually specified in terms related to the use of common nails (the basis for the tables in the *UBC*), but this vast amount of nailing is now commonly done with machine-driven equipment where rated values are acceptable by local codes.

Detail A (Fig. 52.13). This shows the typical front wall condition at the solid wall. The girder, pilaster, pilaster pier, and widened footing are seen in the background. A wood ledger is bolted to the masonry wall to receive the edge of the plywood deck. The deck is nailed to the ledger and the ledger is bolted to the wall to transfer the shear load from the roof diaphragm to the wall.

For the reinforced masonry wall the code requires a minimum vertical spacing of solid horizontal reinforced bond courses. In addition to the minimum spacing, these would be used at the top of the wall, the bottom of the header, and the location of the canopy and roof edge bolting to the wall.

Detail B (Fig. 52.13). This shows the foundation edge at the front, which is essentially similar to that for the wood

structure. The sill bolts would be replaced by dowels for the masonry wall. The pier would be added below the pilaster to carry the load down to the widened footing.

Detail C (Fig. 52.13). This shows the roof edge condition at the building ends. The wood ledger performs the dual task of providing vertical support for the ends of the joists and transfers the lateral loads from the plywood deck to the masonry wall. Because of the roof slope, the top of the steel channel varies 15 in. from front to rear of the building. The horizontal filled block courses and the cutoff to the narrower parapet would be staggered to accommodate this slope. A somewhat larger than usual cant would be used to cover the jog in the wall to the narrower parapet block.

Detail D (Fig. 52.13). This detail is also essentially similar to that for the wood structure. If a footing of increased width is required, care should be taken to ensure that the centroid of the vertical loads is close to the center of the width of the footing.

Detail E (Fig. 52.14). This shows the typical form of the column-top support for a timber (or laminated) beam. While this type of joint can be developed to resist some bending moment, the requirement in this situation is for vertical support only.

The detail could be developed with a wood column, for which the steel connector is probably available as a catalog item from some supplier. If a steel column is used, the U-shaped beam support would be welded to the top of the column.

The detail here shows the support of a single beam that is continuous over the top of the column. A similar detail would be used if a beam joint occurs at the column top, although unbalanced loadings on the two beams would require more concern for the moment-resisting capability of the joint.

Detail F (Fig. 52.14). This shows the column base detail. For a wood column a common form of steel anchor/connector is one that effectively keeps the wood column from having contact with the concrete foundation. Again, this type of connector is commonly available as a standard

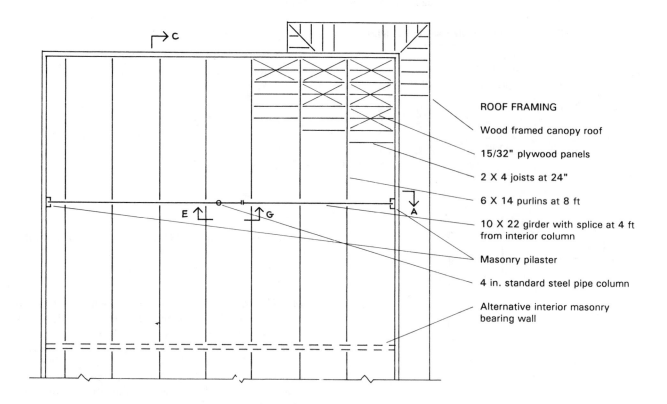

ROOF FRAMING

Wood framed canopy roof

15/32" plywood panels

2 X 4 joists at 24"

6 X 14 purlins at 8 ft

10 X 22 girder with splice at 4 ft from interior column

Masonry pilaster

4 in. standard steel pipe column

Alternative interior masonry bearing wall

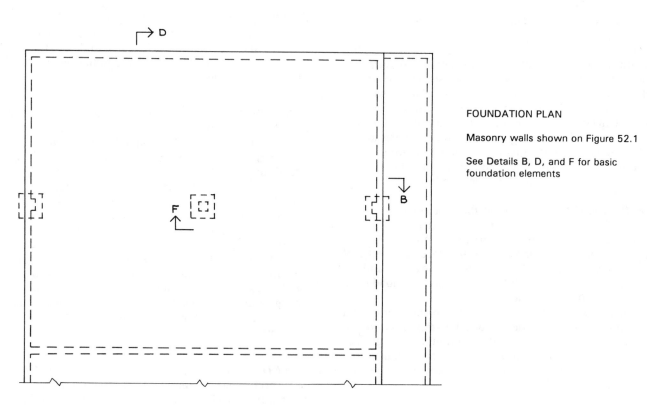

FOUNDATION PLAN

Masonry walls shown on Figure 52.1

See Details B, D, and F for basic foundation elements

FIGURE 52.12. Structural plan.

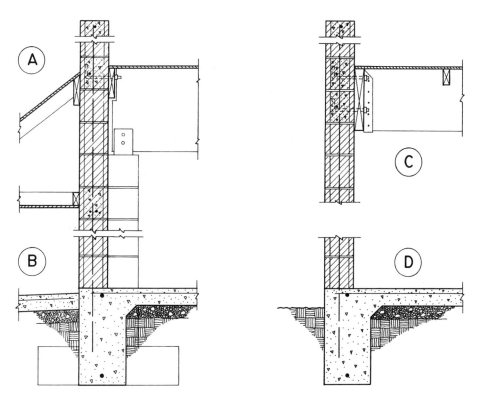

FIGURE 52.13. Construction details.

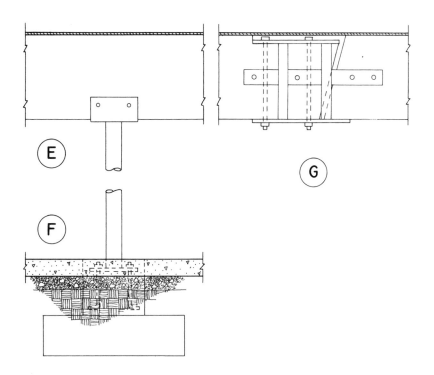

FIGURE 52.14. Details of the interior bearing wall.

hardware item, with a range of forms to accommodate various sizes of columns and magnitudes of vertical load support capacities.

The detail shows the use of a footing and short pier for support of the column, with the concrete floor slab separately developed. This is particularly advantageous if the floor slab is to be cast after the roof structure is erected. Details to be considered here are the desire for construction or control joints in the floor slab for crack control due to shrinkage or thermal change and the desire of providing or avoiding vertical support by the footing for the slab. Soil conditions, floor loadings, and various details of the in-

terior construction can affect the development of these details.

Detail G (Fig. 52.14).   This shows the form of a steel connector for one of the girder splices. In this case the end of the girder on the right is being supported by the end of the girder on the left. Additional concerns may be for the need to transfer tension and/or compression through the joint if the girder is required to perform collector or drag strut functions for lateral loads. For standard timber and laminated member sizes this connector is also available as a standard hardware item.

CHAPTER FIFTY-THREE

Building One, Scheme 4: Concrete and Steel Structure

This scheme accomplishes the same approximate building shape as the previous ones, utilizing exterior bearing walls of precast concrete and a roof framed with steel elements.

53.1. GENERAL CONSIDERATIONS

The precast concrete walls may be custom designed (original engineering design, construction details, and specifications by the building designers), but are more likely to be developed primarily by a specialty contractor or company that does precast concrete work. In the latter case, the contractor is likely to provide engineering design, fabrication of the panels, and field erection. Except for specific dimensions, exposed finishes, and some details, the panels will most likely conform to established practices of the supplier. In regions where use of this form of construction is widespread, such contractors are generally accessible.

The steel roof structure, on the other hand, is likely to be of relatively conventional form; not much related to the concrete bearing walls except in terms of the connection details to the walls at points of support. Thus, the roof structure would be essentially the same if the walls were masonry or of some curtain wall (nonstructural) form with exterior supports developed with steel framing.

The following discussion assumes the preceding situations: with the concrete walls supplied essentially as priority products of a selected manufacturer and a conventional steel-framed roof structure. Basic loads and other criteria are the same as given in Sec. 52.1 for Scheme 3.

Partial plans for the structure are shown in Fig. 53.1 The ground floor plan indicates the layout of the concrete wall units. The roof framing plan indicates the use of a clear-spanning steel girder to achieve the 60-ft span, with a joist and deck system arranged in a manner similar to that for Scheme 1.

Many other framing schemes are possible for the roof, depending somewhat on the perceived need for a clear span. Steel open web joists could be used in a manner similar to that developed with the composite trusses in Scheme 2 (see Fig. 51.4). If interior columns are permis-

sible, the arrangement used for Scheme 3 might be possible (see Fig. 52.1). Some other possible arrangements of the framing with interior supports are shown in Fig. 53.2. Obviously, considerations for architectural planning and any future use of the building will affect the decision between these or other alternatives.

53.2. DESIGN OF THE ROOF

The following construction loads are assumed.

Dead Loads

Roofing: tar and gravel, 6.5 psf.

Insulation: rigid type on top of deck, assume 2 psf.

> Suspended ceiling: gypsum plaster, assume 10 psf total.

> Lights, wiring, ducts, registers: assume average of 3 psf.

> Total dead load: 21.5 psf [1.03 kPa] + the structure.

Roof Deck

A number of roof deck systems may be considered, including those with lightweight concrete or gypsum concrete fill. We use an ordinary single sheet, ribbed deck. With the joists at 6-ft centers, a 20-gauge deck will be adequate. This deck will weigh approximately 2 psf, depending on the specified finish.

The deck must also be considered for seismic load. The choice of the deck itself and the details for its installation must be considered in terms of the required diaphragm shear and the overall diaphragm deflection.

Joists

Several options are possible for the joists. These may be light-rolled, I-shaped sections, open-web steel joists, or cold-formed sections of light-gage sheet steel. With joists at 6-ft centers, the loading will be

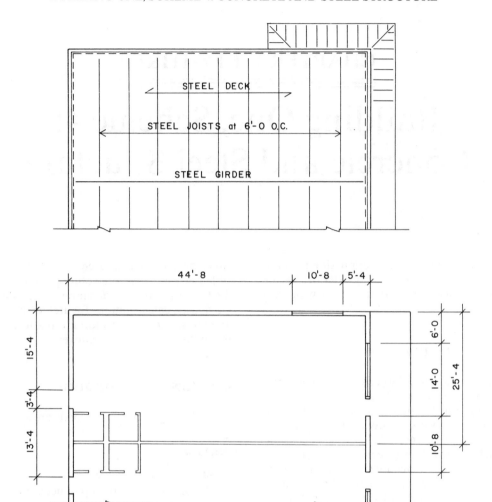

FIGURE 53.1. Building One: plans for the steel-framed roof and masonry wall structure.

DL + LL: 23.5 + 20 = 43.5 psf [2.08 kPa].

Load/joist: 43.5 × 6 = 261 lb/ft [3.81 kN/m] + the joist weight/ft.

For design use: total load 275 = lb/ft [4 kN/m].

Using a total load deflection limit of L/180, the allowable deflection will be

$$\frac{25(12)}{180} = 1.67 \text{ in. [42 mm]}$$

Referring to Table 22.2, we first determine the total load for the joist; thus

$$W = 25 \text{ ft} \times 275 \text{ lb/ft}$$

$$= 6875 \text{ lb} \quad \text{or} \quad 6.875 \text{ kips [30.6 kN]}$$

For this load possible selections are W 12 × 14 or M 12 × 11.8.

Figure 22.6 shows that the 12-in.-deep members are not critical for the deflection limit of $L/180$.

Girder

With the joists at 6-ft centers the actual load on the girder consists of nine joists plus the weight of the girder. Because of the large area supported by the girder, the live load may be reduced to 12 psf. The design load for the girder will thus be

DL + LL: 25.5 + 12 = 37.5 psf [1.8 kPa].

Joists: 9(6)(25)(37.5) = 50,625 lb [225 kN].

Assumed girder weight: 75(60) = 4500 lb.

Total DL + LL: 55,125 lb [245 kN].

The lateral unsupported length is 6 ft (the joist spacing), which should not be critical for this large member. Selection can be made from the load–span tables (Table 4.2) or the maximum moment can be determined and a selection made from either the S-listing tables (Table 22.1) or the graphs in the *AISC Manual* (Ref. 5), which incorporate the consideration for lateral unsupported length. To demonstrate the use of the S-listing tables we first determine the maximum moment as

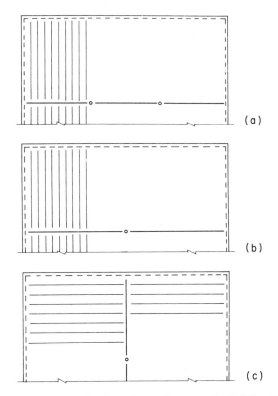

FIGURE 53.2. Plans for alternative roof structures for Building One.

$$M = \frac{WL}{8} = \frac{55.1(60)}{8}$$

$$= 413 \text{ kip-ft } [560 \text{ kN-m}]$$

From the S-listing tables (Table 22.1) the lightest section for this moment is a W 27 × 84 and the next lightest is a W 24 x 94. Both of these have L_c values over 6 ft, indicating no problem with lateral support. If headroom is considered critical, the 24-in.-deep member may be more desirable, although its deflection should be checked as follows.

Actual deflection for the 24-in.-deep beam on the 60-ft span (Fig. 22.6): 3.7 in. [94 mm].

Allowable deflection under total DL + LL:

$$\Delta = \frac{L}{180} = \frac{60(12)}{180} = 4.0 \text{ in. } [102 \text{ mm}]$$

Although the actual deflection is within the specified limit, the dimension of almost 4-in. movement is considerable. Because the live load deflection is the only movement to be experienced after the construction is complete, it may be possible to camber the beam (cold bend it into a permanent upper curvature) by an amount equal to the dead load deflection. Without live load the structure will then be flat, and the construction of interior walls may be able to tolerate the limited deflections due to live load. In any event, the selection of the deeper section will work to the relief of deflection problems.

Columns for the Girder

The end reactions for the joists can probably be borne by the concrete wall panels. However, the magnitude of the end reactions of the girders is most likely too high for the same treatment. This requires the consideration of the development of columns as end supports for the girders.

There are various possible solutions for the development of the whole wall construction on the long sides of the building. With the framing as shown in Fig. 53.1 and the use of some form of column for the girders, the concrete panels on these walls are required to support only the short-span steel deck. However, the openings in the walls must also be developed, with various possible solutions. Header beams may be used with support provided by the structural walls, as was done in Scheme 3. Or, the walls may be developed with a continuous steel framing system, eliminating the need for bearing support by the wall panels on these sides of the building.

An additional consideration in this regard is the development of the walls for shear wall functions, as discussed in Sec. 53.4. This requires that the walls and the roof deck (as the horizontal diaphragm) be attached somehow through the construction for transfer of the lateral forces.

For the support of the girders, two general solutions are possible. The first entails the development of a reinforced concrete column, which may be sitecast in conjunction with the erection and attachment of the precast wall panels. This is more likely to be done if the whole wall structure is basically developed independently of the roof framing.

The second possibility for the column is to use a steel member that is attached to the wall panels only for purposes of the column stability. This solution is most likely to be used in conjunction with the development of a whole independent steel frame for the long walls. This is the case illustrated in the details in Sec. 53.5. In those details, the use of a steel tubular column is shown. This column is considered to be laterally braced by the wall panels and is designed for half the girder load with a column effective length of zero. The result is a relatively small member, which can be relatively easily incorporated into the interior construction finish of the walls.

53.3. DESIGN OF THE WALLS

As discussed previously, the concrete walls are likely to be developed as the products supplied by a specialty contractor or manufacturer. For the design of these walls the building designers are advised to ascertain the availability of local suppliers and to obtain information directly from them for use in planning. Structural details, materials, and construction joints are likely to conform to the suppliers' standards, but finishes, panel dimensions, and some special details at openings or other locations may be custom fit to some of the building designers' preferences.

Guidelines for design and construction detailing with this form of construction are available from various industry agencies, such as the American Concrete Institute, the

Portland Cement Association, and the Prestressed Concrete Institute.

Special concerns for the situation of this building include the following.

Use of the panels to support the ends of the joists; involving some bending induced by the eccentricity of the joist loads as applied through the steel ledger.

General development of the walls to accommodate the openings shown in the plans; involving the attachment of window and door framing, headers over the openings, and so on.

Development of the interior wall surfaces, with provision for attachment of additional interior construction. (See the details in Sec. 53.5.)

Use of the walls for shear wall functions, with special attention to the design of the construction joints to develop the lateral force transfers required.

53.4. DESIGN FOR LATERAL FORCES

The basic lateral-resistive structure for this building is essentially similar to that in the previous schemes, consisting of the use of the roof deck as a horizontal diaphragm and the exterior structural walls as shear walls. The concrete panels have considerable potential for this use, depending on the aspect ratio (height-to-plan length) of the individual panels and the effects of any openings actually occurring within individual panels.

The light-gauge steel roof deck has about the same potential capacity as a wood deck diaphragm. The Steel Deck Institute (SDI) provides some industry standards for design of diaphragms, but specific capacities must usually be obtained for the particular products of individual manufacturers.

As with wood decks, the concerns exist for the attachment of the deck to its supports and the joining of adjacent deck units for the development of the whole roof surface as an integral unit for the diaphragm action. Development of some of the details shown in Sec. 53.5 reflects the need for provision of necessary attachments to develop; the continuity of the diaphragm as well as to transfer forces between the roof and the shear walls.

53.5. CONSTRUCTION DETAILS

Figure 53.3 shows some of the details for the development of the building exterior walls. Development of these details reflects the particular concerns for the enhancement of the interior sides of the walls for this building use. For most industrial or utility buildings in mild climates it is common to omit any interior finish construction and to simply expose the cast panels on the interior.

The details in Fig. 53.3 indicate the development of the exterior walls with a parapet, for which the precast panels are provided as single units from the ground to the top of

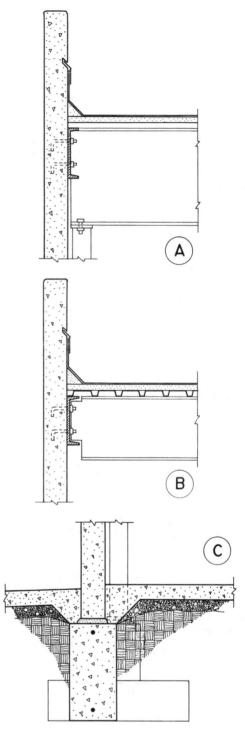

FIGURE 53.3. Construction details for the walls.

the parapet. The roof structure thus runs into the back of the walls. An alternative construction is shown in Fig. 53.4, in which the roof is extended over the tops of the walls, providing an overhang and soffit. This involves some different situations with regard to the development of the vertical

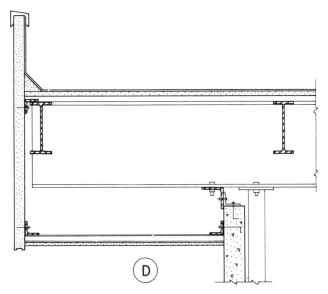

FIGURE 53.4. Construction details for the optional overhang with soffit.

support for the roof structure and the facilitation of the transfer of lateral forces between the roof and the shear walls.

In addition to basic structural concerns, some details we must be considered here are the following.

The type of roofing materials used and how the roofing is attached to a surface that provides a smooth face (the steel deck does not).

How the roof insulation and vapor barrier functions are developed. The deck—if properly sealed at all joints and edges—may function as a vapor barrier, but may not be in the right position to do the job (considering the issue of condensation of moisture within the construction).

How the ceiling and various service elements (ducts, fixtures, piping, etc.) are supported. The details shown here assume support by the joists in most cases. The ceiling is shown as dropped below the girders to permit passage of some elements (ducts, piping, wiring).

CHAPTER FIFTY-FOUR

Building Two: General Considerations

Building Two is a three-story office building, designed as an investment property for speculative rental. As with Building One, there are many possibilities for the construction of this building. However, in any given place at any given time, it will be found that the general planning and the basic construction of such a building vary little from a limited set of choices. The purpose of Chapter 54 is to present the design factors and criteria for the building and discuss the problem of choosing the appropriate construction system.

54.1. THE PROPOSED BUILDING DESIGN

Figure 54.1 presents the initial design scheme for the building in the form of plans and a full building section. In a reasonably rational process of design it is to be expected that the initial design may be modified somewhat in the process of developing the specific details of the building construction as well as the systems for lighting and environmental control. We assume that the plans shown are relatively firmly established, but will discuss some modifications that could improve the structure with certain alternate choices in the construction system.

Although the building exterior is obviously of great importance both architecturally and functionally, we will avoid dealing with the building elevations, except where the choice of certain structural features have a strong relation to the exterior form. We assume, however, that a fundamental requirement for the building is the provision of a significant amount of exterior window surface and the avoidance to long expanses of unbroken solid wall surface. Another assumption—which is reasonably evident in the plans—is that the building is freestanding on the site with all sides having a clear view.

54.2. DESIGN CRITERIA

The following will be assumed as criteria for the building design work:

Building Code: 1991 *UBC* (Ref. 1)
Live loads:
    Roof: *UBC* minimum, Table 23-C.

Floors:
Office: 50 psf [2.39 kPa].
Corridors: 100 psf [4.79 kPa].
Partitions: 20 psf [*UBC* minimum, Sec. 2304(d)] [0.96 kPa].
Wind: map speed, 80 mph; exposure B [129 km/h].
Seismic: zone 3.
Assumed construction loads:
    Floor finish: 5 psf [0.24 kPa].
    Ceilings, lights, ducts: 15 psf [0.72 kPa]
    Walls (average surface weight):
        Interior partitions: 10 psf [0.48 kPa].
        Exterior curtain wall: 15 psf [0.72 kPa].

54.3. STRUCTURAL ALTERNATIVES

Fire codes permitting, the most economical structure for the building will be one that makes the most use of light wood frame construction. It is unlikely that the building would use all wood construction of the type illustrated in Building One, but a mixed system is quite possible. It is also possible to use steel, masonry, or concrete construction and eliminate wood, except for nonstructural uses. In addition to code requirements, consideration must be given to the building owners' preferences and to design criteria or standards for acoustic privacy, thermal control, and so on.

The plan as shown, with 30-ft square bays and a general open interior, is an ideal arrangement for a beam and column system in either steel or reinforced concrete. Other types of systems may be made more effective if some modifications of the basic plans are made. These changes may affect the planning of the building core, the plan dimensions for the column locations, the articulation of the exterior wall, or the vertical distances between the levels of the building.

The general form and basic type of the structural system must relate to both the gravity and lateral force problems. Considerations for gravity require the development of the horizontal spanning systems for the roof and floors and the arrangement of the vertical elements (walls and columns) that provide support for the spanning structure. Vertical elements should be stacked, thus requiring coordinating the plans of the various levels.

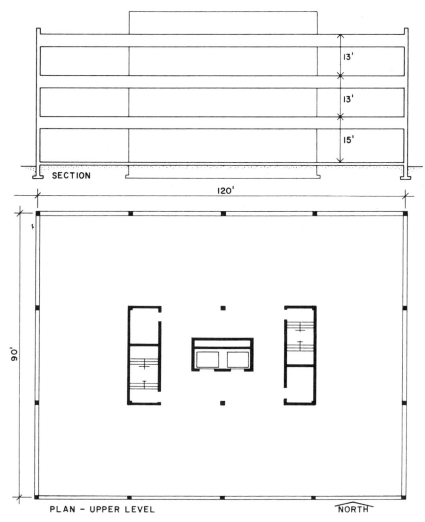

FIGURE 54.1. Building Two.

The most common choices for the lateral bracing system would be the following (see Fig. 54.2).

1. *Core Shear Wall System* (Fig. 54.2a). This consists of using solid walls to produce a very rigid central core. The rest of the structure leans on this rigid interior portion, and the roof and floor constructions outside the core—as well as the exterior walls—are free of concerns for lateral forces as far as the structure as a whole is concerned.

2. *Truss-Braced Core.* This is similar in nature to the shear wall-braced core, and the planning considerations would be essentially similar. The solid walls would be replaced by bays of trussed framing (in vertical bents) using various possible patterns for the truss elements.

3. *Peripheral Shear Walls* (see Fig. 54.2b). This in essence makes the building into a tubelike structure. Because doors and windows must pierce the exterior, the peripheral shear walls usually consist of linked sets of individual walls (sometimes called *piers*).

4. *Mixed Exterior and Interior Shear Walls.* This is essentially a combination of the core and peripheral systems.

5. *Full Rigid-Frame System* (see Fig. 54.2c). This is produced by using the vertical planes of columns and beams in each direction as a series of rigid bents. For this building there would thus be four bents for bracing in one direction and five for bracing in the other direction. This requires that the beam-to-column connections be moment resistive.

6. *Peripheral Rigid-Frame System* (see Fig. 54.2d). This consists of using only the columns and beams in the exterior walls, resulting in only two bracing bents in each direction.

In the right circumstances any of these systems may be acceptable. Each has advantages and disadvantages from both structural design and architectural planning points of view. The braced core schemes were popular in the past, especially for buildings in which wind was the major concern. The core system allows for the greatest freedom in planning the exterior walls, which are obviously of major

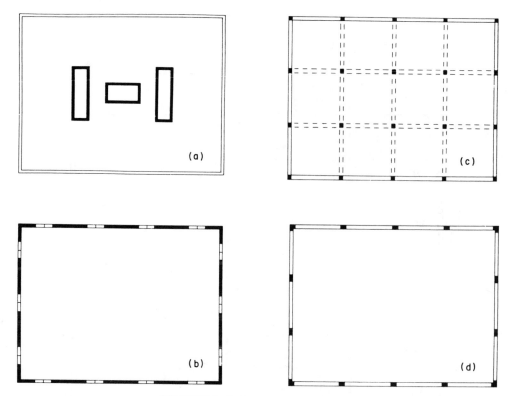

FIGURE 54.2. Options for the lateral bracing.

concern to the architect. The peripheral system, however, produces the most torsionally stiff building—an advantage for seismic resistance.

The rigid frame schemes permit the free planning of the interior and the greatest openness in the wall planes. The integrity of the bents must be maintained, however, which restricts column locations and planning of stairs, elevators, and duct shafts so as not to interrupt any of the column-line beams. If designed for lateral forces, columns

are likely to be large and thus offer more intrusion in the building plan.

Other solutions are also possible, limited only by the creative imagination of designers. In the chapters that follow we will illustrate the design of several possible structures for the building. We do not propose that these are ideal solutions, but merely that they are feasible alternatives. They have been chosen primarily to permit illustrating the design of the elements of the construction.

CHAPTER FIFTY-FIVE

Building Two, Scheme 1: Wood Structure

A structural plan for one of the upper floors of Building Two is shown in Fig. 55.1. The plan indicates the use of bearing walls at the exterior and the building core. Beams are used over window openings in the exterior and to create two rows of support in the interior, providing for three 30-ft-span bays of closely spaced joists. The joists support a short-span structural deck. Two columns are used at the core to provide the necessary open space.

Many options are possible for the development of this system in terms of the materials used for the individual elements of the structure. Chapter 55 demonstrates the design for a structural system that optimizes the use of wood elements. While structurally possible, it is doubtful that fire codes can be satisfied with such a form of construction. The design is illustrated here to show the process for development of the basic elements of such a system; possibly applicable to smaller buildings.

55.1. DESIGN OF THE DECK— JOIST-BEAM SYSTEM

There are a number of options for this system. The spans, the plan layout, the fire code requirements, the anticipated surfacing of floors and ceilings, the need for incorporating elements for wiring, piping, heat and cooling, fire sprinklers, and lighting are all influences on the choice of construction. We assume that the various considerations can be met with a system consisting of a plywood deck, wood joists, and steel beams.

The Deck

The plywood deck will be used for both spanning between the joists to resist gravity loads and as a horizontal diaphragm for distribution of wind and seismic forces to the

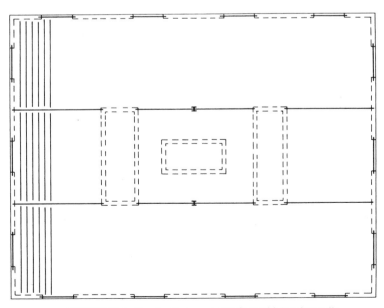

FIGURE 55.1. Structural plan—upper floor with bearing walls.

shear walls. Choice of the deck material, the plan layout of the plywood panels, and the nailing must be done with both functions in mind. The spacing of the joists must also be coordinated with the panel layout. It is also necessary to anticipate floor surfacing in terms of both loading and construction accommodation.

We assume the use of the common size of plywood panel of 4 ft × 8 ft, and a joist spacing of 24 in., which is usually the maximum for floor systems of this type. Office floors will most likely be covered with carpet, thin tile, or wood flooring. The surface of the structural plywood deck, however, is too rough and uneven for these materials, thus resulting in needing some intermediate surfacing, which is most commonly either a second paneling with fiberboard or a thin coat of lightweight structural concrete.

UBC Table 25-S-1 (see Appendix D) indicates that a $\frac{3}{4}$-in.-thick deck may be used with the joist spacing of 24 in. A footnote to the *UBC* table states that the unsupported edges must be blocked or have tongue-and-groove joints if underlayment is not provided. Paneling for floors is quite commonly available with tongue-and-groove joints, but this is not actually required as we will allow for a 2-in. concrete fill.

The panel layout, edge nailing, and the need for blocking must be investigated as part of the design for diaphragm action.

The Joists

It is unlikely that the joists for this span and loading would be solid timber members. As the calculations will show, the required section is quite large and would be very expensive for joists at 24-in. centers. In most regions it is likely that fabricated joists are available that are both lighter and less expensive. For purpose of comparison, however, we will demonstrate the joist design procedure for a solid member.

Economy dictates that a relatively low grade of wood be used for the joist. We will try Douglas fir–larch, No. 1 grade, which is about as high a grade as is feasible. For this grade, the *UBC* yields the following data for sections 2–4 in. thick and 6 in. or more in width:

$$F_b = 1750 \text{ psi (repetitive use) [12.1 MPa]}$$

$$F_v = 95 \text{ psi [0.66 MPa]}$$

$$E = 1,800,000 \text{ psi [12.4 GPa]}$$

Using these data we will design a joist for the 100 psf live load at the corridor. The design loads for a single 30-ft [9.14-m]-span joist are thus as follows.

Live load:
100 psf × 2 ft	= 200 plf [2.92 kN/m]

Dead load:
Carpet + pad	= 5 psf
2-in. concrete fill at 10 lb/in.	
	= 20 psf
$\frac{3}{4}$-in. plywood	= 3 psf
Ceiling, lights, ducts	= 15 psf
Total unit DL	= 43 psf

Load on joist: 43 × 2	= 86 plf	
Estimate joist weight	= 20 plf	
Design DL for joist	= 106 plf [1.55 kN/m]	

The total design load for a joist is thus

$$DL + LL = 106 + 200$$
$$= 306 \text{ plf [4.46 kN/m]}$$

and the maximum bending moment is

$$M = \frac{wL^2}{8} = \frac{306(30)^2}{8}$$
$$= 34,425 \text{ ft-lb [46.7 kN-m]}$$

for which the required section modulus is

$$S = \frac{M}{F_b} = \frac{34,425(12)}{1750}$$
$$= 236 \text{ in.}^3 \text{ [3.87} \times 10^6 \text{ mm}^3\text{]}$$

There is no member in the size range that was assumed that has an S value this high. We must therefore find a new value for S that corresponds to the proper size range. For the "Beams and Stringers" category in the *UBC* the allowable bending stress for grade dense No. 1 is 1550 psi [10.7 MPa]. The corresponding S is thus

$$S = \frac{34,425(12)}{1550}$$
$$= 267 \text{ in.}^3 \text{ [4.38} \times 10^6 \text{ mm}^3\text{]}$$

which may be satisfied with a 6 × 18 ($S = 281$ in.³). This is a mammoth size member for joists at 24 in. on center. If the framing arrangement shown in Fig. 55.1 is to be retained, a fabricated joist will have to be used. One possibility is to use an open-web steel joist with provision for attachment of the plywood deck to its top flange. Using the joist loading determined previously, it will be found from Table 25.1 that a 22K4 joist at 8.0 lb/ft is adequate for both total load capacity and live-load deflection of $L/360$. Other possibilities are a wood-trussed joist, a vertically laminated wood joist, or a wood plus plywood member (wood flanges and plywood web).

Another possibility is to alter the framing system slightly to produce a system similar to that used for the wood roof structure for Building Two. A layout of this type is shown in the partial framing plan in Fig. 55.2a. In this scheme the long-span joists are replaced by purlins at 8-ft centers that support short-span joists at 2-ft centers. As with the roof system for Building One, an advantage with this scheme is that it can be used to provide full edge support for the plywood panels, producing a blocked diaphragm without the need for added blocking.

The purlins for the system in Fig. 55.2a will carry approximately four times the joist load determined for the 2-ft-on-center joists. This would require the use of glue-laminated members or steel wide-flange sections. We will not finish the design of this system, but will proceed with the design assuming the scheme in Fig. 55.1 with fabricated joists.

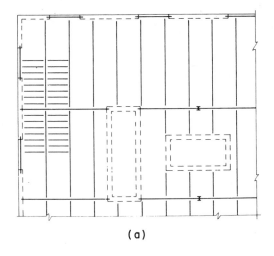

(a)

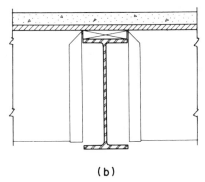

(b)

FIGURE 55.2. Considerations for the floor framing: (a) alternative framing system; (b) detail for the steel beam with wood joists.

The Beams

The large interior beams may be of wood glue-laminated construction or steel wide-flange sections. We consider the latter case first and assume the beam section to be as shown in Fig. 55.2b with a 2-in.-nominal member bolted to the top of the steel section and the joists carried by saddle-type hangers, with the joist tops level with the top of the wood nailer. If open web steel joists or wood-trussed joists are selected, their top chords would be supported on top of the beam, and the detail would be slightly different.

We will design the beam for a live load of 100 psf and a unit dead load of 55 psf, which includes allowances for the weight of the joists, the beams, and any bridging, blocking, and so on. The heaviest-loaded beam is the 30-ft beam that occurs at the building ends. This beam has a total load periphery of 30 ft × 30 ft or 900 ft². The *UBC* allows a reduction of the live load equal to

$$R = 0.08(A - 150)$$
$$= 0.08(900 - 150) = 60\%$$

The reduction is further limited to a maximum of 40% or

$$R = 23.1 \left(1 + \frac{DL}{LL} \right)$$

$$= 23.1 \left(1 + \frac{55}{100} \right) = 35.8\%$$

We will therefore use a 35% reduction, or a design live load of 65 psf.

Using the reduced live load, the total load carried by the 30-ft span beam is

$$W = (65 + 55)(30)(30)$$
$$= 108,000 \text{ lb} \quad \text{or} \quad 108 \text{ kips [480 kN]}$$

and the maximum bending moment is

$$M = \frac{WL}{8} = \frac{108(30)}{8}$$

$$= 405 \text{ kip-ft [549 kN-m]}$$

From Table 22.1, assuming full lateral support for the beam, possible choices are

$W \ 27 \times 84$	$M_R = 426$ kip-ft
$W \ 24 \times 94$	$M_R = 444$ kip-ft
$W \ 18 \times 106$	$M_R = 408$ kip-ft

Deflection of the beam should be a minimum, for the total deflection of the joists must be added to that of the beam for the sag at the middle of the bay. An inspection of Fig. 22.6, shows that either the 27- or 24-in. members will have quite low values of deflection. If a member shallower than 24 in. is desired to allow for passage of ducts, the deflection should be carefully investigated.

Use of the two-span beam at the core will result in considerable reduction of deflection. The shorter spans and reduced loading will result in a section considerably smaller than that required for the 30-ft-span beam.

The short beams that span between the masonry walls at the building edge must be designed to carry the floor structure as well as the wall construction. Selection of the beam section and development of the connection details must be developed as part of the general design of the exterior walls. Some possible details are shown in Sec. 55.5.

For an all-wood structure, the beams would most likely be of glue-laminated construction. The design of such a beam is illustrated in Sec. 52.2. Overall size (width and depth) of the wood beam will usually be in the general range of that for a steel beam.

The Columns

Columns must be provided for the support of the beams. These will occur as freestanding columns for the two-span beam at the core, but will otherwise be built into the wall construction.

To provide for the continuity of the core beams, the columns may be stacked, as shown in Fig. 55.3. This is not the usual procedure for multistory buildings, but may be

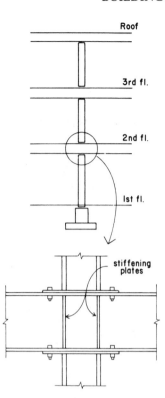

FIGURE 55.3. The interior column.

acceptable as long as the beams are capable of the concentrated compression and the loads are relatively low in magnitude (as is the case for the low-rise building). For the steel wide-flange shape, it is usually necessary to stiffen the web with welded steel plates, as shown in the detail in Fig. 55.3.

Steel columns may also be used with the laminated wood beams, and are most likely to be so for the freestanding columns. At the walls, however, it may be possible to produce concentrated strength by building timber columns into the wall or by using multiples of the wall studs.

The general design of multistory columns is developed more thoroughly in Schemes 2 and 4 for Building Two (Chapters 56 and 58).

55.2. DESIGN OF THE STUD WALLS

The exterior walls and any interior bearing walls will be built essentially as shown for the columns in the upper part of Fig. 55.3. This is the general nature of platform construction for the light wood frame, with story-high wall segments and individual levels of horizontal framing stacked on top of each other. Materials, sizes, and spacing for the studs in each wall segment will be determined by structural design and consideration of the general development of the wall construction.

The exterior walls on the long sides of the buildings support the ends of the 30-ft joists, representing the greater load case for gravity forces. Design of a stud for this situation—including the effects of gravity load and wind pressure on the wall—is illustrated in Sec. 51.3. While larger

than usual studs may be required for the heavier-loaded walls, most walls are likely to be of minimum construction: typically 2 × 4s of stud grade at 16 in. center-to-center spacing.

Columns for the support of the large beams must be incorporated as necessary into the walls, as discussed in Sec. 55.2. The details for achieving this in an acceptable manner may affect the choice of the stud sizes, if a thicker wall is required.

The exterior walls must also be developed as shear walls as discussed in Sec. 55.4.

Some possible details for the stud wall construction are presented in Sec. 55.5.

55.3. DESIGN FOR WIND

Design for lateral force effects includes consideration for wind and seismic forces. Design for wind is based on the requirements of the 1991 edition of the *UBC* (Ref. 1) and an assumed basic wind speed of 80 mph. Seismic requirements are also based on the *UBC* and an assumed seismic zone 3 condition.

It is quite common, when designing for both wind and seismic forces, to have some parts of the structure designed for wind and others for seismic effects. In fact, what is necessary is to analyze for both effects and to design each element of the structure for the condition that produces the greater effect. Thus the shear walls may be designed for seismic effects, the exterior walls and window glazing for wind, and so on.

For wind it is necessary to establish the design wind pressure, defined by the code as

$$p = C_e C_q q_s I$$

where C_e is a combined factor, including concerns for the height above grade, exposure conditions, and gusts. From *UBC* Table 23-G (see Appendix D), assuming exposure *B*:

$$C_e = 0.7 \text{ from 0–20 ft above grade}$$
$$= 0.8 \text{ from 20–40 ft}$$
$$= 1.0 \text{ from 40–60 ft}$$

and C_q is the pressure coefficient. Using the projected area method (method 2) we find from *UBC* Table 23-H (see Appendix D) the following.

For vertical projected area:

$$C_q = 1.3 \text{ up to 40 ft above grade}$$
$$= 1.4 \text{ over 40 ft}$$

For horizontal projected area (roof surface):

$$C_q = 0.7 \text{ upward}$$

The symbol q_s is the wind stagnation pressure at the standard measuring height of 30 ft. From *UBC* Table 23-F the q_s value for a speed of 80 mph is 17 psf.

For the importance factor *I* (*UBC* Table 23-L we use a value of 1.0.

Table 55.1 summarizes the foregoing data for the determination of the wind pressures at the various height zones for Building Two. For the analysis of the horizontal wind effect on the building, the wind pressures are applied and translated into edge loadings for the horizontal diaphragms (roof and floors) as shown in Fig. 55.4. Note that we have rounded off the wind pressures from Table 55.1 for use in Fig. 55.4. Assuming the building to be a total of 122 ft wide in the east–west direction, the forces at the three levels are

$$H_1 = 195 \times 122 = 23{,}790 \text{ lb } [106 \text{ kN}]$$

$$H_2 = 234 \times 122 = 28{,}548 \text{ lb } [127 \text{ kN}]$$

$$H_3 = 227 \times 122 = 27{,}694 \text{ lb } [123 \text{ kN}]$$

and the total wind force at the base of the shear walls is the sum of these loads, or 80,032 lb [356 kN].

The Plywood Shear Walls

The lateral bracing system proposed for this scheme is a combination of the horizontal roof and floor deck diaphragms and the exterior walls sheathed with structural plywood. This makes the structure essentially similar to that for Scheme 2 of Building One. For the wind on the long side of the building the resisting walls are as shown in the wall elevation in Fig. 55.5, consisting of the 92 ft wide walls pierced for the window openings.

The critical section for maximum shear stress in the three story shear wall in Fig. 55.5 occurs through the window openings at the first story. The shear force here is one-

TABLE 55.1.   Design Wind Pressures for Building Two

Height above Average Level of Adjoining Ground (ft)	C_e	C_q	Pressure,[a] p (psf)
0–20	0.7	1.3	15.47
20–40	0.8	1.3	17.68
40–60	1.0	1.4	23.80

[a] Horizontally directed pressure on vertical projected area: $p = C_e \times C_q \times 17$ psf.

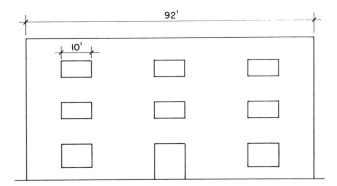

FIGURE 55.5. Elevation of the shear wall.

half the total force previously determined, or 80,032/2 = 40,016 lb. Subtracting the width of the window openings the net section for shear stress is 92 − 30 = 62 ft, and the shear stress is thus 40,016/62 = 645 lb/ft.

From *UBC* Table 25-K-1 (Appendix D), a possible choice is $\frac{15}{32}$-in.-thick Structural I plywood with 10d nails at 3-in. centers. The table footnote indicates that this nailing requires nominal 3-in.-thick studs, but this may already be a choice from the design for gravity plus direct wind pressure.

The upper stories of this wall and the whole wall on the long sides will be less stressed and can probably use a lower-grade plywood, smaller nails, wider nail spacing, or other forms of reduction approaching the minimum required for the plywood sheathing.

Because of the low height-to-length aspect ratio and the considerable dead load of the wall, overturn is not likely to be a problem for this wall or for the building in general.

For horizontal sliding the usual minimum requirement for sill bolts at 6-ft centers would result in approximately 16 bolts for this wall. The bolt load would thus be 40,016/16 = 2501 lb per bolt, which is quite high. This indicates the need for use of a slightly larger-than-average bolt and some closer-than-ordinary spacing, and possibly a sill thicker than the usual 2-in.-nominal size.

55.4.   CONSTRUCTION DETAILS

Details for the construction of the roof and foundation structures for this building will be similar to those shown for Building One. The discussion here is therefore limited to the construction at the typical upper floors of Building Two.

Building finishes, insulation, weather seals, vapor barriers, and other elements will vary by region and climate. However, most of the details for the basic structure will be the same in most parts of the United States. Special concerns for high windstorm areas or areas of high seismic risk may be the single major source for variations.

Floor Framing Plan

Figure 55.6 shows a partial framing plan at one building corner. The section marks on the plan indicate the details shown in Fig. 55.7.

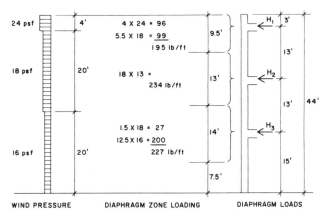

FIGURE 55.4. Investigation for lateral load: (a) development of the wind load; (b) loading of the stair tower.

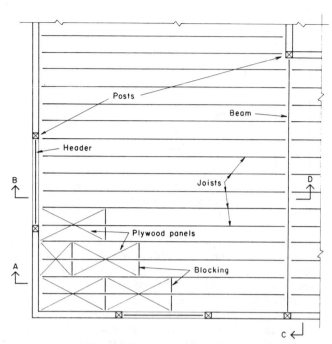

FIGURE 55.6. Framing plan for the typical floor.

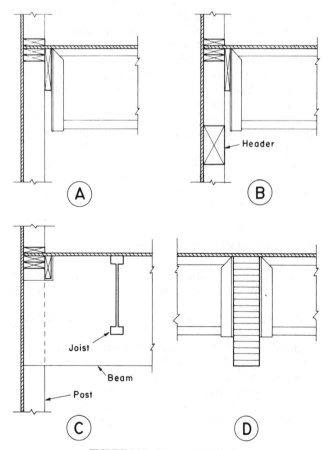

FIGURE 55.7. Construction details.

Note the references on the plan to the columns built into the wall construction at the end of the beams. Indication of these on the plans with the sizes of members to be used is usually sufficient, unless the members do not match the nominal dimensions of the wall studs.

Notes on the plan indicate the use of blocking for the floor deck panels. This may or may not be required for the floor diaphragm action for lateral loads. However, some lateral bracing of the joists is probably required and the support of the panel edges is much more effective for floor usage by this means. Details for the blocking or bridging would depend partly on the form of the joists used. Panels with tongue-and-groove edges might be used to avoid blocking.

Construction Details

Detail A.   This shows the joint between the floor and the solid portion of the exterior wall. This is generally a typical form for multistory construction with the platform framing method.

The floor deck diaphragm is terminated at the outside edge of the stud wall, where the boundary nailing for the diaphragm is required. If the exterior plywood wall sheathing (for the shear wall) is nailed to the same framing manner, the lateral force from the horizontal diaphragm will generally be effectively transferred to the vertical shear wall at this location.

The detail assumes the use of fabricated joists, which are hung at the face of the stud wall. If solid-sawn wood joists are used, they would probably be extended into the stud wall and rest on top of the double-cap plate of the lower-story wall.

Detail B.   This is basically the same as Detail A, except that the joint occurs at the window, and the ends of the joists are supported by the header beam over the wall-opening for the window. Variations on this would relate to the particular relations between the locations of the ceiling and the head of the window and the distance between the top of the floor framing and the level of the ceiling. In this detail the ceiling is suspended some distance and a short stud wall (called a *cripple wall*) is placed on the top of the header. As a result, the joist-to-wall detail is the same as in Detail A.

Detail C.   This shows the joint between the large laminated beam and the exterior wall. The beam is simply extended into the wall cavity and rests on the post that is incorporated into the wall. A standard post cap/beam seat of steel would be used to anchor the beam end to the post inside the wall (see Appendix F).

Detail D.   This shows the joint between the interior beam and the joists. This is simply a variation on the detail shown in Fig. 55.2, with the beam shown here as the glue-laminated timber.

CHAPTER FIFTY-SIX

Building Two, Scheme 2: Steel Structure

A structural framing plan for the upper floors in Building Two is shown in Fig. 56.1, indicating a system using steel beams and columns. For the arrangement shown, the alignment of columns in rows and the continuity of beams on the column lines permits consideration of the development of a *rigid frame* structure for lateral load resistance. (Rigid frame is an old common term; the *UBC* calls this a *moment-resistive space frame*.)

The design of a rigid frame structure is discussed for the sitecast concrete scheme in Chapter 58. For the steel building a truss-braced core will be developed as the vertical element of the lateral load resisting system. Use of the braced core largely relieves the rest of the steel frame from concerns for lateral loads. Thus most columns may be designed for axial compression only and most beams can be designed for simple span conditions.

In fact, the neat arrangement of columns and beams becomes less significant, although it still offers many planning advantages.

56.1. GENERAL CONSIDERATIONS

Figure 56.1 shows a framing system that uses rolled steel beams spaced at some module relating to the column spacing. As shown, the beams are 7.5 ft on center, and the beams that are not on the column lines are supported by column line girders. Thus three-fourths of the beams are supported by the girders and the remainder are supported directly by the columns. The beams in turn support a one-way spanning deck, consisting of formed sheet steel units with a poured-in-place concrete topping.

Within this basic system scheme there are some variables to be considered. The spacing of the beam system is one. The type of deck used and the depth restriction for the beams would influence this choice. Within the 30-ft column module logical possibilities are for 6-, 7.5-, 10-, and 15-ft spacings. For the shallow deck system chosen, the 7.5-ft spacing is reasonable, but not especially evident as a choice.

Orientation of the wide-flange steel columns is another consideration. If a rigid frame is used for lateral load resistance, the system shown might be chosen so that there are the same total number of columns in each direction turned to present their major stiffness (as indicated by I_x).

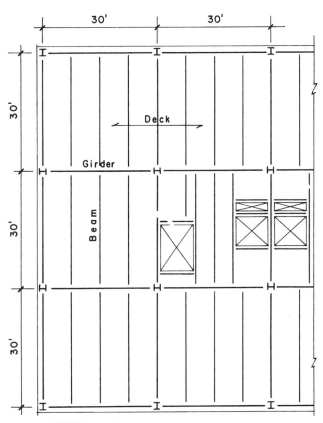

FIGURE 56.1. Steel framing system for Building Two.

This biaxial equalization of total stiffness is somewhat more meaningful for seismic resistance. For wind, the load will be greater in one direction, as the building is oblong in plan.

As shown, location of major openings for duct shafts, elevators, and stairs should be developed so as not to interfere with any of the girders or beams on the column lines. Major plumbing and wiring risers should also be kept off of the column lines. This is essential to the integrity of the vertical column-beam bents for the lateral load resisting system.

A similar system would likely be used for the roof of this building. A light steel frame structure is also assumed for the penthouse structure on the roof that covers the elevators, stairs, and rooftop HVAC equipment.

Some consideration must be given to the fireproofing of the steel frame and deck. We will assume this to be accomplished as follows.

1. Top of the steel deck and exposed faces of the spandrel beams and beams at openings: poured concrete (probably with lightweight aggregate).
2. Exterior columns, interior sides of beams and girders, and the underside of the steel deck: sprayed-on fireproofing.
3. Interior columns: metal lath and plaster.

A36 steel will be used for all steel frame members.

56.2. DESIGN OF THE TYPICAL FLOOR

Several options are possible for the floor deck. In addition to structural concerns, which include gravity loading and diaphragm action for lateral loads, consideration must be given to fireproofing and accommodation of wiring and piping for building services. For office buildings a system that is sometimes used is one that incorporates a two-way wiring distribution network for both power and communication in the structural floor. Figure. 56.2 shows some details and options for this type of system. If the structural floor deck is a concrete slab, either precast or poured in place, the wiring network is usually buried completely in a nonstructural concrete fill on top of the structural slab. If a steel deck is used, it is sometimes possible to use closed cells of the formed deck for wiring in one direction, reducing the wiring in the fill to that for the perpendicular direction only. The latter system is assumed for this design, using a steel deck with a 1.5-in. depth of the corrugations and a fill of 2.5 in. on top of the deck. Total dead weight of this deck combination depends on the thickness of the sheet steel, the profile of the formed deck units, and the unit density of the concrete fill. We will assume the deck to weigh 35 psf [1.68 kPa], which is an average value with lightweight concrete fill. In addition to the deck, we will assume a dead load of 15 psf [0.72 kPa] for the ceiling, lights, ducts, and supported fixtures. Thus the total dead load superimposed on the deck is 50 psf [2.40 kPa].

In addition to this dead load, *UBC* Sec. 2304d requires a load of 20 psf (0.96 kPa] to account for movable partitions. Although this is technically a dead load, its randomness of occurrence makes it similar to live load. Other live load consists of that specified for office buildings. For the typical office areas, the specified load is 50 psf [2.40 kPa], with the extra stipulation that the floor also sustain a concentrated load of 2000 lb [8.90 kN]. For corridors and lobbies, however, the code requires a live load of 100 psf [4.79 kPa]. Since it is not necessary to consider partitions in the corridor or lobby space, the comparison is really between 100 psf in some areas and 70 psf in others. This presents a problem in the building designed to accommodate future rearrangement of the interior, since the exact location of corridors and lobbies cannot be predicted. A conservative approach would be to design for 100-psf live load for the

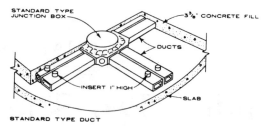

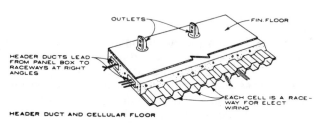

FIGURE 56.2 Details of wiring systems incorporated in floors. Reproduced from *Architectural Graphic Standards* (Ref. 13), with permission of John Wiley & Sons, Inc., New York.

entire floor, ignoring the partition loading. As a compromise we will use the following loadings for the various elements of the structure.

1. For the deck:

    Live load = 100 psf

    Dead load = 50 psf, including the deck weight

2. For the beams that directly support the deck:

    Live load = 80 psf
    (100 psf if combined with partitions)

    Dead load = 70 psf, including partitions
    but not the weight of the beams

3. For girders and columns:

    Live load = 50 psf

    Dead load = 70 psf + weights of
    beams, girders, columns, and walls

Although some generalized information is available for steel decks, it is best to select deck data and determine necessary design data from the product catalogs of manufacturers who service the area of a proposed building site. As shown in the various details presented here, we have used a deck that is quite widely available, although other products are competitive.

The typical beam is a simple span beam supported by ordinary framing connections to the webs of the girders. For the 30-ft-span beam on 7.5-ft centers, the loading is thus:

Beam DL = 7.5(70) = 525 psf + beam weight, say 560 plf

Live-load reduction = 0.08(A − 150) =
$$0.08(225 − 150) = 6\%$$

Live load = 0.94(7.5)(80) = 564 plf

Total unit load on beam = 1124 plf

Total supported load = 1.124(30) = 33.7 kips [150 kN]

For consideration of bending moment only, Table 22.2 yields choices of W 16 × 45, W 18 × 46, or W 21 × 44. Actual selection may be influenced by the depth of the girders or the need for space beneath the beams for ducts or other items. Inspection of Fig. 22.6 will show that the shallowest option beam, the 16-in.-deep section, is not critical for a total load deflection of L/240.

Figure 56.3 shows the loading condition for the girder, assuming only the loads due to the supported beams. While this ignores the effect of the uniformly distributed load of the girder weight, it is reasonable for an approximate design, since the girder weight is a minor loading. For the beam load we determine the following:

Total beam DL = 0.570 × 30 = 17.1 kips.

Live-load reduction = 0.08(A − 150) = 0.08 (675 − 150) = 42%.

(Note that the girder carries three beams, or three-fourths of one full column bay of 900 ft².) The maximum permitted reduction is 40%; thus

total beam live load = 0.60(0.050)(7.5)(30) = 6.75 kips

total load on girder =
$$17.1 + 6.75 = 23.85 \quad \text{say, 24 kips [106 kN]}$$

Using the maximum moment of 360 kip-ft from Fig. 56.3b, a trial section for the girder may be selected from the graphs in Fig. 22.2, assuming a lateral, unsupported length of 7.5 ft (the beam spacing). The lightest section thus determined is a W 24 × 84. It is possible, however, that some other section may be desirable for its increased stiffness or the form of its flanges to facilitate the welded connections. For example, from Table 22.1, it may be observed that there is a deeper beam available at the same weight as the W 24 ×

84; namely, the W 27 × 84. From a comparison of the properties for the moments of inertia for these two beams, as given in Table A.3, it may be determined that the 27-in. beam will have approximately 20%-less deflection; an improvement obtained at no additional cost in terms of weight of steel. However, the additional 3 in. will probably increase the story-to-story height, which becomes a critical issue in terms of additional cost for the exterior walls and other factors that may well rule against the choice.

The spandrel framing (the beams and girders at the building edge) have less gravity loads from the floors, but must carry the building curtain wall. If the wall construction is heavy, they may have a total load similar to that of the interior members. Selection of the spandrel members involves some different considerations from those for the interior members. Deflection may be more critical, as it involves distortions of the curtain wall construction. On the other hand, depth is less restricted, and a deeper and stiffer member may be possible. As is done with the concrete structure in Chapter 58, it may be possible to use a very deep spandrel beam to develop a perimeter bent bracing system, although this is not the scheme chosen for this example.

A consideration for the steel frame is that of the tolerable vertical deflection of the floor and roof systems. This is a complex issue involving considerable judgment and not much factual criteria. Some of the specific considerations are:

The Bounciness of the Floors. This is essentially a matter of the stiffness and fundamental period of vibration of the deck. Use of static deflection limits generally recommended will usually assure reasonable lack of bounce.

Transfer of Bearing to the Curtain Wall and Partitions. The deformations of the frame caused by live gravity loads and wind must be considered in developing the joints between the structure and the nonstructural walls. Flexible gaskets, sliding connections, and so on, must be used to permit the movements caused by these loads as well as those due to thermal expansion and contraction.

Dead-Load Deflection of the Floor Structure. As shown in Fig. 56.4, there is a cumulative deflection at the center

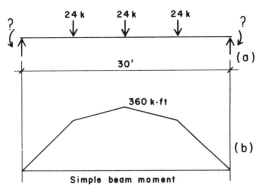

FIGURE 56.3. Approximation of the gravity moments on the girder.

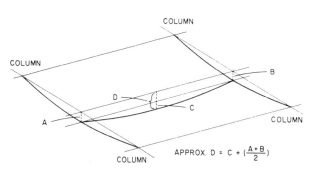

FIGURE 56.4. Deflection of the floor.

of a column bay because of the deflection of the girders plus the deflection of the beams. Within the normally permitted deflections, this could add up to a considerable amount for the 30-ft spans. If permitted to occur, the result would be that the top of the metal deck would be several inches below that at the columns. Because the top of the concrete fill must be as flat as possible, this could produce a much thicker fill in the center of the bays. One means for compensating for this is to specify a camber for the beams approximately equal to the calculated dead-load deflection.

56.3. DESIGN OF THE COLUMNS

With the core bracing system for lateral load resistance, the exterior columns are all basically designed for vertical gravity loads only. With the steel beams having essentially pin connections to the columns (negligible moment transfer capacity), the exterior columns may therefore be designed for axial compression force only.

The interior columns are all involved in core framing conditions (see Fig. 56.1), making the determination of their gravity loads somewhat more complicated. Additionally, some of the core columns will be involved in the development of the trussed bents for lateral loads. Figure 56.5 shows a revised core plan with some additional columns that will be used to develop the trussed bents, which are discussed in Sec. 56.4.

Figure 56.6 is a common form of tabulation used to determine the column loads. For the exterior columns, there are three separate load determinations, corresponding to the three-story-high columns. For the interior columns, the assumption is made that there is a rooftop structure (penthouse) used to house the HVAC and elevator equipment, thus creating a fourth story for these columns.

Figure 56.6 gives tabulations for loads for the two cases of typical exterior columns. It also gives a tabulation for a hypothetical interior column, ignoring the core conditions and assuming a full load periphery of 30 ft × 30 ft; the purpose being to illustrate the procedure for such a column but also to give some approximation for the size selection of the columns on the major column lines.

Organization of Figure 56.6 is designed to facilitate the following determinations.

1. Dead load on the portion of the horizontal structure supported by the column; calculated as an estimated unit load per square foot times the load support periphery. Loads developed in the process of design of the horizontal framing may be used to estimate the unit loads.

2. Live load on the same horizontal periphery.

3. Live load reduction to be used, based on the total periphery; for lower-story columns being the sum of the load peripheries for all the levels that they support.

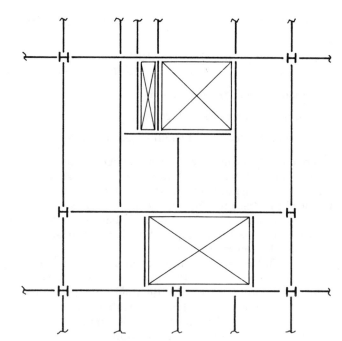

FIGURE 56.5. Plan of revised core with bent columns.

4. Other dead loads supported, such as the estimated weight of the columns and of any walls within the column load periphery that are directly supported.

5. A summation of the total load collected at each story.

6. A design load for each story, using the total accumulation from all the stories supported.

For the entries in Figure 56.6 the following assumptions were made.

Roof unit live load = 20 psf (reducible).

Roof dead load = 40 psf. (Estimated, based on the floor construction.)

Penthouse floor live load = 100 psf (for equipment).

Penthouse floor dead load = 50 psf.

Floor live load = 50 psf (reducible).

Floor dead load = 70 psf (including partitions).

Interior walls weigh 15 psf/ft^2 of wall surface area.

Exterior walls average 25 psf/ft^2 of wall surface area.

Figure 56.7 summarizes the column designs for the three columns for which load tabulations are presented in Figure 56.6. For the pin-connected frame, a K factor of 1.0 is assumed for the columns, and the full-story heights are used as the unbraced column lengths.

Although column loads in the upper stories are quite low, and some small column sizes would be adequate for the loads, a minimum size of 10 in. has been used for the W-shaped columns. There are two considerations for this size choice. First is one involving the form of the horizontal framing members (W shapes in this case) and the type

Level	Load Source	Corner Column 225 ft²			Intermediate Exterior Column 450 ft²			Interior Column 900 ft²		
		DL	LL	Total	DL	LL	Total	DL	LL	Total
P'hse.	Roof							8	5	
Roof	Wall							5		
	Total/level							13	5	
	Design load									18
Roof	Roof	9	5		18	9		36	18	
	Wall	10			10			10		
	Column	3			3			3		
	Total/level	22	5		31	9		49	23	
	Design load			27			40			72
3rd	Floor	16	11		32	23		63	45	
Floor	Wall	10			10			10		
	Column	3			3			3		
	Total/level	51	16		76	32		125	68	
	LL reduction	24%	12		60%	13		60%	27	
	Design load			63			89			152
2nd	Floor	16	11		32	23		63	45	
Floor	Wall	11			11			11		
	Column	4			4			4		
	Total/level	82	27		123	55		203	113	
	LL reduction	42%	16		60%	22		60%	45	
	Design load			98			145			248

FIGURE 56.6. Column load tabulation.

Level	Story	Unbraced Height (ft)	Corner Column		Intermediate Exterior Column		Interior Column	
			Design Load (kips)	Column Choices	Design Load (kips)	Column Choices	Design Load (kips)	Column Choices
Roof								
	3rd	13	27	W10X33	40	W10X33	72	W10X39
3rd Floor								
	2nd	13	63	W10X33	89	W10X33	152	W10X39
2nd Floor			Assumed location of column splice					
	1st	15	98	W10X33	145	W10X33	248	W10X49
1st Floor								

FIGURE 56.7. Column design.

of connections between the columns and the horizontal framing. Regardless of their orientation in plan, the H-shaped columns must usually facilitate the connection of framing on both of their axes. If ordinary beam framing is used, employing connection angles and field erection bolt-ing, a minimum size column is required for practical installation of the angles and bolts.

Figure 56.8 shows the plan layout of a column and horizontal framing at the building corner. For framing members attached to the column flanges, a minimum

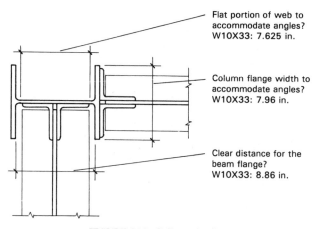

Flat portion of web to accommodate angles? W10X33: 7.625 in.

Column flange width to accommodate angles? W10X33: 7.96 in.

Clear distance for the beam flange? W10X33: 8.86 in.

FIGURE 56.8. Column details.

width of column flange is required—usually at least 6 in. For framing members attached to the column webs, a minimum column depth is required in order to have a sufficient distance of width for the flat portion of the column web to which the angle legs can be attached. The sizes of bolts required, the minimum angle leg width to accommodate the bolts, and the actual thickness of the beam web will determine this required dimension for the column in a specific case. See discussion of steel-bolted connections in Chapter 24.

A second consideration for the column size dimension is the problem of achieving splices in the column, where the building height makes a single piece column impractical. The length of a single column piece that can be handled for transportation and erection depends mostly on the size of the member. A 6-in. W shape will become as flexible as a noodle at a length of 40 ft. A 14-in. W shape, on the other hand, can be handled at some considerable length. The length issue and the chosen column size will generally determine whether a splice is necessary.

If the column can be produced as a single piece (in our case, slightly longer than 40 ft) the splice need not be considered. However, if a splice *is* required, it is generally most easily achieved with two members of the same nominal depth. See discussion of column connections in Chapter 24.

For this design example it is assumed that a 10-in. W shape is the minimum desirable size to comfortably accommodate the framing. It is a marginal case for this size in terms of a one-piece column, so a splice should be anticipated in the design; meaning that any size changes should be made within the range of available 10-in. W shapes. If the steel erector decides that a splice is not necessary, the cost of providing a larger column for upper stories will usually easily be offset by the elimination of the cost of achieving the splice.

With all of these considerations, the trial set of column sizes is given in Figure 56.7. From sizes indicated in Table A.3 a W10 × 33 is selected as the minimum size column; providing a flange width of approximately 8 in. For the final selections, it is assumed that a splice occurs just above

the second floor level. Larger sizes of columns are chosen from Table 23.2 in Part 4, which generally includes all the sizes reasonably useable for columns. It should be noted that Table A.3 presents only a limited number of the available W shapes; the full tabulation being referenced from the tables in the *AISC Manual* (Ref. 5).

56.4. DESIGN FOR WIND

Referring to Fig. 56.5, it may be noted that there are some extra columns in the framing plan at the location of the stairs and the restrooms. These columns are used in conjunction with the regular 30 ft on center columns and the members of the horizontal framing to define vertical planes of framing (called *bents*) for the development of the truss-bracing system shown in Fig. 56.9.

The combination of the vertical and horizontal framing cannot be made resistive to lateral loads unless moment-resisting connections are used (generally flange-to-flange-welded connections for the frame with W shapes). To achieve stability here, the frame is trussed by adding diagonal members. These may be single diagonals, but a system often used is one with double diagonals in the form of X-bracing, as shown in Fig. 56.9. For a simplified design, these diagonals are assumed to work in tension only, which may literally be true if the diagonals are relatively slender and will buckle under a small compression force.

Referring to Fig. 56.9, it may be noted that the bent system provides four bents with X-bracing in each of the two directions of the building's main axes. The individual bents may be approximately designed as single, statically

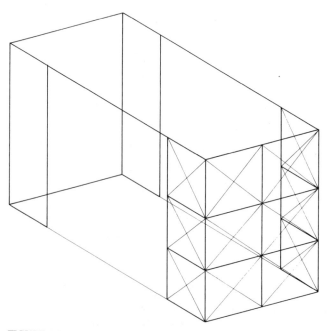

FIGURE 56.9. Development of the core-bracing system for lateral load resistance for Building Two.

determinate, cantilever trusses, loaded at the levels of the horizontal framing by wind forces delivered through the horizontal deck diaphragms. Determination of the wind forces on the building is essentially the same as for Scheme 1. We therefore consider the resistance of the forces indicated in Fig. 55.4b as H_1, H_2, and H_3.

Investigation of one of the braced bents in the long direction of the building plan is shown in Fig. 56.10. The loads at the upper levels of the frame are determined by multiplying the unit loads determined in Fig. 55.4a by the narrow width of the building and dividing by 4; since there are four bents resisting the total wind force. Thus, for example

$$H_1 = 195 \times 92 \div 4 = 4485 \, \text{lb}$$

Analyzed as a cantilever truss, ignoring the compression diagonals, the resulting internal forces in the bent are as shown in Fig. 56.10c. The forces in the diagonals may be used to design tension members, using the usual one third increase in allowable stresses for the wind loads. The compression forces in the vertical members are added to the vertical gravity forces for design of the columns. The principal effects of the vertical tension forces involve the possible necessity to anchor the bents against uplift due to overturning effect if the dead loads do not provide the required counteracting moment of resistance to the overturning moment; essentially similar to the case of the shear wall.

The horizontal forces must be added to the beams in the core framing and an investigation made for the combined bending and axial compression. This can be critical, since beams are ordinarily quite weak on their minor axes (the y-

axis), and it may be practical to add some additional horizontal framing members to reduce the lateral unbraced length of some of these beams.

Design of the diagonal members and of their connections to the frame must be developed with consideration of the form of the frame members and the general form of the wall construction that encloses the steel bents. Figure 56.11 shows some possible details for the diagonals for the bent analyzed in Fig. 56.10. A consideration to be made in the choice of the diagonal members is the necessity for the two diagonals to pass each other in the midheight of a bent level. If the most common truss members—double angles—are used, it will be necessary to use a joint at this crossing, and the added details for the bent are considerably increased.

An alternative to the double angles is to use either single angles or channel shapes. These may cross each other with their flat sides back-to-back without any connection between the diagonals. However, this involves some degree of eccentricity in the member loadings and the connections, so they should be designed conservatively. It should also be carefully noted that the use of single members results in single shear loading on the bolted connections.

56.5. CONSTRUCTION DETAILS

The illustrations in this section show some of the construction details for the building with the steel frame. Since we did not discuss the design of the roof and penthouse, we have shown the structural plans for a typical floor, the first floor, and the basement and foundations only.

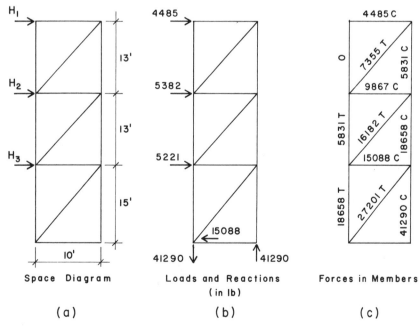

FIGURE 56.10. Investigation of the trussed bents.

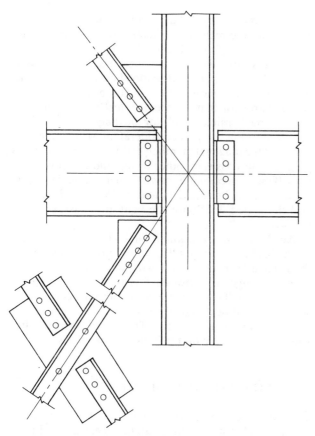

FIGURE 56.11. Details for the braced bent.

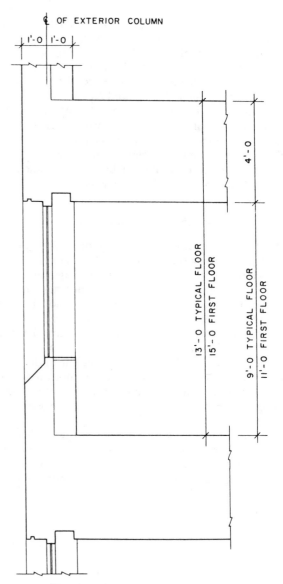

FIGURE 56.12. Profile of the exterior wall.

There are obviously many details that need coordination between the structure and the various architectural elements. We have shown some possibilities for some of the architectural elements of the ceilings, walls, floors, and the exterior curtain wall in order to illustrate the relations that need consideration. Because our principal concern is for the structure, however, we have not shown the complete building construction details.

Building Exterior: General Form

Figure 56.12 shows a profile section of the exterior wall and upper floors. The exterior wall is thickened to incorporate the column, which is centered on the wall. A total distance of 4 ft is provided from the underside of the finished ceiling to the top of the floor above. There are many possible variations of this form and of the general relations between the building skin and the structure. A general discussion of some of the issues is given in Chapter 2. We show in this chapter some details that are not uncommon, but the alternatives are numerable.

Structural Plans

Basement and Foundation Plan (Fig. 56.13). The partial plan in Fig. 56.13 shows details for typical conditions

where a partial basement is used. Although the building did not indicate the existence of a basement, we have shown one here to illustrate some possible construction for this condition. The plan also indicates the use of a series of intermediate piers between the building columns. These piers are for the support of a concrete framed floor structure at the first-floor level. Although a foundation wall is shown at the building edge, the exterior columns are shown with individual square footings. This issue was discussed in the design of the foundations for buildings in Sec. 41.6. If the edge walls are quite short, the plan as shown is probably preferred.

The plan indicates the use of a concrete slab on grade for the basement floor. No intermediate piers are used

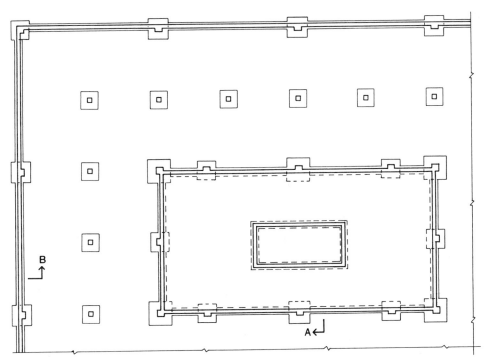

FIGURE 56.13. Foundation plan with grade walls and a partial basement.

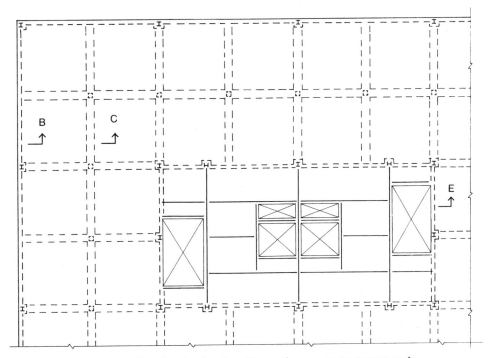

FIGURE 56.14. First-floor framing plan with spanning concrete structure on grade.

here, and it is assumed that the first-floor structure over this area is supported by a structure similar to that for the upper floors.

First-Floor Framing Plan (Fig. 56.14). The first-floor area over the basement is shown with a structure similar to that for the typical upper level floors. At the unexcavated por-

tions the first-floor structure consists of a concrete slab and beam system poured directly over fill. The typical beam is formed as shown in detail C (Fig. 56.17). This network of beams on 15-ft centers is supported by the basement and grade walls and by a series of intermediate piers. The typical slab is a two-way slab on edge supports with beams cast with the slab.

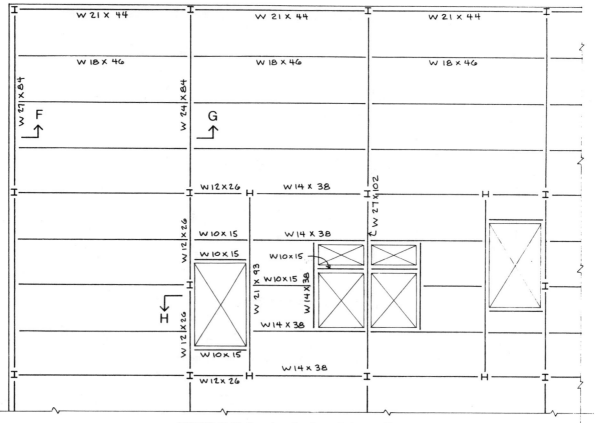

FIGURE 56.15. Framing plan for typical upper floor.

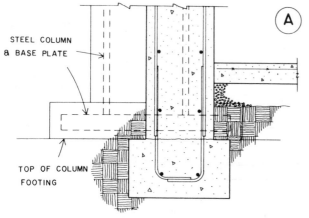

FIGURE 56.16. Foundation detail.

These two systems interface at the interior basement walls, as shown in detail E (Fig. 56.18). The concrete fill is continuous over both systems. A special detail required is the support for the steel beams at the concrete walls, as shown in details D and E (Fig. 56.18). This consists of a pocket in the wall with some erection bolts and a bearing plate for the end of the beam.

Typical Floor Framing Plan: Upper Levels (Fig. 56.15). The typical floor consists of the steel deck placed over the net-

work of beams that provide supports at 7.5-ft centers as well as at the building edges and at the edges of large openings. Some of the considerations in the detailing of this system are as follows:

1. *Attachment of the Steel Deck.* The deck units are normally welded to the steel beams in the valleys of the deck corrugations. These connections must transfer the lateral loads from the floor, acting as a diaphragm, to the steel beams, which in turn transfer the loads to the column-beam bents. Details for this attachment and load ratings for the connections and deck are usually provided by the deck manufacturers.

2. *Support for the Hung Ceiling, Lights, Ducts, and So On.* The usual method of support for the ceiling is by wires that are installed through holes in the steel deck. Wire sizes and spacing and details for installation will depend on the type of ceiling and the specific type of deck used. Lighting units, ceiling registers, and small ducts may be supported as part of the ceiling construction, using the same wire hangers. Heavy ducts and equipment elements of the HVAC system will usually be supported by brackets or hangers attached directly to the beams.

3. *Fireproofing of the Floor Construction.* As discussed previously, the system used here consists of utilizing the concrete fill on top of the deck to protect the upper surface of the deck as well as the faces of beams at the building edge and the edges of openings. The underside of the deck

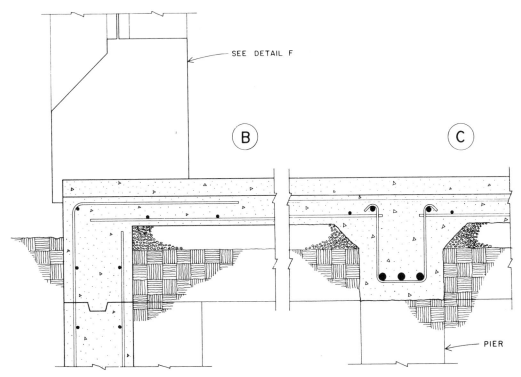

FIGURE 56.17. Details of the first-floor structure.

and the remaining exposed surfaces of beams are protected by sprayed-on fireproofing.

4. *Support for Interior Walls.* Planning of this building envisions two basic types of interior walls—permanent and demountable. Permanent walls are limited to those in the core and would consist of masonry or metal-framed plastered partitions. The weight of these plus their permanency would dictate that the floor structure provide both vertical and lateral support as part of the permanent structural system. Construction and finish of these walls would be influenced by desired architectural details, as well as by considerations of structural design, required code fire rating, acoustic separation, and similar factors.

Demountable walls may consist of a variety of constructions. Some of them may consist of masonry or plastered partitions, although the design of the floor deck should consider this if such is the case. It is expected, however, that most walls will consist of some relatively light construction, including possibly the use of some patented, modular system of relocatable units. The choice of the ceiling system and its detailing would need to consider the necessity for providing attachment and lateral support for whatever walls are anticipated.

The framing plans shown are abbreviated for clarity. Complete construction drawings would include sufficient information to establish the exact location of all beams, to establish the elevations of beam tops, to indicate required camber of beams, and to identify the type of connection for each beam. Some of this information may also be provided

in details or schedules and be referred to by notes or symbols on the plans.

Construction Details

The locations for the details described in this section are indicated on the plans in the preceding section.

Detail A (Fig. 56.16). This section shows the relations between the basement floor, the basement wall, and the steel column and its footing. The basement floor is a paving slab over compacted fill. Necessity for a moisture barrier under the slab and moisture penetration-resistant treatment at the slab-to-wall intersection would depend on the groundwater conditions at the site.

The basement wall at this point is a retaining wall, spanning vertically from the basement floor to the first-floor construction. Location of the wall footing was discussed previously in regard to the foundation plan. As shown in the detail, and also discussed previously, the top of the column footing is shown dropped so that the base plate and anchor bolts can be encased below the floor slab.

Detail B (Fig. 56.17). This shows the general condition at the building edge adjacent to the unexcavated lower level. To accommodate the rental areas on the first floor the first-story wall is assumed to be similar in detail to that for the typical floor, as shown in detail F (Fig. 56.18).

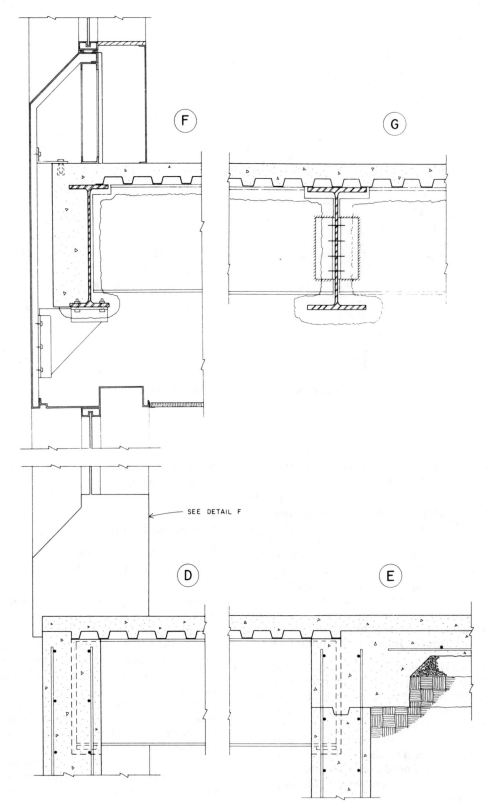

SEE DETAIL F

FIGURE 56.18. Details of the steel floor structures.

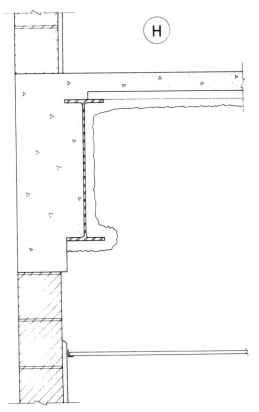

FIGURE 56.19. Framing at the core walls.

The first-floor structural slab, although poured over fill, is a spanning slab and is supported vertically by the grade wall. In cold climates there should be some insulation for the floor at the building edge and a thermal break between the floor slab and the exterior wall.

Detail C (Fig 56.17). This shows the typical beam for the unexcavated portion of the first floor. If the fill material is reasonably cohesive, the lower stem of this beam may be excavated by trenching as shown. The reinforcing in the beam requires the usual 3-in. cover as for footings.

The piers for these beams (midway between columns) would probably be poured with column forms before the fill is placed, with the pour stopped at the level of the bottom of the beams and the pier vertical reinforcing extending into the beam.

Detail D (Fig. 56.18). This shows the edge condition that would exist if the basement extended to the building edge. The first-story wall is assumed to be similar to that for detail B (Fig. 56.18). The steel beams are shown as sup-

ported by the wall, with a pocket, bearing plate, and erection anchor bolts.

Some consideration should be given to the transfer of horizontal force between the top of the wall and the floor construction if the basement wall is a retaining wall.

Detail E (Fig. 56.18). This shows the intersection between the two types of floor construction at the top of the basement wall. Because the concrete fill is continuous over both structures, the tops of the steel deck and the structural concrete slab are at the same level.

The key at the top of the wall pour should be adequate to provide for the lateral force due to the retained fill. The top of the wall pour is dropped to the level of the bottom of the concrete beams to allow the bottom reinforcing in the beams to extend over the supporting wall.

Detail F (Fig. 56.18). This demonstrates the typical spandrel condition at the upper floors. The metal framed window wall is shown centered on the column line with the finished face of the spandrel brought out flush with the finished face of the column. Although not shown in detail, it is assumed that the building skin at the spandrels and columns consists of insulated units with an exterior metal facing. These units are shown supported by brackets from the floor and the spandrel beam. The window wall units rest on a short steel stud wall with a wide sill brought out to the finished face of the column. The space under these sills may house HVAC units, if such a system is used. Lateral support must also be provided for the top of the window units. One way to achieve this would be to add some additional elements to the bracket that is attached to the bottom of the spandrel beam.

These details are of major concern in the architectural design and are subject to considerable variation without significant change in the basic structural system for the building.

Detail G (Fig. 56.18) This illustrates a typical beam-to-girder connection using standard double angle connectors. As shown, the angles are typically welded to the beams in the shop and field connected to the girders with bolts.

Detail H (Fig. 56.19). This gives one possibility for the floor edge condition at the large openings for the stair, elevators, and duct shafts. Although the section has concrete block for the wall, other materials may be used if a thinner wall is desired. One variation would be to stop the concrete closer to the steel beam and to run the wall past the face of the concrete.

CHAPTER FIFTY-SEVEN

Building Two, Scheme 3: Alternative Steel Structure

A possible alternative for the development of the horizontal framing structure for Building Two is presented in Chapter 57.

57.1. GENERAL CONSIDERATIONS

A framing plan for the typical upper floor of Building Two is shown in Fig. 57.1. By comparison with the plan in Fig. 56.1, it may be observed that the principal change relates to

the use of a closer spacing for the joists that directly support the deck. As noted on the plan, the intention here is to use open web steel joists for the joists and steel joist girders for the members that support the joists on the interior.

There are various possibilities for development of the spandrels at the building edge and the core framing for this system. In Fig. 57.1 it is assumed that the spandrels and the core framing are essentially the same as in the system developed in Chapter 56: using all W-shaped elements for the frame.

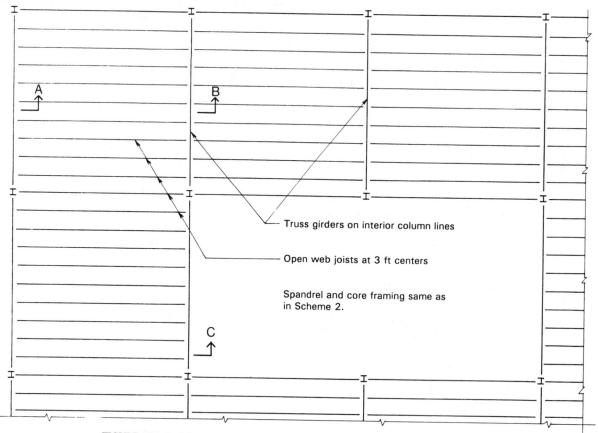

Truss girders on interior column lines

Open web joists at 3 ft centers

Spandrel and core framing same as in Scheme 2.

FIGURE 57.1 Building Two, Scheme 3: partial plan of framing for the typical upper floor.

There are also various alternatives for development of a lateral bracing system with this scheme. One possibility, however—the one assumed here—is the use of the same core trussed bents as in scheme 2 in Chapter 56.

Although considerable use is thus made of the same W-shaped framing as in scheme 2, approximately 80% of the floor and roof surfaces would be supported by the alternative structure in this scheme. Although somewhat more applicable to longer spans and lighter loads (roofs mainly), this system is frequently used for the situation described here.

One potential advantage of using the truss-form joists and girders is the degree of open nature of the interstitial space between the ceilings and the floor or roof above. This space can be used with much greater freedom for the necessary ducts, piping, and wiring than one generated with solid web beams.

57.2. DESIGN OF THE JOISTS

The open web steel joists may be designed by the same basic process as that used for the W-shaped joists in Chapter 56. These, as well as the joist girders, will be provided as proprietary products by some manufacturer, who may also contract or franchise the erection of the system. Determination of specific members will be made from the catalogs of the standard products of such a supplier, who will also supply considerable information for suggested construction details and construction specifications.

Using the data from this case, a joist for this situation is designed in Example 2 in Sec. 25.2 in Part 4. This design produces an answer reflecting the lightest member obtainable from Table 25.1. While this represents some efficiency for economy, a major concern with these light structures is often that of dynamic deflection (bounciness) when they are used for floors. We therefore recommend consideration of the use of the deepest feasible member for the situation without excessive increase in steel weight or compromise of the interstitial space. In general, increasing the depth or height of the trusses will produce reductions in both static deflection (sag) and dynamic deflection (bounce).

57.3. DESIGN OF THE JOIST GIRDERS

The joist girders will also be supplied by the supplier of the joists. Again, although there are industry standards for design and general form, the specific products of the individual supplier must be determined. The pattern of the truss members is relatively fixed and relates to the spacing of the joists. To achieve a reasonable proportion for the panel units of the trusses, the dimension for the depth of the truss girders should be approximately the same as that of the joist spacing. This relates to the achievement of the total dimension for the interstitial space (from underside of the ceiling to top of the finish floor) and to the desired story height (floor-to-floor).

The design of a truss girder for this situation is illustrated in Fig. 57.2. The depth of the girder assumed there is 3.0 ft, which should probably be considered as a *minimum* depth for this structure. Any additional height that can be obtained will probably reduce the amount of steel used and reduce deflections.

57.4. CONSTRUCTION DETAILS

Figure 57.3 shows some details for the construction of the trussed system. The deck shown here is essentially the same as that in scheme 2 with the W-shaped framing. It may be possible to use a lighter gauge deck because of the reduced span in this scheme. However, the design of the decks for horizontal diaphragm actions must also be considered.

There is a slight increase in the overall height of the structure in this system due to the fact that the joist ends sit on top of their supports; whereas the W-shaped joists in scheme 2 have their tops level with the supporting girders.

For the joist girder:

Use 40% LL reduction with LL of 50 psf.
Then, LL = 0.60(50) = 30 psf, and the total joist load is

(30 psf)(3 ft c/c)(30 ft span) = 2700 lb, or 2.7 kips

For DL, add partitions of 20 psf to other DL of 40 psf

(60 psf)(3 ft c/c)(30 ft) = 5400 lb

+ joist weight at (10 lb/ft)(30 ft) = 300 lb

Total DL = 5400 + 300 = 5700 lb, or 5.7 kips

Total load of one joist on the grider:

DL + LL = 2.7 + 5.7 = 8.4 kips

Girder specification for choice from manufacturer's load tables:
(See Ref. 11)

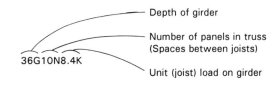

FIGURE 57.2. Design summary for the joist girder.

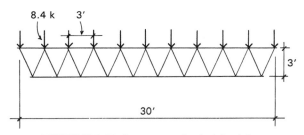

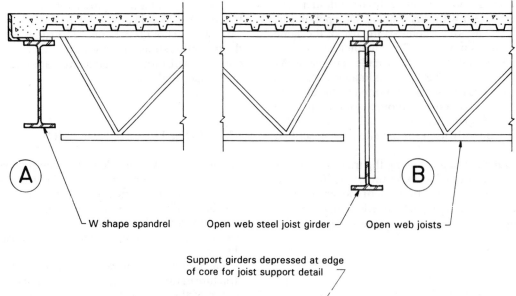

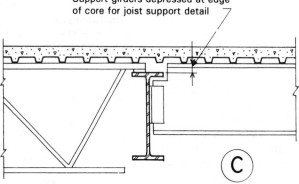

FIGURE 57.3. Details for the trussed floor framing system.

This typically adds only 2.5 in., but this is a carefully monitored dimension in design of multistory construction.

With the relatively closely spaced open web joists, the ceiling may be hung from the joists without requiring major spanning elements in the ceiling structure. Com-

pare this with the situation in scheme 2 where the deck-supporting beams are 7.5 ft on center.

In scheme 2 the ceiling is more likely to be supported by the deck; an option here also, but not so much required to keep the ceiling structure light.

CHAPTER FIFTY-EIGHT

Building Two, Scheme 4: Concrete Structure

Use of sitecast concrete structures for Building Two are discussed in Chapters 58 and 59. In Chapter 58 a solution is developed with the basic intent of using an exposed concrete structure; that is, one in which some parts of the basic structure are exposed to view. This principally involves some additional considerations for development of surface finishes and some possible greater attention to detailing of the form of members.

58.1. GENERAL CONSIDERATIONS

The structure for this scheme consists of a conventional, sitecast concrete system, using columns for vertical supports and slab and beam systems for the horizontal structures for the roof and upper floors. The continuity of the concrete casting and the extensions and anchorages of steel reinforcement tend to produce a natural, rigid frame-type structure and continuous, multiple-span horizontal framing. As a result, the usual form of lateral bracing is a rigid frame (moment-resisting space frame in *UBC* language).

This is a highly indeterminate structure for all loadings, and its precise engineering design will certainly be done with computer-aided methods; for which considerable software exists—albeit at a price. Here, we will discuss the major design considerations, some architectural design implications, and the use of some simplified hand calculation techniques for an approximate design of the structure.

As for other structures discussed in Part 9, there is considerable material in other parts of the book that can be referenced, and we will make note of these resources.

58.2. DESIGN OF THE SLAB AND BEAM FLOOR STRUCTURE

As shown in Fig. 58.1, the basic floor framing system consists of a series of beams at 10-ft centers that support a continuous, one-way spanning slab and are supported by column line girders or directly by the columns. We will discuss the design of three elements of this system: the continuous slab, the four-span beam, and the three-span spandrel girder.

The design conditions for slab, beam, and girder are indicated in Fig. 58.2. Shown on the diagrams are the positive and negative moment coefficients as given in Chapter 8 of the *ACI Code* (Ref. 6). Use of these coefficients is quite reasonable for the design of the slab and beam. For the girder, however, the presence of the concentrated loads makes the use of the coefficients improper according to the *ACI Code*. But for an approximate design of the girder, their use will produce some reasonable results.

Figure 58.1 shows a section of the exterior wall that demonstrates the general nature of the construction. On the basis of the criteria given in Chapter 51 and the construction shown, we determine the loadings for the slab design as follows:

Floor live load:
  100 psf (at the corridor) [4.79 kPa]
Floor dead load:
  Carpet and pad at 5 psf
  Ceiling, lights, and ducts at 15 psf
  2-in. lightweight concrete fill at 18 psf
  Assumed 5-in.-thick slab at 62 psf
    Total dead load: 100 psf [4.79 kPa]

Bending moments in the slab will be determined as

$$M = CwL^2 = C(200)(9)^2$$
$$= C(16,200) \quad \text{in ft-lb}$$

In this determination C is the moment coefficient shown in Fig. 58.2 for the appropriate location in the slab. The unit load w is for a 1-ft-wide strip of slab; L is the clear span in feet assuming a 12-in.-wide beam. We have chosen the 5-in. slab for an estimate primarily because this is probably the minimum thicknesses that will provide the necessary fire resistance to satisfy the building code. Assuming $\frac{3}{4}$-in. cover for the reinforcing, the effective depth will be approximately 4 in. (with a No. 4 bar).

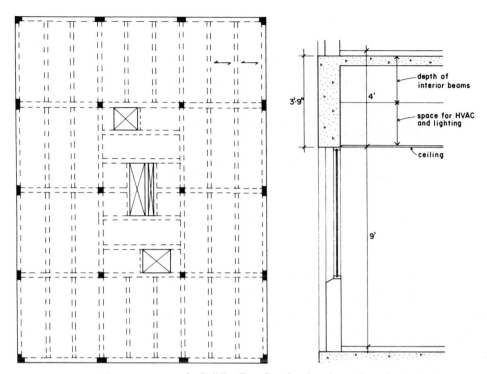

FIGURE 58.1. Concrete structure for Building Two: floor framing plan and exterior wall section.

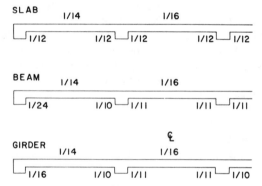

FIGURE 58.2. Moment coefficients for the approximate analysis.

Using the working stress method, the area of steel required per foot of slab width will thus be

$$A_s = \frac{M}{f_s jd} = \frac{C(16,200)(12)}{24,000(0.9)(4)} = 2.25C$$

The slab design is illustrated in Fig. 58.3. Various possible choices are shown for the reinforcing. The most common grade of reinforcing would be that with a yield strength of 60 ksi, which is what we have used for the design. We have thus far not indicated the concrete strength, but will use a fairly low grade with f_c' of 3000 psi.

Flexural and shear stresses in the concrete should be investigated, although the 5-in. slab will be found to be more than adequate for this span and load. It is also well above the recommended minimum thickness for deflection control in the *ACI Code*.

Inspection of the framing plan in Fig. 58.1 reveals that there is a large number of different beams in the structure for the floor with regard to individual loadings and span conditions. Two general types are the beams that carry only uniformly distributed loads as opposed to those that also provide some support for other beams; the latter produce a load condition consisting of a combination of concentrated and distributed loading. We now consider the design of one of the uniformly loaded beams.

The beam that occurs most often in the plan is the one that carries a 10-ft-wide strip of the slab as a uniformly distributed loading, spanning between columns or supporting beams that are 30 ft on center. Assuming the supports to be approximately 12 in. wide, the beam has a clear span of 29 ft and a total load periphery of $29 \times 10 = 290$ ft². Using the *UBC* provisions for reduction of live load,

$$R = 0.08(A - 150)$$

$$= 0.08(290 - 150) = 11.2\%$$

We round this off to a 10% reduction, and, using the loads tabulated previously for the design of the slab, determine the beam loading as follows:

Live load per foot of beam span (with 10% reduction):

0.90(100)(10) = 900 lb/ft   or   0.90 kip/ft [13.1 kN/m]

Slab and superimposed dead load:

100(10) = 1000 lb/ft   or   1.0 kip/ft [14.6 kN/m]

The beam stem weight, estimating a size of 12×20 in. for the beam stem extending below the slab, is

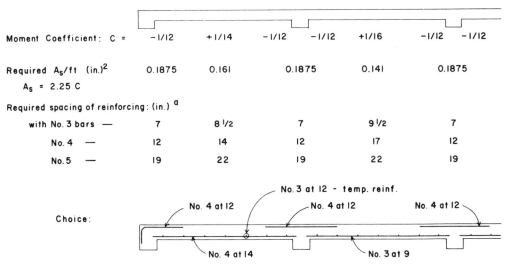

FIGURE 58.3. Design of the continuous slab.

$$\frac{12(20)}{144} \ (150 \ \text{lb/ft}^3)$$

$$= 250 \ \text{lb} \quad \text{or} \quad 0.25 \ \text{kip/ft} \ [3.65 \ \text{kN/m}]$$

The total uniformly distributed load is thus

$$0.90 + 1.0 + 0.25 = 2.15 \ \text{kip/ft} \ [31.35 \ \text{kN/m}]$$

Let us now consider the design of the four-span continuous beam that occurs in the bays on the north and south sides of the building and is supported by the north-south spanning column-line beams that we will refer to as the girders. The approximation factors for design moments for this beam are given in Fig. 58.2 and a summary of the design is presented in Fig. 58.4. Note that the design provides for tension reinforcing only, thus indicating that the beam dimensions are adequate to prevent a critical condition with regard to flexural stress in the concrete. Using the working stress method, the basis for this is as follows.

Maximum bending moment in the beam is

$$M = \frac{wL^2}{10}$$

$$= \frac{2.15(29)^2}{10}$$

$$= 181 \ \text{kip-ft} \ [245 \ \text{kN-m}]$$

Then, for a balanced section, using factors from Table 27.1,

$$\text{required } bd^2 = \frac{M}{R} = \frac{181(12)}{0.204}$$

$$= 10,647 \ \text{in.}^3 \ [175 \times 10^6 \ \text{mm}^3]$$

If b = 12 in.,

$$d = \sqrt{\frac{10,647}{12}} \ = 29.8 \ \text{in.} \ [757 \ \text{mm}]$$

With minimum concrete cover of 1.5 in. on the bars, No. 3 U-stirrups, and moderate-sized flexural reinforcing, this d can be approximately attained with an overall depth of 32 in. This produces a beam stem that extends 27 in. below the slab, and is thus slightly heavier than that assumed previously. Based on this size, we will increase the design load to 2.25 kips/ft for the subsequent work.

Before proceeding with the design of the flexural reinforcing, it is best to investigate the situation with regard to shear to make sure that the beam dimensions are adequate. Using the approximations given in Chapter 8 of the *ACI Code*, the maximum shear is considered to be 15% more than the simple span shear and to occur at the inside end of the exterior spans. We thus consider the following.

The maximum design shear force is

$$V = (1.15) \ \frac{wL}{2} = (1.15) \ \frac{2.25(29)}{2}$$

$$= 37.5 \ \text{kips} \ [167 \ \text{kN}]$$

For the critical shear stress this may be reduced by the shear between the support and the distance of d from the support; thus

$$\text{critical } V = 37.5 \ - \ \frac{29}{12} \ (2.25)$$

$$= 32.1 \ \text{kips} \ [143 \ \text{kN}]$$

Using a d of 29 in., the critical shear stress is

$$v = \frac{V}{bd} = \frac{32,100}{29(12)} = 92 \text{ psi } [634 \text{ kPa}]$$

With the concrete strength of 3000 psi, this results in an excess shear stress of 32 psi, which must be accounted for by the stirrups. The closest stirrup spacing would thus be

$$s = \frac{A_v f_s}{v'b} = \frac{0.22(24,000)}{32(12)}$$

$$= 13.75 \text{ in. } [348 \text{ mm}]$$

Because this results in quite a modest amount of shear reinforcing, the section may be considered to be adequate.

For the approximate design shown in Fig. 58.4, the required area of tension reinforcing at each section is determined as

$$A_s = \frac{M}{f_s jd} = \frac{C(2.25)(29)^2(12)}{24(0.89)(29)}$$

$$= 36.7C$$

Based on the various assumptions and the computations we assume the beam section to be as shown in Fig. 58.5. For the beams the flexural reinforcing in the top, which is required at the supports must pass either over or under the bars in the tops of the girders. Because the girders will carry heavier loadings, it is probably wise to give the girder bars the favored position (nearer the outside for greater value of d) and thus to assume the positions as indicated in Fig. 58.5.

At the beam midspans the maximum positive moments will be resisted by the combined beam and slab section acting as a T-section. For this condition we assume an approximate internal moment arm of $d - t/2$ and may approximate the required steel areas as

$$A_s = \frac{M}{f_s(d - t/2)}$$

$$= \frac{C(2.25)(29)^2(12)}{24(29 - 2.5)} = 35.7C$$

The beams that occur on the column lines are involved in the lateral force resistance actions and are discussed in Sec. 58.5.

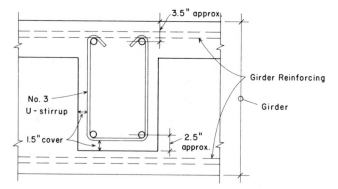

FIGURE 58.5. Section of the interior beam.

Inspection of the framing plan in Fig. 58.1 reveals that the girders on the north–south column lines carry the ends of the beams as concentrated loads at the third points of the girder spans. Let us consider the spandrel girder that occurs at the east and west sides of the building. This member carries the outer ends of the first beams in the four span rows and in addition carries a uniformly distributed load consisting of its own weight and that of the supported exterior wall. The form of the girder and the wall was shown in Fig. 58.1. From the framing plan note that the exterior columns are widened in the plane of the wall. This is done to develop the peripheral bent system, as will be discussed later.

For the spandrel girder we determine the following:

Assumed clear span: 28 ft [8.53 m].

Floor load periphery based on the carrying of two beams and half the beam span load: 15(20) = 300 ft² [27.9 m²].

(*Note:* This is approximately the same total load area as that carried by a single beam, so we will use the live load reduction or 10% as determined for the beam.)

Loading from the beams:

Dead load:
    1.35 kip/ft × 15 ft = 20.35 kips
Live load:
    0.90 kip/ft × 15 ft = 13.50 kips

    Total   = 33.85 kips   say,

                                    34 kips [151 kN]

Moment Coefficient: C =	−1/24	+1/14	−1/10	−1/11	+1/16	−1/11	−1/11
Required Reinforcement: (in.)²							
Top − A_s = 36.7 C	1.53			3.67		3.34	
Bottom − A_s = 35.7 C		2.55			2.23		

Choice:

2 No. 8 (1.58)   4 No. 9 (4.00)   2 No.9 + 2 No. 8 (3.58)

I No. 9 + 2 No. 8 (2.58)   3 No. 8 (2.37)

FIGURE 58.4. Design of the four-span beam.

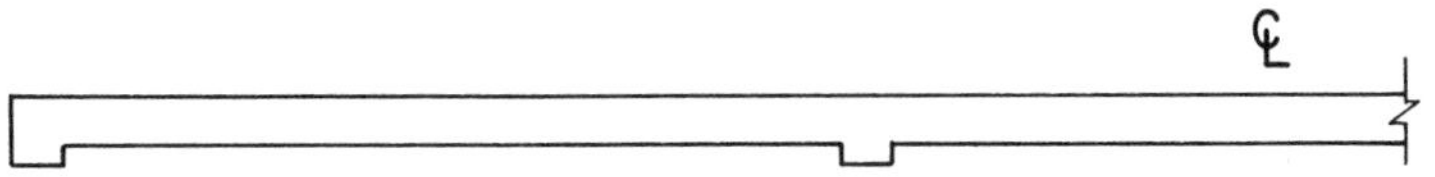

Moment due to distributed load: $M = C w L^2 = C \times 0.8 \times (28)^2 = 627 C$

Coeff – C =	–1/16	+1/14	–1/10	+1/16
M (k‑ft) =	–39.2	+44.8	–62.7	+39.2

Moment due to concentrated load: $M = C P L = C \times 34 \times 28 = 952 C$

Coeff – C =	–1/6	+2/9	–1/3	+1/6
M =	–158.7	+211.6	–317.3	+158.7

Total gravity‑induced moment:

M =	–197.9	+256.4	–380	+197.9

FIGURE 58.6. Gravity moments for the girder.

Uniformly distributed load:
  Spandrel beam weight:
$$\frac{12(45)}{144}(150) = 560 \text{ lb/ft}$$
  Wall assumed at 25 psf:
$$25 \times 9 = \underline{225} \text{ lb/ft}$$
  Total $= 755$ lb/ft   say 0.8 kip/ft [11.7 kN/m]

For the uniformly distributed load approximate design moments may be found using the moment coefficients as was done for the slab and beam. Values for this procedure are given in Fig. 58.2. The *ACI Code* does not permit the use of this procedure for concentrated loads, but we may adapt some values for an approximate design using moments for a beam with the third point loading. Values of positive and negative moments for the third point loading may be obtained from various references, including Refs. 5 and 9.

Figure 58.6 presents a summary of the work for determining the design moments for the spandrel girder under gravity loading. Moment values are determined separately for the two types of load and then added for the total design moment.

We will not proceed further with the girder design at this point, for the effects of lateral loading must also be considered. The moments determined here for the gravity loading will be combined with those from the lateral loading in the discussion in Sec. 58.4.

58.3. DESIGN OF THE COLUMNS FOR GRAVITY LOADS

The four general cases for the columns are (see Fig. 58.7):

The interior column carrying primarily only axial gravity loads.

The intermediate exterior columns on the north and south sides carrying the ends of the interior girders and functioning as members of the peripheral bents for lateral resistance.

The intermediate exterior columns on the east and west sides carrying the ends of the column line beams and functioning as members of the peripheral bents.

The corner columns carrying the ends of the spandrel beams and functioning as the end members in both peripheral bents.

FIGURE 58.7. Framing of the columns and beams.

TABLE 58.1. Column Axial Loads due to Gravity (kips)

Level	Load Source	Corner Column 225 ft²			Intermediate Exterior Column 450 ft²			Interior Column 900 ft²		
		DL	LL	Total	DL	LL	Total	DL	LL	Total
PR	Roof							8	5	
	Wall							5		
	Column									
	Total/level							13	5	
	LL reduction									
	Design load									18
R	Roof	32	5		63	9		126	18	
	Wall	10			10			10		
	Column	8			8			8		
	Total/level	50	5		81	9		157	23	
	LL reduction									
	Design load			55			90			180
3	Floor	32	17		63	34		126	68	
	Wall	10			10			10		
	Column	8			8			8		
	Total/level	100	22		162	43		301	91	
	LL reduction	24%	17		60%	17		60%	36	
	Design load			117			179			337
2	Floor	32	17		63	34		126	68	
	Wall	11			11			11		
	Column	10			10			10		
	Total/level	153	39		246	77		448	159	
	LL reduction	42%	23		60%	31		60%	64	
	Design load			176			277			512

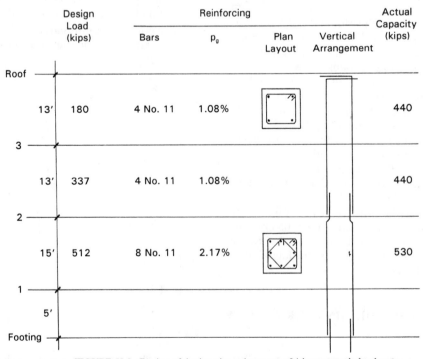

FIGURE 58.8. Design of the interior column as a 24-in square tied column.

A summation of the gravity loads for the columns is given in Table 58.1. In the table we have assumed the existence of a penthouse structure of light metal construction. For the roof we have assumed a dead load equal to that for the floors and a live load of 20 psf. Wall loads for the exterior columns are based on the construction shown in Fig. 58.1. Live load reduction is based on the *UBC* method as illustrated previously for the design of horizontal framing.

Note that in the tabulations in Table 58.1 the load allowed for partitions is assumed as live load, whereas in the tabulation in Fig. 56.6 for Scheme 2 it was treated as dead load. This is done deliberately here to demonstrate the two different methods. This is largely a matter of judgement, although codes may in some cases be more specific. Treating partitions as dead load results in somewhat higher design loads, as live load reductions do not apply. On the other hand, treating them as live load results in some greater potential for unbalanced loading conditions in continuous structures. The difference in design is typically insignificant, so it is mostly an academic difference in methods.

For the interior column it is assumed that a single square size will be used for the three-story-high column. The size would be one that works with minimum reinforcing for the low load at the top story and within the limit for maximum reinforcing for the load at the bottom story. It is necessary, of course, to develop the layout of this column so that it fits into the walls of the building core. The latter may make some shape other than a square more desirable; we have nevertheless assumed a square shape and present a possible design as shown in Fig. 58.8. The design is based on the use of concrete with f_c' of 4000 psi and grade 60 reinforcing. Load capacities have been derived from the load tables in the *CRSI Handbook* (Ref. 9), using 40% of the capacities as determined by strength design; the procedure required for the working stress method is described as the Alternate Design Method in Appendix B of the *ACI Code* (Ref. 6). A minimum eccentricity of 15% of the column dimension has been assumed to allow for some bending caused by the floor and roof framing.

For the intermediate exterior column there are four actions to consider:

1. The vertical compression induced by gravity, as determined in Table 58.1.
2. Bending moment induced by the interior framing that intersects the wall column; the columns are what provides the end moments shown in Fig. 58.3 and 58.4.
3. Bending moments in the plane of the wall induced by unbalanced conditions in the spandrel beams and girders.
4. Bending moments induced by the actions of the peripheral bents in resisting lateral loads.

For the corner column the situation is similar to that for the intermediate exterior column, that is, bending on both axes. The forms of the exterior columns as shown on the plan in Fig. 58.1 have been established in anticipation of the major effects described. Further discussion of these

columns will be deferred, however, until after we have investigated the situations of lateral loading.

58.4. DESIGN FOR LATERAL FORCES

The lateral force resisting systems for the concrete structure are shown in Fig. 58.9. For force in the east–west direction the resistive system consists of the horizontal roof and floor slabs and the exterior bents (columns and spandrel beams) on the north and south sides. For force in the north–south direction the system utilizes the bents on the east and west sides.

If the lateral load is the same in both directions, the stress in the slab (shear in the horizontal diaphragm) is critical for the north–south loading because the slab has less width for resistance to this loading. The loads are not equal, however. Wind force will be greater in the north and

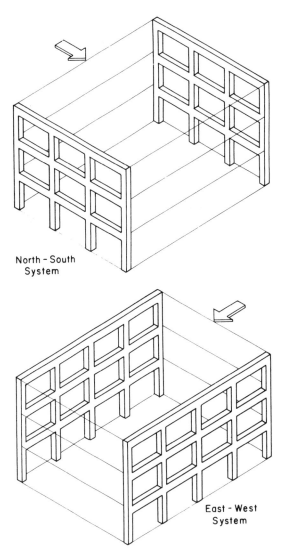

North – South System

East - West System

FIGURE 58.9. The peripheral bent system.

Level	Source of Load	Unit Load (psf)		Load (kips)
Roof	Roof and ceiling	140	120 X 90 X 140 =	1512
	Columns at 0.6 k/ft		0.6 X 6 X 20 =	72
	Window walls	15	400 X 4.5 X 15 =	27
	Interior walls	10	200 X 5 X 10 =	10
	Penthouse + equipment (estimate total load)			25
	Subtotal			1646
Third floor	Floor	140	120 X 90 X 140 =	1512
	Columns		0.6 X 11 X 20 =	132
	Window walls	15	400 X 9 X 15 =	54
	Interior walls	10	200 X 9 X 10 =	18
	Subtotal			1716
Second floor	Floor	140	120 X 90 X 140 =	1512
	Columns		0.6 X 12 X 20 =	144
	Window walls	15	400 X 11 X 15 =	66
	Interior walls	10	200 X 11 X 10 =	22
	Subtotal			1744
Total dead load for base shear				5106

FIGURE 58.10. Dead load for the north–south seismic forces.

Determination of the lateral forces at the upper levels of the structure by redistribution of the total base shear.

Building Level	Gravity Load at Level - w_x (kips)	Height of Level Above Base - h_x (ft)	$(w_x)(h_x)$	F_x (kips)
Roof	1646	41	67,486	167
Third floor	1716	28	48,048	119
Second floor	1744	15	26,160	65
		Total:	141,694	

See text: $F_x = (351/141,694)(w_x h_x) = 0.024772(w_x h_x)$

From this calculation, the bent loads used for design are as shown below.

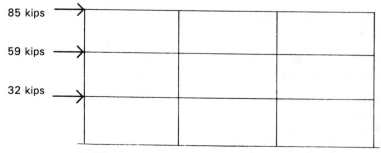

FIGURE 58.11. Redistribution of the seismic forces.

85 kips / 59 kips / 32 kips

south directions because the building has a greater profile in this direction. This makes it even more obvious that this will be the loading critical for the slab in design for wind. However, for seismic load, a true dynamic analysis reveals that the load effect is greater in the east–west direction because the resistive bents are slightly stiffer in this direction. In any event, it is unlikely that the 5-in.-thick slab with properly anchored edge reinforcing at the spandrels will be critically stressed for any loading.

Our considerations for lateral load will be limited to the seismic loading in the north–south direction and to investigations of the effects on the columns and spandrel beams on the east and west sides.

Determination of the Building Weight

Figure 58.10 presents the analysis for determining the building weight to be used for computation of the seismic effects in the north–south direction. In the tabulation we have included the weights of all the walls, which eliminates the necessity for adding the weights of the shear walls in any subsequent analysis of individual walls. Tabulations are done separately for the determination of loads to the three upper diaphragms (eventually producing three forces similar to H_1, H_2, and H_3, as determined for the wind loading). Except for the shear walls, the weight of the lower half of the first-story walls is assumed to be resisted by the first-floor-level construction (assumed to be a concrete structure poured directly on the ground) and is thus not part of the distribution to the rigid bent system.

For the total seismic shear force to the bents, we note from the given data that

Z = 0.3   (zone 3, *UBC* Table 23-I)

I = (standard occupancy, *UBC* Table 23-L)

C = 2.75   [maximum value, *UBC* Sec. 2312(e)1D.2]

R_W = 12   (special concrete frame, *UBC* Table 23-O)

Thus

$$V = \frac{ZIC}{R_W} W = \frac{(0.3)(1.0)(2.75)}{12} W$$

$$= 0.06875 W$$

And using the value from Fig. 58.10 for W, we have

$$V = (0.06875)(5106) = 351 \text{ kips}$$

This total force must be distributed to the roof and upper floors in accordance with the requirements of Sec. 2312(e) of the *UBC*. The force at each level, F_x, is determined from Formula 12-7 as

$$F_x = (V)(w_x h_x)/ \sum_{i=1}^{n} w_i h_i$$

where

F_x = force to be applied at each level x

w_x = total dead load at level x

h_x = height of level x above base of structure

(Notice that F_t has been omitted from the formula because T is less than 0.7 sec.)

The determination of the F_x values is shown in Fig. 58.11.

For an approximate analysis we consider the individual stories of the bent to behave as shown in Fig. 58.12b, with the columns developing an inflection point at their midheight points. Because the columns all move the same distance, the shear load in a single column may be assumed to be equal to the cantilever deflecting load and the individual shears to be proportionate to the stiffnesses of the columns. If the columns are all of equal stiffness in this case, the total load would be simply divided by four. However, the end columns are slightly less restrained as there is a beam on only one side. We will assume the net stiffness of the end columns to be one-half that of the interior columns. Thus the shear force in the end columns will be one-sixth of the load and that in the interior columns one-third

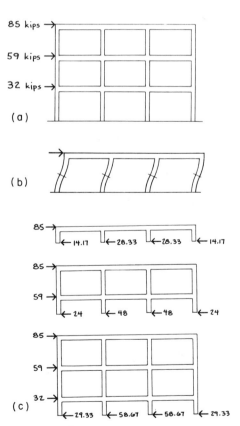

FIGURE 58.12. Considerations of the north–south bents: (a) bent loading; (b) assumed column deformation with midheight inflection; (c) loading for the story shears.

of the load. The column shears for each of the three stories is thus as shown in Fig. 58.12c.

The column shear forces produce moments in the columns. With the column inflection points assumed at midheight, the moment produced by a single shear force is simply the product of the force and half the column height. These moments must be resisted by the end moments in the rigidly attached beams, and the actions are as shown in Fig. 58.13. These effects due to the lateral loads may now be combined with the previously determined effects of gravity loads for an approximate design of the columns and beams.

For the columns, we combine the axial compression forces with any gravity-induced moments and first determine that the load condition without lateral effects is not critical. We may then add the effects of the moments caused by lateral loading and investigate the combined loading condition, for which we may use the one-third increase in allowable stress. Gravity-induced beam moments are taken from Fig. 58.6 and are assumed to induce column moments as shown in Fig. 58.14. The summary of design conditions for the corner and interior column is shown in Fig. 58.15. The design values for axial load and moment and approximate sizes and reinforcing are shown in Fig. 58.16. Column sizes and reinforcing were obtained

from the tables in the *CRSI Handbook* (Ref. 9) using concrete with $f_c' = 4$ ksi and grade 60 reinforcing.

The spandrel beams (or girders) must be designed for the combined shears and moments due to gravity and lateral effects. Using the values for gravity-induced moments from Fig. 58.6 and the values for lateral load mo-

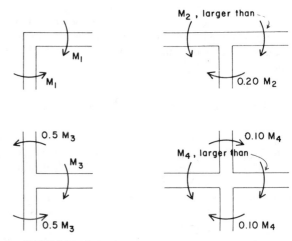

FIGURE 58.14. Assumed gravity moments in the column.

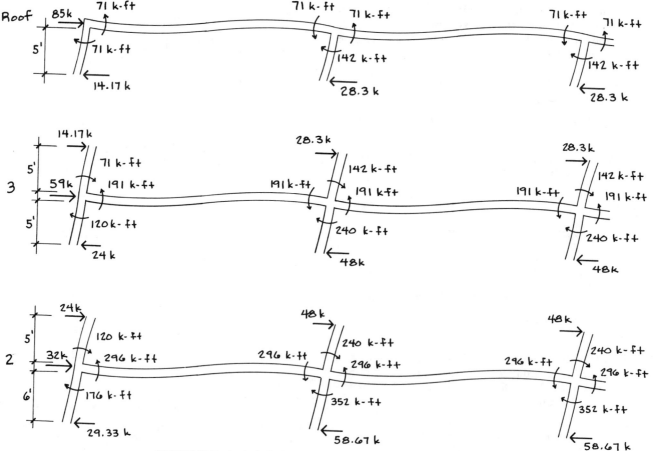

FIGURE 58.13. Analysis for the column and girder moments for lateral load.

	Column	
	Intermediate	Corner

Axial gravity design load (kips) from
Table 58.1:

Third Story	90	55
Second Story	179	117
First Story	277	176

Assumed gravity moment (kip-ft) from
Figures 58.6 and 58.14:

Third Story	60	120
Second Story	40	100
First Story	40	100

Moment from lateral force (kip-ft) from
Figure 58.13:

Third Story	142	71
Second Story	240	120
First Story	352	176

FIGURE 58.15. Summary of design data for the bent columns.

ments from Fig. 58.13, the combined moment conditions are shown in Fig. 58.17a. For design we must consider both the gravity only moment and the combined effect. For the combined effect we use three-fourths of the total combined values to reflect the allowable stress increase of one-third.

Figure 58.17b presents a summary of the design of the reinforcing for the spandrel beam at the third floor. If the construction that was shown in Fig. 58.1 is retained with the exposed spandrel beams, the beam is quite deep. Its width should be approximately the same as that of the

column, without producing too massive a section. The section shown is probably adequate, but several additional considerations must be made as will be discussed later.

For computation of the required steel areas we assume an effective depth of approximately 40 in. and use

$$A_s = \frac{M}{f_s jd} = \frac{M(12)}{24(0.9)(40)} = 0.0139M$$

Because the beam is so deep, it is advisable to use some longitudinal reinforcing at an intermediate height in the section, especially on the exposed face.

Shear design for the beams should also be done for the combined loading effects. The closed form for the shear reinforcing, as shown in Fig 58.17b, is used for considerations of torsion as well as the necessity for tying the compressive reinforcing.

With all of the approximations made, this should still be considered to be a very preliminary design for the beam. It should, however, be adequate for use in preliminary architectural studies and for sizing the members for a dynamic seismic analysis and a general analysis of the actions of the indeterminate structure.

58.5. CONSTRUCTION DETAILS

Development of the construction for Building Two was based on the use of an exposed concrete structure on the building exterior, as shown in Fig. 58.1. It is also possible to use the same basic structural scheme with a completely covered structure. The principal difference would be in the form and detail of the exterior columns and the spandrel beams.

	Intermediate Column					Corner Column				
	Axial Load (kips)	Moment (k-ft)	e (in.)	Column Size (in.)	Reinforcing No. - Size	Axial Load (kips)	Moment (k-ft)	e (in.)	Column Size (in.)	Reinforcing No. - Size
Roof										
	90 X 3/4 = 68	202 X 3/4 = 152	27	20 X 28	6 - 9	55 X 3/4 = 41	191 X 3/4 = 143	42	20 X 24	6 - 10
3										
	179 X 3/4 = 134	280 X 3/4 = 210	19	20 X 28	6 - 9	117 X 3/4 = 88	220 X 3/4 = 165	23	20 X 24	6 - 10
2										
	277 X 3/4 = 208	392 X 3/4 = 294	17	20 X 28	6 - 11	176 X 3/4 = 132	276 X 3/4 = 207	19	20 X 24	6 - 11
1										

FIGURE 58.16. Design of the bent columns.

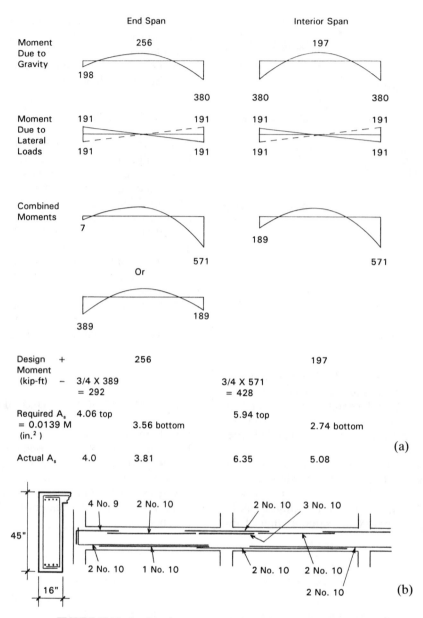

FIGURE 58.17. Combined moments and design for the bent girder.

Construction Illustrated Elsewhere

Some of the drawings shown in the illustration of the work in other chapters are applicable to similar situations that may occur in this building. The following is a description of some of these drawings:

1. Figure 58.1 presents the general form of the framing system for the typical floor. This system could also be used for the roof structure.
2. Figure 58.5 shows the form of the typical interior beam and its reinforcing. This would also be the basic form of the spandrel beam if a covered exterior construction is used for the exterior walls.

3. Figure 56.19 shows a section at the building exterior using a metal and glass exterior curtain wall for the steel frame structure. This wall construction could as well be used with a concrete frame, although the details for attachment of wall elements to the frame would be modified.

Structural Plans

The basic form of the structural framing for the typical floor is shown in Fig. 58.1. If the first-floor structure consists of a simple concrete paving slab poured directly on the ground (called a *slab on grade*), the structural plan for

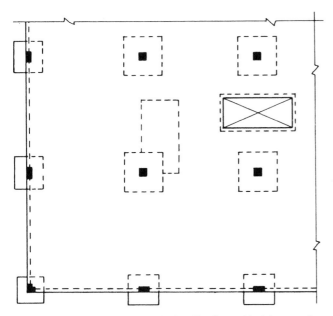

FIGURE 58.18. Partial structural plan: first floor with slab on grade.

the first floor and foundations would be as shown in Fig. 58.18. This assumes the use of ordinary shallow footings for support of the building columns. With the concrete structure, the columns may be extended below the first-floor slab to bear directly on the footings. For taller buildings, in which column axial loads are greater and higher strength values may be used for both the concrete and steel reinforcing in the columns, it may be advisable to use a pedestal between the column and footing, as discussed in Sec. 33.3.

Where ground-level floor conditions do not indicate the advisability of a slab on grade, it may be necessary to use a concrete framed system poured on the ground, as described in Sec. 43.4 This system may use a general plan similar to that for the typical upper floor.

Details at the Exterior Walls

The general form of the exterior wall for the exposed structure is shown in Fig. 58.1. Enlarged details of the concrete structure at the roof and typical upper floor are shown in Fig. 58.19. The detail at the roof indicates the use of a short parapet wall of concrete construction and a lightweight concrete insulating fill with a conventional multilayered membrane roofing.

Figure 58.20 shows a possible detail for a covered structure consisting of precast concrete units for the solid portions of the wall. This is only one of a number of possibilities for development of a covered exterior. As mentioned previously, an all-metal-and-glass exterior, such as

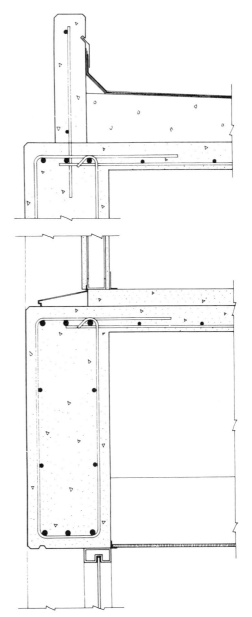

FIGURE 58.19. Details of the exposed concrete structure.

that shown in Fig. 56.20, could also be used with a covered concrete structure.

Details at the Ground Floor

Figure 58.21 shows a section at the edge of the first floor with the exposed concrete structure as shown in Fig. 58.19. The floor paving slab is supported by a short grade wall that is carried down to the level of the top of the column footings. Problems of frost as well as the level at which the column footings must be placed would generally dictate the necessary height for the grade wall.

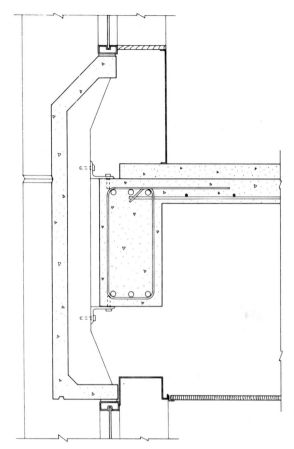

FIGURE 58.20. Alternate spandrel with precast concrete cover units.

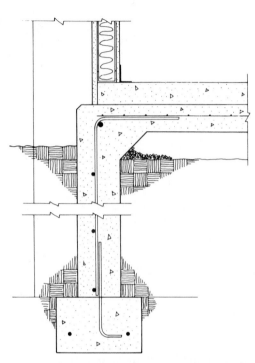

FIGURE 58.21. Details at the first floor.

CHAPTER FIFTY-NINE

Building Two, Scheme 5: Alternative Concrete Structure

Comparison of the structural plans in Figures 55.1, 56.1, and 58.1 will reveal that, despite the use of three different materials, the basic plan form of the systems is essentially the same in character. Basic elements consist of a one-way spanning deck, supported by parallel sets of linear beams, which are in turn supported by linear elements—beams or walls.

The basic deck-beam-girder/wall system is indeed highly useful and can be achieved with a variety of materials. It is generally the most adaptable and accommodating horizontal framing system and the most widely used for building roofs and floors.

With wood and steel, the linear, one-way spanning character of the framing system is quite natural; using the lumber and rolled *shapes*, which come in linear form. With concrete, however, there is no natural form as such, since the material can be cast in just about any shape for which a

mold can be built. Thus the two-way spanning structure is just as natural, as are the three-dimensional forms of domes and shells.

Chapter 59 presents a variation on the concrete structure for Building Two, using a two-way spanning floor system described as a flat slab; a system developed in the early part of the twentieth century and widely used for commercial and industrial buildings.

59.1. GENERAL CONSIDERATIONS

The partial plan in Fig. 59.1 shows the layout of a framing system for Building Two that uses a two-way spanning concrete deck, as described in Chapter 30. The deck is generally of a single thickness of 10 in. with no beam/girder system in general such as that used in Scheme 4. If the

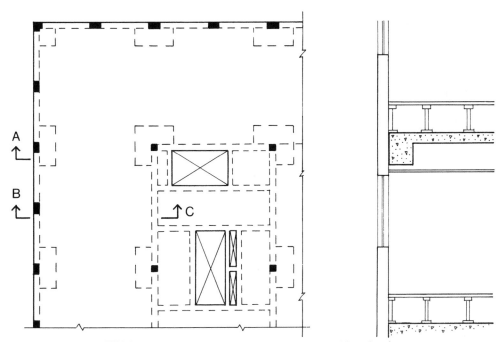

FIGURE 59.1. Building Two, Scheme 5: partial plan of floor framing.

single thickness deck is used throughout with no variation, the structure is described as a *flat plate*. A modification used here is a thickened slab portion around main supporting columns, producing a system described as a *flat slab*.

The presence of the thickened elements (called *drop panels*) serves to define an implied framing network in the slab due to the stiffening that results at the locations of the main column lines (at 30-ft centers in both directions). Thus a strip of stiffened slab exists on the column lines, creating or emulating beam actions on these lines. The slab is thus basically supported for a two-way spanning action within each square bay of the system.

The section in Fig. 59.1 shows the general form of the typical upper-story floor and wall construction. The exterior wall here is a metal and glass curtain wall hung on the outside of the edge of the structure. Floor surfaces are created by a secondary structure consisting of closely spaced, short steel posts and square panels called an *elevated floor* or *access floor*. The principal purpose is to provide a space for easy and frequent modification of the wiring for power and signal wiring communication systems—a process typically ongoing in contemporary computerized offices.

Adding the elevated floor produces a second potential interstitial space. This may result in additional floor-to-floor height with considerable extra cost for more curtain wall and general height of all the interior construction and services. One benefit of the use of the flat slab (or flat plate) system is the possibility of reducing the floor/ceiling sandwich thickness because of the elimination of beams and girders in the general office spaces. By comparison with Scheme 4 with the 32-in.-high beams (and undoubtedly deeper girders, the same dimension for clearance here is only the 10-in. slab. Thus, considering only the beams, the gain in height is 22 in.; more than compensating for the access floor, which is usually about 18 in. high.

Another advantage of the flat slab is its increased thickness, which easily allows for accommodation of the punching character of the post loads from the access floor. A *disadvantage* is the possibility of increased volume and weight of the concrete, since the slab is twice the thickness of the one in Scheme 4.

A possibility with this structure—although not often utilized for this occupancy—is that of exposing the interior columns and the underside of the spanning structure. Fire resistance of the concrete columns and the very thick slab permit this, and the exposed elements are quite simple in form, although generally quite rough in nature as exposed surfaces. The section in Fig. 59.1 and the details in Sec. 59.3 show the exposed structure, although the dotted lines in the sections indicate the possibility for a suspended ceiling of the usual form.

The plan in Fig. 59.1 indicates a core framing system essentially the same as that in Scheme 4 (see Fig. 58.1). This represents a compelling situation for use of the slab and beam system—to accommodate the many openings in the core. The two-way spanning systems have considerable limitation in achieving these discontinuities in the build-

ing plan. The net effect for the two-way portion of the structure is simply to have a continuous rigid support (beam? wall?) along the edges of the core.

The plan in Fig. 59.1 also indicates the use of additional columns at the building exterior. These are used primarily to improve the perimeter bents for lateral load resistance. The net effect is to generally cut the lateral forces to one-half in the exterior columns and to make a less stiff spandrel (basically less deep) a possibility. The section views in Figures 58.1 and 59.1 reflect these differences. The possible reduction of size for the exterior columns is more significant because of their exposure in the plan here, compared with the situations in Schemes 2 and 4. See the details in Sec. 56.5 and 58.5.

59.2. DESIGN OF THE TWO-WAY STRUCTURE

The two-way spanning system shown here works most efficiently with a considerable number of sequential bays in each direction in a larger plan. The three continuous spans on the short side and four continuous spans on the

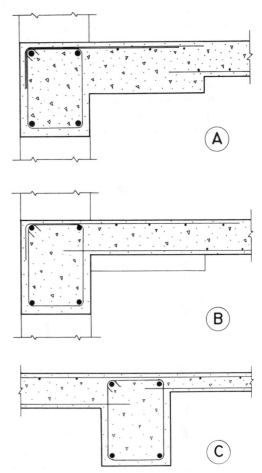

FIGURE 59.2. Construction details, A: section at building edge showing reinforcement in the column strip; B: section through the middle strip; C: section at the edge of the core.

long side of this building permit some realization of this effect, but the system here essentially spans one way for one bay between the core and the perimeter spandrels. However, the use of the drop panels helps to create some of the basic character of the flat slab system. Continuity of the concrete construction between the core and the office area floor also develops some of the nature of the continuous structure.

Proper or not, the system can work here to achieve the basic architectural goals discussed previously. It would, however, be more effective if there were more bays between the core and the perimeter.

This is a highly indeterminate structure and its precise investigation for multiple loading conditions would be extremely laborious. However, approximate and empirical methods for analysis have been used for many years for design of this system and guidelines for their use are still in existence in current design codes.

The basic form of analysis and design is illustrated in Fig. 30.12 and discussed in Sec. 30.4. The two-way system is actually treated as a set of linear strips for determination of sets of reinforcement to be placed on the two major axes of

the system. Thus a "column strip" is visualized as a continuous beam, resulting in a set of steel bars in the top of the slab over the support and a set of steel bars in the bottom in midspan between the supports—just as was done for the continuous one-way slabs and beams. (See Fig. 59.2.) Here, this linear set is simply layered over that in the perpendicular direction to create the two-way structure.

Professional design of this system is now done with computer-aided methods using software based on current code requirements and design practices. An approximate design may be achieved using various tabulated reference materials. One such reference is contained in the *CRSI Handbook* (Ref. 9), which has provided such shortcut design data for many years through several editions of the standard codes. Many actual structures have been fully designed from these tabulations in the recent past. Today, these tabulations still represent sources for quick approximations in early stages of design.

Designs for the columns, spandrel beams, core framing, and the general form and details for the perimeter bents would follow the same processes discussed in Chapter 58 for Scheme 4.

CHAPTER SIXTY

Building Two, Scheme 6: Masonry Wall Structure

Chapters 60 and 61 present options for the construction of the Building Two that involve the use of structural masonry for development of the exterior walls. The walls are used for both vertical bearing loads and lateral shear wall functions. The system in Chapter 60 uses the basic plan as shown for Scheme 2 in Chapter 55, with the masonry walls essentially replacing the wood framed ones and duplicating their functions as developed in that chapter. It is also assumed that the solution developed for the horizontal structures in Scheme 2 are as valid for this scheme, so we will not develop a new system here.

The choice of forms of masonry and details for the construction depend very much on regional considerations (climate, codes, local construction practices, etc.) and on the general architectural design. Major differences occur due to variations in the range in outdoor temperature extremes and the specific critical concerns for lateral forces. The scheme in Chapter 60 proposes a continuous exterior wall construction; a highly questionable situation in cold climates with major concerns for thermal expansion and contraction effects.

60.1. GENERAL CONSIDERATIONS

Figure 60.1 shows a partial elevation of the masonry wall structure and a partial framing plan of the upper floors. As discussed in Chapter 55, the wood construction shown here is most likely not acceptable for fire codes. We are using the example only to demonstrate the general form of the construction; probably only acceptable for smaller buildings by most building codes.

Plan dimensions for structures using CMUs (concrete blocks) must be developed so as to relate to the modular sizes of CMU units. There are a few standard sizes widely used, but individual manufacturers may have some special units or be able to accommodate requests for special shapes or sizes. However, while brick or stone units can be cut to produce precise, nonmodular dimensions, the CMU units generally cannot. Thus the dimensions for the CMU structure itself must be carefully developed to have wall intersections, corners, ends, tops, and openings for windows and doors fall on the fixed modules.

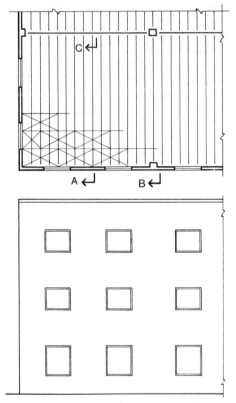

FIGURE 60.1. Building Two, Scheme 6: partial elevation and upper-floor framing plan for the masonry wall structure.

There are various forms of CMU construction. The one shown here is widely used where either windstorm or seismic risk is high. This is described as *reinforced masonry* and is produced to generally emulate reinforced concrete construction, with tensile forces resisted by steel reinforcement that is grouted into the hollow voids in the block construction. This form of construction is described in detail in Chapter 36 in Part 6.

It is assumed here that the three-story-high walls can be developed with block units that produce a nominal 8-in.-thick wall. Combining this with the usual 16-in. single

612

block length and either 6- or 8-in. block height sets up the modular system for the structural layout. We will assume an 8-in. block height and use a running bond wall face pattern that allows for a half-block of 8-in. in wall length. Our three-dimensional module is then 8 in. × 8 in. × 8 in.

For vertical coordination with the module we will assume a story height of 14 ft 8 in. for the first story and 12 ft 8 in. for the upper stories.

Another consideration to be made for the general construction is that involving the relation of the structural masonry to the complete architectural development of the construction, regarding interior and exterior finishes, insulation, incorporation of wiring, and so on.

60.2. DESIGN OF THE TYPICAL FLOOR

As noted previously, the floor framing system here is essentially similar to that for Scheme 2 in Chapter 55. This involves the use of a plywood deck and fabricated joists for the basic framing with supporting interior girders. The girders could be glued-laminated timber, as in Scheme 2, but are shown here as rolled steel shapes. Supports for the steel girders consist of steel columns on the interior and masonry pilaster columns at the exterior walls.

As shown in the details in Sec. 60.5, the exterior masonry walls are used for direct support of the deck and the joists through ledgers bolted and anchored to the interior wall face. With the plywood deck also serving as a horizontal diaphragm for lateral loads, the load transfers for both gravity and lateral forces must be carefully developed in the details for this construction.

The development of the girder support at the exterior wall is a bit tricky with the pilaster columns, which must not only provide support for the individual girders at each level, but also maintain the vertical continuity of the column load from story to story. This is similar to the concept illustrated for the interior columns in Fig. 55.3. There are various options for the development of this detail. The one shown in Detail C in Fig. 60.4 indicates the use of a wide pilaster that virtually straddles the relatively narrow girder. The void created by the girder is thus essentially ignored with the two outer halves of the pilaster bypassing it for a real vertical continuity.

Since the masonry structure in this scheme is used only for the exterior walls, the construction at the building core is free to be developed by any of the general methods shown for other schemes. If the steel girders and steel columns are used here, it is likely that a general steel framing system might be used for most of the core framing.

60.3. DESIGN OF THE MASONRY WALLS FOR GRAVITY LOADS

Buildings much taller than this have been achieved with structural masonry, so the feasibility of the system is well demonstrated. The vertical loads increase in lower stories, so it is expected that some increases in structural capacity will be achieved in lower portions of the walls. The two general means for increasing wall strength are to use thicker CMUs or to increase the amount of core grouting and reinforcement.

It is possible that the usual minimum structure—relating to code minimum requirements for the construction—may be sufficient for the top-story walls, with increases made in steps for lower walls. Without increasing the CMU size, there is considerable range between the minimum and the feasible maximum potential for a wall.

As discussed in Sec. 60.4, it is common to use fully grouted walls (all cores filled) for shear walls. In this scheme that would technically involve using fully grouted construction for *all* of the exterior walls, which might likely rule against the economic feasibility of this scheme. Adding this to the concerns for thermal movements in the long walls might indicate the wisdom of using some control joints to define individual wall segments. This begins to approach the scheme presented in Chapter 61. The scheme could indeed be used here, with nonstructural wall portions filling in between the structural ones, and only the presence of the control joints giving the secret away.

Design of the walls for vertical compression and any combined axial force plus bending is presented in the discussions in Chapter 36. Also discussed there is the design of masonry columns, of which the pilaster is a general form.

60.4. DESIGN FOR LATERAL FORCES

Use of an entire masonry wall as a shear wall, with openings considered as producing the effect of a very stiff rigid frame, is discussed in Sec. 36.4. An example of such a structure is illustrated in that section (see Fig. 36.5). As for gravity loads, the total shear force increases in lower stories. Thus it is also possible to consider the use of the potential range for a wall from minimum construction (defining a minimum structural capacity) to the maximum possible strength with all voids grouted and some feasible upper limit for reinforcement.

The basic approach here is the same as that for the three-story plywood wall discussed in Sec. 55.4. The idea is to design the required wall for each story, using the total shear at that story. In the end, however, the individual story designs must be coordinated for the continuity of the construction. However, it is also possible that the construction itself could be significantly altered in each story if it fits with architectural design considerations.

There are many concerns for the proper detailing of the masonry construction to fulfill the shear wall functions. There are also many concerns for proper detailing to achieve the force transfers between the horizontal framing and the walls. Some of these are addressed in the discussion of the typical details in Sec. 60.5.

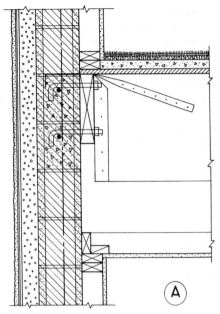

FIGURE 60.2. Detail A.

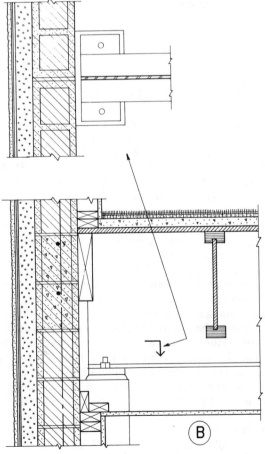

FIGURE 60.3. Detail B.

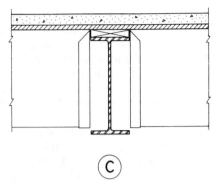

FIGURE 60.4. Detail C.

60.5.   CONSTRUCTION DETAILS

The general form of the framing plan for the upper floor is shown in Fig. 60.1. The location of the details discussed here is indicated by the section marks on that plan.

Detail A (Fig. 60.2).   This shows the general exterior wall construction and the framing of the floor joists at the exterior wall. The wood ledger is used for vertical support of the joists, which are hung from steel framing devices fastened to the ledger. The plywood deck is nailed directly to the ledger to transfer its horizontal diaphragm loads to the wall.

Outward forces on the wall must be resisted by anchorage directly between the wall and the joists. Ordinary hardware elements can be used for this, although the exact details depend on the type of forces (wind or seismic), their magnitude, the details of the joists, and the details of the wall construction. The anchor shown in the detail is really only symbolic.

General development of the construction here shows the use of a concrete fill on top of the floor deck, furred out wall surfacing with batt insulation on the interior wall side, and a ceiling suspended from the joists.

Detail B (Fig. 60.3).   This shows the section and plan details at the joint between the girder and the pilaster. The pilaster unavoidably creates a lump on the inside of the wall in this scheme. Development here relates to the issues discussed in Secs. 60.2 and 60.3.

Detail C (Fig. 60.4)   This shows the use of the steel beam for support of the joists and the deck. After the wood lumber piece is bolted to the top of the steel beam, the attachment of the joists and the deck become essentially the same as they would be with a timber girder.

Attached only to its top, the supported construction does not provide very good lateral support for the steel girder in resistance to torsional buckling. It is advisable, therefore, to use a steel shape that is not too weak on its y-axis, generally indicating a critical concern for lateral unsupported length. Rotation of the girder at the supports is quite well resisted by the encasement in the pilaster column here.

CHAPTER SIXTY-ONE

Building Two, Scheme 7: Alternative Masonry Structure Number One

Chapter 61 presents alternatives for both the exterior masonry walls and the floor construction for Building Two. The general form of the structure is shown in the plan and partial elevation in Fig. 61.1. Instead of the continuous masonry wall, the exterior is developed here with individual masonry piers placed between the window areas. Nonstructural wall construction is used for the wall segments between the piers. The general floor framing is achieved here with precast concrete plank units with voided cores.

The masonry piers and the precast concrete floors are not significantly linked as a system. Indeed, the wall systems and the floor systems shown here and in Chapter 60 could all conceivably be interchanged.

61.1. GENERAL CONSIDERATIONS

The elevation in Fig. 61.1 and the details in Sec. 61.4 show the use of a cast concrete element that is exposed on the exterior face and is used to achieve the joint between the masonry wall and the precast concrete floor elements. In Detail B (Fig. 61.3) this element is shown to be extended to become a spandrel beam for support of the floor edge at the nonstructural portion of the exterior wall.

The interior girders could be developed in a number of ways; one of which is to use a sitecast beam formed and cast with the edge members indicated in Details A and B (Figures 61.2 and 61.3). This is the form shown in option 1 for Detail C (Fig. 61.4). Another option, as shown, is to use a precast concrete member for the girder (option 2 in Fig. 61.4).

These and other concerns for construction and architectural design may affect some decisions for the selection and arrangement of framing elements for the floors.

The walls are developed here as individual, linked piers for both gravity load and lateral load resistance.

61.2. DESIGN OF THE FLOOR STRUCTURE

The precast concrete floor units would be provided by a manufacturer and would most likely be a proprietary prod-

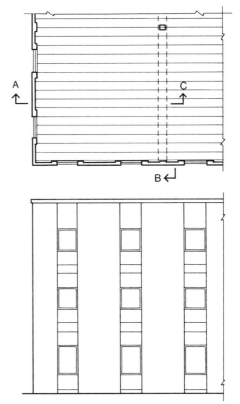

FIGURE 61.1. Building Two, Scheme 7: partial elevation and plan of the typical upper-floor framing.

uct with some standardization of form and dimensions. Because of transportation difficulties, they must be obtained from some local supplier, and thus the range of selection may be limited. The details shown here are based on use of units manufactured by the Flexicore Company.

Most of the engineering design for the precast units is done by agents of the suppliers/erectors of the units. For the building designer, the net effort is generally equivalent to that demonstrated for selection of open web joists, with a load table supplied by the suppliers. The dimensions of the units shown here are generally in keeping with data supplied by the manufacturers.

The concrete topping here is intended for bonding with the precast units, producing a combined structural resistance for flexure and deflection considerations. The sitecast topping will therefore be a structural grade concrete, and its extended use in the development of beams is reasonable.

Development of the building core may be achieved with a wholly sitecast concrete structure or with some use of precast elements. It could also conceivably use some structural masonry walls, as considerable permanent construction is required at stairs, elevators, rest rooms, and vertical duct shafts.

The support of the girder here is essentially the same problem as in Scheme 6 in Chapter 60. This will be most easily accomplished with the sitecast spandrel beams and interior girders, with vertical reinforcement extended through the joint to make a continuity condition for the masonry pilasters. If precast girders are used, the wide pilaster, straddling the girder as in Scheme 6, may be used.

61.3. DESIGN OF THE EXTERIOR WALLS

For resistance to both vertical and lateral forces, the exterior structure consists of the 10-ft-wide masonry piers at intermediate columns and the L-shaped piers at the corners. Vertical loads are transferred to the piers by the concrete planks, the girders, and the spandrel beams at the wall openings. The piers are shown here to be made with CMUs, generally similar in detail to the construction for Scheme 6. The infill walls between piers are developed with a light-gauge steel framing system, similar in detail to ordinary light wood frame construction with wood studs. The infill walls are totally supported—vertically and laterally—by the masonry piers and the concrete spandrel beams.

The masonry walls are generally similar to those in Scheme 6 with regard to resistance of vertical loads, except that there is some less total wall cross section in the horizontal plane. Thus the vertical compression and any bending in the walls is resisted by less wall mass and the stresses will be somewhat higher. Still, most of the discussion in Chapter 60 also applies to the options and considerations for these walls.

For lateral loads, the 10-ft-wide walls would most likely be considered to function as individual piers, only linked by the spandrels. However, if the spandrels are quite deep, the general frame action assumed for Scheme 6 may be more appropriate. We will assume the spandrels to be too flexible to develop this action here and consider the walls to bend independently as three-story-high cantilevers.

With this assumption, a critical concern becomes the overturning moment at the base of the walls. This may actually be mostly a problem for the foundations, requiring a very deep grade beam/wall, as discussed in Sec. 41.4 and illustrated in Fig. 41.8. With the ends of the piers developed as reinforced masonry columns, the ordinary

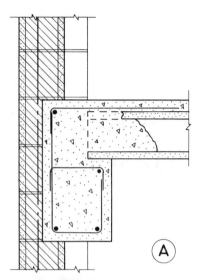

FIGURE 61.2. Detail A.

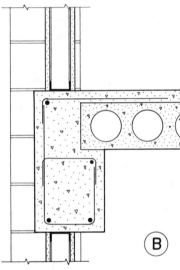

FIGURE 61.3. Detail B.

dowelling of the column reinforcement will most likely be sufficient to anchor the ends of the walls for any required tiedown resistance.

These walls should surely be fully grouted, although the amount of reinforcement may be varied from top to bottom to provide some varied strength for shear (and vertical compression) resistance in the multistory wall.

61.4. CONSTRUCTION DETAILS

The general framing of the floors and the form of the plan for the exterior walls is shown in Fig. 61.1. The locations for

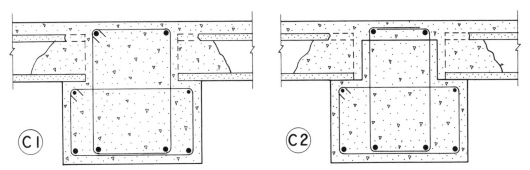

FIGURE 61.4. Detail C.

the details shown here are indicated by the section marks in Fig. 61.1.

Detail A (Fig. 61.2). This shows the section at the masonry pier and the end of the precast plank units. In order to create a more solid bearing for the wall above, it is usually the practice to break out a short portion of the top of the plank units at the voids and use a plug of some kind to permit the cast concrete to fill the ends of the voids, as shown in the details.

Vertical continuity of the wall masonry is achieved by extending the vertical reinforcement through this joint. Some additional vertical bars may be placed in the top of the lower wall as shown (to be bent and cast into the floor topping) to help anchor the floor to the wall for resistance to outward forces on the wall.

Detail B (Fig. 61.3). This shows the detail at the nonstructural wall with the sitecast spandrel beam. The spandrel beam is shown exposed, so that the wall is actually developed as a series of single-story-high wall units; from the top of one spandrel to the underside of the one above. It

would also be possible to develop the wall surface as a continuous one by recessing the face of the spandrel beam.

Detail C (Fig. 61.4).

Option 1. This shows the condition for the sitecast concrete girder as developed in conjunction with the sitecast spandrel members. This is the easiest form of construction for the development of the details, but not the only way to go. A problem here is that of the construction sequence, since the precast plank must be in place before the girder is cast in the same pour as the deck topping. This requires some temporary support for the plank.

Option 2. This shows the use of precast girders. The principal advantage is in eliminating the need for the temporary support of the plank units, since the girder can be placed as soon as the masonry is placed up to the level of its base on the pilaster. The problem—if it is indeed one—is that a wider pilaster, like that in Scheme 6, is probably required to achieve continuity of the vertical columns.

APPENDIX A

Properties of Sections

Computations for structural investigation and design frequently require the use of properties of areas, the areas being those of the cross sections of structural members. The most often used properties are area, centroid, moment of inertia, section modulus, and radius of gyration. This section contains an explanation of these basic properties and tabulations of properties of common geometric shapes (Table A.1). Tabulations are also given for properties of cross sections of standard structural lumber (Table A.2) and common steel shapes (Tables A.3 through A.10). (Tables A.1–A.10 begin on page 624.) Properties for structural lumber are adapted from the *National Design Specification for Wood Construction* (Ref. 7), with permission of the publishers, National Forest Products Association. Properties of steel shapes are adapted from the *AISC Manual* (Ref. 5), with permission of the publishers, American Institute of Steel Construction.

A.1. MASS

The mass of a physical shape may be expressed as the product of a unit density times the number of units of volume of the shape. By this means, mass becomes analogous with the concept of geometry of the physical shape. A shape may be visualized as a force in the form of the weight exerted by the mass. An area may be considered as a mass by conceiving it as having a unit density with the unit being some constant thickness. Thus instead of being an area in the pure mathematical sense, it consists of a shape cut from some thin material, such as cardboard. It is helpful to think of areas in this way in the problems that follow.

A.2. MOMENT OF A PHYSICAL SHAPE

Considered as a mass, and thus as a force (weight), a physical shape may be seen to have the capability of exerting direct force and also of exerting moment. The latter occurs when the shape exerts its mass force other than along a line through its mass center. Referring to the ball on a string shown in Fig. A.1a:

When the string is vertical, that is, it is aligned with the line of action of the weight of the ball, the effect of the ball is to exert only a direct pull on the string.

When the string is other than vertical, there is a moment, or a rotational effect, caused by the component of weight that is perpendicular to the string.

A.3. CENTROID

The concept of centroid (or mass center, or center of gravity) may be visualized in terms of the lack of rotational effect. The mass center is the point about which the sum of the moments of the incremental units of the mass is zero. If defined only geometrically, this point is called the *centroid*. Referring to Fig. A.1b, the shape defined as an area has a moment about the axis *O–O* expressed as

$$\sum M_{O\text{-}O} = \sum wxda = (wA)\bar{x}$$

where

$\sum M_{O\text{-}O}$ = total moment of the area about *O–O*
w = unit density of the area (thickness)
wda = unit weight of the area
wA = total weight of the area
$\bar{x}$ = distance from the axis *O–O* to the center of gravity or centroid of the area

If only the geometric property of the location of the centroid is desired, the unit density w may be dropped from the expression and the centroidal distance can be expressed as

$$\bar{x} = \frac{\sum M_{O\text{-}O}}{A}$$

For an area the precise location of the centroid can be established by determining centroidal distances from two reference axes that are not parallel. Using the customary x- and y-axis references, as shown in Fig. A.1c, the precise location of the centroid for the area shown can be expressed as

$$\bar{x} = \frac{\sum M_{y\text{-}y}}{A} \quad \text{and} \quad \bar{y} = \frac{\sum M_{x\text{-}x}}{A}$$

The following examples illustrate the process for determination of the centroid for an area.

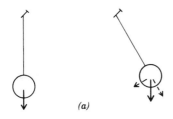

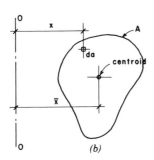

(b)

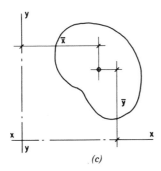

(c)

FIGURE A.1.

The uncut area is considered as a positive area and the hole as a negative area. Thus

$$\text{total } A = A_1 + A_2 = (4 \times 8) - \frac{\pi(1.5)^2}{4}$$

$$= 32 - 1.77 = 30.23 \text{ in.}^2$$

$$\sum M_y = (A_1 \times 2) - (A_2 \times 1.5)$$
$$= 64 - 2.66 = 61.34 \text{ in.}^3$$

$$\sum M_x = (A_1 \times 4) - (A_2 \times 5)$$
$$= 128 - 8.85 = 119.15 \text{ in.}^3$$

$$\bar{x} = \frac{\sum M_y}{A} = \frac{61.34}{30.23} = 2.02 \text{ in.}$$

$$\bar{y} = \frac{\sum M_x}{A} = \frac{119.15}{30.23} = 3.94 \text{ in.}$$

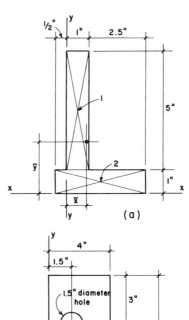

(a)

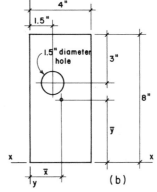

(b)

FIGURE A.2.

Example 1. The shape in Fig. A.2*a* represents the area of a planar figure of uniform density. Find the location of the centroid of the area.

Solution. Consider the area as consisting of two units as shown; the areas and centroids of which are determinable. Then

$$\text{total } A = A_1 + A_2 = 5 + 4 = 9 \text{ in.}^2$$

$$\sum M_y = (A_1 \times 0.5) + (A_2 \times 1.5)$$
$$= 2.5 + 6 = 8.5 \text{ in.}^3$$

$$\sum M_x = (A_1 \times 3.5) + (A_2 \times 0.5)$$
$$= 17.5 + 2 = 19.5 \text{ in.}^3$$

$$\bar{x} = \frac{\sum M_y}{A} = \frac{8.5}{9} = 0.944 \text{ in.}$$

$$\bar{y} = \frac{\sum M_x}{A} = \frac{19.5}{9} = 2.167 \text{ in.}$$

Example 2. Find the location of the centroid of the area shown in Fig. A.2*b*.

Solution. In this case we consider the area to be comprised of two units: the area without the hole and the hole.

A.4. MOMENT OF INERTIA

In certain derivations of relationships involving the behavior of structures a mathematical expression occurs that is related to the geometry of a physical shape. If the shape is a planar area, the expression is of the form

$$\int x^2 \, da$$

where x^2 is the distance of an incremental area da from some fixed reference axis in the plane of the area.

This expression is similar to the one for the moment of an area, except for the second power of x. For this reason it is sometimes called the *second moment of an area*. The more common name given to it in structural engineering is the *moment of inertia*. The symbol commonly used for this property is I.

For the area shown in Fig. A.3a:

$I_x = \int y^2 \, da =$ moment of inertia of the area with respect to the x-axis

$I_y = \int x^2 \, da =$ moment of inertia of the area with respect to the y-axis

$\bar{I}_x = \int (y - \bar{y})^2 \, da =$ moment of inertia of the area with respect to a centroidal axis parallel to the x-axis

For structural investigations, the most used properties of this kind are the values of moment of inertia about centroidal axes. Centroidal I values for common geometric shapes and common structural shapes are given in the tables in this section.

A.5. PARALLEL AXIS THEOREM

A relationship of use in determining the moment of inertia of composite shapes is the parallel axis theorem. The derivation of this relationship is as follows. Referring to Fig. A.3b, we have

$$I_{A-A} = \int s^2 \, da = \int (D + x)^2 \, da$$
$$= \int (D^2 + 2Dx + x^2) \, da$$
$$= \int D^2 \, da + \int 2Dx \, da + \int x^2 \, da$$

We note:

$$\int D^2 \, da = D^2 \int da = A \, D^2$$

$\int 2Dx \, da = 0$ since $\int x \, da =$ moment of area about its centroid, which by definition is zero

$\int x^2 \, da = \bar{I}_{0-0} =$ moment of inertia of the area about the centroidal axis parallel to A–A

and therefore

$$I_{A-A} = \bar{I}_{0-0} + AD^2$$

or, stated differently,

$$\bar{I}_{0-0} = I_{A-A} - AD^2$$

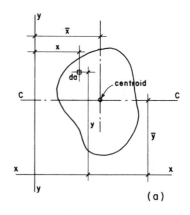

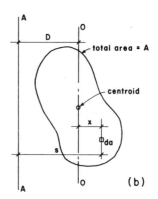

FIGURE A.3.

The main use of the parallel axis theorem is in the transferring of moments of inertia from one axis to another. This is used in finding the moment of inertia of composite shapes for which moments of inertia of component units of the area are determinable. The following example demonstrates this process.

Example 1. Find the moment of inertia of the area shown in Fig. A.4a with respect to the centroidal axis that is parallel to the base of the shape.

Solution. The axis through the base is designated A–A in the figure and the designated centroidal axis is C–C. As the location of the centroid is not given, the first step is to find its location—in this case only with respect to the A–A axis. The process for this is demonstrated in Examples 1. and 2. of Sec. A.3. Thus

$$\bar{y} = \frac{(A_1 \times 4) + (A_2 \times 3) + (A_3 \times 1)}{A_1 + A_2 + A_3} = 2.06 \text{ in.}$$

With the location of the centroid known, the moment of inertia of the individual parts about the C–C axis can be determined with the parallel axis theorem. Referring to Fig. A.4b, the moment of inertia of part 1 is found as follows:

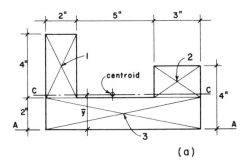

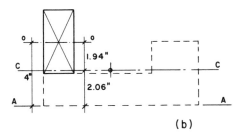

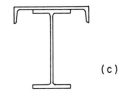

FIGURE A.4.

$$I_{C-C} = I_{O-O} + AD^2$$

in which I_{C-C} is the moment of inertia of the component area of part 1; I_{O-O} is the moment of inertia of part 1 about its own centroidal axis parallel to C–C; and AD^2 is the area of part 1 times the square of the distance from its centroid to the centroid of the whole shape. For the simple rectangular shape of part 1, the moment of inertia about its own centroid is found as (see Table A.1)

$$I_{O-O} = \frac{bd^3}{12} = \frac{2(4)^3}{12} = 10.67 \text{ in.}^4$$

Then for part 1,

$$I_{C-C} = 10.67 + 8(1.94)^2 = 10.67 + 30.11$$
$$= 40.78 \text{ in.}^4$$

and using similar procedures for the other parts, the moment of inertia of the whole figure about the axis C–C is found to be 77.22 in.^{4}.

A procedure similar to that shown in the example can also be used for composite sections such as that shown in Fig. A.4c, consisting of a combination of two standard structural steel rolled shapes. Although the component units are more complex in this case, the areas, the locations

of the centroids, and the centroidal moments of inertia are all given for such shapes in tables of properties in the *AISC Manual* (Ref. 5).

A.6. RADIUS OF GYRATION

In the investigation of buckling due to compression of slender elements, a derived property used to measure the relative slenderness of the elements is the *radius of gyration*. This property is defined as

$$r = \sqrt{\frac{I}{A}}$$

in which:

  r is the radius of gyration, in linear units (inches).
  I is the moment of inertia of the area (sectional shape) about a reference axis, yielding the r for the same axis.
  A is the area of the section.

While the area remains constant for a given shape, the radius of gyration and moment of inertia can be derived for many axes. Of major significance for buckling effects is usually the *least* value of r, derived from the least value of I for the shape. (See the discussion in Sec. A.8 regarding major axes.)

This property is utilized particularly in the investigation of steel compression members, and values for the radius of gyration are thus typically listed in tabulations of properties for rolled steel shapes (see Table A.3).

A.7. POLAR MOMENT OF INERTIA

Figure A.5 shows a planar shape that exists in the x-y plane. As discussed in Sec. A.4, the moment of inertia (second moment) of this shape about the x-axis is defined as the integration (summation) of $x^2 \, da$. Similarly, the moment of inertia about the y-axis is defined as the integration of $y^2 \, da$. A third property, called the *polar moment of inertia*, is defined as the integration of $r^2 \, da$, in which r is the distance of the point of the unit area from the intersection of the three reference axes. To differentiate it from the moments of inertia about the axes in the plane of the shape, the symbol J is usually used for the polar moment of inertia; thus

$$J_z = \int r^2 \, da$$

This property is used primarily in the investigation of torsional effects, in which the z-axis is the longitudinal axis of a member. Another application is in the investigation for rotational (torsional) effects of lateral forces on buildings with unsymmetrical plans, as illustrated in Chapter 51.

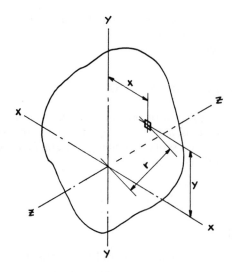

FIGURE A.5.

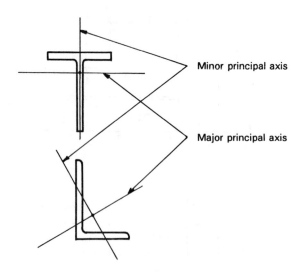

FIGURE A.6.

Using the calculus, it can be shown that

$$J_z = \int r^2 \, da = \int x^2 \, da + \int y^2 \, da = I_x + I_y$$

This is a useful relationship for determination of J values for some shapes.

A.8. PRINCIPAL AXES

For any shape there are two mutually perpendicular axes through the centroid of the shape about which the moments of inertia are, respectively, maximum and mini-

mum. These are called the *principal axes* of the shape (see Fig. A.6). If the shape has an axis of symmetry (*I, T, C, U,* etc.), it will always be a principal axis.

The axis about which the moment of inertia is maximum is called the *major principal axis*, and the axis about which the moment of inertia is least is called the *minor principal axis*. The major principal axis relates to the maximum bending resistance of a member, yielding the maximum values for section modulus (relating to bending stress) and moment of inertia (relating to deflection). On the other hand, the minor principal axis yields the least value for radius of gyration, relating to the maximum potential for buckling.

TABLE A.1. Properties of Geometric Shapes

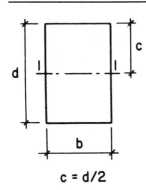

$$A = bd$$
$$I_1 = \frac{bd^3}{12}$$
$$S_1 = \frac{bd^2}{6}$$
$$r_1 = \frac{d}{\sqrt{12}}$$
$$c = d/2$$

$$A = bd$$
$$I_1 = \frac{bd^3}{3}$$
$$S_1 = \frac{bd^2}{3}$$
$$r_1 = \frac{d}{\sqrt{3}}$$
$$c = b$$

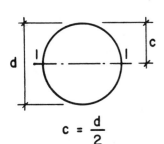

$$A = \frac{\pi d^2}{4}$$
$$I_1 = \frac{\pi d^4}{64}$$
$$S_1 = \frac{\pi d^3}{32}$$
$$r_1 = \frac{d}{4}$$
$$c = \frac{d}{2}$$

$$A = \frac{bd}{2}$$
$$I_1 = \frac{bd^3}{36}$$
$$S_1 = \frac{bd^2}{24}$$
$$r_1 = \frac{d}{\sqrt{18}}$$
$$c = \frac{2d}{3}$$

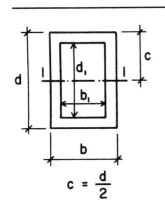

$$A = bd - b_1 d_1$$
$$I_1 = \frac{bd^3 - b_1 d_1^3}{12}$$
$$S_1 = \frac{bd^3 - b_1 d_1^3}{6d}$$
$$r_1 = \sqrt{\frac{bd^3 - b_1 d_1^3}{12 A}}$$
$$c = \frac{d}{2}$$

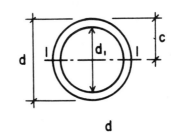

$$A = \frac{\pi (d^2 - d_1^2)}{4}$$
$$I_1 = \frac{\pi (d^4 - d_1^4)}{64}$$
$$S_1 = \frac{\pi (d^4 - d_1^4)}{32 d}$$
$$r_1 = \frac{\sqrt{d^2 - d_1^2}}{4}$$
$$c = \frac{d}{2}$$

A = Area I = Moment of inertia S = Section modulus = $\frac{I}{c}$ r = Radius of gyration = $\sqrt{\frac{I}{A}}$

$S_{rn} = m/F_0$

TABLE A.2. Properties of Structural Lumber[a]

Nominal size b(inches)d	Standard dressed size (S4S) b(inches)d	Area of Section A	Moment of inertia I	Section modulus S	Weight in lb/Ft 35 lb.
1 x 3	3/4 x 2-1/2	1.875	0.977	0.781	0.456
1 x 4	3/4 x 3-1/2	2.625	2.680	1.531	0.638
1 x 6	3/4 x 5-1/2	4.125	10.398	3.781	1.003
1 x 8	3/4 x 7-1/4	5.438	23.817	6,570	1.322
1 x 10	3/4 x 9-1/4	6.938	49.466	10.695	1.686
1 x 12	3/4 x 11-1/4	8.438	88.989	15.820	2.051
2 x 3	1-1/2 x 2-1/2	3.750	1.953	1.563	0.911
2 x 4	1-1/2 x 3-1/2	5.250	5.359	3.063	1.276
2 x 5	1-1/2 x 4-1/2	6.750	11.391	5.063	1.641
2 x 6	1-1/2 x 5-1/2	8.250	20.797	7.563	2.005
2 x 8	1-1/2 x 7-1/4	10.875	47.635	13.141	2.643
2 x 10	1-1/2 x 9-1/4	13.875	98.932	21.391	3.372
2 x 12	1-1/2 x 11-1/4	16.875	177.979	31.641	4.102
2 x 14	1-1/2 x 13-1/4	19.875	290.775	43.891	4.831
3 x 1	2-1/2 x 3/4	1.875	0.088	0.234	0.456
3 x 2	2-1/2 x 1-1/2	3.750	0.703	0.938	0.911
3 x 4	2-1/2 x 3-1/2	8.750	8.932	5.104	2.127
3 x 5	2-1/2 x 4-1/2	11.250	18.984	8.438	2.734
3 x 6	2-1/2 x 5-1/2	13.750	34.661	12.604	3.342
3 x 8	2-1/2 x 7-1/4	18.125	79.391	21.901	4.405
3 x 10	2-1/2 x 9-1/4	23.125	164.886	35.651	5.621
3 x 12	2-1/2 x 11-1/4	28.125	296.631	52.734	6.836
3 x 14	2-1/2 x 13-1/4	33.125	484.625	73.151	8.051
3 x 16	2-1/2 x 15-1/4	38.125	738.870	96.901	9.266
4 x 1	3-1/2 x 3/4	2.625	0.123	0.328	0.638
4 x 2	3-1/2 x 1-1/2	5.250	0.984	1.313	1.276
4 x 3	3-1/2 x 2-1/2	8.750	4.557	3.646	2.127
4 x 4	3-1/2 x 3-1/2	12.250	12.505	7.146	2.977
4 x 5	3-1/2 x 4-1/2	15.750	26.578	11.813	3.828
4 x 6	3-1/2 x 5-1/2	19.250	48.526	17.646	4.679
4 x 8	3-1/2 x 7-1/4	25.375	111.148	30.661	6.168
4 x 10	3-1/2 x 9-1/4	32.375	230.840	49.911	7.869
4 x 12	3-1/2 x 11-1/4	39.375	415.283	73.828	9.570
4 x 14	3-1/2 x 13-1/4	46.375	678.475	102.411	11.266
4 x 16	3-1/2 x 15-1/4	53.375	1034.418	135.66	12.975
5 x 2	4-1/2 x 1-1/2	6.750	1.266	1.688	1.641
5 x 3	4-1/2 x 2-1/2	11.250	5.859	4.688	2.734
5 x 4	4-1/2 x 3-1/2	15.750	16.078	9.188	3.828
5 x 5	4-1/2 x 4-1/2	20.250	34.172	15.188	4.922
6 x 1	5-1/2 x 3/4	4.125	0.193	0.516	1.003
6 x 2	5-1/2 x 1-1/2	8.250	1.547	2.063	2.005
6 x 3	5-1/2 x 2-1/2	13.750	7.161	5.729	3.342
6 x 4	5-1/2 x 3-1/2	19.250	19.651	11.229	4.679
6 x 6	5-1/2 x 5-1/2	30.250	76.255	27.729	7.352
6 x 8	5-1/2 x 7-1/2	41.250	193.359	51.563	10.026
6 x 10	5-1/2 x 9-1/2	52.250	392.963	82.729	12.700
6 x 12	5-1/2 x 11-1/2	63.250	697.068	121.229	15.373
6 x 14	5-1/2 x 13-1/2	74.250	1127.672	167.063	18.047
6 x 16	5-1/2 x 15-1/2	85.250	1706.776	220.229	20.720
6 x 18	5-1/2 x 17-1/2	96.250	2456.380	280.729	23.394
6 x 20	5-1/2 x 19-1/2	107.250	3398.484	348.563	26.068
6 x 22	5-1/2 x 21-1/2	118.250	4555.086	423.729	28.741
6 x 24	5-1/2 x 23-1/2	129.250	5948.191	506.229	31.415
8 x 1	7-1/4 x 3/4	5.438	0.255	0.680	1.322
8 x 2	7-1/4 x 1-1/2	10.875	2.039	2.719	2.643
8 x 3	7-1/4 x 2-1/2	18.125	9.440	7.552	4.405
8 x 4	7-1/4 x 3-1/2	25.375	25.904	14.803	6.168
8 x 6	7-1/2 x 5-1/2	41.250	103.984	37.813	10.026
8 x 8	7-1/2 x 7-1/2	56.250	263.672	70.313	13.672
8 x 10	7-1/2 x 9-1/2	71.250	535.859	112.813	17.318
8 x 12	7-1/2 x 11-1/2	86.250	950.547	165.313	20.964
8 x 14	7-1/2 x 13-1/2	101.250	1537.734	227.813	24.609
8 x 16	7-1/2 x 15-1/2	116.250	2327.422	300.313	28.255
8 x 18	7-1/2 x 17-1/2	131.250	3349.609	382.813	31.901
8 x 20	7-1/2 x 19-1/2	146.250	4634.297	475.313	35.547
8 x 22	7-1/2 x 21-1/2	161.250	6211.484	577.813	39.193
8 x 24	7-1/2 x 23-1/2	176.250	8111.172	690.313	42.839
10 x 1	9-1/4 x 3/4	6.938	0.325	0.867	1.686
10 x 2	9-1/4 x 1-1/2	13.875	2.602	3.469	3.372
10 x 3	9-1/4 x 2-1/2	23.125	12.044	9.635	5.621
10 x 4	9-1/4 x 3-1/2	32.375	33.049	18.885	7.869
10 x 6	9-1/2 x 5-1/2	52.250	131.714	47.896	12.700
10 x 8	9-1/2 x 7-1/2	71.250	333.984	89.063	17.318
10 x 10	9-1/2 x 9-1/2	90.250	678.755	142.896	21.936
10 x 12	9-1/2 x 11-1/2	109.250	1204.026	209.396	26.554
10 x 14	9-1/2 x 13-1/2	128.250	1947.797	288.563	31.172
10 x 16	9-1/2 x 15-1/2	147.250	2948.068	380.396	35.790
10 x 18	9-1/2 x 17-1/2	166.250	4242.836	484.896	40.408
10 x 20	9-1/2 x 19-1/2	185.250	5870.109	602.063	45.026
10 x 22	9-1/2 x 21-1/2	204.250	7867.879	731.896	49.644
10 x 24	9-1/2 x 23-1/2	223.250	10274.148	874.396	54.262
12 x 1	11-1/4 x 3/4	8.438	0.396	1.055	2.051
12 x 2	11-1/4 x 1-1/2	16.875	3.164	4.219	4.102
12 x 3	11-1/4 x 2-1/2	28.125	14.648	11.719	6.836
12 x 4	11-1/4 x 3-1/2	39.375	40.195	22.969	9.570
12 x 6	11-1/2 x 5-1/2	63.250	159.443	57.979	15.373
12 x 8	11-1/2 x 7-1/2	86.250	404.297	107.813	20.964
12 x 10	11-1/2 x 9-1/2	109.250	821.651	172.979	26.554
12 x 12	11-1/2 x 11-1/2	132.250	1457.505	253.479	32.144
12 x 14	11-1/2 x 13-1/2	155.250	2357.859	349.313	37.734
12 x 16	11-1/2 x 15-1/2	178.250	3568.713	460.479	43.325
12 x 18	11-1/2 x 17-1/2	201.250	5136.066	586.979	48.915
12 x 20	11-1/2 x 19-1/2	224.250	7105.922	728.813	54.505
12 x 22	11-1/2 x 21-1/2	247.250	9524.273	885.979	60.095
12 x 24	11-1/2 x 23-1/2	270.250	12437.129	1058.479	65.686
14 x 2	13-1/4 x 1-1/2	19.875	3.727	4.969	4.831
14 x 3	13-1/4 x 2-1/2	33.125	17.253	13.802	8.051
14 x 4	13-1/4 x 3-1/2	46.375	47.34	27.052	11.266
14 x 6	13-1/2 x 5-1/2	74.250	187.172	68.063	18.047
14 x 8	13-1/2 x 7-1/2	101.250	474.609	126.563	24.609
14 x 10	13-1/2 x 9-1/2	128.250	964.547	203.063	31.172
14 x 12	13-1/2 x 11-1/2	155.250	1710.984	297.563	37.734
14 x 14	13-1/2 x 13-1/2	182.250	2767.922	410.063	44.297
14 x 16	13-1/2 x 15-1/2	209.250	4189.359	540.563	50.859
14 x 18	13-1/2 x 17-1/2	236.250	6029.297	689.063	57.422
14 x 20	13-1/2 x 19-1/2	263.250	8341.734	855.563	63.984
14 x 22	13-1/2 x 21-1/2	290.250	11180.672	1040.063	70.547
14 x 24	13-1/2 x 23-1/2	317.250	14600.109	1242.563	77.109
16 x 3	15-1/4 x 2-1/2	38.125	19.857	15.885	9.267
16 x 4	15-1/4 x 3-1/2	53.375	54.487	31.135	12.975
16 x 6	15-1/2 x 5-1/2	85.250	214.901	78.146	20.720
16 x 8	15-1/2 x 7-1/2	116.250	544.922	145.313	28.255
16 x 10	15-1/2 x 9-1/2	147.250	1107.443	233.146	35.790
16 x 12	15-1/2 x 11-1/2	178.250	1964.463	341.646	43.325
16 x 14	15-1/2 x 13-1/2	209.250	3177.984	470.813	50.859
16 x 16	15-1/2 x 15-1/2	240.250	4810.004	620.646	58.394
16 x 18	15-1/2 x 17-1/2	271.250	6922.523	791.146	65.929
16 x 20	15-1/2 x 19-1/2	302.250	9577.547	982.313	73.464
16 x 22	15-1/2 x 21-1/2	333.250	12837.066	1194.146	80.998
16 x 24	15-1/2 x 23-1/2	364.250	16763.086	1426.646	88.533
18 x 6	17-1/2 x 5-1/2	96.250	242.630	88.229	23.394
18 x 8	17-1/2 x 7-1/2	131.250	615.234	164.063	31.901
18 x 10	17-1/2 x 9-1/2	166.250	1250.338	263.229	40.408
18 x 12	17-1/2 x 11-1/2	201.250	2217.943	385.729	48.915
18 x 14	17-1/2 x 13-1/2	236.250	3588.047	531.563	57.422
18 x 16	17-1/2 x 15-1/2	271.250	5430.648	700.729	65.929
18 x 18	17-1/2 x 17-1/2	306.250	7815.754	893.229	74.436
18 x 20	17-1/2 x 19-1/2	341.250	10813.359	1109.063	82.943
18 x 22	17-1/2 x 21-1/2	376.250	14493.461	1348.229	91.450
18 x 24	17-1/2 x 23-1/2	411.250	18926.066	1610.729	99.957
20 x 6	19-1/2 x 5-1/2	107.250	270.359	98.313	26.068
20 x 8	19-1/2 x 7-1/2	146.250	685.547	182.813	35.547
20 x 10	19-1/2 x 9-1/2	185.250	1393.234	293.313	45.026
20 x 12	19-1/2 x 11-1/2	224.250	2471.422	429.813	54.505
20 x 14	19-1/2 x 13-1/2	263.250	3998.109	592.313	63.984
20 x 16	19-1/2 x 15-1/2	302.250	6051.297	780.813	73.464
20 x 18	19-1/2 x 17-1/2	341.250	8708.984	995.313	82.943
20 x 20	19-1/2 x 19-1/2	380.250	12049.172	1235.813	92.422
20 x 22	19-1/2 x 21-1/2	419.250	16149.859	1502.313	101.901
20 x 24	19-1/2 x 23-1/2	458.250	21089.047	1794.813	111.380
22 x 6	21-1/2 x 5-1/2	118.250	298.088	108.396	28.741
22 x 8	21-1/2 x 7-1/2	161.250	755.859	201.563	39.193
22 x 10	21-1/2 x 9-1/2	204.250	1536.130	323.396	49.644
22 x 12	21-1/2 x 11-1/2	247.250	2724.901	473.896	60.095
22 x 14	21-1/2 x 13-1/2	290.250	4408.172	653.063	70.547
22 x 16	21-1/2 x 15-1/2	333.250	6671.941	860.896	80.998
22 x 18	21-1/2 x 17-1/2	376.250	9602.211	1097.396	91.450
22 x 20	21-1/2 x 19-1/2	419.250	13284.984	1362.563	101.901
22 x 22	21-1/2 x 21-1/2	462.250	17806.254	1656.396	112.352
22 x 24	21-1/2 x 23-1/2	505.250	23252.023	1978.896	122.804
24 x 6	23-1/2 x 5-1/2	129.250	325.818	118.479	31.415
24 x 8	23-1/2 x 7-1/2	176.250	826.172	220.313	42.839
24 x 10	23-1/2 x 9-1/2	223.250	1679.026	353.479	54.262
24 x 12	23-1/2 x 11-1/2	270.250	2978.380	517.979	65.686
24 x 14	23-1/2 x 13-1/2	317.250	4818.234	713.813	77.109
24 x 16	23-1/2 x 15-1/2	364.250	7292.586	940.979	88.533
24 x 18	23-1/2 x 17-1/2	411.250	10495.441	1199.479	99.957
24 x 20	23-1/2 x 19-1/2	458.250	14520.797	1489.313	111.380
24 x 22	23-1/2 x 21-1/2	505.250	19462.648	1810.479	122.804
24 x 24	23-1/2 x 23-1/2	552.250	25415.004	2162.979	134.227

[a] Values given for weight are in pounds per linear foot assuming a density of 35 pcf.

TABLE A.3. Properties of Wide-Flange Shapes

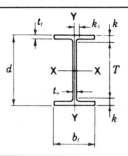

	Area A	Depth d	Web Thickness t_w	Flange		k	Elastic Properties					
				Width b_f	Thickness t_f		Axis X-X			Axis Y-Y		
							I	S	r	I	S	r
Designation	in²	in.	in.	in.	in.	in.	in⁴	in³	in.	in⁴	in³	in.
W 36 × 300	88.3	36.74	0.945	16.655	1.680	2.81	20,300	1110	15.2	1300	156	3.83
× 260	76.5	36.26	0.840	16.550	1.440	2.56	17,300	953	15.0	1090	132	3.78
× 230	67.6	35.90	0.760	16.470	1.260	2.38	15,000	837	14.9	940	114	3.73
× 194	57.0	36.49	0.765	12.115	1.260	2.19	12,100	664	14.6	375	61.9	2.56
× 170	50.0	36.17	0.680	12.030	1.100	2.00	10,500	580	14.5	320	53.2	2.53
× 150	44.2	35.85	0.625	11.975	0.940	1.88	9,040	504	14.3	270	45.1	2.47
× 135	39.7	35.55	0.600	11.950	0.790	1.69	7,800	439	14.0	225	37.7	2.38
W 33 × 241	70.9	34.18	0.830	15.860	1.400	2.19	14,200	829	14.1	932	118	3.63
× 201	59.1	33.68	0.715	15.745	1.150	1.94	11,500	684	14.0	749	95.2	3.56
× 152	44.7	33.49	0.635	11.565	1.055	1.88	8,160	487	13.5	273	47.2	2.47
× 130	38.3	33.09	0.580	11.510	0.855	1.69	6,710	406	13.2	218	37.9	2.39
× 118	34.7	32.86	0.550	11.480	0.740	1.56	5,900	359	13.0	187	32.6	2.32
W 30 × 211	62.0	30.94	0.775	15.105	1.315	2.13	10,300	663	12.9	757	100	3.49
× 173	50.8	30.44	0.655	14.985	1.065	1.88	8,200	539	12.7	598	79.8	3.43
× 124	36.5	30.17	0.585	10.515	0.930	1.69	5,360	355	12.1	181	34.4	2.23
× 108	31.7	29.83	0.545	10.475	0.760	1.56	4,470	299	11.9	146	27.9	2.15
× 99	29.1	29.65	0.520	10.450	0.670	1.44	3,990	269	11.7	128	24.5	2.10
W 27 × 178	52.3	27.81	0.725	14.085	1.190	1.88	6,990	502	11.6	555	78.8	3.26
× 146	42.9	27.38	0.605	13.965	0.975	1.69	5,630	411	11.4	443	63.5	3.21
× 102	30.0	27.09	0.515	10.015	0.830	1.56	3,620	267	11.0	139	27.8	2.15
× 84	24.8	26.71	0.460	9.960	0.640	1.38	2,850	213	10.7	106	21.2	2.07
W 24 × 162	47.7	25.00	0.705	12.955	1.220	2.00	5,170	414	10.4	443	68.4	3.05
× 131	38.5	24.48	0.605	12.855	0.960	1.75	4,020	329	10.2	340	53.0	2.97
× 104	30.6	24.06	0.500	12.750	0.750	1.50	3,100	258	10.1	259	40.7	2.91
× 84	24.7	24.10	0.470	9.020	0.770	1.56	2,370	196	9.79	94.4	20.9	1.95
× 69	20.1	23.73	0.415	8.965	0.585	1.38	1,830	154	9.55	70.4	15.7	1.87
× 55	16.2	23.57	0.395	7.005	0.505	1.94	1,350	114	9.11	29.1	8.30	1.34
W 21 × 147	43.2	22.06	0.720	12.510	1.150	1.88	3,630	329	9.17	376	60.1	2.95
× 122	35.9	21.68	0.600	12.390	0.960	1.69	2,960	273	9.09	305	49.2	2.92
× 101	29.8	21.36	0.500	12.290	0.800	1.56	2,420	227	9.02	248	40.3	2.89
× 83	24.3	21.43	0.515	8.355	0.835	1.56	1,830	171	8.67	81.4	19.5	1.83
× 68	20.0	21.13	0.430	8.270	0.685	1.44	1,480	140	8.60	64.7	15.7	1.80
× 57	16.7	21.06	0.405	6.555	0.650	1.38	1,170	111	8.36	30.6	9.35	1.35

TABLE A.3. (*continued*)

Designation	Area A	Depth d	Web Thickness t_w	Flange Width b_f	Flange Thickness t_f	k	Axis X-X I	S	r	Axis Y-Y I	S	r
	in²	in.	in.	in.	in.	in.	in⁴	in³	in.	in⁴	in³	in.
× 44	13.0	20.66	0.350	6.500	0.450	1.19	843	81.6	8.06	20.7	6.36	1.26
W 18 × 119	35.1	18.97	0.655	11.265	1.060	1.75	2,190	231	7.90	253	44.9	2.69
× 97	28.5	18.59	0.535	11.145	0.870	1.56	1,750	188	7.82	201	36.1	2.65
× 76	22.3	18.21	0.425	11.035	0.680	1.38	1,330	146	7.73	152	27.6	2.61
× 65	19.1	18.35	0.450	7.590	0.750	1.44	1,070	117	7.49	54.8	14.4	1.69
× 55	16.2	18.11	0.390	7.530	0.630	1.31	890	98.3	7.41	44.9	11.9	1.67
× 46	13.5	18.06	0.360	6.060	0.605	1.25	712	78.8	7.25	22.5	7.43	1.29
× 35	10.3	17.70	0.300	6.000	0.425	1.13	510	57.6	7.04	15.3	5.12	1.22
W 16 × 100	29.4	16.97	0.585	10.425	0.985	1.69	1,490	175	7.10	186	35.7	2.51
× 77	22.6	16.52	0.455	10.295	0.760	1.44	1,110	134	7.00	138	26.9	2.47
× 57	16.8	16.43	0.430	7.120	0.715	1.38	758	92.2	6.72	43.1	12.1	1.60
× 45	13.3	16.13	0.345	7.035	0.565	1.25	586	72.7	6.65	32.8	9.34	1.57
× 36	10.6	15.86	0.295	6.985	0.430	1.13	448	56.5	6.51	24.5	7.00	1.52
× 26	7.68	15.69	0.250	5.500	0.345	1.06	301	38.4	6.26	9.59	3.49	1.12
W 14 × 730	215.0	22.42	3.070	17.890	4.910	5.56	14,300	1280	8.17	4720	527	4.69
× 605	178.0	20.92	2.595	17.415	4.160	4.81	10,800	1040	7.80	3680	423	4.55
× 455	134.0	19.02	2.015	16.835	3.210	3.88	7,190	756	7.33	2560	304	4.38
× 370	109.0	17.92	1.655	16.475	2.660	3.31	5,440	607	7.07	1990	241	4.27
× 283	83.3	16.74	1.290	16.110	2.070	2.75	3,840	459	6.79	1440	179	4.17
× 211	62.0	15.72	0.980	15.800	1.560	2.25	2,660	338	6.55	1030	130	4.07
× 159	46.7	14.98	0.745	15.565	1.190	1.88	1,900	254	6.38	748	96.2	4.00
× 120	35.3	14.48	0.590	14.670	0.940	1.63	1,380	190	6.24	495	67.5	3.74
× 90	26.5	14.02	0.440	14.520	0.710	1.38	999	143	6.14	362	49.9	3.70
× 68	20.0	14.04	0.415	10.035	0.720	1.50	723	103	6.01	121	24.2	2.46
× 53	15.6	13.92	0.370	8.060	0.660	1.44	541	77.8	5.89	57.7	14.3	1.92
× 43	12.6	13.66	0.305	7.995	0.530	1.31	428	62.7	5.82	45.2	11.3	1.89
× 34	10.0	13.98	0.285	6.745	0.455	1.00	340	48.6	5.83	23.3	6.91	1.53
× 30	8.85	13.84	0.270	6.730	0.385	0.94	291	42.0	5.73	19.6	5.82	1.49
× 26	7.69	13.91	0.255	5.025	0.420	0.94	245	35.3	5.65	8.91	3.54	1.08
× 22	6.49	13.74	0.230	5.000	0.335	0.88	199	29.0	5.54	7.00	2.80	1.04
W 12 × 336	98.8	16.82	1.775	13.385	2.955	3.69	4,060	483	6.41	1190	177	3.47
× 279	81.9	15.85	1.530	13.140	2.470	3.19	3,110	393	6.16	937	143	3.38
× 210	61.8	14.71	1.180	12.790	1.900	2.63	2,140	292	5.89	664	104	3.28
× 152	44.7	13.71	0.870	12.480	1.400	2.13	1,430	209	5.66	454	72.8	3.19
× 106	31.2	12.89	0.610	12.220	0.990	1.69	933	145	5.47	301	49.3	3.11
× 79	23.2	12.38	0.470	12.080	0.735	1.44	662	107	5.34	216	35.8	3.05
× 58	17.0	12.19	0.360	10.010	0.640	1.38	475	78.0	5.28	107	21.4	2.51

TABLE A.3. (*continued*)

				Flange			Elastic Properties					
				Width	Thickness		Axis X-X			Axis Y-Y		
	Area A	Depth d	Web Thickness t_w	b_f	t_f	k	I	S	r	I	S	r
Designation	in²	in.	in.	in.	in.	in.	in⁴	in³	in.	in⁴	in³	in.
× 50	14.7	12.19	0.370	8.080	0.640	1.38	394	64.7	5.18	56.3	13.9	1.96
× 40	11.8	11.94	0.295	8.005	0.515	1.25	310	51.9	5.13	44.1	11.0	1.93
× 30	8.79	12.34	0.260	6.520	0.440	0.94	238	38.6	5.21	20.3	6.24	1.52
× 22	6.48	12.31	0.260	4.030	0.425	0.88	156	25.4	4.91	4.66	2.31	0.847
× 19	5.57	12.16	0.235	4.005	0.350	0.81	130	21.3	4.82	3.76	1.88	0.822
× 16	4.71	11.99	0.220	3.990	0.265	0.75	103	17.1	4.67	2.82	1.41	0.773
× 14	4.16	11.91	0.200	3.970	0.225	0.69	88.6	14.9	4.62	2.36	1.19	0.753
W 10 × 112	32.9	11.36	0.755	10.415	1.250	1.88	716	126	4.66	236	45.3	2.68
× 88	25.9	10.84	0.605	10.265	0.990	1.63	534	98.5	4.54	179	34.8	2.63
× 60	17.6	10.22	0.420	10.080	0.680	1.31	341	66.7	4.39	116	23.0	2.57
× 45	13.3	10.10	0.350	8.020	0.620	1.25	248	49.1	4.32	53.4	13.3	2.01
× 33	9.71	9.73	0.290	7.960	0.435	1.06	170	35.0	4.19	36.6	9.20	1.94
× 26	7.61	10.33	0.260	5.770	0.440	0.88	144	27.9	4.35	14.1	4.89	1.36
× 19	5.62	10.24	0.250	4.020	0.395	0.81	96.3	18.8	4.14	4.29	2.14	0.874
× 15	4.41	9.99	0.230	4.000	0.270	0.69	68.9	13.8	3.95	2.89	1.45	0.810
× 12	3.54	9.87	0.190	3.960	0.210	0.63	53.8	10.9	3.90	2.18	1.10	0.785
W 8 × 67	19.7	9.00	0.570	8.280	0.935	1.44	272	60.4	3.72	88.6	21.4	2.12
× 48	14.1	8.50	0.400	8.110	0.685	1.19	184	43.3	3.61	60.9	15.0	2.08
× 40	11.7	8.25	0.360	8.070	0.560	1.06	146	35.5	3.53	49.1	12.2	2.04
× 35	10.3	8.12	0.310	8.020	0.495	1.00	127	31.2	3.51	42.6	10.6	2.03
× 31	9.13	8.00	0.285	7.995	0.435	0.94	110	27.5	3.47	37.1	9.27	2.02
× 28	8.25	8.06	0.285	6.535	0.465	0.94	98.0	24.3	3.45	21.7	6.63	1.62
× 24	7.08	7.93	0.245	6.495	0.400	0.88	82.8	20.9	3.42	18.3	5.63	1.61
× 21	6.16	8.28	0.250	5.270	0.400	0.81	75.3	18.2	3.49	9.77	3.71	1.26
× 18	5.26	8.14	0.230	5.250	0.330	0.75	61.9	15.2	3.43	7.97	3.04	1.23
× 15	4.44	8.11	0.245	4.015	0.315	0.75	48.0	11.8	3.29	3.41	1.70	0.876
× 13	3.84	7.99	0.230	4.000	0.255	0.69	39.6	9.91	3.21	2.73	1.37	0.843
× 10	2.96	7.89	0.170	3.940	0.205	0.63	30.8	7.81	3.22	2.09	1.06	0.841
W 6 × 25	7.34	6.38	0.320	6.080	0.455	0.81	53.4	16.7	2.70	17.1	5.61	1.52
× 20	5.87	6.20	0.260	6.020	0.365	0.75	41.4	13.4	2.66	13.3	4.41	1.50
× 15	4.43	5.99	0.230	5.990	0.260	0.63	29.1	9.72	2.56	9.32	3.11	1.46
× 16	4.74	6.28	0.260	4.030	0.405	0.75	32.1	10.2	2.60	4.43	2.20	0.966
× 12	3.55	6.03	0.230	4.000	0.280	0.63	22.1	7.31	2.49	2.99	1.50	0.918
× 9	2.68	5.90	0.170	3.940	0.215	0.56	16.4	5.56	2.47	2.19	1.11	0.905
W 5 × 19	5.54	5.15	0.270	5.030	0.430	0.81	26.2	10.2	2.17	9.13	3.63	1.28
× 16	4.68	5.01	0.240	5.000	0.360	0.75	21.3	8.51	2.13	7.51	3.00	1.27
W 4 × 13	3.83	4.16	0.280	4.060	0.345	0.69	11.3	5.46	1.72	3.86	1.90	1.00

Source: Adapted from data in the *Manual of Steel Construction*, 8th ed., with permission of the publishers, American Institute of Steel Construction.

TABLE A.4. Properties of American Standard (S) Shapes

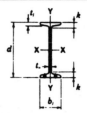

Designation	Area A in²	Depth d in.	Web Thickness t_w in.	Flange Width t_f in.	Flange Thickness b_f in.	k in.	Axis X-X I in⁴	Axis X-X S in³	Axis X-X r in.	Axis Y-Y I in⁴	Axis Y-Y S in³	Axis Y-Y r in.
S 24 × 121	35.6	24.50	0.800	8.050	1.090	2.00	3160	258	9.43	83.3	20.7	1.53
× 106	31.2	24.50	0.620	7.870	1.090	2.00	2940	240	9.71	77.1	19.6	1.57
× 100	29.3	24.00	0.745	7.245	0.870	1.75	2390	199	9.02	47.7	13.2	1.27
× 90	26.5	24.00	0.625	7.125	0.870	1.75	2250	187	9.21	44.9	12.6	1.30
× 80	23.5	24.00	0.500	7.000	0.870	1.75	2100	175	9.47	42.2	12.1	1.34
S 20 × 96	28.2	20.30	0.8000	7.200	0.920	1.75	1670	165	7.71	50.2	13.9	1.33
× 86	25.3	20.30	0.660	7.060	0.920	1.75	1580	155	7.89	46.8	13.3	1.36
× 75	22.0	20.00	0.635	6.385	0.795	1.63	1280	128	7.62	29.8	9.32	1.16
× 66	19.4	20.00	0.505	6.255	0.795	1.63	1190	119	7.83	27.7	8.85	1.19
S 18 × 70	20.6	18.00	0.711	6.251	0.691	1.50	926	103	6.71	24.1	7.72	1.08
× 54.7	16.1	18.00	0.461	6.001	0.691	1.50	804	89.4	7.07	20.8	6.94	1.14
S 15 × 50	14.7	15.00	0.550	5.640	0.622	1.38	486	64.8	5.75	15.7	5.57	1.03
× 42.9	12.6	15.00	0.411	5.501	0.622	1.38	447	59.6	5.95	14.4	5.23	1.07
S 12 × 50	14.7	12.00	0.687	5.477	0.659	1.44	305	50.8	4.55	15.7	5.74	1.03
× 40.8	12.0	12.00	0.462	5.252	0.659	1.44	272	45.4	4.77	13.6	5.16	1.06
× 35	10.3	12.00	0.428	5.078	0.544	1.19	229	38.2	4.72	9.87	3.89	0.980
× 31.8	9.35	12.00	0.350	5.000	0.544	1.19	218	36.4	4.83	9.36	3.74	1.00
S 10 × 35	10.3	10.00	0.594	4.944	0.491	1.13	147	29.4	3.78	8.36	3.38	0.901
× 25.4	7.46	10.00	0.311	4.661	0.491	1.13	124	24.7	4.07	6.79	2.91	0.954
S 8 × 23	6.77	8.00	0.441	4.171	0.426	1.00	64.9	16.2	3.10	4.31	2.07	0.798
× 18.4	5.41	8.00	0.271	4.001	0.426	1.00	57.6	14.4	3.26	3.73	1.86	0.831
S 7 × 20	5.88	7.00	0.450	3.860	0.392	0.94	42.4	12.1	2.69	3.17	1.64	0.734
× 15.3	4.50	7.00	0.252	3.662	0.392	0.94	36.7	10.5	2.86	2.64	1.44	0.766
S 6 × 17.25	5.07	6.00	0.465	3.565	0.359	0.88	26.3	8.77	2.28	2.31	1.30	0.675
× 12.5	3.67	6.00	0.232	3.332	0.359	0.88	22.1	7.37	2.45	1.82	1.09	0.705
S 5 × 14.75	4.34	5.00	0.494	3.284	0.326	0.81	15.2	6.09	1.87	1.67	1.01	0.620
× 10	2.94	5.00	0.214	3.004	0.326	0.81	12.3	4.92	2.05	1.22	0.809	0.643
S 4 × 9.5	2.79	4.00	0.326	2.796	0.293	0.75	6.79	3.39	1.56	0.903	0.646	0.569
× 7.7	2.26	4.00	0.193	2.663	0.293	0.75	6.08	3.04	1.64	0.764	0.574	0.581
S 3 × 7.5	2.21	3.00	0.349	2.509	0.260	0.69	2.93	1.95	1.15	0.586	0.468	0.516
× 5.7	1.67	3.00	0.170	2.330	0.260	0.69	2.52	1.68	1.23	0.455	0.390	0.522

Source: Adapted from data in the *Manual of Steel Construction*, 8th ed., with permission of the publishers, American Institute of Steel Construction.

TABLE A.5. Properties of American Standard Channels

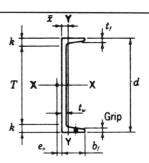

Designation	Area A	Depth d	Web Thickness t_w	Flange Width t_f	Flange Thickness b_f	k	Axis X-X I	Axis X-X S	Axis X-X r	Axis Y-Y I	Axis Y-Y S	Axis Y-Y r	$\bar{x}$
	in²	in.	in.	in.	in.	in.	in⁴	in³	in.	in⁴	in³	in.	in.
C 15 × 50	14.7	15.00	0.716	3.716	0.650	1.44	404	53.8	5.24	11.0	3.78	0.867	0.798
× 40	11.8	15.00	0.520	3.520	0.650	1.44	349	46.5	5.44	9.23	3.37	0.886	0.777
× 33.9	9.96	15.00	0.400	3.400	0.650	1.44	315	42.0	5.62	8.13	3.11	0.904	0.787
C 12 × 30	8.82	12.00	0.510	3.170	0.501	1.13	162	27.0	4.29	5.14	2.06	0.763	0.674
× 25	7.35	12.00	0.387	3.047	0.501	1.13	144	24.1	4.43	4.47	1.88	0.780	0.674
× 20.7	6.09	12.00	0.282	2.942	0.501	1.13	129	21.5	4.61	3.88	1.73	0.799	0.698
C 10 × 30	8.82	10.00	0.673	3.033	0.436	1.00	103	20.7	3.42	3.94	1.65	0.669	0.649
— × 25	7.35	10.00	0.526	2.886	0.436	1.00	91.2	18.2	3.52	3.36	1.48	0.676	0.617
× 20	5.88	10.00	0.379	2.739	0.436	1.00	78.9	15.8	3.66	2.81	1.32	0.692	0.606
× 15.3	4.49	10.00	0.240	2.600	0.436	1.00	67.4	13.5	3.87	2.28	1.16	0.713	0.634
C 9 × 20	5.88	9.00	0.448	2.648	0.413	0.94	60.9	13.5	3.22	2.42	1.17	0.642	0.583
× 15	4.41	9.00	0.285	2.485	0.413	0.94	51.0	11.3	3.40	1.93	1.01	0.661	0.586
× 13.4	3.94	9.00	0.233	2.433	0.413	0.94	47.9	10.6	3.48	1.76	0.962	0.669	0.601
C 8 × 18.75	5.51	8.00	0.487	2.527	0.390	0.94	44.0	11.0	2.82	1.98	1.01	0.599	0.565
× 13.75	4.04	8.00	0.303	2.343	0.390	0.94	36.1	9.03	2.99	1.53	0.854	0.615	0.553
× 11.5	3.38	8.00	0.220	2.260	0.390	0.94	32.6	8.14	3.11	1.32	0.781	0.625	0.571
C 7 × 14.75	4.33	7.00	0.419	2.299	0.366	0.88	27.2	7.78	2.51	1.38	0.779	0.564	0.532
× 12.25	3.60	7.00	0.314	2.194	0.366	0.88	24.2	6.93	2.60	1.17	0.703	0.571	0.525
× 9.8	2.87	7.00	0.210	2.090	0.366	0.88	21.3	6.08	2.72	0.968	0.625	0.581	0.540
C 6 × 13	3.83	6.00	0.437	2.157	0.343	0.81	17.4	5.80	2.13	1.05	0.642	0.525	0.514
× 10.5	3.09	6.00	0.314	2.034	0.343	0.81	15.2	5.06	2.22	0.866	0.564	0.529	0.499
× 8.2	2.40	6.00	0.200	1.920	0.343	0.81	13.1	4.38	2.34	0.693	0.492	0.537	0.511
C 5 × 9	2.64	5.00	0.325	1.885	0.320	0.75	8.90	3.56	1.83	0.632	0.450	0.489	0.478
× 6.7	1.97	5.00	0.190	1.750	0.320	0.75	7.49	3.00	1.95	0.479	0.378	0.493	0.484
C 4 × 7.25	2.13	4.00	0.321	1.721	0.296	0.69	4.59	2.29	1.47	0.433	0.343	0.450	0.459
× 5.4	1.59	4.00	0.184	1.584	0.296	0.69	3.85	1.93	1.56	0.319	0.283	0.449	0.457
C 3 × 6	1.76	3.00	0.356	1.596	0.273	0.69	2.07	1.38	1.08	0.305	0.268	0.416	0.455
× 5	1.47	3.00	0.258	1.498	0.273	0.69	1.85	1.24	1.12	0.247	0.233	0.410	0.438
× 4.1	1.21	3.00	0.170	1.410	0.273	0.69	1.66	1.10	1.17	0.197	0.202	0.404	0.436

Source: Adapted from data in the *Manual of Steel Construction*, 8th ed., with permission of the publishers, American Institute of Steel Construction.

TABLE A.6. Properties of Angles

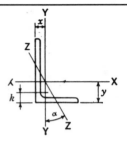

Size and Thickness in.	k in.	Weight lb	Area in²	Axis X-X I in⁴	S in³	r in.	y in.	Axis X-X I in⁴	S in³	r in.	x in.	Axis Z-Z r in.	Tan α
8 × 8 × 1 1/8	1.75	56.9	16.7	98.0	17.5	2.42	2.41	98.0	17.5	2.42	2.41	1.56	1.000
1	1.625	51.0	15.0	89.0	15.8	2.44	2.37	89.0	15.8	2.44	2.37	1.56	1.000
7/8	1.50	45.0	13.2	79.6	14.0	2.45	2.32	79.6	14.0	2.45	2.32	1.57	1.000
3/4	1.375	38.9	11.4	69.7	12.2	2.47	2.28	69.7	12.2	2.47	2.28	1.58	1.000
5/8	1.25	32.7	9.61	59.4	10.3	2.49	2.23	59.4	10.3	2.49	2.23	1.58	1.000
1/2	1.125	26.4	7.75	48.6	8.36	2.50	2.19	48.6	8.36	2.50	2.19	1.59	1.000
8 × 6 × 1	1.50	44.2	13.0	80.8	15.1	2.49	2.65	38.8	8.92	1.73	1.65	1.28	0.543
3/4	1.25	33.8	9.94	63.4	11.7	2.53	2.56	30.7	6.92	1.76	1.56	1.29	0.551
1/2	1.0	23.0	6.75	44.3	8.02	2.56	2.47	21.7	4.79	1.79	1.47	1.30	0.558
8 × 4 × 1	1.50	37.4	11.0	69.6	14.1	2.52	3.05	11.6	3.94	1.03	1.05	0.846	0.247
3/4	1.25	28.7	8.44	54.9	10.9	2.55	2.95	9.36	3.07	1.05	0.953	0.852	0.258
1/2	1.0	19.6	5.75	38.5	7.49	2.59	2.86	6.74	2.15	1.08	0.859	0.865	0.267
7 × 4 × 3/4	1.25	26.2	7.69	37.8	8.42	2.22	2.51	9.05	3.03	1.09	1.01	0.860	0.324
1/2	1.0	17.9	5.25	26.7	5.81	2.25	2.42	6.53	2.12	1.11	0.917	0.872	0.335
3/8	0.875	13.6	3.98	20.6	4.44	2.27	2.37	5.10	1.63	1.13	0.870	0.880	0.340
6 × 6 × 1	1.50	37.4	11.0	35.5	8.57	1.80	1.86	35.5	8.57	1.80	1.86	1.17	1.000
7/8	1.375	33.1	9.73	31.9	7.63	1.81	1.82	31.9	7.63	1.81	1.82	1.17	1.000
3/4	1.25	28.7	8.44	28.2	6.66	1.83	1.78	28.2	6.66	1.83	1.78	1.17	1.000
5/8	1.125	24.2	7.11	24.2	5.66	1.84	1.73	24.2	5.66	1.84	1.73	1.18	1.000
1/2	1.0	19.6	5.75	19.9	4.61	1.86	1.68	19.9	4.61	1.86	1.68	1.18	1.000
3/8	0.875	14.9	4.36	15.4	3.53	1.88	1.64	15.4	3.53	1.88	1.64	1.19	1.000
6 × 4 × 3/4	1.25	23.6	6.94	24.5	6.25	1.88	2.08	8.68	2.97	1.12	1.08	0.860	0.428
5/8	1.125	20.0	5.86	21.1	5.31	1.90	2.03	7.52	2.54	1.13	1.03	0.864	0.435
1/2	1.00	16.2	4.75	17.4	4.33	1.91	1.99	6.27	2.08	1.15	0.987	0.870	0.440
3/8	0.875	12.3	3.61	13.5	3.32	1.93	1.94	4.90	1.60	1.17	0.941	0.877	0.446
6 × 3 1/2 × 3/8	0.875	11.7	3.42	12.9	3.24	1.94	2.04	3.34	1.23	0.988	0.787	0.767	0.350
5/16	0.8125	9.8	2.87	10.9	2.73	1.95	2.01	2.85	1.04	0.996	0.763	0.772	0.352
5 × 5 × 7/8	1.375	27.2	7.98	17.8	5.17	1.49	1.57	17.8	5.17	1.49	1.57	0.973	1.000
3/4	1.25	23.6	6.94	15.7	4.53	1.51	1.52	15.7	4.53	1.51	1.52	0.975	1.000
1/2	1.0	16.2	4.75	11.3	3.16	1.54	1.43	11.3	3.16	1.54	1.43	0.983	1.000
5 × 5 × 3/8	0.875	12.3	3.61	8.74	2.42	1.56	1.39	8.74	2.42	1.56	1.39	0.990	1.000
5/16	0.8125	10.3	3.03	7.42	2.04	1.57	1.37	7.42	2.04	1.57	1.37	0.994	1.000
5 × 3 1/2 × 3/4	1.25	19.8	5.81	13.9	4.28	1.55	1.75	5.55	2.22	0.977	0.996	0.748	0.464
→ 1/2	1.0	13.6	4.00	9.99	2.99	1.58	1.66	4.05	1.56	1.01	0.906	0.755	0.479
3/8	0.875	10.4	3.05	7.78	2.29	1.60	1.61	3.18	1.21	1.02	0.861	0.762	0.486
5/16	0.8125	8.7	2.56	6.60	1.94	1.61	1.59	2.72	1.02	1.03	0.838	0.766	0.489
5 × 3 × 1/2	1.0	12.8	3.75	9.45	2.91	1.59	1.75	2.58	1.15	0.829	0.750	0.648	0.357
3/8	0.875	9.8	2.86	7.37	2.24	1.61	1.70	2.04	0.888	0.845	0.704	0.654	0.364
5/16	0.8125	8.2	2.40	6.26	1.89	1.61	1.68	1.75	0.753	0.853	0.681	0.658	0.368
1/4	0.75	6.6	1.94	5.11	1.53	1.62	1.66	1.44	0.614	0.861	0.657	0.663	0.371

TABLE A.6. (*continued*)

Size and Thickness in.	k in.	Weight lb	Area in²	Axis X-X				Axis Y-Y				Axis Z-Z	
				I in⁴	S in³	r in.	y in.	I in⁴	S in³	r in.	x in.	r in.	Tan α
4 × 4 × 3/4	1.125	18.5	5.44	7.67	2.81	1.19	1.27	7.67	2.81	1.19	1.27	0.778	1.000
5/8	1.0	15.7	4.61	6.66	2.40	1.20	1.23	6.66	2.40	1.20	1.23	0.779	1.000
→ 1/2	0.875	12.8	3.75	5.56	1.97	1.22	1.18	5.56	1.97	1.22	1.18	0.782	1.000
3/8	0.75	9.8	2.86	4.36	1.52	1.23	1.14	4.36	1.52	1.23	1.14	0.788	1.000
5/16	0.6875	8.2	2.40	3.71	1.29	1.24	1.12	3.71	1.29	1.24	1.12	0.791	1.000
1/4	0.625	6.6	1.94	3.04	1.05	1.25	1.09	3.04	1.05	1.25	1.09	0.795	1.000
4 × 3 1/2 × 1/2	0.9375	11.9	3.50	5.32	1.94	1.23	1.25	3.79	1.52	1.04	1.00	0.722	0.750
3/8	0.8125	9.1	2.67	4.18	1.49	1.25	1.21	2.95	1.17	1.06	0.955	0.727	0.755
5/16	0.75	7.7	2.25	3.56	1.26	1.26	1.18	2.55	0.994	1.07	0.932	0.730	0.757
1/4	0.6875	6.2	1.81	2.91	1.03	1.27	1.16	2.09	0.808	1.07	0.909	0.734	0.759
4 × 3 × 1/2	0.9375	11.1	3.25	5.05	1.89	1.25	1.33	2.42	1.12	0.864	0.827	0.639	0.543
3/8	0.8125	8.5	2.48	3.96	1.46	1.26	1.28	1.92	0.866	0.879	0.782	0.644	0.551
5/16	0.75	7.2	2.09	3.38	1.23	1.27	1.26	1.65	0.734	0.887	0.759	0.647	0.554
1/4	0.6875	5.8	1.69	2.77	1.00	1.28	1.24	1.36	0.599	0.896	0.736	0.651	0.558
3 1/2 × 3 1/2 × 3/8	0.75	8.5	2.48	2.87	1.15	1.07	1.01	2.87	1.15	1.07	1.01	0.687	1.000
5/16	0.6875	7.2	2.09	2.45	0.976	1.08	0.990	2.45	0.976	1.08	0.990	0.690	1.000
1/4	0.625	5.8	1.69	2.01	0.794	1.09	0.968	2.01	0.794	1.09	0.968	0.694	1.000
3 1/2 × 3 × 3/8	0.8125	7.9	2.30	2.72	1.13	1.09	1.08	1.85	0.851	0.897	0.830	0.625	0.721
5/16	0.75	6.6	1.93	2.33	0.954	1.10	1.06	1.58	0.722	0.905	0.808	0.627	0.724
1/4	0.6875	5.4	1.56	1.91	0.776	1.11	1.04	1.30	0.589	0.914	0.785	0.631	0.727
3 1/2 × 2 1/2 × 3/8	0.8125	7.2	2.11	2.56	1.09	1.10	1.16	1.09	0.592	0.719	0.660	0.537	0.496
5/16	0.75	6.1	1.78	2.19	0.927	1.11	1.14	0.939	0.504	0.727	0.637	0.540	0.501
1/4	0.6875	4.9	1.44	1.80	0.755	1.12	1.11	0.777	0.412	0.735	0.614	0.544	0.506
3 × 3 × 1/2	0.8125	9.4	2.75	2.22	1.07	0.898	0.932	2.22	1.07	0.898	0.932	0.584	1.000
3/8	0.6875	7.2	2.11	1.76	0.833	0.913	0.888	1.76	0.833	0.913	0.888	0.587	1.000
5/16	0.625	6.1	1.78	1.51	0.707	0.922	0.865	1.51	0.707	0.922	0.865	0.589	1.000
1/4	0.5625	4.9	1.44	1.24	0.577	0.930	0.842	1.24	0.577	0.930	0.842	0.592	1.000
3/16	0.5000	3.71	1.09	0.962	0.441	0.939	0.820	0.962	0.441	0.939	0.820	0.596	1.000
3 × 2 1/2 × 3/8	0.75	6.6	1.92	1.66	0.810	0.928	0.956	1.04	0.581	0.736	0.706	0.522	0.676
1/4	0.625	4.5	1.31	1.17	0.561	0.945	0.911	0.743	0.404	0.753	0.661	0.528	0.684
3/16	0.5625	3.39	0.996	0.907	0.430	0.954	0.888	0.577	0.310	0.761	0.638	0.533	0.688
3 × 2 × 3/8	0.6875	5.9	1.73	1.53	0.781	0.940	1.04	0.543	0.371	0.559	0.539	0.430	0.428
5/16	0.625	5.0	1.46	1.32	0.664	0.948	1.02	0.470	0.317	0.567	0.516	0.432	0.435
1/4	0.5625	4.1	1.19	1.09	0.542	0.957	0.993	0.392	0.260	0.574	0.493	0.435	0.440
3/16	0.5000	3.07	0.902	0.842	0.415	0.966	0.970	0.307	0.200	0.583	0.470	0.439	0.446
2 1/2 × 2 1/2 × 3/8	0.6875	5.9	1.73	0.984	0.566	0.753	0.762	0.984	0.566	0.753	0.762	0.487	1.000
5/16	0.625	5.0	1.46	0.849	0.482	0.761	0.740	0.849	0.482	0.761	0.740	0.489	1.000
1/4	0.5625	4.1	1.19	0.703	0.394	0.769	0.717	0.703	0.394	0.769	0.717	0.491	1.000
3/16	0.5000	3.07	0.902	0.547	0.303	0.778	0.694	0.547	0.303	0.778	0.694	0.495	1.000
2 1/2 × 2 × 3/8	0.6875	5.3	1.55	0.912	0.547	0.768	0.831	0.514	0.363	0.577	0.581	0.420	0.614
5/16	0.625	4.5	1.31	0.788	0.466	0.776	0.809	0.446	0.310	0.584	0.559	0.422	0.620
1/4	0.5625	3.62	1.06	0.654	0.381	0.784	0.787	0.372	0.254	0.592	0.537	0.424	0.626
3/16	0.5000	2.75	0.809	0.509	0.293	0.793	0.764	0.291	0.196	0.600	0.514	0.427	0.631
2 × 2 × 3/8	0.6875	4.7	1.36	0.479	0.351	0.594	0.636	0.479	0.351	0.594	0.636	0.389	1.000
5/16	0.625	3.92	1.15	0.416	0.300	0.601	0.614	0.416	0.300	0.601	0.614	0.390	1.000
1/4	0.5625	3.19	0.938	0.348	0.247	0.609	0.592	0.348	0.247	0.609	0.592	0.391	1.000
3/16	0.5000	2.44	0.715	0.272	0.190	0.617	0.569	0.272	0.190	0.617	0.569	0.394	1.000

Source: Abstracted from the *Manual of Steel Construction*, 8th ed., with permission of the publishers, American Institute of Steel Construction.

TABLE A.7. Properties of Structural Tees[a]

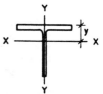

Designation	Area (in.²)	Q_s	Depth (in.)	Stem thickness (in.)	Flange Width (in.)	Flange Thickness (in.)	X–X Axis I (in.⁴)	X–X Axis S (in.³)	X–X Axis r (in.)	X–X Axis y (in.)	Y–Y Axis I (in.⁴)	Y–Y Axis S (in.³)	Y–Y Axis r (in.)
WT4 × 5	1.48	0.735	3.945	0.170	3.940	0.205	2.15	0.717	1.20	0.953	1.05	0.532	0.841
WT4 × 6.5	1.92	—	3.995	0.230	4.000	0.255	2.89	0.974	1.23	1.03	1.37	0.683	0.843
WT4 × 7.5	2.22	—	4.055	0.245	4.015	0.315	3.28	1.07	1.22	0.998	1.70	0.849	0.876
WT5 × 6	1.77	0.793	4.935	0.190	3.960	0.210	4.35	1.22	1.57	1.36	1.09	0.551	0.785
WT5 × 7.5	2.21	0.977	4.995	0.230	4.000	0.270	5.45	1.50	1.57	1.37	1.45	0.723	0.810
WT5 × 8.5	2.50	—	5.055	0.240	4.010	0.330	6.06	1.62	1.56	1.32	1.78	0.888	0.844
WT5 × 9.5	2.81	—	5.120	0.250	4.020	0.395	6.68	1.74	1.54	1.28	2.15	1.07	0.874
WT5 × 11	3.24	0.999	5.085	0.240	5.750	0.360	6.88	1.72	1.46	1.07	5.71	1.99	1.33
WT5 × 13	3.81	—	5.165	0.260	5.770	0.440	7.86	1.91	1.44	1.06	7.05	2.44	1.36
WT5 × 15	4.42	—	5.235	0.300	5.810	0.510	9.28	2.24	1.45	1.10	8.35	2.87	1.37
WT6 × 7	2.08	0.626	5.955	0.200	3.970	0.225	7.67	1.83	1.92	1.76	1.18	0.594	0.753
WT6 × 8	2.36	0.741	5.995	0.220	3.990	0.265	8.70	2.04	1.92	1.74	1.41	0.706	0.773
WT6 × 9.5	2.79	0.797	6.080	0.235	4.005	0.350	10.1	2.28	1.90	1.65	1.88	0.939	0.822
WT6 × 11	3.24	0.891	6.155	0.260	4.030	0.425	11.7	2.59	1.90	1.63	2.33	1.16	0.847
WT6 × 13	3.82	0.767	6.110	0.230	6.490	0.380	11.7	2.40	1.75	1.25	8.66	2.67	1.51
WT6 × 15	4.40	0.891	6.170	0.260	6.520	0.440	13.5	2.75	1.75	1.27	10.2	3.12	1.52
WT6 × 17.5	5.17	—	6.250	0.300	6.560	0.520	16.0	3.23	1.76	1.30	12.2	3.73	1.54
WT7 × 11	3.25	0.621	6.870	0.230	5.000	0.335	14.8	2.91	2.14	1.76	3.50	1.40	1.04
WT7 × 13	3.85	0.737	6.955	0.255	5.025	0.420	17.3	3.31	2.12	1.72	4.45	1.77	1.08
WT7 × 15	4.42	0.810	6.920	0.270	6.730	0.385	19.0	3.55	2.07	1.58	9.79	2.91	1.49
WT7 × 17	5.00	0.857	6.990	0.285	6.745	0.455	20.9	3.83	2.04	1.53	11.7	3.45	1.53
WT7 × 19	5.58	0.934	7.050	0.310	6.770	0.515	23.3	4.22	2.04	1.54	13.3	3.94	1.55
WT7 × 21.5	6.31	0.947	6.830	0.305	7.995	0.530	21.9	3.98	1.86	1.31	22.6	5.65	1.89
WT7 × 24	7.07	—	6.895	0.340	8.030	0.595	24.9	4.48	1.87	1.35	25.7	6.40	1.91
WT7 × 26.5	7.81	—	6.960	0.370	8.060	0.660	27.6	4.94	1.88	1.38	28.8	7.16	1.92
WT8 × 13	3.84	0.563	7.845	0.250	5.500	0.345	23.5	4.09	2.47	2.09	4.80	1.74	1.12
WT8 × 15.5	4.56	0.668	7.940	0.275	5.525	0.440	27.4	4.64	2.45	2.02	6.20	2.24	1.17
WT8 × 18	5.28	0.754	7.930	0.295	6.985	0.430	30.6	5.05	2.41	1.88	12.2	3.50	1.52
WT8 × 20	5.89	0.784	8.005	0.305	6.995	0.505	33.1	5.35	2.37	1.81	14.4	4.12	1.57
WT8 × 22.5	6.63	0.904	8.065	0.345	7.035	0.565	37.8	6.10	2.39	1.86	16.4	4.67	1.57
WT8 × 25	7.37	0.890	8.130	0.380	7.070	0.630	42.3	6.78	2.40	1.89	18.6	5.26	1.59
WT8 × 28.5	8.38	—	8.215	0.430	7.120	0.715	48.7	7.77	2.41	1.94	21.6	6.06	1.60

[a] *Cut from l-shaped elements.*

TABLE A.8. Properties of Double Angles

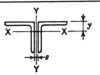

Two equal leg angles

Designation	Wt. per Ft. 2 Angles Lb.	Area of 2 Angles In.²	AXIS X — X				AXIS Y — Y Radii of Gyration Back to Back of Angles, Inches			Q_s* Angles in Contact		Angles Separated	
			I In.⁴	S In.³	r In.	y In.	0	3/8	3/4	F_y = 36 ksi	F_y = 50 ksi	F_y = 36 ksi	F_y = 50 ksi
L 8 x8 x1⅛	113.8	33.5	195.0	35.1	2.42	2.41	3.42	3.55	3.69	—	—	—	—
1	102.0	30.0	177.0	31.6	2.44	2.37	3.40	3.53	3.67	—	—	—	—
⅞	90.0	26.5	159.0	28.0	2.45	2.32	3.38	3.51	3.64	—	—	—	—
¾	77.8	22.9	139.0	24.4	2.47	2.28	3.36	3.49	3.62	—	—	—	—
⅝	65.4	19.2	118.0	20.6	2.49	2.23	3.34	3.47	3.60	—	—	.997	.935
½	52.8	15.5	97.3	16.7	2.50	2.19	3.32	3.45	3.58	.995	.921	.911	.834
L 6 x6 x1	74.8	22.0	70.9	17.1	1.80	1.86	2.59	2.73	2.87	—	—	—	—
⅞	66.2	19.5	63.8	15.3	1.81	1.82	2.57	2.70	2.85	—	—	—	—
¾	57.4	16.9	56.3	13.3	1.83	1.78	2.55	2.68	2.82	—	—	—	—
⅝	48.4	14.2	48.3	11.3	1.84	1.73	2.53	2.66	2.80	—	—	—	—
½	39.2	11.5	39.8	9.23	1.86	1.68	2.51	2.64	2.78	—	—	—	.961
⅜	29.8	8.72	30.8	7.06	1.88	1.64	2.49	2.62	2.75	.995	.921	.911	.834
L 5 x5 x ⅞	54.4	16.0	35.5	10.3	1.49	1.57	2.16	2.30	2.45	—	—	—	—
¾	47.2	13.9	31.5	9.06	1.51	1.52	2.14	2.28	2.42	—	—	—	—
½	32.4	9.50	22.5	6.31	1.54	1.43	2.10	2.24	2.38	—	—	—	—
⅜	24.6	7.22	17.5	4.84	1.56	1.39	2.09	2.22	2.35	—	—	.982	.919
5/16	20.6	6.05	14.8	4.08	1.57	1.37	2.08	2.21	2.34	.995	.921	.911	.834
L 4 x4 x ¾	37.0	10.9	15.3	5.62	1.19	1.27	1.74	1.88	2.83	—	—	—	—
⅝	31.4	9.22	13.3	4.80	1.20	1.23	1.72	1.86	2.00	—	—	—	—
½	25.6	7.50	11.1	3.95	1.22	1.18	1.70	1.83	1.98	—	—	—	—
⅜	19.6	5.72	8.72	3.05	1.23	1.14	1.68	1.81	1.95	—	—	—	—
5/16	16.4	4.80	7.43	2.58	1.24	1.12	1.67	1.80	1.94	—	—	.997	.935
¼	13.2	3.88	6.08	2.09	1.25	1.09	1.66	1.79	1.93	.995	.921	.911	.834
L 3½x3½x ⅜	17.0	4.97	5.73	2.30	1.07	1.01	1.48	1.61	1.75	—	—	—	—
5/16	14.4	4.18	4.90	1.95	1.08	.990	1.47	1.60	1.74	—	—	—	.986
¼	11.6	3.38	4.02	1.59	1.09	.968	1.46	1.59	1.73	—	.982	.965	.897
L 3 x3 x ½	18.8	5.50	4.43	2.14	.898	.932	1.29	1.43	1.59	—	—	—	—
⅜	14.4	4.22	3.52	1.67	.913	.888	1.27	1.41	1.56	—	—	—	—
5/16	12.2	3.55	3.02	1.41	.922	.865	1.26	1.40	1.55	—	—	—	—
¼	9.8	2.88	2.49	1.15	.930	.842	1.26	1.39	1.53	—	—	—	.961
3/16	7.42	2.18	1.92	.882	.939	.820	1.25	1.38	1.52	.995	.921	.911	.834
L 2½x2½x ⅜	11.8	3.47	1.97	1.13	.753	.762	1.07	1.21	1.36	—	—	—	—
5/16	10.0	2.93	1.70	.964	.761	.740	1.06	1.20	1.35	—	—	—	—
¼	8.2	2.38	1.41	.789	.769	.717	1.05	1.19	1.34	—	—	—	—
3/16	6.14	1.80	1.09	.685	.778	.694	1.04	1.18	1.32	—	—	.982	.919
L 2 x2 x ⅜	9.4	2.72	.958	.702	.594	.636	.870	1.01	1.17	—	—	—	—
5/16	7.84	2.30	.832	.681	.601	.614	.859	1.00	1.16	—	—	—	—
¼	6.38	1.88	.695	.494	.609	.592	.849	.989	1.14	—	—	—	—
3/16	4.88	1.43	.545	.381	.617	.569	.840	.977	1.13	—	—	—	—
⅛	3.30	.960	.380	.261	.626	.546	.831	.965	1.11	.995	.921	.911	.834

* Where no value of Q_s is shown, the angles comply with Specification Sect. 1.9.1.2 and may be considered fully effective.

For F_y = 36 ksi: $C'_c = 126.1/\sqrt{Q_s}$

For F_y = 50 ksi: $C'_c = 107.0/\sqrt{Q_s}$

TABLE A.8. (*continued*)

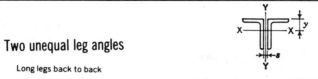

Two unequal leg angles

Long legs back to back

Designation	Wt. per Ft. 2 Angles (Lb.)	Area of 2 Angles (In.²)	AXIS X – X I (In.⁴)	S (In.³)	r (In.)	y (In.)	AXIS Y – Y Radii of Gyration Back to Back of Angles, Inches 0	3/8	3/4	Q_s* Angles in Contact F_y = 36 ksi	F_y = 50 ksi	Angles Separated F_y = 36 ksi	F_y = 50 ksi
L 8 x6 x1	88.4	26.0	161.0	30.2	2.49	2.65	2.39	2.52	2.66	—	—	—	—
3/4	67.6	19.9	126.0	23.3	2.53	2.56	2.35	2.48	2.62	—	—	—	—
1/2	46.0	13.5	88.6	16.0	2.56	2.47	2.32	2.44	2.57	—	—	.911	.834
L 8 x4 x1	74.8	22.0	139.0	28.1	2.52	3.05	1.47	1.61	1.75	—	—	—	—
3/4	57.4	16.9	109.0	21.8	2.55	2.95	1.42	1.55	1.69	—	—	—	—
1/2	39.2	11.5	77.0	15.0	2.59	2.86	1.38	1.51	1.64	—	—	.911	.834
L 7 x4 x 3/4	52.4	15.4	75.6	16.8	2.22	2.51	1.48	1.62	1.76	—	—	—	—
1/2	35.8	10.5	53.3	11.6	2.25	2.42	1.44	1.57	1.71	—	—	.965	.897
3/8	27.2	7.97	41.1	8.88	2.27	2.37	1.43	1.55	1.68	—	—	.839	.750
L 6 x4 x 3/4	47.2	13.9	49.0	12.5	1.88	2.08	1.55	1.69	1.83	—	—	—	—
5/8	40.0	11.7	42.1	10.6	1.90	2.03	1.53	1.67	1.81	—	—	—	—
1/2	32.4	9.50	34.8	8.67	1.91	1.99	1.51	1.64	1.78	—	—	—	.961
3/8	24.6	7.22	26.9	6.64	1.93	1.94	1.50	1.62	1.76	—	—	.911	.834
L 6 x3½x 3/8	23.4	6.84	25.7	6.49	1.94	2.04	1.26	1.39	1.53	—	—	.911	.834
5/16	19.6	5.74	21.8	5.47	1.95	2.01	1.26	1.38	1.51	—	—	.825	.733
L 5 x3½x 3/4	39.6	11.6	27.8	8.55	1.55	1.75	1.40	1.53	1.68	—	—	—	—
1/2	27.2	8.00	20.0	5.97	1.58	1.66	1.35	1.49	1.63	—	—	—	—
3/8	20.8	6.09	15.6	4.59	1.60	1.61	1.34	1.46	1.60	—	—	.982	.919
5/16	17.4	5.12	13.2	3.87	1.61	1.59	1.33	1.45	1.59	—	—	.911	.834
L 5 x3 x 1/2	25.6	7.50	18.9	5.82	1.59	1.75	1.12	1.25	1.40	—	—	—	—
3/8	19.6	5.72	14.7	4.47	1.61	1.70	1.10	1.23	1.37	—	—	.982	.919
5/16	16.4	4.80	12.5	3.77	1.61	1.68	1.09	1.22	1.36	—	—	.911	.834
1/4	13.2	3.88	10.2	3.06	1.62	1.66	1.08	1.21	1.34	—	—	.804	.708
L 4 x3½x 1/2	23.8	7.00	10.6	3.87	1.23	1.25	1.44	1.58	1.72	—	—	—	—
3/8	18.2	5.34	8.35	2.99	1.25	1.21	1.42	1.56	1.70	—	—	—	—
5/16	15.4	4.49	7.12	2.53	1.26	1.18	1.42	1.55	1.69	—	—	.997	.935
1/4	12.4	3.63	5.83	2.05	1.27	1.16	1.41	1.54	1.67	—	.982	.911	.834
L 4 x3 x 1/2	22.2	6.50	10.1	3.78	1.25	1.33	1.20	1.33	1.48	—	—	—	—
3/8	17.0	4.97	7.93	2.92	1.26	1.28	1.18	1.31	1.45	—	—	—	—
5/16	14.4	4.18	6.76	2.47	1.27	1.26	1.17	1.30	1.44	—	—	.997	.935
1/4	11.6	3.38	5.54	2.00	1.28	1.24	1.16	1.29	1.43	—	—	.911	.834
L 3½x3 x 3/8	15.8	4.59	5.45	2.25	1.09	1.08	1.22	1.36	1.50	—	—	—	—
5/16	13.2	3.87	4.66	1.91	1.10	1.06	1.21	1.35	1.49	—	—	—	.986
1/4	10.8	3.13	3.83	1.55	1.11	1.04	1.20	1.33	1.48	—	—	.965	.897
L 3½x2½x 3/8	14.4	4.22	5.12	2.19	1.10	1.16	.976	1.11	1.26	—	—	—	—
5/16	12.2	3.55	4.38	1.85	1.11	1.14	.966	1.10	1.25	—	—	—	.986
1/4	9.8	2.88	3.60	1.51	1.12	1.11	.958	1.09	1.23	—	—	.965	.897
L 3 x2½x 3/8	13.2	3.84	3.31	1.62	.928	.956	1.02	1.16	1.31	—	—	—	—
1/4	9.0	2.63	2.35	1.12	.945	.911	1.00	1.13	1.28	—	—	—	.961
3/16	6.77	1.99	1.81	.859	.954	.888	.993	1.12	1.27	—	—	.911	.834
L 3 x2 x 3/8	11.8	3.47	3.06	1.56	.940	1.04	.777	.917	1.07	—	—	—	—
5/16	10.0	2.93	2.63	1.33	.948	1.02	.767	.903	1.06	—	—	—	—
1/4	8.2	2.38	2.17	1.08	.957	.993	.757	.891	1.04	—	—	—	.961
3/16	6.1	1.80	1.68	.830	.966	.970	.749	.879	1.03	—	—	.911	.834
L 2½x2 x 3/8	10.6	3.09	1.82	1.09	.768	.831	.819	.961	1.12	—	—	—	—
5/16	9.0	2.62	1.58	.932	.776	.809	.809	.948	1.10	—	—	—	—
1/4	7.2	2.13	1.31	.763	.784	.787	.799	.935	1.09	—	—	—	—
3/16	5.5	1.62	1.02	.586	.793	.764	.790	.923	1.07	—	—	.982	.919

* Where no value of Q_s is shown, the angles comply with Specification Sect. 1.9.1.2 and may be considered fully effective.

For F_y = 36 ksi: $C'_c = 126.1/\sqrt{Q_s}$

For F_y = 50 ksi: $C'_c = 107.0/\sqrt{Q_s}$

TABLE A.9. Properties of Steel Pipe

Dimensions				Weight per Foot Lbs. Plain Ends	Properties			
Nominal Diameter In.	Outside Diameter In.	Inside Diameter In.	Wall Thickness In.		A In.^{2}	I In.^{4}	S In.^{3}	r In.
Standard Weight								
½	.840	.622	.109	.85	.250	.017	.041	.261
¾	1.050	.824	.113	1.13	.333	.037	.071	.334
1	1.315	1.049	.133	1.68	.494	.087	.133	.421
1¼	1.660	1.380	.140	2.27	.669	.195	.235	.540
1½	1.900	1.610	.145	2.72	.799	.310	.326	.623
2	2.375	2.067	.154	3.65	1.07	.666	.561	.787
2½	2.875	2.469	.203	5.79	1.70	1.53	1.06	.947
3	3.500	3.068	.216	7.58	2.23	3.02	1.72	1.16
3½	4.000	3.548	.226	9.11	2.68	4.79	2.39	1.34
4	4.500	4.026	.237	10.79	3.17	7.23	3.21	1.51
5	5.563	5.047	.258	14.62	4.30	15.2	5.45	1.88
6	6.625	6.065	.280	18.97	5.58	28.1	8.50	2.25
8	8.625	7.981	.322	28.55	8.40	72.5	16.8	2.94
10	10.750	10.020	.365	40.48	11.9	161	29.9	3.67
12	12.750	12.000	.375	49.56	14.6	279	43.8	4.38
Extra Strong								
½	.840	.546	.147	1.09	.320	.020	.048	.250
¾	1.050	.742	.154	1.47	.433	.045	.085	.321
1	1.315	.957	.179	2.17	.639	.106	.161	.407
1¼	1.660	1.278	.191	3.00	.881	.242	.291	.524
1½	1.900	1.500	.200	3.63	1.07	.391	.412	.605
2	2.375	1.939	.218	5.02	1.48	.868	.731	.766
2½	2.875	2.323	.276	7.66	2.25	1.92	1.34	.924
3	3.500	2.900	.300	10.25	3.02	3.89	2.23	1.14
3½	4.000	3.364	.318	12.50	3.68	6.28	3.14	1.31
4	4.500	3.826	.337	14.98	4.41	9.61	4.27	1.48
5	5.563	4.813	.375	20.78	6.11	20.7	7.43	1.84
6	6.625	5.761	.432	28.57	8.40	40.5	12.2	2.19
8	8.625	7.625	.500	43.39	12.8	106	24.5	2.88
10	10.750	9.750	.500	54.74	16.1	212	39.4	3.63
12	12.750	11.750	.500	65.42	19.2	362	56.7	4.33
Double-Extra Strong								
2	2.375	1.503	.436	9.03	2.66	1.31	1.10	.703
2½	2.875	1.771	.552	13.69	4.03	2.87	2.00	.844
3	3.500	2.300	.600	18.58	5.47	5.99	3.42	1.05
4	4.500	3.152	.674	27.54	8.10	15.3	6.79	1.37
5	5.563	4.063	.750	38.55	11.3	33.6	12.1	1.72
6	6.625	4.897	.864	53.16	15.6	66.3	20.0	2.06
8	8.625	6.875	.875	72.42	21.3	162	37.6	2.76

The listed sections are available in conformance with ASTM Specification A53 Grade B or A501. Other sections are made to these specifications. Consult with pipe manufacturers or distributors for availability.

TABLE A.10. Properties of Structural Tubing

Square

DIMENSIONS				PROPERTIES**			
Nominal* Size	Wall Thickness		Weight per Foot	Area	I	S	r
In.	In.		Lb.	In.^{2}	In.^{4}	In.^{3}	In.
16 x 16	.5000	1/2	103.30	30.4	1200	150	6.29
	.3750	3/8	78.52	23.1	931	116	6.35
	.3125	5/16	65.87	19.4	789	98.6	6.38
14 x 14	.5000	1/2	89.68	26.4	791	113	5.48
	.3750	3/8	68.31	20.1	615	87.9	5.54
	.3125	5/16	57.36	16.9	522	74.6	5.57
12 x 12	.5000	1/2	76.07	22.4	485	80.9	4.66
	.3750	3/8	58.10	17.1	380	63.4	4.72
	.3125	5/16	48.86	14.4	324	54.0	4.75
	.2500	1/4	39.43	11.6	265	44.1	4.78
10 x 10	.6250	5/8	76.33	22.4	321	64.2	3.78
	.5000	1/2	62.46	18.4	271	54.2	3.84
	.3750	3/8	47.90	14.1	214	42.9	3.90
	.3125	5/16	40.35	11.9	183	36.7	3.93
	.2500	1/4	32.63	9.59	151	30.1	3.96
8 x 8	.6250	5/8	59.32	17.4	153	38.3	2.96
	.5000	1/2	48.85	14.4	131	32.9	3.03
	.3750	3/8	37.69	11.1	106	26.4	3.09
	.3125	5/16	31.84	9.36	90.9	22.7	3.12
	.2500	1/4	25.82	7.59	75.1	18.8	3.15
	.1875	3/16	19.63	5.77	58.2	14.6	3.18
7 x 7	.5000	1/2	42.05	12.4	84.6	24.2	2.62
	.3750	3/8	32.58	9.58	68.7	19.6	2.68
	.3125	5/16	27.59	8.11	59.5	17.0	2.71
	.2500	1/4	22.42	6.59	49.4	14.1	2.74
	.1875	3/16	17.08	5.02	38.5	11.0	2.77
6 x 6	.5000	1/2	35.24	10.4	50.5	16.8	2.21
	.3750	3/8	27.48	8.08	41.6	13.9	2.27
	.3125	5/16	23.34	6.86	36.3	12.1	2.30
	.2500	1/4	19.02	5.59	30.3	10.1	2.33
	.1875	3/16	14.53	4.27	23.8	7.93	2.36
5 x 5	.5000	1/2	28.43	8.36	27.0	10.8	1.80
	.3750	3/8	22.37	6.58	22.8	9.11	1.86
	.3125	5/16	19.08	5.61	20.1	8.02	1.89
	.2500	1/4	15.62	4.59	16.9	6.78	1.92
	.1875	3/16	11.97	3.52	13.4	5.36	1.95
4 x 4	.5000	1/2	21.63	6.36	12.3	6.13	1.39
	.3750	3/8	17.27	5.08	10.7	5.35	1.45
	.3125	5/16	14.83	4.36	9.58	4.79	1.48
	.2500	1/4	12.21	3.59	8.22	4.11	1.51
	.1875	3/16	9.42	2.77	6.59	3.30	1.54
3.5 x 3.5	.3125	5/16	12.70	3.73	6.09	3.48	1.28
	.2500	1/4	10.51	3.09	5.29	3.02	1.31
	.1875	3/16	8.15	2.39	4.29	2.45	1.34
3 x 3	.3125	5/16	10.58	3.11	3.58	2.39	1.07
	.2500	1/4	8.81	2.59	3.16	2.10	1.10
	.1875	3/16	6.87	2.02	2.60	1.73	1.13
2.5 x 2.5	.2500	1/4	7.11	2.09	1.69	1.35	.899
	.1875	3/16	5.59	1.64	1.42	1.14	.930
2 x 2	.2500	1/4	5.41	1.59	.766	.766	.694
	.1875	3/16	4.32	1.27	.668	.668	.726

* Outside dimensions across flat sides.
** Properties are based upon a nominal outside corner radius equal to two times the wall thickness.

APPENDIX B

Beam Diagrams and Formulas

Simple cases of beam support and loading conditions occur so frequently that it is common to use derived formulas in structural computations, rather than to follow the full procedure of beam analysis as described in Chapter 9. Such a procedure is required only when lack of symmetry or other factors present complicating relationships. Hand-books such as the *AISC Manual* (Ref. 5) and the *CRSI Handbook* (Ref. 9) have extensive tables that provide critical values for the reactions, shears, moments, and deflections of beams with various support and loading conditions. Examples of these are given in Figs. B.1, B.2, B.3 and B.4.

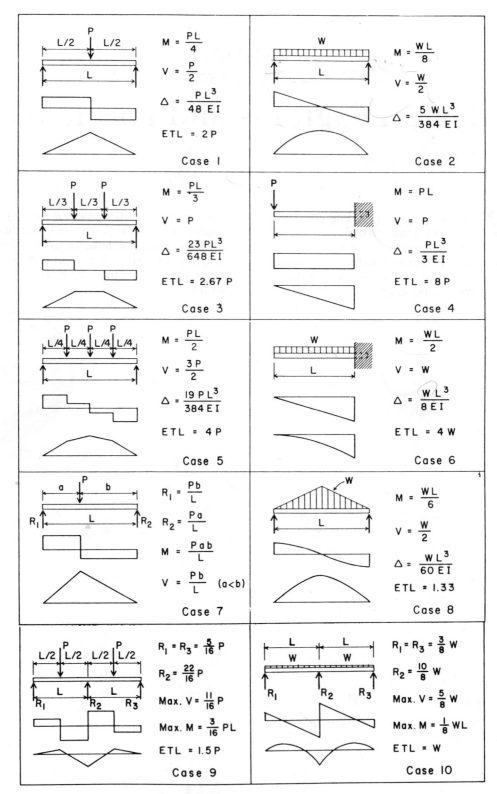

FIGURE B.1. Values for beams with various loading and support conditions.

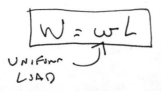

W = TOTAL EQUIV PT LOAD

W = wL

UNIFORM
LOAD

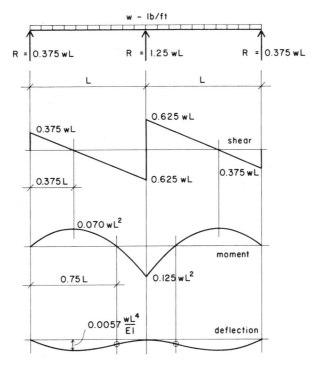

FIGURE B.2. Values for a two-span beam with uniformly distributed load.

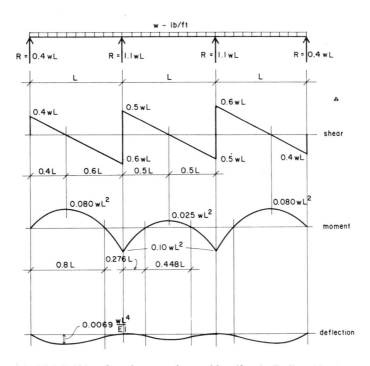

FIGURE B.3. Values for a three-span beam with uniformly distributed load.

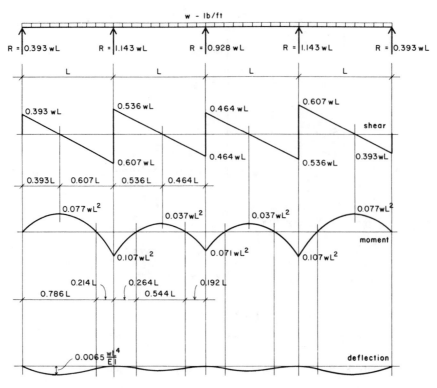

FIGURE B.4. Values for a four-span beam with uniformly distributed load.

Weights of Elements of Building Construction

Table C.1 contains data for use in determining the dead weight of elements of building construction. These are averaged estimates, based on typical construction details and practices. Most values given are for the total weight of the element described, in units of pounds per square foot or kilonewtons per square meter. Weights of materials of variable thickness (plywood, concrete deck, etc.) are given as the weight of a 1-in- or 1-mm-thick element.

TABLE C.1. Weights of Elements of Building Construction

	lb/ft²	kN/m²
Roofs		
3-ply ready roofing (roll, composition)	1	0.05
3-ply felt and gravel	5.5	0.26
5-ply felt and gravel	6.5	0.31
Shingles		
Wood	2	0.10
Asphalt	2–3	0.10–0.15
Clay tile	9–12	0.43–0.58
Concrete tile	8–12	0.38–0.58
Slate, ¼ in.	10	0.48
Fiber glass	2–3	0.10–0.15
Aluminum	1	0.05
Steel	2	0.10
Insulation		
Fiber glass batts	0.5	0.025
Rigid foam plastic	1.5	0.075
Foamed concrete, mineral aggregate	2.5/in.	0.0047/mm
Wood rafters		
2 × 6 at 24 in.	1.0	0.05
2 × 8 at 24 in.	1.4	0.07
2 × 10 at 24 in.	1.7	0.08
2 × 12 at 24 in.	2.1	0.10
Steel deck, painted		
22 ga	1.6	0.08
20 ga	2.0	0.10
18 ga	2.6	0.13
Skylight		
Glass with steel frame	6–10	0.29–0.48
Plastic with aluminum frame	3–6	0.15–0.29
Plywood or softwood board sheathing	3.0/in.	0.0057/mm
Ceilings		
Suspended steel channels	1	0.05
Lath		
Steel mesh	0.5	0.025
Gypsum board, ½ in.	2	0.10
Fiber tile	1	0.05
Drywall, gypsum board, ½ in.	2.5	0.12
Plaster		
Gypsum, acoustic	5	0.24
Cement	8.5	0.41
Suspended lighting and air distribution Systems, average	3	0.15

	lb/ft²	kN/m²
Floors		
Hardwood, ½ in.	2.5	0.12
Vinyl tile, ⅛ in.	1.5	0.07
Asphalt mastic	12/in.	0.023/mm
Ceramic tile		
¾ in.	10	0.48
Thin set	5	0.24
Fiberboard underlay, ⅝ in.	3	0.15
Carpet and pad, average	3	0.15
Timber deck	2.5/in.	0.0047/mm
Steel deck, stone concrete fill, average	35–40	1.68–1.92
Concrete deck, stone aggregate	12.5/in.	0.024/mm
Wood joists		
2 × 8 at 16 in.	2.1	0.10
2 × 10 at 16 in.	2.6	0.13
2 × 12 at 16 in.	3.2	0.16
Lightweight concrete fill	8.0/in.	0.015/mm
Walls		
2 × 4 studs at 16 in., average	2	0.10
Steel studs at 16 in., average	4	0.20
Lath, plaster; see Ceilings		
Gypsum drywall, ⅝ in. single	2.5	0.12
Stucco, ⅞ in., on wire and paper or felt	10	0.48
Windows, average, glazing + frame		
Small pane, single glazing, wood or metal frame	5	0.24
Large pane, single glazing, wood or metal frame	8	0.38
Increase for double glazing	2–3	0.10–0.15
Curtain walls, manufactured units	10–15	0.48–0.72
Brick veneer		
4-in., mortar joints	40	1.92
½-in., mastic	10	0.48
Concrete block		
Lightweight, unreinforced—4 in.	20	0.96
6 in.	25	1.20
8 in.	30	1.44
Heavy, reinforced, grouted—6 in.	45	2.15
8 in.	60	2.87
12 in.	85	4.07

APPENDIX D

===

Building Code Requirements

The material in Appendix D consists of selected reprints from the *UBC* (Ref. 1) and the *City of Los Angeles Building Code* (Ref. 4). While most of this material is similar to that found in any building code, local practices and circumstances produce unique requirements and recommendations in most city, county, and state codes.

Excerpts from the *UBC*, 1991 edition are reprinted with permission of the publisher, International Conference of Building Officials, 5360 South Workmanmill Road, Whittier, CA 90601.

Excerpts from the *City of Los Angeles Building Code*, 1976 edition, are reprinted with permission of the publisher. Building News, Inc., 3055 Overland Avenue, Los Angeles, CA 90034.

TABLE NO. 23-A—UNIFORM AND CONCENTRATED LOADS

USE OR OCCUPANCY		UNIFORM LOAD[1]	CONCEN- TRATED LOAD
Category	**Description**		
1. Access floor systems	Office use	50	2,000[2]
	Computer use	100	2,000[2]
2. Armories		150	0
3. Assembly areas[3] and auditoriums and balconies therewith	Fixed seating areas	50	0
	Movable seating and other areas	100	0
	Stage areas and enclosed platforms	125	0
4. Cornices, marquees and residential balconies		60	0
5. Exit facilities[4]		100	0[5]
6. Garages	General storage and/or repair	100	6
	Private or pleasure-type motor vehicle storage	50	6
7. Hospitals	Wards and rooms	40	1,000[2]
8. Libraries	Reading rooms	60	1,000[2]
	Stack rooms	125	1,500[2]
9. Manufacturing	Light	75	2,000[2]
	Heavy	125	3,000[2]
10. Offices		50	2,000[2]
11. Printing plants	Press rooms	150	2,500[2]
	Composing and linotype rooms	100	2,000[2]
12. Residential[7]		40	0[5]
13. Restrooms[8]			
14. Reviewing stands, grandstands, bleachers, and folding and telescoping seating		100	0
15. Roof decks	Same as area served or for the type of occupancy accommodated		
16. Schools	Classrooms	40	1,000[2]
17. Sidewalks and driveways	Public access	250	6
18. Storage	Light	125	
	Heavy	250	
19. Stores	Retail	75	2,000[2]
	Wholesale	100	3,000[2]

[1]See Section 2306 for live load reductions.
[2]See Section 2304 (c), first paragraph, for area of load application.
[3]Assembly areas include such occupancies as dance halls, drill rooms, gymnasiums, playgrounds, plazas, terraces and similar occupancies which are generally accessible to the public.
[4]Exit facilities shall include such uses as corridors serving an occupant load of 10 or more persons, exterior exit balconies, stairways, fire escapes and similar uses.
[5]Individual stair treads shall be designed to support a 300-pound concentrated load placed in a position which would cause maximum stress. Stair stringers may be designed for the uniform load set forth in the table.
[6]See Section 2304 (c), second paragraph, for concentrated loads.
[7]Residential occupancies include private dwellings, apartments and hotel guest rooms.
[8]Restroom loads shall not be less than the load for the occupancy with which they are associated, but need not exceed 50 pounds per square foot.

TABLE NO. 23-B—SPECIAL LOADS[1]

USE		VERTICAL LOAD	LATERAL LOAD
Category	**Description**	(Pounds per Square Foot unless Otherwise Noted)	
1. Construction, public access at site (live load)	Walkway, see Sec. 4406	150	
	Canopy, see Sec. 4407	150	
2. Grandstands, reviewing stands, bleachers, and folding and telescoping seating (live load)	Seats and footboards	120[2]	See Footnote No. 3
3. Stage accessories (live load)	Gridirons and fly galleries	75	
	Loft block wells[4]	250	250
	Head block wells and sheave beams[4]	250	250
4. Ceiling framing (live load)	Over stages	20	
	All uses except over stages	10[5]	
5. Partitions and interior walls, see Sec. 2309 (live load)			5
6. Elevators and dumbwaiters (dead and live load)		2 x Total loads[6]	
7. Mechanical and electrical equipment (dead load)		Total loads	
8. Cranes (dead and live load)	Total load including impact increase	1.25 x Total load[7]	0.10 x Total load[8]
9. Balcony railings and guardrails	Exit facilities serving an occupant load greater than 50		50[9]
	Other		20[9]
10. Handrails		See Footnote No. 10	See Footnote No. 10
11. Storage racks	Over 8 feet high	Total loads[11]	See Table No. 23-P
12. Fire sprinkler structural support		250 pounds plus weight of water-filled pipe[12]	See Table No. 23-P
13. Explosion exposure	Hazardous occupancies, see Sec. 910		

[1]The tabulated loads are minimum loads. Where other vertical loads required by this code or required by the design would cause greater stresses, they shall be used.
[2]Pounds per lineal foot.
[3]Lateral sway bracing loads of 24 pounds per foot parallel and 10 pounds per foot perpendicular to seat and footboards.
[4]All loads are in pounds per lineal foot. Head block wells and sheave beams shall be designed for all loft block well loads tributary thereto. Sheave blocks shall be designed with a factor of safety of five.
[5]Does not apply to ceilings which have sufficient total access from below, such that access is not required within the space above the ceiling. Does not apply to ceilings if the attic areas above the ceiling are not provided with access. This live load need not be considered as acting simultaneously with other live loads imposed upon the ceiling framing or its supporting structure.
[6]Where Appendix Chapter 51 has been adopted, see reference standard cited therein for additional design requirements.
[7]The impact factors included are for cranes with steel wheels riding on steel rails. They may be modified if substantiating technical data acceptable to the building official is submitted. Live loads on crane support girders and their connections shall be taken as the maximum crane wheel loads. For pendant-operated traveling crane support girders and their connections, the impact factors shall be 1.10.
[8]This applies in the direction parallel to the runway rails (longitudinal). The factor for forces perpendicular to the rail is 0.20 x the transverse traveling loads (trolley, cab, hooks and lifted loads). Forces shall be applied at top of rail and may be distributed among rails of multiple rail cranes and shall be distributed with due regard for lateral stiffness of the structures supporting these rails.
[9]A load per lineal foot to be applied horizontally at right angles to the top rail.
[10]The mounting of handrails shall be such that the completed handrail and supporting structure are capable of withstanding a load of at least 200 pounds applied in any direction at any point on the rail. These loads shall not be assumed to act cumulatively with Item 9.
[11]Vertical members of storage racks shall be protected from impact forces of operating equipment, or racks shall be designed so that failure of one vertical member will not cause collapse of more than the bay or bays directly supported by that member.
[12]The 250-pound load is to be applied to any single fire sprinkler support point but not simultaneously to all support joints.

TABLE NO. 23-C—MINIMUM ROOF LIVE LOADS[1]

ROOF SLOPE	METHOD 1 TRIBUTARY LOADED AREA IN SQUARE FEET FOR ANY STRUCTURAL MEMBER			METHOD 2		
	0 to 200	201 to 600	Over 600	UNIFORM LOAD[2]	RATE OF REDUCTION r (Percent)	MAXIMUM REDUCTION R (Percent)
1. Flat or rise less than 4 inches per foot. Arch or dome with rise less than one eighth of span	20	16	12	20	.08	40
2. Rise 4 inches per foot to less than 12 inches per foot. Arch or dome with rise one eighth of span to less than three eighths of span	16	14	12	16	.06	25
3. Rise 12 inches per foot and greater. Arch or dome with rise three eighths of span or greater	12	12	12	12		
4. Awnings except cloth covered[3]	5	5	5	5	No Reductions Permitted	
5. Greenhouses, lath houses and agricultural buildings[4]	10	10	10	10		

[1]Where snow loads occur, the roof structure shall be designed for such loads as determined by the building official. See Section 2305 (d). For special-purpose roofs, see Section 2305 (e).
[2]See Section 2306 for live load reductions. The rate of reduction r in Section 2306 Formula (6-1) shall be as indicated in the table. The maximum reduction R shall not exceed the value indicated in the table.
[3]As defined in Section 4506.
[4]See Section 2305 (e) for concentrated load requirements for greenhouse roof members.

TABLE NO. 23-F—WIND STAGNATION PRESSURE (q_s) AT STANDARD HEIGHT OF 33 FEET

Basic wind speed (mph)[1]	70	80	90	100	110	120	130
Pressure q_s (psf)	12.6	16.4	20.8	25.6	31.0	36.9	43.3

[1]Wind speed from Section 2314.

TABLE NO. 23-G—COMBINED HEIGHT, EXPOSURE AND GUST FACTOR COEFFICIENT (C_e)[1]

HEIGHT ABOVE AVERAGE LEVEL OF ADJOINING GROUND (feet)	EXPOSURE D	EXPOSURE C	EXPOSURE B
0-15	1.39	1.06	0.62
20	1.45	1.13	0.67
25	1.50	1.19	0.72
30	1.54	1.23	0.76
40	1.62	1.31	0.84
60	1.73	1.43	0.95
80	1.81	1.53	1.04
100	1.88	1.61	1.13
120	1.93	1.67	1.20
160	2.02	1.79	1.31
200	2.10	1.87	1.42
300	2.23	2.05	1.63
400	2.34	2.19	1.80

[1]Values for intermediate heights above 15 feet may be interpolated.

TABLE NO. 23-H—PRESSURE COEFFICIENTS (C_q)

STRUCTURE OR PART THEREOF	DESCRIPTION	C_q FACTOR
1. Primary frames and systems	**Method 1** (Normal force method) Walls: Windward wall	0.8 inward
	Leeward wall	0.5 outward
	Roofs[1]: Wind perpendicular to ridge Leeward roof or flat roof	0.7 outward
	Windward roof less than 2:12	0.7 outward
	Slope 2:12 to less than 9:12	0.9 outward or 0.3 inward
	Slope 9:12 to 12:12	0.4 inward
	Slope > 12:12	0.7 inward
	Wind parallel to ridge and flat roofs	0.7 outward
	Method 2 (Projected area method) On vertical projected area Structures 40 feet or less in height	1.3 horizontal any direction
	Structures over 40 feet in height	1.4 horizontal any direction
	On horizontal projected area[1]	0.7 upward
2. Elements and components not in areas of discontinuity[2]	Wall elements All structures	1.2 inward
	Enclosed and unenclosed structures	1.2 outward
	Open structures	1.6 outward
	Parapets walls	1.3 inward or outward
	Roof elements[3] Enclosed and unenclosed structures Slope < 7:12	1.3 outward
	Slope 7:12 to 12:12	1.3 outward or inward
	Open structures Slope < 2:12	1.7 outward
	Slope 2:12 to 7:12	1.6 outward or 0.8 inward
	Slope > 7:12 to 12:12	1.7 outward or inward
3. Elements and components in areas of discontinuities[2,4,6]	Wall corners[7]	1.5 outward or 1.2 inward
	Roof eaves, rakes or ridges without overhangs[7] Slope < 2:12	2.3 upward
	Slope 2:12 to 7:12	2.6 outward
	Slope > 7:12 to 12:12	1.6 outward
	For slopes less than 2:12 Overhangs at roof eaves, rakes or ridges, and canopies	0.5 added to values above
4. Chimneys, tanks and solid towers	Square or rectangular	1.4 any direction
	Hexagonal or octagonal	1.1 any direction
	Round or elliptical	0.8 any direction
5. Open-frame towers[3,4]	Square and rectangular Diagonal	4.0
	Normal	3.6
	Triangular	3.2
6. Tower accessories (such as ladders, conduit, lights and elevators)	Cylindrical members 2 inches or less in diameter	1.0
	Over 2 inches in diameter	0.8
	Flat or angular members	1.3
7. Signs, flagpoles, lightpoles, minor structures[4]		1.4 any direction

[1]For one story or the top story of multistory open structures, an additional value of 0.5 shall be added to the outward C_q. The most critical combination shall be used for design. For definition of open structures, see Section 2312.
[2]C_q values listed are for 10-square-foot tributary areas. For tributary areas of 100 square feet, the value of 0.3 may be subtracted from C_q, except for areas at discontinuities with slopes less than 7:12 where the value of 0.8 may be subtracted for C_q. Interpolating may be used for tributary areas between 10 and 100 square feet. For tributary areas greater than 1,000 square feet, use primary frame values.
[3]For slopes greater than 12:12, use wall element values.
[4]Local pressures shall apply over a distance from the discontinuity of 10 feet or 0.1 times the least width of the structure, whichever is smaller.
[5]Wind pressures shall be applied to the total normal projected area of all elements on one face. The forces shall be assumed to act parallel to the wind direction.
[6]Discontinuities at wall corners or roof ridges are defined as discontinuous breaks in the surface where the included interior angle measures 170 degrees or less.
[7]Load is to be applied on either side of discontinuity but not simultaneously on both sides.

TABLE NO. 23-I
SEISMIC ZONE FACTOR Z

ZONE	1	2A	2B	3	4
Z	0.075	0.15	0.20	0.30	0.40

The zone shall be determined from the seismic zone map in Figure No. 23-2.

TABLE NO. 23-J
SITE COEFFICIENTS[1]

TYPE	DESCRIPTION	S FACTOR
S_1	A soil profile with either: (a) A rock-like material characterized by a shear-wave velocity greater than 2,500 feet per second or by other suitable means of classification, or (b) Stiff or dense soil condition where the soil depth is less than 200 feet.	1.0
S_2	A soil profile with dense or stiff soil conditions, where the soil depth exceeds 200 feet.	1.2
S_3	A soil profile 70 feet or more in depth and containing more than 20 feet of soft to medium stiff clay but not more than 40 feet of soft clay.	1.5
S_4	A soil profile containing more than 40 feet of soft clay characterized by a shear wave velocity less than 500 feet per second.	2.0

[1]The site factor shall be established from properly substantiated geotechnical data. In locations where the soil properties are not known in sufficient detail to determine the soil profile type, soil profile S_3 shall be used. Soil profile S_4 need not be assumed unless the building official determines that soil profile S_4 may be present at the site, or in the event that soil profile S_4 is established by geotechnical data.

TABLE NO. 23-K
OCCUPANCY CATEGORIES

OCCUPANCY CATEGORIES	OCCUPANCY TYPE OR FUNCTIONS OF STRUCTURE
I. Essential Facilities[1]	Hospitals and other medical facilities having surgery and emergency treatment areas.
	Fire and police stations.
	Tanks or other structures containing, housing or supporting water or other fire-suppression materials or equipment required for the protection of essential or hazardous facilities, or special occupancy structures.
	Emergency vehicle shelters and garages.
	Structures and equipment in emergency-preparedness centers.
	Standby power-generating equipment for essential facilities.
	Structures and equipment in government communication centers and other facilities required for emergency response.
II. Hazardous Facilities	Structures housing, supporting or containing sufficient quantities of toxic or explosive substances to be dangerous to the safety of the general public if released.
III. Special Occupancy Structure	Covered structures whose primary occupancy is public assembly–capacity > 300 persons.
	Buildings for schools through secondary or day-care centers–capacity > 250 students.
	Buildings for colleges or adult education schools–capacity > 500 students.
	Medical facilities with 50 or more resident incapacitated patients, but not included above.
	Jails and detention facilities.
	All structures with occupancy > 5,000 persons.
	Structures and equipment in power-generating stations and other public utility facilities not included above, and required for continued operation.
IV. Standard Occupancy Structure	All structures having occupancies or functions not listed above.

[1]Essential facilities are those structures which are necessary for emergency operations subsequent to a natural disaster.

TABLE NO. 23-L—OCCUPANCY REQUIREMENTS

OCCUPANCY CATEGORY[1]	IMPORTANCE FACTOR I	
	Earthquake[2]	Wind
I. Essential facilities	1.25	1.15
II. Hazardous facilities	1.25	1.15
III. Special occupancy structures	1.00	1.00
IV. Standard occupancy structures	1.00	1.00

[1]Occupancy types or functions of structures within each category are listed in Table No. 23-K and structural observation requirements are given in Sections 305, 306 and 307.
[2]For life-safety-related equipment, see Section 2336 (a).

TABLE NO. 23-M
VERTICAL STRUCTURAL IRREGULARITIES

IRREGULARITY TYPE AND DEFINITION	REFERENCE SECTION
A. **Stiffness Irregularity—Soft Story** A soft story is one in which the lateral stiffness is less than 70 percent of that in the story above or less than 80 percent of the average stiffness of the three stories above.	2333 (h) 3 B
B. **Weight (mass) Irregularity** Mass irregularity shall be considered to exist where the effective mass of any story is more than 150 percent of the effective mass of an adjacent story. A roof which is lighter than the floor below need not be considered.	2333 (h) 3 B
C. **Vertical Geometric Irregularity** Vertical geometric irregularity shall be considered to exist where the horizontal dimension of the lateral force-resisting system in any story is more than 130 percent of that in an adjacent story. One-story penthouses need not be considered.	2333 (h) 3 B
D. **In-plane Discontinuity in Vertical Lateral Force-resisting Element** An in-plane offset of the lateral load-resisting elements greater than the length of those elements.	2334 (g)
E. **Discontinuity in Capacity–Weak Story** A weak story is one in which the story strength is less than 80 percent of that in the story above. The story strength is the total strength of all seismic-resisting elements sharing the story shear for the direction under consideration.	2334 (i) 1

TABLE NO. 23-N
PLAN STRUCTURAL IRREGULARITIES

IRREGULARITY TYPE AND DEFINITION	REFERENCE SECTION
A. **Torsional Irregularity—to be considered when diaphragms are not flexible.** Torsional irregularity shall be considered to exist when the maximum story drift, computed including accidental torsion, at one end of the structure transverse to an axis is more than 1.2 times the average of the story drifts of the two ends of the structure.	2337 (b) 9 E
B. **Reentrant Corners** Plan configurations of a structure and its lateral force-resisting system contain reentrant corners, where both projections of the structure beyond a reentrant corner are greater than 15 percent of the plan dimension of the structure in the given direction.	2337 (b) 9 E 2337 (b) 9 F
C. **Diaphragm Discontinuity** Diaphragms with abrupt discontinuities or variations in stiffness, including those having cutout or open areas greater than 50 percent of the gross enclosed area of the diaphragm, or changes in effective diaphragm stiffness of more than 50 percent from one story to the next.	2337 (b) 9 E
D. **Out-of-plane Offsets** Discontinuities in a lateral force path, such as out-of-plane offsets of the vertical elements.	2334 (g), 2337 (b) 9 E
E. **Nonparallel Systems** The vertical lateral load-resisting elements are not parallel to or symmetric about the major orthogonal axes of the lateral force-resisting system.	2337 (a)

TABLE NO. 23-O—STRUCTURAL SYSTEMS

BASIC STRUCTURAL SYSTEM[1]	LATERAL LOAD-RESISTING SYSTEM—DESCRIPTION	R_w[2]	H[3]
A. Bearing Wall System	1. Light-framed walls with shear panels		
	a. Plywood walls for structures three stories or less	8	65
	b. All other light-framed walls	6	65
	2. Shear walls		
	a. Concrete	6	160
	b. Masonry	6	160
	3. Light steel-framed bearing walls with tension-only bracing	4	65
	4. Braced frames where bracing carries gravity loads		
	a. Steel	6	160
	b. Concrete[4]	4	—
	c. Heavy timber	4	65
B. Building Frame System	1. Steel eccentrically braced frame (EBF)	10	240
	2. Light-framed walls with shear panels		
	a. Plywood walls for structures three stories or less	9	65
	b. All other light-framed walls	7	65
	3. Shear walls		
	a. Concrete	8	240
	b. Masonry	8	160
	4. Concentrically braced frames		
	a. Steel	8	160
	b. Concrete[4]	8	—
	c. Heavy timber	8	65
C. Moment-resisting Frame System	1. Special moment-resisting frames (SMRF)		
	a. Steel	12	N.L.
	b. Concrete	12	N.L.
	2. Concrete intermediate moment-resisting frames (IMRF)[6]	8	—
	3. Ordinary moment-resisting frames (OMRF)		
	a. Steel	6	160
	b. Concrete[7]	5	—
D. Dual Systems	1. Shear walls		
	a. Concrete with SMRF	12	N.L.
	b. Concrete with steel OMRF	6	160
	c. Concrete with concrete IMRF[6]	9	160
	d. Masonry with SMRF	8	160
	e. Masonry with steel OMRF	6	160
	f. Masonry with concrete IMRF[4]	7	—
	2. Steel EBF		
	a. With steel SMRF	12	N.L.
	b. With steel OMRF	6	160
	3. Concentrically braced frames		
	a. Steel with steel SMRF	10	N.L.
	b. Steel with steel OMRF	6	160
	c. Concrete with concrete SMRF[4]	9	—
	d. Concrete with concrete IMRF[4]	6	—
E. Undefined Systems	See Sections 2333 (h) 3 and 2333 (i) 2	—	—

[1]Basic structural systems are defined in Section 2333 (f).
[2]See Section 2334 (c) for combination of structural system.
[3]H—Height limit applicable to Seismic Zones Nos. 3 and 4. See Section 2333 (g).
[4]Prohibited in Seismic Zones Nos. 3 and 4.
[5]N.L.—No limit.
[6]Prohibited in Seismic Zones Nos. 3 and 4, except as permitted in Section 2338 (b).
[7]Prohibited in Seismic Zones Nos. 2, 3 and 4.

TABLE NO. 23-Q
R_w FACTORS FOR NONBUILDING STRUCTURES

STRUCTURE TYPE	R_w
1. Tanks, vessels or pressurized spheres on braced or unbraced legs.	3
2. Cast-in-place concrete silos and chimneys having walls continuous to the foundation.	5
3. Distributed mass cantilever structures such as stacks, chimneys, silos and skirt-supported vertical vessels.	4
4. Trussed towers (freestanding or guyed), guyed stacks and chimneys.	4
5. Inverted pendulum-type structures.	3
6. Cooling towers.	5
7. Bins and hoppers on braced or unbraced legs.	4
8. Storage racks.	5
9. Signs and billboards.	5
10. Amusement structures and monuments.	3
11. All other self-supporting structures not otherwise covered.	4

TABLE NO. 23-P—HORIZONTAL FORCE FACTOR, C_P

ELEMENTS OF STRUCTURES AND NONSTRUCTURAL COMPONENTS AND EQUIPMENT[1]	VALUE OF C_P	FOOTNOTE
I. **Part or Portion of Structure**		
1. Walls including the following:		
a. Unbraced (cantilevered) parapets	2.00	
b. Other exterior walls above the ground floor	0.75	2,3
c. All interior bearing and nonbearing walls and partitions	0.75	3
d. Masonry or concrete fences over 6 feet high	0.75	
2. Penthouse (except when framed by an extension of the structural frame)	0.75	
3. Connections for prefabricated structural elements other than walls, with force applied at center of gravity	0.75	4
4. Diaphragms	—	5
II. **Nonstructural Components**		
1. Exterior and interior ornamentations and appendages	2.00	
2. Chimneys, stacks, trussed towers and tanks on legs:		
a. Supported on or projecting as an unbraced cantilever above the roof more than one half their total height	2.00	
b. All others, including those supported below the roof with unbraced projection above the roof less than one half its height, or braced or guyed to the structural frame at or above their centers of mass	0.75	
3. Signs and billboards	2.00	
4. Storage racks (include contents)	0.75	10
5. Anchorage for permanent floor-supported cabinets and book stacks more than 5 feet in height (include contents)	0.75	
6. Anchorage for suspended ceilings and light fixtures—see also Section 4701 (e)	0.75	4,6,7
7. Access floor systems	0.75	4,9
III. **Equipment**		
1. Tanks and vessels (include contents), including support systems and anchorage	0.75	
2. Electrical, mechanical and plumbing equipment and associated conduit, ductwork and piping, and machinery	0.75	8

[1]See Section 2336 (b) for items supported at or below grade.

[2]See Section 2337 (b) 4 C and Section 2336 (b).

[3]Where flexible diaphragms, as defined in Section 2334 (f), provide lateral support for walls and partitions, the value of C_p for anchorage shall be increased 50 percent for the center one half of the diaphragm span.

[4]Applies to Seismic Zones Nos. 2, 3 and 4 only.

[5]See Section 2337 (b) 9.

[6]Ceiling weight shall include all light fixtures and other equipment or partitions which are laterally supported by the ceiling. For purposes of determining the seismic force, a ceiling weight of not less than four pounds per square foot shall be used.

[7]Ceilings constructed of lath and plaster or gypsum board screw or nail attached to suspended members that support a ceiling at one level extending from wall to wall need not be analyzed provided the walls are not over 50 feet apart.

[8]Machinery and equipment include, but are not limited to, boilers, chillers, heat exchangers, pumps, air-handling units, cooling towers, control panels, motors, switch gear, transformers and life-safety equipment. It shall include major conduit, ducting and piping serving such machinery and equipment and fire sprinkler systems. See Section 2336 (b) for additional requirements for determining C_p for nonrigid or flexibly mounted equipment.

[9]W_p for access floor systems shall be the dead load of the access floor system plus 25 percent of the floor live load plus a 10 psf partition load allowance.

[10]In lieu of the tabulated values, steel storage racks may be designed in accordance with U.B.C. Standard No. 27-11.

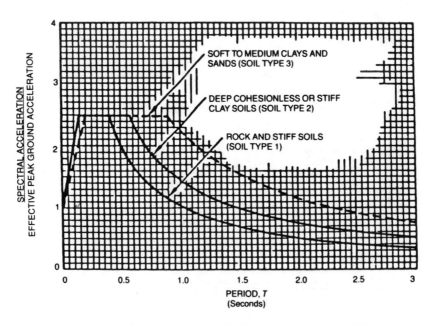

FIGURE NO. 23-3—NORMALIZED RESPONSE SPECTRA SHAPES

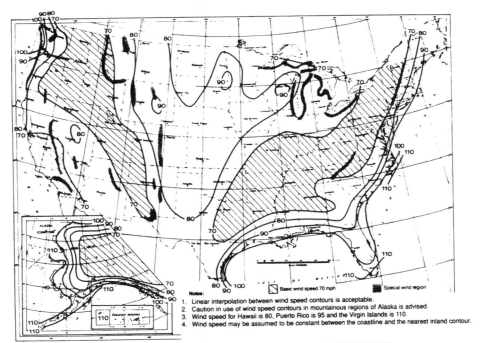

Notes:

1. Linear interpolation between wind speed contours is acceptable.
2. Caution in use of wind speed contours in mountainous regions of Alaska is advised.
3. Wind speed for Hawaii is 80, Puerto Rico is 95 and the Virgin Islands is 110.
4. Wind speed may be assumed to be constant between the coastline and the nearest inland contour.

FIGURE NO. 23-1—MINIMUM BASIC WIND SPEEDS IN MILES PER HOUR

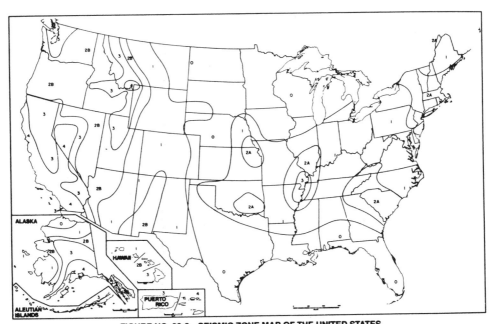

FIGURE NO. 23-2—SEISMIC ZONE MAP OF THE UNITED STATES

For areas outside of the United States, see Appendix Chapter 23.

TABLE NO. 25-A-1—ALLOWABLE UNIT STRESSES—STRUCTURAL LUMBER—(Continued)
Allowable Unit Stresses for Structural Lumber—VISUAL GRADING
(Normal loading. See also Section 2504)

SPECIES AND COMMERCIAL GRADE	SIZE CLASSIFICATION	ALLOWABLE UNIT STRESSES IN POUNDS PER SQUARE INCH							GRADING RULES UNDER WHICH GRADED
		EXTREME FIBER IN BENDING F_b		Tension Parallel to Grain F_t	Horizontal Shear F_v	Compression perpendicular to Grain $F_c\perp$ [21]	Compression Parallel to Grain F_c	MODULUS OF ELASTICITY E [21]	
		Single-member Uses	Repetitive-member Uses						
DOUGLAS FIR – LARCH (Surfaced dry or surfaced green. Used at 19% max. m.c.)									
DOUGLAS FIR – LARCH (North)									
Dense Select Structural	2" to 4" thick 2" to 4" wide	2450	2800	1400	95	730	1850	1,900,000	NLGA, WCLIB, and WWPA (See footnotes 2 through 9, 11, 13, 15 and 16)
Select Structural		2100	2400	1200	95	625	1600	1,800,000	
Dense No. 1		2050	2400	1200	95	730	1450	1,900,000	
No. 1		1750	2050	1050	95	625	1250	1,800,000	
Dense No. 2		1700	1950	1000	95	730	1150	1,700,000	
No. 2		1450	1650	850	95	625	1000	1,700,000	
No. 3		800	925	475	95	625	600	1,500,000	
Appearance		1750	2050	1050	95	625	1500	1,800,000	
Stud		800	925	475	95	625	600	1,500,000	
Construction	2" to 4" thick 4" wide	1050	1200	625	95	625	1150	1,500,000	
Standard		600	675	350	95	625	925	1,500,000	
Utility		275	325	175	95	625	600	1,500,000	
Dense Select Structural	2" to 4" thick 5" and wider	2100	2400	1400	95	730	1650	1,900,000	
Select Structural		1800	2050	1200	95	625	1400	1,800,000	
Dense No. 1		1800	2050	1200	95	730	1450	1,900,000	
No. 1		1500	1750	1000	95	625	1250	1,800,000	
Dense No. 2		1450	1700	775	95	730	1250	1,700,000	
No. 2		1250	1450	650	95	625	1050	1,700,000	
No. 3 and Stud		725	850	375	95	625	675	1,500,000	
Appearance		1500	1750	1000	95	625	1500	1,800,000	
Dense Select Structural	Beams and Stringers [12]	1850	—	1100	85	730	1300	1,700,000	WWPA (See footnotes 2 through 10)
Select Structural		1600	—	950	85	625	1100	1,600,000	
Dense No. 1		1550	—	775	85	730	1100	1,700,000	
No. 1		1350	—	675	85	625	925	1,600,000	
Dense No. 2		1000	—	500	85	730	700	1,400,000	
No. 2		875	—	425	85	625	600	1,300,000	
Dense Select Structural	Posts and Timbers [12]	1750	—	1150	85	730	1350	1,700,000	
Select Structural		1500	—	1000	85	625	1150	1,600,000	
Dense No. 1		1400	—	950	85	730	1200	1,700,000	
No. 1		1200	—	825	85	625	1000	1,600,000	
Dense No. 2		800	—	550	85	730	550	1,400,000	
No. 2		700	—	475	85	625	475	1,300,000	
Selected Decking	Decking	—	2000	—	—	—	—	1,800,000	
Commercial Decking		—	1650	—	—	—	—	1,700,000	
Selected Decking	Decking	—	2150	(Surfaced at 15% max. m.c. and used at 15% max. m.c.)				1,900,000	
Commercial Decking		—	1800					1,700,000	

[2]The design values shown in Table No. 25-A-1 are applicable to lumber that will be used under dry conditions such as in most covered structures. For 2-inch- to 4-inch-thick lumber, the dry surfaced size shall be used. In calculating design values, the natural gain in strength and stiffness that occurs as lumber dries has been taken into consideration as well as the reduction in size that occurs when unseasoned lumber shrinks. The gain in load-carrying capacity due to increased strength and stiffness resulting from drying more than offsets the design effect of size reductions due to shrinkage. For 5-inch and thicker lumber, the surfaced sizes also may be used because design values have been adjusted to compensate for any loss in size by shrinkage which may occur.

[3]Values for F_b, F_t and F_c for the grades of Construction, Standard and Utility apply only to 4-inch widths.

[4]The values in Table No. 25-A-1 for dimension 2 inches to 4 inches are based on edgewise use. Where such lumber is used flatwise, the recommended design values for extreme fiber stress in bending may be multiplied by the following factors:

WIDTH	THICKNESS		
	2"	3"	4"
2 inches to 4 inches	1.10	1.04	1.00
5 inches and wider	1.22	1.16	1.11

Values for decking may be increased by 10 percent for 2-inch decking and 4 percent for 3-inch decking.

(Continued)

[5]When 2-inch- to 4-inch-thick lumber is manufactured at a maximum moisture content of 15 percent and used in a condition where the moisture content does not exceed 15 percent, the design values shown in Table No. 25-A-1 for surfaced dry and surfaced green may be multiplied by the following factors:

EXTREME FIBER IN BENDING F_b	TENSION PARALLEL TO GRAIN F_t	HORIZONTAL SHEAR F_v	COMPRESSION PERPENDICULAR TO GRAIN $F_{c\perp}$	COMPRESSION PARALLEL TO GRAIN F_c	MODULUS OF ELASTICITY E
1.08	1.08	1.05	1.00	1.17*	1.05*

*For redwood use 1.15 for F_c and 1.04 for E.

[6]When 2-inch- to 4-inch-thick lumber is designed for use where the moisture content will exceed 19 percent for an extended period of time, the values shown in Table No. 25-A-1 shall be multiplied by the following factors:

EXTREME FIBER IN BENDING F_b	TENSION PARALLEL TO GRAIN F_t	HORIZONTAL SHEAR F_v	COMPRESSION PERPENDICULAR TO GRAIN $F_{c\perp}$	COMPRESSION PARALLEL TO GRAIN F_c	MODULUS OF ELASTICITY E
0.86	0.84	0.97	0.67	0.70	0.97

[7]When lumber 5 inches thick and thicker is designed for use where the moisture content will exceed 19 percent for an extended period of time, the values shown in Table No. 25-A-1 shall be multiplied by the following factors:

EXTREME FIBER IN BENDING F_b	TENSION PARALLEL TO GRAIN F_t	HORIZONTAL SHEAR F_v	COMPRESSION PERPENDICULAR TO GRAIN $F_{c\perp}$	COMPRESSION PARALLEL TO GRAIN F_c	MODULUS OF ELASTICITY E
1.00	1.00	1.00	0.67	0.91	1.00

[8]Specific horizontal shear values may be established by use of the following tabulation when length of split or size of check or shake is known and no increase in them is anticipated. For California redwood, southern pine, Virginia pine-pond pine, or yellow poplar, the provisions in this footnote apply only to the following F_v values: 80 psi, California redwood; 95 psi, southern pine (KD-15); 90 psi, southern pine (S-dry); 85 psi, southern pine (S-green); 95 psi, Virginia pine-pond pine (KD-15); 90 psi, Virginia pine-pond pine (S-dry); 85 psi, Virginia pine-pond pine (S-green); and 75 psi, yellow poplar.

SHEAR STRESS MODIFICATION FACTOR					
Length of Split on Wide Face of 2" Lumber (nominal):	Multiply Tabulated F_v Value by:	Length of Split on Wide Face of 3" and Thicker Lumber (nominal):	Multiply Tabulated F_v Value by:	Size of Shake* in 3" and Thicker Lumber (nominal):	Multiply Tabulated F_v Value by:
No split	2.00	No split	2.00	No shake	2.00
$^1/_2$ by wide face	1.67	$^1/_2$ by narrow face	1.67	$^1/_6$ by narrow face	1.67
$^3/_4$ by wide face	1.50	1 by narrow face	1.33	$^1/_3$ by narrow face	1.33
1 by wide face	1.33	$1^1/_2$ by narrow face or more .	1.00	$^1/_2$ by narrow face or more . . .	1.00
$1^1/_2$ by wide face or more . .	1.00				

*Shake is measured at the end between lines enclosing the shake and parallel to the wide face.

[9]Stress-rated boards of nominal 1-inch, $1^1/_4$-inch and $1^1/_2$-inch thickness, 2 inches and wider, are permitted the recommended design values shown for Select Structural, No. 1, No. 2, No. 3, Construction, Standard, Utility, Appearance, Clear Heart Structural and Clear Structural grades as shown in the 2-inch- to 4-inch-thick categories herein, where graded in accordance with the stress-rated board provisions in the applicable grading rules.

[10]When decking is used where the moisture content will exceed 15 percent for an extended period of time, the tabulated design values shall be multiplied by the following factors: extreme fiber in bending F_b – 0.79; modulus of elasticity E – 0.92.

[11]Where lumber is graded under the NLGA values shown for Select Structural, No. 1, No. 2, No. 3, and Stud grades are not applicable to 3-inch x 4-inch and 4-inch x 4-inch sizes.

[12]Lumber in the beam and stringer or post and timber size classification may be assigned different working stresses for the same grade name and species based on the grading rules of the specific agency involved. It is therefore necessary that the grading rule agency be identified to properly correlate permitted design stresses with the grademark.

[13]Utility grades of all species may be used only under conditions specifically approved by the building official.

[14]A horizontal shear F_v of 70 may be used for eastern white pine graded under the NHPMA and NELMA grading rules.

[15]Tabulated tension parallel to grain values for species 5 inches and wider, 2 inches to 4 inches thick (and $2^1/_2$ inches to 4 inches thick) size classifications apply to 5-inch and 6-inch widths only, for grades of Select Structural, No. 1, No. 2, No. 3, Appearance and Stud (including dense grades). For lumber wider than 6 inches in these grades, the tabulated F_t values shall be multiplied by the following factors:

GRADE (2 inches to 4 inches thick, 5 inches and wider) ($2^1/_2$ inches to 4 inches thick, 5 inches and wider) (includes "Dense" grades)	Multiply tabulated F_t values by		
	5 inches and 6 inches wide	8 inches wide	10 inches and wider
Select Structural	1.00	0.90	0.80
No. 1, No. 2, No. 3 and Appearance	1.00	0.80	0.60
Stud	1.00		

[16]Design values for all species of Stud grade in 5-inch and wider size classifications apply to 5-inch and 6-inch widths only.

TABLE NO. 25-G—SAFE LATERAL STRENGTH AND REQUIRED PENETRATION OF BOX AND COMMON WIRE NAILS DRIVEN PERPENDICULAR TO GRAIN OF WOOD

SIZE OF NAIL	STANDARD LENGTH (inches)	WIRE GAUGE	PENETRA-TION REQUIRED (inches)	LOADS (Pounds)[1][2][3]	
				Douglas Fir Larch or Southern Pine	Other Species
BOX NAILS					
6d	2	12½	1⅛	51	See U.B.C. Standard No. 25-17
8d	2½	11½	1¼	63	
10d	3	10½	1½	76	
12d	3¼	10½	1½	76	
16d	3½	10	1½	82	
20d	4	9	1⅝	94	
30d	4½	9	1⅝	94	
40d	5	8	1¾	108	
COMMON NAILS					
6d	2	11½	1¼	63	See U.B.C. Standard No. 25-17
8d	2½	10¼	1½	78	
10d	3	9	1⅝	94	
12d	3¼	9	1⅝	94	
16d	3½	8	1¾	108	
20d	4	6	2⅛	139	
30d	4½	5	2¼	155	
40d	5	4	2½	176	
50d	5½	3	2¾	199	
60d	6	2	2⅞	223	

[1]The safe lateral strength values may be increased 25 percent where metal side plates are used.

[2]For wood diaphragm calculations these values may be increased 30 percent. (See U.B.C. Standard No. 25-17.)

[3]Tabulated values are on a normal load-duration basis and apply to joints made of seasoned lumber used in dry locations. See U.B.C. Standard No. 25-17 for other service conditions.

TABLE NO. 25-I—MAXIMUM DIAPHRAGM DIMENSION RATIOS

MATERIAL	HORIZONTAL DIAPHRAGMS Maximum Span-Width Ratios	VERTICAL DIAPHRAGMS Maximum Height-Width Ratios
1. Diagonal sheathing, conventional	3:1	2:1
2. Diagonal sheathing, special	4:1	3½:1
3. Plywood and particleboard, nailed all edges	4:1	3½:1
4. Plywood and particleboard, blocking omitted at intermediate joints	4:1	2:1

TABLE NO. 25-Q—NAILING SCHEDULE

CONNECTION	NAILING[1]
1. Joist to sill or girder, toenail	3-8d
2. Bridging to joist, toenail each end	2-8d
3. 1" x 6" subfloor or less to each joist, face nail	2-8d
4. Wider than 1" x 6" subfloor to each joist, face nail	3-8d
5. 2" subfloor to joist or girder, blind and face nail	2-16d
6. Sole plate to joist or blocking, face nail	16d at 16" o.c.
7. Top plate to stud, end nail	2-16d
8. Stud to sole plate	4-8, toenail or 2-16d, end nail
9. Double studs, face nail	16d at 24" o.c.
10. Doubled top plates, face nail	16d at 16" o.c.
11. Top plates, laps and intersections, face nail	2-16d
12. Continuous header, two pieces	16d at 16" o.c. along each edge
13. Ceiling joists to plate, toenail	3-8d
14. Continuous header to stud, toenail	4-8d
15. Ceiling joists, laps over partitions, face nail	3-16d
16. Ceiling joists to parallel rafters, face nail	3-16d
17. Rafter to plate, toenail	3-8d
18. 1" brace to each stud and plate, face nail	2-8d
19. 1" x 8" sheathing or less to each bearing, face nail	2-8d
20. Wider than 1" x 8" sheathing to each bearing, face nail	3-8d
21. Built-up corner studs	16d at 24" o.c.
22. Built-up girder and beams	20d at 32" o.c. at top and bottom and staggered 2-20d at ends and at each splice
23. 2" planks	2-16d at each bearing
24. Plywood and particleboard:[5]	
Subfloor, roof and wall sheathing (to framing):	
1/2" and less	6d[2]
19/32"-3/4"	8d[3] or 6d[4]
7/8"-1"	8d[2]
11/8"-11/4"	10d[3] or 8d[4]
Combination Subfloor-underlayment (to framing):	
3/4" and less	6d[4]
7/8"-1"	8d[4]
11/8"-11/4"	10d[3] or 8d[4]
25. Panel Siding (to framing):	
1/2" or less	6d[6]
5/8"	8d[6]
26. Fiberboard Sheathing:[7]	
1/2"	No. 11 ga.[8] 6d[3] No. 16 ga.[9]
25/32"	No. 11 ga.[8] 8d[3] No. 16 ga.[9]

[1]Common or box nails may be used except where otherwise stated.

[2]Common or deformed shank.

[3]Common.

[4]Deformed shank.

[5]Nails spaced at 6 inches on center at edges, 12 inches at intermediate supports except 6 inches at all supports where spans are 48 inches or more. For nailing of plywood and particleboard diaphragms and shear walls, refer to Section 2513 (c). Nails for wall sheathing may be common, box or casing.

[6]Corrosion-resistant siding or casing nails conforming to the requirements of Section 2516 (j) 1.

[7]Fasteners spaced 3 inches on center at exterior edges and 6 inches on center at intermediate supports.

[8]Corrosion-resistant roofing nails with 7/16-inch-diameter head and 1 1/2-inch length for 1/2-inch sheathing and 1 3/4-inch length for 25/32-inch sheathing conforming to the requirements of Section 2516 (j) 1.

[9]Corrosion-resistant staples with nominal 7/16-inch crown and 1 1/8-inch length for 1/2-inch sheathing and 1 1/2-inch length for 25/32-inch sheathing conforming to the requirements of Section 2516 (j) 1.

TABLE NO. 25-J-1—ALLOWABLE SHEAR IN POUNDS PER FOOT FOR HORIZONTAL PLYWOOD DIAPHRAGMS WITH FRAMING OF DOUGLAS FIR-LARCH OR SOUTHERN PINE[1]

PLYWOOD GRADE	Common Nail Size	Minimum Nominal Penetration in Framing (in inches)	Minimum Nominal Plywood Thickness (in inches)	Minimum Nominal Width of Framing Member (in inches)	BLOCKED DIAPHRAGMS — Nail spacing at diaphragm boundaries (all cases), at continous panel edges parallel to load (Cases 3 and 4) and at all panel edges (Cases 5 and 6) / Nail spacing at other plywood panel edges — 6	4	2½/2	2	UNBLOCKED DIAPHRAGM — Load perpendicular to unblocked edges and continuous panel joints (Case 1)	Other configurations (Cases 2, 3, 4, 5 and 6)
					6	6	4	3		
STRUCTURAL I	6d	1¼	5/16	2	185	250	375	420	165	125
				3	210	280	420	475	185	140
	8d	1½	3/8	2	270	360	530	600	240	180
				3	300	400	600	675	265	200
	10d[3]	1⅝	15/32	2	320	425	640	730	285	215
				3	360	480	720	820	320	240
C-D, C-C, STRUCTURAL II and other grades covered in U.B.C. Standard No. 25-9	6d	1¼	5/16	2	170	225	335	380	150	110
				3	190	250	380	430	170	125
			3/8	2	185	250	375	420	165	125
				3	210	280	420	475	185	140
	8d	1½	3/8	2	240	320	480	545	215	160
				3	270	360	540	610	240	180
			15/32	2	270	360	530	600	240	180
				3	300	400	600	675	265	200
	10d[3]	1⅝	15/32	2	290	385	575	655	255	190
				3	325	430	650	735	290	215
			19/32	2	320	425	640	730	285	215
				3	360	480	720	820	320	240

[1]These values are for short-time loads due to wind or earthquake and must be reduced 25 percent for normal loading. Space nails 12 inches on center along intermediate framing members.

Allowable shear values for nails in framing members of other species set forth in Table No. 25-17-J of the U.B.C. Standards shall be calculated for all grades by multiplying the values for nails in Structural I by the following factors: Group III, 0.82 and Group IV, 0.65.

[2]Framing at adjoining panel edges shall be 3-inch nominal or wider and nails shall be staggered where nails are spaced 2 inches or $2\frac{1}{2}$ inches on center.

[3]Framing at adjoining panel edges shall be 3-inch nominal or wider and nails shall be staggered where 10d nails having penetration into framing of more than $1\frac{5}{8}$ inches are spaced 3 inches or less on center.

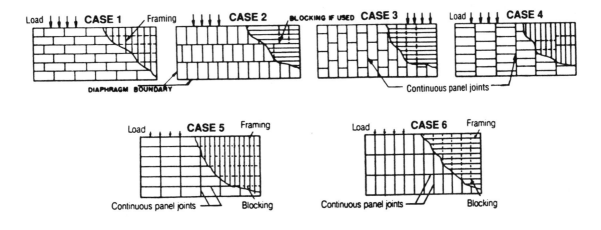

Note: Framing may be oriented in either direction for diaphragms, provided sheathing is properly designed for vertical loading.

TABLE NO. 25-K-1—ALLOWABLE SHEAR FOR WIND OR SEISMIC FORCES IN POUNDS PER FOOT FOR PLYWOOD SHEAR WALLS WITH FRAMING OF DOUGLAS FIR-LARCH OR SOUTHERN PINE[1][4]

PLYWOOD GRADE	MINIMUM NOMINAL PLYWOOD THICKNESS (inches)	MINIMUM NAIL PENETRATION IN FRAMING (inches)	NAIL SIZE (Common or Galvanized Box)	PLYWOOD APPLIED DIRECT TO FRAMING — Nail Spacing at Plywood Panel Edges				NAIL SIZE (Common or Galvanized Box)	PLYWOOD APPLIED OVER 1/2-INCH OR 5/8-INCH GYPSUM SHEATHING — Nail Spacing at Plywood Panel Edges			
				6	4	3	22		6	4	3	22
STRUCTURAL I	5/16	1 1/4	6d	200	300	390	510	8d	200	300	390	510
	3/8	1 1/2	8d	230[3]	360[3]	460[3]	610[3]	10d[5]	280	430	550	730[2]
	15/32	1 1/2	8d	280	430	550	730	10d[5]	280	430	550	730
	15/32	1 5/8	10d[5]	340	510	665	870	—	—	—	—	—
C-D, C-C STRUCTURAL II, plywood panel siding and other grades covered in U.B.C. Standard No. 25-9.	5/16	1 1/4	6d	180	270	350	450	8d	180	270	350	450
	3/8	1 1/4	6d	200	300	390	510	8d	200	300	390	510
	3/8	1 1/2	8d	220[3]	320[3]	410[3]	530[3]	10d[5]	260	380	490	640
	15/32	1 1/2	8d	260	380	490	640	10d[5]	260	380	490	640
	15/32	1 5/8	10d[5]	310	460	600	770	—	—	—	—	—
	19/32	1 5/8	10d[5]	340	510	665	870	—	—	—	—	—
			NAIL SIZE (Galvanized Casing)					NAIL SIZE (Galvanized Casing)				
Plywood panel siding in grades covered in U.B.C. Standard No. 25-9	5/16	1 1/4	6d	140	210	275	360	8d	140	210	275	360
	3/8	1 1/2	8d	130[3]	200[3]	260[3]	340[3]	10d[5]	160	240	310	410

[1]All panel edges backed with 2-inch nominal or wider framing. Plywood installed either horizontally or vertically. Space nails at 6 inches on center along intermediate framing members for 3/8-inch plywood installed with face grain parallel to studs spaced 24 inches on center and 12 inches on center for other conditions and plywood thicknesses. These values are for short-time loads due to wind or earthquake and must be reduced 25 percent for normal loading.

    Allowable shear values for nails in framing members of other species set forth in Table No. 25-17-J of U.B.C. Standards shall be calculated for all grades by multiplying the values for common and galvanized box nails in STRUCTURAL I and galvanized casing nails in other grades by the following factors: Group III, 0.82 and Group IV, 0.65.

[2]Framing at adjoining panel edges shall be 3-inch nominal or wider and nails shall be staggered where nails are spaced 2 inches on center.

[3]The values for 3/8-inch-thick plywood applied direct to framing may be increased 20 percent, provided studs are spaced a maximum of 16 inches on center or plywood is applied with face grain across studs.

[4]Where plywood is applied on both faces of a wall and nail spacing is less than 6 inches on center on either side, panel joints shall be offset to fall on different framing members or framing shall be 3-inch nominal or thicker and nails on each side shall be staggered.

[5]Framing at adjoining panel edges shall be 3-inch nominal or wider and nails shall be staggered where 10d nails having penetration into framing of more than 1 5/8 inches are spaced 3 inches or less on center.

TABLE NO. 25-R-3—SIZE, HEIGHT AND SPACING OF WOOD STUDS[1]

STUD SIZE (Inches)	BEARING WALLS				NONBEARING WALLS	
	LATERALLY UNSUPPORTED STUD HEIGHT[3] (Feet)	SUPPORTING ROOF AND CEILING ONLY	SUPPORTING ONE FLOOR, ROOF AND CEILING	SUPPORTING TWO FLOORS, ROOF AND CEILING	LATERALLY UNSUPPORTED STUD HEIGHT[3] (Feet)	SPACING (Inches)
		SPACING (Inches)				
1. 2 x 3[2]	—	—	—	—	10	16
2. 2 x 4	10	24	16	—	14	24
3. 3 x 4	10	24	24	16	14	24
4. 2 x 5	10	24	24	—	16	24
5. 2 x 6	10	24	24	16	20	24

[1]Utility grade studs shall not be spaced more than 16 inches on center, or support more than a roof and ceiling, or exceed 8 feet in height for exterior walls and load bearing or 10 feet for interior nonload-bearing walls.
[2]Shall not be used in exterior walls.
[3]Listed heights are distances between points of lateral support placed perpendicular to the plane of the wall. Increases in unsupported height are permitted where justified by an analysis.

TABLE NO. 25-S-1—ALLOWABLE SPANS FOR PLYWOOD SUBFLOOR AND ROOF SHEATHING CONTINUOUS OVER TWO OR MORE SPANS AND FACE GRAIN PERPENDICULAR TO SUPPORTS[1][8]

PANEL SPAN RATING[3]	PLYWOOD THICKNESS (Inch)	ROOF[2]				FLOOR MAXIMUM SPAN[4] (in Inches)
		Maximum Span (In Inches)		Load (In Pounds per Square Foot)		
		Edges Blocked	Edges Unblocked	Total Load	Live Load	
1. 12/0	$5/16$	12		135	130	0
2. 16/0	$5/16, 3/8$	16		80	65	0
3. 20/0	$5/16, 3/8$	20		70	55	0
4. 24/0	$3/8$	24	16	60	45	0
5. 24/0	$15/32, 1/2$	24	24	60	45	0
6. 32/16	$15/32, 1/2, 19/32, 5/8$	32	28	55	35[5]	16[6]
7. 40/20	$19/32, 5/8, 23/32, 3/4, 7/8$	40	32	40[5]	35[5]	20[6][7]
8. 48/24	$23/32, 3/4, 7/8$	48	36	40[5]	35[5]	24

[1]These values apply for C-C, C-D, Structural I and II grades only. Spans shall be limited to values shown because of possible effect of concentrated loads.
[2]Uniform load deflection limitations $1/180$ of the span under live load plus dead load, $1/240$ under live load only. Edges may be blocked with lumber or other approved type of edge support.
[3]Span rating appears on all panels in the construction grades listed in Footnote No. 1.
[4]Plywood edges shall have approved tongue-and-groove joints or shall be supported with blocking unless $1/4$-inch minimum thickness underlayment, or $1 1/2$ inches of approved cellular or lightweight concrete is placed over the subfloor, or finish floor is $3/4$-inch wood strip. Allowable uniform load based on deflection of $1/360$ of span is 165 pounds per square foot (psf).
[5]For roof live load of 40 psf or total load of 55 psf, decrease spans by 13 percent or use panel with next greater span rating.
[6]May be 24 inches if $3/4$-inch wood strip flooring is installed at right angles to joists.
[7]May be 24 inches where a minimum of $1 1/2$ inches of approved cellular or lightweight concrete is placed over the subfloor and the plywood sheathing is manufactured with exterior glue.
[8]Floor or roof sheathing conforming with this table shall be deemed to meet the design criteria of Section 2516.

TABLE NO. 25-S-2—ALLOWABLE LOADS FOR PLYWOOD ROOF SHEATHING CONTINUOUS OVER TWO OR MORE SPANS AND FACE GRAIN PARALLEL TO SUPPORTS[1][2]

	THICKNESS	NO. OF PLIES	SPAN	TOTAL LOAD	LIVE LOAD
STRUCTURAL I	$15/32$	4	24	30	20
		5	24	45	35
	$1/2$	4	24	35	25
		5	24	55	40
Other grades covered in U.B.C. Standard No. 25-9	$15/32$	5	24	25	20
	$1/2$	5	24	30	25
	$19/32$	4	24	35	25
		5	24	50	40
	$5/8$	4	24	40	30
		5	24	55	45

[1]Uniform load deflection limitations: $1/180$ of span under live load plus dead load, $1/240$ under live load only. Edges shall be blocked with lumber or other approved type of edge supports.
[2]Roof sheathing conforming with this table shall be deemed to meet the design criteria of Section 2516.

TABLE NO. 25-U-J-1—ALLOWABLE SPANS FOR FLOOR JOISTS—40 LBS. PER SQ. FT. LIVE LOAD

DESIGN CRITERIA: Deflection—For 40 lbs. per sq. ft. live load. Limited to span in inches divided by 360. Strength—Live load of 40 lbs. per sq. ft. plus dead load of 10 lbs. per sq. ft. determines the required fiber stress value.

JOIST SIZE (IN)	SPACING (IN)	Modulus of Elasticity, E, in 1,000,000 psi													
		0.8	0.9	1.0	1.1	1.2	1.3	1.4	1.5	1.6	1.7	1.8	1.9	2.0	2.2
2x6	12.0	8-6 720	8-10 780	9-2 830	9-6 890	9-9 940	10-0 990	10-3 1040	10-6 1090	10-9 1140	10-11 1190	11-2 1220	11-4 1280	11-7 1320	11-11 1410
	16.0	7-9 790	8-0 860	8-4 920	8-7 980	8-10 1040	9-1 1090	9-4 1150	9-6 1200	9-9 1250	9-11 1310	10-2 1360	10-4 1410	10-6 1460	10-10 1550
	24.0	6-9 900	7-0 980	7-3 1050	7-6 1120	7-9 1190	7-11 1250	8-2 1310	8-4 1380	8-6 1440	8-8 1500	8-10 1550	9-0 1610	9-2 1670	9-6 1780
2x8	12.0	11-3 720	11-8 780	12-1 830	12-6 890	12-10 940	13-2 990	13-6 1040	13-10 1090	14-2 1140	14-5 1190	14-8 1230	15-0 1280	15-3 1320	15-9 1410
	16.0	10-2 790	10-7 850	11-0 920	11-4 980	11-8 1040	12-0 1090	12-3 1150	12-7 1200	12-10 1250	13-1 1310	13-4 1360	13-7 1410	13-10 1460	14-3 1550
	24.0	8-11 900	9-3 980	9-7 1050	9-11 1120	10-2 1190	10-6 1250	10-9 1310	11-0 1380	11-3 1440	11-5 1500	11-8 1550	11-11 1610	12-1 1670	12-6 1780
2x10	12.0	14-4 720	14-11 780	15-5 830	15-11 890	16-5 940	16-10 990	17-3 1040	17-8 1090	18-0 1140	18-5 1190	18-9 1230	19-1 1280	19-5 1320	20-1 1410
	16.0	13-0 790	13-6 850	14-0 920	14-6 980	14-11 1040	15-3 1090	15-8 1150	16-0 1200	16-5 1250	16-9 1310	17-0 1360	17-4 1410	17-8 1460	18-3 1550
	24.0	11-4 900	11-10 980	12-3 1050	12-8 1120	13-0 1190	13-4 1250	13-8 1310	14-0 1380	14-4 1440	14-7 1500	14-11 1550	15-2 1610	15-5 1670	15-11 1780
2x12	12.0	17-5 720	18-1 780	18-9 830	19-4 890	19-11 940	20-6 990	21-0 1040	21-6 1090	21-11 1140	22-5 1190	22-10 1230	23-3 1280	23-7 1320	24-5 1410
	16.0	15-10 790	16-5 860	17-0 920	17-7 980	18-1 1040	18-7 1090	19-1 1150	19-6 1200	19-11 1250	20-4 1310	20-9 1360	21-1 1410	21-6 1460	22-2 1550
	24.0	13-10 900	14-4 980	14-11 1050	15-4 1120	15-10 1190	16-3 1250	16-8 1310	17-0 1380	17-5 1440	17-9 1500	18-1 1550	18-5 1610	18-9 1670	19-4 1780

NOTES:

(1) The required extreme fiber stress in bending (F_b) in pounds per square inch is shown below each span.

(2) Use single or repetitive member bending stress values (F_b) and modulus of elasticity values (E) from Tables Nos. 25-A-1 and 25-A-2.

(3) For more comprehensive tables covering a broader range of bending stress values (F_b) and modulus of elasticity values (E), other spacing of members and other conditions of loading, see U.B.C. Standard No. 25-21.

(4) The spans in these tables are intended for use in covered structures or where moisture content in use does not exceed 19 percent.

TABLE NO. 25-U-J-6—ALLOWABLE SPANS FOR CEILING JOISTS—10 LBS. PER SQ. FT. LIVE LOAD
(Drywall Ceiling)

DESIGN CRITERIA: Deflection—For 10 lbs. per sq. ft. live load. Limited to span in inches divided by 240. Strength—Live load of 10 lbs. per sq. ft. plus dead load of 5 lbs. per sq. ft. determines the required fiber stress value.

JOIST SIZE (IN)	SPACING (IN)	Modulus of Elasticity, E, in 1,000,000 psi													
		0.8	0.9	1.0	1.1	1.2	1.3	1.4	1.5	1.6	1.7	1.8	1.9	2.0	2.2
2x4	12.0	9-10 710	10-3 770	10-7 830	10-11 880	11-3 930	11-7 980	11-10 1030	12-2 1080	12-5 1130	12-8 1180	12-11 1220	13-2 1270	13-4 1310	13-9 1400
	16.0	8-11 780	9-4 850	9-8 910	9-11 970	10-3 1030	10-6 1080	10-9 1140	11-0 1190	11-3 1240	11-6 1290	11-9 1340	11-11 1390	12-2 1440	12-6 1540
	24.0	7-10 900	8-1 970	8-5 1040	8-8 1110	8-11 1170	9-2 1240	9-5 1300	9-8 1360	9-10 1420	10-0 1480	10-3 1540	10-5 1600	10-7 1650	10-11 1760
2x6	12.0	15-6 710	16-1 770	16-8 830	17-2 880	17-8 930	18-2 980	18-8 1030	19-1 1080	19-6 1130	19-11 1180	20-3 1220	20-8 1270	21-0 1310	21-8 1400
	16.0	14-1 780	14-7 850	15-2 910	15-7 970	16-1 1030	16-6 1080	16-11 1140	17-4 1190	17-8 1240	18-1 1290	18-5 1340	18-9 1390	19-1 1440	19-8 1540
	24.0	12-3 900	12-9 970	13-3 1040	13-8 1110	14-1 1170	14-5 1240	14-9 1300	15-2 1360	15-6 1420	15-9 1480	16-1 1540	16-4 1600	16-8 1650	17-2 1760
2x8	12.0	20-5 710	21-2 770	21-11 830	22-8 880	23-4 930	24-0 980	24-7 1030	25-2 1080	25-8 1130	26-2 1180	26-9 1220	27-2 1270	27-8 1310	28-7 1400
	16.0	18-6 780	19-3 850	19-11 910	20-7 970	21-2 1030	21-9 1080	22-4 1140	22-10 1190	23-4 1240	23-10 1290	24-3 1340	24-8 1390	25-2 1440	25-11 1540
	24.0	16-2 900	16-10 970	17-5 1040	18-0 1110	18-6 1170	19-0 1240	19-6 1300	19-11 1360	20-5 1420	20-10 1480	21-2 1540	21-7 1600	21-11 1650	22-8 1760
2x10	12.0	26-0 710	27-1 770	28-0 830	28-11 880	29-9 930	30-7 980	31-4 1030	32-1 1080	32-9 1130	33-5 1180	34-1 1220	34-8 1270	35-4 1310	36-5 1400
	16.0	23-8 780	24-7 850	25-5 910	26-3 970	27-1 1030	27-9 1080	28-6 1140	29-2 1190	29-9 1240	30-5 1290	31-0 1340	31-6 1390	32-1 1440	33-1 1540
	24.0	20-8 900	21-6 970	22-3 1040	22-11 1110	23-8 1170	24-3 1240	24-10 1300	25-5 1360	26-0 1420	26-6 1480	27-1 1540	27-6 1600	28-0 1650	28-11 1760

NOTES:

(1) The required extreme fiber stress in bending (F_b) in pounds per square inch is shown below each span.

(2) Use single or repetitive member bending stress values (F_b) and modulus of elasticity values (E) from Tables Nos. 25-A-1 and 25-A-2.

(3) For more comprehensive tables covering a broader range of bending stress values (F_b) and modulus of elasticity values (E), other spacing of members and other conditions of loading, see U.B.C. Standard No. 25-21.

(4) The spans in these tables are intended for use in covered structures or where moisture content in use does not exceed 19 percent.

TABLE NO. 25-U-R-1—ALLOWABLE SPANS FOR LOW- OR HIGH-SLOPE RAFTERS
20 LBS. PER SQ. FT. LIVE LOAD (Supporting Drywall Ceiling)

DESIGN CRITERIA: Strength—15 lbs. per sq. ft. dead load plus 20 lbs. per sq. ft. live load determines required fiber stress. Deflection—For 20 lbs. per sq. ft. live load. Limited to span in inches divided by 240. RAFTERS: Spans are measured along the horizontal projection and loads are considered as applied on the horizontal projection.

RAFTER SIZE (IN)	SPACING (IN)	Allowable Extreme Fiber Stress in Bending F_b (psi).														
		500	600	700	800	900	1000	1100	1200	1300	1400	1500	1600	1700	1800	1900
2x6	12.0	8-6 / 0.26	9-4 / 0.35	10-0 / 0.44	10-9 / 0.54	11-5 / 0.64	12-0 / 0.75	12-7 / 0.86	13-2 / 0.98	13-8 / 1.11	14-2 / 1.24	14-8 / 1.37	15-2 / 1.51	15-8 / 1.66	16-1 / 1.81	16-7 / 1.96
	16.0	7-4 / 0.23	8-1 / 0.30	8-8 / 0.38	9-4 / 0.46	9-10 / 0.55	10-5 / 0.65	10-11 / 0.75	11-5 / 0.85	11-10 / 0.97	12-4 / 1.07	12-9 / 1.19	13-2 / 1.31	13-7 / 1.44	13-11 / 1.56	14-4 / 1.70
	24.0	6-0 / 0.19	6-7 / 0.25	7-1 / 0.31	7-7 / 0.38	8-1 / 0.45	8-6 / 0.53	8-11 / 0.61	9-4 / 0.70	9-8 / 0.78	10-0 / 0.88	10-5 / 0.97	10-9 / 1.07	11-1 / 1.17	11-5 / 1.28	11-8 / 1.39
2x8	12.0	11-2 / 0.26	12-3 / 0.35	13-3 / 0.44	14-2 / 0.54	15-0 / 0.64	15-10 / 0.75	16-7 / 0.86	17-4 / 0.98	18-0 / 1.11	18-9 / 1.24	19-5 / 1.37	20-0 / 1.51	20-8 / 1.66	21-3 / 1.81	21-10 / 1.96
	16.0	9-8 / 0.23	10-7 / 0.30	11-6 / 0.38	12-3 / 0.46	13-0 / 0.55	13-8 / 0.65	14-4 / 0.75	15-0 / 0.85	15-7 / 0.96	16-3 / 1.07	16-9 / 1.19	17-4 / 1.31	17-10 / 1.44	18-5 / 1.56	18-11 / 1.70
	24.0	7-11 / 0.19	8-8 / 0.25	9-4 / 0.31	10-0 / 0.38	10-7 / 0.45	11-2 / 0.53	11-9 / 0.61	12-3 / 0.70	12-9 / 0.78	13-3 / 0.88	13-8 / 0.97	14-2 / 1.07	14-7 / 1.17	15-0 / 1.28	15-5 / 1.39
2x10	12.0	14-3 / 0.26	15-8 / 0.35	16-11 / 0.44	18-1 / 0.54	19-2 / 0.64	20-2 / 0.75	21-2 / 0.86	22-1 / 0.98	23-0 / 1.11	23-11 / 1.24	24-9 / 1.37	25-6 / 1.51	26-4 / 1.66	27-1 / 1.81	27-10 / 1.96
	16.0	12-4 / 0.23	13-6 / 0.30	14-8 / 0.38	15-8 / 0.46	16-7 / 0.55	17-6 / 0.65	18-4 / 0.75	19-2 / 0.85	19-11 / 0.96	20-8 / 1.07	21-5 / 1.19	22-1 / 1.31	22-10 / 1.44	23-5 / 1.56	24-1 / 1.70
	24.0	10-1 / 0.19	11-1 / 0.25	11-11 / 0.31	12-9 / 0.38	13-6 / 0.45	14-3 / 0.53	15-0 / 0.61	15-8 / 0.70	16-3 / 0.78	16-11 / 0.88	17-6 / 0.97	18-1 / 1.07	18-7 / 1.17	19-2 / 1.28	19-8 / 1.39
2x12	12.0	17-4 / 0.26	19-0 / 0.35	20-6 / 0.44	21-11 / 0.54	23-3 / 0.64	24-7 / 0.75	25-9 / 0.86	26-11 / 0.98	28-0 / 1.11	29-1 / 1.24	30-1 / 1.37	31-1 / 1.51	32-0 / 1.66	32-11 / 1.81	33-10 / 1.96
	16.0	15-0 / 0.23	16-6 / 0.30	17-9 / 0.38	19-0 / 0.46	20-2 / 0.55	21-3 / 0.65	22-4 / 0.75	23-3 / 0.85	24-3 / 0.97	25-2 / 1.07	26-0 / 1.19	26-11 / 1.31	27-9 / 1.44	28-6 / 1.56	29-4 / 1.70
	24.0	12-3 / 0.19	13-5 / 0.25	14-6 / 0.31	15-6 / 0.38	16-6 / 0.45	17-4 / 0.53	18-2 / 0.61	19-0 / 0.70	19-10 / 0.78	20-6 / 0.88	21-3 / 0.97	21-11 / 1.07	22-8 / 1.17	23-3 / 1.28	23-11 / 1.39

NOTES:
(1) The required modulus of elasticity (E) in 1,000,000 pounds per square inch is shown below each span.
(2) Use single or repetitive member bending stress values (F_b) and modulus of elasticity values (E) from Tables Nos. 25-A-1 and 25-A-2. For duration of load stress increases, see Section 2504 (c) 4.
(3) For more comprehensive tables covering a broader range of bending stress values (F_b) and modulus of elasticity values (E), other spacing of members and other conditions of loading, see U.B.C. Standard No. 25-21.
(4) The spans in these tables are intended for use in covered structures or where moisture content in use does not exceed 19 percent.

TABLE NO. 25-U-R-14—ALLOWABLE SPANS FOR HIGH-SLOPE RAFTERS, SLOPE OVER 3 IN 12
30 LBS. PER SQ. FT. LIVE LOAD (Light Roof Covering)

DESIGN CRITERIA: Strength—7 lbs. per sq. ft. dead load plus 30 lbs. per sq. ft. live load determines required fiber stress. Deflection—For 30 lbs. per sq. ft. live load. Limited to span in inches divided by 180. RAFTERS: Spans are measured along the horizontal projection and loads are considered as applied on the horizontal projection.

RAFTER SIZE (IN)	SPACING (IN)	Allowable Extreme Fiber Stress in Bending F_b (psi).														
		500	600	700	800	900	1000	1100	1200	1300	1400	1500	1600	1700	1800	1900
2x4	12.0	5-3 / 0.27	5-9 / 0.36	6-3 / 0.45	6-8 / 0.55	7-1 / 0.66	7-5 / 0.77	7-9 / 0.89	8-2 / 1.02	8-6 / 1.15	8-9 / 1.28	9-1 / 1.42	9-5 / 1.57	9-8 / 1.72	10-0 / 1.87	10-3 / 2.03
	16.0	4-7 / 0.24	5-0 / 0.31	5-5 / 0.39	5-9 / 0.48	6-1 / 0.57	6-5 / 0.67	6-9 / 0.77	7-1 / 0.88	7-4 / 0.99	7-7 / 1.11	7-11 / 1.23	8-2 / 1.36	8-5 / 1.49	8-8 / 1.62	8-10 / 1.76
	24.0	3-9 / 0.19	4-1 / 0.25	4-5 / 0.32	4-8 / 0.39	5-0 / 0.47	5-3 / 0.55	5-6 / 0.63	5-9 / 0.72	6-0 / 0.81	6-3 / 0.91	6-5 / 1.01	6-8 / 1.11	6-10 / 1.21	7-1 / 1.32	7-3 / 1.43
2x6	12.0	8-3 / 0.27	9-1 / 0.36	9-9 / 0.45	10-5 / 0.55	11-1 / 0.66	11-8 / 0.77	12-3 / 0.89	12-9 / 1.02	13-4 / 1.15	13-10 / 1.28	14-4 / 1.42	14-9 / 1.57	15-3 / 1.72	15-8 / 1.87	16-1 / 2.03
	16.0	7-2 / 0.24	7-10 / 0.31	8-5 / 0.39	9-1 / 0.48	9-7 / 0.57	10-1 / 0.67	10-7 / 0.77	11-1 / 0.88	11-6 / 0.99	12-0 / 1.11	12-5 / 1.23	12-9 / 1.36	13-2 / 1.49	13-7 / 1.62	13-11 / 1.76
	24.0	5-10 / 0.19	6-5 / 0.25	6-11 / 0.32	7-5 / 0.39	7-10 / 0.47	8-3 / 0.55	8-8 / 0.63	9-1 / 0.72	9-5 / 0.81	9-9 / 0.91	10-1 / 1.01	10-5 / 1.11	10-9 / 1.21	11-1 / 1.32	11-5 / 1.43
2x8	12.0	10-11 / 0.27	11-11 / 0.36	12-10 / 0.45	13-9 / 0.55	14-7 / 0.66	15-5 / 0.77	16-2 / 0.89	16-10 / 1.02	17-7 / 1.15	18-2 / 1.28	18-10 / 1.42	19-6 / 1.57	20-1 / 1.72	20-8 / 1.87	21-3 / 2.03
	16.0	9-5 / 0.24	10-4 / 0.31	11-2 / 0.39	11-11 / 0.48	12-8 / 0.57	13-4 / 0.67	14-0 / 0.77	14-7 / 0.88	15-2 / 0.99	15-9 / 1.11	16-4 / 1.23	16-10 / 1.36	17-4 / 1.49	17-11 / 1.62	18-4 / 1.76
	24.0	7-8 / 0.19	8-5 / 0.25	9-1 / 0.32	9-9 / 0.39	10-4 / 0.47	10-11 / 0.55	11-5 / 0.63	11-11 / 0.72	12-5 / 0.81	12-10 / 0.91	13-4 / 1.01	13-9 / 1.11	14-2 / 1.21	14-7 / 1.32	15-0 / 1.43
2x10	12.0	13-11 / 0.27	15-2 / 0.36	16-5 / 0.45	17-7 / 0.55	18-7 / 0.66	19-8 / 0.77	20-7 / 0.89	21-6 / 1.02	22-5 / 1.15	23-3 / 1.28	24-1 / 1.42	24-10 / 1.57	25-7 / 1.72	26-4 / 1.87	27-1 / 2.03
	16.0	12-0 / 0.26	13-2 / 0.34	14-3 / 0.43	15-2 / 0.53	16-2 / 0.63	17-0 / 0.74	17-10 / 0.85	18-7 / 0.97	19-5 / 1.09	20-1 / 1.22	20-10 / 1.35	21-6 / 1.49	22-2 / 1.63	22-10 / 1.78	23-5 / 1.93
	24.0	9-1 / 0.19	10-9 / 0.25	11-7 / 0.32	12-5 / 0.39	13-2 / 0.47	13-11 / 0.55	14-7 / 0.63	15-2 / 0.72	15-10 / 0.81	16-5 / 0.91	17-0 / 1.01	17-7 / 1.11	18-1 / 1.21	18-7 / 1.32	19-2 / 1.43

NOTES: (1) The required modulus of elasticity (E) in 1,000,000 pounds per square inch is shown below each span.
(2) Use single or repetitive member bending stress values (F_b) and modulus of elasticity values (E) from Tables Nos. 25-A-1 and 25-A-2. For duration of load stress increases, see Section 2504 (c) 4.
(3) For more comprehensive tables covering a broader range of bending stress values (F_b) and modulus of elasticity values (E), other spacing of members and other conditions of loading, see U.B.C. Standard No. 25-21.
(4) The spans in these tables are intended for use in covered structures or where moisture content in use does not exceed 19 percent.

TABLE NO. 29-B—ALLOWABLE FOUNDATION AND LATERAL PRESSURE

CLASS OF MATERIALS[2]	ALLOWABLE FOUNDATION PRESSURE LBS./SQ. FT.[3]	LATERAL BEARING LBS./SQ./FT./FT. OF DEPTH BELOW NATURAL GRADE[4]	LATERAL SLIDING[1]	
			COEF-FICIENT[5]	RESISTANCE LBS./SQ. FT.[6]
1. Massive Crystalline Bedrock	4000	1200	.70	
2. Sedimentary and Foliated Rock	2000	400	.35	
3. Sandy Gravel and/or Gravel (GW and GP)	2000	200	.35	
4. Sand, Silty Sand, Clayey Sand, Silty Gravel and Clayey Gravel (SW, SP, SM, SC, GM and GC)	1500	150	.25	
5. Clay, Sandy Clay, Silty Clay and Clayey Silt (CL, ML, MH and CH)	1000[7]	100		130

[1] Lateral bearing and lateral sliding resistance may be combined.

[2] For soil classifications OL, OH and PT (i.e., organic clays and peat), a foundation investigation shall be required.

[3] All values of allowable foundation pressure are for footings having a minimum width of 12 inches and a minimum depth of 12 inches into natural grade. Except as in Footnote No. 7 below, increase of 20 percent allowed for each additional foot of width or depth to a maximum value of three times the designated value.

[4] May be increased the amount of the designated value for each additional foot of depth to a maximum of 15 times the designated value. Isolated poles for uses such as flagpoles or signs and poles used to support buildings which are not adversely affected by a $^1/_2$-inch motion at ground surface due to short-term lateral loads may be designed using lateral bearing values equal to two times the tabulated values.

[5] Coefficient to be multiplied by the dead load.

[6] Lateral sliding resistance value to be multiplied by the contact area. In no case shall the lateral sliding resistance exceed one half the dead load.

[7] No increase for width is allowed.

TABLE NO. 26-E—ALLOWABLE SERVICE LOAD ON EMBEDDED BOLTS
(Pounds)[1,2]

DIAMETER (In inches)	MINIMUM[3] EMBEDMENT (In inches)	MINIMUM CONCRETE STRENGTH (In psi)		
		SHEAR[4]		TENSION[5]
		2000	3000	2000 to 5000
$^1/_4$	$2^1/_2$	500	500	200
$^3/_8$	3	1100	1100	500
$^1/_2$	4	2000	2000	950
$^5/_8$	4	2750	3000	1500
$^3/_4$	5	2940	3560	2250
$^7/_8$	6	3580	4150	3200
1	7	3580	4150	3200
$1^1/_8$	8	3580	4500	3200
$1^1/_4$	9	3580	5300	3200

[1] Values are for natural stone aggregate concrete and bolts of at least A307 quality. Bolts shall have a standard bolt head or an equal deformity in the embedded portion.

[2] Values are based upon a bolt spacing of 12 diameters with a minimum edge distance of 6 diameters. Such spacing and edge distance may be reduced 50 percent with an equal reduction in value. Use linear interpolation for intermediate spacings and edge margins.

[3] An additional 2 inches of embedment shall be provided for anchor bolts located in the top of columns for buildings located in Seismic Zones Nos. 2, 3 and 4.

[4] Values shown are for work with or without special inspection.

[5] Values shown are for work without special inspection. Where special inspection is provided values may be increased 100 percent.

TABLE NO. 47-I—ALLOWABLE SHEAR FOR WIND OR SEISMIC FORCES IN POUNDS PER FOOT FOR VERTICAL DIAPHRAGMS OF LATH AND PLASTER OR GYPSUM BOARD FRAME WALL ASSEMBLIES[1]

TYPE OF MATERIAL	THICKNESS OF MATERIAL	WALL CONSTRUCTION	NAIL SPACING[2] MAXIMUM (In Inches)	SHEAR VALUE	MINIMUM NAIL SIZE[3,4]
1. Expanded metal, or woven wire lath and portland cement plaster	7/8″	Unblocked	6	180	No. 11 gauge, 1 1/2″ long, 7/16″ head No. 16 gauge staple, 7/8″ legs
2. Gypsum lath	3/8″ Lath and 1/2″ Plaster	Unblocked	5	100	No. 13 gauge, 1 1/8″ long, 19/64″ head, plasterboard blued nail
3. Gypsum sheathing board	1/2″ x 2′ x 8′	Unblocked	4	75	No. 11 gauge, 1 3/4″ long, 7/16″ head, diamond-point, galvanized
	1/2″ x 4′	Blocked	4	175	
	1/2″ x 4′	Unblocked	7	100	
4. Gypsum wallboard or veneer base.	1/2″	Unblocked	7	100	5d cooler or wallboard
			4	125	
		Blocked	7	125	
			4	150	
	5/8″	Unblocked	7	115	6d cooler or wallboard
			4	145	
		Blocked	7	145	
			4	175	
		Blocked Two ply	Base ply 9 Face ply 7	250	Base ply—6d cooler or wallboard Face ply—8d cooler or wallboard

[1]These vertical diaphragms shall not be used to resist loads imposed by masonry or concrete construction. See Section 4714 (b). Values are for short-term loading due to wind. Values must be reduced 25 percent for normal loading. The values for gypsum products must be reduced 50 percent for dynamic loading due to earthquake in Seismic Zones Nos. 3 and 4.
[2]Applies to nailing at all studs, top and bottom plates and blocking.
[3]Alternate nails may be used if their dimensions are not less than the specified dimensions.
[4]For properties of cooler or wallboard nails, see U.B.C. Standard No. 25-17, Table No. 25-17–H.

TABLE NO. 28-B — ALLOWABLE FRICTIONAL & BEARING VALUES FOR ROCK[1]

Type	Friction Coefficient	Allowable Lateral Bearing	
		lbs. per sq. ft.	per ft. Max. Value
Massive Crystalline Bedrock	1.0	4,000	20,000
Foliated Rocks	.8	1,600	8,000
Sedimentary Rocks	.6	1,200	6,000
Soft or Broken Bedrocks	.4	400	2.000

ALLOWABLE FRICTIONAL & LATERAL BEARING VALUES FOR SOILS
Frictional Resistance — Gravels and Sands[1]

Soil Type	Friction Coefficient
Gravel, Well Graded	0.6
Gravel, Poorly Graded	0.6
Gravel, Silty	0.5
Gravel, Clayey	0.5
Sand, Well Graded	0.4
Sand, Poorly Graded	0.4
Sand, Silty	0.4
Sand, Clayey	0.4

1. Coefficient to be multiplied by the Dead Load.

ALLOWABLE FRICTIONAL RESISTANCE
(lbs. per sq. ft.) — Clay and Silt[2]

Soil Type	Loose or Soft	Compact or Stiff
Silt, Inorganic	250	500
Silt, Organic	250	500
Silt, Elastic	200	400
Clay, Lean	500	1000
Clay, Fat	200	400
Clay, Organic	150	300
Peat	0	0

2. Frictional values to be multiplied by the width of footing subjected to positive soil pressure. In no case shall the fricitional resistance exceed ½ the dead load on the area under consideration.

ALLOWABLE LATERAL BEARING PER FT. OF DEPTH BELOW NATURAL GROUND SURFACE
(lbs. per sq. ft.) (Natural Soils or approved compacted fill)

Soil Type	Loose or Soft	Compact or Stiff	Max. Values
Gravel, Well Graded	200	400	8000
Gravel, Poorly Graded	200	400	8000
Gravel, Silty	167	333	8000
Gravel, Clayey	167	333	8000
Sand, Well Graded	183	367	6000
Sand, Poorly Graded	77	200	6000
Sand, Silty	100	233	4000
Sand, Clayey	133	300	4000
Silt, Inorganic	67	133	3000
Silt, Organic	33	67	2000
Silt, Elastic	33	67	1500
Clay, Lean	267	867	3000
Clay, Fat	33	167	1500
Clay, Organic	33	------	500
Peat	0	0	0

GENERAL CONDITIONS OF USE

1. Frictional and lateral resistance of soils may be combined, provided the lateral bearing resistance does not exceed ⅔ of allowable lateral bearing.

2. A ⅓ increase in frictional and lateral bearing values will be permitted to resist loads caused by wind pressure or earthquake forces.

3. Isolated poles such as flag poles or signs may be designed using lateral bearing values equal to two times the tabulated values.

4. Lateral bearing values are permitted only when concrete is deposited against natural ground or compacted fill, approved by the Superintendent of Building.

SEC. 91.2311 — POLES

(a) Design. Flag poles, sign poles, columns or other poles canti-levering from and receiving lateral stability from the ground shall have their lateral support designed in accordance with the following formulas or other methods approved by the Superintendent of Building. Bearing stresses so obtained shall not exceed the values permitted by Section 91.2803 (d).

CASE I — POLES WITH LATERAL RESTRAINT AT THE GROUND SURFACE

$$f = \frac{3.8M}{bd^2}$$

Where:

f = lateral soil pressure in lbs/sq. ft.

M = moment at natural ground surface resulting from applied loads in ft. pounds.

b = diameter of round pole or 1.27 times width of rectangular pole, measured in feet.

R = Reaction capable of taking resultant loads.

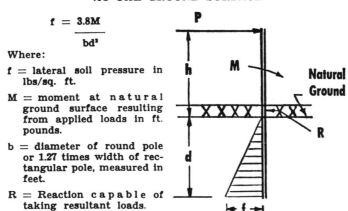

CASE II — POLES WITHOUT LATERAL RESTRAINT AT THE GROUND SURFACE

$$f_1 = \frac{2.85\ P}{bd} + \frac{f_2}{4}$$

$$f_2 = \frac{7.62\ P\ (2h + d)}{bd^2}$$

Where:

f_1 and f_2 = lateral soil pressure in lbs/sq. ft.

b = diameter of round pole or 1.27 times width of rectangular pole, measured in feet.

d = depth of embedment below natural ground in feet (minimum four feet).

h = height of applied lateral load above natural ground measured in feet.

P = lateral force in pounds.

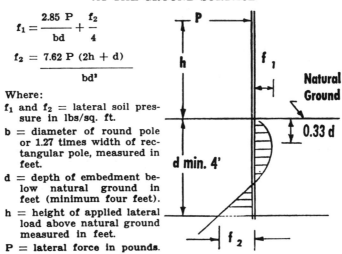

APPENDIX E

Design Data for Masonry

Appendix E contains material that supports various common tasks in the investigation and design of elements of masonry structures. The use of these aids is demonstrated in the example problems in the text.

E.1. WALL REINFORCEMENT

Both vertical and horizontal reinforcement in walls are ordinarily achieved with standard deformed steel bars, ranging typically from size No. 3 to size No. 9. Typical spacings follow modules most common in CMU construction. Table E.1 gives values for the rapid determination of alternative choices once a required amount of steel area has been computed. Areas for required reinforcement are ordinarily computed in units of square inches per foot of wall height or length; this is the unit used for the table entries. Table values are given for typical spacings.

Example.   Determine the alternative choices for reinforcement for a CMU wall with 8-in.-high block with cells at 8 in. on center along the wall length. Required reinforcement is as follows:

Vertical: A_s = 0.130 in.^{2}/ft
Horizontal: A_s = 0.067 in.^{2}/ft

Solution.   Reading from Table E.1, our choices are as follows:

Vertical: No. 3 at 8 (0.165), No. 4 at 16 (0.150), No. 5 at 24 (0.155), No. 6 at 32 (0.165), No. 7 at 48 (0.150)

Horizontal: No. 3 at 16 (0.082), No. 4 at 32 (0.075), No. 5 at 48 (0.077)

Choices respond to the requirement for an 8-in. modular spacing and a code-required maximum spacing of 48 in. Ordinarily, the usual preference is for the widest allowable spacing, permitting the handling of the fewest number of bars for installation.

E.2. FLEXURE IN MASONRY WALLS

Figure E.1 may be used for investigation of bending in masonry walls by the working-stress method. There are three variables in the table, as follows.

TABLE E.1. Average Reinforcement Per Foot in Walls (in^2/ft)[a]

Bar Spacing (in.)	Bar Size						
	No. 3	No. 4	No. 5	No. 6	No. 7	No. 8	No. 9
6	0.220	0.400	0.620	0.880	1.200	1.580	2.000
8	0.165	0.300	0.465	0.660	0.900	1.185	1.500
12	0.110	0.200	0.310	0.440	0.600	0.790	1.000
16	0.082	0.150	0.232	0.330	0.450	0.592	0.750
18	0.073	0.133	0.207	0.293	0.400	0.527	0.667
24	0.055	0.100	0.155	0.220	0.300	0.395	0.500
30	0.044	0.080	0.124	0.176	0.240	0.316	0.400
32	0.041	0.075	0.116	0.165	0.225	0.296	0.375
36	0.037	0.067	0.103	0.147	0.200	0.263	0.333
40	0.033	0.060	0.093	0.132	0.180	0.237	0.300
42	0.031	0.057	0.088	0.126	0.171	0.226	0.286
48	0.027	0.050	0.077	0.110	0.150	0.197	0.250

[a] Table entry = (bar cross section area)(12)/(spacing).

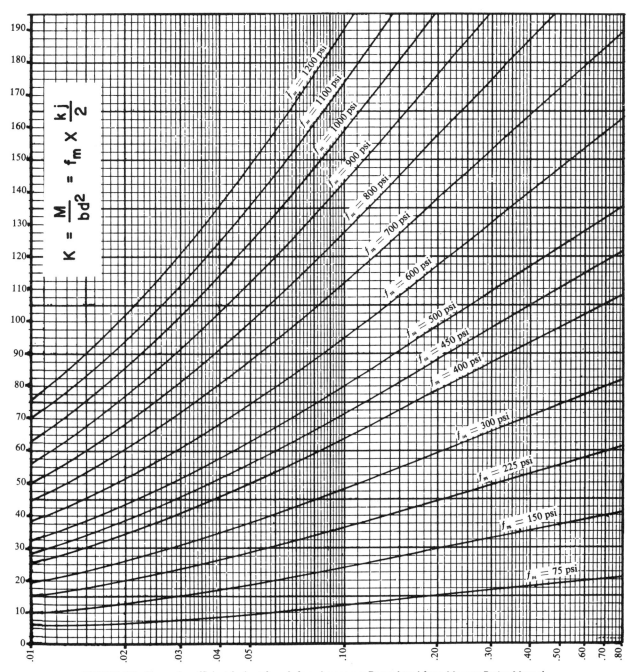

FIGURE E.1. Flexural coefficient k-chart for reinforced masonry. Reproduced from *Masonry Design Manual* (Ref. 7) with permission of the publishers, Masonry Institute of America.

The K factor for bending is plotted vertically on the chart.

The percentage of reinforcement is plotted horizontally on the chart.

The allowable bending stress is represented by the curves on the chart. (Note that the figure uses f_m for the values we give as F_b in this book.)

The chart may be used in a number of ways, a frequent use being the determination of the required reinforcement, as the following example illustrates.

Example 1. A wall of reinforced CMU construction uses 8-in.-nominal units ($t = 7.625$ in.) with $f'_m = 1500$ psi. The wall sustains a wind pressure of 20 psf and spans 16.7 ft vertically. Find the minimum reinforcement, considering only the wind loading. Use steel with $F_y = 40,000$ psi.

Solution. From the data we determine the following:

$$\text{Allowable } F_b = (0.5)(0.33\, f'_m)(1.333)$$
$$= (0.5)(0.33)(1500)(1.333)$$
$$= 333 \text{ psi}$$

(Assume 50% reduction for no inspection; one third increase for wind loading.)

$$\text{Allowable } F_s = 0.5 F_y (1.333)$$
$$= (0.5)(40,000)(1.333)$$
$$= 26,667 \text{ psi}$$

From Table E.2, $n = 40$.

With steel bars in center, $d = t/2 = 3.813$ in.

For the wind load, the wall spans 16.7 ft as a simple span beam. Thus, the maximum bending moment is

$$M = \frac{wL^2}{8} = \frac{(20)(16.7)^2}{8} \times 12$$
$$= 8367 \text{ in.-lb}$$

For the chart

$$K = \frac{M}{bd^2} = \frac{8367}{(12)(3.813)^2} = 48$$

Enter the chart (Fig. E.1) at the left with $K = 48$, proceed to the right to intersect a value of $f_m = 333$ psi, then read at the bottom a value of $np = 0.073$, from which

$$p = \frac{0.073}{n} = \frac{0.073}{40} = 0.001825$$

$$A_s = pbd = (0.001825)(12)(3.813)$$
$$= 0.0835 \text{ in.}^2/\text{ft}$$

From Table E.1, possible choice: No. 5 at 40 in., $A_s = 0.093$ in.^{2}/ft.

Since the example does not deal with the problem of combined stress, we consider the following situation.

TABLE E.2. Balanced Section Properties for Rectangular Masonry Sections with Tension Reinforcement

Reinforcement	f_m (psi)	Modular Ratio $n = E_s/E_m$	$F_b = f_m/3$ (psi)	Balanced Section Properties			
				k	j	K	$p = A_s/bd$
	Without Special Inspection—Code Values Reduced to Half						
Grade 40 $F_y = 40$ ksi	675	44	225	0.333	0.889	33.3	0.00187
	750	40	250	0.333	0.889	37.0	0.00208
	1000	30	333	0.333	0.889	49.3	0.00278
	2000	15	667	0.333	0.889	98.7	0.00556
Grade 60 $F_y = 60$ ksi	675	44	225	0.273	0.909	27.9	0.00128
	750	40	250	0.273	0.909	27.9	0.00142
	1000	30	333	0.273	0.909	41.3	0.00189
	2000	15	667	0.273	0.909	82.7	0.00379
	With Special Inspection—Full Code Values						
Grade 40 $F_y = 40$ ksi	1350	22	450	0.333	0.889	66.6	0.00375
	1500	20	500	0.333	0.889	74.0	0.00416
	2000	15	667	0.333	0.889	89.7	0.00556
	4000	7.5	1333	0.333	0.889	197.0	0.01111
Grade 60 $F_y = 60$ ksi	1350	22	450	0.273	0.909	55.8	0.00256
	1500	20	500	0.273	0.909	62.0	0.00284
	2000	15	667	0.273	0.909	82.7	0.00379
	4000	7.5	1333	0.273	0.909	165.4	0.00758

Example 2. For the wall in Example 1, suppose that the vertical loading combined with the wind loading develops a stress condition such that $f_a/F_a = 0.2$. Then for the combined stress condition.

$$\frac{f_a}{F_a} + \frac{f_b}{F_b} = 1 \quad \text{or} \quad 0.2 + \frac{f_b}{F_b} = 1$$

Thus,

$$\frac{f_b}{F_b} = 0.8$$

or, since stress is proportional to moment, we can say

$$\frac{\text{Loading-induced } M}{\text{Allowable } M} = 0.8$$

Now, if we design for this allowable moment, the stress combination will be acceptable. We therefore determine a design moment as

$$M = \frac{8367}{0.8} = 10{,}459 \text{ in.-lb}, \quad \text{or}$$

$$K = \frac{48}{0.8} = 60$$

Using this K on Fig. E.1, we find $np = 0.155$. Then $p = 0.003875$, $A_s = 0.177$ in.^{2}, and No. 6 at 24 in. (0.220 in.^{2}/ft) is the new required reinforcement.

E.3. STIFFNESS FACTORS FOR MASONRY PIERS

Tables E.3 and E.4 present factors that may be used to determine the relative stiffness of masonry piers. The general use of these factors is for the determination of lateral load distribution to piers of different stiffness that share a single lateral load, as discussed in Sec. 47.2.

DESIGN DATA FOR MASONRY

TABLE E.3. Rigidity Coefficients for Cantilevered Masonry Walls

h/d	Rc	h/d	Rc	h/d	Rc	h/d	Rc	h/d	Rc	h/d	Rc
9.90	.0006	5.20	.0043	1.85	.0810	1.38	.1706	0.91	.4352	0.45	1.4582
9.80	.0007	5.10	.0046	1.84	.0821	1.37	.1737	0.90	.4452	0.44	1.5054
9.70	.0007	5.00	.0049	1.83	.0833	1.36	.1768	0.89	.4554	0.43	1.5547
9.60	.0007	4.90	.0052	1.82	.0845	1.35	.1800	0.88	.4659	0.42	1.6063
9.50	.0007	4.80	.0055	1.81	.0858	1.34	.1832	0.87	.4767	0.41	1.6604
9.40	.0007	4.70	.0058	1.80	.0870	1.33	.1866	0.86	.4899	0.40	1.7170
9.30	.0008	4.60	.0062	1.79	.0883	1.32	.1900	0.85	.4994	0.39	1.7765
9.20	.0008	4.50	.0066	1.78	.0896	1.31	.1935	0.84	.5112	0.38	1.8380
9.10	.0008	4.40	.0071	1.77	.0909	1.30	.1970	0.83	.5233	0.37	1.9098
9.00	.0008	4.30	.0076	1.76	.0923	1.29	.2007	0.82	.5359	0.36	1.9738
8.90	.0009	4.20	.0081	1.75	.0937	1.28	.2044	0.81	.5488	0.35	2.0467
8.80	.0009	4.10	.0087	1.74	.0951	1.27	.2083	0.80	.5621	0.34	2.1237
8.70	.0009	4.00	.0093	1.73	.0965	1.26	.2122	0.79	.5758	0.33	2.2051
8.60	.0010	3.90	.0100	1.72	.0980	1.25	.2162	0.78	.5899	0.32	2.2913
8.50	.0010	3.80	.0108	1.71	.0995	1.24	.2203	0.77	.6044	0.31	2.3828
8.40	.0010	3.70	.0117	1.70	.1010	1.23	.2245	0.76	.6194	0.30	2.4802
8.30	.0011	3.60	.0127	1.69	.1026	1.22	.2289	0.75	.6349	0.29	2.5838
8.20	.0012	3.50	.0137	1.68	.1041	1.21	.2333	0.74	.6509	0.28	2.6945
8.10	.0012	3.40	.0149	1.67	.1058	1.20	.2378	0.73	.6674	0.27	2.8130
8.00	.0012	3.30	.0163	1.66	.1074	1.19	.2425	0.72	.6844	0.26	2.9401
7.90	.0013	3.20	.0178	1.65	.1091	1.18	.2472	0.71	.7019	0.25	3.0769
7.80	.0013	3.10	.0195	1.64	.1108	1.17	.2521	0.70	.7200	0.24	3.2246
7.70	.0014	3.00	.0214	1.63	.1125	1.16	.2571	0.69	.7388	0.23	3.3845
7.60	.0014	2.90	.0235	1.62	.1143	1.15	.2622	0.68	.7581	0.22	3.5583
7.50	.0015	2.80	.0260	1.61	.1162	1.14	.2675	0.67	.7781	0.21	3.7479
7.40	.0015	2.70	.0288	1.60	.1180	1.13	.2729	0.66	.7987	0.20	3.9557
7.30	.0016	2.60	.0320	1.59	.1199	1.12	.2784	0.65	.8201	.195	4.0673
7.20	.0017	2.50	.0357	1.58	.1218	1.11	.2841	0.64	.8422	.190	4.1845
7.10	.0017	2.40	.0400	1.57	.1238	1.10	.2899	0.63	.8650	.185	4.3079
7.00	.0018	2.30	.0450	1.56	.1258	1.09	.2959	0.62	.8886	.180	4.4379
6.90	.0019	2.20	.0508	1.55	.1279	1.08	.3020	0.61	.9131	.175	4.5751
6.80	.0020	2.10	.0577	1.54	.1300	1.07	.3083	0.60	.9384	.170	4.7201
6.70	.0020	2.00	.0658	1.53	.1322	1.06	.3147	0.59	.9647	.165	4.8736
6.60	.0021	1.99	.0667	1.52	.1344	1.05	.3213	0.58	.9919	.160	5.0364
6.50	.0022	1.98	.0676	1.51	.1366	1.04	.3281	0.57	1.0201	.155	5.2095
6.40	.0023	1.97	.0685	1.50	.1389	1.03	.3351	0.56	1.0493	.150	5.3937
6.30	.0025	1.96	.0694	1.49	.1412	1.02	.3422	0.55	1.0797	.145	5.5904
6.20	.0026	1.95	.0704	1.48	.1436	1.01	.3496	0.54	1.1112	.140	5.8008
6.10	.0027	1.94	.0714	1.47	.1461	1.00	.3571	0.53	1.1439	.135	6.0261
6.00	.0028	1.93	.0724	1.46	.1486	0.99	.3649	0.52	1.1779	.130	6.2696
5.90	.0030	1.92	.0734	1.45	.1511	0.98	.3729	0.51	1.2132	.125	6.5306
5.80	.0031	1.91	.0744	1.44	.1537	0.97	.3811	0.50	1.2500	.120	6.8136
5.70	.0033	1.90	.0754	1.43	.1564	0.96	.3895	0.49	1.2883	.115	7.1208
5.60	.0035	1.89	.0765	1.42	.1591	0.95	.3981	0.48	1.3281	.110	7.4555
5.50	.0037	1.88	.0776	1.41	.1619	0.94	.4070	0.47	1.3696	.105	7.8215
5.40	.0039	1.87	.0787	1.40	.1647	0.93	.4162	0.46	1.4130	.100	8.2237
5.30	.0041	1.86	.0798	1.39	.1676	0.92	.4255				

TABLE E.4. Rigidity Coefficients for Fixed Masonry Walls

h/d	R$_f$	h/d	R$_f$	h/d	R$_f$	h/d	R$_f$	h/d	R$_f$	h/d	R$_f$
9.90	.0025	5.20	.0160	1.85	.2104	1.38	.3694	0.91	.7177	0.45	1.736
9.80	.0026	5.10	.0169	1.84	.2128	1.37	.3742	0.90	.7291	0.44	1.779
9.70	.0027	5.00	.0179	1.83	.2152	1.36	.3790	0.89	.7407	0.43	1.825
9.60	.0027	4.90	.0189	1.82	.2176	1.35	.3840	0.88	.7527	0.42	1.874
9.50	.0028	4.80	.0200	1.81	.2201	1.34	.3890	0.87	.7649	0.41	1.924
9.40	.0029	4.70	.0212	1.80	.2226	1.33	.3942	0.86	.7773	0.40	1.978
9.30	.0030	4.60	.0225	1.79	.2251	1.32	.3994	0.85	.7901	0.39	2.034
9.20	.0031	4.50	.0239	1.78	.2277	1.31	.4047	0.84	.8031	0.38	2.092
9.10	.0032	4.40	.0254	1.77	.2303	1.30	.4100	0.83	.8165	0.37	2.154
9.00	.0033	4.30	.0271	1.76	.2330	1.29	.4155	0.82	.8302	0.36	2.219
8.90	.0034	4.20	.0288	1.75	.2356	1.28	.4211	0.81	.8442	0.35	2.287
8.80	.0035	4.10	.0308	1.74	.2384	1.27	.4267	0.80	.8585	0.34	2.360
8.70	.0037	4.00	.0329	1.73	.2411	1.26	.4324	0.79	0.873	0.33	2.437
8.60	.0038	3.90	.0352	1.72	.2439	1.25	.4384	0.78	0.888	0.32	2.518
8.50	.0039	3.80	.0377	1.71	.2468	1.24	.4443	0.77	0.904	0.31	2.605
8.40	.0040	3.70	.0405	1.70	.2497	1.23	.4504	0.76	0.920	0.30	2.697
8.30	.0042	3.60	.0435	1.69	.2526	1.22	.4566	0.75	0.936	0.29	2.795
8.20	.0043	3.50	.0468	1.68	.2556	1.21	.4628	0.74	0.952	0.28	2.900
8.10	.0045	3.40	.0505	1.67	.2586	1.20	.4692	0.73	0.969	0.27	3.013
8.00	.0047	3.30	.0545	1.66	.2617	1.19	.4757	0.72	0.987	0.26	3.135
7.90	.0048	3.20	.0590	1.65	.2648	1.18	.4823	0.71	1.005	0.25	3.265
7.80	.0050	3.10	.0640	1.64	.2679	1.17	.4891	0.70	1.023	0.24	3.407
7.70	.0052	3.00	.0694	1.63	.2711	1.16	.4959	0.69	1.042	0.23	3.560
7.60	.0054	2.90	.0756	1.62	.2744	1.15	.5029	0.68	1.062	0.22	3.728
7.50	.0056	2.80	.0824	1.61	.2777	1.14	.5100	0.67	1.082	0.21	3.911
7.40	.0058	2.70	.0900	1.60	.2811	1.13	.5173	0.66	1.103	0.20	4.112
7.30	.0061	2.60	.0985	1.59	.2844	1.12	.5247	0.65	1.124	.195	4.220
7.20	.0063	2.50	.1081	1.58	.2879	1.11	.5322	0.64	1.146	.190	4.334
7.10	.0065	2.40	.1189	1.57	.2914	1.10	.5398	0.63	1.168	.185	4.454
7.00	.0069	2.30	.1311	1.56	.2949	1.09	.5476	0.62	1.191	.180	4.580
6.90	.0072	2.20	.1449	1.55	.2985	1.08	.5556	0.61	1.216	.175	4.714
6.80	.0075	2.10	.1607	1.54	.3022	1.07	.5637	0.60	1.240	.170	4.855
6.70	.0078	2.00	.1786	1.53	.3059	1.06	.5719	0.59	1.266	.165	5.005
6.60	.0081	1.99	.1805	1.52	.3097	1.05	.5804	0.58	1.292	.160	5.164
6.50	.0085	1.98	.1824	1.51	.3136	1.04	.5889	0.57	1.319	.155	5.334
6.40	.0089	1.97	.1844	1.50	.3175	1.03	.5977	0.56	1.347	.150	5.514
6.30	.0093	1.96	.1864	1.49	.3214	1.02	.6066	0.55	1.376	.145	5.707
6.20	.0097	1.95	.1885	1.48	.3245	1.01	.6157	0.54	1.407	.140	5.914
6.10	.0102	1.94	.1905	1.47	.3295	1.00	.6250	0.53	1.438	.135	6.136
6.00	.0107	1.93	.1926	1.46	.3337	0.99	.6344	0.52	1.470	.130	6.374
5.90	.0112	1.92	.1947	1.45	.3379	0.98	.6441	0.51	1.504	.125	6.632
5.80	.0118	1.91	.1969	1.44	.3422	0.97	.6540	0.50	1.539	.120	6.911
5.70	.0124	1.90	.1991	1.43	.3465	0.96	.6641	0.49	1.575	.115	7.215
5.60	.0130	1.89	.2013	1.42	.3510	0.95	.6743	0.48	1.612	.110	7.545
5.50	.0137	1.88	.2035	1.41	.3555	0.94	.6848	0.47	1.651	.105	7.908
5.40	.0144	1.87	.2058	1.40	.3600	0.93	.6955	0.46	1.692	.100	8.306
5.30	.0152	1.86	.2081	1.39	.3647	0.92	.7065				

Bibliography

The following list contains materials that have been used as references in the development of various portions of the text. Also included are some widely used publications that serve as general references for building design, although no direct use of materials from them has been made in this book. The numbering system is random and merely serves to simplify referencing by text notation.

1. *Uniform Building Code*, 1991 ed., International Conference of Building Officials, 5360 South Workmanmill Road, Whittier, CA 90601. (Called simply the *UBC*.)

2. *American National Standard Minimum Design Loads for Buildings and Other Structures*, American National Standards Institute, 1430 Broadway, New York, NY 10018, 1982.

3. *The BOCA Basic National Building Code/1984*, 9th ed., Building Officials and Code Administrators International, Inc., 4051 Flossmoor Road, Country Club Hills, IL 60477-5795, 1984 (with 1985 supplement). (Called simply the *BOCA Code*.)

4. *City of Los Angeles Building Code*, 1976 ed., Building News, Inc., 3055 Overland Avenue, Los Angeles, CA 90034.

5. *Manual of Steel Construction*, 8th ed., American Institute of Steel Construction, Chicago, IL, 1980. (Called simply the *AISC Manual*.)

6. *Building Code Requirements for Reinforced Concrete*, ACI 318-89, American Concrete Institute, Detroit, MI, 1983. (Called simply the *ACI Code*.)

7. *National Design Specification for Wood Construction*, National Forest Products Association, Washington D. C., 1982.

8. *Timber Construction Manual*, 3rd ed., American Institute of Timber Construction, Wiley, New York, 1985.

9. *CRSI Handbook*, 4th ed., Concrete Reinforcing Institute, Schaumburg, IL, 1982.

10. *Masonry Design Manual*, 3rd ed., Masonry Institute of America, 2550 Beverly Boulevard, Los Angeles, CA 90057, 1979.

11. *Standard Specifications, Load Tables, and Weight Tables for Steel Joists and Joist Girders*, Steel Joist Institute, Suite A, 1205 48th Avenue North, Myrtle Beach, SC 29577.

12. *Steel Deck Institute Design Manual for Composite Decks, Form Decks, and Roof Decks*, Steel Deck Institute, P. O. Box 3812, St. Louis, MO 63122.

13. Charles G. Ramsey and Harold R. Sleeper, *Architectural Graphic Standards*, 8th ed. Wiley, New York, 1988.

14. James Ambrose, *Simplified Design of Building Foundations*, 2nd ed., Wiley, New York, 1988.

15. James Ambrose and Dimitry Vergun, *Simplified Building Design for Wind and Earthquake Forces*, 2nd ed., Wiley, New York, 1990.

16. Jack C. McCormac, *Structural Analysis*, 4th ed., Harper & Row, New York, 1984.

17. S. W. Crawley and R. M. Dillon, *Steel Buildings: Analysis and Design*, 3rd ed., Wiley, New York, 1984.

18. Donald E. Breyer, *Design of Wood Structures*, McGraw-Hill, New York, 1980.

19. Phil M. Ferguson, *Reinforced Concrete Fundamentals*, 4th ed., Wiley, New York, 1979.

20. R. B. Peck, W. E. Hanson, and T. H. Thornburn, *Foundation Engineering*, 2nd ed., Wiley, New York, 1974.

21. Joseph E. Bowles, *Foundation Analysis and Design*, 3rd ed., McGraw-Hill, New York, 1982.

22. R. R. Schneider and W. L. Dickey, *Reinforced Masonry Design*, 3rd ed., Prentice-Hall, Englewood Cliffs, NJ, 1987.

23. Edward Allen, *Fundamentals of Building Construction: Materials and Methods*, 2nd ed., Wiley, New York, 1990.

24. Donald Watson, *Construction Materials and Processes*, 3rd ed., McGraw-Hill, New York, 1986.

25. James M. Fitch, *American Building 2: The Environmental Forces That Shape It*, 2nd ed., rev., Houghton Mifflin, Boston, 1972.

26. Richard D. Rush, *The Building Systems Integration Handbook*, Wiley, New York, 1986.

27. Herman Sands, *Wall Systems: Analysis by Detail*, McGraw-Hill, New York, 1986.

Glossary

The material presented in this glossary constitutes a brief dictionary of words and terms frequently encountered in discussions of the design of building structures. Many of the words and terms have reasonably well-established meanings; in those cases we have tried to be consistent with the accepted usage. In some cases, however, words and terms are given different meanings by different authors or by groups that work in different fields. In these situations we have given the definition as used for this book so that the reader may be clear as to our meaning.

In some cases words or terms are commonly misused with regard to their precise meaning. We have generally used such words and terms as they are broadly understood, but we have given both the correct and popular definitions in some cases.

Words and terms are often used differently in different topic areas. Where such is the case, we have given all the uses, identified by the topic area.

To be clear in its requirements, a legal document such as a building code often defines some words and terms. Reference should be made to such definitions in interpreting such a document.

For a fuller explanation of some of the words and terms given here, as well as definitions not given here, the reader should use the index to find the related discussion in the text.

Abutment. Originally, the end support of an arch or vault. Now, any support that receives both vertical and lateral loading.

Acceleration. The rate of change of the velocity, expressed as the first derivative of the velocity (dv/dt) or as the second derivative of the displacement (d^2s/dt^2). Acceleration of the ground surface is more significant than its displacement during an earthquake because it relates more directly to the force effect. $F = ma$ as a dynamic force.

Active Lateral Pressure. See *Lateral pressure*.

Adequate. Just enough; sufficient. Indicates a quality of bracketed acceptability—on the one hand, not insufficient, on the other hand, not superlative or excessive.

Adobe. Masonry construction that utilizes unburned (fired) clay units.

Aerodynamic. Fluid-flow effects of air, similar to current effects in running water.

Allowable Stress. See *Stress*.

Amplitude. See *Vibration*.

Analysis. Separation into constituent parts. In engineering, the investigative determination of the detail aspects of a particular phenomenon. May be qualitative, meaning a general evaluation of the nature of the phenomenon, or quantitative, meaning the numerical determination of the magnitude of the phenomenon. See also *Synthesis*.

Anchorage. Refers to attachment for resistance to movement; usually a result of uplift, overturn, sliding, or horizontal separation. Tiedown, or holddown, refers to anchorage against uplift or overturn. Positive anchorage generally refers to direct fastening that does not easily loosen.

Angle of Internal Friction (Φ). Property that indicates shear strength in a cohesionless soil.

Aseismic. The correct word for description of resistance to seismic effects. Building design is actually *aseismic* design, although the term *seismic design* is more commonly used.

Assemblage. Something put together from parts. A random, unordered assemblage is called a gathering. An ordered assemblage is called a *system*.

Atterberg Limits. A number of properties and relationships that relate to the identification of cohesive soils. The major tested values are the water contents indicated by the following:

Liquid limit (w_L) is the limit for water content above which the soil displays the character of a liquid with the solid particles carried in suspension.

Plastic limit (w_p) is the water content at the boundary between the physical states of plastic (easily moldable) and solid (not moldable without fracture).

Shrinkage limit (w_s) is the water content at which the soil mass attains its least volume upon drying out.

The numeric difference between the liquid and plastic limits is called the *plasticity index* (I_p) and the values for the liquid limit and the plasticity index are displayed on the *plasticity chart*, which is used to classify the soil as a clay or silt. Soils with high values for both liquid limit and plasticity index are called *fat*; those with low values for both are called *lean*.

Backfill. See *Fill*.

Base Shear. See the *UBC* definition of *base* in Sec. 2312(b).

Basement. Enclosed space within a building that is partly or wholly below the level of the ground surface.

Battering. Describes the effect that occurs when two elements in separate motion bump into each other repeatedly, such as two adjacent parts of a structure during an earthquake. Also called *hammering* or *pounding*.

Beam. A structural element that sustains transverse loading and develops internal forces of bending and shear in resisting loads. Also called a *girder* if large scale; a *joist* if small scale or closely spaced in sets; a *rafter* if used for a roof.

Bearing Foundation. Foundation that transfers loads to soil by direct vertical contact pressure. Usually refers to a *shallow bearing foundation*, which is a foundation that is placed directly beneath the lowest part of the building. See also *Footing*.

Bedrock. The level at which a large rock mass is considered to be solid, stable, and strong. Is typically overlain by masses that are fractured, weathered, or otherwise less stable and strong.

Bending. Turning action that causes change in the curvature of a linear element. Characterized by the development of opposed internal stresses of compression and tension. See also *Moment*.

Bent. A planar framework, or some portion of one, that is designed for resistance to both vertical and horizontal forces in the plane of the frame.

Box System. A structural system in which lateral loads are not resisted by a vertical load-bearing space frame but rather by shear walls or a braced frame.

Braced Frame. Literally, any framework braced against lateral forces. Codes use the term for a frame braced by trussing (triangulation).

Bracing. In structural design usually refers to the resistance to movements caused by lateral forces or by the effects of buckling, torsional rotation, sliding, and so on.

Brittle Fracture. Sudden ultimate failure in tension or shear. The basic structural behavior of so-called brittle materials.

Buckling. Collapse, in the form of sudden sideways deflection, of a slender element subjected to compression.

Buffeting. Wind effect caused by turbulent air flow or by changes in the wind direction that result in whipping, rocking, and so on.

Caisson. See *Pier*.

Calculation. Ordered, rational determination, usually by mathematical methods.

Cavity Wall. A wall built of two or more wythes of masonry units so arranged as to provide a continuous airspace within the wall. The facing and backing, outer wythes, are tied together with noncorrosive ties (e.g., brick or wire).

Centroid. The geometric center of an object, usually analogous to the center of gravity. The point at which the entire mass of the object may be considered to be concentrated when considering moment of the mass.

Clay. General name for soil of very fine grain size and highly cohesive nature. Change in water content dramatically affects consistency, causing change from soft (easily remoldable) to hard (rock-like). *Fat* clays are those with high values for both liquid limit and plasticity index; *lean* clays are those with low values. Major structural property is unconfined compressive strength (q_u).

CMU. Concrete masonry unit; or, good old concrete blocks.

Cohesionless. General lack of cohesiveness; noncohesive. The typical character of clean sands and gravels.

Cohesive. General character of a soil in which the soil particles adhere to each other to produce a nondisintegrating mass. The typical character of fine-grained soils: silts and clays.

Collector. A force transfer element that functions to collect loads from a horizontal diaphragm and distribute them to the vertical elements of the lateral resistive system.

Column. A linear compression member. See also *Pier*.

Compaction. Action that tends to lower the void ratio and increase the density of a soil mass. When produced by artificial means, the degree of compaction obtained is measured in percent with reference to the theoretical minimum volume of the soil.

Compressibllity. The relative resistance of a soil mass to volume change upon being subjected to compressive stress.

Compression. Force that tends to crush adjacent particles of a material together and causes overall shortening of objects in the direction of its action.

Connection. The union or joining of two or more distinct elements. In a structure, the connection itself often becomes an entity. Thus actions of the parts on each other may be visualized in terms of their actions on the connection.

Consistency. The property of a cohesive soil that generally describes its physical state, ranging from soft to hard.

Consolidation. Volume reduction in a soil mass produced by a lowering of the void ratio. The effect resulting from compaction, shrinkage, and so on.

Continuity. Most often used to describe structures or parts of structures that have behavior characteristics influenced by the monolithic, continuous nature of adjacent elements, such as continuous, vertical multistory columns, continuous, multispan beams, and rigid frames.

Core Bracing. Vertical elements of a lateral-bracing system developed at the location of permanent interior walls for stairs, elevators, duct shafts, or rest rooms.

Crawl Space. Space between the underside of the floor construction and the ground surface that occurs when a framed floor is held above the ground but there is no basement.

Creep. Plastic deformation that proceeds with time when certain materials, such as concrete and lead, are subjected to constant, long-duration stress.

Critical Damping. The amount of damping that will result in a return from initial deformation to the neutral position without reversal.

Curb. An edging strip, often occurring at the edge of a pavement; sometimes effects a small change in ground surface elevations on opposite sides of the curb.

Curtain Wall. An exterior building wall that is supported entirely by the frame of the building rather than being self-supporting or load bearing.

Cut. Removal of existing soil deposits during the recontouring (or grading) of the ground surface. See also *Grading*.

Damping. See *Vibration*.

Dead Load. See *Load*.

Deep Foundation. Foundation used to achieve a considerable extension of the bearing effect of a supported structure below the ground surface. Elements most commonly used are *piles* or *piers*.

Deflection. Generally refers to the lateral movement of a structure caused by loads, such as the vertical sag of a beam, the bowing of a surface under wind pressure, or the lateral sway of a tower. See also *Drift*.

Density. See *Mass*.

Design. The conception, contrivance, or planning of a work (verb). The descriptive image (picture, model, etc.) of a proposed work (noun). See also *Synthesis*.

Determinate. Having defined limits; definite. In structures, the condition of having the exact sufficiency of stability externally and internally, therefore being determinable by the resolution of forces alone. An excess of stability conditions produces a structure characterized as indeterminate.

Diaphragm. A surface element (deck, wall, etc.) used to resist forces in its own plane by spanning or cantilevering. See also *Horizontal diaphragm* and *Shear wall*.

Displacement. Movement away from some fixed reference point. Motion is described mathematically as a displacement—time function. See also *Acceleration and Velocity*.

Drag. Generally refers to wind effects on surfaces parallel to the wind direction. Ground drag refers to the effect of the ground surface in slowing the wind velocity near ground level.

Drag Strut. A structural member used to transfer lateral load across the building and into some part of the vertical system. See also *Collector*.

Drift. Lateral movement of one level of a structure with respect to another; may refer to story-by-story movement or the total deflection at the top with respect to the ground.

Dual Bracing System. Combination of a moment-resisting space frame and shear walls or braced frames, with the combined systems designed to share the lateral loads.

Ductile. Describes the load—strain behavior that results from the plastic yielding of materials or connections. To be significant, the plastic strain prior to failure should be considerably more than the elastic strain up to the point of plastic yield.

Dynamic. Usually used to characterize load effects or structural behaviors that are nonstatic in nature. That is, they involve time-related considerations such as vibrations, energy effects versus simple force, and so on.

Earthquake. The common term used to describe sensible ground movements, usually caused by subterranean faults or explosions. The point on the ground surface immediately above the subterranean shock is called the *epicenter*. The magnitude of the energy released at the location of the shock is the basis for the rating of the shock on the *Richter scale*.

Eccentric Braced Flame. Braced frame in which the bracing members do not connect to the joints of the beam-and-column frame, thus resulting in axial force in the braces, but bending and shear in the frame members. Forms include: knee-brace, K-brace, chevron brace (two forms—V-brace and inverted V-brace).

Economy. Thrift; conservation.

Elastic. Used to describe two aspects of stress–strain behavior. The first is a constant stress–strain proportionality, or constant modulus of elasticity, as represented by a straight line form of the stress–strain graph. The second is the limit within which all the strain is recoverable; that is, there is no permanent deformation. The latter phenomenon may occur even though the stress–strain relationship is nonlinear.

Element. A component or constituent part of a whole. Usually a distinct, separate entity.

End-Bearing Pile. See *Pile*.

Energy. Capacity for doing work; what is used up when work is done. Occurs in various forms; mechanical, heat, chemical, electrical, and so on.

Epicenter. See *Earthquake*.

Equalized Settlement. The design method in which a related set of foundations is designed for equal settlement under dead load, rather than for a uniform bearing pressure under total load.

Equilibrium. A balanced state or condition, usually used to describe a situation in which opposed effects neutralize each other to produce a net effect of zero.

Equivalent Static Force Analysis. The technique by which a dynamic effect is translated into a hypothetical (equivalent) static effect that produces a similar result.

Erosion. Progressive removal of a soil mass due to water, wind, or other effects.

Essential Facilities. Building code term for a building that should remain functional after a disaster such as a windstorm or major earthquake; affects establishment of the *I* factor for base shear or design wind pressure.

Excavation. Removal of soil mass to permit construction.

Expansive Soil. Soil that has a tendency to increase in volume because of a change in its water content.

Factored Load. A percentage of the actual service load (usually an increase) used for design by strength methods. See also *Load*.

Failure. The condition of becoming uncapable of a particular function May have partial as well as total connotations in structures. For example, a single connection may fail, but the structure might not collapse because of its ability to redistribute the load.

Fat Clay. See *Clay*.

Fatigue. A structural failure that occurs as the result of a load applied and removed (or reversed) repeatedly through a large number of cycles.

Fault. The subterranean effect that produces an earthquake. Usually a slippage, cracking, sudden strain release, and so on.

Feasible. Capable of being or likely to be accomplished.

Fill. Usually refers to a soil deposit produced by other than natural effects. *Backfill* is soil deposited in the excessive part of an excavation after completion of the construction.

Fines. The portion of the solid particles of a soil sample that have a grain size smaller than the No. 200 sieve (0.003 in. or 0.075 mm).

Fit. As a condition: well matched, adapted, suited, correct, or the right size. Not in conflict.

Flexible. See *Stiffness*.

Flocculent Soil Structure. Soil structure with a high void due to the bonding of soil particles into numerous, small cavelike cells. May be stable and strong under static loading conditions when dry, but will often experience major volume reduction when subjected to saturation or shock.

Flutter. Flapping, vibration type of movement of an object in high wind. Essentially a resonant behavior. See also *Vibration*.

Footing. A shallow, bearing-type foundation element consisting typically of concrete that is poured directly into an excavation.

Force. An effort that tends to change the shape or the state of motion of an object.

Form (Shape). In structures two ideas of form are important. First, the overall form of a structure, such as the profile of an arch. Second, the form of the parts, such as the cross section of the arch rib.

Foundation. The clement, or system of elements, that affects the transition between a supported structure and the ground.

Fracture. A break, usually resulting in actual separation of the material. A characteristic result of tension failure.

Freedom. In structures, this term usually refers to the lack of some type of resistance or constraint. In static analysis, the connections between members and the supports of the structure are qualified as to type, or degree, of freedom. Thus the terms *fixed support*, *pinned support*, and *sliding support* are used to qualify the types of movement resisted. In dynamic analysis the degree of freedom is an important factor in determining the dynamic response of a structure.

Freestanding Wall. See *Wall*.

Frequency. In harmonic motion (bouncing springs, vibrating strings, swinging pendulums, etc.), the number of complete cycles of motion per unit of time. See also *Vibration*.

Friction. Resistance to sliding developed at the contact face between two surfaces.

Friction Pile. See *Pile*.

Frost. The action, or the result, of the freezing of water; usually as condensed water vapor on a surface, or as moisture within a porous material (such as the ground).

Frost Heave. See *Heave*.

Frostline. The level to which freezing of water extends below the ground surface.

Function. Duty; intended use; capability.

Fundamental Period. See *Vibration*.

Gable Roof. Double-sloping roof formed by joined rafters or rigid frames with a ridge or peak at the top. A *gable* is the upper triangular portion of a wall at the end of the roof.

Grade. 1. The level of the ground surface. 2. Rated quality of material used for wood and steel.

Grade Beam. A horizontal element in a foundation system that serves some spanning or load-distributing function.

Grading. Has two usages. 1. Refers to the gradation of size of the solid particles in a sample of soil, which is qualified as *uniform* (major grouping within a limited size range), *well graded* (wide range of sizes with no gaps), or *gap graded* (wide range of sizes with an absence of particles of some sizes). Both uniform and gap-graded soils are considered *poorly graded*. 2. Refers to the general activity of recontouring the ground surface.

Grain. 1. A discrete particle of the material that constitutes a loose material, such as soil. 2. The fibrous orientation of wood.

Gravel. Soil particles consisting of rocks and rock fragments of a size range from $\frac{3}{16}$ to 3 in. in diameter.

Grout. Lean concrete (predominantly water, cement, and sand) used as a filler in the voids of hollow masonry units, under column base plates, and so on.

Grouted Masonry. Masonry of hollow units in which the voids are filled with grout.

Gust. An increase, or surge, of short duration in the wind velocity.

Hammering. See *Battering*.

Header. A horizontal element over an opening in a wall or at the edge of an opening in a roof or floor.

Heave. Upward swelling of the ground surface.

Hertz. Cycles per second.

Hold-Down. See *Anchorage*.

Horizontal Diaphragm. See *Diaphragm*. Usually a roof or floor deck used as part of the lateral bracing system.

Impact. Action of striking or hitting.

Impulse. An impelling force action, characterized by rapid acceleration or deceleration.

Indeterminate. See *Determinate*.

Inelastic. See *Stress–strain behavior*.

Inertia. See *Mass*.

Integration. The bringing into association of distinct but related elements of a system. Thus the plumbing, wiring, ventilating ducts, elevators, and stairs must be integrated with the structure in the building whole.

Interrupted Shear Wall. A shear wall that is not continuous to its foundation.

Intuition. Direct perception, independent of any conscious reasoning process.

Joist. See *Beam*.

Kern Limit. Limiting dimension for the eccentricity of a compression force if tension stress is to be avoided.

Key. A slot or protrusion developed to resist shear, as in the manner of a tongue-and-groove joint.

Lateral. Literally means to the side or from the side. Often used in reference to something that is perpendicular to a major axis or direction. With reference to the vertical direction of the gravity forces, wind, earthquakes, and horizontally directed soil pressures are called *lateral effects*.

Lateral Pressure. Horizontal soil pressure of two kinds: 1. *Active* lateral pressure is that exerted by a retained soil upon the retaining structure. 2. *Passive* lateral pressure is that exerted by soil against an object that is attempting to move in a horizontal direction.

Lean Clay. See *Clay*.

Let-In Bracing. Diagonal boards nailed to studs to provide trussed bracing in the wall plane. In order not to interfere with the surfacing materials of the wall, they are usually notched in, or let in, to the stud faces.

Lintel. A beam placed over an opening in a wall.

Liquid Limit. See *Atterberg limits*.

Live Load. See *Load*.

Load. The active force (or combination of forces) exerted on a structure. *Dead load* is permanent load due to gravity, which includes the weight of the structure itself. *Live load* is any load component that is not permanent, including those due to wind, seismic effects, temperature change, or shrinkage; but the term is most often used for gravity loads that are not permanent. *Service load* is the total load combination that the structure is expected to experience in use. *Factored load* is the service load multiplied by some increase factor for use in strength design.

Macro. Implies upper limits of scale; large, excessive. See also *Micro*.

Masonry Unit. A brick, stone, concrete block, glass block, or hollow clay tile intended to be laid in mortar.

Mass. The dynamic property of an object that causes it to resist changes in its state of motion. This resistance is called *inertia*. The magnitude of the mass per unit volume of the object is called its *density*. Dynamic force is defined by $F = ma$, or force equals mass times acceleration. Weight is defined as the force produced by the acceleration of gravity; thus $W = mg$.

Mat Foundation. A very large bearing-type foundation. When the entire bottom of a building is constituted as a single mat, it is also called a *raft foundation*.

Maximum Density. The theoretical density of a soil mass achieved when the void is reduced to the minimum possible.

Member. One of the distinct elements of an assemblage.

Micro. Implies lower limit of scale. Precise meaning: "very small." See also *Macro*.

Moment. Action tending to produce turning or rotation. Product of a force times a lever arm; gives a unit of force times distance (e.g., foot-pounds). Bending moment causes curvature, torsional moment causes twisting.

Moment of Inertia. The second moment of an area about a fixed line in the plane of the area. A purely mathematical property, not subject to direct physical measurement. Has physical significance in that it can be quantified for any geometric shape and is a measurement of certain structural responses.

Moment-Resisting Space Frame. A vertical load-bearing framework in which members are capable of resisting forces primarily by flexure (*UBC* definition). See also *Rigid Frame*.

Motion. The process of changing position or location. Motion along a line is called *translation*; motion of turning is called *rotation*. The time rate of motion is called *velocity* or *speed*. The time rate of change of velocity is called *acceleration*.

Natural Period. See *Period*.

Noncohesive. See *Cohesionless*.

Normal. 1. The ordinary, usual, unmodified state of something. 2. Perpendicular, such as pressure normal to a surface, stress normal to a cross section, and so on.

Occupancy. In building code language, refers to the use of a building as a residence, school, office, and so on.

Occupancy Importance Factor (I). *UBC* term used in the basic equation for seismic force. Accounts for possible increased concern for certain occupancies.

Optimal. Best; most satisfying. The best solution to a set of criteria is the optimal one. When the criteria have opposed values, there may be no single optimal solution, except by the superiority of a single criterion (e.g., the lightest, the strongest, the cheapest, and so on).

Organic. Refers to material of biological (plant or animal) origin.

Overturn. The toppling, or tipping over, effect of lateral loads.

Parapet. The extension of a wall plane or the roof edge facing above the roof level.

Particle. A minute part. In structures, usually a very small piece of material, slightly bigger than molecular size.

Passive Soil Pressure. See *Lateral pressure*.

P-delta Effect. Secondary effect on members of a frame, induced by the vertical loads acting on the laterally displaced frame.

Pedestal. A short pier or upright compression member. Is actually a short column with a ratio of unsupported height to least lateral dimension of 3 or less.

Penetration Resistance. A standard, field-tested property with significance in cohesionless soils, consisting of the number of blows required to drive a standard sampling device into the soil mass. Generally serves as a direct index of the density and the compressibility of the soil.

Period (of Vibration). The total elapsed time for one full cycle of vibration. For an elastic structure in simple, single-mode vibration, the period is a constant (called the *natural* or *fundamental period*) and is independent of the magnitude of the amplitude, of the number of cycles, and of most damping or resonance effects. See also *Vibration*.

Peripheral Bracing. Vertical elements of a lateral bracing system located at the building perimeter.

Permeability. A measurement of the rate at which water will seep into, or drain out of, a soil mass.

Pier. 1. A short, stocky column with height not greater than three times its least lateral dimension. The *UBC* defines a masonry wall as a pier if its plan length is less than three times the wall thickness. 2. A deep foundation element that is placed in an excavation rather than being driven as a pile. Although it actually refers to a particular method of excavation, the term *caisson* is also commonly used to describe a pier foundation.

Pilaster. An integral portion of the wall that projects on one or both sides and acts as a vertical beam, a column, an architectural feature, or any combination thereof.

Pile. A deep foundation element, consisting of a linear, shaftlike member that is placed by being driven dynamically into the ground. *Friction piles* develop resistance to both downward load and upward (pullout) load through friction between the soil and the surface of the pile shaft. *End-bearing piles* are driven so that their ends are seated in low-lying strata of rock or very hard soil.

Plastic. 1. Usually, a synthetic material of organic origin, including many types of resins, polymers, cellulose derivatives, and casein materials. 2. In structural investigation, the type of stress response that occurs in ductile behavior, usually resulting in considerable, permanent deformation.

Plastic Clay. See *Clay.*

Plastic Hinge. Region where the ultimate moment strength of a ductile member may be developed and maintained with corresponding significant rotation as a result of the local yielding of the material.

Plastic Limit. See *Atterberg limits.*

Plasticity Index. See *Atterberg limits.*

Polar Moment of Inertia. The second moment of an area about a line that is perpendicular to the plane of the area. Of significance in investigation of response to torsion.

Porosity. The percentage of void in a soil mass.

Positive Anchorage. See *Anchorage.*

Pounding. See *Battering.*

Preconsolidation. The condition of a highly compressed soil, usually referring to a condition produced by the weight of soil above on some lower soil strata. May also refer to a condition produced by other than natural causes, such as piling up of soil on the ground surface, vibration, or saturating to dissolve soil bonding.

Pressure. Force distributed over, and normal to, a surface.

Presumptive Soil Pressure. A value for the allowable vertical bearing pressure that is permitted to be used in the absence of extensive investigation and testing. Requires a minimum of soil identification and is usually quite conservative.

Principal Axes. The set of mutually perpendicular axes through the centroid of an area, about which the moments of inertia will be maximum and minimum. Called individually the *major axis* and the *minor axis.*

Radius of Gyration. A defined mathematical property; the square root of the moment of inertia divided by the area. Significant in investigation of the buckling of slender elements subjected to compression.

Raft Foundation. See *Mat foundation.*

Rational. Allowing the application of reasoning.

Reasonable; Sensible. A rational analysis is one that proceeds without recourse to intuition or unwarranted assumptions.

Reaction. Response. In structures, the response of the structure to the loads; the response of the supports to the actions of the structure. *The reactions* usually refers to the components of force developed at the supports.

Reentrant Corner. An exterior corner in a building plan having a form such as that at the junction of the web and flange of a T.

Regular Structure. Structure (building) having no significant discontinuities in plan or in vertical configuration or in its lateral force-resisting systems such as those described for irregular structures. *UBC* definition: Sec. 2312(d)5B.

Reinforce. To strengthen, usually by adding something.

Relative Stiffness. See *Stiffness.*

Resilience. The measurement of the absorption of energy by a structure without permanent deformation or fracture. See *Toughness.*

Resonance. See *Vibration.*

Restoring Moment. The resistance to overturn due to the weight of the affected object.

Retaining Wall. A structure used to brace a vertical cut, or a change in elevation of the ground surface. The term is usually used to refer to a *cantilever retaining wall*, which is a freestanding structure consisting only of a wall and its footing, although basement walls also serve a retaining function.

Richter Scale. A log-based measuring system for evaluation of the relative energy level of an earthquake at its center of origin.

Rigid Bent. See *Rigid frame.*

Rigid Frame. Framed structure in which the joints between the members are made to transmit moments between the ends of the members. Called a *bent* when the frame is planar.

Rigidity. Quality of resistance to deformation. Structures that are not rigid are called *flexible.*

Risk. The degree of probability of loss due to some potential hazard. The risk of an earthquake in a particular location is the basis for the *Z* factor in the *UBC* equation for seismic base shear.

Rock. See *Soil.*

Rotation. See *Motion.*

Runoff. Flowing surface water, usually from rain or melting snow. A major source of surface erosion.

Safety. Relative unlikelihood of failure, absence of danger. The *safety factor* is the ratio of the resisting capacity of a structure to the actual demand on the structure.

Saturation. The condition that exists when the void in the soil is completely filled with water. The *degree of saturation* (S_r) is the expression of the ratio of the volume of water in the soil to the volume of the void, determined in percent. A condition of *partial saturation* exists when the void is only partly filled with water. *Oversaturation*, or *super-saturation*, occurs when the soil contains water in excess of the normal volume of the void, which usually results in some flotation or suspension of the solid particles in the soil.

Scale. A reference of dimensional comparison. A model

may be a scaled reproduction (true in shape and detail) of an object; large scale—as a model of a molecule—or small scale—as a model of a building.

Section. The two-dimensional profile or area obtained by passing a plane through a form. *Cross section* implies a section at right angles to another section or to the linear axis of an object.

Seismic. Pertaining to ground shock. See also *Aseismic*.

Separation. Often used in structural investigation to denote situations in which parts of a structure are made to act independently. Partial separation refers to a selective, controlled situation that allows for some types of interaction while permitting independence of motion in other ways, for example—a connection that provides for vertical support but permits rotation and horizontal movements.

Service Load. See *Load*.

Settlement. The general downward movement of a foundation caused by the loads and the reactions of the supporting soil materials.

Shallow Bearing Foundation. See *Bearing foundation*.

Shear. A force effect that is lateral (perpendicular) to the major axis of a structure, or one that involves a slipping effect, as opposed to a push-pull effect. Wind and earthquake forces are sometimes visualized as shear effects on a building because they are perpendicular to the major vertical (gravity) axis of the building.

Shear Wall. A vertical diaphragm.

Shoring. Bracing; usually refers to the bracing of a cut, a shoreline, or some structure that is in danger from erosion, undercutting, unequal settlements, and so on.

Shrinkage Limit. See *Atterberg limits*.

Sign. Algebraic notation of sense: positive, negative, or neutral. Relates to direction of force—if up is positive, down is negative, to stress—if tension is positive, compression is negative.

Silt. Fine-grained soil in the general size range between sand and clay. Typically, possesses properties that are transitional between those of the purely cohesionless material (sand) and the purely cohesive material (clay).

Simulation. Act of pretending, feigning, or impersonating. In structural investigation, refers to the artificial representation of a structural behavior or loading condition by substitution of another condition.

Site Coefficient (S). A *UBC* term used in the basic equations for base shear to account for the effect of the period of the ground mass under the building.

Slab. A horizontal, planar element of concrete. Occurs as a roof or floor deck in a framed structure. *Slab on grade* refers to a pavement poured directly on the ground surface.

Slenderness. Relative thinness. In structures, the quality of flexibility or lack of buckling resistance is inferred by extreme slenderness.

Soft Story. A story level in a multistory structure in which the lateral stiffness of the story is significantly less than that of stories above. See *UBC* definition in Table 23-M.

Soil. In foundation and soil work a general distinction is made between soil and rock. *Soil* is any material that can be reduced to discrete particles or to a semifluid mass by the actions of water and minor agitation. *Rock* is any material that is generally not affected in its structural properties by variation in water content and that offers considerable resistance to excavation or to separation into discrete particles.

Space Frame. An ambiguous term used variously to describe three-dimensional structures. The *UBC* has a particular definition used in classifying structures for response to seismic effects.

Specific Gravity. A relative indication of density, using the weight of water as a reference. A specific gravity of 2 indicates a density (weight) of twice that of water.

Spectrum. In seismic analysis generally refers to the curve that describes the actual dynamic force effect on a structure as a function of variation in its fundamental period. Response spectra are the family of curves produced by various degrees of damping. This represents the basis for determining the building's general response to seismic force.

Stability. Refers to the inherent capability of a structure to develop force resistance as a property of its form, orientation, articulation of its parts, type of connections, methods of support, and so on. Is not directly related to quantified strength or stiffness, except when the actions involve the buckling of elements of the structure.

Static. The state that occurs when the velocity is zero; thus no motion is occurring. Is generally used to refer to situations in which no change is occurring.

Stiffness. In structures, refers to resistance to deformation, as opposed to strength, which refers to resistance to force. A lack of stiffness indicates a flexible structure. Relative stiffness usually refers to the comparative deformation of two or more structural elements that share a load.

Story Drift. The total lateral displacement occurring in a single story of a multistory structure.

Strain. Deformation resulting from stress. Is usually measured as a percentage of deformation, called *unit strain* or *unit deformation*, and is thus dimensionless.

Strata. Plural of stratum. Levels or layers. Used to describe a soil mass of singular character, typically existing in a layer between other soil masses of different character.

Stratum. See *Strata*.

Strength. Capacity to resist force.

Strength Design. One of the two fundamental design techniques for assuring a margin of safety for a structure. *Stress design*, also called *working stress design*, is performed by analyzing stresses produced by the estimated actual usage loads, and assigning limits for the stresses that are below the ultimate capacity of the materials by some margin. *Strength design*, also called *ultimate strength design*, is performed by multiplying the actual loads by the desired factor of safety (the universal average factor being two) and proceeding to design a structure that will have that load as its ultimate failure load.

Stress. The mechanism of force within the material of a structure; visualized as a pressure effect (tension or compression) or a shear effect on the surface of a unit of the material and quantified in units of force per unit area. *Allowable*, *permissible*, or *working* stress refers to a stress limit that is used in stress design methods. *Ultimate* stress refers to the maximum stress that is developed just prior to failure of the material.

Stress Design. See *Strength design*.

Stress–Strain Behavior. The relation of stress to strain in a material or a structure. Is usually visually represented by a stress–strain graph covering the range from no load to failure. Various aspects of the form of the graph define particular behavioral properties. A straight line indicates an elastic relationship; a curve indicates inelastic behavior. A sudden bend in the graph usually indicates a plastic strain or yield that results in some permanent deformation. The slope of the graph is defined as the modulus of elasticity of the material.

Structure. That which gives form to something and works to resist changes in the form from the action of various forces.

Stud. One of a set of small, closely spaced columns used to produce a framed wall.

Subsidence. Settlement of a soil mass, usually manifested by the sinking of the ground surface.

Surcharge. Vertical load applied at the ground surface or simply above the level of the bottom of a footing. The weight of soil above the bottom of a footing is surcharge for that footing.

Synthesis. The process of combining a set of components into a whole; opposite of analysis.

System. A set of interrelated elements; an ordered assemblage; an organized procedure or method.

Tension. Force action that tends to separate adjacent particles of a material or pull attached elements apart. Produces straightening effects and elongation.

Three-Hinged Structure. An arch, gabled, rafter or rigid-frame structure with two hinged end supports and a third, interior, pin at its peak. Forces on supports due to thermal change are avoided and the structure is statically determinate.

Tiedown. See *Anchorage*.

Torsion. Moment effect involving twisting or rotation that is in a plane perpendicular to the major axis of an element. Lateral loads produce torsion on a building when they tend to twist it about its vertical axis. This occurs when the centroid of the load does not coincide with the center of stiffness of the vertical elements of the lateral load-resisting structural system.

Toughness. The measurement of the total dynamic energy capacity of a structure, up to the point of complete failure. See also *Resilience*.

Translation. Motion of a body along a line, without rotation or turning.

Truss. A framework of linear elements that achieves stability through triangular formations of the elements.

Ultimate Strength. Usually used to refer to the maximum static force resistance of a structure at the time of failure. This limit is the basis for the so-called strength design methods, as compared to the stress design methods that use some established stress limit, called the design stress, working stress, permissible stress, and so on.

Unconfined Compressive Strength. See *Clay*.

Underpinning. Propping up of a structure that is in danger of, or has already experienced, some support failure; notably that due to excessive settlement, undermining, erosion, and so on.

Upheaval. Pushing upward of a soil mass.

Uplift. Net upward force effect; may be due to wind, to overturning moment, or to upward seismic acceleration.

Vector. A mathematical quantity having direction as well as magnitude and sign (sense: + or −). Comparison is made to *scalar* quantities having only magnitude and sense, such as time and temperature. A vector may be graphically represented by an arrow with its length proportional to the magnitude, the angle of its line indicating direction, and the arrowhead representing the sign. See also *Motion*.

Velocity. The time rate of a motion, also commonly called *speed*.

Veneer. A masonry facing that is attached to the backup but not so bonded as to act with it under load, as opposed to Faced Wall.

Vertical Diaphragm. See *Diaphragm*. Also called a *shear wall*.

Vibration. The cyclic, rhythmic motion of a body such as a spring. Occurs when the body is displaced from some neutral position and seeks to restore itself to a state of equilibrium when released. In its pure form it occurs as a harmonic motion with a characteristic behavior described by the cosine form of the displacement–time graph of the motion. The magnitude of linear displacement from the neutral position is called the *amplitude*. The time elapsed for one full cycle of motion is called the *period*. The number of cycles occurring in one second is called the *frequency*. Effects that tend to reduce the amplitude of succeeding cycles are called *damping*. The increase of amplitude in successive cycles is called a *resonant effect*.

Viscosity. The general measurement of the mobility, or free-flowing character, of a fluid or semifluid mass. Heavy oil is viscous (has high viscosity); water has low viscosity.

Visualize. To create a mental image; to make perceptible to the mind.

Void. The portion of the volume of a soil that is not occupied by the solid materials; it is ordinarily filled partly with air and partly with liquid (water, oil, etc.).

Void ratio. The term most commonly used to indicate the amount of void in a soil mass, expressed as the ratio of the volume of the void to the volume of the solids. See also *Porosity*.

Wall. A vertical, planar building element. *Foundation* walls are those that are partly or totally below ground. *Bearing* walls are used to carry vertical loads in direct compression. *Grade* walls are those that are used to achieve the transition between the building that is above the ground and the foundations that are below it; grade is used to refer to the level of the ground surface at the edge of the building. (See also *Grade beam*.) *Shear* walls are those used to brace the building against horizontal forces due to wind or seismic shock. *Freestanding* walls are walls whose tops are not laterally braced. *Retaining* walls are walls that resist horizontal soil pressure.

Weak Story. A story level in a multistory structure in which the lateral strength is significantly less than that of stories above. See *UBC* definition in Table 23-M.

Wind Stagnation Pressure. The reference pressure established by the basic wind speed for the region; used in determining design wind pressures.

Working Stress. See *Stress*.

Working Stress Design. See *Strength design*.

Yield. See *Stress–strain behavior*.

Zone. Usually refers to a bounded area on a surface, such as the ground surface or the plan of a level of a building.

Exercise Problems

The following materials are provided for readers who use this book for individual study or for use as class assignments when the book is used as a text. Answers to problems that involve numerical computation, or otherwise have a single answer, are given following the problems. However, the reader is advised to attempt to work the problems without referring to the answers except for a check at the end of the work. This procedure will serve better to test the reader's understanding of the work.

Problems are grouped by the chapter or part to which the work corresponds. Numbering is given for reference purposes only. In many cases the units used are arbitrary, as the relationships and procedures are more significant; in these cases SI values have been omitted for simplicity.

PART ONE

1. For each of the basic structural materials (wood, steel, concrete, and masonry) list both limitations and advantages in their uses for building structures.

2. For each of the limitations listed in Problem 1, describe what measures (if any) can be taken to overcome them.

3. Find a building that is just beginning to be constructed. Visit the site periodically and photograph the progress of the construction. Photograph the building from the same point on successive visits. Organize the photos or slides obtained into a report on the growth of the building structure.

4. Find a local building that has been built recently. Contact as many people who were involved in the design and construction as possible. Interview them and write a case study report on the design of the structure for the building.

5. *The Make and Break:* This is a classic assignment involving the actual construction of a structure to perform a specific assigned task. The following is an example; endless variations are possible: Design and build a structure to span 4 ft on a simple, horizontal span and to carry a concentrated load at the center of the span. End support is limited to vertical reactions only. Materials for the structure are limited to wood and paper, although any material may be used for connection of parts. The efficiency of the structure on a strength—weight basis is critical. Your structure will be weighed, load tested to destruction, and your score determined using the graph shown in the illustration.

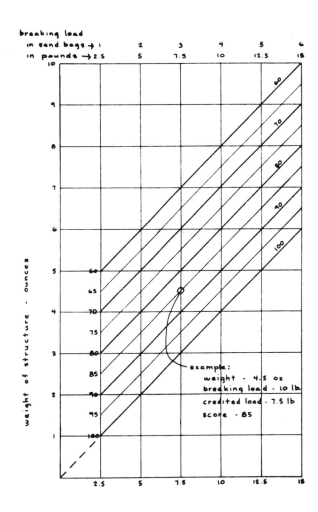

Demonstration Projects: For Assignment or for Classroom Demonstration

6. *Involvement.* Buckling of a linear element as related to slenderness.
   Procedure. Select a slender linear element (strip of wood, plastic, metal) and find its total compression resistance for various increments of length. Start with a ratio of length to thickness of at least 200 for the longest specimen.
   Find. Relation of load capacity to length (or to length-to-thickness ratio.)

7. *Involvement.* Bending resistance related to shape.
   Procedure. Test the bending resistance of a linear element on a single span when subjected to a load at the center of the span. Test elements of the same

material and same total cross-sectional area, but with different shapes and different orientations to the load. Both bending strength (load capacity) and stiffness (deflection) may be tested.
Find. Correlation of bending resistance and shape in beams.

8. *Involvement.* Bending and span.
Procedure. Test a linear element for bending as in Problem 7. Test various specimens of the same material and cross section, but of increasing span length.
Find. Relation of load capacity (and/or deflection) to span length.

9. *Involvement.* Bending resistance and support restraint.
Procedure. Test the bending resistance of a linear element as in Problem 7. Test the same element on the same span, but with three different conditions at the supports, as follows: (a) both ends free to turn, (b) one end clamped to prevent turning, (c) both ends clamped.
Find. Effect of end restraint on load capacity (and/or deflection).

10. *Involvement.* One-way versus two-way spanning. Effect of ratio of span lengths in rectangular two-way spans.
Procedure. Test a thin planar element in bending (a sheet of cardboard, glass, plastic, metal) with a single load at the center of the span. Test specimens with the following shapes and support conditions at the edges: (a) square sheet, two opposite edges supported; (b) square sheet, three edges supported; (c) square sheet, four edges supported; (d) rectangular sheet with the small dimension the same as in (c) and the long dimension a multiple of the short of increasing magnitude in successive specimens: 1.25, 1.5, 1.75, 2.0, 2.5, and so on.
Find. Comparison of effectiveness of one-way and two-way spanning. Relation of panel dimension ratio to effectiveness of two-way action in rectangular panels.

11. *Involvement.* Torsion resistance of various cross-sectional shapes.
Procedure. Test various linear elements of the same material and the same total cross-sectional area, but with different shapes. Fix one end and twist the other end without causing bending. Measure the twisting force for total load capacity or for some constant increment of rotation.
Find. Effectiveness of various cross-sectional shapes in torsion.

12. *Involvement.* Sag ratio of cables.
Procedure. Test a tension element (string, wire, chain) for its total load resistance with various ratios of sag to span. Test by supporting the two ends and loading at the center.
Find. Relation of sag ratio (sag to span) and load capacity.

13. *Involvement.* Rise-to-span ratio in arches.
Procedure. Test a flexible sheet (cardboard, plastic, aluminum) for its resistance to load in arch action. Attach two blocks to a base and bend the sheet to form an arch, kicking against the blocks. Load with a weight at the center of the arch, carefully avoiding any concentrated load effect at the point of load. Test specimens with various span-to-rise ratios.
Find. Mode of failure and effectiveness of arch for various ratios of rise to span.

PART TWO

Chapter 7

14. Using both algebraic and graphic techniques, find the horizontal and vertical components for the force. (a) $F = 100\,lb, \theta = 45°$; (b) $F = 200\,lb, \theta = 30°$; (c) $F = 200\,lb, \theta = 27°$; (d) $F = 327\,lb, \theta = 40°$.

15. Using both algebraic and graphic techniques, find the resultant (magnitude and direction) for the following force combinations: (a) $H = 100\,lb, V = 100\,lb$; (b) $H = 50\,lb, V = 100\,lb$; (c) $H = 43\,lb, V = 61\,lb$; (d) $H = 127\,lb, V = 47\,lb$.

16. Using both algebraic and graphic techniques, find the resultant (magnitude and direction) for the following force combinations: (a) $F_1 = 100\,lb, F_2 = 100\,lb, F_3 = 100\,lb, \theta = 45°$; (b) $F_1 = 100\,lb, F_2 = 200\,lb, F_3 = 150\,lb, \theta = 30°$; (c) $F_1 = 47\,lb, F_2 = 63\,lb, F_3 = 112\,lb, \theta = 40°$; (d) $F_1 = 58\,lb, F_2 = 37\,lb, F_3 = 81\,lb, \theta = 28°$.

17. Using both algebraic and graphic techniques, find the tension in the rope. (a) $x = 0$; (b) $x = 10\,ft$; (c) $x = 5\,ft$; (d) $x = 4\,ft\ 2\,in$.

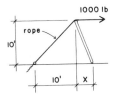

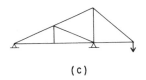

18. Using a Maxwell diagram, find the internal forces in the members of the truss.

21. Find the reactions for the beams. (a) $x = 6$ ft; (b) $x = 4$ ft 7 in.; (c) $x = 5$ ft; (d) $x = 7$ ft 2 in.

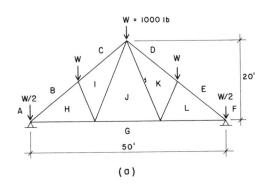

(a)

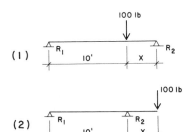

22. Find the magnitude and direction of both reactions. (a) $x = 10$ ft, $H_1 = H_2$; (b) $x = 13$ ft, $H_1 = H_2$; (c) $x = 10$ ft, $H_2 = 0$; (d) $x = 13$ ft, $H_2 = 0$.

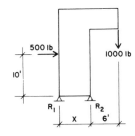

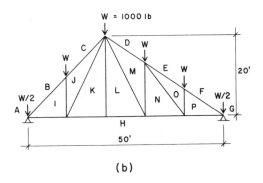

(b)

19. Using the algebraic method of joints, find the internal forces in the members of the truss in Problem 18. Draw the complete separated joint diagram, showing all of the horizontal and vertical components as well as the actual forces in the members.

20. Determine the sense (tension, compression, or zero) of the internal force in each of the members of the truss.

23. Find the resultant of the three forces. Establish the direction of the resultant by finding the coordinates of its intersection with the x-z plane.

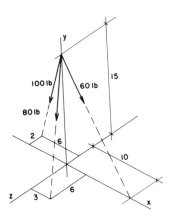

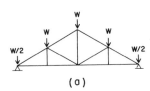

(a)

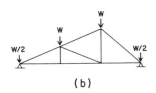

(b)

24. Find the compression force in the struts and the tension force in the wire.

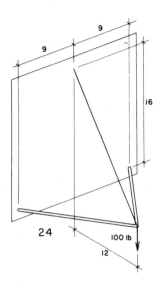

25. Find the tension force in each of the wires.

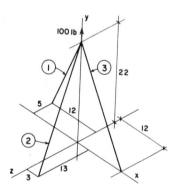

26. Find the resultant and its location with respect to the x- and z-axes.

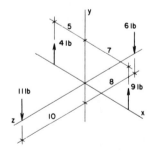

27. Find the tension force in each of the three wires.

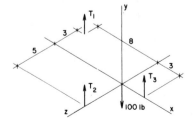

Chapter 8

28. What axial load may be placed on a short timber post whose actual cross-sectional dimensions are 9.5 × 9.5 in. [241.3 mm] if the allowable unit compressive stress in the wood is 1100 psi [7585 kPa]?

29. The allowable bearing capacity of a soil is 8000 psf [383 kPa]. What should be the length of the side of a square footing if the total load (including the weight of the footing) is 240 kips [1067.5 kN]?

30. Determine the minimal cross-sectional area of a steel bar required to support a tensile force of 50 kips [222.4 kN] if the allowable tensile stress is 20 ksi [137.9 MPa].

31. A short square timber post supports a load of 115 kips [511.5 kN]. If the allowable unit compressive stress is 1000 psi [6895 kPa], what nominal-size square timber should be used? (see Table A.2).

32. A joint similar to that shown in Fig. 8.3a is to be used to connect a concrete slab to a supporting concrete wall. What is the minimum dimension for the width of the tongue in the slab if the maximum unit stress in shear is 60 psi [414 kPa] and the load is 2500 lb [11.12 kN] per running foot of the joint?

33. A steel bolt with a diameter of 0.75 in. [19 mm] is to be used in a joint similar to that in Fig. 8.3b. What is the maximum load on the joint if the allowable shear stress in the bolt is 14 ksi [96.5 MPa]?

34. A wood beam has a nominal cross section of 8 × 12 in. If the beam is subjected to a bending moment about its strong axis and the maximum allowable bending stress is 1400 psi [9.65 MPa], what is the maximum bending moment permitted? (See Table A.2 for properties of wood sections.)

35. A solid cylindrical shaft of aluminum is limited to a shear stress in torsion of 8000 psi [55 MPa] and a rotation at the end of 0.5°. Find the maximum allowable torsional moment. (a) Diameter is 3 in. [75 mm], length is 20 in. [500 mm]; (b) diameter is 1.5 in. [38 mm], length is 40 in. [1 m].

36. A hollow cylindrical shaft has an outside diameter of 3 in. [75 mm] and is subjected to a torsional moment of 1600 ft-lb [2170 kN-m]. Find the shear stresses on the inside and outside if the wall thickness is (a) 0.25 in. [6 mm]; (b) 0.625 in. [16 mm].

37. How much will a nominal 8 × 8 Douglas fir post shorten under an axial compression load of 45 kips [200 kN] if the post is 12ft [3.66 m] long? Use E = 1500 ksi (10.34 GPa] for the wood.

38. What force must be applied to a steel bar 1 in. [25 mm] square and 2 ft [610 mm] long to produce an elongation of 0.016 in. [0.4064 mm]?

Chapter 9

39. For each of the beams shown, draw acceptable shear, moment, and deflection diagrams, indicating relationships between the diagrams.

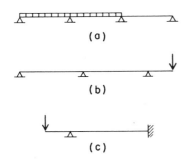

(a)

(b)

(c)

40. For each of the beams shown: find the reactions; draw complete shear and moment diagrams indicating all critical values; sketch the deflected shape; indicate relationships between the diagrams.

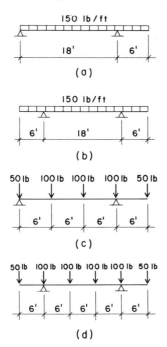

150 lb/ft

18'   6'

(a)

150 lb/ft

6'   18'   6'

(b)

50 lb   100 lb   100 lb   100 lb   50 lb

6'   6'   6'   6'

(c)

50 lb 100 lb 100 lb 100 lb 100 lb 50 lb

6'   6'   6'   6'   6'

(d)

Chapter 10

41. Find the maximum tension in the cable if $T = 10$ kips. (a) $x = y = 10$ ft, $s = t = 4$ ft; (b) $x = 12$ ft, $y = 16$ ft, $s = 8$ ft, $t = 12$ ft.

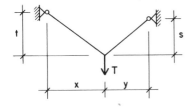

42. A cable is subjected to a load of 1 kip/ft, distributed uniformly with respect to the horizontal span of 120 ft. Find the maximum tension force in the cable at the support if the sag is 16 ft.

43. Select a bridge strand for the cable in Problem 45 for a safety factor of 2. Find the actual length of the cable. What is the magnitude of stretch in the cable due to the applied load?

Chapter 11

44. The compression force at the bottom of a square footing is 40 kips [178 kN] and the bending moment is 30 kip-ft [40.7 kN-m]. Find the maximum soil pressure for footing widths of: (a) 5 ft [1.5 m]; (b) 4 ft [1.2m].

Chapter 12

45. A concrete structure 200 ft long is subjected to a temperature change of $140°$F. Assume a value of $E = 4000$ ksi for the concrete and find (a) the total length change if movement is not prevented; (b) the stress in the concrete if movement is prevented.

46. A 16-in.-diameter round concrete column with $E = 6000$ ksi is reinforced with six 1-in.-diameter round steel rods and sustains an axial compression force of 200 kips. Find the stresses in the steel and the concrete.

47. A 10-in.-diameter round column has an axial compression force of 20 kips. Find the direct compressive stress and the shear stress on a section that is $60°$ from the axis of the column.

48. A roof framing member similar to that shown in Fig. 12.15 consists of a nominal 6×8 with $S_x = 51.6$ in.^{3} and $S_y = 37.8$ in.^{3} and sustains a bending moment of 4 kip-ft in a vertical plane. If the roof slope is $22°$, find the maximum bending stress and the net stress at all four corners of the section.

49. A 16×24-in. column sustains a compression force of 100 kips that is 3 in. eccentric from both major centroidal axes of the section. Find the stress at the four corners of the section and the location of the neutral axis.

50. For a rectangular beam cross section with width equal b and depth equal d, derive an expression for the maximum shear stress in beam action.

51. A beam has an I-shaped cross section with an overall depth of 16 in., web thickness of 2 in., and flanges that are 8 in. wide and 3 in. thick. Compute the critical shear stresses and plot the distribution of shear on the cross section if the beam sustains a shear force of 20 kips.

52. Find the angle at which the block shown in Fig. 12.23 will slip if the coefficient of friction is 0.35.

53. For the block shown, find the value of *P* required to keep the block from slipping if $\phi = 10°$ and $W = 10$ lb.

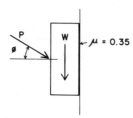

54. For the block shown, find the weight for the block that will result in slipping if $\phi = 15°$ and $P = 10$ lb.

55. For the wall and footing shown, make separate determinations of the possibilities for slipping and overturn. $H = 1000$ lb and $W = 3000$ lb.

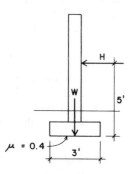

Chapter 13

56. Find the internal forces in the members of the truss (1) using a Maxwell Diagram; (2) using the algebraic beam analogy method.

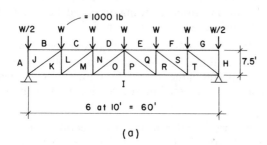

(a)

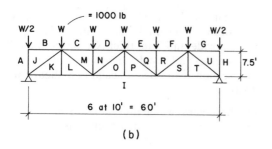

(b)

Chapter 14

57. For the frames shown, find the components of the reactions, draw the free-body diagrams of the whole frame and the individual members, draw the shear and moment diagrams for the members, and sketch the deformed shape of the loaded structure.

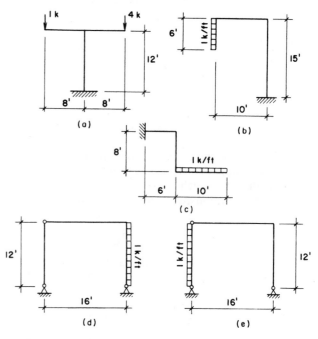

58. Find the components of the reactions for the pinned structure.

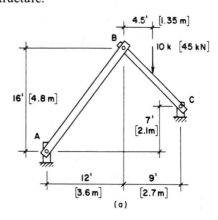

(a)

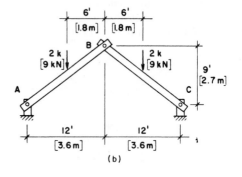

(b)

59. For the beams shown, find the reactions and draw the shear and moment diagrams, indicating all critical values. Sketch the deflected shape and determine the locations of any inflection points not related to the internal pins.

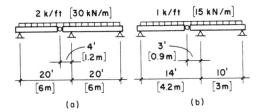

(a)   (b)

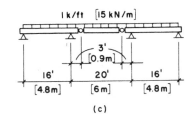

(c)

60. Investigate a three-hinged arch with a span of 80 ft, a rise of 30 ft, and a radius of 41.67 ft for a uniformly distributed loading of 1.2 kips/ft. Use the procedures described in Sec. 16.2.

PART THREE

Chapter 16

61. Determine the design values for bending, shear, and axial tension for the following structural members of Douglas fir–larch: (a) a 10 × 16 beam of dense No. 1 grade; (b) a 2 × 12 joist of No. 1 grade; (c) an 8 × 8 post of No. 1 grade.

Chapter 17

62. For the following beams of Douglas fir–larch No. 1 grade, find the nominal section with the least area. Consider bending, shear, and a maximum deflection of $\frac{1}{240}$ of the span. (a) Span = 20 ft, applied load is 1.2 kips/ft; (b) span is 30 ft, applied load is 0.4 kip/ft; (c) span is 12 ft, applied load is 0.6 kip/ft plus 20 kips at midspan.

63. Using the tables from the *UBC* (Appendix D), select Douglas fir–larch floor joists for the following: (a) No. 1 grade joists, 20-ft span; (b) No. 2 grade joists, 14-ft span; (c) No. 3 grade joists, 12-ft span. Live load is 40 psf, dead load is 10 psf, and deflection is limited top $\frac{1}{360}$ of the span under live load.

64. Using the tables from the *UBC* (Appendix D), select Douglas fir–larch rafters for the following: (a) a flat roof, live load is 20 psf, dead load is 15 psf, live-load deflection is limited to $\frac{1}{240}$ of the span, rafter span is 18 ft, wood is No. 1 grade; (b) same as part (a), except span is 25 ft; (c) 4-in-12 slope, live load is 30 psf, dead load is 7 psf, live-load deflection is limited to $\frac{1}{180}$ of the span, rafter span is 16 ft, wood is No. 2 grade; (d) same as part (c), except span is 25 ft.

Chapter 18

65. For a wood compression member of Douglas fir–larch No. 1 grade, find the allowable axial compression load for (a) a 6 × 8 for lengths of 4, 10, and 18 ft; (b) an 8 × 10 for lengths of 6, 12, and 22 ft.

66. A spaced column of the form shown in Fig 21.5 consists of two 3 × 8 members of Douglas fir–larch, select structural grade. As shown in the figure, L_1 is 12 ft and x is 6 in. Find the axial compression capacity.

Chapter 19

67. A truss joint is made similar to that shown in Fig. 22.3 with 3-in.-nominal lumber and $\frac{5}{8}$-in. plywood. The bottom member is attached with 16 8d nails on each side. Find the limit for the tension force in the member, based on the nail capacity.

68. A three-member tension joint has 2 × 12 outer members and a 4 × 12 middle member (Fig. 22.10*a*). The joint is made with six $\frac{3}{4}$-in. bolts in two rows. Find the capacity of the joint as limited by the bolts and the tension stresses in the members of No. 1 grade Douglas fir–larch.

69. A two-member tension joint consists of 2 × 6 members bolted with two $\frac{3}{4}$-in. bolts in a single row (Fig. 22.10*a*). What is the limit for the tension force based on the bolts?

70. Two outer members, each 2 × 8, are bolted to a middle member consisting of a 3 × 12. The outer members form an angle of 45 with respect to the middle member (Fig. 22.10*c*, $\theta = 45°$). Based on the bolts, what is the maximum compression force that the outer members can transfer to the joint if the connection is made with two $\frac{7}{8}$-in. bolts? Wood is Douglas fir–larch, dense No. 1 grade.

Chapter 20

71. A flitched beam consists of a single 10 × 14 of Douglas fir, select structural grade, and two A36 steel plates, each 0.5 by 13.5 in. [13 × 334 mm]. Compute the magnitude of the single concentrated load this beam will support at the center of a 16-ft [4.8-m] simple span. Neglect the beam weight. Use a value of 22 ksi [152 MPa] for the allowable bending stress in the steel.

PART FOUR

Chapter 22

72. An A36 steel beam is to be used to carry a uniformly distributed load of 30 kips [133 kN] on a simple span of 30 ft [9.15 m]. Ignoring the beam weight and considering only the condition of a maximum allowable bending stress of 24 ksi [165 MPa], select the lightest wide-flange section.

73. Same data as Problem 72, except the lateral unsupported length is 10 ft [3.05 m].

74. Same data as Problem 73, except the maximum deflection is limited to 1.0 in. [25 mm].

75. For each of the beams selected in Problems 72 through 74, find the maximum shear capacity.

76. Compute the maximum allowable reaction with respect to web crippling for a W 14×30 of A36 steel with an end bearing plate length of 8 in. [200 mm].

77. A column load of 81 kips [360 kN] is supported on top of the beam in Problem 79 with a column bearing plate length of 10 in. [250 mm]. Is the beam adequate or are web stiffeners required?

78. A W 14×30 beam with a reaction of 20 kips [89 kN] rests on a brick wall with brick of $f'_m = 1500$ psi and type S mortar. The beam has a bearing length of 8 in. [203 mm] parallel to the length of the beam. If the bearing plate is of A36 steel, determine its dimensions. (*Note*: $k_1 = 0.625$ in.)

79. A wall of brick with f'_m of 1500 psi and type S mortar supports a W 18×55 of A36 steel. The beam reaction is 25 kips [111 kN], and the bearing length along the beam is 9 in. [229 mm] (dimension N). Design the bearing plate of A36 steel. (*Note*: $k_1 = 0.8125$ in.)

80. Using data from Table 22.4, select the lightest steel deck for the following: (a) simple span of 7 ft, total load of 45 psf; (b) two-span condition, span of 8.5 ft, total load of 45 psf; (c) three-span condition, span of 6 ft, total load of 50 psf.

Chapter 23 (*Note*: **Assume all columns unbraced on both axes, $K = 1.0$.**)

81. For the following columns of A36 steel, find the maximum allowable compression load: (a) a W 14×120 with unbraced length of 17 ft 4 in.; (b) a W 12×58 with unbraced length of 11 ft 3 in.; (c) a W 8×31 with unbraced length of 10 ft 6 in.

82. Using the tables in Chapter 23, select the lightest columns for the following: (a) wide-flange shape, unbraced length of 16 ft, load of 250 kips; (b) wide-flange shape, unbraced length of 20 ft, load of 30 kips; (c) steel pipe, unbraced length of 14 ft, load of 80 kips; (d) square tube, unbraced length of 14 ft, load of 80 kips, minimum wall thickness of $\frac{3}{8}$ in.; (e) pair of angles, unbraced length of 10 ft, load of 45 kips.

83. Using the bending factor method, select wide-flange shapes of A36 steel for the following. Assume a K of 1.0 and an unbraced length of 14 ft. (a) Axial compression of 100 kips, bending moment on x-axis of 20 kip-ft; (b) axial compression of 40 kips, bending moment on x-axis of 60 kip-ft; (c) axial compression of 300 kips, bending moments of 100 kip-ft on the x-axis and 40 kip-ft on the y-axis.

84. Design a column base plate of A36 steel for a W 8×31 column that is supported on concrete for which the allowable bearing pressure is 750 psi. Column load is 178 kips.

Chapter 24

85. A bolted connection of the general form shown in Fig. 24.9 is required to transfer a tension force of 200 kips [890 kN], using $\frac{7}{8}$-A490N bolts and plates of A36 steel. The outer plates are 8 in. [200 mm] wide, and the center plate is 12 in. [300 mm] wide. Find the required thicknesses of the plates and the number of bolts needed if the bolts are placed in two rows. Sketch the bolt layout with the necessary dimensions.

86. Design a connection for the data in Problem 85, except that the bolts are $\frac{1}{2}$-in. A325F, the outside plates are 9 in. [275 mm] wide, and the bolts are placed in three rows.

87. A $4 \times 4 \times \frac{1}{2}$-in. angle of A36 steel is to be welded to a plate with E70XX electrodes to develop the full tensile strength of the angle. Using $\frac{3}{8}$-in. fillet welds, compute the design lengths L_1 and L_2, as shown in Fig. 24.21, assuming the development of tension on the entire cross section of the angle.

88. Design the welded connection for the data in Problem 87 assuming that the tension force is developed only by the connected leg of the angle.

Chapter 25

89. Open web steel joists are to be used for a roof with a live load of 25 psf and a dead load of 20 psf (not including joists) on a span of 48 ft. Joists are 4 ft center to center, and the deflection under live load is limited to $\frac{1}{360}$ of the span. Select the lightest joists from the tables in Chapter 25.

90. Same as Problem 89, except the live load is 30 psf, the dead load is 18 psf (not including joists), the span is 44 ft, and the joists are 5 ft center to center.

91. Open web steel joists are to be used for a floor with a live load of 50 psf and a dead load of 45 psf (not including joists) on a span of 36 ft. Joists are 2 ft center to center, and deflection is limited to $\frac{1}{360}$ of the span under live load and to $\frac{1}{240}$ of the span under total load. Select the (a) lightest possible and (b) shallowest possible joist.

92. Same as Problem 91, except the live load is 100 psf, the dead load is 35 psf (not including joists), and the span is 26 ft.

PART FIVE

(*Note:* For all of the following problems, use the working stress method unless otherwise directed.)

Chapter 27

93. Using f_c' = 4 ksi [27.6 MPa] and grade 60 reinforcement with f_y = 60 ksi [414 MPa] and f_s = 24 ksi [165 MPa], design the following members for flexure. Assume 1.5 in. [38 mm] cover for beams and $\frac{3}{4}$ in. [19 mm] cover for slabs with No. 3 U-stirrups in all beams. (a) A balanced rectangular section with tension reinforcement only, M = 120 kip-ft [163 kN-m]; (b) a solid slab, maximum M = 12 k-ft [16.3 kN-m]; (c) a T-section with a 5-in. [125-mm]-thick slab, a beam stem width of 15 in. [380 mm], and M = 200 kip-ft [271 kN-m]; (d) a rectangular section with b = 16 in. [406 mm], h = 30 in. [760 mm], and M = 400 kip-ft [542 kN-m]. (*Note:* (d) requires compressive reinforcement.)

94. Using strength methods, redesign the sections from Problem 93 as follows: (a) same as part (a), except M_u = 180 kip-ft [244 kN-m]; (b) same as part (c), except M_u = 300 kip-ft [407 kN-m].

Chapter 28

95. Using f_c' = 4 ksi [27.6 MPa] and grade 60 reinforcement with f_s = 24 ksi [165 MPa], design single U-stirrup shear reinforcement for a rectangular concrete beam with b = 14 in. [355 mm] and d = 26 in. [660 mm] for a span of 18 ft [5.5 m] and a total uniformly distributed load as follows: (a) 100 kips [445 kN]; (b) 80 kips [356 kN]; (c) 50 kips [222 kN].

96. Design the beams in Problem 95 using strength methods. Assume that the load is one-half live load and one-half dead load.

Chapter 29

97. A concrete beam cantilevers from a column and is reinforced as shown in Fig. 29.2. Using f_c' = 4 ksi [27.6 MPa] and f_y = 40 ksi [276 MPa], determine if full bar development is achieved and whether a hook is required for the following: (a) No. 6 bar; anchorage in the column is 12 in. [300 mm] and extension in the beam is 30 in. [750 mm]; (b) No. 10 bar; anchorage in the column is 18 in. [450 mm] and extension in the beam is 40 in. [1 m].

Chapter 31

98. Using the graphs in Chapter 31, pick the minimum-size square tied column and its reinforcement for the following: (a) load of 100 kips and moment of 25 kip-ft; (b) load of 100 kips and moment of 50 kip-ft; (c) load of 150 kips and moment of 75 kip-ft; (d) load of 200 kips and moment of 100 kip-ft; (e) load of 300 kips and moment of 150 kip-ft.

99. Same as Problem 98, except select the minimum size round tied column and its reinforcement.

Chapter 32

100. Using f_c' = 3 ksi [20.7 MPa] and reinforcement With f_y = 60 ksi, design a bearing wall 14 ft [4.25 m] high to support 10-in. [300-mm]-wide beams that are 12 ft [3.65 m] center to center and have end reaction loads of 30 kips live load [133 kN] and 16 kips dead load [71 kN]. Use strength design methods.

101. Using f_c' = 3 ksi [20.7 MPa] and reinforcement with f_s = 24 ksi [165 MPa], design a basement wall 14 ft [4.25 m] high. Assume no surcharge and an active soil pressure of 35 psf [1.68 kPa]/ft of depth.

Chapter 33

102. Design a square footing for a 14-in. [356-mm] square concrete column with a load of 219 kips [974 kN]. The maximum permissible soil pressure is 3000 psf [144 kN/m²]. Use concrete with f_c' = 3 ksi [20.7 MPa] and reinforcement with grade 40 bars with f_y = 40 ksi [276 MPa] and f_s = 20 ksi [138 MPa].

103. Design a wall footing for the following data: concrete wall thickness is 10 in. [250 mm], load on footing is 12 kips/ft [175 kN/m], maximum soil pressure is 2000 psf [96 kPa]. Use concrete with f_c' = 2 ksi [13.8 MPa] and grade 40 reinforcement with f_y = 40 ksi [276 MPa] and f_s = 20 ksi [138 MPa].

104. An 18-in. [450-mm] square tied column with f_c' = 4 ksi [27.6 MPa] is reinforced with No. 11 bars of grade 60 steel with f_y = 60 ksi [414 MPa]. The column axial load is 260 kips [1156 kN], and the allowable soil pressure is 3000 psf [134 kPa]. Using f_c' = 3 ksi [20.7 MPa] and grade 40 reinforcement with f_y = 40 ksi [276 MPa] and f_s = 20 ksi [138 MPa], design the following: (a) a footing without a pedestal; (b) a footing with a concrete pedestal.

PART SIX

Chapter 36

105. Design a reinforced masonry wall with hollow concrete units (CMUs) for the following data: wall height above grade of 8 ft [2.44 m], top of footing 8 in. [200 mm] below grade, wind load of 15 psf [0.72 kPa], medium weight CMUs of grade N, ASTM C90, f_m' of 1350 psi [9.3 MPa], type S mortar, grade 40 reinforcement with f_y = 40 ksi [276 MPa] and f_s of 20 ksi [138 MPa].

PART SEVEN

Chapter 40

106. Given the value for the dry unit weight and assuming a specific gravity of 2.65, find the following for the samples of sand listed: (1) void ratio; (2) soil unit weight and water content if fully saturated; (3) water content if wet sample weighs 110 pcf [1762 kg/m^3]. Samples: (a) dry weight is 90 pcf [1442 kg/m^3], (b) dry weight is 95 pcf [1522 kg/m^3]; (c) dry weight is 100 pcf [1602 kg/m^3]; (d) dry weight is 105 pcf [1682 kg/m^3].

107. Given the value for the saturated unit weight, and assuming a specific gravity of 2.7, find the following for the samples of clay listed: (1) void ratio; (2) water content of the saturated sample; (3) dry unit weight of the sample. Samples: (a) saturated unit weight is 90 pcf [1442 kg/m^3]; (b) saturated unit weight is 100 pcf [1602 kg/m^3], (c) saturated unit weight is 110 pcf [1762 kg/m^3]; (d) saturated unit weight is 120 pcf [1922 kg/m^3].

108. Given the following data from the grain size analysis, classify the following soil samples according to general type (sand or gravel) and nature of gradation (uniform, well-graded, gap-graded). Samples: (sizes in mm) (a) $D_{10} = 1.00$, $D_{30} = 4.00$, $D_{60} = 50.00$; (b) $D_{10} = 0.20$, $D_{30} = 1.00$, $D_{60} = 2.00$; (c) $D_{10} = 0.10$, $D_{30} = 0.25$, $D_{60} = 0.40$; (d) $D_{10} = 0.06$, $D_{30} = 0.15$, $D_{60} = 7.00$.

109. Given the information listed for each soil sample, identify the soil using the Unified System and estimate values for allowable bearing pressure, lateral passive resistance, and friction resistance using Table 40.4. Samples: (a) grain size analysis—60% retained on No. 4 sieve, only 4% passes No. 200 sieve, $D_{10} = 0.4$, $D_{30} = 3.0$, $D_{60} = 8.0$, penetration resistance—$N = 20$; (b) grain size analysis—80% passes No. 4 sieve, 6% passes No. 200 sieve, $D_{10} = 0.1$, $D_{30} = 0.4$, $D_{60} = 2.0$, penetration resistance—$N = 8$; (c) 60% passes No. 200 sieve, liquid limit is 40%, plasticity index is 10, unconfined compressive strength—$q_u = 1.6$ k/ft^2 [76.6 kPa], strong, musty odor.

Chapter 41 (See also Problems 102 and 103)

110. Select a column footing of reasonably large size from Table 33.2, assume the width in one direction is limited to three-fourths of that given in the table, and design the required oblong rectangular footing. Compare the steel and concrete quantities with those in the table entry.

111. Design a single footing to carry two columns with the loads and center-to-center spacings given. Use

$f_c' = 3$ ksi [20.7 MPa], $f_s = 20$ ksi [138 MPa] allowable soil pressure of 3000 psf [144 kPa]. (a) Column 1—100 kips [445 kN]; column 2—100 kips [445 kN], columns 6 ft [1.83 m] on center, (b) column 1 – 200 kips [890 kN], column 2 – 200 kips [890 kN], columns 10 ft [3.05 m] on center; (c) column 1—100 kips [445 kN] column 2—200 kips [890 kN], columns 8 ft [2.44 m] on center.

112. Design a cantilever footing for the data given. The edge of the footing is limited to 12 in. [300 mm] from the center of column 1 in one direction. Use $f_c' = 3$ ksi [20.7 MPa], $f_s = 20$ ksi [138 MPa], allowable soil pressure of 3000 psf [144 kPa]. (a) Column 1—100 kips [445 kN], column 2—200 kips [890 kN], columns 10 ft [3.05 m] on center; (b) column 1—200 kips, [890 kN], column 2—400 kips [1800 kN], columns 15 ft [4.57 m] on center; (c) column 1—300 kips [1334 kN], column 2—400 kips [1800 kN], columns 18 ft [5.49 m] on center.

113. Design a footing for the masonry wall in Problem 108. Use $f_c' = 3$ ksi [20.7 MPa], $f_s = 20$ ksi [138 MPa], allowable soil bearing pressure of 3000 psf [144 kPa], passive soil resistance of 150 psf/ft of depth [7.18 kPa/m], and a friction coefficient of 0.25. Place the footing so that its bottom is 18 in. [460 mm] below grade and keep the resultant of the forces within the kern limit of the footing.

PART EIGHT

Chapter 48

114. Design a plywood roof deck and plywood shear walls for a building similar to Building A. Data: wind speed of 90 mph, exposure condition B; seismic zone 4; building plan dimensions are 120 ft × 55 ft, end shear walls are 19 ft long, front and rear shear walls are 12 ft long.

115. Same as Problem 114, except walls are concrete hollow masonry units and roof has steel joists and steel deck with lightweight concrete fill.

116. Same as Problem 114, except all-steel frame with X-braced bents.

117. Same as Problem 114, except all-steel frame with north–south rigid frame bents 24 ft center to center.

Answers to Selected Exercise Problems

PART TWO

Chapter 7

14. (a) $F_v = F_h = 70.7$ lb.
 (c) $F_v = 90.8$ lb, $F_h = 178.2$ lb.
15. (a) $R = 141.5$ lb, $\theta = 45°$.
 (c) $R = 74.6$ lb, $\theta = 54.82°$.
16. (a) $R = 41.4$ lb, $\theta = 225°$ (see Fig. 7.23).
 (c) $R = 39.8$ lb, $\theta = 13.06°$.
17. (a) $T = 1414$ lb.
 (c) $T = 942.8$ lb.
21. (a) $R_1 = 37.5$ lb, $R_2 = 62.5$ lb.
 (c) $R_1 = 50$ lb down, $R_2 = 150$ lb up.
22. (a) $R_1 = 1128$ lb, $\theta_1 = 257.2°$, $R_2 = 2115$ lb, $\theta_2 = 96.79°$.
23. $R = 234.9$ lb, $x = 5.91$ ft, $z = 2.89$ ft.
24. $T = 125$ lb, $C = 46.9$ lb.
25. $T_1 = 50.8$ lb, $T_2 = 19.7$ lb, $T_3 = 45.0$ lb.
26. $R = 4$ lb down, $x = 10.75$ ft right, $z = 15.5$ ft left.

Chapter 8

28. 99.275 kips [442 kN].
29. 5.48 ft, or 5 ft 6 in. [1.67 m].
30. 2.5 in.^{2} [1613 mm^2].
31. Required area is 115 in.^{2}, 12 × 12 has 132.25 in.^{2}.
32. 3.47 in. [88 mm].
33. 6.185 kips [27.36 kN].
34. 231.4 kip-in. or 19.3 kip-ft [26.15 kN-m].
35. (a) 13.01 kip-in. [4.56 kN-m], limited by rotation.
 (b) 407 lb-in. [45.95 N-m], limited by rotation.
36. (a) 6995 psi [52.17 MPa], 5829 psi [43.82 MPa].
 (b) 6434 psi [46.13 MPa], 3753 psi [26.45 MPa].
37. 0.077 in. [1.95 mm].
38. 19.33 kips [83.3 kN].

Chapter 10

41. (a) 13.46 kips.
 (b) 9.43 kips.
42. 112.5 kips.

Chapter 11

44. (a) 3.04 ksf [151.5 kPa].
 (b) 5.33 ksf [266.5 kPa].

Chapter 12

45. (a) 1.85 in. [47 mm].

 (b) 3.08 ksi [21 MPa].
46. $f_c = 0.730$ ksi, $f_s = 3.53$ ksi.
47. $f = 0.191$ ksi, $v = 0.110$ ksi.
48. Read on section clockwise, starting with the lower, down-slope corner, $f = +1338$ psi, -387 psi, -1338 psi, $+387$ psi.
49. Using notation as shown on Fig. 12.16: $f_A = -0.2279$ ksi, $f_B = +0.3581$ ksi, $f_C = +0.1627$ ksi, $f_D = +0.7487$ ksi, neutral axis measured from A toward $B = 6.22$ in. and measured from A toward $C = 14.0$ in.
50. $f_v = 1.5$ (V/bd).
51. At neutral axis $f_v = 811.4$ psi; at junction of web and flange $f_v = 699.3$ psi in web and 174.8 psi in flange.
52. 19.3°.
53. 58.45 lb.
54. 0.9927 lb.
55. Will not slide, but will tip.

PART THREE

Chapter 16

61. All stresses in psi: (a) $f_b = 1550$, $f_b = 85$, $f_t = 775$, (b) $f_b = 1500$ or 1750, $f_v = 95$, $f_t = 1000$, (c) $f_b = 1200$, $f_v = 85$, $f_t = 825$.

Chapter 17

62. (a) Shear is critical; 10 × 20 is lightest if critical shear is at d distance from support; otherwise, a 12 × 20 must be used.
 (b) Bending is critical; 8 × 20 is lightest.
 (c) Shear is critical; 12 × 22 is lightest.
63. (a) 2 × 12 at 16.
 (b) 2 × 8 at 12 or 2 × 10 at 16 or 2 × 12 at 24.
 (c) 2 × 8 at 12 or 2 × 10 at 16 or 2 × 12 at 24.
64. (a) 2 × 8 at 12 or 2 × 10 at 16 or 2 × 12 at 24 (from *UBC* Table U-R-1, see Appendix D).
 (b) 2 × 10 at 12 or 2 × 12 at 16 (Table U-R-1).
 (c) 2 × 6 at 12 or 2 × 8 at 16 or 2 × 10 at 24 (Table U-R-13).
 (d) 2 × 10 at 16 (Table U-R-13).

Chapter 18

65. (a) 41 kips, 35 kips, 12.8 kips.
 (b) 71 kips, 65 kips, 27.6 kips.
66. 14.75 kips.

Chapter 19

67. 896 lb.
68. 15,780 lb, limited by bolts.
69. 1660 lb.
70. 3500 lb.

Chapter 20

71. 21.2 kips.

PART FOUR

Chapter 22

72. From Table 22.1—W 18 × 35.
73. From Fig. 22.2—W 16 × 40.
74. From analysis: W 16 × 40 deflects 1.20 in., no good. Use W 21 × 44; deflects 0.75 in.
75. W 18 × 35—77.0 kips, W 16 × 40—70.8 kips, W 21 × 44—104.8 kips.
76. 65.17 kips.
77. 86.6 kips.
78. Approximately 8 in. by 15 in. by 1 in.
80. (a) WR20.
　　(b) WR18.
　　(c) IR22 or WR22.

Chapter 23

81. (a) 630 kips.
　　(b) 306 kips.
　　(c) 157 kips.
82. (a) W 10 × 54.
　　(b) W 6 × 20.
　　(c) 6-in.-nominal-diameter standard pipe.
　　(d) $5 \times 5 \times \frac{3}{8}$.
　　(e) $3\frac{1}{2} \times 2\frac{1}{2} \times \frac{3}{8}$ or $4 \times 3 \times \frac{5}{16}$ (weight the same).
83. (a) W 12 × 65.
　　(b) W 10 × 49.
　　(c) W 14 × 145.

Chapter 24

85. 6 bolts, outer plate—$\frac{5}{8}$ in., inner plate—$\frac{7}{8}$ in.
86. 8 bolts, outer plate—$\frac{5}{8}$ in., inner plate—$\frac{7}{8}$ in.
87. $L_1 = 11$ in., $L_2 = 5$ in.
88. Minimum $L = 4.25$ in. on each side (3.95 in. computed).

Chapter 25

89. 28K7.
90. 28K7.
91. (a) 22K4.
　　(b) 18K7.
92. 18K4.

PART FIVE

Chapter 29

97. (a) Anchorage is adequate, no hooks required.
　　(b) Anchorage is insufficient in both column and beam; standad hook can develop additional required anchorage at both locations.
98. (*Note:* Selections are approximate, especially when the intersection in the graph falls almost on a line.)
　　(a) No. 8—12 × 12, 4 No. 11 bars.
　　(b) No. 14—20 × 20, 4 No. 11 bars (almost No. 13—4 No. 8 bars).
　　(c) No. 16—20 × 20, 8 No. 11 bars (almost No. 15—6 No. 11 bars).
　　(d) No. 18—24 × 24, 6 No. 11 bars.
　　(e) No. 21—30 × 30, 6 No. 10 bars (almost No. 20—24 × 24, 8 No. 14 bars).
99. (*Note:* Selections are approximate, especially when the intersection in the graph falls almost on a line.)
　　(a) No. 8—14 in. diameter, 6 No. 9 bars.
　　(b) No. 14—20 in. diameter, 4 No. 11 bars (almost No. 13—4 No. 8 bars).
　　(c) No. 16—20 in. diameter, 8 No. 11 bars (almost No. 15—6 No. 11 bars).
　　(d) No. 18—24 in. diameter, 6 No. 11 bars.
　　(e) No. 21—30 in. diameter, 6 No. 10 bars (almost No. 20—24 in. diameter, 8 No. 14 bars).

PART SEVEN

Chapter 40

106. (a) $e = 0.838$; saturated weight = 118 pcf [1890 kg/m³]; $w = 22\%$.
107. (a) $e = 2.85$; $w = 105\%$; dry weight = 43.8 pcf [702 kg/m³].
108. (a) Sandy gravel, poorly graded.
109. (a) Sandy gravel, well graded—GW. Allowable bearing pressure is 1500 psf + 200 psf for each foot of surcharge over 1.5. Passive resistance is 300 psf/ft of depth below grade. Friction resistance is 0.6 times the dead load.

Index